MAMMALS
OF THE
NEOTROPICS

T0138241

Mammals of the Neotropics, Volume 1, *The Northern Neotropics: Panama, Colombia, Venezuela, Guyana, Suriname, French Guiana*, by John F. Eisenberg, was published by the University of Chicago Press in 1989.

Mammals of the Neotropics, Volume 2, *The Southern Cone: Chile, Argentina, Uruguay, Paraguay*, by Kent H. Redford and John F. Eisenberg, was published in 1992.

MAMMALS OF THE NEOTROPICS

The Central Neotropics

VOLUME 3

Ecuador, Peru, Bolivia, Brazil

With plates in color and black and white by Fiona A. Reid

John F. Eisenberg

and

Kent H. Redford

with seven contributors

The University of Chicago Press
Chicago and London

The University of Chicago Press, Chicago 60637
The University of Chicago Press, Ltd., London

© 1999 by The University of Chicago
Plates © 1999 by Fiona A. Reid
All rights reserved. Published 1999

25 24 23 22 21 20 19 18   9
ISBN: 0-226-19541-4 (cloth)
ISBN-13: 978-0-226-19542-1 (paper)
ISBN: 0-226-19542-2 (paper)

Library of Congress Cataloging-in-Publication Data
(Revised for vol. 3)

Eisenberg, John Frederick.
    Mammals of the Neotropics

    Vol. 3 by John F. Eisenberg and Kent H. Redford.
    Includes bibliographical references and indexes.
    Contents: v. 1. The northern Neotropics: Panama,
Colombia, Venezuela, Guyana, Suriname, French
Guiana—v. 2. The southern cone: Chile, Argentina,
Uruguay, Paraguay—v. 3. The central Neotropics: Ecuador,
Peru, Bolivia, Brazil
    1. Mammals—South America.   I. Redford, Kent Hub-
bard.   II. Title
QL725.A1E38   1989   599.098   88-27479
ISBN: 0-226-19541-4 (cloth: v. 3)
ISBN: 0-226-19542-2 (paper: v. 3)

⊗ The paper used in this publication meets the minimum
requirements of the American National Standard for Information Sciences—Permanence of Paper for Printed Library Materials, ANSI Z39.48-1992.

To *Karl, Elise, James, Suzanne, Jennifer,* and above all
Genevieve, Elisha, Tana, Jon David, Sofia, and Hugh,
with whom the future rests

Contents

Acknowledgments

This book could have not been prepared without enormous cooperation from many colleagues. Since this is the third and penultimate volume, we do not wish to reiterate those acknowledged in volumes 1 and 2; rather, we highlight the tremendous cooperation we have had from certain individuals in the preparation of volume 3. Our efforts were aided enormously by Michael Carleton and Richard Thorington at the United States National Museum of Natural History, who provided us with a computer printout including localities of all specimens held at that institution. Alceu Rancy at the University of Acre, Brazil, was helpful in many ways. Alfredo Langguth, while at the Museum in Rio de Janeiro, was most cooperative. Special thanks to Tadeu Gomes de Oliviera, who checked and verified the measurements and collection localities for many species within the collection at the Museu de Zoologia da Universidade in São Paulo. He was aided by Alexandre Percequillo. Sidney Anderson at the American Museum of Natural History generously provided us with a printout of his data from Bolivia. Bruce Patterson and the Field Museum of Natural History, Chicago, provided us with computer printouts of specimens held in their collections. Mark Hafner at Louisiana State University generously opened for inspection the Peruvian and Ecuadorian collections in his custody. As always, Michael Mares at the University of Oklahoma was supportive and offered data. James Patton of the Museum of Vertebrate Zoology at Berkeley was always supportive and helpful in many ways.

In the preparation of this volume, we visited the following institutions: Museo Goeldi at Belém, Museo Nacional at Rio de Janeiro, the state museum at Belo Horizonte, and the state museum at Pôrto Alegre in Brazil. In addition we relied on previous and recent visits to the Carnegie Museum in Pittsburgh, the United States National Museum of Natural History in Washington, D.C., the American Museum of Natural History in New York City, the Museum of Vertebrate Zoology at Berkeley, the Museum at Louisiana State University in Baton Rouge, the Museum at the University of Michigan at Ann Arbor, the Field Museum of Natural History in Chicago, the British Museum of Natural History in London, the University of Connecticut Museum collection at Storrs, and of course the Museum of Comparative Zoology at Harvard.

Tangential to our work with these collections were visits to the Ryks Museum at Leiden, the Netherlands, the Museo Nacional, Buenos Aires, and the Museo La Plata. And once again we must acknowledge the fine collection of small cetaceans housed at the Florida Museum of Natural History.

The maps were prepared by Hugo Ochoa and Sigrid James Bonner, and in addition Sigrid prepared numerous line drawings. Dale Johnson helped complete the maps. Once again Fiona A. Reid prepared the color plates. We are indebted to their efforts. Our old friends at the National Zoological Park, Washington, D.C., again offered us data concerning life history parameters. F. J. Leeuwenberg provided useful geographical data on the larger mammals in Brazil. Tadeu Gomes de Oliviera helped greatly with measurements and records for the Carnivora of Brazil. Our thanks.

Particular thanks are due to the Nature Conservancy for assistance with the costs of the color plates.

While working on this volume, we were inspired by several of our former students. These include Rosa Lemos de Sa, Rodrigo Medellin, Joseph Dudley, Jay Malcolm, Alceu Rancy, José Ottenwalder, Damian Rumiz, Tadeu Gomes de Oliviera, Carlos Peres, Theresa Pope, Jody Stallings, Gustavo Fonseca, Claudio Padua, Hugo Ochoa, John Polisar, Rafael Hoogesteijn, Wendy Townsend, Laurence Pinder, Peter Crawshaw, Andreas Navaro, José Fragoso, and Kirsten Silvius. Tia Cordier, Joan de Lorenzo, Sylvia Fussell, and Regina Ventura faithfully prepared the manuscript. Rhoda J. Bryant was invaluable in proofreading all copyedited material. We could not have accomplished this task without her special expertise.

Fiona Reid is grateful to Guy Tudor for helpful comments on sketches and for the use of his extensive photo reference files. She also thanks the curators and other staff members at the Royal Ontario Museum, American Museum of Natural History, British Museum of Natural History, Field Museum, and National Museum of Natural History, who took time out of their busy schedules to help her locate and examine specimens of mammals in their care.

Introduction

Background for the Preparation of This Volume

This volume evolved as a natural consequence of the publication of volumes 1 and 2 of *Mammals of the Neotropics*. After JFE completed volume 1, we thought a second volume dealing with the "southern cone" would be a logical complement. There was of course some overlap in the faunal composition, but volumes 1 and 2 could be used independently. Then by popular demand, particularly from those living and working in Brazil, Bolivia, Peru, and Ecuador, we were asked to prepare volume 3, but this presented some difficulties since most of the mammalian forms described in volumes 1 and 2 are also included in volume 3. To avoid duplication, therefore, we adopted an abbreviated style for the species accounts in this volume. Volume 3 is best used in conjunction with volume 1 or volume 2.

KHR has worked extensively in Bolivia, Brazil, and Paraguay. JFE has worked in the vicinity of Iquitos, Peru, and the areas surrounding Manaus, Rio de Janeiro, Pôrto Alegre, Belo Horizonte, and Acre, Brazil. In addition, he has visited the area of Iguazú from both the Argentine and Brazilian sides. Some of the most productive work JFE did for this volume was conducted near Manaus, at Montes Claros, Minas Gerais, and at the state park Rio Doce, Minas Gerais, Brazil, in 1984 and 1989.

Organization of the Book

This volume is organized in nearly the same way as the previous two. After the introduction and the introductory chapters in part 1, a chapter is devoted to each order. The volume closes with several synthetic discussions of biogeography and ecology. The taxonomic order follows Wilson and Reeder (1993) for orders and families. Within families, the genera are arranged in taxonomic order, and species are arranged alphabetically. To avoid duplication, descriptive material for the higher taxonomic categories is not repeated at the species level. The species accounts are of course the most detailed.

How to Use This Book

As in the previous volumes, we maintain the use of Latin and Greek binomials, but we have inserted common names where they are in current use.

Many of the species accounts for the rodents are provisional, since the taxonomic categories are currently unstable. For most species a distribution map is provided, and map references are included in the literature cited at the end of each chapter. We have included some keys in this volume. Keys have been compiled for a few taxa and regions, but no key to the species of rodents for the entire area is available (Peterson and Pine 1982; Pine 1973; Emmons and Feer 1997; Vizotto and Taddei 1973; Anderson 1997).

Abbreviations are used in many tables and also when referring to collections. The following refer to museums: AMNH, American Museum of Natural History; BA, Museo Argentino de Ciencias Naturales "Bernardino Rivadavia"; BMNH, British Museum of Natural History; CMNH, Carnegie Museum of Natural History; DNPM, Departamento Nacional da Produção Mineral; FMNH, Field Museum of Natural History; LSU, Louisiana State University; MVZ, Museum of Vertebrate Zoology, Berkeley; UConn, University of Connecticut, Storrs; NZP, National Zoological Park; UM, Museum of Zoology, University of Michigan, Ann Arbor; and USNMNH, United States National Museum of Natural History. "PCorps" refers to unpublished data from Peace Corps collections in Paraguay, made available courtesy of Jody Stallings and Mark Ludlow.

Throughout the text, the following abbreviations refer to linear measurements and weights: TL, total length; HB, head and body length; T, tail length; HF, hind foot length; E, ear length as measured from the notch to the tip; FA, forearm length (useful for bats);

Wt, weight of the adult. Linear measurements are usually given in millimeters (mm), and weights are in grams (g) or sometimes kilograms (kg).

Tables and maps portraying the measurements and distribution for the various mammalian taxa are referenced in detail within those chapters that deal with the ordinal taxa. Some data of great significance were derived from computer printouts of localities as recorded by museum curators and their assistants: Peru (LSU, FMNH, USNMNH), Ecuador (LSU), Bolivia (AMNH), and Brazil (FMNH and USNMNH). To verify specimen locations, lists of specimens with location data made trips to museums much more efficient, and for common, wide-ranging forms they helped us establish the limits of a species' range. In addition key references include the following:

General: Voss and Emmons 1996
Bolivia: Anderson 1985, 1997; Anderson et al. 1993; Townsend 1995
Brazil: Alho 1982; Bernardes, Machado, and Rylands 1989; Carvalho 1962; Fonseca et al. 1994, 1996; Mares et al. 1981; Reis and Peracchi 1987; Vieira 1955
Ecuador: Albuja 1982, 1991
Peru: Hutterer et al. 1995; Pacheco et al. 1995; Pearson 1951, 1957; Pulido Capurro 1991; Tovar 1971; Valqui 1995; Woodman et al. 1991

This list is *not* exhaustive; it represents a selection of works for the various countries whose fauna is described in this volume, including efforts at compilations by several pioneering workers during the past forty years.

The mammals of South America are still not well known, and new species are being described every year. This volume attempts to capture our knowledge concerning the mammals found in the "core" of South America at the close of the twentieth century.

Comment on Taxonomy

In preparing volumes 1 and 2 we relied on Honacki, Kinman, and Koeppl (1982) for a standardized nomenclature. In this volume we have tried to follow the scientific names as given in Wilson and Reeder (1993) for chapters 5 through 13. When we depart, we give reasons for doing so. Of course some species have been named since the completion of the chapters in Wilson and Reeder, and some recent revisions depart from the taxonomy proposed in 1993. We have presented alternative schemes where genuine disputes are in progress.

To aid readers we have added an appendix in which we compare key species from volumes 1 and 2 with those of volume 3 where significant changes have occurred in nomenclature and content.

Distribution Maps

The primary base map includes the region of Colombia east of the Andes lying between the Japurá and Putumayo Rivers. Aside from this transgression, the outline follows the political boundaries of Ecuador, Peru, Bolivia, and Brazil. The map was drawn from a curvilinear projection; thus the latitude marks are correct but must not be connected to the longitudinal lines by a line at right angles if one wishes to check coordinates. Depending on the printing reduction, the coordinates will have to be drawn empirically. This warning is meant for contemporary biogeographers who may wish to extrapolate data. It is wise to go back to the original references, since the accuracy is seldom better than one-tenth of a degree. Major rivers are included, and contour lines are given at 500 m and 3,000 m. Smaller derived maps are restricted to eastern Brazil or to Ecuador, Peru, and adjacent Bolivia.

History of Mammal Collecting

We focus this introduction on Brazil, since it is the largest country dealt with in this volume. After Christopher Columbus returned to Spain in the mistaken belief that he had reached Asia, the race was on for claiming territories in the Western Hemisphere. Portugal and Spain became early rivals, and it took intervention by the pope and the Treaty of Tordesillas (1494) to demarcate territories where their Catholic majesties of Spain and Portugal could send their emissaries to develop and Christianize the yet unexplored lands. Most of South America was offered to the Spanish, but Brazil came into being under the influence of the Portuguese—and the race for riches was on. Portugal, through the years, had to struggle against not only the Spanish, but the French, English, and Dutch, who were all attempting to gain a foothold in the New World at the expense of Portugal.

Exploration and active collecting of animal specimens began in earnest in the eighteenth century. Alexandre Ferreira was the first native-born Brazilian to mount successful collecting expeditions in the Amazon region. He studied in Portugal and was dispatched to explore the Amazon during 1783–92. His collections were sent to Portugal to be deposited in the Museu d'Ajuda. Unfortunately, this valuable collection was never described by Portuguese scientists. The Napoleonic Wars intervened, and when Portugal was conquered by France the entire collection was removed to the Natural History Museum in Paris, where much of it was ultimately described by Etienne Geoffroy Saint-Hilaire.

The nineteenth century was dominated by two biologists working in South America: Johann Baptist Ritter von Spix and Karl Friedrich Philipp von Martius.

Their exploits are amply documented by Hershkovitz (1987). Spix was a zoologist and Martius was a botanist. In their expedition, they enlisted the services of Johann Natterer, a formidable collector, whose explorations were published by Pelzeln (1871, 1883) some fifty years after the fact. During this same period one of the crown princes of Austria, Maximilian, prince of Wied-Neuwied, traveled extensively in North and South America, amassing a very important collection (Wied-Neuwied 1826).

Clearly, from the time of discovery through the nineteenth century, with the exception of Alexandre Ferreira, the collection and description of vertebrates was largely in the hands of Europeans, and the type specimens were placed in European museums. Even the British Museum of Natural History (BMNH) played an important role. Although the English never dominated the continent of South America, the BMNH had such an outstanding reputation that collectors in Latin America were often solicited to provide the museum with specimens that were subsequently described in the international scientific literature. Oldfield Thomas was in the enviable position of being on the staff of the BMNH and having sufficient funds to pay collectors. It is no coincidence many of the type specimens of rodents and marsupials for Brazil, Ecuador, and Peru reside in the BMNH.

During the twentieth century, mammal studies in Brazil often were oriented within the realm of applied research. Since many mammals carry ecto- and endoparasites and might transmit microorganisms to humans, it was only logical that surveys were conducted by epidemiologists. Indeed, a great body of literature exists concerning populations of rodents that were potentially harmful to human colonization and agriculture. Accounts of the natural history of mammals were sparse, but notably the efforts of Eladio da Cruz Lima (1945) set a trend for nonapplied natural history. Some seven years later, João Moojen (1952) published his charming essay on the rodents of Brazil.

Paleontology has had a distinguished record as a subject of study in Brazil. The first extensive collection of Pleistocene fossils was developed by Peter Lund in what is now the state of Minas Gerais, Brazil. Lund was a Dane who traveled to Brazil in the early nineteenth century and stayed for the rest of his life. He collected fossils in many of the caves near what is now the city of Belo Horizonte. One of his most famous caves was the cave of Lagoa Santa. His collections were sent to the museum in Copenhagen and were described in the latter part of the nineteenth century by Herluf Winge. The cave collections were remarkable in that they included recently extinct forms as well as extant species. Several extinct species of rodents were described as fossils before being rediscovered as living forms

(e.g., *Carterodon* and *Pseudoryzomys;* Voss and Myers 1991) (see also chapter 4 below).

The consolidation of the Brazilian paleontological material during the past fifty years fell to Paulo Couto (1979). Through his long relationship with George G. Simpson at the American Museum of Natural History, the diverse collections of mammals from Brazilian sites were formally cataloged and organized. The challenge to Brazilian paleontology still remains, but Paulo Couto performed the initial twentieth-century synthesis. Pascual (1996) summarizes the history of mammals on the South American continent.

One feature that amazes us is how soon the process of exploration and synthesis was initiated in South America in contrast to, let us say, Africa. In part it was made easier because the great rivers of South America in the main flow east through flat terrain, so that it is possible to take a boat from the mouth of the Amazon to above Iquitos, Peru, without encountering a single cataract. This ease of access, in spite of sandbars, hostile indigenous inhabitants, and disease, led to an early exploration of the interior. The same comment could well be made about the Mississippi River and its tributaries in North America.

Since the New World was colonized by European empires, it was inevitable that the early explorations and their results would repose across the Atlantic in the museums of the great capitals. Although it is often an annoyance, and an expensive one at that, to travel to Europe to inspect type specimens, the journeys do have their positive side. It is heartening to see the zeal with which collecting and natural history studies are being pursued in Brazil, Peru, Ecuador, and Bolivia. Important collections are currently being assembled at La Paz, Bolivia; Lima, Peru; Quito, Ecuador; and São Paulo, Rio de Janeiro, Belém, Brasília, and Belo Horizonte, Brazil.

Geography

On inspecting a topographic map of South America, one is struck by the position of the Andes Mountains far to the west. Aside from the Brazilian shield and the Guiana shield, there is little topographic relief in the eastern portion of the continent (fig. 1.1). The Andes, deriving from upthrusts perhaps commencing before the Miocene, ultimately influence rainfall and drainage patterns for the east. Before the Andean uplift much of the Amazon basin might have been an inland sea of great magnitude. The continued uplift of the Andes had profound effects on the climate of the continent. Glaciations are manifest on the higher slopes of the Andes, and one assumes that at times of glacial maxima during the Pleistocene these areas were even more dominated by ice. Clearly, at the close of each

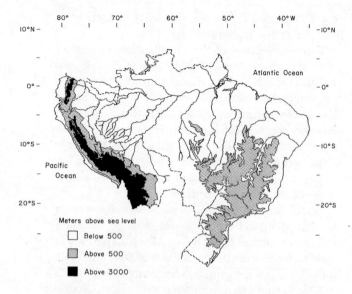

Figure 1.1. Elevation of the landscape in the central portions of South America.

episode of glaciation there must have been tremendous runoff during the final "meltdown." In the closing stages of the most recent glaciation, severe flooding must have occurred off and on from 14,000 B.P. to 8000 B.P. The tropical to subtropical forests we now see on the slopes of the Andes may not have existed 10,000 years ago. Indeed, Liu and Colinvaux (1985) suggest that in Ecuador the lowland flora was more temperate in its character. What did the Amazon basin look like some 10,000 to 14,000 years ago? Frailey et al. (1988) propose that at the last glacial maximum, the Amazon basin was a lake or swamp. Indeed, they go so far as to revive the old idea that the Miocene drainage pattern was not to the Atlantic but to the north and via the Orinoco to the Caribbean. During the last glacial maximum, the great ice fields on the upper plateaus in Peru and Bolivia might have functioned like a huge cool lake. Thus the rainfall patterns in the Amazon basin 12,000 years ago might be quite different from those observed today. In fact Rancy (1991) proposes that the shores of "Lake Amazon" at 11,000 B.P. might well have been dominated by moist savanna grading to dry savanna (see chapter 13).

Rancy's dissertation is provocative in that he documents a Pleistocene faunal assemblage from Rio Branco, Brazil, that is characterized by mammalian forms normally associated with savanna conditions. Glyptodonts, ground sloths, and llamas are found together with genera that normally characterize areas much farther to the south in Argentina. Clearly the surface has only been scratched and much more effort needs to be expended, but the work of Rancy building on the earlier collections of Simpson and Paulo Couto substantiates the idea that the rain forest that now characterizes Acre is of rather recent origin.

In Peru and Bolivia, the eastern and western ranges of the Andes are widely separated, and there are high-altitude plateaus of considerable extent. This area surely was partially glaciated many times, and it should not be surprising to find that the mammalian fauna is somewhat depauperate and that what fauna exists is derived from premontane zones that presumably served as refugia during glacial maxima. This in no way diminishes the importance of these highland endemics; we mean only to say that the survivors now populating the high plateaus form an endemic ecosystem only recently reconstituted with the retreat of the glaciers. The effects of human occupation on these higher plateaus have obviously been dramatic.

Although no one is certain when the first native members of *Homo sapiens* colonized the area, it is quite clear that by 10,000 B.P. the process was actively in motion. Domestication of plants and animals soon followed; thus, when the Spanish explorers first encountered the Inca civilization, the guinea pig (*Cavia*) was semidomesticated as a food source and the llama (*Lama*) had been domesticated as a source of wool and as a pack animal. The controlled exploitation of the vicuna was well developed under the rules laid down by the Inca kings. Arguably, the first sustained yield harvest of a wild mammal was achieved by the Incas before European contact. Indeed, one suspects that if the Europeans had not reached the Western Hemisphere in the fifteenth century, domestication might have gone even further (Wing 1986, 1989).

The savanna formations of South America have been receiving increased attention. The term "savanna," broadly applied, means an area with few trees. Of course there are tropical savannas, temperate savannas, wet savannas, and semiarid savannas. Thus the terminology is a bit confusing. But environments not dominated by trees have been part of the South American landscape at least since the late Oligocene. It should not be surprising, then, that savanna-adapted forms of vertebrates were once widespread. The mammalian fauna of "old" endemic South American mammals dominated until the Pliocene, when there was a general commingling of forms from North America emigrating across the newly created Isthmus of Panama. The faunal exchange was not entirely successful. Many North American forms went extinct during the Pleistocene, as did most of the larger old South American endemics. Many of the major extinctions occurred before *Homo sapiens* appeared on the continent. The structure of these Pleistocene mammal communities will be discussed in chapters 2, 3, and 4.

We have remarked before (Eisenberg 1989) on the complications of making simple statements that more or less cover some generalized features of vegetative formations (fig. 1.2). For example, the term "cerrado" describes an area dominated by a dry, deciduous, trop-

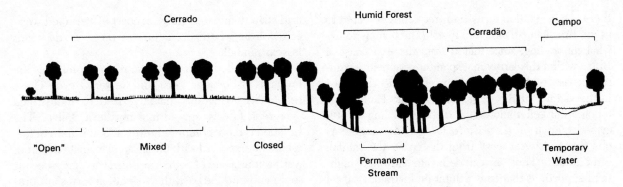

Figure 1.2. Vegetation types along a hypothetical moisture gradient in central Brazil. Note the horizontal heterogeneity.

ical forest in the eastern portion of Brazil. Without pursuing the question too far, let me point out that there are at least three basic subdivisions of the "cerrado" formation: *campo limpo, campo sujo,* and cerrado. Cerrado literally means "closed" and in fact refers to upslope dry, deciduous vegetation where tree crowns almost touch one another. As one proceeds downslope, one can get into *campo sujo,* dominated by an herbaceous understory and small shrubs. Finally, one comes to the *campo limpo,* which is open and almost devoid of woody vegetation and is frequently subject to seasonal flooding. On top of all of this complexity, those rivers that run permanently often have vegetation termed "gallery forest" along their banks. On the high bank—the one that does not normally flood—the forest may in fact support evergreen tree forms that normally would be found in the contiguous forests of Amazonas itself. These gallery forests serve as "ribbons of life" through which species that nor-

mally would be found in evergreen tropical rain forests can penetrate the savanna areas, thus enhancing the biodiversity of the savanna. They may also support refugial populations for the eventual colonization of the savannas should climatic events prove more favorable for their survival (fig. 1.3). This biogeographic circumstance was previously outlined by Eisenberg and Redford (1979) and by Eisenberg (1989).

Given that the continent of South America extends from approximately 10° north latitude to 50° south, the contemporary global wind patterns show considerable variation, which has implications for rainfall on the continent. Added to this is the topographic relief of the Andes set so far to the west. Nevertheless, there are considerable areas of elevation exceeding 500 m that are disjunct from the Andes, including the Guiana highlands and the very dissected and eroded Brazilian shield in the southeastern portion of that country. If we consider eastern Brazil from roughly 10° to 30° S,

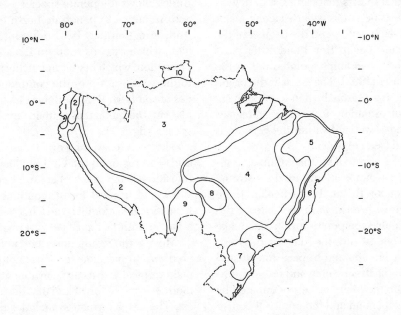

Figure 1.3. Major vegetation forms for central South America outlining zoogeographic regions: (1) Choco region of lowland tropical rain forest; (2) Andean region; (3) Amazonian rain forest; (4) cerrado; (5) caatinga; (6) Atlantic rain forest region; (7) Araucaria forest region; (8) Pantanal; (9) portion of the Gran Chaco; (10) portion of the Grand Savanna.

there is a vegetation formation frequently referred to as the Brazilian Atlantic rain forest. The term is somewhat misleading, since inland from the east coast of Brazil within this latitudinal span there appear a complex series of highlands, almost reaching the coast in the vicinity of Rio de Janeiro and São Paulo. This topographic relief allows for adiabatic cooling and elevated rainfall on the eastern face of the ranges. As one proceeds west away from the coast, the rainfall can diminish. Thus one can delineate extremely high-rainfall portions that truly exhibit multistratal tropical forest, but on the "lee side" rainfall patterns may be highly variable. Nevertheless, this forested strip has obviously been a center of endemism, and many of the mammalian species are distinctive within this region.

Understandably, this portion of Brazil was among the first areas to be colonized by Europeans. It was on the Atlantic coast that early settlements were made, and the impact of European activity has been progressive since the 1500s. As a result, much of the original forest cover has been cut over, exotics have been introduced, and most of the endemic Atlantic rain forest mammals of Brazil are severely threatened (Oliver and Santos 1991). We are concerned with mammals in this volume, but the endangered status of all animal and plant life endemic to the region is clear. To the west of this area in Brazil, one imperceptibly grades into dry deciduous tropical forests of varying formations; these are dealt with in the sections that follow. The dry deciduous forests and associated campos are frequently neglected by environmental groups, yet they exhibit a marvelous flora with many endemics, and they have been and may continue to be refugia for forms adapted to tropical multistratal forests that can exist in dry areas where such forests persist as galleries along major river courses (fig. 1.4). In any event, the neglect of the forgotten forest formations in Brazil and other parts of South America has been amply pointed out in the recent article by Mares (1992). The basin at the origin of the Rio Paraguay is a seasonally flooded area lying mostly in Brazil but extending to parts of Paraguay. This seasonally flooded wetland within the cerrado is termed the Pantanal (see Por 1995).

Within the region covered by this volume, temperate forests begin to appear only in the extreme southeastern portion of Brazil. In particular, there are relict formations of *Araucaria* in the state of Rio Grande do Sul. Genuine temperate formations make their appearance as one ascends the Andes in the west. One passes through premontane tropical forests and finally into zones dominated by pine and oak, eventually crossing over into the high plateaus dominated by the vegetation of the paramo and the puna. Of course the reverse occurs as one descends the western slopes to the Pacific (see Eisenberg 1989). A rather narrow, arid strip dominates the west coast of Peru and western Bolivia, where the failure of the westerly winds lessens rainfall.

The so-called Amazon biotic region dominates the lowlands of Ecuador, Peru, and western Brazil. These vegetation formations, characterized by multistratal evergreen forests, spread into northern Bolivia. The situation is more complex, however. Some areas of forest are interspersed with the savanna formations, often wet savannas; and of course in Bolivia, as one proceeds south and into the east, the vegetation forms are predominantly savanna and thorn scrub. The Pantanal, a large area of wet savanna, is found in south-central Brazil and eastern Bolivia. Its condition derives from the peculiarities of the drainage pattern in that region, which contribute to long periods of standing water. Interposed between the tropical savannas and multistratal rain forests is the semiarid tropical deciduous forest belt—roughly including the caatinga, the cerrado, and the Chaco (see fig. 1.3).

The Amazon rain forest itself exhibits vast horizontal and vertical heterogeneity (fig 1.3). Seasonal flooding results in distinctive forest formations termed the várzea, with faunas different from those found in the forests of terra firma. The disturbance effects of seasonal, variable floods create a dynamic system that will be discussed in greater detail in chapters 20 and 21.

Zoogeography

The region discussed in this volume lies mainly in the tropics, but the elevational gradient imposed by the Andes allows temperate species associated with Chile and Argentina to penetrate north to almost 5° S. The Andean region may be divided into the northern region, which extends south to Ecuador, and the southern region, which begins in northern Peru and extends to the southern tip of the continent. The central Andes of southern Peru to Bolivia include a great central plateau, the puna, that contains many endemic rodent species (fig. 1.1).

The west coast of Peru is semiarid to extremely arid. The region of the Gulf of Guayaquil, Ecuador, is transitional, and much of coastal Ecuador to the north is lowland tropical forest, continuous with the same formation in Colombia (the Chocó) and exhibits faunal affinities with Panama (see Eisenberg 1989).

Most of the region described in this volume is covered with tropical forests (see chapters 20 and 21). The main tropical forests are conveniently subdivided into moist evergreen forest and dry, deciduous tropical forest. The latter formation includes the Gran Chaco of Paraguay, northern Argentina, and south Bolivia extending northeast to the coast of northeastern Brazil.

This dry deciduous forest in Brazil is divisible into the caatinga and cerrado (see chapter 20). The dry tropical forest-savanna complex divides the evergreen tropical forest into two components, the Atlantic rain forest and Amazonas (see chapters 20 and 21).

References

Albuja, L. 1982. *Murciélagos de Ecuador.* Quito: Editorial Escuela Politécnica Nacional.

———. 1991. Mamíferos. In *Lista de vertebrados del Ecuador,* 163–203. Politecnica, 16, no. 3. Quito: Revista de Información Técnico-Científico.

Alho, C. J. R. 1982. Brazilian rodents: Their habitats and habits. In *Mammalian biology in South America,* ed. M. A. Mares and H. H. Genoways, 143–66. Pymatuning Symposia in Ecology 6. Special Publication Series. Pittsburgh: Pymatuning Laboratory of Ecolology, University of Pittsburgh.

Anderson, S. 1985. *Lista preliminar de mamíferos bolivianos.* Cuadernos Academia Nacional de Ciencias de Bolivia, #65. La Paz: Cienc Naturaleza Museo Nacional de Historia Natural Zoología, 3:5–16.

———. 1997. Mammals of Bolivia: Taxonomy and distribution. *Bull. Amer. Mus. Nat. Hist.* 231:1–652.

Anderson, S., B. R. Riddle, T. Yates, and J. Cook. 1993. *Los mamíferos del Parque Nacional Amboró y la región de Santa Cruz de la Sierra, Bolivia.* Special Publications 2. Albuquerque: Museum of Southwest Biology.

Bernardes, A. T., A. B. M. Machado, and A. B. Rylands. 1989. *Fauna brasileira ameaçada de extinção.* Belo Horizonte: Sociedade Editora de Acão.

Carvalho, C. T. de. 1962. Lista preliminar dos mamíferos do Amapa. *Papels Avulsos de Departmento Zoologia: Seção de Agricultura São Paulo, Brasil* 15 (21): 283–97.

Cruz Lima, E. da. 1945. *Mammals of Amazonia.* Vol. 1. *General introduction and primates.* Rio de Janeiro: Museu Paraense Emilio Goeldi de História Natural e Etnografia.

Eisenberg, J. F. 1989. *Mammals of the Neotropics.* Vol. 1. *The northern Neotropics: Panama, Colombia, Venezuela, Guyana, Suriname, French Guiana.* Chicago: University of Chicago Press.

Eisenberg, J. F., and K. H. Redford. 1979. A biogeographic analysis of the mammalian fauna of Venezuela. In *Vertebrate ecology in the northern Neotropics,* ed. J. F. Eisenberg, 31–38. Washington, D.C.: Smithsonian Institution Press.

Emmons, L. H., and F. Feer. 1997. *Neotropical rain-forest mammals: A field guide.* 2d ed. Chicago: University of Chicago Press.

Fonseca, G. A. B. da, G. Herrmann, Y. L. R. Leite, R. A. Mittermeier, A. B. Rylands, and J. L. Patton. 1996. *Lista anotada dos mamíferos do Brasil.* Occasional Papers in Conservation Biology 4. Washington, D.C.: Conservation International.

Fonseca, G. A. B. da, A. B. Rylands, C. M. R. Costa, R. B. Machado, and Y. L. R. Leite, eds. 1994. *Livro vermelho dos mammíferos brasileiros ameaçados de extinção.* Belo Horizonte: Fundação Biodiversitas.

Frailey, C. D., E. L. Lavina, A. Rancy, and J. P. de Souza Filho. 1988. A proposed Pleistocene/Holocene lake in the Amazon basin and its significance to Amazonian geology and biogeography. *Acta Amazonia* 18:119–43.

Hershkovitz, P. 1987. A history of the recent mammalogy of the Neotropical region from 1492 to 1850. *Fieldiana: Zool.,* n.s., 39:11–98.

Honacki, J. H., K. E. Kinman, and J. W. Koeppl, eds. 1982. *Mammal species of the world.* Lawrence, Kans.: Allen Press and Association of Systematics Collections.

Hutterer, R., M. Verhaagh, J. Diller, and R. Podloucky. 1995. An inventory of mammals observed at Panguana Biological Station, Amazonian Peru. *Ecotropica* 1:3–20.

Liu, Kum-bin, and P. A. Colinvaux. 1985. Forest changes in the Amazon basin during the last glacial maximum. *Nature* 318:556–57.

Mares, M. 1992. Tropical deforestation and the myth of Amazonian biodiversity. *Science* 255:976–79.

Mares, M. A., M. R. Willig, K. E. Streilein, and T. E. Lacher Jr. 1981. The mammals of northeastern Brazil, a preliminary assessment (list). *Ann. Carnegie Mus.* 50:81–137.

Moojen, J. 1952. Os roedores do Brasil. *Bibl. Cient. Brasil.* (Rio de Janeiro), ser. A, 2:1–214.

Oliver, W. L. R., and I. B. Santos. 1991. *Threatened endemic mammals of the Atlantic rainforest of southeast Brazil.* Special Scientific Report 4. Jersey: Jersey Wildlife Preservation Trust.

Pacheco, V. H., H. de Macedo, E. Vivar, C. Ascorra, R. Arana-Cardó, and S. Solari. 1995. *Lista anotada de los mamíferos peruanos.* Occasional Papers in Conservation Biology. Washington, D.C.: Conservation International.

Pascual, R. 1996. Late Cretaceous–Recent land mammals: An approach to South American geobiotic evolution. *Mastozool. Neotrop.* 3:133–52.

Paulo Couto, C. de. 1979. *Tratado de paleomastozoologia.* Rio de Janeiro: Academia Brasileira de Ciências.

Pearson, O. P. 1951. Mammals in the highlands of southern Peru. *Bull. Mus. Comp. Zool.* 106 (3): 117–74.

———. 1957. Additions to the mammalian fauna of Peru and notes on some other Peruvian mammals. *Breviora: Mus. Comp. Zool.* 73:1–7.

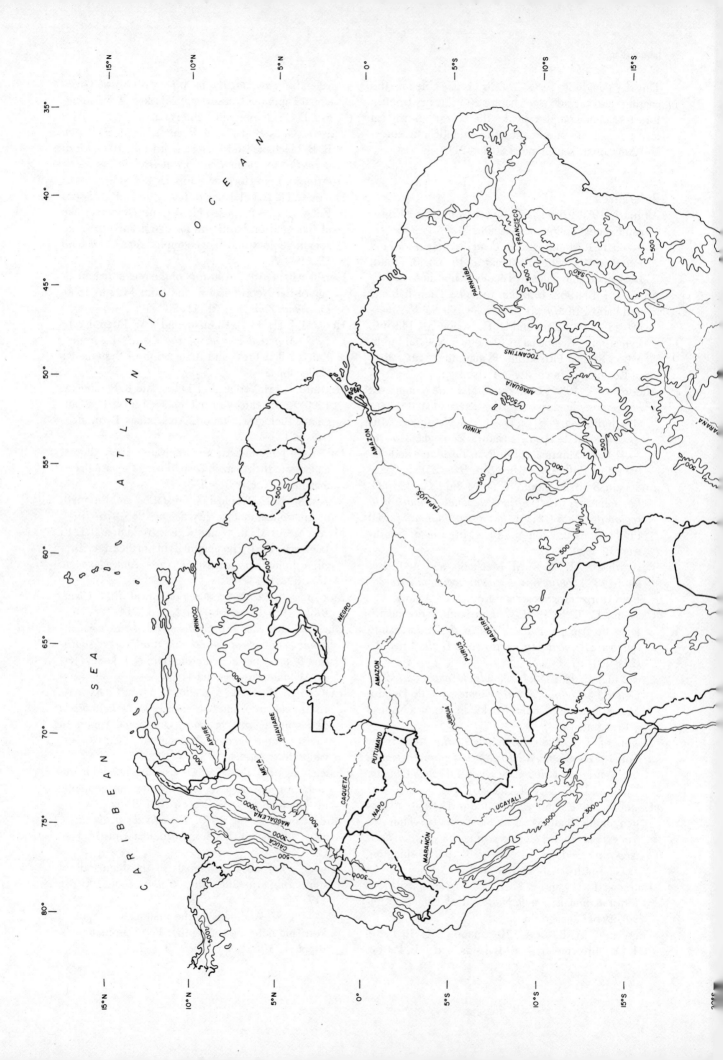

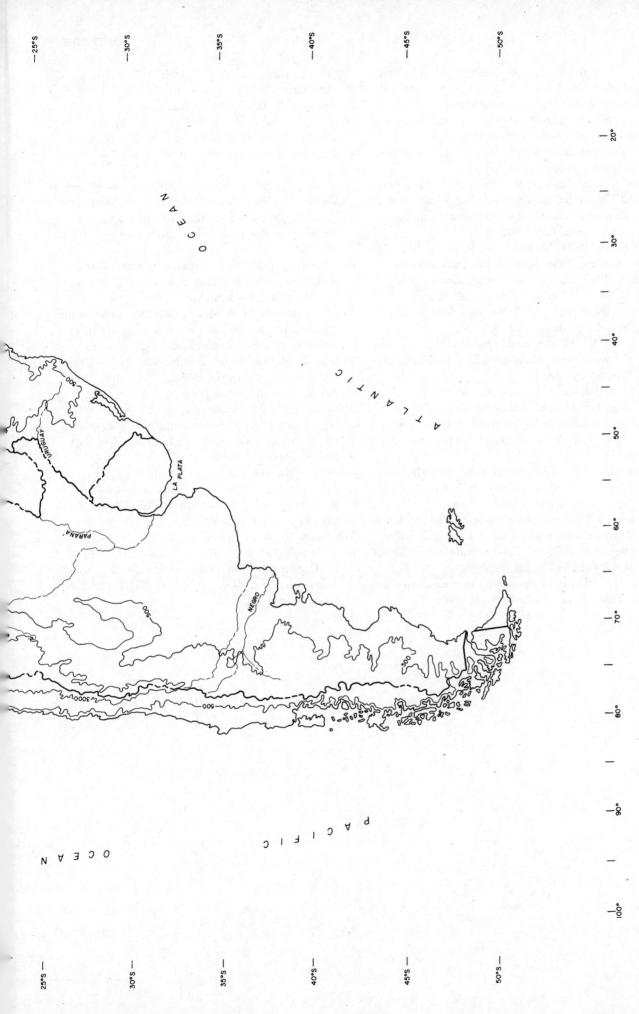

Figure 1.4. Major riverine systems of Central South America.

Pelzeln, A. von. 1871. *Ueber Ornithologie brasiliens: Resultate von Johann Natterer's Reisen in den Jahren 1817 bis 1835.* Vienna: Pichler.

———. 1883. *Brasilische Säugetiere: Resultate von Johann Natterer's Reisen in den Jahren 1817 bis 1835.* Verhandlungen 33. Vienna: Kaiserlich-königlichen zoologisch-botanisch Gesellschaft.

Peterson, N. E., and R. H. Pine. 1982. Chave para identificação de mamíferos da região amazonica brasileira com exceção dos Quirópteros y primates. *Acta Amazonica* 12 (2): 465–482.

Pine, R. H. 1973. Mammals (exclusive of bats) of Belém, Pará, Brazil. *Acta Amazonica* 3:465–82.

Por, F. D. 1995. *The Pantanal of Mato Grosso (Brazil): World's largest wetlands.* Dordrecht: Kluiver.

Pulido Capurro, V. 1991. *El libro rojo de la fauna silvestre del Peru.* Lima: INIAA.

Rancy, A. 1991. Pleistocene mammals and paleoecology of the western Amazon. Ph.D. diss., University of Florida, Gainesville.

Reis, N. R. dos, and A. L. Peracchi. 1987. Quirópteros de região de Manaus, Amazonas, Brasil. *Bol. Mus. Paraense Emilio Goeldi, Sec. Zool.* 3:161–82.

Tovar, A. 1971. Catálogo de mamíferos peruanos. *An. Cient.* 9:18–37.

Townsend, W. 1995. Living on the edge: Sirinó hunting and fishing in lowland Bolivia. Ph.D. diss., University of Florida, Gainesville.

Valqui, M. H. 1995. *Proechimys:* A terrestrial, small rodent community in northeastern Peru. M.S. thesis. Department of Wildlife Ecology and Conservation, University of Florida, Gainesville.

Vieira, C. 1955. Lista remissiva dos mamíferos do Brasil. *Arq. Zool. São Paulo* 8:341–474.

Vizotto, L. Dino, and V. A. Taddei. 1973. Chave para determinação de quirópteros brasileiros. *Bol. Ciênc.* (São José do Rio Preto), n1:72.

Voss, R. S., and L. H. Emmons. 1996. Mammalian diversity in Neotropical lowland rainforests: A preliminary assessment. *Bull. Amer. Mus. Nat. Hist.* 230:1–115.

Voss, R. S., and P. Myers. 1991. *Pseudoryzomys simplex* (Rodentia: Muridae) and the significance of Lund's collections from the caves of Lagoa Santa, Brazil. *Bull. Amer. Mus. Nat. Hist., Biol. Sci.* 206: 414–32.

Wied-Neuwied, M. A. P., Prinz von. 1826. *Beiträge zur Naturgeschichte von Brasilien.* Vol. 2. Weimar: Landes-Industrie-Comtpoir.

Willig, M. R., and M. A. Mares. 1989. Mammals from the caatinga: An updated list and summary of recent research. *Rev. Brasil. Biol.* 49 (2): 361–67.

Wilson, D. E., and D. M. Reeder, eds. 1993. *Mammal species of the world.* 2d ed. Washington, D.C.: Smithsonian Institution Press.

Wing, E. 1986. The domestication of animals in the high Andes. In *High altitude tropical biogeography,* ed. M. Monasterio and F. Vuilleumier. New York: Oxford University Press.

———. 1989. Human use of canids in the central Andes. In *Advances in Neotropical mammalogy,* ed. K. Redford and J. F. Eisenberg, 265–78. Gainesville, Fla.: Sandhill Crane Press.

Woodman, N., R. M. Timm, R. Arana C., V. Pacheco, C. A. Schmidt, E. D. Hooper, and C. Pacheco A. 1991. Annotated list of the mammals of Cuzco, Amazonico, Peru. *Occas. Pap. Mus. Nat. Hist. Univ. Kansas* 145:1–12.

Part 1

Mammalian Faunas
in the
Plio-Pleistocene
of Brazil

2   Isolation and Interchange
A Deep History of South American Mammals

S. David Webb

The study of mammalogy must be placed not only within an appropriate geographic context and an array of environmental conditions, but also in an immense span of geological time. In the case of South American mammals, the time dimension reveals an extraordinarily complex (and improbable) history of physical and biological relationships. Nor is this deep history completely known, for new organisms and new environmental patterns turn up almost annually. Newly discovered evidence of a trans-Amazonian seaway as recent as the middle Miocene now indicates a more divided and more dynamic biotic context than previously imagined for the Amazon basin (Webb 1995).

In this chapter and the next two, three paleontologists review the Cenozoic succession of South American mammal faunas and place them in a broad evolutionary and ecogeographic time frame. Figure 2.1 outlines both the major geological features that helped to shape the South American biota and the immigrant land mammal groups that profoundly affected South American mammal evolution. These immigrants are represented by the classic terminology of Simpson (1940), in which three "faunal strata" denote the successive waves of allochthons that punctuated South American mammal history. Major advances in our understanding of the underlying plate tectonic mechanisms, of the magnitude of the faunal episodes, and of the timing of their action have not altered the utility of Simpson's scheme as the fundamental outline of South American mammal history during the Cenozoic era.

Faunal Stratum 1

Many features of the South American biota go back to the time, more than one hundred million years ago, when South America was still part of the southern supercontinent, Gondwanaland. Although the Cretaceous period obviously predates the age of mammals, its South American terrestrial samples produce some extremely interesting evidence of Mesozoic mammals.

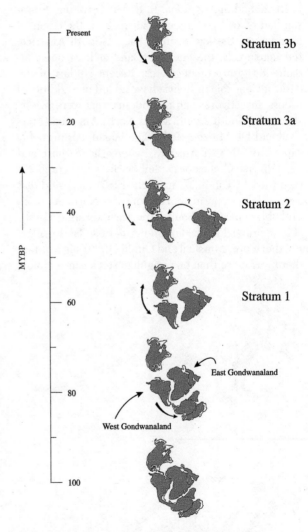

Figure 2.1. Layers of mammalian immigrants in the deep history of South America.

Most remarkable is the recent discovery of an extinct monotreme, clearly dramatizing the southern affiliation with Australia (via Antarctica). Also, the presence of symmetrodonts and a distinct group of multituberculates adds a more cosmopolitan appearance to the

South American Mesozoic mammal fauna than previously known.

Since none of these Mesozoic mammal groups survived, their mention here might be considered esoteric were it not for their geographic implications. The important point geographically is that South American Mesozoic mammals seem to link them both to the Southern Hemisphere and to North America. Monotremes presumably represent an Australian connection, whereas multituberculates and symmetrodonts are probably immigrants from North America. And the same geographic duality continues into the Paleocene. The first faunal stratum of Simpson (1940) is represented by an excellent Paleocene record both in Brazil and in Argentina. In this first stratum, Simpson had recognized three basic stocks as the essential founders of the age of mammals in South America: the marsupials, the edentates, and at least one fundamental group of ungulates. Recent studies of the earliest Cenozoic in Bolivia have added two additional orders, insectivores and pantodonts, that were briefly shared by South America and North America (Marshall and De Muizon 1988). The marsupials probably came from North America, where the earliest and most diverse Cretaceous members are known (e.g., recent work by Cifelli). Gingerich (1985) suggested that several Paleocene mammal groups in North America probably came from South America, indicating a probable reciprocal use of an inter-American land bridge. Furthermore, Marshall and Cifelli (1990) suggest that there was more than one ungulate stock, since almost

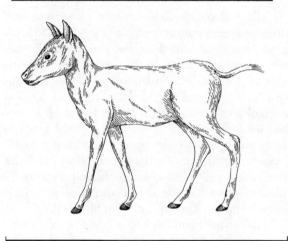

Figure 2.3. Two notable stratum 1 descendants: the litoptern *Thoatherium* (bottom), of early Miocene age, was more precociously monodactyl than even a Pleistocene equid. *Thylacosmilus* (top) was one of the last borhyaenids, with highly elaborate sabers and carnassial teeth.

immediately one finds the following six orders of extinct ungulates: condylarths, litopterns, notoungulates, pyrotheres, astrapotheres, and xenungulates. These diverse elements of the Paleocene fauna imply a far more extensive faunal affiliation with North America than postulated by Simpson (1980).

Faunal Stratum 2

The second faunal stratum of land mammals was more narrowly filtered. Simpson's (1980) work featured ceboid primates and caviomorph rodents as the essential members of stratum 2. Tortoises and phyllostomatid bats may also have reached the isolated continent at that time (Webb 1978). They were present by the beginning of the Deseadan land mammal age, which dates to about 30 million years ago. Recent discovery of an early caviomorph skull in new deposits at Termos del Flaco in Chile suggests a somewhat earlier first appearance, but the exact age requires further clari-

Figure 2.2. Two marsupials derived from stratum 1: *Borhyaena* (top) and *Cladosictis* (bottom) are both early Miocene carnivorous marsupials.

fication (Wyss et al. 1994). The geographic origin of the Oligocene waifs that constitute stratum 2 remains debatable. Neither tropical North America nor equatorial Africa was easily accessible. Most modern systematic studies of rodents and primates favor an African affiliation. This second stratum is the only one that was not reciprocated by South American mammals appearing in North America. That suggests that the land vertebrates of this stratum may have been conducted southward by a "conveyor belt" movement of one or more plates in the Caribbean region during the Oligocene. Such a mechanism would explain the absence of a reciprocal faunal movement (Webb 1985).

Faunal Stratum 3

The third faunal stratum carries with it the story of the great American interchange and comes suddenly only about 2.5 million years ago. There is, however, a prelude to stratum 3, not clearly separated by Simpson (1940), which consists of one or two genera of the family Procyonidae (raccoons). *Cyonasua* of the late Miocene (Huayquerian stage) in South America can be traced to its North American sister genus *Arctonasua*, also of late Miocene age (Baskin 1982). At the same time, more than eight million years ago, two families of South American ground sloths (Mylodontidae and Megalonychidae) reached North America (Webb 1985, 1991). Although this late Miocene episode of reciprocal mammal dispersal was not distinguished by Simpson as a numbered stratum, he was aware that these families had moved before the great American interchange. Now we know that these "heralds" of the great American interchange moved at a much earlier age, it is necessary to distinguish these three taxa as stratum 3a.

By far the most important external influence on South American biotic history since the Paleocene was the great American biotic interchange about 2.5 million years ago. Tectonic forces raised the cordilleran ranges along the Pacific margins of the Americas and ultimately, as part of the same process, formed a land bridge between the continents (now the Isthmus of Panama). The third faunal stratum then walked into South America. It had the most profound bearing on the modern South American mammal fauna not only because it is more recent, but also because it involved a much fuller faunal connection between the two continents. Interactions between the older native fauna and the third faunal stratum from North America were so extensive ecogeographically that they must be considered in some detail. These interactions are the focus of the rest of this chapter and the following two paleontological chapters.

The great American interchange (faunal stratum 3b) introduced seventeen families of land mammals into South America. The diverse groups of herbivores and carnivores that immigrated from North America to South America across the isthmian land bridge had a powerful impact on the South American fauna. Many of the northern taxa successfully spread and diversified in their adopted continent. The details of these diversifications are summarized elsewhere by Pascual and Jaureguizar (1990), by Marshall and Cifelli (1990), and by Webb (1985, 1991). An almost equal number of land mammal families extended their ranges from South America into North America as well, but then they declined rather than increased.

Many families from the north experienced substantial diversification in South America, and one, the Cricetidae, radiated explosively. Among ungulates, Camelidae, Cervidae, Tayassuidae, Equidae, and Gomphotheriidae radiated effectively, even in the face of native forms that may have played comparable adaptive roles (Simpson's "ecological vicars"). The competition between northern and southern ungulates, to the extent that it took place, apparently was settled in favor of the northern taxa (Webb 1976; Simpson 1980). Carnivore families that came with the interchange had no competitors (certainly no mammalian ones), and Canidae, Procyonidae, Felidae, and Mustelidae all evidently exploited new evolutionary opportunities in South America.

Brief History of Cricetidae

By far the most successful family in the third faunal stratum, as measured by its contribution to generic diversity in South America, is the Cricetidae (considered Sigmodontinae in chapter 16). The known record of Cricetidae in South America falls entirely within the past 2.5 million years. Yet the antiquity of the Cricetidae in South America has been repeatedly questioned because of the family's extraordinary diversity and its success in adapting to many environments throughout the continent. For example, one must stand in awe of the remarkable adaptations that Ichthyomyinae demonstrate in Andean streams. How did such unique behavioral and physiological adaptations evolve in just over two million years? Hershkovitz (1972) postulated that cricetids must have arrived in South America by at least the mid-Tertiary to account for their remarkable phylogenetic diversity. This view was countered immediately by paleontologists (Patterson and Pascual 1972), but the debate has continued ever since. In the past two decades paleontologists in both American continents have searched intensely for better records of Cricetidae in critical late Cenozoic deposits.

The evidence of cricetid history in both American continents conforms with that of other members of stratum 3b and can be divided into the following four parts: (1) presence of cricetids in North America during the late Miocene and Pliocene, including several stocks that later appear in South America; (2) absence of cricetids from South America until the latest Pliocene (time of the interchange); (3) presence of small caviomorphs filling roughly equivalent niches in South America during the late Miocene and early Pliocene; and (4) appearance and rapid increase of cricetids in South America during the latest Pliocene and early Pleistocene (at the beginning of the interchange).

1. The early record of cricetids in North America is *Copemys* from the Hemingfordian and more commonly Barstovian (middle Miocene) (Lindsay 1995). By the Clarendonian and Hemphillian (late Miocene), more derived forms such as *Abelmoschomys, Bensonomys, Prosigmodon,* and *Symmetrodontomys* represent probable sister groups of South American genera. The genera broadly resemble oryzomyines and akodontines, but more detailed analyses of the relevant North American fossil cricetids are needed (see Webb 1985; Baskin 1986). The records from Florida, Arizona, and Chihuahua suggest that early differentiation of Neotropical cricetids took place in low temperate latitudes in North America before the interchange. Unfortunately, serious efforts to screenwash for Central American records have been unproductive (Webb and Perrigo 1984).

2. In South America both positive and negative evidence of cricetids must be considered carefully. Because of Hershkovitz's early influence noted above, considerable hope had been held out for discovering mid-Tertiary cricetids. Reig (1978) described some early cricetid records from the base of the cliffs at Mar del Plata in Argentina, referring them to the Montehermosan stage and suggesting that they were from the Miocene epoch. Continuing detailed studies along the coastal cliffs suggests that Reig's specimens actually may have come from the overlying Chapadmalalan stage of the late Pliocene age (Marshall and Sempere 1993). And even for that stage, unfortunately, the absolute time range of the lower part of the Chapadmalal Formation is not known. Above the Chapadmalal, the Uquia Formation definitely contains several cricetid genera, along with other elements of the immigrant fauna. The age of the Uquian is between 2.5 and 1.5 million years (Tonni et al. 1992; Marshall and Sempere 1993).

3. Marshall (1979) introduced a compromise hypothesis in which he postulated cricetids in northern South America during the mid-Miocene. According to this speculative scheme the cricetids were unrecorded heralds of the interchange that "hid" in South America near the equator for about five million years before they reached the south temperate fossil deposits. Marshall had proposed a barrier that restrained the cricetid rodents in equatorial latitudes, although the new evidence for a north-south trans-Amazon seaway suggests that rodents could easily have moved along its western shores. Recently Frailey (1994) has provided direct negative evidence for this hypothesis by describing in the Amazon basin a varied mid-Tertiary rodent fauna, including caviomorph rodents of remarkably small size, but no cricetids. And in the northern Andes, MacFadden, Anaya, and Argollo (1993) dated a rich late Pliocene mammal fauna to a time (between 4.0 and 3.3 million years ago) just before the great American interchange. This Inchasi fauna has other rodents but no cricetids. Meanwhile, Marshall and Sempere (1993) have abandoned the Marshall hypothesis and include Cricetidae as one of the families that walked across the Panamanian land bridge.

4. Taking the fossil evidence at face value, Cricetidae experienced a remarkably rapid radiation. Some of it probably began in low latitudes of North and Central America before the great American interchange, as suggested above. The follow-up in South America was an explosive radiation that spread to all settings throughout the continent. The Uquian records in Argentina, including Reig's material, reflect early stages of the radiation. *Holochilus* from the mid-Pleistocene of Tarija, Bolivia, establishes an early sigmodontine in the Amazon basin (Steppan 1996).

Three Phases of Pleistocene Mammal Evolution

The intricacies of environmental and evolutionary changes that followed the great American interchange may be more readily understood by dividing that history into three successive phases. These are not absolutely distinct, but they help to organize the major patterns of mammalian dispersal and differentiation in South America. The three phases are "savanna corridors," "rain forest corridors" and "latest Pleistocene extinctions."

Savanna Corridors

In the late Pliocene and early Pleistocene, after the establishment of the land bridge, the interchange evidently involved a substantial percentage of the large mammal fauna. The classical studies of the great American biotic interchange emphasized the great number of mammal families that suddenly became widely distributed on both American continents. In North America the Inglis fauna on the west coast of Florida yields a rich land mammal fauna of earliest Pleistocene age. That fauna is remarkable because about 20% of its genera are new immigrants from South America and about 40% represent North American groups that appear at the same time in Argentina (Webb 1976). Thus most of the Inglis mammals represented an inter-American fauna distributed from north temperate into south temperate latitudes. It was also evident from the entire vertebrate fauna that Inglis sampled a coastal scrub habitat, and that the habitat preference of nearly all of the interchange mammals in the fauna was savanna (Webb 1978). This pattern suggests that the interchange was promoted in the late Pliocene and early Pleistocene by a continuous mosaic of open-habitat corridors that crossed through the tropics, especially at high elevations.

Rain Forest Corridors

The middle and late Pleistocene saw the interchange shift emphasis to northward dispersal of lowland equatorial rain forest biota. Continuity of savanna forms through the American tropics was broken in the late Pleistocene. In Central America, for example, such late arriving grazers as *Mammuthus, Bison,* and *Camelops* reached only as far south as Honduras, El Salvador, and Nicaragua (Webb and Perrigo 1984). Similarly in southeastern Brazil, the tropical limits of various south temperate savanna-adapted mammals, such as *Nothrotheriops, Doedicurus,* and *Arctodus,* occur in areas now consisting of cerrado habitats (see chapter 4).

The Amazon large mammal fauna provides a particularly intriguing view of lowland tropical habitats in the late Pleistocene. Most of the large herbivores were mixed feeders, and a few, such as glyptodonts and llamas were almost surely savanna adapted. They may be envisioned as occupying a mobile mosaic of savanna patches among meandering ribbons of gallery forest. The absence of horses may indicate seasonal flooding even in savanna habitats. Seven sites producing such large mammal fauna directly contradict the presence of rain forest refugia widely postulated by studies of modern biota (see chapter 3).

Regional differentiation of many land mammal groups took place in the mid- and late Pleistocene. Cervidae, Canidae, and Cricetidae provide excellent examples of rapid radiations throughout the American tropics in lowland and upland settings. A pattern characteristic of several extinct large mammal groups involves divergence between trans-Andean and cis-Andean taxa. The South American ungulate genera *Toxodon* and *Mixotoxodon,* the giant sloth siblings *Megatherium* and *Eremotherium,* and the Neotropical proboscideans *Notiomastodon* and *Cuvieronius* each represent a south-north vicariance event produced by an Andean barrier that disrupted a more continuous range in the early Pleistocene.

Latest Pleistocene Extinctions

The demise of about 75% of the American tropical mammal fauna at the end of the Pleistocene destroyed a marvelous megafauna, including both North American and South American groups. The loss of vast herds of herbivores greatly reduced the effects of grazing and burning on equatorial forest succession. Even northern families that had succeeded in diversifying during the early part of the interchange—for example, Equidae and Gomphotheriidae—were now totally lost. Other families lost some but not all genera; for example, among the Felidae, only the saber-toothed cat *Smilodon* was lost. A remarkable "Lazarus" genus is the tayassuid *Catagonus,* thought to have become extinct but then discovered still living in the Chaco (Wetzel 1977). All in all, the Pleistocene extinctions had a devastating effect on the large mammal fauna of South America established mainly during the late Pliocene and early Pleistocene active phases of the great American interchange. The net loss was about 75% of the large mammal genera in South America (Webb and Rancy 1996).

The cause of these losses was probably a combination of abrupt climatic change associated with the glacial termination and the impact of human hunting and clearing. In some respects the loss of most large herbivores may have opened opportunities for other small herbivores. The widespread loss of heavy grazers and browsers surely must have radically reduced their earlier impact on opening forests and distributing seeds, although the exact details may never be precisely known.

Conclusion

Although the age of mammals in South America was largely an era of isolation, the rare sets of immigrants, conveyed covertly into the continent by the mysterious modalities of plate tectonics, were supremely important. The three faunal strata punctuated and reoriented the stately progression of endemic land mammal evolution.

In the early studies of Simpson, the outlines of South American mammal evolution were clearly set forth at the family level. In later studies the details were filled in to essentially the generic level. This made it possible to give a more quantitative view of the impact of each successive faunal stratum. Improved chronological control also refined this history. The multilayered origin of South America's modern mammal fauna can be counted in terms of generic percentages. Webb and Marshall (1982) calculated these figures, and they were refined by Marshall and Cifelli (1990) as follows:

Stratum 1 (Paleocene): 17%
Stratum 2 (Oligocene-Miocene): 29%
Stratum 3a: 3%
Stratum 3b: 51%

As might be expected, the most important contributions are the latest. Even so, it is astonishing to recognize how much of the present fauna is recently imported from North and Central America. The improved chronology that places the beginning of the great American interchange at only about 2.5 million years ago makes this conclusion even more surprising. More than half of the mammalian genera now living in South America came from North America since the late Pliocene.

This conclusion sharply focuses the attention of mammalogists on the Quaternary events in South America that brought about such an extraordinary distribution and diversification. The paleontological record of Cricetidae, as shown above, clearly conforms to the standard history of stratum 3b (the great American interchange), but this merely heightens one's need to unravel this family's explosive Quaternary history in South America. Much the same urgency attends the Quaternary records of other interchange families, such as Canidae, Cervidae, and Felidae. The great extinctions at the end of the Quaternary move almost like an antistrophe in a Greek chorus, whereby the loss of many of the larger players simply dramatizes the continued spread and diversification of the smaller players.

The next two chapters represent very important contributions to this urgent effort to comprehend the outlines of mammalian evolution in South America during the Quaternary. It is arresting to realize how regularly one finds a very high degree of difference between the late Quaternary fauna and the Recent fauna in a given place. Rancy (chapter 3) shows that even in the Amazon basin radical change was the norm, and Cartelle (chapter 4) reveals dramatic differences between the fossil and Recent faunas of the cerrado and caatinga. Because the eastern Brazilian cave faunas are so extraordinarily rich, Cartelle has been able to carry their analysis to the species level and to address the origins and geographic movements of many of the modern species.

A special appreciation is owed to our two Brazilian colleagues for their profound contributions to this improving Quaternary record and its analysis. The tropical regions in which they have labored are generally regarded as unproductive at best and dangerous at worst. Raup was reflecting a broad consensus when he reported to the U.S. National Academy of Sciences that "the fossil record in present-day rain forest areas is notoriously poor because of the paucity of good rock exposures from which collections can be made" (Raup 1988, 56). Rancy and Cartelle have overcome these difficulties to produce valuable new evidence of mammalian evolution in the heart of their continent. Their work is leading to a much fuller understanding of the great radiation and redistribution of land mammals that ultimately produced the living fauna of South America.

References

Baskin, J. A. 1982. Tertiary Procyoninae (Mammalia: Carnivora) of North America. *J. Vert. Paleontol.* 2:71–93.

———. 1986. The late Miocene radiations of Neotropoical sigmodontine rodents in North America. *Contrib. Geol. Univ. Wyoming,* spec. pub. 3:287–313.

Frailey, C. D. 1994. Size estimates of small caviomorph rodents from the late Miocene of the Amazon basin with comments on the niche utilization and the cricetid invasion. *J. Vert. Paleontol.* 2:71–93.

Gingerich, P. D. 1985. South American mammals in the Paleocene of North America. In *The great American biotic interchange,* ed. F. G. Stehli and S. D. Webb, 123–37. New York: Plenum Press.

Hershkovitz, P. 1972. The Recent mammals of the Neotropical region: A zoogeographic and ecological review. In *Evolution, mammals, and southern continents,* ed. A. Keast, F. C. Erk, and B. Glass, 311–431. Albany: State University of New York Press.

Lindsay, E. H. 1995. *Copemys* and the Barstovian/ Hemingfordian boundary. *J. Vert. Paleontol.* 15: 357–65.

MacFadden, B. J., F. Anaya, and J. Argollo. 1993. Magnetic polarity stratigraphy of Inchasi: A Pliocene mammal-bearing locality from the Bolivian Andes deposited just before the great American interchange. *Earth Planet. Sci. Letters* 114:229–41.

Marshall, L. G. 1979. A model for paleobiogeography of South American cricetine rodents. *Paleobiology* 5:126–32.

Marshall, L. G., and R. L. Cifelli. 1990. Analysis of changing diversity patterns in Cenozoic land mammal age faunas, South America. *Palaeovertebrata* 19:169–210.

Marshall, L. G., and C. DeMuizon. 1988. The dawn of the age of mammals in South America. *Nat. Geogr. Res.* 4 (1):23–55.

Marshall, L. G., and T. Sempere. 1993. Evolution of the Neotropical Cenozoic land mammal fauna in its geochronologic, stratigraphic, and tectonic context. In *The biotic relationships between Africa and South America,* ed. P. Goldblatt, 329–92. New Haven: Yale University Press.

Pascual, R., M. Archer, E. O. Juareguizar, J. L. Prado, H. Godthelp, and S. J. Hand. 1992. First discovery of monotremes in South America. *Nature* 356: 704–5.

Pascual, R., and E. O. Juareguizar. 1990. Evolving limates and mammal faunas in Cenozoic South America. *J. Hum. Evol.* 19:23–60.

Patterson, B., and R. Pascual. 1972. The fossil mammal fauna of South America. In *Evolution, mammals, and southern continents,* ed. A. Keast, F. C. Erk, and B. Glass, 247–310. Albany: State University of New York Press.

Raup, D. M. 1988. Diversity crises in the geological past. In *Biodiversity*, ed. E. O. Wilson, 51–57. Washington D.C.: National Academy Press.

Reig, O. A. 1978. Roedores cricetidos del Pliocene superior de la Provincia de Buenos Aires (Argentina). *Publ. Mus. Mun. Cienc. Nat. Mar del Plata "Lorenzo Scaglia"* 2 (8): 164–90.

Reig, O. A. 1980. A new fossil genus of South American cricetid rodents allied to *Wiedomys* with an assessment of the Sigmodontinae. *J. Zool.* (London) 192:257–81.

Simpson, G. G. 1940. Mammals and land bridges. *J. Wash. Acad. Sci.* 30:137–63.

———. 1980. *Splendid isolation: The curious history of South American mammals.* New Haven: Yale University Press.

Steppan, S. J. 1996. A new species of *Holochilus* (Rodentia: Sigmodontinae) from the middle Pleistocene of Bolivia and its phylogenetic significance. *J. Vert. Paleontol.* 16:522–30.

Tonni, E. P., M. T. Alberdi, J. L. Prado, M. S. Bargo, and A. L. Cione. 1992. Changes of mammal assemblages in the Pampean region (Argentina) and their relations with the Plio-Pleistocene boundary. *Paleogeogr., Palaeoclimatol., Palaeoecol.* 959:179–94.

Webb, S. D. 1976. Mammalian faunal dynamics of the great American interchange. *Paleobiology* 2: 216–34.

———. 1978. A history of savanna vertebrates in the New World. 2. South America and the great American interchange. *Ann. Rev. Ecol. System.* 9:393–426.

———. 1985. Late Cenozoic dispersals between the Americas. In *The great American biotic interchange,* ed. F. G. Stehli and S. D. Webb, 357–86. New York: Plenum Press.

———. 1991. Ecogeography of the great American interchange. *Paleobiology* 17:226–80.

———. 1995. Biological implications of the middle Miocene Amazon seaway. *Science* 269:361–62.

Webb, S. D., and L. G. Marshall. 1982. Historical biogeography of Recent South American land mammals. In *Mammalian biology in South America,* ed. M. A. Mares, and H. H. Genoways, 39–52. Pymatuning Symposia in Ecology 6. Special Publications Series. Pittsburgh: Pymatuning Lab of Ecology, University of Pittsburgh.

Webb, S. D., and S. C. Perrigo. 1984. Late Cenozoic vertebrates from Honduras and El Salvador. *J. Vert. Paleontol.* 4 (2): 237–54.

Webb, S. D., and A. Rancy. 1996. Late Cenozoic evolution of the Neotropical mammal fauna. In *Evolution and environments in tropical America,* ed. J. Jackson, N. Budd, and A. Coates, 335–58. Chicago: University of Chicago Press.

Wetzel, R. 1977. The Chacoan peccary *Catagonus wagneri* (Rusconi). *Bull. Carnegie Mus. Nat. Hist.* 3:1–36.

Wyss, A. R., J. J. Flynn, M. A. Norell, C. C. Swisher III, M. J. Novacek, M. C. McKenna, and R. Charrier. 1994. Paleogene mammals from the Andes of central Chile: A preliminary taxonomic, biostratigraphic and geochronologic assessment. *Amer. Mus. Novitat.* 3098:1–31.

3

Fossil Mammals of the Amazon as a Portrait of a Pleistocene Environment

Alceu Rancy

The present is surely the key to the past, but one may assert also with good reason, that the past is the key to the present
S. David Webb, 1977

Introduction

During the past two decades, the Amazon has become the focus of intense scientific interest because of its high biodiversity. In the same period the Pleistocene forest refugia hypothesis was developed and was largely accepted as the best way of explaining the high biodiversity of the rain forest biota, with several of the refugia provisionally located within the Amazon (Haffer 1969). Notwithstanding the recommendation by Haffer (1981, 411) that "forest and nonforest refugia should not be identified only on the basis of biological data such as centers of endemism, but rather on the basis of geoscientific data," the Pleistocene forest refugia model was based primarily on the distribution of living plants, reptiles, butterflies, and birds (Whitmore and Prance 1987).

In this context the obvious advantage of studying fossil mammals is that they directly sample past environments. On the other hand, mammals, particularly extant ones, do not precisely represent the structure of the surrounding vegetation in the sites of their deposition. "Classifying living mammals into categories according to the utilization of certain environmental substrates and according to dietary preferences is fraught with difficulty," as Eisenberg testifies (1981, 247).

This chapter deals mainly with the Pleistocene mammalian fauna of the western Amazon (Peru, Ecuador, and Brazil) and its important contribution to understanding the Amazon biotic history (fig. 3.1).

In order to develop reasonable inferences about mammal habitats, and in estimating the environments of Pleistocene mammal assemblages, I will apply four

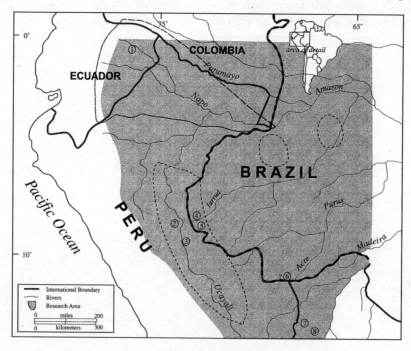

Figure 3.1. Areas of Pleistocene savanna mammals in the western Amazon: (1) Napo River (Ecuador); (2–3) Ucayali River (Peru); (4–5) Juruá River (Brazil); (6) Acre River; (7–8) region of Bolivia.

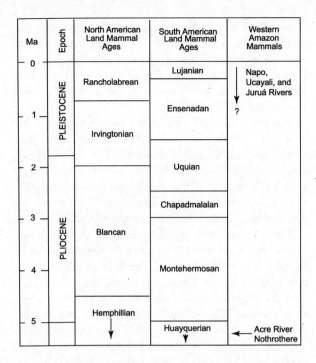

Ma	Epoch	North American Land Mammal Ages	South American Land Mammal Ages	Western Amazon Mammals
0 —	PLEISTOCENE	Rancholabrean	Lujanian	Napo, Ucayali, and Juruá Rivers
1 —		Irvingtonian	Ensenadan	?
2 —	PLIOCENE	Blancan	Uquian	
3 —			Chapadmalalan	
4 —			Montehermosan	
5 —		Hemphillian	Huayquerian	Acre River Nothrothere

Figure 3.2. South American and North American land mammal ages and the late Cenozoic fauna of the western Amazon (after Marshall et al. 1984).

kinds of evidence for each Pleistocene mammal taxon: present ecology of living species or closely related living species; authoritative opinions of other paleontologists based on other sites; morphological adaptations insofar as they suggest a preferred environment (for example, "grazing dentition"); and analogous taxa and analogous assemblages from other continents. Mammal temporal distributions with North and South American equivalents are portrayed in figure 3.2.

The Mammal Palaeofauna

Edentata/Pilosa

The giant ground sloth *Eremotherium* of the Megatheriidae ranged from tropical South America to the southern United States (Kurtén and Anderson 1980; Paula Couto 1979) and is relatively common in Pleistocene sediments of Central America as well as in the state of Florida (USA), as attested by the magnificent collection of the Florida Museum of Natural History.

Eremotherium, along with its sister taxon *Megatherium*, was the largest ground sloth and weighed about four tons. As shown first by Owen (1842), this animal used its strong tail and powerful hind limbs to maintain a bipedal position and grasped tree branches with its robust front claws while browsing. The quadrupedal posture supposedly was used to walk and to forage. Its huge, ever-growing (hypselodont) teeth with two

transverse shearing crests per molar clearly provided an effective chopping mill for leaves and other plant matter.

The huge *Eremotherium* was described from the Juruá River (Brazil) by Paula Couto (1956), Rancy (1981), and Simpson and Paula Couto (1981). Hoffstetter (1952) reported the genus from Napo River (Ecuador), and Marshall et al. (1984) recorded its presence in the area of the Ucayali River (Peru).

The Mylodontidae are well represented in the Pleistocene American fauna, ranging from Patagonia in South America to the state of Washington in the United States (Kurtén and Anderson 1980). Of all the ground sloths, this group is considered to be the most adapted to grazing (Coombs 1983; Salmi 1955). This is particularly evident in the genus *Lestodon*, with its wide symphyseal area, presumably indicating its ability to take large mouthfuls of grass. Such an adaptation is characteristic of relatively indiscriminate grazers such as the white or square-lipped rhinoceros (*Cerathotherium simum;* Owen-Smith 1988). The Mylodontidae in the western Amazon are represented by three (possibly four) genera, from the Juruá and Napo Rivers. *Glossotherium* was described by Paula Couto (1983b), Rancy (1981), and Simpson and Paula Couto (1981). *Lestodon* was cited by Paula Couto (1956), although Cartelle (1989) considered the genus restricted to southern Brazil. Paula Couto (1956, 1983b) and Simpson and Paula Couto (1981) recorded the presence of *Scelidotherium* from the Juruá River. *Mylodon* was reported by Spillmann (1949) from the Napo River, but Hoffstetter (1952) denied its presence in Ecuador and regarded all the materials as *Glossotherium*. Whether or not this fourth genus is confirmed to occur in the western Amazon, the Mylodontidae were well represented in the area, leading to a strong paleoecological inference that there was extensive open landscape in the region during some significant interval of the late Pleistocene (Webb 1978).

Sloths of the family Megalonychidae were described by Simpson and Paula Couto (1981) based on materials (housed at the American Museum of Natural History, New York) from the Juruá River region in Brazil. *Megalonyx* "is known mainly from the Pleistocene of North America, where it ranged as far north as Alaska, but it has also been found in the Amazon basin," as noted by Simpson (1980, 90).

The presence of *Nothropus* in the western Amazon, as supposed by Frailey (1986), would have been extremely important because the genus is closely related to *Nothrotheriops,* a taxon from arid habitats of North America, with direct evidence of its diet from cave dung (Hansen 1978). Further investigation shows that Frailey (1986) misidentified this genus; instead, it has been clearly demonstrated that this nothrothere is a new Miocene (Huayquerian) taxon related to the genus *Pronothrotherium* (Rancy 1991). This latter genus is Pliocene in origin but implies a savanna habitat.

Edentata/Cingulata

The distribution and ecology of the extant Dasypodidae are presented by Wetzel in Montgomery (1985). In the western Amazon, the group is represented today by four species: *Dasypus kappleri, D. novemcinctus, Priodontes maximus,* and *Cabassous unicinctus* (Emmons 1997). As fossils, shelled edentates are represented in the Pleistocene of the western Amazon by eight genera. As noted by Simpson (1980, 87), the Glyptodontidae were characterized by uniformly deep and short skulls and jaws; the presence of a descending spikelike process from the cheekbone below the eye; strong, heavily muscled legs, longer behind than in front; and stout, blunt, almost hooflike claws. All were herbivorous and probably ate coarse, abrasive plants such as siliceous pampas grasses. *Glyptodon* was reported from the Napo and Ucayali Rivers by Spillmann (1949) and from Juruá River by Paula Couto (1983a)

Figure 3.3. Two of the ancient native groups that were probably grazers in the Amazon basin. *Toxodon* (bottom) is a large rhino- or hippolike notoungulate. *Glyptodon* (top) is a tanklike ungulate edentate.

(fig. 3.3). Rancy (1981) reported the presence of *Hoplophorus* in the Ucayali River, and Paula Couto (1983a) recognized that genus from the Juruá River. *Euphractus, Propraopus, Dasypus, Neuryurus,* and *Panocthus* were described from the Juruá River by Paula Couto (1956, 1983a).

Pampatherium is a well-known extinct South American genus, along with its North and South American sister taxon *Holmesina* (Edmund 1987). It has a long cranium and at least eight bilobate molariform teeth in each jaw quadrant. These teeth have flat occlusal surfaces, suggesting that pampatheres ate coarse vegetation and were associated with savanna habitats.

Dasypus has two living species in the Amazon, *D. novemcinctus* and *D. kappleri. D. novemcinctus* is broadly adapted to diverse habitats from thorn scrub and savanna to wet forest, and it exhibits an omnivorous diet together with an enormous geographic range from the southeastern United States to central Argentina (Wetzel 1985). On the other hand, *D. kappleri* is a lowland rain forest animal (Emmons 1997). The presence of *Dasypus* in the western Amazon during the Pleistocene (Paula Couto 1983a) is not a good paleoecological indicator because the genus is adapted to such a wide range of environments.

Euphractus was described by Paula Couto (1983a) based on material from the Juruá River (Brazil). Wetzel (1985, 25) considered its optimal habitats to be "savanna, parkland, forest edge and steppe." The presence of *Euphractus* in the western Amazon, as proposed by Paula Couto (1983a), suggests a synchronous occurrence of savanna in that region during the late Pleistocene.

Notoungulata

The Toxodontidae were large mammals and were widely and abundantly represented in South America. The name of the family refers to the strong curvature of their ever-growing (hypselodont) upper cheek teeth. They "may have been partly amphibious, and they seem to have converged adaptively to a limited extent toward both rhinoceroses and hippopotamuses," according to Simpson (1980, 131) (fig. 3.3).

Toxodon, a common Pleistocene genus in the pampas of Argentina, was reported from the Ucayali River (Peru) by Marshall et al. (1984) and Willard (1966) and was also described from the Juruá River (Brazil) by Paula Couto (1982) and Rancy (1981).

Mixotoxodon was reported from the Juruá River (Brazil) by Paula Couto (1982) and Rancy (1981). The identification of *Mixotoxodon,* a typical Pleistocene genus, by Radambrasil (1976, 68) along the Acre River (Brazil-Peru border), needs to be carefully reexamined because our work in the area has revealed only fossils of Miocene age, and other Notoungulata were common at that time. Note that *Mixotoxodon* was described from Pleistocene sites in Argentina, Venezuela, El Salvador, and Honduras by Marshall et al. (1984), Van Frank (1957), and Webb and Perrigo (1984). The presence of *Mixotoxodon* with large, ever-growing (hypselodont) teeth is evidence that it was adapted to eating abrasive grasses and suggests a continuum of open habitat during the late Pleistocene from Argentina to Central America along the eastern side of the Andes (Webb 1978; Webb and Perrigo 1984).

Proboscidea/Gomphotheriidae

The family Gomphotheriidae dispersed from North America during the time of the great American biotic interchange and then became abundant in the Pleistocene of South America, where it is represented by three genera: *Natiomastodon, Haplomastodon,* and *Cuvieronius* (Patterson and Pascual 1972). The latter two genera occur in the western Amazon. *Haplomastodon* is known from the Juruá, Napo, and Ucayali Rivers, and *Cuvieronius* (= *Cordillerion* Osborn, 1926) occurs in the Napo River (Hoffstetter 1952; Marshall et al. 1984; Rancy 1981; Simpson and Paula Couto 1957, 1981; Spillmann 1949; Willard 1966).

As the name implies, mastodons have cusps shaped like mammary glands on their molar teeth. Their large size and broadly adapted masticatory apparatus indicate a wide-ranging association with forest and savanna-like habitats. On the other hand, Owen-Smith (1988, 99) pointed out that "the extinct mammoths and grazing gomphotheres such as *Stegomastodon,* which from their dentition were adapted for a diet of fine grass leaves, most probably resembled the grazing rhinos in having a relatively efficient hindgut fermentation." Janzen and Martin (1982) noted the important role that gomphotheres may have played in dispersing the seeds of tropical palms such as *Sheelea.*

Simpson and Paula Couto (1957) reviewed the abundant and widespread material of gomphotheres from Brazil and concluded that only a single taxon, *Haplomastodon waringi,* lived in the tropical part of Brazil, including the western Amazon. On the other hand, one molar (DGM-541-M) in the collection of DNPM in Rio de Janeiro suggests the presence of *Cuvieronius* in the upper Juruá River as observed by Llewellyn Ivor Price in his field notes (Paula Couto 1976). The most diagnostic difference between the two genera is the spiral distribution of enamel along the tusks of *Cuverionius;* in the absence of tusk material, positive identification of the latter genus remains dubious.

Artiodactyla

During the Pleistocene in South America, the Camelidae of the Artiodactyla were more widespread than today (Paula Couto 1979). Abundant material

of *Palaeolama,* recovered from caves of the caatinga realm of northeast Brazil, is housed in a very well preserved condition, at the Museu de Paleontología of the Pontífica Universidade Católica de Minas Gerais (Brazil). Extant relatives of this genus, *Lama* and *Vicugna,* extended in historic times along the Andean chain from Colombia southward and onto the plains of Patagonia in the southern extremity of South America; *L. guanicoe* extended into the pampas of Argentina as far eastward as Buenos Aires (Franklin 1982; Mares, Ojeda, and Barquez 1989).

Simpson and Paula Couto (1981) and Rancy (1981) reported both *Vicugna* and *Palaeolama* along the Juruá River (Brazil). Both genera constitute the strongest evidence of drier conditions and savanna-like vegetation including grasses in the region during the late Pleistocene. Hershkovitz (1972, 392) pointed out that "the American camelids are hardy and thrive in pastures and climates where introduced domestic cattle cannot live."

Two species of Tayassuidae, *Tayassu pecari* and *T. tajacu,* live at present in the Amazon basin. The family ranges from the southwestern United States to northern Argentina, almost comparable to the distribution of *Dasypus* (Eisenberg 1989). The presence of the genus *Tayassu,* with relatively brachydont dentition, in the Pleistocene of the western Amazon was reported by Rancy (1981), Simpson and Paula Couto (1981), Spillmann (1949), and Willard (1966). There may be an environmental resemblance between the Pleistocene of the western Amazon and the Chaco of Argentina and Paraguay, where the surviving genera *Tayassu* and *Catagonus,* with its more lophodont dentition, are both important elements of the Chacoan biota.

Perissodactyla

All the extant forms of tapirs seem to be browsers and frugivorous. All are tropical but may range to high elevations. *Tapirus pinchaque* habitually is found at elevations exceeding 1,000 m. On the other hand, *T. indicus* is a lowland species, associated with streams and lacustrine environments in southeast Asian forested habitats. Gallery forests may allow *T. terrestris* to penetrate seasonally arid areas in South America, such as chaco and cerrado (John Eisenberg, pers. comm.). Emmons (1997, 174) pointed out that *T. terrestris* is "found in rainforest, gallery forest, and more open grassy habitats with water and dense vegetation for refuge." Kent Redford (pers. comm.) found evidence of the presence of *T. terrestris* in a dry (500 mm per year) area of the chaco in Paraguay.

Tapirs are well represented as fossils in the Pleistocene of the areas of the Ucayali (Peru) and Juruá Brazil) Rivers (Marshall et al. 1984; Rancy 1981; Simpson and Paula Couto 1981). The presence of *Tapirus* in the western Amazon during the Pleistocene fits, in a broad sense, with indications of savanna with riparian vegetation along watercourses in the area.

Conclusion

Table 3.1 lists the genera of Pleistocene mammals known at present from the western Amazon region, with the inferred habitat and diet for each taxon. Many of these same genera ranged into south temperate parts of Argentina, including Patagonia. On the other hand, figure 3.1 shows that the fossil occurrences directly contradict some of the supposed rain forest refugia, particularly the eastern Peruvian and Napo River refugia of Haffer (1969) and Prance (1973). The broad continuity of large mammals from the Amazon into south temperate regions reinforces that many taxa were savanna adapted.

In a review of the Pleistocene mammals and climate of South America, Pascual and Juareguizar (1990) recognized the strong relations between mammal megafauna and environment during the late Pleistocene. A similar conclusion can be drawn for the late Pleistocene of the western Amazon, based on the abundance of pastoral mammals. The contrast with

present conditions, however, is far more extreme in the Amazon than in the open landscape of Brazil and Argentina.

Radambrasil (1976, 1977) documented Pleistocene landforms in the western Amazon that are characteristic of a dry climate with little or no vegetative cover. Damuth and Fairbridge (1970) independently arrived at a similar conclusion (semiarid climate in the Amazon during the Pleistocene) based on their studies of sediments in the Amazon River delta, which consisted of extensive eolian sands.

The megaherbivores vanished and the last extinction supposedly occurred in the western Amazon during the transition from the late Pleistocene to the early Holocene epochs approximately 10,000 B.P.

Some of the mammals described as fossils in the western Amazon continue to live in the area as survivors of the late Pleistocene extinction. Examples are *Tapirus, Tayassu, Dasypus,* and *Eira.* All of these taxa are widely adapted, ranging from rain forest to deciduous forest, cerrado, chaco, and caatinga.

The western Amazon, during the Pleistocene, was dominated by large and extremely herbivorous edentates and ungulates, such as *Toxodon, Haplomastodon, Eremotherium, Mixotoxodon, Pampatherium, Glyptodon,* and *Palaeolama.* When compared with analogous forms in extant and extinct faunas in other parts of the American continents, these genera are recognized as browsers and grazers and are characteristic of savanna habitat.

By the evidence that arises from the vertebrate fossils, the most probable place to support relict forest vegetation and shelter for forest fauna during the dry times of the Pleistocene was the gallery forest along the edges of the major Amazonian rivers.

Acknowledgments

This work was supported by grants from the Conselho Nacional de Desenvolvimento Científico e Tecnológico–CNPq and Universidade Federal do Acre–UFAC (Brazil). I thank John Eisenberg, David Webb, and Bruce MacFadden for the valuable encouragement I received as a graduate student at the University of Florida. Aulio Gelio Alves de Souza, former rector of Universidade Federal do Acre, Brazil, deserves special recognition by his strong support in the beginning of my scientific career. Thanks to my colleagues acreanos Jean Bocquentin, Jonas Filho, and José Carlos Rodrigues dos Santos for their constant companionship at the Laboratório de Pesquisas Paleontológicas.

Table 3.1   Known Pleistocene Mammals
              of the Western Amazon Region

Genus	Habitat	Diet
Edentata/Pilosa		
Eremotherium	Forest edge/savanna	Grass/browse
Ocnopus	Forest edge/savanna	Grass/browse
Glossotherium	Savanna	Grass/browse
Lestodon	Savanna	Grass/browse
Scelidotherium	Savanna	Grass/browse
Mylodon	Savanna	Grass/browse
Megalonyx	Savanna	Grass/browse
Edentata/Cingulata		
Propraopus	Forest edge/savanna	Omnivore
Dasypus	Forest/savanna	Insectivore
Euphractus	Savanna	Omnivore
Pampatherium	Savanna	Grass
Hoplophorus	Savanna	Grass
Neuryurus	Savanna	Grass
Panocthus	Savanna	Grass
Glyptodon	Savanna	Grass
Notoungulata		
Toxodon	Savanna	Grass/low browse
Mixotoxodon	Savanna	Grass/low browse
Proboscidea		
Cuvieronius	Savanna	Grass/browse/fruit
Haplomastodon	Savanna	Grass/browse/fruit
Perissodactyla		
Tapirus	Forest/savanna	Browse/fruit
Artiodactyla		
Vicugna	Savanna	Grass/low browse
Palaeolama	Savanna	Grass/low browse
Tayassu	Forest/savanna	Omnivore/frugivore
Carnivora		
Eira	Forest	Carnivore

References

Cartelle, C. 1989. Nota prévia sobre um novo Mylodontinae (Edentata, Xenarthra) relacionado com

Mylodon darwini Owen, 1839 do Brasil intertropical. *An. Acad. Brasil. Ciênc.* 61:481.

Coombs, M. C. 1983. Large mammalian clawed herbivores: A comparative study. *Trans. Amer. Phil. Soc.* 73:1–80.

Damuth, J. E., and R. W. Fairbridge. 1970. Equatorial Atlantic deep-sea arkosik sand and ice-age aridity in tropical America. *Geol. Soc. Amer. Bull.* 81:198–206.

Edmund, A. G. 1987. Evolution of the genus *Holmesina* (Pampatheriidade, Mammalia). Texas Memorial Museum, *Pearce-Sellards Series* 45:1–20.

Eisenberg, J. F. 1981. *The mammalian radiations: An analysis of trends in evolution, adaptation, and behavior.* Chicago: University of Chicago Press.

———. 1989. *Mammals of the Neotropics.* Vol. 1. *The northern Neotropics: Panama, Colombia, Venezuela, Guyana, Suriname, French Guiana.* Chicago: University of Chicago Press.

Emmons, L. H., and F. Feer. 1997. *Neotropical rainforest mammals: A field guide.* 2d ed. Chicago: University of Chicago Press.

Frailey, C. D. 1986. *Late Miocene and Holocene mammals, exclusive of the Notoungulata, of the Rio Acre region, western Amazonia.* Contributions in Science 374. Los Angeles: Los Angeles County Museum.

Franklin, W. L. 1982. Biology, ecology, and relationship to man of the South American camelids. In *Mammalian biology in South America,* ed. M. A. Mares and H. H. Genoways, 457–89. Pymatuning Symposia in Ecology 6. Special Publication Series. Pittsburgh: Pymatuning Laboratory of Ecology, University of Pittsburgh.

Haffer, J. 1969. Speciation in Amazonian forest birds. *Science* 165:131–37.

———. 1981. Aspects of Neotropical bird speciation during the Cenozoic. In *Vicariance biogeography: A critique,* ed. G. Nelson and D. E. Rosen, 371–412. New York: Columbia University Press.

Hansen, R. M. 1978. Shasta ground sloth food habits, Rampart cave, Arizona. *Paleobiology* 4:302–19.

Hershkovitz, P. 1972. The Recent mammals of the Neotropical region: A zoogeographic and ecological review. In *Evolution, mammals, and southern continents,* ed. A. Keast, F. C. Erk, and B. Glass, 311–431. Albany: State University of New York Press.

Hoffstetter, R. 1952. Les mammifères pléistocène de la république de l'Equateur. *Mém. Soc. Géol. France* 31 (66): 1–391.

Janzen, D. H., and P. S. Martin. 1982. Neotropical anachronisms: The fruits the gomphotheres ate. *Science* 215:19–27.

Kurtén, B., and E. Anderson. 1980. *Pleistocene mammals of North America.* New York: Columbia University Press.

Mares, M. A., R. A. Ojeda, and R. M. Barquez. 1989. *Guide to the mammals of Salta province, Argentina.* Norman: University of Oklahoma Press.

Marshall, L. G., A. Berta, R. Hoffstetter, R. Pascual, O. A. Reig, M. Bombin, and A. Mones. 1984. Mammals and stratigraphy: Geochronology of the continental mammal-bearing Quaternary of South America. *Palaeovertebrata,* mémoire extraordinaire, 1–76.

Montgomery, G. G., ed. 1985. *The evolution and ecology of armadillos, sloths, and vermilinguas.* Washington, D.C.: Smithsonian Institution Press.

Owen, R. 1842. *Description of the skeleton of an extinct gigantic sloth,* Mylodon robustus, *Owen, with observations on osteology, natural affinities, and probable habits of the megatherioid quadrupeds in general.* London: Royal College of Surgeons of England.

Owen-Smith, R. N. 1988. *Megaherbivores: The influence of very large body size on ecology.* Cambridge: Cambridge University Press.

Pascual, R., and E. O. Jaureguizar. 1990. Evolving climates and mammal faunas in Cenozoic South America. *J. Hum. Evol.* 19:23–60.

Patterson, B., and R. Pascual. 1972. The fossil mammal fauna of South America. In *Evolution, mammals and southern continents,* ed. A. Keast, F. C. Erk, and B. Glass, 247–310. Albany: State of New York University Press.

Paula Couto, C. de. 1956. Mamíferos fósseis do Cenozóico da Amazônia. *Bol. Conselho Nacional de Pesquisas* 3:1–121.

———. 1976. Fossil mammals from the Cenozoic of Acre, Brazil. 1. Astrapotheria. *Anais do XXVIII Congresso Brasileiro de Geologia* 1:236–49.

———. 1979. *Tratado de paleomastozoologia.* Rio de Janeiro: Academia Brasileira de Ciências.

———. 1982. Fossil mammals from the Cenozoic of Acre, Brazil. 5. Notoungulata, Nesodontinae (II), Haplodontheriinae, and Litopterna, Pyrotheria, and Astrapotheria (II). *Iheringia,* ser. geol. 7:5–43.

———. 1983a. Fossil mammals from the Cenozoic of Acre, Brazil. 6. Edentata Cingulata. *Iheringia,* ser. geol., 8:33–49.

———. 1983b. Fossil mammals from the Cenozoic of Acre, Brazil. 7. Miscellanea. *Iheringia,* ser. geol., 8:101–20.

Prance, G. T. 1973. Phytogeographic support for the theory of Pleistocene forest refuges in the Amazon basin, based on evidence from distribution patterns in Caryocaraceae, Chrysobalanaceae, Dichapetalaceae and Lecythidaceae. *Acta Amazonica* 3:5–28.

Radambrasil. 1976. *Levantamento de recursos naturais (geologia, geomorfologia, pedologia, vegetação, uso potencial de terra): Folha SC 19 Rio Branco,* vol. 12. Rio de Janeiro: DNPM.

———. 1977. *Levantamento de recursos naturais (geologia, geomorfologia, pedologia, vegetação, uso potencial da terra): Folha SC 18 Javarí/Contamana,* vol. 13. Rio de Janeiro: DNPM.

Rancy, A. 1981. Mamíferos fósseis do Cenozóico do Alto Juruá-Acre. Master's thesis, Curso de Pós-Graduação em Geociências, Universidade Federal do Rio Grande do Sul, Pôrto Alegre.

———. 1991. Pleistocene mammals and paleoecology of the western Amazon. Ph.D. diss., University of Florida, Gainesville.

Salmi, M. 1955. Additional information on the findings in the *Mylodon* cave at Ultima Esperanza. *Acta Geogr.* 14:313–33.

Simpson, G. G. 1980. *Splendid isolation: The curious history of South American mammals.* New Haven: Yale University Press.

Simpson, G. G., and C. de Paula Couto. 1957. The mastodons of Brazil. *Bull. Amer. Mus. Nat. Hist.* 112:125–90.

———. 1981. Fossil mammals from the Cenozoic of Acre, Brazil. 3. Pleistocene Edentata Pilosa, Proboscidea, Sirenia, Perissodactyla and Artiodactyla. *Iheringia,* ser. geol. 6:11–73.

Spillmann, F. 1949. *Contribución a la paleontología del Peru: Una mamifauna fósil de la región del Rio Ucayali.* Publicaciones del Museo de Historia Natural "Javier Prado," Serie C, Geología y Paleontología 1. San Marcos: Universidad Mayor San Marcos.

Van Frank, R. 1957. A fossil collection from northern Venezuela. 1. Toxodontidae (Mammalia, Notoungulata). *Amer. Mus. Novitat.* 1850:1–38.

Webb, S. D. 1977. A history of savanna vertebrates in the New World. 1. North America. *Ann. Rev. Ecol. System.* 8:355–80.

———. 1978. A history of savanna vertebrates in the New World. 2. South America and the great American interchange. *Ann. Rev. Ecol. System.* 9:393–426.

Webb, S. D., and S. C. Perrigo. 1984. Late Cenozoic vertebrates from Honduras and El Salvador. *J. Vert. Paleontol.* 4 (2): 237–54.

Wetzel, R. M. 1985. Taxonomy and distribution of armadillos, Dasypodidae. In *The evolution and ecology of armadillos, sloths, and vermilinguas,* ed. G. G. Montgomery. Washington, D.C.: Smithsonian Institution Press.

Whitmore, T. C., and G. T. Prance, eds. 1987. *Biogeography and Quaternary history in tropical America.* Oxford Monographs on Biogeography 3. Oxford: Clarendon Press.

Willard, B. 1966. *The Harvey Bassler collection of Peruvian fossils.* Bethlehem, Pa.: Lehigh University Press.

4   Pleistocene Mammals of the Cerrado and Caatinga of Brazil

Castor Cartelle

Introduction

This chapter includes a large Brazilian geographic region mainly occupied by cerrado and caatinga ecosystems, although it does not reach all of the Brazilian states where they can be found. In the northeast region I have included Piauí (PI), Ceará (CE), Rio Grande do Norte (RN), Paraíba (PB), Pernambuco (PE), Alagoas (AL), Sergipe (SE), and Bahia (BA) and have excluded Maranhão; in the southeast region I have included Espírito Santo (ES), Minas Gerais (MG), and Rio de Janeiro (RJ) and have excluded São Paulo (SP); and in the central-west region I have included only Goiás (GO). I excluded São Paulo, Mato Grosso do Sul, Mato Grosso, and Maranhão because I consider all these states transition zones (ecotones). Maranhão is actually pre-Amazon; Mato Grosso and Mato Grosso do Sul are a transition between two differentiated ecosystems: Amazon and Pantanal.

There are no mammal paleontological findings in Maranhão and only a few in Mato Grosso, with no specific reports as yet. Apparently the fauna from Mato Grosso is identical to the fauna of this study area. Although there were important paleontological findings in São Paulo from the late Pleistocene (Ameghino 1907; Carvalho 1952; Paula Couto 1973, 1980), I consider the cerrado and caatinga as a transition or buffer zone in relation to the southern Brazilian fauna, markedly influenced by the southern cone fauna (Argentina and Uruguay).

The study area is approximately 2,140,000 km^2, with 88,000 km^2 of remnant *mata atlântica*. Cerrado is a characteristic vegetation formation with a pronounced dry/wet season (IBGE 1988). It can include many variants such as *cerradão* and *campos*. According to Redford and Fonseca (1986) its main constitutive elements are

1. *Campos:* tree coverage mostly or completely absent
2. *Cerrado* sensu stricto: a few deciduous trees or xeromorphic coverage.
3. *Cerradão:* high-density deciduous tree coverage
4. *Floresta de galeria:* gallery forest

The first three vegetation types spread over 80% of the total area of cerrado. In terms of the surface area this is the second largest Brazilian ecosystem. It covers 25% of the country (Joly 1970). The cerrado trees are tortuous, with thick-barked stems and coriaceous (thick) leaves. There are many endemic forms (Joly 1970).

The weather is warm and semihumid, with 1,100 to 2,000 mm of rainfall each year, but there are divergent subsystems. The rain is not evenly distributed during the year; in general 55% falls during three consecutive months. Temperature also varies greatly, depending on the latitude as well as the altitude. On average we can establish a range of 16–39°C (IBGE 1988).

Caatinga spreads in a semiarid region that includes most of the northeastern region of Brazil and also the northern portion of Minas Gerais. Sometimes it presents a forested pattern with trees as tall as 5 to 12 m. Many kinds of caatingas are recognized, such as: *arbustiva* (shrublike trees), *densa* (densely vegetated), *aberta* (sparsely vegetated), and *agrupada* (vegetated in clusters). The mean rainfall varies from 700–1,000 mm on the seashore to 500–700 mm in the interior. From 50% to 70% of the total yearly rainfall is concentrated in the rainy season (two consecutive months). The rain regime is very irregular, however, with large variation from year to year. Floods can be followed by years of drought. The mean temperature during the year is 26°C (IBGE 1988).

History of Brazilian Paleontology

The work of Peter W. Lund, a Danish paleontologist, in Lagoa Santa, Minas Gerais, was the first systematic study of mammalian paleontology in Brazil. Before his work there were only a few scattered findings from the beginning of the nineteenth century (Spix and Martius

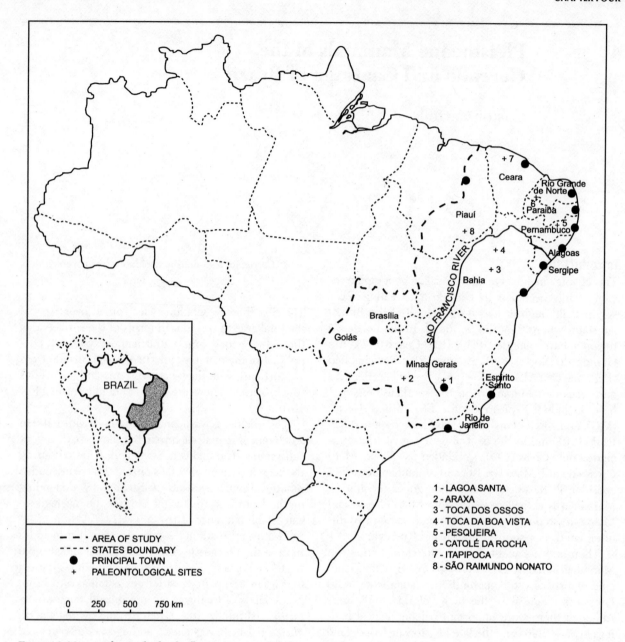

Figure 4.1. Major study sites for fossil collecting in eastern Brazil.

1828; Saint Hilaire 1830). The region Lund chose is a "karst type" area with many caves in the Velhas River valley. This is an affluent of the São Francisco River, which is considered the Brazilian national waterway because of its historical role during the colonization of the Brazilian territory.

Lund worked from 1835 to 1845 and continued to live in the small village of Lagoa Santa until his death in 1880. Between 1845 and 1847 he sent his specimens and notes to Denmark, where they have been kept until the present in the Zoologische Museum of Copenhagen. His findings were used by him and by other Danish researchers such as Boas (1881), Reinhardt (1878), and especially Winge (1888, 1895–96, 1906,

1915), who published a monumental work on late Pleistocene Brazilian mastofauna that is still considered the main reference for this time. After Winge there was a long period when only a few efforts were made (Cope 1885; Holland 1920; Kraglievich 1932).

In the 1940s Carlos de Paula Couto, after studying the southern Brazilian mastofauna (basically related to the Argentine and Uruguayan faunas), started his vast scientific production on this same region. After the 1940s numerous finds were made in the northeast and in Minas Gerais, so Brazilian paleontologists had access to an increasing collection. Most of these finds were in caves associated with limestone beds. Other fossils have been recovered from peculiar places called *caldeirões, cacimbas,* and *tanques.* Such formations present a distinct topographic aspect and origin indicating displacement. According to Oliveira (1989), *caldeirões* are natural depressions in riverbeds caused by turbulence. *Cacimbas* are excavations that were dug in natural depressions by humans in search of subterranean water and were also used to store water during the rainy season. They are very common in arid regions. *Tanques* are the result of natural erosion of crystalline rocks; they are usually deep cavities, oval or elongated, with steep walls as margins. Intermediary fossiliferous layers with sterile basal and top strata can usually be found (Oliveira and Hackspacher 1989). Bigarella, Andrade-Lima, and Riehs (1975) and Oliveira (1989) thought fossils in such environments formed when animals accidentally fell into depressions, since the bones frequently have fractures. I disagree; I have never found complete fossiliferous skeletons in *tanques* unless they were very deep, and I think such deposits were brought there by temporary streams.

We can deduce two things from the observations above: that a set of connected or even isolated caves can represent a reliable sample of the Pleistocene–late Holocene mastofauna from the surrounding region; and that a fluvial network could bring the fauna from a very large territory to the interior of a cave. For example, the Salitre River and its affluents drain an area of 2,000 km^2 in Toca dos Ossos (Ourolândia, Bahia). On the other hand, the fauna collected in *tanques* could be less representative of the regional fauna because of the local drainage area of the temporary streams that supplied them.

The overall faunal list for the cerrado and caatinga is presented in table 4.1. There have been some fundamental changes in the systematics of the Rodentia and Marsupialia since the time of Winge. Preliminary revisions of some taxa were provided by Paula Couto (1946) and by Voss and Myers (1991). I list species rather than just genera because they better reflect the detailed history of this region. This list supersedes paleontological summaries that have been done at genus level (e.g., Marshall et al. 1984). I believe the possible inaccuracies of the species-level analysis are compensated for by the increased precision it provides. Table 4.1 cites recent bibliographical references as well as personal observations.

The age of the Lagoa Santa fauna found by Lund is controversial (Paula Couto 1979; Marshall et al. 1984; Voss and Myers 1991). The main concern is that the Lund excavations did not have a strict stratigraphic control and may include Holocene samples. I address this problem below based on many years of experience in the same caves and similar ones.

In my excavations, the fossils are hardly ever collected under a stalagmitic ground or encrusted by travertine. According to Khoeler (1989), the fossils accumulated under cooler and drier conditions than obtain at present. The carbonate cover would have been formed after the bones were deposited in these caves. Lund frequently referred to the hard "lays" he had to break to reach the fossils, and the results can still be seen in many caves he worked 150 years ago. The material described by Lund and Winge appears synchronous, because in most of the fossiliferous caves we find associated materials of both extinct and living species. These observations are valid for the productive caves in both Minas Gerais and Bahia. The living species that occur with the extinct species in these caves almost surely indicate that they did persist from the Pleistocene into the Holocene.

Most of the fossiliferous beds in Bahia and Minas Gerais are probably synchronous. We can deduce that this vast region experienced the same pluvial factors that caused deposition in many local settings. No older Quaternary beds survived in the region, however, so we cannot know the fauna from earlier periods. The only exception might be the fauna from Toca da Esperança (central Bahia) studied by Lumley et al. (1988), which they estimated as middle Pleistocene by uranium-thorium dating. These authors correlate this fauna with that of Tarija in Bolivia. I reject this hypothesis, and I suspect the date may be marred by contamination of the bones. Faure (1986) and Lalou (1987) also object to this method. Dating errors could derive partly from older limestone adhering to the bones. Their listed species coincide with some local Brazilian intertropical faunas dated from no later than the late Pleistocene.

Biogeography

If we compare the late Pleistocene fauna from Argentina, Uruguay, and Bolivia with the Brazilian fauna, the extinct mammals found in the study areas may be divided into the following groups: intertropical autochthonous species and austral allochthonous species. I consider autochthonous only the species found in the study area (tables 4.2 and 4.4). Some of them may even be endemic. I count as allochthonous those species also reported from Argentina and Uruguay. I disregard the wide-ranging species whose individuals can be found in a large geographical regions (e.g., *Smilodon populator* in the United States, Brazil, and Argentina; Cartelle and Abuhid 1989). These species are listed in tables 4.1 and 4.3.

Intertropical Fauna Extinctions in the Late Pleistocene

I refer to autochthonous and allochthonous species. Most of them are listed in the following sections

Table 4.1   Species List of Late Pleistocene–Holocene Mammals of the Brazilian Cerrado and Caatinga

Species	Ce	Ca	Ma	MG	BA	Other
Marsupialia						
Marmosidae						
Gracilinanus agilis	2	x	x	x	x	
Gracilinanus microtarsus	2			x		
Marmosops incanus	2	—	x	x		
Marmosa grisea				x		
Marmosa pusilla				x		
Marmosa velutina	2	—	x	x		
Micoureus cinereus	2	x	x	x	x	
Monodelphis domestica	2	x	x	x	x	
Caluromyidae						
Caluromys lanatus	3	—	x	x		
Didelphidae						
Philander opossum	2	—	x	x		
Didelphis albiventris	2	x	—	x	x	
Didelphis marsupialis	2	—	x	x		
Lutreolina crassicaudata	2	—	x	x		
Chiroptera						
Emballonuridae						
Peropteryx macrotis					x	
Mormoopidae						
Mormoops aff. *M. megalophylla*					x	
Pteronotus parnelllii					x	x
Phyllostomidae						
Midronycteris megalotis					x	
Tonatia bidens				x	x	
Mimon longifolium				x		
Phyllostomus hastatus				x	x	
Chrotopterus auritus				x	x	x
Glossophaga soricina				x		
Lonchophylla mordax				x		
Anoura caudifer				x		
Carollia perspicillata				x		
Sturnira lilium				x		
Vampyrops lineatus				x		
Chiroderma doriae				x		
Artibeus jamaicensis					x	
Artibeus lituratus				x		
Desmodontidae						
+ ° *Desmodus draculae*						x
Desmodus rotundus				x	x	
Natalidae						
Natalus stramineus						x
Vespertilionidae						
Myotis nigricans				x		
Eptesicus brasiliensis				x	x	
Histiotus velatus				x		
Lasiurus borealis				x		
Lasiurus ega				x		
Molossidae						
Tadarida brasiliensis					x	
Molossops temminckii				x		
Eumops abrasus				x		
Eumops bonariensis				x		
Eumops perotis				x		
Promops nasutus				x		
Molossus molossus					x	
Primates						
Callitrichidae						
Callithrix penicillata	2	—	x	x		
Cebidae						
Callicebus personatus	3	—	x	x		
Alouatta caraya	3	—	—	x	x	

Table 4.1   (continued)

Species	Ce	Ca	Ma	MG	BA	Other
Cebus apella	3	x	x	x	x	
Brachyteles arachnoides	—	—	x	x?		
+ *Brachyteles brasiliensis*				x	x	
Hominidae						
Homo sapiens				x	x	
Edentata						
Megolonichidae						
+ *Nothrotherium maquinense*				x	x	x
+ *"Ocnopus gracilis"*			x			
+ ° *Xenocnus cearensis*						x
Megatheriidae						
+ *Eremotherium laurillardi*				x	x	x
Mylodontidae						
+ *Scelidodon (= Catonyx) cuvieri*				x	x	x
+ ° *Glossotherium* aff. *G. lettsomi*					x	x
+ *Ocnotherium giganteum*				x	x	
+ ° *Mylodopsis ibseni*				x?	x	
Myrmecophagidae						
Myrmecophaga tridactyla	2	x	x	x	x	
Tamandua tetradactyla	2	x	x	x	x	
Dasypodidae						
Tolypeutes tricinctus	2	—	x	—	x	
Euphractus sexcintus	2	x	—	x	x	x
Cabassous unicinctus	1	x	—	x	x	
Dasypus hybridus				x		
Dasypus novemcinctus	2	x	x	x		
Dasypus septemcinctus	2	x	x	x	x	
+ *Propraopus punctatus*				x		x
+ *Propraopus sulcatus*				x	x	
+ *Pampatherium humboldti*				x	x	x
+ *Holmesina paulacoutoi*					x	x
Glyptodontidae						
+ *Hoplophorus euphractus*				x	x	
+ ° *Glyptodon clavipes*				x	x	x
+ ° *Panocthus greslebini*					x	x
+ ° *Panocthus jaguaribensis*						x
Incertae sedis						
+ ° *Valgipes deformis*				x		
Lagomorpha						
Sylvilagus brasiliensis	2	x	x	x		
Rodentia						
Sciuridae						
Sciurus aestuans	3		x	x		
Muridae						
+ *Oryzomys anoblepas*				x		
+ *Oryzomys coronatus*				x		
Oryzomys ratticeps	2			x		
Oryzomys subflavus	2	x	x	x		
Oligoryzomys eliurus	2	x	x	x		
Delomys plebejus				x		
Nectomys squamipes	3	—	x	x		
Rhipidomys mastacalis	2	x	x	x		
+ *Akodon angustidens*				x		
+ *Akodon clivigens*				x		
Akodon cursor	2	x	—	x		
Akodon nigrita	3	—	x	x		
Bolomys lasiurus	2	x	x	x		
+ *Oxymycterus cosmodus*				x		
Oxymycterus roberti	2			x		
Blarinomys breviceps	3	—	x	x		
+ *Juscelinomys (?) talpinus*				x		
Bibimys labiosus	2			x		
Kunsia fronto	2	—	—	x		

Continued on next page

Table 4.1   (continued)

Species	Ce	Ca	Ma	MG	BA	Other
Kunsia tormentosus	2			x		
Calomys callosus	2	x	—	x		
Calomys tener (= laucha)	2	—	—	x		
Holochilus brasiliensis	2	x	—	x		
+ Holochilus molitor				x		
Pseudoryzomys simplex	3	—	—	x		
Erethizontidae						
+ Coendou magnus			x	x?		
Coendou prehensilis	3	—	—	x	x	
Sphiggurus villosus	3	—	x	x	x	
Caviidae						
Galea spixii	1	x	—	x		
Galea flavidens	1	—	—	x		
Cavia aperea	2	—	x	x		
+ Cavia vates				x		
Kerodon rupestris	1	x	—	x	x	
Hydrochaeridae						
Hydrochaeris hydrochaeris	2	x	x	x	x	
+ Neochoerus sulcidens			x	x		
Agoutidae						
+ Agouti laticeps				x		
+ Agouti major				x		
Agouti paca	3	—	x	x	x	x
Dinomyidae						
Tetrastylus walteri				x		
Dasyproctidae						
Dasyprocta azarae	3	—	x	x		
Myocastoridae						
Myocastor coypus				x	x	
Echimyidae						
Proechimys setosus				x		
Euryzygomatomys spinosus?				x		
Clyomys laticeps	1	—	—	x		
Carterodon sulcidens	2	—	—	x		
Trichomys aperoides	2	x	—	x		
Isothrix picta				x		
+ Dicolpomys fossor				x		
Echimys brasiliensis	2	—	—	x		
Kannabateomys amblyonyx				x		
Carnivora						
Canidae						
Lycalopex vetulus	2	—	—	x		
Pseudalopex gymnocercus				x		
Chrysocyon brachyurus	2	—	—	x		
Cerdocyon thous	2	x	x	x	x	
+ Protocyon troglodytes				x	x	x
+ Speothos pacivorus				x		
Speothos venaticus	3	—	x	x	x	
Ursidae						
+ Arctotherium brasiliense				x	x	x
Procyonidae						
Procyon cancrivorus	2	x	x	x	x	
Nasua nasua	2	—	x	x	x	
Mustelidae						
Eira barbara	3	—	x	x	x	
Galictis vittata	2	x	x	x		
Conepatus semistriatus	2	x	x	x	x	
Lontra longicaudis	3	—	x	x		
Pteronura brasiliensis	3	—			x	
Felidae						
Felis concolor	2	x	x	x	x	
Felis jagouaroundi	2	x	x	x	x	
Felis pardalis	3	—	x	x	x	

Table 4.1   (continued)

Species	Ce	Ca	Ma	MG	BA	Other
Felis tigrina	3	—	x	x		
Felis wiedii	3	—	x	x		
Panthera onca	2 ·	x	x	x	x	
+ *Smilodon populator*				x	x	x
Litoptherna						
Macraucheniidae						
+ ° *Xenorhinotherium bahiense*				x?	x	x
Notoungulata						
Toxodontidae						
+ ° *Toxodon platensis*				x?	x	x?
+ ° *Trigonodops lopesi*				x?	x	x
Proboscidea						
+ ° *Haplomastodon waringi*						
Perissodactyla						
Equidae						
+ *Equus (Amerhippus) neogeus*				x	x	x
+ *Hippidion principale*				x	x	x
Tapiridae						
+ *Tapirus cristatellus*				x	x	
Tapirus terrestris	2			x	x	x
Artiodactyla						
Tayassuidae						
Tayassu pecari	2	x	· x	x	x	x
Tayassu tajacu	2	x	x	x	x	x
+ *Brasiliochoerus stenocephalus*				x		
Camelidae						
Lama guanicoe					x	
+ *Palaeolama (= Hemiauchenia) major*				x	x	x
Cervidae						
+ ° *Antifer* sp.				x		
+ ° *Morenelaphus* sp.				x		x
Odocoileus virginianus	2	—	—		x	
Mazama americana	2	x	x	x	x	
Mazama gouazoubira	2	x	x	x	x	x
Blastocerus dichotomus	2	—	—	x		x
Ozotoceros bezoarticus	2	x	—	x	x	x

Note: This table includes the northeast (except Maranhão), southwest (except São Paulo), and central-west (Goiás only) regions of Brazil. We distinguish Minas Gerais (MG), Bahia (BA), and "others." A plus sign indicates extinct species determined through materials collected by Lund. An asterisk after the plus sign indicates an extinct species determined through material not found by Lund in Lagoa Santa. A question mark indicates doubt in relation to the identification validity, species, or identification, or presence in the place marked by x.

The present habitat in the study area is indicated for the living species (except bats), as follows: Ce: cerrado; 1: no forest-dependent species; 2: species that may live in the forest but do not depend on it; 3: forest-dependent species; Ca: caatinga; Ma: *mata atlântica*.

as autochthons or autochthonous, or as allochthons or allochthonous. (See also tables 4.2 and 4.4.)

Autochthonous Fauna

CHIROPTERA

Desmodus rotundus was the only species of the three hematophagous living species found as a fossil in Lagoa Santa and Bahia. I considered *D. draculae* an intertropical autochthon. This species was originally identified in Venezuela (Morgan, Linares, and Ray 1988), but there are also reports from São Paulo (Trajano and Vivo 1991) and Bahia (Cartelle 1992). All *Desmodus* were found in cave beds. The finding from Bahia was articulated with the extinct fauna, but the findings from São Paulo and Venezuela were not. The specimen from São Paulo appears to be recent (a sub-

fossil), but living specimens have not been found in that site region (Ypiranga) (Trajano and Vivo 1991).

PRIMATES

Recently almost complete skeletons of two Atelinae were uncovered at the Toca da Boa Vista (BA). One of them, *Protopithecus brasiliensis*, was previously known only through fragments of humerus and femur in the site Lund studied in Lagoa Santa (Winge 1895–96). Examination of these specimens (Hartwig and Cartelle1996), particularly the skull, indicates that it is related to *Alouatta* and not to *Brachyteles*, as previously suspected (Zingeser 1973). It would have weighed almost twice as much as *B. arachnoides*. The second species (Cartelle and Hartwig 1996) is more closely related to *Ateles* and slightly smaller than

Table 4.2   Composition of the Extinct Intertropical Megafauna of Brazil at the End of the Pleistocene-Holocene in the Study Area

Autochthonous Species (Intertropical)	Allochthonous Species (Primary)
Protopithecus brasiliensis	*Glyptodon clavipes*
Pampatherium humboldti	*Glossotherium* aff. *G. lettsomi*
Holmesina paulacoutoi	+ *Protocyon troglodytes*
Propraopus sulcatus	*Arctotherium brasiliense*
Hoplophorus euphractus	*Toxodon platensis*
Panochtus greslebini	+ *Equus (A.) neogeus*
Panochtus jaguaribensis	+ *Hippidion principale*
Nothrotherium maquinense	*Palaeolama major*
"Ocnopus gracilis"	*Antifer* sp.
Xenocnus cearensis	*Morenelaphus* sp.
Eremotherium laurillardi	
Scelidodon (= *Catonyx*) *cuvieri*	
Ocnotherium giganteum	
Mylodonopsis ibseni	
Valgipes deformis	
Tetrastylus walteri	
Neochoerus sulcidens	
Speothos pacivorus	
Xenorhinotherium bahiense	
Trigonodops lopesi	
Haplomastodon waringi	
Tapirus cristatellus	
Brasiliochoerus stenocephalus	

Note: + indicates possibly allochthonous species.

P. brasiliensis. Both were brachiators and had long, prehensile tails. These forms required closed forest, suggesting the possible expansion of Atlantic forest during the late Pleistocene up to the sites where the fossils were found.

EDENTATA

Megalonychidae

I include in this family the following species: *Nothrotherium maquinense, Ocnopus gracilis, Xenocnus cearensis,* and by convention, *Valgipes deformis,* although this species is poorly known and may be considered *incertae sedis* (Hoffstetter 1954). *N. maquinense* was the first species Lund found in Lagoa Santa. It is considered a member of the Megatheridae by Engelman (1985). Despite this controversy (De Iuliis 1995), I consider it Megalonychidae because of its cranial and limb features. The peculiar shape of its limbs is similar to the living great anteater (*Tamandua bandeira*). It has been reported in São Paulo (Ameghino 1907), Minas Gerais (Reinhardt 1878; Paula Couto 1980), Paraíba (Berqvist 1989), and especially Bahia (Cartelle 1992), where many specimens, including a pregnant female with a very well preserved fetus, were collected in 1993.

I consider *N. maquinense* a peripheral forest tree climber and leaf-eating species. The likely local paleoenvironment and a preliminary analysis of a coprolite containing leaf and fruit tissue deriving from xeromorphic vegetation (Duarte and Souza 1991) corroborate this hypothesis. *O. gracilis* is a taxonomically difficult species. Lund found postcranial skeleton pieces in Lagoa Santa, and a doubtful skull was collected in a limestone cave in Ypiranga, São Paulo (Paula Couto, unpubl. data). *X. cearensis* is also barely known. Paula Couto (1980) considered it an aberrant species of the family Megalonychidae, based on some pieces (especially the astragulus) collected in Ceará. I consider it an endemic species.

Megatheridae

Eremotherium laurillardi is considered a Pan-American species (Cartelle and De Iuliis 1995). There are reports of its occurrence from as far north as New Jersey (New York Museum of Natural History) to Rio Grande do Sul (Toledo 1986). Despite the range, I include it among the intertropical autochthons because of its absence from Uruguay and Argentina. Its large zone of distribution is probably related to its generalized herbivorous feeding habits (grazing and leaf browsing). Many specimens of different ontogenetic ages have been found together by Cartelle and Bohórquez (1982), indicating the gregarious social behavior of this species. Sexual size dimorphism also apparently occurred.

Mylodontidae

The systematics of this family remain poorly defined (Scillato-Yane 1987). I consider Mylodontidae as predominantly grazing and leaf-eating generalized herbivores like *Eremotherium*. *Scelidodon* (= *Catonyx*) *cuvieri* probably includes one of the largest number of synonymous species of any extinct mammal. Hoffstetter (1954) lists twenty-seven synonyms (fifteen of them from Lund), and Cartelle (1992) lists thirty-seven. However, its morphology is clearly known because of its occurrence in Minas Gerais and Bahia. Its cranial morphology and the constant lack of the humeral entepicondylar foramina shows that this species is a peculiar kind of intertropical autochthon clearly differentiated from the other Scelidotheriinae from the southern cone.

Ocnotherium giganteum was described only from two teeth found in Lagoa Santa (Winge 1915, pl. 26). There was some controversy concerning its systematic position. Some authors have identified it as *Glossotherium* (Hoffstetter 1954; Kraglievich 1921). I believe it is a lestodontine because of the presence of divided facets on the calcaneous astragulus within a recently discovered skeleton. It presents an ample diastema between the M_2 and M_3 in what seems to be a unique case among the Edentates (Cartelle 1992). The presence of this lestodontine in Acre (Simpson and Paula Couto 1981) is not certain.

Figure 4.2. *Eremotherium* (top) is one of the largest gravigrade edentates. It ranged northward from the Brazilian cerrado as far as present-day New Jersey. *Macrauchenia* (bottom) is one of the last notoungulates. Best known in Argentina, it is the sister group of the endemic *Xenorhinotherium bahiense.*

Mylodonopsis ibseni has been reported for the first time in Bahia (Cartelle 1991). It is quite similar to *Mylodon darwini,* however, except that its first superior molarlike tooth is atrophied, rather than absent as in *M. darwini.*

Dasypodidae

Four species of extinct armadillos have been identified: *Propraopus punctatus, P. sulcatus, Pampatherium humboldti,* and *Holmesina paulacoutoi.* Paula Couto (1979) considers *P. punctatus* and *P. sulcatus* the same species. I accept Pampatherinae as a valid subfamily (Cartelle and Bohórquez 1985; Cartelle, López de Prado, and Garzon 1989) in disagreement with Edmund (1987) and Engelman (1985). These authors separate *Pampatherium* from the superfamily of the armadillos because of some synapomorphic features of the carapace scutes and the skull. I interpret certain dental features as an adaptation to grazing behavior.

P. humboldti is smaller than *H. paulacoutoi* (Cartelle and Bohórquez 1985). They fall within different genera because of their quite different osteodermal ornamentation and thickness. The geographic distribution of *P. humboldti* could have been much wider than that of *H. paulacoutoi* and must have included Minas Gerais (Winge 1915), Mato Grosso and Piauí (pers. obs.), Ceará (Gomide 1989), Paraíba (Berqvist 1989), and Bahia (Cartelle, López de Prado, and Garzon 1989). Both species must have had similar feeding habits.

Glyptodontoidea

I consider *Panocthus greslebini* and *P. jaguaribensis* autochthonous (endemic) species (Moreira 1971; and Berqvist 1989). *Hoplophorus euphractus* is restricted to Minas Gerais and Bahia (Cartelle 1992).

RODENTIA

Table 4.1 presents some extinct rodents. Some of the species identified may be synonyms of living ones (Paula Couto 1946). A full revision of the material stored in Copenhagen might clarify many problems. Two large species that are well defined in the Lagoa Santa fauna are intertropical autochthons. The big

Tetrastylus species characterizes the Argentine Mio-Pliocene (Paula Couto 1979). *T. walteri* was found in a cave environment in association with the late Pleistocene extinct fauna. It is likely to be a remnant, endemic intertropical species (Paula Couto 1979).

The genus *Neochoerus* had an ample geographic distribution (Moncs 1984). *N. sulcidens* (giant capybara) is a well-defined species with habitats quite different from those of *Hydrochaeris hydrochaeris* (the living capybara). The giant capybara apparently was a specialist grazer of drier habitats. It was probably more vulnerable to predators than the living species. This may be the reason for its early extinction in the cerrado and caatinga.

CARNIVORA

I consider *Speothos pacivorus* endemic because it has been reported only in Lagoa Santa. The identification of *Protocyon troglodytes* is particularly difficult. Two other species of the genus *Protocyon* were identified in South America: *P. orcesi* and *P. scagliarum*. *P. orcesi* was identified based on its mandible and a few other pieces found in Ecuador (Hoffstetter 1952). *P. scagliarum* from Argentina was based on a skull and associated mandible (Kraglievich 1952). Many pieces of *P. troglodytes* have been found in Lagoa Santa, but they were very fragmented (Berta 1988). A complete skeleton of this species has recently been found (Cartelle and Langguth, n.d.). Its dentition exhibits some intraspecific variation relative to the accessory cuspules in the premolars. The proportions of the limbs, which Lund assumed were the same as in *Canis lupus*, are in fact identical to those of *Pseudalopex culpaeus* (= *gymnocercus*). There is some controversy concerning the distinction between *P. troglodytes* and *P. scagliarum*, rejected by Paula Couto (1979) but accepted by Berta (1988). In such a case we might consider the species an allochthon. However, the Argentine specimen belongs to the Lower Pleistocene (Uquian). According to Berta (1988), this species became extinct because of the extinction of its large mammal prey species. I suspect that *P. troglodytes* preyed mainly on the smaller vertebrate species that lived in the cerrado-forest border. The Tayassuidae and Cervidae, for instance, have been found in great quantity in both sites.

LITOPTERNA

Because this group displays great morphological uniformity (Cartelle and Lessa 1988), the scarce and very fragmented pieces found in Brazil were originally identified as *Macrauchenia patachonica*. This is a common species in Argentina and has been also found in Rio Grande do Sul. Subsequently, Cartelle and Lessa (1988) identified *Xenorhinotherium bahiense* as a distinct endemic taxon through the cranial and dental elements of an abundant sample. This species presents the most pronounced regression of the nasal fossae found among the Macraucheniinae.

NOTOUNGULATA

Numerous findings (Cartelle 1992) permitted the clear differentiation between *Trigonodops lopesi* and *Toxodon platensis*. *T. platensis* presents the cranial face (*rostro craniano*), an expanded symphyseal region (*mandíbula sinfisiaria*), and some incisors with an oval section. *T. lopesi* presents a fine mandible including incisors with a triangular section. The species found in this study area may or may not be different from the type of *T. lopesi* found in Acre. At present no morphological features distinguish it from the few teeth from Acre (Paula Couto 1956; Van Franck 1957). Notwithstanding its taxonomic status, *T. lopesi* is here considered an intertropical autochthon and probably a grazer. It is clearly distinct from *T. platensis*.

PROBOSCIDEA

Haplomastodon waringi had a wide distribution in South America. I have considered it an autochthonous species, as I did with *Eremotherium laurillardi*, because it has not been found in the southern cone. It was probably a generalist herbivore, including grass, leaves, and even small scrub branches in its diet.

PERISSODACTYLA

Tapirus terrestris and *T. cristatellus* have been identified in Lagoa Santa, although there has been some controversy concerning the identification of *T. cristatellus* (Paula Couto 1979). Some cranial pieces recently found have confirmed its former identification (Winge 1906). It is little larger than the living species. In addition, when the two species are compared, there are some differences in the shape and orientation of the nasal bones and in the skull morphology. Their dentition is quite similar.

ARTIODACTYLA

Brasiliochoerus (= *Dicotyles*) *stenocephalus* has been reported only from Lagoa Santa. It has a slender (narrow and elongated) skull and is larger than *Tayassu pecari*. Paula Couto (1982) argued that *Dicotyles stenocephalus* should be considered a *nomen nudum*.

Allochthonous Fauna
MYLODONTINAE

The "glossotherids" *qua tale* had been reported only in southern Brazil (Cartelle and Fonseca 1981). As I pointed out above, the specimen Lund found in Lagoa Santa was actually a lestodontine. *Glossotherium lettsomi*, found in Bahia, presents clear sexual dimorphism and great intraspecific variability (Abuhid 1991). The lack of sufficient reports of similar species from Argentina, Uruguay, and Chile prevents us from comparing the two species; perhaps they are the same one. Therefore I decided to consider it an allochthon from the south.

GLYPTODONTOIDEA

Glyptodon clavipes has a wide geographical distribution from Argentina to Ceará in Brazil (Paula Couto 1979). There are reports of *Glyptodon* in Argentina ranging chronologically from the Uquian to the late Pleistocene, whereas in Brazil we have only the late Pleistocene record. Based on these data, I consider it an allochthonous species.

URSIDAE

Arctotherium brasiliense has been reported in Brazil and Bolivia (Paula Couto 1979). In addition to Lagoa Santa (Winge 1895–96; Paula Couto 1960), it was found in Ceará (Trajano and Ferraezzi 1994) and recently in Bahia by Cartelle (1995). I consider it an allochthon, perhaps originating in the Bolivian Andes.

TOXODONTINAE

Toxodon platensis, a species frequently reported in the Lujanian-age deposits of Argentina, has also been recognized in Brazil (São Paulo and Bahia) by Cartelle (1992).

EQUIDAE

South American horses are divided into two groups: Equidae *qua tale* and hippidiformes (Hoffstetter 1952). The first includes many species of *Equus* (*Amerhippus*). There is some controversy concerning hippidiformes. Alberdi (1987) considers *Hippidion* the only genus, whereas MacFadden and Skinner (1979) also include *Onohippidium* within this group.

Since some *Equus* species from the southern cone are not well defined, some authors prefer to call them *Equus* (*Amerhippus*) species (Alberdi et al. 1989). The Brazilian species is considered not well defined by MacFadden and Azzaroli (1987). Some recent findings clarify some aspects relating to its dentition, skull, mandible, and limbs. It is not clearly distinguishable from the other southern species (Boule and Thevenin 1920). Therefore I consider it an allochthon.

Lund described two species from Lagoa Santa using differences in molar size. Based on one skull and many teeth, Cartelle (1992) suggested the only species was *Hippidion principale.* Paula Couto (1979) suggested that the Brazilian species (*H. arcides*) and the Argentine (*H. bonaerense*) species could be the same. The classification of the other southern species presents the same difficulty (Alberdi, Menegaz, and Prado 1987). *H. principale* has nomenclatural primacy. Based on that, I consider the Brazilian hippidiforme a possible allochthon.

CAMELIDAE

Lagoa Santa produced an excellent sample of *Palaeolama major.* According to Cabrera (1935), most of these findings belong to young individuals. Based on pieces from more than fifty individuals, Cartelle (1992) concluded that *P. major* presented a limb proportion similar to that of the other larger "llamas." Thus *P. major* and *P. paradoxa* of Argentina could be synonymous.

According to Cabrera (1935), the morphology and size of the Argentine specimens are quite similar to those found in Bahia. Webb (1974) considered *Hemiauchenia* a valid South American genus based on its longer limbs in relation to the other "llamas." However, the information given by Winge (1906) is distorted because it is based on only a few specimens, mixing both adults and young. Cartelle (1992) recalculated the data from Webb (1974) employing the new material from Bahia. His values were identical to those of the Ecuadorian species (Hoffstetter 1952) and the living species of *Lama* (Webb 1974). Consequently, considering the limb proportions, no adaptive form specific for the lowlands of South America has been confirmed (Hoffstetter 1952). Based on body proportions and morphology, the Brazilian and Argentine specimens are considered conspecifics. *P. major* probably originated in the south. Its presence in Bolivia corroborates this assumption (Marshall et al. 1984).

CERVIDAE

Morenelaphus sp. was found in Minas Gerais and Pernambuco and identified by Vidal (1946) as *Hippocamelus* sp. In both sites the findings consisted of incomplete pieces (Cartelle 1989). The age of the genus ranges from Uquian to Lujanian (Marshall et al. 1984). Magalhães (pers. comm.) has recently identified material from Lagoa Santa in the Museu Nacional do Rio de Janeiro as *Antifer* sp. Both genera have been reported before only in the southern cone, making it reasonable to assume that they originated there.

Some Surviving Fossil Species

The morphology, habits, and habitat of *Tolypeutes tricinctus* are barely known (Wetzel 1985). It has been found in the caatinga of central and northern Bahia as well as in other states with open ecosystems. *Tolypeutes* are also found within the late Pleistocene boundary (Cartelle 1992), suggesting a long occurrence in this region. I have identified its osteoderms from late Pleistocene in Rio Grande do Norte.

Cartelle and Lessa (1988) reported pieces from three individuals of *Myocastor coypus* in Bahia, establishing the extreme northern limit of this species. It has also been reported by Lund in Lagoa Santa (Winge 1888). After the extinction of the megafauna this species remained in its original southern territory.

Cartelle (1994) surprisingly reports *Lama guanicoe* in 1993 in Toca da Boa Vista (Campo Formoso, Bahia). He based his identification on the following reliable pieces: mandible fragment with the symphysis, a superior molar, vertebrae, and limb bones. The unexpected presence of *Lama*, a typical cold-climate species, indicates the presence of allochthonous species from the southern region of the continent.

Odocoileus virginianus occurs in Brazil only in Roraima north of the Rio Negro (Cabrera 1960). Some late Pleistocene postcranial and dental pieces, besides two complete antlers, have been found in Ourolândia (Bahia). This site corresponds to the extreme southern limit of its distribution and indicates that the species is at present withdrawn from its original distribution to the far north of Brazil.

Paleoclimatic Data from the Southern Late Pleistocene

Climatic Implications

Apparently the South American Quaternary weather changes followed the global tendencies. More detailed bio- and chronostratigraphic regional data appear to present slight differences from the Southern Hemisphere. Paula Couto (1975), Bigarella, Andrade-Lima, and Riehs (1975), and Ab'Saber (1977) made global analyses for the Brazilian Pleistocene.

My analysis below is concerned with the mammalian fauna from the late Pleistocene to the early Holocene. The continuing weather changes detected during the Quaternary must have persisted through this short period; peripheral weather has clearly influenced this study area. There are reports of glaciation advances from 13,000 to 10,000 B.P. (Wright 1984; Seltzer 1990). This pattern occurred in Colombia (Schubert 1974), Venezuela (Salgado-Laboriou et al. 1988), Ecuador (Clapperton 1987), and the Peruvian and Bolivian Andes, where there were both glacial advances and retreats (Mercer and Palácios 1977; Wright 1984; Seltzer 1990), and it would have influenced the water level of the Bolivian plateau lakes (ORSTOM 1987).

In the southern Andes (especially in Patagonia) the latest ice advance corresponds to the late glacial period of the Northern Hemisphere (Rabassa and Clapperton 1990). A considerable ice retreat is reported at approximately 10,000 B.P. (Rabassa 1991). However, there is some controversy concerning the ice advance that could have started in that region about 12,000 B.P. (Mercer 1976; Markgraf 1980; Ashworth and Hoganson 1991; Heusser 1987). South American lowlands present great diversity because of their complex geomorphological and climatic evolution (Coltrinari 1992), varying from rain forest to tropical desert (Markgraf 1989). In these areas the weather was cooler and drier before 10,000 B.P. (Markgraf and Bradbury 1982). Paleoclimatic studies are still sparse and not

precise. In short, possibly in the late Pleistocene glaciers covered a larger area of the Andes than today, and the lowlands would also have had a cooler and drier climate (Absy et al. 1989; Siffdine et al. 1991). There are still few conclusions based on data from this study area; the inferred processes are actually guesses.

A group of researchers of the Museo de la Plata (Argentina) have recently carried out some studies concerning the late Pleistocene–early Holocene in the province of Buenos Aires. Some of their conclusions seem to be related to our study areas. A drier and cooler climate occurred in most of the "*bonaerensis*" territory during the late Lujanian compared with the present weather of that region (Tonni 1985; Prado et al. 1987; Alberdi et al. 1989). This situation is probably related to the climatic changes cited above. There was a cooling period in this region at 12,000 B.P. (Fairbridge 1972; Tonni and Fidalgo 1978, 1982). This seems to be the reason for the presence of guanaco within the region at that time (Salemme 1983). Tonni and Politis (1980) believe the further retreat of this species to the south is basically due to climatic changes. After comparing many local late Pleistocene to present faunas, Tonni and Fidalgo (1978) presented the following conclusions about the fauna of the province of Buenos Aires:

1. Species deriving from Patagonia and central Argentina were north and east of their normal distribution zone during the late Pleistocene.
2. In the local studied fauna there were no species derived from subtropical regions. Only the Lujanian local fauna presented some subtropical species. However, all of them were species having a naturally large geographical distribution.
3. The movement of Patagonian and central Argentine fauna to the north and east was due to the decrease of temperature.

The presence of subtropical species is due to the recent postglacial increase in temperature and the decrease of desertification by increased rainfall. Tonni and Laza (1981) presented the same conclusions in relation to the avifauna of Paso Otero (late Pleistocene) in the province of Buenos Aires.

A Model for This Study Area in the Late Pleistocene–Early Holocene

I will not try here to make a global model that could be applied to the entire Pleistocene. Such attempts were made by Ab'Saber (1977) and Brown and Ab'Saber (1979). Neither am I referring to the model of Haffer and Vanzolini concerning refuges (Cerqueira 1982). The analysis of the Brazilian late Pleistocene–Holocene fauna does not allow for the inclusion of those two topics. I insist that when we consider the fauna in question as absolutely synchronous, it reflects essentially a given historical moment.

In the late Pleistocene of the Brazilian intertropical territory there is unquestionable sympatry of autochthonous (intertropical) and allochthonous (from temperate climate) species that has not been discussed. Some authors have not recognized tropical adaptations in species of austral origin (Hoffstetter 1952), but the conclusions of the Argentine researchers seem to support a reasonable explanation for the anomalous sympatry of species, as I indicated above. This is the same climatic development that caused the movement of fauna from higher to lower latitudes, into Brazilian territory. The situation has been described by Prado et al. (1987): "In the late Pleistocene the pampean region presented a biologically inadequate average winter temperature." Webb (1984) noticed a similar effect in North America. Species of *Glossotherium*, *Glyptodon*, *Arctotherium*, *Toxodon*, *Equus (Amerhippus)*, *Hippidion*, *Antifer*, *Morenelaphus*, and *Palaeolama* are examples. *Lestodon* would have moved north up to the state of São Paulo, but not beyond.

The discovery of *Myocastor coypus* in Lagoa Santa and Bahia and of *Lama guanicoe* in Bahia supports such an interpretation. Those two species are not typical of tropical climates. The discovery of allochthonous species in several Brazilian sites seems to support my explanation of the faunal complex during the Brazilian late Pleistocene and early Holocene. The movement of species owing to decreasing temperatures in their original territories turned intertropical Brazil into a refuge for them. It is possible that some species later returned to their original territories.

The paleontological sites mentioned in this studied region can be considered synchronous because they all show the same species (current and extinct). None exhibit a divergent fauna. I have already discussed the only site (central) where the fauna is thought to be from the middle Pleistocene (Lumley et al. 1988) and given my reasons for disagreeing with that chronology. Factors such as rainfall may have caused the deposition of specimens in caves and *tanques* as I described.

Significance of Living Species

Of special significance are species that occurred in the late Pleistocene yet still survive. Redford and Fonseca (1986) noticed that in cerrado regions there is a rich plant endemism as opposed to the low endemism of the mammalian fauna (11% of the species in the area). The surviving species identified as fossils in Lagoa Santa and Bahia indicate a strong dominance of the cerrado, even with the natural caution in the identification made by Winge. Among the current species,

there are representatives of fauna from higher latitudes such as *L. guanicoe* and *M. coypus*. This supports my hypotheses that the presence of an allochthonous fauna in intertropical territory reflects climatic change.

Winge (1895–96) mentioned the presence of "*Canis azarae*" (= *Pseudalopex gymnocercus*) among the Lagoa Santa fossils. He identified the species by comparing two skeletons of the species, "one of which comes from the republic of Argentina" (Winge 1895–96, 108). He also observed that the species has not been found among the current faunas in Lagoa Santa. Walker (1983) notes that the current distribution of *P. gymnocercus* is in grasslands of southern Brazil and Paraguay, northern Argentina, and Uruguay (Winge 1895–96, 943). Such species (at least *L. guanicoe*, *M. coypus*, and *P. gymonocercus*, whose intertropical presence is certain) would have returned to their original territory after the climatic crisis of the late Pleistocene–Holocene, thus disappearing from the intertropical territory. This pattern is similar to what happened with the guanaco in Buenos Aires province. During that period the cerrado would have occupied most of the area currently known as caatinga, which at that time was reduced to habitat islands. This conclusion is supported by the absence of xeric endemics in the recent mammalian fauna of Brazil. *Tolypeutes tricinctus* and *Kerodon rupestris*, considered by some to be xeric forms, are known today to occupy both the cerrado and the caatinga. The *mata atlântica* would have been much larger than today, at least in Bahia, where it expanded toward the west, as can be deduced by the presence of two large atelids in that state (Hartwig and Cartelle 1996).

It is intriguing that Redford and Fonseca (1986) found a low proportion of endemic fauna of open cerrado areas, when we take into consideration the group's extent and age (Ab'Saber 1977; Cerqueira 1982). I postulate that the endemism values for the mammal fauna were higher than they are today. The extinctions at the end of the Pleistocene were more selective in open cerrado, and the better-adapted species were more directly affected. In that vulnerable fauna I include the autochthonous species (twenty-three) and some rodent species studied by Winge that are probably extinct (twelve species?). An accurate review of the material at the Zoologische Museum of Copenhagen would allow us to test this hypothesis more fully. The cerrado endemic fauna in the late Pleistocene would constitute about 35% of the species in the area.

Another hypothesis to explain the current low endemism of the cerrado mammal fauna is that of Redford and Fonseca (1986): the *matas de galeria* that occurred in cerrado may have acted as isolated wet sites so that the regional fauna did not have to adapt to the xeric environment. This explanation alludes to the vegetation type, and I do not think it can be applied to the large grazing mammals that became extinct.

This discussion leads us to a few conclusions about the problem of extinction within the ecosystem as a whole. The analyses of Paula Couto (1975), Bigarella, Andrade-Lima, and Riehs (1975), and Marshall et al. (1984) did not consider the great variablity among the vegetative subdivisions within the cerrado domain. The widely varied plant species of the area, which I consider to be a mosaic of ecosystems, is reflected by the faunal composition of the late Pleistocene. I consider the contents of the caves where Lund collected the Lagoa Santa fossils a local fauna, gathered from an area of a few square kilometers.

If this fauna as a whole is compared with the local fauna from Toca dos Ossos (Bahia), we observe striking differences that reflect the different quantitative composition of the constituent populations. Both have a considerable number of species, and this could reflect the approximate regional faunal composition of both localities in that time frame. The series of caves Lund studied in the Lagoa Santa highlands extend over an area of about 4,000 km^2, whereas the river network (basin) related with Toca dos Ossos (the major force behind the introduction of specimens found as fossils in the cave) was estimated to extend about 2,000 km^2. Concerning the extinct megafauna, I observed that the remains of *Palaeolama major* are numerous in both sites. In Lagoa Santa there are more "glyptodonts" and *Scelidodon cuvieri* specimens. On the other hand, findings of *Equus (Asinus) neogeus, Hippidium principale,* Toxodontinae, *Xenorhinotherium bahiense,* and *Eremotherium laurillardi* are abundant in Toca dos Ossos, whereas these taxa are scarce among Lund's findings from Lagoa Santa. Winge paid little attention to that material, except for *E. (A.) neogeus.* It is possible that the Toca dos Ossos local ecosystem was more open and more abundant in grasslands than the Lagoa Santa ecosystem, where the *mata atlântica* was present.

Three caves in Bahia were studied recently: the already mentioned Toca dos Ossos (town of Ourolândia), Gruta dos Brejões (town of Morro do Chapéu), and Toca da Boa Vista (town of Campo Formoso). Roughly they correspond to the vertices of an imaginary isosceles triangle whose base is 40 km long (the distance between the two first caves), and whose two other sides are 100 km (the distance between the third cave and each of the other two). Table 4.3 shows some of the species found in those caves. In reduced territory the faunas of the three sites show a clearly different ecosystem. The environment of Toca da Boa Vista and Gruta dos Brejões would be predominantly of *mata,* whereas that of Toca dos Ossos would be more open,

Table 4.3   Records of Some Species of the Late Pleistocene-Holocene in Brazilian Caves

	1	2	3
Alouatta sp.	—	—	x
Protopithecus brasiliensis	—	x	—
Eremotherium laurillardi	x	—	x
Glossotherium aff. *G. lettsomi*	x	—	—
Ocnotherium giganteum	x	—	—
Mylodonopsis ibseni	x	—	x
Scelidodon cuvieri	x	x	—
Nothrotherium maquinense	—	x	x
Myrmecophaga tridactyla	—	x	x
Tamandua tetradactyla	x	—	—
Topypeutes tricinctus	x	—	—
Glyptodon clavipes	x	—	—
Pampatherium humboldti	x	—	x
Coendou sp.	x	x	—
Hydrochaeris hydrochaeris	x	—	x
Neochoerus sulcidens	x	—	x
Myocastor coypus	x	—	—
Protocyon troglodytes	—	x	—
Smilodon populator	x	x	—
Toxodon platensis	x	—	—
Trigonodops lopesi	x	—	—
Haplomastodon waringi	x	—	x
Equus (Asinus) neogeus	x	—	—
Hippidion principale	x	—	—
Tayassu tajacu	x	x	x
Lama guanicoe	—	x	—
Palaeolama major	x	—	—
Odocoileus virginianus	x	—	—
Mazama sp.	x	x	x

Note: 1, Toca dos Ossos; 2, Toca da Boa Vista; 3, Gruta dos Brejões.

typical of cerrado. Nevertheless, the presence of *L. guanicoe* and *S. cuvieri* in Toca da Boa Vista indicates the occurrence of open areas.

At some point in history we see the disappearance of a very rich mammal fauna from the study area. The event seems to be post-Lujanian. In the intertropical region the causes for extinction would have acted more slowly than in temperate regions. Many authors have written on the possible causes of extinction in the late Pleistocene, and Borrero (1977) listed the principal ones. The action of humans, supported by Martin (1973; 1984), does not seem to be relevant in Brazil. There is no evidence of any human influence aside from a single sample of a fossil piece of the extinct fauna with marks of human activity (Prous and Guimarães, 1981–82). Neither art (carvings, paintings) nor other archaeological findings have supported this theory. The actions of our indigenous human populations are supportive: there is no evidence that they were responsible for the extinction of any species. And after the western discovery of Brazil (1500 A.D.) their populations were larger than that of "Lagoa Santa men" of the late Pleistocene. I agree with Webb (1984, 192) that "climatic deteriorations became the principal causal hypotheses for the Late Cenozoic large mammal extinctions."

Table 4.4   Distribution of Extinct Autochthonous Species

	CE	BA	PB	PE	MG	PI	RN
Eremotherium laurillardi	x	x	x	x	x	x	x
Scelidodon cuvieri	x	x	x	?	x	x	
Ocnotherium giganteum		x			x		
Mylodonopsis ibseni		x			x	?	
Nothrotherium maquinense	x	x	?			x	
"Ocnopus gracilis"			x		x		
Xenocnus cearensis	x						
Valgipes deformis					x		
Popraopus sulcatus		x			x		
Pampatherium humboldti		x	x		x	x	x
Holmesina paulacoutoi	x	x	x				
Hoplophorus euphractus		x			x	?	
Panocthus greslebini	x	x	x			x	x
Panocthus jaguaribensis	x		x		x		x
Neochoerus sulcidens					x		x
Speothos pacivorus					x		
Xenorhinotherium bahiense	x	x	x	x	?	?	x
Trigonodops lopesi	?	x	x	x	?	?	
Haplomastodon waringi	x	x	x	x	x	x	
Tapirus cristatellus		x			x		
Protopithecus brasiliensis		x		x			x
Brasiliochoerus stenocephalus					x		

Note: The table displays the main occurences by state. CE, Ceará; BA, Bahia; PB, Paraiba; PE, Pernambuco; MG, Minas Gerais; PI, Piaui; RN, Rio Grande do Norte.

As I noted earlier, the model I propose is "punctuated," since there are no faunal temporal sequences (that are synchronous) in the Brazilian sites of the Pleistocene–late Holocene. The fauna identified in these areas exhibits a very diversified set of habitats, yielding a mosaic of microregions that is difficult to analyze. Probably climatic variations shifted the equilibrium in the diversified ecosystem, making it more homogeneous. It is possible that the rainfall regime was changed, extending the dry period and shortening the rainy season. Brown and Ab'Saber (1979) indicated that a probable cause for this change was the influx of the Falkland Current. Vast forests were reduced to *matas ciliares* (riverside forest areas), which became the refuge of surviving species. In the cerrado the primary productivity was reduced or depleted as a result of periodic droughts and natural fires. The caatinga "island" expanded, homogenizing the ecosystem. The chances of survival for the large grazing mammals were reduced, and in fact these forms experienced the greatest rates of extinction. The paleontological findings show that extinction particularly affected the megagrazers and herbivores that inhabited the borders of forests. The surviving megamammals were precisely those that tolerated or depended on forest habitats.

Many living organisms show evidence of periodic growth during their lives. In mammals this is marked by the Park-Harris osseous lines and by dental enamel hypoplasia ("Hipoplasia carencial do esmalte dentário," in the original Portuguese; Ferigolo 1987). The latter is manifested by small orifices organized in rows or diffusely distributed over the enamel surface in furrows of different widths and even by the total absence of enamel in more or less extensive areas (Ainamo and Cotress 1982, cited in Ferigolo 1987).

The enamel hypoplasia is not specific and can be related to several systemic disorders. Possible causes are cyclic alimentary depletion, hyperthermia, metabolic disorders, and other factors. Ferigolo (1987) thinks the most probable cause is cyclic alimentary depletion. The lines that show enamel hypoplasia reflect the deficit in tissue formation by the "ameloblasts." They are permanent because no repair and remodeling process occurs in the enamel. In Brazilian Pleistocene findings I have seen many teeth of species showing "euhipsodontia" (continuous growth). The furrows I interpret as a manifestation of hypoplasia were also found on the cement of the molariforms of the Edentata. Teeth of *Glyptodon clavipes, G. giganteum, Glossotherium* aff. *G. lettsomi, Eremotherium laurillardi, Toxodon platensis, T. lopesi,* and *Neochoerus sulcidens* exhibit this trait, and in some groups up to half of the studied specimens show it. Causes such as metabolic disorders would hardly include so many individuals. It is reasonable to believe that this effect was more correlated with cyclic dietary deficiencies during the dry seasons.

The adverse shift would also have reduced the number and survival of young. It is possible that there was a relation between drying and lowered plant productivity, with negative consequences for the reproductive strategy of large animals. The habitat became less variable in composition and the seasonality more pronounced. The faunas collected in Lagoa Santa and Bahia include many young individuals. As I noted earlier, the endemic fauna of the cerrado open areas was the most damaged during the climate shift. At the same time the Amazon forest expanded, causing the reduction of *Odocoileus virginianus* and the creation of a separate population of *Euphractus sexcinctus* in Suriname and the Brazilian northeast (Wetzel 1985). This mode of extinction was gradual instead of sudden. The main cause, in my opinion, was the climate.

There is a final issue that has not received much attention, perhaps because it is hard to verify. Epidemic, epizootic, and parasitic diseases are not rare in the tropics. Any of these problems could have contributed to the extinction scenario described or could have intensified populational instability.

References

Ab'Saber, A. N. 1977. Os dominios morioclimáticos na América do Sul. *Geomorfológia,* 1–22.

Absy, M. L., T. Van Der Hammen, F. Soubius, H. Suguio, L. Martin, M. Fourmer, and B. Turco. 1989. Data on the history of vegetation and climate in Carajás, eastern Amazonia. In *International symposium on global changes in South America during the Quaternary,* 129–131. Special Publication 1. São Paulo.

Abuhid, V. S. 1991. Sobre um glossotério pleistocênico do Estado da Bahia. Master's thesis, Instituto de Geociéncias da UFRJ), Rio de Janeiro.

Aguirre, A. C. 1971. *O mono* Brachyteles arachnoides. Rio de Janeiro: Academia Brasileira de Ciéncias.

Alberdi, M. T. 1987. La familia Equidae Gray, 1821 (Perissodactyla, Mammalia) en el Pleistoceno de Sudamérica. In *Anales IV Congresso Latinoamericano de Paleontología* (Bolivia), 484–99.

Alberdi, M. T., A. N. Menegaz, and J. L. Prado. 1987. Formas terminales de *Hippidion* (Mammalia, Perissodactyla) de los yacimientos del Pleistoceno tardio-Holoceno de la Patagonia (Argentina y Chile). *Estud. Geol.* 43:107–15.

Alberdi, M. T., A. N. Menegaz, J. L. Prado, and E. Tonni. 1989. La fauna local Quequen Salado-Indio Rico (Pleistoceno tardio) de la provincia de Buenos Aires, Argentina: Aspectos paleoambientales y bioestratigráficos. *Ameghiniana* 25 (3): 225–36.

Alho, J. O. C. 1990. Distribuição da fauna num gradiente de recursos em mosáico. In NOVAIS PINTO, *Cerrado: Caraterização e perspectivas,* 205–54. Brasília: Editora da Universidade de Brasília.

Ameghino, F. 1907. Notas sobre una pequeña colección de huesos de mamíferos procedentes de las grutas calcáreas de Iporanga. *Rev. Mus. Paulista* 7:59–124.

Ashworth, A. C., and J. W. Hoganson. 1991. The magnitude and rapidity of the climate change marking the end of the Pleistocene in the mid-latitudes of South America. In *Abstracts INQUA Congress, 13* (Beijing), 1:11.

Berqvist, L. R. 1989. Os mamíferos pleistocênicos do estado da Paraiba, Brasil, depositados no Museu Nacional, Rio de Janeiro. Master's thesis, Museu Nacional (UFRJ).

Berta, A. 1988. Quaternary evolution and biogeography of the large South American Canidae (Mammalia: Carnivora). *Univ. Calif. Publ. Geol. Sci.* 132: 1–149.

Bigarella, J. J. D. Andrade-Lima, and P. J. Riehs. 1975. Cosiderações a respeito das mudanças paleoambientais na distribuição de algumas espécies vegetais e animais do Brasil. *An. Acad. Brasil. Ciênc.* 47 (suppl.): 411–64.

Boas, J. E. V. 1881. Om en fossil Zebra-Form fra Brasilienes Campos: Med. et Tillaeg om to Arter af Salaegten *Hippidion. Kongelige Danske Videnskabernes Selskabs Skrifter, Naturvidenskabelig og Mathematisk Afdeling,* ser. 6, 1 (5): 305–30.

Borrero, L. A. 1977. La extinción de la megafauna: Su explicación en la Patagonia austral. *Anal. Inst. Patagonia, Punta Arenas* 8:81–93.

Boule, M., and A. Thevenin. 1920. *Mammifères fossiles de Tarija.* Mission Scientifique G. de Crequi-Monfort et E Senechal de la Grange. Paris: Imprimerie Nacional.

Brown, K. S., and N. A. Ab'Saber. 1979. Ice glace forest refuges and evolution in the Neotropics: Paleoclimatological, geomorphological and pedacological data with modern biological endemism. *Palaeoclimas* 5:1–30.

Cabrera, A. 1935. Sobre la osteologia de Palaeolama. *Anal. Mus. Argent. Cienc. Nat. "Bernardino Rivadavia"* 33:283–312.

———. 1960. *Catálogo de los mamíferos de América del Sur.* Vol. 2. Buenos Aires: Museo Argentino de Ciencias Naturales "Bernardino Rivadavia."

Camara, I. G. 1991. *Plano de ação para a Mata Alantica.* São Paulo: Fundação SOS Mata Atlantica.

Cartelle, C. 1989. Sobre uma pequena coleção de restos fósseis de mamíferos do Pleistoceno-Holoceno de Janaúba (MG). In *Anais do XI Congresso Brasileiro de Paleontologia* (Curitiba), 1:635–49.

———. 1991. Um novo Mylodontinae (Edentata, Xenarthra) do Pleistocene final da região intertropical brasileira. *An. Acad. Brasil. Ciênc.* 63 (2): 161–70.

———. 1992. Edentata e megamamíferos herbivoros extintos da Toca dos Ossos (Ourolandia, BA). Doctoral diss., Curso de Pós-Graduação em Morfologia (UFMG), Belo Horizonte.

———. 1994. Presença de *Lama* (Artiodactyla, Camelidae) no Pleistoceno final—Holoceno de Bahia. *Acta Geol. Leopoldensia* 39, 1 (17): 339–410.

————. 1995. A fauna local de mamíferos pleistocênicos da Toca da Boa Vista (Campo Formoso, BA). Thesis, Concurso de Professor Titular, Instituto de Geociências (UFMG), Belo Horizonte.

Cartelle, C., and V. S. Abuhid. 1989. Novos especimes brasileiros de *Smilodon populator* (Lund, 1842), Carnivora, Machairodontinae: Morfologia e conclusoes taxonômicas. In *Anais XI Congresso Brasileiro de Paleontología* (Curitiba), 1:607–20.

Cartelle, C., and G. A. Bohórquez. 1982. *Eremotherium laurillardi* (Lund, 1842). 1. Determinação especifica e dimorfismo sexual. *Iheringia,* ser. geol., 7:45–63.

————. 1985. *Pampatherium paulacoutoi,* uma nova espécie de tatu gigante da Bahia, Brasil (Edentata, Dasypodidae). *Rev. Brasil. Zool.* 2 (4): 229–54.

Cartelle, C., and J. S. Fonseca. 1981. Espécies do género *Glossotherium* no Brasil. In *Anais do XX Congresso Latinoamericano de Paleontología* (Pôrto Alegre), 2:805–18.

————. 1982. Contribuição ao melhor conhecimento da pequena preguiça terricola *Nothrotherium maquinense* (Lund) Lydekker, 1889. *Lundiana* 2: 127–82.

Cartelle, C., and G. De Iuliis. 1995. *Eremotherium laurillardi:* The Panamerican late Pleistocene megatheriid sloth. *J. Vert. Paleontol.* 15:830–41.

Cartelle, C., and W. C. Hartwig. 1996. A new extinct primate from the Pleistocene megafauna of Bahia, Brazil. *Proc. Nat. Acad. Sci.* 93:6405–9.

Cartelle, C., and A. Langguth. n.d. *Protocyon troglodytes* (Lund): Um canideo intertropical extinto. In press.

Cartelle, C., and G. Lessa. 1988. Descrição de um novo genero e espécie de Macraucheniidae (Mammalia, Litopterna) do Pleistoceno do Brasil. *Paulacoutiana* 3:3–26.

Cartelle, C., P. López de Prado, and B. Garzon. 1989. Estudo comparativo dos esqueletos da mão e pé de *Pampatherium humboldti* (Lund, 1839) e *Holmesina paulacoutoi* (Cartelle and Bohórquez, 1985): Edentata Pampatheriinae. In *Anais XI Congresso Brasileiro de Paleontología* (Curitiba), 1:607–14.

Carvalho, A. M. V. de. 1952. Ocorrências de *Lestodon trigonidens* na mamalofauna de Alvares Machado (estado de São Paulo). *Bol. Fac. Filosof., Ciênc. Let.* 134 (Geologia 4): 43–55.

Casamiquela, R. M. 1968. Noticia sobre la presencia de *Glossotherium* (Xenarthra, Mylodontidae) em Chile central. *An. Mus. Hist. Nat. Valparaiso* 1: 59–75.

Cerqueira, R. 1982. South American landscapes and their mammals. In *Mammalian biology in South America,* ed. M. A. Mares and H. H. Genoways,

53–75. Pymatuning Symposia in Ecology 6. Special Publication Series. Pittsburgh: Pymatuning Laboratory of Ecology, University of Pittsburgh.

Clapperton, C. M. 1987. Glacial geomorphology, Quaternary glacial sequence and paleoclimatic inferences in the Ecuadorian Andes. In *International Geomorphology 1986,* ed. V. Gardiner, 843–70. London: Wiley.

Coltrinari, L. 1992. Paleoambientes Quaternários na América do Sul: Primeira aproximação. In *Anais do III Congresso ABEQUA* (Belo Horizonte), 13:42.

Cope, E. D. 1885. A contribution to the vertebrate paleontology of Brazil. *Proc. Amer. Phil. Soc.* 23: 1–21.

De Iuliis, G. 1995. Relationships of the Megatheriinae, Nothrotheriinae and Planosinae: Some skeletal characteristics and their importance for phylogeny. *J. Vert. Palontol.* 14 (4): 577–91.

Duarte, L., and M. M. Souza. 1991. Restos de vegetais conservados em coprólitos de mamíferos (*Palaeolama* sp. *Nothrotherium maquiense* (Lund, Lidekker) na Gruta dos Brejões, BA. *Resumos do XII Congresso Brasileiro de Paleotologia* (São Paulo), 74.

Edmund, A. G. 1987. Evolution of the genus *Holmesina* (Pampatheriidae, Mammalia). *Pearce-Sellards Series* 45:1–20.

Engelmann, F. F. 1985. The phylogeny of the Xenarthra. In *The evolution and ecology of armadillos, sloths, and vermilinguas,* ed. G. G. Montgomery, 51–64. Washington, D.C.: Smithsonian Institution Press.

Fairbridge, R. 1972. Climatology of a glacial cycle. *Quat. Res.* 2:283–302.

Faure, G. 1986. *Principles of isotope geology.* New York: John Wiley.

Ferigolo, J. 1987. *Paleopatologia comparada de Vertebrados: Homem de Lagoa Santa, Homen do Sambaqui da Cabeçuda e mamíferos pleistocênicos.* 2 vols. Pôrto Alegre: Tese de Doutorado. Instituto de Geociências (UFRGS).

Gomide, S. M. M. 1989. Mamiferos pleistocênicos de Itapipoca, Ceará, Brasil, depositados no Museu Nacional. Master's thesis, Museu Nacional (UFRJ), Rio de Janeiro.

Hartwig, W., and C. Cartelle. 1996. A complete skeleton of the giant South American primate *Protopithecus. Nature* (London) 381:307–61.

Heine, K. 1991. Late Quaternary climatic changes of the Southern Hemisphere. In *Abstracts, INQUA Congress 13* (Beijing),. 1:132.

Heusser, C. J. 1987. Quaternary vegetation of southern South America. *Quat. S. Amer. Antarctic Penin.* 5:197–222.

Hoffstetter, R. 1952. Les mammifères pléistocènes

de la république de l'Equateur. *Mém. Soc. Géol. France* 31 (66): 1–391.

———. 1954. Les gravigrades (Edentés Xénarthres) des cavernes de Lagoa Santa (Minas Gerais, Brasil). *Ann. Sci. Nat. Zool. Biol. Anim.* 16 (3): 741–64.

Holland, W. J. 1920. Fossil mammals collected at Pedra Vermelha, Bahia, Brazil, by Gerald A. Waring. *Ann. Carnegie Mus.* 13 (1–2): 224–32.

IBGE. 1988. *Mapa de Vegetação do Brasil.* Rio de Janeiro: Fundação Instituto Brasileiro de Geografía e Estatística.

Joly, A. B. 1970. *Conheça a vegetação brasileira.* São Paulo: Editor Universidade de São Paulo e Poligono.

Khoeler, H. C. 1989. Geomorfologia Cárstica na região de Lagoa Santa, MG. Doctoral diss., Instituto de Geociências (USP), São Paulo.

Kraglievich, G. L. 1952. Un cánido del Ecuartorio de Mar del Plata y sus relaciones con otras formas brasileñas y norteamericanas. *Rev. Mus. Mun. Cienc. Nat. Trad. Mar del Plata* 1 (1): 53–70.

Kraglievich, L. 1921. Estudios sobre los Mylodontinae: Descripción del cráneo y mandibula del *Pseudolestodon mylloides galleni. An. Mus. Nac. Hist. Nat. Buenos Aires* 3: 119–34.

———. 1932. Sobre *Trigonodops lopesi* (Roxo) Kraglievich. *Rev. Soc. Amigos Arq.* 5: 81–89.

———. 1934. Contribución al conocimiento de *Mylodon darwini* Owen y espécies afines. *Rev. Mus. La Plata* 34: 255–92.

Lalou, C. 1987. Deséquilibres radioctifs dans la famille de liuranium. In *Géologie de la pré-histoire,* ed. J. C. Miskovsky, 1073–85. Paris: Géopré.

Langguth, A. 1975. Ecology and evolution in the South American canids. In *The wild canids: Their systematics, behavioral ecology, and evolution,* ed. M. W. Fox, 192–206. New York: Van Nostrand Reinhold.

Lumley, H., M. A. Lumley, M. C. Beltráo de, Y. Yokoyama, J. Lareyrie, J. Danon, S. Delibrias, C. Falgueres, and J. L. Bischoff. 1988. Découverte d'outils taillés associés à des faunes du Pléistocène moyen dans la Toca da Esperança état de Bahia, Brésil. *C.R. Acad. Sci. Paris,* ser. 2, 306: 241–47.

Lund, P. W. 1839. Blik pae Brasiliens Dyravenden för Bidste Jerasmyaelining, Andes Afhandling: Patterdyrene. *Kongelige Danske Videnskabernes Seskbas Naturvidenskabelige og Mathematiske Afhandlinger* 5: 61–144.

Lund, P. W., and C. Paula Couto, C. 1950. *Memórias sobre a paleontología brasileira: Revistas e comentadas por Carlos de Paula Couto.* Rio de Janerio: Instituto Nacional de Livro.

Lydekker, R. 1887. *Catalogue of the fossil Mammalia in the British Museum (Natural History),* part 5. London: British Museum of Natural History.

MacFadden, B., and A. Azzaroli. 1987. Cranium of *Equus insulatus* (Mammalia, Equidae) from the middle Pleistocene of Tarija, Bolivia. *J. Vert. Paleontol.* 7 (3): 325–34.

MacFadden, B., and M. F. Skinner. 1979. Diversification and biogeography of the one-toed horses *Onohippidium* and *Hippidion, Postilla* 175: 1–10.

Markgraf, V. 1980. Paleoclimatic reconstruction of the last 15,000 years in subantarctic and temperate regions in Argentina. *Mem. Mus. Hist. Nat.* vol. 27.

———. 1989. Palaeoclimates in Central and South America since 18,000 BP based on pollen and lake-level records. *Quat. Sci. Rev.* 8 (1): 1–24.

Markgraf, V., and J. P. Bradbury. 1982. Holocene climatic history of South America. In *Chronostratigraphic subdivision of the Holocene,* ed. J. Mangerud, H. J. B. Birks, and K. D. Hager, 40–45. *Striae* 16: 40–45.

Marshall, L. G., A. Berta, R. Hoffstetter, R. Pascual, O. A. Reig, M. Bombim, and A. Mones. 1984. Mammals and stratigraphy: Geochronology of the continental mammal-bearing Quaternary of South America. *Palaeovertebrata,* mémoire extraordinaire, 1–76.

Martin, P. S. 1973. The discovery of America. *Science* 179: 969–74.

———. 1984. Prehistoric overkill: The global model. In *Quaternary extinctions: A prehistoric revolution,* ed. P. S. Martin and R. G. Klein, 354–403 Tucson: University of Arizona Press.

Mercer, C. H., and M. O. Palácios. 1977. Radiocarbon dating of the last glaciation in Peru. *Geology* 5: 600–604.

Mercer, J. H. 1976. Glacial history to southernmost South America. *Quat. Res.* 6: 125–66.

Mones, A. 1984. Estudios sobre la familia Hydrochaeridae. 14. Revision sistematica. *Senckenbergiana Biol.* 65 (1–2): 1–17.

Moreira, L. E. 1971. Os gliptodontes do nordeste du Brasil. *An. Acad. Brasil. Ciênc.* 43 (suppl.): 529–52.

Morgan, G. S., O. J. Linares, and C. S. Ray. 1988. New species of fossil vampire bats (Mammalia, Chiroptera, Desmontidae) from Florida and Venezuela. *Proc. Biol. Soc. Wash.* 101 (4): 912–28.

Oliveira, D. L. D. 1989. Consideraçoes sobre a emprego da terminologia da "Formação cacimbas" e caldeiroes para os tanques fossiliferos do nordeste do Brasil. In *Anais do XI Congresso Brasileiro de Paleontología* (Curitiba), 1: 535–39.

Oliveira, D. L. D., and P. C. Hackspacher. 1989. Genese e provável idade dos tanques fossilíferos de São Rafael-RN. In *Anais do XI Congresso Brasileiro de Paleontología* (Curitiba), 1: 541–49.

ORSTOM. 1987. Séminaire "Paléolacs et paléolacs et

paléoclimats en Amérique Latine et en Afrique" (20.000 BP–Actuel). *Géodynamique* 2 (2): 93–167.

Paula Couto, C. de. 1946. Atualização da nomenclatura genéruca e específica usada por Herluf Winge em "E Museo Lundii." *Est. Brasil. Geol.* 1 (3): 59–80.

———. 1947. Contribuição para a estudo de *Hoplophorus euphractus* Lund, 1839. *Summa Brasil. Geol.* 1 (4): 33–53.

———. 1956. Mamíferos fósseis do Cenozóico da Amazônia. *Bol. Conselho Nacional de Pesquisas* 3:1–121.

———. 1960. Um urso extinto do Brasil. *Bol. Soc. Brasil. Geol.* 9 (1): 5–27.

———. 1973. Edentados fósseis de São Paulo. *An. Acad. Bras. Ciênc.* 45 (2): 261–75.

———. 1975. Mamíferos fósseis do Quaternário do sudeste brasileiro. In International Symposium on the Quaternary, *Boletim Paraense de Geologia* (Curitiba), ed. J. J. Bigarella and B. R. D. Baker.

———. 1979. *Tratado de paleomastozoologia.* Rio de Janeiro: Academia Brasileira de Ciências.

———. 1980. Pleistocene mammals from Minas Gerais and Bahia, Brazil. In *Actas del 2º Congreso Argentino de Paleontologia y Bioestratigrafia y 1ᵒʳ Congreso Latinamericano de Paleontología* (Buenos Aires), 3:193–209.

———. 1981. On an extinct peccary from the Pleistocene of Minas Gerais. *Iheringia,* ser. geol., 6: 75–76.

———. 1982. Fossil mammals from the Cenozoic of Acre, Brazil. 5. Notoungulata, Nesodontinae (II), Haplodontheriinae, and Litopterna, Pyrotheria, and Astrapotheria (II). *Iheringia,* ser. geol. 7: 5–43.

Prado, J. L., N. Menegaz, E. P. Tonni, and M. C. Salemme. 1987. Los mamíferos de la fauna local Paso Otero (Pleistocene tardio), provencia el Buenos Aires: Aspectos paleontología y bioestratigráficos. *Ameghiniana* 24 (3–4): 217–33.

Prous, A., and C. L. Guimarães. 1981–82. Recentes descobertas sobre os mais antigos caçadores de Minas Gerais e da Bahia. *Arq. Mus. Hist. Nat.* 6–7:23–31.

Rabassa, J. 1991. Evidence of Quaternary global changes in Tierra del Fuego, southernmost South America. In *Abstracts INQUA Congresso, 13* (Beijing), 1:297.

Rabassa, J., and C. M. Clapperton. 1990. Quaternary glaciations of the southern Andes. *Quat. Sci. Rev.* 9 (1): 153–74.

Redford, K., and G. A. B. da Fonseca. 1986. The role of gallery forest in the zoogeography of the cerrado's non-volant mammalian fauna. *Biotropica* 18 (2): 126–35.

Reinhardt, J. 1878. Kaempedovendryr-Slaegten. *Coelodon. Videnskabernes Selskabs Skrifter, Raekke, Naturvidenskabelige og Mathematiske Afhandlinger* 12 (3): 257–349.

Saint Hilaire, A. 1830. *Voyage dans les provinces de Rio de Janeiro et de Minas Gerais.* 2:314–15.

Salemme, M. C. 1983. Distribución de algunas espécies de mamíferos en el nordeste de la provincia de Buenos Aires durante el Holoceno. *Ameghiniana* 20 (1–2): 81–94.

Salgado-Laboriou, M. L., V. Rull, C. Schubert, and S. Valastro. 1988. The establishment of vegetation after late Pleistocene deglaciation of the Páramo de Miranda, Venezuelan Andes. *Rev. Paleobot. Palynol.* 55:1–17.

Schubert, C. 1974. Late Pleistocene Mérida glaciation. *Boreas* 3:147–52.

Scillato-Yane, G. 1977. Octomylodontinae nueva subfamilia de Mylodontidae (Edentata, Tardigrada): Descripción del cráneo y mandíbula de *Octomylodon robertoscagliai* n. sp. procedentes de la Formación Arroyo Chasicó (edad Chasiquense, Plioceno Temprano) del sur de la provincia de Buenos Aires (Argentina): Algunas consideraciones y sistemáticas sobre los Mylodontoidea. *Publ. Mus. Mun. Ciênc. Nat.* 2 (5): 123–40.

Seltzer, S. O. 1990. Recent glacial history and paleoclimate of the Peruvian-Bolivian Andes. *Quat. Sci. Rev.* 9 (1): 137–52.

Siffdine, A., F. Frohllich, J. L. Helice, B. Turcq, K. Cuguio, and F. Soubies. 1991. Detritic fluxes in an Amazon lake: A record of climatic fluctuations during the past 60,000 years. In *Abstracts INQUA Congress, 13* (Beijing), 1:331.

Simpson, G. G., and Paula Couto, C. de. 1981. Fossil mammals from the Cenozoic of Acre, Brazil. 3. Pleistocene Edentata Pilosa, Proboscidea, Sirenia, Perissodactyla and Artiodactyla. *Iheringia,* ser. geol., 6:11–73.

Spix, J. B. von, and C. F. P. von Martius. 1828. *Reise in Brasilien.* Munich.

Trajano, E., and H. Ferrarezzi. 1994. A fossil bear from northeastern Brazil with a phylogenetic analysis of the South American extinct Tremarctinae (Ursidae). *J. Vert. Paleonol.* 14 (4): 552–61.

Toledo, P. M. 1986. Descrição do sincrânio de Eromotherium laurillardi (Lund, 1842): Taxonomia e paleobiogeografia. Master's thesis, Instituto de Geociências da UFRGS, Pôrto Alegre.

Tonni, E. P. 1985. The Quaternary climate in the Buenos Aires province through the mammals. *Acta Geocriogen.* 3:113–21.

Tonni, E. P., and F. Fidalgo. 1978. Consideraciones sobre los cambios climáticos durante el Pleistocene tardio—reciente en la provincia de Buenos Aires:

Aspectos ecológicos y zoográficos relacionados. *Ameghiniana* 15 (1–2): 234–53.

———. 1982. Geología y paleontología de los sedimentos del Pleistoceno en el area de Punta Hermengo (Miramar, prov. de Buenos Aires, Re. Argentina): Aspectos paleoclimáticos. *Ameghiniana* 19 (1–2): 79–108.

Tonni, E. P., and J. H. Laza. 1981. Las aves de la fauna local Paso Otero (Pleistoceno tardio) de la provincia de Buenos Aires: Su significación ecológica, climática y zoogeográfica. *Ameghiniana* 17 (4): 313–21.

Tonni, E. P., and G. G. Politis. 1980. La distribución del guanaco (Mammalia, Camelidae) en la provincia de Buenos Aires durante el Pleistoceno tardío y Holoceno: Los factores climáticos como causas de su retracción. *Ameghiniana* 17 (1): 53–66.

Trajano, E., and M. Vivo. 1991. *Desmodus draculae* Morgan, Linares and Ray, 1988, reported for southeastern Brasil, with paleoecological comments (Phyllostomidae, Desmontinae). *Mammalia* 55 (3): 456–59.

Van Frank, R. 1957. A fossil collection from northern Venezuela. 1. Toxodontidae (Mammalia, Notoungulata). *Amer. Mus. Novitat.* 1850: 1–38.

Vidal, N. 1946. Contribuição ao conhecimento da paleontología do nordeste brasileiro: Noticia sôbre a descoberta de vertebrados pleistocênicos no municipio de Pesqueira em Pernambuco. *Bol. Mus. Nac.*, n.s., geol., 6: 1–11.

Voss, R. S., and P. Myers. 1991. *Pseudoryzomys simplex* and the significance of Lund's collections from the caves of Lagoa Santa, Brazil. *Bull. Amer. Mus. Nat. Hist.* 206: 414–32.

Walker, E. P. 1983. *Mammals of the world.* 4th ed., Vol. 2. Baltimore: Johns Hopkins University Press.

Webb, S. D. 1974. Pleistocene llamas of Florida with a brief review of the Lamini. In *Pleistocene mammals of Florida,* ed. S. D. Webb, 170–213. Gainesville: University of Florida Press.

———. 1984. Ten million years of mammal extinctions in North America. In *Quaternary extinctions: A prehistoric revolution,* ed. P. S. Martin and R. G. Klein, 189–210. Tucson: University of Arizona Press.

Wetzel, R. M. 1985. Taxonomy and distribution of Recent Xenarthra (= Edentata). In *The evolution and ecology of armadillos, sloths, and vermilinguas,* ed. G. G. Montgomery, 23–50. Washington, D.C.: Smithsonian Institution Press.

Winge, H. 1888. Jorfundne of nulevende Gnawers (Rodentia) fra Lagoa Santa, Minas Geraes, Brasilien. *E. Mus. Lundii* 1 (3): 1–200.

———. 1895–96. Jorfundne of nulevende Aber (Primates)-Rovdyr (Carnivora) fra Lagoa Santa, Minas Geraes, Brasilien. *E. Mus. Lundii* 2 (2): 1–187.

———. 1906. Jordgundne og nulevende Hoydyr (Ungulata) fra Lagoa Santa, Minas Geraes, Brasilien. *E. Mus. Lundii* 3 (1): 1–239.

———. 1915. Jordgundne og nulavende Gumlere (Edentata) fra Lagoa Santa, Minas Geraes, Brasilian. *E. Mus. Lundii* 3 (2): 1–321.

Wright, I. E. 1984. Late Pleistocene glaciation and climate around the Junín plain, central Peruvian Andes. *Geogr. Ann,* 65A: 35–43.

Zingeser, M. R. 1973. Dentition of *Brachyteles arachnoides* with reference to Alouttine and Ateline affinities. *Folia Primatol.* 20: 351–90.

Part 2

The Contemporary Mammalian Fauna

5   The New World Marsupials
(Didelphimorphia, Paucituberculata, and Microbiotheria)

Diagnosis and Comments on Reproduction of the Extant Metatheria

In volume 1 of *Mammals of the Neotropics* (Eisenberg 1989), several alternative classifications of marsupials were discussed. In volume 2 (Redford and Eisenberg 1992), a cladogram suggesting marsupial relationships was presented. Gardner (1993) places the family Didelphidae in the order Didelphimorphia, the family Caenolestidae within the order Paucituberculata, and the family Microbiotheriidae within the order Microbiotheria. This arrangement reflects current thinking concerning the antiquity of the marsupial lineages and a revision of the content of the Metatheria (raised to an infraclass within the subclass Theria by Szalay 1993). In Szalay's scheme *Dromiciops* (see Redford and Eisenberg 1992) is allied with the Old World marsupials within the cohort Australidelphia. Szalay (1993) proposes that the extant New World marsupials belong to the cohort Ameridelphia but relegates the Didelphimorphia to subordinal status and the Paucituberculata to infraordinal status within the suborder Glirimetatheria (Szalay 1994).

The dentition of marsupials is heterodont, with easily distinguished incisors, canines, premolars, and molars, and the number of teeth often exceeds the basic eutherian number of 44. All New World marsupials have 46–50 teeth. The bony palate is fenestrated, a diagnostic character for most species when compared with the extant eutherians. The auditory bullae are formed principally by the alisphenoid. Epipubic bones are present in all extant species and are usually diagnostic for the order (monotremes share this character). The brain structure differs from that of eutherian mammals in that the corpus callosum is lacking, but the anterior commissure is well developed.

Marsupial young have a brief intrauterine development and are born in an extremely undeveloped state. At birth the forelimbs are well developed, and the newborn crawls to the teat area, where it attaches to a nipple and remains for four to seven weeks. In many New World marsupials the teat attachment phase terminates when the female deposits the young in a nest, where she nurses them for another three to seven weeks until they are fully weaned. The teat area in many, but not all, species is enclosed in a pouch. Marsupial reproduction has been reviewed by Tyndale-Biscoe and Renfree (1987). The biology of New World forms is reviewed in the volume edited by Hunsaker (1977a,b) and in Eisenberg (1989) and Redford and Eisenberg (1992).

Distribution

At present, marsupials in the Eastern Hemisphere are confined to the Australian region, and they have undergone an extensive adaptive radiation in Australia and New Guinea. Some species have colonized the Celebes and the Moluccas. In the Western Hemisphere, the major adaptive radiation of marsupials occurred in South America. Most species are now found in South America, although many have currently extended their ranges into Central America. One species, *Didelphis virginiana,* has colonized temperate North America, perhaps with a Central American origin.

History and Classification

The oldest known marsupials are found in the mid-Cretaceous of North America. Cretaceous records from South America are late Cretaceous, but the radiation, as exemplified by the fossil record in South America, suggests that remains of earlier marsupials may await discovery (Marshall, DeMuison, and Sige 1983). Throughout the Tertiary in South America marsupials radiated into insectivore, frugivore, and carnivore niches (see Marshall 1982a,b,c and Streilein 1982a for an extended account). Fossil marsupials first appear in Australia during the Oligocene, but clearly they must have arrived on the island continent before it was so distant from Antarctica. The recent discovery of fossil marsupials in the peninsula of Antarctica suggests that before the glaciation, when Antarctica, Australia, and South America were closer together, Antarctica may have served as the bridge for the earliest marsupials' transit from South America to Australia (Woodburne and Zinmeister 1982).

The old order "Marsupialia" was classically divided into thirteen families, eight of which are extant. Ride (1964) proposed that the Marsupialia be divided into four orders. This view has been refined, and in this volume we follow the classification developed by Kirsch and Calaby (1977) and Alpin and Archer (1987) as modified by Gardner (1993). Szalay (1994) and others have attempted to link an extant genus of South American marsupial (*Dromiciops australis*) with the major Australian radiation. Hershkovitz (1992a) reviews the evidence and concurs that *Dromiciops* is a survivor of an ancient, unique South American radiation but is not necessarily allied closely with the Australian marsupials. Reference to Szalay's work has been made in the first section of this chapter. More recent efforts have reinforced the notion that New World marsupial lineages are phylogenetically older than previously recognized (Kirsch and Palma 1995).

Kielan-Jaworowska and Nessov (1990) propose a radical but logical extension of the content of the taxon Marsupialia. They offer evidence that many extinct forms assigned to the taxon Deltatheroidea were in fact "marsupials." This view supports the suggestion made by Van Valen (1966, 1974).

Order Didelphimorphia
FAMILY DIDELPHIDAE
Opossums, Comadrejas, Cuicas

Diagnosis
The dental formula tends to be conservative, with 50 teeth typical: I 5/4, C 1/1, P 3/3, M 4/4. The postorbital bar is not developed (fig. 5.1). The pouch may

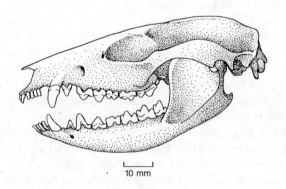

10 mm

Figure 5.1. Skull of *Didelphis marsupialis*.

be either absent as in the genera of mouse opossums (in the broadest sense, p. 57), *Monodelphis*, and *Metachirus*, or well developed as in *Didelphis*, *Philander*, *Chironectes*, and during lactation in some species of *Caluromys*. There are five digits on each foot, and the hind foot has an opposable thumb. Members of this family have a long rostrum, large, naked ears, and a tail that is often highly prehensile and usually almost hairless for at least the distal two-thirds. *Glironia* is exceptional in having its tail furred almost to the tip. This family exhibits a wide range of sizes. Some species may be as small as 80 mm in head and body length, while the larger species may reach 1,020 mm in total length. Pelage color is highly variable, ranging from gray to deep, rich brown. Some species exhibit a banding pattern of the dorsal pelage. Karotypic data have been summarized by Reig et al. (1977) and Seluja et al. (1984).

Comment
Certain morphological characters of *Caluromys*, *Caluromysiops*, and *Glironia* lend credence to the tendency to place them in their own subfamily (Caluromyinae) (Creighton 1984; Reig, Kirsch, and Marshall 1987). Based on genetic studies, Patton, Dos Reis, and Silva (1996) confirm that *Caluromys* is distinct from the other didelphids, but *Glironia* is not closely allied with *Caluromys*.

Distribution
Species of this family occur widely over the Neotropics and occupy almost every habitat type except extremely high elevations and the extreme desert areas surrounding the Gulf of Venezuela, southwestern Peru, northwestern Chile, and the extreme temperate south.

Natural History
Most members of this family are nocturnal (some species of *Monodelphis* may be diurnal or crepuscular). All occupy omnivore, insectivore, frugivore/insectivore, or carnivore feeding niches. Many members are strongly arboreal and seldom come to the ground, especially those long-tailed forms that are specialized for multistratal tropical rain forests.

Marsupials are characterized by a unique mode of reproduction. All forms studied in the Western Hemisphere have only a yolk-sac placenta, and the shell membrane remains intact during the intrauterine growth phase. The young are born after a brief period of gestation, generally thirteen to fourteen days for the South American species. On passing out of the reproductive tract, the young climb unassisted to the nipple area, where each grasps a teat in its mouth and remains attached to it for a varying time during early growth and development. The mother transports the entire litter attached to her teats for approximately the first five to six weeks of their development. The young then detach and begin a nest phase where the mother does not continuously transport them but returns to the nest to nurse. The teat area may be enclosed in a pouch, or no pouch may be present (e.g., *Marmosa, Marmosops, Gracilinanus, Micoureus, Thylamys, Metachirus*, and *Monodelphis*).

Eisenberg and Wilson (1981) measured the cranial capacities of several species of didelphid marsupials. Comparing their cranial capacities with those of eutherian mammals, they found that several didelphid species (e.g., *Monodelphis* and *Didelphis*) had extremely small brains, but many others had brain sizes near the norm for small eutherian mammals. The wide variety of relative brain sizes within the family Didelphidae strongly suggests that we should reexamine generalizations concerning members' behavioral capacities based on studies of *Didelphis*.

By comparing the relative proportions of limb segment length, trunk length, and tail length, we can classify didelphid marsupials into groups reflecting different locomotor adaptations. Terrestrial adaptations include a shortened tail (e.g., *Monodelphis*), and rapid terrestrial locomotion involves increasing the relative mass of the hind leg (e.g., *Metachirus*). Arboreal adaptation is reflected by a long tail with a proportionately high mass and relatively large, powerful forelimbs (Grand 1983).

McNab (1978) studied the energetics of Neotropical marsupials and concluded that, in contrast to most Australian species, their basal metabolic rates fall in line with rates measured for many insectivorous and frugivorous eutherian mammals. It is true that no didelphid marsupial has a high metabolic rate, but it is wrong to say that they have lower metabolic rates than comparable-size eutherians specialized for similar diets.

The community ecology of didelphids has been receiving increasing attention. Fleming (1972) studied three sympatric species in Panama. Charles-Dominique (1983) has completed an extensive comparative study of didelphid ecology in French Guiana. O'Connell (1979) investigated a didelphid community in northern Venezuela and has analyzed reproductive performance, habitat preference, and substrate utilization. Atramentowicz (1986, 1988), Julien-Lafferière and Atrementowicz (1990), and Julien-Lafferière (1991) have made similar surveys for the marsupials of French Guiana, as have Fonseca and Kierulff (1988) in Brazil and Fleck and Harder (1995) in Peru.

Communication mechanisms were reviewed by Eisenberg and Golani (1977). The natural history of didelphid marsupials has been reviewed in an article by Streilein (1982a) and in books by Hunsaker (1977a), Eisenberg (1989), Redford and Eisenberg (1992), and Collins (1973).

SUBFAMILY DIDELPHINAE
Tribe Didelphini
The use of tribes under the subfamily Didelphinae follows Alpin and Archer (1987).

Genus *Chironectes* Illiger, 1811
Chironectes minimus (Zimmerman, 1780)
Yapok, Water Opossum, Cuica de Agua

Measurements

	Mean	Min.	Max.	N
TL	641.0	592	710	14
HB	289.1	259	350	
T	351.9	310	386	
HF	64.5	60	69	10
E	26.5	24	29	
Wt	386.7	550	650	3

Redford and Eisenberg 1992.

Description
The dental formula is I 5/4, C 1/1, P 3/3, M 4/4. The ears are moderately large, naked, and rounded. In addition to the usual distribution of facial hairs, the yapok has supernumerary facial bristles that are stout and long. The tail, which is longer than the head and body, is round and powerful, naked along most of its length, and coarsely scaled. In males the scrotum is clearly visible because of its mustard-colored fur and is pulled into a small fold of skin when the animal is in the water. The female when lactating has a well-developed backward-opening pouch and is able to seal it when in the water. The forefeet have expanded fingertips, lack claws, and are used with great dexterity. The hind feet, webbed to the end of the toes, are used in swimming (Marshall 1978c; Mondolfi and Medina Padilla 1957; Oliver 1976).

The peculiar color pattern and webbed feet make this, the only semiaquatic opossum, unmistakable. It has short, dense, water-repellent fur, generally grayish white marbled with deep brown. The muzzle, the top of the head, and a band extending through the eye to below the ear are deep blackish brown. The dorsum is marbled gray or black, with rounded black areas coming together along the midline and extending from the top of the head to the base of the tail and expanding laterally into four broad transverse patches placed over the shoulders, center of the back, loins, and rump. The chin, chest, and belly are pure white, and the tail is black proximally and yellowish terminally.
Chromosome number: 2n = 22; FN 20 (Reig et al. 1977).

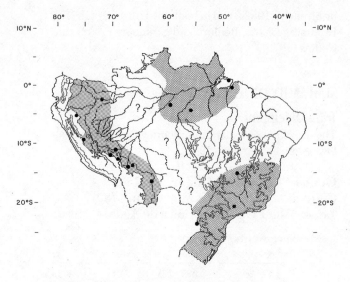

Map 5.1. Distribution of *Chironectes minimus*.

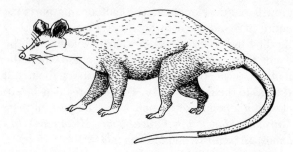

Figure 5.2. Sketch of *Didelphis marsupialis* highlighting general anatomical features.

Distribution

The yapok is confined mostly to tropical and subtropical areas, from southernmost Mexico and Central America through Colombia, Venezuela, the Guianas, Ecuador, Peru, Paraguay, and along the eastern side of Brazil to northeastern Argentina (Marshall 1978c) (map 5.1).

Life History and Ecology

Litter size in *Chironectes minimus* is one to five, with two to three most common. Animals from Venezuela have four or five nipples. The female keeps the young in her pouch when she swims. In captivity one female had her first estrous cycle at ten months of age (Marshall 1978c; Mondolfi and Medina Padilla 1957; Oliver 1976).

The yapok is confined to areas of permanent water such as streams or rivers, usually within a forest. It is an excellent swimmer and diver, paddling with its hind feet and using its tail as a rudder. The den is usually a subterranean cavity, reached through a hole in the stream bank just above water level. Within this cavity it builds a nest; in captivity animals have been observed to transport nesting material with their tails.

Yapoks are carnivorous, eating small fish, crabs, crustaceans, insects, and frogs. Prey is captured with either the front feet or the mouth (Marshall 1978c; Mondolfi and Medina Padilla 1957; Oliver 1976).

Genus *Didelphis* Linnaeus, 1758
Large American Opossum, Rabipelado, Comadreja, Gambá

Description, Taxonomy, and Comments on Behavior

This is the large opossum commonly encountered over most of North and South America. The dental formula is I 5/4, C 1/1, P 2/3, M 4/4. Head and body length ranges from 325 to 500 mm and the tail from 255 to 535 mm. Animals may weigh up to 5.5 kg. The pouch is well developed in the female. The pelage consists of two hair types, a dense underfur and long guard hairs that are generally white tipped, giving a shaggy appearance. The almost naked tail is darkly pigmented from the base for approximately one-third of its length; the distal portion is generally white to pink (fig. 5.2). The forelimbs tend to be black, and the face is white to yellow. Depending on the species, the basic dorsal color may be gray with varying amounts of black; melanistic forms are known from several parts of its range.

Recent research on *D. virginiana* suggests that the sternal gland of the male has social significance for females. Since sternal glands are prominent in male didelphids, more research needs to be directed toward explaining their function in olfactory communication (Holmes 1992). *D. virginiana* can form stable, linear dominance hierarchies in captivity. Males appear to be subservient to females (Holmes 1991).

There are three widely recognized species: *Didelphis albiventris*, *D. marsupialis*, and *D. virginiana*. Cerqueira (1985), however, recognizes *D. aurita* as a distinct species. *D. virginiana* ranges from Massachusetts in the United States to Costa Rica (Gardner 1973). Its behavior has been studied in detail by McManus (1970). Life history strategies for *D. marsupialis* and *D. virginiana* are reviewed by Sunquist and Eisenberg (1993).

Distribution

The genus is distributed from the northeastern United States to Patagonia in Argentina. The species *D. virginiana* has been introduced to the western United States, where it occurs from southern British Columbia to California.

Didelphis albiventris Lund, 1840
White-eared Opossum, Comadreja Común, Overa

Description

Measurements ($n = 14$): TL 763.5; T 372.9; HF 59.6; E 54.1; Wt 1.56 kg (Redford and Eisenberg

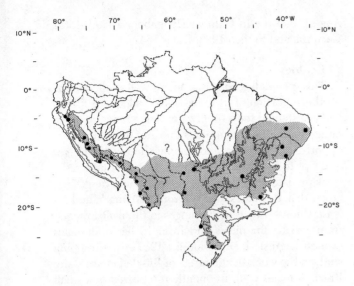

Map 5.2. Distribution of *Didelphis albiventris*.

1992). The white-eared opossum closely resembles its congener *D. marsupialis*, but adults are clearly separable by ear color. As in the black-eared opossum, there is considerable variation in body color within a population, with some individuals considerably darker than others. Another character that appears to separate the congeners is the length of the furred portion of the tail: *D. albiventris* has fur extending several centimeters up from the base, whereas the tail of *D. marsupialis* is mostly unfurred. Care must be taken in identifying juvenile *Didelphis*, because the white ear color of *D. albiventris* is much less pronounced in young animals (Varejão and Valle 1982; pers. obs.). Chromosome number: $2n = 22$; FN = 20 (Reig et al. 1977).

Distribution

The white-eared opossum inhabits the subtropical and pampean zones from Mato Grosso state in Brazil south to central Argentina and also in the Andean zones from Venezuela to western Argentina. In southern South America *D. albiventris* is found throughout Paraguay and Uruguay, and in Argentina it occurs south to about Río Negro (about 40° S) (Cerqueira 1985; Olrog 1979) (map 5.2).

Life History

In Uruguay nine females were found to have an average of 9.4 young, with a range of 4 to 12. In northeastern Argentina the mean number of young per litter was recorded as 7.1, with one or two litters produced between August and January. A study of opossum reproduction in northeastern Brazil concluded that mature males are continuously fertile and that females become fertile with the onset of rains. However, Nogueira (1988, 1989) indicates that male genital systems demonstrate subtle seasonal changes. That males

can reproduce opportunistically was also recorded in a study of this species in Minas Gerais, Brazil, and may well apply to the Chaco as well (Barlow 1965; Cerqueira 1984; Crespo 1982; González 1973; Rigueira et al. 1987). The testicles of young males descend at the time of weaning (Fonseca and Nogueira 1991).

Ecology

Didelphis albiventris is a habitat generalist, apparently found everywhere except at extreme altitudes, in very dry areas, and in dense woodland or forest. It is mainly terrestrial, though it can climb well. In Misiones province, Argentina, stomachs were found to contain large numbers of worms together with ants, small birds, eggshells, and vegetation. In northern Argentina it feeds on ripening grapes and other fruit. A study in Tucumán, Argentina, found an average minimum home range of 0.57 ha for six animals (Barlow 1965; Cajal 1981; Crespo 1982; Fonseca, Redford, and Pereira 1982; Mares 1973; Mares, Ojeda, and Kosco 1981; UM).

Comment

Didelphis albiventris is a senior name for *D. azarae*, but see Cerqueira (1985).

Didelphis aurita Wied-Neuwied, 1826

Comment

Didelphis aurita, as noted by Gardner (1993), is now listed as a full species in accordance with Cerqueira (1985). According to the current interpretation, those forms assigned to *D. marsupialis* from southeastern Brazil (map 5.3), southeastern Paraguay, and northeastern Argentina in fact could be assigned to the species *D. aurita*, originally proposed over 150 years ago (Cerqueira 1985). Although *D. azarae* was proposed in 1824

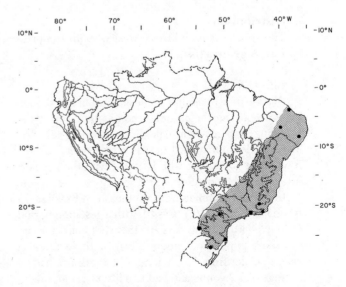

Map 5.3. Distribution of *Didelphis aurita*.

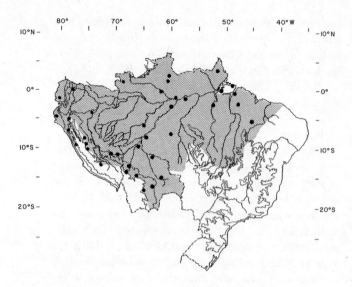

Map 5.4. Distribution of *Didelphis marsupialis*.

by Temminck, this binomial was applied to *D. albiventris* for over a century. Measurements of *D. aurita* from Minas Gerais Brazil (*n* = 9) were HB 355–73; T 355–77; HF 57; E 51; Wt 939–1,159 g (Stallings 1988b).

Didelphis marsupialis Linnaeus, 1758
Black-eared Opossum, Comadreja Grande, Mucura, Gambá

Description
Measurements: TL 716–930; T 356–465; Wt 750–2,450 g (Venezuela). *Didelphis marsupialis* is similar in appearance to *D. albiventris* but has darker dorsal pelage. The black ears separate it from *D. albiventris*. As in the white-eared opossum, there is considerable color variation within a population (Crespo 1974; pers. obs.).

Distribution
The species is found from Mexico south to northeastern Argentina (map 5.4).

Life History
Black-eared opossums are usually solitary, though two or more may be encountered together during the breeding season when males actively court females. The female builds a leaf nest in a tree cavity or burrow. Litter size varies with latitude, with the smallest litters near the equator.

In captivity, animals from southeastern Brazil had a mean litter size of 10.7 (*n* = 3), with a gestation period of fourteen to fifteen days. In the wild, animals from the same population produce two or three litters a year. Females from Rio de Janeiro state, Brazil, had an average of 7.2 young and had two litters a year, one in

August and one in October (Eisenberg 1989; Motta, Carreira, and Franco 1983; Davis 1947).

Ecology
This species has been reasonably well studied in the northern portion of its range. Given adequate shelter and a sustained food supply, the home range of a lactating female may be rather stable, but the animals are opportunistic feeders and readily shift home ranges to adapt to fluctuating resources. In Venezuela, radio tracking data indicate that extended home ranges are larger than estimates derived from trap, mark, and release studies. Mean home ranges in the llanos ranged from 123 ha for males and from 16 ha for females (Sunquist, Austed, and Sunquist 1987). In a trapping study in the wet Atlantic forest of Rio de Janeiro state, Brazil, females were frequently recaptured in a small area, but males were recaptured very infrequently. In southeastern Brazil, *D. albiventris* can occur in microsympatry with *D. marsupialis* (C. M. C. Valle, pers. comm.), though in one study *D. marsupialis* was more typically found in moister habitats. *D. marsupialis* may be more sensitive to human disturbance than its congener (Cerqueira 1985; Davis 1945; Stallings 1988a; Sunquist and Eisenberg 1993).

Genus *Lutreolina* Thomas, 1910
Lutreolina crassicaudata (Desmarest, 1804)
Little Water Opossum, Thick-tailed Opossum, Comadreja Colorada

Measurements

	Mean	Min.	Max.	N
TL	574.1	466	781	8
HB	289.4	197	378	17
T	281.9	221	336	18
HF	43.8	35	54	
E	26.3	2	38	16
Wt	432.6	176	800	14
	642.0	455	1,100	9

Redford and Eisenberg 1992.

Description
The long weasel-like body and short, dense reddish or yellowish fur make this opossum distinctive. The tail is quite thick, naked at the tip and somewhat prehensile; it is furred for about 30% to 50% of its length. The ears are short and rounded and project only slightly above the fur. The limbs and feet are short and stout. There is disagreement about the extent of pouch development, but if it does occur, it is slight. Some adults are only half the size of others. Males are often larger than females, suggesting a long period of continued growth in males.

There is considerable variation in coat color, but the upperparts are generally a rich, soft yellow buff or dark

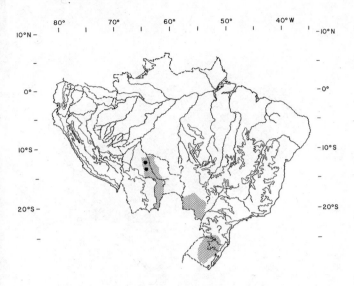

Map 5.5. Distribution of *Lutreolina crassicaudata*.

brown, and the underparts are reddish ochraceous or pale to dark brown. Most of the tail is black, though the tip is often gray. The face is devoid of eye-rings or prominent markings, but the cheeks and chin may be lighter than the rest of the head. It appears that coat color may vary with diet and climate (Barlow 1965; Marshall 1978a; Ximénez 1967).
Chromosome number: $2n = 22$; FN 20 (Reig et al. 1977).

Distribution
Two apparently distinct populations of this species exist, one known only from a few specimens in Guyana and Venezuela, the other occurring east of the Andes from Bolivia and southern Brazil south to central Argentina. The distribution of *L. crassicaudata* in Argentina may not currently connect across the center of the country (Barlow 1965; Lemke et al. 1982; Marshall 1978a; Olrog 1976; Ximénez 1967) (map 5.5).

Life History and Ecology
Very little information is available on this interesting opossum. One litter was recorded as seven and another as eleven (Marshall 1978a; Ximénez 1967). This mesic-adapted animal is found along areas of permanent water. *L. crassicaudata* is somewhat weasel-like in shape and apparently in habits as well. It can swim, and it climbs well. It is reported to be nocturnal, preying on small vertebrates, fishes, and insects. One stomach contained remnants of mollusk shells and sand, and this species has been caught in traps baited with mice. In captivity *Lutreolina* will eat insects and fruit and kills birds and mammals up to the size of *Microcavia*. A brief trapping study found that two animals had an average home range of 800 m^2 (Cajal 1981)

(Barlow 1965; Marshall 1978a; Olrog 1976; Ximénez 1967).

Genus *Philander* Tiedemann, 1808
Gray Four-eyed Opossum, Chucha Mantequera, Zorro de Cuatrojos

Description
Weights of adults range from 300 to 800 g (see species accounts). These opossums have a dorsal pelage ranging from gray to black contrasting with a buff to cream venter. The base of the tail is furred. The most distinctive feature is a white to yellow spot over each eye. Hershkovitz (1997) reviews in detail the morphology and behavior of this genus.

Comment
Philander has sometimes been referred to *Metachirops*, but see Gardner (1993).

Philander andersoni (Osgood, 1913)

Description
Measurements: HB 223–307; T 255–332; HF 34–51; E 35–42; Wt 225–425 g (Emmons and Feer 1997). Similar to *P. opossum*, but slightly larger. The middorsum is black, contrasting with gray sides. The fur at the base of the tail is not as extensive as in *P. mcilhennyi* (Patton and da Silva 1997). The feet are black, and the venter is buff.

Distribution
The type locality is "Yurimaguas," Loreta, Peru. The total range of the species is poorly defined, but it apparently ranges in the Amazonian portions of Ecuador, Peru, and adjacent Brazil (map 5.6). It may occur with *P. opossum* (Fleck and Harder 1995).

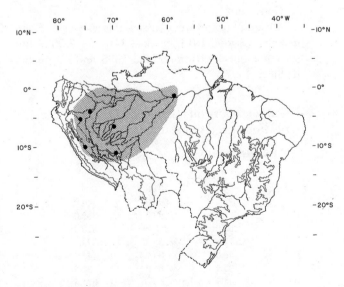

Map 5.6. Distribution of *Philander andersoni*.

Philander frenata (Olfers, 1818)

Description

Patton and da Silva (1997) recognize *P. opossum frenata* as a distinct species. This is a large dark gray form with a cream venter. Their conclusions are based on a comparison of six hundred base pairs of the light strand of the mitochondrial cytochrome 6 gene for six taxa of *Philander*.

Distribution

The southeastern coast of Brazil appears to be the range of the species. Limits of the range remain to be determined but include the state of Bahia south to Paraná and to the west including portions of Minas Gerais and Goiás.

Philander mcilhennyi Gardner and Patton, 1972

Description

Measurements of the type: HB 579; T 298; HF 43; E 39 (Gardner and Patton 1972). This species in uniformly almost black with the exception of the white spots above the eyes, the white to buff patches on the cheeks, and the small buff patches in front of the ears. The base of the tail is furred for about 20% of its length. The tips of the hairs on the forelimbs and hind limbs may be silver. The distal third of the scaly tail is white (see also description in Hershkovitz 1997).

Distribution

The type locality is Balta (10°08′ S, 17°13′ W), Río Curanja at 300 m, department of Loreto, Peru. Known from the Amazonian region of Peru and adjacent portions of Brazil (not mapped).

Comment

Most authorities would assign this species to *P. andersoni* (Osgood 1913), as does Gardner (1993), but see comments by Hutterer et al. 1995, and Patton and da Silva (1997).

Philander opossum (Linnaeus, 1758)
Gray Four-eyed Opossum, Guaiki, Cuíca

Measurements

	Mean	Min.	Max.	N
TL	558.6	437.0	620.0	12
	525.6	500.0	546.0	5
HB	265.6	202.0	302.0	12
T	293.0	235.0	320.0	12
	277.8	253.0	299.0	5
HF	42.0	39.0	46.0	5
E	35.2	34.0	36.0	5
Wt	444.4	256.3	674.5	13

Redford and Eisenberg 1992.

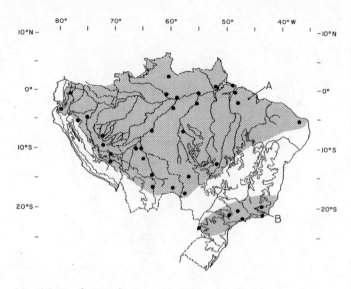

Map 5.7. Distribution of two species of *Philander:* (a) *Philander opossum;* (b) *P. frenata.*

Description

The most striking characters of this opossum are a sharply defined white spot above each eye (as in *Metachirus*) and a bicolored tail with the dark proximal part usually sharply separated from the much shorter white distal part. The fur is short, soft, and woolly. The dorsum is dark gray grizzled with white; usually the central part is darker than the sides. The color is more blackish and less brownish than in *Metachirus*. Another pale spot is present in front at the base of the ears. The cheeks are whitish, and the large, rounded ears are whitish with black rims. The venter is cream, and the tail is furred for about 15% to 25% of its length (Husson 1978; pers. obs.).

Distribution

Philander opossum is distributed from southern Mexico south to northern Argentina, and in Ecuador, eastern Peru, Bolivia, and Brazil (Crespo 1974; Gardner 1993; Massoia 1970) (map 5.7).

Life History

In Nicaragua the average litter size of twenty-one females was 6, with a range of 3 to 7; in Suriname the number of pouch young averaged 2.8 (range 1–5; $n = 6$ females); in southeastern Brazil the average litter size of seven females was 4.5; and in northeastern Argentina litter size ranged from 4 to 6. In Misiones province, Argentina, breeding occurs from August to February.

Philander has a lower mean litter size than does *Didelphis*. Compared with *Caluromys*, *Philander* has a higher reproductive rate. Sexual maturity is attained earlier, often at less than seven months, and the rearing phase of a *Philander* female is somewhat shortened. The phase of teat attachment by the young is as short as sixty days, and the nest phase lasts eight to fifteen days before dispersal (Charles-Dominique 1983; Davis 1947; Husson 1978; Phillips and Knox Jones 1969). The male genitalia have been described in great detail by Ribeiro and Nogueira (1983, 1990).

Ecology

Philander uses clicks, chirps, and hisses in communication, comparable to the sounds noted for *Caluromys* and *Marmosa*. It has a relatively large brain, approximating the condition of *Caluromys* (Eisenberg and Wilson 1981). In French Guiana *Philander* has a home-range pattern characteristic of didelphids, with broad overlap among the ranges of neighboring adults. There is no clear-cut defense of a territory; home-range stability depends on the availability of adequate resources. Apart from mating, contact among adult animals is minimal (Charles-Dominque et al. 1981).

In southeastern Brazil *P. opossum* was found most commonly in moist areas, but it wanders through nearly all vegetation types. In French Guiana the stomach contents of four individuals contained 85% animal matter and 15% fruit and seeds (Charles-Dominique 1983; Davis 1945, 1947). The water balance of *Philander* suggests it is an obligate inhabitant of mesic environments (Fonseca and Cerqueira 1991).

Tribe Marmosini
(includes *Gracilinanus, Marmosa, Marmosops, Micoureus, Monodelphis,* and *Thylamys*)
Mouse Opossums

The use of the tribe Marmosini follows Alpin and Archer (1987). These authors also place *Metachirus* in a separate tribe. The use of tribes is somewhat unsatisfactory, but molecular data do not indicate a strong relationship between *Metachirus* and *Didelphis* (Patton, Dos Reis, and Silva 1996), and *Didelphis* and *Philander* are closely related but not close to the smaller genera. On the other hand, the Marmosini may comprise at least three clades. For example, Kirsch and Palma (1995) suggest that *Marmosa, Micoureus,* and *Monodelphis* group together and separate from *Gracilinanus, Thylamys, Lestodelphys,* and *Marmosops*. However, *Marmosops* and *Gracilinanus* are a distinct "sister pair." The data from Patton, Dos Reis, and Silva (1996) are in agreement.

Description of the Marmosini

The dental formula is I 5/4, C 1/1, P 3/3, M 4/4. The mouse opossums are highly variable in size and include some forty-seven species in five genera. Head and body lengths vary from 60 mm to over 200 mm, depending on the species, and the tail ranges from 100 to 281 mm and generally exceeds the head and body in length. The ear is usually slightly shorter than the hind foot. The tail is fully prehensile, and if its base is haired it is always for less than one-third the length of the tail. There are no white spots on the face; the eye has dark brown hairs around it so that from the front one sees a contrasting mask. Dorsal pelage can vary from red brown to gray, and the venter is usually paler, varying from tan to cream.

Distribution

The Marmosini are widely distributed from southern Veracruz, Mexico, to central Argentina. Different species often occupy discrete altitudinal zones, and some species are found at over 3,000 m.

Natural History, Identification, and Systematics

The mouse opossums exhibit an interesting adaptive radiation, with many species apparently having narrow habitat requirements while others tolerate a wide range of habitat types. Climbing ability varies from species to species; some are strongly arboreal and seldom come to the ground, while others show a more scansorial tendency (O'Connell 1979; Handley 1976). The female does not develop a pouch. The teats are arranged in varying symmetrical patterns in the posterior ventral area, and the number of nipples is variable both within and between species. For the genus, the nipples range from nine to nineteen (Osgood 1921; Tate 1933); litter size is very high in some species. The young are born in an extremely undeveloped state after a thirteen- to fourteen-day gestation period and remain attached to the teats for almost the first thirty days of life (Eisenberg and Maliniak 1967).

Creighton (1984) has proposed that *Marmosa* may be divided into at least four genera, whereas Reig, Kirsch, and Marshall (1987) would group the species into three genera. The recent classification by Gardner and Creighton (1989) is shown in table 5.1. Under the new scheme *Marmosa elegans* and allies would become *Thylamys* Gray 1943. Species of this genus have a woolly pelage and rounded premaxillae and store fat in their tails. Whereas most species of *Marmosa* have the teats arranged in a circular pattern, in members of the genus *Thylamys* the teats form the bilaterally symmetrical rows common in eutherians.

The genus *Micoureus* Lesson, 1842, is distinctive in that most of the species are large, with head and body length exceeding 125 mm. The tail vastly exceeds the head and body, averaging at least 1.3 times as long. The scales on the tail are rhomboid and coarse, with four-

Table 5.1   Classification of *Marmosa* into Five Genera

Micoureus	*M. incanus*
M. alstoni	*M. invictus*
M. cinereus	*M. neblina*
M. constantiae	*M. noctivagus*
M. regina	*M. parvidens*
Marmosa	*M. paulensis*
M. andersoni	*Thylamys*
M. canescens	*T. elegans*
M. lepida	*T. macrura*
M. mexicana	*T. pallidior*
M. murina	*T. pusillus*
M. robinsoni	*T. velutinus*
M. rubra	*Gracilinanus*
M. tyleriana	*G. aceramarcae*
M. xerophila	*G. agilis*
Marmosops	*G. dryas*
M. cracens	*G. emiliae*
M. dorothea	*G. marica*
M. fuscatus	*G. microtarsus*
M. handleyi	
M. impavidus	

Source: Gardner and Creighton 1989.

teen to sixteen rows per centimeter. The tail is never bicolored but is often whitish or mottled distally. The fur of the body extends at least 5 cm on the proximal portion of the tail. The tympanic bullae of the skull are small, and the postorbital processes are prominent.

The genus *Marmosops* Matschie, 1916, also has rhomboid and spirally arranged or annular scales on the tail, but they are much finer, with twenty-two to twenty-eight per centimeter. The bullae are small, but the postorbital processes are lacking. The lower canines are premolariform, which is distinctive.

The genus *Marmosa* Gray, 1821, is now subdivided into the *murina* group, or *Marmosa* in the strict sense, and the *microtarsus* group; the latter could be elevated to the genus *Grymaeomys* Burmeister, 1854, or *Gracilinanus* as proposed by Gardner and Creighton, 1989, and is acknowledged by Hershkovitz (1992b). In the *microtarsus* group the tail scales are square and very fine grained, as many as forty per centimeter, and are annular rather than spiral. The tail tends to be weakly bicolored, and the bullae are large. Pectoral mammae are present.

The *murina* group has tail scales that are rhomboid and coarse, at about fifteen to twenty per centimeter, and spiral as in *Micoureus*. Although fur may extend on the proximal portion of the tail, it never exceeds 2.5 cm. The premaxillaries exhibit an acute outline. Postorbital processes are present and prominent.

The southern and high-altitude species of mouse opossums often encounter extreme cold for brief periods. Some may be able to enter torpor for a few days. Fat storage on the body and in the tail may ameliorate nutritional stress when little food is available. McNab (1978, 1982) has suggested that small marsupials may be near their energy limits when adapting to temperate climates. Their relatively long lactation and the obligate teat attachment phase of the young may place severe constraints on successfully rearing a litter. It is no wonder that as marsupials reach temperate climates fat storage becomes a necessity.

Comment

In accordance with Gardner and Creighton (1989), the former species of *Marmosa* will be discussed under their new generic scheme. New species of this group may well await description (Husband 1992).

Genus *Gracilinanus* Gardner and Creighton, 1989

Description

To quote from Gardner and Creighton (1989), "Species of *Gracilinanus* are distinguishable from those of *Marmosops* by the shape and arrangement of tail scales and bristles . . . , the non-premolariform lower canine, and the presence of pectoral mammae. They can be separated from species of *Thylamys* by the absence of seasonal fat deposits in the tail (thus non-incrassate), absence of densely granular central palmar and plantar surfaces, and relatively longer digits and broader interdigital pads on manus and pes. *Gracilinanus* can be distinguished from *Marmosa* and *Micoureus* by the lack of postorbital processes on the frontals and by the annular arrangement of minute scales on the tail." Hershkovitz (1992b) amplifies the diagnosis and description of the species within this genus.

Distribution

This lowland, southern genus is recorded from Brazil, Bolivia, Peru, Paraguay, Argentina, Venezuela, and Ecuador.

Gracilinanus aceramarcae (Tate, 1931)

Description

Measurements: HB 83; T 112; HF 16; E 16 (Tate 1933). The dorsum is reddish brown; the pelage is "long and lax." The tail is not conspicuously long and is also reddish brown.

Distribution

A specimen is known from southern Peru (Limbani Valley, department of Puno), at 2,700 m (J. L. Patton, pers. comm.). The type locality is Río Aceramarca (a tributary of the Río Unduavi), Yungas, Bolivia (not mapped).

Gracilinanus agilis (Burmeister, 1854)
Marmosa Rojiza

Measurements

	Mean	Min.	Max.	N
TL	242.4	226	259	8
HB	100.9	91	109	
	75.4	70	84	13
T	141.5	128	162	8
	106.9	95	120	13
HF	17.4	16	19	8
	14.4	13	15	13
E	21.4	20	23	8
	15.9	15	18	13
Wt	27.0	23	34	8
	14.9	12	19	10

Redford and Eisenberg 1992.

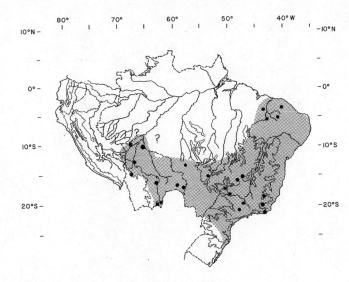

Map 5.8. Distribution of *Gracilinanus agilis*.

Description

Measurements from Brazil: HB 72–94; T 119–26; HF 16; E 18–19 (Tate 1933). Dorsal color varies from dusty brown to gray brown. As originally understood and described by Tate (1933), variation in the color of the dorsal pelage was duly noted, and four subspecies were proposed (plate 1). After noting the color, Tate said, "the fur lies smooth and close and is very uniform in length." The tail is relatively shorter than in *M. microtarsus*.

This medium-sized mouse opossum (note differences between samples) has a very long bicolored tail equaling 140% of its head and body length. The ears are very large. The fur is close and smooth and uniform in length. The body color varies from dull, dusty brown to grayish brown, lighter on the sides and cream on the venter. The face is paler, with the cream of the belly extending well onto the cheeks, and adults have distinct black eye-rings (Massoia and Fornes 1972; Tate 1933; pers. obs.).

Distribution

Brazil, Paraguay, northern Argentina, and Uruguay, through eastern Bolivia to Peru (map 5.8).

Life History and Ecology

This mouse opossum is reported to have up to twelve young. Females lack a true pouch, and the teats remain hidden when the female is not lactating (Massoia and Fornes 1972; Nitikman and Mares 1987).

G. agilis is a characteristic inhabitant of the gallery forests of southern South America but has broad habitat tolerance. It has been caught under fallen trunks, in tree holes, and in moist woodland. It is reported to be an adept climber, and nests made of vegetation have been found 1.6 m off the ground. One such nest contained seven individuals. In eastern Paraguay it has usually been captured in vegetation but sometimes has been caught on the ground. In captivity these mouse opossums eat ground meat and drink a lot. They are frequently found in *Tyto* owl pellets.

G. agilis is found throughout the Brazilian cerrado, usually associated with mesic areas such as gallery forests. In one gallery forest it was found at a biomass of 126 g/ha (Massoia and Fornes 1972; Nitikman and Mares 1987; UM).

Gracilinanus dryas (Thomas, 1898)

Description

In this very small species the head and body length ranges from 90 to 100 mm, tail length from 130 to 150 mm. Weight averages about 18 g. The dorsum is dark reddish brown, with the venter pale cream to pale brown.

Range and Habitat

This species occurs in the mountains of western Venezuela, eastern Colombia, and perhaps to Ecuador. It is associated with moist forest habitat at elevations over 2,000 m (not mapped) (see also Eisenberg 1989).

Natural History

This small species forages both arboreally and terrestrially. It is found in cloud forest over most of its range but may be more abundant in second growth (Aagaard 1982). In the mountains of northern Venezuela it is replaced by the equally small *Marmosa marica*.

Gracilinanus emiliae (Thomas, 1909)

Description

Measurements: HB 75; T 142; HF 15; E 16 (Tate, 1933). Husson (1978) recorded specimens from Suriname resembling the type with measurements HB 60;

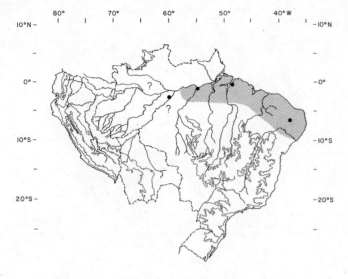

Map 5.9. Distribution of *Gracilinanus emiliae*.

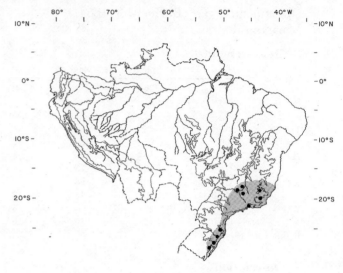

Map 5.10. Distribution of *Gracilinanus microtarsus*.

T 112. The dorsum is chestnut brown, contrasting with a cream venter. The feet are pale dorsally. The tail is extremely long. The type specimen is a male not fully adult. The ears are large.

Distribution

The type locality is listed as Pará, Brazil. If the specimens recorded in Husson (1978) are correctly assigned, then the range could well be from northeastern Brazil to the Guianas (map 5.9).

Gracilinanus microtarsus (Wagner, 1842)

Description

Measurements: TL 221–56; HB 80–112; T 95–130; HF 15; E 18, Wt 31 g (Redford and Eisenberg 1992). The tail is equal to 118% of the head and body length. The long pelage is usually rough or shaggy looking because of the numerous guard hairs. This comparatively bright colored species has a dorsum ranging from tawny to russet. The face is distinctly paler than the body, and the vibrissae are well developed and rather long. *G. microtarsus* is more vividly colored than *G. agilis* and has a rougher coat (Tate 1933).

Distribution

G. microtarsus is known only from southeastern Brazil and from Misiones province in Argentina (Davis 1947; Massoia 1980; Vieira 1949) (map 5.10).

Ecology

In moist Atlantic forest in eastern Brazil *G. microtarsus* was regularly caught on the ground as well as in trees, in both virgin and second-growth forest, though its morphology suggests extensive arboreality. It will probably be found to be confined to areas of moderate to high rainfall (Davis 1947; Stallings 1988b; Tate 1933).

Genus *Marmosa* Gray, 1821
Mouse Opossum

Description

The dental formula is I 5/4, C 1/1, P 3/3, M 4/4. The tail scales are rhomboid but rather coarse. Although the tail scales spiral, the count averages fifteen to twenty scales per centimeter. There are no white spots on the face; the eye has dark brown hairs around it so that when viewed from the front the effect is of a contrasting mask, a character shared with other "mouse opossums."

Distribution

The genus is widely distributed from southern Veracruz, Mexico, to northern Argentina. Different species often occupy discrete altitudinal zones.

Natural History

Marmosa robinsoni females exhibit ovarian senescence after fourteen to sixteen months of age (Barnes and Barthold 1969; pers. obs.). Given sexual maturity at six months, this senescence restricts breeding to eight to ten months following maturity. It is doubtful if this species can breed more than once under more stringent climatic regimes where a long annual drought occurs (Eisenberg 1988). All species studied to date seem to be nocturnal. Most appear to eat fruit when it is available and feed heavily on invertebrates (Hunsaker 1977a; O'Connell 1983).

Social life is rudimentary in *Marmosa*, and aside from a courting pair or a female and her dependent young, adults forage and nest alone. Communication between same-sex adults is agonistic. An open-mouth threat and hissing constitute a common defensive posture. Young animals chirp when detached from the teats, and the mother will return and push them under her body so they can reach the teat area. Males courting females approach and produce clicking sounds. Males mark by rubbing the chin, chest, or cheeks on the substrate. Both sexes urine mark and drag their cloaca on the substrate, leaving chemical traces that presumably coordinate reproductive behavior (Eisenberg and Golani 1977).

Marmosa andersoni Pine, 1972

Description
Measurements of the type: T 181; HF > 20; E 16.9. The tail exceeds the head and body in length. Pine (1972) notes that the postorbital processes of the skull are quite large. This is a medium-sized species with prominent black eye-rings. The dorsal fur is gray but tipped with brown. The venter is cream, but the hairs are gray at the base.

Distribution
Known only from the type locality: Peru, Cuzco, Hda Villa Carmen (Cosnipata), 12°52′ S, 71°15′ W at 600 m (map 5.11).

Marmosa lepida (Thomas, 1888)

Description
This small mouse opossum has a total length of 208 to 262 mm and an average head and body length of 108 mm, with the tail 150 mm and the hind foot 16 to 19 mm. The dorsum is a dark reddish brown, and the venter is cream or dirty white.

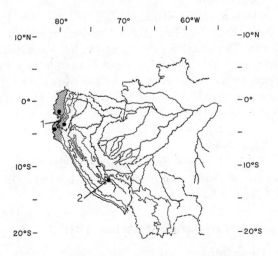

Map 5.11. Distribution of two species of *Marmosa*: (1) *Marmosa robinsoni*; (2) *M. andersoni*.

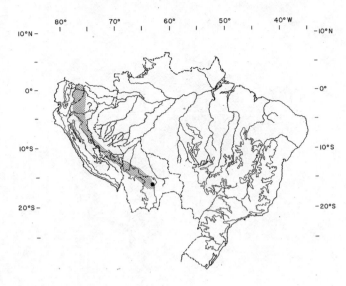

Map 5.12. Distribution of *Marmosa lepida*.

Range and Habitat
This species is widely distributed from eastern Peru, Ecuador, and Bolivia extending east to Suriname. It may penetrate extreme southern Colombia from Ecuador, but its presence there has not been confirmed (map 5.12).

Marmosa murina (Linnaeus, 1758)
Marmosa Ratón

Measurements

	Mean	Min.	Max.	N
TL	275.29	221	358	14
T	161.71	133	212	14
HF	18.75	14	25	14
E	20.67	20	23	9
Wt	26.00	13	44	5

Redford and Eisenberg 1992; Eisenberg 1989.

Description
Total length averages 332 mm (220–358), with head and body length 110 to 146 mm; the tail is longer than the head and body (140–212 mm) (Mount Duida, Venezuela). *Marmosa murina* has a pale buff dorsal pelage contrasting with cream underparts. The black facial mask is prominent, and the tail may be faintly bicolored.
Chromosome number: $2n = 14$; FN = 24.

Range and Habitat
This species is distributed from northeastern Brazil west to Ecuador and Peru and north to adjacent parts of Venezuela and the Guianas. In Venezuela it extends through the north coast range to the Maracaibo basin. It is found at elevations below 1,300 m and is strongly associated with moist habitats and tropical evergreen forest (map 5.13).

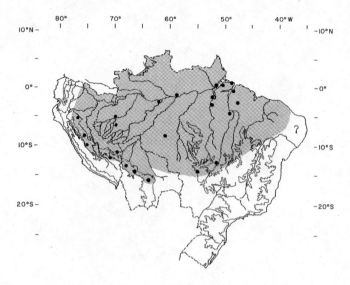

Map 5.13. Distribution of *Marmosa murina*.

Natural History

This mouse opossum is arboreal, nocturnal, and insectivorous, but it is versatile in its habitat exploitation and is frequently trapped on the ground, sometimes near human dwellings. The female is tolerant of the male only during estrus; copulation may last several hours, and gestation takes thirteen days. The litter size averages 5.8, and the teat number is eleven. The species' behavior in captivity was analyzed by Eisentraut (1970). Maternal behavior is similar to that described for *Micoureus;* the female constructs a leaf nest by transporting nesting material with her prehensile tail. Young are weaned at about 12 g body weight.

Marmosa robinsoni Bangs, 1898

Description

External measurements show a rather wide range: HB 110–57; T 135–83; HF 18–23; E 20–24. Males tend to be considerably larger than females. A large male may have a total length of over 350 mm, with the tail exceeding 200 mm. The dorsum is light brown, and the venter tends to be white. Dark coloration around the eyes gives the effect of a little black mask. Chromosome number: $2n = 14$; FN $= 24$.

Range and Habitat

The species' range extends from Panama across northern Colombia and Venezuela; it is not recorded from the Guianas. From southern Panama it extends along the Pacific coast of Columbia through western Ecuador. This opossum has a broad habitat tolerance and has been taken at from sea level to 1,260 m. Most specimens in Venezuela were taken at below 500 m. It is widely tolerant of seasonally dry, deciduous habitats and ranges down into the llanos of Colombia and Venezuela (map 5.11).

Natural History

This species appears to forage for fruit and insects both on the ground and in vines and trees. In the llanos of Venezuela it frequently shelters under the fronds of the palm *Copernicia tectorem.* Home-range size is influenced by flooding in the llanos. August (1981) estimated a mean home range of 0.36 ha in the dry season and 0.17 ha in the wet season when flooding occurs.

Its reproductive biology has been studied by Fleming (1973) in Panama and by O'Connell (1979) in Venezuela. The onset of reproduction is influenced by the rainy season. Captive studies of reproduction include Eisenberg and Maliniak (1967) and Barnes and Barthold (1969). A male courts a female by approaching and emitting clicks. While copulating the male maintains a grip with his prehensile tail wrapped around a branch. Copulation lasts several hours. Gestation time is thirteen to fourteen days. The female builds a leaf nest in a suitable cavity, and while rearing young she may show extreme site fidelity. From birth to five weeks of age, she carries the young continuosly. They remain attached to the teats until they are approximately twenty-one days old, then they may begin to detach and crawl on the female's venter. Their eyes open at thirty-nine to forty days, and they remain in the nest while the female forages alone. The young begin to venture out of the nest at forty to fifty days of age. They are fully weaned at sixty-five days and shortly thereafter disperse and establish solitary nest sites. The female has fifteen teats. In Panama, litter size ranged from six to thirteen (Fleming 1973). In the llanos of Venezuela, litter sizes ranged from thirteen to fifteen, with an average size of fourteen (O'Connell 1983).

Marmosa rubra Tate, 1931

Description

Measurements: HB 145–50; T 195–210; HF 22–24; E 18–21 (Tate 1933). The dorsum is reddish brown. The venter is pale cinnamon, but some specimens have a distinct stripe about 1 cm wide from the throat to the chest, bordered by paler hairs. An eyering is present but reduced behind the eye. The tail is bicolored, and the feet are dusky dorsally.

Distribution

Lowland eastern Ecuador, probably the Río Napo basin and thence south into Amazonian Peru (Tate 1933) (map 5.14).

Genus *Marmosops* Matschie, 1916

Description

The lower canines are premolariform. The size range is intermediate (see p. 57). Very small rhomboid

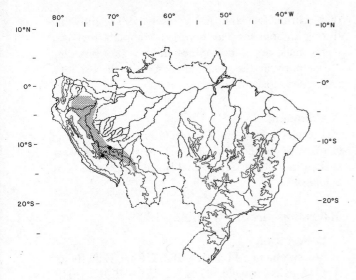

Map 5.14. Distribution of *Marmosa rubra*.

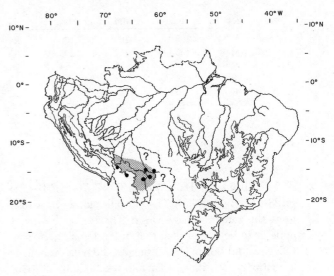

Map 5.15. Distribution of *Marmosops dorothea*.

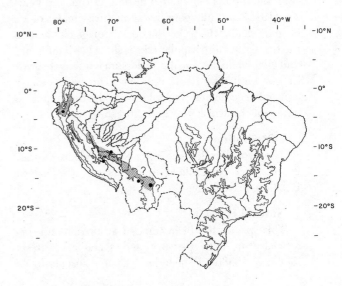

Map 5.16. Distribution of *Marmosops impavidus*.

scales are present on the tail. A scale count would yield from twenty-two to twenty-eight per linear centimeter. The tympanic bullae are not enlarged, and the postorbital processes are lacking. Mustrangi (1995) provides a comprehensive review of the genus.

Distribution
The genus is widely distributed in lowland, tropical South America (see species accounts).

Marmosops dorothea (Thomas, 1911)

Description
Measurements: HB 150; T 158; HB 18; E 20. The dorsum is brown and the venter cream. The bases of the ventral hairs are gray. The species is based on two female specimens.

Distribution
The Yungas, Bolivia (Tate 1933), type locality is listed as Bolivia, La Paz, Río Solocame, 67° W, 16° S (map 5.15).

Comment
Tate (1933) lists this form as a subspecies of *M. noctivaga*.

Marmosops impavidus (Tschudi, 1844)

Description
This medium-sized species has a total length of about 300 mm (280–320); HB 125–50; T 155–87; HF 19–22; E 21–23. Weights range from 36 to 45 g (department of Puno, Peru, MVZ). The dorsum is dark brown, blending to gray brown on the sides; the midventer is cream separated by a gray-based, buff-tipped lateral area separating the cream venter from the

brown dorsum. The black eye-rings are prominent, and the tail is distinctly bicolored.

Range and Habitat
Marmosops impavidus occurs from the montane portions of Panama to the mountains of western Venezuela and then is distributed south to Ecuador, Peru, Bolivia, and western Brazil. This species prefers moist evergreen forests. It is arboreal, but often taken on the ground, and occurs at lowland elevations up to 2,300 m (map 5.16).

Marmosops incanus (Lund, 1840)

Measurements

	Sex	Mean	Min.	Max.	N
HB	F	137.5	110	162	22
	M	138.0	95	192	46

	Sex	Mean	Min.	Max.	N
T	F	183.4	163	199	22
	M	195.1	162	296	45
HF	F	19.8	16	24	22
	M	21.5	18	26	46
E	F	25.9	23	29	22
	M	27.5	24	32	46
Wt	F	58.6	44	73	22
	M	66.0	35	130	46

Stallings 1988b; Brazil.

Description

Measurements: HB 120–35; T 147–70; HF 19–21; E 23–27 (Tate 1933). The dorsal color is brownish gray; the venter is snow white. The fur of the males appears to be longer than that of the female. Persson, Oliveira, and Lorini (1992) suggest the pelage changes at maturity according to age and sex. Stallings (1988b) also noted that adult males had a distinctive pelage color but did not elaborate. Oliveira, Lorini, and Persson (1992) carefully note that at maturity males of *M. incanus* exhibit the following dorsal pelage characters: the fur of the interscapular area is short (4–9 mm) and dull gray, while the remaining dorsum is a shiny brownish gray with hairs 10–12 mm long. The adult females are similar to juveniles in color, with the dorsal pelage uniformly brownish gray and short (5–8 mm).

Distribution

The type locality, is Lagoa Santa, Minas Gerais, Brazil; possibly confined to forests of southeastern Brazil (map 5.17).

Natural History

This species has been trapped arboreally and terrestrially, in both primary and secondary forests (Stallings 1988b). It includes fruit and insects in its

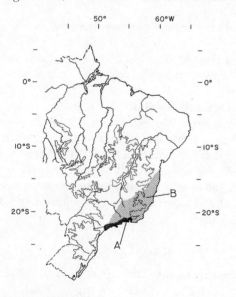

Map 5.17 Distribution of two species of *Marmosops:* (a) *Marmosops paulensis;* (b) *M. incanus.*

diet. Lorini, Oliveira, and Persson (1994) note that during an annual cycle there is a complete turnover in the male population. Since there is one breeding season per year, they conclude that the males exhibit a nearly semelparous breeding strategy.

Comment

Lorini, Oliveira, and Persson (1992) and Mustrangi (1995) discuss the sympatry of two species of *Marmosops* in southeastern coastal Brazil (see *Marmosops paulensis*).

Marmosops neblina (Gardner, 1989)

Description

Measurements: TL 291; T 163; HF 20; E 24 (holotype). Means for eight males: TL 277.8; T 152.4; HF 19.5; E 21.8; means for five females: TL 267.8; T 152; HF 17.4; E 20.4. Gardner (1989) originally placed *M. neblina* as a subspecies of *M. impavidus*. The dorsum is dark brown to the furred base of the tail. The toes of the forefeet are clothed with white hair tufts. The tail is brown above and not distinctly bicolored. There is a conspicuous black eye patch. The venter is pale brown with a white chin, and there may be a white chest patch.

Distribution

The type locality is Cerro Neblina, territory of Amazonas, Venezuela. May extend to Roraima, Brazil (see Mustrangi and Patton 1997) (not mapped).

Marmosops noctivagus (Tschudi, 1844)

Description

Three specimens from Itajubá, Brazil, measured TL 265–80; T 150–70; HF 17; E 22; Wt 35–40 g (Belém). This medium-sized species has a brownish gray dorsum grading to a tan brown on the face and flanks. The venter is creamy white extending to the chin and cheeks. The black eye "mask" is prominent. The tail is bare for its entire length (plate 1).

Distribution

Amazonian portions of Ecuador, Peru, and Bolivia extending to western Brazil (map 5.18).

Marmosops parvidens (Tate, 1931)

Measurements

	Mean	Min.	Max.	N
TL	235.67	225	248	6
T	136.67	129	145	6
HF	16.33	15	18	6
E	18.83	15	20	6

Location: Venezuela, Bolivia, Colombia (AMNH).

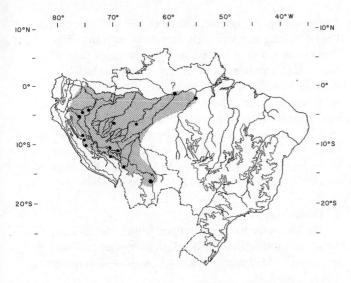

Map 5.18. Distribution of *Marsmosops noctivagus*.

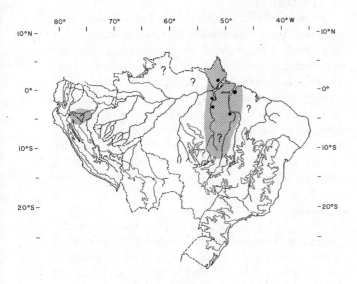

Map 5.19. Distribution of *Marsmosops parvidens*.

Description

This species rarely is over 250 mm in total length (227–93), and the tail greatly exceeds the head and body length. The dorsum is pale brown to chocolate brown, the venter is gray to white. Handley notes that the coat has a shaggy appearance that distinguishes it from *Marmosops cracens* (Handley and Gordon 1979).

Range and Habitat

The range of this species appears to be highly fragmented. It occurs in the Guianas, Venezuela, Colombia, and south through Amazonian Brazil to Peru. In Venezuela it occurs at elevations below 1,000 m, but Pine (1981) has records up to 2,000 m. It is strongly associated with moist habitats (map 5.19).

Natural History

This species appears to forage both in the trees and on the ground, and it prefers moist tropical forest. Embryo counts vary from six to seven; teat number is seven (Pine 1981). Malcolm (1990) reports densities of 14.4/km^2 near Manaus, Brazil.

Marmosops paulensis (Tate, 1931)

Description

Average measurements for adults are males: TL 260–370; Wt 20–70 g; females: TL 240–320; Wt 20–50 g. Additional measurements are females: HB 108–53; T 150–212; HF 19–24; E 20–27; males HB 94–150; T 145–81; HF 17–22; E 19–23 (Mustrangi 1995; Mustrangi and Patton 1977). Generally smaller than *M. incanus*. Ears are smaller than in *M. incanus* (<25 mm). Dorsum is dark brownish gray with a reddish wash, and venter is creamy white. The pelage is long (10 mm). Eye-rings are dark and prominent. Hind feet are covered with white hairs. Tail is bicolored with the last 2–3 cm white.

Distribution

Tate (1933) considered these forms to be the "coastal representative of *M. incana*," Rio de Janeiro to Paraná, but Mustrangi (1995) clearly separates this taxon from *M. incanus*. The two species can occur in sympatry (map 5.17).

Comment

Formerly considered a subspecies of *M. incanus* (see Mustrangi and Patton 1997).

Genus *Micoureus* Lesson 1842

Description

This genus is distinctive for its large body size (an adult usually is over 150 mm in head and body length). The tail generally exceeds the head and body length. The species of this genus are highly arboreal (see species accounts).

Distribution

The genus extends from Veracruz, Mexico, to northern Argentina.

Micoureus constantiae (Thomas, 1904)

Measurements

	Mean	Min.	Max.	N
TL	367.5	330	400	11
HB	161.7	140	180	
T	205.7	190	220	
HF	23.5	21	27	
E	25.8	23	31	
Wt	90.0	1		

Redford and Eisenberg 1992.

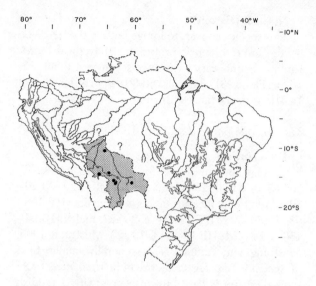

Map 5.20. Distribution of *Micoureus constantiae.*

Description

The tail of this large murine opossum is about 125% the length of the head and body, proportionally slightly shorter than in *M. demerarae.* It is gray for most of its length, but white on top for its final 20% and on the bottom for its final 40%. The dorsum is gray to gray brown; the venter is some shade of cream to yellow. There is frequently a thin line of bright tannish orange separating the dorsal and ventral colors. The black eye-rings are well developed (Tate 1933; Miranda-Ribeiro 1936; pers. obs.).

Distribution

This mouse opossum is known from Mato Grosso, Brazil, western Bolivia, and northwestern Argentina (map 5.20).

Life History

Females are reported to have fifteen inguinal mammae (Tate 1933).

Micoureus demerarae (Thomas, 1905)

Measurements

	Mean	Min.	Max.	N
TL	388.0	270.0	460.0	16
HB	168.6	120.0	200.0	
T	219.4	150.0	260.0	
HF	25.1	22.5	29.5	8
E	27.5	25.0	30.0	2
Wt	109.9	56.0	194.0	36 m
	99.1	53.0	230.0	28 f

Redford and Eisenberg 1992.

Description

This is one of the largest mouse opossums in South America and is readily identifiable by its size. In southeastern Brazil males are considerably larger than

females. The ears are large to moderate and are gray; the tail is long and strikingly bicolored, with the distal portion cream or yellowish and the proximal portion black. Hair extends about 30 mm along the tail. The dorsum is some shade of gray, the sides are lighter, and the venter varies from buff to yellowish. The white of the venter extends to the entire chin and the lower cheeks. The eye is surrounded by a distinct dark eye-ring (plate 1). The reddish dorsal coloration of live individuals fades soon after death (Husson 1978; Massoia 1972; Stallings 1988b; Tate 1933; pers. obs.).

Distribution

M. demerarae has been trapped from Colombia and Venezuela south through Brazil to northeastern Paraguay and in Misiones province, Argentina (Eisenberg 1989) (map 5.21).

Life History

Females of this species do not have a pouch, and they have up to eleven teats. This mouse opossum has been studied in captivity, where it makes a nest with material it carries in its mouth. When detached from the teat the young utters a repetitive chirping cry, inducing the female to approach, grasp it with her forepaws, and push it under her venter, whereupon it reattaches to the nipple. In Venezuela breeding is tied to rainfall, with no reproduction during the winter dry season (Beach 1939; Eisenberg 1989; O'Connell 1979; Stallings 1988b).

Ecology

This species forages both arboreally and on the ground. In Venezuela it was trapped 47% of the time on the ground and 53% in trees and bushes. In Minas Gerais, Brazil, it was found in brushy and forested habitats, though almost always off the ground. An ex-

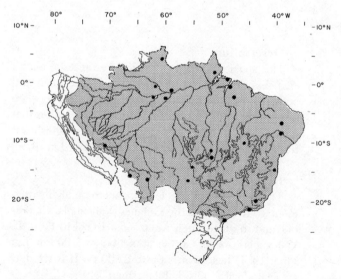

Map 5.21. Distribution of *Micoureus demerarae.*

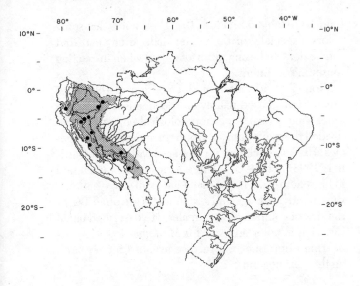

Map 5.22. Distribution of *Micoureus regina*.

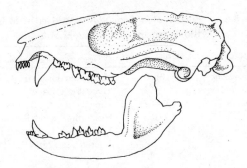

Figure 5.3. Skull of *Monodelphis* sp.

amination of three stomachs showed only insects. It is nocturnal and constructs open, arboreal nests. From the limited data from southern South America, it appears that this species lives mostly in rich subtropical forest, though it has been captured arboreally in thorn forest in eastern Paraguay (Massoia 1972; Stallings 1988a,b). Malcolm (1990) reports densities of 4.8 km^2 near Manaus, Brazil.

Comment

Changed by Gardner (1993) from *Marmosa cinerea* (Temminck, 1824) to *Micoureus demerarae* (Thomas, 1905), since "*cinerea* is preoccupied by *D. cinerea* Goldfuss, 1812."

Micoureus regina (Thomas, 1898)

Description

Measurements: HB 230; T 220; HF 30; E25. The teeth are unusually small (Tate 1933). This is a rather large species resembling *M. demerarae*. The dorsum is cinnamon brown, and the tail is uniform in color. The tips of the dorsal hairs are reddish yellow with a gray base. The venter is reddish cinnamon. The ear is of modest size. The basal portion of the tail is furred for only 1.5 cm.

Distribution

Colombia, Ecuador, Peru, Bolivia, and western Brazil (map 5.22).

Genus *Monodelphis* Burnett, 1830
Short-tailed Opossum

Description

The dental formula is I 5/4, C 1/1, P 3/3, M 4/4 (fig. 5.3). Head and body length ranges from 71 to 179

mm, and tail length from 45 to 65 mm. In some species the adult male may be 30% larger than the adult female. The tail is approximately half the head and body length, and this characteristic immediately distinguishes the short-tailed opossums from any other opossums within their range. Color varies widely depending on the species, from gray to chestnut brown. Some species have black dorsal stripes (plate 1). Chromosome number: typically $2n = 18$.

Distribution

The genus *Monodelphis* is widespread from southwestern Panama to Argentina.

Life History and Ecology

Some species of this genus may be both nocturnal and diurnal. Their activity is predominantly terrestrial and crepuscular. The litter size may be high in some species, and Pine, Dalby, and Matson (1985) suggest that *Monodelphis dimidiata* may exhibit a form of reproduction similar to that of the Australian *Antechinus stuartii*. In this pattern the males die shortly after a sharp seasonal reproductive period and the adult females, after rearing their litters, usually do not survive to reproduce in the following year. The data of Pine, Dalby, and Matson (1985) are provocative, and appropriate field and laboratory studies could shed some light on this proposed reproductive pattern.

Especially in Brazil, Peru, and Bolivia, the numerous species of the genus *Monodelphis* appear to occupy the small terrestrial, insectivore, omnivore niche. Rodents of the genera *Oxymycterus* and *Akodon* are clearly providing them with some trophic competition. One wonders, given the absence of the classical "Insectivora," to what extent rodents sustain competition with the "old endemic," insectivorous marsupials. Gardner (1993) recognizes eight species of *Monodelphis* that could occur in southeastern Brazil. Five of these are rather restricted in their distribution and are poorly represented in collections. There is little information on their ecology and current status. This area of Brazil was colonized early on by Europeans, and exotic introductions of the domestic cat (*Felis catus*) and

Table 5.2   Species of *Monodelphis* viewed at the Museo, Rio de Janeiro, and the Field Museum

Name	HB (mm)	Description (as Labeled)
M. americana	≈101–5	Three bold dorsal stripes
M. iheringi	≈92	Three dorsal stripes
M. rubida	125–37	Three faint dorsal stripes
M. scalops	≈133	Head, rump, and tail rufous
M. sorex	110–30	Feet reddish; head, neck, and forequarters gray. No dorsal stripes
M. theresa	≈77	One dorsal stripe
M. unistriata	˜≈140	A single chestnut dorsal stripe

Note: These observations are in conformity with Emmons and Feer 1997.

European rodents (*Mus* and *Rattus*) have no doubt had a negative impact on the native small mammals. Studies on small mammals described for the states of Espírito Santo, São Paulo, Rio de Janeiro, and Paraná are urgently needed.

Comment

Monodelphis americana is reputed to extend from southeastern Brazil to French Guiana (Honacki, Kinman, and Koeppl 1982), but we cannot confirm the French Guiana record. Husson (1978) recognizes only *M. brevicaudata* in Suriname. Pine and Handley (1984) recognize *M. americana* as a valid species and concur with Reig et al. (1977) in separating *M. orinoci* from *M. brevicaudata*. Wilson and Reeder (1993) list *M. americana* for eastern Brazil. The systematics of the genus *Monodelphis* are still in flux, but field studies offer great promise (Pine, Dalby, and Matson 1985).

Pine (1976) recognizes four color patterns within the genus *Monodelphis*. Species exhibiting similar patterns need not be closely related. Within the area covered in this volume, four distinct color patterns may be found (see table 5.2). Pine (1973) also notes that some colors, especially of the venter, fade after death but can be detected using ultraviolet light when inspecting museum specimens.

Gomez (1991) attempted a revision of the genus *Monodelphis* based mainly on specimens from Brazil, employing a variety of morphological characters. He divided the genus into six groups including fifteen species. His nomenclature is compared with that of Gardner (1993) in table 5.4. Three forms remain unnamed in his scheme. He noted that adult skull shape and size vary greatly with age and sex. Females are only 70% of the size of comparable-age males. He described clinal variation in color and size for *M. domestica*. His pioneering effort suggests further avenues for research on this vexing genus.

Monodelphis adusta (Thomas, 1897)

Description
Measurements: 108 mm; T 60; HF 17; E 14; Wt 35 g. The tail is only slightly prehensile. The upperparts are dark brown, darkest along the dorsal midline. In some specimens a dorsal midline stripe is discernible, but this is not invariant. The throat is gray and the venter darker gray. The midline of the venter is marked by a buff white stripe from the pectoral region to the teat area.

Range and Habitat
This species is found from southwestern Panama and west of the western cordillera of the Andes in Colombia, then to Ecuador, Peru, and Bolivia (map 5.23).

Natural History
The habits of this species are very poorly known. It is assumed to be terrestrial and apparently feeds on

Table 5.3   Some Measurements for *Monodelphis*

	Source	HB	T	HF	E
M. americana	a	87–111	42–51	14.5–16	11–12
	b	101–5	45–55	16–18	15
M. dimidiata	a	119–51	66–88	15.5–17	7.1–11
M. iheringii	a	87	47	13	10
M. rubida°	a	123	53	16	13
	b	125–37	56–60	15–18	12
M. scalops	a	123	53	16	13
	b	133	71		8
M. sorex°°	a	106–8	55–62	15.5	5.3
	b	110–30	65–85	13–16	10–13
M. theresa	a	97	47	17	14.5
	b	—	—	—	—
M. unistriata	a	140	65	15.4	5.5
	b	140	˜60	—	—

[a] Miranda-Ribiero 1936.
[b] Emmons and Feer 1997.
° includes *M. umbristriata*.
°° includes *M. henselii*.

Table 5.4   Nomenclature for the Brazilian Species
of *Monodelphis*

Gomez 1991	Gardner 1993
Group 1	
M. domestica	*M. domestica*
Group 2	
M. glirina	*M. brevicaudata*
M. touan	*M. brevicaudata*
Monodelphis sp. 1	?
Group 3	
M. emiliae	*M. emiliae*
Monodelphis sp. 2	?
Group 4	
M. brevicaudis = *hensi*	*M. sorex?*
M. dimidiata	*M. dimidiata*
Group 5	
M. brevicaudata	*M. brevicaudata*
M. rubida	*M. rubida*
Monodelphis sp. 3	?
Group 6	
M. americana (incl. *M. unistriata*)	*M. americana*
M. iheringi	*M. iheringi*
M. scalops	*M. scalops*
M. umbristriata	*M. rubida*

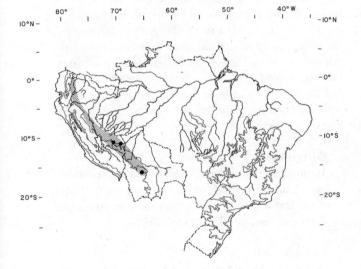

Map 5.23. Distribution of *Monodelphis adusta*.

terrestrial invertebrates, fruits, and small vertebrates (Eisenberg 1989).

Monodelphis americana (Muller, 1776)
Three-striped Short-tailed Opossum,
Colicorto Estriado

Measurements

	Mean	Min.	Max.	N	Loc.
TL	153	142	159	3	Br
HB	107	100	111		
T	46	42	48		

Vieira 1949.

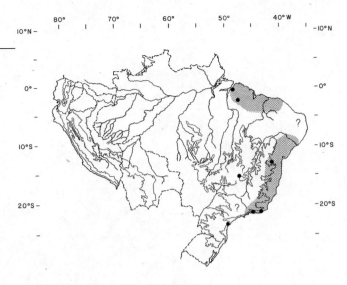

Map 5.24. Distribution of *Monodelphis americana*.

Description
This distinctive short-tailed opossum has three black dorsal stripes, the middle one starting at the nose and extending to the base of the short tail, with the flanking ones shorter. The venter is heavily washed with orange (pers. obs.).
Chromosome number: $2n = 18$; FN = 22 (Langguth and Lima 1988).

Distribution
Monodelphis americana is distributed from Brazil possibly to northern Argentina (Wilson and Reeder 1993; Massoia 1980) (map 5.24).

Ecology
In the Atlantic forest of Brazil this species was found to make nests in the forks of trees or in bushes. In the Brazilian cerrado it was found to be terrestrial and active during the day (Davis 1947; Nitikman and Mares 1987).

Monodelphis brevicaudata (Erxleben, 1777)

Description
This short-tailed mouse opossum averages 54 g in weight (67–95 g): TL 170–242; T 60–90; HF 18–24; E 16–21 (Suriname, CMNH); TL 173–210; T 61–83; HF 15–20; E 10–13 (Auyan Tepui, Venezuela, AMNH). It is gray brown above, grading to reddish on the sides. The venter varies from pale brown with a violet cast to cream, but the color fades in a dead specimen. The proximal portion of the tail is furred for about 20 mm.
Chromosome number: $2n = 18$; FN = 30.

Range and Habitat
This species occurs in eastern Colombia, Venezuela, the Guianas, and south to Brazil. It tolerates a

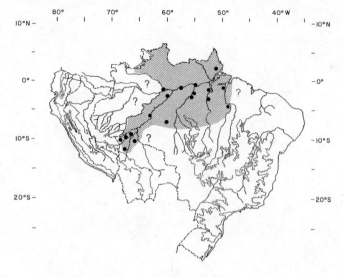

Map 5.25. Distribution of *Monodelphis brevicaudata*.

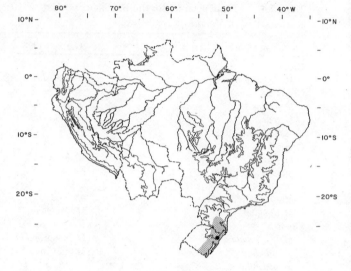

Map 5.26. Distribution of *Monodelphis dimidiata*.

variety of habitat types, occurring at up to 1,200 m elevation. Although it frequents multistratal tropical evergreen forests, it also may be found in edge habitats around clearings. It is less abundant in dry deciduous forests (map 5.25).

Natural History

This species is predominantly terrestrial and crepuscular in its habits. Up to seven young are born in a single litter. In northern Venezuela the breeding season extends from May through August. Malcolm (1990) reports densities of 12.9/km² near Manaus Brazil. The behavior patterns and breeding biology are probably similar to those of *M. domestica,* a species well studied in captivity (Trupin and Fadem 1982; Streilein 1982b).

Comment

Monodelphis touan is probably a junior synonym of *M. brevicaudata.* Reig et al. (1977) separated *M. orinoci* from *M. brevicaudata,* suggesting that species of *Monodelphis* near the Orinoco are distinct from the northern forms, but this contention is not supported by Gardner (1993).

Monodelphis dimidiata (Wagner, 1847)
Eastern Short-tailed Opossum, Colicorto Pampeano

Measurements

	Mean	Min.	Max.	N
TL	164.3	114.0	231	23
HB	99.6	55.0	151	16
T	55.8	37.0	80	
HF	14.8	11.0	27	17
E	10.9	7.1	14	
Wt	51.5	40.0	84	7

Redford and Eisenberg 1992.

Description

This short-tailed opossum resembles a small weasel (*Mustela rixosa*) in general body form. The ears are very short. The tail, thick at the base and sparsely haired, is dark gray above and yellowish below. In animals from Uruguay the short, dense fur is ash colored from the center of the forehead back over the entire dorsum. Animals from Argentina have an olive brown dorsum, rufous sides, and a bright orange tan venter. On the sides of the head, flanks, and legs the fur is yellowish orange, paling on the venter (Barlow 1965; Miranda-Ribeiro 1936; pers. obs.).

Distribution

Monodelphis dimidiata is found in southeastern Brazil, through Uruguay to the pampean region of Argentina, and as far west as Salta province, Argentina (Wilson and Reeder 1993) (map 5.26).

Life History

In Buenos Aires province, Argentina, litters were found in December–January, with a maximum of sixteen young. Litter sizes ranging from eight to fourteen also have been reported. Neonates weighed between 0.08 and 0.11 g (*n* = 3). Breeding is done only by young of the previous year. There is great sexual dimorphism, and it has been suggested that this species may be semelparous like some of the Australian species of *Antechinus* (Pine, Dalby, and Matson 1985).

Monodelphis domestica (Wagner, 1842)
Gray Short-tailed Opossum, Colicorto Gris

Measurements

	Mean	Min.	Max.	N
TL	212.3	178	270	18
HB	143.2	123	179	

	Mean	Min.	Max.	N
T	69.1	46	91	
HF	17.7	14	22	
E	19.8	14	25	17
Wt	71.4	58	95	11

Redford and Eisenberg 1992.

Description

This short-tailed opossum is a more or less uniform grayish brown dorsally with a whitish venter washed with orange (plate 1). Some individuals have olive speckling on the dorsum and are lighter on the sides (Fadem et al. 1982; Olrog and Lucero 1981; pers. obs.).

Distribution

Monodelphis domestica is distributed in eastern and central Brazil, Bolivia, and Paraguay (Myers and Wetzel 1979; Wilson and Reeder 1993) (map 5.27).

Life History

Animals of this species obtained from the caatinga of Brazil were used to start a captive breeding colony. Extensive work on colony animals has made *M. domestica* one of the best studied of the South American small opossums. In captivity this species breeds throughout the year, and some females produce four litters a year. The female builds a compact, complicated nest, carrying nesting material with her tail.

Gestation lasts fourteen or fifteen days; young are born at about 0.10 g; litter size is three to fourteen, with an average of seven; and the estrous cycle is twenty-eight days. Young are attached to the nipple for about two weeks and then enter a nest phase. The female does not have a pouch but will transport young on her back. Young eat solid food at four to five weeks, can be separated from the female at seven weeks, and can

reproduce at fifteen months. In captivity males often weigh considerably more than females. Bergallo and Cerqueria (1994) confirmed in a field study that females cease growing at sexual maturity but males continue to grow.

In the Brazilian caatinga *M. domestica* reproduces through much of the year; it has at least two litters a year and maybe up to six. The shortest time between pregnancies is seven to eight weeks, and lactation lasts six to eight weeks. Age of first reproduction is five to seven months, and litters range from 6 to 11 with an average of 8.4. This reproductive pattern is probably the same in the Chaco. In Bahia reproduction is seasonal. Litter sizes range from 2 to 16 with a mean of 7.9 (Bergallo and Cerqueria 1994). This species does not exhibit a near-semelparous mode of reproduction (Fadem et al. 1982; Streilein 1982a,b; Unger 1982).

Ecology

Primarily found in xeric situations, this species has been trapped in grassy areas, brush piles, and among jumbled rocks in a dry riverbed. It is also tolerant of man-made clearings. It is an accomplished predator, feeding primarily on invertebrates (Myers and Wetzel 1979; Streilein 1982a).

Monodelphis emiliae (Thomas, 1912)

Description

Measurements: HB 120–58; T 50–70; HF 17–24; E 14–18; Wt 60 g (Emmons and Feer 1997; Pine and Handley 1984). The head and rump are reddish brown while the middorsum is gray with white-tipped hairs, lending a grizzled appearance. The venter is a light orange. The postorbital process of the skull is robust. This character and others are shared with *M. americana* and *M. sorex* (Pine and Handley 1984).

Distribution

Found discontinuously south of the Amazon in eastern Peru and central Brazil. It may possibly occur in sympatry with *M. brevicaudata* (map 5.28).

Monodelphis henseli (Thomas, 1888)

Comment

Pine, Dalby, and Matson (1985) propose that *M. henseli* be included in *M. sorex*.

Monodelphis iheringi (Thomas, 1888)

Description

The head and body length is about 90 mm (see table 5.2). This small species of *Monodelphis* has three brown dorsal stripes; it appears similar to *M. americana*.

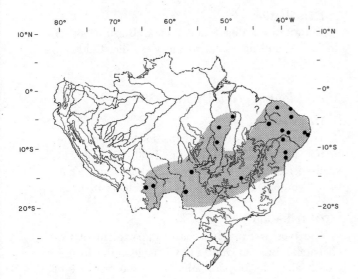

Map 5.27. Distribution of *Monodelphis domestica*.

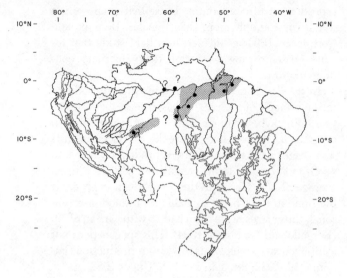

Map 5.28. Distribution of *Monodelphis emiliae*.

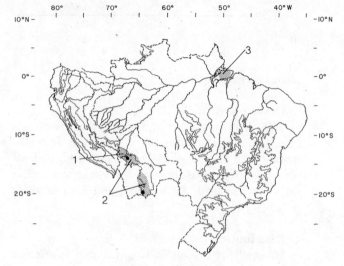

Map 5.30. Distribution of three species of *Monodelphis*: (1) *Monodelphis osgoodi*; (2) *M. kunsi*; (3) *M. maraxina*.

Map 5.29. Distribution of *Monodelphis iheringi*.

Distribution

The type locality is listed as Rio Grande de Sul (see Gardner 1993). Found in southeastern Brazil (map 5.29).

Comment

Considered by some authors to be a subspecies of *M. americana*, but see Pine (1977).

Monodelphis kunsi (Pine, 1975)

Description

Measurements: TL 113; T 42; HF 12; E 12. A very small, short-tailed opossum, it resembles a short-tailed shrew (*Blarina*). The canines are not exaggerated in length; the dorsum is brown without stripes; the venter is a "dirty white." The tail is slightly bicolored. The small size distinguishes this form from all other species of *Monodelphis* (Anderson 1982).

Distribution

Known only from two sites in Bolivia and two in Brazil, but could occur in Peru and northern Argentina (map 5.30).

Monodelphis maraxina Thomas, 1923

Description

A large species of *Monodelphis* with measurements of TL 213; T 79. The tail is relatively hairless and scaly. The dorsum is gray, resembling *M. domestica*. Its status is reviewed by Pine (1979, 1980).

Distribution

Known from Ilha Marajó, Pará, Brazil, 0°44′ S, 48°31′ W, and an island to the west Ilha Caldeirão (map 5.30).

Monodelphis osgoodi Doutt, 1938

Description

Two specimens exhibited the following measurements: TL 154–58; HB 94–96; T 60–62; E 9. This small species has a cinnamon brown dorsum and a pale brown venter. The bases of the ventral hairs are gray.

Distribution

The type is from Incachaco, department of Cochabamba, Bolivia. Doutt (1938) suggests this form is closely allied to *M. peruvianus* from the province of Huánuco, Peru (map 5.30).

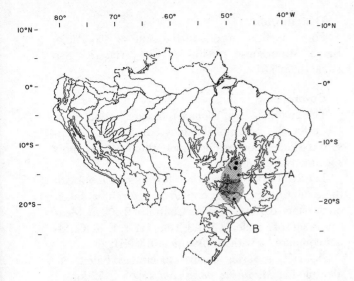

Map 5.31. Distribution of two species of *Monodelphis*: (a) *Monodelphis rubida*; (b) *M. unistriata*.

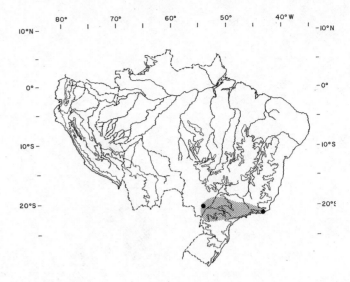

Map 5.32. Distribution of *Monodelphis scalops*.

Comment

M. peruvianus is now considered a synonym of *M. adusta*. Originally described as *M. peruvianus osgoodi*.

Monodelphis rubida (Thomas, 1899)

Description

Measurements: HB 125–37; T 56–60; HF 15–18; E 12. The dorsum is a reddish brown bearing three faint brown stripes (see table 5.2). The venter is gray. The tail is not bicolored.

Distribution

The species has been recorded from central eastern Brazil including the states of Goiás, Minas Gerais, and São Paulo (map 5.31).

Comment

Formerly included in *M. umbristriata*, but see Pine (1976). Pine, Dalby, and Matson (1985) review the history of *M. umbristriata*, but the topic needs further study. Gardner (1993) includes *umbristriata* in *M. rubida*. Pine and Handley (1984, 242) suggested *umbristriata* be included in *M. rubida*.

Monodelphis scalops (Thomas, 1888)

Description

Measurements: HB 133; T 71; E 8 (Emmons and Feer 1997; Pine and Abravaya 1978). The basic dorsal color is reddish brown, but white hairs on the anterior dorsum yield a grizzled effect. The venter is gray. The rostrum is long and slender.

Distribution

Southeastern Brazil and perhaps to adjacent Argentina (map 5.32).

Monodelphis sorex (Hensel, 1872)

Description

Miranda-Ribeiro (1936) offers measurements for *M. sorex*: HB 72–73 HB; T 41–46. These measurements could be referable to immature specimens. If we include *M. henseli* within *M. sorex*, then the range of adult measurements could fall: HB 106–30; T 55–65; HF 16; E 10–13, as noted in Emmons and Feer (1997).

This species has a very dark brown dorsum, very lightly washed with rufous. The venter is a lighter shade, washed with yellow. On the venter the hair is sparse (pers. obs.).

Distribution

Although the species is thought to be confined to southeastern Brazil, specimens have been recorded from Paraguay and from Misiones province, Argentina (map 5.33).

Comment

M. henseli is often considered a junior synonym (Wilson and Reeder 1993). Some species labeled *M. sorex* in collections may be improperly assigned.

Monodelphis theresa Thomas, 1921

Description

A specimen at the Museo, Rio de Janeiro, had a head and body length of 77 mm. The reddish brown dorsum exhibited a faint median dorsal stripe (see table 5.2). Miranda-Ribeiro (1936) offers measurements as HB 97; T 47; HF 17; E 14.5. He notes that this species typically has three faint dorsal stripes.

Map 5.33. Distribution of *Monodelphis sorex*.

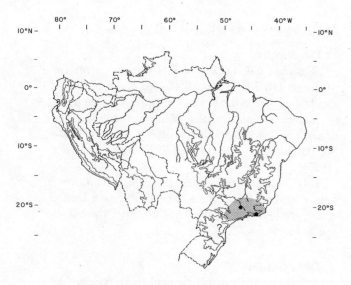

Map 5.34. Distribution of *Monodelphis theresa*.

Distribution

The type locality is Rio de Janeiro, Brazil (map 5.34).

Comment

The status of this species is obscure. It is occasionally included under *M. americana*, but this policy seems questionable. Its relation to *M. unistriata* remains to be determined.

Monodelphis unistriata (Wagner, 1842)

Description

Measurements: HB 135; T 62 (see table 5.2). The dorsum is reddish mixed with gray with a single dark dorsal median stripe; the venter is reddish. The tail is sparsely haired.

Distribution

Found in the portions of the state of São Paulo, Brazil. Waterhouse (1846–48) lists "Ytarare," São Paulo (map 5.31).

Comment

The systematics of *Monodelphis* are hazy at best. We offer the provisional tables 5.2–5.4. Some unsolved riddles concerning unidentified *Monodelphis* specimens are as follows:

1. LSU 14019. A very dark brown specimen with three faint dorsal stripes and an orange venter. Taken on the slope of the Cordillera Central at 2,400 m. Measurements of a male: HB 149; T 49; HF ?; E 14. Gardner compared it with *M. theresa* at BMNH.

2. MVZ, a series of five specimens tentatively identified as *M. adusta peruviana*, taken at 72 km in northeastern Paucartambo (km marker 152), department of Cuzco, Peru, at 1,460 m. Measurements: TL 139–63; T 54–67; HF 13–14; E 10–11; Wt 12–23 g. The dorsum and tail are very dark brown, almost black, with the venter only slightly paler.

3. MVZ, a single specimen similar to those above, caught at Huampanai, Río Cenapa, department of Amazonas, Peru. A male measured HB 164; T 66; HF 19; E 11; Wt 22 g. The color is brown, but not as dark as the highland series.

Genus *Thylamys* Gray, 1943

Description

The premaxillae of the skull are rounded. The dorsal pelage is woolly, and most species can store fat in their tails. There is no pouch, and the nipples of the female are not organized in a circular pattern but are in two bisymmetrical lines on the venter. For comments on species validity see Palma (1995, 1997). A key to the species of this genus is included in Palma (1997).

Distribution

This genus occupies the more temperate portions of South America. It occurs from Peru to Brazil and south to Argentina. Palma and Yates (1998) review the status of the species and distributions.

Thylamys elegans (Waterhouse, 1839)
Marmosa Elegante, Yaca

Measurements

	Mean	Min.	Max.	N
TL	239.8	221.0	256.0	19
HB	114.1	106.0	121.0	
T	125.7	115.0	142.0	
HF	15.3	13.5	16.0	
E	22.5	21.0	24.6	
Wt	28.9	18.5	41.0	15

Redford and Eisenberg 1992.

Description

Thylamys elegans is a medium-sized mouse opossum whose tail, only slightly longer than the head and body, is finely haired throughout and frequently incrassated. The fur is very dense and soft, and the ears are large and naked. The color is generally grayish or light brownish, with lighter sides, a pure white, yellowish white, or gray white venter, and conspicuous black eye-rings that extend toward the nose. There is considerable variation in body color throughout the range (Osgood 1943; pers. obs.).

Distribution

This wide-ranging mouse opossum is found in central coastal Peru and most of Chile, from sea level to at least 2,500 m. Except for one report eighty years ago, there are no verifiable specimens of *T. elegans* from south of the Bío-bío River in Chile (Miller 1980). Anderson (1997) follows Palma (1995) in restricting *T. elegans* to Chile and Argentina. Bolivian specimens are referable to *T. pallidior* and *T. venustus* (map 5.35).

Life History

Females are reported to have nineteen nipples and up to fifteen young, though eight to twelve is a more common number. In Chile they reproduce from September to March, during which time two litters can be produced. There is little or no pouch development (Mann 1978; Schneider 1946; Thomas 1927).

Ecology

Thylamys elegans nests in various places: under rocks or roots, in trees and in cane patches, in rocky embankments, and in holes in the ground made by

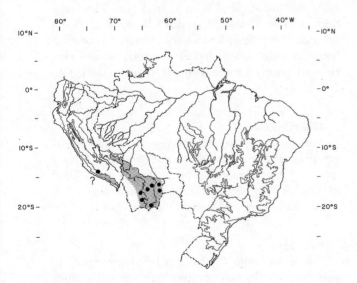

Map 5.35. Distribution of *Thylamys elegans* (more properly referred to as *T. venustus;* see text and Anderson 1997).

Cavia. This species stores fat, especially in the base of the tail, and animals hibernate during the winter.

The species has wide habitat tolerance: in Fray Jorge National Park in Chile it has been trapped in wet forest, brush, and riverine scrub. It is a typical element of the matorral scrub in Chile. In all habitats it seems to occur at fairly low densities. In Tucumán two animals were found to have an average home range of 289 m^2.

Thylamys elegans can be captured both arboreally and terrestrially. It is primarily an arthropod eater (arthropods and insect larvae made up 90% by volume of the diet in one study), though it also eats fruit, small vertebrates, and probably even carrion (Bruch 1917; Cajal 1981; Glanz 1977; Mann 1978; Mares, Ojeda, and Kosco 1981; Meserve 1981; Roig 1971; Schamberger and Fulk 1974; Schneider 1946; Simonetti, Yañez, and Fuentes 1984; Thomas 1926).

Thylamys karimii (Petter, 1968)

Description

Measurements: HB 95; T 72; HF 11; E 19, similar to *Marmosa* (= *Thylamys*) *pallidior*, but the tail is reputed to be nonprehensile. Pine, Bishop, and Jackson (1970) challenged the validity of the species. A full discussion in given in Julien-Laferrière (1994).

Distribution

The type specimen is from Karimi à Exu, Pernambuco, Brazil, 7°30' S, 39°45' W (not mapped).

Thylamys macrura (Olfers, 1818)

Description

This small mouse opossum is known only from a few specimens. One specimen from UM measured TL 275; T 140; HF 17; E 25; Wt 54 g. One dried skin had a head and body length of 120 mm and a tail length of 155 mm. It has short, soft velvety fur, mouse gray above with a rufous tone on the shoulders and a pure white to creamy white venter. The eyes are clearly ringed with black, the ears are large and leafy, and the long tail is naked except for its basal centimeter. The tail is gray above for one-third to one-half its length, and the terminal half and underside are white (Goeldi 1894; Thomas 1894; pers. obs.).

Distribution

The species is found in Paraguay and possibly adjacent Brazil (map 5.36).

Comment

T. macrura is listed as a full species by Gardner (1993). *T. grisea* Desmarest, 1827, is listed as a synonym.

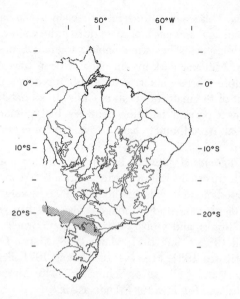

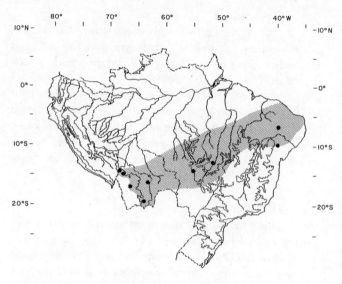

Map 5.36. Distribution of *Thylamys macrura*.

Map 5.38. Distribution of *Thylamys pusilla*.

Thylamys pallidior (Thomas, 1902)

Description
Measurements in Bolivia: HB 84–99; T 104–13; H 14–15; E 20–21 (Tate 1933). This opposum has a pale cinnamon brown dorsum with a white to cream venter. It resembles *T. elegans* but is smaller. The eyerings (mask) are very faint.

Distribution
The species occurs from eastern Bolivia to northern Argentina (map 5.37).

Comment
Possibly closely related to *Thylamys bruchi* (Redford and Eisenberg 1992), but see *Thylamys karimii*.

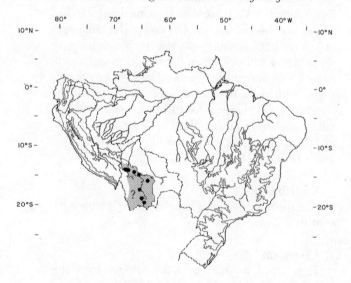

Map 5.37. Distribution of *Thylamys pallidior* (see comments in text).

Thylamys pusilla (Desmarest, 1804)
Marmosa Común

Measurements

	Mean	Min.	Max.	N
TL	197.9	170.0	235.0	26
HB	94.3	75.0	120.0	
T	103.6	90.0	134.0	
HF	12.6	8.0	15.0	
E	21.7	18.0	24.1	
Wt	18.3	12.1	29.5	12

Redford and Eisenberg 1992.

Description
The tail is equal to about 110% of the head and body length and is strongly bicolored. In some specimens the throat gland is well developed. The fur is thick and fine. Dorsally *T. pusilla* is brownish gray, and ventrally it is yellowish to white. The ventral color extends out onto the legs and cheeks, and the black eyerings are poorly defined (Olrog and Lucero 1981; Tate 1933).

Distribution
This mouse opossum is widely distributed in Argentina, southwestern Bolivia, and Paraguay. In Jujuy province, Argentina, it has been trapped up to 3,500 m (Daciuk 1974; Honacki, Kinman, and Koeppl 1982; Olrog 1959; Olrog and Lucero 1981) (map 5.38).

Ecology
Within its range *T. pusilla* is frequently found in very dry areas. In Salta province, Argentina, it inhabits much drier areas than its congener *T. elegans* and has been trapped on rocky xeric hillsides, in dry thorn

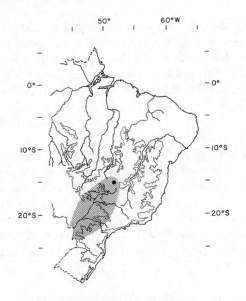

Map 5.39. Distribution of *Thylamys velutinus.*

scrub, and along watercourses in dense vegetation. In the Paraguayan Chaco this mouse opossum was frequently trapped in and near thorn forest. One burrow inhabited by this species was under a cactus at 1,950 m, and in Salta *T. pusilla* was observed active even when there was snow on the ground (Lucero 1983; Mares 1973; Mares, Ojeda, and Kosco 1981; Olrog 1979).

Thylamys velutinus (Wagner, 1842)

Description
Vieira and Palma (1996a,b) offer a series of measurements: HB 141–212 (x = 173.3); T 73–85 (x = 78). The short tail may serve as a fat storage organ.

Distribution
The type locality is "Ypanema," São Paulo, Brazil (not mapped). Vieira and Palma (1996a) find it to be distributed in the cerrado and caatinga of east-central Brazil (map 5.39).

Comment
The status of this taxon in Brazil and Bolivia is being reviewed by Palma (1995).

Ecology
In the cerrado of Brazil *T. velutinus* has been found to reach densities of 0.55/ha. It appears to be omnivorous, and 75% of its diet is animal in origin including 44% arthropods (Vieira and Palma 1996b). Home ranges based on recaptures were estimated at 2.28 ha for males and 1.7 ha for a female (Vieira and Palma 1996a,b).

Tribe Metachirini

Genus *Metachirus* Burmeister, 1854
Metachirus nudicaudatus (E. Geoffroy, 1803)
Brown Four-eyed Opossum, Cuica Común

Measurements

	Sex	Mean	Min.	Max.	N
HB	F	222.9	150	265	50
	M	233.9	170	300	35
T	F	298.0	178	363	51
	M	307.7	227	373	35
HF	F	41.4	34	47	51
	M	43.7	35	52	35
E	F	35.7	31	43	50
	M	35.4	28	40	35
Wt	F	235.9	91	345	51
	M	281.5	102	480	35

Stallings 1988b; Brazil.

Description
This medium-sized opossum is often mistaken for *Philander,* the gray four-eyed opossum, but the difference in body color makes *Metachirus* unmistakable. The tail is considerably longer than the head and body (118%) and has fur extending only 5–25 mm over its base, much less than in *Philander.* The scaly part of the tail is black and white but not as sharply demarcated as in *Philander.* There is no pouch. The dorsum is grayish brown, and the head has a striking color pattern: a dark band extends from the tip of the snout through the eyes and along the base of the ears, reaching a point midway between the ears and on some individuals stretching several centimeters past them. The dark color forms eye-rings. The ears are large, rounded, and unlike those of *Philander,* entirely dark. The venter is white or cream (Husson 1978; pers. obs.).
Chromosome number: $2n$ = 14; FN = 24 (Reig et al. 1977).

Distribution
This widespread species is found from Nicaragua through most of lowland tropical South America to Paraguay and northeastern Argentina. It has been recorded from extreme northeastern Brazil (Husband et al. 1992; Crespo 1950; Massoia 1980) (map 5.40).

Life History and Ecology
Litter size ranges from one to nine, with a mean of five. The average teat number for a female is nine (Osgood 1921). *M. nudicaudatus* is omnivorous; its diet includes fruits, small vertebrates, and invertebrates. It appears to have a narrow feeding niche when compared with the sympatric *Didelphis aurita* (Santori, Astua de Moraes, and Cerqueira 1996). In Minas Gerais state, Brazil, this species was caught in all forested habitats and found to be strongly terrestrial

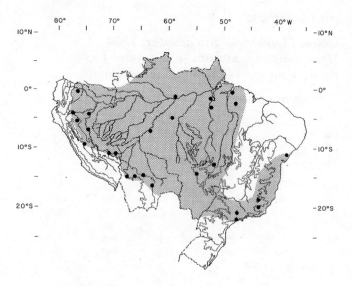

Map 5.40. Distribution of *Metachirus nudicaudatus*.

(Stallings 1988a,b). Malcolm (1990) reports densities of 25.6/km² near Manaus, Brazil.

SUBFAMILY CALUROMYINAE
Genus *Caluromys* J. A. Allen, 1900
Woolly Opossum, Zorrito de Palo

Description
The dental formula of *Caluromys* is I 5/4, C 1/1, P 3/3, M 4/4. The first upper premolar is very small, situated directly behind the canine, and there is a distinct gap between the first premolar and the much larger second premolar. The fur of the body is very thick, hence the common name woolly opossum. The dorsal pelage varies from brown to gray brown, and a brown stripe runs from the muzzle to between the ears. The tail, furred for more than one-third of its length, is extremely long and fully prehensile. There are three recognized species.

Distribution
The genus is distributed from southern Veracruz in Mexico to Paraguay. The strong arboreal adaptations of this genus limit its distribution to areas of moist forest.

Life History and Ecology
The woolly opossums are distinctive when compared with other members of the family Didelphidae. Their litter size tends to be somewhat reduced, and they can live longer in captivity. *Caluromys* is typified by a comparatively large encephalization quotient (Eisenberg and Wilson 1981). A pattern of more extended maternal care is suggested by the studies of Charles-Dominique (1983) and Charles-Dominique et al. (1981). Trapping records and radiotelemetry indicate that these strongly arboreal animals come to the ground infrequently. They tend to be mixed feeders,

eating the pulp of various fruits and supplementing this diet with nectar, invertebrates, and small vertebrates. Vocalizations of *Caluromys* are similar to those of other didelphids. A hiss is employed with an open-mouth threat during defensive behavior. Click sounds are made during male-female encounters and courtship and may grade into chirps (Eisenberg, Collins, and Wemmer 1975).

Caluromys derbianus and *Caluromys lanatus* appear not to overlap geographically, but *Caluromys philander* can occur with *C. lanatus*. The latter species is much larger than *C. philander,* and there is a strong suggestion that there is ecological segregation in feeding habits, indirectly indicated by the size differences (Eisenberg and Wilson 1981; Charles-Dominique et al. 1981).

Caluromys derbianus (Waterhouse, 1841)

Description
External measurements exhibit a wide range. Total length ranges from 760 to 587 mm, of which the tail may be 490–395 mm. The ear averages 40 mm and the hind foot 35 mm. Mean weight for six adults from Panama was 295 g. The basic dorsal pelage is a lighter brown than that of *C. lanatus;* otherwise they are similar in color.
Chromosome number: $2n = 14$; FN = 24.

Range and Habitat
This species occurs from western Ecuador to Veracruz, Mexico, and apparently does not pass over the eastern cordillera of the Andes. To the east it is replaced by *Caluromys lanatus* (map 5.41). This species appears to be confined to lower elevations of moist evergreen tropical rain forests.

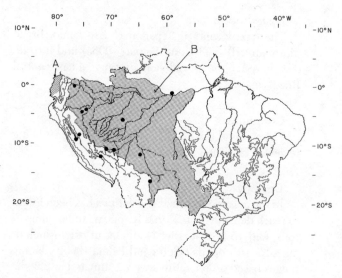

Map 5.41. Distribution of two species of *Caluromys:* (a) *Caluromys derbianus;* (b) *C. lanatus.*

Natural History

This nocturnal, strongly arboreal species rarely descends to the ground. It forages in the upper strata of mature rain forest, sheltering in tree cavities and supplementing its diet of ripe fruit with almost any invertebrate or small vertebrate it can capture. It is not known to exhibit torpor. The average litter size is three (Bucher and Hoffmann 1980), and the interval between estrous periods is twenty-eight days (Bucher and Fritz 1977). In Panama breeding commences with the onset of the dry season, generally in late January or early February (Enders 1966). Biggers (1967) recorded a similar seasonality in Nicaragua. Sexual maturity is attained at seven to nine months (Collins 1973).

Caluromys lanatus (Illiger, 1815)
Woolly Opossum, Cuica Lanosa

Measurements

	Mean	Min.	Max.	N
TL	661.0	602	702	18
HB	273.3	201	319	
T	387.7	341	440	
HF	42.2	30	48	16
E	34.9	30	40	
Wt	3,201.0			

Redford and Eisenberg 1992.

Description

The tail is long and heavily furred over 50% of its length, the fur gradually becoming thinner, with the distal 30% of the tail naked. All digits have well-developed pads and claws. This opossum has long, woolly fur that is light brown dorsally and darker, grading to orangish, on the shoulders, the front legs, and the top of the head. The venter is lighter. The ears are uniformly darkish, and there are indistinct orangish eye-rings. The face is grayish white with a dark brown stripe originating at the nose and passing between the eyes to fade into orangish on top of the head (pers. obs.).
Chromosome number: $2n = 14$; FN $= 24$ (Reig et al. 1977).

Distribution

C. lanatus is distributed from Colombia and Venezuela south to northeastern Argentina. In southern South America it is found only in eastern Paraguay and the Argentine province of Misiones (map 5.41).

Life History and Ecology

In captivity estrus occurs at an average interval of twenty-eight days (Bucher and Fritz 1977). A specimen from Paraguay was obtained on a branch 10 m above the ground, and all indications are that this spe-

cies is highly arboreal. In Venezuela all specimens taken were associated with multistratal evergreen tropical forests. This species is typical of rain forest and is at the extreme portion of its range in southern South America (Eisenberg 1989; Handley 1976).

Caluromys philander (Linnaeus, 1758)

Description

Measurements: HB 160–240; T 250–350; HF 32–39; E 30–35. In Suriname, head and body length ranges from 258 to 245 mm, while the tail ranges from 362 to 317 mm. One specimen recorded in Suriname attained a weight of 350 g, but most specimens are about 250 g. *C. philander* is usually smaller in Venezuela than in Suriname, with a mean weight of 170 g ($n = 3$). This opossum has a brown dorsum and a gray head. A stripe of brown extends from the nose to between the eyes. The venter is orange to gray. The basal portion of the tail is furred.
Chromosome number: $2n = 14$; FN $= 24$.

Range and Habitat

The species ranges from northern Venezuela to Brazil. In Venezuela, Handley (1976) found *C. philander* to be highly arboreal; it was rarely taken on the ground and was strongly associated with moist habitats. Although most specimens were taken in multistratal evergreen tropical forests, the species is adaptable, and some were caught in orchards. *C. philander* can range up to 1,600 m in elevation (map 5.42).

Natural History

C. philander is the best-studied species of the genus in the field (Atramentowicz 1982; O'Connell 1979; Charles-Dominique et al. 1981; Charles-Dominique

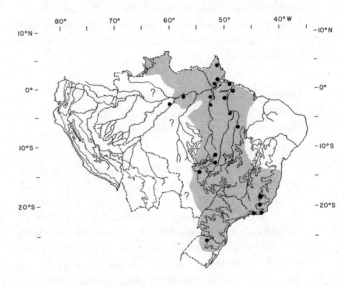

Map 5.42. Distribution of *Caluromys philander*.

1983). It is nocturnal and strongly arboreal, concentrating its activity in the upper canopy of a multistratal forest. It often shelters in tree cavities, where it constructs a nest of dead leaves. Although its diet includes ripe fruit, gum, nectar, small vertebrates, and invertebrates, it feeds primarily on fruit. It is not known to exhibit torpor.

In French Guiana litter size is about four, but O'Connell (1979) recorded six young in Venezuela. Gestation is assumed to be about fourteen days. The young remain in the mother's pouch for approximately eighty days, then enter a nest phase for another thirty days. During the nest phase the mother returns from her nocturnal forays to nurse. Age criteria for developing young in captivity are noted in detail by D'Andrea, Cerqueria, and Hingst (1994). Young disperse from the natal nest at about 130 days of age. In French Guiana a female does not breed until she is approximately one year old. Maternal care is prolonged in *Caluromys philander*. A female can produce three litters a year, but if there is a seasonal scarcity of food she will probably not rear more than one litter each year (Atramentowicz 1982).

Charles-Dominique (1983) describes a nightly home-range area of 0.3 to 1.0 ha. Home-range stability depends on food availability. Interactions among adults tend to be agonistic except when a male is courting a female. There seems to be no strong territorial defense, and home ranges of adjacent adult females may show some overlap. Males' ranges broadly overlap each other and the ranges of several females. No permanent social grouping persists except the mother-young unit. Vocalizations during encounters have been analyzed by Charles-Dominique (1983). Both young and adults produce clicks, and clicking by suckling young apparently helps maintain maternal activity. Agonistic vocalizations include hisses and grunts. A distress scream is produced upon extreme disturbance. *Caluromys* will actively attempt to bite predators or human handlers.

Genus *Caluromysiops* Sanborn, 1951
Caluromysiops irrupta Sanborn, 1951
Black-shouldered Opossum

Description
Head and body length exceeds 200 mm, and the tail is longer than the head and body: TL 570; T 310; HF 67; E 37 (Peru). This species resembles the woolly opossum in that the tail is haired for over two-thirds of its length, but its color pattern is quite different—light gray on the dorsum and face with a distinctive black shoulder. The black markings on the shoulder join at the midline to form either a single dorsal stripe or a double stripe slightly off center of the midline.

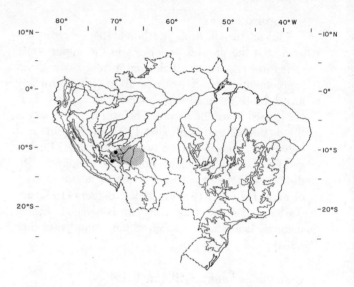

Map 5.43. Distribution of *Caluromysiops irrupta*.

Range and Habitat
The species is known from multistratal tropical evergreen forests in the Amazonian portion of Peru (map 5.43) and was reported from the extreme southeast of Colombia by Simonetta (1979). Simonetta offers a photograph, but there is no museum specimen to confirm this record. Vivo and Gomèz (1989) have recorded this species from western Brazil.

Natural History
Little is known concerning the behavior and ecology of this rare opossum. Although its feeding habits are unknown from the wild, specimens in captivity show rather individual dietary preferences, taking a wide range of fruits as well as small rodents. Judging from its body proportions and behavior in captivity, this species is probably highly arboreal. Captives have lived more than seven years (Collins 1973; Izor and Pine 1987).

Genus *Glironia* Thomas, 1912
Glironia venusta Thomas, 1912
Bushy-tailed Opossum

Description
The dental formula is I 5/4, C 1/1, P 3/3, M 4/4. Head and body length ranges from 160 to 205 mm; the tail exceeds the head and body, ranging from 195 to 225 mm (Marshall 1978b). This genus is characterized by the almost fully haired tail, though the tip is lightly haired and presumably retains some prehensile ability. The dorsal pelage is brown to gray, and the face is adorned by two black lines extending from the snout through the eye to the ear.

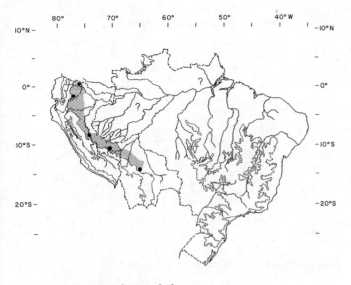

Map 5.44. Distribution of *Glironia venusta*.

Table 5.5   Distribution of the Known Caenolestidae

Species	Location
Caenolestes caniventer Anthony, 1921[a]	Eastern cordillera, Ecuador
Caenolestes condorensis Albuja and Patterson, 1996	Cordillera Cóndor, Ecuador
Caenolestes convelatus Anthony, 1924	Western cordillera, Ecuador
Caenolestes fuliginosus (Tomes, 1863)	Southern Ecuador to Peru
Caenolestes obscurus Thomas, 1895	Colombia and Venezuela
Caenolestes tatei Anthony, 1923	Eastern cordillera, Ecuador
Lestoros Oehser, 1934	Peru
Lestoros inca (Thomas, 1917)	Peru
Rhyncholestes Osgood, 1924	Chile
Rhyncholestes raphanurus Osgood, 1924	Chile

Note: Albuja and Patterson 1996 describe *Caenolestes condorensis* and provide a key to all the species of *Caenolestes* they recognize. They give full species rank to the newly described form as well as *C. caniventer*, *C. convelatus*, and *C. fuliginosus*. In their scheme *C. obscurus* and *C. tatei* are included within *C. fuliginosus*.
[a] Full references in Honacki, Kinman, and Koeppl 1982 or in Wilson and Reeder 1993.

Distribution

Specimens have been taken in the lowland rain forests of western Amazonas in Ecuador, east-central Brazil, Peru, and Bolivia (map 5.44).

Natural History

Only a few specimens exist in museums. The species is assumed to be arboreal and similar in its habits to *Caluromys* (Marshall 1978b).

Order Paucituberculata

FAMILY CAENOLESTIDAE

Rat Opossums

Diagnosis

This family contains three genera and seven species. The dental formula is I 4/3, C 1/1, P 3/3, M 4/4. Rat opossums are easily distinguished from members of the family Didelphidae in that the first lower incisors are enlarged and project forward. The remaining incisors are very reduced and show a single cusp

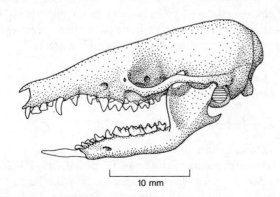

10 mm

Figure 5.4. Skull of *Caenolestes obscurus*.

(fig. 5.4). Members of the Caenolestidae are ratlike in appearance, with head and body length ranging from 90 to 135 mm while the tail ranges from 65 to 135 mm. The rostrum is long and narrow, and the eyes are small. The tail is covered with short hairs.

Distribution

The species of this family are confined to the high montane paramos of the Andes or the moist, temperate lowlands of Chile and Argentina. They range from western Colombia to southern Chile.

There are five recognized species within the genus *Caenolestes*, found in the montane areas of Colombia, Venezuela, and Ecuador to central Peru. They are confined to alpine meadows in the Andes. There is one species in the genus *Lestoros* confined to central Peru.

Natural History

Little material is available concerning the behavior of these animals in the field. The family has an ancient history, being recognizable in the Eocene of South America, and underwent an adaptive radiation in the Oligocene with numerous species that subsequently became extinct (Marshall 1980). The extant forms are relicts in the high montane meadows of the Andes and in the temperate lowlands of Chile (Osgood 1924). In these habitats they use runways, where they feed on small invertebrates.

Genus *Caenolestes* Thomas, 1895

Caenolestes caniventer Anthony, 1921

Description

Measurements of the type: TL 256; T 127; HF 26.5. The dorsum is very dark brown; the dorsal hairs are dark gray at the base and darker at the tips, with a scattering of white-tipped hairs. The venter is lighter with the hair tips whitish. The tail is only faintly bicolored. Anthony, in his original description, believed this form to be distinct from *C. fuliginosus* and *C. obscurus* (plate 1).

Distribution

The type is known from El Chiral, El Oro province, Ecuador, taken at 1,600 m. The species appears to be distributed in southern Ecuador on the western cordillera (not mapped).

Canenolestes condorensis (Albuja and Patterson, 1996)

Description

The dentition is typical for the genus; the canines are large and massive. Measurements: TL 260; T 130; HF 30; E 18; Wt 48 g. This newly discovered species is the largest known caenolestid. The skull is massive, with the premaxillary bones extending posteriorly past the anterior margin of P2. The dorsal pelage is fuscous, contrasting with the drab venter.

Distribution

The type specimen is from the Cordillera del Cóndor, province of Morona-Santiago, Ecuador, at 3°27'05" S; 78°21'39" W, at 2,080 m (Albuja and Patterson 1996).

Caenolestes convelatus Anthony, 1924

Description

In *C. convelatus* the antorbital vacuity is in the shape of a crescent, situated at the margin of the nasal and maxillary bones. In all other species the antorbital is comma shaped and bounded by the nasal maxillary and frontal bones (Albuja and Patterson 1996).

Distribution

The type locality is Pichincha, Ecuador, at 7,000 ft (not mapped).

Caenolestes fuliginosus (Tomes, 1863)
Ratón Comadreja

Measurements

	Mean	Min.	Max.	N
TL	235.75	232	242	4
T	118.25	114	120	4
HF	23.00	22	24	4
E	14.00	14	14	4

Location: Colombia (FMNH).

Description

Head and body length ranges from 93 to 135 mm and tail length from 93 to 127 mm. The dorsal pelage is dense and soft. Although the tail is haired, the hairs

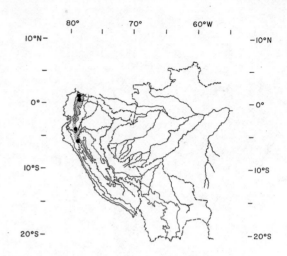

Map 5.45. Distribution of *Caenolestes* sp. (see text).

are sparse. There is no pouch. The rostrum is elongated, and the ears and eyes are greatly reduced in size. The dorsum is dark brown tending toward lighter brown on the venter.

Range and Habitat

Distributed from the Andes of western Venezuela to the Andes of northern Colombia and perhaps south to Ecuador and Peru. Specimens are generally taken at above 2,300 m (map 5.45).

Natural History

This species is adapted to cool, wet habitats. The animals use runways at interfaces between grassland and scrub forest. When disturbed they hiss and open their mouths wide. In captivity they feed on insects, earthworms, and baby rats. When moving rapidly they bound, with forelimbs and hind limbs striking the ground alternately. The tail is not prehensile but serves as a prop when they sit upright. The female has four teats not enclosed in a pouch (Kirsch and Waller 1979).

Barkley and Whittaker (1984) studied *C. fuliginosus* in Peru, where lepidopteran larvae, centipedes, and arachnids made up 75.5% of the diet by volume. They conclude that *Caenolestes* is an opportunistic feeder that prefers invertebrate larvae but will eat vertebrate prey, fruit, and other vegetable matter.

Comment

Includes *C. obscurus.*

Caenolestes tatei Anthony, 1923

Description

Measurements of the type: TL 213; T 117; HF 22. As Anthony describes the species, it is similar to *C. fuliginosus* and smaller than *C. caniventer.* The dorsum is a dark gray to almost black. The tips of some dorsal hairs are dark brown. The tail is not bicolored. The venter is somewhat lighter, but the transition from the dorsal color is gradual. Anthony emphasizes the darker dorsal color of *C. tatei*. This form is not recognized as a species in Wilson and Reeder (1993).

Distribution

The type locality is Molleturo, Azuay province, Ecuador, at 2,288 m. It is believed to be confined to the western Andes of southern Ecuador.

Comment

C. caniventer and *C. tatei* have been noted in "near" microsympatry in the eastern cordillera of the Andes in Ecuador (Barnett 1991). Barnett suggests that *Akodon* sp. and *Cryptotis montivaga* may be potent scramble competitors for insect prey when they occur with *Caenolestes*. Gardner (1993) considers *C. tatei* a junior synonym of *C. fuliginosus*.

Genus *Lestoros* Oehsor, 1934

Lestoros inca (Thomas, 1917)
Peruvian "Shrew" Opossum

Description

The dental formula is I 4/3, C 1/1, P 3/3, M 4/4. The canine is double-rooted. Head and body length range from 90 to 120 mm. The tail only slightly exceeds the head and body length, ranging from 96 to 135 mm. Weight ranges from 25 to 32 g (Kirsch and Waller 1979). The basic dorsal color is dark brown, and the venter is a lighter shade.

Distribution

The species is found at high altitudes (2,100 to 3,600 m) in moist habitats of the southern Peruvian Andes (map 5.46).

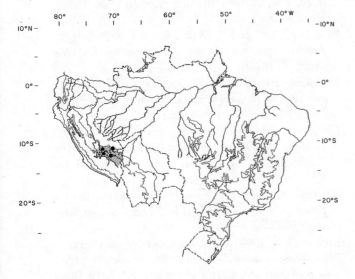

Map 5.46. Distribution of *Lestoros inca*.

Natural History

According to Kirsch and Waller (1979), these animals are terrestrial and nocturnal in their habits. They appear to be invertebrate feeders and similar to *Caenolestes* in their ecology.

Comment

Lestoros has been subsumed in *Caenolestes* by some authors.

References

• References used in preparing distribution maps

Aagaard, E. M. J. 1982. Ecological distribution of mammals in the cloud forests and paramos of the Andes, Mérida, Venezuela. Ph.D. diss., Colorado State University.

Albuja, L. V., and B. D. Patterson. 1996. A new species of northern shrew-opossum (Paucituberculata: Caenolestidae) from the Cordillera del Cóndor, Ecuador. *J. Mammal.* 77:41–53.

Alpin, K. P., and M. Archer. 1987. Recent advances in marsupial systematics with a new syncretic classification. In *Possums and opossums: Studies in evolution,* ed. M. Archer, xv–lxxii. Chipping Norton, N.S.W., Australia: Surrey Beatty.

• Anderson, S. 1982. *Monodelphis kunsi. Mammal. Species* 190:1–3.

———. 1997. Mammals of Bolivia: Taxonomy and distribution. *Bull. Amer. Mus. Nat. Hist.* 231:1–652.

Atramentowicz, M. 1982. Influence du milieu sur l'activité locomotrice et la reproduction de *Caluromys philander* (L.). *Rev. Ecol. (Terre et Vie)* 36:373–95.

———. 1986. Dynamique de population chez trois marsupiaux didelphides de Guyane. *Biotropica* 18: 136–49.

———. 1988. La frugivorie opportuniste de trois marsupiaux didelphides de Guyane. *Rev. Ecol. (Terre et Vie)* 43:47–57.

August, P. 1981. Population ecology and community structure of small mammals in northern Venezuela. Ph.D. diss., Boston University.

Barkley, L. J., and J. O. Whittaker Jr. 1984. Confirmation of *Caenolestes* in Peru with information on diet. *J. Mammal.* 65:328–30.

Barlow, J. C. 1965. Land mammals from Uruguay: Ecology and zoogeography. Ph.D. diss., University of Kansas.

Barnes, R. D., and S. W. Barthold. 1969. Reproduction and breeding behaviour in an experimental colony of *Marmosa mitis* Bangs (Didelphidae). *J. Reprod. Fert.* 6 (suppl.): 477–92.

Barnett, A. P. 1991. Records of the grey-bellied shrew opossum *Caenolestes caniventer* and *C. tatei*. *Mammalia* 55 (3): 443–45.

Beach, F. A. 1939. Maternal behavior of the pouch-less marsupial *Marmosa cinerea*. *J. Mammal.* 20: 315–22.

Bergallo, H. G., and R. Cerqueria. 1994. Reproduction and growth of the opossum *Mondelphis domestica*. *J. Zool. (London)* 232:551–64.

Biggers, J. D. 1967. Notes on the reproduction of the wooly opossum *Caluromys derbianus* in Nicaragua. *J. Mammal.* 48:678–80.

• Bruch, C. 1917. La comadrejita *Marmosa elegans*. *Rev. Jardin Zool. Buenos Aires* 13 (51, 52): 208–12.

Bucher, J. E., and H. I. Fritz. 1977. Behavior and maintenance of the wooly opossum (*Caluromys*) in captivity. *Lab. Anim. Sci.* 27 (6): 1007–12.

Bucher, J. E., and R. S. Hoffmann. 1980. *Caluromys derbianus*. *Mammal. Species* 140:1–4.

Cajal, J. L. 1981. Estudios preliminares sobre el área de acción en marsupiales (Mammalia-Marsupialia). *Physis,* sec. C, 40 (98): 27–37.

Cerqueira, R. 1984. Reproduction de *Didelphis albiventris* dans le nord-est du Brésil (Polyprotodontia, Didelphidae). *Mammalia* 48 (1): 95–104.

———. 1985. The distribution of *Didelphis* in South America (Polyprotodontia, Didelphidae). *J. Biogeogr.* 12:135–45.

Charles-Dominique, P. 1983. Ecology and social adaptions in didelphid marsupials: Comparison with eutherians of similar ecology. In *Advances in the study of mammalian behavior,* ed. J. F. Eisenberg and D. G. Kleiman, 395–422. Special Publication 7. Shippensburg, Pa.: American Society of Mammalogists.

Charles-Dominique, P., M. Atramentowicz, M. Charles-Dominique, H. Gérard, A. Hladik, C. M. Hladik, and M. F. Prévost. 1981. Les mammifères frugivores arboricoles nocturnes d'une forêt guyanaise: Interrelations plantes-animaux. *Rev. Ecol. (Terre et Vie)* 35:342–435.

Collins, L. R. 1973. *Monotremes and marsupials: A reference for zoological institutions.* Washington, D.C.: Smithsonian Institution Press.

• Contreras, J. R. 1982. Mamíferos de Corrientes. 1. Nota preliminar sobre la distribución de algunas especies. *Hist. Nat.* 2 (10): 71–72.

Creighton, G. K. 1984. Systematic studies on opossums (Didelphidae) and rodents (Cricetidae). Ph.D. diss., University of Michigan, Ann Arbor.

• Crespo, J. A. 1950. Nota sobre mamíferos de Misiones nuevos para Argentina. *Comun. Inst. Nac. Invest. Cienc. Nat., Mus. Argent. Cienc. Nat. "Bernardino Rivadavia," Zool.* 1 (14): 1–14.

• ———. 1974. Comentarios sobre nuevas localidades para mamíferos de Argentina y de Bolivia. *Rev. Mus. Argent. Cienc. Nat. "Bernardino Rivadavia," Zool.* 11 (1): 1–31.

———. 1982. Ecología de la comunidad de mamíferos del Parque Nacional Iguazú, Misiones. *Rev Argent. Cienc. Nat. "Bernardino Rivadavia," Ecol.* 3 (2): 45–162.

Daciuk, J. 1974. Notas faunísticas y bioecológicas de Península Valdés y Patagonia. 12. Mamíferos colectados y observados en la Península Valdés y zona litoral de los Golfos San José y Nuevo (provincia de Chubut, república Argentina). *Physis,* sec. C, 33 (86): 23–39.

D'Andrea, P. S., R. Cerqueria, and E. D. Hingst. 1994. Age estimation of the gray four-eyed opossum, *Philander opossum*. *Mammalia* 58:283–91.

Davis, D. E. 1945. The annual cycle of plants, mosquitoes, birds and mammals in two Brazilian forests. *Ecol. Monogr.* 15:244–95.

———. 1947. Notes on the life histories of some Brazilian mammals. *Bol. Mus. Nac., Zool.* 76:1–8.

Doutt, K. 1938. Two new mammals from South America. *J. Mammal.* 19:100–101.

Eisenberg, J. F. 1988. Reproduction in polyprotodont marsupials and similar-sized eutherians with a speculation concerning the evolution of litter size in mammals. In *Evolution of life histories of mammals,* ed. M. S. Boyce, 291–311. New Haven: Yale University Press.

———. 1989. *Mammals of the Neotropics.* Vol. 1. *Mammals of the northern Neotropics: Panama, Colombia, Venezuela, Guyana, Suriname, French Guiana.* Chicago: University of Chicago Press.

Eisenberg, J. F., L. R. Collins, and C. Wemmer. 1975. Communication in the Tasmanian devil (*Sarcophilus harrisii*) and a survey of auditory communication in the Marsupialia. *Z. Tierpsychol.* 37: 379–99.

Eisenberg, J. F., and I. Golani. 1977. Communication in Metatheria. In *How animals communicate,* ed. T. Sebeok, 575–99. Bloomington: Indiana University Press.

Eisenberg, J. F., and E. Maliniak. 1967. Breeding the murine opossum *Marmosa* in captivity. *Int. Zoo Yearb.* 7:78–79.

Eisenberg, J. F., and D. E. Wilson. 1981. Relative brain size and demographic strategies in didelphid marsupials. *Amer. Nat.* 118:1–15.

Eisentraut, M. 1970. Beitrag zur Fortpflanzungsbiologie der Zwergbeutelratte *Marmosa murina* (Didelphidae, Marsupialia). *Z. Säugetierk.* 35:159–73.

Emmons, L. H., and F. Feer. 1997. *Neotropical rainforest mammals: A field guide.* 2d ed. Chicago: University of Chicago Press.

Enders, R. K. 1966. Attachment, nursing, and survival of young in some didelphids. In *Comparative biology of reproduction in mammals,* ed. J. W. Rowlands, 195–203. New York: Academic Press.

Fadem, B. H., G. L. Trupin, E. Maliniak, J. L. Vande-

Berg, and V. Hayssen. 1982. Care and breeding of the gray, short-tailed opossum (*Monodelphis domestica*). *Lab. Anim. Sci.* 32 (4): 405–9.

Fleck, D. W., and J. D. Harder. 1995. Ecology of marsupials in two Amazonian rain forests in northeastern Peru. *J. Mammal.* 76:809–18.

Fleming, T. H. 1972. Aspects of the population dynamics of three species of opossums in the Panama Canal Zone. *J. Mammal.* 53:619–23.

———. 1973. The reproductive cycles of three species of opossums and other mammals in the Panama Canal Zone. *J. Mammal.* 54:439–55.

Fonseca, C. C. and J. C. Nogueira. 1991. Morphological aspects of the testicular descent in the white-belly opossum *Didelphis albiventris* (Marsupialia). *Zool. Jb. Anat.* 121:115–26.

Fonseca, G. A. B. da, and M. C. M. Kierulff. 1988. Biology and natural history of Brazilian Atlantic forest small mammals. *Bull. Florida State Mus., Biol. Sci.* 34:99–152.

Fonseca, G. A. B. da, K. H. Redford, and L. A. Pereira. 1982. Notes on *Didelphis albiventris* (Lund, 1841) of central Brazil. *Ciênc. Cult.* 34 (10): 1359–62.

Fonseca, S. D., and R. Cerqueira. 1991. Water and salt balance in a South American marsupial, the gray four-eyed opossum (*Philander opossum*). *Mammalia* 55:421–32.

• Fornes, A., and E. Massoia. 1965. Micromamíferos (Marsupialia y Rodentia) recolectados en la localidad Bonaerense de Miramar. *Physis* 25 (69): 99–108.

Gardner, A. L. 1973. *The systematics of the genus Didelphis (Marsupialia: Didelphidae) in North and Middle America.* Special Publications of the Museum 4. Lubbock: Texas Tech University Press.

———. 1989. Two new mammals from southern Venezuela and comments on the affinities of the highland fauna of Cerro de la Neblina. In *Advances in Neotropical mammalogy*, ed. K. Redford and J. F. Eisenberg, 411–424. Gainesville, Fla.: Sandhill Crane Press.

———. 1993. Didelphimorphia. In *Mammal species of the world*, ed. D. E. Wilson and D. M. Reeder, 15–23. Washington, D.C.: Smithsonian Institution Press.

Gardner, A. L., and G. K. Creighton. 1989. A new generic name for Tate's (1933) *Microtarsus* group of South American mouse opossums (Marsupialia: Didelphidae). *Proc. Biol. Soc. Wash.* 102 (1): 3–7.

Gardner, A. L., and J. L. Patton. 1972. New species of *Philander* (Marsupialia: Didelphidae) and *Mimon* (Chiroptera: Phyllostomidae) from Peru. *Occas. Papers Mus. Zool., Louisiana State Univ.* 43:1–12.

• Glanz, W. E. 1977. Small mammals. In *Chile-California Mediterranean scrub atlas: A comparative analysis*, ed. N. J. W. Thrower and D. E. Bradbury,

232–37. Stroudsburg, Pa.: Dowden, Hutchinson and Ross.

• Goeldi, E. A. 1894. Critical gleanings on the Didelphidae of the Serra dos Orgãos, Brazil. *Proc. Zool. Soc. London* 1894:457–67.

Gomez, N. F. 1991. Revisão sistematica do gênero *Monodelphis* (Didelphidae: Marsupialia). Master's thesis, Instituto Biociências, Universidade de São Paulo, Brazil.

González, J. C. 1973. Observaciones sobre algunos mamíferos de Bopicuá (dpto. de Río Negro, Uruguay). *Comun. Mus. Mun. Hist. Nat. Río Negro, Uruguay* 1 (1): 1–14.

Grand, T. I. 1983. Body weight: Its relationship to tissue composition, segmental distribution of mass, and motor function. 3. The Didelphidae of French Guyana. *Austral. J. Zool.* 31:299–312.

Handley, C. O., Jr. 1976. Mammals of the Smithsonian Venezuelan project. *Brigham Young Univ. Sci. Bull., Biol. Ser.* 20 (5): 1–90.

Handley, C. O., Jr., and L. K. Gordon. 1979. New species of mammals from northern South America: Mouse opossums, the genus *Marmosa*. In *Vertebrate ecology in the northern Neotropics*, ed. J. F. Eisenberg, 65–71. Washington, D.C.: Smithsonian Institution Press.

Harder, J. D. 1992. Reproductive biology of South American marsupials. In *Reproductive biology of South American vertebrates*, ed. W. C. Hamlett, 211–28. New York: Springer-Verlag.

Hershkovitz, P. 1992a. Ankle bones: The Chilean opossum *Dromiciops gliroides* Thomas, and marsupial phylogeny. *Bonn. Zool. Beitrag.* 43:181–213.

———. 1992b. The South American gracile mouse opossums, genus *Gracilinanus* Gardner and Creighton, 1989: A taxonomic review with notes on general morphology and relationships. *Fieldiana: Zool.*, n.s., 70:1–56.

Hershkovitz, P. 1997. Composition of the family Didelphidae Gray, 1821 (Didelphoidea: Marsupialia) with a review of the morphology and behavior of the included four-eyed pouched opossums of the genus *Philander* Tiedermann, 1801. *Fieldiana: Zool.*, n.s., 86:1–103.

Holmes, D. J. 1991. Social behavior in captive Virginia opossums, *Didelphis virginiana*. *J. Mammal.* 72 (2): 402–10.

———. 1992. Sternal odors as cues for social discrimination by female Virginia opossums, *Didelphis virginiana*. *J. Mammal.* 73 (2): 286–91.

• Honacki, J. H., K. E. Kinman, and J. W. Koeppl, eds. 1982. *Mammal species of the world.* Lawrence, Kans.: Allen Press and Association of Systematics Collections.

Hunsaker, D., ed. 1977a. *The biology of marsupials.* New York: Academic Press.

———. 1977b. The ecology of New World marsupials. In *The biology of marsupials,* ed. D. Hunsaker, 95–156. New York: Academic Press.

Husband, T. P. 1992. A new mouse opossum found in northeast Brazil. *Mammalia* 56:297–98.

Husband, T. P., G. D. Hobbs, C. N. Santos, and H. J. Stillwell. 1992. First record of *Metachirus nudicaudatus* for northeast Brasil. *Mammalia* 56:298–99.

Husson, A. M. 1978. *The mammals of Suriname.* Leiden: E. J. Brill.

Hutterer, R., M. Verhaagh, J. Diller, and R. Podloucky. 1995. An inventory of mammals observed at Panguana Biological Station, Amazonian Peru. *Ecotropica* 1:3–20.

Izor, R. J., and R. Pine. 1987. Notes on the black shouldered opossum. *Fieldiana: Zool.,* n.s., 39:117–24.

Julien-Lafferrière, D. 1991. Organisation du peuplement de marsupiaux en Guyane française. *Rev. Ecol. (Terre et Vie)* 46:125–44.

• ———. 1994. Catalogue des types de marsupiaux de Museum National d'Histoire Naturelle, Paris. *Mammalia* 58:1–39.

Julien-Lafferrière, D., and M. Atramentowicz. 1990. Feeding and reproduction of three didelphid marsupials in two Neotropical forests (French Guiana). *Biotropica* 22:404–15.

Kielan-Jaworowska, Z., and L. A. Nessov. 1990. On the metatherian nature of the Deltatheroida, a sister group of the Marsupialia. *Lethaia* 23:1–10.

Kirsch, J. A. W., R. E. Bleiweiss, A. W. Dickerman, and O. A. Reig. 1993. DNA/DNA hybridization studies of carnivorous marsupials. 3. Relationships among species of *Didelphis* Didelphinae). *J. Mammal. Evol.* 1:75–97.

Kirsch, J. A. W., and J. H. Calaby. 1977. The species of living marsupials: An annotated list. In *The biology of the marsupials,* ed. B. Stonehouse and D. Gilmore, 9–26. London: Macmillan.

Kirsch, J. A. W., and R. E. Palma. 1995. DNA/DNA hybridization studies of carnivorous marsupials. 5. A further estimate of relationships among opossums (Didelphidae). *Mammalia* 59 (3): 403–25.

Kirsch, J. A. W., R. Pine, and R. G. Van Gelder. 1982. Didelphidae. In *Mammal species of the world,* ed. J. H. Honacki, K. E. Kinman, and J. W. Koeppl, 18–26. Lawrence, Kansas: Allen Press and Association of Systematics Collections.

Kirsch, J. A. W., and P. F. Waller. 1979. Notes on the trapping and behavior of the Caenolestidae. *J. Mammal.* 60:390–95.

Langguth, A., and J. Fernando S. Lima. 1988. The karyotype of *Monodelphis americana. Rev. Nordeste Biol.* 6 (1): 1–5.

• Lemke, T. O., A. Cadena, R. H. Pine, and J. Hernandez-Camacho. 1982. Notes on opossums, bats, and rodents new to the fauna of Colombia. *Mammalia* 46:225–34.

Lorini, M. L., J. A. Oliveira, and V. G. Persson. 1992. Status taxonômico clas formas de *Marmosa* incluidas na seção *incana* de Tate. In *Resumos do XIX Congresso Brasileiro de Zoologia,* 154.

———. 1994. Annual age structure and reproduction patterns in *Marmosa incana* (Lund, 1841). *Z. Säugetierk.* 59:65–73.

• Lucero, M. M. 1983. Lista y distribución de aves y mamíferos de la provincia de Tucumán. Ministerio de Cultura y Educación, Fundación Miguel Lillo, *Miscelánea* 75:5–53.

McManus, J. J. 1970. The behavior of captive opossum *Didelphis marsupialis virginiana. Amer. Midl. Nat.* 84:144–69.

McNab, B. K. 1978. The comparative energetics of Neotropical marsupials. *J. Comp. Physiol.* 25:115–28.

———. 1982. The physiological ecology of South American mammals. In *Mammalian biology in South America,* ed. M. A. Mares and H. H. Genoways, 187–208. Pymatuning Symposia in Ecology 6. Special Publication Series. Pittsburgh: Pymatuning Laboratory of Ecology, University of Pittsburgh.

Malcolm, J. R. 1990. Estimation of mammalian densities in a continuous forest north of Manaus. In *Four Neotropical rainforests,* ed. A. H. Gentry, 339–57. New Haven: Yale University Press.

Mann, G. 1978. *Los pequeños mamíferos de Chile.* Guyana: Zoología 40. Santiago: Universidad de Concepción.

Mares, M. A. 1973. Climates, mammalian communities and desert rodent adaptations: An investigation into evolutionary convergence. Ph.D. diss., University of Texas at Austin.

Mares, M. A., R. A. Ojeda, and M. P. Kosco. 1981. Observations on the distribution and ecology of the mammals of Salta province, Argentina. *Ann. Carnegie Mus.* 50 (6): 151–206.

Marshall, L. G. 1978a. *Lutreolina crassicaudata. Mammal. Species* 91:1–4.

———. 1978b. *Glironia venusta. Mammal. Species* 107:1–3.

———. 1978c. *Chironectes minimus. Mammal. Species* 109:1–6.

———. 1980. The systematics of the South American marsupial family Caenolestidae. *Fieldiana: Geol.,* n.s., 5:1–145.

———. 1982a. Calibration of the age of mammals in South America. *Geobios,* memo. spec., 6:427–37.

———. 1982b. Evolution of South American Marsupialia. In *Mammalian biology in South America,* ed. M. A. Mares and H. H. Genoways, 251–72. Pymatuning Symposia in Ecology 6. Special Publication Series. Pittsburgh: Pymatuning Laboratory of Ecology, University of Pittsburgh.

———. 1982c. Systematics of the South American marsupial family Microbiotheriidae. *Fieldiana: Geol.*, n.s., 10:1–75.

Marshall, L. G., C. DeMuizon, and B. Sige. 1983. Late Cretaceous mammals (Marsupialia) from Bolivia. *Geobios* 16:739–45.

Massoia, E. 1970. Contribución al conocimiento de los mamíferos de Formosa con noticias de los que habitan zonas viñaleras. *IDIA* 276:55–63.

• ———. 1972. La presencia de *Marmosa cinerea paraguayana* en la república Argentina, provincia de Misiones (Mammalia-Marsupialia-Didelphidae). *Rev. Invest. Agropecuarias, INTA* (Buenos Aires), ser. 1, *Biol. Prod. Anim.* 9 (2): 63–70.

• ———. 1973. Observaciones sobre el género *Lutreolina* en la república Argentina (Mammalia-Marsupialia-Didelphidae). *Rev. Invest. Agropecuarias, INTA* (Buenos Aires), ser. 1. *Biol. Prod. Anim.* 10 (1): 13–20.

• ———. 1980. Mammalia de Argentina. 1. Los mamíferos silvestres de la provincia de Misiones. *Iguazú* 1 (1): 15–43.

• Massoia, E., and A. Fornes. 1972. Presencia y rasgos etoecológicos *Marmosa agilis chacoenis* Tate en las provincias de Buenos Aires, entre Ríos y Misiones (Mammalia-Marsupialia-Didelphidae). *Rev. Invest. Agropecuarias, INTA* (Buenos Aires), ser. 1, *Biol. Prod. Anim.* 9 (2): 71–81.

Meserve, P. L. 1981. Resource partitioning in a Chilean semi-arid small mammal community. *J. Anim. Ecol.* 40:747–57.

• Miller, S. D. 1980. Human influences on the distribution and abundance of wild Chilean mammals: Prehistoric–present. Ph.D. diss., University of Washington, Seattle.

Miranda-Ribeiro, A. 1936. Didelphia ou mammalia-ovovivipara. *Rev. Mus. Paulista, São Paulo* 20:245–427.

Mondolfi, E., and G. Medina Padilla. 1957. Contribución al conocimiento del "Perrito de Agua" (*Chironectes minimus* Zimmerman). *Mem. Soc. Cient. Nat. La Salle* 17:140–55.

Moore, H. D. M. 1992. Reproduction in the gray short-tailed opossum, *Monodelphis domestica*. In *Reproductive biology of South American vertebrates*, ed. W. C. Hamlett, 229–41. New York: Springer-Verlag.

Motta, M. de F. D., J. C. de A. Carreira, and A. M. R. Franco. 1983. A note on reproduction of *Didelphis marsupialis* in captivity. *Mem. Inst. Oswaldo Cruz* 78 (4): 507–9.

Mustrangi, M. A. 1995. Phylogeography of *Marmosops* (Marsupialia; Didelphidae) in the Atlantic forest in Brazil and the phylogenetic relationships of the Atlantic rainforest and Amazonian species in the genus. Ph.D. diss., University of California, Berkeley.

Mustrangi, M. A., and J. Patton. 1997. Phylogeography and systematics of the slender mouse opossum, *Marmosops* (Marsupialia, Didelphidae). *Univ. Calif. Publ. Zool.* 130:1–86.

• Myers, P., and R. M. Wetzel. 1979. New records of mammals from Paraguay. *J. Mammal.* 60:638–41.

Nitikman, L. Z., and M. A. Mares. 1987. Ecology of small mammals in a gallery forest of central Brazil. *Ann. Carnegie Mus.* 56:75–95.

Nogueira, J. C. 1988. Anatomical aspects and biometry of the male genital system of the white-belly opossum *Didelphis albiventris* Lund, 1841, during the annual reproductive cycle. *Mammalia* 52: 233–42.

———. 1989. Reprodução de gambá, *Didelphis albiventris. Ciênc. Hoje* 9 (53): 8–9.

O'Connell, M. A. 1979. Ecology of didelphid marsupials from northern Venezuela. In *Vertebrate ecology in the northern Neotropics,* ed. J. F. Eisenberg, 73–87. Washington, D.C.: Smithsonian Institution Press.

———. 1983. *Marmosa robinsoni. Mammal. Species* 203:1–6.

Oliveira, J. A. de, M. L. Lorini, and V. G. Persson. 1992. Pelage variation in *Marmosa incana* with notes on taxonomy. *Z. Säugetierk.* 57:129–36.

Oliver, W. L. R. 1976. The management of yapoks (*Chironectes minimus*) at Jersey Zoo, with observations on their behaviour. In *Jersey Wildlife Preservation Trust, Thirteenth Annual Report,* 32–36. Jersey: Trinity.

Olrog, C. C. 1959. Notas mastozoológicas. 2. Sobre la colección del Instituto Miguel Lillo. *Acta Zool. Lilloana* 17:403–19.

———. 1976. Sobre mamíferos del noroeste argentino. *Acta Zool. Lilloana* 32:5–12.

• ———. 1979. Los mamíferos de la selva húmeda, Cerro Calilegua, Jujuy. *Acta Zool. Lilloana* 33: 9–14.

• Olrog, C. C., and M. M. Lucero. 1981. *Guía de los mamíferos argentinos.* Tucumán, Argentina: Ministerio de Cultura y Educación, Fundación Miguel Lillo.

Osgood, W. H. 1913. New Peruvian mammals. *Field Mus. Nat. Hist., Zool. Ser.* 12:93–100.

———. 1921. A monographic study of the American marsupial *Caenolestes. Field Mus. Nat. Hist., Zool. Ser.* 14:1–162.

———. 1924. Review of living caenolestids with description of a new genus from Chile. *Field Mus. Nat. Hist., Zool. Ser.* 14 (2): 165–73.

————. 1943. The mammals of Chile. *Field Mus. Nat. Hist., Zool. Ser.* 30:1–268.

Palma, R. E. 1995. The karyotypes of two South American mouse opossums of the genus *Thylamys* from the Andes and eastern Paraguay. *Proc. Biol. Soc. Wash.* 108:1–5.

————. 1997. *Thylamys elegans. Mammal. Species* 572:1–4.

Palma, R. E., and T. L. Yates. 1998. Phylogeny of southern South American mouse opossums (*Thylamys*, Didelphidae) based on allozyme and chromosomal data. *Zeit. Säugetierk.* 63:1–15.

Patton, J. L., S. F. Dos Reis, and M. Nazareth F. da Silva. 1996. Relationships among didelphid marsupials based on sequence variation in the mitochondrial cytochrome 6b gene. *J. Mammal. Evol.* 3: 3–29.

Patton, J. L., and M. Nazareth F. da Silva. 1997. Definition of species of pouched four-eyed opossums (Didelphidae, *Philander*). *J. Mammal.* 78:90–102.

Persson, V. G., J. A. Oliveira, and M. L. Lorini. 1992. Componentes de variação de pelage em *Marmosa incana* (Lund 1841). In *Resumes XIX Congresso Brasileiro de Zoologia,* 152.

Petter, F. 1968. *Mammalia* 32:313–16.

Phillips, C. J., and J. Knox Jones Jr. 1969. Notes on reproduction and development in the four-eyed opossum, *Philander opossum*, in Nicaragua. *J. Mammal.* 50:345–49.

Pine, R. 1972. A new subgenus and species of murine opossum (genus *Marmosa*) from Peru. *J. Mammal.* 53:279–82.

————. 1973. Anatomical and nomenclatural notes on opossums. *Proc. Biol. Soc. Wash.* 86:391–402.

————. 1976. *Monodelphis umbristriata* is a distinct species of opossum. *J. Mammal.* 57:785–87.

————. 1977. *Monodelphis iheringi* (Thomas) is a recognizable species of Brazilian opossum. *Mammalia* 41:235–37.

————. 1980. Taxonomic notes on "*Monodelphis dimidiata itatiayae* (Miranda-Rabiero)," *Monodelphis domestica* (Wagner) and *Monodelphis maraxina* (Thomas). *Mammalia* 43:495–99.

————. 1981. Reviews of the mouse opossums *Marmosa parvidens* Tate and *Marmosa invieta* Goldman with description of a new species. *Mammalia* 45:55–70.

Pine, R. H., and J. P. Abravaya. 1978. Notes on the Brazilian opossum *Monodelphis scalops* (Thomas). *Mammalia* 52:379–82.

Pine, R. H., I. R. Bishop, and R. L. Jackson. 1970. Preliminary list of mammals of the Xavantina/Cachimbo expedition. *Trans. Roy. Soc. Trop. Med. Hyg.* 64:668–70.

Pine, R. H., P. L. Dalby, and J. O. Matson. 1985. Ecology, postnatal development, morphometrics, and

taxonomic status of the short-tailed opossum, *Monodelphis dimidiata*, an apparently semelparous annual marsupial. *Ann. Carnegie Mus.* 54 (6): 195–231.

Pine, R. H., and C. O. Handley Jr. 1984. A review of the Amazonian short-tailed opossum *Monodelphis emiliae. Mammalia* 48:239–45.

Redford, K. H., and J. F. Eisenberg. 1992. *Mammals of the Neotropics. Vol. 2. The southern cone: Chile, Argentina, Uruguay, Paraguay.* Chicago: University of Chicago Press.

• Reig, O. A. 1964. Roedores y marsupiales del partido de General Pueyrredón y regiones adyacentes (provincia de Buenos Aires, Argentina). *Pub. Mus. Mun. Cienc. Nat. Mar del Plata* 1 (6): 203–24.

Reig, O. A., A. L. Gardner, N. O. Bianchi, and J. L. Patton. 1977. The chromosomes of the Didelphidae (Marsupialia) and their evolutionary significance. *Biol. J. Linnean Soc.* 9:191–216.

Reig, O. A., J. A. W. Kirsch, and L. G. Marshall. 1987. Systematic relationships of the living and Neocenozoic American "opossum-like" marsupials. In *Possums and opossums: Studies in evolution,* ed. M. Archer, 1–89. Chipping Norton, N.S.W., Australia: Surrey Beatty.

Ribeiro, M. G. and J. C. Nogueira. 1983. Aspectos anatômicos do sistema genital masculino da cuíca *Philander opossum* (Linnaeus, 1758) Didelphidae-Marsupialia. *Separata da Revista, Lundiana,* 2: 57–69.

————. 1990. The penis morphology of the four-eyed opossum *Philander opossum. Anat. Anz., Jena* 171: 65–72.

Ride, W. D. L. 1964. A review of Australian fossil marsupials. *J. Proc. Roy. Soc. West Austral.* 47:97–131.

Rigueira, S. E., C. M. de Carvalho Valle, J. B. M. Varejão, P. V. de Albuquerque, and J. C. Noguleira. 1987. Algumas observações sobre o ciclo reprodutivo anual de fêmeas do gambá *Didelphis albiventris* (Lund, 1841) (Marsupialia, Didelphidae) em populações naturais no estado de Minas Gerais, Brasil. *Rev. Brasil. Zool., São Paulo* 4 (2): 129–37.

Roig, V. G. 1971. La presencia de estados de hibernación en *Marmosa elegans* (Marsupialia-Didelphidae). *Acta Zool. Lilloana* 28:5–12.

Santori, R. T., D. Astua de Moraes, and R. Cerqueira. 1996. Diet composition of *Metachirus nudicaudatus* and *Didelphis aurita* in southeastern Brazil. *Mammalia* 59:511–16.

Schamberger, M., and G. Fulk. 1974. Mamíferos del Parque Nacional Fray Jorge. *Idesia* (Chile) 3:167–79.

Schneider, C. O. 1946. Catálogo de los mamíferos de la provincia de Concepción. *Bol. Soc. Biol. Concepción* 21:67–83.

Seluja, G. A., M. V. DiTomaso, M. Brun-Zorilla, and

H. Cordoso. 1984. Low karyotypic variation in two didelphids. *J. Mammal.* 65:702–7.

Simonetta, A. M. 1979. First record of *Caluromysiops* from Colombia. *Mammalia* 43:247–48.

Simonetti, J. A., J. L. Yañez, and E. R. Fuentes. 1984. Efficiency of rodent scavengers in central Chile. *Mammalia* 48:608–9.

Stallings, J. R. 1988a. Small mammal communities in an eastern Brazilian park. Ph.D. diss., University of Florida, Gainesville.

———. 1988b. Small mammal inventories in an eastern Brazilian park. *Bull. Florida State. Mus.* 34:153–200.

• Streilein, K. E. 1982a. Behavior, ecology, and distribution of the South American marsupials. In *Mammalian biology in South America*, ed. M. A. Mares and H. H. Genoways, 231–50. Pymatuning Symposia in Ecology 6. Special Publications Series. Pittsburg: Pymatuning Laboratory of Ecology, University of Pittsburgh.

———. 1982b. The ecology of small mammals in the semiarid Brazilian caatinga. 3. Reproductive biology and population ecology. *Ann Carnegie Mus.* 51 (13): 251–69.

Sunquist, M. E., S. N. Austad, and F. Sunquist. 1987. Movement patterns and home range in the common opossum (*Didelphis marsupialis*). *J. Mammal.* 68:173–76.

Sunquist, M. E., and J. F. Eisenberg. 1993. Reproductive strategies of female *Didelphis*. *Bull. Fla. Mus. Nat. Hist., Biol. Sci.* 36:109–40.

Szalay, F. S. 1993. Metatherian taxon phylogeny. In *Mammal phylogeny: Placentals*, ed. F. S. Szalay, M. J. Novacek, and M. C. McKenna, 216–42. New York: Springer-Verlag.

———. 1994. *Evolutionary history of the marsupials and an analysis of osteological characters*. New York: Cambridge University Press.

Tate, G. H. H. 1933. A systematic revision of the marsupial genus *Marmosa*. *Bull. Amer. Mus. Nat. Hist.* 66:1–250.

• Thomas, O. 1888. Diagnoses of four new species of *Didelphis*. *Ann. Mag. Nat. Hist.*, ser. 6, 1:158–59.

———. 1894. On *Microureus griseus* Desm., with the description of a new genus and species of Didelphidae. *Ann. Mag. Nat. Hist.*, ser. 6, 14:194–88.

———. 1926. Two new mammals from north Argentina. *Ann. Mag. Nat. Hist.*, ser. 9, 17 (99): 311–13.

———. 1927. On a further collection of mammals made by Sr. E. Budin in Neuquén, Patagonia. *Ann. Mag. Nat. Hist.*, ser. 9, 19 (114): 650–58.

Trupin, G. L., and B. H. Fadem. 1982. Sexual behaviour of the gray short-tailed opossum (*Monodelphis domestica*). *J. Mammal.* 63:409–14.

Tyndale-Biscoe, C. H., and M. B. Renfree. 1987. *Reproduction in marsupials*. Cambridge: Cambridge University Press.

Unger, K. L. 1982. Nest building behavior of the Brazilian bare-tailed opossum *Monodelphis domestica*. *J. Mammal.* 63:150–62.

Van Valen, L. 1966. Deltatheridia, a new order of mammals. *Bull. Amer. Mus. Nat. Hist.* 132:1–126.

———. 1974. *Deltatheridium* and marsupials. *Nature* (London) 248:165–66.

Varejão, J. B. M., and C. M. C. Valle. 1982. Contribuições ao estudo da distribução geográfica das espécies do gênero *Didelphis* no estado de Minas Gerais Brasil. *Lundiana* 2:5–55.

• Vieira, C. 1949. Xenartros e marsupiais do estado de São Paulo. *Arq. Zool. São Paulo* 7:325–62.

Vieira, E. M., and A. R. T. Palma. 1996a. Natural history of *Thylamys velutinus* (Marsupialia, Didelphidae) in central Brazil. *Mammalia* 60:481–84.

———. 1996b. Area de vida e hábitos alimentares de *Thylamys velutinus* (Marsupialia, Marmosidae) no Brasil central. In *Resumos do XXI Congresso Brasileiro de Zoologia*, 222. Pôrto Alegre.

Vivo, M. de, and N. F. Gomès. 1989. First record of *Caluromysiops irrupta* Sanborn, 1951 (Didelphidae) from Brazil. *Mammalia* 53:310–11.

Waterhouse, G. R. 1846–48. *A natural history of the Mammalia*. Reprint New York: Arno Press, 1979.

Wilson, D., and D. M. Reeder, eds. 1993. *Mammal species of the world*. Washington, D.C.: Smithsonian Institution Press.

Woodburne, M. O., and W. J. Zinmeister. 1982. Fossil land mammals from Antarctica. *Science* 218:284–86.

Ximénez, A. 1967. Contribución al conocimiento de *Lutreolina crassicaudata* (Desmarest, 1804) y sus formas geográficas (Mammalia-Didelphidae). *Commun. Zool. Mus. Hist. Nat. Montevideo* 9 (112): 1–7.

6   Order Xenarthra
(Edentata)

Diagnosis

The name Xenarthra is used here because we concur with Emry (1970) that the Palaenodonta are not closely related or ancestral to the Xenarthra. Rose and Emry (1993) reinforce our conviction concerning the unique status of this taxon. In spite of the older ordinal name Edentata, not all members lack teeth; however, the tooth number is often reduced and tooth structure is simplified. Among the existing families the incisors are absent as well as the canines, but the two-toed sloth has a presumptive premolar that is caniniform. The premolars and molars, when present, do not have enamel. The tympanic bone is ring shaped in many forms. All forms have trunks reinforced with broadened ribs and accessory zygapophyses on the thoracic and lumbar vertebrae, hence the ordinal name Xenarthra (strange joints). Many of the living forms show great specialization for vastly different ways of life, making a simple diagnosis for the order difficult.

Distribution

The extant edentates are entirely confined to the New World. One species, the nine-banded armadillo (*Dasypus novemcinctus*) has extended its range in recent times to the southeastern United States and the southern portions of the Great Plains. Other southern species, if they range across the Panamanian land bridge, generally do not extend farther north than southern Mexico. In the south of Argentina three species of armadillos range to Patagonia, and *Zaedyus pichiy* ranges to the Strait of Magellan and thence to Chile (see Redford and Eisenberg 1992).

History and Classification

Xenarthrans appear to have originated in South America and to have undergone their major adaptive radiation there (Reig 1981). Although the suborder Palaenodonta is first noted in the Paleocene of North America, this is now believed to be either an early offshoot from South American stock or an entirely separate taxon (Rose and Emry 1993). The more recent xenarthrans first appear in the Paleocene of South America. These specimens are believed to belong to the armadillo family, the Dasypodidae. Six families of the Xenarthra are now extinct, including the giant ground sloths of the families Mylodontidae and Megatheriidae (Patterson and Pascual 1972). The glyptodonts (Glyptodontidae) were somewhat similar to armadillos in having bony scutes forming a turtle-like carapace, but they were an herbivorous, independent offshoot that first appeared in the Eocene and persisted to the Pleistocene (Gillette and Ray 1981). The glyptodonts, as well as some of the giant ground sloths, crossed from South America to North America in the late Miocene or early Pliocene at the completion of the Panamanian land bridge. They subsequently became extinct in North America at the close of the Pleistocene (Webb and Marshall 1982; see chapters 2, 3, and 4 this volume).

The biology of the Xenarthra is admirably reviewed in the volume edited by Montgomery (1985a). Captive maintenance has been summarized by Merrett (1983). The extant xenarthrans are included in four families: the Myrmecophagidae, the Bradypodidae, the Megalonychidae, and the Dasypodidae.

Comment

The genera *Bradypus* and *Choloepus* were formerly subsumed under Bradypodidae, but see Gardner (1993) and Webb (1985).

FAMILY MYRMECOPHAGIDAE
Anteaters

Diagnosis

Teeth are totally lacking, and the rostrum of the skull is extremely elongated (fig. 6.1). The long tongue,

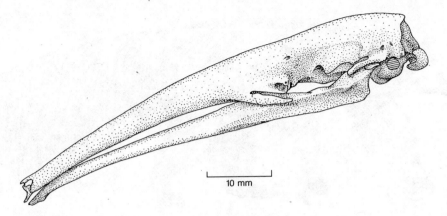

Figure 6.1. Skull of *Myrmecophaga*.

extended for feeding on ants and termites, is coated with a copious sticky saliva produced from greatly enlarged submaxillary glands. The forefeet have an enlarged third digit with a strong claw, and the other digits are somewhat reduced or absent. The hind feet have four or five digits. The body is covered with hair; the tail is long and also haired to varying degrees.

Distribution

Species of anteaters occur from southern Mexico to northern Argentina. Ranges for the species are reviewed by Wetzel (1985a).

Natural History

Members of this family are highly adapted for feeding on ants and termites. The strong claws are used to open the nests, and the long, extensible tongue aids in feeding. Females produce only a single young at a time, and parental care is highly developed. The young is frequently carried on the mother's back when she changes resting areas or shortly before weaning, when it accompanies her to feeding sites. Relative to their body size anteaters have a low metabolic rate, apparently an adaptation for feeding on ants and termites, a ubiquitous resource of low nutritional quality (McNab 1982; Redford 1987a,b).

Genus *Cyclopes* Gray, 1821
Cyclopes didactylus (Linnaeus, 1758)
Pygmy or Silky Anteater

Description

This is the smallest of the anteaters; total length ranges from 360 to 450 mm, and the tail is 180 to 262 mm. Weight is rarely more than 400 g. The color is golden brown, with some of the dorsal hairs silver tipped. The pelage is very dense. The tail is fully prehensile, and two claws on each forepaw are greatly enlarged (plate 2).

Range and Habitat

Cyclopes occurs from extreme southern Mexico south to Brazil and perhaps to Paraguay. It is generally confined to moist multistratal tropical evergreen forests and is absent from savanna areas (map 6.1).

Natural History

This arboreal specialist has an extremely prehensile tail. It is almost totally nocturnal and rarely descends to the ground, occurring where the forest canopy is continuous (Sunquist and Montgomery 1972). A single young is born after an unknown gestation period. This species may occur at a density of 0.77/ha in Panama. Females are strictly spaced with little home-range overlap, but the home range of a single male may overlap those of several females (Montgomery 1983). Best and Harada (1985) note that these animals are opportunistic feeders on ants and seem to track the relative abundance of their prey, taking ubiquitous species of ants more often than rare species.

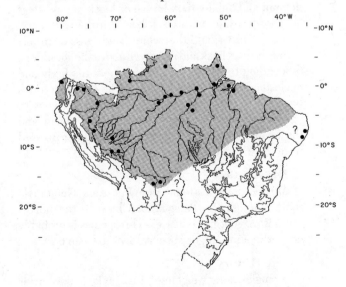

Map 6.1. Distribution of *Cyclopes didactylus*.

Figure 6.2. Sketch of *Myrmecophaga tridactyla*.

Genus *Myrmecophaga* Linnaeus, 1758
Myrmecophaga tridactyla Linnaeus, 1758
Giant Anteater, Oso Hormiguero, Yuru Mi

Measurements

	Mean	Min.	Max.	N
HB	1,265.5	1,100.0	2,000.0	16
T	734.0	600.0	900.0	
HF	165.0	150.0	180.0	13
E	46.7	35.0	50.0	9
Wt	32.9 kg	22.0 kg	39.0 kg	5
	32.1 kg	26.4 kg	36.4 kg	12 m
	29.2 kg	25.5 kg	31.8 kg	12 f

Redford and Eisenberg 1992.

Description

The giant anteater is one of the most distinctive mammals in South America, with its large size, long tail with long, coarse hair, and greatly elongated snout. The ears and eyes are very small. The front claws, particularly the third, are greatly enlarged and carried curled back as the animal walks on its "wrists" (fig. 6.2). The dorsum and tail are dark brown or black; the forelegs are mostly white, with black bands at the wrists and above the claws; and thin white bands pass from just below the ears back and up to well above the shoulders, descending to where the forelegs meet the body and enclosing a broad band of black. The tail, which can be nearly as long as the body, is uniformly brown with very coarse hair. There is little sexual dimorphism, and the phallus of the male must be manually expressed from the urogenital sinus.

Distribution

Myrmecophaga is found from Belize and Guatemala south through the Paraguayan Chaco to the northernmost Argentine provinces. This species is probably now extinct in Uruguay (Wetzel 1982a) (map 6.2).

Life History

A single young weighing 1.1 to 1.6 kg is born after a gestation period of 183–90 days. Its eyes open after about six days, and the lactation period is six to eight weeks or longer. The young is carried on its mother's back for about six to nine months, though it may be left in a "nest" while the female feeds. In captivity the age of first reproduction varies from 2.5 to 4.0 years; animals breed throughout the year, and the interbirth interval can be as low as nine months. In northeastern Argentina this species is reported to reproduce between September and March, though in central Brazil young are seen throughout the year. Giant anteaters can live at least sixteen years in captivity (Bickel, Murdock, and Smith 1976; Byrne 1962; Crespo 1982; Hardin 1976; Merrett 1983; Shaw, Machado-Neto, and Carter 1987; Smielowski, Stanislawski, and Taworksi 1981).

Ecology

Giant anteaters are found in a large variety of habitats from tropical forest to the xeric Chaco. They appear to be most abundant in open vegetation formations with an abundance of ants and termites and can be active throughout the day and night depending

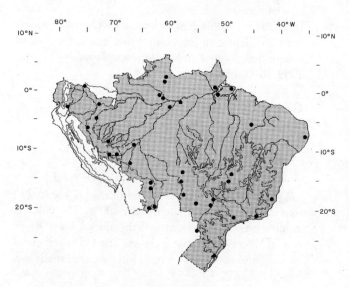

Map 6.2. Distribution of *Myrmecophaga tridactyla*.

on temperature and rainfall. In southeastern Brazil *Myrmecophaga* occurred at a density of between 1.3 and 2.0 per square kilometer. In south-central Brazil female giant anteaters have average home ranges of 3.7 km² ($n = 4$), while male home ranges averaged 2.7 km² ($n = 4$)—not a statistically significant difference. In the Venezuelan llanos home ranges were much larger, with one report of a 25 km² home range.

Myrmecophaga appears to have very poor eyesight and fairly poor hearing, relying on smell to locate prey. Ants and termites compose most of its diet. The distribution of the social insect fauna available to a giant anteater appears to dictate to what extent it feeds on ants versus termites. It finds insects either by rooting with its nose or by digging into nests or mounds with its large, powerful front claws. The giant anteater is a very selective feeder, visiting many termite mounds or ant nests in the course of feeding. Feeding seems to end whenever the soldier caste of ants or termites arrives at the breach to defend the nest, generally by noxious chemical secretions or by biting.

Giant anteaters are solitary and, except during the breeding season, seem to ignore one another. Adults probably have no serious predators; the major sources of mortality are humans and fire. The long, coarse hair is highly flammable, and anteaters are frequently found burned to death after a severe grass or forest fire (Eisenberg 1989; Montgomery 1985b; Montgomery and Lubin 1977; Redford 1985a, 1987b; Schmid 1938; Shaw, Machado-Neto, and Carter 1987; Shaw, Carter, and Machado-Neto 1985).

Genus *Tamandua* Gray, 1825

Description

The tamandua is easily separated from the giant anteater by its smaller size, its coloration, and the shape of its tail. Most tamanduas in southern South America are golden brown with a black vest covering the dorsum and venter, crossing the shoulders in a black band, but on some individuals the vest may be greatly reduced or even absent. The tail is prehensile and sparsely haired, the snout is considerably less elongated than that of the giant anteater, and the ears are proportionally longer. There is little sexual dimorphism, and the phallus of the male must be expressed manually from the urogenital sinus. The claws on the front feet are enlarged but are not as long proportionally as those of *Myrmecophaga* (see plate 2) (Wetzel 1975, 1985a).

Tamandua mexicana (Saussure, 1860)

Description

The head and body length is approximately 563 mm; tail length, 544 mm; weight, 3.2 to 5.4 kg. Over

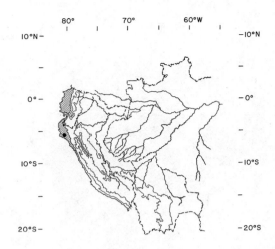

Map 6.3. Distribution of *Tamandua mexicana*.

most of its range *T. mexicana* exhibits the black vest markings described in the genus account (see plate 2).

Range and Habitat

The species is distributed from southern Veracruz, Mexico, and south through Colombia, west of the eastern Andean cordillera to Peru (map 6.3). It is present in the Maracaibo basin in Venezuela but is replaced to the east in Venezuela and Colombia by *T. tetradactyla*. It occurs at low elevations in a variety of habitats from gallery forests to multistratal tropical evergreen forests.

Life History and Ecology

This species was studied extensively at Barro Colorado Island, Panama (Montgomery 1985b). Home ranges averaged 25 ha, and daily movements averaged about 1,000 m. Both ants and termites were included in the diet, with termites predominating at 67%. Activity periods were both diurnal and nocturnal. Although the newborn is carried by the mother, as it matures the young is often left in a shelter while the mother forages alone.

Tamandua tetradactyla (Linnaeus, 1758)
Southern Tamandua, Oso Melero, Caguare

Measurements

	Mean	Min.	Max.	N
TL	1,002.4	905.0	1,047.0	8
HB	590.3	522.0	635.0	
	640.5	594.0	692.0	6
T	412.1	370.0	436.0	8
	458.5	425.0	498.0	6
HF	88.3	57.0	105.0	8
	94.5	80.0	105.0	6
E	46.5	40.0	50.0	8
	46.6	41.0	51.0	6
Wt	5.12 kg	3.8 kg	8.5 kg	5
	6.20 kg	4.9 kg	7.0 kg	5

Redford and Eisenberg 1992.

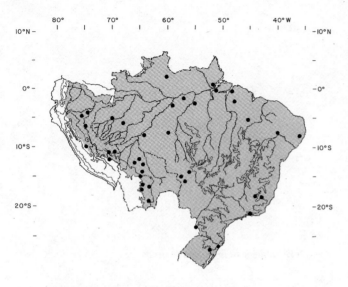

Map 6.4. Distribution of *Tamandua tetradactyla*.

Description

See account for genus. *T. tetradactyla* is uniformly golden over much of its range in eastern Venezuela and Colombia (see Wetzel 1985a), but black-vested golden and melanistic individuals may occur throughout Brazil (plate 2).

Distribution

Tamandua tetradactyla is found from Venezuela south through Paraguay to northern Uruguay and the northern Argentine provinces of Santa Fe, Chaco, Salta, and Jujuy; in the southern part of its range the "vest" is prominent (map 6.4).

Life History

The southern tamandua produces a single young after a gestation period of about 160 days. The young is carried on the mother's back or left in a nest. When it is old enough, it accompanies its mother as she forages (Crespo 1982; Meritt 1975; Montgomery and Lubin 1977).

Ecology

Tamandua tetradactyla is found in a wide variety of habitats, including tropical forest, dry scrub forest, and open grassland. With its prehensile tail it is an excellent climber, though it is often found well away from trees. The animal shelters in tree hollows as well as abandoned holes in the ground such as those made by armadillos.

This medium-sized anteater feeds on both ants and termites, depending on the available prey. There is extensive individual variation in prey choice. The young, feeding with its mother, apparently learns her food preference, which may account for observed individual preferences as adults. There is also intrapopulation

variation in activity cycles; some individuals are diurnal and others are nocturnal.

Tamanduas have been reported as prey for ocelots and jaguars, and the young are probably vulnerable to other felines and foxes. They are killed by humans for their meat as well as for their tough hide (Mares, Ojeda, and Kosco 1981; Montgomery 1985b; Montgomery and Lubin 1977; Redford 1983).

FAMILY BRADYPODIDAE
Three-toed Sloths

Diagnosis

The exact identification of the tooth types remains in some doubt; thus the dental formula is usually listed as 5/4–5. The teeth are cylindrical and grow throughout life (fig. 6.3). Head and body length ranges from 400 to 800 mm and weight from 2.25 to 5.50 kg; the stout tail is approximately 68 mm long. The neck contains eight or nine cervical vertebrae. The forelimbs are slightly longer than the hind limbs, and each forefoot has three long claws. The hair is long and rather stiff; individual hairs may support algae and thereby have a blue-green color. Adult males are about the same size as females; their distinctive black dorsal patch distinguishes them from females, which lack the patch. In addition to algal symbionts, moths have been recorded in the fur of members of this family (Waage and Best 1985).

Distribution

Three-toed sloths extend from eastern Honduras across South America to northern Argentina.

Natural History

Three-toed sloths are active both diurnally and nocturnally. They are highly specialized as selective browsers on the leaves of trees; the stomach is compartmentalized, and fermentative reduction occurs in

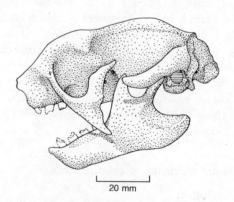

20 mm

Figure 6.3. Skull of *Bradypus variegatus*.

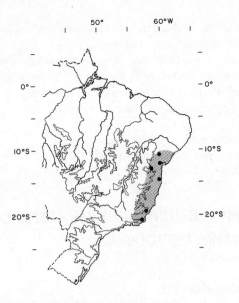

Map 6.5. Distribution of *Bradypus torquatus*.

both the stomach and the intestines. Three-toed sloths are characterized by a low metabolic rate and a low core body temperature, apparently an adaptation for feeding on leaves that are abundant but have a low nutrient content (McNab 1978). They locomote slowly while suspended beneath a tree branch by all four limbs. A single young is produced after a gestation period of approximately six months.

Genus *Bradypus* Linnaeus, 1758
Three-toed Sloth, Perezoso de Tres Dedos
Bradypus torquatus Desmarest, 1816
Maned Sloth

Description
In general, this sloth looks like *B. variegatus*. The dorsum is grizzled light brown, but from the back of the neck to the shoulders there is a middorsal patch of long black hair, hence the common English name "maned sloth." The mane is absent in infants and juveniles, whose dorsum is white to pale brown (plate 2).

Distribution
Currently this sloth occurs in remnant patches of the Brazilian Atlantic rain forest in the states of Bahia, Espírito Santo, and Rio de Janeiro (map 6.5).

Natural History
Its diet is similar to that of *B. variegatus*. The two species have overlapping ranges, but where one species is abundant, the other is rare. They may be active at any hour in the twenty-four-hour cycle, but are most active from 0900 to 2100 hours. They can survive in secondary forests, but they prefer tall trees supporting vines. Home ranges average 2 ha, but animals often

shift to new ranges in the rainy season. The single young weighs about 300 g at birth (Pinder 1993).

Bradypus tridactylus Linnaeus, 1758
Pale-throated Sloth

Description
This species is similar in size and proportions to *B. variegatus*. Head and body length averages 500 mm and tail length 31 to 75 mm. The throat is white to buff; the dorsal color varies but frequently exhibits a marked speckled pattern, immediately distinguishing it from the basic brown to yellow brown of *B. variegatus*.

Range and Habitat
This species is found to the east of the Andes and south of the Orinoco River, although it crosses at the delta. It occurs in south-central Venezuela, eastward through the Guianas, and south across the Amazon and the Río Negro to Brazil (map 6.6) (Wetzel 1982a).

Natural History
The natural history patterns of this sloth have been described by Beebe (1925). Its habits are very similar to those described for *B. variegatus*.

Bradypus variegatus Schinz, 1825
Brown-throated Sloth

Description
External measurements are recorded under the diagnosis for the family. Weight averages 4.3 kg. The coat consists of long, coarse hairs overlying a short, dense underfur. The face is white with a brown stripe on each side enclosing the eye, the throat is brown, and the body varies from pale brown to yellowish. There is considerable variation in the pattern of lighter spots or

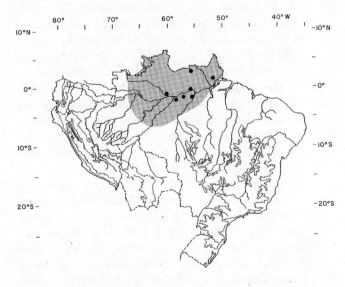

Map 6.6. Distribution of *Bradypus tridactylus*.

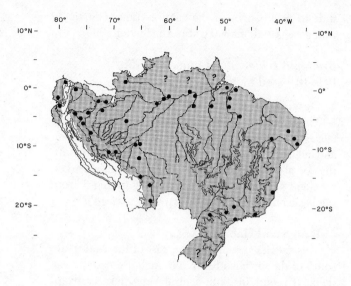

Map 6.7. Distribution of *Bradypus variegatus*.

splotches on the back. On the middorsum of the male the hair is extremely short, generally black surrounded by a yellow band (see plate 2).

Distribution

Bradypus variegatus is distributed from Honduras south to northernmost Argentina (Massoia 1980; Onelli 1913; Wetzel and Avila-Pires 1980; Wetzel 1982a, 1985a; FMNH) (map 6.7).

Natural History

A single young is born after a gestation period of between 120 and 180 days and is nutritionally dependent on the female for about four weeks. It remains with its mother for about six months, after which it occupies a portion of its natal home range (Montgomery and Sunquist 1978). Green (1989) made a remarkable observation of agonistic behavior between two male three-toed sloths in Costa Rica. He saw a male ascend a *Cecropia* tree and attack the "resident" by striking with the forelimbs. The resident fought back, emitting high-pitched screams, and the intruder withdrew.

The sloth is strictly arboreal, regularly descending to the ground only to defecate. Because they are strictly folivorous, sloths are found only in areas where most trees keep their leaves, though they are reported to be able to survive subfreezing temperatures for a short period.

In some areas sloths can be important components of the mammalian biomass: on Barro Colorado Island (Panama) they made up an estimated 70% of the arboreal mammalian biomass. In Panama animals were found to occupy home ranges averaging 1.61 ha. They are active both day and night and favor some species of trees over others, with this preference varying between individuals. Young sloths show the same preference for food and tree species as their mothers, learned as

they are carried around on their mothers' backs. One of the criteria used in choosing trees is a crown that receives abundant sun: sloths behaviorally thermoregulate while resting by moving in and out of the sun (Luederwaldt 1918; Montgomery and Sunquist 1975, 1978; Sunquist and Montgomery 1973).

Comment

Bradypus variegatus includes *B. infuscatus* and *B. boliviensis* (Gardner 1993).

FAMILY MEGALONYCHIDAE (CHOLOEPIDAE)

SUBFAMILY CHOLOEPINAE
Two-toed Sloths

Diagnosis

This subfamily was formerly subsumed under the Bradypodidae, but behavioral and anatomical differences are so profound that it deserves a higher rank of its own. Webb (1985) places the two-toed sloths within the family Megalonychidae, a group that has a rich fossil record. Nomenclature for the tooth types remains in doubt; the dental formula is given as 5/4−5. The anterior teeth in both the upper and lower jaw are distinctly caniniform (see fig. 6.4), and a wide gap separates the upper caniniform tooth from the remaining teeth. The short neck has six cervical vertebrate, the forefoot has only two claws, and the tail is absent or vestigial. The sexes are not easily distinguished by either size or color, and the genitalia must be examined manually by opening the cloaca (McCrane 1966). The color varies from dark brown to pale yellow (see plate 2I), and algal symbionts in the fur have been reported for this family.

Distribution

The species of two-toed sloths occur from northern Nicaragua to the Amazon region of Peru, Bolivia, and Brazil.

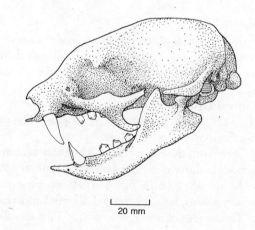

20 mm

Figure 6.4. Skull of *Choloepus hoffmanni*.

Natural History

Highly specialized for arboreal life, the two-toed sloth exhibits adaptations similar to those of *Bradypus* in locomoting while suspended from a branch (Goffart 1971). In conformity with its leaf-eating habits, it has a relatively low metabolic rate. A single young is produced at two- to three-year intervals (see species accounts).

Genus *Choloepus* Illiger, 1811
Perezoso de Dos Dedos, Prequiça-real

Comment

See subfamily account. Wetzel (1982; pers. comm.) indicated a disjunct distribution for *Choloepus hoffmanni* and possible sympatry with *C. didactylus* in the upper Amazon regions including portions of Brazil, Peru, and possibly Ecuador. At the time of his terminal illness Wetzel was convinced there was a genuine possibility of a third species of *Choloepus* in this region of upper Amazonas distinct from *C. didactylus* and *C. hoffmanni*, but the puzzle was not resolved before his death.

Choloepus didactylus (Linnaeus, 1758)

Description

This species of two-toed sloth is somewhat larger than *C. hoffmanni*. Head and body length ranges from 600 to 860 mm, the vestigial tail from 14 to 15 mm, and weight from 4.0 to 8.4 kg. The fur is generally tan to buffy brown.

Range and Habitat

The species occurs east of the Andes in southern Colombia, Venezuela, and along both banks of the Amazon in Brazil to the Amazon basin of Colombia, Ecuador, and Peru (map 6.8).

Natural History

The natural history of *Choloepus didactylus* has not been studied in the wild but is assumed to be similar to that of *C. hoffmanni*. In an extensive recovery operation in French Guiana following flooding from river dam construction, the adult sex ratio was found to be 0.74 males to one female (*n* = 315) (E. Taube, pers. comm.). A single young is born after a gestation period of approximately eleven months. The young is dependent on the female for four to five months and is carried until it reaches approximately 15% of the mother's body weight (Eisenberg and Maliniak 1985).

Choloepus hoffmanni Peters, 1859

Description

Head and body length ranges from 540 to 700 mm, and weight averages 5.7 kg. This species is usually pale yellow with brownish limbs (plate 2).

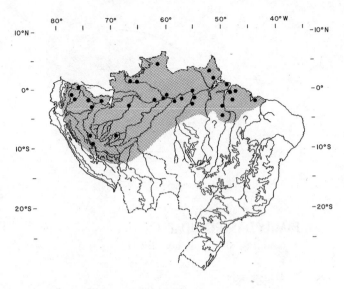

Map 6.8. Distribution of *Choloepus didactylus*.

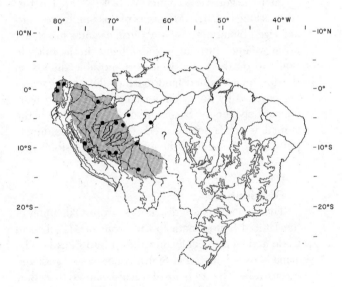

Map 6.9. Distribution of *Choloepus hoffmanni*.

Range and Habitat

From northern Nicaragua, its range extends west of the Andes to Ecuador (map 6.9). East of the Andes it occurs in Ecuador, Peru, and Brazil. Salazar, Redford, and Stearman (1990) have recorded *C. hoffmanni* as far south as 16°47' at the base of the Andes in Bolivia. Le Pont and Desjeux (1992) have also reordered the species in Bolivia.

Natural History

Two-toed sloths are nocturnal. Although specialized for feeding on leaves, they eat a considerable amount of fruit. They have a larger home range than the three-toed sloth, often 2–3 ha (Montgomery and Sunquist 1975, 1978). In the wild, adult females seem to vastly outnumber adult males, suggesting that there is some spacing among males and perhaps competition for access to females.

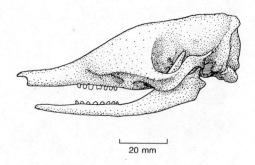

20 mm

Figure 6.5. Skull of *Dasypus* sp.

FAMILY DASYPODIDAE
Armadillos, Quirquinchos, Tatús

Diagnosis

The dental formula is highly variable and will be given in the generic accounts (Wetzel 1985b). The rostrum is long (fig. 6.5). Members of this family are characterized by numerous bony dermal scutes in regular arrangements, forming movable bands in the midsection and on the tail as well as immovable shields on the forequarters and hindquarters. The bony scutes are covered by horny epidermal scales. Sparse hairs appear between the bands and on the underside of the animal, which is not armored. There is no size dimorphism, but the penis of the male is readily visible anterior to the rectum.

Distribution

The armadillos are distributed from Oklahoma in the United States south to the Strait of Magellan in Chile and were recently introduced into Florida. The armadillos originated in South America and radiated extensively. The rich fossil record shows that they entered North and Central America during the late Miocene and Pliocene as the Panamanian land bridge neared completion (Webb and Marshall 1982; Webb 1985).

Natural History

The dermal scutes apparently are an antipredator defense. Armadillos' dietary habits vary greatly depending on the species. Some, such as *Cabassous*, feed almost exclusively on ants and termites; others (*Euphractus*) eat fruits, invertebrates, and small vertebrates. All have extremely well-developed claws and are capable burrowers. Members of the genus *Dasypus* frequently forage on the surface, whereas others, such as *Cabassous*, are adapted for a fossorial existence.

The feeding habits of the armadillos have been reviewed by Redford (1985b). Species adapted for a fossorial life and specialized for feeding on ants and

termites (e.g., *Priodontes* and *Cabassous*) have very low metabolic rates (McNab 1982). The distribution of most armadillos is strongly governed by the mean low temperature of an area, since they have a high thermal conductance and readily lose body heat at low temperatures. Though burrowing protects an individual from the extremes of ambient temperatures, only four species of armadillos have significantly occupied the warmer portions of the temperate zone (McNab 1982). Further information on this widely varying group is given in the species accounts.

Genus *Cabassous* McMurtrie, 1831
Naked-tailed Armadillo, Cabasú

Description

The four species within this genus differ little in external morphology except for size (Wetzel 1980). The number of teeth is highly variable, ranging between 7/8 and 10/9. There are no teeth in the premaxillary bone. Head and body length ranges from 300 to 490 mm and the tail from 90 to 200 mm. The snout is very short and broad, the ears are moderately large and funnel shaped, and the eyes are extremely small. The forefeet have five claws, the middle one extremely large and sickle shaped. The dorsal plates are arranged in transverse rows for the entire length of the body. The slender tail is distinctive, with either no armor or small, widely spaced thin plates; hence the common name naked-tailed armadillo. The tail alone serves to distinguish this genus from all other armadillos (see plate 3).

Distribution

The genus *Cabassous* is distributed from southern Mexico south to northern Argentina. It prefers rather moist habitats with well-drained, loose soil (Wetzel 1980).

Life History

The species of *Cabassous* are specialized for feeding on ants and termites. They are strongly fossorial and may forage underground. The female gives birth to a single young.

Cabassous centralis (Miller, 1899)

Description

Head and body length is approximately 340 mm; tail length, 130 to 180 mm; weight, 2.0 to 3.5 kg. This is the smaller of the two species occurring in northern South America.

Range and Habitat

Cabassous centralis occurs from southern Mexico, through Panama, to Ecuador west of the eastern cordillera of the Andes (map 6.10).

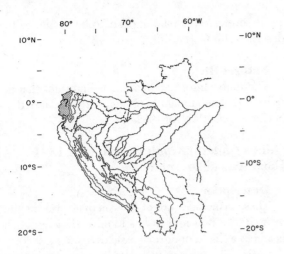

Map 6.10. Distribution of *Cabassous centralis.*

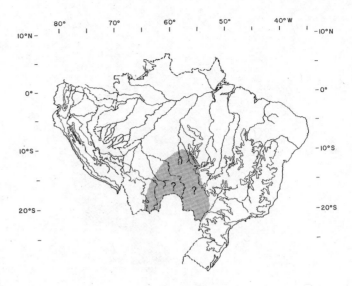

Map 6.11. Distribution of *Cabassous chacoensis.*

Natural History

Little is known concerning the behavior of *Cabassous* in the field. Apparently it is nocturnal and specialized for feeding on terrestrial ants and termites. Its tongue can be extruded to a great length when feeding. It is highly specialized for burrowing, and a great deal of its activity may take place underground. The few observations that have been made were of animals digging into rotten logs, apparently seeking termites.

If the animal is disturbed while foraging on the surface, it usually tries to burrow into the ground to escape. When disturbed it produces a low buzz or growl, which is startlingly loud, since many armadillos make very soft sounds (Wetzel 1982a). A single young is produced after an unknown gestation period.

Cabassous chacoensis Wetzel, 1980
Chacoan Naked-tailed Armadillo,
Tatú-ai Menore

Description

Measurements of two specimens were HB 300–306; T 90–96; HF 61; E 14–15. *C. chacoensis* is the smallest of the species of *Cabassous* and can be distinguished by its ears, which are strikingly smaller than in any other species in this genus and have a unique fleshy expansion of their anterior margins (Wetzel 1980).

Distribution

This species is found in the Gran Chaco of northwestern Argentina, in western Paraguay, in southeastern Bolivia, and possibly into adjacent areas of Brazil (Wetzel 1980) (map 6.11).

Ecology

This species appears to be confined to the xeric Chaco of Argentina and Paraguay and adjoining coun-

tries. Nothing is known of its natural history. When handled, males emit a loud grunt, but females remain silent (Meritt 1985; Wetzel 1982b).

Cabassous tatouay (Desmarest, 1804)
Greater Naked-tailed Armadillo

Measurements

	Mean	Min.	Max.	N
HB	457.8	410.0	490.0	5
T	179.0	150.0	200.0	
HF	82.2	80.0	86.0	
E	41.7	40.0	44.0	
Wt	5.35 kg	3.4 kg	6.4 kg	3

Redford and Eisenberg 1992.

Description

Cabassous tatouay is rather large compared with *C. unicinctus.* It is further separable from *C. unicinctus* by the size of its ear: *C. tatouay* has a much larger, funnel-shaped ear that extends well above the top of the head (Wetzel 1980).

Distribution

This species is found from Brazil south through Uruguay, to southeastern Paraguay, and to northern Argentina (Ximénez and Achaval 1966; Wetzel 1982a) (map 6.12).

Ecology

C. tatouay appears to be a species of open areas of vegetation, but as with all species of *Cabassous,* its highly fossorial habits make information on this species scarce. In southeastern Brazil, *C. tatouay* dug single-entrance burrows only into active termite mounds and was never found to reuse a burrow (Carter and Encarnação 1983).

Map 6.12. Distribution of *Cabassous tatouay*.

Cabassous unicinctus (Linnaeus, 1758)
Cabasú de Orejas Largas

Description

This is one of the larger species of the genus, as specimens from Venezuela and Suriname show: head and body length may range from 347 to 445 mm and tail length from 165 to 200 mm. In external appearance it is similar to *C. tatouay*, and identity can easily be determined by the ear.

Range and Habitat

Cabassous unicinctus ranges to the east of the Andes in Colombia, Venezuela, and the Guianas and in the east south to the Brazilian shield, and in the west to northern Peru. There is one record from Acre, Bra-

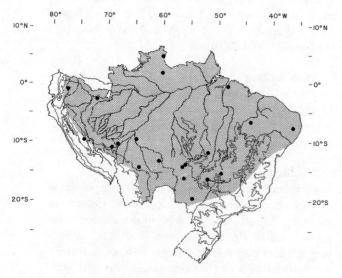

Map 6.13. Distribution of *Cabassous unicinctus*.

zil. Its preferred habitat appears to be similar to that described for *C. centralis* (map 6.13).

Natural History

The sparse information available indicates that this species' diet and foraging strategies are similar to those of *C. centralis* (Wetzel 1982a).

Genus *Chaetophractus* Fitzinger, 1871
Hairy Armadillo, Peludo

Description

These armadillos are of intermediate size, ranging from 200 to 400 mm in head and body length. The head has a shield of dermal ossicles that almost covers the nose. The dorsum is made distinctive by a linear series of about eighteen bands of dermal scutes, the middle seven or eight bands being flexible. The venter is haired (see plate 3). Sometimes confused with *Euphractus*, members of the genus *Chaetophractus* have longer ears (see species accounts).

Distribution

The current distribution of the genus includes western Bolivia to Paraguay and thence to central Argentina. Outlying populations have been recorded from southern Peru and northern Chile.

Chaetophractus nationi (Thomas, 1894)
Andean Hairy Armadillo, Quirquincho Andino

Description

Chaetophractus nationi is intermediate in size between *C. villosus* and *C. vellerosus*. One specimen from Bolivia measured HB 268; HF 52; E 30. The hair on the carapace varies from tan to buffy white, is up to 72 mm long, and can be sparse or thick (Wetzel 1985b).

Distribution

The distribution of this species is poorly known because it has frequently been confused with *C. vellerosus*. It is found in the puna of Bolivia and extends into the altiplano of Chile (Cabrera 1958; Mann 1978; Tamayo and Frassinetti 1980; Wetzel 1985b) (map 6.14).

Life History

One or two young are born in the summer (Mann 1978).

Ecology

This armadillo inhabits high-altitude areas with brushy vegetation. It deposits fat seasonally and hibernates during the winter. It is omnivorous, preferring insects and other invertebrates (Mann 1978).

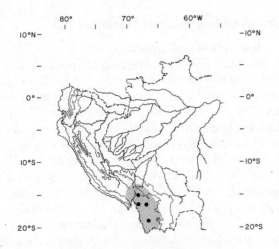

Map 6.14. Distribution of *Chaetophractus nationi*.

Comment

More specimens are needed to either support the validity of this species or establish it as a high-altitude subspecies of *C. vellerosus* (Wetzel 1985b).

Cook, Cáceres, and Miranda (1991) note that *C. nationi* ($2n = 62$) is karyotypically separable from *C. villosus* ($2n = 60$), but comparable data for *C. vellerosus* are lacking.

Chaetophractus vellerosus (Gray, 1865)
Small Hairy Armadillo, Small Screaming Armadillo, Quirquincho Chico, Piche Llorón

Measurements

	Mean	Min.	Max.	N
TL	376	328	400	76 m
	368	265	419	71 f
T	114	84	131	76 m
	112	77	138	71 f
HF	49	44	53	76 m
	48	31	56	71 f
E	28	22	31	76 m
	27	22	31	71 f
Wt	860	543	1,329	76 m
	814	257	1,126	71 f

Greegor 1974.

Description

This species is the smallest and slenderest of the hairy armadillos and differs from other *Chaetophractus* and from *Zaedyus pichiy* in having much longer ears. The hair on the dorsum is usually tan (pers. obs.; Wetzel 1985b).

Distribution

Chaetophractus vellerosus is found in the Gran Chaco of Bolivia, western Paraguay, and Argentina (Carlini and Vizcaíno 1987; Wetzel 1982a) (map 6.15).

Ecology

This armadillo is found primarily in xeric environments from low to high altitudes and does not occur in rocky areas, where burrow construction is impossible. It is usually found where yearly rainfall is between 200 and 600 mm, but a population exists in eastern Buenos Aires province, Argentina, where the annual rainfall is 1,000 mm.

It is nocturnal in the summer and diurnal in the winter and can go for long periods without drinking water. Burrows range from 8 to 15 cm in diameter, can be several meters long, may have multiple entrances, and are often at the bases of shrubs. One animal uses several burrows throughout its home range. Apparently no nest is built in the burrow. The entrance is usually sealed when an animal is inside. The minimum home-range area for one individual was 3.4 ha.

When not in the burrow, *C. vellerosus* spends most of its time foraging. Cassini (1993) reports that during capture experiments this species does not forage randomly but is systematic. Its diet varies seasonally. During the summer the major food item was insects (46% by volume; $n = 48$ stomachs), whereas during the winter it was plant material (50.7% by volume; $n = 36$ stomachs), especially pods from the *Prosopis* tree. A significant percentage of the diet is vertebrates (27.7% by volume in summer; 13.9% by volume in winter), including anurans, lizards, birds, and the mice *Eligmodontia typus* and *Phyllotis griseoflavus*. It ingests a great deal of sand while foraging, and sand can compose 50% of the volume of a single stomach. In the winter males and females are up to 10% heavier than during the summer because they have a 1–2 cm layer of subcutaneous fat. When handled, this species frequently emits loud cries of protest (Carlini and Vizcaíno 1987; Crespo 1944; Greegor 1974, 1975,

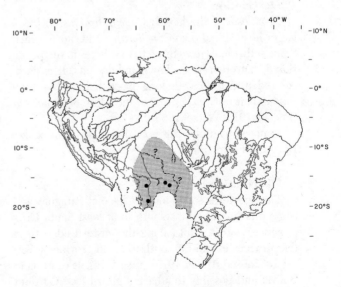

Map 6.15. Distribution of *Chaetophractus vellerosus*.

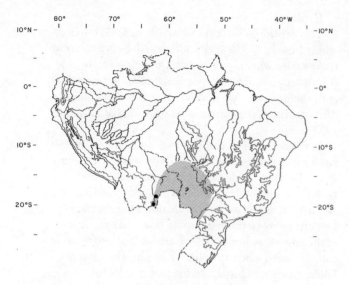

Map 6.16. Distribution of *Chaetophractus villosus*.

1980a,b, 1985; Mares, Ojeda, and Kosco 1981; Myers and Wetzel 1979).

Chaetophractus villosus (Desmarest, 1804)
Larger Hairy Armadillo, Quirquincho Grande

Measurements

	Mean	Min.	Max.	N
TL	436.7	386.0	486.0	10
HB	291.1	261.0	344.0	
T	145.6			
HF	61.7	57.0	66.0	9
E	24.0	22.0	31.0	10
Wt	1.32 kg	1.0 kg	1.40 kg	4
	3.42 kg	3.2 kg	3.65 kg	2

Redford and Eisenberg 1992.

Description

C. villosus is the largest *Chaetophractus*. It has a larger head shield than *C. vellerosus* and shorter ears. Some individuals have three to four holes in the pelvic shield that open to shallow glandular pits. It is dark-colored armadillo with long black hair rather than the tan hair of *C. vellerosus* or the white of *C. nationi* (Pocock 1913; Wetzel 1985b).
Chromosome number: $2n = 60$; FN = 90 (Benirschke, Low, and Ferm 1969).

Distribution

This species is found in the Chaco of Paraguay and Argentina (in Argentina south to at least Santa Cruz province) and in Chile along the eastern edge from the province of Bío-bío south to Aisén province (Wetzel 1982a; see Atalah 1975) (map 6.16). It extends to Bolivia and possibly to adjacent Brazil (see Gardner 1993).

Genus *Chlamyphorus* Harlan, 1825

Description and Taxonomy

The genus *Burmeisteria* is included in this account. There are two extant species. These extremely small armadillos have a head and body length usually less than 150 mm. The dorsal plates are almost free from the body, being attached at the head and loosely along the spine and at the pelvis. These animals are strongly fossorial and appear to occupy an almost molelike niche. The tail is short, and the rump is abruptly squared off, so it looks like a plug when the animal is burrowing (see plate 3). The small size, large claws on the forefeet, reduced ear, and diminutive eye render these armadillos so distinctive that they cannot readily be confused with any other species.

Distribution

The genus is at present confined to the more xeric portions of Bolivia, Paraguay, and northern Argentina. *C. truncatus* of Argentina is discussed in Redford and Eisenberg 1992.

Chlamyphorus (*Burmeisteria*) *retusus* (Burmeister, 1863)
Chacoan Fairy Armadillo, Greater Fairy Armadillo, Pichiciego Grande

Description

Chlamyphorus retusus is distinguishable from *C. truncatus* by its larger size. One specimen from Paraguay measured TL 158.5; HB 116; T 42.5; HF 39; E 4.5 (PCorps). In addition, the species has the carapace completely attached to the skin of the back and head, not underlain by fur. The head shield is wider, extending laterally and ventrally to the level of the eye, and the tail is not spatulate at the tip (see plate 3). The dorsum is golden yellowish, and there is short gray white hair on the belly (Myers and Wetzel 1979; Wetzel 1985b; pers. obs.).
Chromosome number: $2n = 64$ (Benirschke, Low, and Ferm 1969).

Distribution

Chlamyphorus retusus is found in the Gran Chaco of southeastern Bolivia, western Paraguay, and northwestern Argentina (Wetzel 1982a) (map 6.17).

Genus *Dasypus* Linnaeus, 1758
Long-nosed Armadillo, Cachicame, Mulita, Tatú

Description

The rostrum is long (see fig. 6.5), and the dental formula is 7-9/7–9. The dark brown carapace comprises scapular and pelvic shields, with six to eleven movable

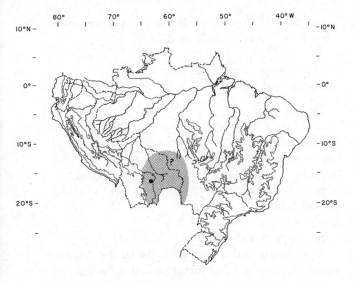

Map 6.17. Distribution of *Chlamyphorus retusus*.

bands separating the two shields. The ears are long
and have no scales or scutes. The long tail, generally
exceeding 55% of the head and body length, tapers to
a slender tip. The proximal two-thirds of the tail is covered with rings, each formed by two or more rows of
scales and scutes. The forefoot bears four long claws,
the longest two on the second and third digits. The
hind foot bears five claws, with the longest on the third
digit (plate 3).

Distribution

The genus *Dasypus* is distributed from the south-
central United States to the Río Negro in Argentina.

Life History

Species of *Dasypus* are insectivorous but will opportunistically eat small vertebrates and fruit. They
construct burrows and transport nesting material
such as grass and leaves by raking it under the body,
pressing it against the abdomen with the forefeet, and
hopping to the burrow on their hind legs (Eisenberg
1961).

Females are monovular but produce litters of two
to eight. The young are genetically identical, since the
zygote cleaves to produce several blastocysts before
implantation (see *D. novemcinctus*) (Redford 1985b,
1987a).

Dasypus hybridus (Desmarest, 1804)
Southern Lesser Long-nosed Armadillo,
Mulita Orejuda

Measurements

	Mean	Min.	Max.	N
TL	459.5	397.0	498.0	15
T	168.4	132.0	191.0	
HF	67.3	55.0	75.0	

E	25.1	23.0	28.0	
Wt	1.5 kg	1.09 kg	2.04 kg	14

Redford and Eisenberg 1992.

Description

This small *Dasypus* usually has seven movable
bands. It has smaller ears but a larger body than *D.
septemcinctus*. The very lightly haired carapace and
tail are dull brownish gray, and the belly is grayish to
pink (Barlow 1965; Wetzel and Mondolfi 1979).
Chromosome number: $2n = 64$ (Beirschke, Low, and
Ferm 1969).

Distribution

Dasypus hybridus is found from eastern Paraguay,
eastern Argentina, southern Brazil, and Uruguay west
through northern Argentina (Wetzel 1982a) (map
6.18).

Life History

Embryo counts have ranged from seven to twelve,
with eight the most common number of viable embryos (Crespo 1982; Galbreath 1985).

Ecology

Dasypus hybridus is an animal of the grasslands.
It apparently digs burrows only in grassland or other
very open vegetation, unlike *D. novemcinctus*, whose
burrows are frequently found in the forest. Burrows
are about 2 m long, with a single entrance less than
25 cm in diameter. Burrows are usually in sandy soils
on flat or gently sloping ground, in banks, under rocks,
or among the roots of trees.

This armadillo feeds much like *D. novemcinctus*,
moving rapidly across the ground, snuffling constantly,
and digging shallow foraging holes. In Uruguay it often

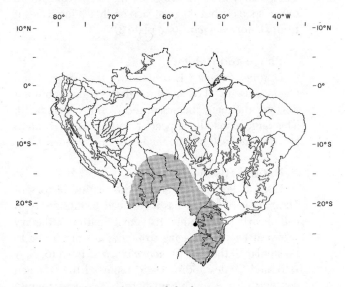

Map 6.18. Distribution of *Dasypus hybridus*.

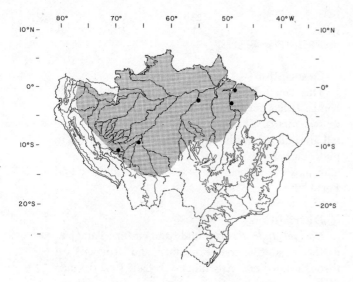

Map 6.19. Distribution of *Dasypus kappleri*.

digs into ant and termite nests, and one stomach contained mostly ants and termites as well as Orthoptera, Lepidoptera, other invertebrates, and the remains of a small rodent (Barlow 1965; Olrog 1979).

Dasypus kappleri Krauss, 1862
Greater Long-nosed Armadillo

Description

There are seven to nine teeth in both the upper and lower jaws. Head and body length ranges from 510 to 575 mm, tail length from 325 to 483 mm, and weight from 8.5 to 10.5 kg. There are seven to eight movable bands in the midsection between the shields covering the forequarters and the hindquarters, and large projecting scales or spurs are arranged in transverse rows on the proximal posterior surface of the hind leg. These scales on the hind legs are themselves diagnostic of the species, but its great size easily distinguishes it from *D. novemcinctus* (see plate 3).

Range and Habitat

Dasypus kappleri is confined to the Amazon and Orinoco basins. It occurs through the Guianas but not north of the main channel of the Orinoco (map 6.19). This large armadillo appears to prefer forested areas and does not extend into the northern savannas.

Natural History

Little is known concerning the natural history of *Dasypus kappleri*. It has a reduced litter size, typically bearing two young. Its activity patterns, dietary preferences, and foraging strategies are presumed to be similar to those of *D. novemcinctus*. Barreto, Barreto, and D'Alessandro (1985) reported the stomach contents of one specimen where 14% was vertebrate and 86% invertebrate material. The species constructs

burrows in well-drained soil and does not appear to be gregarious (Wetzel and Mondolfi 1979).

Dasypus novemcinctus Linnaeus, 1758
Nine-banded Armadillo

Description

Head and body length ranges from 395 to 573 mm and the tail from 290 to 450 mm. This is the second largest species of the genus. Weight ranges from 3.2 to 4.1 kg. The number of movable bands varies from eight to ten, but in northern South America the modal number is nine.

Range and Habitat

The nine-banded armadillo has the largest range of any armadillo species, occurring from the central and southern United States to approximately 32° S (map 6.20). It occupies a wide range of habitat types at low elevations, from dry deciduous forests to multistratal tropical evergreen forests, and even ranges into the semiarid llanos of Venezuela and Colombia and the caatinga of Brazil.

Natural History

Depending on season and temperature, the armadillo may be active during both day and night, but it avoids extremes of temperature and is generally inactive at midday. The long claws permit the animal to dig efficiently, and it constructs its own dwelling burrows. It transports nesting material by raking leaves under the body and then, clasping them with its forepaws, hopping to the burrow on its hind legs (Eisenberg 1961). It forages by scratching in the forest litter and probing with its long, sensitive nose. Nine-banded armadillos eat a wide variety of foods. Insect larvae and ants can predominate, but their diets also include

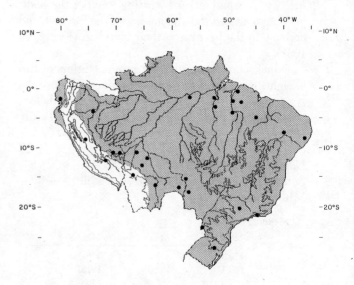

Map 6.20. Distribution of *Dasypus novemcinctus*.

small vertebrates and some plants (Taber 1954; Redford 1985b, 1987b).

Dasypus novemcinctus has a delayed implantation in the northern part of its range, so the time from insemination to birth can exceed 240 days, but the true gestation is approximately 70 days. Identical quadruplets are typically produced at birth, indicating that one egg divides into four separate blastocysts. Armadillos tend to forage alone, but there is considerable home-range overlap. Though several individuals may use the same burrow, they may be members of a family group. Home-range size in *Dasypus* varies considerably depending on the carrying capacity of the habitat and may be as small as 3.4 ha. At lower carrying capacities it can exceed 15 ha (Wetzel and Mondolfi 1979). It will eat virtually any food it can find and ingest. In South America this species feeds to a greater extent on ants and termites. These armadillos have very poor eyesight and locate food by smell. Most food items are probably taken from the ground surface or just beneath it. The strong claws are used to dig short, triangular feeding holes.

Dasypus novemcinctus is heavily hunted throughout its range for its delicate white meat. Because of its relatively large litter size and tolerance of humans, it can probably coexist with humans in rural areas (Barlow 1965; Eisenberg 1989; Galbreath 1982; González and Ríos 1980; Humphrey 1974; Redford 1987a,b; Wetzel and Mondolfi 1979).

Dasypus pilosus (Fitzinger, 1856)

Description
One male measured TL 575; T 252; HF 72; E 43; Wt 1,293 g (LSU). Body form and proportions are as found in the typical *Dasypus*, but the dorsal surface is covered with light brown hairs that extend through minute pores in the scutes.

Distribution
Found at high elevations in north-central Peru. Specimens at LSU derive from the department of Huánuco, Bosque Zapetqoch above Acomayo (see also Wetzel 1985a) (map 6.21).

Dasypus septemcinctus Linnaeus, 1758
Brazilian Lesser Long-nosed Armadillo, Mulita Común

Measurements

	Mean	Min.	Max.	N
HB	260.5	240.0	305.0	8
T	147.5	125.0	170.0	7
HF	60.0	45.0	72.0	2
E	30.9	30.0	38.0	7
Wt	1.63 kg	1.45 kg	1.8 kg	2

Redford and Eisenberg 1992.

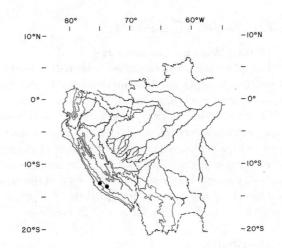

Map 6.21. Distribution of *Dasypus pilosus*.

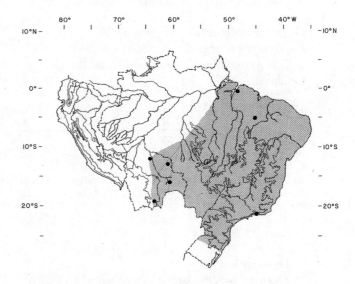

Map 6.22. Distribution of *Dasypus septemcinctus*.

Description
Dasypus septemcinctus is the smallest species of *Dasypus*. It does, however, have longer ears than *D. hybridus*. It has six or seven movable bands. Its carapace is dark, with little of the yellow found on *D. novemcinctus* (Redford, n.d.; Wetzel 1985b).

Distribution
This species is found from eastern Amazonian Brazil south to the northern Argentine provinces (Wetzel 1982a) (map 6.22).

Ecology
Dasypus septemcinctus is apparently a grassland species, although in southeastern Brazil it is reported to prefer gallery forests. It frequently expands burrows dug by other species of armadillos. In captivity young animals build nests at low temperatures (Block 1974; T. S. Carter, unpubl. data; Wetzel 1985b).

106

Dasypus yepesi Vizcaino, 1995

Description and Commentary

Dasypus mazzai was described from northwestern Argentina by Yepes in 1933. Subsequently *D. mazzai* was relegated to a junior synonym of *D. novemcinctus* (Wetzel and Mondolfi 1979). This inclusion was based on conclusions by Hamlett, who had studied the holotype, but the designated paratypes apparently indicate a new species so named by Vizcaino. The new species has 7–9 movable bands with ears 44%-54% of the cranial length. It appears to be intermediate in measurements between *D. novemcinctus* and *D. hybridus*.

Distribution

Currently known from Jujuy and Salta provinces, Argentina, but may extend to Bolivia (not mapped).

Genus *Euphractus* Wagler, 1830

Euphractus sexcinctus (Linnaeus, 1758)
Yellow Armadillo, Six-banded Armadillo, Gualacate

Measurements

	Mean	Min.	Max.	N
TL	616.4	556.0	700.0	13
HB	395.7	341.0	445.0	12
T	220.2	200.0	255.0	
HF	83.5	80.0	89.0	13
E	35.2	24.0	41.09	12
Wt	3.95 kg	3.0 kg	5.9 kg	6
	4.68 kg	3.2 kg	6.5 kg	14

Redford and Eisenberg 1992.

Description

This species is the largest of the "hairy" armadillos. At the anterior margin of the scapular shield there is no movable band. There are two to four holes in the pelvic shield. The carapace is characteristically yellow or tan and sparsely covered with pale hair (Redford and Wetzel 1985) (plate 3, fig. 6.6).
Chromosome number: $2n = 58$; FN $= 102$ (Benirschke, Low, and Ferm 1969; Jorge, Meritt, and Benirschke 1977).

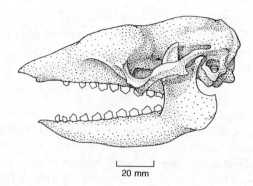

20 mm

Figure 6.6. Skull of *Euphractus sexcinctus*.

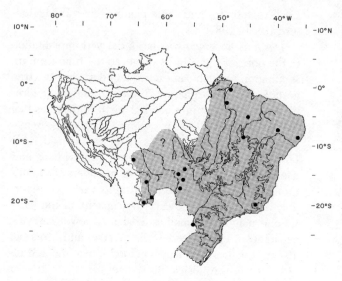

Map 6.23. Distribution of *Euphractus sexcinctus*.

Distribution

Euphractus sexcinctus is found in the savannas of Suriname and adjacent Brazil and south to Uruguay, eastern and western Paraguay, and Argentina south to Buenos Aires province (Redford and Wetzel 1985; Wetzel 1982a) (map 6.23).

Life History

Litter size ranges from one to three, and litters include both sexes. In captivity the female builds a nest before giving birth, and young weighing 95–116 g are born after a gestation period of sixty to sixty-four days. The eyes open after twenty-two to twenty-five days, and the young reach sexual maturity at nine months. Two pregnant females were found in September and October in central Brazil and another in January in Uruguay (Barlow 1965; Gucwinska 1971; Kühlhorn 1954; Redford and Wetzel 1985; Sanborn 1930).

Ecology

The yellow armadillo is most commonly found in savannas, other open vegetation formations, and forest edges. It appears to use higher, drier habitats, though in Uruguay it is reported as most common in ecotonal areas, especially near streams. In the Brazilian Pantanal, in all vegetation types the biomass of *Euphractus* was estimated at 18.8 kg/km^2.

This species is largely diurnal, though occasionally it is active at night. It is a good digger and builds burrows with a single inverted U-shaped entrance. Unlike many other species of armadillos, it frequently reuses burrows. Captive animals are reported to mark the corners of their cages with secretions from the pelvic shield scent gland, so this gland is probably used to mark burrows.

Like other hairy armadillos, *Euphractus* stores fat, and animals in captivity can weigh up to 11 kg. It is omnivorous and eats a broad range of animal and plant foods including carrion, small vertebrates, insects, bromeliad fruits, tubers, and palm nuts. Plant material can compose a significant proportion of the diet. In captivity the yellow armadillo will kill and eat large rats; they are very inefficient predators, however, unable to effect a killing bite. *Euphractus* are active, alert animals with poor eyesight, and they locate food by smell. Unlike many armadillos, they run to escape and bite when handled (Barlow 1965; Carter and Encarnação 1983; Redford 1985b; Redford and Wetzel 1985; Roig 1969; Schaller 1983).

Genus *Priodontes* F. Cuvier, 1825

Priodontes maximus Kerr, 1792
Giant Armadillo, Tatú Carrera, Tatú-guaca

Measurements

	Mean	Min.	Max.	N
HB	895.5	832.0	960.0	4
T	528.2	510.0	550.0	
HF	190.7	185.0	200.0	3
E	53.8	45.0	60.0	5
Wt	26.8 kg	18.7 kg	32.3 kg	

Redford and Eisenberg 1992.

Description

The giant armadillo is unmistakable because of its size. The only armadillos it can be confused with are those of the genus *Cabassous*. *Priodontes* is much larger, however, with a darker shell, sharply marked laterally by a buffy border, and a well-armored tail. Like *Cabassous* it has a rounded, blunt muzzle, a carapace with many narrow bands, and large scimitar-shaped foreclaws, the third of which is greatly enlarged (Redford, n.d.).
Chromosome number: $2n = 50$; FN = 76 (Benirschke, Low, and Ferm 1969).

Distribution

Priodontes is found east of the Andes from Colombia and Venezuela south to Paraguay and northern Argentina (Wetzel 1982a) (map 6.24).

Life History and Ecology

The litter size is one or two (Krieg 1929). The mode is apparently one. Giant armadillos range over much of South America and are found in tropical forest and open savanna. They are largely nocturnal and highly fossorial and therefore are seldom seen. They are extremely powerful diggers, and their very large foraging and sleeping holes are unmistakable. Burrows tend to be clumped and are usually found in active or dead termite mounds. Individuals usually remain in their burrows for at least twenty-four hours and may stay there several days.

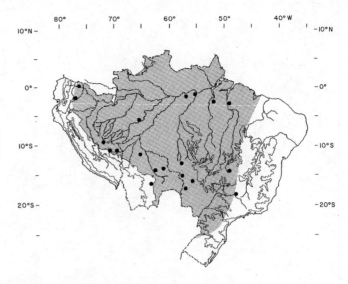

Map 6.24. Distribution of *Priodontes maximus*.

Priodontes is probably the most myrmecophagous of the armadillos and has been recorded as eating virtually nothing but ants and termites. Unlike other armadillos, *Priodontes* often destroys a termite mound while feeding (Barreto, Barreto, and D'Alessandro 1985; Carter 1983; Carter and Encarnaçõ 1983; Redford 1985b).

Genus *Tolypeutes* Illiger, 1811

Description

This small armadillo has a head and body length of about 250 mm. The tail is shorter than the head and body. The dorsal carapace is extremely hard and has three movable bands. Given the scutes on the head and the way the head and tail fit together when the animal is attacked, it can roll into an impregnable ball (fig. 6.7). This attribute makes it vulnerable to human predation, however (Redford and Eisenberg 1992). The animals walk on their toes, in part because of the long claws and reduction or fusion of the toes. Their gait looks somewhat like that of a ballet dancer on pointe.

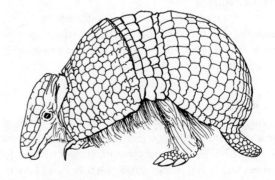

Figure 6.7. Sketch of *Tolypeutes*.

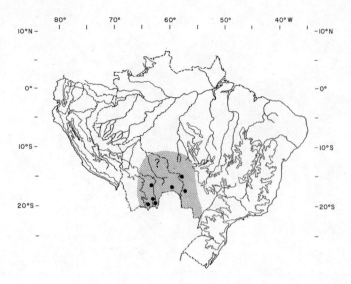

Map 6.25. Distribution of *Tolypeutes matacus*.

Tolypeutes matacus (Desmarest, 1804)
Southern Three-banded Armadillo,
Quirquincho Bola

Measurements

	Mean	Min.	Max.	N
HB	250.7	218	273.0	17
T	63.7	60	80.0	15
HF	42.3	38	47.0	15
E	22.8	21	32.0	
Wt	1.1 kg	1 kg	1.15 kg	10

Redford and Eisenberg 1992.

Description

Tolypeutes is probably the most distinctive of the armadillos and is unmistakable because of its small size, its hard, inflexible carapace, and its habit of curling up. Members of *Tolypeutes* are the only armadillos that can roll into a ball (plate 3) (pers. obs.). The second, third, and fourth toes of the hind foot are fused. The claws on the fused toes are hooflike.

Distribution

This species is found from southeastern Bolivia south through the Paraguayan Chaco to the Argentine province of Santa Cruz (Wetzel 1982a, as amended by O. Pearson, pers. comm.) (map 6.25).

Life History

Tolypeutes matacus gives birth to a single young after a gestation period of 120 days. The young opens its eyes after about twenty-two days and suckles for approximately ten weeks during October and January (Krieg 1929; Meritt 1971; Sanborn 1930).

Ecology

Tolypeutes matacus prefers dry vegetation formations and is abundant in the most xeric parts of the Paraguayan Chaco. It apparently does not dig its own burrows. In fact, unlike most other armadillos except *Euphractus*, it will often run when chased rather than dig. *T. matacus* can be active throughout the day and night, though its major activity peaks are probably dictated by temperature and rainfall. It seems to feed primarily from the ground surface, occasionally digging shallow foraging holes. It is largely myrmecophagous but will take other soft-bodied invertebrates and fruit (Redford 1985b; Sanborn 1930; Schaller 1983; Bulkovic, Caziani, Protomoastro 1995).

Tolypeutes tricinctus (Linnaeus, 1758)
Tatú-bola

Description

Characters as noted under the description for the genus. Head and body length is about 300 mm, with a tail of 65 mm. There are five toes on the forefoot, distinguishing it from *T. matacus*, which has only four (pers. obs.).

Distribution

Before 1980 only six museum specimens were known, all from east-central Brazil. Recently specimens have been taken from southwestern Bahia (G. Fonseca, pers. comm.). Current data are included in Santos (1994) and extend the range. Cardoso da Silva and Oren (1993) report recently discovered specimens from 13°45'S, 44°28'W and from 14°17'S, 43°20' W in Brazil. They comment that this species may be considered endemic to the cerrado and caatinga of Brazil (map 6.26).

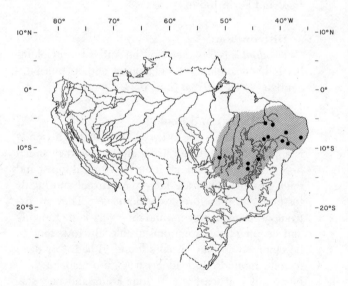

Map 6.26. Distribution of *Tolypeutes tricinctus*.

References

• References used in preparing distribution maps

• Atalah G., A. 1975. Presencia de *Chaetophractus villosus* (Edentata, Dasypodidae) nueva especie para la región de Magallanes, Chile. *Anal. Inst. Patagonia, Punta Arenas* 6 (1–2): 169–71.

Barlow, J. C. 1965. Land mammals from Uruguay: Ecology and zoogeography. Ph.D. diss., University of Kansas.

Barreto, M., P. Barreto, and A. D'Alessandro. 1985. Colombian armadillos: Stomach contents and infection with *Trypanosoma cruzi*. *J. Mammal.* 66:188–93.

Beebe, W. 1925. The three-toed sloth, *Bradypus cuculliger* Wagler. *Zoologica* 7:1–67.

Benirschke, K., R. J. Low, and V. H. Ferm. 1969. Cytogenetic studies of some armadillos. In *Comparative mammalian cytogenetics*, ed. K. Benirschke, 330–45. New York: Springer-Verlag.

Best, R. C., and A. Y. Harada. 1985. Food habits of the silky anteater (*Cyclopes didactylus*) in the central Amazon. *J. Mammal.* 66:780–81.

Bickel, C. L., G. K. Murdock, and M. L. Smith. 1976. Hand-rearing a giant anteater *Myrmecophaga tridactyla* at Denver Zoo. *Int. Zoo Yearb.* 16:195–98.

Block, J. A. 1974. Hand-rearing seven-banded armadillos *Dasypus septemcinctus* at the National Zoological Park, Washington. *Int. Zoo Yearb.* 14: 210–14.

• Bulkovic, M. L., S. M. Caziani, and J. Protomastro. 1995. Food habits of the three banded armadillo (Xenarthra, Dasypodidae) in the dry Chaco, Argentina. *J. Mammal* 76:1199–1204.

Byrne, P. S. 1962. Giant ant-eaters born in zoo. *J. British Guiana Mus. Zoo Roy. Agric. Commerc. Soc.* 36:28–29.

Cabrera, A. 1958. *Catálogo de los mamíferos de América del Sur*. Vol. 1. Buenos Aires: Museo Argentino de Ciencias Naturales "Bernardino Rivadavia."

Cardoso da Silva, J. M., and D. C. Oren. 1993. Observations on the habitat and distribution of the Brazilian three-banded armadillo *Tolypeutes tricinctus*, a threatened caatinga endemic. *Mammalia* 57:149–52.

• Carlini, A. A., and S. F. Vizcaíno. 1987. A new record of the armadillo *Chaetophractus vellerosus* (Gray, 1865) (Mammalia, Dasypodidae) in the Buenos Aires province of Argentina: Possible causes for the disjunct distribution. *Stud. Neotrop. Fauna Environ.* 22 (1):53–55

Carter, T. S. 1983. The burrows of the giant armadillos, *Priodontes maximus* (Edentata: Dasypodidae). *Säugetierk. Mitt.* 31:47–53.

Carter, T. S., and C. D. Encarnação. 1983. Character-
istics and use of burrows by four species of armadillos in Brazil. *J. Mammal.* 64:103–8.

Cassini, M. H. 1993. Searching strategies within food patches in the armadillo *Chaetophractus vellerosus*. *Anim. Beh.* 46:400–402.

Cook, J., F. Cáceres, and C. Miranda. 1991. Cariotipo del quirquincho (*Chaetophractus nationi*). *Ecol. Bolivia* 18:21–27.

• Crespo, J. A. 1944. Contribución al conocimiento de la ecología de algunos dasipódidos (Edentata) argentinos. *Rev. Argent. Zoogeogr.* 4 (1–2): 7–16.

• ———. 1982. Ecología de la comunidad de mamíferos del Parque Nacional Iguazú, Misiones. *Rev. Mus. Argent. Cienc. Nat. "Bernardino Rivadavia," Ecol.* 3 (2): 45–162.

Eisenberg, J. F. 1961. The nest-building behavior of armadillos. *Proc. Zool. Soc. London* 137:322–24.

———. 1989. *Mammals of the Neotropics*. Vol. 1. *The northern Neotropics: Panama, Colombia, Venezuela, Guyana, Suriname, French Guiana*. Chicago: University of Chicago Press.

Eisenberg, J. F., and E. Maliniak. 1985. Maintenance and reproduction of the two-toed sloth *Choloepus didactylus* in captivity. In *The evolution and ecology of armadillos, sloths, and vermilinguas*, ed. G. G. Montgomery, 327–32. Washington, D. C.: Smithsonian Institution Press.

Emry, R. J. 1970. A North American Oligocene pangolin and other additions to the Pholidota. *Bull. Amer. Mus. Nat. Hist.* 142 (6): 457–510.

Galbreath, G. J. 1982. Armadillo, *Dasypus novemcinctus*. In *Wild mammals of North America: Biology, management, and economics*, ed. J. A. Chapman and G. A. Feldhamer, 71–79. Baltimore: Johns Hopkins University Press.

———. 1985. The evolution of monozygotic polyembryony in *Dasypus*. In *The evolution and ecology of armadillos, sloths, and vermilinguas*, ed. G. G. Montgomery, 243–46. Washington, D.C.: Smithsonian Institution Press.

Gardner, A. L. 1993. Xenarthra. In *Mammal species of the world*, ed. D. E. Wilson and D. M. Reeder, 63–68. Washington, D.C.: Smithsonian Institution Press.

Gillette, D. D., and C. E. Ray. 1981. *Glyptodonts of North America*. Smithsonian Contributions to Paleobiology 40. Washington, D.C.: Smithsonian Institution Press.

Goffart, M. 1971. *Function and form in the sloth*. Oxford: Pergamon Press.

González, J. C., and C. Ríos. 1980. Refugios epideos del "tatú" *Dasypus n. novemcinctus* Linne (Mammalia: Dasypodidae). *Res. J. C. Nat. Montevideo* 1:129–30.

Greegor, D. H., Jr. 1974. Comparative ecology and

distribution of two species of armadillos, *Chaeto-phractus vellerosus* and *Dasypus novemcinctus*. Ph.D. diss., University of Arizona.

———. 1975. Renal capabilities of an Argentine desert armadillo. *J. Mammal.* 56:626–32.

———. 1980a. Preliminary study of movements and home range of the armadillo, *Chaetophractus vellerosus. J. Mammal.* 61:334–35.

———. 1980b. Diet of the little hairy armadillo, *Chaetophractus vellerosus*, of northwestern Argentina. *J. Mammal.* 61:331–34.

———. 1985. Ecology of the little hairy armadillo *Chaetophractus vellerosus*. In *The evolution and ecology of armadillos, sloths, and vermilinguas*, ed. G. G. Montgomery, 397–405. Washington, D.C.: Smithsonian Institution Press.

Green, H. W. 1989. Agonistic behavior by three-toed sloths, *Bradypus variegatus. Biotropica* 21 (4): 369–72.

Gucwinska, H. 1971. Development of six-banded armadillos *Euphractus sexcinctus* at Wrocław Zoo. *Int. Zoo Yearb.* 11:88–89.

Hamlet, G. W. D. 1939. Identity of *Dasypus septemcinctus* Linnaeus with notes on some related species. *J. Mammal.* 20:328–36.

Hardin, C. J. 1976. Hand-rearing a giant anteater *Myrmecophaga tridactyla* at Toledo Zoo. *Int. Zoo Yearb.* 15:199–200.

Honacki, J. H., K. E. Kinman, and J. W. Koeppl, eds. 1982. *Mammal species of the world.* Lawrence, Kans.: Allen Press and Association of Systematics Collections.

Humphrey, S. R. 1974. Zoogeography of the nine-banded armadillo (*Dasypus novemcinctus*) in the United States. *BioScience* 24 (8): 457–62.

Jorge, W., D. A. Meritt Jr., and K. Benirschke. 1977. Chromosome studies in Edentata. *Cytobiosis* 18: 157–72.

Krieg, H. 1929. Biologische Reisestudien in Süd-amerika. 9. Gürteltiere. *Z. Morph. Ökol. Tiere* 14 (1): 166–90.

Kühlhorn, F. 1954. Säugetierkundliche Studien aus Süd-Mattogrosso. 2. Edentata, Rodentia. *Säugetierk. Mitt.* 2:66–72.

Le Pont, F., and P. Desjeux. 1992. Présence de *Choeloepus hoffmanni* dans les Yungas do la Paz, Bolivia. *Mammalia* 56:404–85.

Luederwaldt, H. 1918. Observações sobre a preguica (*Bradypus tridactylus* L.) em liberdade en cativeiro. *Rev. Mus. Paulista* 10:795–812.

McCrane, M. P. 1966. Birth, behavior and development of a hand reared two-toed sloth. *Int. Zoo Yearb.* 6:153–63.

McNab, B. 1978. Energetics of arboreal folivores: Physiological problems and ecological consequences of feeding on a ubiquitous food supply. In *The ecology of arboreal folivores*, ed. G. G. Montgomery, 153–62. Washington, D.C.: Smithsonian Institution Press.

———. 1982. The physiological ecology of South American mammals. In *Mammalian biology in South America*, ed. M. Mares and H. Genoways, 187–207. Pymatuning Symposia in Ecology 6. Special Publication Series. Pittsburgh: Pymatuning Laboratory of Ecology, University of Pittsburgh.

• Mann, G. 1978. *Los pequeños mamíferos de Chile.* Guyana: Zoología 40. Santiago: Universidad de Concepción.

• Mares, M. A., R. A. Ojeda, and M. P. Kosco. 1981. Observations on the distribution and ecology of the mammals of Salta province, Argentina. *Ann. Carnegie Mus.* 50 (6): 151–206.

• Massoia, E. 1980. Mammalia de Argentina. 1. Los mamíferos silvestres de la provincia de Misiones. *Iguazú* 1 (1): 15–43.

Meritt, D. A., Jr. 1971. The development of the La Plata three-banded armadillo *Tolypeutes matacus* at Lincoln Park Zoo, Chicago. *Int. Zoo Yearb.* 11: 195–96.

———. 1975. The lesser anteater *Tamandua tetradactyla* in captivity. *Int. Zoo Yearb.* 15:41–45.

———. 1985. Naked-tailed armadillos *Cabassous* sp. In *The evolution and ecology of armadillos, sloths, and vermilinguas*, ed. G. G. Montgomery, 389–91. Washington, D.C.: Smithsonian Institution Press.

Merrett, P. K. 1983. *Edentates.* La Villiaze, Saint Andrew's, Guernsey: Zoological Trust of Guernsey.

Montgomery, G. G. 1983. *Cyclopes didactylus*. In *Costa Rican natural history*, ed. D. H. Janzen, 461–63. Chicago: University of Chicago Press.

———. 1985a. Montgomery, G. G., ed. 1985. *The evolution and ecology of armadillos, sloths, and vermilinguas.* Washington, D.C.: Smithsonian Institution Press.

———. 1985b. Movements, foraging and food habits of the four extant species of Neotropical vermilinguas (Mammalia; Myrmecophagidae). In *The evolution and ecology of armadillos, sloths, and vermilinguas*, ed. G. G. Montgomery, 365–77. Washington, D.C.: Smithsonian Institution Press.

Montgomery, G. G., and Y. D. Lubin. 1977. Prey influences on movements of Neotropical anteaters. In *Proceedings of the 1975 Predator Symposium*, ed. R. L. Philips and C. Jonkel, 103–31. Missoula: Montana Forest and Conservation Experiment Station, University of Montana.

Montgomery, G. G., and M. E. Sunquist. 1975. Impact of sloths on Neotropical forest energy flow and nutrient cycling. In *Tropical ecological systems*, ed. F. B. Golley and E. Medina, 69–98. New York: Springer-Verlag.

———. 1978. Habitat selection and use by two-toed

and three-toed sloths. In *The ecology of arboreal folivores,* ed. G. G. Montgomery, 329–59. Washington, D.C.: Smithsonian Institution Press.

• Myers, P., and R. M. Wetzel. 1979. New records of mammals from Paraguay. *J. Mammal.* 60:638–41.

• Oliver, W. L. R., and I. B. Santos. 1991. *Threatened endemic mammals of the Atlantic forest region of southeast Brasil.* Special Scientific Report 4. Jersey: Jersey Wildlife Preservation Trust.

• Olrog, C. C. 1979. Los mamíferos de la selva húmeda, Cerro Calilegua, Jujuy. *Acta Zool. Lilloana* 33:9–14.

• Onelli, C. 1913. Biología de algunos mamíferos argentinos. *Rev. Jardín Zool.* 9:77–142.

Patterson, B., and R. R. Pascual. 1972. The fossil mammal fauna of South America. In *Evolution, mammals, and southern continents,* ed. A. Keast, F. C. Erk, and B. Glass, 247–310. Albany: State University of New York Press.

Pinder, L. 1993. Body measurements, karyotype, and birth frequencies of maned sloth (*Bradypus torquatus*). *Mammalia* 57:43–48.

Pocock, R. I. 1913. Dorsal glands in armadillos. *Proc. Zool. Soc. London* 72:1099–1103.

• Redford, K. H. 1983. Lista preliminar de mamíferos do Parque Nacional das Emas. *Brasil Florestal* 55:29–32.

———. 1985a. Feeding and food preferences in captive and wild giant anteaters (*Myrmecophaga tridactyla*). *J. Zool.* (London) 205:559–72.

———. 1985b. Food habits of armadillos (Xenarthra: Dasypodidae). In *The evolution and ecology of armadillos, sloths, and vermilinguas,* ed. G. G. Montgomery, 429–37. Washington, D.C.: Smithsonian Institution Press.

———. 1987a. Dietary specialization and variation in two mammalian insectivores. *Rev. Chil. Hist. Nat.* 59 (2): 201–8.

———. 1987b. Patterns of ant and termite eating in mammals. *Current Mammal.* 1:349–400.

• Redford, K. H. n.d. Sinopse dos edentados (tamanduás, preguiças etatús) do Brasil central. Unpublished manuscript.

Redford, K. H., and J. F. Eisenberg. 1992. *Mammals of the Neotropics.* Vol. 2. *The southern cone: Chile, Argentina, Uruguay, Paraguay.* Chicago: University of Chicago Press.

Redford, K. H., and R. M. Wetzel. 1985. *Euphractus sexcinctus. Mammal. Species* 252:1–4.

Reig, O. A. 1981. *Teoría del origen y desarrollo de la fauna de mamíferos de América del Sur.* Monografie Naturae. Mar del Plata, Argentina: Museo Municipal de Ciencias Naturales Lorenzo Scaglia.

Roig, V. G. 1969. Termoregulación en *Euphractus sexcinctus* (Mammalia, Dasypodidae). *Physis* 29 (78): 27–32.

Rose, K. D., and R. J. Emry. 1993. Relationships of Xenarthra, Pholidota, and fossil "edentates": The morphological evidence. In *Mammal Phylogeny: Placentals,* ed. F. S. Szalay, M. J. Novacek, and M. C. McKenna, 81–102. New York: Springer-Verlag.

• Salazar B., J., K. H. Redford, and A. M. Stearman. 1990. El perezoso de dos dedos de Hoffmann (*Choloepus hoffmanni* Peters, 1859; Megalonychidae) en Bolivia. *Mus. Nac. Hist. Nat. (Bolivia) Comunicación* 9:18–21.

Sanborn, C. C. 1930. Distribution and habits of the three-banded armadillo (*Tolypeutes*). *J. Mammal.* 11:61–68.

• Santos, I. B. 1994. *Tolypeutes tricinctus.* In *Livro vermelho dos mamíferos brasilieros ameaçados de extinção,* ed. G. A. B. da Fonseca, A. B. Rylands, C. M. R. Costa, C. M. Machado, and Y. L. R. Leite, 25–32. Belo Horizonte, Brazil: Fundacão Biodiversitas.

Schaller, G. B. 1983. Mammals and their biomass on a Brazilian ranch. *Arq. Zool. São Paulo* 31 (1): 1–36.

Schmid, B. 1938. Psychologische Beobachtungen und Versuche an einem jungen männlichen Ameisenbären (*Myrmecophaga tridactyla* L.). *Z. Tierpsychol.* 2:117–26.

Shaw, J. H., T. S. Carter, and J. C. Machado-Neto. 1985. Ecology of the giant anteater *Myrmecophaga tridactyla* in Serra da Canastra, Minas Gerais, Brazil: A pilot study. In *The evolution and ecology of armadillos, sloths, and vermilinguas,* ed. G. G. Montgomery, 379–84. Washington, D.C.: Smithsonian Institution Press.

Shaw, J. H., J. Machado-Neto, and T. S. Carter. 1987. Behavior of free-living anteaters (*Myrmecophaga tridactyla*). *Biotropica* 19:255–59.

Smielowski, J., P. Stanislawski, and T. Taworski. 1981. Breeding the giant anteater. *Int. Zoo News* 28, 5 (174): 1–6.

Sunquist, M. E., and G. G. Montgomery. 1972. Activity pattern of a translocated silky anteater (*Cyclopes didactylus*). *J. Mammal.* 54:782.

———. 1973. Activity patterns and rate of movement of two-toed and three-toed sloths. *J. Mammal.* 54:946–54.

Taber, F. W. 1954. Contributions on the life history and ecology of the nine-banded armadillo. *J. Mammal.* 26:211–26.

• Tamayo, M., and D. Frassinetti. 1980. Catálogo de los mamíferos fósiles y vivientes de Chile. *Mus. Nac. Hist. Nat., Chile* 37:323–99.

Vizcaino, S. F. 1995. Identificación específica de las "mulitas" genero *Dasypus* L. (Mammalia, Dasypodidae), del noroeste Argentina. Descripción de una nueva especie. *Mastozool. Neotrop.* 1 (2): 5–13.

Waage, J. K., and R. C. Best. 1985. Arthropod associates of sloths. In *The evolution and ecology of ar-*

madillos, sloths, and vermilinguas, ed. G. G. Montgomery, 297–311. Washington, D.C.: Smithsonian Institution Press.

Webb, S. D. 1985. The interrelationships of tree sloths and ground sloths. In The evolution and ecology of armadillos, sloths, and vermilinguas, ed. G. G. Montgomery, 105–12. Washington, D.C.: Smithsonian Institution Press.

Webb, S. D., and L. G. Marshall. 1982. Historical biogeography of Recent South American land mammals. In Mammalian biology in South America, ed. M. A. Mares and H. H. Genoways, 39–52. Pymatuning Symposia in Ecology 6. Special Publication Series. Pittsburgh: Pymatuning Laboratory of Ecology, University of Pittsburgh.

• Wetzel, R. M. 1975. The species of Tamandua Gray (Edentata, Myrmecophagidae). Proc. Biol. Soc. Washington 88:95–112.

• ———. 1980. Revision of the naked-tailed armadillos, genus Cabassous. Ann. Carnegie Mus. 49: 323–57.

• ———. 1982a. Systematics, distribution, ecology and conservation of South American edentates. In Mammalian biology in South America, ed. M. Mares and H. Genoways, 345–75. Pymatuning Symposia in Ecology 6. Special Publication Series. Pittsburgh: Pymatuning Laboratory of Ecology, University of Pittsburgh.

———. 1982b. The mammals of the Chaco of Paraguay. Nat. Geog. Soc. Res. Rpts. 14:679–84.

• ———. 1985a. The identification and distribution of Recent Xenarthra (= Edentata). In The evolution and ecology of armadillos, sloths, and vermilinguas, ed. G. G. Montgomery, 5–21. Washington, D.C.: Smithsonian Institution Press.

• ———. 1985b. Taxonomy and distribution of armadillos, Dasypodidae. In The evolution and ecology of armadillos, sloths, and vermilinguas, ed. G. G. Montgomery, 23–46. Washington, D.C.: Smithsonian Institution Press.

• Wetzel, R. M., and F. D. de Avila-Pires. 1980. Identification and distribution of the Recent sloths of Brazil (Edentata). Rev. Brasil. Biol. 40 (4): 831–36.

• Wetzel, R. M., and E. Mondolfi. 1979. The subgenera and species of long-nosed armadillos, genus Dasypus L. In Vertebrate ecology in the northern Neotropics, ed. J. F. Eisenberg, 43–63. Washington, D.C.: Smithsonian Institution Press.

Wilson, D. E., and D. M. Reeder, eds. 1993. Mammal species of the world. 2d ed. Washington, D.C.: Smithsonian Institution Press.

• Ximénex, A., and F. Achaval. 1966. Sobre la presencia en el Uruguay del tatú de Rabo Molle, Cabassous tatouay (Desmarest) (Edentata-Dasypodidae). Comun. Zool. Mus. Hist. Nat. Montevideo 9 (109): 1–5.

7   Order Insectivora

Diagnosis and History of Classification

In this volume the order Insectivora is used in accordance with Hutterer (1993). The higher-level taxonomy of the "insectivores" has been modified several times in the past forty years. The inclusion of the major families in one order has repeatedly been challenged. The order Insectivora has been defined in a restricted sense, excluding the tenrecs (Tenrecidae), golden moles (Chrysochloridae), tree shrews (Tupaiidae), and elephant shrews (Macroscelidea), which have been considered to be within ordinal ranks in their own right (see also McKenna 1975; Eisenberg 1981). Although MacPhee and Novacek (1993) recognize the unity of the lipotyphlan insectivores, they believe the characters they have analyzed do not indicate a special relationship between the Chrysochloridae and Tenrecidae. As a result the Tenrecidae are included with the "soricoids." We reject this conclusion based on the behavioral research on *Solenodon* and the tenrecs summarized in Eisenberg and Gould (1966, 1970). In addition, the latest publication by MacPhee (1994) casts further doubt on the stability of "insectivore" taxonomy at higher levels. The problem will require further study.

This diagnosis applies only to the hedgehogs (Erinaceidae), moles (Talpidae), shrews (Soricidae), and solenodons (Solenodontidae). Modern insectivores are characterized by a rather small body size. Members of the order as so defined have long, pointed snouts and relatively small brains dominated by large olfactory bulbs. The tympanic bone is annular, with no auditory bullae. The jugal bone is absent, and the zygomatic arch is reduced or absent. The teeth have well-defined cusps, and the molars and premolars retain many characters reminiscent of conservative placentals. In the Western Hemisphere only the moles, solenodons, and shrews are now represented. The solenodons are Antillean in their distribution. The moles are strongly adapted for a fossorial life, with forefeet and claws en-

Table 7.1   Alternative Classifications for the Insectivora

Simpson 1945	McKenna 1975; Eisenberg 1981
Order Insectivora	Order Tenrecomorpha
Family Tenrecidae	Family Tenrecidae
	Subfamily Tenrecinae
Family Potomogalidae	Subfamily Potomogalinae
Family Chrysochloridae	Family Chrysochloridae
	Order Erinaceomorpha
Family Erinaceidae	Family Erinaceidae
	Order Soricomorpha
Family Soricidae	Family Soricidae
Family Solenodontidae	Family Solenodontidae
Family Talpidae	Family Talpidae
	Order Macroscelidea
Family Macroscelididae	Family Macroscelididae

larged for digging. Although shrews are also specialized to some extent for digging in loose soil, the development of musculature and shortening of the humerus so characteristic of the moles are not morphological specializations of shrews. Shrews have a small to minute body size, and many species have moderately long tails. The pelage is soft and velvety. The eyes are very reduced, as is the external ear. Further diagnostic features will be discussed at the family level.

History and Distribution

At present the Insectivora, as defined for this volume, are distributed in North America, South America, Eurasia, and Africa. Members of the order did not cross into South America until the completion of the Panamanian land bridge in the Pliocene. The Recent families of insectivores originated on the northern continents. The hedgehogs, family Erinaceidae, first appeared in the Eocene and remain conservative in many of their morphological features. The shrews (Soricidae) and moles (Talpidae) appeared in the early Oligocene.

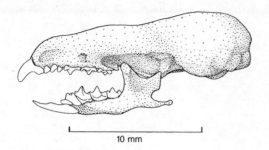

Figure 7.1. Skull of *Cryptotis parva*.

FAMILY SORICIDAE
Shrews, Musaranas

Diagnosis
The first upper incisors are long and hooked, with a cusp projecting ventrally at the base, and the first lower incisors are directed forward, producing a pincerlike grip (fig. 7.1). The skull is narrow, and there is no zygomatic arch. Instead of a tympanic bulla there is a bony ring. The rostrum is long and pointed. The eyes are extremely small, and though pinnae are present, they are reduced. Normally there are five toes on the forepaws and hind feet (fig. 7.2).

Distribution
Members of the family Soricidae are found in Eurasia, Africa, North America, and northern South America.

Natural History
Shrews are specialized for feeding on small invertebrates in the litter of the forest floor. They forage beneath the leaf litter, and their activity extends into the day as well as the night. Shrews are represented in the Neotropics by a single genus, *Cryptotis*.

Genus *Cryptotis* Pomel, 1848
Least Shrew, Musarana de Cola Corta

Description
The dental formula is I 3/1, C 1/1, P 2/1, M 3/3. The total length is less than 130 mm, and the tail does not exceed 50% of the head and body length. The external ear is greatly reduced. There are some twelve species within the genus, and four occur in the area covered by this volume.

Distribution
The genus is distributed from the northeastern portion of North America south through Panama, and in the Andes to Peru (map 7.1).

Natural History
This genus is insectivorous and forages terrestrially in leaf litter. The natural history of the northern species *C. parva* is summarized in Broadbrooks (1952), Conaway (1958), and Whitaker (1974).

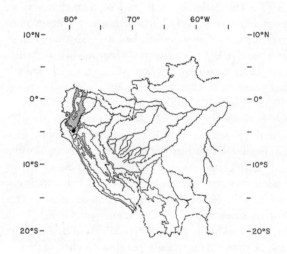

Map 7.1. Distribution of *Cryptotis*.

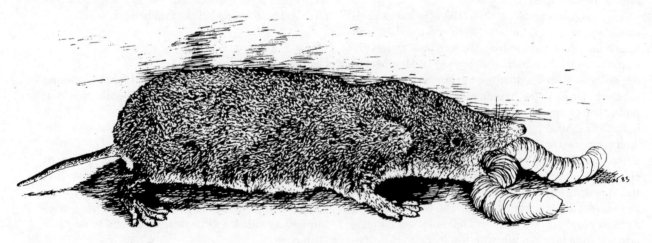

Figure 7.2. Sketch of *Cryptotis parva*.

Table 7.2   Classification of the Lipotyphla

Order Lipotyphla
Suborder Chrysochloromorpha
Family Chrysochloridae
Suborder Erinaceomorpha
Family Erinaceidae
Suborder Soricomorpha
Family Soricidae
Family Talpidae
Family Solenodontidae
Family Tenrecidae

Source: MacPhee and Novacek 1993.

Table 7.3   Key to the Species of *Cryptotis* Occurring in Northern South America

1	Venter (pale buff to buff) markedly paler than dorsum (gray brown)	*C. thomasi*
1′	Venter only slightly if at all paler than dorsum...	2
2	Pelage pale gray both above and below	*C. montivaga*[a]
2′	Pelage dark brown or black above and below..	3
3	Size large (palatal length 9.2 to 10 mm).......	*C. squamipes*
3′	Size small (palatal length 8.3 to 8.6 mm)...........	*C. avia*

Note: Does not include *C. colombiana,* which is related to the Central American *C. nigriscens* group. *C. colombiana* is known from Antioquia of the Colombian central cordillera of the Andes (Woodman and Timm 1993).
Source: Adapted from Choate and Fleharty 1974.
[a]To date recorded from Ecuador and extreme north-central Peru.

Comment on Systematics

All the species of least shrews are extremely similar in external appearance. Identification often requires skull characters, but when the geographical location is known it is possible to assign a specific identification with some certainty, within the limits of literature (Choate 1970). Keys to the species are provided by Choate and Fleharty (1974), and the Central American forms are reviewed in Woodman (1992) and Woodman and Timm (1993). The following accounts summarize data for some of the South American species that may extend to Ecuador and Peru (see also table 7.2). Vivar, Pacheco, and Valqui (1997) recognize *C. equatoris* from southern Ecuador as distinct from *C. thomasi* and include *C. osgoodi* within *C. equatoris.*

Cryptotis peruviensis Vivar, Pacheco, and Valqui, 1997

Description

Measurements of one specimen are HB 63.2; T 30.5; HF 14; E vestigial; Wt 9 g. This specimen from Ashitas is slightly smaller than the LSU specimen (LSU 26887). The latter could be assigned to *Cryptotis montivaga.* The "Ashitas" specimen (see below) was taken only 60 km south of the LSU example. The teeth

are lightly pigmented, in contrast to those of *C. thomasi* and *C. squamipes.* The dorsum is dark gray, with the venter slightly paler (M. Valqui, pers. comm.; see also Vivar, Pacheco, and Valqui 1997).

Distribution

Known only from the capture locality: an elfin forest at 3,150 m on the La Palma–Sallique Trail in "Las Ashitas" in Chontali, department of Cajamarca, Peru (not mapped).

Cryptotis montivaga (Anthony, 1921)

Description

Measurements of the type: TL 112; T 31; HF 15. Measurements from Barnett (1992): HB 65–85; T 28–35; HG 12–14; Wt 9–14 g. ($n = 9$). In size and form this species resembles a typical *Cryptotis.* The dorsum is dark gray, with the venter slightly lighter. The tail is slightly paler on the ventral side. The gray dorsal pelage is distinctive.

Distribution

The type was collected from Bestion, Azuay province, Ecuador, at 3,100 m.

Natural History

Barnett (1992) studied *C. montivaga* in southern Ecuador. Analysis of stomach contents demonstrated beetles, spiders, and caterpillars as prey. Specimens were consistently trapped in habitats with closed, continuous vegetation at elevations ranging from 2,800 to 3,100 m. A litter size of two and two litters per year were inferred from the trapping data.

Cryptotis squamipes (J. A. Allen, 1912)

Description

This large species of *Cryptotis* has a palatal length of 9.2 to 10 mm and a total length greater than 106 mm. The dorsum and venter do not contrast in color, varying from dark brown to black.

Range and Habitat

This shrew apparently replaces *C. nigrescens* in the southern portion of the western cordillera of the Colombian Andes.

Cryptotis thomasi (Merriam, 1897)

Description

Measurements are: TL 118–29; T 35–39; HF 14–17; E 6–8; Wt 11.7–15.9 g (Venezuela). Aagaard (1982) found this species to vary in total length from 117 to 131 mm; weight ranged from 9 to 15 g. The dorsum is dark brown, almost black, and the venter is lighter than the dorsum.

Range and Habitat

The species is found in the Andes of Venezuela and presumably in the higher montane areas of the Andes rimming the western side of Lake Maracaibo and then south to an undetermined limit.

Natural History

Aagaard (1982) found this species moderately abundant at between 3,500 and 4,100 m in the Venezuelan Andes. It was the most common catch in the paramos, and within this habitat breeding takes place in the wet season (April to December). These observations may well be restricted to *C. meridensis* Thomas, 1898, as reconstituted by Hutterer (1993).

References

• References used in preparing distribution maps

Aagaard, E. M. J. 1982. Ecological distribution of mammals in the cloud forests and paramos of the Andes, Mérida, Venezuela. Ph.D. diss., Colorado State University.

Barnett, A. A. 1992. Notes on the ecology of *Cryptotis montivaga* Anthony, 1921 (Insectivora, Soricidae), a high-altitude shrew from Ecuador. *Mammalia* 56 (4): 588–92.

Broadbrooks, H. E. 1952. Nest and behavior of a short-tailed shrew, *Cryptotis parva. J. Mammal.* 33: 241–43.

• Cabrera, A. 1958. *Catálogo de los mamíferos de América del Sur.* Vol. 1. Buenos Aires: Museo Argentino de Ciencias Naturales "Bernardino Rivadavia."

• Choate, J. R. 1970. Systematics and zoogeography of Middle America shrews of the genus *Cryptotis. Univ. Kansas Publ. Mus. Nat. Hist.* 19:195–317.

Choate, J. R., and E. D. Fleharty. 1974. *Cryptotis goodwini. Mammal. Species* 44:1–3.

Conaway, C. H. 1958. Maintenance, reproduction, and growth of the least-shrew in captivity. *J. Mammal.* 39:507–12.

Eisenberg, J. F. 1981. *The mammalian radiations: An analysis of trends in evolution, adaptation, and behavior.* Chicago: University of Chicago Press.

Eisenberg, J. F., and E. Gould. 1966. The behavior of *Solenodon paradoxus* in captivity with comments on the behavior of other Insectivora. *Zoologica* 51 (1): 49–58.

———. 1970. The tenrecs: A study in mammalian behavior and evolution. *Smithsonian Contrib. Zool.* 27:1–137.

• Hall, E. R. 1981. *The mammals of North America.* 2d ed., 2 vols. New York: John Wiley.

• Handley, C. O., Jr. 1976. Mammals of the Smithsonian Venezuelan project. *Brigham Young Univ. Sci. Bull., Biol. Ser.* 20 (5): 1–90.

• Honacki, J. H., K. E. Kinman, and J. W. Koeppl, eds. 1982. *Mammal species of the world.* Lawrence, Kans.: Allen Press and Association of Systematics Collections.

• Hutterer, R. 1993. Order Insectivora. In *Mammal species of the world,* ed. D. E. Wilson and D. M. Reeder, 89–130. Washington, D.C.: Smithsonian Institution Press.

McKenna, M. C. 1975. Toward a phylogenetic classification of the Mammalia. In *Phylogeny of the primates,* ed. W. P. Luckett and F. S. Szalay, 21–46. New York: Plenum Press.

MacPhee, R. D. E. 1994. Morphology, adaptations and relationships of *Plesiorycteropus,* and a diagnosis of a new order of eutherian mammals. *Bull. Amer. Mus. Nat. Hist.* 220:1–214.

MacPhee, R. D. E., and M. J. Novacek. 1993. Definition and relationships of Lipotyphla. In *Mammal phylogeny: The placentals,* ed. F. S. Szalay, M. J. Novacek, and M. C. McKenna, 13–31. New York: Springer-Verlag.

Simpson, G. G. 1945. The principles of classification and a classification of the mammals. *Bull. Amer. Mus. Nat. Hist.* 85:1–350.

Vivar, E., V. Pacheco, and M. Valqui. 1997. A new species of *Cryptotis* (Insectivora: Soricidae) from northern Peru. *Amer. Mus. Novitat.* 3202:1–15.

Whitaker, J. O., Jr. 1974. *Cryptotis parva. Mammal. Species* 43:1–8.

Woodman, N. 1992. Biogeographical and evolutionary relationships among Central American small-eared shrews of the genus *Cryptotis.* Ph.D. diss., University of Kansas, Lawrence.

Woodman, N., and R. M. Timm. 1993. Intra- and interspecific variation in the *Cryptotis nigrescens* species complex of small-eared shrews (Insectivora: Soricidae). *Fieldiana, Zool.,* n.s., 74:1–30.

8   Order Chiroptera (Bats, Murciélagos, Morcegos)

Diagnosis

This order includes the only true flying mammals. Some other mammals glide, but the bats really fly. A wing membrane extends from each side of the body and hind leg to the forearm, where it is supported by the fingers, which are elongated to create a large surface area for the membrane's support. There may be an additional membrane between the hind legs, sometimes enclosing the tail. The clavicle is well developed. The hind leg has become rotated to support the wing membrane, so that the knee is directed laterally and backward. The sternum is usually keeled for attachment of the massive pectoral muscles that are used in the power stroke during flying (see fig. 8.1).

In his classifications of bats Miller (1907) used the morphology of the humerus as an indicator of specialization for flight. Family diagnoses often refer to the relative development of the greater and lesser tuberosities (trochiter and trochin) on the proximal end of the humerus.

Distribution

At present bats are found on all continents except Antarctica. The families Pteropodidae (= Pteropidae), Rhinopomatidae, Nycteridae, Megadermatidae, Rhinolophidae, Hipposideridae, Craseonycteridae, Mystacinidae, and Myzopodidae are confined to the Old World (Koopman 1993 includes the Hipposideridae within the Rhinolophidae). The families Thyropteridae, Natalidae, Furipteridae, Mormoopidae, Phyllostomidae (= Phyllostomatidae), and Noctilionidae are confined to the New World. The families Emballonuridae, Vespertilionidae, and Molossidae are worldwide in their distribution. Bats show their greatest species richness and numbers in the subtropical and tropical regions of the world.

History, Classification, and Literature

Bats are poorly represented in the early fossil record. Their small size and delicate bones apparently reduce

Figure 8.1. External anatomy exemplified by a phyllostomine bat: (a) antebrachial membrane; (b) interfemoral membrane; (c) nose leaf; (d) tragus.

the probability of preservation. The earliest known bats are from the Eocene of North America (Jepsen 1970) and Europe (Schaal and Ziegler 1992). These fossils clearly indicate that the bats were completely volant; thus the early ancestors of bats showing pre-flight adaptations are as yet unknown or unrecognized in the fossil record. The earliest records of the phyl-lostomid bats are from the Miocene of Colombia.

The bats are typically divided into two suborders, the Megachiroptera and the Microchiroptera. The megachiropterans are entirely Old World and are rep-resented by a single family, the Pteropodidae. These are the Old World fruit bats, specialized for feeding on fruits, pollen, and nectar. Their ecological equivalent in the New World tropics is the family Phyllostomidae. The latter family, together with sixteen other families, is included in the Microchiroptera (see fig. 8.2).

The megachiropterans are chiefly distinguished from the microchiropterans by the following charac-ters: the second finger of a megachiropteran is capable

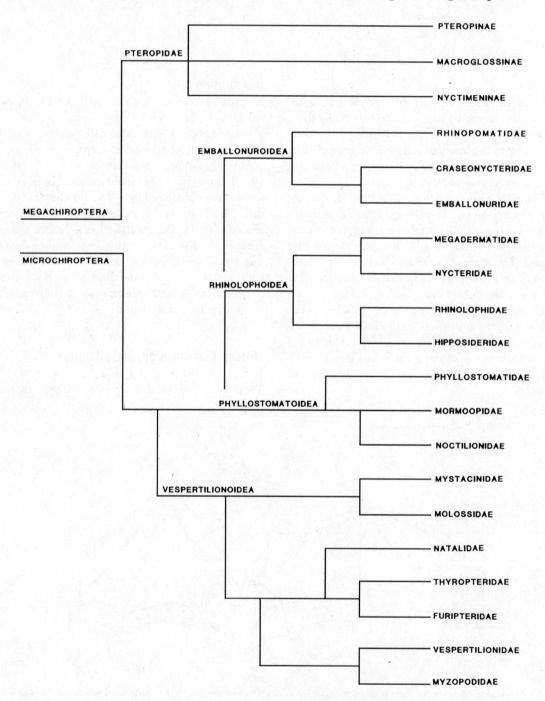

Figure 8.2. Cladogram indicating the relationships of the families of the Chiroptera (Eisenberg 1981).

Table 8.1   Key to the Families of New World Bats

1   Nose leaf or conspicuous folds or plates of skin on chin present 2
1' Nose leaf and folds or plates of skin absent 3
2   Folds of skin on chin; no nose leaf present . . . . Mormoopidae
2' Nose leaf present; incisors 2/2, 2/1, or 2/0 . . . . Phyllostomidae[a]
3   Tail perforates uropatagium and is enclosed in a sheath for half its length; upper lip is not noticeably cleft;
    claws of hind feet are not noticeably long; most species have glandular sacs in the wing or tail (see text). . . . . Emballonuridae
3' Tail and other characters not as above 4
4   Conspicious suckers (adhesive disks) on forefeet and hind feet . . . . Thyropteridae
4' Sucker disks absent from hind feet 5
5   Tail usually is longer than head and body but is completely enclosed in an interfemoral membrane; ears funnel
    shaped; tragus triangular. . . . . Natalidae
5' Tail and other characters not as above 6
6   Thumb (first finger) very reduced and enclosed in a membrane or absent . . . . Furipteridae
6' First finger present 7
7   Upper lip divided; tail shorter than interfemoral membrane but perforates the uropatagium; claws of hind feet
    very long . . . . Noctilionidae
7' Not as above 8
8   Tail extends beyond the interfemoral membrane; incisors 1/2–3; tooth number 28–32 . . . . Molossidae
8' Not as above; incisors 1–2/2–3, upper incisors widely spaced at base; tooth number 28–38 . . . . Vespertilionidae

[a]The nose leaf is vastly modified in *Desmoda, Centurio,* and *Sphaeronycteris* (see text). The Desmodontinae have a postcanine dental formula less than 4/4, and the upper incisors are bladelike (see text).

of some independent movement and generally has a claw; and the postorbital process of the skull is well developed, helping to support the very large eyes that are typical of megachiropterans.

In the microchiropterans the second finger lacks a claw and is not capable of independent movement. Usually a median fold of skin termed the tragus projects into the pinna, or external ear. All microchiropterans show specializations of the auditory nerves and larynx that correlate with their highly evolved ability to echolocate, which is used in orientation and prey capture. Only one genus of the Megachiroptera, *Rousettus,* uses echolocation for orientation. Bats of this genus produce the echolocating pulse by tongue clicks, not from the larynx as do the Microchiroptera.

The standard reference works for the Chiroptera are Wimsatt (1970 a,b, 1977) and Kunz (1982). Methods for studying the ecology and behavior of bats are reviewed in Kunz (1988). A key to the South American families is given in table 8.1, and table 8.2 portrays a matrix of morphological features. A key to the bats of Brazil is offered by Vizotto and Taddei (1973). Goodwin and Greenhall (1961) include keys to the bats of Trinidad and Tobago, while Linares (1987) offers a more recent compendium for Venezuela. Husson (1962) presents a very useful treatise for Suriname. Anderson (1997) includes keys in his treatise on Bolivia, and Brosset and Charles-Dominique (1990) have performed a similar task for French Guiana. The preceding works also contain information applicable to Brazil, Peru, and Ecuador. In a similar fashion the works of Barquez, Giannini, and Mares (1993), Ximénez (1969), and Acosta y Lara (1950) will be useful to workers in Bolivia and South Brazil. Albuja V.

(1982) has prepared a very useful handbook for the bats of Ecuador. Lists including bat distributions have been compiled for parts of lowland Peru (Tuttle 1970; Woodman et al. 1991); Bolivia (Anderson 1985, 1991); and Brazil (Willig and Mares 1989; Taddei 1975, 1979). Ruschi (1951–54) published twenty-one papers on the bats of Espíritu Santo. Emmons and Feer (1997) offer an excellent field guide with keys to the genera of bats of the lowland, moist Neotropics.

The natural history of New World bats has been described in numerous publications. Useful summaries are contained in Goodwin and Greenhall (1961), Husson (1962, 1978), and Baker, Knox Jones, and Carter (1976, 1977, 1979). Kunz (1982) contains much useful information on the ecology of the Chiroptera. Koopman (1982) has summarized distribution patterns for the South American species, and Baker et al. (1982) have summarized the data on karyotypes. Kunz (1988) has edited a useful volume dealing with methods for studying bats in the wild and in captivity.

Volumes 1 and 2 of *Mammals of the Neotropics* (Eisenberg 1989; Redford and Eisenberg 1992) offered brief summaries of bat natural history. To avoid duplication we will not repeat some of the published material. We will mention useful papers on work carried out within the geographical area encompassed by this volume. The species accounts also include other recent and regional references.

Bats are important components of the ecosystems they occupy; their numerical abundance and and their role in terrestrial ecosystems are underappreciated. Frugivorous bats are organized into guilds (Bonaccorso 1979; Humphrey and Bonaccorso 1979; Humphrey, Bonaccorso, and Zinn 1983) and are important

Table 8.2   Comparison of Morphological Features of New World Bat Families and Subfamilies

	Nose Leaf Present	Chin or Cheek Flaps Present	Stripes on Dorsum	Stripes on Face	Tail Vestigial or Absent	Uropatagium Reduced	Uropatagium Virtually Absent	Sacs in Antebrachial Membrane	Glandular Sac at Base of Tail
Emballonuridae	−	−	+/−	−	−	−	−	+/−	+/−
Emballonurinae	−	−	+/−	−	−	−	−	+/−	−
Diclidurinae	−	−	−	−	−	−	−	−	+
Noctilionidae	−	−	−	−	−	−	−	−	−
Mormoopidae	−	+	−	−	−	−	−	−	−
Phyllostomidae	+/−	−	+/−	+/−	+/−	+/−	+/−	−	−
Phyllostominae	+	−	−	−	−	+/−	−	−	−
Glossophaginae	+	−	−	−	+	+	+	−	−
Carolliinae	+	−	−	−	−	+	−	−	−
Sturnirinae	+	−	−	−	+	+	+	−	−
Stenoderminae	+/−	+/−	+/−	+/−	+/−	+/−	+/−	−	−
Desmodontinae	+[a]	−	−	−	+	+	+	−	−
Natalidae	−	−	−	−	−	−	−	−	−
Furipteridae	−	−	−	−	−	−	−	−	−
Thyropteridae	−	−	−	−	−	−	−	−	−
Vespertilionidae	−	−	−	−	−	−	−	−	−
Molossidae	−	−	−	−	−	−	−	−	−

	Tufts of Fur on Forearms	Tail Perforates Interfemoral Membrane	Dorsal-Surface of Uropatagium Densely Haired	Dorsal Surface of Uropatagium Sparsely Haired	Back Covered with Naked Skin	Thumb Reduced or Absent	Tail Projects beyond Uropatagium	Tail > HB	Suckers on Forefeet and Hind Feet
Emballonuridae	+/−	+/−	−	+/−	−	+/−	−	−	−
Emballonurinae	+/−	+/−	−	+/−	−	−	−	−	−
Diclidurinae	−	+	−	−	−	+/−	−	−	−
Noctilionidae	−	+	−	−	−	−	−	−	−
Mormoopidae	−	+/−	−	+/−	+/−	−	−	−	−
Phyllostomidae	−	+/−	−	+/−	−	−	−	−	−
Phyllostominae	−	+/−	−	+/−	−	−	−	−	−
Glossophaginae	−	+/−	−	+/−	−	−	−	−	−
Carolliinae	−	+	−	+/−	−	−	−	−	−
Sturnirinae	−	−	−	+/−	−	−	−	−	−
Stenoderminae	−	−	−	+/−	−	−	−	−	−
Desmodontinae	−	−	−	+/−	−	−	−	−	−
Natalidae	−	−	−	−	−	−	−	+	−
Furipteridae	−	−	−	−	−	+	−	−	−
Thyropteridae	−	−	−	−	−	−	−	−	+
Vespertilionidae	−	−	+/−	+/−	−	−	−	−	−
Molossidae	−	−	−	−	−	−	+	−	−

Note: +, character present; −, character absent; +/−, character variable.
[a] Highly modified.

seed dispersal agents (Reis and Guillaumet 1983). Co-existence among species exploiting similar resources in the same community is an active area of research (Marinho-Filho 1991), and bats' activity patterns often may reflect avoidance of competition (Marinho-Filho and Sazima 1989). That some species of plants have coevolved with bats serving as pollinators is amply documented by Sazima and Sazima (1975). The ecology of cavernicolous bats in Brazil has been explored by Trajano (1984).

Useful reviews arranged topically include activity patterns (Brown 1968; Erkert 1982); brain and behav-ior (Eisenberg and Wilson 1978); chromosomal numbers and evolution (Baker, Genoways, and Seyfarth 1981; Baker et al. 1982; Gardner 1977b); echolocation (Brown, Brown, and Grinnell 1983; Fenton 1982; Fuzessery and Pollak 1984; Gould 1977; Griffin 1958; Griffin, Webster, and Michael 1960; Novick 1977; Novick and Dale 1971); feeding (Ayala and D'Alessandro 1973; Carvalho 1961; Fleming 1982; Freeman 1979; Gardner 1977a; Howell 1974; Sazima 1976; Whitaker and Findley 1980; Wilson 1973b); morphometries (Findley 1993; Findley and Wilson 1982; Myers 1978; Ralls 1976; Vaughan 1959; Willig 1983);

ontogeny (Kleiman and Davis 1979; Phillips 1971; Tuttle and Stevenson 1982); physiological ecology (McNab 1969, 1982); and reproduction (Arata and Vaughan 1970; Fleming, Hooper, and Wilson 1972; Kleiman and Davis 1979; Myers 1977; Taddei 1976; Wilson 1973a, 1979).

Finally, phyllostomid bats may be useful indicators of forest disturbance, with some species of the subfamily Phyllostomatinae being especially sensitive (Fenton et al. 1992). The effect of human predation on bats has only recently been addressed (Setz and Sazima 1987).

The range maps have been shaded, with minor modifications, in conformity with Koopman (1982). In some cases we have indicated recent changes deriving from new collections. The measurements attributed to the USNMNH for the most part are taken with permission from unpublished data assembled by C. O. Handley Jr. and Katherine Ralls. Taxonomy follows Koopman (1993) unless otherwise noted.

FAMILY EMBALLONURIDAE
Sac-winged Bats, Sheath-tailed Bats

Diagnosis

A distinguishing but not unique feature of this family of bats is that the tail is enclosed in a sheath of the interfemoral membrane and the tip perforates the membrane's upper surface to lie free upon it; hence the name sheath-tailed bats. The Noctilionidae also exhibit this feature. The dental formula is highly variable in the family, from I 2/3, C 1/1, P 2/2, M 3/3 in *Emballonura* to I 1/2, C 1/1, P 2/2, M 3/3 in *Taphozous*. The premaxillary bones do not meet anteriorly. Many species have an opening on the antebrachial membrane over a gland field, and scent is produced during some of their wing-waving displays. Other species (the diclidurines) have a gland field surrounding the tail in the uropatagium. During resting, the third finger exhibits a reflexed proximal phalanx. Although the trochiter is well developed, it is smaller than the trochin and does not articulate with the scapula (Sanborn 1937) (see table 8.3 and figs. 8.3 and 8.4A–C).

Distribution

This family is widely distributed in both the Old World and New World tropics and includes some twelve genera and forty-seven species. In the Neotropics it extends from Sonora, Mexico, south through the Isthmus of Panama to southern Brazil.

Natural History

These bats are entirely nocturnal, and most are specialized for feeding on insects. Strategies of insect capture are variable, but many species seem to be adapted for aerial pursuit (see Bradbury and Emmons 1974; Bradbury and Vehrencamp 1976a,b, 1977a,b). The usual number of young is one.

Table 8.3   Key to the Common Adult Emballonuridae of Northern South America

1	Glandular sacs not present in either the interfemoral membrane or the antebrachial membrane	2
1'	Sacs present in either the interfemoral membrane or the antebrachial membrane	4
2	Tufts of whitish hairs on the forearms; muzzle pointed; two faint stripes on the dorsum	*Rhychonycteris naso*
2'	Not as above	3
3	Dorsal pelage tawny	*Centronycteris maximiliani*
3'	Dorsal pelage smoky gray to black	*Cyttarops alecto*
4	Glandular sac present in the interfemoral membrane (Diclidurinae)	5
4'	Glandular sac present in the antebrachial membrane of males	8
5	Head and body length exceeds 75 mm; forearm exceeds 65 mm; color pale gray	*Diclidurus ingens*
5'	Head and body length less than 75 mm	6
6	Dorsal pelage light brown; head and body length less than 60 mm	*Diclidurus isabellus*
6'	Not as above; head and body length greater than 60 mm	7
7	Dorsal pelage white; wings yellow; head and body length averages greater than 69 mm	*Diclidurus albus*
7'	Dorsal pelage pale; head and body length greater than 65, less than 69 mm	*Diclidurus scutatus*
8	Three upper incisors present; sac in the antebrachial membrane long, extending from the bend of the elbow to the anterior edge of the antebrachial membrane	*Cormura brevirostris*
8'	Fewer than three upper incisors	9
9	Distal portion of wings white; ears connected at the base by a membrane	*Peropteryx leucoptera*
9'	Not as above	10
10	Two longitudinal white stripes present on the dorsum (*Saccopteryx*)	11
10'	No dorsal stripes (*Peropteryx*)	13
11	Head and body length exceeds 45 mm	*Saccopteryx bilineata*
11'	Head and body length less than 45 mm	12
12	Head and body length less than 45 mm; hind foot less than 8 mm	*Saccopteryx canescens*
12'	Head and body less than 45 mm; hind foot exceeds 8 mm	*Saccopteryx leptura*
13	Head and body length exceeds 55 mm	*Peropteryx kappleri*
13'	Head and body length less than 55 mm	*Peropteryx macrotis*

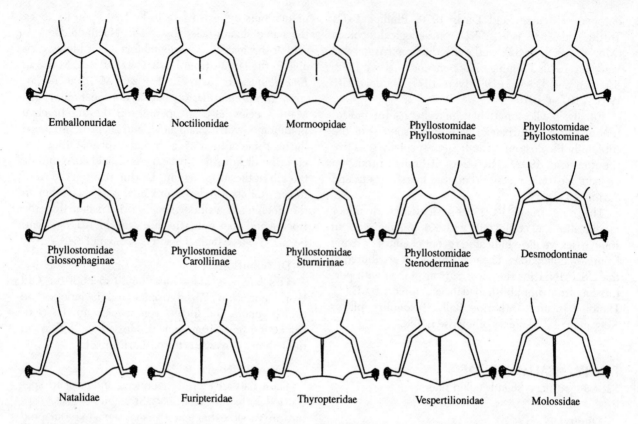

Emballonuridae	Noctilionidae	Mormoopidae	Phyllostomidae Phyllostominae	Phyllostomidae Phyllostominae
Phyllostomidae Glossophaginae	Phyllostomidae Carolliinae	Phyllostomidae Sturnirinae	Phyllostomidae Stenoderminae	Desmodontinae
Natalidae	Furipteridae	Thyropteridae	Vespertilionidae	Molossidae

Figure 8.3. Variation in tail length and interfemoral membrane structure in the families and subfamilies of bats (after Husson 1962).

SUBFAMILY EMBALLONURINAE

Genus *Rhynchonycteris* Peters, 1867

Rhynchonycteris naso (Wied-Neuwied, 1820)
Sharp-nosed Bat, Murciélago de Trompa

Description

The dental formula is I 1/3, C 1/1, P 2/2, M 3/3. Head and body length ranges from 37 to 43 mm. In Venezuela, total length averaged 56.5 mm for males and 59.2 mm for females; tails made up 15.4 mm and 16.8 mm of the total length. Weight ranged from 3.8 to 3.9 g (see Eisenberg 1989). The muzzle is elongated and pointed. There are no conspicuous wing sacs, so characteristic of other bat genera in this subfamily. The dorsum is grizzled yellow, and the venter is paler. The dorsal surfaces of the forearms bear tufts of whitish hairs; this character is diagnostic (plate 4). Chromosome number: $2n = 22$; FN = 36.

Range and Habitat

This species ranges from southern Veracruz, Mexico, to southeastern Brazil and is widely distributed at low elevations (map 8.1). It is almost always associated with moist areas near multistratal evergreen forests, generally at elevations below 500 m.

Natural History

These bats tend to roost in small, single-species colonies of about ten to twenty-four, on tree trunks or large tree branches (often over water), in tree cavities, or in rock caves. When roosting they are often aligned in vertical rows with individuals about 10 mm apart. Colonies range from three to forty-five individuals. Several males occur in a roosting group, and according to Plumpton and Knox Jones (1992) males appear to form harems; however, some authors dispute this claim. Harem defense may be seasonal. These bats are aerial insectivores (Husson 1978; Goodwin and

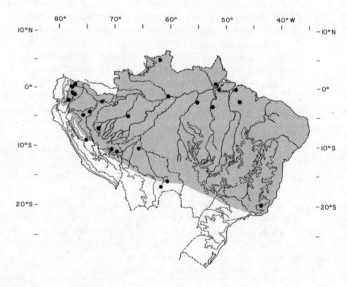

Map 8.1. Distribution of *Rhynchonycteris naso*.

Figure 8.4A. Some external features of the Chiroptera (part 1): (a) proximal flexure of emballonurid phalanges; (b) scent glands of *Peropteryx;* (c) face of *Noctilio;* (d) nose leaf; (e) face of *Carollia;* (f) face of *Peropteryx macrotis;* (g) head of *Glossophaga soricina;* (h) tail membrane of *Sturnira;* (i) face of *Artibeus jamaicensis.*

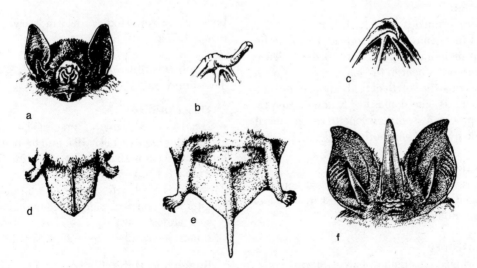

Figure 8.4B. Some external features of the Chiroptera (part 2): (a) *Desmodus rotundus* face; (b) thumb of *Desmodus;* (c) absence of thumb in Furipteridae; (d) furred interfemoral membrane of *Lasiurus;* (e) free-tailed bat, *Molossus* (relation of tail to uropatagium); (f) long nose leaf of *Lonchorhina.*

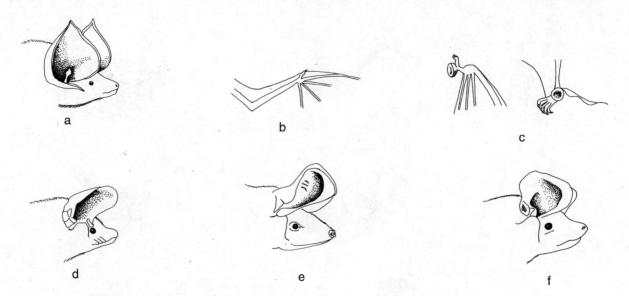

Figure 8.4C. Some external features of the Chiroptera (part 3): (a) ear and face of *Natalus;* (b) thumb of *Furipterus;* (c) suction disks of *Thyroptera;* (d) ear and face of *Nyctinomops;* (e) ear and face of *Eumops;* (f) ear and face of *Molossus.*

Greenhall 1961), and they tend to feed over water, flying only a short distance above the surface.

Females bear a single young, and parturition may be asynchronous. Young are large at birth and fly independently at one week of age, although they stay close to the mother (Bradbury and Emmons 1974). Adult size is attained within fourteen days postpartum. The bat's coloration provides some camouflage, since it resembles a lichen, but it is preyed on by hawks (*Buteo*), falcons (*Falco*), and herons (*Leucophoyx*) (Plumpton and Knox Jones 1992).

Genus *Saccopteryx* Illiger, 1811
White-lined Bat, Murciélago Rayado

Description
The dental formula is I 1/3, C 1/1, P 2/2, M 3/3. Members of this genus possess a glandular sac in the antebrachial membrane, quite close to the forearm, which is more prominent in the male. A pair of white dorsal stripes runs the length of the body on either side of the median axis. The skull is highly domed, and the postorbital process is well developed. The premaxillary bones do not meet anteriorly (see fig. 8.5).

Distribution
The genus is broadly distributed from southern Mexico to southeastern Brazil and west to southeastern Peru and Bolivia.

Natural History
The sac-winged bats tend to have preferred roosts that may include tree trunks or cave walls. They are aerial insectivores. Mating and foraging systems have

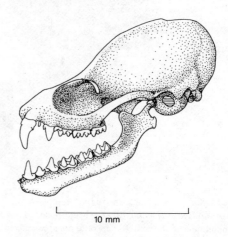

Figure 8.5. Skull of *Saccopteryx bilineata.*

been extensively studied by Bradbury and Vehrencamp (1976a,b, 1977a,b).

Saccopteryx bilineata (Temminck, 1838)
Two-lined Bat, Murciélago de Listas

Description
Head and body length averages 53.4 mm for males and 55.3 for females; tail is 19.6 mm for males, 19.6 mm for females; ear is 16.6 mm for males, 16.2 for females (see table 8.17). Weights range from 8.5 to 9.3 g. The dorsal fur is black with distinctive white lines running longitudinally on each side of the midline; the venter is paler, almost gray.
Chromosome number: $2n = 26$; FN = 36.

Range and Habitat
This species ranges from Colima, Mexico, south to southeastern Brazil (map 8.2). The bats usually forage

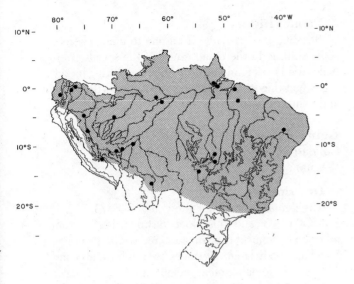

Map 8.2. Distribution of *Saccopteryx bilineata*.

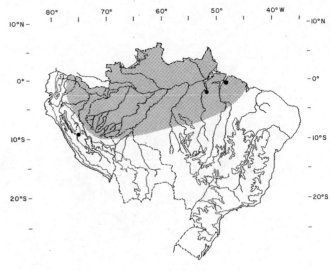

Map 8.3. Distribution of *Saccopteryx canescens*.

near streams and in moist areas; they prefer multistratal evergreen forests but may forage in clearings and rarely range above 500 m elevation.

Natural History

Roosting colonies average about a dozen individuals. These bats may roost with other species of bats in hollow trees or caves. Males defend harems; they have well-developed scent glands in their antebrachial membranes and emit scent while flapping their wings in ritualized combat (Bradbury and Emmons 1974).

Saccopteryx canescens Thomas, 1901

Description

Head and body length averages 38.6 mm for males, 44 mm for females; weights average 3.2 g for males and 3.5 g for females (see table 8.17). This species is much smaller than *S. bilineata* but nearly the same size as *S. leptura*. The hind foot of *S. canescens* is considerably shorter than that of *S. leptura*, being 7.2 mm for both sexes, whereas the tail is slightly longer in *S. leptura*. Tail length averages 16.1 mm for males and 17.6 mm for females; forearms are 35.9 mm for males and 38.6 mm for females. The fur extends onto the wing membranes both above and below for about a third of the distance from the body to the elbow. The dorsal surface of the interfemoral membrane is furred to the distal part of the tail. The basic color of the dorsal pelage varies from gray brown to brown and has a grizzled appearance. The two white longitudinal stripes may be indistinct in some specimens, and the striped pattern is not as bold as in *S. bilineata*. The venter is only slightly lighter than the dorsum.
Chromosome number: $2n = 24$; FN = 38.

Range and Habitat

This species is distributed from eastern Colombia across northern South America to northern Brazil and Peru (map 8.3). It frequents multistratal evergreen forests as well as open fields adjacent to them and occurs at low elevations.

Natural History

This small bat has not been the subject of any detailed investigations. Often found near streams and moist areas and in clearings, it co-occurs over much of its range with *S. leptura*, a better-studied species, but exactly how they partition resources is poorly understood. In Venezuela *S. canescens* commonly occurs below 150 m elevation; *S. leptura* can be taken at up to 500 m (Handley 1976).

Saccopteryx gymnura Thomas, 1901

Description

This is the smallest species of the genus, since the forearm ranges from 33 to 35 mm. *S. leptura* has a range from 37 to 43 mm and *S. canescens* from 36 to 41 mm. The dorsum is a light brown, but the pale dorsal stripes are visible. The venter is slightly paler.

Distribution

Apparently confined to the eastern Amazon basin of Brazil (map 8.4).

Saccopteryx leptura (Schreber, 1774)
Murciélago de Listas y de Cola Corta

Description

Head and body length averages 43.8 mm for males and 45 mm for females. Weight averages 4.4 g for

126

CHAPTER EIGHT

Map 8.4. Distribution of *Saccopteryx gymnura* (based on Koopman 1982).

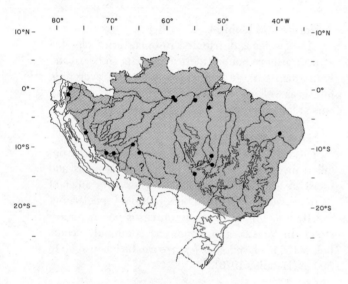

Map 8.5. Distribution of *Saccopteryx leptura*.

males and 5.7 g for females (see table 8.17). In general, *S. leptura* is heavier and slightly larger than *S. canescens*, but both species are smaller than *S. bilineata*. The scent sac opening on the dorsal side of the antebrachial membrane is quite large in the male. The upperparts of *S. leptura* are brown. The two longitudinal stripes are usually distinct, but when the pelage is worn they may not be sharply set off.
Chromosome number: $2n = 28$; FN = 38.

Range and Habitat
This species extends from Chiapas, Mexico, south through the Isthmus to eastern Brazil and adjacent Peru (map 8.5). It occurs in moist habitats and is strongly associated with multistratal evergreen forests at elevations below 500 m.

Natural History
Roosting groups range from one to nine. There is some shifting in the composition of the social group; males tend to defend individual females during breeding, when they exist as monogamous pairs (Bradbury and Vehrencamp 1977a).

Genus *Cormura* Peters, 1867
Cormura brevirostris (Wagner, 1843)
Wagner's Sac-winged Bat

Description
The dental formula for the genus is I 3/3, C 1/1, P 2/2, M 3/3, and this alone distinguishes it from other emballonurid bats encountered within its range. Head and body length averages 54 mm for males and 54.2 mm for females; mean weights are 8.6 g for males and 10.2 g for females (see table 8.17). The sac in the antebrachial membrane opens via a rather long aperture from near the anterior edge of the membrane to the elbow. The wing membrane extends to the ankle, and this character alone separates the genus from *Peropteryx*. The upperparts are buffy brown to black with no longitudinal dorsal stripes; the venter does not contrast with the dorsum.
Chromosome number: $2n = 22$; FN = 40.

Range and Habitat
The genus *Cormura* is widely distributed from Nicaragua south through the Isthmus to Amazonian Ecuador, Peru, and Brazil (map 8.6). These rare bats are associated with streams and moist areas, preferably multistratal tropical evergreen forests. Most specimens in Venezuela were taken below 500 m elevation (Handley 1976).

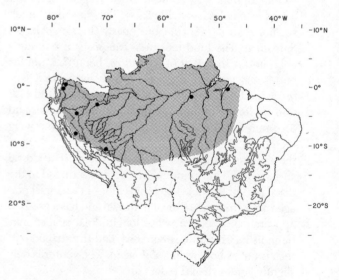

Map 8.6. Distribution of *Cormura brevirostris*.

Natural History

This rare bat has not been the subject of a detailed field study. Specimens have been collected singly, roosting in hollow fallen trees or under the projecting ends of fallen trees (Handley 1976).

Genus *Peropteryx* Peters, 1867
Peters's Sac-winged Bat

Description

The dental formula is I 1/3, C 1/1, P 2/2, M 3/3. Head and body length for the genus ranges from 45 to 55 mm, and tail length is 12 to 18 mm. Adults weigh from 4 to 10 g. The opening to the gland in the wing is near the anterior edge of the antebrachial membrane, thus distinguishing the genus from *Saccopteryx* (fig. 8.4A).

Distribution

The genus is distributed from southern Mexico to southeastern Brazil.

Natural History

These sac-winged bats may roost in shallow caves or rock crevices as well as in the hollows of dead trees. They are aerial insectivores.

Peropteryx kappleri Peters, 1867

Description

This is the largest species of the genus. Head and body length averages 58.2 mm for males and 60.9 mm for females, with forearm length 48.3 mm for males and 50.3 mm for females (see table 8.17). Weight averages 7.7 g for males and 8.5 g for females. There are two color phases, ranging from light brown to dark

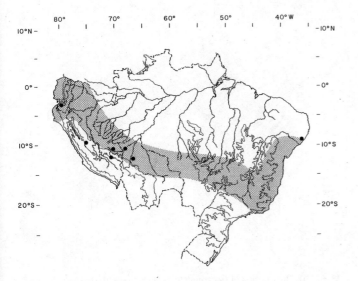

Map 8.7. Distribution of *Peropteryx kappleri*.

brown, but in both cases the underparts are slightly lighter. There is no white line on the dorsum.

Range and Habitat

This species ranges from southern Veracruz across northern South America to eastern Brazil, Ecuador, and Peru (map 8.7). It tolerates dry situations, and though it prefers evergreen forest, it does forage considerably in open fields. It may range to moderate elevations of 850 m in Venezuela (Handley 1976; Husson 1978).

Natural History

Roosting colonies are small, ranging from one to seven, and roosts are in caves and boulder crevices. The mating system is based on a monogamous pair, and the male defends his female against intruding males (Bradbury and Vehrencamp 1977a).

Peropteryx leucoptera Peters, 1867

Description

This species was formerly assigned to the genus *Peronymus*. It is approximately the size of *Peropteryx kappleri*, with forearm length ranging from 40 to 44 mm in males and 42 to 47 mm in females. Means for seven specimens were TL 59; HB 56; T 13; E 16; FA 43 (Suriname, CMNH; Husson 1962). The upperparts are reddish brown to dark brown with the underparts slightly paler, and the distal portions of the wings are white. In addition to the wing coloration, the best distinguishing feature for this species is that the ears tend to be joined at the bases by a very low membrane. In *Peropteryx kappleri* the ears are entirely separate (see fig. 8.3).
Chromosome number: $2n = 48$; FN = 62.

Range and Habitat

This species ranges from Peru and eastern Brazil north to the Guianas (map 8.8). It is found at low elevations. Ochoa-G. (1984b) recorded two specimens from Amazonas territory, Venezuela.

Natural History

Very little is known concerning the natural history of this species; it is assumed to be an aerial insectivore like other members of this family. Roosts are in hollow trees or in hollow rotten logs on the ground.

Peropteryx macrotis (Wagner, 1843)

Description

In this account *Peropteryx trinitatis* is considered a subspecies of *P. macrotis*. In general the Trinidadian subspecies is slightly smaller (see table 8.17). Males

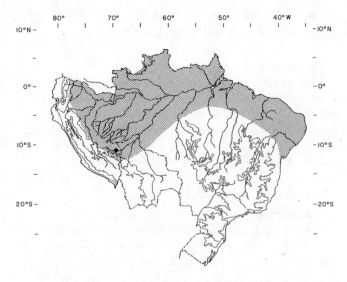

Map 8.8. Distribution of *Peropteryx leucoptera* (based in part on Koopman 1982).

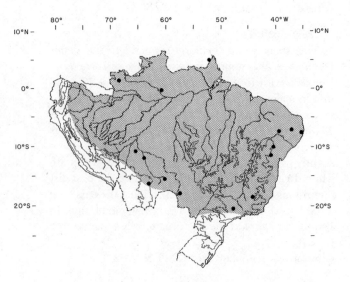

Map 8.9. Distribution of *Peropteryx macrotis*.

range from 56.5 to 60.5 mm in total length and females from 64.4 to 62.2 mm. Head and body length ranges from 46.4 to 47.2 mm in males and 49.37 to 51.09 mm in females (see table 8.17). The dorsal fur varies from buffy brown to reddish brown to very dark brown, and the venter is only slightly paler (plate 4).
Chromosome number: $2n = 26$; FN = 48.

Range and Habitat

This species ranges from Oaxaca, Mexico, southward to Peru and eastward to Paraguay and southeastern Brazil (map 8.9). It prefers moist sites and multistratal evergreen forest interspersed with savannas and generally does not occur above 800 m elevation in Venezuela (Handley 1976).

Natural History

This bat roosts in limestone and coral caves, sometimes with other bat species. Roosting groups may be as large as six individuals. It is entirely insectivorous, apparently feeding on the wing (Goodwin and Greenhall 1961). In the Brazilian caatinga this species roosts in rock piles and culverts and in Paraguay in limestone caves and crevices (Myers and Wetzel 1983; Willig 1983). In the caatinga of Brazil, this bat has been found roosting in colonies of up to ten individuals, always with only one male, probably in small harems (Willig 1983).

Genus *Centronycteris* Gray, 1838
Centronycteris maximiliani (Fischer, 1829)

Description

The dental formula is I 1/3, C 1/1, P 2/2, M 3/3. Head and body length averages 52 mm (range 50–62 mm), tail length is 18 to 23 mm, and forearm length averages 45 mm. Measurements for one specimen from Suriname were TL 71; HB 47; T 24; HF 7; E 18; Wt 15 g (CMNH). The long, lax dorsal pelage is tawny, with no dorsal stripes, and the venter does not contrast with the dorsum. This small bat does not have a sac in the antebrachial membrane.
Chromosome number: $2n = 28$; FN = 48.

Range and Habitat

This species occurs from southern Veracruz, southwest of the central cordillera of the Andes, to Ecuador and thence across Brazil, reappearing in Venezuela and the Guianas (map 8.10). It prefers lower elevations.

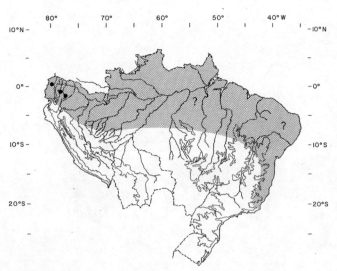

Map 8.10. Distribution of *Centronycteris maximiliani* (based in part on Koopman 1982).

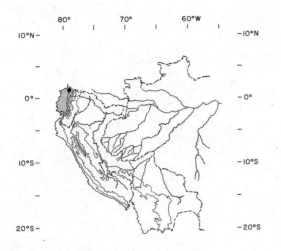

Map 8.11. Distribution of *Balantiopteryx infusca*.

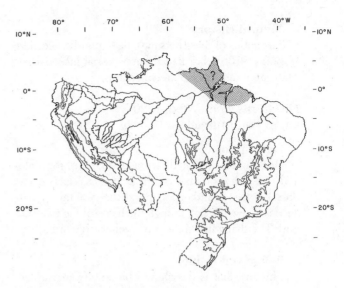

Map 8.12. Distribution of *Cyttarops alecto*.

Natural History

This bat tends to roost in hollow trees. It appears to be a slow flier and has a rather regular pattern of foraging in its home range, a feature shared with other emballonurids.

Genus *Balantiopteryx* Peters, 1867
Balantiopteryx infusca (Thomas, 1897)

Description

The dental formula is I 1/3, C 1/1, P 2/2, M 3/3. Total length averages 59 mm: HB 42; T 27; E 11; FA 41 (Albuja 1982). The wing sac is centrally placed in the antebrachial membrane. The dorsum is dark brown to dark gray with the venter paler, and dorsal stripes are absent.

Range and Habitat

The genus has been recorded from Central America, but on the southern continent the species *B. infusca* is known from the lowlands of western Ecuador but may range to southwestern Colombia (map 8.11).

Natural History

The species *B. infusca* is poorly known. *B. io* roosts in caves, and colonies in Costa Rica can consist of several hundred individuals (Arroyo-Cabrales and Knox Jones 1988).

SUBFAMILY DICLIDURINAE

Genus *Cyttarops* Thomas, 1913
Cyttarops alecto Thomas, 1913

Description

The dental formula is I 1/3, C 1/1, P 2/2, M 3/3. Head and body length ranges from 50 to 55 mm, tail from 20 to 25, and forearm from 45.8 to 47.2 mm. Females are slightly larger that males. There are no ante-brachial glands, and no gland is associated with the tail as in the genus *Diclidurus*. The ultimate phalanx of the thumb is free of the propatagium. The nostrils open through short, diverging tubes (Starrett 1972). The long, soft fur is almost black, but is grayer on the lower back and shoulders, and the venter is gray to black.

Range and Habitat

This rare bat is known disjunctly from Costa Rica, Guyana, and near Belém, Brazil (map 8.12).

Natural History

This bat is found in lowland forests roosting under palm fronds in groups of one to four. It is insectivorous (Starrett 1972).
Chromosome number: $2n = 32$; FN = 60.

Genus *Diclidurus* Wied, 1820
Ghost Bat

Description

The dental formula is I 1/3, C 1/1, P 2/2, M 3/3. Head and body length ranges from 50 to 80 mm, tail length from 15 to 25 mm, and forearm length from 45 to 66 mm. Species of this genus are unique in having white or whitish gray pelage. The base of the hair tends to be slate color, and in the white species the wings are depigmented. These bats do not have a wing sac, but where the tail penetrates the interfemoral membrane there is a small pouch and associated glandular tissue. The four species found in this region appear to be graded in size, with *Diclidurus ingens* the largest and *D. isabellus* the smallest.

Distribution

The genus is distributed from southern Mexico across northern South America to Brazil.

Natural History

The habits of these bats have been poorly recorded; Wilson (1973b) classified them as aerial insectivores. They are relatively rare in collections.

Diclidurus albus Wied, 1820
Ghost Bat, Murciélago Blanco

Description

Head and body length averages 70.4 mm for males and 71.4 mm for females. Only *D. ingens* is larger. The dorsal pelage is rather long, and the general appearance is of a white bat, though the bases of the hairs are gray. The flying membranes are yellowish white.

Range and Habitat

This rare bat is distributed broadly from Nayarit, Mexico, south through the Isthmus to eastern Brazil (map 8.13). No specimens have been taken in Venezuela above 850 m. We may conclude that it is a low elevation bat. It prefers moist areas and tolerates man-made clearings.

Natural History

In Trinidad this bat has been seen roosting between the leaves of tall coconut palms. It is not colonial and appears to roost individually. It is known to feed on insects (Goodwin and Greenhall 1961).

Diclidurus ingens Hernandez-Camacho, 1955

Description

This is the largest species of the genus. One female collected in Venezuela had a head and body length of 80.6 mm, with a forearm length of 71.4 mm (see table 8.17). Size alone should distinguish this rare species from other members of the genus.

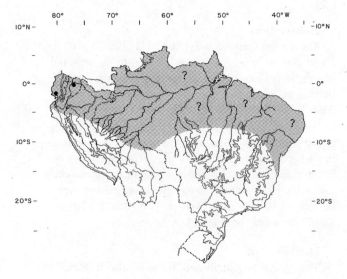

Map 8.13. Distribution of *Diclidurus albus*.

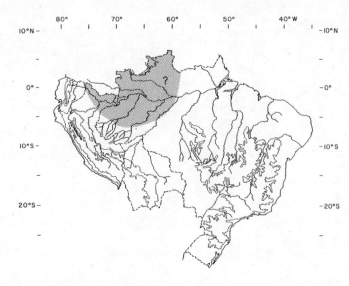

Map 8.14. Distribution of *Diclidurus ingens*.

Range and Habitat

The species ranges from southeastern Colombia and adjacent Venezuela to the Guianas, extending south to northwestern Brazil (map 8.14). It is associated with multistratal evergreen forests and prefers moist areas. In Venezuela specimens have been taken below 200 m in elevation.

Diclidurus isabellus (Thomas, 1920)

Description

This is the smallest species within the genus in the areas covered by this volume. Head and body length ranges from 59.6 to 59.3, and the bat weighs from 13.3 to 15.6 g. Males and females are nearly the same size. The dorsum is whitish but grades to pale brown posteriorly.

Range and Habitat

This species occurs from northwestern Brazil into southern Venezuela (map 8.15).

Natural History

This bat is strongly associated with wet habitats and multistratal evergreen forests. In Venezuela it was not collected above 155 m (Handley 1976).

Diclidurus scutatus Peters, 1869

Description

The species is larger than *D. isabellas* but smaller than *D. albus*. Head and body length averages 68 mm for males and 65 mm for females; weight averages 13.6 g for males and 12.9 g for females (see table 8.17). The male is clearly larger than the female. Where the

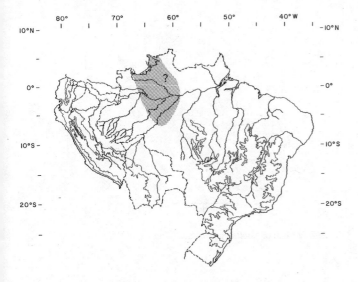

Map 8.15. Distribution of *Diclidurus isabellus.*

tail emerges from the membrane there are two distinct glandular pouches, one immediately adjacent to the tip of the tail, the other separated by a distinct ridge. The second pouch extends almost to the end of the inter-femoral membrane.

Range and Habitat
This species is distributed in the Amazonian portions of Peru, Brazil, adjacent Venezuela and the Guianas (map 8.16). It is strongly associated with moist habitats and broadly tolerant of human activity. In Venezuela it occurs at 99–850 m elevation (Handley 1976).

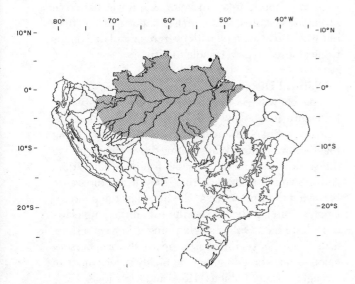

Map 8.16. Distribution of *Diclidurus scutatus.*

FAMILY NOCTILIONDAE
Bulldog Bats, Murciélagos Pescadores

Diagnosis
This family includes only one genus and two species. The dental formula is I 2/1, C 1/1, P 1/2, M 3/3. There is no prominent nose leaf, and the upper lip is medially divided (see fig. 8.4A). The tail is considerably shorter than the interfemoral membrane, and its tip protrudes through the membrane on the dorsal side. There are three phalanges on the third finger, and the claws of the feet are greatly enlarged. This suite of characters distinguishes the family from other Neotropical bats.

Distribution
This family is distributed from Sinaloa, Mexico, to northern Argentina.

Natural History
Noctilio leporinus is known as the fishing bat because it flies near the surface of water and seizes minnows in its claws. The other species, *N. albiventris,* is not specialized for fishing but feeds primarily on aquatic insects.

Noctilio albiventris Desmarest, 1818
Lesser Bulldog Bat, Pescador Menor

Description
Noctilio albiventris, in its basic morphology, is similar to *N. leporinus* but much smaller. Males average 74.9 mm in head and body length, while females average 71.6 mm. Weight for males averages 28.7 g and that for females 23 g (see table 8.17). The dorsum is brown (though some specimens may be light brown) or bright red on the head and shoulders; the venter is paler, ranging from gray to dark orange. Chromosome number: $2n = 34$; FN = 58–62.

Range and Habitat
Noctilio albiventris ranges from Nicaragua south to northern Argentina (map 8.17). It prefers lower elevations; in Venezuela bats were not taken above 300 m. It is strongly associated with streamside habitats and tolerates a wide variety of vegetation types, using areas modified for croplands as well as evergreen forests.

Natural History
This bat often feeds over water and appears to eat primarily aquatic insects (Whitaker and Findley 1980). Prey may be caught aerially or snatched from the surface with the hind feet. There is an early feeding phase at dusk, followed by a second activity peak after midnight. The bats often roost in hollow trees near water.

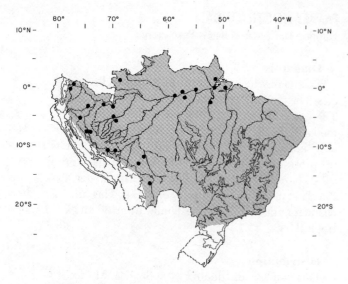

Map 8.17. Distribution of *Noctilio albiventris*.

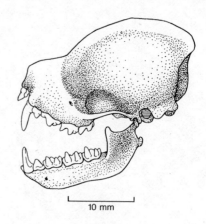

Figure 8.6. Skull of *Noctilio leporinus*.

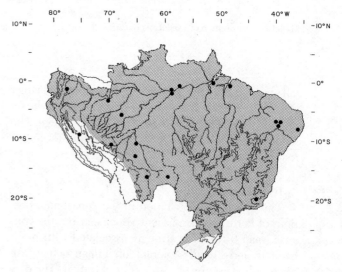

Map 8.18. Distribution of *Noctilio leporinus*.

Timing of reproduction may be strongly influenced by insect abundance. Only a single young is born each year. The litter size is one, and there is probably only one litter per year. Young do not fly until thirty-five to forty-four days of age and have a long period of maternal attachment before weaning, which occurs at seventy-five to ninety days (mean = 80.5). The young eats solid food for the first time at about forty-five days and is fed masticated food from the mother's cheek pouches. When juveniles first leave the roost they fly with their mothers and continue to nurse even after learning how to fly. The females nurse only their own young, which they recognize by vocal signals (Brown, Brown, and Grinnell 1983; Hood and Pitocchelli 1983). The natural history has been reviewed in detail by Hood and Pitoccheli (1983).

Noctilio leporinus (Linnaeus, 1758)
Fishing Bat, Murciélago Pescadore

Description
The dental formula is I 2/1, C 1/1, P 1/2, M 3/3 (fig. 8.6). In this large bat, males' head and body length averages 96.9 mm, while that of females averages 91.3 mm. Weight for males averages 67 g, and that for females 56 g (see table 8.17). As with *N. albiventris*, the male is larger, unlike members of many other Neotropical bat families. The dorsum is orange brown. A pale median dorsal stripe is evident. The venter ranges from gray to orange.
Chromosome number: $2n = 34$; FN = 58–62.

Range and Habitat
The fishing bat extends from Sinaloa, Mexico, to northern Argentina (map 8.18). It is usually found below 200 m elevation and is strongly associated with streams or moist areas. In Venezuela it was netted in savannas, pastures, and marshes about half the time but was also associated with evergreen and deciduous tropical forest types (Handley 1976).

Natural History
This bat will roost in colonies of up to seventy-five, frequently in a hollow tree. Foraging tends to be individualistic, but occasionally two bats may be seen flying together over water. While fishing they fly very close to the surface and can detect small fish by echolocation. As fish swim close to the surface they distort the water, betraying their presence as the bat scans the pond for ripples. The bat then seizes the fish in the long claws of its hind feet. Only a single young is born at a time, usually one per year. Females apparently form nursery colonies, and breeding may be highly synchronous in strongly seasonal habitats (Hood and Knox Jones 1984; Goodwin and Greenhall 1961; Willig 1983).

FAMILY MORMOOPIDAE

Diagnosis

This family was formerly included in the Phyllostomidae. It was defined at the family rank and reclassified by J. D. Smith (1972). Although almost all members of the family Phyllostomidae have a prominent nose leaf, the mormoopid bats do not possess such a structure. The lips are expanded and adorned with flaps and folds, and there are often bristlelike hairs on the sides of the lips. The nostrils are incorporated into the expanded upper lip and often have ridges and protuberances above or between them. The trochiter is well developed but smaller than the trochin and does not articulate with the scapula. Two genera are recognized by Smith, *Pteronotus* and *Mormoops*. *Pteronotus* is divisible into three subgenera, *Phyllodia*, *Chilonycteris*, and *Pteronotus*. Identifications are outlined in table 8.4. External features are illustrated in Eisenberg (1989). The skull of *Pteronotus* is portrayed in figure 8.7.

Distribution

This family is confined to the New World tropics, ranging from southern Mexico to northeastern Brazil. The Antilles have been extensively colonized by species of this family.

Natural History

These bats have been classified as aerial insectivores by Wilson (1973b). They primarily roost in caves or tunnels. A single young is born after gestation period of at least sixty days. The echolocation pulses of mormoopid bats are distinctive and separable from those of the phyllostomids. The high-frequency sounds mormoopids produce during echolocation resemble the sounds emitted by rhinolophids more

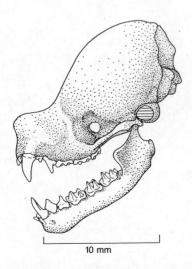

Figure 8.7. Skull of *Pteronotus parnellii*.

10 mm

Table 8.4   Key to the Common Mormoopidae
             of South America

1	Hairless skin of wings meets in the midline of the back (subgenus *Pteronotus*) 2
1'	Skin of wings does not meet in the midline of of back; back presents the usual furred appearance 3
2	Head and body length greater than 60 mm . . . . *Pteronotus suapurensis*
2'	Head and body length less than 60 mm . . . . *Pteronotus davyi*
3	Ears joined by flap of skin extending over the crown of the head; chin leaf prominent . . . . *Mormoops megalophylla*
3'	Ears not connected by skin flap 4
4	Head and body length greater than 59 mm . . . . *Pteronotus parnellii*
4'	Head and body length less than 59 mm . . . . *Pteronotus personatus*

Note: See family account.

closely than they resemble those of any of the phyllostomids. The individual pulses are long and loud, without major frequency modulation, and have three harmonics that contribute qualitatively different echoes, allowing the bat to discriminate two-dimensional cues while echolocating (Fuzessery and Pollak 1984). Sound is emitted from the mouth, not through the nose as in phyllotomids.

Genus *Pteronotus* Gray, 1838

Subgenus *Phyllodia*
Pteronotus parnellii (Gray, 1843)
Mustached Bat

Description

The dental formula is I 2/2, C 1/1, P 2/3, M 3/3 (see fig. 8.7). This is one of the largest species of the genus. Males' head and body length averages 71.7 mm while that of females is 70.4 mm, and weight averages 20.4 g for males and 19.6 g for females (see table 8.17). The dorsum ranges from dark to light brown depending on age and molt. This description includes *Pteronotus* (= *Chilonycteris*) *rubiginosa*.
Chromosome number: $2n = 38$; FN = 60.

Range and Habitat

This species ranges from southern Sonora, Mexico, south through the Isthmus and across the northern Neotropics to Brazil. It is found at up to 1,500 m altitude in Venezuela, but in the main it occurs below 500 m. It generally lives in moist areas but tolerates both multistratal evergreen forest and dry deciduous forest (Handley 1976) (map 8.19).

Natural History

These bats often roost in caves and may co-occur with other species of mormoopids and phyllostomids. In Venezuela they breed once a year. Mating takes

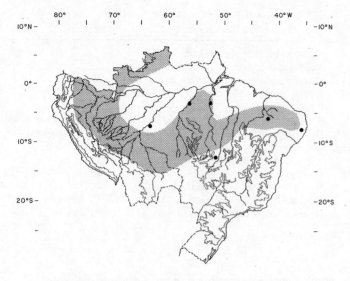

Map 8.19. Distribution of *Pteronotus parnellii*.

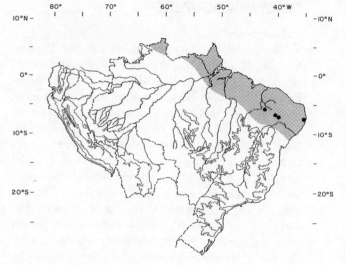

Map 8.20. Distribution of *Pteronotus davyi*.

place in January, when both sexes roost together, and the young are born in May when insects are most abundant. They feed mostly on Lepidoptera and Coleoptera. Insects are found by echolocation, but the echolocation pulse is unique in that the major portion of the call is not frequency modulated. Apparently the bats determine the distance and direction of their prey from the Doppler shift in pitch of the echo. This echolocation system is convergent with that of the Old World families Hipposideridae and Rhinolophidae (Herd 1983; Novick 1977).

Subgenus *Pteronotus*
Naked-backed Bat

Description
This subgenus is easily separated from the other mormoopids because the skin of the wing meets in the midline, giving a naked-backed appearance. There are two species, *Pteronotus davyi* and *Pteronotus gymnonotus* (= *P. suapurensis*).

Pteronotus davyi Gray, 1838

Description
The dental formula is I 2/2, C 1/1, P 2/3, M 3/3. This is the smallest species of the naked-backed bats. Head and body length averages 56.4 mm for males and 57.9 mm for females; weight averages 9.3 g for males and 9.6 g for females (see table 8.17). The dorsal fur is dark brown, and the venter is somewhat paler. Chromosome number: $2n = 38$; FN = 60.

Range and Habitat
This species is distributed from southern Sonora, Mexico, south to northwestern Peru and northeastern Brazil. In Venezuela it most frequently occurs below

500 m elevation. It shows a broad tolerance for habitat types; most specimens in Venezuela occur in dry areas. The bats frequently occur in dry deciduous thorn forests, but they range into multitratal wet evergreen forest (Handley 1976) (map 8.20).

Natural History
This species prefers to roost in damp caves, often with several other species including *Pteronotus parnellii* and various phyllostomid bats.

Pteronotus gymnonotus Natterer, 1843
Greater Naked-backed Bat

Description
This species is similar in color to *Pteronotus davyi* but considerably larger. Head and body length aver-

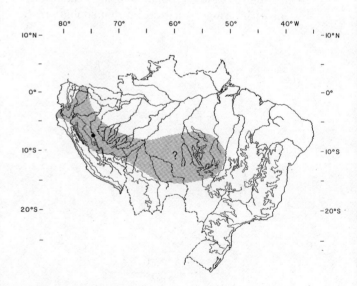

Map 8.21. Distribution of *Pteronotus gymnonotus* (based in part on Koopman 1982).

ages 64.3 mm for males and 64 mm for females, and weight averages 12.6 g for males and 13.9 for females. Chromosome number: $2n = 38$; FN = 60.

Range and Habitat

This species ranges from southern Veracruz, Mexico, south to Peru, and across northern South America to the Guianas and southwestern Brazil (map 8.21).

Natural History

No specific observations have been recorded on this species other than that it generally occurs below 400 m in Venezuela and prefers dry deciduous forests. Note that Smith (1972) synonymized *P. suapurensis* with *P. gymnonotus*.

Subgenus *Chilonycteris*
Pteronotus personatus (Wagner, 1843)
Mustached Bat

Description

The species resembles *Pteronotus parnellii* in morphology, color, and dentition but is smaller. With its haired back, it cannot be confused with *P. davyi* or *P. gymnonotus*. Head and body length for males averages 53.5 mm; females are larger, with an average length of 58 mm. Weight averages 8 g for males and 7 g for females (see table 8.17). Chromosome number: $2n = 38$; FN = 60.

Range and Habitat

This species ranges from northern Colombia south through Peru and east to Brazil and Suriname. Its range continues northward from northern Colombia to southern Mexico, reaching southern Sonora. It generally occurs at elevations below 400 m and tolerates

both multistratal evergreen forests and dry deciduous tropical forests (Handley 1976) (map 8.22).

Genus *Mormoops* Leach, 1821
Mormoops megalophylla (Peters, 1864)
Leaf-chin Bat

Description

The dental formula is I 2/2, C 1/1, P 2/3, M 3/3. Males have an average head and body length of 68 mm, while females are slightly larger at 69.6 mm (see table 8.17). Weights range from 17.2 g for males and 16.5 g for females (see table 8.17). This reddish brown bat has a highly ornamented face with a prominent chin leaf. A flap of skin connecting the ears is diagnostic and separates this genus from other mormoopid bats.
Chromosome number: $2n = 38$; FN = 62.

Range and Habitat

This species ranges from southern Texas through the Isthmus of Panama and across northern Colombia and northern Venezuela. It has a disjunct southern population in Ecuador and western Peru. It has colonized many of the Caribbean islands. It is frequently found in dry deciduous thorn forests but also forages in moist areas, primarily below 400 m elevation (map 8.23).

Natural History

Although it exploits dry deciduous forest, this species roosts in moist caverns or mine shafts. It may occur in colonies of up to twenty, often with other species of mormoopids and phyllostomids (Goodwin and Greenhall 1961). It appears to feed on moths. Nursing females roost separately from males and nonreproductive females (Rezsutek and Cameron 1993).

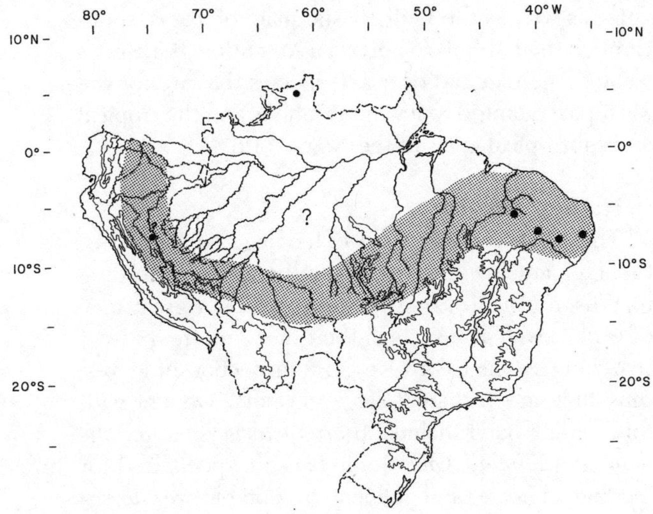

Map 8.22. Distribution of *Pteronotus personatus*.

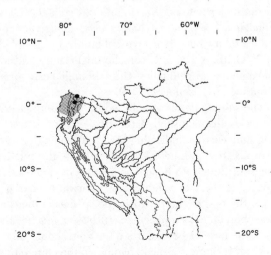

Map 8.23. Distribution of *Mormoops megalophylla*.

FAMILY PHYLLOSTOMIDAE
American Leaf-nosed Bats

Taxonomy and Identification

This diverse family of bats is divided here into seven subfamilies. We have retained the Sturnirinae as distinct from the Stenoderminae. Sturnirinae are subsumed in the Stenoderminae by Knox Jones and Carter (1976). True vampire bats are considered a subfamily, that is, the Desmodontinae of the Phyllostomidae. The Phyllonycterinae and Brachyphyllinae are endemic to the Greater and Lesser Antilles and are not covered by this volume.

Baker, Hood, and Honeycutt (1989) propose that the family Phyllostomidae may be properly divided into three subfamilies: the Desmodontinae, Phyllostominae, and Vampyrinae. They employed multiple data sets, including morphology of the reproductive tract, albumin immunological data, and chromosonal banding data. Their proposal would have the Desmodontinae intact. The Vampyrinae would include the larger, carnivorous genera of the classical Phyllostominae (*Vampyrum, Trachops,* and *Chrotopterus*). The newly divided Phyllostominae would include all other genera divided into three tribes: Phyllostomini, Glossophagini, and Stenodermatini. The Stenodermatini would include the Carolliinae and Sturnirinae with the stenodermatines. Unfortunately their scheme would have the genera *Micronycteris* and *Macrotus* as *incertae sedis.* The subfamilies in this volume follow Koopman (1993).

Pacheco and Patterson (1991) argue persuasively that the genus *Sturnira* is monophyletic, lending credence to the earlier tendency to raise the group to a subfamily in full recognition that it is closely allied to the Stenodermatinae. Koopman (1993) is conservative in maintaining most of the older subfamilial categories, although the sturnirines are retained within the Stenodermatinae. Koopman does split the Glossophaginae, separating *Lionycteris, Lonchophylla,* and *Platalina* into their own subfamily Lonchophyllinae (Griffiths 1982). Given the uncertainty and for convenience, we will follow the subfamilial categories as outlined in Eisenberg (1989).

The subfamilies as so outlined in Eisenberg (1989) are easily distinguished (table 8.6). If the muzzle is long and narrow, indicating a mixed-feeding or carnivorous strategy, one need only estimate the relative length of the ear to break specimens clearly into two groups. If the ear, when pressed forward, does not reach the end of the rostrum, the species belongs to the subfamily Glossophaginae. If the ear reaches the tip of the nose or beyond, then the chin should be inspected. If there is a single papilla on the chin, either standing alone or surrounded by smaller papillae, the species belongs to the subfamily Carolliinae; if there is a Y-or V-shaped shield on the chin, it belongs to the subfamily Phyllostominae.

If the bat in hand has a short, broad muzzle, one should immediately inspect the face; if it bears white stripes, the species is referable to the subfamily Stenoderminae. Only the genera *Pygoderma, Ectophylla, Ametrida, Sphaeronycteris,* and *Centurio* lack the white facial stripes within this subfamily, though they are faint in some species of *Artibeus.* Once one eliminates these five genera, then if there are no white facial stripes, no interfemoral membrane is present, and the foot is rather hairy, the species in question probably belongs to the subfamily Stenoderminae (see table 8.5). This "in-the-hand" key should be crosschecked with Knox Jones and Carter (1976), which includes a complete key to the subfamilies and genera.

Diagnosis

For this description the subfamily Desmondontinae (the true vampire bats) and some stenodermines are aberrant, since the nose leaf is very reduced, but their morphology is so distinctive that they cannot be mistaken for other species of bats (see Desmodontinae, below). The New World leaf-nosed bats are characterized by a skull without a postorbital process. The third finger has three complete bony phalanges. The most distinctive feature is the nose leaf that uniquely defines almost all members of this family. (See Arita 1990 for a discussion of morphological variations.) The tooth number is highly variable from 26 to 34, reflecting in part the diverse feeding specializations exhibited by the family as a whole.

Distribution

This family is confined to the Western Hemisphere and is distributed from southern California and Arizona, as well as the Gulf coastal plain of Texas, south through the Isthmus to northern Argentina. It includes some 51 genera and over 140 species that are for the most part adapted to lower elevations and the tropical and subtropical areas of the New World.

Natural History

These bats show a variety of feeding specializations, though many feed on fruit (Carvalho 1961). According to Wilson (1973b), most genera include a great deal of fruit as well as insects in their diet, and their insect searching seems to involve close inspection of leaves; thus they are termed foliage gleaners. Genera with this feeding habit include *Micronycteris, Tonatia, Mimon,* and *Phylloderma.* Some bats are specialized for feeding on nectar and pollen: most notable are *Glossophaga, Anoura,* and *Lionycteris.* Two genera belonging to the subfamily Desmodontinae are specialized

Table 8.5   Key to the Subfamilies and Genera of Phyllostomidae

Subfamilies

1   Single upper incisor and upper canine enlarged and bladelike . . . . Desmodontinae
1′   Upper incisor(s) and canine not enlarged and bladelike 2
2   Nose leaf rudimentary, without distinct upright process; tail present . . . . Phyllonycterinae
2′   Nose leaf usually well developed; tail absent if nose leaf rudimentary 3
3   Tongue elongate, with conspicuous bristlelike papillae on anterodorsal surface; first upper premolar usually
     distinctly separated from canine and rarely in contact with second upper premolar (first upper premolar
     sometimes in contact with canine in *Monophyllus*, but distinctly separated from second upper premolar) . . . . Glossophaginae
3′   Tongue not elongate, lacking conspicuous bristlelike papillae; first upper premolar in contact with canine and
     usually with second upper premolar 4
4   Zygomatic arch incomplete . . . . Carolliinae
4′   Zygomatic arch complete 5
5   Molars dilambdodont (distinct W-shaped pattern of lophs on occlusal surface) . . . . Phyllostominae
5′   Molars lacking dilambdodont pattern . . . . Stenoderminae

Phyllostominae

1   One lower incisor 2
1′   Two lower incisors 4
2   Two lower premolars . . . . *Mimon*
2′   Three lower premolars (second small to minute) 3
3   Second lower premolar crowded to lingual side of tooth row, first and third lower premolars usually in contact . . . . *Chrotopterus*
3′   Second lower premolar not crowded from tooth row, first and third lower premolars not in contact . . . . *Tonatia*
4   Two lower premolars . . . . *Phyllostomus*
4′   Three lower premolars (second sometimes crowded to lingual side of tooth row) 5
5   Rostrum as long as braincase . . . . *Vampyrum*
5′   Rostrum shorter than braincase 6
6   Second lower premolar large, but smaller than first and third premolars 7
6′   Second lower premolar small to minute, much smaller than first and third premolars 8
7   Auditory bullae large, greatest diameter much exceeding distance between them . . . . *Macrotis*
7′   Auditory bullae small, greatest diameter less than distance between them . . . . *Micronycteris*
8   Second lower premolar displaced lingually from tooth row; first and second lower premolars in contact or
     nearly so 9
8′   Second lower premolar not displaced lingually from tooth row; first and second lower premolars usually not
     in contact 10
9   Greatest length of skull less than 20 mm . . . . *Macrophyllum*
9′   Greatest length of skull more than 20 mm . . . . *Trachops*
10   Dorsal profile of rostrum strongly convex; deep depression present between orbits . . . . *Lonchorhina*
10′ Dorsal profile of rostrum not convex; no depression between orbits . . . . *Phylloderma*

Glossophaginae

1   Permanent lower incisors lacking 2
1′   Two pairs of permanent lower incisors, usually well developed 8
2   Premolars 3/3 . . . . *Anoura*
2′   Premolars 2/3 3
3   Molars 2/2 . . . . *Lichonycteris*
3′   Molars 3/3 4
4   Pterygoids highly modified, expanded at base and inflated in appearance; pterygoid wings long and in contact,
     or nearly so, with auditory bullae 5
4′   Pterygoids normal, not expanded at base or inflated in appearance, pterygoid wings short and not in contact
     with auditory bullae 7
5   First and second upper incisors separated by distinct gap; upper premolars low, barely exceeding height of
     molars . . . . *Choeroniscus*
5′   First and second upper incisors in contact, or nearly so; upper premolars distinctly higher than molars 6
6   Rostrum distinctly longer than postrostral part of cranium; upper molars essentially equal in size, all with a
     distinct metastyle . . . . *Musonycteris*
6′   Rostrum about equal in length to postrostral part of cranium; third upper molar somewhat smaller than first
     two and lacking a distinct metastyle . . . . *Choeronycteris*
7   Upper molars lacking mesostyle; lower molars long and narrow; known only from Middle America . . . . *Hylonycteris*
7′   Mesostyle present on all upper molars; lower molars only moderately compressed; known only from Brazil
     and Venezuela . . . . *Scleronycteris*
8   Molars 2/2 . . . . *Leptonycteris*
8′   Molars 3/3 9
9   Zygomatic arch complete, first upper incisor not markedly enlarged and spatulate 10

Table 8.5   (*continued*)

Subfamilies

9′ Zygomatic arch incomplete, first upper incisor enlarged and spatulate. 11

10 Evident gap between upper premolars and between them and adjacent teeth; tail relatively long and extending beyond posterior border of uropatagium. . . . . *Monophyllus*

10′ Upper premolars usually in contact and filling space between canine and first molar; tail short and not extending beyond posterior border of uropatagium . . . . *Glossophaga*

11 Rostrum elongate, longer than postrostal part of cranium; postcanine maxillary teeth reduced in size and with evident gaps between them . . . . *Platalina*

11′ Rostrum not elongate, no longer than postrostral part of cranium; postcanine maxillary teeth of normal size; last premolars and molars in contact or nearly so 12

12 First upper premolar smaller than second and laterally compressed. . . . . *Lonchophylla*

12′ First upper premolar essentially same size as second, not laterally compressed (triangular in outline) . . . . *Lionycteris*

Carolliinae

1 Tail present; upper premolars essentially equal in size. . . . . *Carollia*

1′ Tail absent; first upper premolar much smaller than second . . . . *Rhinophylla*

Stenoderminae (includes Sturnirinae)

1 Molars 2/2. 2

1′ Molars 2/3 or 3/3 7

2 Upper dental arcade semicircular, rostrum less than half as long as braincase . . . . *Centurio*

2′ Upper dental arcade not semicircular, rostrum more than half as long as braincase 3

3 Rostrum inflated, nearly cuboid. . . . . *Pygoderma*

3′ Rostrum not inflated or cuboid 4

4 Posterior margin of external nares with marked, lyre-shaped emargination . . . . *Chiroderma*

4′ Posterior margin of external nares lacking lyre-shaped emargination 5

5 Second upper molar markedly larger than first; upper premolars separated from each other and adjacent teeth by evident gaps. . . . . *Ectophylla* (part)

5′ Second upper molar essentially equal in size to, or smaller than, first; no gaps between anterior upper cheek teeth. 6

6 Posterior margin of external nares more or less straight; second upper molar much smaller than first and differing in form. . . . . *Artibeus* (part)

6′ Posterior margin of external nares broadly V-shaped; second upper molar resembling first in size and form . . . . *Vampyressa* (part)

7 Molars 2/3. 8

7′ Molars 3/3. 12

8 Palate short, posterior border having deep U-shaped emargination that reaches level of first molar . . . . *Ariteus*

8′ Palate long, posterior boarder having shallow emargination that falls far short of level of tooth row 9

9 First upper incisor markedly bifid, less than twice size of second incisor . . . . *Artibeus* (part)

9′ First upper incisor not bifid or only weakly so, more than twice size of second incisor 10

10 Second upper molar noticeably larger than first; upper premolars separated from each other and from adjacent teeth by evident gaps . . . . *Ectophylla* (part)

10′ Second upper molar equal to or smaller than first, no gaps between anterior upper cheek teeth. 11

11 Incisors 2/1 or 2/2; height of first incisor greater than height of first premolar, greatest length of skull less than 22 mm . . . . *Vampyressa* (part)

11′ Incisors 2/2; height of first incisor much less than height of first premolar; greatest length of skull more than 24 mm . . . . *Vampyrodes*

12 Upper dental arcade expanded laterally to form semicircular arc 13

12′ Upper dental arcade not expanded laterally, U-shaped in occlusal view 14

13 Orbital space wider than long; interorbital constriction less than 5 mm . . . . *Ametrida*

13′ Orbital space longer than wide; interorbital constriction more than 5 mm. . . . . *Sphaeronycteris*

14 Palate short, posterior palatal emargination reaching level of first upper molar. 15

14′ Palate of medium length or long, posterior border variously emarginate but never to level of tooth row. 17

15 Palatal emargination broadly V-shaped. . . . . *Phyllops*

15′ Palatal emargination deeply U-shaped 16

16 Well-developed V-shaped ridge from sagittal crest to anterior margin of orbits, forming deep rostral depression . . . . *Stenoderma*

16′ V-shaped ridge from sagittal crest to anterior margin of orbits lacking, rostrum normal. . . . . *Ardops*

17 Upper molars distinctly grooved longitudinally, the first two subquadrate in outline and lacking well-developed cusps; first upper incisor approximately half as high as canine. . . . . *Sturnira*

17′ Upper molars lacking longitudinal groove, the first two not subquadrate in outline and possessing well-developed cusps; first upper incisor much less than half as high as canine 18

18 First upper incisor less than twice size of second and resembling it in shape; upper incisors in contact and filling space between canines. 19

Table 8.5   (*continued*)

Subfamilies

18′ First upper incisor more than twice size of second and differing from it in shape; evident gaps present
between upper incisors. 20
19   First upper incisor deeply bifid; M3, if present, minute and peglike. . . . . *Artibeus* (part)
19′ First upper incisor not bifid; M3 relatively large and well developed . . . . *Enchisthenes*
20   Crowns of first upper incisors parallel, deeply bifid; lower incisors in contact . . . . *Uroderma*
20′ Crowns of first upper incisors converge distally, not deeply bifid; lower incisors separated by distinct gaps . . . . *Vampyrops*

Phyllonycterinae

1   Tail not extending beyond edge of uropatagium. . . . . *Brachyphylla*
1′   Tail extending beyond edge of uropatagium 2
2   Zygomatic arch complete; second and third lower molars distinctly cuspidate. . . . . *Erophylla*
2′   Zygomatic arch incomplete; second and third lower molars not distinctly cuspidate. . . . . *Phyllonycteris*

Desmodontinae

1   First lower incisors in contact; interfemoral membrane with distinct fringe of moderately long hairs . . . . *Diphylla*
1′   First lower incisors not in contact; interfemoral membrane without fringe of hair 2
2   Lower incisors not bifid; wing white from middle of proximal phalanx to tip. . . . . *Diaemus*
2′   Lower incisors bifid; wing usually pigmented to tip (if white tipped, white does not extend proximally
to first phalanx). . . . . *Desmodus*

Source: Modified from Knox Jones and Carter 1976.

Table 8.6   Field Key to the Subfamilies of the Phyllostomidae

1   Nose leaf rudimenentary, face not wrinkled; calcar absent, wing attaches near the knee joint; upper incisors
broader than canines . . . . Desmodontinae
1′   Nose leaf prominent or face exhibits prominent wrinkles; upper incisors smaller than canines 2
2   Muzzle long and narrow 3
2′   Muzzle short and broad 5
3   Ear when pressed forward does not reach the nose tip. . . . . Glossophaginae
3′   Ear when pressed forward reaches the tip of the nose or beyond 4
4   Single papilla on chin or large central papilla on chin surrounded by smaller papillae . . . . Carolliinae
4′   Y- or V-shaped shield on chin . . . . Phyllostominae
5   White facial stripes and often median dorsal stripe present . . . . Stenoderminae
5′   No white facial stripes 6
6   Face naked, wrinkled, or naked with hornlike structure on nose 8
6′   Face not naked, no wrinkles 7
7   Interfemoral membrane absent . . . . Sturnirinae
7′   Interfemoral membrane present . . . . *Ametrida*
8   Face exhibits extreme wrinkling, fold of skin on throat. . . . . *Centurio*
8′   Face with folds, hornlike structure on nose; fold of skin on throat . . . . *Sphaeronycteris*

for feeding on blood, including *Diphylla* and *Desmodus*. Some bats are active carnivores, preying on lizards, birds, and small mammals; these include *Phyllostomus*, *Trachops*, *Chrotopterus*, and *Vampyrum*. *Artibeus*, *Carollia*, and *Sturnira* are specialized for frugivory, as are many stenodermine species (Gardner 1977a).

Because many phyllostomid bats are foliage gleaners, nectarivores, or frugivores, their food does not require precise localization by echolocation, in contrast to that of the aerial insectivores. The echolocating pulse of phyllostomids generally is of low amplitude, rather brief, and highly frequency modulated. Because of these acoustical characters, these bats are sometimes referred to as "whispering bats."

Most species of phyllostomids produce a single young. Some are highly seasonal in the timing of births, especially in areas subject to seasonal aridity. Birth often occurs at a time of maximum fruit or insect abundance (Wilson 1979).

An analysis of bat communities in Panama by Bonaccorso (1979) and Humphrey and Bonaccorso (1979) indicates that abundance varies considerably depending on the habitat. Some species are specialists in second-growth habitats, others are specialized for feeding along streams or creeks, and still others are specialists in mature forests. Regardless of the habitat preference, only three or four species make up the most abundant forms. Over 70% of all bats netted in a mature forest in Panama may belong to just four spe-

Table 8.7   Key to the Common Genera of the Phyllostominae

1	Single lower incisor on each side, 2/1	2
1'	At least two lower incisors	4
2	Head and body length greater than 80 mm	*Chrotopterus*
2'	Head and body length less than 80 mm	3
3	Nose leaf rather long, greater than 14 mm	*Mimon*
3'	Nose leaf rather short, less than 14 mm	*Tonatia*
4	Head and body length greater than 130 mmp; no tail; interfemoral membrane vastly reduced	*Vampyrum*
4'	Head and body length less than 130 mm	5
5	Nose leaf equals tragus in length (greater than 25 mm)	*Lonchorhina*
5'	Nose leaf does not equal tragus in length	6
6	Wartlike protrusions evident on both upper and lower lips; head and body length greater than 75 mm	*Trachops*
6'	Wartlike protrusions not evident on upper lip	7
7	Only two lower premolars present	*Phyllostomus*
7'	Three lower premolars present	8
8	Head and body length greater than 80 mm	*Phyllostomus*
8'	Head and body length less than 80 mm	9
9	Tail extends to end of uropatagium	*Macrophyllum*
9'	Tail extends to middle of uropatagium	10
10	One upper incisor on each side	*Barticonycteris*
10'	Two upper incisors on each side	*Micronycteris*

cies: *Artibeus jamaicensis, Artibeus lituratus, Glossophaga soricina,* and *Carollia perspicillata.*

Within any habitat type, species of bats belonging to the same trophic category seem to avoid direct competition by specializing for different vertical strata or different foods. It appears from the work of Gardner (1977a) and Bonaccorso (1979) that feeding categories are not as easily demarcated as previously thought. Wilson's trophic categorization into frugivores, foliage gleaners, and so on has been refined in Bonaccorso's publication and further refined in the paper by Humphrey, Bonaccorso, and Zinn (1983). Many of the so-called frugivores are highly opportunistic feeders, eating insects at one time of year and switching to fruit at another. Opportunism seems more characteristic of most medium-size phyllostomids (Humphrey, Bonaccorso, and Zinn 1983).

The biology of phyllostomid bats has been elegantly summarized in three volumes edited by R. J. Baker, J. Knox Jones Jr., and D. C. Carter (1976, 1977, 1979). A key to the subfamilies of the Phyllostomidae (as defined for this volume) is included in table 8.5. Chromosomal data included in the species accounts are from Baker et al. (1982).

SUBFAMILY PHYLLOSTOMINAE

Diagnosis

This diverse subfamily exhibits a wide range of sizes, including the largest Neotropical bat, *Vampyrum spectrum.* The rostrum is long, and when pressed forward the ears reach the tip of the nose or extend beyond. The interfemoral membrane shows little reduction in the insectivorous forms (*Micronycteris*), but

in other genera it may be greatly reduced. External features are illustrated in figures 8.3 and 8.4A. A key to the genera appears in table 8.7. The genera *Chrotopterus, Vampyrum,* and *Trachops* clearly are closely related (see chapter 1).

Distribution

The subfamily as so defined in this volume is widespread in suitable habitat from the western Gulf Coast and southwestern United States to northern Argentina.

Genus *Micronycteris* Gray, 1866
Little Leaf-nosed Bat

Description

The dental formula is usually I 2/2, C 1/1, P 2/3, M 3/3 (*Barticonycteris* has a single pair of upper incisors) (fig. 8.8). The tail extends only to approximately

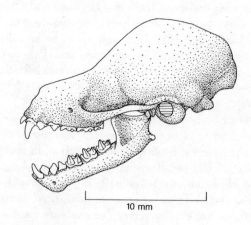

10 mm

Figure 8.8. Skull of *Micronycteris megalotis.*

the middle of the interfemoral membrane. These leaf-nosed bats are small, with the forearm less than 41 mm long. The head and body length ranges from 42 to 52 mm, and tail length is about 12 mm. In some species the ears are connected by a band of skin, which can aid in species recognition. The dorsum is usually some shade of brown, with buff below. Sanborn (1949) offered a benchmark review of the genus. Keys to the species are included in Medellin, Wilson, and Navarro (1985) and in Eisenberg (1989).

Medellin, Wilson, and Navarro (1985) subdivide the genus *Micronycteris* into five subgenera: *Micronycteris* includes *M. hirsuta, M. megalotis, M. minuta,* and *M. schmidtorum.* These species all have the ears connected by a notched band of skin. The subgenus *Lampronycteris* includes *M. brachyotis;* the subgenus *Neonycteris* includes *M. pusilla;* the subgenus *Trinycteris* includes *M. nicefori;* the subgenus *Barticonycteris* is reserved for *M. daviesi;* and *Glyphonycteris* includes *M. sylvestris* and *M. behni.* We follow their designations. Ten species of *Micronycteris* are known to occur within the range of this volume.

Distribution
The genus is distributed from central Mexico south across northern South America to Brazil.

Natural History
These small leaf-nosed bats are mainly insectivorous. Although they occasionally take fruit, they are not highly specialized for a frugivorous diet but are foliage gleaners (Gardner 1977a).

Comment
Simmons (1996), in the process of describing a new species, *Micronycteris sanborni,* has noted that *M. microtis,* considered a subspecies of *M. megalotis* by Sanborn and followed by Koopman (1993), may well be a full species, and thus *M. mexicana,* considered a subspecies of *M. megalotis,* may be either a subspecies of *M. microtis* or a distinct species of its own.

Micronycteris behnii (Peters, 1865)

Description
The ears are not connected by a high, notched band; the fourth metacarpal is the shortest and the fifth is the longest. In these characters it resembles *M. sylvestris* but has a longer forearm than the latter, exceeding 42.5 mm.

Habitat and Range
The species inhabits the forested regions of west-central Brazil to Amazonian Peru (map 8.24).

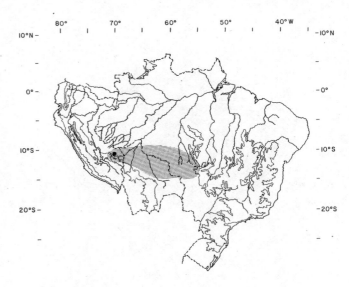

Map 8.24. Distribution of *Micronycteris behnii.*

Micronycteris brachyotis (Dobson, 1879)
Yellow-throated Bat

Description
This bat is of intermediate size: TL 57–75; T 7–14; HF 10–18; E 12–19; FA 39–43; Wt 9–15 g (Medellin, Wilson, and Navarro L. 1985). There is no band connecting the ears over the top of the head, a character it shares with *M. silvestris* and *M. nicefori.* The dorsum is light brown, with the venter slightly paler. Chromosome number: $2n = 32$; FN = 60.

Range and Habitat
This species ranges from Oaxaca, Mexico, through the Isthmus, across northern South America and south to Amazonian Brazil. In northern Venezuela most specimens were caught below 150 m. It is strongly associated with moist evergreen habitats (Handley 1976) (map 8.25).

Natural History
This bat appears to roost in the hollow trunks of trees. A colony may contain up to ten individuals. One male may occur with nine females, suggesting a polygynous mating system (Medellin, Wilson, and Navarro L. 1985). Although it takes some fruits, insects make up the bulk of its diet. Bonaccorso (1979) reports that it is most commonly caught 3–12 m off the ground in Panama.

Micronycteris (Barticonycteris) daviesi
(Hill, 1964)

Description
Included by Koopman (1993) within *Micronycteris,* this species is certainly closely allied, but the dental

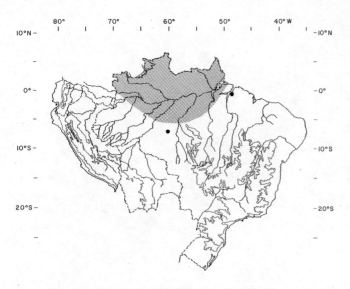

Map 8.25. Distribution of *Micronycteris brachyotis*.

formula is I 1/2, C 1/1, P 2/3, M 3/3. Head and body length ranges from 69 to 84 mm and the forarm from 54 to 58 mm. Measurements of one specimen from Suriname (CMNH) were TL 77; T 7; HF 16; E 31; Wt 8 g. The medium-sized ears are not connected by a band of skin. The dark brown dorsal pelage is long and lax; the venter is gray brown.
Chromosome number: $2n = 28$; FN = 52.

Range and Habitat

This lowland bat has a disjunct range. Specimens have been taken from Costa Rica, eastern Colombia, the Guianas, Ecuador, southeastern Peru, and northeast Brazil (map 8.26).

Micronycteris hirsuta (Peters, 1869)

Description

This is the largest species of *Micronycteris:* TL 78; HB 61–64; T 14–17; HF 11–13; E 25–27; FA 43–44; Wt 12–14 g. The ears are connected by a band over the top of the head that is low in profile and does not have a notch, which distinguishes *M. hirsuta* from *M. megalotis* and *M. minuta*. The dorsum is gray brown, scarcely contrasting with the venter.
Chromosome number: $2n = 28–30$; FN = 32.

Range and Habitat

This species ranges from Honduras south through the Isthmus to Amazonian Peru and Brazil. It prefers lower elevations and concentrates its activity near streams or moist areas (map 8.27).

Natural History

This bat will nest in hollow trees and under bridges, often with other species of bats. The timing of birth may correspond with abundance of the insects and fruits it feeds on.

Micronycteris megalotis (Gray, 1842)

Description

Head and body length averages 43.8 mm for males and 44.6 mm for females, and weight averages 5 g for males and 5.7 g for females. The ears of this pale brown bat are connected by a band that is high and notched in the center (plate 5). *M. microtis* is considered a junior synonym of *M. megalotis*.
Chromosome number: $2n = 40$; FN = 68.

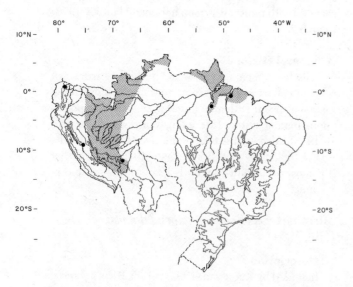

Map 8.26. Distribution of *Micronycteris daviesi*.

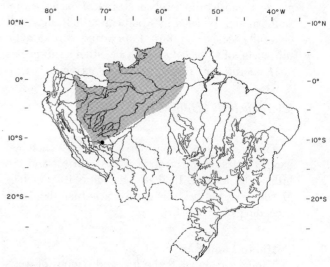

Map 8.27. Distribution of *Micronycteris hirsuta*.

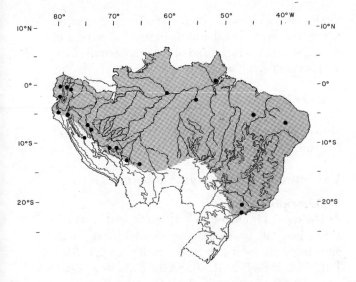

Map 8.28. Distribution of *Micyonycteris megalotis.*

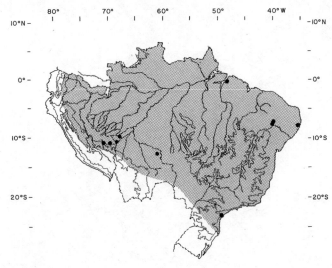

Map 8.29. Distribution of *Micronycteris minuta.*

Range and Habitat

This species ranges from southern Tamaulipas in Mexico south to Peru, east across Colombia and Venezuela to French Guiana and thence south to Brazil. This species does not occur above 800 m elevation in Venezuela. It is broadly tolerant of both multistratal evergreen forests and dry thorn forests (map 8.28).

Natural History

This bat forage near streams, and it roosts in hollow trees, logs, caverns, or even houses in groups of up to twelve. It is a mixed feeder, taking fruits when in season as well as insects.

Micronycteris minuta (Gervais, 1856)

Description

Head and body length for males averages 47.4 mm; females are larger, averaging 49 mm. Average weight for males is 6.9 g and for females 7.2 g (see table 8.17). A band connects the ears over the head but is not raised as in *M. megalotis,* and a deep notch divides the band into equal halves. Though this bat is pale brown, the light basal color of the dorsal hairs may yield the effect of a buff dorsal midline stripe.
Chromosome number: $2n = 28$; FN = 50.

Range and Habitat

This species ranges from Nicaragua south to Amazonian Brazil and Peru. It is broadly distributed over northern South America. In Venezuela, most specimens occur below 500 m. It will forage in man-made clearings and dry deciduous forests as well as in multistratal evergreen forest (map 8.29).

Natural History

Fleming, Hooper, and Wilson (1972) found 76% insects and 24% fruit by volume in the stomachs of specimens taken in Costa Rica and Panama. The bats roost in hollow trees and caves, often with other species (Goodwin and Greenhall 1961).

Micronycteris nicefori Sanborn, 1949

Description

This is the smallest of the species not exhibiting a connecting band between the ears. Males average 55 mm in head and body length, and females average 58 mm. Weight averages 7.9 g for males and 9.2 for females (see table 8.17). A reddish cast to the dorsal pelage is evident (plate 5).
Chromosome number: $2n = 28$; FN = 52.

Range and Habitat

This species occurs from Nicaragua to northern Colombia, then eastward across Venezuela to Amazonian Brazil and south from Colombia into Amazonian Peru. It is strongly associated with both multistratal tropical evergreen forests and dry uplands (Handley 1976) (map 8.30).

Natural History

Groups of about twelve individuals roost in hollow trees in company with other species (Goodwin and Greenhall 1961).

Micronycteris pusilla Sanborn, 1949

Description

Similar to *M. nicefori* in color and size. The ears are not connected by a notched band of skin. The fourth

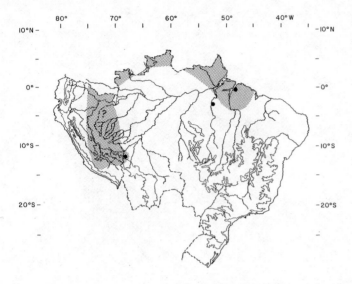

Map 8.30. Distribution of *Micronycteris nicefori*.

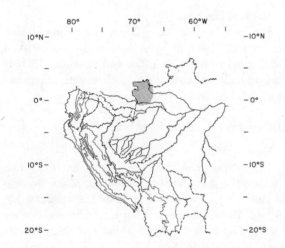

Map 8.31. Distribution of *Micronycteris pusilla*.

metacarpal is the shortest, and the third metacarpal is the longest. By these characters it resembles *M. nicefori*, but the ears are rounded in *M. pusilla* and the nose leaf is blunt, whereas in *M. nicefori* the nose leaf and ears are pointed.

Range and Habitat

Found from northwestern Brazil to adjacent Colombia (map 8.31).

Micronycteris sanborni Simmons, 1996

Description

This small species of *Micronycteris* measures TL 55.5–65; T 12–14; HF 8–9; E 19–21; FA 32–34; Wt 6.3 g ($n = 6$). The maxillary tooth row at 5.76 mm, which is shorter than *M. megalotis*, unambiguously distinguishes it. White ventral fur extends anteriorly to the throat, and a small patch of white fur occurs on the ventral surface of the uropatagium at the base of the tail. There is a high band of skin between the ears, deeply notched in the middle.n

Distribution

Currently known from the Chapada do Araripe plateau in northeastern Brazil; found in Ceará and Pernambuco states.

Micronycteris schmidtorum Sanborn, 1935

Description

The species is immediately characterized by its relatively long tail, which averages about 16 mm. Mean measurements for males and females are TL 64.8; T 16, 16.2; HF 11; E 21, 23; FA 34, 35.8. Weights average 7.6 g for males and 7.2 g for females. The dorsal pelage is a very pale brown, and the venter is pale gray to almost white.
Chromosome number: $2n = 38$; FN = 66.

Range and Habitat

This species occurs from southern Mexico to Peru and Brazil. It is strongly associated with moist habitats when foraging, although it occurs in both multistratal tropical evergreen forest and dry thorn forest. It frequently roosts in tree holes (map 8.32).

Micronycteris sylvestris (Thomas, 1896)

Description

Total length ranges from 53 to 73 mm; T 9–14; HF 10–12; E 20–22 (Suriname, CMNH). Weight averages 7.2 g for males and 10.3 for females (see table 8.17). The dorsum is dark brown with a frosted cast because the tips of the dorsal hairs are white. There is no

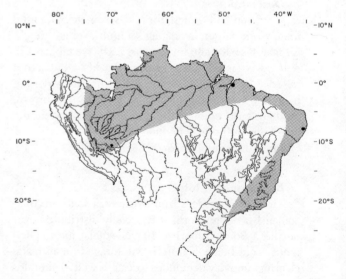

Map 8.32. Distribution of *Micronycteris schmidtorum*.

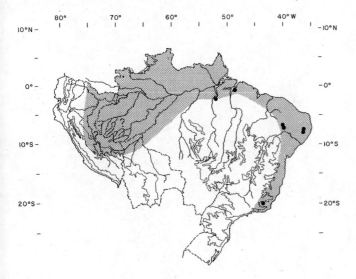

Map 8.33. Distribution of *Micronycteris sylvestris* (based in part on Koopman 1982).

band connecting the ears. The species may be distinguished from *M. nicefori* by its relatively longer tail. It has been placed in its own genus, *Glyphonycteris*, but here *Glyphonycteris* is considered at best a subgenus. Chromosome number: $2n = 22$; FN = 40.

Range and Habitat
This species occurs from Veracruz, Mexico, south through the Isthmus to Peru and southeastern Brazil. It prefers low elevations and tropical evergreen forest habitat (map 8.33).

Natural History
The bats roost in hollow trees in groups of up to seventy-five and tolerate other species (Goodwin and Greenhall 1961).

Genus *Lonchorhina* Tomes, 1863
Sword-nosed Bat, Murciélago de Espada

Description
The dental formula is I 3/3, C 1/1, P 3/3, M 2/3 (Hernandez-Camacho and Cadena-G. 1978). These medium-sized bats have a forearm length exceeding 40 mm. Head and body length ranges from 52 to 60 mm, and the tail nearly equals the head and body in length. The nose leaf is exceedingly long, almost equaling the length of the ears, which exceed 30 mm (fig. 8.4B). The large ear, with a prominent tragus, and the narrow, long nose leaf clearly distinguish this genus from any other New World bat. The fur is a light reddish brown, and the venter color is not set off from that of the dorsum, although it is paler in some specimens.

Distribution
The genus is broadly distributed from southern Veracruz, Mexico, to southeastern Brazil. These bats tend to roost in colonies, and *L. aurita* has been found in groups of over five hundred.

Lonchorhina aurita Tomes, 1863
Description
This is the second largest species within the genus. Males average 60.9 mm in head and body length, and females average 61.1 mm. Forearm length ranges from 48 to 55 mm (see table 8.17), and weight averages 14.3 g for males and 15 g for females. The bat has a medium brown dorsum with a gray brown venter. Chromosome number: $2n = 32$; FN = 60.

Range and Habitat
This species is broadly distributed from southern Veracruz, Mexico, to southeastern Brazil. It is strongly associated with moist habitats and is most frequently encountered in multistratal tropical forests. In Venezuela it was not taken above 1,000 m elevation (map 8.34) (Handley 1976).

Natural History
This bat is strongly insectivorous, highly specialized for the aerial pursuit of insects. Gardner (1977a) records some plant material in its diet, but plants are a minor component in its feeding. It roosts in caves in colonies of twenty to twenty-five (Goodwin and Greenhall 1961).

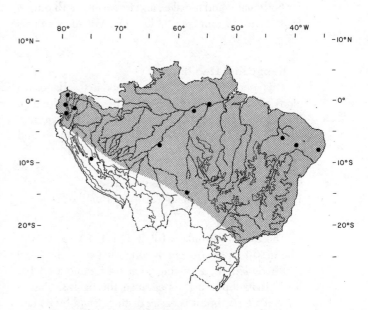

Map 8.34. Distribution of *Lonchorhina aurita*.

Lonchorhina fernandez Ochoa and Ibañez, 1982

Description

This is the smallest species of the genus: TL 90.23; HF 9; E 19 (*n* = 33) (Ochoa-G. and Ibañez 1982). The dorsum is brown and the venter is similar.

Range and Habitat

The species is known from the type locality, 40–50 km northeast of Puerto Ayacucho, Amazonas territory, Venezuela, but may extend farther south to Brazil (not mapped).

Lonchorhina marinkellei Hernandez-Camacho and Cadena-G. 1978

Description

This species is similar in appearance to *L. aurita* but smaller. Head and body length averages 52.3 mm for both males and females, and forearm length ranges from 41 to 44 mm (see table 8.17). Weight averages 8.7 g.

Range and Habitat

The species is known to occur in the extreme southeast of Colombia, and the description was based on a small number of specimens. However, reports of this species from French Guiana raise the possibility that it may occur in Brazil (not mapped).

Lonchorhina orinocensis Linares and Ojasti, 1971

Description

This species is similar in appearance to *L. aurita* but smaller. Head and body length averages 52.3 mm for both males and females, and forearm length ranges from 41 to 44 mm (see table 8.17). Weight averages 8.7 g.

Range and Habitat

This species appears to be confined in the Ilanos regions of western Venezuela and east-central Colombia. Strongly associated with savanna habitats, it apparently roosts in rock crevices during the day. It may extend into Brazil or Peru.

Genus *Macrophyllum* Gray, 1838
Long-legged Bat, Falso Vampiro Portilargo

Description

The dental formula is I 2/2, C 1/1, P 2/3, M 3/3. The hind feet are extremely long relative to the head and body length, averaging some 14.5 mm (see table 8.17). In its superficial appearance, this bat is similar to *Micronycteris,* but it is easily distinguished by its tail, which extends to the edge of the uropatagium. The dorsum is sooty brown, and the venter is paler brown.

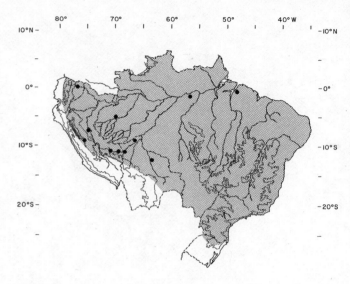

Map 8.35. Distribution of *Macrophyllum macrophyllum.*

Macrophyllum macrophyllum (Schinz, 1821)

Description

Head and body length averages 44.7 mm for males and 48.5 mm for females; weight averages 7 g for males and 7.2 g for females (see table 8.17). The dorsum is dark brown and the venter is paler (plate 5). Chromosome number: $2n = 32$; FN = 56

Range and Habitat

This species ranges from Tabasco, Mexico, through the Isthmus and over most of northern South America through Brazil to the south. It forages near streams or in other moist areas and, though tolerant of deciduous scrub forest, is most abundant in multistratal tropical evergreen forest (map 8.35).

Natural History

This species roosts singly or in very small groups and has been found in caves. In Central America there appears to be only a single breeding period during the late dry season. It is largely insectivorous, and Gardner (1977a) suggests it may specialize on aquatic insects.

Genus *Tonatia* Gray, 1827
Round-eared Bat

Description

The dental formula is I 2/1, C 1/1, P 2/3, M 3/3. Within the subfamily Phyllostominae only three genera have the single lower incisor: *Tonatia, Mimon,* and *Chrotopterus.* These large-eared bats have an average-sized nose leaf. The tail is short, but the uropatagium is retained. The dorsal pelage is dark to light brown, and the venter is slightly paler. Superficially, species of *Tonatia* resemble *Micronycteris;* they may be distinguished from *Mimon,* which has an extremely long nose leaf (see plate 5). *Chrotopterus* cannot be confused with either *Mimon* or *Tonatia* because of its extremely large size.

Distribution

The genus *Tonatia* is distributed from Guatemala south through the Isthmus to extreme southeastern Brazil.

Natural History

This genus appears to be frugivorous, although insects are a significant part of its diet seasonally. Species of *Tonatia* typically bear a single young, but in some parts of the range there are two birth peaks, suggesting that individual females may conceive twice in a given year.

Tonatia bidens (Spix, 1823)
Spix's Round-eared Bat

Description

This is the largest species of the genus *Tonatia*. Head and body length averages 76 mm for males and 75.4 mm for females. Average weight for males is 26.3 g and for females 24.7 g. The ears are rounded proximally, hence the common name round-eared bat. A light-colored median stripe occurs on the forehead in some populations. Upperparts vary from tawny to blackish brown, and underparts tend to be paler. Chromosome number: $2n = 16$; FN = 20.

Range and Habitat

This species is widely distributed from southern Guatemala to southeastern Brazil. It has been taken in both moist sites and dry deciduous sites. All specimens taken in Venezuela were below 200 m elevation. Specimens have been found roosting with other species in hollow trees (Handley 1976) (map 8.36). Williams,

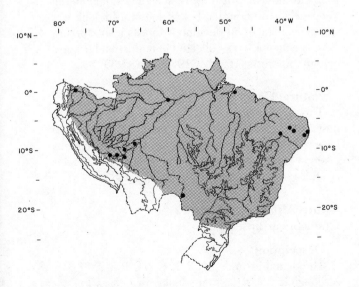

Map 8.36. Distribution of *Tonatia bidens* (includes *T. saurophila*).

Willig, and Reid (1995) recognize *Tonatia saurophila* as distinct from *T. bidens*. Anderson (1997) comments on the situation. Map 8.36 does not discriminate the separation.

Natural History

Two females from Peru were found to contain two embryos each (Gardner 1976). In Paraguay *T. bidens* has been collected over isolated ponds in thorn scrub and over a stream in high tropical forest. Specimens have been found roosting with other species in hollow trees. Insect remains and fruit pulp were found in their stomachs (Eisenberg 1989; Myers and Wetzel 1983; Myers, White, and Stallings 1983). Recently the species has been documented to feed on small birds. Although only slightly smaller than *Trachops cirrhosis*, these bats were not recorded as feeding on vertebrates until the observations by Martuscelli (1995). He reports that they kill small birds and carry them to the roosting site to eat them.

Tonatia brasiliense (Peters, 1866)

Description

This is the smallest species of the genus in South America. Head and body length averages 54.8 mm for males and 55.7 mm for females. Weight averages 10.2 g for males and 10.6 g for females (see table 8.17). The species is easily distinguished from the larger *T. bidens*. *T. brasiliense* includes *T. minuta*, *T. nicaraguae*, and *T. venezuelae* (Honacki, Kinman, and Koeppl 1982). Chromosome number: $2n = 30$; FN = 56.

Range and Habitat

Distributed from southern Veracruz, Mexico, south to Peru and eastward across northern Brazil (map 8.37). In northern Venezuela this species was common below 500 m elevation. It is strongly associated with streamside habitats and other moist areas but can range into deciduous forests. The preferred habitat appears to be multistratal evergreen forest, though it is broadly tolerant of man-made clearings (Handley 1976).

Natural History

In Trinidad this species has been seen roosting in abandoned termite nests (Goodwin and Greenhall 1961).

Tonatia carrikeri (J. A. Allen, 1910)
Allen's Round-eared Bat

Description

Measurements: TL 66–76; T 14–15; HF 12–16; E 22–25; FA 45–50 ($n = 7$; Colombia, CMNH). One male from Venezuela had a head and body length of

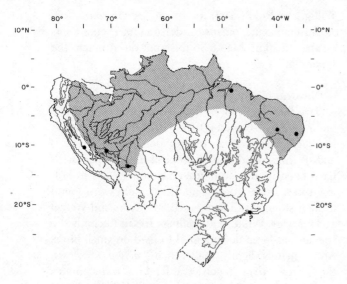

Map 8.37. Distribution of *Tonatia brasiliense*.

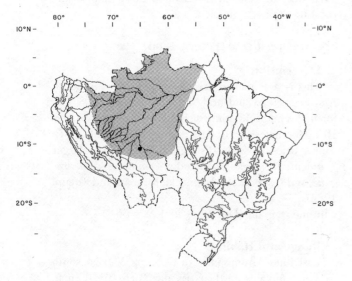

Map 8.38. Distribution of *Tonatia carrikeri*.

72 mm, a forearm of 44 mm, and a weight of 22 g. It approaches *T. schulzi* in size, but the color pattern is distinctive since the brown dorsum and sides contrast with the white venter.

Chromosome number: $2n = 26$; FN $= 46$.

Range and Habitat

Thus far this species has been described in southeastern Colombia, the extreme southern part of Amazonas territory in Venezuela, Suriname extending to north-central Brazil, and adjacent to Peru. The Venezuelan specimens were taken at elevations below 155 m in multistratal tropical evergreen forests near streams (map 8.38).

Natural History

This species is associated with forested habitats. It may invade llanos habitats via gallery forests. It uses hollow termite nests as roosting sites. Roosting groups range from five to twelve. It apparently feeds mainly on arthropods (McCarthy, Gardner, and Handley 1992).

Tonatia schulzi Genoways and Williams, 1980

Description

This newly discovered species is intermediate in size: TL 68–78; T 11–13; HF 13–14; E 27–29 ($n = 2$; Suriname, CMNH).

Chromosome number: $2n = 28$: FN $= 36$ (Genoways and Williams 1980).

Range and Habitat

The species is known only from a restricted locality in Suriname, 38°48′ N, 56°08′ W (Genoways and Williams 1980), but may extend to adjacent Brazil (not mapped).

Tonatia silvicola (d'Orbigny, 1836)

Description

Head and body length averages 73.3 mm for males and 71.5 for females, with weights averaging 30.6 g for males and 24.1 g for females. This bat is approximately the same size as *T. carrikeri* and *T. bidens*. It lacks the median white stripe on the head characteristic of *T. bidens,* and it is distinguishable from *T. carrikeri* by its much larger ear (plate 5). In *T. carrikeri*, the ear of one specimen was only 30 mm long, but in *T. silvicola* ear lengths average 36.6 mm for males and 37.2 mm for f males (see table 8.17).

Chromosome number: $2n = 34$; FN $= 60$.

Range and Habitat

The species ranges from Honduras south to Bolivia and most of Brazil. Most specimens obtained in Venezuela were found at below 460 m. Although it is occasionally taken in deciduous forests near streams, most specimens were collected in multistratal tropical evergreen forests (Handley 1976) (map 8.39).

Natural History

This species occasionally roosts in hollow termite nests. Fleming, Hooper, and Wilson (1972) recorded insects from the stomachs of twenty-two specimens taken in Panama. In the Brazilian caatinga this species is a foliage-gleaning insectivore (Willig 1983).

Genus *Mimon* Gray, 1847
Gray's Spear-nosed Bat

Description

The dental formula is I 2/1, C 1/1, P 2/2, M 3/3. Although it shares the same dental formula as *Tonatia* and *Chrotopterus*, *Mimon* is easily distinguishable be-

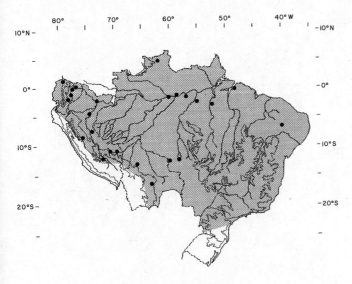

Map 8.39. Distribution of *Tonatia silvicola*.

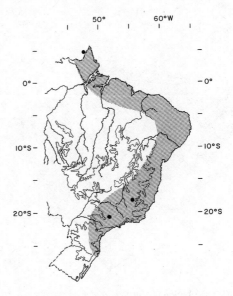

Map 8.40. Distribution of *Mimon bennettii*.

cause of its much larger nose leaf. The fur is long and woolly. Color markings vary according to the species but are always some shade of brown.

Distribution
The genus *Mimon* is distributed from southern Veracruz, Mexico, to southern Brazil.

Natural History
Gardner (1977a) reports that the species of *Mimon* feed on fruits and insects.

Mimon bennettii (Gray, 1838)

Description
Total length ranges from 85 to 95 mm: T 20–25; HF 15–17; E 36–38 ($n = 4$; Suriname, CMNH). This bat typically has no striping on the back, which distinguishes it from *M. crenulatum*. *M. bennettii* has a pale brownish dorsum, and there are small whitish patches behind the ears. For this discussion *M. cozumelae* is considered a junior synonym of *M. bennettii*. Chromosome number: $2n = 30$: FN = 56.

Range and Habitat
Mimon bennettii occurs from southern Veracruz, Mexico, to northern Colombia, then in the north coastal region of Venezuela to the Guianas, and follows the coast of Brazil to southeastern Brazil (map 8.40).

Natural History
This species is tolerant of a variety of habitats but is associated with multistratal tropical forest although capable of invading gallery forests in dry zones. Insects and small vertebrates are included in its prey as well as fruit. Roosting sites include caves and hollow logs. It shares roosting sites with up to ten other species, but usually only one or two at any given site (Ortega and Arita 1997).

Mimon crenulatum (E. Geoffroy, 1810)

Description
Head and body length of males averages 57.9 mm, while females are slightly larger at 58.5 mm. Weight averages 12.8 g for males and 12 g for females (see table 8.17). In this species there is generally a white line down the center of the back. The basic color of the dorsum is a bright mahogany brown, sometimes grading to blackish brown (plate 5). The underparts are rusty to gray.
Chromosome number: $2n = 32$; FN = 60.

Range and Habitat
This species occurs from Campeche in Mexico south over most of northern South America, including northern Peru and Brazil (map 8.41). It was not taken above 600 m in elevation in northern Venezuela. Although it ranges into dry deciduous forests, it prefers multistratal tropical evergreen forests. It frequently forages in natural openings or man-made fields, and it roosts in hollow tree trunks and buildings (Handley 1976).

Genus *Phyllostomus* Lacépède, 1799
Spear-nosed Bat

Description
The dental formula is I 2/2, C 1/1, P 2/2, M 3/3 (see fig. 8.9). These are medium to large bats with forearms exceeding 50 mm in length. The tail is short, and the interfemoral membrane is reduced in size. There is a glandular throat sac that is well developed in males but vestigial in females. The lower lip bears a V-shaped

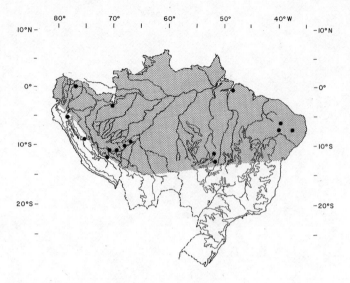

Map 8.41. Distribution of *Mimon crenulatum.*

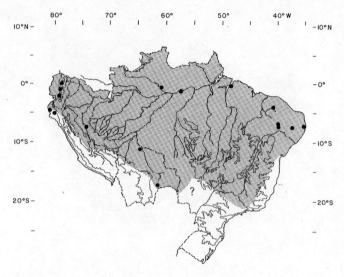

Map 8.42. Distribution of *Phyllostomus discolor.*

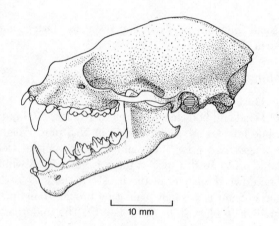

Figure 8.9. Skull of *Phyllostomus hastatus.*

groove edged with small warts. The dorsal color ranges from light to dark brown.

Distribution
Species of this genus range from southern Veracruz, Mexico, to southern Brazil.

Natural History
The species *P. hastatus* is one of the largest New World bats. Although all species of this genus eat fruit, the larger species are also carnivorous. *P. hastatus* preys on small rodents and other bats.

Phyllostomus discolor Wagner, 1843

Description
This is one of the smallest species of the genus; forearm length averages about 61 mm. In males, head and body length averages 83.6 mm, and in females it is 84.2 mm. Weight averages 38.2 g for males and 35.4 g for females (table 8.17). The dorsal pelage is brown,

but the white tips of the hairs convey a mottled effect. The venter is grayish to yellowish white.
Chromosome number: $2n = 32$; FN $= 60$.

Range and Habitat
P. discolor occurs from southern Mexico south across northern South America to northern Bolivia and southeastern Brazil. In Venezuela most specimens were taken below 500 m elevation (map 8.42). The species is broadly tolerant of both dry and wet habitats, occurring in deciduous tropical forest and multistratal tropical evergreen forest.

Natural History
This species appears to be mainly frugivorous but eats insects and pollen seasonally. It does not exhibit predatory behavior like *P. hastatus* (Gardner 1977a). The bats roost in hollow tree trunks in groups of twenty-five; in breeding roosts the sex ratio may be one male to twelve females, suggesting harem formation (Willig 1983).

Phyllostomus elongatus (E. Geoffroy, 1810)

Description
This bat is somewhat larger than *P. discolor;* forearms average 66 mm for males and females. Males average 80.4 mm in head and body length, while females average 82.5 mm. Weight for males and females averages 42 g and 40 g, respectively (table 8.17). The dorsal and ventral pelage is a uniform dark brown (plate 5).
Chromosome number: $2n = 32$; FN $= 58$.

Range and Habitat
P. elongatus is confined to South America to the east of the Andes, ranging across northern South America

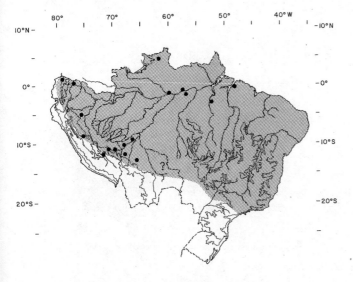

Map 8.43. Distribution of *Phyllostomus elongatus.*

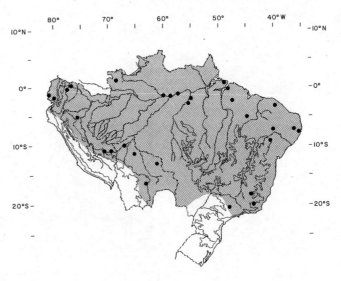

Map 8.44. Distribution of *Phyllostomus hastatus.*

to southeastern Brazil (map 8.43). In Venezuela it is occasionally taken in drier habitats near streams, but it strongly prefers multistratal tropical evergreen forests. In Venezuela most specimens occur below 350 m elevation (Handley 1976).

Natural History

This species roosts in tree cavities. Apparently it is strongly disposed to feed on fruits, but it may also take nectar and pollen (Gardner 1977a).

Phyllostomus hastatus (Pallas, 1767)
Vampiro de Lanza

Description

This is one of the largest New World bats (see fig. 8.9). The venter and dorsum are colored similarly, ranging from dark brown to reddish brown. Chromosome number: $2n = 32$; FN = 58.

Range and Habitat

This species ranges from Honduras south through the Isthmus to Bolivia and southeastern Brazil (map 8.44). In Venezuela it usually occurs below 500 m elevation, but some specimens have taken at as high as 1,394 m. It tolerates a variety of habitat types including deciduous forests, man-made clearings, and multistratal tropical evergreen forests (Handley 1976).

Natural History

This species roosts opportunistically in caves and buildings and under palm leaves, forming both small groups and colonies exceeding five hundred individuals. In addition to being frugivorous this species preys actively on lizards, rodents, bats, and other small vertebrates. Even within a colony, males will defend

groups of females and form temporary harems of thirty females per male. A single young is born (Mc-Cracken and Bradbury 1977, 1981; Tuttle 1970; Willig 1983).

Phyllostomus latifolius (Thomas, 1901)

Description

This species is smaller than *P. elongatus:* TL 91–95; T 13–17; HF 14–17; E 27–29; FA 56–59; Wt 25–30 g ($n = 4$; CMNH). The dorsum and venter are brown. Chromosome number: $2n = 32$; FN = 58.

Range and Habitat

The species is confined to northwestern Brazil and adjacent portions of Guyana and Colombia (map 8.45). Little is known concerning this bat.

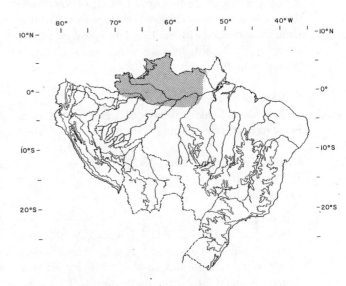

Map 8.45. Distribution of *Phyllostomus latifolius.*

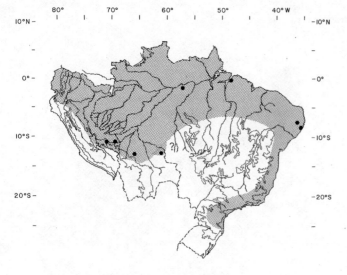

Map 8.46. Distribution of *Phylloderma stenops*.

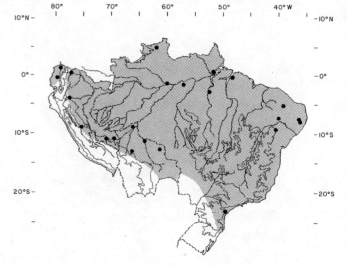

Map 8.47. Distribution of *Trachops cirrhosus*.

Genus *Phylloderma* Peters, 1865
Phylloderma stenops Peters, 1865

Description

There is a single species in South America, *P. stenops*. The dental formula is I 2/2, C 1/1, P 2/3, M 3/3. *Phylloderma* has one more lower premolar than does *Phyllostomus*. Head and body length ranges from 85 to 120 mm, and forearm length from 67 to 80 mm (see table 8.17). Males possess a glandular throat sac. The dorsum is brown, but the pale bases of the dorsal hairs are noticeable on the shoulders. The venter is gray, and the tips of the wing membranes may show depigmentation.

Range and Habitat

This species occurs from Honduras south to Brazil, but it appears to be absent from dry deciduous forest over much of southeastern Brazil. In Venezuela this species was taken below 206 m (map 8.46). It is strongly associated with multistratal tropical evergreen forests but is broadly tolerant of man-made clearings (Handley 1976).

Natural History

This mixed feeder will take fruit as well as insects. Females bear a single young.

Genus *Trachops* Gray, 1847
Trachops cirrhosus (Spix, 1823)
Fringe-lip Bat

Description

The dentition is I 2/2, C 1/1, P 2/3, M 3/3. The genus is typified by robust size. Head and body length averages 77.5 mm for males and 78.2 mm for females. Weight averages 34.2 g for males and 32.9 g for females (table 8.17). The lips, both upper and lower, are characteristic because of the great number of wartlike protrusions (plate 5). The upper parts vary from dark brown to cinnamon; the underparts are a dull brown, contrasting slightly with the dorsum.

Range and Habitat

This bat is distributed from southern Mexico south through the Isthmus and ranges broadly over the tropical portions of South America (map 8.47). In Venezuela it is found below 500 m elevation. It is strongly associated with tropical evergreen forest but occurs in regions of dry deciduous forest near moist habitats.

Natural History

This species tends to roost in caves and hollow trees. The colonies are small (fewer than six individuals). There is some evidence that the young associate with a parent for a considerable time. Although they eat insects, these bats are active predators and also feed on lizards, other bats, and frogs. In Panama, Tuttle, Taft, and Ryan (1982) report that populations of some frog species have been under considerable selection to produce calls that render them less conspicuous to the ears of these predators (Tuttle and Ryan 1981).

Genus *Chrotopterus* Peters, 1865
Chrotopterus auritus (Peters, 1856)
Peters's Woolly False Vampire Bat

Description

The dental formula is I 2/1, C 1/1, P 2/3, M 3/3. This bat's large size and single lower incisor easily distinguish it from other phyllostomines. Forearm length ranges from 76 to 80 mm. A glandular throat pouch is conspicuous in the male, reminiscent of some species of *Phyllostomus*. The hair of the dorsum is long and

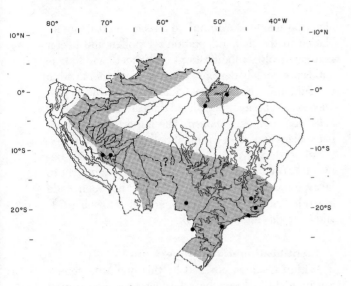

Map 8.48. Distribution of *Chrotopterus auritus*.

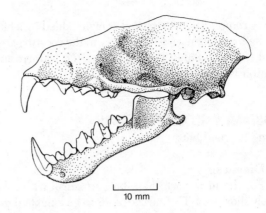

Figure 8.10. Skull of *Vampyrum spectrum*.

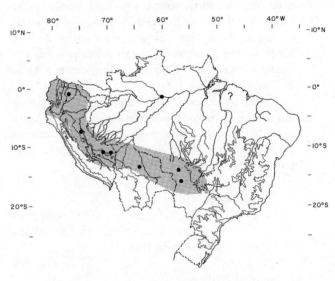

Map 8.49. Distribution of *Vampyrum spectrum*.

soft, from light to dark brown with a gray aspect. The venter tends to be more grayish.

Range and Habitat

C. auritus ranges from southern Mexico south through the Isthmus to southeastern Brazil (map 8.48) but appears to be absent over much of the Amazon region. In Venezuela, specimens were taken below 500 m. The species is strongly associated with multistratal tropical evergreen forests.

Natural History

This bat feeds on insects and fruit but also preys actively on other vertebrates (Sazima 1978). Medellin (1988) reports heavy predation on small rodents in Mexico. A single young is born after a gestation exceeding one hundred days. The bats roost in caves and hollow trees (Taddei 1976). Colonies vary in size from one to seven. Where *Chrotopterus* co-occurs with *Vampyrum*, *Chrotopterus* may take smaller vertebrate prey (Medellin 1989).

Genus *Vampyrum* Rafinesque, 1815
Vampyrum spectrum (Linnaeus, 1758)
False Vampire Bat

Description

The dental formula is I 2/2, C 1/1, P 2/3, M 3/3, separating this large bat from *Chrotopterus* (see fig. 8.10). This is the largest bat in the New World tropics, easily distinguished from any other species. Measurements for males and females, respectively, are TL 140, 158; HF 32, 38; E 46, 48; FA 105, 107; Wt 169, 199 g (USNM). It has no tail, and the interfemoral membrane is vastly reduced. The dorsum is reddish brown and the venter somewhat paler (plate 5).
Chromosome number: $2n = 30$; FN = 56.

Range and Habitat

This species is distributed from southern Mexico to Peru, Bolivia, and southwestern Brazil. It appears to be absent over much of the Amazon region of Brazil (map 8.49). It occurs in both dry deciduous forest and tropical evergreen forest but is strongly associated with moist habitats. It occurs up to 1,032 elevation in Venezuela.

Natural History

Our knowledge of the natural history of this interesting bat species has been greatly augmented by Vehrencamp, Stiles, and Bradbury (1977). It roosts in very small groups in hollow tree trunks and is apparently monogamous. Males have been observed to capture prey and bring it back to the roosting area, apparently to feed the female and dependent young. These bats are known to feed on fruit and small vertebrates, and there is reason to believe that individuals may specialize on certain types of vertebrates. In the Vehrencamp study one male actively preyed on the groove-billed ani (*Crotophagus*); it attacked at night, killed by a bite to the brain, then carried the prey to the roosting area (Vehrencamp, Stiles, and Bradbury 1977; Navarro and Wilson 1982).

SUBFAMILY GLOSSOPHAGINAE
Long-tongued Bats

Diagnosis
For the most part, these bats are small with relatively short ears. The muzzle tends to be long, and the tongue is extremely long when extruded and has papillae on the tip. The tongue specialization and long rostrum correlate with nectar and pollen feeding. In conjunction with the tongue specialization, and to permit extrusion of the long tongue, in several genera the lower incisors are absent. Dental evolution in this subfamily is discussed by Phillips (1971). (See fig. 8.4A for anatomical details.)

For the purposes of this volume the Glossophaginae are considered in the broad, classical sense, but Koopman (1993) separates the genera *Lionycteris*, *Lonchophylla*, and *Platalina* into their own subfamily Lonchophyllinae according to Griffiths (1982).

Distribution
Members of this subfamily are distributed from the southwestern United States through the Isthmus of Panama to southeastern Brazil.

Natural History
This subfamily of bats exhibits adaptations for feeding on pollen and nectar. Indeed, some species of the genera *Leptonycteris* and *Glossophaga* have been shown to be important pollinators of certain plant species. On the other hand, pollen and nectar are not always available year round, so insects and fruit are included in the diet. Adaptation for pollen and nectar feeding involves modifications of the teeth and tongue and is not equally developed among the genera within the subfamily. For example, *Glossophaga* and *Leptonycteris* are far less specialized than are *Choeroniscus*, *Scleronycteris*, and *Anoura*. The importance of these bat species and others as plant pollinators and dispersers of seeds as well as the degree of coevolution between certain plant species and bat species has yet to be explored, although an excellent start has been made in studies of the genera *Leptonycteris* (Howell 1974) and *Carollia* (Fleming 1988).

Identification of the Genera
Within the area covered by this book the genera can be identified only by reference to a combination of external and tooth characters (see table 8.8). Some genera are readily distinguished in the hand. As the group has specialized for nectar and pollen feeding, there has been a tendency to lose the lower incisors and reduce the molars to 2/2. Tooth number can vary from 34 in *Glossophaga* and *Lonchophylla* to a low of 26 in *Lichonycteris*. By inspecting the lower incisors one can immediately decide between two groups, those with two lower incisors per side (*Glossophaga*, *Lonchophylla*, *Lionycteris*, and *Leptonycteris*) and the other genera that have no lower incisors. Among the four genera with two lower incisors per side, *Leptonycteris* distinctively exhibits a rudimentary tail and a vastly reduced interfemoral membrane. The genus *Lionycteris* has a low number of teeth (30), since it has reduced its molars to 2/2 per side.

The genera with no lower incisors are more difficult to separate, but *Anoura* is distinctive because of its loss of the interfemoral membrane and its rudimentary tail (see table 8.17).

Table 8.8   Field Key to the Genera of Common "Glossophagine" Bats of South America

1	Lower incisors present	2
1'	Lower incisors absent	5
2	Size large (head and body length greater than 75 mm); interfemoral membrane vastly reduced; tail rudimentary	*Leptonycteris*
2'	Not as above	3
3	Only two upper and lower molars per side	*Lionycteris*
3'	Three upper and lower molars per side	4
4	Upper incisors about the same length	*Glossophaga*
4'	Upper incisors unequal in length	*Lonchophylla*
5	Interfemoral membrane absent or vastly reduced; tail absent or rudimentary	*Anoura*
5'	Interfemoral membrane present; tail present but may be quite short	6
6	Only two upper and lower molars per side	*Lichonycteris*
6'	Three upper and lower molars per side	7
7	Interfemoral membrane somewhat abbreviated; first and second incisors separated by a distinct gap; pterygoids expanded at base, pterygoid wings in contact with the bullae	*Choeroniscus*
7'	Interfemoral membrane moderate to well developed; pterygoids normal	8
8	Upper molars lacking mesostyle	*Hylonycteris*
8'	Upper molars with mesostyle	*Scleronycteris*

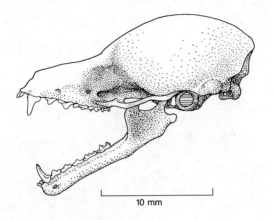

Figure 8.11. Skull of *Glossophaga soricina*.

Genus *Glossophaga* E. Geoffroy, 1818
Long-tongued Bat, Murciélago Nectario

Description
The dental formula is I 2/2, C 1/1, P 2/3, M 3/3 (see fig. 8.11). These are small bats, with head and body length ranging from 48 to 65 mm and forearm length from 32 to 42 mm. The interfemoral membrane is reduced but clearly visible, and the tail is less than half the length of the interfemoral membrane. Color varies from dark reddish brown to gray brown. The dorsal hairs are bicolored, brownish at tips and gray at the base.

Distribution
The genus is distributed from Mexico across South America to southeastern Brazil and extreme northeastern Argentina.

Natural History
These bats feed on insects, fruits, pollen, nectar, and flower parts.

Glossophaga commissarisi Gardner, 1962

Description
The lower incisors are very small. Head and body length ranges from 43 to 60 mm; HF 9–13; E 11–16; FA 33–36.

Range and Habitat
The species is found in western Mexico from Sinaloa south through Panama to western Colombia and then occurs in eastern Ecuador, northeastern Peru, and adjacent Brazil) (map 8.50).

Natural History
Stomach contents indicate the species feeds on moths and fruit. Foraging is most frequent before midnight (Webster and Knox Jones 1993).

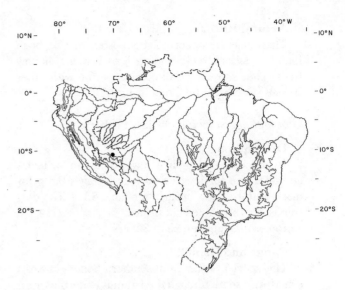

Map 8.50. Distribution of *Glossophaga commissarisi*.

Glossophaga longirostris Miller, 1898

Description
This is the largest of the three species that occur in South America. Head and body length averages 61.6 mm for males and 61.4 mm for females; weight averages 12.8 g for males and 12.9 g for females (see table 8.17). The dorsum is light brown.
Chromosome number: $2n = 32$; FN = 60.

Range and Habitat
This species is distributed in northeastern Colombia south to Ecuador and across northern Venezuela to Guyana and north-central Brazil. In Venezuela most specimens were taken below 500 m elevation (map 8.51). This bat is adapted to tropical dry deciduous forest habitats; it is the common nectar-feeding bat in the llanos.

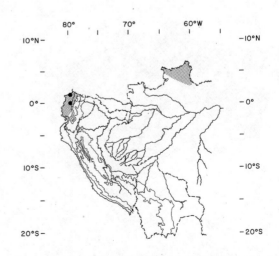

Map 8.51. Distribution of *Glossophaga longirostris*.

Natural History

These bats are opportunistic roosters, having been found in caverns, rocks, and crevices as well as hollow trees. They roost in small colonies, often with other species (Goodwin and Greenhall 1961).

Glossophaga soricina (Pallas, 1766)

Description

Head and body length averages 53.9 mm for males and 54.9 mm for females; weight averages 9.5 g for males and 9.4 g for females (see table 8.17). The dorsum is dark brown to light brown. Chromosome number: $2n = 32$; FN = 60.

Range and Habitat

This species occurs from southern Sonora in western Mexico south through the Isthmus to northeastern Argentina and southeastern Brazil (map 8.52). It prefers moist situations and is rarely found in dry areas. It tolerates man-made clearings and croplands, but in undisturbed areas it prefers multistratal tropical evergreen forests. In northern Venezuela most specimens were caught below 500 m elevation, but it can range to 1,560 m (Handley 1976).

Natural History

This species forms maternity colonies in shelters such as caves and hollow trees. Several hundred females and their young can roost together (Willig 1983), but Goodwin and Greenhall (1961) report colonies of twelve to sixteen in Trinidad. Normally a single young is born. The time of reproduction can be strongly seasonal in habitats with pronounced rainfall cycles, but females are polyestrous and can bear two or three young per year. Though fully capable of hovering in flight while taking nectar from flowers, this bat is also

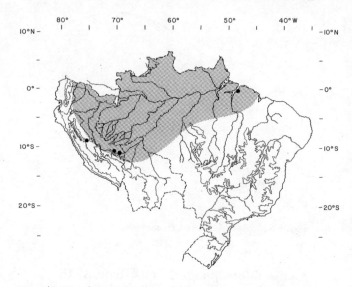

Map 8.53. Distribution of *Lionycteris spurrelli*.

to some extent a foliage-gleaning insectivore (Howell 1983; Alvarez et al. 1991). Activity periods are bimodal, just before dawn and just after dusk. Sazima, Fabian, and Sazima (1982) note that *G. soricina* actively pollinates *Luchea* sp. in southern Brazil. During the period of nectar production by *Agave*, individual bats will defend the plants against conspecifics to ensure an exclusive supply of nectar and pollen (Lemke 1984).

Genus *Lionycteris* Thomas, 1913
Lionycteris spurrelli Thomas, 1913

Description

The dental formula is I 2/2, C 1/1, P 2/3, M 2/2. Head and body length averages 53.5 mm for males and 54.8 mm for females; weight averages 8.7 g for males and 8.9 g for females (see table 8.17).

Range and Habitat

The species' range extends from Central America to northern South America. It occurs from 135 to 1,400 m elevation in Venezuela. It prefers moist areas for foraging and is strongly associated with multistratal tropical evergreen forest (Handley 1976) (map 8.53).

Natural History

These bats roost in caves and crevices. The diet has not been recorded but is probably similar to that of *Lonchophylla*.

Genus *Lonchophylla* Thomas, 1903

Description

The dental formula is I 2/2, C 1/1, P 2/3, M 3/3. Head and body length ranges from 45 to 60 mm and the short tail from 8 to 10 mm; weight ranges from 6 to 14 g. Species of this genus are graded in size, with

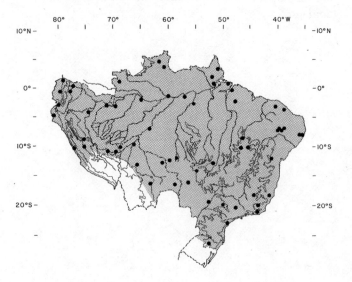

Map 8.52. Distribution of *Glossophaga soricina*.

L. thomasi the smallest, *L. mordax* intermediate, and *L. robusta* the largest. The uropatagium is well developed. The dorsum is rusty to dark brown, the venter somewhat paler (plate 6).

Distribution

The genus is distributed from Nicaragua to Brazil. Some species show an extremely disjunct range.

Natural History

These bats frequently roost in caves. They are specialized for feeding on flowers and are strongly implicated in the pollination of night-blooming plants such as *Bauhinia rata*. In common with other members of the subfamily, they are not confined to pollen and nectar but feed on insects and fruit as well (Sazima, Vizotto, and Taddei 1978; Gardner 1977a).

Lonchophylla bokermanni Sazima, Vizotto, and Taddei, 1978

Description

This species is characterized by its large size, with a forearm measurement of 38.7–41.3 mm. The maxillary tooth row exceeds 7.8 mm. The head and body length is 60.5 to 63.5 mm.

Distribution

The type locality is Minas Gerais, Brazil, at Jaboficatubas, Serra do Cipo. The species is apparently confined to southeastern Brazil (map 8.54).

Lonchophylla dekeyseri Taddei, Vizotti, and Sazima, 1983

Description

This medium-sized species has a forearm measuring 34.7 to 37.7 mm and a relatively short skull at 22–

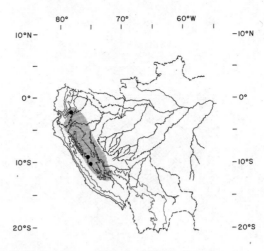

Map 8.55. Distribution of *Lonchophylla handleyi*.

22.6 mm. The maxillary tooth row is short, from 7.5 to 7.6 mm. The short tooth row in combination with the short skull distinguishes this species. A key to *Lonchophylla* is offered with the description of this species (Taddei, Vizotto, and Sazima 1983).

Distribution

Type locality is given as eight kilometers north of Brasília, Federal District, Brazil. It appears to be confined to southeastern Brazil (not mapped).

Lonchophylla handleyi Hill, 1980

Description

The species is similar to *L. robusta*, but the second upper premolar exhibits a reduced posterior and anterior basal cusps (styles) as slightly longer covdylar basal length (26.9–29.2 mm) compared with *L. robusta*. The forearm ranges from 44.9 to 47.9 mm compared with lengths for *L. robusta* of 39.7–44.5 mm.

Distribution

The type locality is 3°7′ S and 18°12′ W. The range encompasses southern Colombia, Ecuador, and adjacent Peru (map 8.55).

Lonchophylla hesperia G. M. Allen, 1908

Description

The species is Intermediate in size, with forearm length 36–40.6 mm, candylar-basal length 24.5–26.1 mm, and a rather elongated rostrum. The dorsum is pale brown, and the venter is pale brown to gray.

Distribution

The type locality is listed as Tumbes, Zorritos, Peru. Total range may be northwestern Peru and adjacent Ecuador (map 8.56).

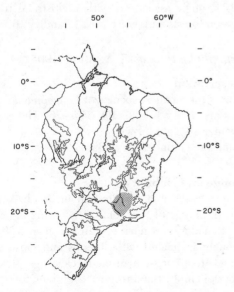

Map 8.54. Distribution of *Lonchophylla bokermanni*.

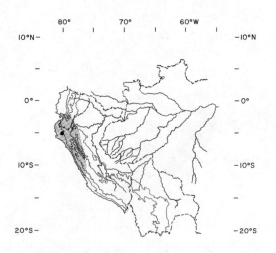

Map 8.56. Distribution of *Lonchophylla hesperia*.

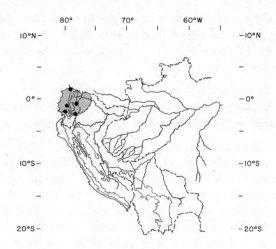

Map 8.58. Distribution of *Lonchophylla robusta*.

Lonchophylla mordax Thomas, 1903

Description

Head and body length ranges from 55 to 58 mm, placing this species between *L. robusta* and *L. thomasi*. The rostrum is considerably longer than in *L. thomasi*. It rarely exceeds 9 g in weight. A sample from the caatinga of Brazil yielded the following measurements: males, TL 64.8; T 10.0; HF 9.0; E 14.6 FA 34.7; females, TL 66.1; T 10.2; HF 9.2; E 14.8; FA 35.1 (Willig 1983). The dorsum is reddish brown, the venter paler.

Range and Habitat

This species, according to Koopman (1982), has a disjunct distribution. It is recorded from Costa Rica through the Isthmus of Panama and in western Colombia (map 8.57). Specimens referable to *L. mordax* are also described from east-central Brazil. No specimens have been recorded in areas between these two populations.

Lonchophylla robusta Miller, 1912

Description

This is the largest species of the genus. Head and body length averages 69.4 mm for males and 69.3 mm for females; weight averages 14.3 g for males and 13.7 g for females (see table 8.17). The dorsum and venter are light brown.
Chromosome number: $2n = 28$; FN = 50.

Range and Habitat

This species ranges from southern Nicaragua through the Isthmus to western Colombia and northwestern Venezuela and thence to Ecuador (map 8.58). It tolerates elevations of 75 to 1,135 m in Venezuela and is strongly associated with multistratal tropical evergreen forests and moist areas (Handley 1976).

Lonchophylla thomasi J. A. Allen, 1904

Description

This is the smallest species of *Lonchophylla*. Measurements: HB 50–54.5; FA 35.5–36. The dorsum and venter are light brown.
Chromosome number: $2n = 30–32$; FN = 34–38.

Range and Habitat

This is a disjunct distribution in South America within the range of this volume; it occurs in eastern Peru, Ecuador, and northern Brazil (map 8.59) and is strongly associated with streams and moist areas. Although tolerant of man-made clearings, this species prefers multistratal tropical evergreen forests. All specimens in Venezuela were caught below 850 m elevation (Handley 1976).

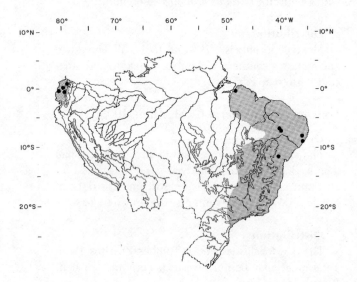

Map 8.57. Distribution of *Lonchophylla mordax*.

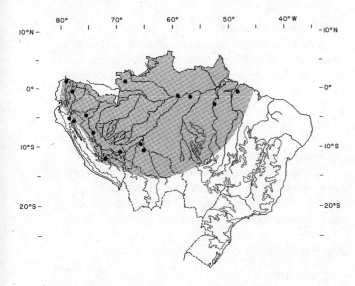

Map 8.59. Distribution of *Lonchophylla thomasi*.

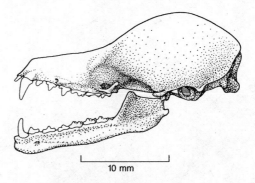

Figure 8.12. Skull of *Anoura geoffroyi*.

Genus *Platalina* Thomas, 1928
Platalina genovensium Thomas, 1928

Description
The type measured HB 72; T 9; FA 46. Additional specimens extend the forearm measurements to 48 mm and yield a weight of 47 g. The basic color is pale brown with the venter not demarcated. The inner upper incisor teeth are broad and spatulate. The muzzle is long, and the tongue is not only elongate but supplied with papillae (Jimenez and Pefauer 1982).

Distribution
Known only from five localities in western, coastal Peru (map 8.60).

Natural History
Apparently this bat is a pollen and nectar feeder. It has been noted to shelter in caves. It feeds on cacti

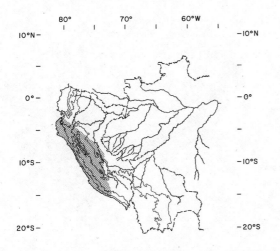

Map 8.60. Distribution of *Platalina genovensium*.

when they are in bloom. Pregnant females were captured in September (Graham 1987; Jimenez and Pefauer 1982). Sahley and Baraybar (1996) report that this bat can be adversely affected during El Niño events. Decreased fruit and flower production by the colmnar cactus (*Weberbauerocereus*) during prolonged droughts resulted in emigration or death within colonies of this species at Arequipa, Peru.

Genus *Anoura* Gray, 1838
Tailless Bat, False Vampire, Hocicuda

Description
The dental formula is I 2/0, C 1/1, P 3/3, M 3/3 (see fig. 8.12). There are six species of this genus; one, *A. latidens*, has recently been described by Handley from Venezuela and is similar in size to *A. caudifer*. Where three species co-occur, in general they are graded in size, ranging in head and body length from 50 to 90 mm and in forearm length from 34 to 48 mm. The tail is rudimentary or absent, and the interfemoral membrane is greatly reduced or virtually absent (plate 6).

Distribution
The genus is distributed from Sinaloa in western Mexico south through the Isthmus of Panama to southeastern Brazil. Species of *Anoura* appear to be absent from the central Amazon region.

Natural History
Anoura feeds on fruit, nectar, pollen, and insects (Gardner 1977a).

Anoura caudifer (E. Geoffroy, 1818)

Description
This is one of the smaller species of *Anoura*. Measurements: TL 65–66; T 4.6; HF 12.0; E 14.4–14.7; FA 36.2–36.3; Wt 9.9–10.3 g (see table 8.17). The upperparts are dark brown, the venter paler brown. Chromosome number: $2n = 30$; FN = 56.

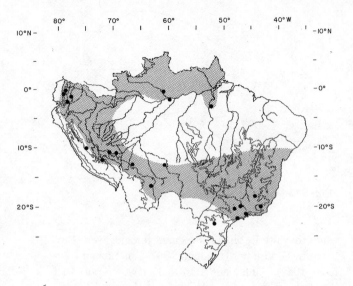

Map 8.61. Distribution of *Anoura caudifer*.

Range and Habitat

A. *caudifer* is distributed over northern South America; it is generally absent from the Amazon region but broadly distributed over north-central Bolivia east to southeastern Brazil (map 8.61). In Venezuela it occurs from 500 to 1,500 m elevation. It is strongly associated with streams within multistratal tropical evergreen forest (Handley 1976).

Anoura cultrata Handley, 1960

Description

This is the largest species of the genus. The male is considerably larger than the female. Mean measurements for males and females, respectively, are TL 79, 71; T 4, 4.5; HF 14.5, 14.5; E 16, 17; FA 42, 40; Wt 20.4, 14.9 g (*n* = 4; USNM). The dorsal pelage is dark brown to black.
Chromosome number: $2n = 30$; FN = 54.

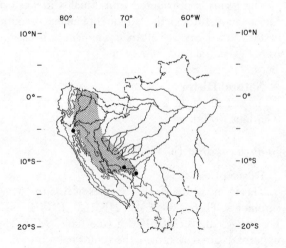

Map 8.62. Distribution of *Anoura cultrata*.

Range and Habitat

A. *cultrata* ranges from Costa Rica to northwestern Venezuela and in the western part of Colombia, south to Amazonian Peru (map 8.62). It occurs at modest elevations from 200 to 1,870 m in Venezuela, and its preferred habitat is moist premontane forest. It roosts in caves (Tamsitt and Nagorsen 1982).

Natural History

This species is known to feed on fruits, pollen, nectar, and insects. It may use foliage gleaning to capture insects. Seasonal breeding tied to rainfall is strongly implied from the data (Tamsitt and Nagorsen 1982).

Anoura geoffroyi Gray, 1838

Description

This species is slightly larger than A. *caudifer*. Head and body length averages 69 mm for males and 68 mm for females; weight averages 15 g for males and 13 g for females. This species lacks a tail, which immediately sets it off from A. *caudifer* and A. *cultrata*, both of which have tiny tails, but in all species of the genus the interfemoral membrane is distinctively reduced. The dorsum is dull brown blending to gray on the shoulders, and the venter is gray.
Chromosome number: $2n = 30$; FN = 56.

Range and Habitat

This species is occurs from Sinaloa in western Mexico and Tamaulipas in eastern Mexico through the Isthmus, across northern South America, through Peru and Bolivia to east-central Brazil (map 8.63). It appears to be absent from most of the Amazon region. In Venezuela most specimens were taken below 1,500 m. The species is broadly tolerant of man-made

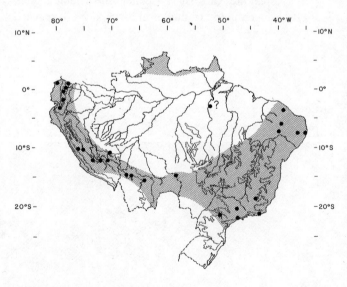

Map 8.63. Distribution of *Anoura geoffroyi*.

clearings and is strongly associated with moist areas and multistratal tropical evergreen forest (Handley 1976).

Natural History

Anoura geoffroyi is seasonally monestrous, with births at the beginning of the dry season and lactation during the flowering of *Pseudobombax*. The young fly when the forearm length equals 90% of adult length (Baumgarten and Vieira 1994).

A. geoffroyi has been implicated in the pollination of certain night-blooming plants, including *Eperua falcata*. This species roosts in caves in groups of up to fifty. Seasonally, sexes can be segregated or roost in mixed-sex colonies. This bat is highly insectivorous at certain seasons of the year (Gardner 1977a; Wilson 1979).

Anoura latidens Handley, 1984

Description

This species was described by Handley (1984). Males average 64.4 mm in head and body length, while females average 66.2 mm. Weight averages 14.5 g for males and 14.8 g for females.

Range and Habitat

Thus far this species has been found widely distributed in Venezuela. Strongly associated with moist areas and multistratal tropical evergreen forests, in Venezuela it tolerates elevations from 50 to 2,240 m. It may extend to Brazil (not mapped).

Genus *Lichonycteris* Thomas, 1895

Lichonycteris obscura Thomas, 1895

Description

The dental formula is I 2/0, C 1/1, P 2/3, M 2/2. Head and body length averages 46–55 mm and forearm length 30–33.9 mm. The upperparts are uniform dark brown and the underparts slightly darker (Hall 1981). *L. degener* is a common synonym according to Gardner (1976), but Honacki, Kinman, and Koeppl (1982) retain the separation, while Koopman (1993) includes *L. degener* in *L. obscura*.
Chromosome number: $2n = 24$; FN = 44.

Range and Habitat

This species occurs from Honduras south through the Isthmus of Panama, through western Venezuela to northern Peru (map 8.64). It has been sporadically reported across northern Venezuela, Guyana, and Suriname in moist evergreen forests.

Natural History

This bat visits flowers and probably feeds on nectar, pollen, and insects (Gardner 1977a).

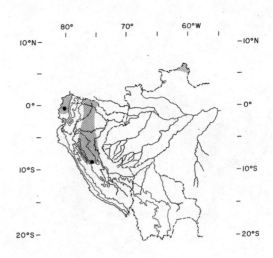

Map 8.64. Distribution of *Lichonycteris obscura*.

Comment

Includes *Lichonycteris degener*.

Genus *Choeroniscus* Thomas, 1928

Hog-nosed Bat

Description

The dental formula is I 2/0, C 1/1, P 2/3, M 3/3. Head and body length ranges from 50 to 55 mm, with the tail averaging about 12 mm. Forearm length ranges from 32 to 38 mm. The pelage is usually dark to light brown above, with the venter paler to nearly the same color.

Distribution

Species of this genus are distributed from Sinaloa, Mexico, and south in western Mexico to Central America. In South America, species are distributed mostly in the northern parts. Some species tolerate rather arid habitats.

Choeroniscus godmani (Thomas, 1903)

Description

Measurements: HB 53; T 6–9; HF 9–10; FA 32–34; Wt 7–8 g (see table 8.17). The muzzle is exceptionally long. This, together with the small nose leaf and relatively small ears, serves to identify the species. Coloration is uniformly dark brown on the dorsum and paler on the venter.
Chromosome number: $2n = 20$; FN = 32 or 36.

Range and Habitat

This species occurs from Sinaloa in western Mexico south through Central America and across northern South America. In Venezuela it ranges from 2 to 350 m elevation and is strongly associated with moist habitats and multistratal tropical evergreen forest. It may fre-

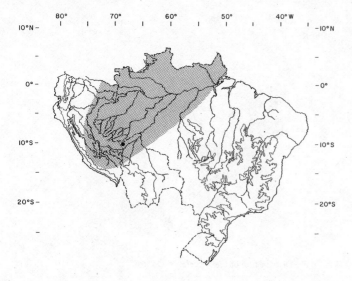

Map 8.65. Distribution of *Choeroniscus intermedius.*

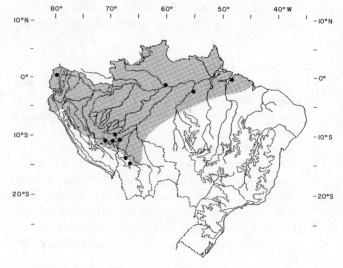

Map 8.66. Distribution of *Choeroniscus minor.*

quent orchards (Handley 1976). Within the range covered by this volume, it could extend to Roraima, Brazil (not mapped).

Natural History

These bats roost in groups of twelve to twenty-four. They feed on fruit, pollen, nectar, and insects.

Choeroniscus intermedius (J. A. Allen and Chapman, 1893)

Description

Body proportions are approximately the same as for *C. minor.* Measurements of one female were TL 69; T 9; HF 10; E 13 (Suriname, CMNH). This very dark bat has an exceptionally long muzzle. Chromosome number: $2n = 20$; FN = 36.

Range and Habitat

This species occurs in the Guianas and extreme eastern Venezuela and ranges south into northern Brazil and Peru (map 8.65).

Natural History

These bats roost in tree hollows in groups of up to eight (Goodwin and Greenhall 1961).

Choeroniscus minor (Peter, 1868)

Description

In spite of the name, this species is larger than *C. godmani.* Its measurements are TL 69–71; T 9; HF 9–10; E 12–13; Wt 10 g (Suriname, CMNH) (see also table 8.17). The dorsum and venter are light brown.

Range and Habitat

The species is confined to South America, including the southern parts of Venezuela, the Guianas, south-

eastern Colombia, Ecuador, Peru, and northern Brazil (map 8.66). It is associated with moist areas in multistratal tropical evergreen forests (Handley 1976).

Choeroniscus periosus Handley, 1966

Description

This medium-sized species measures TL 62; T 10; FA 41.2 (Handley 1966). The dorsum is brown and the venter lighter.

Range and Habitat

This species has been reported from the extreme southwest of Colombia to Ecuador (not mapped).

Genus *Scleronycteris* Thomas, 1912
Scleronycteris ega Thomas, 1912

Description

The dental formula is I 2/0, C 1/1, P 2/3, M 3/3. Measurements of the type specimen are HB 57; T 6; FA 35. The dorsum is dark brown, the venter light brown. This species is very poorly known.

Range and Habitat

This bat is known only from extreme southern Venezuela and could extend to adjacent portions of Brazil (map 8.67). In Venezuela the single specimen collected was caught in multistratal tropical evergreen forest at 135 m elevation.

SUBFAMILY CAROLLIINAE

Diagnosis

The muzzle is long and narrow but not as pronounced as in the Glossophaginae. The ears are relatively longer than in the Glossophaginae; when laid

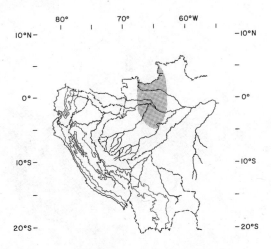

Map 8.67. Distribution of *Scleronycteris ega* (according to Koopman 1982).

Table 8.9   Key to the Carolliinae

1	Short tail present (*Carollia*)	2
1′	Short tail absent (*Rhinophylla*)	5
2	Fur of dorsum long, forearm sparsely haired	*C. subrufa*
2′	Forearm well haired	3
3	Tail less than 7 mm	*C. brevicauda*
3′	Tail greater than 7 mm	4
4	Head and body length less than 60 mm	*C. castanea*
4′	Head and body length greater than 60 mm ...	*C. perspicillata*
5	Head and body length greater than 50 mm	*R. alethina*
5′	Head and body length less than 50 mm	*R. pumilio*

forward they reach almost to the tip of the nose. The naked pad on the chin with the large central O-shaped wart is diagnostic. No facial stripes are present. The tail is extremely reduced or absent, but the interfemoral membrane is still present, though much less extensive than in genera having longer tails (plate 6). External features are shown in figures 8.3 and 8.4A, and a key is given in table 8.9.

Distribution

These bats range from Sinaloa, Mexico, south through the Isthmus across South America to extreme southeastern Brazil and northeastern Argentina.

Genus *Carollia* Gray, 1838
Short-tailed Bat, Leaf-nosed Bat,
Murciélago, Colicorto

Description

The dental formula is I 2/2, C 1/1, P 2/2, M 3/3 (see fig. 8.13). Head and body length ranges from 48 to 65 mm. The tail, which is lacking in the genus *Rhinophylla*, ranges from 3 to 14 mm. The dorsum is dark brown to reddish brown, and the venter is similar. Pine (1972) has written a monograph on the genus.

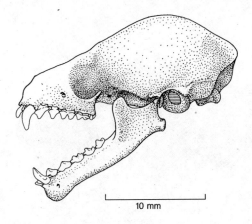

Figure 8.13. Skull of *Carollia perspicillata*.

Distribution

The genus is distributed from southern Sinaloa, Mexico, to northeastern Argentina and southeastern Brazil.

Natural History

These bats are generalized feeders, eating fruits, flowers, and insects. They use a variety of roosting sites, including caves, hollow trees, and crevices.

Identification

Owen, Schmidly, and Davis (1984) have developed criteria for the field identification of the species of *Carollia*.

Carollia brevicauda (Schinz, 1821)

Description

Measurements: TL 64.5; T 6.2; HF 14.1; E 19.7; FA 38.6; Wt 12.5–13.7 g (see table 8.17). This species is characterized by an extremely short tail, averaging less than 6.7 mm, but *C. subrufa* also has a short tail (see below). Although the head and body length is similar to that of *C. perspicillata*, the forearm is considerably shorter, averaging 38.5 mm for males and 38.6 mm for females (see table 8.17). The dorsum and venter are approximately the same shade of medium brown.
Chromosome number: $2n = 20$–21; FN = 36.

Range and Habitat

This species ranges from southern Veracruz, Mexico, south through the Isthmus over most of tropical South America to central Brazil (map 8.68). It is strongly associated with moist habitats for foraging and typically prefers multistratal tropical evergreen forests. It is broadly tolerant of a wide range of elevations. Roosts are found in caves, rock crevices, houses, and under the leaves of *Musa*. In northern Venezuela it occurs at from 24 to 2,147 m (Handley 1976).

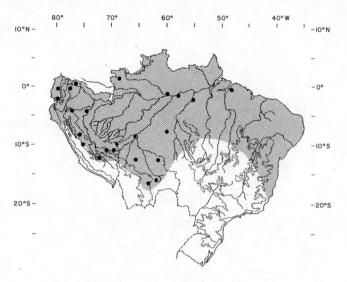

Map 8.68. Distribution of *Carollia brevicauda*.

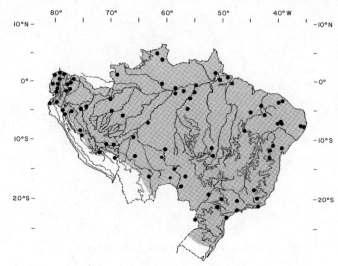

Map 8.70. Distribution of *Carollia perspicillata*.

Carollia castanea H. Allen, 1890

Description

Measurements: TL 63–64; T 9–10; HF 12–14; FA 36–37; Wt 14.6 g. This is the smallest of the three species occurring in South America (see table 8.17). It is similar in appearance to *C. brevicauda*.
Chromosome number: $2n = 20$–22; FN = 36–38.

Range and Habitat

This species is distributed from Honduras south through the Isthmus in the western portion of South America to Bolivia (map 8.69). It penetrates eastward in western Venezuela to the Guianas but is absent over a vast portion of Brazil. In Venezuela it is strongly associated with multistratal tropical evergreen forests, and all specimens taken there were below 460 m elevation (Handley 1976).

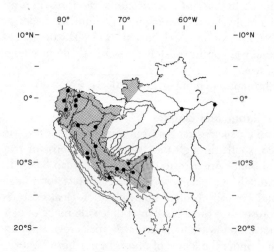

Map 8.69. Distribution of *Carollia castanea*.

Carollia perspicillata (Linnaeus, 1758)

Description

Measurements: HB 56; T 10.4; HF 14.4; E 20.7; FA 40.8; Wt 17.4 g. This species has roughly the same head and body length and color as *Carollia brevicauda*, but the forearms of males and females average 41 mm, and weights average 17 g for males and 16 g for females. The relatively longer tail, averaging approximately 10 mm, distinguishes *C. perspicillata* from *C. brevicauda* (see table 8.17).
Chromosome number: $2n = 20$–21; FN = 36.

Range and Habitat

This species is distributed from southern Veracruz south through the Isthmus, across northern South America and south through Amazonian Peru, Bolivia, and Brazil (map 8.70). In Venezuela it forages near moist areas, being taken most frequently in multistratal tropical evergreen forests. It has a wide altitude tolerance, ranging to 1,260 m elevation in Venezuela (Handley 1976).

Natural History

This is one of the best-studied species of the genus. *Carollia* depends on the fruit of *Piper* for most of its diet, but it also gleans foliage for insects. Males defend small groups of females in a harem breeding system. The bats roost in hollow trees, caves, crevices, and other moist places in colonies of up to one hundred. Over most parts of their range they have two birth peaks; a female thus produces two young annually (Fleming 1983). Interbirth intervals range from 115 to 173 days, and gestation is approximately 115 to 120 days. The young are born in an advanced state with the eyes open. The newborn remains more or less continuously attached to the mother for the first fourteen days of life. Young weigh approximately 5 g at birth (Kleiman and Davis 1979).

The short-tailed bat is characteristic of lower elevations and is found in both tropical evergreen and deciduous forests. There is some indication that in certain parts of its range *C. perspicillata* may migrate seasonally. It will roost almost anywhere that is dark or shaded and provides some protection, and it can be found singly or in roosts of up to 1,000 individuals.

C. perspicillata is primarily frugivorous and can be harmful to fruit crops in some areas. In many areas the fruit of *Piper* is a major food source, although insects are also gleaned from foliage. Males defend small groups of females and thus have a harem breeding system. The definitive work on the natural history and ecology of this bat has been done by Fleming (1983, 1988; see also Charles-Dominique 1991; Porter 1979; Cloutier and Thomas 1992).

Genus *Rhinophylla* Peters, 1865

Description
The dental formula is I 2/2, C 1/1, P 2/2, M 3/3. This genus is similar to *Carollia* but is immediately distinguishable by its lack of a tail. The interfemoral membrane is small but still conspicuous, not nearly as reduced as that of *Sturnira*. The dorsum and venter are medium brown to very dark brown.

Distribution
The genus is distributed solely in South America, ranging over the northern portion with *R. pumilio* extending into south-central Brazil.

Natural History
These bats are believed to be primarily frugivorous.

Rhinophylla alethina Handley, 1966

Description
The dental formula is I 2/2, C 1/1, P 2/2, M 3/3. Head and body length ranges from 55 to 58 mm; HF 11; FA 35–37; Wt 12–16 g. The dorsum is blackish, shading to brownish black on the rump; the underparts are paler.

Range and Habitat
The species is known from Río Rapaso, Valle, Colombia; Pasco, Peru; Ecuador; and Brazil (map 8.71).

Rhinophylla fischerae Carter, 1966

Description
This small species has a head and body length ranging from 47 to 54 mm, and the forearm is 33 to 36 mm.

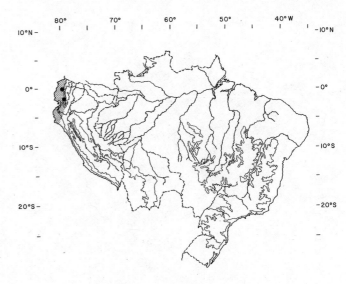

Map 8.71. Distribution of *Rhinophylla alethina*.

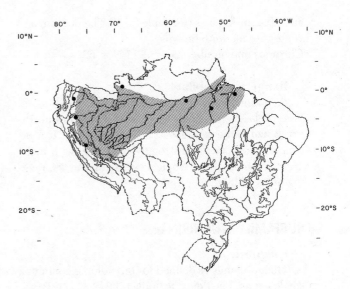

Map 8.72. Distribution of *Rhinophylla fischerae*.

The interfemoral membrane has a fringe of stiff hairs. The dorsal pelage varies from gray brown to reddish brown.
Chromosome number: $2n = 34$; FN = 56.

Range and Habitat
Known from sixty-one miles southeast of Pucallpa, Ucayali, Peru, and thence to Belém, Brazil. The species may extend to southern Colombia. It was taken at 180 m elevation (map 8.72).

Rhinophylla pumilio Peters, 1865

Description
Head and body length averages 48.3 mm for males and 50 mm for females; weight averages 9.4 g for males

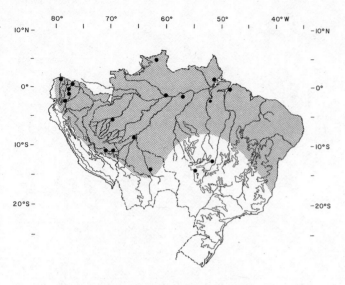

Map 8.73. Distribution of *Rhinophylla pumilio*.

Table 8.10   Key to the Common Species
             of the Sturnirinae

1   High-elevation species occurring at
     1,500 to 2,500 m 2
1′   Low-elevation species, generally
     occurring below 1,000 m 4
2   Head and body length less than
     60 mm . . . . *Sturnira erythromos*
2′   Head and body length greater than 60 mm 3
3   Head and body length greater than
     66 mm, only two lower molars . . . . *S. bidens*
3′   Head and body length greater than
     66 mm, three lower molars . . . . *S. bogatensis*
4   Head and body length less than 65 mm . . . . *S. lilium*
4′   Head and body length greater than 65 mm 5
5   Head and body length less than 75 mm 6
5′   Head and body lenght greater than 75 mm 7
6   Ears greater than 20 mm . . . . *S. tildae*
6′   Ears less than 20 mm . . . . *S. ludovici*
7   Head and body length 85 to 90 mm . . . . *S. magna*
7′   Head and body length greater than 95 mm . . . *S. aratathomasi*

and 10.4 for females (see table 8.17). The dorsum and venter are medium brown.
Chromosome number: $2n = 36$; FN = 62.

Range and Habitat

R. pumilio ranges across southern Colombia and Venezuela, the Guianas, Amazonian Peru, and northern Brazil to the central east coast (map 8.73). In Venezuela this species was strongly associated with moist areas and multistratal tropical evergreen forests, ranging up to 1,400 m elevation.

SUBFAMILY STURNIRINAE

Diagnosis

This subfamily as defined for this volume contains a single genus. The dental formula is I 2/2, C 1/1, P 2/2, M 3/2–3 (see fig. 8.14). These bats are tailless, and the interfemoral membrane is so reduced as to be inconspicuous, but the remnant fringe is well haired (see fig. 8.3). The ears are relatively short. Many species

bear patches of stiff, yellowish hair on the shoulder glands. The dorsum varies from pinkish buff to dark brown, with the underparts usually considerably paler (plate 6). There is little size dimorphism within this subfamily. A key is given in table 8.10, and keys for all species may be found in Davis (1980).

Distribution

The genus is distributed from southern Sinaloa in western Mexico and southern Tamaulipas in eastern Mexico south through the Isthmus to northern Argentina and Uruguay. Many species occur at modest to relatively high elevations.

Natural History

These bats are mainly frugivorous and possibly feed on pollen and nectar (Gardner 1977a). A single young is born annually. They often exist at rather high altitudes extending to 2,000 m at 10° N.

Comment

Koopman (1993) subsumes *Sturnira* within the Stenodermatinae. Pacheco and Patterson (1991) confirm monophylly, but the group is closely allied to the stenodermatines.

Genus *Sturnira* Gray, 1842
See subfamily account.

Sturnira aratathomasi Peterson and Tamsitt, 1968
Yellow-shouldered Bat

Description

This large, rare species exceeds 100 mm in total length, and the forearm may reach 60 mm. It may be distinguishable from *S. magna* in that its inner upper

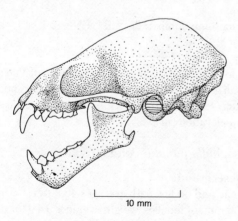

10 mm

Figure 8.14. Skull of *Sturnira ludovici*.

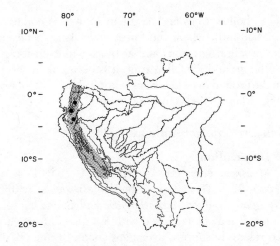

Map 8.74. Distribution of *Sturnira bidens*.

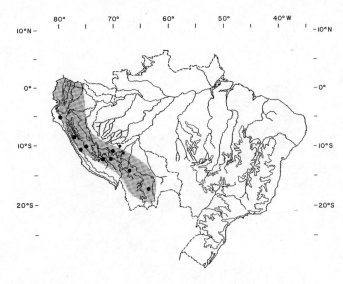

Map 8.75. Distribution of *Sturnira bogotensis* (includes *Sturnira oporaphilum*).

incisors are elongated and centrally pointed and not in contact terminally (Tamsitt and Hauser 1985). The dorsal pelage is darkish brown.

Range and Habitat

The species is apparently confined to the Pacific side of the Andes from southern Colombia to Ecuador (not mapped). It is known to feed on fruit (Thomas and McMurray 1974).

Sturnira bidens Thomas, 1915

Description

Measurements: HB 64; HF 14; E 15.7; FA 39; Wt 17 g (see table 8.17). This bat differs from other species of *Sturnira* in possessing only two lower molars. The dorsum looks smoky gray because the brown-tipped hairs have gray bases; the venter is brown. Chromosome number: $2n = 30$; FN = 56.

Range and Habitat

The species is distributed in the foothills of the Andes from western Venezuela south to Peru (map 8.74). In Venezuela it ranges between 2,550 and 2,640 m elevation and is strongly associated with cloud forest. It seems broadly tolerant of both moist and drier sites.

Natural History

In central Peru breeding occurs in July, with pregnant females taken in August. A bimodal breeding pattern has been noted in Venezuela. The bats are known to use caves as roosting sites. Ripe fruits are taken in season, and insects are rare to absent in the diet (Molinari and Soriano 1987).

Sturnira bogotensis Shamel, 1927

Description

This species is larger than *S. bidens,* measuring TL 65–67; HF 14–15.5; E 17–17.5; FA 43–43.3;

Wt 19.6 g (n = 4; USNM). Forearms average approximately 43 mm. The coloration is similar to that of *S. bidens* (includes *S. oporaphilum,* but see Anderson, Koopman, and Creighton 1982).

Range and Habitat

This species co-occurs with *S. bidens* in the Andes from western Venezuela south to Bolivia (map 8.75). In Venezuela specimens were collected between 2,107 and 2,640 m elevation. It is strongly associated with moist tropical evergreen or cloud forest (Handley 1976).

Sturnira erythromos (Tschudi, 1844)

Description

Measurements: HB 57; HF 12.6–13; 15.6; FA 38.7–39.4; Wt 14.4; 12.0 g. This is the smallest species in the genus; the sexes are nearly equal in size. It can be distinguished from *S. bogotensis* by its small size. The coloration is similar to that of *S. bidens,* with the dorsum smoky gray and the venter brown.

Range and Habitat

This species occurs in the premontane areas of the Andes in Venezuela, Colombia, and south to northwest Argentina (map 8.76). In Venezuela most specimens were taken at between 1,000 and 2,500 m elevation. It prefers moist habitats of multistratal evergreen forest or cloud forest.

Sturnira lilium (E. Geoffroy, 1810)

Description

Measurements: HB 60.6–61.3; HF 13.6–13.9; E 16.5–16.7; FA 40.6–41; Wt 18.4–20.4 g. There is little

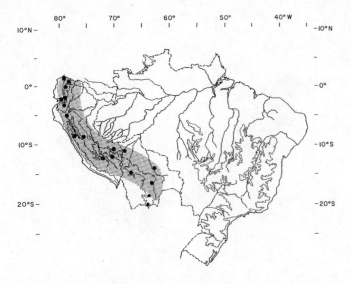

Map 8.76. Distribution of *Sturnira erythromos*.

ing in hollow trees and caves. In Jujuy it is common to 2,600 m or where the forest ends. This species is exclusively frugivorous and at times will even forage on the ground for fallen fruit; it is occasionally taken on the ground in rodent traps (Crespo 1982; Eisenberg 1989; Mares, Ojeda, and Kosco 1981).

Sturnira ludovici Anthony, 1924

Description
Measurements: HB 71; HF 16; F 18.5; F 46.6; Wt 25–27 g. This is one of the largest species within the genus; only *S. aratathomasi* and *S. magna* are larger. It may be confused with *S. tildae*, but its ears are relatively shorter, averaging about 19 mm. Basically this is a brown bat with the venter not contrasting, but the bases of the dorsal hairs are pale.

Range and Habitat
This species ranges from southern Sinaloa in western Mexico and southern Tamaulipas in eastern Mexico south through the Isthmus to northern South America. It extends to Ecuador and across Venezuela to Guyana (map 8.78). It ranges from sea level to 2,240 m elevation in Venezuela, but most specimens are taken below 1,500 m. It forages in moist areas and mainly occurs in multistratal tropical evergreen forests, though it is occasionally taken in dry deciduous forests.

Comment
Over much of the northern range of *Sturnira lilium*, *S. ludovici* co-occurs. The species are easily separable by size. In the southern part of South America, *S. ludovici* is replaced by the similar-sized *S. tildae* and co-occurs with *S. lilium*.

sexual size dimorphism. Yellow shoulders are very pronounced in males; the dorsum is brown, the venter lighter.

Range and Habitat
This species is widely distributed from southern Sonora in western Mexico and Tamaulipas in eastern Mexico south through the Isthmus, over all of South America to northern Argentina (map 8.77). In Venezuela it ranges from sea level to 1,982 m elevation. At higher elevations this species is replaced by *S. bogotensis* and *S. erythromos*.

Natural History
One of the most common bats where it occurs, *S. lilium* seems to be confined to moister habitats, roost-

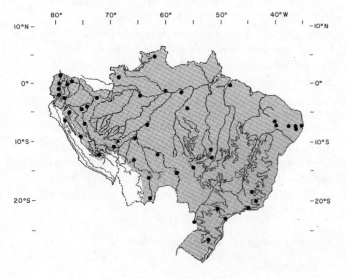

Map 8.77. Distribution of *Sturnira lilium*.

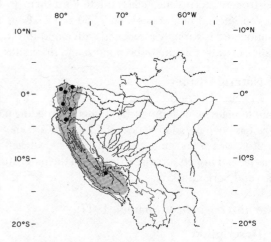

Map 8.78. Distribution of *Sturnira ludovici*.

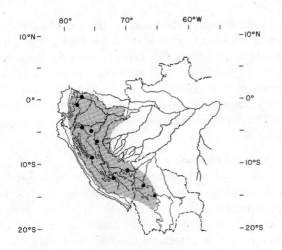

Map 8.79. Distribution of *Sturnira magna*.

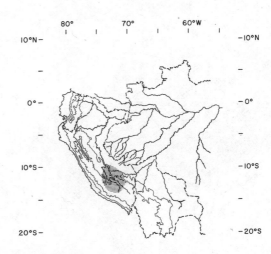

Map 8.80. Distribution of *Sturnira nana*.

Sturnira luisi Davis, 1980

Comment

The species is described from Costa Rica but may extend to Ecuador and northeast Peru. It may be confused with *S. ludovici* (Davis 1980).

Sturnira magna de la Torre, 1966

Description

This species is similar to *S. aratathomasi*, but the inner upper incisors are broad and almost in contact terminally. The head and body length ranges from 85 to 90 mm and the forearm from 56 to 57 mm. The dorsal pelage is yellow brown to gray brown; the venter is pale yellow, brown, or gray, and the bases of the hairs are very pale.

Range and Habitat

The species is known from the Río Manito, Iquitos, Peru, thence south to Bolivia and north to extreme southeastern Colombia (map 8.79). It has a wide altitude tolerance, ranging from 200 to 2,300 m in Peru. Moist lowland and premontane forests are preferred. It is assumed to be a frugivore (Tamsitt and Hauser 1985).

Sturnira nana Gardner and O'Neill, 1971

Description

S. nana is a very small species with the forearm measuring 34.2–35.7 mm. Measurements of holotype: TL 51; HF 10; E 13. The zygomatic arch is thin or incomplete, a character it shares with *S. bidens*. The dorsum is a dark grayish brown; the venter is slightly lighter in color. There is no indication of shoulder glands.

Distribution

The species is found in southeastern Peru. Type locality: Peru, Ayacucho, Huanhuachayo, 1,660 m (map 8.80).

Sturnira oporaphilum

See *S. bogatensis*.

Sturnira tildae de la Torre, 1959

Description

Head and body length averages 69.6 mm; HF 16.5; E 20.5; FA 47.7; Wt 26.5 g. There is little sexual size dimorphism, but considerable variation in size among individuals can be noted (see Marinkelle and Cadena 1971). The ears are slightly larger in proportion to head and body length than in *S. ludovici*. The bases of the dorsal hairs are very light while the tips are brown, giving the general effect of a variable dorsal coloration of brown washed with white. The venter is pale brown.

Range and Habitat

This species is entirely South American and occurs in the southeastern portion of Colombia, southern Venezuela, the Guianas, Amazonian Peru, and across much of Brazil (map 8.81). In Venezuela specimens were taken below 1,165 m. It is strongly associated with moist habitats and multistratal tropical evergreen forests. This species of *Sturnira* replaces *S. ludovici* at lower elevations over much of the South American continent.

SUBFAMILY STENODERMINAE

Diagnosis

There is a tendency within the subfamily to reduce the number of molars to two lower molars per side.

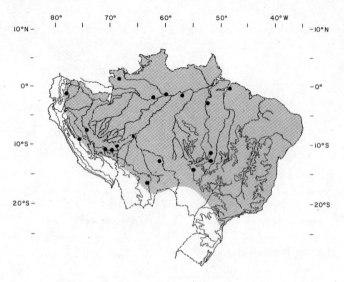

Map 8.81. Distribution of *Sturnira tildae*.

These are medium-sized to small bats. The tail is rudimentary or absent, but the interfemoral membrane is present and supported by the calcar bone on the heel. Although there may be varying degrees of reduction in the interfemoral membrane, the reduced size is not as pronounced as one finds in the genera *Anoura* and *Sturnira*. The muzzle is short, in some cases very short and broad, which apparently correlates with an increase in specialization for feeding on fruit. Many species of the subfamily have white facial stripes, generally two pairs from the nose to just above the eye and from the corner of the mouth to the ear. The dorsum frequently has a median white stripe. Facial stripes are absent in the genera *Pygoderma*, *Ametrida*, *Sphaeronycteris*, and *Centurio*. A key to the genera (exclusive of *Sturnira*) is given in table 8.11. Anatomical details are portrayed in figures 8.3 and 8.4A. An extended review of tent making by bats worldwide is presented by Kunz et al. (1994).

Natural History

"Tent making" by bats involves biting leaves by several techniques that ultimately fold the leaf to form a temporary shelter (Timm 1987). Thirteen species from five genera of stenodermatines have been identified to date as practicing this "art" (see table 8.14). Many species exhibiting this trait may be nearly obligate tentmakers, but *Artibeus jamaicensis* is surely not (Handley, Wilson, and Gardner 1991). The tent provides shelter from the sun and rain, and apparently makes the roosting site difficult for predators to identify. An extended review of tent making by bats worldwide is presented by Kunz et al. (1994).

Genus *Uroderma* Peters, 1865
Tent-making Bat

Description

The dental formula is I 2/2, C 1/1, P 2/2, M 3/3 (fig. 8.15). Head and body length ranges from 60 to 62.5 mm. The white facial stripes are extremely pronounced. The basic color of the dorsum is gray brown to brown, and the venter is usually paler.

Table 8.11   Field Key to the Genera of the Stenoderminae

1	Face naked; nose leaf very reduced 2
1'	Face furred; nose leaf prominent 3
2	Face very wrinkled, fold of skin under chin .... *Centurio*
2'	Face not wrinkled; hornlike protuberance above nostrils of male; fold of skin under the chin prominent in male .... *Sphaeronycteris*
3	Interfemoral membrane furred on dorsal side; nasal bones absent .... *Chiroderma*
3'	Interfemoral membrane not furred 4
4	Fur whitish, wings yellow .... *Ectophylla alba*
4'	Fur colored 5
5	No striping on face or dorsum 6
5'	Stripes usually present on face and/or dorsum 7
6	Single white spot on each shoulder, dermal outgrowths on cheeks .... *Ametrida*
6'	Single white spot on each shoulder, no dermal outgrowths on face .... *Pygoderma*
7	Inner upper incisor slightly higher than outer incisor 8
7'	Inner upper incisor much higher than outer 10
8	Basal portion of nose leaf bilobate .... *Uroderma*
8'	Basal portion of nose leaf not bilobate 9
9	Inner upper incisor bifid .... *Artibeus*
9'	Inner upper incisor entire .... *Enchisthenes*
10	Upper incisor trilobate .... *Platyrrhinus*
10'	Not as above 11
11	Stripes on face and back prominent; incisors with cutting edge entire .... *Vampyrodes*
11'	No dorsal stripe .... *Vampyressa*

Note: See also table 8.5.

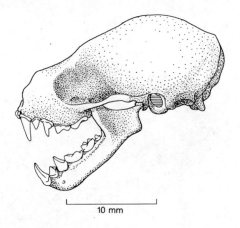

Figure 8.15. Skull of *Uroderma bilobatum*.

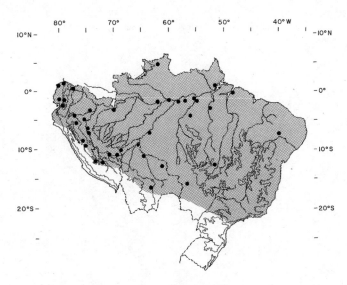

Map 8.82. Distribution of *Uroderma bilobatum*.

Distribution

The genus occurs from southern Veracruz, Mexico, south to the main continent, extending to southeastern Brazil.

Natural History

Species of this genus roost under palm fronds or banana leaves. They often bite through the ribs of fronds and cause the leaf to collapse on itself, thereby providing a shelter. For this reason they are called tent-making bats, although the trait is not exclusive to the genus.

Uroderma bilobatum Peters, 1866

Description

Head and body length ranges from 54 to 61 mm. In Venezuela specimens averaged about 60.8 mm. The sexes are nearly identical in size. The facial stripes are very pronounced in *U. bilobatum,* and there is a white stripe down the midline of the back. The ear edge is pigmented yellowish white.
Chromosome number: $2n = 42, 44, 38$; FN = 48, 50.

Range and Habitat

This species occurs from southern Veracruz, Mexico, south through the Isthmus to southeastern Brazil and west to Peru and Ecuador and on to Bolivia and Brazil (map 8.82). It is found widely over all tropical areas of South America. It generally occurs below 1,000 m elevation. It tolerates man-made clearings, and though it is strongly associated with multistratal tropical evergreen forest, it also occurs in drier situations (Handley 1976).

Natural History

These bats typically cut the surface of palm fronds with a series of bites, causing the leaf to bend in half and form a shelter. They may roost in small colonies (up to ten) of both sexes. They are strongly frugivorous but include insects in their diet (Goodwin and Greenhall 1961; Timm and Lewis 1991).

Netting data suggest a bimodal activity rhythm. A single young is produced at each parturition, and the timing of reproduction is synchronized with local conditions of rainfall and subsequent fruiting by trees, vines, and shrubs. Females usually bear two young per year (Baker and Clark 1987).

Uroderma magnirostrum Davis, 1968

Description

This species is somewhat larger than *U. bilobatum.* Head and body length ranges from 58 to 65 mm (see table 8.17). The facial stripes are less conspicuous than in *U. bilobatum.* The ear tends to be uniformly colored in this species, whereas in *U. bilobatum* the ear is edged with yellowish white.
Chromosome number: $2n = 36$; FN = 60 or 62.

Range and Habitat

This species ranges from Oaxaca, Mexico, south through the Isthmus to central Brazil (map 8.83). In Venezuela it ranges below 1,000 m; most specimens were taken below 500 m. It is often associated with moist habitats. It makes use of open areas and man-made clearings and seems less tolerant of arid habitats than *U. bilobatum* (Handley 1976).

Natural History

Females form roosting colonies when they bear their young, and the sexes tend to roost separately during the rearing season. In Panama the young are born from February through April (Wilson 1979).

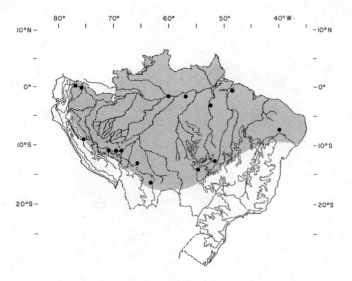

Map 8.83. Distribution of *Uroderma magnirostrum*.

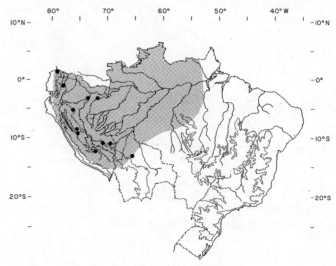

Map 8.84. Distribution of *Platyrrhinus brachycephalus*.

Genus *Platyrrhinus* Saussure, 1860
White-lined Bat

Description
The dental formula is I 2/2, C 1/1, P 2/2, M 3/3 (see fig. 8.16). Head and body length ranges from 48 to 98 mm. The basic color of the dorsum is dark brown to almost black, and the white or gray dorsal stripe is prominent, extending from the ears to the tail membrane (see plate 6). There are distinct white facial stripes.
Chromosome number: $2n = 30$; FN = 56.

Distribution
The genus is distributed from southern Mexico through the Isthmus to northeastern Argentina and Uruguay. Within its range it is divisible into a number of different species.

Natural History
White-lined bats are basically frugivorous. They roost in small groups of three to ten in leafy tangles,

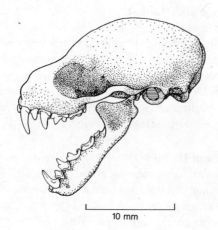

10 mm

Figure 8.16. Skull of *Platyrrhinus helleri*.

tree hollows, or caves. Reproduction usually coincides with the onset of the rainy season and varies locally.

Comment
Vampyrops Peters, 1865, has now been replaced. The systematics of this genus have been reviewed by Alberico (1990). Koopman (1993), following Gardner and Farrell (1990), resurrects *Platyrrhinus* Saussure, 1860, as the generic name with priority.

Platyrrhinus brachycephalus (Rouk and Carter, 1972)

Description
This is one of the smaller species, only slightly larger than *P. helleri*, and the sexes are nearly the same size. Head and body length averaged 63 mm, FA 39.8 mm, and weight 15.5 g for three specimens from Suriname (CMNH). The dorsum is light to medium brown, the venter pale brown. There is one bold line on the side of the face; the second line is fainter. The median dorsal stripe is conspicuous.

Range and Habitat
The species is confined to northern South America in the Amazonian portion of Brazil, the Guianas, southern Venezuela, Colombia, and Amazonian Peru. It could extend to adjacent Brazil (map 8.84). Strongly associated with multistratal tropical evergreen and moist sites, in Venezuela it was taken between 175 and 375 m (Handley 1976).

Platyrrhinus dorsalis (Thomas, 1900)

Description
Head and body length ranges from 69 to 80 mm and forearm length from 44.6 to 50.1 mm. One female specimen from Colombia had the following measure-

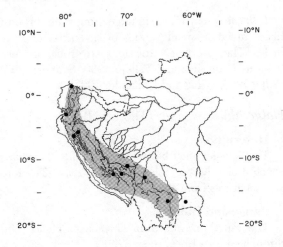

Map 8.85. Distribution of *Platyrrhinus dorsalis*.

ments: TL 80; HF 13; E 20; FA 47 (CMNH). This species is intermediate in size between *P. helleri* and *P. vittatus*. The middorsal stripe is prominent, as are the two stripes on each side of the face. The dorsum is dark brown and the venter gray brown.

Range and Habitat

This species occurs from the extreme eastern portion of Panama west of the Andes, and southward through Colombia to Bolivia (map 8.85).

Platyrrhinus helleri (Peters, 1867)

Description

Average measurements: HB 56; HF 12; E 17; FA 38; Wt 13–14 g. This is the smallest species in the genus, and males and females are nearly the same size. The dorsum is brown and the venter gray; a dorsal stripe and facial stripes are present (plate 6).

Range and Habitat

The species ranges from southern Mexico through the Isthmus, broadly across northern South America and the western portion of Brazil (map 8.86). In Venezuela it is found below 1,000 m elevation and prefers moist habitats, but it can range into dry deciduous forest. It tolerates man-made clearings.

Natural History

These bats roost in pairs high in the crowns of trees. There is evidence that they are strongly frugivorous (Gardner 1977a). The species may be very abundant in favorable habitat but can be netted in large numbers only by setting nets in the canopy (Handley 1967).

Platyrrhinus infuscus Peters, 1881

Description

This large species is exceeded in size only by *P. vittatus*. Its measurements are TL 97–101; HF 15–16; E 21–23; FA 55–57 (CMNH). The dorsum is buff to light brown with the venter paler. The dorsal and facial stripes are faint.

Range and Habitat

The species is found in the Amazon region of Peru north to Colombia and south to Bolivia (Gardner and Carter 1972) (see map 8.87).

Platyrrhinus lineatus (E. Geoffroy, 1810)
Vampiro de Franjas Blancas

Description

This small, dark species has forearms measuring 41 to 48 mm. One male specimen from Colombia measured TL 41; HF 12; E 13; FA 41 (CMNH). A sample from Paraguay measured HB 69.5; HF 13; FA 46.4; Wt 26.7 g. The dorsal pelage varies from light to blackish

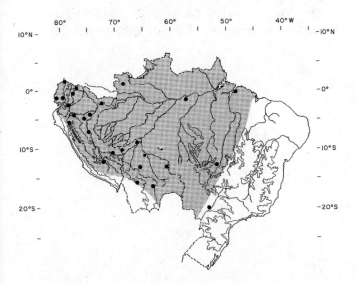

Map 8.86. Distribution of *Platyrrhinus helleri*.

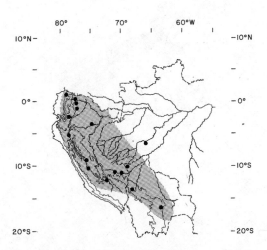

Map 8.87. Distribution of *Platyrrhinus infuscus*.

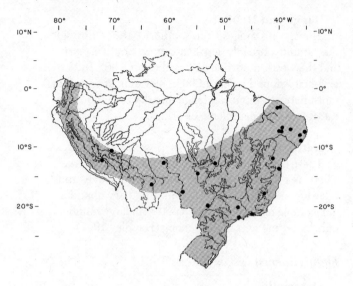

Map 8.88. Distribution of *Platyrrhinus lineatus*.

brown with a prominent white stripe; the venter is a paler shade of brown. The four facial stripes are buff and indistinct. Includes *P. nigellus*.

Distribution
Found from Colombia to Peru, then across northern Paraguay to Argentina and southern Brazil, then north in eastern Brazil to the Guianas (map 8.88).

Natural History
Females with one and two embryos have been collected. A single neonate weighed 8.3 g. In São Paulo state, Brazil, pregnant females were caught in all months except April, whereas in northeastern Brazil there was a bimodal distribution of pregnancy (Fornes and Massoia 1966; González and Vallejo 1980; Willig and Hollander 1987).

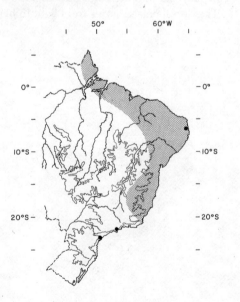

Map 8.89. Distribution of *Platyrrhinus recifinus*.

P. lineatus may forage in large groups, but in roosts males maintain small harems of seven to fifteen females. They are frugivorous, insectivorous, and nectivorous (González and Vallejo 1980; Sanborn 1955; Willig 1983; Willig and Hollander 1987).

Platyrrhinus recifinus (Thomas, 1901)

Description
This species may be conspecific with *P. lineatus*, which it closely resembles (see *P. lineatus*). The forearm measures from 40 to 42 mm.

Range and Habitat
The range extends from Guyana east and south through eastern Brazil (map 8.89).

Platyrrhinus vittatus (Peters, 1860)

Description
Measurements: HB 86–91; HF 17–18; E 23–24; FA 57–60. This is the largest species of the genus. The dorsal pelage is a dark blackish brown. The lower facial stripes are poorly developed, but the upper facial stripe and dorsal stripe are buffy and contrast sharply with the basic dorsal color.

Range and Habitat
This species occurs from Costa Rica south through the Isthmus over much of western South America, with the exception of Brazil (map 8.90). In Venezuela it occurred up to 2,119 m elevation and was strongly associated with moist habitats and multistratal evergreen forest (Handley 1976).

Genus *Vampyrodes* Thomas, 1900
Vampyrodes caraccioli (Thomas, 1889)
Great Stripe-faced Bat

Description
The dental formula is I 2/2, C 1/1, P 2/2, M 2/2; the reduction in number of molars distinguishes this

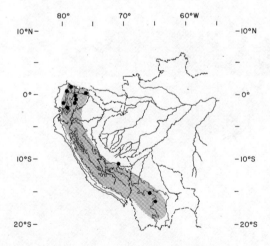

Map 8.90. Distribution of *Platyrrhinus vittatus*.

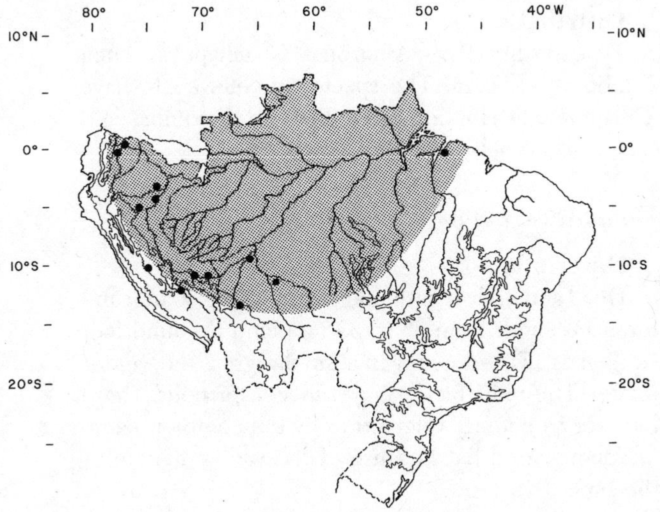

Map 8.91. Distribution of *Vampyrodes caraccioli*.

genus from *Platyrrhinus* and *Uroderma*. Measurements: HB 72–75; HF 16; E 21–22; FA 51–53; Wt 27–30 g. The dorsal pelage varies from brownish to gray brown, and the venter is paler. There are four white facial stripes, and a white line extends from the head down the dorsal midline.
Chromosome number: $2n = 30$; FN = 56.

Range and Habitat

This species occurs from southern Veracruz, Mexico, south through the Isthmus and over most of northern South America to Amazonian Peru and northern Brazil (map 8.91). In northern Venezuela it occurs below 1,000 m elevation and is strongly associated with multistratal tropical evergreen forest.

Natural History

These bats may be found roosting in small groups (two to four) under the branches of shrubs. They are frugivores (Goodwin and Greenhall 1961).

Individuals have been noted to feed on bananas, papaya, and figs. They fly to fruit trees at sunset and feed intermittently throughout the night. They rarely use the same day roost for more than two days. They are usually found in moist habitats below 600 m elevation (Willis, Willig, and Knox Jones 1990).

Genus *Vampyressa* Thomas, 1900
Yellow-eared Bat

Description

The dental formula is variable: I 2/2, C 1/1, P 2/2, M 2/2 or I 2/1, C 1/1, P 2/2, M 2/2. Head and body length ranges from 43 to 65 mm. The dorsum varies from smoky gray to pale brown or dark brown, and no dorsal stripe is usually present. The white facial stripes are prominent in some species but lacking in others.

Table 8.12 Key to the Known Species and Subspecies of the Genus *Vampyressa*

1	One pair of lower incisors	*V. bidens*
1′	Two pairs of lower incisors	2
2	M_3 present	*V. melissa*
2′	M_3 absent	3
3	M_2 approximately as long as wide, with high anterior and posterior cusps	4
3′	M_2 longer than wide, with lower anterior and posterior cusps	5
4	Forearm in full adults more than 34 mm	*V. pusilla pusilla*
4′	Forearm in full adults less than 34 mm	*V. pusilla thyone*
5	Forearm in full adults more than 34 mm, greatest length of skull more than 20 mm	*V. nymphaea*
5′	Forearm in full adults less than 34 mm, greatest length of skull less than 20 mm	*V. brocki*

Source: After Peterson 1968.

The ears are typified by a yellow margin. The genus has been reviewed by Gardner (1977b) and Peterson (1968). For a key to the genus see table 8.12.

Distribution

The genus occurs from southern Mexico south through the Isthmus to northeastern Brazil and adjacent portions of Amazonian Peru to Argentina. One species, *V. pusilla*, extends its range to southeastern Brazil but is absent from the Amazon region.

Natural History

The species of this genus are believed to be largely frugivorous. Reproduction is seasonally timed by the onset of the rains (Gardner 1977a; Wilson 1979).

Vampyressa bidens (Dobson, 1878)

Description

Only one pair of lower incisors is present. Measurements: HB 50–52; HF 11–11.3; E 17–17.2; FA 35.4–35.8; Wt 11–12 g. The dorsum is dark brown and the venter grayish brown. In contrast to some species of the genus, a faint forsal line is present, and there are four facial stripes.
Chromosome number: $2n = 26$; FN = 48.

Range and Habitat

This species is confined solely to South America and is found in the Amazon basin (map 8.92). In Venezuela most specimens were taken below 500 m elevation, and the species was strongly associated with multistratal tropical evergreen forest (Handley 1976).

Vampyressa brocki Peterson, 1968

Description

This is a small species, similar in size to *V. pusilla:* TL 49–51; HF 9–10; E 13–15; Wt 7–8 g ($n = 4$; Suri-

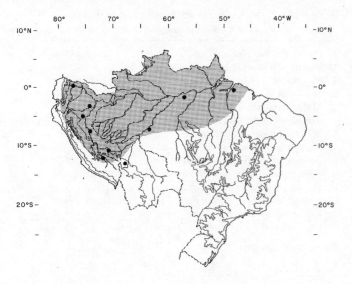

Map 8.92. Distribution of *Vampyressa bidens*.

name, CMNH). The dorsum is light brown and the venter gray, and there is no dorsal stripe. The facial stripes are conspicuous and demarcate a dark stripe from the nose leaf on each side of the eye to the ear. Chromosome number: $2n = 24$; FN = 44.

Range and Habitat

The type specimen was taken in Guyana at 2°50′ N, 58°55′ W. Since the original description, specimens have been taken from Suriname and southeastern Colombia (Peterson 1968; Genoways and Williams 1979). It may extend to Brazil (not mapped).

Vampyressa melissa Thomas, 1926

Description

Very similar to *Vampyressa pusilla*, but besides two pairs of lower incisors, it also has the M_3 (third upper molar).

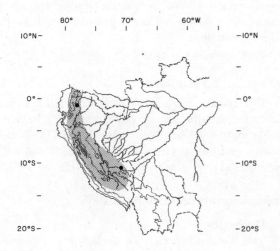

Map 8.93. Distribution of *Vampyressa melissa*.

Distribution

Type locality: Peru, Amazonas, Chachapoyas, Pucu Tambo at 1,480 m. The species is believed to have a disjunct distribution from Peru to Colombia and thence to French Guiana (map 8.93).

Vampyressa nymphaea Thomas, 1909

Description

Head and body length ranges from 55 to 60 mm and forearm length from 36 to 39 mm, and the hind foot is 13 mm. The dorsum is a smoky gray, the venter lighter. The white facial stripes are conspicuous. These bats are frequently referred to by the common name of yellow-eared bat because of the yellow margin of the ears.
Chromosome number: $2n = 26$; FN = 48.

Range and Habitat

The species is found from southern Nicaragua south through the Isthmus to extreme western Colombia extending to western Ecuador. It has not been taken east of the Andes (not mapped).

Vampyressa pusilla (Wagner, 1843)
Little Yellow-eared Bat

Description

Measurements: HB 46–47; HF 10; E 14.8; FA 31; Wt 7.7–7.9 g. The dorsum is light brown, and the facial stripes are not conspicuous, though the upper facial stripe is usually more boldly marked than the lower. The venter is light gray, and the throat region may be yellowish.
Chromosome number: variable, $2n = 18, 24$; FN = 20–22.

Range and Habitat

The species is distributed from southern Mexico south through the Isthmus over most of northern South America, then in the Amazonian portion of Peru and Bolivia to Argentina. There is a disjunct population in southeastern Brazil (map 8.94). In Venezuela it occurs up to 1,537 m, but most specimens were taken below 500 m. It is strongly associated with moist habitats and multistratal evergreen forest (Handley 1976).

Natural History

Females probably exhibit bimodal polyestry (Lewis and Wilson 1987). Individuals of this species were captured in Paraguay at the edge of a small clearing in tropical forest. In Venezuela this bat is strongly associated with moist habitats and multistratal evergreen forest. It is entirely frugivorous and has been described as a common fig specialist (Handley 1976; Lewis and Wilson 1987; Myers and Wetzel 1983).

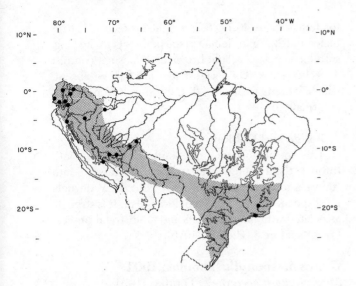

Map 8.94. Distribution of *Vampyressa pusilla*.

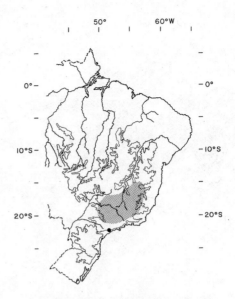

Map 8.95. Distribution of *Chiroderma doriae*.

Genus *Chiroderma* Peters, 1860
Large-eyed Bat

Description
The dental formula is I 2/2, C 1/1, P 2/2, M 2/2. The skull is unusual in that the nasal bones are absent. Head and body length ranges from 55 to 77 mm, and the forearm ranges from 37 to 53 mm. The upperparts are usually lighter. The presence of facial stripes varies from species to species, and the dorsal stripe may be faint or absent.
Chromosome number: all species show $2n = 26$; FN $= 48$.

Distribution
The genus is distributed from Sinaloa in western Mexico and southern Veracruz in eastern Mexico south through the Isthmus to Brazil.

Natural History
The species of *Chiroderma* are frugivorous. Reproduction is seasonal and timed by the onset of the rains.

Chiroderma doriae Thomas, 1891

Description
Similar to *C. villosum* but larger, with the forearm measuring 49.5–55.5 mm.

Distribution
Found in the states of Minas Gerais and São Paulo, Brazil (map 8.95).

Chiroderma salvini Dobson, 1878

Description
Head and body length averages 72.1 mm for males and 73.2 mm for females; HF 14; FA 48; weight averages 27.2 g for males and 29.1 g for females (see

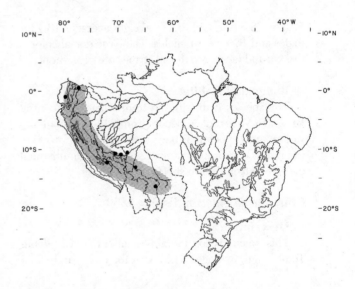

Map 8.96. Distribution of *Chiroderma salvini*.

table 8.17). The facial and dorsal stripes are very conspicuous in this species.

Range and Habitat
The species is distributed from Sinaloa in western Mexico and southern Veracruz south through the Isthmus to northern Venezuela, western Colombia, and Amazonian Peru to Bolivia (map 8.96). It occurs over a range of elevations, from 611 to 2,240 m in Venezuela, and prefers moist habitats and multistratal tropical evergreen forest (Handley 1976).

Chiroderma trinitatum Goodwin, 1958

Description
This is the smallest species of the genus. Males average 54.8 mm in total length and females 56.7 mm;

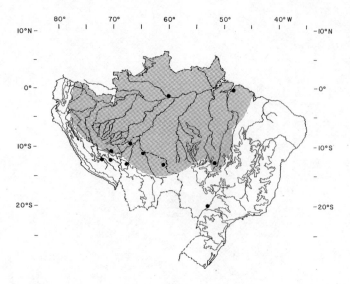

Map 8.97. Distribution of *Chiroderma trinitatum*.

FA 38–38.5 (see table 8.17). Weight averages 13 g for males and 13.9 g for females. The basic dorsal color is brown, and facial and dorsal stripes are prominent.

Range and Habitat

The species is distributed from Panama to central Brazil. In Venezuela it occurs from 24 to 1,032 m elevation; most specimens have been taken below 500 m (map 8.97). It prefers moist habitats and multistratal evergreen tropical forest (Handley 1976).

Chiroderma villosum Peters, 1860

Description

This species is considerably smaller than *C. salvini*. Total length averages 64.57 mm for males and 67.55

for females; FA 43.7–45.7; weight averages 21 g for males and 22.9 g for females (see table 8.17). The long, soft dorsal pelage is light brown and does not contrast with the venter. The facial stripes are faint or absent, which, with its small size, distinguishes it from *C. salvini* (plate 6).

Range and Habitat

This species occurs from southern Mexico south through the Isthmus, over much of northern South America to western Brazil (map 8.98). In Venezuela most specimens were taken below 500 m. It is strongly associated with moist habitats and multistratal tropical evergreen forest (Handley 1976).

Genus *Mesophylla* Thomas, 1901
Mesophylla macconnelli Thomas, 1901

Description

The dental formula is I 2/2, C 1/1, P 2/2, M 2/2. Head and body length averages 42.8 mm for males and 44.1 mm for females; there is no external tail. Weight averages 6.5 g for males and 6.6 g for females and forearm 30.7 to 30.9 (see table 8.17). The dorsum is a dull brownish white, darkening to brown on the lower back. This species is markedly different from the white *Ectophylla alba,* once considered congeneric. Chromosome number: $2n = 21$–22; FN = 20.

Range and Habitat

The species ranges from southern Costa Rica through the Isthmus over much of northern South America and northeastern Brazil (map 8.99). In Venezuela it is common below 500 m elevation. It prefers moist habitats and multistratal tropical evergreen forest (Handley 1976).

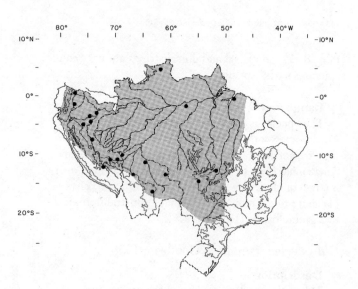

Map 8.98. Distribution of *Chiroderma villosum*.

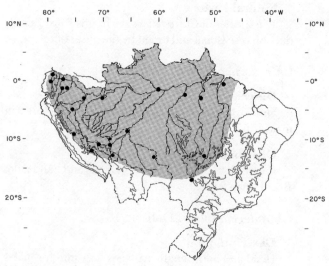

Map 8.99. Distribution of *Mesophylla macconnelli*.

Natural History

In common with many phyllostomids, this bat is seasonally polyestrous. It is associated with lowland rain forests and will construct a leaf "tent" for roosting (Foster 1992). Females with nursing young tend to roost alone. The diet is mainly fruit (Kunz and Pena 1992).

Comment

The taxonomic status of *M. macconnelli* has been a puzzle. Kunz and Pena (1992) present the most useful summary.

Genus *Artibeus* Leach, 1821
Fruit-eating Bat

Description

The dental formula is I 2/2, C 1/1, P 2/2, M 2-3/2-3 (fig. 8.17). Head and body length ranges from 53 to 100 mm and forearm length from 35 to 76 mm. The species of this genus present an array of sizes. The interfemoral membrane is rather narrow. In general the dorsum is brownish to black, and the underparts are usually paler. The dorsal line, so characteristic of most stenodermine bats, is absent. Four whitish facial stripes may be present, but this varies from species to species.

Chromosome number: all species show $2n = 30-31$; FN = 56.

Distribution

Species of *Artibeus* range from Sinaloa in western Mexico and Tamaulipas in eastern Mexico south over most of tropical South America to northern Argentina and southeastern Brazil.

Natural History

Artibeus contains some fourteen species (at least two more are yet to be formally described). Within any community they are usually graded in size, presum-

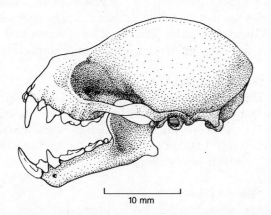

Figure 8.17. Skull of *Artibeus jamaicensis.*

ably reflecting specialization for different-sized foods. These bats are primarily frugivores, and one species (*A. jamaicensis*) has been studied in some detail (see species account). Most species exhibit a seasonal bimodal polyestrous pattern, with births occurring late in the dry season and in the midportion of the rainy season where rainfall is seasonal (Wilson 1979).

Comment on Taxonomy

Some years ago the late Randolph Peterson confided to me (JFE) that the genus *Artibeus* was a "taxonomic mess." His warning was heeded by Handley and Owen. The result is a ferment with respect to the taxonomic status of the genus. In 1989 Handley proposed that *Artibeus fimbriatus* Gray include *A. perspicillatus, A. grandis,* and *A. lituratus* in part. Furthermore, *A. jamaicensis* Leach, 1821, includes *Artibeus lobatus.* Handley goes on to include *A. davisi* in *Artibeus fuliginosus* (Handley 1989). He notes that if *A. fuliginosus* has been correctly applied to the blackish forms of Amazonas it may be predated by *A. obscurum* Schinz, 1821. In 1991 Handley proposed the view that *Phyllostoma planirostre* Spix, 1823, is in fact a composite of *Artibeus planirostris* and *A. fimbriatus.*

Owen (1987), using a phenetic approach, was able to divide the genus *Artibeus* into two groups. He proposed a new genus *Dermanura* to unite *Artibeus andersoni, A. cinerea, A. phaeotis, A. tolteca, A. glauca, A. watsoni,* and *A. azteca* separately from the remaining species of species of *Artibeus. A. concolor* was ambiguous in its status. In 1991 Owen reviewed two characters of a larger sample and concluded that *A. concolor* cannot be assigned to either *Dermanura* or *Artibeus.* He concluded that *A. concolor* should be assigned to its own genus *Koopmania.* Marques-Aguiar (1995) offers a furthur discussion of the systematics of *Artibeus.* Given this introduction, I must say that in presenting the species in the old genus *Artibeus* we will recognize some of the new designations. Synonyms will be noted to aid the reader. We tend to follow Koopman (1993). Subgenera are given in parentheses following the genus *Artibeus.*

Artibeus (Artibeus) amplus Handley, 1987

Description

This new species, described by Handley in 1987, is very similar to *A. literatus.* It is one of the largest species of the genus. The validity of this taxon is accepted in Wilson and Reeder (1993). Head and body length averages 91.5 mm for males and 90.3 mm for females; weight averages about 60 g. The skull is longer and narrower than in *A. jamaicensis.* The lower edge of the nose leaf is fused to the facial skin. The wings are never white tipped (see also Lim and Wilson 1993).

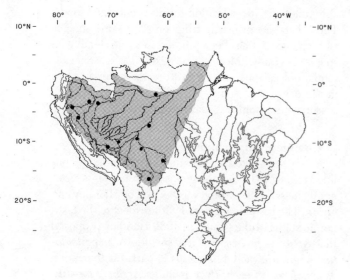

Map 8.100. Distribution of *Artibeus andersoni.*

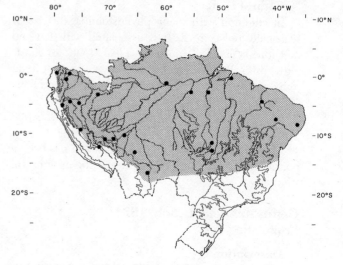

Map 8.101. Distribution of *Artibeus cinereus* (but see Anderson 1997 for Bolivia).

Range and Habitat

The species was diagnosed from specimens taken in Venezuela. It has been recorded from the "northern foothills of the Colombian Andes, . . . the Venezuelan Andes . . . and the vicinity of Cerro Duida and the low southeastern mountains of Est. Bolívar in southeastern Venezuela. It probably occurs in adjacent parts of Guyana and Brazil as well" (Handley 1987, 164–65) (not mapped).

Artibeus (Dermanura) andersoni Osgood, 1916

Description and Comments

The species was previously considered a subspecies of *A. cinereus,* but see Koopman (1978).

Distribution

This bat is found from western Brazil to Ecuador and Peru.

Comment

Koopman (1982) suggests that *Artibeus andersoni* could extend to southeastern Colombia and the eastern Guianas, but I cannot record a specimen (map 8.100).

Artibeus (Dermanura) cinereus (Gervais, 1856)

Description

This is one of the smaller species of the genus, and some workers include *Artibeus andersoni* within this taxon. Measurements: HB 55; HF 11.5; E 16.7; FA 40; Wt 13 g (see table 8.17). Both dorsum and venter are medium brown, and the throat may be pale brown. The white facial stripes are prominent.

Range and Habitat

This species ranges from southern Veracruz, Mexico, through Panama and across most of northern

South America to central Brazil. In Venezuela it ranges from 1,000 to 2,000 m elevation (map 8.101). It prefers multistratal tropical evergreen forest and in general occurs above 1,000 m. Where it co-occurs with *A. concolor,* the latter species generally replaces *A. cinereus* at lower elevations.

Natural History

These bats roost in small groups, usually in trees such as palms (Goodwin and Greenhall 1961).

Artibeus (Koopmania) concolor Peters, 1865

Comment

The species was previously assigned to the genus *Dermanura* Owen, 1988, thence elevated to *Koopmania* Owen, 1991.

Description

Measurements: HB 61–64; HF 11.5; E 19; FA 47–48; Wt 18–20 g (see table 8.17). This species is smaller than *A. jamaicensis* but larger than *A. cinereus;* females are larger than males. The coloration is similar to that of *A. cinereus,* but the facial stripes are indistinct.

Range and Habitat

This species is found in southern Colombia, Venezuela, the Guianas, and northern Brazil (map 8.102). In Venezuela specimens were caught in moist areas from 100 to 1,000 m elevation, but the vast majority were found below 500 m (Handley 1976).

Natural History

Apparently breeding follows patterns of pulsed primary productivity with its range. Although taken in the dry deciduous forests of Brazil, the species may

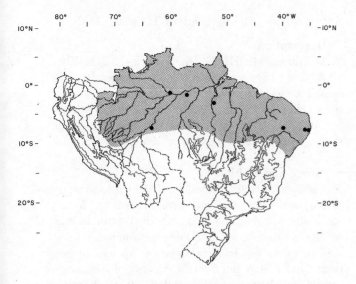

Map 8.102. Distribution of *Artibeus concolor.*

be closely tied to more mesic riverine habitats. It is assumed to be primarily frugivorous (Acosta and Owen 1993).

Artibeus (Artibeus) fimbriatus Gray, 1838

Description

Measurements: TL 95.2; T 18; HF 23.5; E 65; Wt 54 g (Redford and Eisenberg 1992). This very large *Artibeus* has long, lax, silky fur. It is sooty brown above, a little paler below, with white-tipped hairs giving it a frosted appearance. There is a wide median band of dorsal hairs; the face is blackish with facial stripes at least faintly indicated. The ears are blackish. The wings are usually white tipped. The legs and interfemoral membranes are distinctly haired (Handley 1989).

Distribution

A. fimbriatus occurs from Rio de Janeiro state, Brazil, south along the coast to eastern Paraguay and adjacent Argentina at altitudes from sea level to 530 m (Handley 1989) (map 8.103).

Comment

The type is recognized, but it may be confused with *A. lituratus* (see Handley 1989) (see also *A. lituratus* in this volume).

Artibeus (Artibeus) fraterculus Anthony, 1924

Description

The dorsal hairs are faintly bicolored, light brown on the tips and pale brown at the base. The general effect is a light brown dorsum washed with white. The venter is a gray brown, and the facial stripes are prominent. There is no dorsal stripe. Five specimens from Peru measured TL 74–83; T 0; HF 14–18; E 17–20; FA 48–50; Wt 36–43 g (LSU).

Distribution

This bat has apparently been collected only in the lowland rain forests of western Ecuador and Peru (map 8.104).

Artibeus fuliginosus Gray, 1838

See *Artibeus obscurus.*

Artibeus glaucus Thomas, 1893

Description

Measurements: HB 50; HF 11; E 17.4; FA 38; Wt 10.5 g (see table 8.17). This is one of the smallest species of the genus (Handley 1987). The facial stripes and dorsal pelage are similar to those of *A. jamaicensis.*

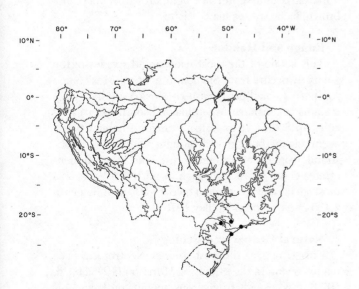

Map 8.103. Distribution of *Artibeus fimbriatus.*

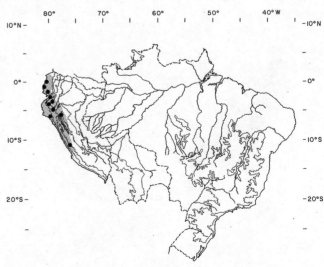

Map 8.104. Distribution of *Artibeus fraterculus.*

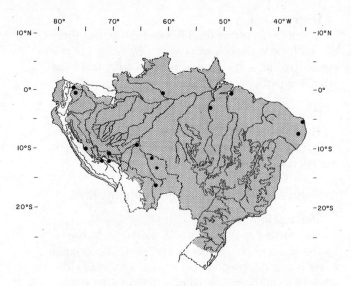

Map 8.105. Distribution of *Artibeus glaucus* (includes *A. gnomus*).

Range and Habitat

The species has been taken in Venezuela, and I (JFE) have seen similar specimens from Suriname (CMNH). Handley has the following comments: "Distribution—The Amazon Basin and bordering regions, from northern Amazonas Territory (14 km SSE Pto. Ayacucho) and northern Bolívar State (28 km SE El Manteco) in Venezuela and northern Guyana to Pará (Belém) and Mato Grosso (Serra do Roncador), Brazil and Loreto (Santa Rosa), Peru" (Handley 1987, 167) (map 8.105).

Comment

Koopman (1993) subordinates *Artibeus gnomus* Handley (1987) into *Artibeus glaucus*.

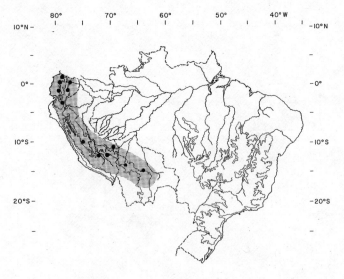

Map 8.106. Distribution of *Artibeus hartii*.

Artibeus (Enchisthenes) hartii Thomas, 1892

Description

Measurements: HB 60–61; HF 12–12.2; E 15.8; FA 39.2; Wt 18 g (see table 8.17). This species does not exhibit noticeable sexual size dimorphism. The dorsum is dark brown, the venter paler; it is exceptional for *Artibeus* in that it has a prominent white dorsal stripe. White facial stripes are present but faint.

Range and Habitat

The species is distributed from Nayarit in western Mexico and southern Tamaulipas in eastern Mexico south through Panama to northern Venezuela and western Colombia, south to Peru and Amazonian Bolivia (map 8.106). In Venezuela this species ranges from 2 to 2,250 m, but the vast majority of specimens were taken between 1,000 and 2,250 m elevation (Handley 1976). It is strongly associated with moist habitats and multistratal tropical evergreen forest, but it does penetrate cloud forest, and some specimens were taken in dry deciduous forest.

Natural History

These bats are often captured while foraging over water and are associated with moist, multistratal tropical evergreen forest. They feed almost exclusively on small figs. They are vulnerable to predation by owls (Arroyo-Cabrales and Owen 1997).

Artibeus (Artibeus) jamaicensis Leach, 1821

Description

Males average 80.1 mm in total length and females 81.8 mm; weight averages 40.4 g for males and 43.2 g for females (see table 8.17). There is considerable geographic variation in size (see Handley 1987). The upperparts and venter are light brown to very dark brown; facial stripes may be faint.

Range and Habitat

This is one of the most widespread species of the genus. It occurs from Sinaloa in western Mexico and southern Tamaulipas in eastern Mexico south through Panama to northern Argentina and southeastern Brazil (map 8.107). It tolerates a range of habitat types, occurring in both dry deciduous forest and multistratal tropical evergreen forest, and it even penetrates cloud forest. Although in northern Venezuela most specimens were taken below 1,500 m, it ranges to 2,135 m elevation (Handley 1976).

Natural History and Ecology

This is one of the best-studied Neotropical bats, mainly owing to the efforts of Morrison (1978a,b,c,d, 1979). It is strongly frugivorous, feeding on *Ficus* (Au-

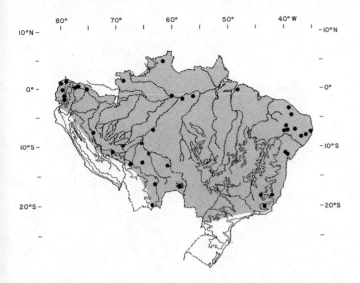

Map 8.107. Distribution of *Artibeus jamaicensis*.

Table 8.13   Numerical Densities (Estimated) of Common Bats from Barro Colorado Island, Panama

Uroderma bilobatum	$50.6/\text{km}^2$
Artibeus lituratus	$33.0/\text{km}^2$
Chiroderma villosum	$24.2/\text{km}^2$
Carollia perspicillata	$22.0/\text{km}^2$
Vampyrodes caraccioli	$21.5/\text{km}^2$
Phyllostomus discolor	$18.2/\text{km}^2$
Carollia castanea	$11.0/\text{km}^2$
Micronycteris hirsuta	$9.9/\text{km}^2$
Vampyressa pusilla	$9.9/\text{km}^2$
Vampyressa nymphaea	$9.9/\text{km}^2$

Source: Based on capture data from Handley, Wilson, and Gardner 1991.
Note: Densities are approximate. *A. jamaicensis* not included; see text.

gust 1981), and its reproduction is closely timed with maximum abundance of fruit (Fleming 1971). Breeding colonies of up to twenty-five involve harem defense by the male (Kunz, August, and Burnett 1983). The mating system is strongly polygynous. During lactation, females have a daytime roost, and when moving to forage they deposit their babies in crèches near the feeding tree. Bright moonlight strongly inhibits the flight of these bats. It is suspected that owls may be significant predators at night and snakes and the bat falcon, *Falco rufigularis,* during the day. When foraging the bats fly in small groups; if an individual is caught it will emit distress calls, inducing "mobbing" behavior by fellow flock members (August 1979). These bats occasionally construct tents by biting the midribs of large leaves, causing the leaves to fold over (Foster and Timm 1976) (see also table 8.14). They also roost in hollow trees and well-lighted caves (Goodwin and Greenhall 1961). According to Handley, Wilson, and Gardner (1991), there are roughly fifty-six species of bats recorded from Barro Colorado Island, Panama. The island itself is approximately 15 km^2 in size. During their research effort they took thirty-nine species, but they were netting for fruit bats and thus the insectivorous forms are underrepresented. I believe, however, that the thirty-nine species constitute a fair sample for the frugivore community. According to their efforts, *A. jamaicensis* is estimated to occur at a density of approximately 200 individuals/km^2. This includes not only the island itself but adjacent areas. *Artibeus jamaicensis can* exist on a daily intake of roughly 7 ± 2 fruits of *Ficus insipida.* Thus approximately 1,400 *F. insipida* fruits/day/km^2 are consumed by this species alone. It must be appreciated, however, that *A. jamaicensis* coexists with other guilds of frugivorous bats (see table 8.13).

Table 8.13 is taken in part from table 1–2 in Handley, Wilson, and Gardner (1991). Utilizing column 1 of their table (mean catch/species night), if we assume equal probability of capture, it follows that the ten most common species after *A. jamaicensis* would exist at the density suggested in table 8.13. The remaining seven frugivorous species of bats exist at < 9/km^2 but > 5.0/km^2. In the list in table 8.13, the carnivorous and mixed feeding, as well as insectivorous species, are excluded.

Thus, if we consider the biomass of frugivorous bats, the seven species that are most common exist at 63/km^2. The eleven species tabulated, including *A. jamaicensis,* in total exist at 410/km^2, giving a grand total for the frugivorous group of 473/km^2. If we assume a mean weight/individual of $\approx$ 30 g, then the total biomass for the frugivorous bats on Barro Colorado Island is roughly 14.19 kg/km^2.

Now let us consider the impact on fruits. If we consider *only* species of the genus *Ficus,* then the production of figs on the island is $\approx$ 100 kg dry mass of fruit/ha/yr. The *Ficus* consumed by bats (and here we will only consider the genus *Artibeus* that includes three species) would result in a predation level of about 28 kg (dry mass) of *Ficus*/ha/yr. If all frugivorous bats of the canopy consume 40 kg (dry mass) of fruit/ha/yr, then this group of bats is taking just under 50% of the fruit productivity; but note that other mammalian predators are present, including the kinkajou (*Potos*) and the howler monkey (*Alouatta*). These two species alone may consume an additional 20 kg (dry mass) of *Ficus*/ha/yr. In this series of estimates, predation of insects and birds on *Ficus* is not considered. It will be appreciated, however, that when birds and mammals eat fruits they disperse the seeds. Thus the loss of the fleshy portion of the fruit to the consumers is offset by the benefits of seed dispersal. *Artibeus jamaicensis* without a doubt is one of the best studied species of phyllostomatid bats (Handley, Wilson, and

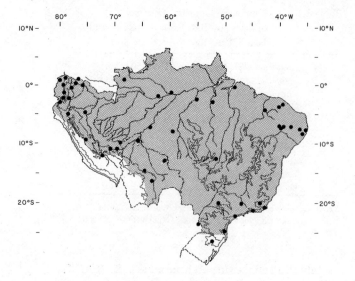

Map 8.108. Distribution of *Artibeus lituratus*.

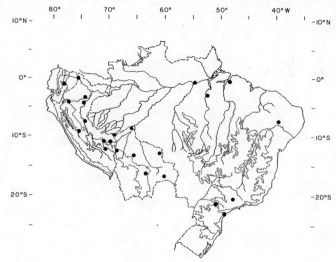

Map 8.109. Distribution of *Artibeus obscurus* (includes *A. fuliginosus*).

Gardner 1991), matched only by the studies on *Carollia perspicillata* conducted by Fleming (1988). These two treatises are benchmark studies of tropical frugivorous bats.

Artibeus (Artibeus) lituratus (Olfers, 1818)

Description

Measurements: HB 89–91; HF 19–20; E 24; FA 68–69; Wt 65–68 g (see table 8.17). This is one of the largest species of *Artibeus*. The dorsum and venter are light brown, and the facial stripes are conspicuous.

Range and Habitat

This species occurs from Sinaloa in western Mexico and southern Tamaulipas in eastern Mexico south through Panama over most of the South American continent, to Argentina and Brazil (map 8.108). Together with *A. jamaicensis,* this is one of the most common species of fruit bats. In Venezuela it mainly occurs below 500 m, but some specimens may be taken up to 2,000 m elevation. Although it prefers moist areas and multistratal evergreen forest, it tolerates man-made clearings.

Natural History

Handley (1967) noted that in Brazil this species forages high in the canopy. The bats roost in well-lighted caves and in palm trees. Colonies may be as large as twenty-five individuals (Goodwin and Greenhall 1961). Reproduction is controlled by rainfall and subsequent primary productivity (Sazima 1989). In an urban forest in Brazil this species fed on eleven species of plants. They will construct "tents" by biting the midrib of leaves (Zortea and Chiarello 1994). A total of thirteen species of plants were utilized for their fruits in Brazil, with from five to nine different species uti-

lized in any month (Galletti and Morellato 1994). Selective feeding on leaves has also been noted for this species (Zortea and Mendes 1993).

Artibeus (Artibeus) obscurus Gray 1838

Description

Five specimens from Peru measured TL 70–73; T 0: HF 13–17; E 18–22; FA 50–58; Wt 29–35 g (LSU). The dorsal pelage presents a uniform dark brown appearance, and though the venter is a paler brown, it does not contrast. Facial and dorsal striping patterns are faint or lacking. Handley (1989) suggests this taxon may be subordinate to *A. obscurum* Schinz.

Range and Habitat

The species ranges south of the Orinoco in Venezuela and into the adjacent portions of the Guianas and northern Brazil. In the west it occurs in southern Colombia south to Bolivia (map 8.109). Most specimens taken in Venezuela were caught below 500 m elevation. It is associated with moist habitats and multistratal tropical evergreen forest (Handley 1976).

Artibeus (Dermanura) phaeotis (Miller, 1902)

Description

This small species has a head and body length of 51 to 60 mm; forearm length is 35.2 to 41.8 mm. A single specimen from Veracruz measured TL 59; HF 10; E 17 (CMNH). The dorsum and venter are brown, and the facial stripes are prominent.

Range and Habitat

This species occurs from southern Sinaloa in western Mexico and southern Veracruz in eastern Mexico south through Panama, to Colombia and Ecuador

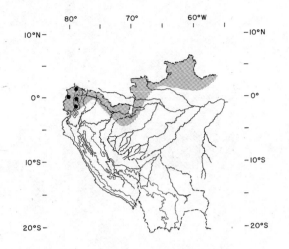

Map 8.110. Distribution of *Artibeus phaeotis*.

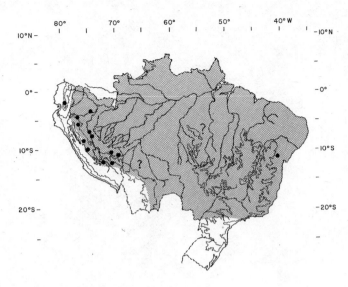

Map 8.111. Distribution of *Artibeus planirostris.*

(map 8.110). The status of this species in Venezuela still needs to be clarified pending a future publication by Handley.

Natural History
Though primarily frugivorous, this bat will feed on insects and pollen. Strongly associated with lowland forests, it constructs a "tent" using banana leaves as described for several other species of the stenodermines (Timm 1985).

Artibeus (Artibeus) planirostris (Spix, 1823)

Description
This heavy-bodied, medium-sized bat is similar in size to *A. lituratus*. Five specimens from Peru exhibited the following measurements: HB 88–91; HF 15–17; E 21–25; Wt 42–65 g (LSU). It has ashy brown dorsal coloration, with the light gray of the hair bases showing on the shoulders. Dorsally there is faint grizzling, and this is more pronounced on the venter. The facial stripes are faint. The wings are dark and the ears lighter (plate 6).

Distribution
Artibeus planirostris is found from Colombia south to northern Argentina and eastern Brazil (Barquez 1987) (map 8.111).

Comment
Some authors use the genus *Cynomops* for this species.

Artibeus (Dermanura) toltecus (Saussure, 1860)

Comment
Although the species may occur in Ecuador, it has not been recorded (see Eisenberg 1989).

Genus *Pygoderma* Peters, 1863
Pygoderma bilabiatum (Wagner, 1843)

Description
The dental formula is I 2/2, C 1/1, P 2/2, M 2/2. Measurements: HB 58.4; HF 10.5; E 21; FA 40–43; Wt 21–22 g (see Redford and Eisenberg 1992). The dorsum is pale brown varying almost to black; the venter is grayish brown. There is a white spot on each shoulder near the wing, but this bat lacks facial stripes and a dorsal stripe. Males have large subdermal glands surrounding the eye (Meyers 1981). In the male the fur of the chest is very thin or absent.
Chromosome number: $2n = 30-31$; FN = 56.

Range and Habitat
This species has been described from southern Brazil, Bolivia, Argentina, and Paraguay. Recently specimens have been collected in Suriname (map 8.112) (Webster and Owen 1984). This bat prefers tropical forest habitats.

Natural History
In Paraguay this species exhibits two marked birth peaks during the fall and winter. It is believed to feed on fruit pulp (Myers 1981). The litter size is one, and in Paraguay it reproduces in fall and winter (Myers 1981). This species has been collected in mature tropical forest, subtropical forest, and second growth bordering forest (Carvalho 1961; Olrog 1967).

Genus *Ametrida* Gray, 1847
Ametrida centurio Gray, 1847

Description
The dental formula is I 2/2, C 1/1, P 2/2, M 3/3. Measurements: HB 41–47; HF 10.6–11.3; E 14.2–

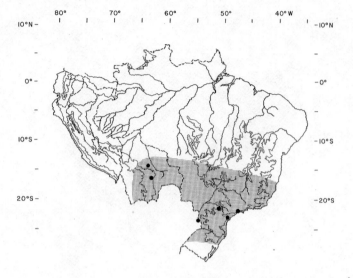

Map 8.112. Distribution of *Pygoderma bilabiatum*.

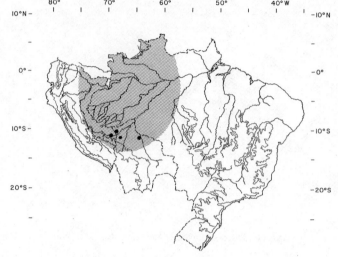

Map 8.114. Distribution of *Sphaeronycteris toxophyllum*.

15.0; FA 26–32; Wt 8–9.5 g. Males average 8 g, and females average 10 to 13 g (see table 8.17). The pronounced sexual size dimorphism formerly led to the small males' being given the name *A. minor*. The rostrum is very short. There are some dermal outgrowths on the sides of the face, though they are less extensive than in *Centurio*, and the nose leaf is prominent. The dorsum is light to dark brown, and there is a single gray patch on each shoulder.
Chromosome number: $2n = 30–31$; FN = 56.

Range and Habitat

The species is distributed east of the Andes in Venezuela, across the Guianas, and south into north-central Brazil (map 8.113). In Venezuela specimens were taken at up to 2,150 elevation, but most were taken below 1,500 m. These bats are strongly associ-

ated with moist areas and multistratal tropical evergreen forest (Handley 1976). They are frugivores.

Genus *Sphaeronycteris* Peters, 1882
Sphaeronycteris toxophyllum Peters, 1882

Description
The dental formula is I 2/2, C 1/1, P 2/2, M 3/3. Measurements: HB 56–58; HF 12; E 15.5–16; FA 37.6–40; Wt ≈ 17 g (see table 8.17). The dorsum is cinnamon brown. There is some fleshy outgrowth on the side of the face, reminiscent of *Centurio* (plate 6). The male has a hornlike growth on the forehead, but in the female the structure is rudimentary. A fold of skin is present under the chin, and in the male it can be rolled over the face like that described for *Centurio* (see Eisenberg 1989).
Chromosome number: $2n = 28$; FN = 52.

Range and Habitat
This species occurs from Venezuela and eastern Colombia, east of the Andes, south to Amazonian Peru and northwestern Brazil (map 8.114). In Venezuela specimens were taken at up to 2,240 m. They may follow gallery forest into dry habitats but are usually associated with multistratal tropical evergreen forest, though they tolerate man-made clearings.

SUBFAMILY DESMODONTINAE

Diagnosis
This subfamily contains three species that range in head and body length from 65 to 90 mm. The dental formula for the three species is highly variable and will be discussed in the species accounts, but the incisors and canines are specialized for cutting, and the cutting edges form a V (fig. 8.18). They use these teeth

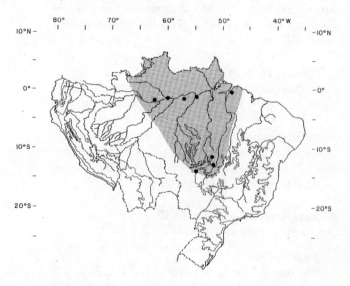

Map 8.113. Distribution of *Ametrida centurio*.

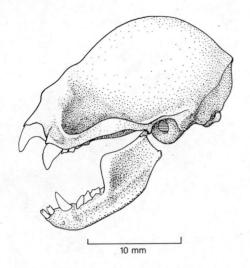

Figure 8.18. Skull of *Desmodus rotundus*.

to make incisions in their prey, for these are the true vampire bats. The premolars and molars are greatly reduced, indicating less selection for their retention, since these animals have specialized for their peculiar dietary habits. There is no tail. The nose leaf is reduced to U-shaped fleshy pads that surround each nostril. The dorsum is generally some shade of brown.

Distribution

Members of this subfamily occur from southern Sonora and southeastern Texas south across the continent of South America to central Argentina, Uruguay, and northern Chile.

Natural History

The three species of this subfamily are specialized for feeding on the blood of warm-blooded vertebrates. The saliva contains an enzyme that retards coagulation, and they lap the blood after making an incision. They can walk on their feet and thumbs and usually crawl on the body of their prey before feeding. Vampire bats are of medical importance because they have been implicated in transmitting rabies.

Genus *Desmodus* Wied-Neuwied, 1826
Vampiro, Mordedor
Desmodus rotundus (E. Geoffroy, 1810)
Common Vampire Bat

Description

The dental formula is I 1/2, C 1/1, P 2/3, M 0/0 (fig. 8.18). Head and body length ranges from 75 to 90 mm and the forearm from 50 to 63 mm (see table 8.17). Measurements: HB 81–86; HF 17–18; E 16–18; FA 61–66; Wt ≈ 41 g. The weight is highly variable, since these bats become extremely distended after feeding. The thumb is rather long and has a distinct pad at the base (fig 8.4B). The dorsum is gray brown, and the underparts are somewhat paler (plate 6).

Chromosome number: $2n = 28$; FN = 52.

Range and Habitat

The species is distributed from southern Sonora and Tamaulipas, Mexico, south through the Isthmus to central Argentina and Uruguay. It rarely ranges above 1,000 m elevation (map 8.115). It tolerates a broad variety of cover types, being found in multistratal evergreen forest, man-made clearings, pastures, and deciduous tropical forest (Handley 1976).

Natural History

This species appears to feed almost exclusively on the blood of mammals. It is prone to feed on livestock and thus becomes a pest. It is implicated in transmitting the rabies virus, and the livestock losses may be considerable. The single young is born after a 110-day gestation period. These bats may roost in hollow trees, though they seem to prefer caves, and colonies may be as large as five hundred. Vampire bats often share roosting places with other species. The bats forage in small groups that may be families. They are quite able to run over surfaces, an adaptation they employ when selecting a place to bite on ungulates such as the domestic cow or white-tailed deer (Turner 1975; Schmidt and Manske 1973).

Over much of its range the common vampire can apparently breed throughout most of the year. The litter size is usually one, the gestation approximately seven months, and the neonatal weight 5–7 g. Young are suckled up to three hundred days, and there is a gradual switch from milk to blood, the young being fed regurgitated blood. After about four months the young accompanies its mother to prey, drinking blood from her feeding site, and after about five months growth is largely complete. In a Costa Rican population the

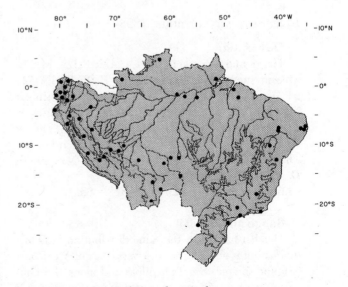

Map 8.115. Distribution of *Desmodus rotundus*.

interbirth interval was ten months and the infant mortality rate was 54% (Barlow 1965; Crespo 1982; Greenhall, Joermann, and Schmidt 1983; Langguth and Achaval 1972; Wilkinson 1985a,b).

D. rotundus is found in many different habitats and at elevations up to 1,500 m in Jujuy province, Argentina, and 2,000 m in Chile, but apparently it prefers tropical and subtropical woodland and open grasslands for foraging. The bats commonly forage 5–8 km from the roost, but distances of up to 20 km have been recorded.

In different areas, different roost sites are selected. In northern Argentina vampires roost in tree trunks, especially *Chorisia* and *Caesalpinia,* and in Chile they also roost in sea caves. Throughout their range they exhibit a preference for mine shafts, old wells, and abandoned buildings. The colonies are usually small (twenty to one hundred), but colonies up to about five thousand have been observed. Vampires commonly share roosts with other species; over their entire range, forty-five species of bats have been recorded as roosting with vampires.

In a Costa Rican population the social unit is a group of eight to twelve females. Males fight for access to the top of the preferred female roosting site. Males fight for access to the top of the preferred female roosting site. Males that gain this position mate preferentially with the females in that roost. Within the colony there is extensive sharing of blood by regurgitation, and bats that fail to obtain a blood meal are fed by successful roost mates. Food sharing is based both on degree of relatedness and on opportunity for reciprocation. There is also extensive allogrooming, which takes place preferentially between related bats. Male offspring leave the maternal group and females remain (see Redford and Eisenberg 1992; Wilkinson 1984, 1985a,b, 1986).

Diaemus youngi (Jentink, 1893)

Description

The dental formula is I 1/2, C 1/1, P 1/2, M 2/1. Measurements: HB 84; HF 18–19; E 18–19; FA 52–54; Wt 31–32 g (see table 8.17). This species is similar in size to *D. rotundus.* Head and body length averages approximately 83 mm, and weight averages about 35 g. The thumb is shorter than that of *D. rotundus.* The tips of the wings are white, which easily separates *D. youngi* from *D. rotundus.*
Chromosome number: $2n = 32$; FN = 60.

Range and Habitat

This bat has nearly the same distribution as *D. rotundus* but is absent from the west coast of Mexico. It is found in southern Tamaulipas and along the Gulf Coast south to Central America, and from there it

Map 8.116. Distribution of *Diaemus youngi.*

ranges throughout South America to southeastern Brazil and northern Argentina (map 8.116). It prefers moist areas but forages in dry deciduous forest as well as in multistratal evergreen forest. In Venezuela specimens were taken below 500 m elevation (Handley 1976).

Natural History

Like *D. rotundus,* this species feeds on the blood of higher vertebrates. Glands in the mouth produce a sharp odor in the male, but their exact function in communication is poorly understood. The bats nest in hollow trees, in colonies of up to thirty (Goodwin and Greenhall 1961).

Genus *Diphylla* Spix, 1823
Diphylla ecaudata Spix, 1823
Vampiro de Doble Escudo

Description

The dental formula is I 2/2, C 1/1, P 1/2, M 2/2. The outer lower incisor is fan shaped and has seven tiny lobes; it may be important in grooming the fur. Measurements: HB 73–79; HF 16–18; E 15–16; FA 50–52; Wt 25–26 g (see table 8.17). These bats are smaller than *Desmodus.* Head and body length averages 73.2 mm for males and 78.7 mm for females; weight averages 25.9 g for males and 25.2 g for females. The dorsum is medium brown, and the venter is nearly the same shade.
Chromosome number: $2n = 32$; FN = 60.

Range and Habitat

This species ranges from the Gulf Coast of Texas south in eastern Mexico to Central America, and thence to Colombia, Venezuela, Peru, Ecuador, Amazonian Bolivia, and Brazil south of the Amazon (map

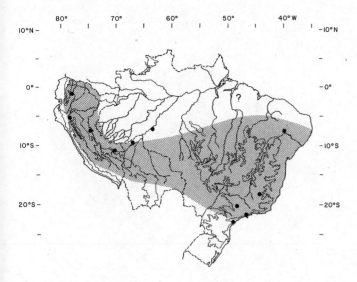

Map 8.117. Distribution of *Diphylla ecaudata*.

8.117). In Venezuela it was collected at elevations up to to 1,537 m. It tolerates a variety of habitats, including farms, multistratal evergreen forest, and dry deciduous forest (Handley 1976).

Natural History

This species apparently is specialized for feeding on the blood of birds and can be a serious pest to poultry farmers. It has been found roosting in caves, generally in colonies smaller than those typical of *D. rotundus* and as a monospecific unit. It rests in small groups (one to three) in caves or hollow trees (Greenhall, Schmidt, and Joermann 1984).

SUPERFAMILY VESPERTILIONOIDEA

These bats appear to be the most specialized for flight, based on the anatomy of the pectoral girdle. The greater tuberosity of the humerus is enlarged, and on the upstroke of the wingbeat it articulates with the scapula, thus mechanically terminating the upswing (Miller 1907) (see figs. 8.1 and 8.11). For an amplified discussion see Eisenberg (1989) and Redford and Eisenberg (1992).

FAMILY NATALIDAE

Diagnosis

This family includes a single genus, *Natalus*. The dental formula is I 2/3, C 1/1, P 3/3, M 3/3. The tail is very long and completely enclosed in the interfemoral membrane, and it nearly equals or exceeds the head and body length (fig. 8.3). The thumb is short and almost enveloped in the antebrachial membrane, and the third phalanx of the third finger remains cartilagi-

nous even in the adult. Adult males have a glandlike structure in the center of the forehead. The ears have a peculiar funnel shape (see fig. 8.4C).

Distribution

This family is confined to the Western Hemisphere. It is distributed from northern Mexico to northern South America and eastern Brazil and has speciated in the Antilles.

Natural History

These bats are aerial insectivores and find their prey through echolocation. They roost in caves or tunnels, sometimes with other cavernicolous bats, and the colonies may be very large.

Genus *Natalus* Gray, 1838
Funnel-eared Bat
Natalus stramineus Gray, 1838

Description

Head and body length is approximately 50 mm; the tail is longer than the head and body, averaging about 55 mm, and forearms range from 35 to 45 mm. Two females from Colombia measured TL 92, 101; T 45, 53; FH 9, 16; E 13, 17; Wt 4.3, 6.2 g. Over most of its range there are two color phases. The dorsum may be either buffy in the light phase or reddish brown in the dark phase. The underparts are always paler (plate 4).

Range and Habitat

This species ranges from Sonora in western Mexico and Tamaulipas in eastern Mexico south to western Panama (map 8.118). Koopman (1982) gives the range as extending across northern South America into eastern Brazil.

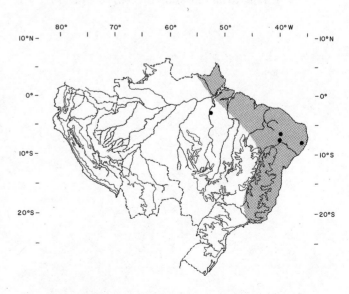

Map 8.118. Distribution of *Natalus stramineus*.

Natural History
Linares (1972) reports on the traditional use of caves as roosting sites as setting the stage for the evolution of local variation in measurements and pelage color.

FAMILY FURIPTERIDAE
Smoky Bats

Diagnosis
The thumb is extremely reduced or absent. The claw, if present, is minute and functionless (see fig. 8.4C). There are two genera, *Furipterus* and *Amorphochilus*.

Distribution
This family is confined to the New World tropics. *Furipterus* ranges from Costa Rica across northern South America to eastern Brazil. *Amorphochilus* occurs on the Pacific coast of South America.

Genus *Amorphochilus* Peters, 1877
Amorphochilus schnablii Peters, 1877

Description
The dental formula is I 2/3, C 1/1, P 2/3, M 3/3. Measurements: TL 76.1; HB 48.9; T 27.2; HF 10; E 14.7; FA 36.8. This small bat has long, light gray fur, lightly washed with brown. The tail membrane is light brown, and the wings and ears are brownish gray (pers. obs.).

Distribution
This poorly known bat is found from Isla Puno in Ecuador south along the coast to northern Chile (Mann 1978; Tamayo and Frassinetti 1980) (map 8.119).

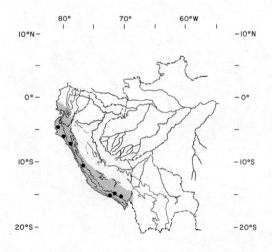

Map 8.119. Distribution of *Amorphochilus schnablii*.

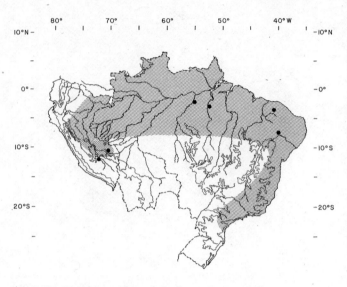

Map 8.120. Distribution of *Furipterus horrens*.

Ecology
In Chile, *A. schnablii* is found along the coast near the mouths of rivers. It feeds on small dipterans and lepidopterans, and it roosts in cracks in rocks along the arid northern coast (Mann 1978).

Genus *Furipterus* Bonaparte, 1837
Furipterus horrens (F. Cuvier, 1828)

Description
Measurements: TL 62; T 24.5; HF 7.9; E 11.1; FA 35.3; Wt 3.1 g. In this small bat, head and body length averages 37.5 mm. The dorsum varies from brownish gray to almost black, and the venter is somewhat paler (plate 4).

Range and Habitat
This species is distributed from Costa Rica south across northern South America to eastern Brazil (map 8.120). In northern Venezuela it was strongly associated with moist habitats. All specimens were taken at below 150 m elevation (Handley 1976).

Natural History
This insectivorous bat roosts in caves. Its natural history is very poorly known, and it is infrequently collected. Uieda, Sazima, and Storti-Filho (1980) report the following observations from Brazil. Females are slightly larger than males. The mammary glands of the female are situated posteriorly on the abdomen; thus when the mother is hanging upside down the suckling young is oriented with its head up. It appears to be insectivorous, and the remains of Lepidoptera have been found in its stomach.

FAMILY THYROPTERIDAE
Disk-winged Bats, Murciélagos con Mamantones

Diagnosis
This distinctive family has suction disks at the bases of the thumbs and on the soles of the feet (see fig. 8.4C). Only two other taxa of bats have such suction disks, the Myzopodidae, confined to the Island of Madagascar, and some vespertilionids including *Eudiscops* from Southeast Asia.

Distribution
This family is confined to the New World and distributed from Chiapas, Mexico, south through the Isthmus, across much of northern South America to eastern Brazil.

Genus *Thyroptera* Spix, 1823
Thyroptera discifera (Lichtenstein and Peters, 1855)

Description
The dental formula is I 2/3, C 1/1, P 3/3, M 3/3. Head and body length ranges from 37 to 47 mm, tail length from 24 to 33 mm, and forearm length from 31 to 35 mm. One male measured TL 73; T 36; HF 6; E 12; Wt 6 g. The calcar has a single cartilaginous projection extending into the posterior border of the interfemoral membrane. This species differs from *T. tricolor* in that the venter is only slightly paler than the light brown dorsum.

Range and Habitat
This species is known from Nicaragua and from scattered localities in northern South America (map 8.121). The exact limits of its distribution are unknown.

Natural History
These bats have been found roosting communally under a dead banana leaf. Groups consist of adults of both sexes and young (Wilson 1978).

Thyroptera laveli Pine, 1993

Description
This is a large species with its tail extending prominently beyond the interfemoral membrane. Measurements: FA 38.8–40.7 (Pine 1993). Dorsal pelage is brown with the individual hairs unicolor. The ventral hairs are bicolor, brown at the base and buff at the tips.

Distribution
The species is known from several localities in Loreto Peru: 4°21′ S, 71°58′ W (not mapped).

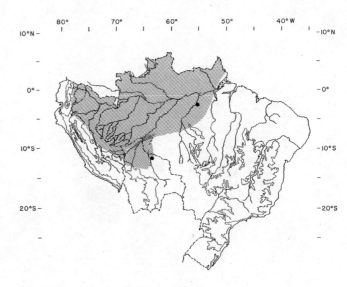

Map 8.121. Distribution of *Thyroptera discifera*.

Thyroptera tricolor Spix, 1823
Murciélago Tricolor de Brasil

Description
Measurements: TL 73; HB 40; T 32; HF 7; E 13; FA 36–37. This small bat has a total length averaging 72.6 mm; the tail averages about 32.5 mm, and weight averages about 4.6 g. There is little difference in size between the sexes. The calcar has two cartilaginous projections extending onto the posterior border of the interfemoral membrane. The dorsum is reddish brown to almost black, but the roots of the hairs are pale. The venter contrasts strongly, being either white or yellowish.

Range and Habitat
This species occurs from southern Mexico across South America including Peru and Bolivia to eastern Brazil (map 8.122). It is strongly associated with moist habitats. In northern Venezuela all specimens were collected below 850 m elevation (Handley 1976).

Natural History
This species is an aerial insectivore. The bats form small colonies, rarely exceeding nine individuals, that show stability over time although roosting sites are changed frequently. Their specialized roosting habits, inside rolled *Heliconia* leaves, may limit colony size. The suction disks allow them to cling to the smooth surface of leaves (Findley and Wilson 1974; Wilson and Findley 1977).

FAMILY VESPERTILIONIDAE

Diagnosis
These bats are highly specialized for flight. The trochiter of the humerus is much larger than the trochin

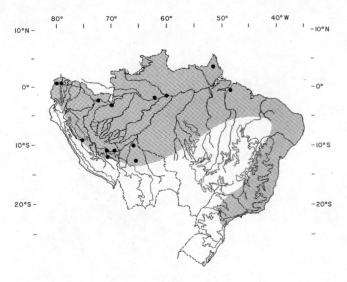

Map 8.122. Distribution of *Thyroptera tricolor*.

and has a surface of articulation with the scapula more than half as large as the glenoid fossa. The tail is rather long and extends to the edge of a wide interfemoral membrane. These bats are of medium to small, with forearm length ranging from 24 to 90 mm. The nostrils and lips are relatively unmodified, and there are no decorative facial folds. The dental formula is highly variable and will be discussed under the genera. A key to the genera is given in table 8.15.

Distribution

This family is worldwide in its distribution except for Antarctica and New Zealand.

Natural History

Vespertilionid bats are usually specialized as aerial insectivores, although some show specializations for feeding on fish (e.g., *Pizonyx*). Bats of this family roost in caves, hollow trees, or other sheltered areas. These are the common bats of the north temperate zone, but they also occupy the tropics. In the temperate zone they are specialized for winter hibernation.

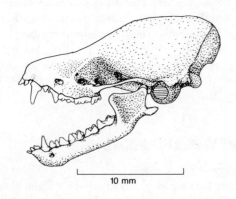

Figure 8.19. Skull of *Myotis nigricans*.

Echolocation in the capture of flying insects has been well studied in certain species of this family (Griffen 1958; Gould 1977).

Genus *Myotis* Kaup, 1829
Little Brown Bat

Description

The dental formula is usually I 2/3, C 1/1, P 3/3, M 3/3 (fig. 8.19). Head and body length ranges from 35 to 80 mm, tail length from 40 to 60 mm, and forearm length from 29 to 68 mm. The dorsum is usually some shade of brown, and the underparts are somewhat paler. The Neotropical species were reviewed by LaVal (1973a).

Distribution

The genus is worldwide in its distribution except for the Arctic and Antarctic regions.

Myotis albescens (E. Geoffroy, 1806)

Description

Measurements: TL 80–86; T 35–37; HF 9; FA 34–35; Wt 7.3–7.6 g. This species is similar in size to *M. nigricans* but lighter in color. The dorsum is brown, but the dorsal hairs are black at the base and white at the tip, giving a frosted appearance. The venter is paler.

Range and Habitat

The species is distributed from southern Veracruz, Mexico, through the Isthmus and across most of South America to northern Argentina and Uruguay (map 8.123). This species prefers lower elevations than does *M. nigricans*; in Venezuela specimens were all taken below 155 m. It tolerates both dry deciduous for-

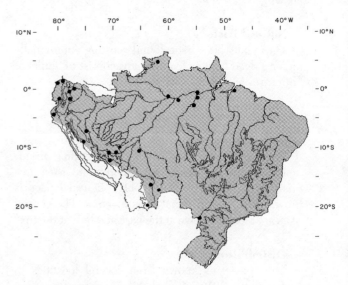

Map 8.123. Distribution of *Myotis albescens*.

est and humid multistratal tropical evergreen forest (Handley 1976).

Natural History

In Paraguay the first births occur in October, followed by copulation and a second pregnancy. Some females may even breed a third time. The gestation period is three months or slightly less, though it may be a little shorter for the second and third pregnancies. The litter size is always one, and lactation lasts about one month. Males probably breed in their first year, and females can store sperm (Myers 1977).

Although a few specimens of this species are known from above 1,500 m, most have been taken below 500 m. In Venezuela specimens were taken in dry deciduous and moist evergreen forests. In Uruguay *M. albescens* is found in open woodlands and forages in open grasslands. It roosts in inhabited buildings, walls, crevices, rocks, and trees, typically near fast moving streams. In Paraguay it is apparently dependent on human habitations, where it typically roosts in large groups. Most likely this bat does not hibernate but does undergo daily torpor (Barlow 1965; González 1973; LaVal 1973a; Myers 1977).

Myotis atacamensis (Lataste, 1892)

Description

Measurements: TL 89; HB 49; T 39; HF 8; FA 38. This is the smallest of the Neotropical *Myotis* species. It has a tiny skull and long, blond fur. The sparse fur extends distally on the dorsum of the uropatagium to a point halfway from the knee to the ankle. Fur is sparse or absent on the rest of the membranes, which are pale brown (LaVal 1973a; Miller and Allen 1928).

Distribution

M. atacamensis is known only from southern Peru and northern Chile. In Chile it extends south to the Elqui River (Tamayo and Frassinetti 1980) (map 8.124).

Ecology

This bat is found from the coastal deserts from sea level to an altitude of 2,400 m (Tamayo and Frassinetti 1980).

Myotis keaysi J. A. Allen, 1914

Description

Measurements: TL 83; HB 47; T 36; HF 8; E 12; FA 36; Wt 5.3 g. This is one of the larger species of *Myotis*. Males tend to be slightly larger than females; head and body length averages 48.4 mm for males and 44.6 mm for females. The dorsum is a very dark brown, and the venter is paler.

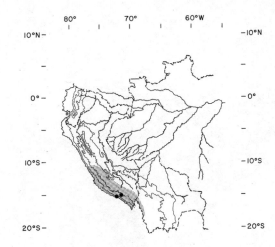

Map 8.124. Distribution of *Myotis atacamensis*.

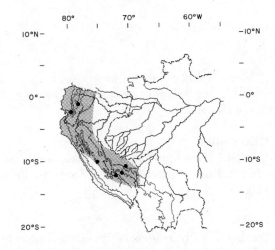

Map 8.125. Distribution of *Myotis keaysi*.

Range and Habitat

The species is distributed from southern Veracruz, Mexico, through Panama to northern Venezuela and western Colombia, following the foothills of the Andes through Peru (map 8.125). This high-altitude bat usually occurs above 1,000 m elevation in Venezuela and can be found up to 2,000 m. It prefers moist habitats and montane tropical humid forest (Handley 1976).

Natural History

Little is known concerning the biology of this bat. Clearly it is highly adapted for semiarid habitats, and it is probably a relict on its present range with the waning of arid habitats in northern South America since the close of the last glaciation.

Myotis levis (I. Geoffroy, 1824)
Murciélago Común

Description

Measurements: TL 88; HB 52; T 37; HF 8; E 14; Wt 4.6 g. This pretty little bat is mottled chestnut to

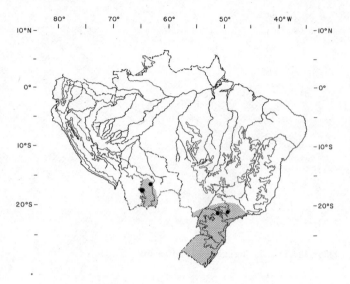

Map 8.126. Distribution of *Myotis levis*.

tan. There is a dark cast to the dorsal color because of the dark hair bases showing through. The belly is gray washed with white.

Distribution

M. levis is found in central Argentina, southeastern Brazil, Paraguay, and Uruguay (Koopman 1993) (map 8.126).

Ecology

This bat is locally common in Uruguay and adjacent Brazil, where it forages primarily over open country, in open woodland or over streams. It characteristically roosts in colonies numbering in the thousands (Barlow 1965).

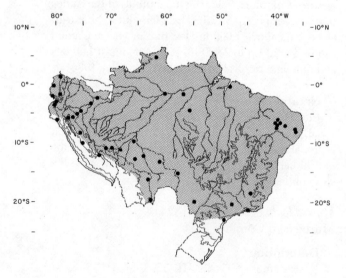

Map 8.127. Distribution of *Myotis nigricans*.

Myotis nigricans (Schinz, 1821)

Description

Measurements: TL 78–80; HB 41–44; T 36–37; HF 8; E 12–13; FA 33–35; Wt 4–5 g. In northern Venezuela males are slightly larger than females. Head and body length averages 43.4 mm for males and 41.2 mm for females; forearm length averages 34.56 mm for males and 33.4 mm for females; weight ranges from 3 to 5.5 g. The dorsal color varies geographically from light to dark brown, and the venter is nearly the same shade.

Range and Habitat

M. nigricans occurs from southern Mexico across most of South America to northern Argentina (map 8.127). Although it ranges to 2,240 m elevation in Venezuela, the vast majority of specimens were taken below 1,200 m. It tolerates a wide range of vegetation types, occurring in dry deciduous forest, and is found near human habitations.

Natural History

This is one of the best studied of the tropical species of *Myotis*. These bats tend to roost in sheltered areas; females rearing young form separate groups. In Panama this species breeds in December and early January. Gestation is approximately sixty days, and a birth peak occurs in February. The birth is followed by a postpartum estrus and a repeat of the cycle, with a second birth peak in April–May and another in August. Reproductive activity then declines, to be initiated again in December. Reproductive periods coincide with the time of maximum insect abundance. When feeding, females leave their young in large groups termed crèches. The young achieve adult size at approximately five to six weeks of age (Wilson 1971; Wilson and LaVal 1974).

In Paraguay breeding appears to occur most of the year, though most females copulate in May to December, with the first birth peak occurring in late October. The second period occurs between late December and early January. There may be a third peak for some females. The gestation period is slightly less than three months, but there may be a delay in implantation of the blastocyst. Litter size is one, and the lactation period is about one month. Males can breed year round, the females store sperm, and both males and females can breed at about four months of age.

Young remain attached to their mothers for the first two or three days and then are left behind in large groups or crèches when the mothers leave to fly and feed. Adult weight is reached by week two and flight begins in week three. Some individuals are known to reach seven years of age in the wild (Myers 1977; Wilson and LaVal 1974).

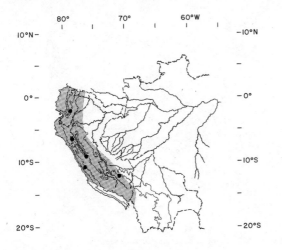

Map 8.128. Distribution of *Myotis oxyotus*.

Map 8.129. Distribution of *Myotis riparius*.

M. nigricans occurs in all the vegetation zones within its large range but appears to be uncommon over 1,200 m. In Uruguay it shows a distinct preference for subtropical woodland. In Paraguay this species is dependent on human habitations for roosts and forages above man-made cattle ponds. Males are often found roosting singly outside the breeding season whereas females tend to occur in groups. One study found evidence of stable groups and perhaps harems (Barlow 1965; LaVal 1973a; Myers 1977).

Myotis oxyotus (Peters, 1867)

Description
Measurements: TL 84.5; T 39.3; HF 8.8; E 13.9; FA 36.7; Wt 5.5 g. In this large species, head and body length averages 45.5 mm for males and 45.6 mm for females. The forearm averages about 36 mm, and weight generally exceeds 5 g. The dorsum is brown but has a slight frosted appearance in some specimens. The ventral hairs have brown bases with gray tips, creating a mottled effect.

Range and Habitat
The species occurs from Costa Rica south through Panama to Venezuela and in the west to Peru (map 8.128). These large bats are adapted to high elevations; in northern Venezuela specimens were taken between 800 and 2,110 m elevation. They prefer multistratal montane tropical forest and wet habitats (Handley 1976).

Myotis riparius Handley, 1960

Description
Measurements: TL 80; T 37; HF 8; E 12–13; FA 35; Wt 4.4 g. Males and females are nearly the same size, averaging about 42.7 mm in head and body length, with forearms approximately 35 mm. The venter and

dorsum are dark brown (plate 4). The species may be confused with *M. simus*. The dorsal hairs of *M. riparius* are bicolored, whereas in *M. simus* the hairs are monochromatic (see LaVal 1973a for a comparison).

Range and Habitat
This species is distributed from Honduras to South America, ranging widely over the continent; the southernmost distribution is in Uruguay extending to Rio Grande do Sul, Brazil (Gonzáles and Fabian 1995) (map 8.129). These bats generally range at lower elevations; although in northern Venezuela some specimens were taken at up to 1,000 m, 90% of all specimens collected were taken below 200 m. Broadly tolerant of human activities, they may be found in croplands, dry deciduous forests, and multistratal tropical evergreen forest (Handley 1976).

Myotis ruber (E. Geoffroy, 1806)
Murciélago Ruber

Description
Measurements: HB 49.8; T 41.4; HF 8.8; E 15; FA 40.2; Wt 7.3 g. *M. ruber* has reddish, monocolored fur of medium length, thick and silky. Fur is thick on the basal third of the dorsal surface of the uropatagium and extends one-third to one-half the distance from the knee to the ankle. Within its known range this species is larger than *M. nigricans*, *M. albescens*, and *M. riparius*. Of these, only *M. riparius* is similar in color, but it is much smaller (LaVal 1973a).

Distribution
This species has a fairly limited range, being known only from southeastern Brazil, Paraguay, and northeastern Argentina (Honacki, Kinman, and Koeppl 1982) (map 8.130).

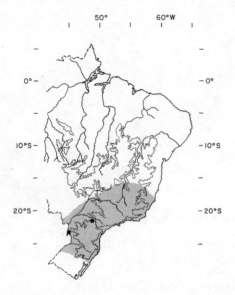

Map 8.130. Distribution of *Myotis ruber*.

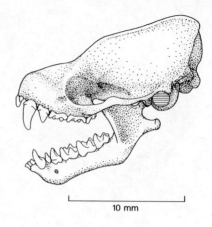

Figure 8.20. Skull of *Eptesicus brasiliensis*.

Myotis simus Thomas, 1901

Description

Measurements: TL 82–97; T 35–39; HF 10; E 14; FA 37–39; Wt 8 g. This medium-sized species is similar to *M. riparius* in size and proportions, with forearm length about 35 mm. The dorsum is orange to cinnamon brown, and the fur is short and woolly.

Range and Habitat

The species is confined to the lowland rain forests of the Amazon basin; it reaches extreme southeastern Colombia and extends south to Bolivia, Paraguay, and northeastern Argentina (map 8.131).

Genus *Eptesicus* Rafinesque, 1820
Big Brown Bat, Murciélago con Orejas de Ratón

Description

The dental formula is I 2/2, C 1/1, P 1/1, M 3/3. The genus is distinguishable from *Myotis* by the reduction in number of premolars (see fig. 8.20). Head and body length ranges from 35 to 75 mm and forearm length from 28 to 55 mm. In spite of the common name, there is a considerable size range among the species of the genus. The dorsum is usually dark brown to almost black, and the venter is paler.

Distribution

This genus contains approximately thirty species and is worldwide in its distribution except for the Arctic and Antarctic regions.

Eptesicus brasiliensis (Desmarest, 1819)

Description

Measurements: TL 95–97; T 41–43; HF 10; E 15; FA 40–41; Wt 9–10 g. The dorsum and venter are dark brown. In this account *Eptesicus andinus* is considered a junior synonym of *E. brasiliensis*. The species is slightly larger than *E. furinalis* but considerably smaller than *E. fuscus*. Females are only slightly larger than males.

Range and Habitat

The species ranges through montane regions of southern Mexico, through the Isthmus, and across much of South America to Uruguay (map 8.132). In northern Venezuela these bats occur below 1,000 m elevation. They prefer moist habitats and multistratal tropical evergreen forest, though they will forage in man-made clearings (Handley 1976).

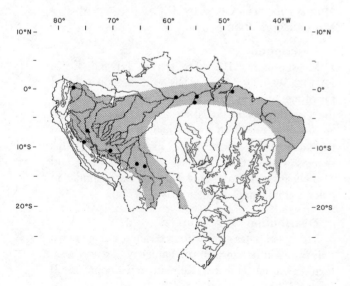

Map 8.131. Distribution of *Myotis simus*.

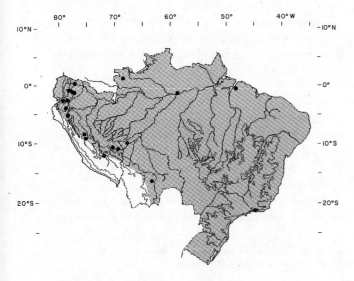

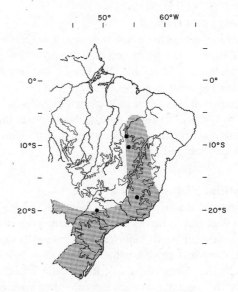

Map 8.132. Distribution of *Eptesicus brasiliensis* (includes *E. andinus*).

Map 8.133. Distribution of *Eptesicus diminutus*.

Eptesicus diminutus Osgood, 1915

Description
This is the smallest species of the genus in the northern Neotropics. Two specimens measured TL 91–91.9; T 37; HF 9; E 15; FA 35.5 (Venezuela, USNM). A series from Paraguay measured TL 87.8; T 33.5; HF 7.0; E 13.6; FA 34.4; Wt 5.6 g. The dorsum is brown, as is the venter, but in some specimens the venter is washed with gray (plate 4).

Range and Habitat
The species is distributed from Venezuela discontinuously through eastern Brazil to central Argentina. In northern Venezuela it was collected at 100 m elevation in a mixed habitat of grassland and deciduous tropical forest (Handley 1976). In Paraguay it roosts in houses and in trees (Myers, White, and Stallings 1983) (map 8.133).

Eptesicus furinalis (D'Orbigny, 1847)

Description
Measurements: TL 93–97; T 37–39; HF 9–10; E 16. In this account *E. montosus* is considered a junior synonym of *E. furinalis*. This bat is smaller than *E. brasiliensis;* head and body length averages 52 mm and the forearm 38 mm. Females are only slightly larger than males in linear measurements but are about 1.5 g heavier; average weights range from 7.5 to 9 g. The dorsum and venter are dark brown.

Range and Habitat
This species occurs from Nayarit in western Mexico and Tamaulipas in eastern Mexico south through the Isthmus over most of South America to northern

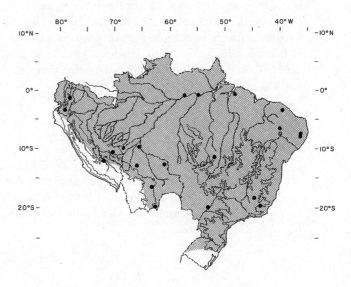

Map 8.134. Distribution of *Eptesicus furinalis* (includes *E. montosus*).

Argentina (map 8.134). It is generally found at higher altitudes up to 1,580 m and prefers moist habitats and montane tropical forest or multistratal evergreen forest.

Eptesicus fuscus (Beaubois, 1796)

Description
This is the largest species of the genus and is easily distinguishable from the other species by size alone. Measurements for two males averaged TL 117.5; T 48; HF 12.5; E 17; FA 56.6 (Venezuela, USNM). Head and body length averages 70 mm for males and 84 mm for females; weight may reach 26 g in females. The dorsum is dark brown with a slightly paler venter.

Range and Habitat

This bat occurs from Saskatchewan and Quebec in North America south through the Isthmus to the montane portions of Colombia, Venezuela, and northern Brazil. A high-elevation species in the tropical portion of its range, it was taken in northern Venezuela from 1,260 to 1,524 m. It prefers moist habitats, cloud forest, and premontane tropical evergreen forest (Handley 1976) (not mapped).

Natural History

Although this is one of the best-studied bats in North America, little is known concerning the details of its biology in South America. This bat roosts during the day in small groups, usually in a hollow tree or a cave. In the northern part of its range it hibernates during the winter, sometimes with other species of bats but never in large groups of its own species. In the northern part of its range two or three young may be born, and they attain adult size within sixty days. These bats are strongly insectivorous, taking a wide variety of flying insects. In the north, beetles constitute up to 50% of their diet and they seldom take moths (Barbour and Davis 1969).

Eptesicus innoxius (Gervais, 1841)

Description

Two specimens measured TL 90–91; T 36–38; HF 8–9; E 10–13; FA 38.9–39.1. This small chocolate brown bat has a gray dorsum and faint white lines on the venter. The wings are dark (Villa-R. and Villa Cornejo 1969; pers. obs.).

Map 8.135. Distribution of *Eptesicus innoxius*.

Distribution

Eptesicus innoxius is found in Ecuador and Peru. It does extend to Argentina (Villa-R. and Villa Cornejo 1969) (map 8.135).

Ecology

The specimens from Argentina were caught in early evening with a net across a small stream in mixed forest (Villa-R. and Villa Cornejo 1969).

Comment

Barquez (1987) suggests that the Argentine specimens of *E. innoxius* should be referred to *E. furinalis*.

Genus *Histiotus* Gervais, 1856
Murciélago Orejudo

Description

Head and body length ranges from 54 to 70 mm and the forearm from 42 to 52 mm. This genus resembles *Eptesicus*. However, *E. fuscusi* with a total length of 117 mm will have an ear approximately 17 mm long, whereas a comparable-sized *Histiotus* will have an ear 31 mm long. The upperparts of the body are generally light brown or grayish brown; the underparts are slightly paler (plate 4).

Distribution

This genus is confined to the southern parts of South America.

Natural History

These aerial insectivores are found in small colonies (three to seventeen individuals) inside natural cavities or buildings.

Histiotus alienus Thomas, 1916

Description

The ears are not connected by a conspicuous membrane as in *H. montanus*. They are oval and extremely long (29 mm or longer). The species is similar to *H. velatus*, but the longer ear should distinguish it.

Distribution

This bat occurs from southeastern Brazil to Uruguay (map 8.136).

Histiotus macrotus (Poeppis, 1835)

Description

Measurements: HB 57; T 48; HF 9; E 32; FA 49; Wt 12 g. This medium-sized vespertilionid has very large ears connected at the base by a membrane. When folded down, the ears pass the nose, a character that separates this species from the shorter-eared

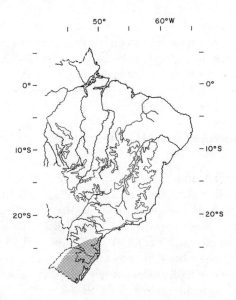

Map 8.136. Distribution of *Histiotus alienus* (from Koopman 1982).

H. montanus. The dorsum is light brown, and the venter is whitish gray (Mann 1978; Osgood 1943; pers. obs.). This species was identified as *H. montanus* by Mares, Ojeda, and Kosco (1981) (see also Ojeda and Mares 1989).

Distribution

Histiotis macrotus is found in Chile, southern Bolivia, western and southern Argentina, and southern Peru (Olrog 1959; Tamayo and Frassinetti 1980) (map 8.137).

Natural History

In southern Argentina reproduction seems to be well synchronized. All females examined bore fetuses in only the right uterine horn (Pearson and Pearson

1989). This species has been found roosting in roofs and in cracks in deep mines (Mann 1978; Pearson and Pearson 1989).

Histiotus montanus (Philippi and Landbeck, 1861)
Murciélago Oregón Chico

Description

Mean measurements for three males were TL 108.7; T 49; HF 10.3; E 31: FA 45.7 (Venezuela, USNM). Measurements from a sample in Uruguay were HB 61.7; T 47.3; HF 8.4; E 21.4; FA 45. The dorsum is medium brown and the venter somewhat paler.

Range and Habitat

The species ranges at high elevations in the Andean portions of Venezuela and Colombia south through Ecuador and Peru to Bolivia and Argentina. In Venezuela it occurred from 1,498 to 2,101 m (map 8.138).

Natural History

In Argentine Patagonia pregnant bats are found between August and November. Unlike that of the sympatric *Myotis chiloensis,* their breeding is not tightly synchronized. Implantation is always in the right horn of the uterus, litter size is one, and age of first reproduction is approximately one year (Mann 1978; Pearson and Pearson 1989).

This wide-ranging bat is most commonly found in dense woodlands, and also savannas. In Chile it demonstrates a remarkable tolerance for cold temperatures. *H. montanus* is generally found in small colonies in mines, caves, roofs, and tree holes (Barlow 1965; Greer 1966; Mann 1978; Tamayo and Frassinetti 1980; Villa-R. and Villa Cornejo 1969).

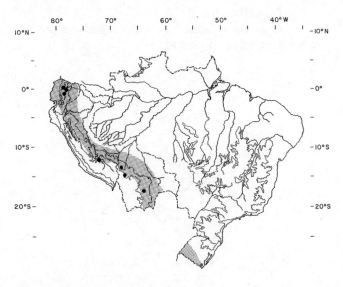

Map 8.138. Distribution of *Histiotus montanus.*

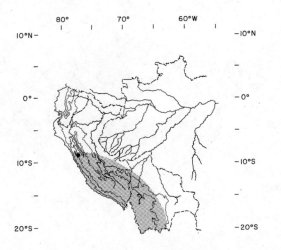

Map 8.137. Distribution of *Histiotus macrotus.*

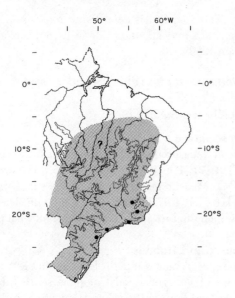

Map 8.139. Distribution of *Histiotus velatus*.

Histiotus velatus (I. Geoffroy, 1824)
Murciélago Oregón Tropical

Description
Measurements: HB 62.7; T 42.5; HF 8.8; E 23.2; Wt 13 g. This bat's large ears are not connected at the base. It is grayish to tannish dorsally, with the gray at the base of the hairs prominent. The venter is appreciably lighter than the dorsum, and in some specimens it is white (pers. obs.).

Distribution
Histiotus velatus is found in southern Brazil, Uruguay, Paraguay, and northeastern Argentina (Massoia 1980; Thomas 1898) (map 8.139).

Life History
In southern Brazil this species begins to reproduce in September (Peracchi 1968).

Ecology
Histiotus velatus in southern Brazil roosts in roofs in groups averaging twenty-three (range 12–30, $n = 6$); there are always more males than females, but the composition of sexes at peak reproduction is not recorded (Peracchi 1968).

Genus *Rhogeessa* H. Allen, 1866
Little Yellow Bat

Description
The dental formula is I 1/3, C 1/1, P 1/2, M 3/3. This genus is distinguishable from *Myotis* and *Eptesicus* by its single upper incisor (see fig. 8.21). Head and body length ranges from 37 to 50 mm, and forearm length ranges from 25 to 34 mm. The color is yellowish brown or light brown dorsally and paler below. The genus was reviewed by LaVal (1973b).

Distribution
This New World genus has species ranging from Mexico to Brazil.

Rhogeessa tumida H. Allen, 1866

Description
Measurements: HB 41–42; T 29; HF 6–7; E 12–13; FA 27–28; Wt 3.5–4.0 g. This is an extremely small bat with an unadorned face. It is light yellow brown on the dorsum with a slightly paler venter.

Range and Habitat
This species occurs from southern Tamaulipas, Mexico, south through the Isthmus and across much of northern South America (map 8.140). This form prefers moist habitats. It is broadly tolerant of man-made clearings and is the common species of this genus taken below 570 m in appropriate habitat.

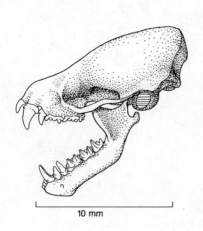

10 mm

Figure 8.21. Skull of *Rhogeessa tumida*.

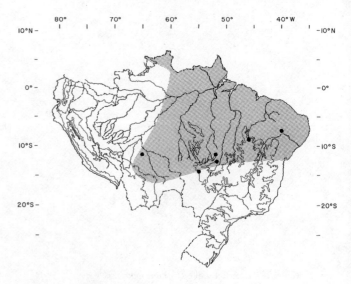

Map 8.140. Distribution of *Rhogeessa tumida*.

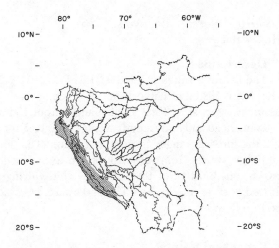

Map 8.141. Distribution of *Tomopeas ravus* (from Koopman 1982).

Natural History

These bats roost in hollow trees, and colonies may be large. They are aerial insectivores, and individuals appear to have established hunting routes. Females may bear up to two young (LaVal 1973b).

Genus *Tomopeas* Miller, 1900

Comment

Sudman, Barkley, and Hafner (1994) place *Tomopeas* as a subfamily of the Molossidae. Koopman (1993) leaves it within the Vespertilionidae as the sole genus in the subfamily Tomopeatinae.

Tomopeas ravus Miller, 1900

Description

Measurements: TL 73–85; T 34–45; FA 31–35; Wt 2–3.5 g. The dorsal color is pale brown contrasting with a cream buff venter. The tragus of the ear is blunt. The long tail is distinctive, being almost equal to or greater than the head and body length.

Distribution

This bat is taken from sea level to 1,000 m elevation in the semiarid areas of coastal northwestern Peru (map 8.141).

Natural History

Roosts are under granite boulders. Graham (1987) provides evidence that single young are born from May to September.

Genus *Lasiurus* Gray, 1831

Description

Species of this genus have a considerable range in size. Head and body length ranges from 50 to 90 mm, tail length from 40 to 75 mm, and the forearm from 37 to 57 mm. Weight ranges from 3 to 6 g. The distinguishing feature for this group is that the dorsal sur-

face of the interfemoral membrane is well haired for at least half its length and usually for its entire length (see fig. 8.4B).

Distribution

These bats are confined to the New World and have colonized the Galápagos and the Hawaiian Islands. They are distributed from Canada to Argentina.

Natural History

Bats of the genus *Lasiurus* have been studied in North America, but little is known concerning their habits in South America. In the north they exhibit seasonal migrations and are aerial insectivores. *L. borealis* has two or three young, in contrast to the single young characteristic for most of the Chiroptera.

Lasiurus blossevillii (Lesson and Carmot, 1826)

Description

The forearm length is 35.8–39.7 in males and 38.6–40.6 in females. Koopman (1993) considers *L. blossevillii* a junior synonym of *L. borealis,* but Morales and Bickham (1995) offer evidence that the name *L. blossevillii* may be correctly applied to most specimens described as *L. borealis* from the southeastern United States and Central and South America.

Distribution

As redefined, the range could include the southeastern United States, Central America, and South America (see account for *L. borealis*).

Lasiurus borealis (Müller, 1776)

Description

The dental formula is I 1/3, C 1/1, P 2/2, M 3/3 (see fig. 8.22). Measurements: TL 101; T 9.4; E 11–12; FA 39–40; Wt 6.4 g. Head and body length averages 51.8 mm for males and 51.1 mm for females, and forearm length averages 38.6 mm for males and 40.5 for

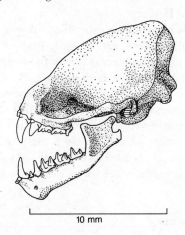

10 mm

Figure 8.22. Skull of *Lasiurus borealis.*

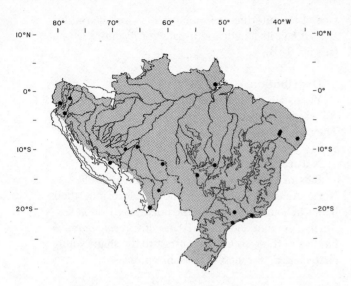

Map 8.142. Distribution of *Lasiurus borealis* (South American map specimens may be referred to *L. blossevillii*; see text).

females. Weight exceeds 6 g. The dorsum is bright red or rust, and the venter is only slightly paler. Males tend to be brighter in color than females. There is usually a buffy patch on the front of each shoulder. The interfemoral membrane is furred on its entire upper surface.

Range and Habitat

This species was formerly thought to range from southern Canada through the Isthmus to the southern tip of the South American continent in suitable habitats (map 8.142). The account that follows is a generalized account applicable to *L. blossevillii*. In northern South America *Lasiurus* is strongly associated with moist habitats and multistratal tropical evergreen forest. In Venezuela specimens were taken at below 155 m; at higher elevations it is replaced by *L. cinereus* (Handley 1976).

Natural History

Much is known concerning the biology of this bat in the northern parts of its range. In North America it tends to migrate and hibernates in the winter. When not hibernating it often roosts on the bark of trees. Its behavior in tropical climates is poorly understood. In North America approximately 26% the diet of this aerial insectivore can consist of moths. Females may have up to five young, although three is the usual number (Barbour and Davis 1969). These bats are solitary and do not congregate in nurseries (Shump and Shump 1982a).

Comment

See *L. blossevillii*.

Lasiurus cinereus (Beauvois, 1796)
Hoary Bat

Description

The dental formula is I 1/3, C 1/1, P 2/2, M 3/3. This bat is distinguishable from all other species of *Lasiurus* by its size. Total length ranges from 134 to 140 mm, head and body length averages about 85 mm, and the forearm ranges from 46 to 55 mm. The dorsal pelage waries from yellowish brown to mahogany brown, but the tips of the hairs are white, giving a frosted or silver appearance. The underparts are whitish to yellow.

Range and Habitat

This species occurs from southern Canada south through the Isthmus in suitable habitat to southern Argentina (map 8.143). It is strongly associated with higher elevations in South America; in Venezuela it was taken at up to 1,456 m elevation. It prefers moist areas, though it has been taken in dry deciduous forest as well as multistratal tropical evergreen forest (Handley 1976).

Natural History

In the Northern Hemisphere this bat is migratory and hibernates. Little is known concerning its natural history in tropical regions. This species tends to be solitary. The female has an average litter size of two in the northern parts of its range. It is an aerial insectivore taking a wide variety of flying insects; moths seem to predominate in the diet of the temperate zone races (Shump and Shump 1982b).

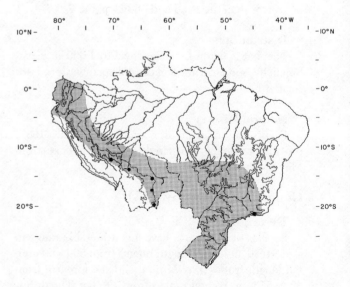

Map 8.143. Distribution of *Lasiurus cinereus*.

Lasiurus ebenus Fazzolari-Corrêa, 1994

Description

Clearly related to *L. borealis* group, this species is distinctive with its black coloration. The cranium exhibits a short rostrum and a weakly developed sagittal crest; the first upper premolar is present, and there is a double-rooted fourth upper premolar. Measurements: TL 115; T 58; HF 8; FA 45.7. This species differs from *L. blossevillii* in both color and its larger size (Fazzolari-Corrêa 1994).

Distribution

The type locality is Brazil, eastern São Paulo, Parque Estadual da Ilha do Cardoso (20°05′ S; 47°59′ W) (not mapped).

Lasiurus (Dasypterus) ega (Gervais, 1856)
Southern Yellow Bat, Murciélago Leonado

Description

Measurements: TL 118–26; T 50–52; HF 9–10; E 19; FA 45–48; Wt 12.3 g. This species is easily separated from the other southern South American species of *Lasiurus* because of its pale tan color with a grayish wash both ventrally and dorsally. It has pale ears, a pale tail membrane, and dark wing membranes.

Distribution

L. ega is distributed from the southern United States south to Argentina and Uruguay (map 8.144). In southern South America it is found commonly in Paraguay and Uruguay and in Argentina to about 35° S. There are occasional unsubstantiated records of its appearance in Chile (Barlow 1965; Honacki, Kinman, and Koeppl 1982; Myers and Wetzel 1983; Tamayo and Frassinetti 1980).

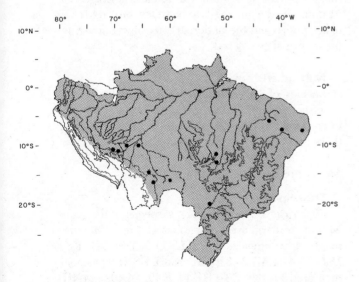

Map 8.144. Distribution of *Lasiurus ega*.

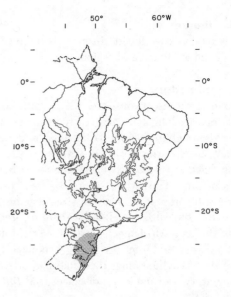

Map 8.145. Distribution of *Lasiurus egregius* (from Koopman 1982).

Lasiurus egregius (Peters, 1871)
Big Red Bat

Description

This species is similar to *L. borealis* in coloration and dental characters, but it is distinguishable from all other red species of *Lasiurus* by its large size: total length is 127 mm and forearm 48 mm (the type).

Range and Habitat

The species occurs in extreme southeastern Panama, but the extent of its range into northwestern Colombia is unknown. It has been described from French Guiana and Brazil (map 8.145).

Comment

The species' relation to *Lasiurus ebenus* is unclear.

FAMILY MOLOSSIDAE

Free-tailed Bats, Mastiff Bats, Murciélagos de Cola de Ratón

Diagnosis

The family contains ten genera and about eighty species. Head and body length ranges from 40 to 130 mm, tail from 14 to 80 mm, and forearm from 27 to 85 mm. The fibula is well developed, supporting the lower leg. In these bats the tail extends beyond the edge of the interfemoral membrane, and this single character is diagnostic for the entire family, hence the common name "free-tailed bat" (see fig. 8.4B).

Distribution

This family is worldwide in its distribution but is mainly confined to tropical regions.

Natural History

These aerial insectivores are swift, high fliers, and their wings are long and narrow. Freeman (1981) has made a thorough morphometric analysis of the family. The jaw size covaries positively with body size, suggesting that there is great specialization for certain size classes of prey. Thus when several species co-occur they tend to present an array of sizes that probably reflects this specialization. Beetles seem to predominate in the diet. Most species nest in caves, tunnels, or hollow trees, and some are strongly colonial. In the northern parts of their range these bats may be seasonally migratory over moderate distances (see *Tadarida brasiliensis*).

Identification of the Genera

Within the area covered by this volume there are seven genera of free-tailed bats: *Mormopterus, Molossus, Molossops, Tadarida, Promops, Eumops,* and *Nyctinimops*. Should the conclusions of Sudman, Barkley, and Hafner (1994) be substantiated, *Tomopeas* may be added to this family. The species span an array of sizes, with the forearm ranging from a minimum of 25 mm to a maximum of 80 mm. All the very large free-tailed bats with forearms longer than 60 mm belong to the genus *Eumops*, but this genus also includes some very small species such as *E. nanus*. Species of the genus *Molossops* are all quite small, and *Promops* and *Tadarida* are intermediate in size. Thus size alone does not discriminate among the genera, and one must pay close attention to dental characteristics and the shape of the ear. In all species of free-tailed bats the ears are very stout and somewhat flattened, projecting forward and laterally. Whether the ears are joined in the center of the forehead is an important diagnostic character. The peculiar shape of the ear has suggested that the ear itself may act as an airfoil and provide some lift to the head as the animal flies rapidly forward.

Although not readily visible in a live specimen, the number of premolars easily separates the genera into two groups. If the premolar formula is 1/2, you have in hand either *Molossops* or *Molossus*. If the premolar formula is 2/2, it could be *Nyctinomops, Tadarida, Eumops,* or *Promops*. An inspection of the lower incisors usually separates out *Molossus*, for which the incisor formula is 1/1; for all other genera the formula is 1/2 (but see *Molossops neglectus* and *M. temminckii*). Having eliminated *Molossus*, one should inspect the ears. If the ears are separate at the midline and do not connect with a fold of skin, the specimen belongs to the genus *Molossops*. If the ears are not separate but

are joined or almost joined, inspect the upper lip. If the upper lip is grooved and separated into two parts as in a rodent, and if folds of skin from the ears almost meet at the midline, the specimen belongs to the genus *Tadarida* or *Nyctinomops*. If there is no groove and the ears are completely fused across the forehead unambiguously, one should then look at the ears from the side. If they are very large and sculpted in a unique fashion, reaching almost to the end of the snout, the species belongs to the genus *Eumops* (see fig. 8.4C). If the ears are not so sculpted and clearly do not reach to the end of the snout, the bat belongs to the genus *Promops*. A key from Freeman (1981) is included in table 8.16. For aid in identifying *Mormopterus* see the generic account.

The question of subgenera is vexing, but there are clear morphological criteria to aid in recognition. *Neoplatymops* (a subgenus of *Molossops* by Koopman 1993) has a highly flattened body and head, in addition to a unique series of skin tubercules visible on the forearm. *Nyctinomops* (often listed as a subgenus of *Tadarida*) has the ears joined on the forehead by a skin band; the ears are quite long and extend beyond the tip of the rostrum when pressed forward. The lips are wrinkled. *Cynomops* is recognized as a subgenus of *Molossops* by Koopman (1993). Species referable to this taxon have rounded, short ears (see Emmons and Feer 1997 for other characters).

Genus *Molossops* Peters, 1865
Malaga's Free-tailed Bat

Description

The dental formula is I 1/2, C 1/1, P 1/2, M 3/3 (but see *M. neglectus*). Head and body length ranges from 40 to 95 mm, and the tail shows a corresponding range from 14 to 30 mm. The forearm averages 28–51 mm. The dorsum tends to be yellow brown to chocolate brown, and the underparts usually contrast, being gray or slate colored.

Natural History

Species of *Molossops* are specialized for feeding on insects. Some are highly gregarious and may live in colonies of up to seventy-five. In forested habitats these bats tend to roost in hollow trees.

Molossops abrasus (Temminck, 1827)

Description

M. abrasus is one of the largest species of the genus, with the forearm averaging about 41.4 mm. Measurements of two females were TL 111.5; T 34; HF 13; E 18.5; FA 41.4; Wt 24.8 g (Venezuela, USNH); one male measured TL 136; T 37; HF 14; E 19; FA 40 (CMNH). A series from Paraguay (*n* = 7) measured HB 83; T 38;

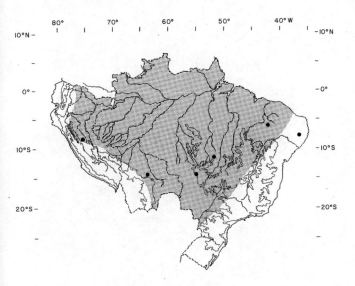

Map 8.146. Distribution of *Molossops abrasus*.

HF 11; E 19; FA 48; Wt 33 g. The dorsum is dark reddish brown, and the venter is lighter. For this account *M. brachymeles* is included in *M. abrasus*.

Range and Habitat

This species is distributed in the Amazonian portions of Venezuela and Colombia, west through the Guianas, and broadly over much of Amazonian Brazil and adjacent portions of Peru, Bolivia, and Paraguay. In northern Venezuela specimens were taken below 150 m in association with mature tropical evergreen forest (map 8.146).

Molossops aequatorianus (Abrera, 1917)

Description

Listed as *Cabreramops* by Albuja (1982). The dental formula is I 1/2, P 1/2. Measurements: HB 50–52; T 29–31; HF 7–7.6; E 14; FA 35.9–37.6.

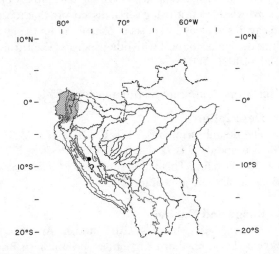

Map 8.147. Distribution of *Molossops aequatorianus*.

Distribution

Known from Babahoyo, Los Ríos province, Ecuador, with a second record from Peru (map 8.147).

Molossops greenhalli (Goodwin, 1958)

Description

This species is only slightly smaller than *M. abrasus*. Measurements for two specimens were TL 94–105; T 34; HF 8–10; E 15–16; FA 35; Wt 18.9 g (Venezuela, USNMNH; Linares and Kiblinski 1969). The dorsum is dark brown, the venter paler.
Chromosome number: $2n = 34$.

Range and Habitat

The species' range extends from southern Nayarit in Mexico south along the Pacific coast to Honduras and thence through the Isthmus to the Caribbean coast of northern South America. It has been taken in the extreme northeastern portions of Brazil (map 8.148). In Venezuela it occurs at low elevations in association with multistratal tropical evergreen forest (Handley 1976). It has been found roosting in colonies of fifty to seventy-five (Goodwin and Greenhall 1961).

Molossops (Neoplatymops) mattogrossensis Vieira, 1942

Description

Measurements: TL 75–80; T 24–25; HF 7–8; E 13–14; FA 29–30; Wt 7–7.5 g. This is one of the smallest species of *Molossops* within the range covered by this volume. As with most species of this genus, the male is larger than the female. This bat is frequently placed in the genus *Neoplatymops* because of its very

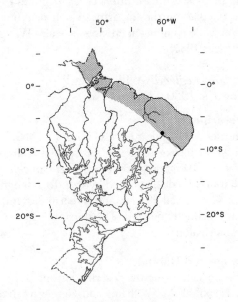

Map 8.148. Distribution of *Molossops greenhalli*.

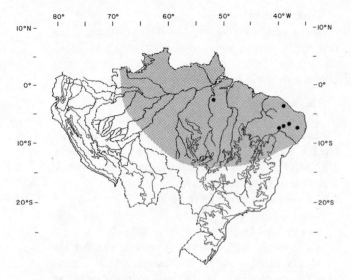

Map 8.149. Distribution of *Molossops mattogrossensis*.

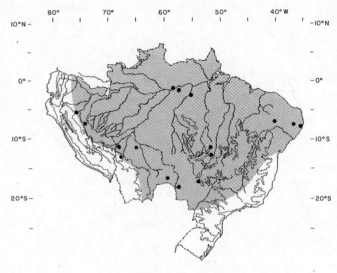

Map 8.150. Distribution of *Molossops planirostris*.

flattened skull (Willig and Knox Jones 1985). The upperparts are brown, but the white bases of the hairs yield a pale brown cast. The venter is white to gray, and the ears are dark brown.
Chromosome number: $2n = 48$; FN = 60.

Range and Habitat

This species occurs in the Amazonian portions of Venezuela, in Colombia, and south into Brazil (map 8.149). In Venezuela it was taken below 200 m elevation in tropical evergreen forest. It has a habit of feeding over ponds and streams (Handley 1976).

Natural History

This bat is an aerial insectivore (Willig 1985). It roosts in rock crevices, often close to ground level. It has papillae on the forearms, which may increase traction while climbing in crevices (see also Willig and Knox Jones 1985).

Molossops neglectus Williams and Genoways, 1980

Description

The dental formula is I 1/1, C 1/1, P 1/2, M 3/3. The single lower incisor separates this species from other members of the genus except for *M. temminckii,* from which it may be distinguished by size. Total length averages 89 mm, tail 29 mm, and forearm 35 mm. The ears are distinctly pointed. The dorsum is brown and the venter paler.

Range and Habitat

The species is newly described from Suriname, 5°25′ N, 55°03′ W (Williams and Genoways 1980b). It may extend to extreme northeastern Brazil (not mapped).

Molossops planirostris (Peters, 1865)
Dog-faced Bat, Moloso Hocico Aplanado

Description

Measurements: HB 53–56; T 25–26; E 13; FA 31–32; Wt 7–9 g. This very small molossid is pale brown dorsally with the white at the bases of the hairs showing through. Ventrally it is white on the chin and chest, with the white continuing in a medial line to the base of the tail. The rest of the venter is colored like the dorsum.

Distribution

Molossops planirostris ranges from Panama to northern Argentina (Myers and Wetzel 1983; Olrog and Barquez 1979) (map 8.150).

Natural History

This species has a litter size of one. *Molossops planirostris* is a fast-flying aerial insectivore that roosts in cavities in trees and posts in colonies of up to eight. It never seems to roost with other species (Vizotto and Taddei 1976; Willig 1983).

Molossops temminckii (Burmeister, 1854)

Description

The dental formula is I 1/1, C 1/1, P 1/2, M 3/3. Measurements: TL 72–74; T 22–26; HF 7–8; E 13; FA 30; Wt 5.5 g. This bat is similar to *M. mattogrossensis* in size and appearance.

Range and Habitat

The species is broadly distributed in Amazonian Bolivia, Peru, northern Argentina, and southern Brazil. It may extend to southeastern Colombia (map 8.151).

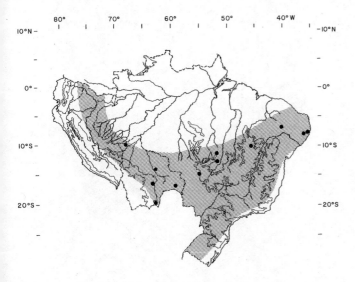

Map 8.151. Distribution of *Molossops temminckii*.

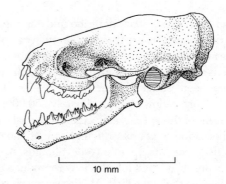

Figure 8.23. Skull of *Tadarida brasiliensis*.

Natural History

Molossops temminckii has a litter size of one. It is widespread in the cerrado and caatinga of Brazil (Mares et al. 1981; Vizotto and Taddei 1976). In Paraguay most individuals were caught over ponds in thorn scrub or at the edge of clearings in thorn scrub, and in Salta province, Argentina. it is the most common bat in chaco vegetation when free water is present. It may be seen flying before sunset and roosts in roofs, hollows of trees, and fence posts (Myers and Wetzel 1983).

Genus *Tadarida* Rafinesque, 1814
Free-tailed Bat

Description

The dental formula is I 1/2–3, C 1/1, P 2/2, M 3/3. This genus contains approximately thirty-five species, which exhibit considerable variation in body size. Head and body length ranges from 45 to 100 mm, the tail from 20 to 60 mm, and the forearm from 27 to 65 mm. Males possess a small glandular throat sac. The dorsum is generally reddish brown to black, and the venter tends to be somewhat paler.

Distribution

This genus is distributed in the subtropical and tropical portions of both the Old World and the New World.

Tadarida brasiliensis (I. Geoffroy, 1824)
Murciélago de Cola de Ratón

Description

The dental formula is I 1/3, C 1/1, P 2/2, M 3/3 (see fig. 8.23). Measurements: HB 60–63; T 32–33; HF 10; FA 43–45; Wt 11–12 g. In this intermediate-sized species of the genus, head and body length aver-

ages 62.3 mm for males and 60.6 mm for females. The dorsal pelage ranges from dark gray to dark brown.

Range and Habitat

This species is distributed from California east to Florida and southward through Mexico through western South America to Argentina and Chile (map 8.152). Its range then extends through Uruguay to the south Atlantic coast of Brazil. In Venezuela this tolerant species may range up to 2,107 m elevation. It utilizes a wide variety of habitats.

Natural History

This is one of the best-studied bats and one of the first in North America to be examined as a potential vector of rabies. In North America this species shows seasonal migratory movements of several hundred kilometers; populations in southern Texas will migrate to Mexico for the winter (Davis, Herreid, and Short 1962). This species is highly adapted for feeding on small moths, and practically all its food is captured on the wing. The bat roosts in caves in parts of their range but tolerates human dwellings as roosting sites. Davis,

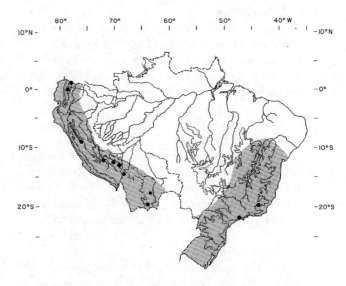

Map 8.152. Distribution of *Tadarida brasiliensis*.

Herreid, and Short (1962) describe in detail the ecology of this species where it uses caves in Texas. When they form colonies in caves the bats may number in the hundreds of thousands. At Carlsbad Caverns the population has been estimated from as high as 8.7 million in 1936 to a current low of approximately 250,000. Subpopulations of this species may exhibit seasonal torpor, but hibernation is not as profound as that shown by many of the north temperate vespertilionids. A single young is produced, and females may reproduce in the year following their own birth. While rearing young the females form nursery colonies spatially separate from male groups; the young are not carried by the mothers when they feed but are left in the crèche. Mothers can recognize the call of their own young within the crèche (Wilkins 1989).

Tadarida espiritosantensis (Ruschi, 1951)

This species is considered a synonym of *N. laticaudatus.*

Genus *Nyctinomops* Miller, 1902

Description

Nyctinomops has been raised to a full genus by many authors (see Koopman 1993). The incisor number is 1/2; the second phalanx of the fourth digit of the wing is shorter than the same phalanx in *Tadarida brasiliensis.* In species of this group the ears meet at the midline, though the point of intersection may be deeply notched, whereas in *T. brasiliensis* there is a slight gap in the fold of skin between the ears. This subgenus includes members distributed from the southwestern United States south across most of South America to northern Argentina.

Nyctinomops aurispinosus (Peale, 1848)

Description

This is one of the largest species of the genus *Nyctinomops,* with forearm length ranging from 47.8 to 54.2 mm. Males are slightly larger than females. The dorsum is wood brown to russet, and the venter is somewhat paler (see Knox Jones and Arroyo-Cabrales 1990).

Range and Habitat

This species occurs from Sinaloa in western Mexico and southern Tamaulipas in eastern Mexico south through the Isthmus and across most of northern South America, extending to Peru, Bolivia, and southeastern Brazil (see Ochoa-G. 1984a) (map 8.153).

Natural History

Females appear to bear one young annually, like most molossids. This bat shelters in caves; its diet in-

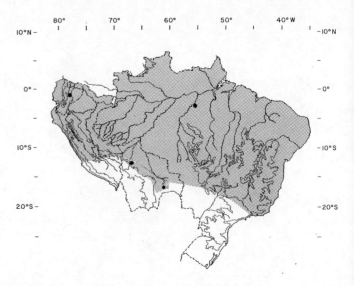

Map 8.153. Distribution of *Nyctinomops aurispinosus.*

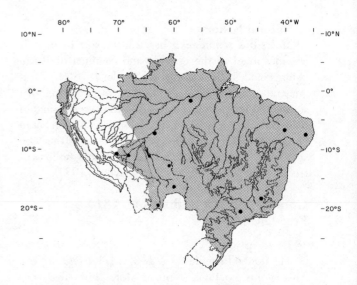

Map 8.154. Distribution of *Nyctinomops laticaudatus.*

cludes soft-bodied insects. Owls appear to be a major predator (Knox Jones and Arroyo-Cabrales 1990).

Nyctinomops laticaudatus (E. Geoffroy, 1805)

Description

Measurements: HB 67.8; T 42.3; HF 10.4; E 20.5; FA 44.6; Wt 14.5 g. This account includes *N. europs, N. gracilis,* and *N. yucatanica.* This small species of *Nyctinomops* shows little sexual size dimorphism. The upperparts are deep brown, and the bases of the hairs are white. The venter is pinkish buff.

Range and Habitat

This species extends from southern Tamaulipas in Mexico south across northern South America to Argentina and eastern Brazil (map 8.154). It was col-

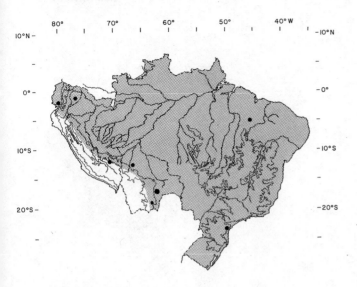

Map 8.155. Distribution of *Nyctinomops macrotis*.

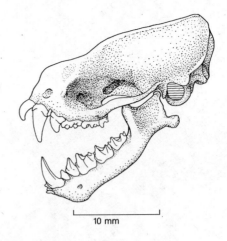

Figure 8.24. Skull of *Eumops glaucinus*.

lected in Venezuela at elevations below 350 m. Although frequently foraging over streams and other moist sites, it is broadly tolerant of both dry deciduous forest and multistratal tropical evergreen forest (Handley 1976). It roosts in colonies of up to fifty.

Nyctinomops macrotis (Gray, 1839)

Description

This is the largest species of *Nyctinomops* in the New World. Average measurements are TL 180; T 54; HF 12; E 28; FA 58–64. The tail of this large molossid extends a good 25 mm beyond the interfemoral membrane. The ears are large and join basally at the midline. The dorsum varies from reddish brown to black, and the basal portion of each hair is nearly white. The fur is shiny (plate 4).

Range and Habitat

This species extends from southern Utah and Colorado in the United States south through Mexico; it occurs intermittently throughout the Isthmus and then is broadly distributed across South America east of the Andes to Uruguay and northern Argentina (map 8.155). It prefers rocky outcrops and tolerates a variety of habitat types.

Natural History

This bat roosts in crevices in rocks, generally in small groups. Only a single young is produced each year, and the females form small nursery colonies. This species is associated with rocky outcrops in arid to semiarid habitats. They specialize in feeding on large moths, but orthopterans and hemipterans are included in the diet. They are powerful fliers and range widely each night from roosting sites (Milner, Jones, and Knox Jones 1990).

Genus *Eumops* Miller, 1906
Bonneted Bat, Mastiff Bat

Description

The dental formula is I 1/2, C 1/1, P 2/2, M 3/3 (see fig. 8.24). There are eight species, showing an extreme range in size. Head and body length may vary between 40 and 130 mm, and the tail ranges from 35 to 80 mm. The large ears are rounded and usually connected across the head at the base. Males of some species have throat sacs that produce glandular secretions when in breeding condition. Eger (1977) has written a monograph on the genus.

Distribution

Species of this genus are distributed from southern California and Florida in the United States south through Central America to southern South America as far as northern Argentina.

Eumops auripendulus (Shaw, 1800)

Description

A medium-sized species of the genus, *Eumops auripendulus* has a forearm length from 56 to 63 mm. One female in Venezuela had a total length of 130.15 mm, of which 43 mm was the tail. A male from Trinidad measured TL 141; T 45; HF 15; E 23; Wt 32.5 g. The upperparts are dark reddish brown, the underparts slightly paler. The bases of the dorsal hairs are buffy.

Range and Habitat

This species is distributed from southern Veracruz, Mexico, south through the Isthmus and across South America to Paraguay, Argentina, and southeastern Brazil (map 8.156). In northern Venezuela it prefers dry deciduous thorn forests with rocky outcroppings.

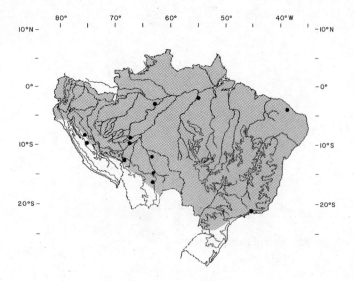

Map 8.156. Distribution of *Eumops auripendulus*.

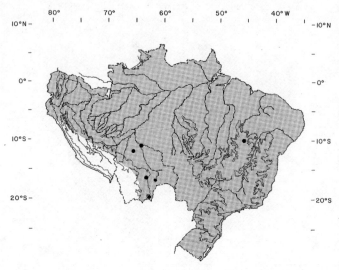

Map 8.157. Distribution of *Eumops bonariensis*.

Eumops bonariensis (= *E. nanus*) (Peters, 1874)

Description

Measurements: TL 99–101; T 33–35; HF 9; E 19–20; FA 43–44; Wt 12.6 g. This is generally the smallest species of the genus taken over northern South America. *E. nanus* is included in *E. bonariensis*.

Range and Habitat

This species is found from southern Veracruz south through the Isthmus, across most of South America, to northern Argentina (map 8.157). In northern Venezuela it was found at extremely low elevations and was strongly associated with dry deciduous tropical forest.

Eumops dabbenei Thomas, 1914

Description

Because of its large size this species can be confused only with *E. perotis*. Measurements of a female taken in northern Venezuela were total length 177 mm; head and body length 120 mm; forearm 79 mm.

Range and Habitat

This species occurs from western Venezuela sporadically through central Brazil to northern Argentina (map 8.158). In northern Venezuela it was taken in dry deciduous thorn forest, near water (Handley 1976).

Eumops glaucinus (Wagner, 1843)

Description

Measurements: TL 144–49; T 51; HF 13–14; E 27–28; FA 60–62; Wt ll g. This bat is smaller than *E. perotis* but larger than *E. auripendulus* and *E. maurus*. The sexes are nearly alike in size. The upperparts are dark cinnamon brown, the venter slightly paler.

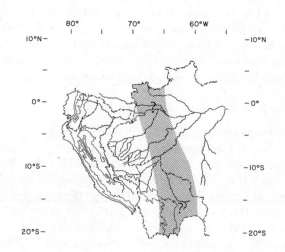

Map 8.158. Distribution of *Eumops dabbenei* (from Koopman 1982).

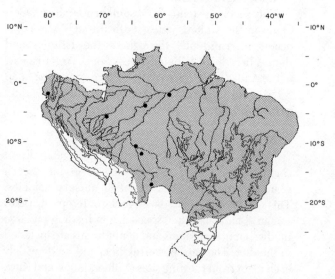

Map 8.159. Distribution of *Eumops glaucinus*.

Range and Habitat

This species occurs from central Mexico and Florida south through the Isthmus of Panama and the island of Cuba, and thence over most of South America to Paraguay, northwestern Argentina, and southeastern Brazil (map 8.159). In northern Venezuela it was taken below 600 m elevation and generally was found foraging in the vicinity of streams, swamps, and lagoons. It is strongly associated with moist habitats and multistratal tropical evergreen forest. It roosts in trees (Handley 1976).

Natural History

This species roosts in colonies, and during the breeding season one male may associate with a group of females. Tree cavities are preferred roosts. Colony size is small, with an average of about fifteen. These bats are active predators on flying insects, including up to 58% Coleoptera (Best, Kiser, and Rainey 1997).

Eumops hansae Sanborn, 1932

Description

This small species of *Eumops* exhibits rather pronounced sexual size dimorphism. Males have forearm lengths from 40 to 41.6 mm, and females have forearm lengths of approximately 37 mm. Head and body length for a typical female is approximately 71 mm, and males have a head and body length of about 75 mm. The color is darker than that of *E. nanus*, but the species are similar in size (plate 4). *E. amazonicus* is considered a junior synonym of *E. hansae*.

Range and Habitat

This species occurs from southern Costa Rica south through the Isthmus across northern South America to Guyana; it occurs sporadically in Amazonian Brazil. In Venezuela specimens were taken at below 155 m elevation in moist multistratal tropical evergreen forest (not mapped).

Eumops perotis (Schinz, 1821)
Moloso de Orejas Anchas

Description

This large bat has a forearm ranging from 72 to 82 mm in length. Average measurements are TL 171; T 55; HF 17; E 40. The dorsum ranges from dark gray to brownish gray, and the venter barely contrasts. The species *E. trumbulli* is considered a junior synonym of *E. perotis*.

Range and Habitat

This species ranges from central California south into western Mexico. It is sparsely distributed in Central America but occurs widely over most of South America south to northern Argentina (map 8.160).

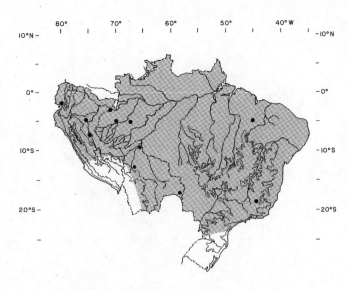

Map 8.160. Distribution of *Eumops perotis*.

Natural History

Like all members of the Molossidae, this large bat has very narrow wings. As a result, it has difficulty taking flight from flat surfaces and generally prefers crevices in rock outcroppings for daytime roosting sites, where it can make a vertical drop of several feet before launching into flight. This bat has been extensively studied in the southwestern United States by Vaughan (1959). Females form maternity colonies during the breeding season. A single young is born each year. The species, at least in the southern parts of the United States, feeds extensively on Hymenoptera (Eger 1977).

Genus *Molossus* E. Geoffroy, 1805

Description

The dental formula is I 1/1, C 1/1, P 1/2, M 3/3. Head and body length ranges from 50 to 95 mm, tail length from 20 to 70 mm, and the forearm from 33 to 60 mm. The general dorsal color is reddish brown to dark chestnut brown. It has been remarked that several species have two color phases: dark brown to almost black, and a lighter brown phase.

Distribution

This New World genus is distributed from northern Mexico south to northern Argentina and Uruguay.

Molossus ater E. Geoffroy, 1805
Black Mastiff Bat

Description

Measurements: TL 130; T 44; HF 13; E 17; FA 48–49; Wt 30 g. This species is rather large for the genus. The upperparts are a rich russet brown to blackish,

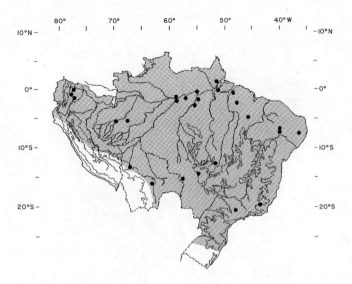

Map 8.161. Distribution of *Molossus ater* (includes *M. rufus;* see Anderson 1997).

with the underparts only slightly paler. The basal portions of the hairs are lighter.

Range and Habitat

Molossus ater occurs from Sinaloa in western Mexico and Nuevo León in eastern Mexico south through the Isthmus over much of South America to northern Argentina (map 8.161). In northern Venezuela this species was taken below 1,180 m; most specimens were found below 500 m elevation. It is broadly tolerant of dry deciduous forest, man-made clearings, and multistratal tropical evergreen forest (Handley 1976).

Natural History

This species tends to roost in hollow trees or human dwellings and generally prefers moist areas. Colony size may reach fifty (Goodwin and Greenhall 1961). In the Brazilian Amazon region females are repro-

ductively active through most of the year. Gestation is estimated to be from two to three months (Marques 1986). Rasweiler (1992) summarizes important data on placentation and the physiology of reproduction.

Molossus bondae J. A. Allen, 1904

Description

Measurements: TL 105–12; T 39–41; HF 11–12; E 12–14; FA 39–41; Wt 17–19 g. This species is intermediate in size. The forearm length in Venezuela averaged 41 mm for males and 39 mm for females. The dorsum is dark brown to reddish brown, the venter paler.

Range and Habitat

This species is distributed from Costa Rica south through the Isthmus across northern South America extending to Ecuador (map 8.162). In Venezuela specimens were taken below 600 m and in association with dry deciduous tropical forest in coastal areas.

Molossus molossus (Pallas, 1766)

Description

Measurements: TL 100–102; T 36; HF 10; E 12–13; FA 38–39; Wt 13–15 g. The dorsum is gray brown to dark brown, the venter pale brown (plate 4). This species includes *M. coibensis. M. aztecus* may prove to be a subspecies of *M. molossus*.

Range and Habitat

This species ranges from Central America south across almost all of South America well into northern Argentina and Uruguay (map 8.163). It is the common small species of *Molossus* taken over an enormous range. In northern Venezuela most specimens were

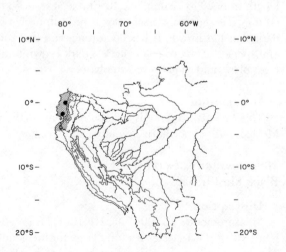

Map 8.162. Distribution of *Molossus bondae.*

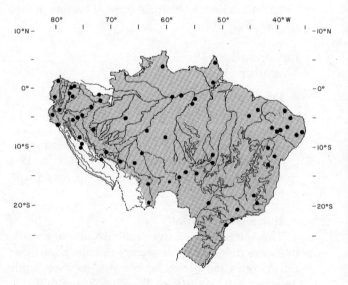

Map 8.163. Distribution of *Molossus molossus.*

taken below 900 m elevation. It is broadly tolerant of both dry deciduous forest and moist tropical multistratal evergreen forest.

Natural History

Although this bat forages near streams, it roosts in hollow trees and in human dwellings. In drier areas it frequently feeds near ponds and may be the dominant species taken in mist nets set near pond sites (Handley 1976). Haussler, Moller, and Schmidt (1981) conducted a field and captive study on the growth and development of this species. Reproduction takes place at the onset of the rainy season, and females form nursery colonies for rearing the young. Although the young hang as a compact group while the mothers leave the cave to forage, each female can identify her own by its cry. The young begin thermoregulation at twenty days of age; they nurse until sixty-five days of age, and the forearm reaches adult size at sixty days.

These bats are entirely insectivorous and specialize on beetles (Myers and Wetzel 1983; Fabian and Marques 1989).

Genus *Mormopterus* Peters, 1865

Description

The dental formula is 1/2, 1/1, 1-2/2, 3/3. The jaw bears a strong coronoid process, and the third upper molar is quite large. These small molossid bats have entirely separate, erect ears.

Distribution

As a genus the taxon has a scattered distribution including parts of Africa, Southeast Asia, and South America.

Mormopterus kalinowskii (Thomas, 1893)

Description

The generic status of this species is in some doubt since all the other congeners are Old World in their distribution. I (JFE) examined six specimens from Cerrola Diego south of Motupé, state of Lambayeque, Peru. Measurements were TL 79–88; T 29–35; HF 5–7; E 13–17; FA 35–38; Wt 6–7 g (LSU). The dorsum is light brown, barely contrasting with the pale gray brown venter.

Distribution

Known only from central Peru and northern Chile (map 8.164).

Mormopterus phrudus Handley, 1956

Description

The dental formula is 1/2, 1/1, 2/2, 3/3. Measurements: HB 50–51, T 29–32, HF 8; E 13–14 ($n = 2$).

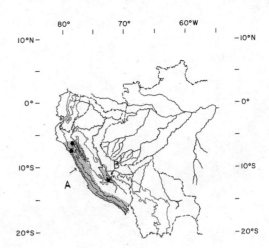

Map 8.164. Distribution of (a) *Mormopterus kalinowskii* (b) *Mormopterus phrudus*.

Closely resembles *M. kalinowskii* and *Tadarida* sp. Only two specimens were available to Handley.

Distribution

Type locality: San Miguel Bridge, Urubamba River, Machu Picchu, Cuzco, Peru, at 2,000 m (map 8.164).

Genus *Promops* Gervais, 1855

Description

The dental formula is I 1/2, C 1/1, P 2/2, M 3/3. In these intermediate-sized molossid bats, head and body length ranges from 60 to 90 mm, tail length from 45 to 75 mm, and forearm from 43 to 63 mm. The dorsum is drab brown to glossy black, with the venter paler.

Distribution

Species of this genus are distributed from Central America south to Paraguay. There are two species within the range covered by this volume.

Promops centralis Thomas, 1915

Description

This is the largest species of the genus. Head and body length averages about 89 mm, and the forearm exceeds 50 mm. One female from Venezuela measured TL 139; T 50; HF 13; E 18; FA 52.5. The upperparts are dark brown to glossy black, the underparts slightly paler.

Range and Habitat

This species occurs from the Yucatán peninsula in Mexico south through the Isthmus across northern South America and in western Colombia and Peru through Bolivia to Paraguay and northern Argentina (map 8.165). It is absent over much of Brazil. It roosts in colonies of up to six individuals under palm fronds (Goodwin and Greenhall 1961).

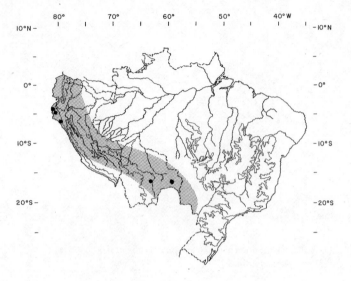

Map 8.165. Distribution of *Promops centralis*.

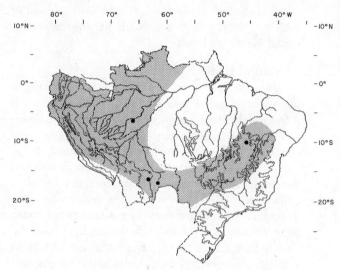

Map 8.166. Distribution of *Promops nasutus*.

Promops nasutus (Spix, 1823)

Description
This bat is slightly smaller than *P. centralis:* TL 126–41; T 57–61; HF 12–13; E 13–17; FA 48.5; Wt 19.9 g (*n* = 2 females; Venezuela, USNMNH).

Range and Habitat
This species is confined to the continent of South America, being found in the Amazonian portions of Venezuela, Colombia, Peru, and Bolivia and sporadically distributed from south-central Brazil to northern Argentina (map 8.166). Specimens in Venezuela were taken at below 1,000 m in wet habitat and multistratal tropical evergreen forest. Like other tropical molossids, it roosts in hollow trees.

Natural History
Little is known concerning the life history of this bat..It produces a single young, is insectivorous, and shelters in natural cavities. Reproduction in seasonal habitats is synchronized by rainfall (Sazima and Uieda 1977).

Table 8.14   List of Phyllostomid Tent-Making Bats

Artibeus andersoni	*Uroderma bilobatum*
Artibeus cinereus	*Uroderma magnirostrum*
Artibeus glaucus	*Ectophylla alba*
Artibeus gnomus	*Vampyressa nymphaea*
Artibeus jamaicensis	*Vampyressa pusilla*
Artibeus phaeotis	*Mesophylla macconnelli*
Artibeus toltecus	*Rhinophylla pumilio*
Artibeus watsoni	

Source: Foster and Timm 1976; Foster 1992; Timm 1987; Kunz et al 1994.

Table 8.15   Key to the Common Genera of the
Vespertilionidae of South America

1	Dorsal surface of interfemoral membrane furred at least halfway to the distal edge	*Lasiurus*
1'	Dorsal surface of interfemoral membrane naked	2
2	Only a single upper incisor on each side	*Rhogeessa*
2'	Two or more upper incisors per side	3
3	Only a single premolar per side	4
3'	Two or more premolars per side	*Myotis*
4	Ears 25% of head and body length	*Histiotus*
4'	Ears 12% or less of head and body length	*Eptesicus*

References
• References used in preparing distribution maps

Acosta, C. E., and R. D. Owen. 1993. *Koopmania concolor. Mammal. Species* 429:1–3.

• Acosta y Lara, E. F. 1950. Quirópteros del Uruguay. *Comun. Zool. Mus. Hist. Nat. Montevideo* 3 (58): 1–71.

• Aellen, V. 1970. Catalogue raisonné des chiroptères de la Colombie. *Rev. Suisse Zool.* 77 (1): 1–37.

Alberico, M. 1990. Systematics and distribution of the genus *Vampyrops* (Chiroptera: Phyllostomidae) in north western South America. In *Vertebrates in the tropics,* ed. G. Peters and R. Hutterer, 345–54, Bonn: Museum Alexander Koenig.

Albuja V., L. 1982. *Murciélagos de Ecuador.* Quito: Escuela Politechnica Nacional.

Allen, G. M. 1908. Notes on Chiroptera. *Bull. Mus. Comp. Zool.* 52 (3): 25–62.

Alverez, J., M. R. Willig, J. Knox Jones Jr., and W. D. Webster. 1991. *Glossophaga soricina. Mammal. Species* 379:1–7.

Table 8.16   Key to the Genera and Subgenera of New World Molossid Bats of South America

1a. Ears usually widely separated, basisphenoid pits usually shallow or nonexistent.
  2a. Palate without anterior emargination, no wrinkles on lips.
    3a. Tooth number 26 (I 1/1, C 1/1, P 1/2, M 3/3), M^3 V-shaped, 2d phalanx of digit 4 16–20% of total length of digit 4. Basisphenoid pits shallow, slight cusp on M^3, size very small (GSL near 13 mm) . . . . *Molossops (Molossops)*
    3b. Tooth number 28 or 30, M^3 either V- or N-shaped, 2d phalanx variable.
      4a. Tooth formula 1/2(1), 1/1, 1/2, 3/3 = 28 (26 possibly), 2d phalanx of digit 4 near 7.5% (6.9–8.5) of digit 4, M^3 V-shaped . . . . *Molossops (Cynomops)*
      4b. Tooth formula 1/2, 1/1, 2/2, 3/3 = 30, 2d phalanx of digit 4 near 12% of digit 4, M^3 N-shaped . . . . *Molossops (Neoplatymops)*
  2b. Palate notched anteriorly, coronoid process usually high and curved posteriorly, lips not wrinkled or slightly wrinkled or covered with spines. Skull not extremely flattened, 28–30 teeth, M^3 complete N-shape. Tooth formula 1/2, 1/1, 1/2, 3/3 . . . . New World *Mormopterus*
1b. Ears not widely separated but just joining to form a V-shaped valley or well joined over the nose by a band; basisphenoid pits usually medium deep to very deep.
  5a. No anterior emargination of palate.
    6a. PM^3 absent . . . . *Molossus*
      PM^3 vestigal, domed palate . . . . *Promops*
    6b. Lips skirtlike with many fine wrinkles, PM^3 present, ears often large and forward facing, M^3 variable but usually always with posterior commissure . . . . *Eumops*
  5b. Anterior emargination of palate either slight or deep, lips either deeply wrinkled or skirtlike.
    7a. 2d phalanx of digit 4 9% or more of digit 4 length. M^3 usually well-developed N-shape, deeply notched anterior palate, PM^3 usually well developed, 2d phalanx of digit 4 usually 9–12% of digit 4 . . . . *Tadarida*
    7b. 2d phalanx between 3% and 5% of digit 4, narrow rostrum and deeply notched anterior palate . . . . *Nyctinomops*

Source: Adapted from Freeman 1981.

Table 8.17   Average External Measurements (mm) and Weights (g) of Some Neotropical Chiroptera

	Sex	N	TL	HB	T	HF	E	FA	Wt
Emballonuridae									
Rhynchonycteris naso	m	46	56.48	41.09	15.39	7.18	13.20	36.87	3.80
	f	38	59.18	42.34	16.84	7.45	13.63	38.57	3.99
Saccopteryx bilineata	m	44	72.98	53.39	19.59	11.76	16.60	46.27	8.58
	f	99	74.93	55.33	19.59	11.65	16.24	47.74	8.82
Saccopteryx canescens	m	8	54.75	38.63	16.13	7.25	12.63	35.93	3.23
	f	8	61.63	44.00	17.63	7.25	13.75	38.68	3.47
Saccopteryx leptura	m	12	58.83	43.75	15.08	8.42	13.92	37.98	4.47
	f	21	61.14	45.05	15.90	8.33	14.33	39.48	5.71
Cormura brevirostris	m	6	67.83	54.00	13.83	10.33	14.50	45.73	8.62
	f	5	69.80	54.20	15.60	8.80	16.20	46.48	10.15
Peropteryx kappleri	m	14	71.93	58.21	13.71	10.86	18.07	48.30	7.73
	f	242	74.67	60.96	13.71	11.42	18.00	50.28	8.54
Peropteryx macrotis	m	47	60.49	47.21	13.28	8.57	15.30	41.94	5.14
	f	76	64.43	49.37	15.07	9.01	15.94	44.88	6.09
Peropteryx m. trinitatis	m	26	56.58	46.38	10.19	8.35	12.46	38.57	3.61
	f	11	62.18	51.09	11.09	8.64	12.45	41.81	4.42
Diclidurus albus	m	8	90.50	70.38	20.13	12.50	16.63	62.45	14.93
	f	14	91.93	71.43	20.50	12.57	16.93	64.38	16.25
Diclidurus ingens	f	3	103.67	80.67	23.00	13.00	20.67	71.43	—
Diclidurus isabellus	m	21	80.38	59.67	20.71	12.19	17.71	57.93	13.36
	f	6	81.33	59.33	22.00	12.83	17.67	59.10	15.63
Diclidurus scutatus	m	2	86.00	68.00	18.00	10.00	14.50	53.10	13.60
	f	10	83.40	65.10	18.30	10.40	13.80	56.22	12.94
Noctilionidae									
Noctilio albiventris	m	65	91.88	74.88	17.00	18.20	23.70	59.72	28.71
	f	146	88.26	71.55	16.71	17.14	23.03	58.64	22.99
Noctilio leporinus	m	26	127.42	96.88	30.54	33.65	28.00	85.12	67.72
	f	37	120.03	91.27	28.76	30.62	28.11	83.36	55.87

Continued on next page

Table 8.17   (continued)

	Sex	N	TL	HB	T	HF	E	FA	Wt
Mormoopidae									
Pteronotus davyi	m	55	78.69	56.44	22.25	11.75	17.87	46.52	9.38
	f	49	80.51	57.96	22.55	11.00	18.08	47.19	9.06
Pteronotus parnellii	m	127	95.13	71.77	23.44	15.06	23.69	59.95	20.42
	f	118	93.88	70.44	23.44	15.14	23.56	60.45	19.60
Pteronotus personatus	m	2	70.50	53.50	17.00	12.00	17.50	44.55	8.00
	f	2	75.00	58.00	17.00	11.00	16.00	45.00	6.90
Pteronotus suapurensis	m	18	87.89	64.33	23.56	12.94	18.94	51.13	12.62
	f	15	88.00	64.07	23.93	12.88	19.06	51.87	13.98
Mormoops megalophylla	m	46	94.57	68.07	26.50	12.17	15.35	55.41	17.21
	f	15	95.53	69.60	25.93	12.53	16.40	55.06	16.55
Phyllostomidae									
Phyllostominae									
Micronycteris brachyotis	m	1	68.00	57.00	11.00	11.00	17.00	38.60	10.60
	f	1	70.00	59.00	11.00	13.00	16.00	40.00	—
Micronycteris hirsuta	m	2	78.00	64.00	14.00	11.50	25.50	44.15	14.00
	f	3	78.00	61.00	17.00	13.33	27.00	43.60	12.43
Micronycteris megalotis	m	24	56.41	43.78	12.63	9.96	21.85	32.13	5.09
	f	68	58.32	44.68	13.64	10.07	21.69	33.25	5.71
Micronycteris microtis	m	23	57.74	43.35	12.39	10.26	22.04	33.64	5.97
	f	17	60.35	46.12	14.24	10.47	22.59	34.97	6.64
Micronycteris minuta	m	25	59.52	47.40	12.12	11.56	22.32	34.96	6.86
	f	25	61.24	49.08	12.12	11.58	21.88	35.11	7.20
Micronycteris nicefori	m	41	64.02	54.98	9.05	13.00	18.37	37.51	7.90
	f	39	67.49	58.03	9.46	13.38	18.49	38.84	9.20
Micronycteris schmidtorum	m	2	64.00	48.00	16.00	11.00	23.00	35.80	7.60
	f	5	64.80	48.60	16.20	11.00	21.00	34.60	7.23
Micronycteris sylvestris	m	2	61.50	47.50	14.00	12.50	21.50	37.95	7.25
	f	1	65.00	50.00	15.00	11.00	22.00	41.00	10.30
Lonchorhina aurita	m	61	113.92	60.97	59.89	14.55	32.66	50.97	14.34
	f	40	115.03	61.10	53.93	14.48	32.18	51.48	15.07
Lonchorhina orinocensis	m	29	102.34	52.28	50.07	11.00	30.76	42.16	8.59
	f	24	103.08	52.38	50.71	11.00	30.71	42.37	8.97
Macrophyllum macrophyllum	m	4	86.00	44.70	41.90	14.53	18.83	35.10	7.08
	f	11	93.38	48.54	44.85	15.00	18.77	35.89	7.15
Tonatia bidens	m	9	97.89	76.00	21.10	16.80	33.50	57.33	26.25
	f	7	95.00	75.43	19.57	16.86	31.14	57.04	24.65
Tonatia brasiliense	m	24	65.50	54.83	10.67	11.96	24.58	36.15	10.18
	f	21	65.43	55.71	9.71	12.09	24.86	36.06	10.61
Tonatia carrikeri	m	1	92.00	72.00	20.00	17.00	30.00	44.00	22.10
Tonatia silvicola	m	15	92.80	73.33	19.47	16.93	36.60	52.56	30.64
	f	20	92.20	71.50	20.70	17.25	37.20	51.91	24.13
Mimon crenulatum	m	33	83.15	57.88	25.53	12.58	26.00	49.18	12.85
	f	32	83.94	58.50	25.44	12.47	26.41	49.49	12.03
Phyllostomus discolor	m	128	97.70	83.57	14.13	17.29	23.36	60.91	38.15
	f	107	98.23	84.24	13.99	17.06	23.18	60.86	35.35
Phyllostomus elongatus	m	52	103.17	80.37	22.76	18.85	30.35	66.35	41.08
	f	55	104.78	82.47	22.31	18.22	29.96	66.36	40.06
Phyllostomus hastatus	m	8	131.09	109.75	21.34	22.56	32.36	84.24	104.54
	f	12	126.92	107.78	19.14	22.49	31.86	82.82	87.55
Phylloderma stenops	m	9	113.71	—	17.71	21.24	27.67	70.92	50.63
Trachops cirrhosus	m	125	97.02	77.54	19.47	19.89	34.36	60.06	34.20
	f	122	97.15	78.16	18.98	19.80	34.65	60.84	32.86
Chrotopterus auritus	m	24	107.58	97.48	9.38	25.50	45.50	76.53	65.06
	f	9	111.67	103.43	10.00	25.67	47.56	79.47	68.92
Vampyrum spectrum	m	3	140.00	140.00	0.00	32.67	46.33	105.17	· 169.43
	f	1	158.00	158.00	0.00	38.00	48.00	107.40	199.70
Glossophaginae									
Glossophaga longirostris	m	236	68.30	61.62	6.47	12.30	15.90	37.78	12.81
	f	209	68.91	61.36	7.61	11.97	15.69	38.24	12.91

Table 8.17   (*continued*)

	Sex	N	TL	HB	T	HF	E	FA	Wt
Glossophaga soricina	m	221	61.67	53.96	7.71	11.14	14.29	34.27	9.46
	f	218	62.82	54.94	7.91	11.14	14.52	34.97	9.35
Lionycteris spurrelli	m	48	61.52	53.50	8.02	11.19	13.35	34.59	8.66
	f	77	63.30	54.73	8.57	11.35	13.29	35.55	8.89
Lonchophylla robusta	m	13	77.31	68.38	8.92	13.23	16.92	42.75	14.34
	f	9	78.44	69.33	9.11	13.00	16.67	42.85	13.72
Lonchophylla thomasi	m	6	58.83	51.33	7.50	10.17	15.00	31.84	7.08
	f	7	56.86	49.14	7.71	10.14	14.86	31.73	5.93
Anoura caudifer	m	36	65.81	62.61	5.50	12.03	14.36	36.64	10.98
	f	31	64.55	61.65	4.72	12.06	14.48	36.39	10.00
Anoura cultrata	m	2	79.00	75.00	4.00	14.50	16.50	41.90	20.35
	f	2	71.00	66.50	4.50	14.50	17.00	40.50	14.95
Anoura geoffroyi	m	21	65.86	65.86	—	13.14	15.41	42.63	15.80
	f	33	65.12	65.12	—	13.21	15.52	42.86	14.50
Anoura latidens	m	45	64.44	64.44	—	12.93	15.33	42.91	14.53
	f	46	66.20	66.20	—	12.65	14.65	43.15	14.81
Choeroniscus godmani	m	4	59.75	53.75	6.00	9.50	8.25	31.87	7.27
	f	9	64.78	55.67	9.11	9.89	12.13	33.98	7.61
Choeroniscus minor	m	1	69.00	61.00	8.00	10.00	13.00	34.60	8.20
	f	2	70.50	61.00	9.50	11.00	12.50	35.30	8.30
Leptonycteris curasoae	m	97	81.35	81.35	—	16.85	17.47	53.37	26.33
	f	86	79.02	79.02	—	17.03	17.22	52.55	24.93
Scleronycteris ega	m	1	—	—	—	—	—	35.20	—
Carolliinae									
Carollia brevicauda	m	66	65.35	58.75	6.69	13.88	19.49	38.48	14.57
	f	83	65.19	58.54	6.65	13.96	19.37	38.63	12.93
Carollia castanea	m	4	63.75	53.75	10.00	13.50	17.75	36.50	14.55
	f	5	62.60	53.80	8.80	12.20	18.60	35.73	—
Carollia perspicillata	m	547	68.23	58.21	10.04	14.39	20.79	40.91	17.04
	f	655	69.29	58.43	9.85	14.33	20.63	41.14	15.90
Rhinophylla pumilio	m	12	48.33	48.33	—	10.33	16.33	34.71	9.36
	f	26	50.04	50.04	—	10.77	15.81	35.14	10.41
Sturnirinae									
Sturnira bidens	m	3	64.33	64.33	—	14.00	15.33	39.63	17.20
	f	13	64.92	64.92	—	14.77	16.31	40.64	16.30
Sturnira bogotensis	m	2	67.00	67.00	—	15.50	17.50	43.05	19.60
	f	2	65.50	65.50	—	14.00	17.00	43.30	—
Sturnira erythromos	m	40	58.23	58.23	—	13.40	15.83	39.50	15.11
	f	32	58.34	58.34	—	12.94	15.78	38.91	15.07
Sturnira lilium	m	76	61.30	61.30	—	13.88	16.66	41.03	20.43
	f	89	60.61	60.61	—	13.62	16.48	40.56	18.87
Sturnira ludovici	m	90	70.80	70.80	—	16.39	19.27	47.21	26.77
	f	48	70.92	70.92	—	16.13	18.88	46.82	23.98
Sturnira tildae	m	67	69.66	69.66	—	16.63	20.37	47.78	26.16
	f	89	69.47	69.47	—	16.40	20.22	47.67	25.46
Stenoderminae									
Uroderma bilobatum	m	167	60.92	60.92	—	12.50	18.16	41.68	16.05
	f	215	60.73	60.73	—	12.60	18.14	41.73	17.19
Uroderma magnirostrum	m	52	61.37	61.37	—	12.38	17.90	42.54	15.92
	f	103	62.48	62.48	—	12.71	17.66	43.34	17.83
Platyrrhinus aurarius	m	17	76.06	76.06	—	16.06	21.88	51.92	33.64
	f	15	78.27	78.27	—	16.47	21.53	53.22	34.68
Platyrrhinus brachycephalus	m	3	62.67	62.67	—	11.33	17.00	39.80	15.50
•*Platyrrhinus dorsalis*	—	45	78.0	—	—	12.5	—	46.5	—
Platyrrhinus helleri	m	159	57.05	57.05	—	11.85	17.19	37.54	13.01
	f	181	57.60	57.60	—	12.08	17.39	37.92	14.33
Platyrrhinus umbratus	m	81	69.85	69.85	—	13.52	18.67	44.20	23.35
	f	79	71.61	71.61	—	13.80	18.66	44.58	25.56
Platyrrhinus vittatus	m	5	86.80	86.80	—	17.40	23.20	57.24	—
	f	3	89.67	89.67	—	18.33	24.33	59.77	—

Continued on next page

Table 8.17　　(*continued*)

	Sex	N	TL	HB	T	HF	E	FA	Wt
Vampyrodes caraccioli	m	4	71.75	71.75	—	15.75	21.25	51.06	26.88
	f	15	74.50	74.50	—	16.13	21.67	53.24	29.95
Vampyressa bidens	m	57	50.14	50.14	—	11.04	17.19	35.42	11.04
	f	48	52.10	52.10	—	11.23	17.08	35.81	12.21
Vampyressa pusilla	m	18	46.72	46.72	—	10.16	15.00	31.37	8.22
	f	62	47.76	47.76	—	10.31	14.95	31.86	9.43
Chiroderma salvini	m	9	72.11	72.11	—	13.78	18.78	48.67	27.22
	f	13	73.23	73.23	—	14.23	19.23	48.19	29.10
Chiroderma trinitatum	m	13	54.85	54.85	—	11.92	17.46	38.85	13.09
	f	46	56.70	56.70	—	11.91	17.70	38.48	13.98
Chiroderma villosum	m	46	64.57	64.57	—	13.37	19.15	44.21	21.00
	f	198	67.55	67.55	—	13.72	19.34	45.80	22.91
Ectophylla macconnelli	m	16	42.81	42.81	—	10.25	15.56	30.15	6.39
	f	47	44.06	44.06	—	9.91	15.45	30.97	6.56
Artibeus amplus	m	10	91.50	91.50	—	20.10	23.10	69.12	60.03
	f	16	90.31	90.31	—	19.75	23.13	69.43	—
Artibeus cinereus	m	109	54.01	54.01	—	11.46	16.77	39.74	12.62
	f	101	53.99	53.99	—	11.74	16.95	39.98	13.05
Artibeus concolor	m	18	59.61	59.61	—	12.33	18.17	46.93	18.21
	f	69	64.09	64.09	—	11.91	18.88	48.33	19.96
• *Artibeus fimbriatus*	—	22	95.2	—	18.0	23.5	14.6	—	54.0
Artibeus fuliginosus	m	105	75.00	75.00	—	16.53	23.01	58.94	35.34
	f	50	77.24	77.24	—	17.26	23.48	60.16	36.61
Artibeus gnomus	m	20	48.95	48.95	—	11.05	17.25	37.50	10.16
	f	23	49.78	49.78	—	11.00	17.26	37.45	10.49
Artibeus hartii	m	39	60.33	60.33	—	12.13	15.92	39.18	17.37
	f	36	61.00	61.00	—	12.00	15.78	39.18	17.26
Artibeus jamaicensis	m	294	80.11	80.11	—	17.57	22.23	59.57	40.41
	f	344	81.83	81.83	—	17.96	22.77	60.45	43.16
Artibeus lituratus	m	161	89.43	89.43	—	19.96	23.77	68.82	62.94
	f	188	91.14	91.14	—	20.30	24.19	69.59	68.15
• *Artibeus planirostris*	—	11	85.0	—	17.6	24.5	25.6	52.4	—
Ametrida centurio	m	67	42.01	42.01	—	10.63	14.27	25.58	8.01
	f	69	47.94	47.94	—	10.94	15.23	31.87	12.60
Sphaeronycteris toxophyllum	m	31	56.55	56.55	—	11.55	15.84	37.66	14.91
	f	93	57.89	57.89	—	11.92	16.38	40.14	14.73
Centurio senex	f	5	55.40	55.40	—	14.80	17.00	42.22	18.70

Desmodontinae

	Sex	N	TL	HB	T	HF	E	FA	Wt
Desmodus rotundus	m	71	74.51	74.51	—	16.93	18.75	—	—
	f	63	80.57	80.57	—	17.68	19.33	—	—
Desmodus youngi	m	11	83.36	83.36	—	17.91	18.18	51.65	36.27
	f	4	84.00	84.00	—	19.50	18.75	53.48	32.20
Diphylla ecaudata	m	6	73.17	73.17	—	15.67	15.17	49.90	25.98
	f	3	78.67	78.67	—	17.67	15.67	51.80	25.17

Natalidae

	Sex	N	TL	HB	T	HF	E	FA	Wt
Natalus tumidirostris	m	89	100.69	48.73	52.96	10.06	15.42	29.43	6.23
	f	42	101.40	49.95	51.45	9.95	14.95	39.41	6.29

Furipteridae

	Sex	N	TL	HB	T	HF	E	FA	Wt
Furipterus horrens	m	6	62.00	37.50	24.50	7.92	11.08	35.28	3.08

Thyropteridae

	Sex	N	TL	HB	T	HF	E	FA	Wt
Thyroptera tricolor	m	6	72.67	40.17	32.50	6.67	12.50	36.43	4.68
	f	4	72.50	40.25	32.35	7.00	12.50	37.63	4.50

Vespertilionidae

	Sex	N	TL	HB	T	HF	E	FA	Wt
Myotis albescens	m	27	79.11	44.48	34.63	8.96	14.00	34.18	5.59
	f	48	81.08	45.64	35.47	9.00	14.02	34.71	5.72
• *Myotis chiloensis*	—	15	86.3	—	38.4	8.0	12.3	37.3	6.96
Myotis keaysi	m	8	85.00	48.38	36.63	8.25	12.63	36.00	5.25
	f	14	81.50	44.64	36.86	8.36	12.57	36.17	5.25
• *Myotis levis*	—	13	88.3	51.8	36.5	7.8	14.1	—	4.6
Myotis nesopolus	m	12	79.92	42.33	37.58	7.67	12.58	31.80	3.50
	f	6	80.33	43.00	37.33	7.50	13.17	32.49	—

Table 8.17   (continued)

	Sex	N	TL	HB	T	HF	E	FA	Wt
Myotis nigricans	m	45	78.98	43.38	35.60	8.02	13.02	34.56	4.66
	f	46	77.54	41.17	36.37	8.02	12.46	33.41	4.48
Myotis oxyotus	m	4	84.00	45.50	38.50	9.00	14.00	36.93	5.37
	f	5	85.60	45.60	40.00	8.40	13.80	36.40	5.60
Myotis riparius	m	3	79.00	42.67	36.33	8.00	12.67	34.78	4.33
	f	13	81.38	42.85	38.54	8.15	12.92	35.33	4.43
• *Myotis ruber*	—	11	91.2	49.8	41.4	8.8	15.0	40.2	7.3
• *Myotis simus*	f	4	97.3	—	38.8	10.0	14.3	38.7	8.0
Eptesicus andinus	m	3	94.00	54.67	39.33	10.00	14.00	41.63	8.73
	f	10	100.30	57.40	42.90	10.20	14.30	43.74	10.19
Eptesicus brasiliensis	m	12	94.67	54.42	40.25	10.00	14.42	40.05	8.90
	f	32	97.53	54.53	43.00	9.81	14.94	40.93	9.89
Eptesicus diminutus	f	1	91.90	54.00	37.00	9.00	15.00	35.50	—
Eptesicus furinalis	m	8	92.50	52.00	40.50	9.00	13.25	38.35	7.53
	f	6	92.33	54.00	38.33	13.83	39.03	8.08	54.00
• *Eptesicus furinalis*	m	20	92.8	—	36.8	9.4	16.4	—	—
	f	29	97.0	—	38.6	9.7	16.2	—	—
Eptesicus fuscus	m	2	117.50	69.50	48.00	12.50	17.00	56.63	—
	f	1	132.00	84.00	48.00	11.00	19.00	50.30	26.30
Eptesicus montosus	m	8	89.25	52.25	37.00	9.25	14.38	38.63	7.67
	f	21	92.19	54.86	37.33	9.86	14.62	39.16	9.08
Histiotus sp.	m	3	108.67	59.67	49.00	10.33	31.00	45.57	—
• *Histiotus macrotus*	—	9	105.4	57.1	48.3	9.3	31.7	48.7	12.0
Rhogeessa minutilla	m	24	72.17	40.54	31.63	6.71	14.38	27.33	3.51
	f	47	72.36	41.23	31.13	6.85	14.23	27.39	3.78
Rhogeessa tumida	m	8	70.50	41.13	29.39	6.29	12.38	27.70	3.47
	f	14	71.29	42.36	28.93	6.79	12.79	28.42	4.00
Lasiurus borealis	m	5	100.80	51.80	49.00	9.30	10.70	38.64	6.43
• *Lasiurus borealis*	—	27	96.5	51.3	45.1	8.0	9.4	39.1	10.4
	f	2	102.5	51.5	51.0	9.5	11.5	40.5	—
Lasiurus cinereus	f	3	138.00	77.00	61.00	12.33	17.00	54.33	19.70
• *Lasiurus cinereus*	—	11	126.9	70.4	55.7	10.7	12.4	53.3	18.8
Lasiurus ega	m	9	114.44	61.11	53.33	10.44	17.56	45.43	11.87
	f	8	120.50	65.75	54.75	10.88	17.75	46.15	13.51
• *Lasiurus ega*	m	31	118.3	—	50.1	9.8	18.7	45.1	10.0
	f	32	126.1	—	51.7	10.5	19.0	47.6	10.0
Molossidae									
Molossops abrasus	f	2	111.50	77.50	34.00	13.00	18.50	41.40	54.80
• *Molossops abrasus*	—	14	121.1	82.9	37.6	11.2	19.5	48.1	33.0
Molossops greenhalli	m	1	105.00	80.00	25.00	10.00	16.00	35.00	18.90
Molossops paranus	f	2	87.50	61.00	26.50	8.50	16.00	32.00	11.65
Molossops planirostris	m	18	89.56	62.22	27.33	10.00	16.22	33.54	14.19
	f	52	83.02	57.48	25.54	9.64	16.02	32.02	11.33
• *Molossops planirostris*	m	15	—	55.7	25.2	—	13.3	31.6	10.5
	f	15	—	52.9	25.6	—	12.7	30.7	—
Molossops mattogrossensis	m	10	78.90	53.50	25.40	7.82	13.55	29.81	7.52
	f	6	74.67	51.00	23.67	7.83	14.33	29.78	7.15
• *Molossops temminckii*	m	48	73.8	—	26.0	7.6	13.0	30.0	—
	f	50	72.3	—	22.9	7.5	13.1	29.7	—
Nyctinomops brasiliensis	m	3	96.33	63.33	33.00	9.67	20.00	44.57	12.43
	f	5	94.00	60.60	33.40	10.60	19.20	43.42	11.04
Nyctinompos gracilis	m	81	96.63	59.19	37.44	10.04	19.48	41.97	9.76
	f	99	94.45	57.95	37.51	10.52	19.52	41.69	9.73
Tadarida laticaudata	m	1	108.00	67.00	41.00	11.00	23.00	45.00	14.60
	f	1	102.00	60.00	42.00	10.00	18.00	44.30	13.80
Eumops amazonicus	f	1	100.00	71.00	29.00	9.00	18.00	37.50	15.40
Eumops auripendulus	f	4	100.50	86.75	43.40	15.00	22.40	57.35	23.10
• *Eumops auripendulus*	—	10	136.8	83.3	53.0	11.3	21.1	64.4	—
• *Eumops bonariensis*	m	27	101.2	—	35.2	9.7	19.9	44.2	12.6
	f	40	99.0	—	33.8	9.3	19.2	43.6	12.6
Eumops dabbenei	f	1	177.00	120.00	57.00	21.00	32.00	79.10	64.70
Eumops glaucinus	m	15	136.93	88.93	48.00	14.40	28.80	58.78	35.33
	f	49	136.14	88.00	48.14	14.41	27.77	58.56	33.00

Continued on next page

Table 8.17   (continued)

	Sex	N	TL	HB	T	HF	E	FA	Wt
Eumops nanus	m	4	89.00	58.00	31.00	9.00	18.00	37.72	7.35
	f	7	91.14	60.43	30.71	8.43	18.71	38.11	6.95
•*Eumops perotis*	—	16	169.2	—	56.4	17.3	40.1	77.9	64.0
Molossus ater	m	107	125.21	79.36	45.85	13.69	16.56	48.29	32.10
	f	160	120.69	76.29	44.42	13.48	16.00	47.36	26.45
Molossus aztecus	m	26	92.65	59.73	32.92	10.73	13.58	35.27	15.61
	f	34	88.44	55.35	33.09	10.51	13.74	34.84	12.48
Molossus bondae	m	4	111.50	71.00	40.50	12.50	14.25	41.05	18.80
	f	15	104.47	65.93	38.53	11.20	12.27	39.23	17.00
Molossus molossus	m	78	102.14	65.73	36.41	10.26	13.26	39.38	15.85
	f	167	100.17	64.49	35.71	9.94	13.11	38.66	13.55
Molossus sinaloae	f	4	130.25	82.50	47.75	12.75	15.50	49.45	24.23
Promops centralis	f	1	139.00	89.00	50.00	13.00	18.00	52.50	—
•*Promops centralis*	—	10	132.7	79.6	53.1	12.2	17.3	53.7	24.1
Promops nasutus	f	2	133.50	74.50	59.00	12.50	15.00	48.40	19.90
•*Promops nasutus*	—	10	118.5	71.0	49.0	7.7	14.2	47.7	12.3
•*Nyctinomops laticaudatus*	—	12	110.1	67.8	42.30	10.4	20.5	44.6	14.5

Source: Data from USNMNH Venezuela collection, courtesy of K. Ralls and C. Handley Jr., and from Redford and Eisenberg 1992 [•].

Note: TL, total length; HB, head and body length; T, tail length; E, ear length as measured from the notch to the tip; FA, forearm length; Wt, adult weight (mean weights and measurements).

Anderson, S. 1985. *Lista preliminar de mamíferos bolivianos.* Cuadernos Academia Nacional de Ciencias de Bolivia 65. La Paz: Naturaleza Museo Nacional de Historia Natural Zoología.

•———. 1991. A brief history of Bolivian chiroptology and new records of bats. *Bull. Amer. Mus. Nat. Hist.* 206:138–44.

———. 1997. Mammals of Bolivia: Taxonomy and distribution. *Bull. Amer. Mus. Nat. Hist.* 231:1–652.

Anderson, S., K. F. Koopman, and G. K. Creighton. 1982. Bats of Bolivia: An annotated checklist. *Amer. Mus. Novitat.* 2750:1–24.

Arata, A. A., and J. B. Vaughan. 1970. Analyses of the relative abundance and reproductive activity of bats in southwestern Colombia. *Caldasia* 10:517–28.

Arita, H. T. 1990. Noseleaf morphology and ecological correlates in phyllostomid bats. *J. Mammal.* 71:36–47.

———. 1995. Description of noseleaves of phyllostomid bats using Fourier analysis. *Rev. Mex. Mastozool.* 1:59–68.

Arita-Watanabe, H. T. 1992. Ecology and conservation of cave communities in Yucatán, Mexico. Ph.D. diss., University of Florida, Gainesville.

Arroyo-Cabrales, J., and J. Knox Jones Jr. 1988. *Balantiopteryx io* and *Balantiopteryx infusca. Mammal. Species* 313:1–3.

Arroyo-Cabrales, J., and R. D. Owen. 1997. *Enchisthenes hartii. Mammal. Species* 546:1–4.

August, P. V. 1979. Distress calls in *Artibeus jamaicensis.* In *Vertebrate ecology in the northern Neotropics,* ed. J. F. Eisenberg, 151–59. Washington, D.C.: Smithsonian Institution Press.

———. 1981. Fig fruit consumption by *Artibeus jamaicensis* in the llanos of Venezuela. *Biotropica* (Reprod. Bot. Suppl.) 13:70–76.

Ayala, S. C., and A. D'Alessandro. 1973. Insect feeding of some Colombian fruit-eating bats. *J. Mammal.* 54:266–67.

Baker, R. J., and C. Clark. 1987. *Uroderma bilobatum. Mammal. Species* 279:1–4.

Baker, R. J., H. H. Genoways, and P. A. Seyfarth. 1981. Additional chromosomal data for bats (Mammalia: Chiroptera) from Suriname. *Ann. Carnegie Mus.* 50 (12): 333–43.

Baker, R. J., M. W. Haiduk, L. W. Robbins, A. Cadena, and B. F. Koop. 1982. Chromosomal studies of bats and their implications. In *Mammalian biology in South America,* ed. M. A. Mares and H. H. Genoways, 303–44. Pymatuning Symposia in Ecology 6. Special Publications Series. Pittsburgh: Pymatuning Laboratory of Ecology, University of Pittsburgh.

Baker, R. J., C. S. Hood, and R. L. Honeycutt. 1989. Phylogenetic relationships and classification of the higher categories of the New World bat family Phyllostomidae. *Syst. Zool.* 38:228–38.

Baker, R. J., J. Knox Jones Jr., and D. C. Carter, eds. 1976. *Biology of bats of the New World family Phyllostomatidae, part 1.* Special Publications of the Museum 10. Lubbock: Texas Tech Press.

———. 1977. *Biology of bats of the New World family Phyllostomatidae, part 2.* Special Publications of the Museum 13. Lubbock: Texas Tech Press.

———. 1979. *Biology of bats of the New World family Phyllostomatidae, part 3.* Special Publications of the Museum 16. Lubbock: Texas Tech Press.

Barbour, R. W., and W. H. Davis. 1969. *Bats of America*. Lexington: University Press of Kentucky.

• Barlow, J. C. 1965. Land mammals from Uruguay: Ecology and zoogeography. Ph.D. diss., University of Kansas.

Barquez, R. M. 1983a. Una nueva localidad para la distribución de *Peropteryx macrotis macrotis* (Wagner) (Chiroptera: Emballonuridae). *Hist. Nat.* 3 (21): 185–86.

• ———. 1983b. Breves comentarios sobre *Molossus molossus* (Chiroptera: Molossidae) de Bolivia. *Hist. Nat.* 3 (18): 169–73.

• ———. 1984a. Morfometría y comentarios sobre la colección de murciélagos de la Fundación Miguel Lillo, familias Emballonuridae, Noctilionidae, Mormoopidae, Phyllostomatidae, Furipteridae, Thyropteridae (Mammalia, Chiroptera). *Hist. Nat.* 3 (25): 213–23.

• ———. 1984b. Significativa extension del rango de distribución de *Diaemus youngii* (Yentink, 1893) (Mammalia, Chiroptera, Phyllostomidae). *Hist. Nat.* 4 (7): 67–68.

• ———. 1987. Los murciélagos de Argentina. Ph.D. diss., Facultad de Ciencias Naturales e Instituto Miguel Lillo. Universidad Nacional de Tucumán.

———. 1988. Notes on identity, distribution, and ecology of some Argentine bats. *J. Mammal.* 69: 873–76.

Barquez, R. M., N. P. Giannini, and M. A. Mares. 1993. *Guide to the bats of Argentina*. Norman: University of Oklahoma Press.

Barquez, R. M., and R. Ojeda. 1992. The bats of the Argentine chaco. *Ann. Carnegie Mus.* 61:239–61.

Barriga-Bonilla, E. 1965. Estudios mastozoológicos Colombianos. 1. Chiroptera. *Caldasia* 9:241–68.

Baumgarten, J. E., and E. M. Vieira. 1994. Reproductive seasonallity and development of *Anoura geoffroyi* (Chiroptera: Phyllostomidae) in central Brazil. *Mammalia* 58:415–22.

• Bergmans, W. 1979. A record from Surinam of the bat *Chiroderma trinitatum* Goodwin, 1958 (Mammalia, Chiroptera). *Zool. Med. Ed.* 54 (22): 313–17.

Best, T., M. W. Kiser, and P. Freeman. 1996. *Eumops perotis*. *Mammal. Species* 534:1–8.

Best, T., W. M. Kiser, and J. C. Rainey. 1997. *Eumops glaucinus*. *Mammal. Species* 551:1–6.

Bonaccorso, F. J. 1979. Foraging and reproductive ecology in a Panamanian bat community. *Bull. Florida State Mus. (Biol. Sci.)* 24 (4): 359–408.

Bradbury, J. W. 1977. Social organization and communication. In *Biology of bats*, vol. 3, ed. W. A. Wimsatt, 1–72. New York: Academic Press.

Bradbury, J. W., and L. Emmons. 1974. Social organization of some Trinidad bats. 1. Emballonuridae. *Z. Tierpsychol.* 36:137–83.

Bradbury, J. W., and Vehrencamp, S. L. 1976a. Social organization and foraging in emballonurid bats. 1. Field studies. *Behav. Ecol. Sociobiol.* 1:337–81.

———. 1976b. Social organization and foraging in emballonurid bats. 2. A model for the determination of group size. *Behav. Ecol. Sociobiol.* 1:383–404.

———. 1977a. Social organization and foraging in emballonurid bats. 3. Mating systems. *Behav. Ecol. Sociobiol.* 2:1–17.

———. 1977b. Social organization and foraging in emballonurid bats. 4. Parental investment patterns. *Behav. Ecol. Sociobiol.* 2:19–30.

• Brosset, A., and P. Charles-Dominique. 1990. The bats from French Guiana: A taxonomic, faunistic and ecological approach. *Mammalia* 54:509–60.

• Brosset, A., and G. Dubost. 1967. Chiroptères de la Guyane française. *Mammalia* 31 (4): 583–94.

Brown, J. H. 1968. Activity patterns of some Neotropical bats. *J. Mammal.* 49:754–57.

Brown, P. E., T. W. Brown, and A. D. Grinnell. 1983. Echolocation, development, and vocal communication in the lesser bulldog bat, *Noctilio albiventris*. *Behav. Ecol. Sociobiol.* 13:287–98.

• Cabrera, A. 1958. *Catálogo de los mamíferos de América del Sur.* Vol. 1. Buenos Aires: Museo Argentino de Ciencias Naturales "Bernardino Rivadavia."

• Carter, D. C. 1966. A new species of *Rhinophylia* (Mammalia, Chiroptera, Phyllostomatidae) from South America. *Proc. Biol. Soc. Washington* 79: 235–38.

Carvalho, C. T. 1961. Sobre los hábitos alimentares de phillostomídeos (Mammalia, Chiroptera). *Rev. Biol. Trop.* 9:53–60.

Charles-Dominique, P. 1991. Feeding strategy and activity budget of the frugivorous bat *Carollia perspicillata* in French Guiana. *J. Trop. Ecol.* 7: 243–56.

Cloutier, D., and D. W. Thomas. 1992. *Carollia perspicillata*. *Mammal. Species* 417:1–9.

• Crespo, J. A. 1947. Comentarios sobre neuvas localidades para mamíferos de Argentina y de Bolivia. *Rev. Mus. Argent. Cienc. Nat. "Bernardino Rivadavia," Zool.* 11 (1): 1–31.

• ———. 1982. Ecología de la comunidad de mamíferos del Parque Nacional Iguazú, Misiones. *Rev. Mus. Argent. Cienc. Nat. "Bernardino Rivadavia," Ecol.* 3 (2): 45–162.

Davis, R. B., C. F. Herreid, and H. Short. 1962. Mexican free-tailed bats in Texas. *Ecol. Monogr.* 32: 311–46.

• Davis, W. B. 1966. Review of South American bats of the genus *Eptesicus*. *Southwest. Nat.* 11 (2): 245–74.

———. 1980. New *Sturnira* from Central and South

America with a key to currently recognized species. *Occas. Pap. Mus. Texas Tech Univ.* 70:1–5.

• de la Torre, L. 1966. New bats of the genus *Sturnira* (Phyllostomidae) from the Amazonian lowlands of Peru and the Windward Islands, West Indies. *Proc. Biol. Soc. Washington* 79:267–72.

• de la Torre, L., and A. Schwartz. 1966. New species of *Sturnira* (Chiroptera: Phyllostomidae) from the Islands of Guadeloupe and Saint Vincent, Lesser Antilles. *Proc. Biol. Soc. Washington* 79:297–304.

Eger, J. L. 1977. Systematics of the genus *Eumops* (Chiroptera: Molossidae). *Roy. Ontario Mus. Life Sci. Contrib.* 110:1–69.

Eisenberg, J. F. 1981. *The mammalian radiations: An analysis of trends in evolution, adaptation, and behavior.* Chicago: University of Chicago Press.

———. 1989. *Mammals of the Neotropics.* Vol. 1. *The northern Neotropics: Panama, Colombia, Venezuela, Guyana, Suriname, French Guiana.* Chicago: University of Chicago Press.

Eisenberg, J. F., and D. E. Wilson. 1978. Relative brain size and feeding strategies in the Chiroptera. *Evolution* 32 (4): 740–51.

• Emmons, L. H., and F. Feer. 1997. *Neotropical rainforest mammals: A field guide.* 2d ed. Chicago: University of Chicago Press.

Erkert, H. G. 1982. Ecological aspects of bat activity rhythms. In *Ecology of bats,* ed. T. Kunz, 201–42. New York: Plenum.

Fabian, M. E., and R. V. Marques. 1989. Contribução ao conhecimento da biologia reprodutiva de *Molossus molossus* (Pallas, 1766). *Rev. Brasil. Zool.* 6 (4): 603–10.

Fazzolari-Corrêa, S. 1994. *Lasiurus ebenus,* a new vespertilionid bat from southeastern Brazil. *Mammalia* 58:119–23.

Fenton, M. B. 1982. Echolocation, insect hearing and feeding ecology of insectivorous bats. In *Ecology of bats,* ed. T. Kunz, 261–86. New York: Plenum.

Fenton, M. B., L. Acharya, D. Audet, M. B. C. Hickey, C. Merriman, M. K. Obrist, D. M. Syme, and B. Adkins. 1992. Phyllostomid bats as indicators of habitat disruption in the Neotropics. *Biotropica* 24:440–46.

Findley, J. S. 1993. *Bats: A community perspective.* Cambridge: Cambridge University Press.

Findley, J. S., and D. E. Wilson. 1974. Observations on the Neotropical disk-winged bat *Thyroptera tricolor* Spix. *J. Mammal.* 55:562–71.

———. 1982. Ecological significance of chiropteran morphology. In *Ecology of bats,* ed. T. Kunz, 243–60. New York: Plenum.

Fleming, T. H. 1971. *Artibeus jamaicensis:* Delayed embryonic development in a Neotropical bat. *Science* 171:402–4.

———. 1982. Foraging strategies of plant visiting bats. In *Ecology of bats,* ed. T. Kunz, 287–326. New York: Plenum.

———. 1983. *Carollia perspicillata* (murciélago candelaro, lesser short-tailed fruit bat). In *Costa Rican natural history,* ed. D. H. Janzen, 457–58. Chicago: University of Chicago Press.

———. 1988. *The short-tailed fruit bat.* Chicago: University of Chicago Press.

Fleming, T. H., E. T. Hooper, and D. E. Wilson. 1972. Three Central American bat communities: Structure: reproductive cycles, and movement patterns. *Ecology* 53 (4): 555–69.

Fornes, A. 1964. Consideraciones sobre *Eumops abrasus* y *Tadarida molossa* (Mammalia, Chiroptera, Molossidae). *Acta Zool. Lilloana* 20:171–75.

Fornes, A., and E. Massoia. 1966. *Vampyrops lineatus* (E. Geoffroy), nuevo género y especie para la república Argentina (Chiroptera, Phyllostomidae). *Physis* 26 (71): 181–84.

Foster, M. S. 1992. Tent roosts of Macconnelli bat (*Vampyressa macconnelli*). *Biotropica* 24:447–54.

Foster, M. S., and R. M. Timm. 1976. Tent making by *Artibeus jamaicensis* with comments on plants used by bats for tents. *Biotropica* 8:265–69.

Freeman, P. W. 1979. Specialized insectivory: Beetle-eating and moth-eating molossid bats. *J. Mammal.* 60:467–79.

———. 1981. A multivariate study of the family Molossidae: Morphology, ecology, evolution. *Fieldiana: Zool.* 7.

Fuzessery, Z. M., and G. D. Pollak. 1984. Neural mechanisms of sound localization in an echolocating bat. *Science* 225:725–28.

Galletti, M., and L. P. C. Morellato. 1994. Diet of the large fruit bat in forest fragments in Brazil. *Mammalia* 58:661–64.

• Gardner, A. L. 1976. The distributional status of some Peruvian mammals. *Occas. Pap. Mus. Zool. Louisiana State Univ.* 48:1–18.

———. 1977a. Feeding habits. In *Biology of bats of the New World family Phyllostomatidae, part 2,* ed. R. J. Baker, J. Knox Jones Jr., and D. C. Carter, 293–350. Special Publications of the Museum 13. Lubbock: Texas Tech Press.

———, 1977b. Chromosomal variation in *Vampyressa* and a review of chromosomal evolution in the Phyllostomidae. *Syst. Zool.* 26:300–318.

• Gardner, A. L., and D. C. Carter. 1972. A review of the Peruvian species of *Vampyrops* (Chiroptera: Phyllostomatidae). *J. Mammal.* 53:72–82.

Gardner, A. L., and C. S. Farrell. 1990. Comments on the nomenclature of some Neotropical bats. *Proc. Biol. Soc. Wash.* 103:501–8.

• Gardner, A. L., and J. P. O'Neill. 1969. The taxonomic status of *Sturnira bidens* with notes on its karyotype and life history. *Occas. Pap. Mus. Zool. Louisiana State Univ.* 38:1–8.

———. 1971. A new species of *Sturnira* from Peru. *Occas. Pap. Mus. Zool. Lousianaa State Univ.* 42:1.

• Gardner, A. L., and J. L. Patton. 1972. New species of *Philander* (Marsupialia: Didelphidae) and *Mimon* (Chiroptera: Phyllostomidae) from Peru. *Occas. Pap. Mus. Zool. Louisiana State Univ.* 43:1–12.

• Genoways, H. H., and S. L. Williams. 1979. Records of bats (Mammalia: Chiroptera) from Suriname. *Ann. Carnegie Mus.* 48:323–35.

• ———. 1980. A new species of bat of the genus *Tonatia* (Mammalia: Phyllostomatidae). *Ann. Carnegie Mus.* 49 (14): 203–11.

• Genoways, H. H., S. L. Williams, and J. A. Groen. 1981. Noteworthy records of Surinamese mammals. *Ann. Carnegie Mus.* 50 (11): 319–32.

• González, J. C. 1973. Observaciones sobre algunos mamíferos de Bopicuá (dpto. de Río Negro, Uruguay). *Comun. Mus. Mun. Hist. Nat. Río Negro, Uruguay* 1 (1): 1–14.

González, J. C., and M. Fabian. 1995. Uno nueva espécie de Murciélago para estado de Rio Grande do Sul, Brazil: *Myotis riparius. Comun. Mus. Sci. Tecnol. PUCRS, ser. Zool.* 8:55–60.

González, J. C., and S. Vallejo. 1980. Notas sobre *Vampyrops lineatus* (Geoffroy), del Uruguay (Phyllostomidae, Chiroptera). *Comun. Zool. Mus. Hist. Nat. Montevideo* 10 (144): 1–8.

Goodwin, G. G. 1942. A summary of recognizable species of *Tonatia*, with descriptions of two new species. *J. Mammal.* 23:204–9.

———. 1963. American bats of the genus *Vampyressa* with the description of a new species. *Amer. Mus. Novitat.* 2125:1–24

• Goodwin, G. G., and A. M. Greenhall. 1961. A review of the bats of Trinidad and Tobago. *Bull. Amer. Mus. Nat. Hist.* 122 (3): 187–302.

Gould, E. 1977. Echolocation and communication. In *Biology of bats of the New World family Phyllostomatidae, part 2,* ed. R. J. Baker, J. Knox Jones Jr., and D. C. Carter, 247–80. Special Publications of the Museum 13. Lubbock: Texas Tech Press.

Graham, G. L. 1987. Seasonality of reproduction in Peruvian bats. *Fieldiana: Zool.,* n.s., 39:173–86.

Greenhall, A. M., G. Joermann, and U. Schmidt. 1983. *Desmodus rotundus. Mammal. Species* 202:1–6.

Greenhall, A. M., U. Schmidt, and G. Joermann. 1984. *Diphylla ecaudata. Mammal. Species* 227:1–3.

Greenhall, A. M., and W. A. Schutt Jr. 1996. *Diaemus youngi. Mammal. Species* 534:1–8.

Greer, J. K. 1966. Mammals of Malleco province, Chile. *Publ. Mus., Michigan State Univ., Biol. Ser.* 3 (2): 49–152.

Griffin, D. R. 1958. *Listening in the dark.* New Haven: Yale University Press.

Griffin, D. R., F. A. Webster, and C. R. Michael. 1960. The echolocation of flying insects by bats. *Anim. Behav.* 8:141–54.

Griffiths, T. A. 1982. Systematics of the New World nectar feeding bats based on the morphology of the lingual and hyoid regions. *Amer. Mus. Novitat.* 2742:42.

• Hall, E. R. 1981. *The mammals of North America.* 2d ed., 2 vols. New York: John Wiley.

• Handley, C. O., Jr. 1966. Descriptions of new bats (*Choeroniscus* and *Rhinophylla*) from Colombia. *Proc. Biol. Soc. Washington* 79:83–88.

———. 1967. Bats of the canopy of an Amazonian forest. *Atas Simp. Biota Amazônica, Zool.* 5:211–15.

• ———. 1976. Mammals of the Smithsonian Venezuelan project. *Brigham Young Univ. Sci. Bull., Biol. Ser.* 20 (5): 1–90.

• ———. 1984. New species of mammals from northern South America: A long-tongued bat, genus *Anoura* Gray. *Proc. Biol. Soc. Washington* 97 (3): 513–21.

———. 1987. New species of mammals from northern South America: Fruit-eating bats, genus *Artibeus* Leach. *Fieldiana: Zool.,* n.s. 39:163–72.

———. 1989. The *Artibeus* of Gray 1838. In *Advances in Neotropical mammalogy,* ed. K. H. Redford and J. F. Eisenberg, 443–68. Gainesville: Sandhill Crane Press.

———. 1991. The identity of *Phyllostoma planirostre* Spix, 1823. In Contributions to mammalogy in honor of Karl F. Koopman, ed. T. A. Griffiths and D. Kingener, 12–17. *Bull. Amer. Mus. Nat. Hist.* 206:1–432.

Handley, C. O., Jr., D. E. Wilson, and A. L. Gardner, eds. 1991. Demography and natural history of the common fruit bat, *Artibeus jamaicensis,* on Barro Colorado Island, Panama. *Smithsonian Contrib. Zool.* 511:1–173.

Harrison, D. L. 1975. *Macrophyllum macrophyllum. Mammal. Species* 62:1–3.

Haussler, U., E. Moller, and U. Schmidt. 1981. Zur Haltung und Jugendentwicklung von *Molossus molossus. Z. Säugetierk.* 46:337–51.

• Herd, R. M. 1983. *Pteronotus parnellii. Mammal. Species* 209:1–5.

• Hernandez-Camacho, J., and A. Cadena-G. 1978. Notas para la revisión del género *Lonchorhina* (Chiroptera, Phyllostomidae). *Caldàsia* 12 (57): 199–251.

• Hershkovitz, P. 1949. Mammals of northern Colombia, preliminary report no. 5: Bats (Chiroptera). *Proc. U.S. Nat. Mus.* 99:429–54.

• Hill, J. E. 1964. Notes on bats from British Guiana

with the description of a new genus and species of Phyllostomidae. *Mammalia* 28 (4): 553–72.

———. 1980. A note on *Lonchophylla* from Ecuador and Peru with the description of a new species. *Bull. Brit. Mus. (Nat. Hist.) Zool.* 38:233–36.

• Honacki, J. H., K. E. Kinman, and J. W. Koeppl, eds. 1982. *Mammal species of the world.* Lawrence, Kans.: Allen Press and Association of Systematics Collections.

• Honeycutt, R. L., R. J. Baker, and H. H. Genoways. 1980. Chromosomal data for bats (Mammalia: Chiroptera) from Suriname. *Ann. Carnegie Mus.* 49 (16): 237–50.

Hood, C. S., and J. Knox Jones Jr. 1984. *Noctilio leporinus. Mammal. Species* 216:1–7.

Hood, C. S., and J. Pitocchelli. 1983. *Noctilio albiventris. Mammal. Species* 197:1–5.

Hooper, E. T., and J. H. Brown. 1968. Foraging and breeding in two sympatric species of Neotropical bats, genus *Noctilio. J. Mammal.* 49:310–12.

Howell, D. J. 1974. Bats and pollen: Physiological aspects of chiropterophily. *Comp. Biochem. Physiol.* 48A:263–76.

———. 1983. *Glossophaga soricina.* In *Costa Rican natural history,* ed. D. H. Janzen, 472–74. Chicago: University of Chicago Press.

Humphrey, S. R., and F. J. Bonaccorso. 1979. Population and community ecology. In *Biology of bats of the New World family Phyllostomatidae, part 3,* ed. R. J. Baker, J. Knox Jones Jr., and D. C. Carter, 406–41. Special Publications of the Museum 16. Lubbock: Texas Tech Press.

Humphrey, S. R., Bonaccorso, F. J., and T. L. Zinn. 1983. Guild structure of surface-gleaning bats in Panama. *Ecology* 64 (2): 284–94.

• Husson, A. M. 1958. Notes on the Neotropical leaf-nosed bat *Sphaeronycteris toxophyllum* Peters. *Arch. Neerland. Zool.* 13 (suppl. 1): 114–19.

• ———. 1962. *The bats of Suriname.* Zoologische Verhandelingen 58. Leiden: E. J. Brill.

• ———. 1978. *The mammals of Suriname.* Leiden: E. J. Brill.

• Ibañez, C. 1979. Nuevos datos sobre *Eumops dabbenei* Thomas, 1914. *Doñana-Acta Vert.* 4 (2): 248–52.

Iudica, C. A. 1994. Role of a bat community in the rejuvenation process of a forest after human disturbance in northwestern Argentina. M.S. thesis, University of Florida, Gainesville.

———. 1995. Frugivoria en murciélagos el frutero (*Sturnira lilium*) en las Yungas de Jujuy, Argentina. In *Investigación conservación y desarrollo en selvas subtropicales de Montana,* ed. A. D. Brown and H. R. Grau, 123–18. Buenos Aires: Projecto de Desarrollo Agroforestal/LIEY.

Jepsen, G. L. 1970. Bat origins and evolution. In *Biology of bats,* vol. 1, ed. W. A. Wimsatt, 1–64. New York: Academic Press.

Jimenez, M. P., and J. Pefauer. 1982. Aspectos sistemáticos y ecológicos de *Platalina genovensium. Actas Congr. Latinoamer. Zool.* 8:707–18.

Kleiman, D. G., and T. M. Davis. 1979. Ontogeny and maternal care. In *Biology of bats of the New World family Phyllostomatidae, part 3,* ed. R. J. Baker, J. Knox Jones Jr., and D. C. Carter, 387–402. Special Publications of the Museum 16. Lubbock: Texas Tech Press.

Knox Jones, J., Jr., and J. Arroyo-Cabrales. 1990. *Nyctinomops aurispinosus. Mammal. Species* 350:1–3.

Knox Jones, J., Jr., and D. C. Carter. 1976. Annotated checklist, with keys to subfamilies and genera. In *Biology of bats of the New World family Phyllostomatidae, part 1,* ed. R. J. Baker, J. Knox Jones Jr., and D. C. Carter, 31–37. Special Publications of the Museum 10. Lubbock: Texas Tech Press.

• Koopman, K. 1978. Zoogeography of Peruvian bats. *Amer. Mus. Novitat.* 2651:1–33.

• ———. 1982. Biogeography of the bats of South America. In *Mammalian biology in South America,* ed. M. A. Mares and H. H. Genoways, 273–302. Pymatuning Symposia in Ecology 6. Special Publication Series. Pittsburgh: Pymatuning Laboratory of Ecology, University of Pittsburgh.

• ———. 1993. Chiroptera. In *Mammal species of the world,* ed. D. E. Wilson and D. M. Reeder, 137–242. Washington, D.C.: Smithsonian Institution Press.

———. 1994. Chiroptera: Systematics. In *Handbuch der Zoologie,* vol. 8, *Mammalia,* 1–217. Berlin: Walter de Gruyter.

Kunz, T. H., ed. 1982. *Ecology of bats.* New York: Plenum.

———. 1988. *Ecological and behavioral methods for the study of bats.* Washington, D.C.: Smithsonian Institution Press.

Kunz, T. H., P. V. August, and C. D. Burnett. 1983. Harem social organization in cave roosting *Artibeus jamaicensis* (Chiroptera: Phyllostomidae). *Biotropica* 15 (21): 133–38.

Kunz, T. H., M. S. Fujita, A. P. Brooke, and, G. F. McCracken. 1994. Convergence in tent architecture and tent making behavior among Neotropical bats. *J. Mammal. Evol.* 2:57–78.

Kunz, T. H., and I. M. Peña. 1992. *Mesophylla macconnelli. Mammal. Species* 405:1–5.

Kurta, A., and R. H. Baker. 1990. *Eptesicus fuscus. Mammal. Species* 356:1–10.

Langguth, A., and F. Achaval. 1972. Notas ecológicas sobre el vampiro *Desmodus rotundus rotun-*

dus (Geoffroy) en el Uruguay. *Neotrópica* 18 (55): 45–53.

LaVal, R. K. 1973a. A revision of the Neotropical bats of the genus *Myotis*. *Bull. Nat. Hist. Mus., Los Angeles County* 15:1–53.

———. 1973b. *Systematics of the genus* Rhogeessa. Occasional Papers 19. Lawrence: Museum of Natural History, University of Kansas.

Lemke, T. O. 1984. Foraging ecology of the long-nosed bat, *Glossophaga soricina*. *Ecology* 65:538–48.

• Lemke, T. O., A. Cadena, R. H. Pine, and J. Hernandez-Camacho. 1982. Notes on opossums, bats, and rodents new to the fauna of Colombia. *Mammalia* 46 (2): 225–34.

• Lemke, T. O., and J. R. Tamsitt. 1979. *Anoura culturata* from Colombia. *Mammalia* 43:579–81.

Lewis, S. E., and D. E. Wilson. 1987. *Vampyressa pusilla. Mammal. Species* 292:1–5.

Lim, B. K., and D. E. Wilson. 1993. Taxonomic status of *Artibeus amplus* in northern South America. *J. Mammal.* 74:521–30.

• Linares, O. J. 1972. Studies in the bat *Natalus stramineus* of Venezuelan caves, with special reference to variation and isolation. *Bol. Soc. Venez. Espeleol.* 3 (3): 231–33.

• ———. 1987. *Murciélagos de Venezuela.* Caracas: Cuardenas Lagoven, Premio Nacional de Perodismo.

• Linares, O. J., and P. Kiblinski. 1969. Note on a new locality and karyotype of *Molossops greenhalli* from Venezuela. *J. Mammal.* 50 (4): 831–32.

• Linares, O. J., and J. Ojasti. 1971. Una nueva especie de murciélago del género *Lonchorhina* (Chiroptera: Phyllostomidae) del sur de Venezuela. *Nov. Cient., Ser. Zool.* 36:1–8.

• McCarthy, T. J. 1983. Comments on the first *Tonatia carrikeri* from Colombia. *Lozania* (Bogotá) 40: 1–5.

McCarthy, T. J., A. L. Gardner, and C. O. Handley Jr. 1992. *Tonatia carrikeri. Mammal. Species* 407:1–4.

McCracken, G. F., and J. W. Bradbury. 1977. Paternity and genetic heterogeneity in the polygynous bat *Phyllostomus hastatas. Science* 198:303–6.

———. 1981. Social organization and kinship in the polygynous bat *Phyllostomus hastatus. Behav. Ecol. Sociobiol.* 8:11–34.

McNab, B. K. 1969. The economics of temperature regulation in Neotropical bats. *Comp. Biochem. Physiol.* 31:227–68.

———. 1982. Evolutionary alternatives in the physiological ecology of bats. In *Ecology of bats*, ed. T. H. Kunz, 151–200. New York: Plenum.

Mann, G. 1978. *Los pequeños mamíferos de Chile.* Guyana: Zoología 40. Santiago: Universidad de Concepción.

Mares, M. A., R. A. Ojeda, and M. P. Kosco. 1981. Observations on the distribution and ecology of the mammals of Salta province, Argentina. *Ann. Carnegie Mus.* 50 (6): 151–206.

Mares, M. A., M. R. Willig, K. E. Streilein, and T. E. Lacher Jr. 1981. The mammals of northeastern Brazil, a preliminary assessment (list). *Ann. Carnegie Mus.* 50:81–137.

Marinho-Filho, J. S. 1991. The co-existence of two frugivorous bat species and phenology of their food plants in Brazil. *J. Trop. Ecol.* 7:59–67.

Marinho-Filho, J. S., and I. Sazima. 1989. Activity patterns of six phyllostomid bat species in southeastern Brazil. *Rev. Brasil. Biol.* 49 (3): 777–82.

• Marinkelle, C. J., and A. Cadena. 1971. Remarks on *Sturnira tildae* in Colombia. *J. Mammal.* 52 (1): 235–37.

• ———. 1972. Notes on bats new to the fauna of Colombia. *Mammalia* 36:50–58.

Marques, S. A. 1986. Activity cycle, feeding and reproduction of *Molossus ater* (Chiroptera: Molossidae) in Brazil. *Bol. Mus. ParaenseEmilio Goeldi, Zool.* 2 (2): 159–79.

Marques-Aguiar, S. A. 1995. A systematic review of *Artibeus* Leach, 1821 (Mammalia: Chiroptera) with some phylogenetic inference. *Bol. Mus. Paraense Emilio Goeldi, Zool.* 10:3–84.

Martuscelli, P. 1995. Avian predation by the round-eared bat (*Tonatia bidens*, Phyllostominae) in the Brazilian Atlantic forest. *J. Trop. Biol.* 11:461–64.

Massoia, E. 1980. Mammalia de Argentina. 1. Los mamíferos silvestres de la provincia de Misiones. *Iguazú* 1 (1): 15–43.

Medellin, R. A. 1988. Prey of *Chrotopterus auritus*, with notes on feeding behavior. *J. Mammal.* 69: 841–44.

———. 1989. *Chrotopterus auritus. Mammal. Species* 343:1–5.

Medellin, R. A., D. E. Wilson, and D. Navarro L. 1985. *Micronycteris brachyotis. Mammal. Species* 251: 1–4.

Miller, G. S., Jr. 1907. The families and genera of bats. *U.S. Nat. Mus. Bull.* 57:1–282.

Miller, G. S., Jr., and G. M. Allen. 1928. The American bats of the genera *Myotis* and *Pizonyx. U.S. Nat. Mus. Bull.* 144:175–213.

Milner, J., C. Jones, and J. Knox Jones Jr. 1990. *Nyctinomops macrotis. Mammal. Species* 351:1–4.

Molinari, J., and P. J. Soriano. 1987. *Sturnira bidens. Mammal. Species* 276:1–4.

Morales, J. C., and J. W. Bickham. 1995. Molecular systematics of the genus *Chiroptera* (Vespertilionidae) based on restriction maps of the mitochondrial ribosomal genes. *J. Mammal.* 76:730–49.

Morrison, D. W. 1978a. Lunar phobia in a Neotropical

fruit bat, *Artibeus jamaicensis* (Chiroptera: Phyllostomidae). *Anim. Behav.* 26:852–55.

———. 1978b. Influence of habitat on the foraging distances of the fruit bat, *Artibeus jamaicensis. J. Mammal.* 59 (3): 622–24.

———. 1978c. On the optimal searching strategy for refuging predators. *Amer. Nat.* 112:925–34.

———. 1978d. Foraging ecology and energetics of the frugivorous bat, *Artibeus jamaicensis. Ecology* 59 (4): 716–23.

———. 1979. Apparent male defense of tree hollows in the fruit bat, *Artibeus jamaicensis. J. Mammal.* 60 (1): 11–15.

Myers, P. 1977. Patterns of reproduction of four species of vespertilionid bats in Paraguay. *Univ. Calif. Pub. Zool.* 107:1–41.

———. 1978. Sexual dimorphism in size of vespertilionid bats. *Amer. Nat.* 112 (986): 701–11.

———. 1981. Observations on *Pygoderma bilabiatum. Z. Säugetierk.* 46:146–51.

• Myers, P., and R. M. Wetzel. 1983. Systematics and zoogeography of the bats of the Chaco Boreal. *Misc. Publ. Mus. Zool., Univ. Michigan* 165:1–59.

Myers, P., R. White, and J. Stallings. 1983. Additional records of bats from Paraguay. *J. Mammal.* 64: 143–45.

Navarro, L. D., and D. E. Wilson. 1982. *Vampyrum spectrum. Mammal. Species* 184:1–4.

Novick, A. 1977. Acoustic orientation. In *Biology of bats,* vol. 3, ed. W. A. Wimsatt, 74–287. New York: Academic Press.

Novick, A., and B. A. Dale. 1971. Foraging behavior in fishing bats and their insectivorous relatives. *J. Mammal.* 52:817–18.

• Nowak, R. M. 1994. *Walker's "Bats of the World."* Baltimore: John Hopkins University Press.

• Ochoa-G., J. 1980. Lista y comentarios ecológicos de las especies de murciélagos (Mammalia-Chiroptera) en la ciudad de Maracay y el Parque Nacional "Henri Pittier" (Rancho Grande), Aragua, Venezuela. Thesis, Universidad Central de Venezuela, Facultad de Agronomía, Instituto de Zoología Agricola, Maracay.

• ———. 1984a. Presencia de *Nyctinomops aurispinosa* en Venezuela. *Acta Cient. Venez.* 35:147–50.

• ———. 1984b. Nuevo hallazgo de *Peronymus leucopterus* en Venezuela. *Acta Cient. Venez.* 35: 160–61.

• Ochoa-G., J., and C. Ibañez. 1982. Nuevo murciélago del género *Lonchorhina* (Chiroptera: Phyllostomidae). *Mem. Soc. Cienc. Nat. La Salle* 42 (118): 145–59.

• Ojasti, J., and O. J. Linares. 1971. Adiciones a la fauna de murciélagos de Venezuela con notas sobre las especies del género *Diclidurus. Acta Biol. Venez.* 7:421–41.

• Ojeda, R. A., and R. M. Barquez. 1978. Contribución al conocimiento de los quirópteros de Bolivia. *Neotrópica* 24 (71): 33–38.

Ojeda, R. A., and M. A. Mares. 1989. *A biogeographic analysis of the mammals of Salta province, Argentina.* Special Publications of the Museum 27. Lubbock: Texas Tech University.

Olrog, C. C. 1959. Notas mastozoológicas. 2. Sobre la colección del Instituto Miguel Lillo. *Acta Zool. Lilloana* 17:403–19.

———. 1967. *Pygoderma bilabiatum,* un murciélago nuevo para la fauna argentina (Mammalia, Chiroptera, Phyllostomidae). *Neotrópica* 13:104.

Olrog, C. C., and R. M. Barquez. 1979. Dos quirópteros nuevos para la fauna argentina. *Neotrópica* 25 (74): 185–86.

Ortega, J., and H. T. Arita. 1997. *Mimon bennettii. Mammal. Species* 549:1–4.

Osgood, W. H. 1916. New mammals from Brazil and Peru. *Field Mus. Nat. Hist. Publ. Zool. Ser.* 10:212.

———. 1943. The mammals of Chile. *Field Mus. Nat. Hist. Zool. Ser.* 30:1–268.

Owen, J. G., D. J. Schmidly, and W. B. Davis. 1984. A morphometric analysis of three species of *Carollia* from Middle America. *Mammalia* 48:85–93.

Owen, R. D. 1987. *Phylogenetic analysis of the bat subfamily Stenodermatinae.* Special Publications of the Museum 26. Lubbock: Texas Tech University Press.

———. 1991. The systematic status of *Dermanura concolor* (Peters, 1865) with a description of a new genus. In Contributions to mammalogy in honor of Karl F. Koopman, ed. T. A. Griffiths and D. Klingener, 18–25. *Bull. Amer. Mus. Nat. Hist.* 206:1–432.

Owen, R. D., and W. D. Webster. 1983. Morphological variation in the Ipanema bat, *Pygoderma bilabiatum,* with description of a new subspecies. *J. Mammal.* 64:146–49.

Pacheco, V., and B. D. Patterson. 1991. Phylogenetic relationships of the New World bat genus *Sturnira* (Chiroptera: Phyllostomatidae). *Bull. Amer. Mus. Nat. Hist.* 206:101–21.

Paradiso, J. L. 1967. A review of the wrinkle-faced bats (*Centurio senex*) with a description of a new subspecies. *Mammalia* 31:595–604.

Pearson, O. P. and A. K. Pearson. 1989. Reproduction of bats in southern Argentina. In *Advances in Neotropical mammalogy,* ed. K. H. Redford and J. F. Eisenberg, 549–66. Gainesville, Fla.: Sandhill Crane Press.

Peracchi, A. L. 1968. Sobre os hábitos de *Histiotus velatus* (Geoffroy, 1824) (Chiroptera, Vespertilionidae). *Rev. Brasil. Biol.* 28 (4): 469–73.

Peterson, R. L. 1968. A new bat of the genus *Vampyressa* from Guyana, South America, with a brief systematic review of the genus. *Roy. Ontario Mus., Life Sci. Contrib.* 73:1–17.

• Peterson, R. L., and J. R. Tamsitt. 1968. A new species of bat of the genus *Sturnira* (Family Phyllostomatidae) from northwestern South America. *Roy. Ontario Mus., Life Sci. Contrib.* 12:1–8.

Phillips, C. J. 1971. *The dentition of glossophagine bats: Development, morphological characteristics, variation, pathology, and evolution.* Miscellaneous Publications 54. Lawrence: Museum of Natural History, University of Kansas.

Pine, R. H. 1972. *The bats of the genus* Carollia. Technical Monograph 8. College Station: Agricultural Experiment Station, Texas A&M University.

———. 1993. A new species of *Thyroptera* Spix from the Amazonia basin of northeastern Peru. *Mammalia* 57:213–25.

• Pirlot, P. 1965. Chiroptères de l'est du Venezuela et delta de l'Orenoque. *Mammalia* 29:375–89.

• ———. 1967. Nouvelle récolte de chiroptères dans l'ouest du Venezuela. *Mammalia* 31:260–74.

Plumpton, D. L., and J. Knox Jones Jr. 1992. *Rhynchonycteris naso. Mammal. Species* 413:1–5.

Porter, F. L. 1979. Social behavior in the leaf-nosed bat, *Carollia perspicillata*. 2. Social communication. *Z. Tierpsychol.* 50:1–8.

Ralls, K. 1976. Mammals in which females are larger than males. *Quart. Rev. Biol.* 51:245–76.

Rasweiler, J. J. 1992. Reproductive biology of the female black mastif bat *Molossus ater*. In *Reproductive biology in South American vertebrates*, ed. W. C. Hamlett, 262–82. New York: Springer-Verlag.

Redford, K., and J. F. Eisenberg. 1992. *Mammals of the Neotropics.* Vol. 2. *The southern cone: Chile, Argentina, Uruguay, Paraguay.* Chicago: University of Chicago Press.

Reis, N. R. dos, and J. L. Guillaumet. 1983. Les chauves-souris frugivores de la région de Manous et leur rôle dans la dissémination des espèces végétates. *Rev. Ecol. (Terre et Vie)* 38:147–68.

Reis, N. R. dos, and A. L. Peracchi. 1987. Quirópteros de região de Manaus, Amazonas, Brasil. *Bol. Mus. Paraense Emilio Goeldi, Sec. Zool.* 3:161–82.

Reis, S. F. dos. 1989. Biología reproductiva de *Artibeus lituratus* (Olfers, 1818). *Rev. Brasil. Biol.* 49 (2): 369–72.

Rezsutek, M., and G. Cameron. 1993. *Mormoops megalophylla. Mammal. Species* 448:1–5.

Ruschi, A. 1951–54. [Series of twenty-one papers on the bats of Espíritu Santo]. *Bol. Mus. Biol. Prof. Mella Leitão (Esp. Sant. Zool.)*, vols. 1–22.

Sahley, C. T., and L. E. Baraybar. 1996. Natural history of the long-snouted bat, *Platalina genovensium* (Phyllostomidae: Glossophaginae) in southwestern Peru. *Vida Silvestre Neotrop.* 5 (2): 101–9.

Sanborn, C. C. 1937. American bats of the subfamily Emballonurinae. *Field Mus. Nat. Hist. Zool. Ser.* 29 (24): 321–54.

• ———. 1941. Descriptions and records of Neotropical bats. In Papers on mammalogy, published in honor of Wilfred Hudson Osgood. *Field Mus. Nat. Hist. Zool. Ser.* 27:371–87.

———. 1949. Bats of the genus *Micronycteris. Fieldiana: Zool.* 31:215–33.

———. 1955. Remarks on the bats of the genus *Vampyrops. Fieldiana: Zool.* 37:403–13.

Sazima, I. 1976. Observations on the feeding habits of phyllostomatid bats (*Carollia, Anoura,* and *Vampyrops*) in southeastern Brazil. *J. Mammal.* 57: 381–82.

———. 1978. Vertebrates as food items of the wooly false vampire, *Chrotopterus auritus. J. Mammal.* 59:617–18.

Sazima, I., and W. Uieda. 1977. O morcego *Promops nasutus* do sudeste Brasileiro. *Ciênc. Cult.* (São Paulo) 29 (3): 312–14.

Sazima, I., L. D. Vizotto, and V. A. Taddei. 1978. Uma nova espécie de *Lonchophylla* da serra do cipó, Minas Gerais, Brasil (Mammalia, Chiroptera, Phyllostomidae). *Rev. Brasil. Biol.* 38 (1): 81–89.

Sazima, M., M. E. Fabian, and I. Sazima. 1982. Polinização do *Luehea* (Tiliaceae) por *Glossophaga soricina. Rev. Brasil. Biol.* 42 (3): 505–13.

Sazima, M., and I. Sazima. 1975. Quiropterofilia em *Lafoensia pacari* St. Hill. (Lythraceae), na Serra do Cipo, Minas Gerais. *Ciênc. Cult.* (São Paulo) 27 (4): 405–16.

Schaal, S., and W. Ziegler, eds. 1992. *Messel: An insight into the history of life and of the earth.* Oxford: Clarendon Press.

Schmidt, U., and U. Manske. 1973. Jugendentwicklung der Vampirfledermause (*Desmodus rotundus*). *Z. Säugetierk.* 38:14–33.

Setz, E. Z. F., and I. Sazima. 1987. Bats eaten by Nambiquara Indians in western Brazil. *Biotropica* 19 (2): 190.

Shump, K. A., Jr., and A. U. Shump. 1982a. *Lasiurus borealis. Mammal. Species* 183:1–6.

———. 1982b. *Lasiurus cinereus. Mammal. Species* 185:1–5.

Simmons, N. 1996. A new species of *Micronycteris* from northeast Brazil with comments on phylogenetic relationships. *Amer. Mus. Novitat.*, no. 3158.

Smith, J. D. 1972. *Systematics of the chiropteran family Mormoopidae.* Miscellaneous Publications 56. Lawrence: Museum of Natural History, University of Kansas.

• Smith, J. D., and H. H. Genoways. 1974. Bats of Margarita Island, Venezuela, with zoogeographic comments. *Bull. So. Calif. Acad. Sci.* 73 (2): 64–79.

Snow, J. L., J. Knox Jones Jr., and W. D. Webster. 1980. *Centurio senex. Mammal. Species* 138:1–3.

Starrett, A. 1972. *Cyttarops alecto. Mammal. Species* 13:1–2.

Sudman, P. D., L. J. Barkley, and M. S. Hafner. 1994. Familial affinity of *Tomopeas ravus* based on protein electrophoretic and cytochrome b sequence data. *J. Mammal.* 75:365–77.

Taddei, V. A. 1975. Phyllostomidae (Chiroptera) do norte-ocidental do estado de São Paulo. 2. Glossophaginae; Carolliinae; Sturnirinae. *Ciênc. Cult.* (São Paulo) 27 (7): 723–34.

———. 1976. The reproduction of some Phyllostomatidae from the northwestern regions of the state of São Paulo. *Bol. Zool. Univ. São Paulo* 1:313–30.

———. 1979. Phyllostomidae (Chiroptera) do norte-ocidental do estado do São Paulo. 3 Stenodermatinae. *Ciênc. Cult.* (São Paulo) 31:900–914.

Taddei, V. A., L. D. Vizotto, and I. Sazima. 1983. Uma nova espécie de *Lonchophylla* do Brasil e chave para identificação des espécies do gênero. *Ciênc. Cult.* (São Paulo) 35:625–29.

Tamayo, M., and D. Frassinetti. 1980. Catálogo de los mamíferos fósiles y vivientes de Chile. *Mus. Nac. Hist. Nat., Chile* 37:323–99.

Tamsitt, J. R., and C. Hauser. 1985. *Sturnira magna. Mammal. Species* 240:1–4.

Tamsitt, J. R., and D. Nagorsen. 1982. *Anoura culturata. Mammal. Species* 179:1–5.

• Tamsitt, J. R., and D. Valdivieso. 1963. Records and observations on Colombian bats. *J. Mammal.* 44: 168–80.

• Tate, G. H. H. 1939. The mammals of the Guiana region. *Bull. Amer. Mus. Nat. Hist.* 76 (5): 151–229.

• ———. 1947. A list of the mammals collected at Rancho Grande, in a montane cloud forest of northern Venezuela. *Zoologica* 32:65–66.

Thomas, M. E., and D. N. McMurray. 1974. Observations on *Sturnira aratathomasi* from Colombia. *J. Mammal.* 55:834–36.

Thomas, O. 1891. *Ann. Mus. Civ. Stor. Nat. Genova,* ser. 2, 10:881.

• ———. 1898. On the small mammals collected by Dr. Borelli in Bolivia and northern Argentina. *Boll. Mus. Zool. Anat. Comp.* 13 (315): 1–4.

———. 1926. The Stephan Lewis expedition. 2. Mammals collected Tarija dpt. so. Bolivia. *Ann. Mag. Nat. Hist.,* ser. 9, 17:318–28.

Timm, R. M. 1985. *Artibeus phaeotis. Mammal. Species* 235:1–6.

———. 1987. Tent construction by bats of the genera *Artibeus* and *Uroderma. Fieldiana: Zool.,* n.s., 39: 187–212.

Timm, R. M., and S. E. Lewis. 1991. Tent construction and use by *Uroderma bilobatum* in coconut palms (*Cocos nucifera*) in Costa Rica. In Contributions to mammalogy in honor of Karl F. Koopman, ed. T. A. Griffiths and D. Klingener, 251–60. *Bull. Amer. Mus. Nat. Hist.* 206:1–432.

• Trainer, M., and J. L. Berthier. 1984. Trois chauves-souris nouvelles pour la Guyane française. *Mammalia* 48:303.

Trajano, E. 1984. Ecologia de populacões de morcegagos cavernicales em uma regino cartistica do sudeste do Brasil. *Rev. Brasil. Zool.* 2 (5): 255–320.

Turner, D. C. 1975. *The vampire bat.* Baltimore: Johns Hopkins University Press.

• Tuttle, M. D. 1970. Distribution and zoogeography of Peruvian bats with comments on natural history. *Univ. Kansas Sci. Bull.* 49:45–86.

Tuttle, M. D., and M. Ryan. 1981. Bat predation and the evolution of frog vocalizations in the Neotropics. *Science* 214:677–78.

Tuttle, M. D., and D. Stevenson. 1982. Growth and survival of bats. In *Ecology of bats,* ed. T. H. Kunz, 105–50. New York: Plenum.

Tuttle, M. D., L. Taft, and M. Ryan. 1982. Evasive behavior of a frog in response to bat predation. *Anim. Behav.* 30:393–97.

Uieda, W., I. Sazima, and A. Storti-Filho. 1980. Aspectos da biologia do morcego *Furipterus horrens. Rev. Brasil. Biol.* 40 (1): 59–66.

• Valdivieso, D. 1964. La fauna quiróptera del departamento de Cudinamarca, Colombia. *Rev. Biol. Trop.* 12:19–45.

Vaughan, T. 1959. *Functional morphology of three bats:* Eumops, Myotis, *and* Macrotus. Miscellaneous Publications 12. Lawrence: Museum of Natural History, University of Kansas.

Vehrencamp, S. L., F. G. Stiles, and J. W. Bradbury. 1977. Observations on the foraging behavior and avian prey of the Neotropical carnivorous bat, *Vampyrum spectrum. J. Mammal.* 158:469–78.

Vieria, C. O. DaCunha. 1942. Ensaio monographico sobre os quirópteros do Brasil. *Arq. Zool. São Paulo* 3:219–471.

• Villa-R., B., and M. Villa Cornejo. 1969. Algunos murciélagos del norte de Argentina. In *Contributions in mammalogy,* ed. J. Knox Jones Jr., 409–29. Miscellaneous Publications 51. Lawrence: Museum of Natural History, University of Kansas.

Vizotto, L. Dino, and V. A. Taddei. 1973. Chave para determinação de quirópteros brasileiros. *Bol. Ciênc.* (São José do Rio Preto) 1:1–72.

———. 1976. Notas sobre *Molossops temminckii temminckii* e *Molossops planirostris* (Chiroptera: Molossidae). *Naturalia* 2:47–59.

Webster, W. D., and J. Knox Jones Jr. 1982a. *Artibeus aztecus. Mammal. Species* 178:1–3.

———. 1982b. *Artibeus toltecus. Mammal. Species* 178:1–3.

———. 1993. *Glossophaga commissarisi. Mammal. Species* 446:1–4.

• Webster, W. D., and W. B. McGillivray. 1984. Additional records of bats from French Guiana. *Mammalia* 48:463–65.

Webster, W. D., and R. D. Owen. 1984. *Pygoderma bilabiatum. Mammal. Species* 220:1–3.

Whitaker, J. O., and J. S. Findley. 1980. Foods eaten by some bats from Costa Rica and Panama. *J. Mammal.* 61:540–44.

Wilkins, K. T. 1989. *Tadarida brasiliensis. Mammal. Species* 331:1–10.

Wilkinson, G. S. 1984. Reciprocal food sharing in the vampire bat. *Nature* 308 (5955): 181–84.

———. 1985a. The social organization of the common vampire bat. 1. Pattern and cause of association. *Behav. Ecol. Sociobiol.* 17:111–21.

———. 1985b. The social organization of the common vampire bat. 2. Mating system, genetic structure, and relatedness. *Behav. Ecol. Sociobiol.* 17:123–34.

———. 1986. Social grooming in the common vampire bat, *Desmodus rotundus. Anim. Behav.* 34:1880–89.

• Williams, D. F. 1978. Taxonomic and karyologic comments on small brown bats, genus *Eptesicus* from South America. *Ann. Carnegie Mus.* 47 (16): 361–83.

Williams, D. F., and M. A. Mares. 1978. Karyologic affinities of the South American big-eared bat, *Histiotus montanus* (Chiroptera: Vespertilionidae). *J. Mammal.* 59:844–46.

• Williams, S. L., and H. H. Genoways. 1980a. Additional records of bats (Mammalia: Chiroptera) from Suriname. *Ann. Carnegie Mus.* 49 (15): 213–36.

• ———. 1980b. A new species of bat of the genus *Molossops* (Mammalia: Molossidae). *Ann. Carnegie Mus.* 49 (25): 487–98.

Williams, S. L., M. R. Willig, and F. A. Reid. 1995. Reviews of the *Tonatia bidens* complex with descriptions of two new subspecies. *J. Mammal.* 76:612–26.

Willig, M. R. 1983. Composition, microgeographic variation, and sexual dimorphism in caatinga and cerrado bat communities from northeastern Brazil. *Bull. Carnegie Mus. Nat. Hist.* 23:1–131.

———. 1985. Ecology, reproductive biology, and systematics of *Neoplatymops mattogrossensis* (Chiroptera: Molossidae). *J. Mammal.* 66:618–28.

Willig, M. R., and R. R. Hollander. 1987. *Vampyrops lineatus. Mammal. Species* 275:1–4.

• Willig, M. R., and J. Knox Jones Jr. 1985. *Neoplatymops mattogrossensis. Mammal. Species* 224:1–3.

• Willig, M. R., and M. Mares. 1989. Mammals of the caatinga: An updated list and summary of recent research. *Rev. Brasil. Biol.* 49 (2): 361–67.

Willis, K., M. R. Willig, and J. Knox Jones Jr. 1990. *Vampyrodes caraccioli. Mammal. Species* 359:1–4.

Wilson, D. E. 1971. The ecology of *Myotis nigricans* on Barro Colorado Island, Panama. *J. Zool.* 163:1–13.

———. 1973a. Reproduction in Neotropical bats. *Period. Biol.* 75:215–17.

———. 1973b. Bat faunas: A trophic comparison. *Syst. Zool.* 22:14–29.

• ———. 1978. *Thyroptera discifera. Mammal. Species* 104:1–3.

———. 1979. Reproductive patterns. In *Biology of bats of the New World family Phyllostomatidae, part 3,* ed. R. J. Baker, J. Knox Jones Jr., and D. C. Carter, 317–78. Special Publications of the Museum 16. Lubbock: Texas Tech Press.

• Wilson, D. E., and J. S. Findley. 1977. *Thyroptera tricolor. Mammal. Species* 71:1–3.

Wilson, D. E., and R. K. LaVal. 1974. *Myotis nigricans. Mammal. Species* 39:1–3.

Wilson, D. E., and D. M. Reeder, eds. 1993. *Mammal species of the world.* 2d ed. Washington, D.C.: Smithsonian Institution Press.

Wimsatt, W. A. 1970a. *Biology of bats.* Vol. 1. New York: Academic Press.

———. 1970b. *Biology of bats.* Vol. 2. New York: Academic Press.

———. 1977. *Biology of bats.* Vol. 3. New York: Academic Press.

Woodman, N., R. M. Timm, R. Arana C., V. Pacheco, C. A. Schmidt, E. D. Hooper, and C. Pacheco A. 1991. Annotated checklist of the mammals of Cuzco, Amazonico, Peru. *Occas. Pap. Mus. Nat. Hist. Univ. Kansas* 145:1–12.

• Ximénez, A. 1969. Dos nuevos géneros de quirópteros para el Uruguay (Phyllostomidae-Molossidae). *Comun. Zool. Mus. Hist. Nat. Montevideo* 10 (125): 1–8.

• Ximénez, A., A. Langguth, and R. Praderi. 1972. Lista sistemática de los mamíferos del Uruguay. *Anal. Mus. Nac. Hist. Nat. Montevideo* 7 (5): 1–45.

Zortea, M., and A. G. Chiarello. 1994. Observations on the big fruit bat *Artibeus lituratus* in an urban forest reserve in southeast Brazil. *Mammalia* 58:665–69.

Zortea, M., and S. L. Mendes. 1993. Folivory in the big fruit eating bat, *Artibeus lituratus* in eastern Brazil. *J. Trop. Ecol.* 9:117–20.

Diagnosis

If we exclude the fossil Plesiadapiformes, the order Primates comprises a group of species that exhibit rather unspecialized physical characteristics. Except for the extremely terrestrially adapted forms, most have flexible digits and retain five fingers and five toes. Only in some brachiating forms is the thumb lost. The shoulder joint is freely movable, and the radius and ulna remain unfused. The orbits of the skull are directed forward, and a strong postorbital bar separates the eye socket from the temporal fossa. The braincase is relatively large, and in the evolutionary history of this order one can see a trend toward progressive enlargement of the cerebral hemispheres and cerebellum (Hill 1957, 1960). For a comprehensive review see Martin (1990).

Distribution

The Recent distribution of this order, exclusive of humans (*Homo sapiens*), is in both the Old World and New World tropics except for Australasia. In the Old World some members of the genera *Macaca* and *Rhinopithecus* extend their distributions into the temperate zone, but in the Western Hemisphere distributions are currently confined more or less between 23° north and 24° south latitude.

History and Classification

The earliest representatives of the order Primates can be distinguished in the Paleocene, when the first fossils appear in North America. By the time of the Eocene, lemurlike primates are recognizable from North America and Europe. In some manner the early prosimians of the Eocene made their way to Asia and Africa, with one or two stocks transiting to Madagascar, where they became isolated and underwent an adaptive radiation. The New World primates, the Ceboidea, had an origin independent from that of the Old World primates and became established on what was then the island continent of South America, where they underwent an extensive adaptive radiation. The earliest ceboid primates have been found in the Oligocene strata of South America. The Cercopithecoidea and the Hominoidea had an Old World origin and underwent an adaptive radiation roughly parallel to the radiations in South America. In the New World tropics there are currently only three families extant, the Callitrichidae, the Cebidae, and the recently immigrant family Hominidae (Szalay and Delson 1979). Hill (1957, 1960, 1962) summarizes anatomical data for the New World primates. Hershkovitz (1977) has provided the latest discussion of anatomy and phylogeny. Martin (1990) provides a splendid overview of evolution and the place of New World primates within the whole order.

Natural History of New World Primates

In the past the New World primates received less attention from naturalists than their Old World cousins. Carpenter (1934) pioneered the study of New World primates with his work on *Alouatta palliata* on Barro Colorado Island, Panama. After a hiatus of some twenty years the behavior and ecology of Neotropical primates began to be explored by Moynihan (1964, 1976), and subsequently a number of long-term studies have focused on primate community ecology (Hladik and Hladik 1969; Hladik et al. 1971; Klein and Klein 1975; Izawa 1976; Mittermeier and van Roosmalen 1981; Terborgh 1983). The ecology and behavior of New World primates is summarized in the edited volumes by Coimbra-Filho and Mittermeier (1981); Mittermeier et al. (1988); and Norconk, Rosenberger, and Garber (1996).

With the exception of *Aotus*, the New World primates are diurnal. All species are highly arboreal, but they exhibit a variety of locomotor adaptations (Erikson 1963). Dietary specializations among New World primates are diverse, ranging from insect and fruit feeding by the tamarins to fruit and foliage feeding by *Alouatta*. Summaries of New World primate ecology and behavior are included in the volume edited by Smuts et al. (1987).

Primates typically give birth to a single young, but the species of the genera *Callithrix, Saguinus, Leontopithecus,* and *Cebuella* usually produce twins. Gestation is long relative to body size, and although the young are furred and have their eyes open at birth, there is a long period of dependency on the parents. Primates are sociable animals, and group sizes vary from nuclear families (four to seven) up to troops of mixed age and sex classes (twenty-four to fifty). The costs and benefits of group foraging have been analyzed by Terborgh (1983).

The communication system of New World primates involves the basic sensory inputs: tactile, visual, gustatory, auditory, and olfactory. The use of chemical signals is much more important in the ceboids than in the cercopithecoids. Specialized glandular areas on the chest or the genitalia are often used in marking, and chemical marking movements may be highly ritualized (Moynihan 1976; Eisenberg 1977; Epple 1973).

In the past twenty years the research effort directed toward New World primates has vastly increased. There are excellent field guides and summaries of research. When I (JFE) went to Mexico in 1960, the literature on primate distribution and ecology was rudimentary. It is hard for me to comprehend how much has been accomplished in the past thirty-five years, not only in Mexico, but in the vast areas of the southern continent. The meticulous work of C. R. Carpenter in Panama during the 1930s stands as a benchmark. The work of Cruz Lima during the late 1930s and early 1940s in Brazil is of equal stature. Let us not forget the tough collectors of an earlier time who first described the primates of the New World. Their names are immortalized in the species' full binomials.

Comment on Taxonomy

There is a certain amount of nomenclatural instability when one considers the order Primates. Recent explorations in the Amazon region reveal new taxa. The basic problem concerns the level of classification. Are new forms truly species, or are they subspecies? Could we combine forms under subgenera, or are some groupings true genera? Throughout this chapter we will attempt to present alternative interpretations of classification. Clearly, in some taxa (e.g., *Saguinus* and *Callithrix*) there are dramatic variations in color pattern and the length of the pelage. The problem of variation in color, crests, and vibrissae has been dis-

cussed by Hershkovitz (1968). Jacobs, Larson, and Cheverud (1995) have recently tested aspects of the metachromism hypothesis and confirmed many of its predictions. Their results demonstrate that *Saguinus* is a monophyletic genus composed of two major clades: one containing the *S. nigricollis* group of Hershkovitz (1977) and the other clade containing the remaining species.

Oscillations in climate resulting in wet and dry periods during the past 100,000 years surely have contributed to the isolation of primate stocks. Equally important has been the waxing and waning of river systems as well as the natural meander patterns of the Amazonian rivers, given the flat terrain (see Ayres and Clutton-Brock 1992). Then too, though primates are not known to be swimmers, there are notable exceptions, and all Neotropical forms can probably swim to some extent. Genetic variability within defined species is well documented (see Pope 1995 for a review). In any event the New World primates demonstrate a remarkable variety, and those of us who have had the privilege of studying them have been rewarded by their beauty and adaptability. Rylands (1995) offers a species list that reflects the trend toward separating allopatric congeners into full species. In addition to this valuable contribution, he attempts to assess the current population status for each form. Auriccho (1995) published an excellent field guide to the primates of Brazil. Groves (1993) serves as a basic reference on nomenclature and distribution.

FAMILY CALLITRICHIDAE

Marmosets and Tamarins, Titis, Sagüis

Diagnosis

These small primates are highly adapted for an arboreal life, but they have nonprehensile tails. Rather than bearing nails at the ends of the digits, they have secondarily acquired clawlike nails. Head and body length ranges from 150 to 370 mm. The tail can be shorter or longer than the head and body. There are five Recent genera, all occurring in the area dealt with in this volume: *Callithrix, Saguinus, Cebuella, Leontopithecus,* and *Callimico*. The dental formula is I 2/2, C 1/1, P 3/3, M 2/2 (see figs. 9.1–9.3), with the exception of *Callimico*, which retains the third molar for a total of 26 teeth. Based on this characteristic and other skull features, *Callimico* has often been assigned to its own family (Hill 1957; Hershkovitz 1977). The standard reference for the family is Hershkovitz (1977).

Distribution

Species of this family do not occur north of the Panama–Costa Rica border area but range south to

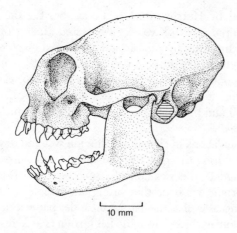

Figure 9.1. Skull of *Saguinus geoffroyi.*

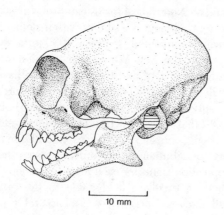

Figure 9.2. Skull of *Cebuella pygmaea.*

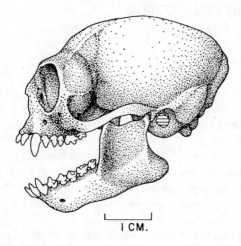

Figure 9.3. Skull of *Callithrix argentata.*

Paraguay. They are mainly confined to tropical forests and do not reach extremely high elevations.

Natural History

All species of this family are strictly diurnal. They have a varied diet including fruits, invertebrates, and in some cases small vertebrates. The genus *Cebuella,*

in common with the genus *Callithrix,* is highly adapted for gouging tree bark to supplement its diet with sap (see Power 1996). With the exception of the uniparous *Callimico,* all species characteristically give birth to two young. Field data indicate that strong pair-bonds exist between an adult male and an adult female during the rearing of the young. Although the existence of two adult males within a group has implied a polyandrous mating system, this interpretation may be premature (see species accounts). The male shares in the early rearing by transporting the young; the time this carrying behavior begins is variable from species to species, being as early as the first day in *Cebuella* and as late as fourteen days of age in some species of *Saguinus.* Although *Callimico* has a single young, it shows the same degree of male parental care. In both rearing patterns and patterns of affiliation there is a strong tendency toward monogamy.

The animals shelter in tree cavities, in tangles of vines, or under the overhanging crowns of palms (see chapters in Kleiman 1977; Mittermeier et al. 1988; Rylands 1993). Group size usually varies from four to seven. Apparently only the adult pair reproduces in any family group (Kleiman 1976), but see species accounts. The older offspring may assist the parents in rearing younger infants by carrying the babies and by contributing food (Hoage 1982). Communication involves an intricate combination of chemical, acoustic, and visual signals (Oppenheimer 1977). The ecology and behavior of this family have been summarized in articles by Kleiman (1977), Mittermeier et al. (1988), and Rylands (1993).

Comment on the Taxonomy of the Callitrichidae

The nonhuman primates of the Western Hemisphere have been the subject of intense scrutiny over the past twenty years. Older names have been "resurrected" (*Saguinus tripartitus*), new taxa have been described (e.g., *Leontopithecus caissara*), and former subspecies have been elevated to full species status. *Callimico goeldii,* a species recognized by all, has shifted from its own family Callimiconidae to the Callitrichidae. The family Callitrichidae has shifted to a new spelling from the older Callithricidae (Groves 1993). Forms from inshore islands have received full species status (e.g., *Alouatta coibensis*). Although genuine new taxa (about four) have been discovered since 1986, the major changes have resulted from elevating former subspecies to full species. The primary genera where these taxonomic decisions have had a major effect on the numbers of species are just three: *Callithrix* (as revised by Vivo 1991), *Aotus* (as revised by Hershkovitz 1983), and *Callicebus* (as revised by Hershkovitz 1990). The results of these taxonomic decisions will be reviewed in subsequent sections. It is useful to note that in Honacki, Kinman, and Koeppl (1982) the New World primate species were subsumed under three families that included sixteen genera and forty-seven species. Groves (1993) groups the New World primates into two families that include fifteen genera and eighty-three species. Note that of the eighty-three "species," only seven were formally described as new taxa from 1983 to 1990 (about three more new species have been proposed since). Obviously, new mammalian species will continue to be discovered in South America, but the alarming jump from forty-seven species in 1982 to eighty-three in 1993 mainly reflects new interpretations and our changing concept of the species rather than the discovery of forms "new to science."

When studying color variation in *Callithrix* and *Saguinus*, Hershkovitz (1968) proposed the theory of "metachromism." In seeking an explanation for the evolutionary change in mammalian tegumentary colors, he realized that social, sexual, and predatory selection were involved and that the highly "visually" adapted primates may be predisposed to select mates based on coat color and hair adornments.

Hershkovitz's key hypotheses concerned the orderly, irreversible loss of pigment within "chromogenetic fields." The key concept here is that genetic drift (coupled with social selection) could fix phenotypes departing from the primitive agouti brown by various degrees of bleaching. Thus nearly white fur could represent the end point of geographic variation in a series of near allopatric morphs deriving ultimately from a highly pigmented form. Thus his genus *Callithrix* comprised just three species, each composed of a series of dramatic color "morphs" more or less allopatric in their distribution.

The superspecies concept may be invoked: when a series of allopatric populations can be identified whose components are distinguishable morphologically by superficial characters, but in almost all aspects of their biology the component taxa are similar, then the taxa can be collectively referred to as a "superspecies." There are many precedents—for example, *Peromyscus* (Rodentia) and *Papio* (Primates).

Although many putative species of the Callitrichidae are currently allopatric, such geographic separation may be recent. The possibility for hybridization producing fertile offspring is a valid question to be explored by captive breeding. Coimbra-Filho, Pissinatti, and Rylands (1993) report the results of their hybridization experiments in Rylands (1993). They have been able to produce many fertile hybrids deriving from crosses between putative species. They conclude in part: "Forest destruction, prevalent in the remains of geographic ranges of these marmosets, may result in isolated hybrid or partially hybrid populations which along with founder effects and genetic drift may cause

rapid 'speciation,' or at least the production of localized, uniformly distinct, true breeding phenotypes." Alternative classifications are included in table 9.1.

Genus *Callithrix* Erxleben 1777

Description

The dental formula is I 2/2, C 1/1, P 3/3, M 2/2. The lower canines are short, and together *Calllithrix* and *Cebuella* constitute the "short-tusked" group of the Callitrichidae (according to Hershkovitz 1977). These small primates have a body weight from 400 to 600 g. The hair color patterns are highly variable (see table 9.2 and plate 7). The tail is nonprehensile and fully haired; the ears may be naked or haired to varying degrees. Except for the hallux, the digits bear claws. The tail usually exceeds the head and body in length. Clearly their morphology implies an arboreal adaptation.

Distribution

The genus is represented predominantly in Brazil, but it extends to Paraguay. This is a low-altitude group, rarely found above 700 m elevation. Within the semiarid areas of Brazil, species may extend their range along the gallery forests of the larger, permanent rivers.

Natural History

The marmosets are diurnal and arboreal. They forage on the ground and will also move from one isolated grove of trees to another by crossing open ground. The female typically produces fraternal twins at birth. The father, or associated subadults, helps carry the young for varying periods after their birth.

Their food consists of fruit, invertebrates, small vertebrates, and sap and gum extracted from trees by gouging wounds in the bark to the cambium. Flowers and associated pollen and nectar are also included in the diet. Home ranges average 35 ha, of which about 30 ha is defended.

The typical group size is from two to seven; the group usually contains one reproductive female, and subadults and adult males help carry the young. Ferrari, Carrêa, and Coutinho (1996) report that *Callithrix aurita* frequently has two reproductive females per troop. Aggregations of several groups in feeding trees have been recorded, yielding temporary group counts of over thirty (Stevenson and Rylands 1988; Rylands 1993).

Identification and Taxonomy of Marmosets

Table 9.1 presents alternative classifications for the Callitrichidae. Table 9.2 presents character matrices for the species of *Saguinus*. Tables 9.3 and 9.4 offer standard measurements for *Callithrix*, *Saguinus*,

Table 9.1   Alternative Classification for the Callitrichidae (Callithricidae)

Groves 1993	Mittermeier et al. 1988	Hershkovitz 1977	Emmons and Feer 1990
Callitrichidae	Callitrichidae	Callimiconidae	Callitrichidae
Callimico goeldii	*Callimico goeldii*	*Callimico goeldii*	*Callimico goeldii*
		Callitrichidae	
Callithrix	*Callithrix*	*Callithrix*	*Callithrix*
C. argentata	*C. argentata* (incl.	*C. argentata*	*C. argentata*
(incl. *C. emiliae*)	*leucippe, argentata*	(as in Mittermeir	
	and *melanura* as subsp.)	et al. 1988)	
C. aurita	*C. jacchus* group	*C. j. aurita*	*C. j. aurita*
C. flaviceps	*C. jacchus* group	*C. j. flaviceps*	*C. j. flaviceps*
C. geoffroyi	*C. jacchus* group	*C. j. geoffroyi*	*C. j. geoffroyi*
C. humeralifer	*C. humeralifera* group (incl.	*C. humeralifera*	*C. humeralifera*
	humeralifera, chrysoleuca,		
	and *intermedius*)		
C. jacchus	*C. jacchus* group	*C. jacchus*	*C. jacchus*
C. kuhlii	*C. jacchus* group	*C. jacchus*	*C. jacchus*
C. penicillata	*C. jacchus* group	*C. jacchus*	*C. jacchus*
C. pygmaea	*Cebuella pygmaea*	*Cebuella pygmaea*	*Cebulla pygmaea*
Leontopithecus	*Leontopithecus*	*Leontopithecus*	*Leontopithecus*
L. caissara	n.a.	n.a.	n.a.
L. chrysomelas	*L. chrysomelas*	(in *L. rosalia*)	(in *L. rosalia*)
L. chrysopygus	*L. chrysopygus*	(in *L. rosalia*)	(in *L. rosalia*)
L. rosalia	*L. rosalia*	*L. rosalia*	*L. rosalia*
Saguinus	*Saguinus*	*Saguinus*	*Saguinus*
S. bicolor	*S. bicolor* group	*S. bicolor*	*S. bicolor*
S. fuscicollis	*S. nigrocollis* group	*S. fuscicollis*	*S. fuscicollis*
S. geoffroyi	*S. oedipus* group	*S. oedipus*	*S. geoffroyi*
S. imperator	*S. mystax* group	*S. imperator*	*S. imperator*
S. inustus	*S. inustus* group	*S. inustus*	*S. inustus*
S. labiatus	*S. mystax* group	*S. mystax*	*S. mystax*
S. leucopus	*S. oedipus* group	*S. oedipus*	*S. leucopus*
S. midas	*S. midas* group	*S. midas*	*S. midas*
S. mystax	*S. mystax* group	*S. mystax*	*S. mystax*
S. nigricollis	*S. migricollis* group	*S. nigricollis*	*S. nigricollis*
S. oedipus	*S. oedipus* group	*S. oedipus*	*S. oedipus*
S. tripartitus	(in *S. fuscicollis*)	(in *S. fuscicollis*)	*S. tripartitus*
			(cf. Thorington)

Note: n.a. = not applicable; form described subsequent to the authors' classification.

Callimico, and *Leontopithecus.* Species identification will require referring to the descriptions, plates, and tables. Emmons and Feer (1997) present very useful field characters.

Callithrix argentata (Linnaeus, 1766)

Description

For measurements see table 9.3. Marmosets of this species group have naked ears. The basic coat color ranges from a pale brown dorsum with darker brown limbs and a dark brown tail (plate 7). *C. argentata leucippe* is a very pale form, almost white, with the tail and four paws pale orange (Hershkovitz 1977).

Comment

Groves (1993) includes *C. emiliae* within *C. argentata.* Mittermeir, Rylands, and Coimbra-Filho (1988) include *C. leucippe* and *C. melanura* as subspecies of *C. argentata.* Vivo (1991) considers *C. leucippe* and

C. melanura distinct species (see table 9.3). The species (subspecies) of this taxon share the characters listed above.

Distribution

This species group is found south of the Amazon and east of the Rio Madeira in Brazil. The western boundary of the range is poorly known, but it extends south to Bolivia (map 9.1).

Callithrix humeralifer (E. Geoffroy, 1812)

Description

For measurements see table 9.3. This species group is characterized by hairy ears, but the ears are not extremely tufted as in the *C. jacchus* group. Hershkovitz considered this grouping to exhibit the same depigmentation trend outlined for *C. argentata.* In *C. h. humeralifer* the trunk is black but "frosted," and the forearms are paler. The tail is banded and the ear tufts

Table 9.2 Matrix of Characters for the Callitrichidae (Excluding *Callimico*)

Taxon	Ear Tuft Present	Pinnae Bare	Tail Uniformly Colored	Lower Canines "Short"	"White" Races Present
Callithrix argentata	−	+	+	+	+
Callithrix humeralifer	+	−	(±)	+	+
Callithrix jacchus	+	−	−	+	−
Saguinus sp.	−	+	+	−	+

Taxon	"White" Races Present	Cheeks Hairy	Saddle Present	Muzzle White	Basic Dorsum Dark	Typical Venter "White"
Saguinus bicolor	−	−	−	−	−	(+)
Saguinus fuscicollis	+	+	+	(+)	−	(−)
Saguinus geoffroyi	−	(−)	−	−	−	+
Saguinus imperator	−	+	−	(−)	−	−
Saguinus inustus	−	(−)	−	(+)	+	−
Saguinus labiatus	−	+	−	+	(+)	−
Saguinus leucopus	−	(−)	−	−	−	+
Saguinus midas	−	+	−	−	+	−
Saguinus mystax	−	+	−	+	+	−
Saguinus nigricollis	−	+	−	(+)	+	−
Saguinus oedipus	−	(−)	−	−	−	+
Saguinus tripartitus	−	+	−	(±)	−	−

Note: Parentheses mean character is not uniformly expressed in all species.

Table 9.3 Measurements of the Species of *Callithrix* as Noted by Vivo (1991)

Taxa	Total Length n	Total Length Range	Tail n	Tail Range	Hind Foot n	Hind Foot Range	Ear n	Ear Range
Callithrix argentata	55	450–612	73	200–380	68	50–72	29	(?)19–35
Callithrix aurita	7	520–80	9	280–350	8	68–72	8	20–30
Callithrix chrysoleuca	27	502–655	27	298–395	27	50–74		
Callithrix flaviceps	2	540–60	3	273–335	3	65–70	3	20–27
Callithrix geoffroyi	18	450–565	27	215–350	26	60–75	23	25–30
Callithrix humeralifer	30	530–640	34	310–70	34	58–65	7	25–31
Callithrix jacchus	24	470–635	34	175–350	34	50–65	32	(?)15–30
Callithrix leucippe	14	527–650	14	290–360	14	40–65	4	(?)16–29
Callithrix melanura	6	540–80	9	315–40	9	62–76	4	25–30
Callithrix penicillata	17	500–555	99	240–390	98	40–71	95	(?)19–31

Adult specimens of *Callithrix* "rough out" at TL ≈ 600; HB ≈ 260; T ≈ 340; HF ≈ 60; E ≈ 30; Wt ≈ < 500 g.
Vivo 1991 gives the following measurements for *Callithrix intermedius* (mm): HB 240; T 320; HF 68; E 32.

Table 9.4 Measurements of the Species of *Saguinus*, *Callimico*, and *Leontopithecus* as Noted in Hershkovitz 1977

Taxon	Head and Body n	Head and Body $\bar{x}$	Tail n	Tail $\bar{x}$	Hind Food n	Hind Food $\bar{x}$	Ear n	Ear $\bar{x}$
Saguinus bicolor	28	237	28	380	28	72	4	29
Saguinus fuscicollis	96	222	95	322	94	66	62	32
Saguinus imperator	7	248	7	370	7	70	5	28
Saguinus inustus	11	233	11	366	10	72	—	—
Saguinus labiatus	8	261	8	387	9	73	—	—
Saguinus leucopus	31	240	30	383	30	74	32	27
Saguinus midas	101	234	101	379	104	69	15	32
Saguinus mystax	22	257	22	390	22	73	20	29
Saguinus nigricollis	8	223	9	338	6	68	7	28
Saguinus oedipus	149	237	147	368	151	72	139	23
Callimico goeldii	18	224	18	302	19	74	15	23
Leontopithecus rosalia	17	261	16	370	14	77	6	25

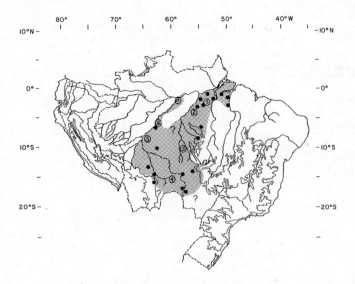

Map 9.1. Distribution of *Callithrix argentata* (nos. 1–7 refer to some named forms within this group as noted in table 9.1; number is in center of range): (1) *C. a. argentata*; (2) *C. a. leucippe*; (3) *C. a. emiliae*; (4) *C. a. melanura*; (5) *C. a. nigriceps*; (6) *C. a. chrysoleuca*; (7) *C. a. mauesi*.

are white (plate 7). In *C. h. chrysoleuca* the trunk and head are pale gold to white and the tail banding is indistinct or absent. *C. h. intermedius* (if a valid taxon) is intermediate.

Distribution

This species group occurs south of the Amazon between the Rio Tapajós and the Rio Roosevelt (see map 9.2).

Comment

Mittermier et al. (1988) include *C. humeralifera*, *C. chrysoleuca*, and *C. intermedius* within the "*humeralifera* group." *C. chrysoleuca* is considered a separate species by Vivo (1991).

Callithrix jacchus (Linnaeus, 1758)

Description

For measurements see table 9.3. These marmosets have conspicuous white tufts on their ears; the tail is usually marked by whitish bands alternating with wider dark bands. The dorsum may also exhibit a banded effect with symmetrical groups of white-tipped hairs. The hairs of the face however, can vary from almost white (e.g., *C. j. geoffroyi*) to deeply pigmented except around the eyes (e.g., *C. j. jacchus*) (see plate 7).

Distribution

These marmosets are found south of the Amazon from within Brazil, including a wide variety of habitats (see maps 9.3–9.7).

Comment

Mittermier, Rylands, and Coimbra (1988) include *C. aurita*, *C. flaviceps*, *C. geoffroyi*, *C. jacchus*, *C. kuhlii*, and *C. penicillata* in the "*jacchus* group." Vivo (1991) recognizes *C. aurita*, *C. flaviceps*, *C. geoffroyi*, *C. jacchus*, and *C. penicillata* as full species.

Callithrix kuhlii (Wied-Neuwied, 1826)

Description

This species is similar to *Callithrix penicillata*. The ears are tufted, but the tufts are dark. The overall coloring is brown, and no pale morphs have been noted. The tail is dark brown.

Comment

Although this form was *not* considered a species by Hershkovitz (1977), it was recognized as such by Mittermeier et al. (1988) but *not* by Vivo (1991). It

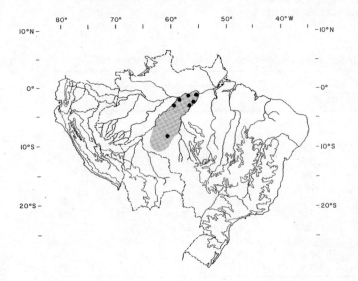

Map 9.2. Distribution of *Callithrix humeralifera*.

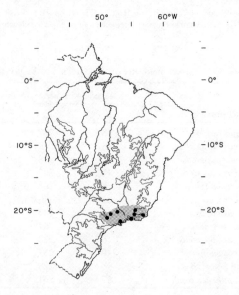

Map 9.3. Distribution of *Callithrix aurita*.

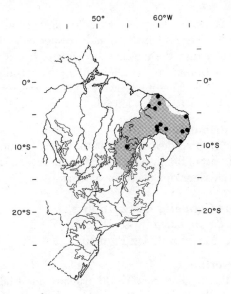

Map 9.6. Distribution of *Callithrix jacchus*.

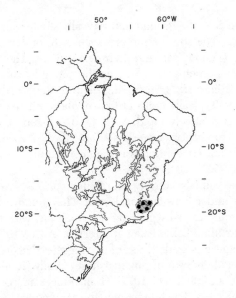

Map 9.4. Distribution of *Callithrix flaviceps*.

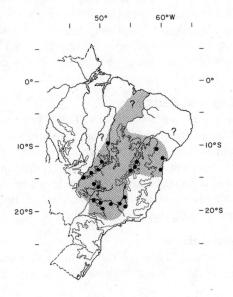

Map 9.7. Distribution of *Callithrix penicillata*.

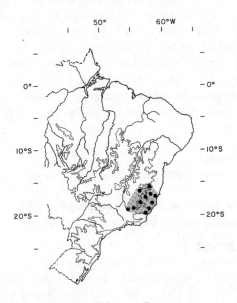

Map 9.5. Distribution of *Callithrix geoffroyi*.

could be subsumed under *C. j. penicillata,* thus being a tufted-ear marmoset of the *C. jacchus* group, but the relationships are in doubt.

Distribution

The type locality is listed as Bolivia, north of Río Belmonte; it is now recognized as distributed between the Río de Contas and Río Jequitinhonha in south-western Bolivia (not mapped).

Callithrix mauesi Schwartz and Ayres, 1992

Description

Measurements: $n = 7$; HB 198–226; T 339–56; HF 59–64; E 30–37; Wt 315–405 g. This species is possibly allied to the *C. humeralifera* group but is similar to *C. argentata* and its allies. The ear tufts are

"erect," in contrast to the long, drooping tufts of the typical *C. humeralifera*. *C. mauesi* is darker in overall color. The dorsum exhibits a banded pattern deriving from alteration of dark brown and buff bands. A more complete description is included in Mittermeier, Schwartz, and Ayres (1992).

Distribution

The species is found on the west bank of the Rio Maués-Acú, directly across the town of Maués, Amazonas state, Brazil (not mapped).

Callithrix nigriceps Ferrari and Lopes, 1992

Description

Measurements: HB 206.3; T 319.3; HF 65.3; E 29.3; Wt 370 g ($n = 4$). This group is clearly allied to the *C. argentata* group but differs in the following respects: The trunk is short, and the species seems to be stouter than the typical *C. argentata*. It is a bare-eared marmoset but is rather dark both ventrally and dorsally. The dorsum is brown with orange russet hind limbs. The face is thinly haired.

Distribution

The group is described from the state of Rondônia, Brazil, east of the Jiparaná and Madeira Rivers. It may extend to the Aripuanã River (not mapped).

Genus *Cebuella* Gray, 1866
Cebuella pygmaea (Spix, 1823)
Pygmy Marmoset, Titi Pigmeo

Description

In this smallest New World primate, unmistakable owing to its diminutive size, total length rarely exceeds 350 mm, and head and body length is approximately 200 mm. The tail is shorter than the head and body and is nonprehensile. A typical weight is 70 g. There are no conspicuous tufts on the head. The tricolored hairs give a definite salt-and-pepper pattern to the dorsal pelage. The skull is portrayed in figure 9.2.

Range and Habitat

The pygmy marmoset is distributed from southern Colombia through Amazonian Peru and west into Brazil (map 9.8). Throughout its range it coexists with several species of the genus *Saguinus*. Its distribution is limited to multistratal tropical evergreen forests at lower elevations. It may be locally abundant, but it is dependent on certain tree species for sap.

Natural History

The natural history of this form is summarized in Soini (1988). This strongly arboreal species rarely

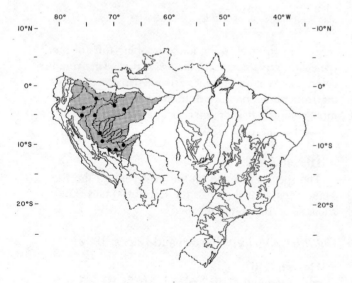

Map 9.8. Distribution of *Cebuella pygmaea*.

descends to the ground and typically shelters in tree cavities. The social structure appears to be based on a monogamous pair and their descendants. The male actively participates in rearing the young by carrying them from birth, transferring them to the female for nursing. Typically twins are born, and their development is relatively rapid. Solid food is taken at from twenty to twenty-seven days of age. Earliest recorded age at first conception was eighteen months (Eisenberg 1977).

The animals actively search for various invertebrates under bark and on leaves. They remove bark from some trees by gouging out small pits with their incisors, then lick the sap at intervals. A tree so used may show hundreds of pits of varying ages that are visited regularly by a pygmy marmoset family (Ramirez, Freese, and Revilla C. 1977). Home ranges may center on such trees. The home range for a troop may be less than 1 ha, and densities can reach 5.6 animals per ha in favorable habitat (Castro and Soini 1977). Developmental studies and captive behavior have been studied by Christen (1974), and vocalizations have been analyzed in detail by Snowden and Cleveland (1980). The birdlike chittering and tweeting sounds of pygmy marmosets span a broad spectrum of frequencies. Parts of their vocal repertoire are out of the range of human hearing.

Comment

Listed as *Callithrix pygmaea* by Groves (1993).

Genus *Saguinus* Hoffmannsegg, 1807
Tamarin, Pinche, Titi

Description

The lower canine teeth are rather large, in contrast to those of *Callithrix* and *Cebuella* (see figs. 9.1–9.3). Approximately eleven species are assigned to this genus, and eight occur within the area covered by this volume. These squirrel-sized monkeys exhibit a variety of color patterns, and tufts of hair often ornament their heads and faces. Head and body length rarely exceeds 370 mm, the tail is slightly longer than the head and body, and weight averages about 500 g. The heavily furred tail is nonprehensile. Clawlike nails are present on the digits. Measurements are included in table 9.4.

Distribution

The genus is distributed from eastern Brazil to Panama, but the distribution is discontinuous since no members of the Callitrichidae have been recorded from east-central Colombia and northern Venezuela. In accordance with their arboreal adaptations, species of *Saguinus* occur only in forested areas. However, some seem to be adapted to second growth and may be found in very scrubby habitats (e.g., *Saguinus oedipus geoffroyi*).

Although *Saguinus* does co-occur with *Callimico* and with *Cebuella*, the pygmy marmoset, tamarins do not co-occur with *Callithrix*. By and large, in the area covered by this volume *Saguinus* occurs in the western region of the Amazon basin, although extending north to Panama. The golden-handed tamarin, *Saguinus midas*, extends its range from northeastern Brazil to the Guianas.

Natural History

The tamarins may be distinguished from the true marmosets (genera *Cebuella* and *Callithrix*) by their long lower canines and unspecialized incisors (see figs. 9.1–3). Although some species do feed on tree exudates, apparently the tamarins do not habitually gouge the bark of trees to obtain sap. Tamarins are generalized frugivore-insectivores and are strictly diurnal. They catch and kill small vertebrates, including lizards and birds. As in other members of the family Callitrichidae, the typical social unit consists of a bonded pair and their offspring. Strong participation in care of the young by the male and older juveniles is characteristic. Two northern species have been the subject of long-term field studies, *Saguinus oedipus geoffroyi* by Dawson (1977) and *Saguinus o. oedipus* by Neyman (1977), while Terborgh and associates have studied *Saguinas imperator* and *S. fuscicollis* in Peru. Rylands (1993) provides an excellent summary of the recent studies of this genus.

Identification of Tamarins

Following Hershkovitz (1977), it is possible to subdivide the tamarins into three groups: the hairy-faced tamarins, where the cheeks and sides of the face are haired; the mottled-faced tamarins, with sparse hair on the cheeks; and the bare-faced tamarins, with a nearly naked face. Table 9.2 offers a character matrix.

Among the hairy-faced tamarins *Saguinas midas* is distinctive in that it does not have a white muzzle. The hands and feet are reddish north of the Amazon, but to the south the hands and feet are black. The white-mouth tamarins include four species or species groups. *Saguinus imperator* possesses a truly white mustache on the upper lip (plate 7). *Saguinus labiatus* has white hairs around the mouth, a white spot on the nape, and a dark brown dorsum, which may have a grizzled appearance on the lower back; the venter is a reddish (chestnut) brown. *Saguinus mystax* is similar to *S. labiatus* but lacks the white spot on the nape, and the venter is a dark brown not contrasting with the dorsal color. *Saguinus nigricollis* possesses the white muzzle, but the dorsum has two distinct colors. Either the forelimbs and anterior portion of the dorsum are dark brown contrasting with the reddish brown hind-

quarters or the shading is reversed in saturation, with the forelimbs and anterior dorsum olive brown and the hindquarters and lower back dark brown. In its typical form *Saguinus fuscicollis* has white hairs around the mouth contrasting with a dark brown head. Typically the species exhibits a salt-and-pepper saddle contrasting with a dark face and reddish arms and legs.

Hershkovitz (1977) portrays thirteen subspecies of *S. fuscicollis* and again invokes metachromism as an "umbrella" for describing the variation. The *S. fuscicollis* complex ranges from the typical pattern of color described above to a nearly white form, *S. f. melanoleucus*. One subspecies, *S. f. tripartitus*, has recently been elevated to full species status (Thorington 1988). Metachromism as a hypothesis must be properly evaluated by appropriately designed experiments. The "saddle" as a coat color character holds for all subspecies except for the "bleached" variant *S. f. melanoleucus* (see species accounts below).

Within the mottled-faced tamarin group there is a single species, *Saguinus inustus,* which resembles the white-mouthed hairy-faced tamarins but does not have a sharply contrasting white muzzle, although there is often a white patch between the upper lip and the nostrils and in some races the cheeks may be whitish. It also does not have the salt-and-pepper saddle but is uniformly brown.

The bare-faced tamarins include four species: *Saguinus oedipus, S. geoffroyi, S. leucopus,* and *S. bicolor*. The first three are distributed in northwestern Colombia and Panama, while *S. bicolor* is Brazilian.

Saguinus bicolor (Spix, 1823)

Description

There are three recognized subspecies. All have in common a naked face and head. The tail is dark brown dorsally and reddish ventrally. The ears are not haired. The three subspecies *S. bicolor, S. martinsi,* and *S. ochraceous* exhibit varying degrees of bleaching. *Saguinus bicolor bicolor* is white on the forepart of the body with pale brown hindquarters. *Saguinus b. ochraceous* is the most depigmented form, so that the entire dorsum is reduced to a light brown.

Distribution

The species is found north of the Amazon River east of the junction between the Amazon and the Rio Negro (map 9.9).

Natural History

Egler (1992) has studied *S. bicolor* near Manaus, Brazil. Groups of tamarins used low-stature trees and consumed fruits, flowers, and exudates of over twenty tree species. The home range of one group included roughly 12 ha.

Map 9.9. Distribution of *Saguinus bicolor.*

Saguinus fuscicollis (Spix, 1823)
Saddle-backed Tamarin

Description

This average-sized tamarin is characterized by a white muzzle and a conspicuous black-and-white (salt-and-pepper) saddle on the middorsum (plate 7), but as noted above, numerous subspecies exhibit varying degrees of "bleaching" in their coat color (Hershkovitz 1977).

Range and Habitat

The species is distributed in the extreme southern part of Colombia and from there south into Amazonian Peru to Bolivia and adjacent portions of Brazil (map 9.10). It is confined to multistratal tropical evergreen forests.

Natural History

The field data indicate that this marmoset lives in small groups, probably based on a nuclear family. At certain times of the year several families may congregate in fruiting trees, then separate into their original groupings. As one of the best-studied species in captivity, it has been the subject of numerous behavioral investigations by Epple (1973, 1975). The gestation period is approximately 140 days; the male begins to carry the young within ten days of their birth. The feeding ecology has been analyzed by Terborgh (1983). Insects and fruits form the important components of the diet, in common with other species of *Saguinus.*

In Peru *S. fuscicollis* forms mixed-species foraging groups with *S. imperator.* These associations are long term and involve subtle differences in feeding patterns, such as in vertical distribution while foraging. Apparently the mixed-species troop can more effectively defend the foraging area. Foraging together by sympatric species may be much more common than previously thought in areas where more than one species of *Saguinus* occurs (Terborgh 1983).

Saguinus imperator (Goeldi, 1907)

Description

Immediately recognizable by the white hairs on the upper lip which droop downward to exhibit a true mustache in the cultural context of western European male adornment. The dorsum looks gray banded, and the tail is a reddish brown (plate 7).

Distribution

As determined at present, the range is confined to the Amazonian portion of southeast Peru and adjacent western Brazil (see map 9.11).

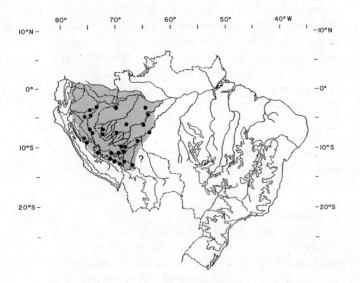

Map 9.10. Distribution of *Saguinus fuscicollis.*

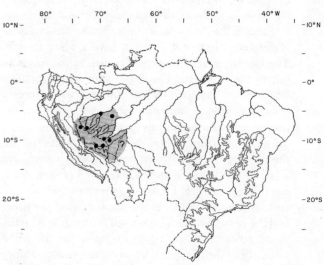

Map. 9.11. Distribution of *Saguinus imperator.*

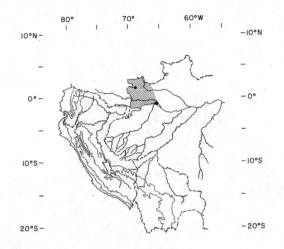

Map 9.12. Distribution of *Saguinus inustus*.

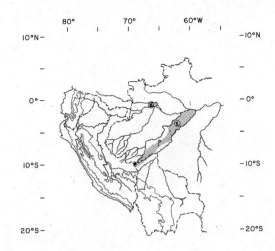

Map 9.13. Distribution of (1) *Saguinus l. labiatus;* (2) *S. l. thomasi.*

Natural History

Terborgh (1983) describes the ecology of this species in southeast Peru. In his study area, *S. imperator* forms a polyspecies association with adjacent groups of *S. fuscicollis.* Mutual recognition of calls and individual identities between members of the two species is documented. In its feeding habits and behavior this species is a typical tamarin.

Saguinus inustus Schwartz, 1951

Description

Saguinus inustus is approximately the same size as *S. fuscicollis.* It does not have a distinctive white muzzle or the contrasting saddle and thereby is easily distinguished from the other two hairy-faced tamarins that range near it.

Range and Habitat

The species occurs from southeastern Colombia into adjacent parts of northwestern Brazil (map 9.12).

Natural History

There are few natural history data concerning this species. Presumably its reproductive behavior and social structure conform to the pattern described for *Saguinus fuscicollis.*

Saguinus labiatus (E. Geoffroy, 1812)

Description

The white muzzle is prominent in all subspecies. The head is dark brown to almost black. The dorsum varies in color from very dark brown to varying degrees of a grizzled effect deriving from white tips on the hairs. The venter is a reddish brown contrasting with the dorsum. The nape usually exhibits a white patch (see plate 7).

Distribution

This species has a disjunct distribution south of the Amazon, occurring in east-central Peru and then in central Brazil (map 9.13).

Saguinus midas (Linnaeus, 1758)
Red-handed Tamarin

Description

This tamarin lacks the white muzzle. The subspecies north of the Amazon has reddish hands and feet.

Range and Habitat

The species is broadly distributed from Guyana east of the Río Essequibo through French Guiana and south into Brazil. It is tolerant of both multistratal tropical evergreen forests and second-growth forests (map 9.14).

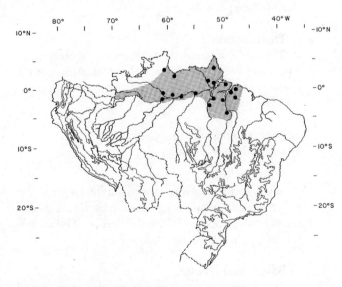

Map 9.14. Distribution of *Saguinus midas.*

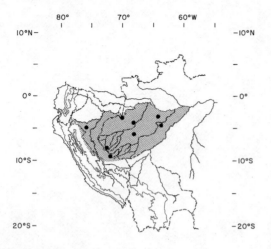

Map 9.15. Distribution of *Saguinus mystax*.

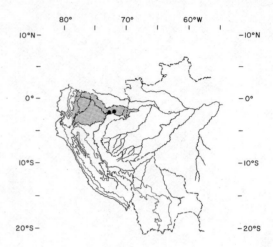

Map 9.16. Distribution of *Saguinus nigricollis*.

Natural History

This species has received attention in the publication by Mittermeier and van Roosmalen (1981). The red-handed tamarin occurs from lower montane rain forests to lowland rain forests. It utilizes edge habitats, especially where forest and savanna intersperse. It does not typically frequent the high emergent canopy, but rather forages in the lower crowns or the understory, eating both fruits and seeds. Seeds compose a large part of its diet, supplemented with insects. In its grouping tendencies and reproductive habits it does not appear to deviate from the typical tamarin pattern.

Saguinus mystax (Spix, 1823)

Description

The muzzle is white, the head and dorsum are usually brown, often quite dark, and the venter is brown. The subspecies *S. m. pileatus* is somewhat paler on the dorsum (see plate 7).

Distribution

The species occurs in the Amazonian portions of Brazil and Peru. It could extend to Colombia, but the current situation in that region restricts our understanding (map 9.15).

Saguinus nigricollis (Spix, 1823)
Black-and-Red Tamarin

Description

Saguinus nigricollis is approximately the same size as *S. fuscicollis* but is distinguishable by the absence of the salt-and-pepper saddle on the dorsum. The muzzle is white, as in *S. fuscicollis* (plate 7).

Range and Habitat

This species is distributed in northwestern Brazil but extends into the extreme southeastern tip of Co-

lombia and also crosses the Río Putumayo into the extreme west of the southwestern portion of Colombia to the east of the Andes (map 9.16). It is confined to multistratal tropical evergreen forests.

Natural History

In captivity *Saguinus nigricollis* is very similar to *S. fuscicollis* in reproduction and behavior. There have been no long-term field studies of this species (Izawa 1976).

Saguinus tripartitus (Milne Edwards, 1878)

Description

Although most workers considered this form a subspecies of *S. fuscicollis*, Thorington (1988) raised it to a full species (see *S. fuscicollis* account). The muzzle is white and the area over each eye is also white. The "saddle" on the rump exhibits a light agouti brown, the shoulders are orange, but the head and neck are dark brown to black. The forefeet and hind feet are reddish (plate 7).

Distribution

The type locality is listed as Río Napo, Ecuador. Thorington implies a distribution east of the Río Curaray on the Brazil-Colombia border, where it comingles with *S. fuscicollis* (map 9.17).

Genus *Callimico* Miranda-Ribeiro, 1911
Callimico goeldii (Thomas, 1904)
Goeldi's Marmoset, Calimico

Description

This species differs from all other members of the family Callitrichidae in possessing three molars. In the region under consideration it is the only all-black tamarin. The hair on its head is rather long, approxi-

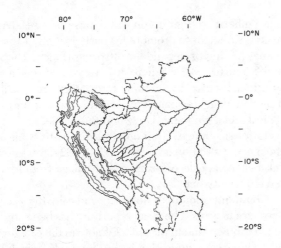

Map 9.17. Distribution of *Saguinus tripartitus*.

mating a crest, but does not contrast with the basic black body color.

Range and Habitat

This tamarin is widely distributed in the upper Amazon basin, extending to Colombia and possibly to Bolivia. (map 9.18).

Natural History

Goeldi's marmoset has been studied extensively in captivity by Heltne, Turner, and Wolhandler (1973) and in the field by Pook (Heltne, Wojeik, and Pook 1981; Pook and Pook 1981). It appears that *Callimico* prefers low-stature strata. It tends to forage in the understory of multistratal tropical evergreen forests, eating small fruits, berries, and insects. Group size can be up to eight. The female typically has one young, in marked contrast to other members of the family Callitrichidae. The female transports the infant for the first two weeks of life, and thereafter the male participates. Gestation takes between 153 and 159 days.

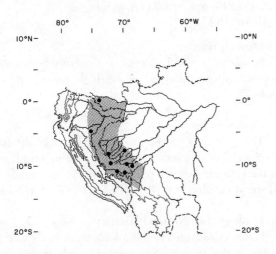

Map 9.18. Distribution of *Callimico goeldii*.

This species is never very abundant. Pook estimated a density of 0.25 groups per km^2. Home range for the group of eight studied by Pook was nearly 60 ha. Activity patterns indicate a range from crepuscular to diurnal. *Callimico* can occur not only with the pygmy marmoset but with other species of tamarins. While foraging, *Callimico* is occasionally seen in association with *S. fuscicollis,* following behind at a considerable distance.

Genus *Leontopithecus* Lesson, 1840
Lion Tamarin
Leontopithecus caissara Lorini and Persson, 1990

Description

This species is clearly a member of the genus *Leonopithecus,* but it differs in color pattern from all other taxa in the genus in that the face is black, as are the hairs of the cheeks, head, and throat. The forefeet and hind feet are also black, as is the tail. The body is a reddish gold. Coimbra-Filho (1990) has suggested it could be considered a subspecies of *L. chrysopygus.*

Distribution

Known only from Superagui Island, Brazil (map 9.19).The location is an offshore, borderline island but is considered within Paraná state.

Leontopithecus rosalia (Linnaeus, 1766)

Comment on Taxonomy

The four color variants of this genus are discussed as subspecies except for *L. caissara.* We recognize as

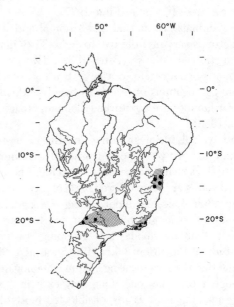

Map 9.19. Distribution of four species of *Leontopithecus:* triangle, *L. rosalia;* square, *L. chrysopygus;* closed circle, *L. chrysomelas;* open circle, *L. caissara.*

subspecies *L. rosalia, L. chrysomelas,* and *L. chryso-pygus.* The fourth form, *L. caissara,* has been described from an island off the coast of Paraná state, Brazil (Lorini and Persson 1990). A discussion of the subspecific status of the first three forms is included in Hershkovitz (1977). The three well-studied forms will be discussed separately (see also Kleiman 1981; Kleiman, Hoage, and Green 1988).

Leontopithecus rosalia rosalia (Linnaeus, 1766)
Golden Lion Tamarin

Description

The lion tamarins are the largest species of the Callitrichidae. Measurements are HB 200–336; T 315–400; weights for captives average 710 (600–800) g (Kleiman 1981; Kleiman, Hoage, and Green 1988). The subspecies *L. r. rosalia* is reddish gold to almost white in basic body color. The terminal portion of the tail and hands may be brown (see plate 8).

Distribution

This form occurred in the Atlantic rain forests of Espírito Santo and Rio de Janeiro, Brazil. It is currently confined to relict forest patches in Rio de Janeiro state (map 9.19).

Natural History

These tamarins formerly occurred at lower elevations in the Atlantic rain forest. They now occur in second-growth remnant patches of rain forest in Rio de Janeiro state. They are omnivores, including fruit, insects, and small vertebrates in their diet. Although they shelter in clumps of vines, they also use natural cavities in trees (Coimbra-Filho 1977).

The animals are diurnal and arboreal and live in small groups varying from two to eight. They may exhibit territorial defense (Peres 1989). Two or more groups may congregate in a large fruiting tree, presenting aggregations of fifteen to twenty individuals. Nuclear families are territorial (Kleiman, Hoage, and Green 1988).

The typical social grouping consists of a mated pair and one or two sets of offspring. The father and older siblings share in carrying the young. The newly weaned young are often offered parts of prey captured by the adult male and older siblings (Hoage 1982). Polyandrous social groupings occur, but they are less common than the modal pair grouping. In a two-year study of this taxon Baker (1991) was able to determine that the basic rearing system is monogamous. Although more than one male may exist in a rearing group, one male is responsible for most of the conceptions. Apparently subordinate males that may in fact immigrate into a breeding group do not sire offspring initially; the possibility that the immigrant

may "inherit" the group on the demise of a resident male is sufficient to promote his participation in communal rearing of dependent offspring (Baker, Dietz, and Kleiman 1993). Where two reproductively mature females exist in the same troop, they are usually a mother-daughter pair. This is a rare variant on the typical social grouping (Dietz and Baker 1993).

Litter size ranges from one to three, with twins as the mode. Gestation lasts from 125 to 132 days. Intervals between peaks of sexual activity range from fourteen to twenty-one days.

Communication does not differ radically from that in other tamarins. The vocal repertoire has been studied in captivity and in the wild. Of note in the category of auditory communication is the "long call," which is involved in territorial display (see also Kleiman 1981; Kleiman, Hoage, and Green 1988).

Leontopithecus rosalia chrysomelas (Kuhl, 1820)

Description

Size is similar to *L. r. rosalia,* but the basic body color is black with a reddish gold color on the head, rump, thighs, and proximal third of the tail (see plate 7).

Distribution

This form occurred in the lowland and premontane tropical forests of the southeastern portion of Bahia state, Brazil. At present its range is restricted to relict patches of forest in the former range (map 9.19).

Ecology and Natural History

Rylands (1982) made observations on this form at Una Bahia. In habits, feeding, and grouping tendencies, it resembles *L. r. rosalia.* He noted that *L. chrysomelas* may form polyspecific feeding associations with *Callithrix kuhlii.*

Leontopithecus rosalia chrysopygus (Mikan, 1823)

Description

Similar in size to *L. r. rosalia.* The basic body color is black with reddish gold on the thighs and rump (see plate 7).

Distribution

This subspecies formerly occurred west of the Rio Tietê in São Paulo state, Brazil. It is currently confined to the extreme western portion of the former range at Morro do Diablo State Park (Padua 1993) (map 9.19).

Ecology and Natural History

Padua (1993) conducted a long-term study of this taxon in the state of São Paulo, Brazil. Padua's results suggest that the ecology and behavior of *L. r.*

chrysopygus is similar to that described for *L. r. rosalia*. His study confirms the versatility of *L. r. chrysopygus* with respect to habitat use and foraging, since it can exist in very modified secondary growth as well as in mature multistratal forests.

FAMILY CEBIDAE

Diagnosis

The dental formula is I 1/1, C 2/2, P 3/3, M 3/3 (see fig. 9.4). The posterior molar may be missing in a small proportion of individuals within a larger population (15% for *Ateles*). There are eleven Recent genera, including some twenty-nine species. Five genera (*Cebus, Ateles, Alouatta, Brachyteles,* and *Lagothrix*) have prehensile tails. All species are small to medium

in size (see tables 9.6 and 9.7). Alternative schemes for classification are included in table 9.5 (see genus accounts).

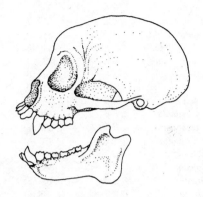

Figure 9.4. Skull of *Cebus apella*.

Table 9.5   Classification of the Family Cebidae according to Groves 1993

Taxon	Comments[a]	Taxon	Comments[a]
Cebidae		Callicebinae	
Alouattinae		*Callicebus*	
Alouatta		C. brunneus	Same
A. belzebul	Same	C. caligatus	Same
A. caraya	Same	C. cinerascens	—? doubtful
A. coibensis	Same	C. cupreus	Same
A. fusca	Same	C. donacophilus	Same
A. palliata	Same	C. dubius	—(in C. cupreus)
A. pigra	Same	C. hoffmansi	—(in C. moloch)
A. sara	—	C. modestus	Same
A. seniculus	Same	C. moloch	Same
Aotinae		C. oenanthe	Same
Aotus		C. olallae	Same
A. azarae	Same; red-necked group	C. personatus	Same
A. brumbacki	Same; gray-necked group	C. torquatus	Same
A. hershkovitzi	—	Cebinae	
A. infulatus	Same; red-necked group	*Cebus*	
A. lemurinus	Same; gray-necked group	C. albifrons	Same
A. miconax	Same; red-necked group	C. apella	Same
A. nancymaae	Same; red-necked group	C. capucinus	Same
A. nigriceps	Same; red-necked group	C. olivaceus	Same
A. trivirgatus	Same; gray-necked group	*Saimiri*	
A. vociferans	Same; gray-necked group	S. boliviensis	Same
Atelinae		S. oerstedii	Same
Ateles		S. sciureus	Same
A. belzebuth	Same	S. ustus	Same
A. chamek	S. s. paniscus[b]	S. vanzolinii	? S. boliviensis
A. fusciceps	Same	Pitheciinae	
A. geoffroyi	Same	*Cacajao*	
A. marginatus	S. s. belzebuth[b]	C. calvus	Same
A. paniscus	Same	C. melanocephalus	Same
Brachyteles		*Chiropotes*	
B. arachnoides	Same	C. albinasus	Same
Lagothrix		C. satanus	Same
L. flavicauda	Same	*Pithecia*	
L. lagothricha	Same	P. aequatorialis	Monachus group
		P. albicans	Monachus group
		P. irrorata	Monachus group
		P. monachus	Monachus group
		P. pithecia	Pithecia group

[a]According to Mittermeier, Rylands, and Coimbra-Filho 1988.

[b]S. s. = subspecies.

Table 9.6   Some Measurements for the Smaller Species of Cebidae

Taxon	Sex	TL	HB	T	HF	E	Wt	Source
Aotus nancymaae	♀♀	—	312 (7)	371 (7)	93 (7)	29 (7)	—	Hershkovitz 1983
Callicebus brunneus	♂♂	—	317 (7)	397 (7)	90 (7)	31 (7)	845 g	Hershkovitz 1990
Callicebus donacophilus	♂♂	—	311 (6)	411 (6)	91 (5)	34 (6)	—	Hershkovitz 1990
Callicebus donacophilus	♀♀	—	340 (5)	440 (4)	89 (6)	36 (5)	—	Hershkovitz 1990
Callicebus hoffmanni	♂♂	—	322 (14)	453 (14)	94 (12)	33 (4)	—	Hershkovitz 1990
Callicebus hoffmanni	♀♀	—	316 (16)	465 (15)	95 (15)	30–	—	Hershkovitz 1990
Callicebus moloch	♂♂	—	333 (19)	466 (18)	93 (17)	34–37	850–1,200 g	Hershkovitz 1990
Callicebus moloch	♀♀	—	331 (22)	448 (21)	91 (22)	29 (4)	700–1,020 g	Hershkovitz 1990
Callicebus personatus	♂♂	—	380 (5)	508 (5)	111 (4)	33 (4)	1,270 g (5)	Hershkovitz 1990
Callicebus personatus	♀♀	—	356 (7)	485 (6)	104 (7)	34 (4)	1,378 g (6)	Hershkovitz 1990
Callicebus torquatus	♀♀	—	425 (4)	460 (4)	—	—	—	
Pithecia albicans	♂♂	—	404–10	416–40	—	—	—	Hershkovitz 1979
Pithecia albicans	♀♀	—	380 (6)	428 (6)	—	—	—	Hershkovitz 1979
Pithecia hirsuta	♂♂	—	416 (12)	463 (12)	—	—	—	Hershkovitz 1979
Pithecia hirsuta	♀♀	—	408 (5)	432 (5)	—	—	—	Hershkovitz 1979
Pithecia pithecia	♂♂	—	350 (7)	397 (7)	—	—	—	Hershkovitz 1979
Chiropotes albinasus	♂♂	—	455–60	360–80	110–20	—	—	Hershkovitz 1985
Chitopotes satanas	♂♂	—	409 (6)	422 (6)	127 (6)	—	—	Hershkovitz 1985
Chiropotes satanas	♀♀	—	390 (7)	422 (7)	126 (7)	—	—	Hershkovitz 1985
Cacajao calvus	♀♀	—	410	100	110	—	—	Hill 1960

Table 9.7   Some Measurements for the Larger Species of Cebidae

Taxon	Sex	Measurements (cm)		Wt (kg)	n	Reference[a]
		HB	T			
Alouatta seniculus	♂	52.3	63	6.5	10	(1)
	♀	46.8	59	4.5	4	(1)
Alouatta palliata	♂	50.6	66.5	7.8	15	(1)
	♀	47.5	65.1	6.6	15	(1)
Ateles fusciceps	♂	59	63	—	1	(2)
	♀	51	75	—	1	(2)
Ateles belzebuth	♂	64	81	—	1	(2)
	♀	63.5	69.5	—	1	(2)
Lagothrix lagotricha	♂	46.5	64.5	5–10	7	(3)
	♀	47.7	57.7	5–6.5	7	(3)

[a](1) Thorington, Rudran, and Mack 1979; (2) Eisenberg 1976; (3) Fooden 1963.

Distribution

The species of this family occur from southern Veracruz, Mexico, to the gallery forests on the upper reaches of the Río Paraná in Argentina. All are adapted to multistratal tropical evergreen forests, but the genera *Alouatta, Aotus,* and *Cebus* have adapted to the semideciduous forests of the more xeric portions of northern and southern South America.

Natural History

All members of this family are diurnal except *Aotus,* which is the only nocturnal New World primate. Most species are adapted to a diet of fruit, seeds, small vertebrates, and invertebrates. *Alouatta* is exceptional in that it generally includes 50% foliage in its diet. The species of *Aotus, Callicebus,* and *Pithecia* show monogamous tendencies. All females of this family usually bear a single young.

History of the Cebidae

The earliest fossil cebids are recorded from the Oligocene. *Branisella* is the oldest recorded genus. By the Miocene, resemblances to modern forms are recognizable: *Neosaimira* resembles *Saimiri,* and *Stirtonia* resembles *Alouatta.* The history of the New World primates to date suggests early adaptation for arboreal niches and a modest body size. Although *Protopithecus* was described by Lund in 1838 from the Pleistocene, recent discoveries of more complete skulls and postcranial elements have revealed some surprises. *Protopithecus brasiliensis* is now known to have reached an estimated 25 kg in body weight (Hartwig and Cartelle

1996). The newly discovered *Caipora bambuiorum* from the Pleistocene of Brazil suggest a body weight of 20 kg (Cartelle and Hartwig 1996; Hartwig 1995). The structure of primate communities in the Pleistocene of eastern Brazil must have been astonishingly diverse.

Genus *Aotus* Illiger, 1811
Night Monkey, Mono de la Noche, Coatá

Taxonomy
During the past thirty years most workers have recognized the single species *Aotus trivirgatus,* but cytogenetic research has indicated that the genus *Aotus* is divisible into populations having different chromosomal morphology (deBoer 1974). Up to nine karyotypes have been described, but it is unknown if hybridization is impossible among them all. Natural hybrids have been found in Colombia (Yunis, Torres de Caballero, and Ramirez 1977). Honacki, Kinman, and Koeppl (1982) listed one species for the genus. Eisenberg (1989) followed Hershkovitz (1983) in the listing of species for the "northern rim" of South America. Wilson and Reeder (1993) list ten species for *Aotus* following Hershkovitz. Mittermeier et al. (1988) follow Hershkovitz but group the species into the red-necked and gray-necked groups, as do Emmons and Feer (1997) and Corbet and Hill (1991), the latter authors do not use the same terminology, but the intent is there. Baer, Weller, and Kakoma (1994) offer the most recent discussion of the problem. We will discuss the taxa within the context of a simplified dichotomy, although table 9.5 contains the taxonomy according to Groves (1993).

Description
These small ceboid primates have nonprehensile tails. Head and body length ranges from 240 to 375 mm and tail length from 220 to 418 mm. Weight averages 700 g (Wright 1981). The dorsal color varies depending on the species, ranging from gray to red brown; the venter can be bright orange to white. There is usually a white patch shaped like a half moon above each eye, and black stripes are generally evident on the head—a median stripe and two lateral ones that come together at the top (plate 8).

All species of *Aotus* in northern South America belong to the gray-neck group. The pelage of the sides of the neck is grayish agouti to brownish agouti, corresponding to the color of the sides. The interscapular area may have a group of long hairs approximating a small crest. The fur of the pectoral area often separates along a distinct midventral line, where the chest gland is situated (Hershkovitz 1983).

In contrast, the more southern red-neck group of species exhibits a reddish throat and neck and has orange at the base of the tail. The inner surfaces of the forelimbs and hind limbs may have a reddish cast. These forms are distributed south of the Amazon and include *Aotus miconax, A. nigriceps, A. nancymaae, A. azarae,* and *A. infulatus.*

Distribution
Aotus ranges from Panama to northern Argentina, but within this latitudinal range there are areas where it is completely absent. It is noticeably absent from the llanos of Venezuela and does not appear to be present in the Guianas. The night monkey occurs in a variety of forest types and can occur at up to 3,200 m elevation in Colombia.

Natural History
Baer, Weller, and Kakoma (1994) have summarized much information in their edited volume. Night monkeys are primarily frugivorous but take some insects. *Aotus* generally lives in small family groups based on a monogamous pair-bond. Apparently several family groups can share a large fruiting tree while foraging. In suitable old-growth habitat the animals may shelter in tree holes. In good habitats they may exist at densities of forty per km^2 (Wright 1981).

In Panama they are strictly nocturnal, generally emerging from their tree holes shortly after dark and returning at dawn. In northern Argentina they may be more diurnal. Home range for a family group appears to be rather small, perhaps less than 4 ha. The male transports the infant briefly during its first two weeks of life, and thereafter he carries it most of the time. The behavior patterns of *Aotus* have been described by Moynihan (1964) and Wright (1981).

When moving through the forest at night the animals produce "gulp" sounds, usually in bursts of two to five, perhaps as a contact note. Their alarm call, which has been termed a sneeze-grunt, sounds like a metallic click, usually produced in bursts of two or three, and serves as a warning signal (Moynihan 1964).

Puertas, Aquino, and Encarnacion (1992) studied *A. vociferans* in Peru. They found that the diet included 83% fruits and 17% flowers and nectar. Fruits of *Ficus* and *Inga* were the most frequently used. The authors report cases of scramble competition with *Potos flavus, Didelphis,* and *Bassaricyon gabbii.*

The Gray-necked Group
Aotus lemurinus (I. Geoffroy, 1984)

Description
The interscapular crest is absent. The chest is orange or buffy, but the color does not extend to the inner sides of the limbs.
Chromosome number: $2n = 55$ or 56.

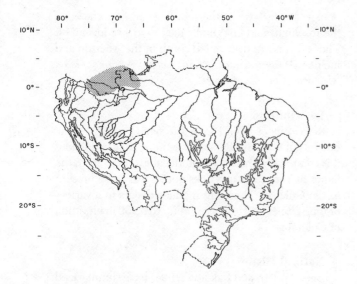

Map 9.20. Distribution of *Aotus trivirgatus*.

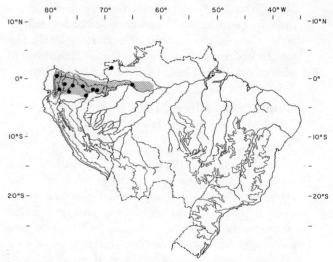

Map 9.21. Distribution of *Aotus vociferans*.

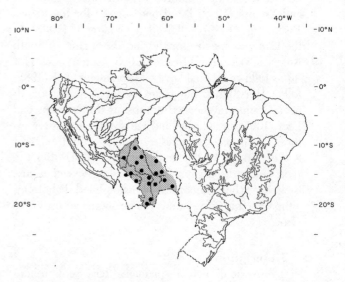

Map 9.22. Distribution of *Aotus azarae*.

Range and Habitat

The species ranges from Panama to Colombia west of the eastern cordillera of the Andes and could extend to western, coastal Ecuador. In Venezuela it is confined to the Maracaibo basin (not mapped).

Aotus trivirgatus (Humboldt, 1811)

Description

Aotus trivirgatus is similar to *A. lemurinus* in lacking the interscapular crest or whorl. It differs from *A. lemurinus* in that the orange or buff ventral color extends to the inner sides of the limbs.

Distribution

The species ranges through Amazonian Venezuela and adjacent portions of Brazil (map 9.20).

Aotus vociferans (Spix, 1823)

Description

The interscapular hairs are organized in a whorl, distinguishing this species from *A. brumbacki*. Chromosome number: $2n = 46$.

Range and Habitat

The species range east of the Andes in southern Colombia and in adjacent portions of Amazonian Brazil, Peru, and Ecuador (map 9.21).

The Red-necked Group
Aotus azarae (Humboldt, 1811)

Description

In this monkey the upperparts are grayish; the tail is reddish on the dorsal surface, as is the nape. Chromosome number: $2n = 49$ or 50.

Distribution

The species is mainly confined to lowland eastern Bolivia but extends to southeast Peru, western Paraguay, and adjacent north-central Argentina (see species account in Redford and Eisenberg 1992) (map 9.22).

Aotus infulatus (Kuhl, 1820)

Description

Characters are typical for the red-necked species but are generally darker in hue than in *A. azarae*. This form may be conspecific with the latter.

Distribution

The species is found in central Brazil to the south of the Amazon River and east of the Rio Tapajós (see map 9.23).

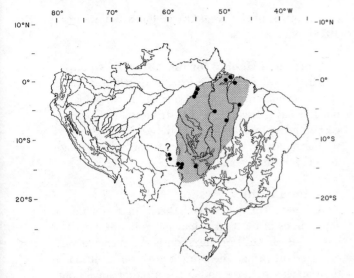

Map 9.23. Distribution of *Aotus infulatus*.

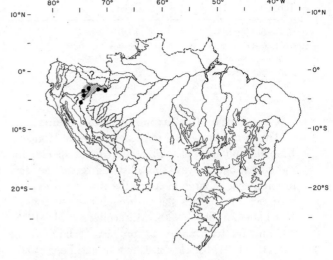

Map 9.25. Distribution of *Aotus nancymaae*.

Map 9.24. Distribution of *Aotus miconax*.

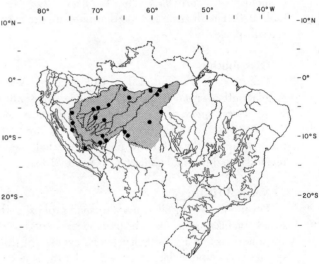

Map 9.26. Distribution of *Aotus nigriceps*.

Aotus miconax Thomas, 1927

Distribution
A. miconax is found in Peru from the Río Ucayali to south of the Río Marañón (see map 9.24).

Comment
The status of this as a species is uncertain.

Aotus nancymaae Hershkovitz, 1983

Description
The species is a typical red-necked form, but the chromosome number differs.
Chromosome number: $2n = 54$.

Distribution
This monkey is found in a restricted area of Loreto department, Peru (map 9.25).

Aotus nigriceps Dollman, 1909

Description
See comment under the red-necked group, p. •••.

Distribution
The species occurs mainly in Brazil south of the Amazon extending to eastern Peru (map 9.26).

Genus *Callicebus* Thomas, 1903
Titi Monkey, Sahui de Collar,
Titi Zocayo, Zogue

Description
Titi monkeys are small, weighing less than a kilogram, with a head and body length of 287 to 390 mm. The tail exceeds the head and body length by about 20% and is nonprehensile (see plate 7).

Table 9.8   The Species of *Callicebus*

Callicebus brunneus	*Callicebus moloch*
Callicebus caligatus	*Callicebus oenathe*
Callicebus cupreus	*Callicebus olallae*
Callicebus donacophilus	*Callicebus personatus*
Callicebus modestus	*Callicebus torquatus*

Comment on the Taxonomy of the Genus

This genus includes three species groups. Honacki, Kinman, and Koeppl (1982) listed three species for the genus, but Groves (1993) lists thirteen. The shift in number is in large measure the result of elevating subspecies to species status through the revisions of Hershkovitz (1988, 1990). Mittermeier et al. (1988) recommend ten species as probably valid. Corbet and Hill (1991) list eleven. Emmons and Feer (1997) consider only three taxa at the species level. We will discuss the named taxa under three species groups: *C. moloch, C. torquatus,* and *C. personatus* (see table 9.8). The distribution maps are broken down to the subspecies.

Distribution

The genus is distributed from Colombia and Venezuela south over much of Brazil to Bolivia. The southern species, *C. personatus,* is isolated, *C. torquatus* can co-occur with the third species, *C. moloch.* In areas of macrosympatry the latter two species are usually segregated into different habitat types (Kinzey 1981).

Natural History

Titi monkeys typically travel in small groups, and social organization seems to be based on a bonded pair and their various-aged offspring. The ecology of the species varies and will be discussed under the species accounts. Extensive captive studies have been carried out with *C. moloch* by Mason (1971). Field studies on *C. moloch* were done by Mason (1968) and Robinson (1977; 1979a,b; 1982). *C. torquatus* has been studied in the field by Defler (1983b) and by Kinzey and his associates (Kinzey 1977, 1981; Kinzey and Gentry 1978).

Callicebus rarely comes to the ground. *C. moloch* appears to forage in the understory, whereas *C. torquatus* seems to be more specialized for feeding in the crowns of trees. A single young is born after a gestation period of approximately 160 days.

The "*Moloch*" Group

Description

This species group represents the smallest members of the genus *Callicebus.* Mean head and body length for *C. moloch* is 334 mm, with the tail 429 mm long. There is virtually no sexual dimorphism in this species. It is typically brown dorsally with a reddish venter, without the distinctive lateral facial striping so characteristic of *C. torquatus* (see Jones and Anderson 1978) (plate 8).

Range and Habitat for the "*Moloch*" Group

The *C. moloch* group is broadly distributed within the Amazonian portions of Brazil and Peru, but in the north it has a discontinuous distribution in Colombia east of the eastern cordillera of the Andes. In the more southern parts of its distribution, *C. moloch* is strongly associated with riverine habitats.

Natural History

Although *C. moloch* is basically a frugivore, at some times of the year it may eat young leaves. The basic unit of social organization is the pair and various-aged offspring. The animals are strictly diurnal in their activity patterns and utilize a rather small home range; 3–5 ha seems average. They seek out a tangle of vines in a preferred sleeping tree when retiring at night. This species has a rather rich vocal repertoire. A bonded male and female will produce a duet calling sequence that has several elements, including bellows and pants. Countercalling between adjacent social groups is involved in territorial defense (Robinson 1977, 1979a,b).

Callicebus brunneus (Wagner, 1842)

Distribution

The species ranges through southeastern Peru to adjacent Bolivia and Brazil (map 9.27).

Callicebus caligatus (Wagner, 1842)

Distribution

The species is found in western Brazil and adjacent northern Peru (map 9.28).

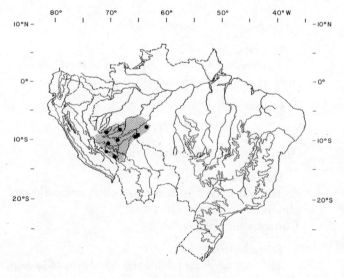

Map 9.27. Distribution of *Callicebus brunneus.*

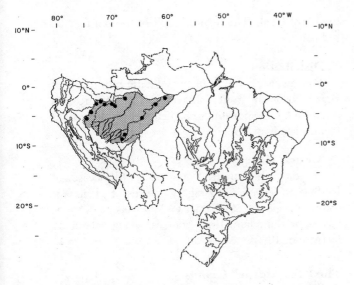

Map 9.28. Distribution of *Callicebus caligatus*.

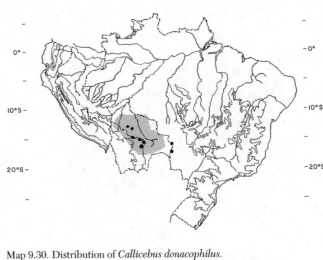

Map 9.30. Distribution of *Callicebus donacophilus*.

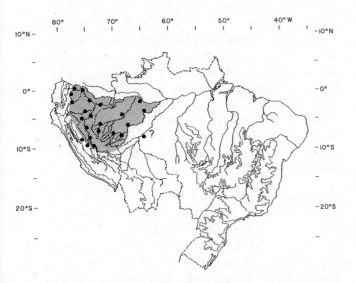

Map 9.29. Distribution of *Callicebus cupreus* (includes *C. dubius*).

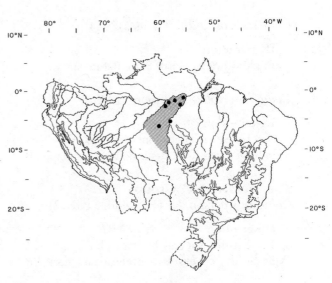

Map 9.31. Distribution of *Callicebus hoffmanni*.

Callicebus cinerascens (Spix, 1823)

Distribution

The form is found in the Rio Madeira basin. Its specific status is doubted (not mapped).

Callicebus cupreus (Spix, 1823)

Distribution

The species is found south of the Amazon from Río Purus to Río Ucayali, Peru (map 9.29).

Callicebus donacophilus (d'Orbigny, 1836)

Distribution

The species ranges through west-central Bolivia to Paraguay (map 9.30).

Callicebus dubius Hershkovitz, 1988

Distribution

The species is found in Brazil between the Rio Purus and the Rio Madeira (map 9.29).

Callicebus hoffmanni Thomas, 1908

Distribution

The species is found in Brazil south of the Amazon between the Rio Madeira and the Rio Tapajós (map 9.31).

Callicebus moloch (Hoffmannsegg, 1807)

Distribution

The species' range is in central Brazil between the Rio Tapajós and the Rio Araguaia (map 9.32).

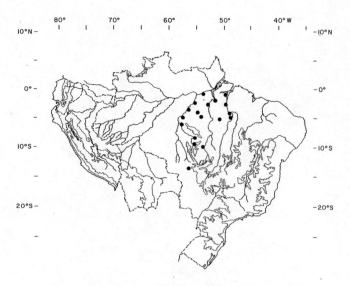

Map 9.32. Distribution of *Callicebus moloch*.

Callicebus oenanthe Thomas, 1924

Distribution
Río Mayo valley, north Peru (not mapped).

Callicebus olallae Lönnberg, 1939

Distribution
The species is known only from the type locality: 5 km from Santa Rosa, Beni, Bolivia (not mapped).

The *"Personatus"* Group
Callicebus personatus (E. Geoffroy, 1812)

Description
The dorsum is a gray brown with a lighter venter. Typically the cheeks and forehead are dark brown

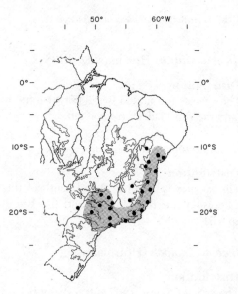

Map 9.33. Distribution of *Callicebus personatus*.

to almost black. The terminal half of the tail may be reddish.

Distribution
The species is found in the Atlantic rain forest of Brazil, formerly from Bahia to Rio de Janeiro and thence inland to portions of São Paulo state and Minas Gerais (see map 9.33).

Natural History
The diet is mainly fruit and seeds with occasional feeding on leaves. Over eighty species of plants are eaten, but species of Myrtaceae, Moraceae, and Sapotaceae predominate. Home ranges average about 24 ha (Müller 1996).

The *"Torquatus"* Group
Callicebus torquatus (Hofmannsegg, 1807)
Sahui de Collar

Description
This is the largest species within the genus, with an average head and body length of approximately 339 mm and a tail length of 460 mm. The pelage is dark brown to black, with white facial markings. In some parts of its range there may be a tendency toward a reduction in melanin deposits, resulting in a buff color, but this color variation is rare. Typically the animals are dark and easily distinguishable from *C. moloch* (plate 8).

Range and Habitat
This species is distributed in the forested parts of the Orinoco and Amazon drainages. *C. torquatus* is associated with tall, mature forests, generally in well-drained ground. Kinzey and Gentry (1978) noted that it often frequents forest on white sand soils, which occur in nutrient-deficient areas. When in association with white sand soils and the so-called black-water forests, *C. torquatus* may stay near streams (map 9.34).

Natural History
This apparently monogamous species travels in small groups (Defler 1983b). Its home ranges and day ranging seem to be greater for *C. moloch*, perhaps because *C. torquatus* is adapted to black-water forests with a reduced carrying capacity. Long-term studies on *C. torquatus* indicate that a family's home range is approximately 29 ha and is vigorously defended against neighboring groups. *C. torquatus* eats only a small quantity of leaves; its diet is almost entirely fruit and seeds. The natural history of this species has been described in detail by Kinzey (1981) and Kinzey and Gentry (1978). Defler (1994) notes that *C. torquatus* is not confined to white sand soils and associated vegetation.

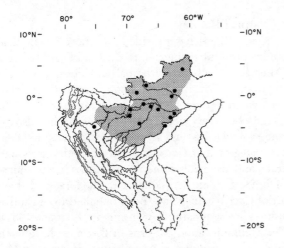

Map 9.34. Distribution of *Callicebus torquatus*.

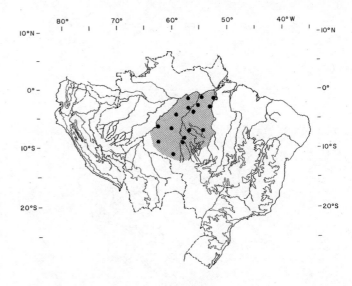

Map 9.35. Distribution of *Chiropotes albinasus*.

Genus *Chiropotes* Lesson, 1840
Bearded Saki, Cuxiú

Description
The skull exhibits the typical cebid dental formula, but the males are strongly size dimorphic and the canines are quite enlarged. Weight ranges from 2.6 to 3.2 kg; HB is 370–460 mm, and the tail ranges from 370 to 463 mm. The tail is nonprehensile. The color is variable, and some species are color dimorphic (see species accounts).

Distribution
The genus is found in Amazonian Brazil, the Guianas, and southern Venezuela (van Roosmalen, Mittermeier, and Milton 1981).

Natural History
Species of the genus are arboreal and quite social. Troop size is variable but can be as large as thirty. They are frugivores and significant seed predators. Home ranges of large troops may include an area of 1 km². The bearded sakis appear to be limited to multistratal, evergreen forests (Ayres 1981; van Roosmalen, Mittermier and Milton 1981).

Chiropotes albinasus (I. Geoffroyi and Deville, 1848)

Description
General features are described under the generic accounts. Three specimens measured HB 390–400; T 405–10; HF 117–23; E 24–30. One specimen weighed 2.3 kg (Belém). This species is characterized by a dorsum dominated by medium-length jet black hair. The face is dominated by a white blaze, generally extending from the midline of the face and covering both nostrils. The venter is darkly pigmented and sparsely haired.

Distribution
The species is apparently confined to north-central Amazonian Brazil (see map 9.35).

Chiropotes satanas (Hoffmannsegg, 1807)
Bearded Saki, Cuxiú

Measurements

	Mean	Min.	Max.	N	Loc.	Source
TL				10	Brazil	Belém
HB	423.2	370.0	460.0			
T	420.0	370.0	450.0			
HF	124.8	117.0	148.0			
E	26.5	18.0	30.0			
Wt	2,502.0	2,350.0	3,250.0			

Description
This medium-sized primate weighs from 2.4 to 3.2 kg. Head and body length ranges from 327 to 480 mm, with a tail length of 370 to 436. Males may slightly exceed females in size, but there is no marked color dimorphism. Although the shoulders may be reddish, the coat is basically black, and there are no obvious white facial markings. The tail is heavily furred and nonprehensile. The hair on the male's head is organized into two rather rounded, dense masses overlying the enlarged temporal muscles, and there is a conspicuous beard. The female has a small beard and much less development of the temporal swellings on the head (plate 9).

Range and Habitat
The bearded saki is found from the Guianas and southern Venezuela south on both sides of the Amazon to the mouth of the river. It generally frequents only mature rain forests (map 9.36).

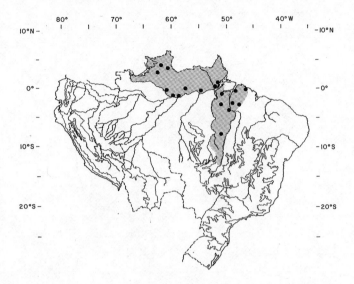

Map 9.36. Distribution of *Chiropotes satanas*.

Natural History

Only three extended field studies have been carried out (van Roosmalen, Mittermeier, and Milton 1981; Ayres 1981; Norconk 1996). *Chiropotes* tends to forage in the upper strata of mature rain forests. Typically this species is found in large bands of up to twenty-five. Several males are present in a band at any one time, and there is no tendency toward monogamy. In Suriname the diet consists primarily of fruits and seeds; van Roosmalen noted that they eat fruits from a wide variety of plant families; thus they are true seed predators (van Roosmalen, Mittermeier, and Fleagle 1988). The feeding trees are widely dispersed, and a troop's daily movements may exceed 3 km. One troop in Suriname had a home range nearly 100 ha in extent. A single young is born after a gestation period of approximately five months (Hick 1973). Detailed analyses of communication patterns and vocalizations have yet to be performed for this species.

Genus *Cacajao* Lesson, 1840
Uacari

Description

These monkeys are typical cebids with respect to skull morphology; however, the adult males are larger than the females and have very large upper canines. The nonprehensile tail is relatively short, and this feature distinguishes the genus from all other Neotropical primates (see species accounts).

Distribution

The genus is confined to southwestern Venezuela, southeastern Colombia, and northwestern Brazil.

Natural History

Ayres (1986) has provided the most recent field study of this intriguing genus. Like most cebids, these primates are strongly arboreal and diurnal. Groups of twenty to thirty are common, and in *C. melanocephalus* troops approximating one hundred individuals have been reported. They feed on fruits but also eat some young leaves and invertebrates. The limited field data suggest that *Cacajao* feeds on both leaves and fruits and also takes some invertebrates in the course of foraging. Some data concerning the foraging behavior of *Cacajo calvus* have been obtained from captive studies in a seminatural environment (Fontaine 1981). The communication patterns of *C. calvus* have been analyzed for captives in a seminatural environment by Fontaine (1981), who noticed antiphonal vocalization bouts between members of the same group. A single young is born at a time, and the gestation period is imperfectly known but is probably five and one-half months. The rather large canine teeth permit them to break open tough-hulled fruits or nuts. They are usually found in gallery forests associated with permanent streams or rivers, although such areas are seasonally flooded. Aquino (1995) summarizes recent conservation efforts.

Cacajao calvus (I. Geoffroyi, 1847)

Description

Two females from Rio Javari, Brazil, exhibited the following measurements: TL 430–40; T 170–80; HF 130–40; E 28–30 (Belém). This distinctive, short-tailed primate is covered with a reddish dorsal pelage. The hairs are very long over the shoulders creating a capelike effect. The neck may have a pale orange to yellow cast. The face and forehead are naked, and in a living specimen they have a reddish cast. The venter is sparsely haired. This is the description for *C. c. rubicundus*. Some subspecies to be listed have gray fur, and various degrees of pigment "dilution" can be noted over the range (see plate 9).

Distribution

The species occurs in the northwest of Brazilian Amazonas and adjacent portions of Peru (map 9.37).

Natural History

Cacajao calvus calvus inhabits swamp forests and often lives in small bands of five to six individuals. Leaves and fruit are included in the diet (Ayres 1986). In upland forests in Peru *Cacajao c. rubicundus* can be found in troops of twenty or more. The troops can contain several males of adult stature, but the details of their social life are poorly known.

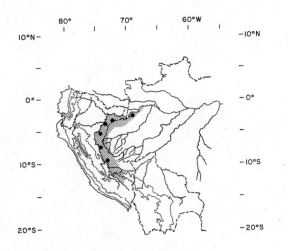

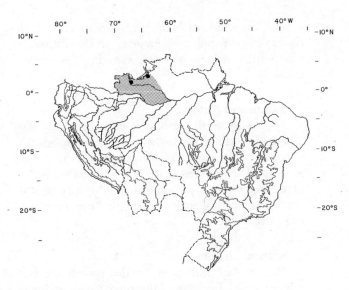

Map 9.37. Distribution of *Cacajao calvus*.

Cacajao melanocephalus (Humboldt, 1812)
Black Uakari, Uakari Negro

Description
This genus is composed of two species, but only the black uakari, *C. melanocephalus*, occurs in the northern Neotropics. These are medium-sized primates, with adult males weighing 3.5 to 4.1 kg. Females are slightly smaller, weighing about 500 g less than males. The uakari cannot be mistaken for any other New World primate because it has an extremely short, nonprehensile tail, rarely exceeding one-third of the head and body length. Whereas *C. calvus* has a red to pale yellow coat, the coat of *C. melanocephalus* tends to be black; however, there may be varying degrees of light brown on the hind limbs, tail, and lower back.

Range and Habitat
This species is confined to the Amazon basin and penetrates into the Amazonian portions of southern Venezuela and Colombia (map 9.38). In common with *Callicebus torquatus*, it appears that the genus *Cacajao* is adapted to black-water river systems.

Natural History
C. melanocephalus lives in rather large groups of fifteen to twenty-five animals; more than one adult-sized male may be present in a troop. The animals are strictly diurnal. Boubli (1994) has found groups exceeding one hundred in Pico Neblina, Brazil. He has noted that this population is found in upland forest at over 500 m elevation.

Genus *Pithecia* Desmarest, 1804
Saki Monkey, Parauacu

Description
Head and body length averages 300 to 480 mm, with a tail length of 255 to 545 mm. Weight ranges

Map 9.38. Distribution of *Cacajao melanocephalus*.

from 1.4 to 2.25 kg. These medium-sized primates have rather long, thick hair and very well haired, nonprehensile tails. Size dimorphism is negligible, but in some species the sexes are strongly dimorphic in color pattern.

According to the revision by Hershkovitz (1979) there are four species. Groves (1993) lists five species. We will deal with these taxa as two species groups: the *monachus* group and the *pithecia* group, following Mittermeier, Rylands, and Coimbro-Filho (1988).

Distribution
The genus has a rather wide distribution in the Amazon basin.

Natural History
Saki monkeys are difficult to maintain in captivity and have received little attention from fieldworkers. The most recent summary of their natural history is contained in Buchanan, Mittermeier, and van Roosmalen (1981). Some species (*P. pithecia*) are found in small groups, suggesting monogamy. Other species (*P. monachus*) may form groups of ten or twelve. Further remarks on natural history are included in the species accounts.

The *"Pithecia"* Group
Pithecia pithecia (Linnaeus, 1766)
White-faced Saki

Description
In this species the sexes are strongly color dimorphic. The adult male is black except for a white face and throat. The female is basically an agouti brown on the dorsum with lighter underparts. White stripes extend from below the eye to the corner of the

mouth. The female of *P. pithecia* is very similar to both the male and female of *P. hirsuta,* but the distinctive color of the male immediately identifies this species (plate 9).

Range and Habitat

The white-faced saki has its main center of distri-bution north of the Amazon and west through the Guianas to Venezuela. It occurs sporadically in the state of Bolívar, Venezuela, but is broadly distributed through the Amazon portion of Venezuela. It is confined to mature multistratal tropical evergreen forests (map 9.39).

Natural History

The standard references for this species are Buchanan (1978), Buchanan, Mittermeier, and van Roosmalen (1981), and Norconk (1996). The animal is found from lowland forests to forests of moderate elevation. It rarely if ever enters seasonally flooded riverine forests. In Suriname it tends to forage in the understory or lower part of the canopy. It is primarily frugivorous but eats some leaves on occasion. Although this species will take insects in captivity, there are no precise data on insect consumption in the field. The saki tends to live in small family groups, apparently based on a monogamous pair. In adequate habitat population densities may reach seven per km². Day ranging has not been studied to any great extent; Buchanan estimates home ranges at 4–10 ha. A single young is born after a gestation period of 163 to 176 days. Vocalizations have received a preliminary analysis by Buchanan (1978). Long, loud calls to promote group spacing are modified in form but are similar to those of the howler monkey though lower in amplitude.

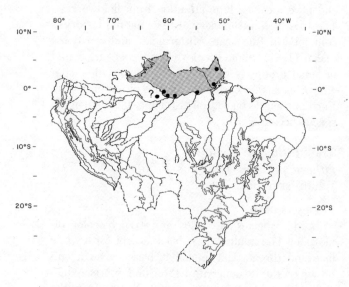

Map 9.39. Distribution of *Pithecia pithecia.*

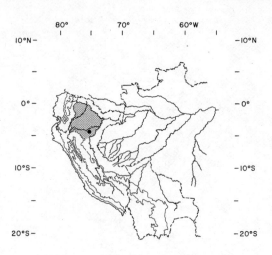

Map 9.40. Distribution of *Pithecia aequatorialis.*

The "*Monachus*" Group
Pithecia aequatorialis Hershkovitz, 1987

Distribution

The species is found from Napo province, Ecuador, to Terate, Peru (map 9.40).

Pithecia albicans Gray, 1860

Description

In proportion and hair pattern this species is typical for the genus. The dorsal pelage is black from between the shoulders across the back extending to the sides until the pelvis. The black pattern extends to the tail for its entire length. The top of the head, the arms, and the sides from the rib cage down are a whitish brown. The arms and the sides from the rib cage forward are covered with a whitish, grizzled fur. The venter is sparsely haired and orange (Belém).

Distribution and Natural History

The species ranges on the south bank of the Amazon River between the Rio Juruá and the Rio Purus (map 9.41). Peres (1993) found that it feeds on seeds to a great extent.

Pithecia hirsuta Spix, 1823

Description

Hershkovitz (1979) separated this species from *P. monachus,* but Groves (1993) subsumes it under *P. monachus. P. hirsuta* is dark agouti brown with pale hands and feet, and its underparts and beard are blackish. The sexes are not strongly dimorphic in color or in size.

Range and Habitat

This species occurs in the western Amazon basin. It typically prefers multistratal tropical evergreen forests (not mapped).

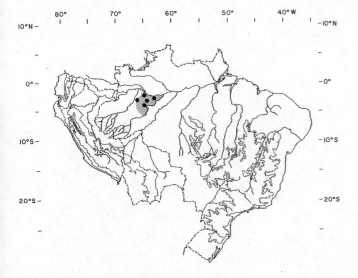

Map 9.41. Distribution of *Pithecia albicans*.

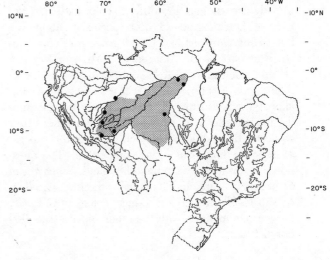

Map 9.42. Distribution of *Pithecia irrorata*.

Natural History

The natural history of this species is not as well known as that of *P. pithecia*. The field observations we have suggest that it lives in small family groups, feeding in the middle regions of the multistratal forest. It appears to be frugivorous, diurnal, and quite similar to *P. pithecia* in its general behavior patterns. No evidence of territorial defense has been noted (Happel 1982).

Pithecia irrorata Gray, 1842

Description

The species is similar in appearance to *P. monachus*.

Distribution

According to Groves (1993), this form occurs south of the Amazon in southwestern Brazil extending to adjacent Peru and Bolivia (map 9.42).

Pithecia monachus (E. Geoffroy, 1812)

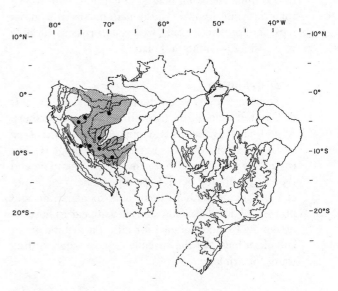

Map 9.43. Distribution of *Pithecia monachus*.

Description

The long dorsal hairs exhibit a very uniform pattern of coloration; they are tipped with black and cream, yielding a grizzled appearance. The tail is heavily furred and does not contrast with the dorsum. The facial hairs are cream colored, as are the short hairs on the hands. The venter is characterized by pale, short hairs. The hair on the head is short and often has a yellow cast, thus contrasting with the dorsum.

Distribution

The species is found in the western Amazon region of Brazil, Colombia, Ecuador, and Peru (map 9.43).

Comment on Taxonomy

Groves (1993) recognizes the species status of *Pithecia aequatorialis* Hershkovitz (1987) as occurring in Loreto department, Peru. This form was included in *P. monachus* by Hershkovitz in 1979. Groves synonymizes *P. vanzolinii* with *P. irrorata* Gray, 1842, which occurs south of the Amazon in southwestern Brazil and adjacent Peru and Bolivia (See maps 9.42 and 9.43). In the account, Groves comments that these forms may be subspecies of *P. monachus*.

Genus *Cebus* Erxleben, 1777
Capuchin, Sapajou, Capuchino

Taxonomy

Cebus apella is distinct and co-occurs with at least one of the remaining three species. The species *Cebus olivaceus*, *C. albifrons*, and *C. capucinus* are all similar in body size, though they differ in color pattern. Over most of their range these three species are allopatric, but *Cebus albifrons* and *C. olivaceus* occur in sympatry in part of the Amazonian portion of Venezuela and again in Venezuela south of the eastern cordillera (see Eisenberg 1989). This condition of near allopatry suggests that the three species are very similar in their mode of habitat exploitation; where they co-occur, they could possibly either be in the process of interbreeding or in strong competition. The question remains unclear. On the other hand, *Cebus apella* co-occurs over much of its range with either *C. olivaceus* or *C. albifrons*. The feeding ecology of *C. apella* reduces competition with the other species (see Terborgh 1983).

Description

Head and body length is 305 to 565 mm, the tail 300 to 560 mm. The adult male is larger and more robust than the adult female. These medium-sized primates have semiprehensile tails. Color patterns are variable (see species accounts and plate 9).

Distribution

There are four recognized species. *Cebus apella* is distributed from northern South America to northern Argentina. This species is sympatric with *C. olivaceus* and *C. albifrons* over many parts of its range. *Cebus capucinus* is confined to northwestern Colombia and then is distributed north to southern Nicaragua. Species of this genus have been able to penetrate and utilize dry deciduous forests. They appear to have an extremely broad tolerance for different habitat types and elevations, but they usually require access to free water for drinking.

Natural History

Capuchin monkeys typically live in moderate-sized bands, usually with a single adult male dominant over all younger males in the troop. They eat a variety of fruits and seeds and a modest amount of both vertebrates and invertebrates. In habitats supporting dry deciduous forests the troop can forage extensively on the ground, turning the leaf litter to find small vertebrates and invertebrates. Species of the genus are most versatile in their use of habitat and range of diet. In the Amazon, capuchin monkeys are often found with the squirrel monkey (*Saimiri*). It appears that *Cebus* is the focal species, with the squirrel monkey moving along in association. It is speculated that this association confers some antipredator benefit on the squirrel monkey, but is of no special benefit to *Cebus*. Freese and Oppenheimer (1981) and Robinson and Janson (1987) summarize the biology of this taxon.

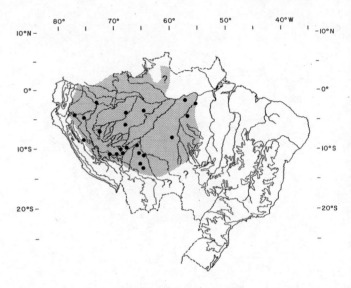

Map 9.44. Distribution of *Cebus albifrons*.

Cebus albifrons (Humboldt, 1812)
Brown Pale-fronted Capuchin

Description

The weight averages 2.3 kg. The color pattern is brown dorsum with white circling the facial region and white on the upper forearms; the cap tends to be darker brown.

Range and Habitat

The species occurs from southern Colombia and southern Venezuela south through the Amazon basin to northern Bolivia (map 9.44). It occupies a wide range of forest types, including dry deciduous tropical forests and multistratal evergreen forests.

Natural History

Like other species of this genus, *Cebus albifrons* is omnivorous, taking fruit, seeds, vertebrates, and invertebrates. Group sizes range from seven to thirty, with fifteen as a mode. Generally there is only one adult male per troop, but instances of an adult male with a satellite subadult have been recorded. Densities in good habitat range from twenty-four to forty-five per km^2. The animals are diurnally active and generally have a rather large home range: 60–70 ha has been recorded. The foraging ecology has been analyzed by Terborgh (1983). In breeding biology, development, and vocalizations this species is similar to *C. capucinus*.

Cebus apella (Linnaeus, 1758)
Brown Capuchin, Macaco-Prego

Description

This robust species averages about 2.5 kg in weight; males tend to be some 800 g heavier than females.

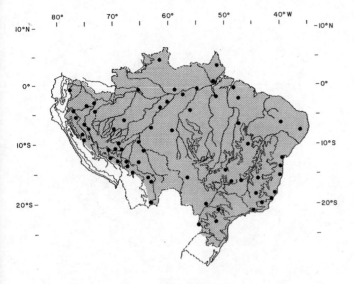

Map 9.45. Distribution of *Cebus apella*.

Map 9.46. Distribution of *Cebus capucinus*.

The cap on top of the head is composed of short, dark, erect hairs that in males form ridges on either side of the crown, and it contrasts sharply with the light brown body (plate 9).

Range and Habitat

Cebus apella is broadly distributed from southern Colombia, Venezuela, and the Guianas to northern Argentina. It occurs on Margarita Island, but this disjunction from the continental range suggests it was originally introduced by Amerindians (map 9.45). It is broadly tolerant of a variety of forest types, from semideciduous to multistratal tropical rain forests.

Natural History

The diet includes fruit, palm nuts, seeds, and a considerable quantity of insects (Terborgh 1983). Group size ranges from five to forty, and population density varies from six to thirty-five per km^2. This species is strictly diurnal and, depending on carrying capacity, can show a home range of 25–40 ha. A single young is born after a gestation period of approximately 160 days. Young males apparently do not become sexually mature until at least seven years of age. Females may conceive in their fourth year. The mating system is polygamous. Generally a single adult male is dominant over all other males in the troop. Communication patterns in this species have been described for captives. The standard review for the behavior of the genus is in Freese and Oppenheimer (1981).

Cebus capucinus (Linnaeus, 1758)
White-throated Capuchin, Carita Blanca

Description

This capuchin is similar in size to *C. albifrons*, but its color is distinctive. The dorsum and hindquarters are black, the chest and forearms are a sharply contrasting white, and the pale face is surrounded by white with a black cap. This species is easily distinguished from all others.

Range and Habitat

The species occurs from northwestern Ecuador north to Honduras. It may be found in multistratal tropical evergreen forests to dry deciduous forests (map 9.46).

Natural History

The standard reference for the behavior and ecology of this species is Oppenheimer (1968). This diurnal primate has a broad diet, eating an enormous variety of fruits, flowers, and invertebrates. It forages over the entire vertical range of the forest, including the ground, and exhibits a versatility unmatched by any other New World primate. Group sizes average about fifteen. Usually only one adult male is present in a troop. They can exist at population densities of five to fifty per km^2. Home ranges vary from 32 to 80 ha. Reproduction is similar to that described for *C. apella*. Communication patterns have been summarized by Freese and Oppenheimer (1981).

Cebus kaapori Queiroz, 1992

Description

This form is clearly allied to *Cebus olivaceus* (= *nigrivittatus*) complex; Queiroz (1992) notes the disjunct distribution and suggests full species status. In his description only one specimen is known, and the measurements are in line with those developed for *C. olivaceus*.

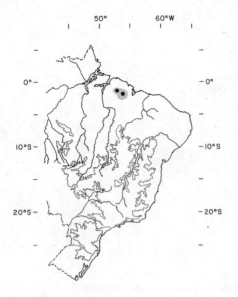

Map 9.47. Distribution of *Cebus kaapori*.

Map 9.48. Distribution of *Cebus olivaceus*.

Distribution

The type is described from the municipality of Ca-rutupera, Maranhão, Brazil (map 9.47).

Cebus olivaceus (= *nigrivittatus*)
Schomburgk, 1848
Weeper Capuchin, Cai Capuchino, Mico Común
(Venezuela)

Description

This species is similar to *C. albifrons* in size and color: light brown on the dorsum, pale buff on the forearms and the hair around the face; but where *C. albifrons* is white, *C. olivaceus* tends to be buff.

Range and Habitat

C. olivaceus is distributed in northern South America from northwestern Venezuela across through the Guianas south to the Amazon basin (map 9.48).

Natural History

Diurnally active, this species is an omnivore that feeds both on the forest floor and at various levels of the forest. Troops generally include a single adult male that is responsible for most of the breeding. Group foraging has been analyzed in detail by Robinson (1981, 1986). While foraging, the dominant male and dominant adult females tend to be at the center of the troop. Juvenile males tend to be at the front or toward the rear. A strict hierarchy among females and among males determines priority of access to preferred food items. Troop size ranges from ten to thirty-five. In the dry deciduous forests of northern Venezuela the home range may exceed 100 ha. In such a habitat home ranges of neighboring troops overlap considerably, and

no territorial behavior is demonstrable. In reproduction, development, and age at sexual maturation this species is similar to *Cebus apella*.

Genus *Saimiri* Voight, 1831
Squirrel Monkey, Titi (Venezuela),
Mico de Chiero

Description

Head and body length of adult females averages about 280 mm, while that of males averages 300 mm; the tail is 350 to 430 mm. Adult females weight from 500 to 750 g, while males range from 700 to 1,100 g. These small primates have long, nonprehensile tails. Color patterns are somewhat variable, but in general most of the body is greenish yellow. The throat, face, and ears are usually white, and the muzzle is black, contrasting sharply. The venter and undersides of the limbs may be white or light yellow. Usually the tip of the tail is black (plate 9).

Distribution

S. oerstedii is found in Costa Rica and Panama. The other species are found in the Amazonian portions of Brazil, Colombia, Ecuador, Peru, Bolivia, and Venezuela and in the Guianas.

Comment on Taxonomy

Hershkovitz (1984) recognizes four species divisible into two species groups. Thorington (1985) recognizes only two species. Groves (1993) recognizes five. Mittermeier, Rylands, and Coimbra-Filho (1988) recognize four species as outlined in table 9.5. Early on it was recognized that the white markings over the eyes assumed two major forms: a roman arch ô ô and

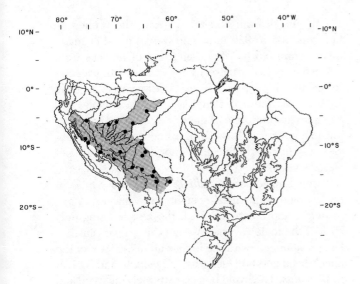

Map 9.49. Distribution of *Saimiri boliviensis*.

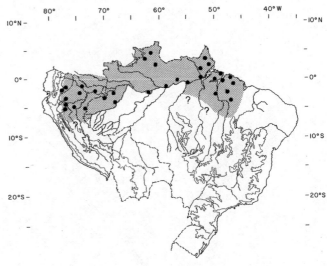

Map 9.50. Distribution of *Saimiri sciureus*.

a gothic arch ô ô. These characters were used as a preliminary dichotomy. As will be noted, the rough morphological characters (gothic type and roman type) still hold, with *S. boliviensis* included in the latter and *S. sciureus, S. oerstedii,* and *S. ustus* included in the former category.

Saimiri boliviensis (I. Geoffroy and de Blainville, 1834)

Description
This taxon is the classic "roman" type, with circular white eye rings and rather hairy ears that are white and contrast with the dark head (see plate 9).

Distribution
The species is broadly distributed in the western Amazon basin in southwestern Brazil, adjacent to Peru and Bolivia (map 9.49).

Comment
Saimiri vanzolinii may be synonymized with this species (Hershkovitz 1987).

Saimiri sciureus (Linnaeus, 1758)
Squirrel Monkey

Description
A classic "gothic" face with white eye markings in the shape of a gothic arch (see genus account).

Range and Habitat
This species occurs widely throughout the Amazonian portions of South America (map 9.50). It is absent from the coastal forests of eastern Brazil. Broadly tolerant of a variety of habitat types, it can occur in multistratal evergreen tropical forests as well as mangroves and secondary forests.

Natural History
Saimiri sciureus primarily feeds on insects and fruit, though it occasionally takes some leafy material. Its foraging ecology has been analyzed by Terborgh (1983). Group size depends on the carrying capacity of the habitat. Groups as small as seven or eight have been noted in relict forest patches, and over one hundred were counted in one troop in the Amazon basin. Several adult males may occur in a troop; the ratio of adult females to adult males averages about four to one. Population density estimates in good habitats range from nineteen to thirty-one animals per km^2. They are diurnally active, and home-range size depends on the carrying capacity of the habitat. Home ranges of adjacent troops appear to overlap extensively. A single young is born after a gestation period of approximately 165 days. Behavior patterns and vocalizations have been extensively studied in the laboratory; Baldwin and Baldwin (1972) provide a useful overview of the material. Rosenblum and Cooper (1968) summarize the anatomical, ecological, and captive maintenance data.

Ploog, Hopf, and Winter (1967) described the vocalization system. The basic vocalization repertoire can develop without imitative learning. Vocalizations can be classified into contact-seeking, fearful, aggressive, and warning calls. Within a single category there are many variants that express different intensities of motivation. A similar functional classification for New World primate vocalizations was developed by Eisenberg (1976).

Saimiri ustus (I. Geoffroy, 1843)

Description
This monkey is distinctive for its nearly naked ears; otherwise it is similar to *S. sciureus*.

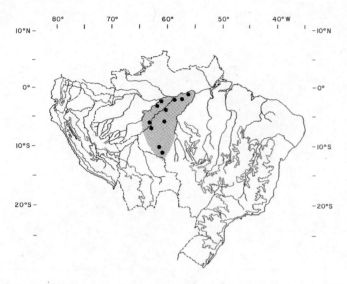

Map 9.51. Distribution of *Saimiri ustus*.

Distribution

The species is found between the Purus and Tapajós Rivers to the south of the Amazon River in Brazil (map 9.51).

Genus *Alouatta* Lacépède, 1799
Howler Monkey, Aullador, Araguato, Guariba

Description

Eight species are recognized within this genus by Groves (1993), and six occur in the region covered by this book. Members of this genus are among the largest of the New World primates. Head and body

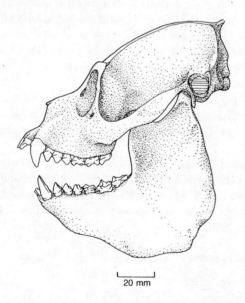

20 mm

Figure 9.5. Skull of *Alouatta* sp.

length ranges from 559 to 915 mm, the prehensile tail from 585 to 920 mm. The weight of adult males ranges from 6.5 to 7.8 kg, and that of females from 4.5 to 6.6 kg. Color patterns are variable from species to species. *A. palliata* is black, *A. seniculus* is reddish. In *A. caraya* the males are black and the females and juveniles are buff, one of the few cases of color dimorphism among the New World primates. Some subspecies of *A. fusca* exhibit color dimorphism also (see species accounts).

Adult females of *A. palliata* weigh about 84% of the average adult male's weight; the value for *A. seniculus* is more extreme, with the female being only 69% of the male's average weight. Size dimorphism is less pronounced, with the female attaining 94% of the male's head and body length in *A. palliata* and 90% in *A. seniculus*. The hyoid bone is extremely enlarged and ossified and no doubt influences the low fundamental frequency of the long call (Thorington, Rudran, and Mack 1979; Sekulic 1982).

Distribution

The genus is the most widely distributed among the New World primates, with a range extending from Vera Cruz, Mexico, to northern Argentina. The species tolerate a range of habitat types varying from semideciduous tropical forests to multistratal tropical evergreen forests (Eisenberg 1979).

Natural History

The howler monkeys are aptly named because they characteristically produce loud roaring bouts at daybreak. These roars have been implicated in intratroop spacing behavior, but they may also be directed at potential conspecific invaders of an occupied home range. Males roar toward alien males, and females roar when they encounter alien females. Both sexes can join in combined choruses. For a full account of the function of roaring in *A. seniculus,* see Sekulic (1982) and Whitehead (1989, 1995). Howler monkeys are organized into troops of both sexes, and the number of adult males in a troop varies from species to species. *A. palliata* seems to exhibit the greatest male-male tolerance and the highest incidence of multiple-male troops; *A. seniculus* tends to exhibit a one-male troop structure in many parts of its range. In favorable habitat, howler troops may have rather small home ranges, in part because they are the most folivorous New World primate and are not totally dependent on ripe fruit for nourishment. Up to 50% of the annual diet may be composed of young leaves. The foraging ecology of *A. palliata* has been amply documented by Milton (1980). A single young is born after a gestation period of 185 days (Crockett and Eisenberg 1987).

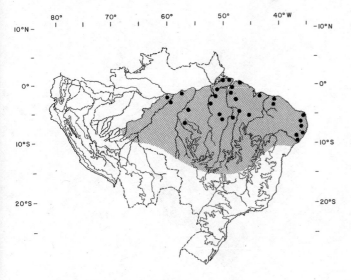

Map 9.52. Distribution of *Alouatta belzebul*.

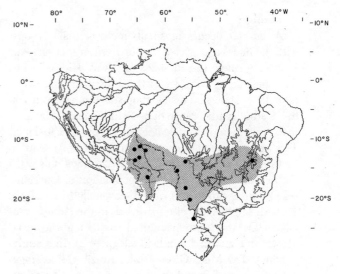

Map 9.53. Distribution of *Alouatta caraya*.

Alouatta belzebul (Linnaeus, 1766)

Description

This species' basic body color is very dark brown to black. The reddish hands and feet contrast sharply with the body color (plate 10). The terminal portion of the tail may often be reddish (Ayres and Milton 1981).

Distribution

The species is found south of the Amazon in north-central to northeastern Brazil and formerly to the Atlantic coast (Coimbra-Filho and Rylands 1995) (map 9.52).

Alouatta caraya (Humboldt, 1812)
Black Howlers, Mono Aullador Negro, Carayá

Measurements

	Mean	Min.	Max.	N	Loc.
TL	1,711.5	1,075	1,260	12 m	A, Br, P
	1,043.3	991	1,100	8 f	
HB	549.8	500	610	12 m	
	483.8	460	510	8 f	
T	591.2	540	650	12 m	
	559.5	526	600	8 f	
HF	142.5	140	145	2 m	
	140.0			1 f	
E	41.0			1 m	
Wt	6,875	4,600	9,800	14 m	
	4,845	3,100	7,800	10 f	

Redford and Eisenberg 1992.

Description

Like other members of the genus *Alouatta*, *A. caraya* demonstrates extreme sexual dimorphism in size, but color also varies with sex. Males are all black and weigh about 6.5 kg, while females are yellowish brown and weigh about 4.5 kg; infants are a reddish buff. In captivity young males change pelage color at between thirty-two and forty months; in the wild they change at about fifty-four months. In Argentina females never exceed 6.5 kg while some males weigh 9.5 kg (Jones 1983; Pope 1966; Thorington, Ruiz, and Eisenberg 1984).

Distribution

A. caraya is confined to southern Brazil, Paraguay, Bolivia, and northern Argentina. In eastern Paraguay this howler monkey occurs commonly in forested habitats, whereas in the Paraguayan Chaco it generally occupies the gallery forests along the Río Paraguay and Río Pilcomayo as well as the small rivers that drain the Chaco. *A. caraya* can also be found sporadically throughout the more xeric areas of the contiguous forested Chaco. In Bolivia it occupies the southern portion extending to south-central Brazil (Cabrera 1939; Olrog 1984; Stallings 1984; Stallings and Mittermeier 1983; UM) (map 9.53).

Life History

Females give birth to a single young, rarely twins, after a gestation period of about 187 days. The birthweight of a single young born in captivity was 113 g. In northeastern Argentina and southwestern Brazil young are born between August and October, and in north-central Argentina the birth peak takes place at the onset of the dry season. Females stop growing at sexual maturity, though males apparently continue growing for several years. Juvenile males maintain female coloration until about 4.5 years of age and about 5 kg weight. In the wild animals may live at least twenty years (Crespo 1982; Miller 1930; Pope 1966; Shoemaker 1979; Thorington, Ruiz, and Eisenberg 1984).

Ecology

A. caraya occurs in small, multiple-male troops with the males age graded. In northern Argentina near Puerto Bermejo the adult sex ratio was 0.56, with infants making up 6%-14% of the individuals and juveniles 16%-21%. The mean troop size ranged from 7.2 to 8.9 adults with a range of 3–19 individuals. The animals were found at an ecological density of 130/km². This agrees with another study in northern Argentina that found nine troops to average 7.2 individuals with a range of 3–15. In an island population in Corrientes province, group size varied seasonally from 2.5 to 10, with single males also recorded. In the Paraguayan Chaco the largest group counted was 10, and the average troop contained 7 individuals ($n = 2$). In a study conducted in Mato Grosso state, Brazil, the average group size was 7.2, and an average group had 1.8 adult males, 0.6 subadult males, 2.4 females, 0.8 yearlings, and 1.6 young.

Within the past seventy-five years yellow fever swept through the range of this species, and it is not known how it may have affected its distribution and abundance (Crespo 1982; Piantanida et al. 1984; Rathbun and Gache 1980; Schaller 1983; Stallings 1984; Thorington, Ruiz, and Eisenberg 1984; Rumiz 1990).

Alouatta fusca (E. Geoffroy, 1812)
Mono Aullador Rufo

Measurements

	Mean	Min.	Max.	N
TL	1,131	1,035	1,220	5 m
	1,000	970	1,022	4 f
HB	559	535	585	5 m
	468	450	490	4 f
T	572	485	670	
	532	515	570	
HF	130			1 m
E	25			1 m

Redford and Eisenberg (Belém).

Description

In southern South America the subspecies *A. fusca clamitans* exhibits color dimorphism; adult males are bright reddish with a golden tint, and most adult females are brown (Cordeiro da Silva 1981; Kinzey 1982).

Distribution

A. fusca is found only in southeastern and south-central Brazil and in the Argentine province of Misiones. Another population of *A. fusca* has been reported from Bolivia, but it is not clear whether this is based on accurate identification of specimens (Crespo 1954) (map 9.54).

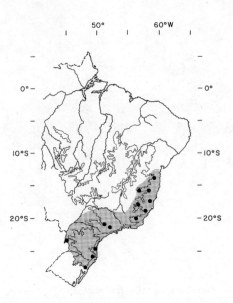

Map 9.54. Distribution of *Alouatta fusca*.

Ecology

This howler prefers moister forests than *A. caraya* and is found in the Brazilian Atlantic forest, the southern araucaria forests of Brazil, and into the tropical forest of Misiones province, Argentina. In the Brazilian state of São Paulo group size ranged from 2 to 22, with an average of 5.76 ($n = 25$), and the mean population density was 80.9/km² (± 32.5). The average group consisted of 1.76 adult males and 2.3 adult females. In this araucaria forest 70.6% of the feeding observations were on leaves and 29.4% on fruit. In a forest in Rio Grande do Sul *A. fusca* was found eating figs (*Ficus*), leaves, *Inga* fruit, and *Cecropia* fruit. In the southern parts of the species range *A. fusca clamitans* exploits forests containing stands of *Araucaria angustifolia*. The monkeys may use this southern conifer seasonally by feeding on seeds in addition to plant parts of other associated trees such as *Mimosa scabrella* and *Pithecoctenium echinatum* (Biedzicki and Ades 1996). Like *A. caraya*, this species was adversely affected by yellow fever (Chitolina and Sander 1981; Cordeiro da Silva 1981; Crespo 1974).

Comment

In older literature this species is often listed as *A. guariba*.

Alouatta palliata (Gray, 1849)
Mantled Howler Monkey

Description

This species is the largest by weight within the genus; in Panama average weight for males was 7.8 kg. Mean head and body length for adult males on Barro Colorado Island is 506 mm; the comparable value for

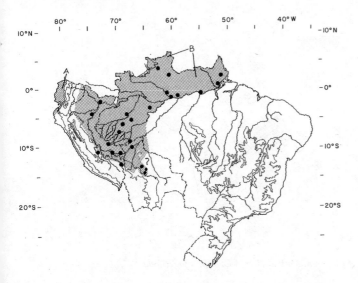

Map 9.55. Distribution of two species of *Alouatta:* (a) *A. palliata;* (b) *A. seniculus.*

females is 475 mm. The tail exceeds the head and body length, and males average 665 mm while females average 651 mm. As can be seen from these figures, the female's tail is proportionately longer than the male's. The basic color is black. The scrotal skin is white in the male, and the vulvar area is white in the female. A fringe of chestnut hairs runs laterally from the arm to the groin, hence the common name mantled howler. The development of the mantle varies widely throughout the range of the species, but some trace can usually be discerned even in those populations where it is reduced.

Range and Habitat

Alouatta palliata occurs from southern Veracruz, Mexico, through Panama and in Colombia west of the western cordillera of the Andes to western Ecuador. It is found in dry deciduous tropical forests as well as multistratal tropical evergreen forests (map 9.55).

Natural History

All species of howler monkeys eat a great deal of leafy material. Because they are not completely dependent on fruit, their daily range and home range tend to be small for a primate of this size class. Howlers rarely descend to the ground, but they will do so in areas where they must cross open country to move from one isolated grove of trees to the next. They are strictly diurnal and characteristically produce loud, roaring calls at dawn. Depending on the carrying capacity of the habitat and the stability of the troop, troop sizes can range from three to forty-four. Fourteen is average. In some habitats *Alouatta palliata* can reach high densities; seventy animals per km^2 is not

uncommon. In Costa Rica, where it co-occurs with *Ateles* and *Cebus,* it can represent 69% of the total primate biomass; in Colombia, where it co-occurs with three other primate species, it can represent 44% (Eisenberg 1979). *Alouatta palliata* has been the subject of long-term studies in both Costa Rica and Panama; see Carpenter (1934), Froehlich, Thorington, and Otis (1981), Milton (l978, 1980), and Glander (1975, 1981).

Aloutta sara Elliot, 1910

Comment

Most workers considered this form a subspecies of *Aloutta seniculus,* but in accordance with Minezawa et al. (1985), Groves (1993) raised it to a full species. Confined to the Sara area of Santa Cruz province, Bolivia.

Alouatta seniculus (Linnaeus, 1766)
Red Howler Monkey, Guariba, Mono Colorado

Description

There is great variation in size over the range of this species. A population studied in Venezuela exhibited a mean head and body length of 523 mm for males and 468 mm for females (Thorington, Rudran, and Mack 1979). Average tail lengths for the two sexes were 630 mm and 590 mm. The basic body color is a chestnut brown in some parts of its range.

Range and Habitat

The species is broadly distributed east of the western cordillera of the Andes in Colombia, south to Bolivia, and east to the Guianas (map 9.55). It is found in a range of habitats from dry deciduous forest to multistratal tropical rain forest.

Natural History

This species has been intensively studied in Venezuela by Braza, Alvarez, and Azcarate (1983), Gaulin and Gaulin (1982), Neville (1972), Rudran (1979), Sekulic (1981), Crockett (1984), Crockett and Pope (1993), and Pope (1992). In the llanos of Venezuela the animals depend on leaves for approximately 40% of their annual diet. Fruit makes up over half the diet, and this species is an important disperser of seeds (Juliot 1994). The mean troop size there is 8.9. There is a pronounced tendency for only one adult male to be present in a given troop, but some troops have two adult males, and the average for the Venezuela population was 1.65 males per troop. In an established troop with more than one adult male, usually one male sires most offspring (Pope 1990, 1992). Variations in troop size and composition as a function of population size have been analyzed using a long-term data set derived form studies in Venezuela by Crockett (1996). Soini (1992) studied this species in Peru. Heterosexual groups ranged from 2 to 11 individuals with a mean of 5.5. Adult sex ratios averaged 1.3 males to 1.8 females. Population densities of 36 animals per km^2 were noted in preferred habitats. Howlers were observed to swim across streams. Home ranges for two groups were 6 ha and 9 ha. Leaves and leaf buds constituted 53% of the annual diet; fruits (50%) and flowers (6%) made up the rest.

At high population densities in Venezuela, alien males will actively attempt to enter a troop and depose the dominant male. During such times of social unrest infant mortality increases. If the alien male can establish himself within the troop, he usually kills infants less than six months old (Rudran 1979; Crockett and Sekulic 1984). Although the size of the home range varies depending on the quality of the habitat, there is usually a correlation between troop size and home-range size. In Venezuela the average home range is approximately 6 ha. In areas of good habitat in the llanos of Venezuela, howler populations can reach a density of 150 animals per km^2. The biology of this species is reviewed in Crockett and Eisenberg (1987).

Genus *Lagothrix* E. Geoffroy, 1812
Woolly Monkey, Macaco Barrigudo

Description
The genus *Lagothrix* contains two species. Head and body lengths range from 390 to 580 mm, while body weights range from 5 to 10 kg. The tail is longer than the head and body. Although males exceed females in total length, females have relatively longer tails (Ramirez 1988). The tail is fully prehensile, and the coat is very dense. The dorsal coat varies from gray brown to dark brown.

Distribution
The species is found in Amazonian portions of Colombia, Ecuador, Peru, and Brazil.

Lagothrix flavicauda (Humboldt, 1812)
Yellow-tailed Woolly Monkey

Description
Measurements are similar to those noted for *L. lagothricha* (see below). The yellow-tailed woolly monkey has a dark brown dorsum, with a white to buff mouth. The posterior lateral ventral surface of the tail is yellow.

Distribution
The species is found in montane rain forest on the eastern slope of the Andes in Peru. It may extend to Ecuador (Ramirez 1988) (map 9.56).

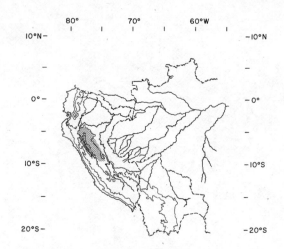

Map 9.56. Distribution of *Lagothrix flavicauda*.

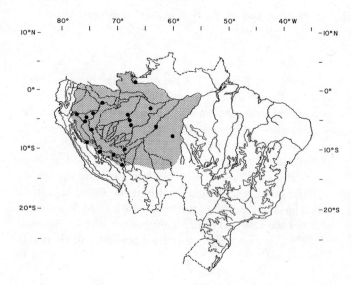

Map 9.57. Distribution of *Lagothrix lagothricha*.

Lagothrix lagothricha (Humboldt, 1812)
Woolly Monkey, Caparro, Barrigudo

Description
This species is one of the largest New World primates by weight. Head and body length ranges from 558 to 686 mm and the prehensile tail from 600 to 720 mm. Adult males may weight over 10 kg and females 8 kg. Color varies from gray to brown, and the head is generally a darker brown than the body. The fur is dense and short, giving the animal a woolly appearance. The face is deeply pigmented.

Range and Habitat
This monkey is distributed in the western part of the Amazon basin from Peru to southern Colombia. In southern Colombia it occurs to the east of the Andes and in several isolated areas in the Magdalena River basin (map 9.57).

Natural History

The woolly monkey has been little studied in the field but intensively studied in captivity (Williams 1968). From the field data we have available it appears that it feeds primarily on fruit and seeds supplemented by leaves (Ramirez 1980). Its consumption of leafy material, however, does not approach that of the howler monkey and is probably less than 20% of its diet. The time budget has been analyzed by Defler (1995). Troop size ranges between twelve and seventy. Depending on the carrying capacity, the home range can exceed 500 ha. Peres (1996) summarizes the ecology of *L. lagothricha* and concludes that home-range size varies with group size. Clearly carrying capacity is also related to differences in primary productivity of different habitats. A large troop of about forty-five animals may range over an area of 10 to 11 km^2. A smaller troop of twenty to twenty-four may have a home range of 1.21 km^2. Troops contain more than one adult male, but the exact nature of the dominance structure within a free-ranging troop has yet to be worked out. Troops may divide into subgroups while foraging, but overall troop cohesion is higher than observed in *Ateles* (Peres 1996). The vocalizations of the woolly monkey are extremely complicated; males' loud, long calls have a complex syllabic structure. There are similarities between the calls of the woolly monkey and the spider monkey (Eisenberg 1976). A single young is born after a gestation period of seven and a half months (see also Ramirez 1988).

Genus *Ateles* E. Geoffroy, 1806
Spider Monkey, Mono Araña,
Marimonda, Coatá

Taxonomy

Since the revision by Kellogg and Goldman (1944), the genus has not received a comprehensive taxonomic treatment. Although there has been a tendency by some workers to recognize only a single species within the genus, we firmly believe that the genus can be subdivided into at least five valid species. Some credence is offered to this viewpoint by the chromosomal studies of Heltne and Kunkel (1975) and Kunkel, Heltne, and Borgaonkar (1980), where karyotypic variants are demonstrable.

Description

Spider monkeys are among the largest New World primates in linear measurements. Adult head and body length can range from 420 to 660 mm, and tail length ranges from 744 to 880 mm. A large adult male can weigh up to 11 kg. The limbs are long and slender, and the tail is extremely long and highly prehensile. Coat color is highly variable and will be discussed under the species accounts. The large size and long limbs immediately distinguish this genus form the other larger New World monkeys in the northern Neotropics, for example, the howler monkey and the woolly monkey (plate 10).

Distribution

The species of *Ateles* range from southern Veracruz, Mexico, to Bolivia. They are adapted to multistratal tropical evergreen forests, although in the drier parts of Costa Rica and Bolivia they are known to occupy semideciduous forests. The species of this genus tend to be allopatric.

Natural History

The spider monkeys are by and large frugivorous, although they may eat young leaves. Typically they forage in the middle to upper strata of the forest and rarely descend to the ground. When they do travel on the ground they walk bipedally. Troop size is variable, and there is a marked tendency for troops to break up into foraging subgroups, then reconstitute at preferred sleeping sites. The species is strongly diurnal, and a troop utilizes a rather large home range. Subgroups communicate through loud, long calls that are easily distinguished from the deeper, more monotonic roars of the howler monkey.

A single young is born after gestation period of seven and a half months. The female carries the young until it is approximately ten months old, and thereafter she carries it at times until it is fourteen months of age. The most cohesive social unit within a troop of spider monkeys appears to be a female and her dependent young. The period of dependency is long, and the interbirth interval averages about thirty months. Spider monkeys show one of the lowest rates of recruitment of all New World monkeys and thus are vulnerable to human predation and slow to recover from catastrophic events. The standard references for behavior and ecology are Carpenter (1935), Dare (1974), Eisenberg and Kuehn (1966), Eisenberg (1973, 1976), Klein (1972), and van Roosmalen (1985).

Ateles belzebuth E. Geoffroy, 1806
Long-haired Spider Monkey, Marimonda,
Mono Frontino

Description

A large male may have a head and body length of 640 mm; males slightly exceed females in head and body length, but the female's tail is proportionately longer. The basic color is brown; the venter tends to be lighter, but the exact coloration is variable over the entire range of the species. The most distinguishing feature is a white diadem on the forehead.

268

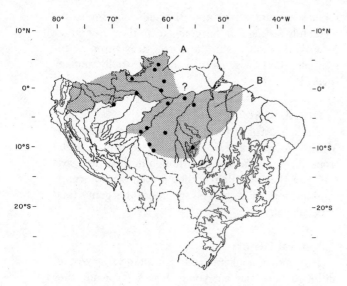

Map 9.58. Distribution of two species of *Ateles:* (a) *A. belzebuth;* (b) *A. marginatus.*

Range and Habitat

In the main this species is distributed in the north-eastern portion of the Amazon basin. It occurs in southern Colombia, in Amazonian Venezuela, and in relict populations in the mountains of northern Venezuela extending to Brazil (Mondolfi and Eisenberg 1979) (map 9.58). This species is replaced in the Guianas by *A. paniscus.* It typically inhabits multistratal evergreen forests.

Natural History

This species received intensive field study by Klein (1972) and by Klein and Klein (1977). In ecology, behavior, and reproduction it resembles *A. geoffroyi.* Troops range from ten to thirty animals and frequently break into subgroups of one to seven. Home ranges

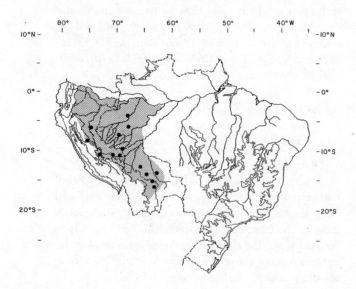

Map 9.59. Distribution of *Ateles chamek.*

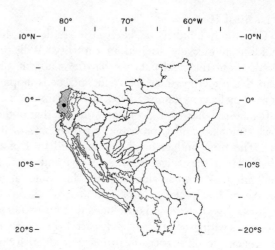

Map 9.60. Distribution of *Ateles fusciceps.*

vary from 2.6 to 3.9 km^2 (Klein and Klein 1977). Consumption of fleshy fruits ranged seasonally from 69% to 90%. Young leaves were an alternative food source (Castellanos and Chanin 1996).

Ateles chamek (Humboldt, 1812)

Mapped separately (map 9.59); see account for *A. paniscus.*

Ateles fusciceps Gray, 1866
Brown-headed Spider Monkey

Description

This species shows considerable variation in coat color over its entire range. A dark brown cap and brown back contrasting with a red venter are characteristic in the southern parts of its range, but in the area of Panama and northwestern Colombia the most frequent color is entirely black. The all-black form is described subspecifically as *Ateles fusciceps robustus* and extends from the Darién region of Panama west of the Andes through extreme western Colombia.

Range and Habitat

This species extends from southern Panama to Ecuador, west of the Andes. It typically inhabits multistratal tropical evergreen forests (map 9.60).

Natural History

There are few field data available for this species, but it has been intensively studied in captivity. Such data as we have indicate that it closely approximates *A. geoffroyi* in its basic natural history patterns. The standard reference for this species in captivity is Eisenberg (1976).

Ateles marginatus E. Geoffroy, 1809

Description

Basically similar to *Ateles belzebuth* with a whitish patch on the forehead. Cruz Lima (1945) considered it a distinct species (plate 10). Cruz Lima also grappled with the problem of loss of color and a series of bleached forms when he studied a series of clinal allopatric forms. Since so many local extinctions have occurred in the past hundred years in Brazil, his comments with respect to *Ateles* are worth looking into.

Distribution

If this form proves to be a valid species according to Groves (1993), it occurs in Brazil south of the Amazon between the Rio Tapajós and the Rio Tocantins (see map 9.58).

Ateles paniscus (Linnaeus, 1758)
Black Spider Monkey

Description

This dark-colored species is almost indistinguishable from *A. fusciceps robustus* when captives from disparate origins are held together, but it may easily be identified in the field by its geographical location. In the north, *A. paniscus* is basically black with a pigmented face. In western Amazonas the face is pink.

Range and Habitat

The species ranges from Guyana eastward and south to the Amazon River (map 9.61).

Natural History

In its behavior and ecology, this species strongly resembles *A. geoffroyi.* In Peru field studies confirm the tendency for troops to subdivide into small foraging units (Symington 1987). The species has been studied by Mittermeier and van Roosmalen (1981) and van Roosmalen (1980, 1985) in Suriname. Troop sizes range from ten to thirty. Home ranges may exceed 3.3 km^2. In Suriname, *Ateles paniscus* prefers mature

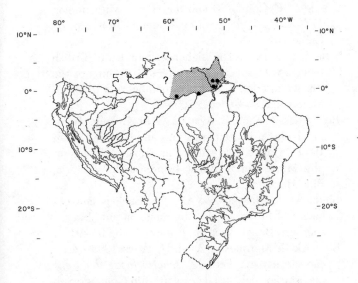

Map 9.61. Distribution of *Ateles paniscus.*

rain forests and does not exploit edge habitats. It typically forages in the upper part of the canopy. Its diet consists largely of fruit, with leaves making up less than 6%. In common with other species of *Ateles,* this monkey is strongly persecuted by humans as a game animal. Over much of its range, *Ateles* thus has become rare or has been exterminated in areas of prolonged human settlement.

Comment

Groves (1993) breaks this species into *Ateles chamek* (Humboldt, 1812) with the black face and *Ateles paniscus* (Linnaeus, 1758) with the pink face. This is a question that requires further study.

Genus *Brachyteles* Spix and Martius, 1823
Woolly Spider Monkey, Muriqui
Brachyteles arachnoides (E. Geoffroy, 1806)

Description

In this largest New World primate, adult males can attain weights of 15 kg and females can weigh 12 kg. Head and body length exceeds 800 mm; the tail generally is equal to the head and body length. The tail is prehensile, and the distal ventral surface is naked. The coat is dense and woolly, presenting a light brown appearance contrasting with the naked face rimmed with pale hairs; the face is brown, but conspicuous areas of depigmentation allow individual identifications (plate 10).

The adult males have very large testes, and the scrotum is clearly visible. The adult females have a conspicuous clitoris, a feature shared with the spider monkeys *Ateles* (Nishamura et al. 1988).

Distribution

This primate apparently occurred in the premontane forests of eastern Brazil from Bahia state south through Espírito Santo, Rio de Janeiro, and Minas Gerais, to São Paulo. Fonseca et al. (1994) consider the species to be much more adaptable with respect to forest type. At present, fewer than five hundred individuals remain in only the latter four states (Fonseca 1985). Cartelli (pers. comm.) reports a Pleistocene fossil from Bahia considerably larger in body size (map 9.62).

Natural History

Aguirre (1971) concluded that the muriqui preferred mature, climax rain forest. The surviving remnants of this species exhibit an unsuspected adaptability in that they cling to remnant second-growth patches of the Brazilian Atlantic rain forest with a surprising tenacity. The woolly spider monkey feeds on fruit, flowers, and seeds, but it can and does include considerable leaves in its diet (Milton 1984).

The muriqui currently exists in forest patches. It

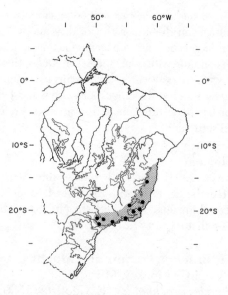

Map 9.62. Distribution of *Brachyteles arachnoides.*

should surprise no one that conflicting data have been assembled concerning the behavior of the species studied in different places (Nishamura et al. 1988). Indeed, we may have two valid subspecies within the fragmented range of this taxon (Lemos de Sa et al. 1993). This species lives in groups of eighteen to twenty-five, and their composition represents all age and sex classes. When several males occupy the same troop, there is evidence that an estrous female will accommodate more than one male. Polyandry? Time will tell. A single infant is born after a gestation period exceeding 210 days. The interbirth interval may be over twenty-four months. The communication system exhibits a close relation to that of *Ateles* (Nishamura et al. 1988; Fonseca 1985; Milton 1984; Valle et al. 1984; Strier 1992). Pope (1995) concludes from her analyses of allozymes that the species *B. arachnoides*, although currently fragmented, could be divided into two subspecies.

References

• References used in preparing distribution maps
• Aguirre, A. C. 1971. *O mono* Brachyteles arachnoides. Rio de Janeiro: Academia Brasileira de Ciências.
Albuja, L. 1994. Nuevos registros de *Saguinus tripartitus* en la Amazonia ecuatoriana. *Neotrop. Primates* 2 (2): 8–10.
• Allen, J. A. 1916. Mammals collected on the Roosevelt Brazilian expedition, with field notes by Leo E. Miller. *Bull. Amer. Mus. Nat. Hist.* 34:559–610.
Aquino, R. 1995. Conservación de *Cacajao calvus ucayalii* en la Amazonia peruana. *Neotrop. Primates* 3 (2): 40–42.

Auriccho, O. 1995. *Primatas do Brasil.* São Paulo: Terras Brasília Editora.
Ayres, J. M. 1981. Observacões sobre a ecologia e o comportamento dos cuxiús (*Chiropotes albinasus* e *Chiropotes satanas,* Cebidae: Primates). Conselho Nacional de Desenvolvimento Cientifico e Tecnologico (CNPq) Instituto Nacional de Pesquisas da Amazonia (INPA) Fundação Universidade do Amazonas (DUA).
———. 1986. Uakaris and Amazonian flooded forest. Ph.D. diss., Cambridge University, Cambridge.
Ayres, J. M., and T. H. Clutton-Brock. 1992. River boundaries and species range size in Amazonian primates. *Amer. Nat.* 140:531–37.
• Ayres, J. M., and K. Milton. 1981. Levantamento de primates e habitat no Rio Tapajós. *Bol. Mus. Paraense Emilio Goeldi, Zool.,* n.s., 111:1–11.
Baer, J. F., R. E. Weller, and I. Kakoma, eds. 1994. *Aotus: The owl monkey.* New York: Academic Press.
Baker, A. J. 1987. Emigration in wild groups of golden lion tamarins (*Leontopithecus rosalia*). *Int. J. Primatol.* 8:500.
———. 1991. Evolution of the social system of the golden lion tamarin (*Leontopithecus rosalia*): Mating system, group dynamics and cooperative breeding. Ph.D. diss., University of Maryland, College Park.
Baker, A. J., J. M. Dietz, and D. G. Kleiman. 1993. Behavioral evidence for monopolization of paternity in multi-male groups of golden lion tamarins. *Anim. Behav.* 46:1091–1103.
Baldwin, J. D., and J. I. Baldwin. 1972. The ecology and behavior of squirrel monkeys (*Saimiri oerstedii*) in a natural forest in western Panama. *Folia Primatol.* 18:161–84.
———. 1981. The squirrel monkeys, genus *Saimiri.* In *Ecology and behavior of Neotropical primates,* ed. A. F. Coimbra-Filho and R. A. Mittermeier, 277–325. Rio de Janeiro: Academia Brasileira de Ciências.
• Beebe, W. 1919. The higher vertebrates of British Guiana with special reference to the fauna of the Bartica district. 7. List of Amphibia, Reptilia, and Mammalia. *Zoologica* 2:205–38.
Bernardes, A. T., A. B. Rylands, C. M. C. Valle, R. M. Machado, A. F. Coimbra-Filho, and L. R. B. Fisher. 1989. Primate field studies in Brazil: A bibliography. Minas Gerais, Brazil: Imprensa Universitária, Belo Horizonte.
Biedzicki de Marques, A. A. 1996. Ecology and behavior of brown howlers in araucaria pine forests, southern Brazil. *Neotrop. Primates* 4 (3): 90–91.
Biedzicki de Marques, A. A., and C. Ades. 1996. Variação sazonar ma dieta de *Alouatta fusca clamitans* ma estação ecológica de Aracuri, Rio Grand do Sul.

Resumos do XXI Congresso Brasileiro de Zoologia (Pôrto Alegre), 217.

Boubli, J. P. 1994. The black uakari in the Pico Neblina National Park. *Neotrop. Primates.* 2 (3): 11–12.

Braza, F., F. Alvarez, and T. Azcarate. 1983. Feeding habits of the red howler monkeys (*Alouatta seniculus*) in the llanos of Venezuela. *Mammalia* 47 (2): 205–14.

Buchanan, D. B. 1978. Communication and ecology of pithecine monkeys, with special reference to *Pithecia pithecia*. Ph.D. diss., Wayne State University.

Buchanan D. B., R. A. Mittermeier, and M. G. M. van Roosmalen. 1981. The saki monkeys, genus *Pithecia*. In *Ecology and behavior of Neotropical primates*, ed. A. F. Coimbra-Filho and R. A. Mittermeier, 391–417. Rio de Janeiro: Academia Brasileira de Ciências.

Cabrera, A. 1939. Los monos de la Argentina. *Physis* 16:3–29.

Carpenter, C. R. 1934. A field study of the behavior and social relations of howling monkeys. *Comp. Psychol. Monogr.* 10 (2): 1–168.

———. 1935. Behavior of red spider monkeys in Panama. *J. Mammal.* 16:171–80.

Cartelle, C., and W. C. Hartwig. 1996. A new extinct primate among the Pleistocene megafauna of Bahia, Brazil. *Proc. Nat. Acad. Sci.* 93:6405–9.

Castellanos, H., and P. Chanin. 1996. Seasonal differences in food choice and patch preference of long-haired spider monkeys (*Ateles belzebuth*). In *Adaptive radiations of Neotropical primates*, ed. M. A. Norconk, A. L. Rosenberger, and P. A. Garber, 451–66. New York: Plenum Press.

Castro, R., and P. Soini. 1977. Field studies on *Saguinus mystax* and other callitrichids in Amazonian Peru. In *The biology and conservation of the Callitrichidae*, ed. D. G. Kleiman, 73–78. Washington, D.C.: Smithsonian Institution Press.

Chitolina, O. P., and M. Sander. 1981. Contribuição ao conhecimento da alimentação de *Alouatta guariba clamitans* Cabrera, 1940 em habitat natural no Rio Grande do Sul (Cebidae, Alouattinae). *Iheringia*, ser. zool., 59:37–44.

Christen, A. 1974. Fortpflanzungsbiologie und Verhalten bei *Cebuella pygmaea* and *Tamarin tamarin*. *Z. Tierpsychol.*, suppl. 14.

• Coimbra-Filho, A. F. 1977. Natural shelters of *Leontopithecus rosalia* and some ecological implications (Callitrichidae, Primates). In *The biology and conservation of the Callitrichidae*, ed. D. G. Kleiman, 79–89. Washington, D.C.: Smithsonian Institution Press.

Coimbra-Filho, A. F. 1990. Sistemática, distribução geográfica e situação actual do símios brasileiros. *Rev. Brasil. Biol.* 50 (4): 1063–79.

Coimbra-Filho, A. F., and R. A. Mittermeier, eds. 1981. *Ecology and behavior of Neotropical primates*, vol. 1. Rio de Janeiro: Academia Brasileira de Ciências.

Coimbra-Filho, A. F., A. Pissinatti, and A. B. Rylands. 1993. Experimental multiple hybridism and natural hybrids among *Callithrix* species from eastern Brazil. In *Marmosets and tamarins: Systematics, behavior and ecology*, ed. A. B. Rylands. 95–120. Oxford: Oxford University Press.

• Coimbra-Filho, A. F., and A. B. Rylands. 1995. On the geographic distribution of the red-handed howling monkey. *Alouatta belzebul*, in north-east Brazil. *Neotrop. Primates* 3:176–79.

Corbet, G. B., and J. E. Hill. 1991. *A world list of mammalian species.* Oxford: Oxford University Press.

Cordeiro da Silva, E. 1981. A preliminary survey of brown howler monkeys (*Alouatta fusca*) at the Cantareira Reserve (São Paulo, Brazil). *Rev. Brasil. Biol.* 41 (4): 897–909.

• Crespo, J. A. 1954. Presence of the reddish howling monkey (*Alouatta guariba clamitans* Cabrera) in Argentina. *J. Mammal.* 35:117–18.

• ———. 1974. Comentarios sobre nuevas localidades para mamíferos de Argentina y de Bolivia. *Rev. Mus. Argent. Cienc. Nat. "Bernardino Rivadavia,"* *Zool.* 11 (1): 1–31.

———. 1982. Ecología de la comunidad de mamíferos del Parque Nacional Iguazú, Misiones. *Rev. Mus. Argent. Cienc. Nat. "Bernardino Rivadavia," Ecol.* 3 (2): 45–162.

Crockett, C. M. 1984. Emigration by female red howler monkeys and the case for female competition. In *Female primates: Studies by women primatologists*, ed. M. Small, 159–73. New York: Alan R. Liss.

———. 1996. The relation between red howler monkey troop size and population growth in two habitats. In *Adaptive radiations of Neotropical primates*, ed. M. A. Norconk, A. L. Rosenberger, and P. A. Garber, 489–510. New York: Plenum Press.

Crockett, C. M., and J. F. Eisenberg. 1987. Howlers: Variations in group size and demography. In *Primate societies*, ed. B. B. Smuts, D. L. Cheney, R. M. Seyfarth, R. W. Wrangham, and T. T. Struhsaker, 54–68. Chicago: University of Chicago Press.

Crockett, C. M., and T. R. Pope. 1993. Consequences of sex differences in dispersal for juvenile red howler monkeys. In *Juvenile primates: Life history, development and behavior*, ed. M. E. Pereira and L. A. Fairbanks. New York: Oxford University Press.

Crockett, C. M., and R. Sekulic. 1984. Infanticide in red howler monkeys (*Alouatta seniculus*). In *Infanticide: Comparative and evolutionary perspectives*, ed. G. Hausfater and S. Hrdy. Hawthorne, N.Y.: Aldine.

Cruz Limada. *Mammals of Amazonia.* Vol. 1. *General introduction and primates.* Rio de Janeiro: Museu Paraense Emilio Goeldi de História Natural e Etnografia.

Dare, R. 1974. The social behavior and ecology of spider monkeys, *Ateles geoffroyi,* on Barro Colorado Island, Panama. Ph.D. diss., University of Oregon.

Dawson, G. A. 1977. Composition and stability of social groups of the tamarin, *Saguinus oedipus geoffroyi,* in Panama: Ecological and behavioral implications. In *The biology and conservation of the Callitrichidae,* ed. D. G. Kleiman, 23–37. Washington D.C.: Smithsonian Institution Press.

deBoer, L. B. M. 1974. Cytotaxonomy of the Platyrrhini. *Genen en Phaenen* 17:1–115.

• Defler, T. R. 1983a. A remote park in Colombia. *Oryx* 17:1–115.

———. 1983b. Some population characteristics of *Callicebus torquatus* in eastern Colombia. *Lozania* (Bogotá) 38:1–8.

———. 1994. *Callicebus torquatus* is not a white-sand specialist. *Amer. J. Primatol.* 33:149–54.

———. 1995. The time budget of a group of wild woolly monkeys (*Lagothrix lagothricha*). *Int. J. Primatol.* 16:107–20.

Dietz, J. M., and A. J. Baker. 1993. Polygamy and female reproductive success in golden lion tamarins, *Leontopithecus rosalia. Animal. Behav.* 46:1067–78.

Egler, S. 1992. Feeding ecology of *Saguinus bicolor* in a relict forest in Manaus, Brazilian Amazonia. *Folia Primatol.* 59:61–76.

Eisenberg, J. F. 1973. Reproduction in two species of spider monkeys, *Ateles fusciceps* and *A. geoffroyi. J. Mammal.* 54:955–57.

———. 1976. Communication mechanisms and social integration in the black spider monkeys, *Ateles fusciceps robustus* and related species. *Smithsonian Contrib. Zool.* 213:1–108.

———. 1977. Comparative ecology and reproduction in New World primates. In *The biology and conservation of the Callitrichidae,* ed. D. G. Kleiman, 13–22. Washington, D.C.: Smithsonian Institution Press.

———. 1979. Habitat, economy and society: Some correlations and hypotheses for the Neotropical primates. In *Primate ecology and human origins,* ed. I. S. Bernstein and E. O. Smith, 215–62. New York: Garland Press.

———. 1989. *Mammals of the Neotropics.* Vol. 1. *The northern Neotropics: Panama, Colombia, Venezuela, Guyana, Suriname, French Guiana.* Chicago: University of Chicago Press.

Eisenberg, J. F., and R. E. Kuehn. 1966. The behavior of *Ateles geoffroyi* and related species. *Smithsonian Misc. Coll.* 151 (8): 1–63.

Emmons, L. H., and F. Feer. 1990. *Neotropical rainforest mammals: A field guide.* Chicago: University of Chicago Press.

———. 1997. *Neotropical rainforest mammals: A field guide.* 2d ed. Chicago: University of Chicago Press.

Epple, G. 1973. The role of pheromones in the social communication of marmoset monkeys. *J. Reprod. Fertil.* 19 (suppl.): 447–54.

———. 1975. The behavior of marmoset monkeys. In *Primate behavior,* ed. L. A. Rosenblum, 4:195–239. New York: Academic Press.

Erikson, G. E. 1963. Brachiation in the New World monkeys. In *The primates,* ed. J. Napier and N. A. Barnicot, 135–64. Symposium 10. London: Zoological Society of London.

Ferrari, S. F. 1993. Ecological differentiation in the Callitrichidae. In *Marmosets and tamarins: Systematics, behavior and ecology,* ed. A. B. Rylands, 314–28. Oxford: Oxford University Press.

Ferrari, S. F., H. Katia M. Corrêa, and P. E. G. Coutinho. 1996. Ecology of the "southern" marmosets (*Callithrix aurita* and *C. flaviceps*). In *Adaptive radiations of Neotropical primates,* ed. M. A. Norconk, A. L. Rosenberger, and P. A. Garber, 157–71. New York: Plenum Press.

Ferrari, S. F., and M. A. Lopes. 1992. A new species of marmoset, genus *Callithrix* Erxleben, 1777 (Callitrichidae, Primates) from western Brazilian Amazonia. *Goeldiana Zool.* 12:1–13.

• Fonseca, G. A. B. da. 1985. The vanishing Brazilian Atlantic forest. *Biol. Conserv.* 34:17–34.

Fonseca, G. A. B. da, A. B. Rylands, C. M. R. Costa, R. B. Machado, and Y. L. R. Leite, eds. 1994. *Livro vermelho dos mammíferos brasileiros ameaçados de extinçõ.* Belo Horizante: Fundação Biodiversitas.

Fontaine, R. 1981. The uakaris, genus *Cacajao.* In *Ecology and bahavior of Neotropical primates,* ed. A. F. Coimbra-Filho and R. A. Mittermeier, 443–93. Rio de Janeiro: Academia Brasileira de Ciências.

Fooden, J. 1963. A revision of the woolly monkeys (genus *Lagothrix*). *J. Mammal.* 44:213–47.

Freese, C., and J. Oppenheimer. 1981. The capuchin monkeys, genus *Cebus.* In *Ecology and behavior of Neotropical primates,* ed. A. F. Coimbra-Filho and R. A. Mittermeier, 331–90. Rio de Janeiro: Academia Brasileira de Ciências.

Froehlich, J. W., R. W. Thorington Jr., and J. S. Otis. 1981. The demography of howler monkeys (*Alouatta palliata*) on Barro Colorado Island, Panama. *Int. J. Primatol.* 2 (3): 207–36.

Garber, P. 1993. Feeding ecology and behavior of the

genus *Saguinus*. In *Marmosets and tamarins: Systematics, behavior and ecology,* ed. A. B. Rylands, 273–89. Oxford: Oxford University Press.

Gaulin, S. J. C., and C. K. Gaulin. 1982. Behavioral ecology of *Alouatta seniculus* in Andean cloud forest. *Int. J. Primatol.* 3:1–32.

Glander, K. E. 1975. Habitat and resource utilization: An ecological view of social organization in the mantled howling monkey. Ph.D. diss., University of Chicago.

———. 1981. Reproduction and population growth in free-ranging mantled howling monkeys. *Amer. J. Phys. Anthropol.* 53:25–36.

Goldizen, A. 1987. Tamarins and marmosets: Communal care of offspring. In *Primate societies,* ed. B. B. Smuts, D. L. Cheney, R. M. Seyfarth, R. W. Wrangham, and T. T. Struhsaker, 34–43. Chicago: University of Chicago Press.

Green, K. M. 1978. Neotropical primate censusing in northern Colombia. *Primates* 19:537–50.

• Groves, C. P. 1993. Order Primates. In *Mammal species of the world,* ed. D. E. Wilson and D. M. Reeder, 243–78. Washington, D.C.: Smithsonian Institution Press.

• Hall, E. R. 1981. *The mammals of North America.* 2d ed., 2 vols. New York: John Wiley.

• Handley, C. O., Jr. 1976. Mammals of the Smithonian Venezuelan project. *Brigham Young Univ. Sci. Bull., Biol. Ser.* 20 (5): 1–90.

Happel, R. E. 1982. Ecology of *Pithecia hirsuta* in Peru. *J. Human Evol.* 11:581–90.

Hartwig, W. C. 1995. A giant New World monkey from the Pleistocene of Brazil. *J. Hum. Evol.* 28:189–95.

Hartwig, W. C., and C. Cartelle. 1996. A complete skeleton of the giant South American primate *Protopithecus. Nature* (London) 381:307–11.

Heltne, P. G., and L. Kunkel. 1975. Taxonomic notes on the pelage of *Ateles paniscus paniscus, A. p. chamek,* and *A. fusciceps rufiventris. J. Med. Primatol.* 4:83–102.

Heltne, P. G., D. C. Turner, and J. Wolhandler. 1973. Maternal and paternal periods in the development of infant *Callimico goeldii. Amer. J. Phys. Anthropol.* 3:555–60.

Heltne, P. G., J. F. Wojeik, and A. G. Pook. 1981. Goeldi's monkey, genus *Callimico.* In *Ecology and behavior of Neotropical primates,* ed. A. F. Coimbra-Filho and R. A. Mittermeier, 169–209. Rio de Janeiro: Academia Brasileira de Ciências.

• Hernandez-Camacho, J., and R. W. Cooper. 1976. The nonhuman primates of Colombia. In *Neotropical primates: Field studies and conservation,* ed. R. W. Thorington Jr. and P. G. Heltne, 35–69. Washington, D.C.: National Academy of Sciences.

• Hershkovitz, P. 1949. Mammals of northern Colombia, preliminary report no. 4: Monkeys (Primates), with taxonomic revisions of some forms. *Proc. U.S. Nat. Mus.* 98:323–427.

———. 1968. Metachromism or the principle of evolutionary change in mammalian tegumentary colors. *Evolution* 22:556–75.

• ———. 1977. *Living New World monkeys (Platyrrhini).* Vol. 1. Chicago: University of Chicago Press.

• ———. 1979. The species of sakis, genus *Pithecia* (Cebidae, Primates), with notes on sexual dichromatism. *Folia Primatol.* 31:1–22.

• ———. 1982. Subspecies and geographic distribution of black-mantle tamarins *Saguinus nigricollis* Spix (Primates: Callitrichidae). *Proc. Biol. Soc. Wash.* 95 (4): 647–56.

• ———. 1983. Two new species of night monkeys, genus *Aotus* (Cebidae, Platyrrhini): A preliminary report on *Aotus* taxonomy. *Amer. J. Primatol.* 4 (3): 209–43.

• ———. 1984. Taxonomy of squirrel monkeys genus *Saimiri:* A preliminary report. *Amer. J. Primatol.* 7:155–210.

• ———. 1985. A preliminary review of the South American bearded saki monkeys genus *Chiropotes* (Cebidae; Platyrrhini), with a description of a new subspecies. *Fieldiana: Zool.,* n.s., 27:1–46.

• ———. 1987a. The taxonomy of South American sakis, genus *Pithecia:* A preliminary report. *Amer. J. Primatol.* 12:387–468.

• ———. 1987b. Uacaris, New World monkeys of the genus *Cacajao:* A preliminary taxonomic review. *Amer. J. Primatol.* 12:1–53.

———. 1988. Origin, speciation and distribution of South American titi monkeys, genus *Callicebus. Proc. Acad. Nat. Sci. Philadelphia* 140:240–72.

• ———. 1990. Titis, New World monkeys of the genus *Callicebus:* A preliminary taxonomic review. *Fieldiana: Zool.,* n.s., 55:1–109.

Hick, U. 1973. Wir sind umgezogen. *Z. Kölner Zoo* 16 (4): 127–45.

Hill, W. C. O. 1957. *Primates.* Vol. 3. Edinburgh: Edinburgh University Press.

———. 1960. *Primates.* Vol. 4. Edinburgh: Edinburgh University Press.

———. 1962. *Primates.* Vol. 5. Edinburgh: Edinburgh University Press.

Hladik, A., and C. M. Hladik. 1969. Rapports trophiques entre végétation et primates dans la forêt de Barro Colorado (Panama). *Terre et Vie* 23: 25–117.

Hladik, C. M., A. Hladik, J. Bousset, P. Valdebouze, G. Viroben, and J. Delort-Laval. 1971. Le régime alimentaire des primates de l'île de Barro-Colorado (Panama): Résultats des analyses quantitatives. *Folia Primatol.* 16:85–122.

Hoage, R. J. 1982. Social and physical maturation in captive lion tamarins, *Leontopithecus rosalia rosalia* (Primates: Callitrichidae). *Smithsonian Contrib. Zool.* 354:1–56.

Honacki, J. H., K. E. Kinman, and J. W. Koeppl, eds. 1982. *Mammal species of the world.* Lawrence, Kans.: Allen Press and Association of Systematics Collections.

• Husson, A. M. 1978. *The mammals of Suriname.* Leiden: E. J. Brill.

Izawa, J. 1976. Group sizes and compositions of monkeys in the upper Amazon basin. *Primates* 17: 367–99.

Jacobs, S. C., A. Larson, and J. M. Cheverud. 1995. Phylogenetic relationships of coat color among tamarins (genus *Saguinus*). *Syst. Biol.* 44 (4): 515–32.

Janson, C. H. 1988. Food competition in the brown capuchin monkey *Cebus apella.* In Food competition in primates, symposium held at the XI Congress of the International Primatological Society, Göttingen. *Behaviour* 105 (1–2): 53–76.

Jones, C., and S. Anderson. 1978. *Callicebus moloch. Mammal. Species* 112:1–5.

Jones, C. B. 1983. Social organization of captive black howler monkeys (*Alouatta caraya*): "Social competition" and the use of non-damaging behavior. *Primates* 24:25–39.

• Juliot, C. 1994. Frugivory and seed dispersal by red howler monkeys: Evolutionary aspects. *Rev. Ecol. (Terre et Vie)* 49 (4): 331–41.

Kellogg, R., and E. A. Goldman. 1944. Review of the spider monkeys. *Proc. U.S. Nat. Mus.* 96 (3186): 1–45.

Kinzey, W. G. 1977. Diet and feeding behavior of *Callicebus torquatus.* In *Primate ecology.* ed. T. H. Clutton-Brock, 127–51. London: Academic Press.

———. 1981. The titi monkeys, genus *Callicebus.* In *Ecology and behavior of Neotropical primates,* ed. A. F. Coimbra-Filho and R. A. Mittermeier, 241–76. Rio de Janeiro: Academia Brasileira de Ciências.

———. 1982. Distribution of primates and forest refuges. In *Biological diversification in the tropics,* ed. G. T. Prance, 455–82. New York: Columbia University Press.

Kinzey, W. G., and A. H. Gentry. 1978. Habitat utilization in two species of *Callicebus.* In *Primate ecology: Problem oriented field studies,* ed. R. W. Sussman, 89–100. New York: John Wiley.

Kleiman, D. G. 1976. Monogamy in mammals. *Quart. Rev. Biol.* 52:39–69.

———. 1977. *The biology and conservation of the Callitrichidae.* Washington, D.C.: Smithsonian Institution Press.

———. 1981. *Leontopithecus rosalia. Mammal. Species* 148:1–7.

Kleiman, R. T., Hoage, R. J., and K. Green. 1988. The lion tamarins. In *Ecology and behavior of Neotropical primates,* ed. R. A. Mittermeier, A. B. Rylands, A. Coimbra-Filho, and G. A. B. da Fonseca, 2:299–347. Washington, D.C.: World Wildlife Fund.

Klein, L. L. 1972. Ecology and social organization of the spider monkey *Ateles belzebuth.* Ph.D. diss., University of California, Berkeley.

Klein, L. L., and D. B. Klein. 1975. Social and ecological contrasts between four taxa of Neotropical primates. In *Socioecology and psychology of primates,* ed. R. H. Tuttle, 59–85. The Hague: Mouton.

———. 1977. Feeding behavior of the Colombian spider monkey. In *Primate ecology,* ed. T. H. Clutton-Brock, 153–82. London: Academic Press.

• Kunkel, L. M., P. G. Heltne, and D. S. Borgaonkar. 1980. Chromosomal variation and zoogeography in *Ateles. Int. J. Primatol.* 1 (3): 223–32.

Lemos de Sa, R., T. R. Pope, T. T. Struhsaker, and K. E. Glander. 1993. Sexual dimorphism in the canine length of woolly spider monkeys. *Int. J. Primatol.* 14 (5): 755–63.

Lemos da Sa, R., and K. B. Strier. 1992. Preliminary comparisons of forest structure and use by two isolated groups of woolly spider monkeys, *Brachyteles arachnoides. Biotropica* 24:455–59.

Lorini, M. L., and V. G. Persson. 1990. Nova espécie de *Leontopithecus* Lesson, 1840 do sul do Brasil. Rio de Janeiro: *Bol. Mus. Nac.,* n.s., *Zool.* 338:1–14.

• ———. 1994. Status of field research on *Leontopithecus caissara:* The black-faced lion tamarin project. *Neotrop. Primates* 2 (suppl.): 52–55.

Martin, R. D. 1990. *Primate origins and evolution.* Princeton: Princeton University Press.

Mason, W. A. 1968. Use of space by *Callicebus* groups. In *Primates: Studies in adaptation and variability,* ed. P. C. Jay, 200–216. New York: Holt, Rinehart and Winston.

———. 1971. Field and laboratory studies of social organization in *Samiri* and *Callicebus. Primate Behav.* 2:107–38.

Miller, F. W. 1930. Notes on some mammals of southern Mato Grosso, Brazil. *J. Mammal.* 11:10–22.

Milton, K. 1978. Behavioral adaptations to leaf eating in the mantled howler monkey. In *The ecology of arboreal folivores,* ed. G. G. Montgomery, 535–50. Washington, D.C.: Smithsonian Institution Press.

———. 1980. *The foraging strategy of howler monkeys.* New York: Colombia University Press.

———. 1984. Habitat, diet and activity patterns of free-ranging woolly spider monkeys (*Brachyteles arachnoides* E. Geoffroy, 1806). *Int. J. Primatol.* 5:491–514.

Minezawa, M., M. Harada, O. C. Jordan, and C. J. Valdivia Borda. 1985. Cytogenetics of the Bolivian endemic red howler monkeys (*Alouatta seniculus sara*): *Accessory chromosomes and Y-autosome translocation related numerical variations.* Kyoto Univ. Overseas Res. Rep. New World Monkeys 5: 7–16.

• Mittermeier, R. A., A. B. Rylands, and A. Coimbra-Filho. 1988. Systematics. In *Ecology and behavior of Neotropical primates,* ed. R. A. Mittermeier, A. B. Rylands, A. Coimbra-Filho, and G. A. B. da Fonseca, 2:13–75. Washington. D.C.: World Wildlife Fund.

• Mittermeier, R. A., A. B. Rylands, A. Coimbra-Filho, and G. A. B. da Fonseca, eds. 1988. *Ecology and behavior of Neotropical primates,* Vol. 2. Washington, D.C.: World Wildlife Fund.

• Mittermeier, R. A., M. Schwartz, and J. M. Ayres. 1992. A new species of marmoset, genus *Callithrix* Erxleben 1777 (Callitrichidae, Primates), from the Rio Maués region, state of Amazonas, central Brazilian Amazonia. *Goeldiana Zool.* 14:1–17.

Mittermeier, R. A., and M. G. M. van Roosmalen, 1981. Preliminary observations on habitat utilization and diet in eight Suriname monkeys. *Folia Primatol.* 36:1–39.

Mondolfi, E., and J. F. Eisenberg. 1979. New records for *Ateles belzebuth hybridus* in northern Venezuela. In *Vertebrate ecology in the northern Neotropics,* ed. J. F. Eisenberg, 93–96. Washington, D.C.: Smithsonian Institution Press.

Moynihan, M. 1964. Some behavior patterns of platyrrhine monkeys. 1. The night monkey (*Aotus trivirgatus*). *Smithsonian Misc. Coll.* 146 (5): 1–84.

———. 1970. Some behavior patterns of platyrrhine monkeys. 2. *Saguinus geoffroyi* and some other tamarins. *Smithsonian Contrib. Zool.* 28:1–77.

———. 1976. *The New World primates.* Princeton: Princeton University Press.

Muckenhirn, N. A. 1967. The behavior and vocal repertoire of *Saguinus oedipus.* Master's thesis, University of Maryland.

• Muckenhirn, N. A., B. Mortensen, S. Vessey, C. E. Fraser, and B. Singh. 1976. *Report on a primate survey in Guyana.* Washington, D.C.: Panamerican Health Organization.

Müller, K. H. 1996. Diet and feeding ecology of masked titis (*Callicebus personatus*). In *Adaptive radiations of Neotropical primates,* ed. M. A. Norconk, A. L. Rosenberger, and P. A. Garber, 383–402. New York: Plenum Press.

Neville, M. K. 1972. The population structure of red howler monkeys. *Alouatta seniculus* in Trinidad and Venezuela. *Folia Primatol.* 17:56–86.

Neville, M. K., K. E. Glander, F. Braza, and A. B. Ry-

lands. 1988. The howling monkeys genus *Alouatta.* In *Ecology and behavior of Neotropical primates,* ed. R. A. Mittermeier, A. B. Rylands, A. Coimbra-Filho, and G. A. B. da Fonseca, 2:349–453. Washington, D.C.: World Wildlife Fund.

Neyman, P. F. 1977. Aspects of the ecology and social organization of free-ranging cotton-top tamarins (*Saguinus oedipus*) and the conservation status of the species. In *The biology and conservation of the Callitrichidae,* ed. D. G. Kleiman, 39–71. Washington, D.C.: Smithsonian Institution Press.

• Nishamura, A., G. A. B. da Fonseca, R. A. Mittermeier, A. L. Young, K. Strier, and C. M. Valle. 1988. The muriqui. In *Ecology and behavior of Neotropical primates,* ed. R. A. Mittermeier, A. B. Rylands, A. Coimbra-Filho, and G. A. B. da Fonseca, 2:577–610. Washington, D.C.: World Wildlife Fund.

Norconk, M. A. 1996. Seasonal variation in the diets of white-faced and bearded sakis in Guri Lake, Venezuela. In *Adaptive radiations of Neotropical primates,* ed. M. A. Norconk, A. L. Rosenberger, and P. A. Garber, 403–26. New York: Plenum Press.

Norconk, M. A., A. L. Rosenberger, and P. A. Garber, eds. 1996. *Adaptive radiations of Neotropical primates.* New York: Plenum Press.

Olrog, C. C. 1984. El mono caraya. In *Mamíferos fauna argentina,* vol. 1. Buenos Aires: Centro Editor de America Latina.

Oppenheimer, J. R. 1968. Behavior, ecology of the white-faced monkey *Cebus capucinus* on Barro Colorado Island, Panama. Ph.D. diss., University of Illinois.

———. 1977. Communication in New World monkeys. In *How animals communicate,* ed. T. A. Sebeok, 851–89. Bloomington: Indiana University Press.

Padua, C. V. 1993. The ecology, behavior and conservation of the black lion tamarin (*Leontopithecus chrysopygus* Mikan, 1823). Ph.D. diss., University of Florida, Gainesville.

Peres, C. A. 1989. Costs and benefits of territorial defense in wild golden lion tamarins (*Leontopithecus rosalia*). *Behav. Ecol. Sociobiol.* 25:227–33.

———. 1993. Notes on the ecology of the buffy saki monkey (*Pithecia albicans* Gray, 1860): A canopy seed predator. *Amer. J. Primatol.* 31:129–40.

———. 1996. Use of space, spatial group structure and foraging group size of gray woolly monkeys (*Lagothrix lagotricha cana*) at Urucu, Brazil. In *Adaptive radiations of Neotropical primates,* ed. M. A. Norconk, A. L. Rosenberger, and P. A. Garber, 467–88. New York: Plenum Press.

Piantanida, M., S. Puig, N. Nani, F. Rossi, L. Cavanna, S. Mazzucchelli, and A. Gil. 1984. Introducción al estudio de la ecología y etología del mono aullador

(*Alouatta caraya*) en condiciones naturales. *Rev. Mus. Argent. Cienc. Nat. "Bernardino Rivadavia," Ecol.* 3 (3): 163–92.

Ploog, D., S. Hopf, and P. Winter. 1967. Ontogenese des Verhaltens von Totenkopf-Affen (*Saimiri sciureus*). *Psychol. Forsch.* 31:1–41.

Pook, A. G., and G. Pook. 1981. A field study of the socioecology of Goeldi's monkey *Callimico goeldii* in northern Bolivia. *Folia Primatol.* 35:288–312.

Pope, B. 1966. The population characteristics of howler monkeys (*Alouatta caraya*) in northern Argentina. *Amer. J. Phys. Anthropol.* 24:361–70.

Pope, T. 1990. The reproductive consequences of male cooperation in the red howler monkey. *Behav. Ecol. Sociobiol.* 27:439–46.

———. 1992. The influence of dispersal patterns and mating system on genetic differentiation within and between populations of the red howler monkey (*Alouatta seniculus*). *Evolution* 46 (4): 1112–28.

———. 1995. Socioecology, population fragmentation, and patterns of genetic loss in endangered primates. In *Conservation genetics: Case histories from nature*, ed. J. C. Avise and J. L. Hamrick, 119–59. New York: Chapman Hall.

Power, M. L. 1996. The other side of callitrichine gumivory: Digestibility and nutritional value. In *Adaptive radiations of Neotropical primates*, ed. M. A. Norconk, A. L. Rosenberger, and P. A. Garber, 97–110. New York: Plenum Press.

Puertas, P., R. Aquino, and E. Encarnacion. 1992. Uso de alimentos y competición entre el mono nocturno *Aotus vociferans* y otros mamíferos, Loreto, Peru. *Folia Amazonia* 4:135–44.

Queiroz, H. L. 1992. A new species of capuchin monkey, genus *Cebus* Erxleben, 1777 (Cebidae: Primates) from eastern Brazilian Amazonia. *Goeldiana Zool.* 15:1–13.

Ramirez, M. F. 1980. Grouping patterns of the woolly monkey, *Lagothrix lagothricha* at the Manu National Park, Peru. *Amer. J. Phys. Anthropol.* 52:269.

———. 1988. The woolly monkeys, genus *Lagothrix*. In *Ecology and behavior of Neotropical primates*, ed. R. A. Mittermeier, A. B. Rylands, A. Coimbra-Filho, and G. A. B. da Fonseca, 2:539–75. Washington, D.C.: World Wildlife Fund.

Ramirez, M. F., C. H. Freese, and J. Revilla C. 1977. Feeding ecology of the pygmy marmoset, *Cebuella pygmaea*, in northeastern Peru. In *The biology and conservation of the Callitrichidae*, ed. D. G. Kleiman, 91–104. Washington, D.C.: Smithsonian Institution Press.

Rathbun, G. B., and M. Gache. 1980. Ecological survey of the night monkey, *Aotus trivirgatus*, in Formosa province, Argentina. *Primates* 21:211–19.

Redford, K. H., and J. F. Eisenberg. 1992. *Mammals of the Neotropics*. Vol. 2. *The southern cone: Chile, Argentina, Uruguay, Paraguay*. Chicago: University of Chicago Press.

Robinson, J. 1977. The vocal regulation of spacing in the titi monkeys, *Callicebus moloch*. Ph.D. diss., University of North Carolina.

———. 1979a. An analysis of the organization of vocal communication in the titi monkey *Callicebus moloch*. *Z. Tierpsychol.* 49:381–405.

———. 1979b. Vocal regulation of use of space by groups of titi monkeys *Callicebus moloch*. *Behav. Ecol. Sociobiol.* 5:1–15.

———. 1981. Spatial structure in foraging groups of wedge-capped capuchin monkeys *Cebus nigrivittatus*. *Anim. Behav.* 29:1036–56.

———. 1982. Vocal systems regulating within-group spacing. In *Primate communication*, ed. C. T. Snowden, C. H. Brown, and M. Petersen, 94–116. Cambridge: Cambridge University Press.

———. 1986. Seasonal variation in use of time and space by the wedge-capped monkey *Cebus olivaceus*: Implications for foraging theory. *Smithsonian Contrib. Zool.* 431:1–60.

Robinson, J. G., and C. H. Janson. 1987. Capuchins, squirrel monkeys and atelines. In *Primate societies*, ed. B. B. Smuts, D. L. Cheney, R. M. Seyfarth, R. W. Wrangham, and T. T. Struhsaker, 69–72. Chicago: University of Chicago Press.

Rosenblum, L. A., and R. W. Cooper, eds. 1968. *The squirrel monkey*. New York: Academic Press.

Rothe, H., and K. Darms. 1993. The social organization of marmosets: A critical evaluation of recent concepts. In *Marmosets and tamarins: Systematics, behavior and ecology*, ed. A. Rylands, 176–99. Oxford: Oxford University Press.

Rudran, R. 1979. The demography and social mobility of a red howler (*Alouatta seniculus*) population in Venezuela. In *Vertebrate ecology in the northern Neotropics*, ed. J. F. Eisenberg, 107–26. Washington, D.C.: Smithsonian Institution Press.

Rumiz, D. 1990. *Alouatta caraya* population density and demography. *Amer. J. Primatol.* 21:279–94.

Rylands, A. B. 1982. The behaviour and ecology of three species of marmosets and tamarins in Brazil. Ph.D. diss., Cambridge University.

———. 1989. Sympatric Brazilian callitrichids: The black-tufted-ear marmoset, *Callithrix kuhlii*, and the golden-headed lion tamarin, *Leontopithecus chrysomelas*. *J. Hum. Evol.* 8 (7): 679–95.

———. 1993. *Marmosets and tamarins: Systematics, behavior and ecology*. Oxford: Oxford University Press.

———. 1995. A species lists for the New World primates (Platyrrhini): Distribution by country, endemism and conservation status according to the Mace-Land system. *Neotrop. Primates* 3 (suppl.): 113–60.

Schaller, G. B. 1983. Mammals and their biomass on a Brazilian ranch. *Arq. Zool. São Paulo* 31 (1): 1–36.

Sekulic, R. 1981. The significance of howling in the red howler monkeys *Alouatta seniculus*. Ph.D. diss., University of Maryland.

———. 1982. Daily and seasonal patterns of roaring and spacing in four red howler *Alouatta seniculus* troops. *Folia Primatol.* 39:22–48.

Shoemaker, A. H. 1979. Reproduction and development of the black howler monkey *Alouatta caraya* at Columbia Zoo. *Int. Zoo Yearb.* 19:150–55.

Smuts, B. B., D. L. Cheney, R. M. Seyfarth, R. W. Wrangham, and T. T. Struhsaker, eds. 1987. *Primate societies.* Chicago: University of Chicago Press.

Snowden, C. T. 1993. A vocal testimony of the callitrichids. In *Marmosets and tamarins: Systematics, behavior and ecology,* ed. A. B. Rylands, 78–94. Oxford: Oxford University Press.

Snowden, C. T., and J. Cleveland. 1980. Individual recognition of contact calls by pygmy marmosets. *Anim. Behav.* 28:717–27.

Soini, P. 1988. The pygmy marmoset, genus *Cebuella.* In *Ecology and behavior of Neotropical primates,* ed. R. A. Mittermeier, A. B. Rylands, A. Coimbra-Filho, and G. A. B. da Fonseca, 2:79–129. Washington, D.C.: World Wildlife Fund.

• ———. 1992. Ecología del coto mono (*Alouatta seniculus*) en el Río Pacaya, Reserva Nacional Pacaya-Samiria, Peru. *Folia Amazonia* 4:103–18.

Stallings, J. R. 1984. Status and conservation of Paraguayan primates. M.A. thesis, University of Florida.

Stallings, J. R., and R. A. Mittermeier. 1983. The black-tailed marmoset (*Callithrix argentata melanura*), record from Paraguay. *Amer. J. Primatol.* 4:159–63.

• Stevenson, M., and, A. Rylands. 1988. The marmosets, *Callithrix.* In *Ecology and behavior of Neotropical primates,* ed. R. A. Mittermeier, A. B. Rylands, A. Coimbra-Filho, and G. A. B. da Fonseca, 2:131–222. Washington, D.C.: World Wildlife Fund.

Strier, K. B. 1992. *Faces in the forest: The endangered muriqui monkeys of Brazil.* Oxford: Oxford University Press.

Symington, M. A. 1987. Food competition and foraging party size in the black spider monkey (*Ateles paniscus chamek*). *Behaviour* 105:117–34.

Szalay, F. S., and E. Delson. 1979. *The evolutionary history of primates.* New York: Academic Press.

• Tate, G. H. H. 1939. The mammals of the Guiana region. *Bull. Amer. Mus. Nat. Hist.* 76 (5): 151–229.

Terborgh, J. 1983. *Five New World primates: A study in comparative ecology.* Princeton: Princeton University Press.

Thorington, R. W., Jr. 1985. The taxonomy and distribution of squirrel monkeys (*Saimiri*). In *Handbook of squirrel monkey research,* ed. L. A. Rosenblum and C. I. Coe, 1–33. New York: Plenum Press.

———. 1988. Taxonomic status of *Saguinus tripartitus* (Milne-Edwards, 1878). *Amer. J. Primatol.* 15: 367–71.

Thorington, R. W., Jr., R. Rudran, and D. Mack. 1979. sexual dimorphism of *Alouatta seniculus* and observation on capture techniques. In *Vertebrate ecology in the northern Neotropics,* ed. J. F. Eisenberg, 97–106. Washington, D.C.: Smithsonian Institution Press.

Thorington, R. W., Jr., J. C. Ruiz, and J. F. Eisenberg. 1984. A study of a black howling monkey (*Alouatta caraya*) population in northern Argentina. *Amer. J. Primatol.* 6:357–66.

Valle, C. M. C., I. B. Santos, M. C. Alves, and C. A. Pinto. 1984. Algumas observações sobre o comportamento do mono (*Brachyteles arachnoides*). In *A Primatologia no Brasil,* ed. T. de Mello, 271–83. Brasília: Sociedade Brasileira de Primatologie.

van Roosmalen, M. G. M. 1980. Habitat preferences, diet, feeding strategy and social organization of the black spider monkey (*Ateles p. paniscus* Linnaeus, 1758) in Suriname. Report, Rijksinstituut voor Natuurbeheer, Arnhem (the Netherlands).

———. 1985. Habitat preferences, diet, feeding strategy and social organization of the black spider monkey (*Ateles paniscus paniscus* Linnaeus, 1758) in Suriname. *Acta Amazonica* 15 (3–4): 1–237.

van Roosmalen, M. G. M., and L. L. Klein. 1988. The spider monkey, genus *Ateles.* In *Ecology and behavior of Neotropical primates,* ed. R. A. Mittermeier, A. B. Rylands, A. Coimbra-Filho, and G. A. B. da Fonseca, 2:455–537. Washington, D.C.: World Wildlife Fund.

van Roosmalen, M. G. M., R. A. Mittermeier, and J. G. Fleagle. 1988. Diet of the northern bearded saki (*Chiropotes satanus chiropotes*) *Amer. J. Primatol.* 12 (1): 11–35.

van Roosmalen, M. G. M., R. A. Mittermeier, and K. Milton. 1981. The bearded sakis, genus *Chiropotes.* In *Ecology and behavior of Neotropical primates,* ed. A. F. Coimbra-Filho and R. A. Mittermeier, 419–42. Rio de Janeiro: Academia Brasileira de Ciências.

• Vieira, C. 1945. Sobre uma coleção de mamíferos de Mato Grosso. *Arq. Zool. (São Paulo)* 4:395–430.

Vivo, M. de. 1985. On some monkeys from Rondônia,

Brazil (Primates: Callithricidae, Cebidae). *Pap. Avulsos Zool. São Paulo* 4:1–31.

———. 1991. *Taxonomia de* Callithrix *Erxleben, 1777.* Belo Horizonte, Brazil: Fundação Biodiversitas.

Whitehead, J. M. 1989. The effect of the location of a simulated intruder on responses to long-distance vocalizations of mantled howling monkeys, *Alouatta palliata palliata. Behaviour* 108 (1–2): 73–103.

———. 1995. Vox Aloattinae: A preliminary survey of acoustic characteristics of long distance calls of howling monkeys. *Int. J. Primatol.* 16:121–44.

Williams, L. 1968. *Man and monkey.* Philadelphia: J. B. Lippincott.

Wilson, D. E., and D. M. Reeder, eds. 1993. *Mammal species of the world.* 2d ed. Washington, D.C.: Smithsonian Institution Press.

Wright, P. 1981. The night monkeys, genus *Aotus.* In *Ecology and behavior of Neotropical primates,* ed. A. F. Coimbra-Filho and R. A. Mittermeier, 211–40. Rio de Janeiro: Academia Brasileira de Ciências.

Yunis, E., O. M. Torres de Caballero, and C. Ramirez. 1977. Genus *Aotus* Q- and G-band karyotypes and natural hybrids. *Folia Primatol.* 27:165–77.

Plates

Plate 1: (a) *Gracilinanus agilis;* (b) *Micoureus demerarae (cinerea);* (c) *Marmosops noctivagus;* (d) *Glironia venusta;* (e) *Monodelphis americana;* (f) *Monodelphis domestica;* (g) *Monodelphis brevicaudata;* (h) *Caenolestes caniventer*

Plate 2: (a) *Bradypus torquatus*; (b) *Bradypus variegatus*; (c) *Choloepus hoffmanni*; (d) *Cyclopes didactylus*; (e) *Tamandua tetradactyla* (blonde phase); (f) *Tamandua tetradactyla* (vested phase)

Plate 3: (a) *Chaetophractus* sp.; (b) *Euphractus sexcinctus*; (c) *Chlamyphorus retusus*; (d) *Dasypus. hybridus*; (e) *D. novemcinctus*; (f) *Cabassous unicinctus*; (g) *Tolypeutes matacus*; (h) *Tolypeutes* rolled into a ball; (i) *Priodontes maximus*

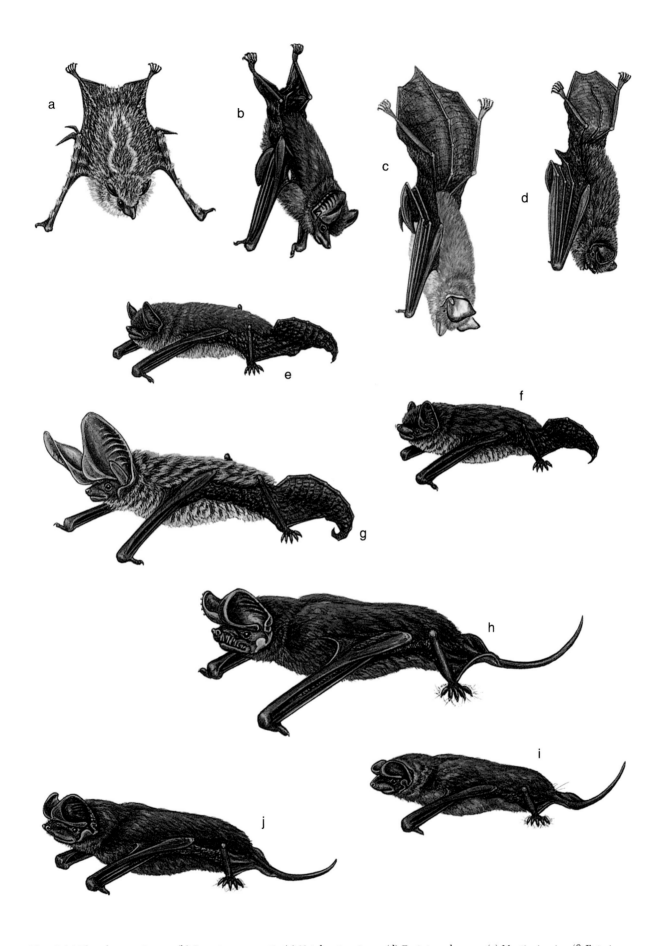

Plate 4: (a) *Rhynchonycteris naso;* (b) *Peropteryx macrotis;* (c) *Natalus stramineus;* (d) *Furipterus horrens;* (e) *Myotis riparius,* (f) *Eptesicus diminutus;* (g) *Histiotus montanus;* (h) *Nyctinomops macrotis;* (i) *Molossus molossus;* (j) *Eumops hansae*

Plate 5: (a) *Micronycteris nicefori;* (b) *M. megalotis;* (c) *Phyllostomus elongatus;* (d) *Macrophyllum macrophyllum;* (e) *Mimon crenulatum;* (f) *Vampyrum spectrum;* (g) *Trachops cirrhosus;* (h) *Tonatia silvicola*

Plate 6: (a) *Lonchophylla robusta*; (b) *Anoura caudifer*; (c) *Desmodus rotundus*; (d) *Sphaeronycteris toxophyllum*; (e) *Sturnira tildae*; (f) *Carollia perspicillata*; (g) *Platyrrhinus helleri*; (h) *Chiroderma villosum*; (i) *Artibeus planirostris*

Plate 7: Faces of (a) *Callithrix penicillata*; (b) *C. jacchus*; (c) *C. geoffroyi*; (d) *C. humeralifer humeralifer*; (e) *C. humeralifer chrysoleuca*; (f) *C. argentata melanura*; (g) *C. argentata argentata*; (h) *C. argentata leucippe*; (i) *Saguinus fuscicollis lagonotus*; (j) *S. fuscicollis weddelli*; faces of (k) *Saguinus nigricollis*; (l) *S. mystax*; (m) *S. labiatus*; (n) *S. imperator*; (o) *S. tripartitus*

Plate 8: (a) *Callicebus personatus*; (b) *C. moloch brunneus*; (c) *C. m. donacophilus*; (d) *C. m. cupreus*; (e) *Leontopithecus chrysomela*; (f) *L. rosalia*; (g) *L. chrysopygus*; (h) *Callicebus torquatus*; (i) *Aotus nigriceps*; (j) *A. vociferans*

Plate 9: (a) *Pithecia monachus;* three faces of *Saimiri;* (b) *S. boliviensis;* (c) *S. sciureus;* (d) *S. ustus;* (e) *Cebus albifrons;* (f) *C. apella;*
(g) *Chiropotes satanas;* (h) *C. albinasus;* (i) *Cacajao melanocephalus;* (j) *Cacajao calvus rubicundus*

Plate 10: (a) *Lagothrix lagotricha*; (b) *L. flavicauda*; (c) *Ateles marginatus*; (d) *Alouatta belzebul*; (e) *Brachyteles arachnoides*

Plate 11: (a) *Sciurus pyrrhinus*; (b) *S. aestuans*; (c) *Microsciurus flaviventer*; (d) *Sciurus stramineus*; (e) *S. igniventris*; (f) *S. spadiceus*; (g) *Chaetomys subspinosus*; (h) *Sphiggurus insidiosus*; (i) *Myoprocta acouchy* (green phase)

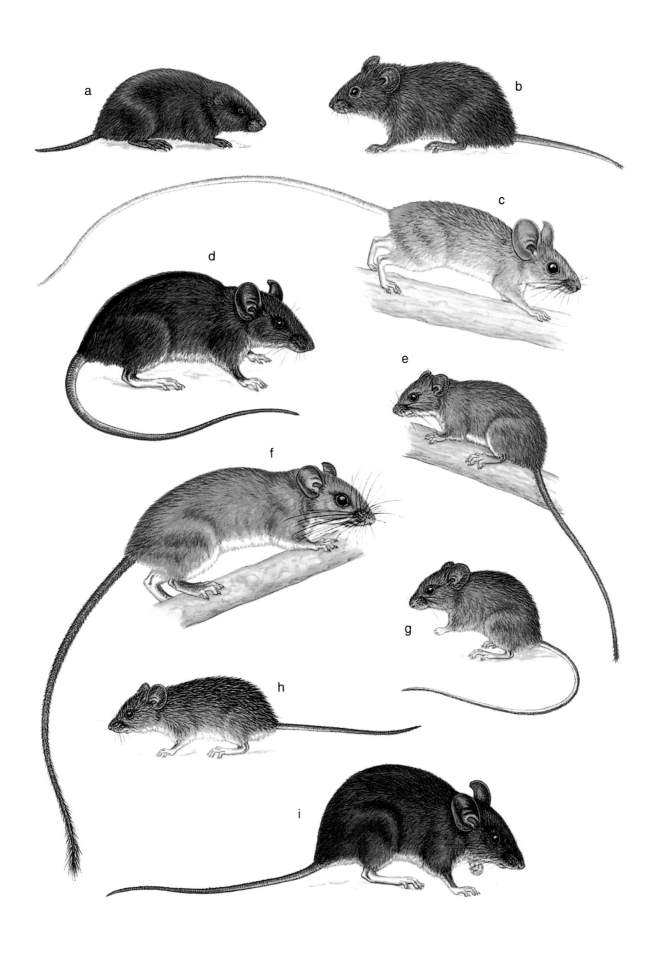

Plate 12: (a) *Blarinomys breviceps*; (b) *Bolomys lasiurus*; (c) *Wiedomys pyrrhorhinus*; (d) *Delomys dorsalis*; (e) *Oecomys bicolor*;
(f) *Rhipidomys mastacalis*; (g) *Oligoryzomys flavescens*; (h) *Neacomys tenuipes*; (i) *Oryzomys megacephalus*

Plate 13: (a) *Carterodon sulcidens;* (b) *Echimys armatus;* (c) *Dactylomys dactylinus;* (d) *Lonchothrix emiliae;* (e) *Mesomys stimulax;* (f) *Isothrix bistriata;* (g) *Isothrix picta;* (h) *Proechimys cuvieri;* (i) *Thrichomys apereoides*

Plate 14: (a) *Sotalia fluviatilis;* (b) *Pontoporia blainvillei;* (c) *Inia geoffrensis*

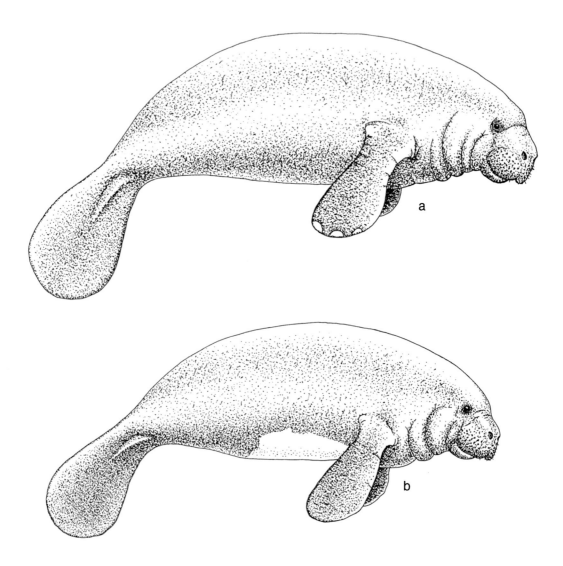

Plate 15: (a) *Trichechus manatus;* (b) *Trichechus inunguis*

Plate 16: (a) *Speothos venaticus;* (b) *Atelocynus microtis;* (c) *Pseudalopex (Lycalopex) vetulus;* (d) *Cerdocyon thous;* (e) *Pseudalopex culpaeus;* (f) *Chrysocyon brachyurus*

Plate 17: (larger scale) (a) *Potos flavus;* (b) *Bassaricyon gabbii;* (c) *Nasua nasua;* (d) *Galictis vittata;* (smaller scale) (e) *Felis pardalis;* (f) *Pteronura brasiliensis;* (g) *Panthera onca.* Both scales = 10 centimeters.

Plate 18: (a) *Tayassu tajacu*; (b) *T. pecari*; (c) *Lama guanicoe*; (d) *Vicugna vicugna*; (e) *Tapirus pinchaque* and young; (f) *Tapirus terrestris*

Plate 19: (a) *Mazama americana;* (b) *Hippocamelus antisensis;* (c) *Odocoileus virginianus;* (d) *Ozotoceros bezoarticus;*
(e) *Blastocerus dichotomus*

Diagnosis

The order Carnivora includes most extant terrestrial mammals specialized for predation on other vertebrates. Many members are omnivores, and most are terrestrial or scansorial, except the semiaquatic otters. Dental formulas are somewhat variable, with reduction in tooth number pronounced in the Felidae, and will be presented under the family accounts. The canine is conical and prominent. The modern Carnivora are usually defined based on the major carnassial shearing effect of the molariform teeth; the upper fourth premolar shears against the lower first molar. The digestive system shows no extreme modifications, and the cecum is usually small. The brain is relatively large and the braincase somewhat inflated, with the tympanic bullae enlarged and hemispheric. Of the eight extant families, five occur in South America.

We treat the Pinnepedia separately, since we consider them monophyletic, and thus discuss them in a separate chapter (see Wyss and Flynn 1993).

Distribution

The Recent distribution includes all continents except Antarctica and Australia; however, the dog (*Canis familiaris*) was introduced to Australia by aboriginal humans about 4,000 to 7,000 B.P., possibly later than its introduction into the Western Hemisphere. Members of this order tolerate a variety of habitats, extending to the extreme climatic conditions of the Arctic and the alpine zones of high mountains.

History and Classification

In Paleocene times, the Creodonta first appeared and radiated into the Eocene. Some creodonts persisted until the early Pliocene, but in general the original radiation was replaced by the Fissipedia, which first appeared in the middle Paleocene as the family Miacidae, known from North America, Europe, and Asia.

The modern families of carnivores are recognizable in the early Oligocene. Members of the order Carnivora were originally absent from South America and Australia; the carnivore niches there were filled by a parallel radiation within the order Marsupialia, resulting in the Dasyuridae (Australia) and the Borhyaenidae and Didelphidae (South America). The first true carnivores to enter South America appeared in the Miocene as an early raccoonlike form (Linares 1981). The remaining carnivores currently found there entered the continent via the Panamanian land bridge during the late Pliocene. Standard references for the biology of the Carnivora are Ewer (1973) and Gittleman (1989, 1996). Gomes de Oliviera (n.d.) provides a useful summary for the carnivores of Brazil.

FAMILY CANIDAE

Diagnosis

The typical dental formula for New World species is I 3/3, C 1/1, P 4/4, M 2/3, but some variation is shown in the number of molars. In the genus *Speothos*, molar number varies 1-2/2 (figs. 10.1 and 10.3). The canines are long and prominent. The molars have small crushing surfaces, and the first lower molar and upper third

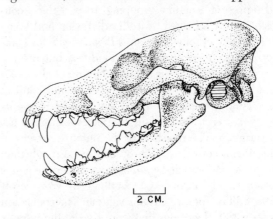

Figure 10.1. Skull of *Pseudalopex culpaeus*.

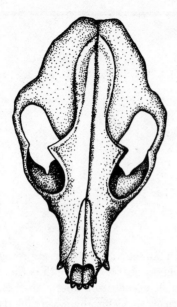

Figure 10.2. Skull of *Pseudalopex (Lycalopex) vetulus.*

premolar provide the carnassial shearing surfaces. The rostrum of the skull is long. Members of this family are adapted for a cursorial gait. They walk on their toes, and some reduction in toe number is demonstrable. There are generally five toes on the forefoot and four on the hind foot, but on the forefoot one toe is reduced to a dewclaw. The claws are nonretractile.

Comment on Taxonomy

Wozencroft (1993) employs *Pseudalopex* as the valid generic name for the South American foxes assigned to the genus *Dusicyon* in volume 2 of this work (Redford and Eisenberg 1992). *Dusicyon australis* (Kerr, 1792) is retained as the binomial for the Falkland Islands fox. Berta (1987) notes that *D. australis* had conservative cranial features suggesting affinities with the earliest "*Canis*" that entered South America. *Chrysocyon, Speothos,* and *Atelocynus* are rather derived, but the "*Pseudalopex*" foxes share many features and represent a natural grouping. In part these conclusions are confirmed by Tedford, Taylor, and Wang (1995). They employ the genus *Pseudalopex* for *gymnocercus, sechurae, griseus,* and *culpeus* but reserve the genus *Lycalopex* for *vetulus.* Zunino et al. (1995) consider the smaller foxes of Argentina and confirm the distinctiveness of *culpeus* but consider *griseus* and *gymnocercus* to be a single species exhibiting clinal variation. The question of the putative *griseus* in Chile is not considered. These authors also subsume the species *culpeus* and *gymnocercus* under the genus *Lycalopex.* The taxonomy as outlined by Wozencroft (1993) will be followed in this volume. The speciation events within the canids that were Recent immigrants to South America undoubtedly occurred quite rapidly

and without a fundamental change in chromosome number (Vitullo and Zuleta 1992).

Distribution

Species of this family occur on all continents except Antarctica. They were introduced by early humans in New Guinea and Australia.

Genus *Atelocynus* Cabrera, 1940
Atelocynus microtis (Sclater, 1883)
Small-eared Dog, Zorro de Orejas Cortas

Description

This genus contains a single species, *A. microtis,* a medium-sized canid with an extremely long rostrum and very small external ears. Head and body length ranges from 700 to 1,000 mm and the tail from 250 to 350 mm; the hind foot is about 140 mm and the ear 56 mm. The color is basically dark brown, but white hairs scattered on the dorsal pelage create a grizzled appearance and dark hairs in the midline give the impression of a dorsal stripe. The tail is black. The venter is paler, and in some specimens there are small white patches on the throat and in the groin.

Measurements of four specimens from Caranja, Ucayalidepartment, Peru, were TL 882–1,073; T 276–336; HF 126–45; E 52–63; Wt 6.5–7.5 kg (LSU). The species is immediately distinguishable from all other canids in the northern Neotropics by ear size, body size, and coat color (plate 16).

Range and Habitat

Atelocynus is known from scattered localities in the Amazon basin, having been recorded from Colombia, Bolivia, Ecuador, Peru, and Brazil. It is found in tropical forests (Hershkovitz 1961; Berta 1986) (map 10.1).

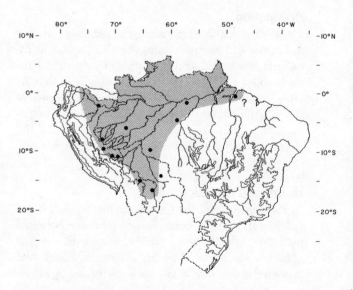

Map 10.1. Distribution of *Atelocynus microtis.*

Natural History

The fragmentary information we have indicates that the small-eared dog inhabits multistratal tropical evergreen forests and gallery forests. When seen, it is usually solitary. Although pairs are tolerant in captivity, there is no strong expression of contact behavior, and one gains the impression that this species is one of the least gregarious of the South American canids. The anal gland of the male produces a marked, musky odor. Definitely a predator on small mammals, it does not appear to be a group hunter (Peres 1992). The vocal repertoire is limited, in contrast to the rich variety of contact notes shown by the closely related bush dog, *Speothos* (Hershkovitz 1961; pers. obs.).

Cerdocyon thous (Linnaeus, 1766)
Crab-eating Fox, Zorro del Monte

Description

Measurements: TL 952–81; HB 640; T 312–16; E 71–74; Wt 5.6–6 kg. This fox is nearly the same size as *Pseudalopex gymnocercus* but slightly more robust, and the fur is shorter and coarser (plate 16). The snout is a little shorter than that of *P. gymnocercus*. The dorsum is grizzled grayish brown with black conspicuous along the middorsum, and there may be black over the shoulders and varying amounts on the hind legs and forearms. In some specimens there is a reddish cast on the cheeks. The lips, the ears, the feet, and the top of the tail are all black; the venter is grayish white. There is a good deal of individual variation, with some individuals much darker than others (Barlow 1965; Berta 1982; Stains 1975; pers. obs.).

Distribution

C. thous is distributed from Venezuela and Colombia south to Uruguay and Paraguay and in northern Argentina south to the province of Entre Ríos (Crespo 1984; Wilson and Reeder 1993) (map 10.2).

Life History

In captivity animals become sexually mature at about nine months and breed twice a year. Litter size ranges from three to six ($n = 4$); young are born weighing 120–60 g; gestation averages fifty-six days (range fifty-two to fifty-nine); and pups open their eyes at fourteen days. During the first thirty days only milk is consumed; until ninety days the pups take both solid food and milk; and after ninety days they are fully weaned. At about five months the family group breaks up. In Venezuela pregnant females are found throughout the year, but pups are most commonly seen in February and March (Berta 1982; Brady 1978).

Ecology

This fox occupies a variety of habitats from savanna to woodland, and up to at least 2,000 m elevation. In

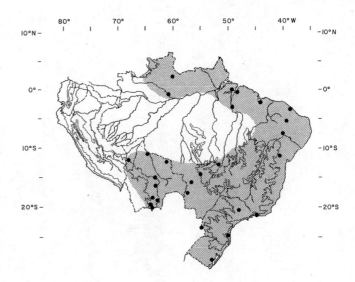

Map 10.2. Distribution of *Cerdocyon thous*.

Uruguay it prefers forest and edge areas, separating itself from *Pseudalopex gymnocercus*, which is found in more open country. However, in northeastern Argentina it is found in all habitat types.

In the Venezuelan llanos *C. thous* is nocturnal and pair-bonded, with individuals usually seen in pairs. Larger groups always consist of adults with their young. Adults travel extensively within relatively small home ranges ranging from 0.6 to 0.9 km^2. Partners may locate one another by a characteristic high-pitched whistle (Brady 1979; Sunquist, Sunquist, and Daneke 1989). MacDonald and Courtney (1996) report that mature, nonbreeding offspring may remain in adult territories, and new pairs formed by dispersing young may establish themselves near their natal territory.

C. thous is omnivorous: in Venezuela 104 stomachs were found to contain (by abundance) small mammals 26.0%, fruits 24%, amphibians 13.0%, insects (mostly orthopterans) 11.0%, reptiles 10.0%, and birds 9.0%. In the low llanos crustaceans, principally crabs, are a very common food during the wet season. In Uruguay one stomach was found to contain palm fruit, one skink, and two grasshoppers; in Paraguay one stomach had four snails, one frog, and two crabs; in Misiones (Argentina) one stomach had 40% plant material, including mushrooms, one *Dasypus* sp., insects, and frogs; while in southern Brazil two stomachs contained grasshoppers, *Astrocarpus*, and *Syzygium* seeds. Four stomachs from Pantanal foxes contained frogs, fruit, crabs, beetles, birds, and fish (Barlow 1965; Berta 1982; Bisbal and Ojasti 1980; Brady 1979; Coimbra-Filho 1966; Crespo 1982; Langguth 1975; Montgomery and Lubin 1978; Olrog 1979; Schaller 1983; Sunquist, Sunquist, and Daneke 1989).

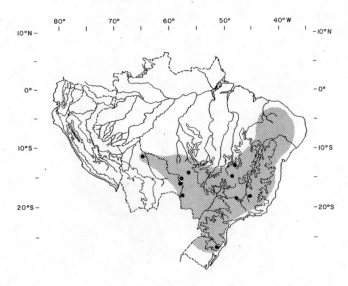

Map 10.3. Distribution of *Chrysocyon brachyurus*.

Genus *Chrysocyon* Smith, 1839
Chrysocyon brachyurus (Illiger, 1815)
Maned Wolf, Aguara Guazu, Lobo de Crin

Description
Measurements: TL 1,491; HB 1,030; T 398; HF 295; E 168; Wt 23.3–24.2 kg. The maned wolf is unmistakable with its long, thick buff red pelage, long legs, and large ears (plate 16). It is the largest canid in South America, standing about 90 cm at the shoulder. Its legs are black, as is part of the erectile mane. There is white on the pinnae as well as under the chin and on the tip of the tail (Dietz 1981; Kleiman 1972).

Distribution
C. brachyurus is found from the lowlands of Bolivia south through central Brazil into Paraguay, Argentina, and northeastern Brazil (Dietz 1984) (map 10.3).

Life History
The maned wolf is monestrous, with matings in the Southern Hemisphere occurring between April and June and most litters born from June to September. The gestation period is sixty-two to sixty-six days, and the litter size averages 2.2 (range 1–5; *n* = 25). The pups are born weighing about 350 g and open their eyes at between eight and nine days (Brady and Ditton 1979; Dietz 1981, 1984).

Ecology
Chrysocyon is an animal of grassland and scrub habitats. In a comprehensive field study in central Brazil maned wolves spent 34% of their time in open grassland, 43% in cerrado, and 24% in forest.

Individual adults are strictly territorial against like-sexed adults, and the species is facultatively monogomous, with a long-term pair-bond between the male and the female, though animals are usually seen alone. There is a good deal of scent marking, and mates often communicate over long distances using a high-amplitude roar-bark. In the wild males do not care for the young but are primarily territory defenders. Females regurgitate food for the young, and in captivity males do so as well.

Maned wolves are omnivores, consuming small vertebrates and invertebrates and large quantities of fruit, particularly fruit of *Solanum lycocarpum*. In central Brazil small rodents were found in 77.9% of the scats examined (*n* = 68), birds in 17.6%, armadillos in 8.8%, *Solanum* fruit in 82.4%, and grass in 2.9%. In central Brazil, out of 740 scats examined 28% of the volume consisted of small mammals, 2.3% of birds, and 57.6% of *Solanum lycocarpum* fruit (Brady and Ditton 1979; Dietz 1981, 1984; Rasmussen and Tilson 1984; Schaller 1983; Schaller and Tarak, unpubl.; UConn).

Genus *Pseudalopex* Burmeister, 1856
South American Fox, Zorro

Description
There has been much controversy concerning the taxonomy of the South American canids. The six species of *Pseudalopex* are medium-sized canids ranging from the small *P. sechurae* (HB 445 mm) to the large, robust *P. culpaeus* (body length >1,000 mm). The dorsum exhibits some pattern of brown, often speckled with black. The venter is usually some shade of cream to white (see species accounts).

Distribution
Depending on one's taxonomic preferences, the genus can have a broad or a restricted range. The members as so defined in this volume are South American and are to be found in almost all major habitat categories from puna to cerrado.

Natural History
The food habits of the species included in *Pseudalopex* are reviewed in Marquet et al. (1993).

Pseudalopex culpaeus (Molina, 1792)
Zorro Colorado

Description
Measurements: TL 1,102–1,173; T 395–440; FH 163–76; E 84–99; Wt 6.7–8.1 kg (Redford and Eisenberg 1992). After *Chrysocyon*, this is the largest of the South American canids. There is quite a bit of variation in color and size, but the overall impression is of a large, broad-headed fox with a considerable

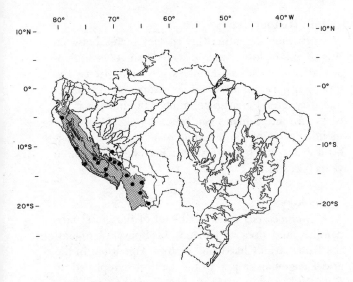

Map 10.4. Distribution of *Pseudalopex culpaeus.*

amount of orange on its coat. It ranges from 700 to 900 mm body length over its range from the mountains of southern Colombia to Chile. In Chile foxes of this species generally are reddish on the sides, with a middorsal region usually showing a brownish stripe. The muzzle and head are broad. Foxes from Argentina are also highly variable in color. The dorsum is black and gray, with an almost spotted to agouti effect. The amount of black on the tail is highly variable. There is a distinct reddish wash on the face, cheeks, limbs, and sometimes tail. The venter is cream to pale orange. There is a mutant form with a much more pronounced orange color to its coat (plate 16) (Crespo 1962; Fuentes and Jaksić 1979).

Distribution

P. culpaeus is distributed along the Andes from Colombia to southermost South America (Crespo 1984; Wozencroft 1993) (map 10.4).

Life History

In central Argentina males produce sperm from June to the middle of October and the females are monestrous, ovulating from early August to October. The gestation period is fifty-five to sixty days, the number of embryos per female averages 5.2 ($n = 6$), and litters are produced from October to December. Lactation lasts about two months. Females are fertile before one year of age. In Chile births were recorded in November, and the litter size ranged from three to five (Crespo and Carlo 1963; Housse 1953). The biology of the species has recently been reviewed by Novaro (1997).

Ecology

P. culpaeus is found in many kinds of habitats up to at least 4,500 m. Most are arid or semiarid. It does penetrate the dense, subantarctic forests of the Patagonian Andes. In Neuquén province, Argentina, this species is found in all available habitats but is most common in areas with abundant cover, where it occurs in densities of about 0.72 per km^2.

P. culpaeus is largely nocturnal and in some areas shows seasonal movements in seeking its prey of sheep and hares. In Peru two stomachs contained lizards, birds, and small mammals, and animals were found to be active both day and night. Over much of its range in the high Andes rodents predominate in the diet. In Chile this species eats lizards, small birds, mice, and the mouse opossoms (*Marmosa*). Its food habits have been well studied and found to vary seasonally and geographically. In most areas rodents make up a major portion of the diet; principal prey species include *Octodon degus, Abrocoma bennetti, Reithrodon,* and *Akodon olivaceus.* Where livestock is common, carrion is a frequent component of the diet, and in some areas the bird *Chloephaga picta* is a major food item. The introduced hare (*Lepus*) is commonly taken; most of the individuals eaten (83.3%) were juveniles, suggesting that nest predation is the main way these foxes catch hares. *P. culpaeus* also takes rabbits out of snares. Plant material, especially fruit, was commonly eaten in some areas.

This species is important as a furbearer, although less so than a smaller species of *Dusicyon:* between 1976 and 1979, 32,000 skins were exported from Argentina (Crespo 1975; Crespo and Carlo 1963; Jaksić, Schlatter, and Yáñez 1980; Jaksić and Yáñez 1983; Jaksić, Yáñez, and Rau 1983; Langguth 1975; Lucero 1983; Mares and Ojeda 1984; Pearson 1951; Romo 1995: Simonetti 1986; Yáñez and Jaksić 1978; Yáñez and Rau 1980).

Pseudalopex griseus (Gray, 1837)
Zorro Gris; Chilla

Description

Measurements: TL 868; HB 557; T 314; HF 122; E 72; Wt 3.9 kg. *P. griseus* is a small gray fox with a blackish middorsal stripe and a black tail tip, and in Chilean specimens a well-marked black chin. The venter is cream, and the underside of the tail is mixed pale tawny and black. In some specimens there is a faint reddish cast. On the island of Chiloé (Chile) and the adjacent mainland foxes of this species are almost black and have been considered a separate species (Medel, Jiménez, and Jaksić 1990) (see chapter 18). In Chile this species ranges in body length from 520 to 670 mm, with the largest animals in the north (34° N) and the smallest in the south (54° S). This is the opposite of the pattern seen in *P. culpaeus* (Fuentes and Jaksić 1979; Osgood 1943).

Distribution

P. griseus is found along the Andes from at least Salta province south to Río Negro province, Argentina, and then across the whole of Chile and Argentina south to Tierra del Fuego. In the north it may extend to Peru and Bolivia (Crespo 1984). Since the validity of applying the name *P. griseus* to the smaller fox of Argentina is questioned (Zunino et al. 1995), the Bolivian specimens are considered *P. gymnocercus* (not mapped).

Comment

P. griseus is considered conspecific with *P. gymnocereus* by Zunino et al. (1995) based on an analysis of Argentine material. See family account.

Pseudalopex gymnocercus (G. Fischer, 1914)
Zorro Pampa, Graxaima

Description

Measurements: TL 920–1,010; HB 591; T 319–45; HF 127–52; E 80–86; Wt 3.9–5.6 kg. This species is grayish on the dorsum and grayish mixed with yellowish on the sides. The ears and parts of the neck are reddish, as are the lower parts of legs. The underparts are markedly white. The tail is full and a mixture of grayish, reddish, and black. The dorsum is similar to that of *Cerdocyon thous*, but the dorsal midline black color is broken by white longitudinal speckles (Barlow 1965; pers. obs.).

Distribution

P. gymnocercus is found in eastern Bolivia, southern Brazil, western Paraguay, and the eastern provinces of Argentina north of Río Negro province (Crespo 1984) (map 10.5).

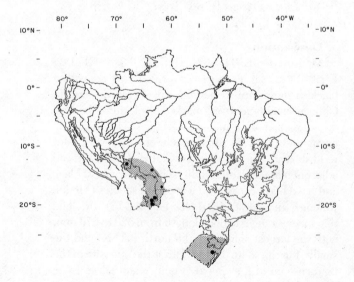

Map 10.5. Distribution of *Pseudalopex gymnocercus*.

Life History

In central Argentina this species mates between August and October, and pups are born between September and October. Females are monestrous; the gestation period is between approximately fifty-eight and sixty days; the average number of embryos is 3.4 (*n* = 72); and females can breed by eight to twelve months. Most of the breeding is done by the youngest females, and few individuals live longer than three years (Crespo 1971).

Ecology

P. gymnocercus is an inhabitant of the grasslands, pampas, and open woodlands of Bolivia, Uruguay, the Paraguayan Chaco, and eastern Argentina. In the arid Patagonian steppe to the south it is replaced by *P. griseus*. *P. gymnocercus* occurs with *Cerdocyon thous* over a good portion of its range, but it seems to prefer more open habitats.

Brooks (1992) estimates densities of five to twelve individuals per 10 km^2 in the Chaco of Paraguay. *P. gymnocercus* is nocturnal where it is hunted. Like other *Pseudalopex* species, it is omnivorous, and plant food, especially fruit, makes up one-quarter of the total diet (*n* = 230 stomachs). In La Pampa province (Argentina) the most important food is mammals, particularly the young of the hare, *Lepus*, which was found in 33% of the stomachs. Next most important was the mouse *Graomys griseoflavus*. Sheep carrion was found in fewer than 20% of the samples. Birds, particularly tinamous of the family Tinamidae, were found in a third of the stomachs. Insects and reptiles were of less importance (Barlow 1965; Crespo 1971, 1975, 1982; Langguth 1975).

Pseudalopex sechurae Thomas, 1900

Description

Measurements of type specimen were HB 580; T 300; HF 120; E 60. Adult weight averages 2.2 kg. This is a small fox with a short, harsh fur. It is pale in color, with relatively large ears. The general color of the dorsum is a grizzled iron gray; the underfur is grayish and the venter a fulvous white. This fox is convergent in form with the better-known *Vulpes velox* of North America.

Distribution

The species is known from southwestern Ecuador and coastal northwestern Peru (map 10.6).

Natural History

The species is apparently adapted to the arid Pacific coast of southern Ecuador and northern Peru. Its habitat is characterized by high diel temperatures (28°C)

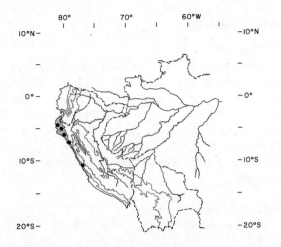

Map 10.6. Distribution of *Pseudalopex sechurae*.

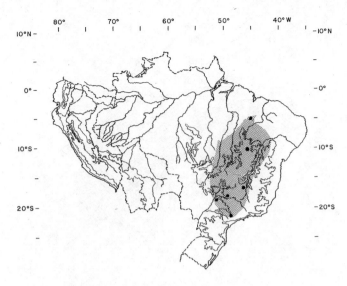

Map 10.7. Distribution of *Pseudalopex vetulus*.

and low rainfall (<300 mm per annum). Its diet apparently consists primarily of plant material and invertebrates (Asa and Wallace 1990).

Pseudalopex (Lycalopex) vetulus (Lund, 1842) Raposinha

Description

This is a very small fox. In the original description the measurements are HB 500 and T 300, comparable in size to *P. sechurae* of Peru. Lund notes that in body proportions it resembles *Vulpes velox* of North America. The temporal ridges on the top of the skull are somewhat lyre shaped, joining at the back of the skull much as in *Urocyon* (fig. 10.2). The teeth are somewhat reduced in size. The dorsal color is a gray brown, darker in the midline, and there is a dark patch at the base of the tail. This fox is rare in collections (plate 16).

Distribution

The species is apparently confined to the cerrado and caatinga vegetational zones of Brazil (map 10.7).

Natural History

Seasonally, this fox feeds on invertebrates, especially termites (J. Dalponte, pers. comm.).

Genus *Speothos* Lund, 1839
Speothos venaticus (Lund, 1842)
Bush Dog, Zorro Vinagre, Cachorro do
Mata Vinagre

Description

Measurements: TL 810; HB 700, T 110; HF 110; E 45. This canid is unmistakable because of its short ears, very short legs, and uniform dark pelage (plate 16). It is stocky with a short tail, small, rounded ears, and a thick muzzle. The fur is dark brown, grading to a more golden color on the neck, with no distinctive markings. The venter is darker than the dorsum (Kleiman 1972; pers. obs.).

Distribution

S. venaticus is found from Panama south through Paraguay to northern Argentina (Crespo 1974b; Hill and Hawkes 1983; Honacki, Kinman, and Koeppl 1982; Wilson and Reeder 1993) (map 10.8).

Life History

In captivity gestation is approximately sixty-five days, and three to five is the usual litter. Pups open their eyes between eight and seventeen days and eat solid food at between thirty-eight and seventy-one days. The male provisions the female and young throughout the rearing phase. Pups share food and show a higher social tolerance than many other canids. The vocal repertoire is very rich, and individuals foraging as a group use a short whine as a contact call (Biben 1982a,b, 1983; Brady 1981; Drüwa 1977).

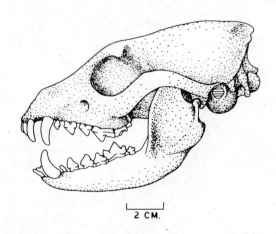

Figure 10.3. Skull of *Speothos venaticus*.

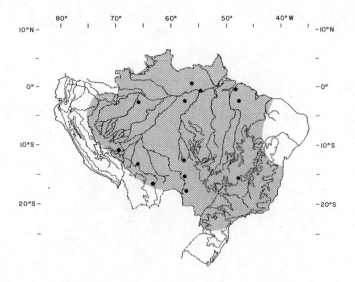

Map 10.8. Distribution of *Speothos venaticus*.

Ecology

Bush dogs in the wild are very poorly studied. They are apparently found only in moist forest, although this species has been seen in central Brazil where the only forest was gallery forest.

All evidence from captivity indicates a strong pair-bond between an adult male and an adult female (Drüwa 1977). Young animals that remain with their parents after attaining maturity do not reproduce. Apparently the dominant female suppresses the expression of estrus in her daughters. Unlike most other canids, this species uses short-distance contact calls, and this combined with the high degree of social tolerance shown in pups suggests that bush dogs may be social. This is corroborated by two observations in Argentina and Paraguay, where a male and a female were trapped together and a male and three females were collected from the same burrow. In both cases the animals were adults. The animals urine mark in a unique fashion, standing on the forelegs and micturating on a spot higher than would have been possible had they stayed on all fours (Crespo 1974a; Eisenberg 1989; Hill, unpubl.; Kleiman 1972).

Aquino and Puertas (1997) observed groups of *Speothos* on three separate occasions in Peru, one group of four individuals and two groups of three, which appeared to be "family groups." They found a den site in a large cavity in a fallen tree trunk. Examination of feces led them to conclude that *Proechimys*, *Dasyprocta*, and *Nasua* were included in the diet. Peres (1992) reaffirms the group hunting activities of this species. It has been suggested that bush dogs, with their short legs, specialize on large caviomorph rodents such as the paca (*Agouti paca*), which they chase into burrows. In Mato Grosso, Brazil, a bush dog was seen to capture a paca out in the open (Deutsch 1983).

FAMILY URSIDAE

Diagnosis

These large to medium-sized carnivores have a dental formula of I 3/3, C 1/1, P 4/4, M 2/3. They are powerfully built, with short ears and tail, and weigh up to 500 kg. There is no reduction in toe number, with five on each foot, and the animals have a plantigrade posture, bearing weight on the full sole of the foot. Bears are immediately distinguishable from all other carnivores.

Distribution

The Recent distribution of this family includes all the major continents except Africa, Australia, and Antarctica.

Genus *Tremarctos* Gervais, 1855
Tremarctos ornatus (F. Cuvier, 1825)
Spectacled Bear, Oso

Description

In this medium-sized bear, weight may reach 200 kg. The basic color pattern is uniform dark brown to black, the face distinctively marked with varying degrees of white stripes. In their most complete form these white markings encircle the eyes, hence the vernacular name spectacled bear.

Range and Habitat

The genus *Tremarctos* once was distributed from southern California to the southeastern United States and through Central America. Since the Pleistocene it exists only as a relict in South America and is confined to premontane and montane habitats in the Andes and adjacent foothills. The present range extends from Panama through Peru to Bolivia. It has become extinct in western Argentina within the past fifty years (Brown and Rumiz 1988) (map 10.9).

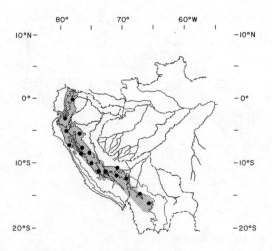

Map 10.9. Distribution of *Tremarctos ornatus*.

Natural History

The spectacled bear is a generalized omnivore, eating a variety of fruits, nuts, small vertebrates, and invertebrates. In Peru it may feed heavily on bromeliad hearts when fruit is unavailable. When climbing trees to reach fruits, it pulls small branches toward itself, creating a platform that will support its weight. This is not a true "nest." Daily resting places may be in cavities at the bases of trees or in rock caves. When it occurs near human populations, it may raid crops such as maize (Peyton 1980). Goldstein (1991) reports predation on immature livestock in Venezuela, but the impact is minimal.

In foraging the bears tend to move as solitary individuals except for females accompanied by young. The litter size averages two, born after seven months' gestation. This is not a true gestation period, however; a delayed implantation phase occurs, as in the temperate zone bears of the Northern Hemisphere. The true gestation period is probably about sixty-five days (Mondolfi 1971; Dathe 1967; Eisenberg 1989).

FAMILY PROCYONIDAE

Diagnosis

The molars of this family tend to be broad and do not exhibit the shearing edge so characteristic of canids and felids. In this feature Procyonidae resemble the bear family (Ursidae) (fig. 10.4). The dental formula is usually I 3/3, C 1/1, P 4/4, M 2/2, except in the genus *Potos*, where the number of premolars is 3/4. Members of the family are medium to small in size with a total length ranging from 60 to 135 cm. There is no toe reduction, and they are either plantigrade or semiplantigrade.

Distribution

If one excludes from the procyonids the Asiatic lesser panda, *Ailurus*, and the giant panda, *Ailuropoda*, then the current distribution of the family is confined to the New World. Raccoons and their allies are dis-

tributed from Canada to Argentina. They are adapted to a variety of habitats; however, they generally occur only where there is tree cover.

Genus *Bassaricyon* J. A. Allen, 1876

Note on Systematics and Taxonomy

In Central America this genus is represented by a single species, *B. gabbii*. Although two other species have been described with extremely restricted ranges in Costa Rica and Panama, these two species, *B. lasius* and *B. pauli*, may be conspecific with *B. gabbii*. *B. alleni* has been described from Ecuador and Peru but is usually synonymized with *B. gabbii*. *Bassaricyon beddardi* has been described from Guayana and possibly adjacent Brazil.

Bassaricyon gabbii J. A. Allen, 1876
Olingo

Description

This genus is typified by its rather small ears and extremely long tail. Adult head and body length ranges from 370 to 420 mm; tail length is 380 to 432 mm; the ear averages 27 mm. The tail exceeds the head and body length and often exhibits banding, though the pattern and degree are variable over the range of the species and in some specimens the banding is very indistinct. The basic color of the dorsum is honey brown. The olingo is lively and quick in its movements. Only the longer muzzle quickly identifies a young animal until you have it in the hand. The tail banding and the lack of prehensility distinguish the olingo from the kinkajou (plate 17).

Range and Habitat

The genus *Bassaricyon* ranges from southern Nicaragua to the Amazon basin. It is absent from the llanos of Colombia and Venezuela and appears to be confined to areas of multistratal tropical evergreen forests below 2,000 m. Except for one record from Bolívar, Venezuela (E. Mondolfi, pers. comm.), it is absent from the eastern portion of northern South America, but in the west it extends west of the Andes through Colombia and Ecuador and then east of the Andes in Peru and Bolivia (map 10.10).

Natural History

The olingo is a nocturnal, arboreal form that has been little studied in the wild; what we know of its behavior derives from captive studies. The standard reference is Poglayen-Neuwall and Poglayen-Neuwall (1965). Its diet includes fruits, invertebrates, and small vertebrates. There is some suggestion that it is more carnivorous than the kinkajou (*Potos*). Adults seem to forage singly, and it is believed to be less social than *Potos;* like-sexed adults are usually incompatible in captivity. The female bears a single young after seventy-

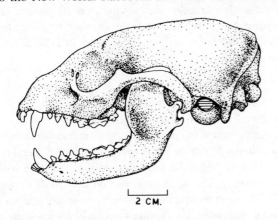

2 CM.

Figure 10.4. Skull of *Procyon cancrivorus*.

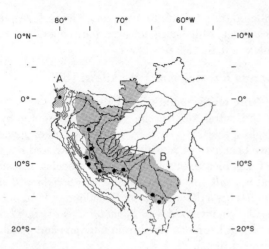

Map 10.10. Distribution of (a) *Bassaricyon gabbii;* (b) *Bassaricyon alleni.*

three to seventy-four days' gestation. The olingo shelters in hollow trees.

Genus *Nasua* Storr, 1780

Description

The tail is nearly equal to the head and body in length; head and body length ranges from 430 to 700 mm and the tail from 420 to 680 mm. The ears are short, the nostrum is narrow, and the snout is mobile. The dorsal pelage is variable, ranging from pale brown to reddish; in adult males, the shoulder region may show yellow or white hairs. The tail is banded with alternate yellow and brown markings, and the eyes are bordered by a mask that varies from reddish to brown.

Range and Habitat

The genus is distributed from the southwestern United States to Argentina. Coatis appear to be absent from the llanos of Venezuela. They are broadly tolerant of a variety of habitat types, exploiting dry deciduous forests to multistratal tropical evergreen forests.

Comment

Wozencroft (1993) presents two species, *N. narica* and *N. nasua. N. narica* is thought to extend from the southeastern United States (Arizona and New Mexico) south through Central America to the northwest coast of Colombia. Over the rest of the range in South America, the species is recognized as *N. nasua,* but the limits of the range of both species in coastal, western Ecuador is poorly understood.

Nasua narica (Linnaeus, 1766)
Coati, Pisote, Quati

Description

See Eisenberg (1989). Superficially similar to *N. nasua.*

Distribution and Comment

The species ranges from southern Arizona and New Mexico south through Central America to Colombia. The southern limit of its range is poorly documented (not mapped).

Gompper (1995) provides a recent summary of the biology of this species.

Nasua nasua (Linnaeus, 1766)
Coati, Pisote

Description

Measurements: TL 999; HB 550; T 464; HF 93; E 50; Wt 3.2–4.9 kg. The coati is unmistakable with its long snout and ringed tail, which is usually held perpendicular to the body (plate 17). Individuals from eastern Paraguay are a grizzled orange and black dorsally and orange tan on the sides, chin, and venter. The tail has pronounced but ill-defined black and orange tan bands. The ears are short, are heavily furred, and have a white spot. There is white on the face. The feet are dark and equipped with long, strong claws. Individuals from other populations may have more black and less orange with the face less marked with white (pers. obs.).

Distribution

Nasua narica is recognized as a distinct species in Mexico (and the adjacent United States) and Central America. *N. nasua* is found from southern Colombia south to Paraguay and thence to northern Argentina, where it has been recorded from Santa Fe and Salta provinces. It has also been introduced onto Isla Robinson Crusoe off the coast of Chile (Redford and Eisenberg 1992) (map 10.11).

Life History

Young are born after seventy-seven days' gestation in an arboreal nest constructed by the female. Litter size ranges from one to six. During the first part of their life young remain in the nest. After two or three weeks they begin to accompany their mother as she rejoins other females with young. In northeastern Argentina coatis reproduce from October to February, and the litter size is three to six (Crespo 1982; Eisenberg 1989).

Ecology

These animals are found in many vegetation types ranging from thorn scrub to moist tropical forest. They occur in groups ranging form one to as many as twenty. In Panama *N. narica* has been well studied. Several females form permanent bands and forage together with their young. The males join the female bands during the breeding season, but at other times of the year they forage solitarily. This form of social life is apparently also characteristic for *N. nasua.*

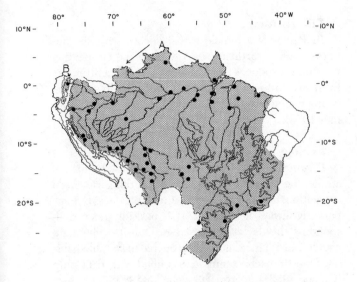

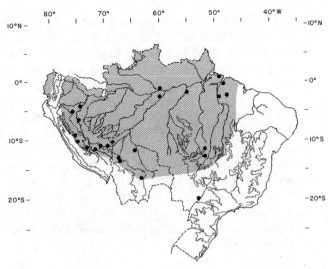

Map 10.11. Distribution of (a) *Nasua nasua;* (b) *Nasuella olivacea.*

Map 10.12. Distribution of *Potos flavus.*

Coatis are diurnal and scansorial. They are omnivorous, foraging by probing the forest litter with their sensitive snouts. They eat a great deal of fruit at certain seasons of the year. They are also opportunistic predators of vertebrates. Two stomachs from Brazilian animals contained fish, snakes, crabs, milipedes, sowbugs, spiders, slugs, cicadas, beetles, ants, and termites, and they also feed on palm nuts and figs (Crespo 1982; Eisenberg 1989; Pine, Miller, and Schamberger 1979; Kaufmann 1962; Russell 1981, 1983; Schaller 1983; Gompper 1994).

Genus *Nasuella* Hollister, 1915
Nasuella olivacea (Gray, 1865)
Mountain Coati, Coati Oliva

Description
This small coati is half the size of *N. nasua.* Head and body length ranges from 360 to 390 mm, and the tail ranges from 200 to 240 mm. The dorsum is gray brown. In body proportions and tail banding *N. olivacea* resembles the common coati, but it is easily distinguished by its small size.

Distribution
Nasuella is found in montane habitats from northern Colombia and Venezuela to Ecuador. It may extend to Peru, but I have found no validated museum specimens. It is a high-altitude specialist, preferring forested habitats at elevations over 2,000 m (Aagaard 1982) (map 10.11).

Natural History
The mountain coati has not been the subject of detailed study, and little is known concerning its ecology and behavior. What fragmentary data are available suggest that its natural history is similar to that

of the lowland coati, and confirmation awaits further fieldwork.

Genus *Potos* E. Geoffroy and F. G. Cuvier, 1795
Potos flavus (Schreber, 1774)
Kinkajou, Cuchicuchi

Description
The kinkajou is unique among the Procyonidae in having a prehensile tail. Head and body length averages 500 mm; tail length is about 450 mm. The range of measurements for four specimens form Colombia was HB 433–54; T 424–68; HF 87–98; E 40–42. Weight averages about 3 kg. The head is rounded, and the muzzle is short but pointed. The teeth are low crowned. The basic dorsal pelage is honey brown, and the fur is short and dense (plate 17).

Range and Habitat
The kinkajou is distributed from Veracruz, Mexico, to southern Brazil. It is associated with forested habitats, in keeping with its strong arboreal adaptations (map 10.12).

Natural History
Kinkajous specialize on fruits and arboreal vertebrates as their primary food items. Their ability to manipulate objects with their forepaws is well documented (McClearn 1992). Charles-Dominique et al. (1981) monitored their foraging behavior and found that they are important dispersers of the seeds of *Ficus, Virola,* and *Inga,* which are ingested with the fruit pulp and pass unharmed through the digestive tract. Redford and Stearman (1993) report that seasonally they may feed on termites in addition to fruits. Often several kinkajous may be found eating in the same tree.

In good habitat, densities can reach fifty-nine per km^2 (Walker and Cant 1977). Julien-Lafferrière (1993) has confirmed much of the informaiton above with a radio-tracking study. In Panama *Potos flavus* was studied by radiotelemetry. Three individuals, two males and one female, exhibited considerable home range overlap (35% between a male and female). Ranges of 8.2 to 53 ha were recorded. The authors noted allogrooming and frequent contact during nocturnal movements (Kays and Gittleman 1995). A single young is born after a rather long gestation of 112 to 120 days; social bonds between mother and offspring may persist for a considerable time. This species has been studied extensively in captivity (Poglayen-Neuwall 1962, 1966; Forman 1985).

Genus *Procyon* Storr, 1780
Procyon cancrivorus (F. Cuvier, 1798)
Crab-eating Raccoon, Osito Lavador, Mayuato

Description
Measurements: TL 1,003; HB 678; T 323; HF 142; E 54; Wt 8.8 kg. This unmistakable procyonid resembles its northern conspecific, *P. lotor*, but has longer legs and shorter fur. The dorsum is agouti black and yellow with yellowing on the sides and venter. The tail is well furred and striped. The face has a prominent dark mask across the eyes.

Distribution
P. cancrivorus is found from Costa Rica south through eastern and western Paraguay, Uruguay, and into the northern Argentine provinces at least as far south as Santa Fe province (Barlow 1965; Honacki, Kinman, and Koeppl 1982; BA, UConn, UM) (map 10.13).

Life History
In Paraguay a female was collected with six embryos, and in northeastern Argentina the crab-eating raccoon reproduces from May to July with a litter size from two to four. The development schedule for the young is similar to that for *P. lotor* (Crespo 1982; Löhmer 1976).

Ecology
P. cancrivorus is found in many habitats, for example, the gallery forest of the llanos, the xeric chaco vegetation, and the moist forest of Amazonas, but apparently always near water. It is basically nocturnal, spending the day asleep in trees. Densities fluctuate widely, but 1/km^2 is not unreasonable in suitable habitat. One stomach examined was filled with fruit pulp (Crespo 1982; Lucero 1983; Schaller 1983).

FAMILY MUSTELIDAE
Weasel Family

Diagnosis
This group consists of small to medium-sized carnivores, generally showing a reduction in molar number. Dental formula for most South American forms is I 3/3, C 1/1, P 3/3, M 1/2. The otters (*Lutra*) do not show the reduction in premolar number, and their formula is 4/3 (see figs. 10.5 and 10.6). The body tends to be elongate with relatively short legs and usually a long tail. The family is classically divided into five subfamilies, three of which occur in South America: the subfamilies Mustelinae (the weasels), Mephitinae (the skunks), and Lutrinae (the otters).

Distribution
Species of the weasel family occur on all continents except Antarctica and Australia. The family has a northern origin, since mustelids first appear in the Oligocene in North America and Europe. Species of the Mustelidae entered South America in the late Plio-

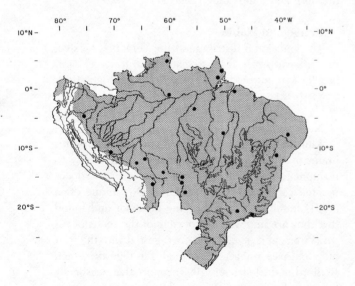

Map 10.13. Distribution of *Procyon cancrivorus*.

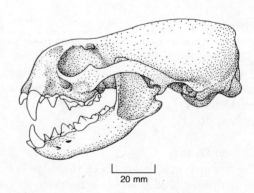

Figure 10.5. Skull of *Lontra* sp.

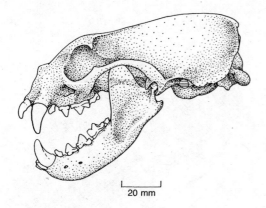

Figure 10.6. Skull of *Eira barbara*.

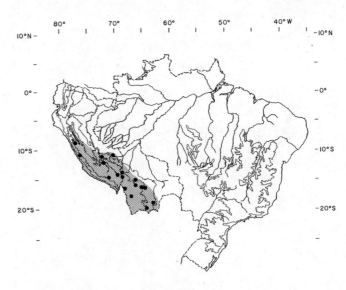

Map 10.14. Distribution of *Conepatus chinga*.

cene and rapidly occupied the small carnivore ecological niches.

Genus *Conepatus* Gray, 1837
Hog-nosed Skunk, Mapurite

Description
The hog-nosed skunks have a typical pelage pattern, being black on the sides and venter with either a strongly contrasting white dorsum or two white dorsal stripes. The length of the head and body ranges from 300 to 490 mm and the tail from 160 to 210. Weights may reach 4 kg. The muzzle is bare and the nose pad is large and flat. The claws of the forepaws are well developed.

Distribution
The genus *Conepatus* is distributed from the southwestern United States to southern Argentina. It is broadly tolerant of many habitat types (Kipp 1965).

Conepatus chinga (*rex*) (Molina, 1792)
Zorrino Común, Zorro Andino

Description
Measurements: TL 518; HB 170–288; T 230–377; 59–71; 24–28; Wt 1.5–2.2 kg. *C. chinga* is blackish to brownish black with a varying amount of white hair. The white stripes vary in length from ones that extend the length of the body and onto the base of the tail to no stripes at all and only a few white hairs on the head. In Uruguay the male is larger than the female and much heavier (Van Gelder 1968; pers. obs.).

Distribution
C. chinga is found from Bolivia south through Uruguay and western Paraguay into Argentina at least as far south as Neuquén and throughout Chile (Honacki, Kinman, and Koeppl 1982; Osgood 1943) (map 10.14).

Life History
In Uruguay this species is apparently monestrous with a protracted breeding season (Barlow 1965).

Ecology
This skunk is found in many habitats from the Paraguayan Chaco into the precordillerean steppe. In Tucumán province (Argentina) it occurs at least to 3,000 m. In Uruguay *C. chinga* prefers open country and has apparently become more abundant as woodland has been replaced by grassland. It digs its own burrow as well as using burrows dug by other animals. It is nocturnal and eats mostly beetles, orthopterans, and spiders. The stomach of one individual collected in Paraguay contained ten frogs (Barlow 1965; Lucero 1983; UM).

Conepatus semistriatus (Boddaert, 1784)

Description
Two white dorsal stripes contrasting with a black body immediately distinguish this species from other mustelids within its range.

Range and Habitat
Conepatus semistriatus is distributed from the Yucatán, Mexico, through northern Colombia into northern Venezuela and thence to Peru and eastern Brazil (map 10.15).

Natural History
The hog-nosed skunk is nocturnal and forages for invertebrates and small vertebrates by digging in the soil. Home ranges of 18 to 53 ha have been recorded in Venezuela (Sunquist, Sunquist, and Daneke 1989). It uses the burrows of armadillos as shelters, but it can

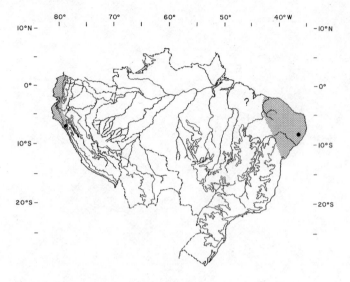

Map 10.15. Distribution of *Conepatus semistriatus*.

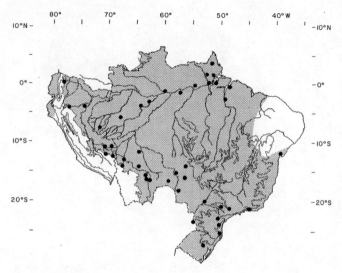

Map 10.16. Distribution of *Eira barbara*.

also construct its own. Four to five young are born after a gestation period of approximately sixty days. Aside from the female-young unit, hog-nosed skunks tend to forage singly. Strong-smelling secretions from the anal glands are employed in self-defense (Mondolfi 1973).

Genus *Eira* H. Smith, 1842
Eira barbara (Linnaeus, 1758)
Tayra, Hurón Mayor

Description
Measurements: TL 959; HB 613; T 334; HF 90; E 36; Wt 3.9 kg. This large mustelid is unmistakable. It has a slender body, and its long legs are equipped with strong claws. The coat color is variable. In general the animals are brown to black with varying amounts of white or tan on the head and throat extending onto the chest. In Panama specimens are usually completely black. In Argentina animals have very dark bodies, with the head and shoulders much lighter brown and a conspicuous cream blaze on the chest. Its black eyes shine blue green at night (Kaufmann and Kaufmann 1965; pers. obs.).

Distribution
The tayra is found from southern Mexico south to Paraguay and northern Argentina, where it is found at least as far south as Catamarca province (Redford and Eisenberg 1992; Lucero 1983) (map 10.16).

Life History
In captivity a female comes into estrus six months after giving birth. After a gestation period of sixty-three to seventy days a litter of two is produced. Young weigh about 100 g and open their eyes at thirty-eight

and forty-one days of age. In captivity a female produced young for the first time at twenty-three months of age (Poglayen-Neuwall 1978; Vaughn 1974).

Ecology
Tayras are active both during the day and at night but are primarily diurnal. They are usually encountered as solitary individuals or family groups (Kaufmann and Kaufmann 1965). In the Venezuelan llanos a female with nursing young had a home range of 9 km^2 through which she traveled extensively (Sunquist, Sunquist, and Daneke 1989). Konecny (1989) recorded home ranges of 16 km^2 for an adult female and 24.4 km^2 for an adult male.

Tayras are found predominantly in forest, though they will range away from trees. They are excellent climbers and are often seen high in trees. Although often termed a frugivore, this large mustelid can be an active predator. Individuals have been seen killing iguanas and chasing agoutis and monkeys (*Saimiri* and *Saguinus midas*) (Galef, Mittermeier, and Bailey 1976). In Belize scats contained (by occurrence; n = 31) 68% fruit and 58% arthropods; small vertebrates were taken with equal frequency. The arboreal spiny rat was the most commonly taken vertebrate in Venezuelan gallery forests (Konecny 1989; Sunquist, Sunquist, and Daneke 1989).

Genus *Galictis* Bell, 1826
Grison, Hurón, Furão

Description
This genus traditionally included three species (*G. cuja*, *G. allamandi*, and *G. vittata*) distributed from southern Veracruz in Mexico to Argentina. Most authorities consider consider the species *G. allamandi* to

be a junior synonym of *G. vittata* (Wozencroft 1993). The genus *Grison* is a synonym for *Galictis*. Head and body lengths are from 475 to 550 mm. The tail is short, usually less than 150 mm. Animals may weigh up to 3.2 kg. The limbs, throat, and venter are black. A white stripe from the forehead to the shoulders demarcates the ventral color from the dorsum. The dorsum is a grizzled salt-and-pepper pattern (plate 17).

Galictis cuja
Grison, Hurón Menor

Description
Measurements: TL 543; HB 387; T 154; HF 56; E 23; Wt 1.6 kg. *G. cuja* is smaller than *G. vittata* and resembles a large weasel in proportions, but it is more robust and has a striking black-and-white pattern. The face, sides, and underparts are black and sharply delineated from the yellowish gray dorsum. On the head a white stripe extends across the forehead and down the sides of the neck. There is quite a lot of variation in the amount of lighter dorsal coloration as well as in the size and placement of the white stripes (Barlow 1965; Osgood 1943; pers. obs.).

Distribution
G. cuja is found from southern Peru, Paraguay, southern Brazil, to central Chile, and throughout Argentina (Honacki, Kinman, and Koeppl 1982; J. A. Iriarte, pers. comm.; Texera 1974b; BA) (map 10.17).

Ecology
The grison is found in a great range of habitats that have water and good cover to at least 3,500 m. It is also found in the xeric Chaco of Paraguay. In Peru one specimen was collected in the afternoon with a stomach full of mice and a lizard. In Patagonia (Valdez peninsula) grisons were seen in a group of three and a group of four, possibly mothers and young (Barlow 1965; Lucero 1983; Osgood 1943; Pearson 1957; Andrew Taber, pers. comm.).

Galictis vittata (Schreber, 1776)
Grison, Hurón

Description
Animals weigh up to 3.2 kg. Head and body length ranges from 475 to 550 mm. The limbs, throat, and venter are black, and the black extends to the face. A white stripe from the forehead to the shoulders separates the dorsum from the venter. The dorsum has a grizzled "salt and pepper" pattern.

Range and Habitat
Galictis vittata occurs from southern Veracruz, Mexico, to Brazil, where it is replaced by *G. cuja*

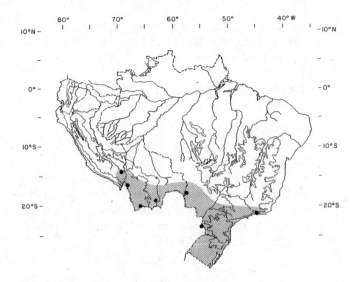

Map 10.17. Distribution of *Galictis cuja*.

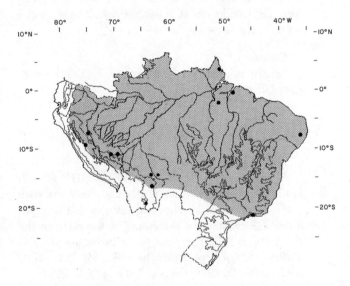

Map 10.18. Distribution of *Galictis vittata*.

(Krumbiegel 1942). It usually occurs below 1,200 m but is broadly tolerant of the vegetation cover, being found in both dry deciduous forest and multistratal tropical rain forest (map 10.18).

Natural History
Grisons are active in the early morning and late afternoon and at night. Although they can climb, they generally forage terrestrially (Kaufmann and Kaufmann 1965). They can eat a wide range of foods but are primarily predators on reptiles, small birds, and small mammals. One radio-collared adult female had a home range of 4.2 km^2 in the llanos of Venezuela (Sunquist, Sunquist, and Daneke 1989). Grisons frequently shelter in burrows, often using abandoned armadillo burrows. They may be seen in groups of three or four,

probably a mother and her young. The gestation period is thirty-nine days, and litter size averages two (Miles Roberts, pers. comm.).

Genus *Lontra* Schreber, 1777
Freshwater Otter, Perro da Agua, Lontra

Description
The dental formula is 3/3, 1/1, 4/3, 1/2 (see fig. 10.5).

A cylindrical body form, very short ears, webbed feet, and stout vibrissae characterize this aquatic carnivore. The tail is thick at the base, tapering to a point. The body contours are smooth. There are approximately twelve species worldwide.

Distribution
The genus is distributed in freshwater and coastal areas of North and South America, most of Africa, Europe, Asia, and parts of Southeast Asia. Three species occur in southern South America.

Comment
Lontra is recorded by Wozencroft (1993) rather than *Lutra* Brunnich, 1771.

Lontra felina (Molina, 1792)
Lobito Marino, Gatuna

Description
Measurements: TL 870–910; T 340; HF 90–97; E 15; Wt 4.5 kg. This small otter has a uniform dark coffee color on the dorsum and a venter that is sometimes lighter than the dorsum. The top edge of the rhinarium is flat, in contrast to the biconcave top rhinarium edge of *L. provocax.* Juveniles are darker than adults (Cabello 1978; Osgood 1943; Sielfeld 1983).

Distribution
L. felina is found from coastal Peru south along the whole Chilean coast in isolated populations to the

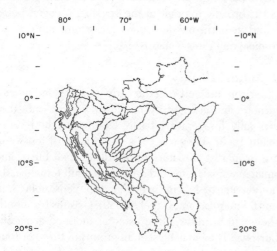

Map 10.19. Distribution of *Lontra felina.*

Strait of Magellan, where it is found in south coastal Argentina as well (map 10.19). The original range has decreased because of hunting (Melquist 1984; Olrog 1950; Torres, Yáñez, and Cattan 1979).

Life History
In the southern part of the range there seems to be a birth peak between September and October. The gestation period is sixty to sixty-five days, and the litter size is two. Young are born in earthen dens or rock crevices and stay with the female for approximately ten months (Cabello 1978, 1983; Sielfeld 1983).

Ecology
L. felina is apparently found virtually only in marine waters, although there are many reports of this species swimming up rivers. Apparently limited to places on the coast with rocky outcroppings, *L. felina* is therefore found in disjunct populations all down the Chilean coast. In the northern part of its range it occurs in small, scarce populations because of a paucity of suitable habitat. In areas where it occurs, densities vary from 0.04 to 10 per km of coastline. The areas where *L. felina* is found are characterized by heavy seas and strong winds. It stays within about 500 m of the coast and occurs mostly solitarily, although groups of three have been seen.

These otters usually sleep ashore and give birth in dens in the rocks. They will shelter in caves at water level. Most of the aquatic activity takes place in the afternoon. Food dives last fifteen to forty-five seconds, and the prey is usually eaten in the water. On Chiloé Island 73% of the feeding observations were on shellfish and 27% on fish. They have also been reported as eating freshwater prawns, crabs, cuttlefish, and a little algae (Castilla 1982).

Darwin reported this species to be very abundant in the last century. Since that time heavy hunting for fur has greatly reduced the numbers of *L. felina* (Allen 1905; Brownell 1978; Cabello 1978, 1983; Melquist 1984; Sielfeld 1983; Van Zyll de Jong 1972).

Lontra longicaudis (Olfers, 1818)
Lobito Común, Guairão

Description
Measurements: TL 1,053; HB 513; T 540; HF 120; E 19; Wt 5.8 kg. This otter is a lustrous grayish brown dorsally and slightly lighter on the venter. The tip of the muzzle and the mandible are usually blazed with yellowish white. The nose pad distinguishes this species from *L. provocax* (Barlow 1965).

Distribution
L. longicaudis is broadly distributed to the south from the Pacific coast riverine drainages of southern

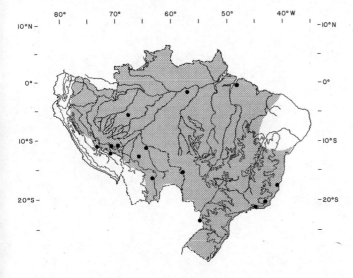

Map 10.20. Distribution of *Lontra longicaudis.*

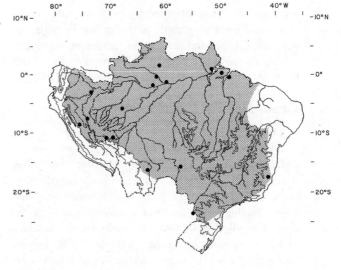

Map 10.21. Distribution of *Pteronura brasiliensis.*

Mexico, including both Atlantic and Pacific systems, to Uruguay and to the major rivers of Paraguay and Argentina (Wilson and Reeder 1993; Massoia 1976; Melquist 1984) (map 10.20).

Life History

In northeastern Argentina *L. longicaudis* reproduces in the spring, with a litter size of two or three. In central Brazil a female had three embryos (Crespo 1982; AMNH).

Ecology

L. longicaudis is found in areas that meet its habitat requirements: ample riparian vegetation along permanent streams and lakes and a readily available year-round food source (Spinola and Vaughan 1995). The animals shelter in terrestrial burrows that they themselves dig. In Jujuy province, Argentina, they are found to at least 3,000 m elevation. Hunting has considerably reduced their abundance and distribution. These otters eat fish as well as mollusks and crustaceans (Crespo 1982; Griva 1978; Massoia 1976; Melquist 1984; Olrog 1979).

Comment

L. platensis is considered a synonym.

Genus *Pteronura* Gray, 1837
Pteronura brasiliensis (Gmelin, 1758)
Giant River Otter, Lontra Gigante, Arirai

Description

Measurements: TL 1625; T 440; Wt 24–29 kg. This otter is unmistakable because of its size. Its long tail becomes dorsoventrally flattened in its distal half. The broad head is flattened, and the eyes are large.

The ears are small and round, and the feet are large and well webbed. The fur, which is thick, glossy, and short, looks shiny chocolate black when wet, drying to brown. The lip, chin, throat, and chest may be spotted with creamy white, which may be almost absent in some animals or expanded to form a bib in others (plate 17) (Duplaix 1980).

Distribution

P. brasiliensis is found in the major river systems of South America east of the Andes as far south as northern Argentina (Honacki, Kinman, and Keoppl 1982; Massoia 1976; Melquist 1984) (map 10.21).

Life History

In captivity the gestation period was recorded as between sixty-five and seventy days, with litters ranging from two to five. Otterlets were born at about 200 g and opened their eyes at thirty-one days. In northeastern Argentina this species reproduces in spring and summer, and litter sizes of two to three have been noted. In Suriname litters of one to three are born at the beginning of the dry season (Autuori and Deutsch 1977; Crespo 1982; Duplaix 1980).

Ecology

These large, diurnal otters seem to prefer slow-moving creeks and rivers. In Suriname, where they have been well studied, they move seasonally as water levels and fish populations change. Otters occur in groups of up to ten, composed of a pair-bonded male and female and related individuals. This group tends to occupy one stretch of river and scent mark certain areas, as well as using common latrines. This species is highly vocal and is easily recognized by its "barks." A large portion of time is spent ashore grooming. Cubs are born in a den attended by both parents. Otters travel and hunt in groups, pursuing fish by sight. In Suriname characoid fish were preferred (Duplaix 1980). The natural history of the species in Brazil has been covered in a monograph by Schweizer (1992).

Genus *Mustela* Linnaeus, 1758

Description

Species of the genus *Mustela* exhibit a wide range of head and body lengths (114 to 228 mm). The tail is usually less than half the head and body length. Sexual size dimorphism is typical, with females generally much smaller than males, but apparently not in *M. felipei*. The soles of the feet are usually furred, but two South American species, *M. africana* and *M. felipei*, have naked soles. The dorsum is brown and the venter white or cream. The tail tip is often black.

Distribution

The genus is widely distributed in the temperate zones of North America, Europe, and Asia. Three species occur in South America: *M. africana*, *M. felipei*, and *M. frenata* (Izor and de la Torre 1978).

Mustela africana Desmarest, 1818

Description

One male specimen measured HB 300; T 240; HF 58; E 27 (Belém). This species is a typical weasel in body form and proportion. Chestnut brown dominates the dorsum and the flanks. The venter and medial parts of limbs are cream, and this color extends to the neck and chin. Some specimens exhibit a blaze of chestnut brown in the midline of the venter. The tail is uniformly brown

Distribution

The species is sparsely distributed in the Amazon basin of Brazil, Ecuador, and Peru (map 10.22).

Mustela felipei Izor and de la Torre, 1978

Description

Head and body length ranges from 217 to 220 mm, and the tail averages 107 mm. The baculum is trifid, a character shared with *M. africana* (Izor and Peterson 1985). The face, dorsum, and tail are dark brown, and the venter is pale yellow from the chin to the groin. The plantar surfaces are naked.

Range and Habitat

The species is known from only two localities in the Cordillera Central of Colombia: San Augustín, Huila, and Popayán, Cauca, between 1,700 and 2,700 m, but it could extend to Ecuador (Izor and de la Torre 1978) (not mapped).

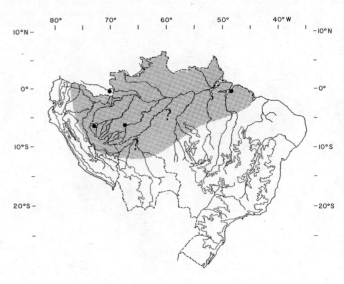

Map 10.22. Distribution of *Mustela africana*.

Mustela frenata Lichtenstein, 1831
Long-tailed Weasel

Description

Head and body length ranges from 180 to 220 mm, and the tail is 50% to 60% of the head and body length. The plantar surfaces are furred. The dorsum is brownish, the venter is cream to white, and the distal third of the tail is usually black. In the northern part of the range the dorsal coat changes to white in winter, but in the southern parts, including South America, there is no seasonal change.

Range and Habitat

This species is distributed from southern British Columbia in Canada to northern South America. It has been taken in scattered localities of Colombia and Venezuela (not mapped). It is associated with the premontane forests of the Andes and with the forests of the Guyana highlands in southern Venezuela and extends to Ecuador, Peru, and possibly Brazil. It prefers wooded habitats and tolerates a wide range of elevations (Hall 1951) (map 10.23).

Natural History

The long-tailed weasel is a specialist predator on ground-nesting birds and rodents. It does not typically forage in trees, although it can climb. This species has been studied in some detail in the North American part of its range; little is known concerning the natural history of the southern subspecies. In the north temperate zone, the long-tailed weasel mates in the summer and exhibits delayed implantation so that the young are born the following spring. The actual gestation time, however, is twenty-three to twenty-four days when the time of delay is subtracted. Litter size may be high, up to seven in North America. Although the

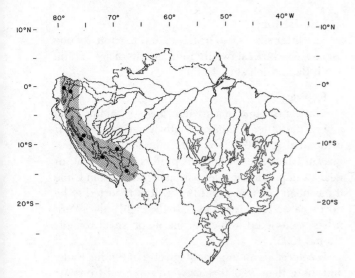

Map 10.23. Distribution of *Mustela frenata*.

mother and young may forage as a social unit during the weaning phase, adults typically are solitary except during the breeding season (Hall 1951). Seasonality of breeding in the tropics has not been investigated. The biology of this species has been reviewed by Sheffield and Thomas (1997).

FAMILY FELIDAE
Cats

Diagnosis
The tooth formula is extremely modified in this family. Most notably, the number of molars is reduced. The dental formula typically is I 3/3, C 1/1, P 2-3/2, M 1/1 (figs. 10.7 and 10.8). The auditory bulla is inflated and subdivided into two chambers. Five toes are retained on the forefoot and four on the hind foot. The claws are retractile except in the African cheetah (*Acinonyx*). The eyes are directed forward, and the rostrum is rather short.

Distribution
The Recent distribution includes all continents except Australia and Antarctica (where they were introduced). Members of the family Felidae are distinguishable in the Oligocene. The early cats appear in North America, Eurasia, and later in Africa. They are placed in the subfamily Nimravinae. Subsequently one could recognize two divergent groups, the true cats and the saber-toothed cats. The latter are frequently placed in their own subfamily, the Machairodontinae. Members of the Felidae entered South America in the Pliocene and rapidly replaced the large carnivorous birds (*Phorusrhachidae*) that then occupied the medium to large cusorial carnivore niche (Vuilleumier 1985). Gomes de Oliveira (1992) has recently summa-

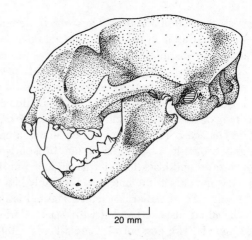

Figure 10.7. Skull of *Felis (Leopardus) pardalis*.

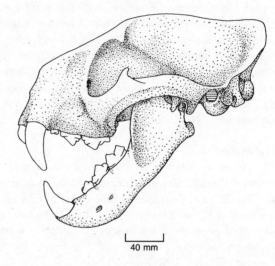

Figure 10.8. Skull of *Panthera onca*.

rized the status of the South American cats. Guix (n.d.) has attempted a similar evaluation for Brazil.

Comment on Taxonomy
Wozencroft (1993) has elevated previously recognized subgenera of *Felis* to full generic status. In this account *Felis* is employed at the generic level with the generic nomenclature according to Wozencroft given in parentheses. Although the smaller Neotropical felids exhibit at least three different lineages, it is not at all clear that the current use of subgenera reflects phylogenetic affinity. *Felis tigrina* presents genuine problems (Johnson et al. 1996).

Felis (Oncifelis) colocolo Molina, 1810
Gato de Pajonal

Description
Measurements: TL 855; HB 571; T 279; HF 123; E 53; Wt 3 kg. This account includes *F. pajeros*. This

robust cat, the size of a bobcat (*Lynx rufus*), demonstrates great variation in coloration. In Chile the dorsum and sides are mottled reddish brown and gray with faint banding. The venter is white with black stripes, and the forelimbs have bold black stripes separated with white. In Argentina and adjacent Bolivia, the striping on the forelimbs and hind limbs is the only constant character. The dorsum can be almost a uniform gray brown to a basic gray to tan, broken with brown rosettes and black lines in the middorsum. In quite a few individuals the hair is considerably longer than on other South American felids. The tail is full and well haired, usually banded with black (Barlow 1965; Osgood 1943; pers. obs.). Garcia-Perea (1994) considers *F. colocolo* to be separable under a full genus as *Lynchailurus*. She further offers evidence to support a subdivision of *F. colocolo* into three full species. Following her analysis the coastal races of Chile would be referable to *L. colocolo*. The Andean forms of Peru, Chile, and Argentina (including three forms in Patagonia) would be named *L. pajeros*. The specimens from Uruguay and Brazil could be named *L. braccatus*. This revision recognizes preexisting names formerly relegated to subspecific status.

Distribution

F. colocolo is distributed from Peru south to southern Brazil through Uruguay and in western Paraguay into Chile from approximately Coquimbo to Chillán. In Argentina *F. colocolo* is found in the northwestern and central portions of the country south to at least Río Negro province (Tamayo and Frassinetti 1980; Ximénez 1961; BA, UM) (map 10.24).

Silviera (1995) reviews the distribution of *Felis colocolo braccatus* in Brazil and confirms its presence in the semiopen habitats of the cerrado, Matto Grosso, and Pantanal.

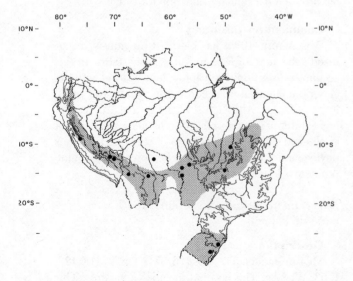

Map 10.24. Distribution of *Felis (Oncifelis) colocolo.*

Life History

The litter size is two or three, and in captivity one female gave birth at twenty-four months of age (Eaton 1984; Rabinovich et al. 1987).

Ecology

This cat is found in a greater range of habitats than any of the other southern South American felids. It is found above 5,000 m in the Andes and in the Paraguayan Chaco. In Uruguay it prefers low areas in or near swamps and marshes with tall grass, and in Chile it has been taken in cloud forest. It is reported to be nocturnal and highly arboreal. In the highlands of Peru it has been noted to feed primarily on small rodents (Romo 1995).

F. colocolo is an important species in the fur trade, and more than 78,000 skins were exported from Argentina between 1976 and 1979 (Barlow 1965; Mares and Ojeda 1984; Pearson 1951; Schamberger and Fulk 1974; Wolffsohn 1908).

Comment

It may be possible to split *colocolo* into three species: *Lynchailurus colocolo; L. pajeros;* and *L. braccatus.* This has been proposed by Garcia-Perea (1994). All are previously named forms. The Brazilian-Uruguayan form, *L. braccatus,* is morphologically distinct and disjunct.

Felis (Oncifelis) geoffroyi d'Orbigny and Gervais, 1844
Geoffroy's Cat, Gato Montés

Description

Measurements: TL 781–956; HB 503; T 277–353; HF 100–135; E 50–56; Wt 2.4–5.2 kg. *F. geoffroyi* is a distinctive small, lithe spotted cat. Its basic body color is gray to reddish brown, with black spots, not rosettes, as the predominant pattern. The ground color lightens on the sides. On the sides, neck, and legs these spots can join to form indistinct parallel stripes. It is the size of a very large domestic cat but has a shorter tail and a longer, more flattened head. The tail is the same basic color as the dorsum and is narrowly banded with black stripes. Melanistic individuals are not unusual (Barlow 1965; Ximénez 1975; pers. obs.).

Distribution

Geoffroyi's cat is found from Bolivia across southern Brazil and into the Paraguayan Chaco, throughout Uruguay and in Argentina south to the southernmost part of the continent, where it enters southern Chile (Melquist 1984; Texera 1974a; Ximénez 1973; UConn) (map 10.25).

Life History

In captivity *F. geoffroyi* has a gestation period of seventy-four to seventy-six days and a litter size of one

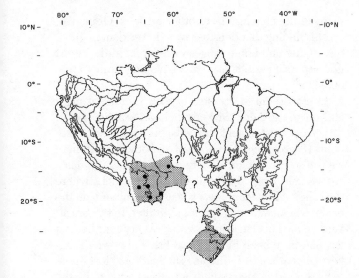

Map 10.25. Distribution of *Felis (Oncifelis) geoffroyi*.

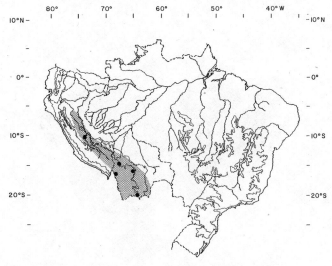

Map 10.26. Distribution of *Felis (Oreailurus) jabcobita*.

to three. Kittens are born weighing 65 g and open their eyes at about twelve days. In Uruguay this cat has one litter per year of two or three kittens (Scheffel and Hemmer 1975; Ximénez 1975).

Ecology

F. geoffroyi is primarily a cat of subtropical and temperate regions of South America. It ranges from sea level to 3,300 m in Bolivia. In Uruguay preferred habitats are open woodland, brushy areas, open savannas, and marshes. It apparently does not occupy subtropical rain forest or the southern coniferous forest, being replaced in the latter habitat by *F. guigna*. In southern Chile this species had modest home ranges. Average range for five males was 9.21 km². Two adult females exhibited ranges of 5.2 km² and 2.3 km² (Johnson and Franklin 1991).

This primarily nocturnal hunter preys on small rodents and birds. Stomachs have been found to contain *Cavia, Ctenomys, Phyllotis, Lepus,* and the birds *Myiopsitta* and *Nothura.* One stomach had five *Oryzomys,* two *Noctura,* and an unidentified passeriform. *F. geoffroyi* is the most important species of spotted cat in the skin trade of the southern cone. More than 340,000 skins were exported from Argentina between 1976 and 1979 (Barlow 1965; Gibson 1899; Melquist 1984; Ojeda and Mares 1984; Ximénez 1973, 1975; UConn).

Felis (Oreailurus) jacobita Cornalia, 1865
Andean Cat, Gato Andino

Description

Measurements: TL 1,047; HB 606; T 441; HF 124; E 31; Wt 4 kg. This medium-sized cat resembles a small snow leopard (*Panthera uncia*) with its long, luxurious fur. The dorsal pelage is pale gray, spotted, and

transversely striped with blackish or brownish. The underparts are white. There are two faint middorsal stripes on the neck and three lines on the forehead. The feet are well furred except for the pads. The tail is long and banded with about seven rings, although some may be very faint. It is larger than *F. colocolo* and *F. guigna* and lacks the spinal crest of long hair seen on many specimens of *F. colocolo.* The legs are banded with two or three bars of black (Osgood 1943; Pearson 1957; pers. obs.).

Distribution

F. jacobita is found in the Andes from west-central Peru through southwestern Bolivia to northeastern Chile and northern Argentina (Honacki, Kinman, and Koeppl 1982; Melquist 1984; Pearson 1957) (map 10.26).

Ecology

This rare felid is an inhabitant of the high Andes. It seems to be confined to treeless, rocky, semiarid, and arid portions of the Andes above 3,000 m. In Peru an individual was collected at 5,100 m in a barren region of rocks and bare ground with scattered clumps of bunchgrass and small bushes. Another individual was observed in similar habitat in the Argentine province of Tucumán at 4,250 m (Cabrera 1961; Melquist 1984; Pearson 1957; Scrocchi and Halloy 1986).

Felis (Leopardus) pardalis Linnaeus, 1758
Ocelot, Gato Onza, Ocelote

Description

Measurements: TL 1,097; HB 742–822; T 355–75; HF 160; E 54–55; Wt 7–8.8 kg. The ocelot is the largest of the small spotted cats. The upper parts of

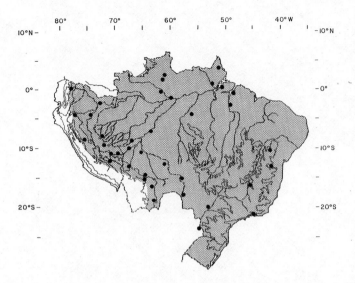

Map 10.27. Distribution of *Felis (Leopardus) pardalis*.

the body are grayish to cinnamon colored, with black markings forming streaks on the neck or elongated spots on the body. The tail exhibits incomplete banding. Spotting and banding extends to the dorsal surfaces of the limbs and onto the venter. The basic color of the venter is white with occasional black spots (plate 17).

Distribution

F. pardalis is found from the southern United States to Paraguay and northern Argentina. In Argentina it has never been known south of Entre Ríos province (Honacki, Kinman, and Koeppl, 1982; Melquist 1984; BA) (map 10.27).

Life History and Ecology

In captivity females reproduce from 2.5 years to at least 12 years of age. In Venezuela the litter size averages 1.6 with a range of 1 to 3. In northeastern Argentina ocelots reproduce from October to January, and litters range from 2 to 3 (Crespo 1982; Eaton 1984; Mondolfi 1986).

Ocelots are found in a wide range of habitats from thorn forest to tropical moist forest. Animals are solitary except when the female is being courted or has cubs. In Venezuela movements of radio-tagged ocelots were variable, with adult females traveling about half as far during a night as adult males. The larger home ranges of males encompassed the smaller ranges of adult females. Animals were usually nocturnal though sometimes active during the day. In Belize ocelots were most active in the early morning and late evening. Home ranges there were almost entirely within areas of second-growth vegetation. Crawshaw (1995) studied the ocelot in Iguazú National Park and confirmed

many of the conclusions of other workers. Of interest was his finding that densities were lower than in previous studies and home ranges were large. In this forest nondispersing adults used areas ranging from 15.6 to 50.9 km^2.

Ocelots seem to do most of their hunting at night and on the ground. In Venezuela an examination of fifteen ocelot stomachs showed a predominance of rodents (*Proechimys, Dasyprocta,* and others), but there were also armadillos, a sloth, marsupials, an iguana, snakes, and frogs. In the llanos of Venezuela, where the cane rat *Zygodontomys* is seasonally abundant, the ocelot will include this species in its diet. In the Brazilian Pantanal ocelots can be active twenty-four hours a day and have been found to eat fish and howler monkeys (*Alouatta*). In Belize an examination of forty-nine scats showed *D. marsupialis* and *P. opossum* to be the most commonly taken prey. In Amazonian Peru larger rodents and marsupials predominate. *Mazama americana, Tamandua,* and *Dasypus* were also occasionally taken (Emmons 1988; Konecny 1989; Ludlow and Sunquist 1987; Mondolfi 1986; Schaller 1983; Schaller, Quigley, and Crawshaw 1984; Sunquist, Sunquist, and Daneke 1989). Sunquist (1992) emphasizes in his review of reproduction that the ocelot with its low litter size (one) is more vulnerable to exploitation than comparable-sized Northern Hemisphere cats such as *Lynx rufus*. Murray and Gardner (1997) present the most comprehensive review of the biology of this species.

Felis (Leopardus) tigrina Schreber, 1775
Little Spotted Cat, Gato Tigre, Chivi

Description

Measurements: TL 753; HB 484; T 269; HF 104; E 43; Wt 2.2 kg. Excluding *Felis guigna*, this is the smallest of the spotted cats. It is often confused with *Felis wiedii*. It is more gracile than *F. wiedii,* and the eyes and ears are relatively larger, the tail is relatively shorter, and the snout is narrower. The ground color is dark tawny. On the neck are several stripes, extending down the center of the back as irregular broken bands. Along the sides are elongate rosettes, with the ground color darker within the rosettes. The venter is whitish between dark spots. In northern Venezuela melanistic morphs have been recorded in lowland and premontane tropical forest (Cabrera 1961; Eisenberg 1989; pers. obs.).

Distribution

The little spotted cat is found in lowland and premontane tropical forest from Costa Rica south into northern Argentina, but the exact distribution in eastern Brazil is poorly documented (Honacki, Kinman, and Keoppl 1982; Melquist 1984; Gomes de Oliveira 1992) (map 10.28).

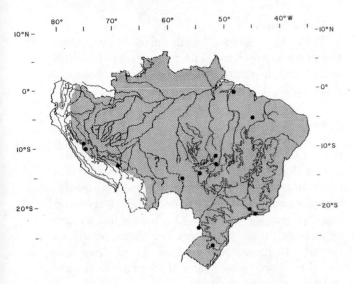

Map 10.28. Distribution of *Felis (Leopardus) tigrina*.

Life History

In captivity the gestation period of *F. tigrina* is between sixty-two and seventy-four days and the litter size is one to four, though one is the mode (Leyhausen and Falkena 1966).

Ecology

In Venezuela this species is strictly a forest dweller and shows a preference of humid evergreen forests up to 3,000 m. It is generally a ground dweller but climbs with ease. In Costa Rica it was taken at 3,500 m. Stomachs have been found to contain small rodents, a shrew, and a small passerine bird (Gardner 1971; Mondolfi 1986).

Comment

Without a doubt, this species will be split into at least two species. The eastern Brazilian form will surely be distinct, as suggested by Leyhausen (1963). In Brazil south of the Amazon, *"tigrina"* is a form disjunct from the northern *F. tigrina*. The upper premolars as noted by Husson (1978) separate all *"tigrina"* from *"wiedii"* regardless of the ultimate taxonomic designation.

Felis (Leopardus) wiedii Schinz, 1821
Margay, Gato Pintado, Gato Brasileiro

Description

Measurements: TL 893; HB 529; T 364; HF 115; E 42.3; Wt 3.2 kg. *F. wiedii* is smaller than *F. geoffroyi* with softer fur and a much longer tail. The postorbital bars of the skull almost fuse with the zygomatic arch. The fur is longer than in many of the other spotted cats, thick and very soft. The tail is proportionately longer in the margay than in the ocelot. This feature,

together with the fact that the margay tends to be smaller, distinguishes them. Young margays, however, are markedly similar in appearance to young ocelots. The margay has the same basic ground color of the dorsum, from grayish to cinnamon, strongly marked with bands and spots on the dorsum, and incomplete black stripes in a transverse pattern on the tail. The venter is white (Barlow 1965; Cabrera 1961; Eisenberg 1989; Osgood 1914).

Distribution

The margay was found from southern United States to Uruguay and northern Argentina to about 24° S (Cabrera 1961; Honacki, Kinman, and Koeppl 1982; Melquist 1984; Ojeda and Mares 1989) (map 10.29).

Life History

In captivity margay kittens are born weighing about 165 g, open their eyes at between eleven and sixteen days, and are weaned at about fifty days of age. In northeastern Argentina they reproduce between July and August and have a litter size of one (Crespo 1982; Petersen and Petersen 1978).

Ecology

F. weidii is an animal of tropical and subtropical forests. In Venezuela the margay is more specific in its habitat requirements than the ocelot, being found in humid lowland forests as well as in premontane moist forest and cloud forest. In Belize Konecny (1989) recorded a home range of 14.7 km^2 for a female and 31.2 km^2 for a subadult male. Crawshaw (1995) recorded a home range of 43.2 km^2 for a male and noted both diurnal and nocturnal activity. The margay is much more arboreal than the ocelot, and unlike

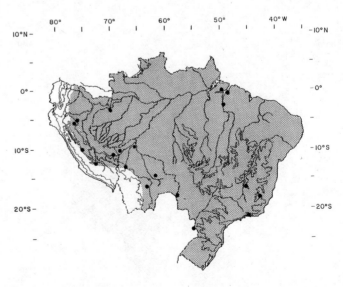

Map 10.29. Distribution of *Felis (Leopardus) wiedii*.

other cats it can pronate and supinate its hind foot. Thus, when descending a tree the hind foot can rotate around the ankle so that this cat can hang vertically, much like a squirrel. In Belize margays always rested above the ground, escaped through the trees, and had arboreal rodents as the most frequently taken prey (*n* = 27 scats). Stomachs have also been found to contain squirrels. In Panama its most frequent prey was *Dasyprocta* and *Proechimys* (Barlow 1965; Enders 1935; Konecny 1989; Leyhausen 1963; Mondolfi 1986).

Felis (Herpailurus) yagouaroundi E. Geoffroy, 1803
Jaguarundi, Gato Eyra, Gato Moro, Gato Mourisco

Description
Measurements: TL 1021; HB 641; T 381; HF 112; E 37; Wt 2.6 kg. The jaguarundi is one of the most distinctive of the cats because of its lithe, mustelid-shaped body and its lack of spots. A medium-sized cat, *F. yagouaroundi* has a flattened head, relatively small ears, and relatively short legs. Its pelt has no spots or stripes at any age and comes in a variety of colors from reddish through gray to black. Different colored animals can be found in the same population (Cabrera 1961; Mondolfi 1986; Zapata 1982; pers. obs.).

Distribution
F. yagouaroundi ranges from southern Texas south across the South American continent to Argentina. It has been collected in both eastern and western Paraguay and in Argentina as far south as Río Negro province (Honacki, Kinman, and Koeppl 1982; BA, UM) (map 10.30).

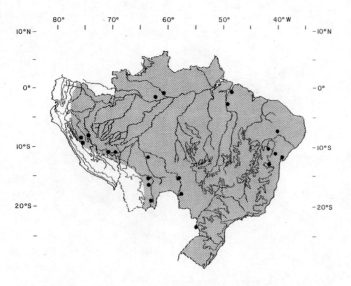

Map 10.30. Distribution of *Felis (Herpailurus) yagouaroundi*.

Life History
In captivity gestation length ranges from seventy-two to seventy-five days. Litter sizes ranges from one to four, and the age of first reproduction is about two years (Hulley 1976).

Ecology
In Venezuela the jaguarundi is found in a great variety of habitats including semiarid thorn forest, deciduous forest, swampy grassland, and moist tropical forest. In southern South America it appears to inhabit a similar range of habitats. Konecny (1989) recorded large home ranges in Belize. One female had a range of 20.1 km², while two males had ranges of 88.3 to 100 km². In Iguazú National Park, Brazil, Crawshaw (1995) recorded home ranges of 19.6 km² for a male and 7.0 km² for a female.

F. yagoauroundi is reported as being active both during the day and at night, but in Belize it is essentially diurnal. It is primarily terrestrial and consumes a large variety of small vertebrates. Manzani and Montiero-Filho (1989) report birds, small mammals, fish, and reptiles in the diet of this species. An examination of sixteen stomachs from Venezuela revealed a predominance of rodents, birds, and lizards, with an armadillo and several rabbits as well. In Belize small mammals were the major diet, principally *Sigmodon* taken from old fields. Other prey items included *Didelphis*, small birds, arthropods, and some fruit (Mondolfi 1986; Zapata 1982).

Genus *Panthera* Oken, 1816
Panthera onca (Linnaeus, 1758)
Jaguar, Yaguar, Yaguareté, Tigre Americano, El Tigre

Description
Measurements: TL 1,882–2,072; HB 1,337–1,472; T 600; Wt 61–119 kg. The jaguar is unmistakable because of its large size, powerful build, and glossy spotted coat (plate 17). This is the largest of the American cats; it has a robust build and a relatively short tail, short, rounded ears, and large feet. The female is almost always smaller than the male. Jaguars from the Pantanal of Brazil are reportedly some of the largest in South America. The ground color of the dorsum ranges from light ochraceous buff to golden tawny; the ventral surfaces and the inner surfaces of the legs are white. There is a great deal of variation in the rosette spotting. Melanistic forms have been reported for Suriname and Brazil (Almeida 1976; Cabrera 1961; Nelson and Goldman 1933).

Distribution
The jaguar was formerly distributed from the southwestern United States to northern Argentina, but it

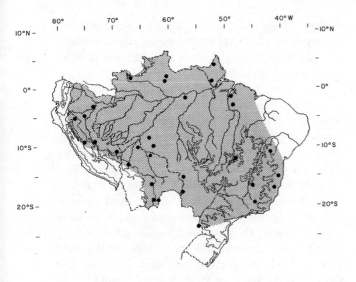

Map 10.31. Distribution of *Panthera onca*.

has been exterminated from much of its range. It persists in eastern Brazil within fragments of the Atlantic rain forest (Arra 1974; Carman 1984; Guix 1992; Ojeda and Mares 1982 (map 10.31). Jorgenson and Redford (1993) offer provocative arguments concerning the past distribution of the jaguar and human influences on its abundance.

Life History

In captivity jaguars can come to estrus every month. The gestation period ranges from 101 to 105 days, the litter size is two, and cubs are born weighing 970 g. They open their eyes within three days of birth, eat solid food at 75 days, and cease nursing at about 157 days of age. In Venezuela females are aseasonally polyestrous and have litters averaging two (range one to four). In northeastern Argentina jaguars reproduce between March and July and have litters of two or three. Cubs may stay with their mothers for a year and a half before dispersing (Almeida, n.d.; Crespo 1982; Mondolfi and Hoogesteijn 1986; Stehlik 1971).

Ecology

Jaguars occur in a wide variety of habitats but seem to require abundant cover, water, and sufficient prey. They are found in the xeric chaco as well as in tropical forest. In the Brazilian Pantanal, where jaguars have been studied, females range over at least 25–38 km² and males over an area more than twice as large. The ranges of females overlap, and the ranges of males include those of several females. However, it seems that the land tenure system of jaguars may vary with prey density. Jaguars are solitary except when females are in heat and when a female is accompanied by cubs.

Crawshaw (1995) studied the jaguar in Iguazú National Park with the aid of radiotelemetry. Within the subtropical forest the home ranges were relatively large, and the prey taken weighed less on average than in the Pantanal. Nondispersing adults had home ranges of 70 to 86.5 km².

Jaguars are most active after dusk and before dawn, though they may be active throughout the day. The loud, repeated hoarse cough (like someone sawing wood) seems to be an important form of long-distance communication. An animal may stay at a kill for several days, and kills are sometimes covered with vegetation.

In different areas jaguars take different prey items in different frequencies. They are remarkably catholic in their diet. In Venezuela jaguars have been noted to eat caiman eggs, tortoises, porcupines, armadillos, collared peccaries, and brocket deer. Where cattle are common, calves can become a major part of the diet. In the Pantanal, of thirty-six stomachs examined, twelve contained cattle, five capybaras, four white-lipped peccaries, four collared peccaries, and two caimans. Other prey included *Tamandua* anteaters, howler monkeys, sloths coatis, tapirs, giant anteaters, marsh deer, jabiru storks, rheas, fish, and land tortoises. There appears to be variation between areas in favored foods, which may be due to preferences passed from female to offspring. In Belize armadillos, pacas, and brocket deer accounted for 70% of identified prey. The range and density of this species have both been negatively affected by human hunting (Almeida 1976, n.d.; Hoogesteijn and Mondolfi 1992; Mondolfi and Hoogesteijn 1986; Rabinowitz and Nottingham 1986; Schaller 1983; Schaller and Crawshaw 1980; Schaller, Quigley, and Crawshaw 1984; Schaller and Vasconcelos 1978).

Genus *Puma* Jardine, 1834
Puma concolor (Linnaeus, 1771)
Puma, León Americano

Description

Measurements: TL 1,554–1,699; HB 1,080; T 571–619; HF 220–36; E 82–85; Wt 23.6–50.4 kg. The puma is unmistakable because of its large size, monocolored coat, and lithe build. There are several color phases including reddish, dark brown red, gray, and tawny. The tail is long and often dark tipped. The ears are short and rounded, the lateral muzzle and backs of the ears are also often black. The chin, median muzzle, and ventral area are creamy white. Young are spotted and striped. The average size of the puma may vary greatly over its geographic range (Currier 1983; Sanborn 1954; pers. obs.).

Distribution

The puma ranges from northern Canada south to southern Chile and Argentina, although it is extinct or very rare over much of its former range. In southern

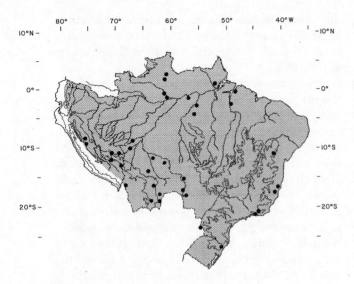

Map 10.32. Distribution of *Puma concolor.*

South America it is found in Paraguay, Uruguay, Argentina south to Tierra del Fuego, and all of Chile (Cabrera 1961; Currier 1983; Ximénez 1972) (map 10.32).

Life History

The gestation period is eighty-two to ninety-six days; the litter size varies from one to six; young are born weighing about 400 g; and adult weight is attained at about two to four years of age. In northern Argentina pumas reproduce from May to July and have a litter of two to three cubs (Crespo 1982; Currier 1983).

Ecology

Puma concolor is found in a wide range of habitats from the moist tropical forests of northeastern Argentina to above the tree line in the Andes. Tracks have been recorded above 5,800 m in southern Peru. In North America *P. concolor* is polygamous, and individuals have stable home ranges. Animals are solitary except for females with cubs. In the Brazilian Pantanal pumas were active throughout the day and night and used the same areas as jaguars, but the species seemed to avoid each other. In this area the home range of one male was 32 km².

In Panama, pumas preyed on collared peccaries, *Mazama* and *Odocoileus* deer, pacas, agoutis, *Proechimys*, iguanas, and snakes. In southwestern Brazil they have been recorded killing *Ozotoceros* and *Mazama* deer and a *Tamandua* anteater, as well as a rhea. In the Paraguayan Chaco, one puma stomach contained a peccary, a rabbit, and a *Tolypeutes* armadillo. An examination of twenty-one scats from the same area showed *Dasypus* armadillos in seven, collared peccaries in three, *Mazama gouazoubira* deer in four,

rabbits in two, coatis in one, and *Tamandua* anteaters in four. In Venezuela the cats prey on *Mazama odocoileus, Dasyprocta,* and *Dasypus* in addition to other small game (Bisbal 1986).

In Patagonia pumas prey on European hares (*Lepus*) and guanacos (*Lama*), and young guanacos are especially vulnerable. Pumas hunt from elevated, concealed positions and kill with a throat bite that crushes the trachea. In an extensive study of 409 puma scats from Torres del Paine National Park, Chile, 92% of all prey items were mammalian—mostly European hares. Guanacos represented the next most frequent prey item. Yearling and juvenile guanacos were taken more frequently than expected (Cajal and López 1987; Currier 1983; Enders 1935; Iriarte 1988; Miller 1930; Pearson 1951; Schaller and Crawshaw 1980; Schaller, Quigley, and Crawshaw 1984; Stallings, unpubl.; Wilson 1984).

Comment

Puma tracks are usually smaller than those of the jaguar, but tracks of a very large puma can be distinguished from a small jaguar's because the toes radiate like spokes from a wheel around a smaller pad instead of pointing forward from a larger pad (Almeida 1976).

References

• References used in preparing distribution maps
• Aagaard, E. M. J. 1982. Ecological distribution of mammals in the cloud forests and paramos of the Andes, Mérida, Venezuela. Ph.D. diss., Colorado State University.
• Allen, J. A. 1905. Mammalia of southern Patagonia: Reports of the Princeton University expeditions to Patagonia, 1896–1899. *Zoology* 3:1–5, 151–61, 164.
•———. 1916. New mammals collected on the Roosevelt Brazilian expedition. *Bull. Amer. Mus. Nat. Hist.* 35:523–30.
•———. 1919. Notes on the synonymy and nomenclature of the smaller spotted cats of tropical America. *Bull. Amer. Mus. Nat. Hist.* 41:341–419.
Almeida, A. E. d'A. 1976. *Jaguar hunting in the Mato Grosso.* London: Stanwill Press.
———. n.d. Some feeding and other habits, measurements and weights of *Panthera onca pallustris,* the jaguar of the "Pantanal" region of Mato-Grosso and Bolivia. Photocopy.
Aquino, R., and P. Puertas. 1997. Observations of *Speothos venaticus* (Canidae: Carnivora) in its natural habitat in Peruvian Amazonia. *Z. Säugetierk.* 62:117–18.
• Arra, M. A. 1974. Distribución de *Leo onca* (L) en Argentina (Carnivora, Felidae). *Neotrópica* 20 (63): 156–58.

Asa, C. S., and M. P. Wallace. 1990. Diet and activity patterns of the Sechuran desert fox. *J. Mammal.* 71 (1): 69–72.

Atalah, G. A., W. Sielfeld K., and C. Venegas C. 1980. Antecedentes sobre el nicho trófico de *Canis g. griseus* Gray 1836 en Tierra del Fuego. *Anal. Inst. Patagonia, Punta Arenas* 11:259–71.

Autuori, M. P., and L. A. Deutsch. 1977. Contribution to the knowledge of the giant Brazilian otter, *Pteronura brasiliensis* (Gmelin 1788), Carnivora, Mustelidae. *Zool. Gart.*, n.s. 47:1–8.

• Barlow, J. C. 1965. Land mammals from Uruguay: Ecology and zoogeography. Ph.D. diss., University of Kansas.

Berrie, P. M. 1978. Home range of young female Geoffroy's cat in Paraguay. *Carnivore* 1 (1): 132–33.

Berta, A. 1982. *Cerdocyon thous. Mammal. Species* 186:1–4.

———. 1984. The Pleistocene bush dog *Speothos pacivorus* (Canidae) from the Lago Santo caves, Brazil. *J. Mammal.* 65 (4): 549–59.

———. 1986. *Atelocyon microtis. Mammal. Species* 256:1–3.

———. 1987. Origin diversification and zoogeography of the South American canids. *Fieldiana: Zool.*, n.s., 39:455–72.

Biben, M. 1982a. Object play and social treatment of prey in bush dogs and crab-eating foxes. *Behaviour* 79:201–11.

———. 1982b. Ontogeny of social behavior related to feeding in the crab-eating fox (*Cerdocyon thous*) and the bush dog (*Speothos venaticus*). *J. Zool. Soc. London* 196:207–16.

———. 1983. Comparative ontogeny of social behavior in three South American canids, the maned wolf, crab-eating fox and bush dog: Implications for sociality. *Anim. Behav.* 31:814–26.

Bisbal, F. 1986. Food habits of some Neotropical carnivores in Venezuela. *Mammalia* 50:329–39.

———. 1987. The carnivores of Venezuela: Their distribution and the ways they have been affected by human activities. M.S. thesis, University of Florida.

———. 1989. Distribution and habitat association of the carnivores in Venezuela. In *Advances in Neotropical mammalogy,* ed. K. H. Redford and J. F. Eisenberg, 339–62. Gainesville, Fla.: Sandhill Crane Press.

Bisbal, F., and J. Ojasti. 1980. Nicho trófico del zorro *Cerdocyon thous* (Mammalia, Carnivora). *Acta Biol. Venez.* 10 (4): 469–96.

Brady, C. A. 1978. Reproduction, growth and parental care in crab-eating foxes *Cerdocyon thous* at the National Zoological Park, Washington. *Int. Zoo Yearb.* 18:130–34.

———. 1979. Observations on the behavior and ecology of the crab-eating fox, *Cerdocyon thous.* In *Vertebrate ecology in the northern Neotropics,* ed. J. F. Eisenberg, 16–71. Washington, D.C.: Smithsonian Institution Press.

———. 1981. Vocal repertoires of the bush dog *Speothos venaticus,* crab-eating fox *Cerdocyon thous,* and maned wolf *Chrysocyon brachyurus. Anim. Behav.* 29:649–69.

Brady, C. A., and M. K. Ditton. 1979. Management and breeding of maned wolves *Chrysocyon brachyurus* at the National Zoological Park, Washington. *Int. Zoo Yearb.* 19:171–76.

Brooks, D. M. 1992. Notes on group size, density, and habitat association of the pampas fox (*Dusicyon gymnocercus*) in the Paraguayan Chaco. *Mammalia* 56:314–16.

• Brown, A. D., and D. I. Rumiz. 1988. Habitat and distribution of the spectacled bear (*Tremarctos ornatus*) in the southern limit of its range. In *Proceedings of the First International Symposium on the Spectacled Bear.* Chicago: Lincoln Park Zoo.

• Brownell, R. L. 1978. Ecology and conservation of the marine otter *L. felina.* In *Otters,* ed. N. Duplaix, 104–6. Proceedings of the First Working Meeting of the Otter Specialist Group. Gland, Switzerland: IUCN.

• Cabello, C. C. 1978. La nutria de mar *L. felina* en la Isla de Chiloé. In *Otters,* ed. N. Duplaix, 108–18. Proceedings of the First Working Meeting of the Otter Specialist Group. Gland, Switzerland: IUCN.

• ———. 1983. *La nutria de mar en la Isla de Chiloé.* Corporación Nacional Forestal, Ministero de Agricultura, Boletín Técnico 6. Santiago: República de Chile.

• Cabrera, A. 1961. Los félidos vivientes de la república Argentina. *Rev. Mus. Argent. Cienc. Nat. "Bernardino Rivadavia," Zool.* 6 (5): 161–247.

Cajal, J. L., and N. E. López. 1987. El puma como depredador de camélidos silvestres en las Reserva San Guillermo, San Juan, Argentina. *Rev. Chil. Hist. Nat.* 60:87–91.

• Carman, R. L. 1984. Límite austral de la distribución del tigre o yaguareté (*Leo onca*) en los siglos XVIII y XIX. *Rev. Mus. Argent. Cienc. Nat. "Bernardino Rivadavia," Zool.* 13 (1–60): 293–96.

• Castilla, J. C. 1982. Nuevas observaciones sobre conducta, ecología y densidad de *Lutra felina* (Molina 1782) (Carnivora: Mustelidae) en Chile. *Mus. Nac. Hist. Nat. Publ. Ocas.* 38:197–206.

Charles-Dominique, P., M. Atramentowicz, M. Charles-Dominique, H. Gerard, A. Hladik, C. M. Hladik, and M. F. Prévost. 1981. Les mammifères frugivores arboricoles nocturnes d'une forêt guyanaise: Interrelations plantes-animaux. *Rev. Ecol. (Terre et Vie)* 35:341–435.

Chehebar, C. E. 1985. A survey of the southern river otter *Lutra provocax* Thomas in Nahuel Huapi National Park, Argentina. *Biol. Conserv.* 32:299–307.

Clutton-Brock, T., G. B. Corbet, and M. Hills. 1976. A review of the family Canidae with a classification by numerical methods. *Bull. British Mus. (Nat. Hist.), Zool.* 29:117–99.

Coimbra-Filho, A. F. 1966. Notes on the reproduction and diet of Azara's fox *Cerdocyon thous azarae* and the hoary fox *Dusicyon vetulus* at Rio de Janeiro Zoo. *Int. Zoo Yearb.* 6:168–69.

Crawshaw, P. 1995. Comparative ecology of the ocelot (*Felis paradalis*) and jaguar (*Panthera onca*) in a protected subtropical forest in Brazil and Argentina. Ph.D. diss., University of Florida, Gainesville.

Crawshaw, P. G., Jr., and H. B. Quigley. 1991. Jaguar spacing, activity and habitat use in a seasonally flooded environment in Brazil. *J. Zool.* (London) 223:357–70.

Crespo, J. A. 1962. Una mutación de pelaje en el zorro colorado, *Dusicyon culpaeus* (Molina) (Mammalia: Carnivora). *Neotrópica* 8 (27): 115–16.

———. 1971. Ecología del zorro gris *Dusicyon gymnocercus antiquus* (Ameghino) en la provincia de la Pampa. *Rev. Mus. Argent. Cienc. Nat. "Bernardino Rivadavia," Ecol.* 1 (5): 147–205.

• ———. 1974a. Comentarios sobre nuevas localidades para mamíferos de Argentina y de Bolivia. *Rev. Mus. Argent. Cienc. Nat. "Bernardino Rivadavia," Zool.* 11 (1): 1–31.

• ———. 1974b. Incorporación de un género de canidos a la fauna de Argentino (fam. Canidae: *Speothos venaticus* [Lund] 1943). *Comun. Mus. Argent. Cienc. Nat. "Bernardino Rivadavia," Zool.* 4 (6): 37–39.

———. 1975. Ecology of the pampas gray fox and the large fox (*culpeo*). In *The wild canids*, ed. M. W. Fox, 179–91. New York: Van Nostrand Reinhold.

• ———. 1982. Ecología de la comunidad de mamíferos del Parque Nacional Iguazú, Misiones. *Rev. Mus. Argent. Cienc. Nat. "Bernardino Rivadavia," Ecol.* 3 (2): 45–162.

———. 1984. Los zorros. *Fauna Argent.* 52:1–32.

Crespo, J. A., and J. M. de Carlo. 1963. Estudio ecológico de unapoblación de zorros colorados *Dusicyon culpaeus culpaeus* (Molina) en el oeste de la provincia de Neuquén. *Rev. Mus. Argent. Cienc. Nat. "Bernardino Rivadavia," Ecol.* 1 (1): 1–55.

Currier, M. J. P. 1983. *Felis concolor. Mammal. Species* 200:1–7.

Dalponte, J. n.d. Diet of the hoary fox (*Dusicyon vetulus*) in Matto Grosso, central Brazil. In press.

Dathe, H. 1967. Bemerkungen zur Aufzucht von Brillenbären. *Zool. Gart.* 34:105–33.

Defler, T. R. 1980. Notes on interactions between the tayra (*Eira barbara*) and the white fronted capuchin (*Cebus albifrons*). *J. Mammal.* 51:156.

Deutsch, L. 1983. An encounter between bush dog (*Speothos venaticus*) and paca (*Agouti paca*). *J. Mammal.* 64:532–33.

Dietz, J. M. 1981. Ecology and social organization of the maned wolf (*Chrysocyon brachyurus*). Ph.D. diss., Michigan State University.

———. 1984. *Ecology and social organization of the maned wolf* (Chrysocyon brachyurus). Smithsonian Contributions to Zoology 392. Washington, D.C.: Smithsonian Institution Press.

Drüwa, P. 1977. Beobachtungen zur Geburt und natürlichen Aufzucht von Waldhunden (*Speothos venaticus*) in der Gefangenschaft. *Zool. Gart.*, n.s., 47:109–37.

Duplaix, N. 1980. Observations on the ecology and behavior of the giant river otter *Pteronura brasiliensis* in Suriname. *Rev. Ecol. (Terre et Vie)* 34:496–620.

———. 1982. Contribution à l'écologie et à l'ethologie de *Peteronura brasiliensis* Gmelin 1788 (Carnivora, Lutrinae): Implications évolutives. Ph.D. diss., Université de Paris—Sud, Centre d'Orsay.

Eaton, R. L. 1984. Survey of smaller felid breeding. *Zool. Gart.*, n.s., 54 (1–2): 101–20.

Eisenberg, J. F. 1989. *Mammals of the Neotropics.* Vol. 1. *The northern Neotropics: Panama, Colombia, Venezuela, Guyana, Suriname, French Guiana.* Chicago: University of Chicago Press.

Emmons, L. H. 1988. A field study of ocelots (*Felis pardalis*) in Peru. *Rev. Ecol. (Terre et Vie)* 43:133–57.

———. 1989. Jaguar predation on chelonians. *J. Herpetol.* 23:311–14.

Enders, R. K. 1935. Mammalian life histories from Barro Colorado Island, Panama. *Bull. Mus. Comp. Zool. (Harvard)* 78 (4): 385–502.

Ewer, R. F. 1973. *The carnivores.* Ithaca, N.Y.: Cornell University Press.

Ferrari, S. F., and M. A. Lopes. 1992. A note on the behaviour of the weasel *Mustela* cf. *africana* (Carnivora, Mustelidae), from Amazonas, Brazil. *Mammalia* 56 (3): 482–83.

Ford, L. S., and R. S. Hoffmann. 1988. *Potos flavus. Mammal. Species* 321:1–9.

Forman, L. 1985. Genetic variation in two procyonids: Phylogenetic, ecological and social correlates. Ph.D. diss., New York University, New York.

Fuentes, E. R., and F. M. Jaksić. 1979. Latitudinal size variation of Chilean foxes: Tests of alternative hypotheses. *Ecology* 60:43–47.

Fuller, T. K., W. E. Johnson, W. L. Franklin, and K. A. Johnson. 1987. Notes on the Patagonian hog-nosed skunk (*Conepatus humboldtii*) in southern Chile. *J. Mammal.* 68:864–67.

Galef, B. G., Jr., R. A. Mittermeier, and R. C. Bailey. 1976. Predation by the tayra (*Eira barbara*). *J. Mammal.* 57:760–61.

Garcia-Perea, R. 1994. The pampas cat group (genus *Lynchailurus* Severtzov, 1858) (Carnivora: Felidae): A systematic and biogeographic review. *Amer. Mus. Novitat.* 3096:1–35.

Gardner, A. L. 1971. Notes on the little spotted cat, *Felis tigrina oncilla* Thomas, in Costa Rica. *J. Mammal.* 52:464–65.

Gibson, E. 1899. Field-notes on the wood-cat of Argentina (*Felis geoffroyi*). *Proc. Zool. Soc. London* 1899:928–29.

Gittleman, J. L., ed. 1989. *Carnivore behavior, ecology, and evolution*, vol. 1. Ithaca, N.Y.: Cornell University Press.

———, ed. 1996. *Carnivore behavior, ecology and evolution*, vol. 2. Ithaca, N.Y.: Cornell University Press.

Goldstein, I. 1991. Spectacled bear predation and feeding behavior on livestock in Venezuela. *Stud. Neotrop. Fauna Environ.* 26:231–35.

Gomes de Oliveira, T. 1992. Ecology and conservation of Neotropical felids. M.S. thesis, University of Florida, Gainesville.

———. 1994. *Neotropical cats: Ecology and conservation.* São Luis Maranhão, Brazil: EDUFMA.

———. n.d. *Carnivora of Brazil.* In press.

Gompper, M. E. 1994. The behavorial ecology and genetics of a coati (*Nasua narica*) population in Panama. Ph.D. diss., University of Tennessee, Knoxville.

———. 1995. *Nasua narica. Mammal. Species* 487: 1–10.

• Greer, J. K. 1966. Mammals of Malleco province, Chile. *Publ. Mus., Michigan State Univ., Biol. Ser.* 3 (2): 49–152.

Griva, E. E. 1978. El programa de cría y preservación de *L. platensis* en Argentina. In *Otters*, ed. N. Duplaix, 86–103. Proceedings of the First Working Meeting of the Otter Specialist Group. Gland, Switzerland: IUCN.

Guix, J. C. 1992. El jaguar en la pluvisilva atlantica de Brasil. *Vida Silvestre* (Madrid) 71:32–37.

———. n.d. Cat communities in six areas of the state of São Paulo, southeastern Brazil, with notes on their feeding habits. In press.

• Hall, E. R. 1951. *American weasels.* Publication 4. Lawrence: Museum of Natural History, University of Kansas.

• Hershkovitz, P. 1958. A synopsis of the wild dogs of Colombia. *Nov. Colombianas, Mus. Hist. Nat. Univ. Cauca,* 3:157–61.

• ———. 1961. On the South American small-eared zorro *Atelocynus microtis* Sclater (Canidae). *Fieldiana: Zool.* 39:505–23.

Hill, K., and K. Hawkes. 1983. Neotropical hunting among the Ache of eastern Paraguay. In *Adaptive responses of native Amazonians*, ed. R. B. Hames and W. T. Vickers, 139–88. New York: Academic Press.

Honacki, J. H., K. E. Kinman, and J. W. Koeppl, eds. 1982. *Mammal species of the world.* Lawrence, Kansas: Allen Press and Association of Systematics Collections.

Hoogesteijn, R., and E. Mondolfi. 1992. *The jaguar.* Caracas: Armitano Editores.

Housse, R. P. R. 1948. Las zorras de Chile o chacales americanos. *Rev. Univ.* 34:33–56.

———. 1953. *Animales salvajes de Chile en su clasificación moderna.* Santiago: Ediciones de la Universidad de Chile.

Hulley, J. T. 1976. Maintenance and breeding of captive jaguarundis (*Felis yagouaroundi*) at Chester Zoo and Toronto. *Int. Zoo Yearb.* 16:120–22.

• Husson, A. M. 1978. *The mammals of Suriname.* Leiden: E. J. Brill.

Iriarte, J. A. 1988. Feeding ecology of the Patagonian puma (*Felis concolor*) in Torres del Paine National Park, Chile. M.A. thesis, University of Florida, Gainesville.

Iriarte, J. A., W. Franklin, W. Johnson, and K. Redford. 1990. Biogeographic variation of food habits and body size of the America puma. *Oecologia* 85: 185–90.

Izor, J. R., and L. de la Torre. 1978. A new species of weasel (*Mustela*) from the highlands of Colombia with comments on the evolution and distribution of South American weasels. *J. Mammal.* 59:92–102.

Izor, J. R., and N. E. Peterson. 1985. Notes on South American weasels. *J. Mammal.* 66 (4): 788–89.

Jaksić, F. M., R. P. Schlatter, and J. L. Yáñez, 1980. Feeding ecology of central Chilean foxes, *Dusicyon culpaeus* and *Dusicyon griseus. J. Mammal.* 61: 254–60.

Jaksić, F. M., and J. L. Yáñez. 1983. Rabbit and fox introductions in Tierra del Fuego: History and assessment of the attempts at biological control of the rabbit infestation. *Biol. Conserv.* 26:367–74.

Jaksić, F. M., J. L. Yáñez, and J. R. Rau. 1983. Trophic relations of the southernmost populations of *Dusicyon* in Chile. *J. Mammal.* 64:693–97.

Johnson, J. P., and W. L. Franklin. 1991. Feeding and spatial ecology of *Felis geoffroyi* in southern Patagonia. *J. Mammal.* 78:815–20.

Johnson, W. E., P. A. Dratch, J. S. Martenson, and S. J. O'Brien. 1996. Resolution of recent radiation within three evolutionary lineages of Felidae using mitochondrial restriction fragment length polymorphism variation. *J. Mammal. Evol.* 3 (2): 97–120.

Jorgensen, J. P., and K. H. Redford. 1993. Humans and big cats as predators in the Neotropics. In *Mammals as predators,* ed. N. Dunstone and M. L. Gorman, 367–90. Oxford: Clarendon Press.

Julien-Lafferrière, D. 1993. Radio-tracking observations on ranging and foraging patterns by kinkajous (*Potos flavus*) in French Guiana. *J. Trop. Ecol.* 9: 19–32.

Kaufmann, J. F., and A. Kaufmann. 1965. Observations of the behavior of tayras and grisons. *Z. Säugetierk.* 30:146–55.

Kaufmann, J. H. 1962. Ecology and social behavior of the coati, *Nasua narica,* on Barro Colorado Island, Panama. *Univ. Calif. Publ. Zool.* 60:95–222.

Kays, R. W., and J. L. Gittleman. 1995. Home range size and social behavior of kinkajouos (*Potos flavus*) in Panama. *Biotropica* 27:530–34.

• Kipp, H. 1965. Beitrag zur Kenntnis der Gattung *Conepatus* Molina, 1782. *Z. Säugetierk.* 30:193–256.

Kleiman, D. G. 1972. Social behavior of the maned wolf (*Chrysocyon brachyurus*) and bush dog (*Speothos venaticus*): A study in contrast. *J. Mammal.* 53:791–806.

Kleiman, D. G., and J. F. Eisenberg. 1973. A comparison of canid and felid social systems from an evolutionary perspective. *Anim. Behav.* 21:637–59.

Knapik, D. P. 1984. Notes on bear reproductive behavior with specific reference to spectacled bears (*Tremarctos ornatus*) at the Calgary Zoo. Apprenticeship paper, Calgary Zoological Society, Alberta.

Konecny, M. J. 1989. Movement patterns and food habits of four sympatric carnivore species in Belize, Central America. In *Advances in Neotropical mammalogy,* ed. K. H. Redford and J. F. Eisenberg, 243–64. Gainesville, Fla.: Sandhill Crane Press.

Krumbiegel, I. 1942. Die Säugetiere der Südamerika-expeditionen Prof. Dr. Kriegs. 17. Hyrare und Grisons (*Tayra* und *Grison*). *Zool. Anz.* 139:81–108.

Langguth, A. 1975. Ecology and evolution in the South American candids. In *The wild canids: Their systematics, behavioral ecology, and evolution,* ed. M. W. Fox, 192–206. New York: Van Nostrand Reinhold.

Leyhausen, P. 1963. Über südamerikanische Pardelkatzen. *Z. Tierpsychol.* 20:627–40.

Leyhausen, P., and M. Falkena. 1966. Breeding the Brazilian ocelot-cat *Leopardus tigrinus* in captivity. *Int. Zoo Yearb.* 6:176–82.

• Linares, O. J. 1968. El perro de monte, *Speothos venaticus* (Lund) en el norte de Venezuela (Canidae). *Mem. Soc. Cient. Nat. La Salle* 27:83–86.

———. 1981. Tres nuevos carnívoros prociónidos fósiles de Mioceno de Norte y Sudamérica. *Ameghiniana* 18:113–21.

Löhmer, R. 1976. Zur Verhaltensontogenese bei *Procyon cancrivorus cancrivorus* (Procyonidae). *Z. Säugetierk.* 41:42–58.

Lucero, M. M. 1983. Lista y distribución de aves y mamíferos de la provincia de Tucumán. Ministerio de Cultura y Educación, Fundación Miguel Lillo, *Miscelánea* 75:5–53.

Ludlow, M. E. and M. E. Sunquist. 1987. Ecology and behavior of ocelots in Venezuela. *Nat. Geog. Res.* 3 (4): 447–61.

Lumpkin, S. 1986. Spectacular spectacled bears. *Zoo-Goer* 15 (4): 4–6.

MacDonald, D. W., and O. Courtney. 1996. Enduring social relationships in a population of crab-eating zorros *Cerdocyon thous* in Amazonian Brazil. *J. Zool.* (London) 239:329–55.

Manzani, P. R., and E. L. A. Montiero-Filho. 1989. Notes on the food habits of the jaguarundi *Felis yagouaroundi. Mammalia* 53:659–60.

Mares, M. A., and R. A. Ojeda. 1984. Faunal commercialization and conservation in South America. *BioScience* 34 (9): 580–84.

Marquet, P. A., L. C. Contreras, J. C. Torres-Mura, S. I. Silva, and F. M. Jaksić. 1993. Food habits of *Pseudalopex* foxes in the Atacama desert, pre-Andean ranges and the high-Andean plateau of northernmost Chile. *Mammalia* 57:130–35.

• Massoia, E. 1976. Mammalia. In *Fauna de agua dulce de la república Argentina,* vol. 44, *Mammalia.* Buenos Aires: Fundación para la Educación, la Ciencia y la Cultura.

———. 1982. *Dusicyon gymnocercus lordi,* una nueva subespecie del "zorro gris grande" (Mammalia Carnivora Canidae). *Neotrópica* 28 (80): 147–52.

McCarthy, T. J. 1992. Notes concerning the jaguarundi cat (*Herpailurus yagouroundi*) in the Caribbean lowlands of Belize and Guatemala. *Mammalia* 56 (2): 302–6.

McClearn, D. 1992. Locomotion, posture, and feeding behavior of kinkajous, coatis, and raccoons. *J. Mammal.* 73:245–61.

Medel, R. G., J. E. Jiménez, and F. M. Jaksić. 1990. Discovery of a continental population of the rare Darwin's fox, *Dusicyon fulvipes* (Martin, 1837) in Chile. *Biol. Conserv.* 51:71–77.

Melquist, W. E. 1984. Status survey of otters (Lutrinae) and spotted cats (Felidae) in Latin America. Completion report to IUCN (contract 9006).

Melquist, W. E., and M. G. Hornocker. 1983. Ecology of river otters in west central Idaho. *Wildl. Monogr.* 83:1–60.

Meserve, P. L., E. J. Shadrick, and D. A. Kelt. 1987.

Diets and selectivity of two Chilean predators in the northern semi-arid zone. *Rev. Chil. Hist. Nat.* 50:93–99.

Miller, F. W. 1930. Notes on some mammals of southern Mato Grosso, Brazil. *J. Mammal.* 11:10–22.

Mondolfi, E. 1970. Las nutrias o perros de aqua. *Defensa Nat.* 1 (1): 24–26, 47.

———. 1971. El oso frontino (*Tremarctos ornatus*). *Defensa Nat.* 1 (2): 31–35.

———. 1973. El mapurite, un animal beneficioso. *Defensa Nat.* 2 (6): 37–41.

———. 1986. Notes on the biology and status of the small wild cats in Venezuela. In *Cats of the world: Biology, conservation, and management,* ed. S. Douglas Miller and D. D. Everett, 125–46. Washington, D.C.: National Wildlife Federation.

Mondolfi, E., and R. Hoogesteijn. 1986. Biology and status of the jaguar in Venezuela. In *Cats of the world: Biology, conservation, and management,* ed. S. D. Miller and D. D. Everett, 85–124. Washington, D.C.: National Wildlife Federation.

Montgomery, G. G., and Y. D. Lubin. 1978. Social structure and food habits of crab-eating fox (*Cerdocyon thous*) in Venezuelan llanos. *Acta Cient. Venez.* 29:382–83.

Müller, P. 1988. Beobachtungen zur Fortpflanzungsbiologie von Brillenbären (*Tremarctos ornatus* F. Cuvier, 1825). *Zool. Gart.* 1:9–21.

Murray, J. L., and G. L. Gardner. 1997. *Leopardus pardalis. Mammal. Species* 548:1–10.

Nelson, E. W., and E. A. Goldman. 1933. Revision of the jaguars. *J. Mammal.* 14:221–40.

Novaro, A. 1997. *Pseudalopex culpaeus. Mammal. Species* 558:1–8.

Ojeda, R. A., and M. A. Mares. 1982. Conservation of South American mammals: Argentina as a paradigm. In *Mammalian biology in South America,* ed. M. A. Mares and H. H. Genoways, 505–22. Pymatuning Symposia in Ecology 6. Special Publication Series. Pittsburgh: Pymatuning Laboratory of Ecology, University of Pittsburgh.

Ojeda, R. A., and M. A. Mares. 1984. La degradación de los recursos naturales y la fauna silvestre en Argentina. *Incerciencia* 9 (1): 21–26.

Ojeda, R. A., and M. A. Mares. 1989. *A biogeographic analysis of the mammals of Salta province, Argentina.* Special Publications of the Museum 27. Lubbock: Texas Tech University.

Olmos, F. 1993. Notes on the food habits of Brazilian "caatinga" carnivores. *Mammalia* 57:126–30.

Olrog, C. C. 1950. Notas sobre mamíferos y aves del archipiélago de Cabo de Hornos. *Acta Zool. Lilloana* 9:509–11.

———. 1979. Los mamíferos de la selva húmeda, Cerro Calilegua, Jujuy. *Acta Zool. Lilloana* 33:9–14.

Osgood, W. H. 1914. Mammals of an expedition across northern Peru. *Field Mus. Nat. Hist. Zool. Ser.* 10 (12): 143–85.

• Osgood, W. H. 1943. The mammals of Chile. *Field Mus. Nat. Hist., Zool. Ser.* 30:1–268.

• Pearson, O. P. 1951. Mammals in the highlands of southern Peru. *Bull. Mus. Comp. Zool.* 106 (3): 117–74.

• ———. 1957. Additions to the mammalian fauna of Peru and notes on some other Peruvian mammals. *Breviora: Mus. Comp. Zool.* 73:1–7.

Peres, C. A. 1992. Observations on hunting by the small eared dog (*Atelocynus microtis*) and bush dogs (*Speothos venaticus*) in central-western Amazonia. *Mammalia* 55:635–39.

Petersen, M. K., and M. K. Petersen. 1978. Growth rates and other post-natal developmental changes in margays. *Carnivore* 1 (1): 87–92.

Peyton, B. 1980. Ecology, distribution and food habits of spectacled bears *Tremarctos ornatus* in Peru. *J. Mammal.* 6:639–52.

• Pine, R. H., S. D. Miller, and M. L. Schamberger. 1979. Contributions to the mammalogy of Chile. *Mammalia* 43:339–76.

Poglayen-Neuwall, I. 1962. Beitrage zu einem Ethogramm des Wickelbären (*Potos flavus* Schreber). *Z. Säugetierk.* 27:1–44.

———. 1966. On the marking behavior of the kinkajou (*Potos flavus* Schreber). *Zoologica* 51:137–41.

———. 1975. Copulatory behavior, gestation, and parturition of the tayra *Eira barbara. Z. Säugetierk.* 40:176–89.

———. 1978. Breeding, rearing and notes on the behaviour of tayras *Eira barbara* in captivity. *Int. Zoo Yearb.* 18:134–40.

Poglayen-Neuwall, I., and I. Poglayen-Neuwall. 1965. Gefangenschaftsbeobachtungen an Makibären (*Bassaricyon gabbii* Allen, 1876). *Z. Säugetierk.* 30:321–66.

Porton, D., D. G. Kleiman, and M. Rodden. 1987. Seasonality of bush dog reproduction and the influence of social factors on the estrous cycle. *J. Mammal.* 68:867–71.

Quigley, H. B., and P. G. Crawshaw Jr. 1992. A conservation plan for the jaguar *Panthera onca* in the Pantanal region of Brazil. *Biol. Conserv.* 61:149–57.

Rabinovich, J., A. Capurro, P. Folgarait, T. Kitzberger, G. Krammer, A. Novaro, M. Puppo, and A. Travaini. 1987. Estado del conocimiento de 12 especies de la fauna silvestre argentina e valor comercial. Paper presented at the second workshop on Elaboración de propuestas de investigación orientada al manejo de la fauna silvestre de valor comercial, Buenos Aires.

Rabinowitz, A. R., and B. G. Nottingham Jr. 1986. Ecology and behaviour of the jaguar (*Panthera onca*) in Belize, Central America. *J. Zool.* (London) 210:149–59.

Rasmussen, J. L., and R. L. Tilson. 1984. Food provisioning by adult maned wolves (*Chrysocyon brachyurus*). *Z. Tierpsychol.* 65:346–52.

Redford, K. H,, and J. F. Eisenberg. 1992. *Mammals of the Neotropics*. Vol. 2. *The southern cone: Chile, Argentina, Uruguay, Paraguay*. Chicago: University of Chicago Press.

Redford, K. H., and A. M. Stearman. 1993. Notas sobre la biología de tres procyonidos simpatricos bolivianos (Mammalia, Procyonidae). *Ecol. Bolivia* 21:35–44.

Romo, M. C. 1995. Food habits of the Andean fox (*Pseudalopex culpeus*) and note on the mountain cat (*Felis colocolo*) and puma (*Felis concolor*) in Río Abiseo Park, Peru. *Mammalia* 59 (3): 35–344.

Russell, J. K. 1981. Exclusion of adult male coatis from social groups: Protection from predation. *J. Mammal.* 62 (1): 206–8.

———. 1983. Altruism in coati bands: Nepotism or reciprocity? In *Social behavior of female vertebrates*, ed. S. K. Wasser, 263–90. New York: John Wiley.

Sanborn, C. C. 1954. Weights, measurements, and color of the Chilean forest puma. *J. Mammal.* 35: 126–28.

Schaller, G. B. 1983. Mammals and their biomas on a Brazilian ranch. *Arq. Zool. São Paulo* 31 (1): 1–36.

Schaller, G. B., and P. G. Crawshaw Jr. 1980. Movement patterns of jaguar. *Biotropica* 12:161–68.

Schaller, G. B., H. B. Quigley, and P. G. Crawshaw. 1984. Biological investigations in the Pantanal, Mato Grosso, Brazil. *Nat. Geogr. Soc. Res. Repts.* 17:777–92.

Schaller, G. B., and J. M. C. Vasconcelos. 1978. Jaguar predation on capybara. *Z. Säugertierk.* 43:296–310.

Schamberger, M., and G. Fulk. 1974. Mamíferos del Parque Nacional Fray Jorge. *Idesia* (Chile) 3:167–79.

Scheffel, W., and H. Hemmer. 1975. Breeding Geoffroy's cat *Leopardus geoffroyi salinarum* in captivity. *Int. Zoo Yearb.* 15:152–54.

Schweizer, J. 1992. *Ariranhas no Patanal: Ecologia e comportamento da* Pternoura brasiliensis. Curitiba, Brazil: Edibran.

• Scrocchi, G. J., and S. P. Halloy. 1986. Notas sistemáticas, ecológicas, etológicas y biogeográficas sobre el gato andino *Felis jacobita* Cornalia (Felidae, Carnivora). *Acta Zool. Lilloana* 38 (2): 157–70.

Seymour, K. 1989. *Panthera onca. Mammal. Species* 340:1–9.

Sheffield, S. R., and H. H. Thomas. 1997. *Mustela frenata. Mammal. Species* 570:1–9.

Sielfeld, W. 1983. *Mamíferos marinos de Chile*. Santiago: Ediciones de la Universidad de Chile.

Silviera, L. 1995. Notes on the distribution and natural history of the pampas cat, *Felis colocolo*, in Brazil. *Mammalia* 59:284–87.

Simonetti, J. A. 1986. Human-induced dietary shift in *Dusicyon culpaeus. Mammalia* 50 (3): 406–8.

Spinola, R. M., and C. Vaughn. 1995. Abundancia relationale y actividad de marcaje de la nutria neotropical (*Lutra longicaudis*) en Costa Rica. *Vida Silvestre Neotrop.* 4 (1): 38–45.

Stains, H. J. 1975. Distribution and taxonomy of the Canidae. In *The wild canids*, ed. M. W. Fox, 3–26. New York: Van Nostrand Reinhold.

Stehlik, J. 1971. Breeding jaguars *Panthera onca* at Ostrava Zoo. *Int. Zoo Yearb.* 11:116–18.

Strahl, S. D., J. L. Silva, and I. R. Goldstein. 1992. The bush dog (*Speothos venaticus*) in Venezuela. *Mammalia* 56 (1): 9–13.

Sunquist, M. E. 1992. The ecology of the ocelot: The importance of incorporating life history traits into conservation plans. In *Felinos de Venezuela*, 117–27. Caracas: MARNR.

Sunquist, M. E., and G. G. Montgomery. 1973. Arboreal copulation by coatimundi (*Nasua narica*). *Mammalia* 37:517–18.

Sunquist, M. E., F. Sunquist, and D. E. Daneke. 1989. Ecological separation in a Venezuelan llanos carnivore community. In *Advances in Neotropical mammalogy*, ed. K. H. Redford and J. F. Eisenberg, 197–232. Gainesville, Fla.: Sandhill Crane Press.

• Tamayo, M., and D. Frassinetti. 1980. Catálogo de los mamíferos fósiles y vivientes de Chile. *Mus. Nac. Hist. Nat. Chile* 37:232–99.

Tate, G. H. H. 1939. The mammals of the Guiana region. *Bull. Amer. Mus. Nat. Hist.* 76 (5): 151–229.

Tedford, R., B. F. Taylor, and X. Wang. 1995. Phylogeny of the Caninae (Carnivora Canidae): The living taxa. *Amer. Mus. Novitat.* 3146:1–37.

• Texera, W. A. 1974a. Nuevos antecedentes sobre mamíferos de Magallanes. 1. La distribucion del gato montes (*Felis geoffroyi leucobapta*) (Mammalia: Felidae) en la region de Magallanes. *Anal. Inst. Patagonia, Punta Arenas* 5 (1–2): 189–92.

• ———. 1974b. Nuevos antecedentes sobre mamíferos de Magallanes. 3. El quique (*Galictis cuja cuja*) (Mammalia: Mustelidae), una nueva adición a la fauna mamal de Magallanes, Chile. *Anal. Inst. Patagonia, Punta Arenas* 5 (1–2): 195–98.

Torres, D., J. Yáñez, and P. Cattan. 1979. Mamíferos marinos de Chile: Antecedentes y situación actual. *Biol. Pesquera, Chile* 11:49–81.

Van Gelder, R. G. 1968. The genus *Conepatus* (Mammalia, Mustelidae): Variation within a population. *Amer. Mus. Novitat.* 2322:1–37.

Van Zyll de Jong, C. G. 1972. *A systematic review of the Nearctic and Neotropical river otters (genus Lutra, Mustelidae, Carnivora).* Life Sciences Contribution 80. Toronto: Royal Ontario Museum.

Vaughn, R. 1974. Breeding the tayra (*Eira barbara*) at Antelope Zoo, Lincoln. *Int. Zoo Yearb.* 14:120–22.

• Vieira, C. 1945. Sobre uma coleção de mamíferos de Mato Grosso. *Arq. Zool. (São Paulo)* 4:395–430.

Vitullo, A. D., and G. A. Zuleta. 1992. Cytogenetics and the fossil record: Confluent evidence for speciation without chromosomal change in South American canids. *Z. Säugetierk.* 57:248–50.

Vuilleumier, F. 1985. Fossil and Recent avifaunas and the interamerican interchange. In *The great American biotic interchange,* ed. F. G. Stehli and S. D. Webb. New York: Plenum Press.

Walker, F. L., and J. G. H. Cant. 1977. A population survey of kinkajous (*Potos flavus*) in a seasonally dry tropical forest. *J. Mammal.* 58:100–102.

Wilson, D. E., and D. M. Reeder, eds. 1993. *Mammal species of the world.* 2d ed. Washington, D.C.: Smithsonian Institution Press.

Wilson, P. 1984. Puma predation on guanacos in Torres del Paine National Park, Chile. *Mammalia* 48 (4): 515–22.

Wolffsohn, J. A. 1908. Contribuciones a la mamalogía chilena. 1. Sobre el *Felis colocolo,* Mol. *Rev. Chil. Hist. Nat.* 12:165–72.

• Wozencroft, W. C. 1993. Order Carnivora. In *Mammal species of the world,* ed. D. E. Wilson and D. M. Reeder, 279–348. Washington, D.C.: Smithsonian Institution Press.

Wyss, A. R., and J. J. Flynn. 1993. Phylogenetic analy-sis and definition of Carnivora. In *Mammal phylogeny: Placentals,* ed. F. S. Szalay, M. J. Novacek, and M. C. McKenna, 32–52. New York: Springer-Verlag.

• Ximénez, A. 1961. Nueva subespecie del gato pajero en el Uruguay. *Comun. Zool. Mus. Hist. Nat. Montevideo* 5 (88): 1–8.

• ———. 1972. Notas sobre félidos neotropicales. 4. *Puma concolor* spp. en el Uruguay. *Neotrópica* 18 (55): 37–39.

• ———. 1973. Notas sobre félidos neotropicales. 3. Contribución al conocimiento de *Felis geoffroyi* d'Orbigny y Gervais, 1844 y sus formas geográficas (Mammalia, Felidae). *Pap. Avulsos Zool. São Paulo* 27 (3): 31–43.

———. 1975. *Felis geoffroyi. Mammal. Species* 54: 1–4.

• Ximénez, A., and E. Palermo. 1971. Confirmación de la presencia de *Felis wiedii wiedii* Schinz (Carnivora, Felidae) en el Uruguay. *Bol. Soc. Zool. Uruguay* 1:7–10.

Yañez, J., and F. Jaksić. 1978. Rol ecológico de los zorros (*Dusicyon*) en Chile central. *Anal. Mus. Hist. Nat. Valparaiso* 11:105–12.

Yañez, J., and J. Rau. 1980. Dieta estacional de *Dusicyon culpaeus* (Canidae) en Magallanes. *Anal. Mus. Hist. Nat. Valparaiso* 13:189–91.

• Zapata, A. R. P. 1982. Sobre el yaguarundi o gato eira, *Felis yagouaroundi ameghinoi* Holmberg y su presencia en la provincia de Buenos Aires, Argentina. *Neotrópica* 28 (80): 165–70.

Zunino, G. E., O. B. Vaccaro, M. Canevari, and A. L. Gardner. 1995. Taxonomy of the genus *Lycalopex* in Argentina. *Proc. Biol. Soc. Wash.* 108:729–47.

11   Order Carnivora (Pinnipedia) Seals, Sea Lions, and Walruses

Comment on Taxonomy

Wozencroft (1993) subsumes the seals,, sea lions, and walruses under the Carnivora. His taxonomic intent reflects the evidence presented in part by King (1964) suggesting that the pinnipeds are polyphyletic. Eisenberg (1989) and Redford and Eisenberg (1992) presented the Pinnipedia as a separate order. We felt at the time that the Pinnipedia, although closely related to the rest of the Carnivora, were at least monophyletic, a view supported by Sarich (1969) and by Berta, Ray, and Wyss (1989). Since the pinnipeds are embedded in the Carnivora but are surely monophyletic (Wyss and Flynn 1993), they are treated here in their own chapter.

Diagnosis

The body form of pinnipeds is strongly modified for swimming. The hind feet and forefeet are in the form of flippers, with the digits completely enclosed in the integument. The tail is short, and the body is spindle shaped. The teeth are usually simplified in structure, often peglike or conical (fig 11.1), but the feeding specialization of such forms as *Lobodon* yields a modified form of the molars and premolars.

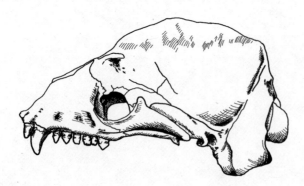

Figure 11.1. Skull of *Zalophus californicus*.

Distribution

Seals and their relatives are found worldwide in the temperate to cold oceans close to continental shorelines or polar ice. The monk seals (*Monachus*) are exceptional in being tropical but are still bound to land. A few species are landlocked in subarctic lakes. They reach their greatest species diversity in the temperate to Arctic and Antarctic latitudes and are generally absent from tropical waters except for the monk seals (King 1956).

History and Classification

The pinnipeds are first detected in the middle Miocene, already extremely specialized for an aquatic existence. It has been suggested that the Pinnipedia may have had a dual origin and thus may be an artificial assemblage, since the ancestors of the true seals (Phocidae) could have been derived from mustelid precursors whereas the eared seals (Otariidae) may have derived from an ancestral form that also gave rise to modern canids and ursids. In contrast, evidence presented by Berta, Ray, and Wyss (1989) and by Sarich (1969) supports a monophyletic origin for the group. The extant pinnipeds are usually grouped into three families: the Otariidae or eared seals, including the fur seals and sea lions; the Odobenidae, including the extant walruses; and the Phocidae or "earless" seals. This last group is characterized by loss of the external ear (pinna) and by hind limbs incapable of rotating forward, thus limiting movement on land. The biology of the Pinnipedia is summarized by King (1964, 1983).

Comment

The South American species of this order are considered in volumes 1 and 2 of this work (Eisenberg 1989; Redford and Eisenberg 1992). Only five species, all from the family Otariidae, regularly occur in the coastal waters of the countries included in this vol-

Table 11.1   Pinnipedia Recorded from the Coastal
               Areas of Brazil, Peru, and Ecuador
               (Including the Galápagos Islands)

Taxon	Ecuador	Galápagos	Peru	Brazil
Family Otariidae				
Arctocephalus				
australis	—	—	x	—
Arctocephalus				
galapagoensis	—	x	—	—
Arctocephalus				
tropicalis	—	—	—	x
Arctocephalus				
gazella	—	—	—	x
Otaria byronia	?	—	x	x
Zalophus				
californicus	?	x	?	—
Hydrurga				
leptonyx	—	—	—	x

Sources: Vaz-Ferreira 1982; Riedman 1990; Majluff and Trillmich 1981; Rosas et al. 1994; Drehmer, Borsatto, and Rosenau 1996; Borges Martins et al. 1996.
Note: Coastal records of "strays" from southeastern Brazil also included in Redford and Eisenberg 1992.

Table 11.2   Species of Pinnipedia Recorded
               for South America

Taxon	Description Reference
Family Otariidae	
Arctocephalus australis	This volume
Arctocephalus galapagoensis	This volume
Arctocephalus gazella	Redford and Eisenberg 1992
Arctocephalus philippii	Redford and Eisenberg 1992
Arctocephalus tropicalis	This volume
Otaria byronia	This volume
Zalophus californicus	This volume
Family Phocidae	
Hydrurga leptonyx	Redford and Eisenberg 1992
Leptonychotes weddelli	Redford and Eisenberg 1992
Lobodon carcinophagus	Redford and Eisenberg 1992
Mirounga leonina	Redford and Eisenberg 1992
Monachus tropicalis	Eisenberg 1989
Ommatophoca rossi	Redford and Eisenberg 1992

ume (see table 11.1). Two species, *Arctocephalus galapagoensis* and *Zalophus californicus,* are associated with the Galápagos archipelago. Species accounts for the two Galápagos forms and three species commonly found off the coasts of Peru and Brazil are included in the following sections.

FAMILY OTARIIDAE
Lobo Marinho

Diagnosis
The dental formula is I 3/2, C 1/1, P 4/4, M 1-2/ (fig 11.2). There is a wide range of size within the fam-

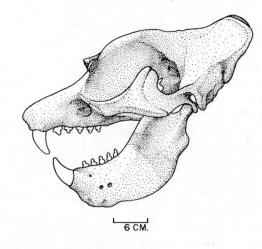

Figure 11.2. Skull of *Otaria byronia.*

ily, from 60 kg to over 1,000 kg. Total length ranges from 1.5 to 3.5 m. Species of this family are usually strongly dimorphic in size, with the males much larger than the females. The body is fusiform, highly adapted for aquatic life, and the limbs are modified into flippers, with well-developed nails. The hind limbs and forelimbs can be brought under the body for partial support. The external ear (pinna) is present but small.

Distribution
Distribution is coastal or insular. The family is broadly distributed from the northeast coast of Asia southward along the Pacific coast of North America and South America, extending in the Atlantic to the coast of Brazil. Populations also occur off southern Australia and New Zealand to the coast of southern Africa and in islands of the extreme southern Indian Ocean.

Natural History
Fur seals and sea lions spend most of their lives at sea, returning to land to breed and to rear their young. Sites for breeding and rearing are usually on isolated offshore islands or inaccessible rocky coasts. These animals feed primarily on fish and cephalopods. A single young is produced after a gestation period of approximately eleven months. The time spent ashore varies widely. Polar populations spend only a short time on land (usually less than a month), while populations in the lower latitudes may spend several months on land and have a more protracted rearing phase. The males are strongly polygynous; they usually stake out a favorable beach area and defend it against other males. Females usually arrive ashore later than males and take up residence in a male's territory (see species accounts).

Genus *Arctocephalus* E. Geoffroy and F. Cuvier, 1826

Description

Arctocephalus is strongly sexually dimorphic in size; males may be 2.7 m in head and body length and weigh over 700 kg, while females are more gracile, rarely exceeding 1.9 m with a maximum weight of 120 kg. The basic dorsal color is a reddish to dark brown. Breeding males have thick fur on the neck, and its white tips yield a frosted appearance.

Distribution

This genus ranges in the Southern Hemisphere following the cold Humboldt current to the Galápagos Islands in South America.

Life History and Ecology

The Southern Hemisphere fur seals are at sea much of their lives but return to isolated insular beaches to breed and give birth. After giving birth the female usually nurses the young for about a week, then she must return to the sea to feed. The nursing phase is of variable duration. The female may suckle the young for four to twelve months after birth, but she alternates feeding bouts with time ashore nursing. A similar rearing pattern is shown by the sea lion (see species accounts).

Arctocephalus australis (Zimmermann, 1783)
Lobo de los Pelos

Description

Adult males are much larger than females; they reach 1.9 to 3 m in total length and weigh more than 159 kg, while females reach 1.4 to 2 m and 48.5 kg. The snout is pointed and is larger than that of *A. flavescens*.

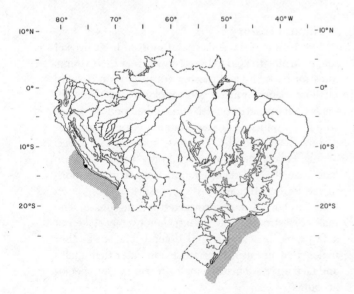

Map 11.1. Distribution of *Arctocephalus australis*.

The rhinarium is globular with nasal openings pointed forward, distinguishing it from *A. philippii*, which has the openings pointed down. This large marine mammal has small external ears and short, thick pelage composed of relatively stiff outer hairs and soft, dense underfur. Dorsally it is dark coffee brown with gray tints, and ventrally it has cinnamon tints. The pups are born black and gain adult coloration after about three months (Sielfeld 1983; Vaz-Ferreira 1979a).

Distribution

Arctocephalus australis is found on the coasts of South America from Rio de Janeiro (Brazil) around the tip of the continent and up the west coast to Lima, Peru. It is also found on the Juan Fernández Islands, Galápagos Islands, and Malvinas Islands. It does not necessarily breed throughout the total range (Daciuk 1974; Miller et al. 1983; Scheffer 1958) (map 11.1).

Life History

The birth season is November and December, with mating postpartum. One pup is born, weighing 3–5 kg, and lactation lasts six to twelve months. The age of first reproduction is probably four to five years; pups swim before two months of age (Housse 1952; Sielfeld 1983; Trillmich and Majluf 1981; Vaz-Ferreira 1979a).

Ecology

Males of *A. australis* are polygynous, setting up territories in November, and in some areas they defend an average of fifteen females and do not eat during the entire reproductive period. There are also groups of nonreproductive bachelor males. Breeding groups are normally established in rocky places. The male-to-female sex ratio in breeding rookeries in some areas varies from 1:1 to 1:13, averaging 1:6.5. In Peru this species breeds in huge sea caves. Females coming in from the ocean call to attract their pups, then smell them to confirm identity. Harcourt (1992) describes the mortality of fur seal pups deriving from reduced abundance of prey (fish) available to lactating females. The shift in abundance was caused by a well-documented El Niño event.

In some parts of its range *A. australis* is a krill (*Euphausia*) feeder, but it is reported to take a broad range of other food items including fish and squid. It feeds over wide areas of the continental shelf and beyond. No definite migration is shown, but seasonally there is wide seaward dispersion (King 1983; Sielfeld 1983; Trillmich and Majluf 1981; Vaz-Ferreira 1979a).

Arctocephalus galapagoensis Heller, 1904

Description

The postcanine teeth are small and lack cusps. This small fur seal has a condylar-basal skull measurement of 201–10 (male) and 171–86 (female) in adults. There is only slight sexual dimorphism in size. The muzzle is short and pointed. The occipital crest is weakly developed.

Distribution

The species is apparently confined to the Galápagos archipelago and was greatly reduced by hunting in the nineteenth century. Colonies are currently found on Genovesa, Santiago, Isabela, and Santa Cruz Islands (not mapped).

Life History

The ecology and behavior of these nonmigratory seals were described by Orr (1973). They prefer rocky beaches with caves where they can shelter from the sun. They feed at night, usually at a shallow depth. After giving birth a female will come into estrus in about eight days. The young are nursed for a variable time and are usually weaned at eighteen to twenty-six months of age (Trillmich 1986a). The Galápagos subspecies of *Zalophus* is much more abundant and better known (Clark 1975).

Arctocephalus tropicalis (Gray, 1872)
Subantarctic Fur Seal

Description

This species is strongly dimorphic in size, with adult males averaging 180 cm in length and having an average weight of 165 kg while females average 145 cm with a mean weight of 55 kg. The dorsum is variable in color, dark gray to brown or almost black on the dorsum and sides. The throat and chest are yellow grading to a brown venter. The male has longer hairs on the neck region and a crest of hair on the head.

Distribution

The species is known to breed on Tristan da Cunha, Gough, Prince Edward, Saint Paul, and New Amsterdam Islands of the South Atlantic. Stray animals have been recorded from the coasts of South Africa, Uruguay, and Brazil (González et al. 1994).

Natural History

Young are born during November and December after fifty-one weeks of gestation. Lactation lasts about seven months. Mating occurs three to seven days after birth. Fish, cephalopods, and euphausids are included in the diet (Bonner 1979).

Genus *Otaria* Peron, 1816
Otaria byronia (Blainville, 1820)
Lobo de un Pelo

Description

This large marine mammal has a large head, a short, broad snout, small external ears, and coarse pelage

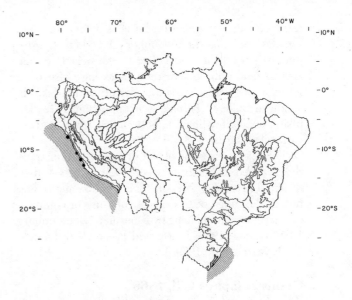

Map 11.2. Distribution of *Otaria byronia*.

with no underfur. Adult males can reach 2.56 to 3.5 m in length and at least 300 to 340 kg in weight, while the much smaller females reach 2 to 2.5 m and at least 144 kg. The males are much more robust than the females, which lack a mane. The color is highly variable: often adult males are dark brown to orange with a mane of a paler shade and the belly shading to a lighter yellow. The females are dark to light brown, and the pups are black (Osgood 1943; Sielfeld 1983; Vaz-Ferreira 1979b).

Distribution

Otaria byronia is widely distributed in South America, from northern Peru down the length of Chile, up the coast of Argentina and Uruguay to southeastern Brazil. It is also found on offshore islands like the Malvinas (Majluf and Trillmich 1981; Scheffer 1958) (map 11.2).

Life History

Newborn male pups are 82 cm long and weigh 14.2 kg, while females measure 79 cm and weigh 11.5 kg. In Uruguay most births are from the end of December to the middle of January. Pups suckle for six to twelve months, and a female may nurse both a pup and a yearling. Females mate postpartum. In Chile births are from September to March, depending on the latitude; lactation lasts about ten months, and young enter the water at about two months of age (Sielfeld 1983; Vaz-Ferreira 1979b).

Ecology

Males are polygynous, and in Uruguay they establish territories from the end of November, on sand or gravel beaches where the male-to-female ratio may be as high as 1:15. In Chile males set up breeding territories in September on beaches or in caves, and females come late in September. Outside the reproductive period these seals live in the pelagic ocean, visiting land only to sun. Individuals have also been found a long way up freshwater rivers and in lakes in southern Chile. *O. byronia* is an opportunistic feeder, taking small schooling fish, bottom fish, squid, and occasional seabirds.

This species was the object of an intense fur trade that reached its height between 1920 and 1954, then declined because of overkill. There has been a reduction in the range of this species because of commercial harvesting and killing by fishermen (Brandenburg 1938; Housse 1952; Iriarte and Jaksić 1986; Sielfeld 1983; Vaz-Ferreira 1979b).

Genus *Zalophus* Gill, 1866
Zalophus californicus (Lesson, 1828)

Description
This species is characterized by marked sexual size dimorphism. Males range from 200 to 250 cm in head and body length, while females range from 150 to 200 cm. The basic body color is dark brown, but in living specimens reflectance varies depending on the moisture content of the fur.

Distribution
The species is basically coastal and occurs from Japan and Korea to the Pacific coast of North America south to Nayarit, Mexico, and thence to the Galápagos archipelago (not mapped).

Life History
Courtship and rearing of pups take place on sandy beaches of offshore islands. The adult males defend terrestrial territories in order to monopolize mating of several adult females. The Galápagos population demonstrates a birth peak in August to October, but births can occur in any month. A single pup is the mode. The population of the Galápagos is nonmigratory. They tend to feed during the day, and nursing females return daily to their young. The young are weaned at about eleven to twelve months of age. After giving birth a female will come into estrus within three weeks (Trillmich 1986b; Kooyman and Trillmich 1986). Females can recognize their own offspring within a "crèche" of young (Eibl-Eibesfeldt 1955; Peterson and Bartholomew 1967).

References

• References used in preparing distribution maps

Berta, A., C. E. Ray, and A. R. Wyss. 1989. Skeleton of the oldest known pinniped *Enaliarctos mealsi*. *Science* 244:60–62.

Bonner, W. N. 1979. Subantarctic fur seal. In *Mammals in the seas*, 2:52–54. FAO Fisheries Series 5. Rome: Food and Agriculture Organization of the United Nations.

Borges Martins, M., D. Damielewicz, L. R. do Oliviera, P. H. Ott, and L. Susin. 1996. Registros de pinepedes para o litoral norte do Rio Grande do Sul, Brasil (1991–1995). Abstract 1191. *Resumos do XXI Congresso Brasileiro de Zoologia* (Pôrto Alegre).

Brandenburg, F. G. 1938. Notes on the Patgonian sea lion. *J. Mammal.* 19:44–47.

Clark, T. W. 1975. *Arctocephalus galapagoensis*. *Mammal. Species* 64:1–2.

• Daciuk, J. 1974. Notas faunísticas y bioecológicas de Península Valdés y Patagonia. 12. Mamíferos colectitativo y observados en la Península Valdés y zona litoral de los Golfos San José y Nuevo (provincia de Chubut, república Argentina). *Physis*, sec. C, 33 (86):23–39.

Drehmer, C. J., E. S. Borsatto, and M. Rosenau. 1996. Ações antropicas sobre Otarideos (Carnivora, Pinnipedia) ma costa do Rio Grande do Sul, Brasil. *Resumos do XXI Congresso Brasileiro de Zoologia* (Pôrto Alegre).

Eibl-Eibesfeldt, I. 1955. Ethologische studien om Galapagos: Seclowen *Zalophus* Wollebacki. *Z. Tierpsychol.* 12:286–303.

Eisenberg, J. F. 1989. *Mammals of the Neotropics*. Vol. 1. *The northern Neotropics: Panama, Colombia, Venezuela, Guyana, Suriname, French Guiana*. Chicago: University of Chicago Press.

• González, J. C., A. Saralegui, E. M. Gonzalez, and R. V. Ferreira. 1994. La presencia de *Arctocephalus tropicalis* en Uruguay. *Comun. Mus. Ciênc. Tecnol.*, Pôrto Alegre, Sér. Zool. 7:205–10.

Harcourt, R. 1992. Factors affecting early mortality in the South American fur seal (*Arctocephalus australis*) in Peru. *J. Zool.* (London) 227:417–38.

Honacki, J. H., K. E. Kinman, and J. W. Koeppl, eds. 1982. *Mammal species of the world*. Lawrence, Kans.: Allen Press and Association of Systematics Collections.

• Housse, R. P. R. 1952. Mamíferos de Chile orden de los pinípedos. *Acad. Chil. Cienc. Nat.* 17, Anal., Rev. Univ. (Santiago) 37:163–75.

Iriarte, J. A., and F. M. Jaksić. 1986. The fur trade in Chile: An overview of seventy-five years of export data (1910–1984). *Biol. Conserv.* 38:243–53.

King, J. E. 1956. The monk seals, genus *Monachus*. *Bull. British Mus. Nat. Hist.* (*Zool.*) 3:203–56.

• ———. 1964. *Seals of the world*. London: British Museum of Natural History.

———. 1983. *Seals of the world*. Rev. ed. London: British Museum of Natural History.

Kooyman, G. L., and F. Trillmich. 1986. Diving behavior of Galapágos sea lions. In *Fur seals: Maternal strategies on land and at sea,* ed. R. L. Gentry and G. L. Kooyman, 209–19. Princeton: Princeton University Press.

Majluf, P., and F. Trillmich. 1981. Distribution and abundance of sea lions (*Otaria byronia*) and fur seals (*Arctocephalus australis*) in Peru. *Z. Säugetierk.* 46:384–93.

• Miller, S. D., J. Rottman, K. J. Raedeke, and R. D. Taber. 1983. Endangered mammals of Chile: Status and conservation. *Biol. Conserv.* 25:335–52.

Orr, R. T. 1973. Galápagos fur seal (*Arctocephalus galapagoensis*). In *Proceedings of the working specialists on threatened and depleted seals of the world,* 124–28. Gland, Switzerland: IUCN.

Osgood, W. H. 1943. The mammals of Chile. *Field Mus. Nat. Hist. Zool. Ser.* 30:1–268.

Peterson, R. S., and G. A. Bartholomew. 1967. *The natural history and behavior of the California sea lion.* Special Publication 1. Shippensburg, Pa.: American Society of Mammalogists.

Redford, K. H., and J. F. Eisenberg. 1992. *Mammals of the Neotropics.* Vol. 2. *The southern cone: Chile, Argentina, Uruguay, Paraguay.* Chicago: University of Chicago Press.

Riedman, M. 1990. *The pinnipeds: Seals, sea lions and walruses.* Berkeley: University of California Press.

Rosas, F. C. W., L. C. Capistrano, A. P. Di Beneditto, and R. Ramos. 1992. *Hydrurga leptonyx* recovered from the stomach of a tiger shark captured off the Rio de Janeiro coast, Brazil. *Mammalia* 56 (1): 153–55.

Rosas, F. C. W., M. C. Pinedo, M. Marmontel, and M. Haimovici. 1994. Seasonal movements of the South American sea lion (*Otaria flavescens* Shaw) off the Rio Grande do Sul coast, Brazil. *Mammalia* 58:51–59.

Sarich, V. M. 1969. Pinniped phylogeny. *Syst. Zool.* 18:416–22.

Scheffer, V. B. 1958. *Seals, sea lions and walruses: A review of the Pinnipedia.* Stanford: Stanford University Press.

• Sielfeld, W. 1983. *Mamíferos marinos de Chile.* Santiago: Ediciones de la Universidad de Chile.

Trillmich, F. 1986a. Attendance behavior of Galápagos fur seals. In *Fur seals: Maternal strategies on land and at sea,* ed. R. L. Gentry and G. L. Kooyman, 168–85. Princeton: Princeton University Press.

————. 1986b. Attendance behavior of Galápagos sea lions. In *Fur seals: Maternal strategies on land and at sea,* ed. R. L. Gentry and G. L. Kooyman, 196–208. Princeton: Princeton University Press.

Trillmich, F., and P. Majluf. 1981. First observation on colony structure, behavior, and vocal repertoire of the South American fur seal (*Arctocephalus australis* Zimmermann, 1783) in Peru. *Z. Säugetierk* 46:310–22.

• Vaz-Ferreira, R. 1979a. South American sea lion. In *Mammals in the seas,* 2:9–12. FOA Fisheries Series 5. Rome: Food and Agriculture Organization of the United Nations.

• ————. 1979b. South American fur seal. In *Mammals in the seas,* 2:34–36. FOA Fisheries Series 5. Rome: Food and Agriculture Organization of the United Nations.

————. 1982. *Arctocephalus australis* Zimmerman, South American fur seal. In *Mammals of the seas,* 4:497–508. FAO Fisheries Series 5. Rome: United Nations.

Wilson, D. E., and D. M. Reeder, eds. 1993. *Mammal species of the world,* 2d ed. Washington, D.C.: Smithsonian Institution Press.

Wozencroft, W. C. 1993. Order Carnivora. In *Mammal species of the world,* ed. D. E. Wilson and D. M. Reeder, 279–348. Washington, D.C.: Smithsonian Institution Press.

Wyss, A. R., and J. J. Flynn. 1993. Phylogenetic analysis and definition of Carnivora. In *Mammal phlogeny: Placentals,* ed. F. Szalay, M. J. Novacek, and M. McKenna, 32–52. New York: Springer-Verlag.

12   Order Cetacea
(Whales, Dolphins, and Their Allies:
Baleias, Botos, and Golfinhos)

Diagnosis

Except for the Sirenia, the Cetacea exhibit the most complete adaptation for aquatic existence shown within the class Mammalia (see plate 14). The body is spindle shaped, hair is absent except for a few bristles around the lips, and the skin is smooth. The forelimbs are modified into paddles; the digits cannot be detected externally. Hindlimbs are absent, and the tail is flattened dorsoventrally and extended laterally to produce two pointed flukes. The nasal passages open through either a single or a double aperture on top of the head. The teeth are absent or extremely modified, exhibiting no cusp pattern in the adult. When teeth are present all are very similar in shape within the jaw of any species (homodont dentition); it is often impossible to distinguish molars, premolars, and canines. Some standard references on the order's biology and taxonomy are Slijper (1962), Norris (1966), Gaskin (1982), Minasian, Balcomb, and Foster (1984), and Hershkovitz (1966).

Distribution

The order is found worldwide in oceans and seas; some small species ascend rivers or are adapted for a permanent freshwater existence.

History and Classification

Cetaceans are classically divided into three suborders, the Archaeoceti (extinct), the Odontoceti, and the Mysticeti. Whether the last two lineages descended from a common ancestor is still open to debate. It is enough to say that the Archaeoceti first appear in the fossil record in the middle Eocene and may have given rise to the extant families. The oldest odontocete fossils are from the late Eocene, and mysticete fossils first appeared in the mid-Oligocene. Odontocetes have teeth and are thus often referred to as "toothed" whales. These forms usually feed on fish and cepha-

lopods (see species accounts). The mysticetes do not have teeth as adults, but there are modified epidermal derivatives hanging from the roof of the mouth in longitudinal plates called baleen (see fig. 12.1), which serve as a straining device. Mysticetes are specialized for feeding on various small crustaceans (krill) that they obtain by swimming through swarms of the small shrimplike creatures with their mouths partly open, periodically expelling excess water and swallowing the retained crustaceans after scraping them off the baleen plates with the tongue. Baleen whales also eat fish (see species account).

Natural History

The great whales and dolphins are entirely aquatic and never come to land unless they are accidentally stranded, washed ashore during storms, or seizing prey from the edge of a beach or ice floe (*Orca*). Usually only a single young is born, after an extended gestation period, delivered at sea in a highly advanced state. Odontocetes are specialized for feeding on fish or pelagic squid. Mysticetes, as noted before, are specialized for feeding on planktonic crustaceans. Many of the great mysticete whales make enormous migrations to and away from the poles following plankton abundance. During their short growing season, the polar regions have high productivity and thus are favor-

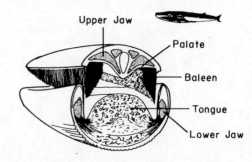

Figure 12.1. Schematic drawing of baleen plates and their position in the mouth of *Balaenoptera* (modified from Slijper 1962).

ite feeding grounds for mysticete whales during their summer movements. Although some whales are specialized for tropical waters, most cetacean taxa show their greatest species richness in the temperate and Arctic latitudes. Gaskin (1982) is a useful reference on the behavior and ecology of cetaceans. Eisenberg (1986) has reviewed the behavioral convergence demonstrable when terrestrial mammals are compared with cetaceans. Many of the smaller, toothed whales have been studied in aquariums. Some years ago it was recognized that small delphinids could locate objects in the water by emitting high-frequency "sonar pulses." The whole subject of underwater echolocation caught the imagination of scientists, and the pioneer results are included in Kellogg (1961).

Comment

The whales, porpoises, and dolphins have been fully discussed in volume 1 and 2 of this work (Eisenberg 1989; Redford and Eisenberg 1992). Table 12.1 lists those species recorded from the coasts of Brazil, Peru, and Ecuador. The biology of *Inia* has recently been reviewed by Best and da Silva (1993). We include here only three species accounts: *Inia geoffroyensis*, *Sotalia fluviatilis*, and *Pontoporia blainvillei*. These are either riverine or inshore species commonly found in the South American waters and, in the case of the freshwater forms, Amazonia.

FAMILY PLATANISTIDAE (INIIDAE)

River Dolphins

Diagnosis

The teeth are numerous and cone shaped, and the skull is distinctive (fig. 12.2). These small dolphins have an extremely long, slender beak that is sharply marked off from the rather bulging forehead. The dorsal fin is low. The eye tends to be very reduced in size, and in some genera it is not detectable externally without dissection.

Distribution

These freshwater dolphins have a very disjunct distribution, being found in the Ganges and Indus Rivers of India, in the Yangtze Kiang of China, in the La Plata River and its major tributaries of Argentina and Uruguay, and in the Amazon and Orinoco river systems.

Inia geoffrensis (de Blainville, 1817)
Amazon River Dolphin, Bouto, Boto Vermelho

Description

The long, slender beak is distinctive. The teeth are simple, single crowned, and number up to 132, with

Table 12.1 Marine Cetaceans Recorded from the Coastal Areas of Ecuador, Peru, and Brazil

Taxon	Ecuador	Peru	Brazil
Balaenidae			
Eubalaena australis			X
Eubalaena glacialis	?		
Balaenopteridae			
Balaenoptera acutorostrata	X	X	X
Balaenoptera borealis	?		
Balaenoptera edeni	X	X	X
Balaenoptera musculus	?	X	X
Balaenoptera physalus	?	X	X
Megaptera novaeangliae	?	X	X
Neobalaenidae			
Caperea marginata	—	—	X
Delphinidae			
Delphinus delphis	X	X	
Feresa attenuata	?	X	—
Globicephala macrorhynchus	?	X	—
Grampus griseus	X	X	X
Lagenodelphis hosei[a]	?	X	—
Lagenorhynchus obscurus	—	X	X
Lissodelphis peronii	—	—	X
Orcinus orca	—	—	X
Peponocephala electra	—	—	X
Pseudorca crassidens	?	X	X
Sotalia fluviatilis[b]	—	—	X
Stenella attenuata	X	X	X
Stenella clymene	—	—	X
Stenella coeruleoalba	X	X	—
Stenella frontalis	—	—	X
Stenella longirostris	X	X	—
Steno bredanensis	—	—	X
Tursiops truncatus	X	X	X
Phocoenidae			
Phocoena spinipinnis	?	X	—
Physeteridae			
Physeter catodon	X	X	X
Kogia breviceps	—	X	X
Kogia simus	—	—	X
Platanistidae			
Pontoporia blainvillei[b]	—	—	X
Ziphiidae			
Berardius arnuxii	—	—	X
Mesoplodon densirostris	—	—	X
Mesoplodon peruvianus	?	X	—
Ziphius cavirostris	—	—	X

Note: Species descriptions are included in Eisenberg 1989 and Redford and Eisenberg 1992. Cetacean strandings are chance events, and no doubt this table will expand in the years to come.

[a] *Lagenodelphis hosei* was recently reviewed by Jefferson and Leatherwood 1994.

[b] Not strictly marine, but *Sotalia* is mainly found in freshwater (see text).

20 to 30 on each side of the upper and lower jaws. The dorsal fin begins in the midback and extends as a ridge to the tail. Length ranges from 2 to 3 m and weight from 90 to 150 kg. The coloration is variable and related to age. Young dolphins have a grayish black dorsum and a pink venter, whereas in older individuals the upperparts become pink (plate 14).

Table 12.2   Coastal Records for Brazilian Cetaceans

Taxon	State	Taxon	State
Balaenopteridae		Delphinidae	
Balaenoptera		*Delphinus delphis*	RJ, SP, SC, RS
acutorostrata	Pba, RJ, SP, RS	*Stenella attenuata*	Northeast coast
Balaenoptera borealis	Pba, ES, RJ	*Stenella coeruleoalba*	RS
Balaenoptera edeni	Pba, RJ, Pn	*Stenella frontalis*	SP, RJ, SC
Balaenoptera musculus	Pba, RJ, RS	*Stenella longirostris*	Pb, RJ, SP
Balaenoptera physalus	Pba, RJ	*Steno bredanensis*	Pm, RJ, SC
Megaptera novaeangliae	RN, Pba, ES, RJ, SP, Pn, SC, RS	*Tursiops truncatus*	RN, Pba, ES, RJ, SP, SC, RS
Balaenidae		*Sotalia fluviatilis*	Pa, Pba, Ba, RN, Se, Pm, ES, RJ,
Eubalaena australis	Ba, RJ, SP, Pn, SC, RS		SP, Pn, SC
Physeteridae		*Peponocephla electra*	Ba
Physeter macrocephalus	RN, Pba, Pm, Se, RJ, SP, SC, RS	*Pseudorca crassidens*	Pba, RJ, SC, RS
Kogia breviceps	RJ, SP, RS	*Orcinus orca*	Pba, RJ, SP, SC, RS
Kogia simus	RS	*Grampus griseus*	SC
Ziphiidae		*Globicephala*	
Hyperoodon planifrons	SC, RS	*macrorhynchus*	SP
Mesoplodon densirostris	SC, RS	*Globicephala melas*	SP, RS
Ziphius cavirostris	Pba, SP	Phocoenidae	
		Phocoena spinipinnis	SC, RS

Source: Pinedo, Rosas, and Marmontel 1992; Hetzel and Lodi 1993.
Note: Platanistidae are not included.
Abbreviations: Pba, Paraíba; Pa, Pará; RN, Rio Grande do Norte; Ba, Bahia; Pm, Pernambuco; Se, Sergipe; ES, Espírito Santo; RJ, Rio de Janiero; SC, Santa Catarina; SP, São Paulo; Pn, Paraná; RS, Rio Grande do Sul.

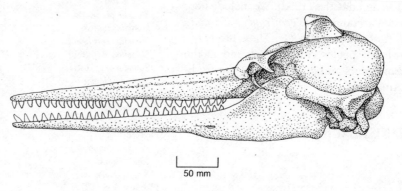

50 mm

Figure 12.2. Skull of *Inia geoffrensis*.

Range and Habitat

The species occurs in both the Amazon and the Orinoco river systems and is the common dolphin sighted in the upstream portions of these rivers (Pilleri and Pilleri 1982) (map 12.1).

Natural History

The Amazon river dolphin is specialized for feeding on fish. Much of the water it forages in is extremely turbid, and it undoubtedly uses echolocation to detect its prey (Layne and Caldwell 1964). Groups of up to half a dozen may be encountered. Group size varies in part because dolphin schools may break up and forage in small groups when the river water is high, and the dolphins follow their fish prey into the flooded forest. At low water dolphins occasionally become trapped in lagoons and *canos* until the next high water (Trebbau

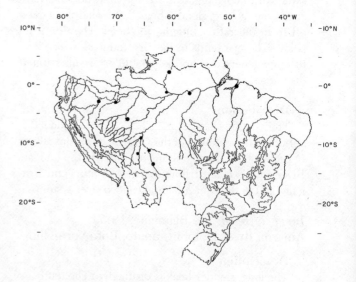

Map 12.1. Distribution of *Inia geoffrensis*.

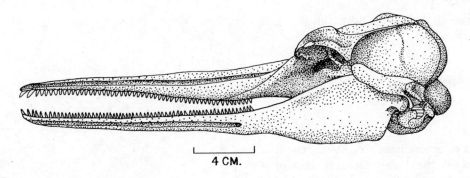

Figure 12.3. Skull of *Pontoporia blainvillei*.

1975; Trebbau and van Bree 1974). Reproduction in the Amazon is seasonal and correlated with the fluctuation in water levels (Best and da Silva 1984). Gestation lasts 10.5 months, with a two-year interval between calves (Brownell 1984). There appears to be a strong bond between the mother and the single young. Defler (1983) reported feeding associations between *Inia* and *Pteronura* in Colombia.

Best and da Silva (1993) have compiled a rather useful summary of the biology of *Inia*. They confirm the distribution in the Rio Madeira extending well into Bolivia. The season of birth is tied to the initial recession of high water, thus occurring from May to June over much of the Amazon basin. Lactation may last more than a year.

Seasonal migrations of groups has been noted in the major river systems, corresponding to seasonal shifts in the fish populations the dolphins prey on.

Genus *Pontoporia* Gray, 1846
Pontoporia blainvillei (Gervais and d'Orbigny, 1844)

Measurements

	Mean	Min.	Max.	N	Loc.
TL	1.33 m	1.22 m	1.47 m	21 m	U
	1.53 f	1.37 f	1.71 f	28 f	
Wt	28.20 kg	19.90 kg	32.20 kg	4 m	
	40.30 kg	29.90 kg	52.10 kg	5 f	

Redford and Eisenberg 1992.

Description
This small dolphin is easily recognized by its elongated body and long, thin beak (see fig. 12.3). It has a disproportionately small head and fifty or more teeth. The distinct triangular dorsal fin is on the midback and extends as a ridge to the flukes. The paddle-shaped flippers are rounded at the tips. It is pale brown dorsally, shading to lighter brown ventrally, but some individuals are nearly white (Minasian, Balcomb, and Foster 1984).

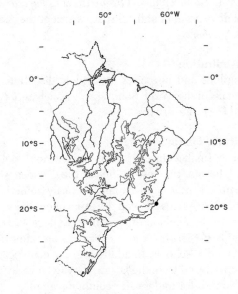

Map 12.2. Distribution of *Pontoporia blainvillei*.

Distribution
Pontoporia blainvillei is found from the Valdés peninsula, Chubut province, Argentina, to Espírito Santo state in Brazil and well into the Río de la Plata in shallow waters (Borobia and Geise 1984; Ximénez, Langguth, and Praderi 1972) (map 12.2).

Life History
The birth season is from December to January. Gestation lasts 10.5 months, and lactation lasts about nine months. Females become sexually mature at between 1.34 m and 1.37 m and males at above 1.4 m (about two to three years of age) (Brownell 1984).

Ecology
This dolphin is found in salt water near the coast and up into the Río de la Plata. It is found in the river from October to April and then seems to migrate north to the Brazilian coast. It apparently feeds at or near the bottom, and examination of nine stomachs revealed mostly fish and squid (Fitch and Brownell 1971;

Minasian, Balcomb, and Foster 1984; Ximénez, Lang-guth, and Praderi 1972).

FAMILY DELPHINIDAE
Dolphins (Delfíns) and Porpoises

Diagnosis
Composed of approximately eighteen genera and sixty-two species, this family includes most of the smaller, toothed whales. The beak is variable in its expression, being prominent in *Tursiops* and virtually absent in *Globicephala, Grampus,* and *Faresa.* The dorsal fin is prominent in all species except the Asiatic genus *Neomeris.*

Distribution
Dolphins and porpoises are widely distributed in the oceans of the world. Some species, such as *Sotalia,* ascend freshwater rivers.

Natural History
These small cetaceans feed on fish and pelagic squid. They are the best studied of the Cetacea, since they can be kept in captivity and are easily trained. Research with captives allowed the experimental demonstration of their echolocating ability (Kellogg 1961).

Highly gregarious, these cetaceans are almost always found in schools. In addition to the echolocation pulses produced from the larynx, they use a wide variety of sounds for underwater communication.

Sotalia fluviatilis (Gervais and De Ville, 1853)
Tucuxi

Description
This is the smallest of the dolphins, averaging 1.9 m. The dorsum ranges from gray to blackish, and the venter tends to be lighter. *S. guianensis* is included in *S. fluviatilis* (plate 14).

Range and Habitat
S. fluviatilis is found within the lower Amazon river system extending coastally to Rio Grande do Sul, Brazil (Magnusson, Best, and da Silva 1980). *Sotalia guianensis* is usually the specific name given to the form found in the coastal waters of Venezuela and the Guianas. The exact relationship of the two named forms remains to be determined (map 12.3).

Natural History
Species of the genus *Sotalia* travel in small groups of up to ten individuals. This coastal dolphin ascends the Orinoco and Amazon Rivers and has been recorded from Lago de Maracaibo (Casinos, Bisbal, and Boher 1981). In the Amazon they may occur far up-

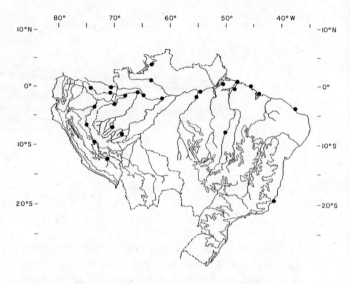

Map 12.3. Distribution of *Sotalia fluviatilis.*

stream, but in the Orinoco they are nearer the delta (Layne 1958). Feeding habits have been recorded for marine populations of this species taken from waters off southeastern Brazil. Both squid and fish were recorded, apparently taken from various depths (Borobia and Barros 1989; da Silva and Best 1996).

References
• References used in preparing distribution maps

Best, R. C., and V. M. F. da Silva. 1984. Preliminary analysis of reproductive parameters of the boutu, *Inia geoffrensis,* and the tucuxi, *Sotalia fluviatilis,* in the Amazon River system. In *Reproduction in whales, dolphins and porpoises,* ed. W. F. Perrin, R. L. Brownell Jr., and D. P. DeMaster, 361–72. Special Issue 6. Cambridge: International Whaling Commission.

———. 1993. *Inia geoffrensis. Mammal. Species* 426: 1–8.

Borobia, M., and N. B. Barros. 1989. Notes on the diet of marine *Sotalia fluviatilis. Marine Mammal Sci.* 5 (4): 395–99.

Borobia, M., and L. Geise. 1984. Registro de *Pontoporia blainvillei* (Cetacea, Plantanistidae) no Espírito Santo, Brasil. In *Primera reunion de trabajo de expertos en mamíferos acuáticos de América del Sud* (resúmenes) 8. Buenos Aires.

Brownell, R. L., Jr. 1984. Review of reproduction in platanistid dolphins. In *Reproduction in whales, dolphins and porpoises,* ed. W. F. Perrin, R. L. Brownell Jr., and D. P. DeMaster, 149–61. Special Issue 6. Cambridge: International Whaling Commission.

Casinos, A., F. Bisbal, and S. Boher. 1981. Sobre tres

ejemplares de *Sotalia fluviatilis* del Lago de Maracaibo (Cetacea-Delphinidae). *Publ. Dept. Zool. Univ. Barcelona* 7:93–96.

Defler, T. R. 1983. Associations of the giant river otter (*Pteronura brasiliensis*) with fresh-water dolphins (*Inia geoffrensis*). *J. Mammal.* 64 (4): 692.

Eisenberg, J. F. 1986. Dolphin behavior and cognition: Evolutionary and ecological aspects. In *Dolphin cognition and behavior: A comparative approach,* ed. R. Buhr, R. Schusterman, J. Thomas, and F. Wood, 261–70. Hillsdale, N.J.: Lawrence Erlbaum.

————. 1989. *Mammals of the Neotropics.* Vol. 1. *The northern Neotropics: Panama, Colombia, Venezuela, Guyana, Suriname, French Guiana.* Chicago: University of Chicago Press.

Fitch, J. E., and R. L. Brownell Jr. 1971. Food habits of the franciscana *Pontoporia blainvillei* (Cetacea: Platanistidae) from South America. *Bull. Marine Sci.* 21:626–36.

Gaskin, D. E. 1982. *The ecology of whales and dolphins.* London: Heinemann.

Geise, L., and M. Borobia. 1987. New Brazilian records for *Kogia, Pontoporia, Grampus,* and *Sotalia* (Cetacea, Physeteridae, Platanistidae, and Delphinidae). *J. Mammal.* 68 (4): 873–78.

Hershkovitz, P. 1966. Catalog of living whales. *Bull. U.S. Nat. Mus.* 246:1–259.

Hetzel, B., and L. Lodi. 1993. *Baleias, botoe e golfinhos: Guia de identifica cão para o Brasil.* Rio de Janeiro: Editora Nova Frontera.

Jefferson, T. A., and S. Leatherwood. 1994. *Lagenodelphis hosei. Mammal. Species* 470:1–5.

Kellogg, W. N. 1961. *Porpoises and sonar.* Chicago: University of Chicago Press.

Layne, J. N. 1958. Observations on freshwater dolphins in the upper Amazon. *J. Mammal.* 39:1–22.

Layne, J. N., and D. K. Caldwell. 1964. Behavior of the Amazon dolphin *Inia geoffrensis* in captivity. *Zoologica* 49:81–108.

Leatherwood, S., F. S. Todd, J. A. Thomas, and F. T. Awbrey. 1982. Incidental records of cetaceans in southern seas, January and February 1981. *Repts. Int. Whaling Comm.* 32:515–20.

Magnusson, W. E., R. C. Best, and V. M. F. da Silva. 1980. Numbers and behaviour of Amazonian dolphins, *Inia geoffrensis* and *Sotalia fluviatilis fluviatilis,* in the Rio Solimões, Brasil. *Aquatic Mammals* 8 (1): 27–32.

• Martuscelli, P., M. Milanelo, and F. Olmos. 1995. First record of Arnoux's beaked whale (*Bearardius arnuxii*) and southern right whale dolphin (*Lissodelphis peronii*) from Brazil. *Mammalia* 59:274–75.

Minasian, S. M., K. Balcomb III, and L. Foster. 1984. *The world's whales.* Washington, D.C.: Smithsonian Institution Press.

Norris, K. S., ed. 1966. *Whales, dolphins, and porpoises.* Berkeley: University of California Press.

Perrin, W. F., E. D. Mitchell, J. G. Mead, D. K. Caldwell, and P. J. H. van Bree. 1981. *Stenella etymene,* a rediscovered tropical dolphin of the Atlantic. *J. Mammal.* 62:583–98.

Pilleri, G., and O. Pilleri. 1982. *Zoologische expedition zum Orinoco and Brazo Casiquiare* 1981. Ostermundigen, Switzerland: Hirnanatomischen Institutes.

Pinedo, M. C., F. W. Rosas, and M. Marmontel. 1992. *Cetáceos e pinípedes do Brasil: Uma revisão dos registros e guia para identificação das espécies.* Manaus, Brazil: UNEP/FUA.

Piñero, M. E., and H. P. Castello. 1975. Sobre "ballenas piloto," *Globicephala melaena edwardi* (Cetacea, Delphinidae) varadas en Isla Trinidad, provincia de Buenos Aires. *Rev. Mus. Argent. Cienc. Nat. "Bernardino Rivadavia," Zool.* 12 (2): 13–24.

Redford, K. H., and J. F. Eisenberg. 1992. *Mammals of the Neotropics.* Vol. 2. *The southern cone: Chile, Argentina, Uruguay, Paraguay.* Chicago: University of Chicago Press.

• Reyes, J. C., J. G. Mead, and K. Van Waerebeck. 1991. A new species of beaked whale *Mesoplodan pervuianus* sp. n. from Peru. *Marine Mammal Sci.* 7 (1): 1–24.

• Silva, F. 1994. *Mamíferos silvestres — Rio Grande do Sul.* 2d ed. Pôrto Alegre: Fundação Zoobotanica do Rio Grande do Sul.

Silva, M. F. da, and R. C. Best. 1996. *Sotalia fluviatilis. Mammal. Species* 527:1–7.

Slijper, E. J. 1962. *Whales.* London: Hutchinson.

Trebbau, P. 1975. Measurements and some observations on the freshwater dolphin *Inia geoffrensis* in the Apure River. *Zool. Gart.,* n.s., 45:153–67.

Trebbau, P., and P. J. H. van Bree. 1974. Notes concerning the freshwater dolphin *Inia geoffrensis* in Venezuela. *Z. Säugetierk.* 39:50–57.

Ximénez, A., A. Langguth, and R. Praderi. 1972. Lista sistemática de los mamíferos del Uruguay. *Anal. Mus. Nac. Hist. Nat. Montevideo* 7 (5): 1–45.

13   Order Sirenia
(Sea Cows)

Diagnosis

These large, smooth-skinned aquatic mammals have no hind limbs. The tail has flukes and is whalelike in the Dugongidae but paddlelike in the Trichechidae. The forelimbs are paddlelike. There is no dorsal fin. The head is large and rounded, with the nostrils on the upper surface of the snout. The rostrum is deflected ventrally (fig. 13.1), most pronounced in the Dugongidae. Rough, horny plates on the palate help to abrade the herbaceous vegetation the animals feed on. The cecum is large, apparently adapted for digestive fermentation.

Distribution

There are two extant families, the Trichechidae and the Dugongidae. The dugongs are found in the Indian Ocean and in the western Pacific from the Philippines to northern Australia. The Trichechidae occur in a disjunct distribution along the coast and rivers of West Africa and in coastal parts of the Caribbean as well as in the Amazon and Orinoco drainage systems (Bertram and Bertram 1973).

History and Classification

The sirenians, or sea cows, were much more diverse in the past. They are known from the Eocene of North Africa, Jamaica, and Florida. The group was widespread in the Miocene but is now extremely restricted in its distribution, occurring only in tropical waters. The two families, the Dugongidae and the Trichechidae, seem to have diverged in the Eocene and thus represent two ancient lineages (Domning 1982).

FAMILY TRICHECHIDAE

Manatees, Manatis

Diagnosis

Functional incisors and canines are absent, and the cheek teeth are numerous and variable in number, since they are replaced sequentially from the rear (fig. 13.1). The teeth are low crowned, with two crests. The manatee may reach 4.5 m in length and weigh over 600 kg. The tail is paddle shaped (plate 15). The upper lip is deeply divided, and the animal can move each side of the lips independently while feeding.

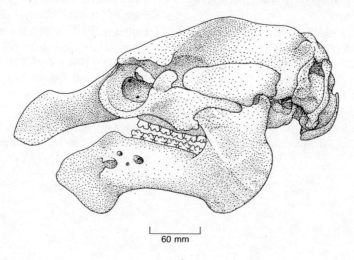

60 mm

Figure 13.1. Skull of *Trichechus manatus.*

Distribution

The three species of this family show a disjunct distribution. Whereas *Trichechus senegelensis* is found in the rivers of West Africa, the other two species are mainly Neotropical. In South America the Caribbean manatee is confined to riverine areas and coastal zones around the Caribbean and thence coastally to northern Brazil. It extends its range up the Orinoco to the first cataract (Mondolfi 1974). The Amazonian manatee is distributed within the Amazon and reaches the extreme southern portions of Venezuela and Colombia and thence the great rivers of Ecuador, Peru, Brazil, and Bolivia (Domning 1982).

Genus *Trichechus* Linnaeus, 1758
Trichechus inunguis (Natterer, 1883)
Amazonian Manatee, Peixe-boi

Description

This species is easily distinguished from the Caribbean manatee, *T. manatus*, by the absence of nails on the flippers. It is slightly smaller (maximum length 2.8 m) and frequently has whitish patches on the venter (plate 15).

Range and Habitat

The species is confined to the Amazon river system and is allopatric with *T. manatus* (map 13.1).

Natural History

Montgomery, Best, and Yamakoshi (1981) radio-tracked a manatee for several days. Rate of movement was 2.6 km per day. They provide a list of food plants and emphasize the use of aquatic vegetation near lake edges and of floating vegetation. In feeding habits, social tendencies, and presumably other attributes of natural history, *T. inunguis* parallels *T. manatus*. It makes frequent use of the "floating meadows" in the Amazon drainage, where *Paspalum repens* is a dominant aquatic plant. It also eats true water grasses such as *Panicum* and *Echinochloa* (Marmol 1976; Husar 1977). A complete review of the biology of the species is included in Rosas (1994).

Trichechus manatus Linnaeus, 1758
Caribbean Manatee, Manati

Description

This is the largest Neotropical manatee, reaching 4.5 m. Three nails are visible on each flipper (plate 15). The body is a dull gray.

Range and Habitat

The range extends from the southeastern United States, especially Florida, disjunctly to southeastern Mexico. The species formerly occurred in bays

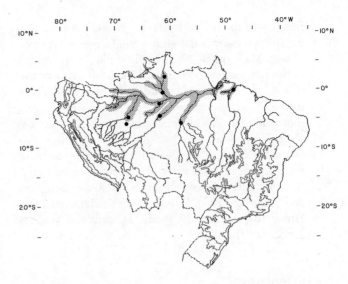

Map 13.1. Distribution of *Trichechus inunguis.*

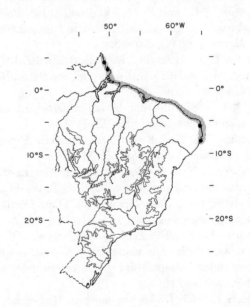

Map 13.2. Distribution of *Trichechus manatus.*

throughout the Greater Antilles and is distributed in coastal areas of eastern Central America and across northern South America (map 13.2). It has recently been exterminated over much of its range.

Natural History

Manatees are almost entirely herbivorous, feeding on aquatic vegetation, including leaves of mangroves, sea grass, and water hyacinths. They may browse on plants hanging over the water if they can reach them. They tend to move singly except for males courting females or females followed by dependent young. A single calf is born after a gestation period of thirteen months. It is dependent on the mother for a considerable time, so interbirth intervals may be three and a half years or more. Interbirth interval may also be a function of the habitat's carrying capacity. Sexual maturity apparently is not attained until five or six years of age. Marmontel (1992, 1995) studied bone annuli in the tympanic region of dead Florida manatees and was able to develop a life table calibrated by skulls of individuals of known age at death. The results suggest that at least in Florida female manatees can conceive at three years of age, thus giving birth at four plus years. Maximum age at death in her sample was more than fifty years. Members of a group maintain contact by underwater vocalizations. (See also Bengtson and Fitzgerald 1985; Eisenberg 1982; Hartman 1979; Husar 1978; Moore 1956; and Schevill and Watkins 1965.)

References

• References used in preparing distribution maps

Bengston, J. L., and S. M. Fitzgerald. 1985. Potential role of vocalizations in West Indian manatees. *J. Mammal.* 66 (4): 816–18.

Bertram, G. C. L., and C. K. R. Bertram. 1973. The modern Sirenia: Their distribution and status. *Biol. J. Linn. Soc.* 5:297–338.

Best, R. C. 1983. Apparent dry season fasting in Amazonian manatee. *Biotropica* 15:61–64.

Domning, D. P. 1982. Evolution of manatees: A speculative history. *J. Paleontol.* 56:599–619.

Eisenberg, J. F. 1982. Moderator's remarks. In *West Indian manatee in Florida,* ed. R. L. Brownell and K. Ralls, 66. Tallahassee: Florida Department of Natural Resources.

Hartman, D. S. 1979. *Ecology and behavior of the manatee* Trichechus manatus *in Florida.* Special Publication 5. Shippensburg, Pa.: American Society of Mammalogists.

Husar, S. L. 1977. *Trichechus inunguis. Mammal. Species* 72:4.

———. 1978. *Trichechus manatus. Mammal. Species* 93:5.

Marmol, B. A. E. 1976. Informe preliminar sobre las plantas que sirven de alimento al manati de la Amazonia (*Trichechus inunguis*). *Resum. I. Congr. Nac. Bot.* (Lima, Peru), 31–32.

Marmontel, M. 1992. Age and reproductive parameter estimates in female Florida manatees. In *Interim report of the technical workshop on manatee population biology,* ed. T. J. Oshea, B. Ackerman, and H. F. Percival. Manatee Population Research 10. Washington, D.C.: U.S. Fish and Wildlife Service.

———. 1995. Age and reproduction in female Florida manatees. In *Population biology of the Florida manatee,* ed. T. J. O'Shea, B. Ackerman, and H. F. Percival. Information and Technology Report 1, August 1995. Washington, D.C.: U.S. Department of the Interior, National Biological Survey.

• Mondolfi, E. 1974. Taxonomy, distribution and status of the manatee in Venezuela. *Mem. Soc. Cienc. Nat. La Salle* 34 (97): 5–23.

Montgomery, G. G., R. C. Best, and M. Yamakoshi. 1981. A radio-tracking study of the Amazonian manatee *Trichechus inunguis. Biotropica* 13:81–85.

Moore, J. C. 1956. Observations of manatees in aggregations. *Amer. Mus. Novitat.* 1811:124.

Reynolds, J. E., III. 1981. Behavior patterns in the West Indian manatee, with emphasis on feeding and diving. *Florida Scientist* 44 (4): 233–42.

Rosas, F. C. W. 1994. Biology conservation and status of the Amazonian manatee, *Trichechus inunguis. Mammal Rev.* 24 (2): 49–59.

Schevill, W. E., and W. A. Watkins. 1965. Underwater calls of *Trichechus. Nature* 205:373–74.

Thornback, J., and M. Jenkins. 1982. *Threatened mammalian taxa of the Americas and the Australian zoogeographic region.* Gland, Switzerland: IUCN.

14   Order Perissodactyla
(Odd-toed Ungulates)

Diagnosis

These ungulates characteristically have an enlarged middle digit on both the forefeet and the hind feet. The major weight-bearing axis is on the middle digit, in contradistinction to the Artiodactyla, where the major toes are reduced to two and the axis of weight passes between them. Among the extant Perissodactyla, toe reduction reaches its most extreme form in the horse family (Equidae), where there is only one functional toe. The digestive system does not exhibit an extreme modification of the stomach, but the cecum, a blind pouch at the union of the small and large intestines, is enlarged and serves as a fermentation chamber. Tooth reduction is not pronounced, and the incisors are retained. Horns in extant forms, if they are developed at all, occur as epidermal derivatives and are situated in the midline of the nasal bones (e.g., Rhinocerotidae).

Distribution

The Recent distribution includes South America, Africa, Europe, and Asia. Recently extinct in North America (ca. 8000 B.P.), the order but was reintroduced by humans in the form of the domestic horse and ass (Equidae).

History and Classification

There are three extant families, the Rhinocerotidae, the Equidae, and the Tapiridae. The last one, the tapirs, is the only extant family in South America, although the horse has been reintroduced by Europeans. This order has had a long evolutionary history, first appearing in the Eocene. It radiated rapidly in Asia and North America to include twelve families by the end of the Eocene, but their diversity diminished until only four families existed in the late Miocene.

FAMILY TAPIRIDAE
Genus *Tapirus* Brunnich, 1772
Tapirs, Dantas

Diagnosis

The dental formula is I 3/3, C 1/1, P 4/3–4, M 3/3 (fig. 14.1). Head and body length averages 2 m; the short tail is less than 10 cm long (plate 18). The nasal bones of the skull are short, and the animal has a distinct proboscis formed from the nostrils and upper lip, which overhangs the lower lip. The forefeet bear four toes, although a vestige of the thumb can be detected on dissection. The hind foot has three toes.

Distribution

Species of this family were once widely distributed in North America and Asia. Tapirs crossed into South America during the Pliocene, and they persist today in South and Central America and in Southeast Asia. Their distribution in Colombia was reviewed by Hershkovitz (1954).

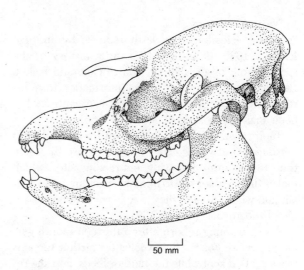

50 mm

Figure 14.1. Skull of *Tapirus terrestris*.

Natural History

The family has been reviewed in Eisenberg, Groves, and MacKinnon (1989). All living species are browsers and frugivores. Tapirs tend to confine much of their activity to the hours of darkness, but they may move around during the day, especially where they are not hunted. They are excellent swimmers and generally have a wallowing area in their home range. They show great fidelity to trails and thus are vulnerable to human predation. Tapirs tend to defecate at special loci near water. Communal use of trails and defecation sites permits a loose communication system within a population. Densities can reach 0.8 animals per km^2 in good habitat (Eisenberg and Thorington 1973). A single young is born after a thirteen-month gestation period. The coat of the young is characterized by horizontal yellow stripes on a brown background. The current status of the genus is summarized in Brooks, Bodmer, and Matola (1997).

Tapirus bairdii (Gill, 1865)
Central American Tapir, Baird's Tapir

Description

The largest of the three South American species, with a shoulder height of 120 cm, this tapir may reach a weight of over 300 kg. It can be distinguished from the common tapir by the absence of a neck crest and by its uniformly short, dull brown coat.

Range and Habitat

The species is distributed west of the western cordillera of the Andes from Ecuador to Veracruz, Mexico (Hershkovitz 1954). It occurs in a variety of habitats, including dry deciduous forest as well as multistratal tropical evergreen forest (map 14.1).

Natural History

This species has been vastly reduced by hunting. Enders (1935) described the natural history of the species from observations on Barro Colorado Island. Leopold (1959) provided useful information from his researches in Mexico. Reproduction in captivity has been analyzed by Alvarez del Toro (1966).

Tapirs tend to move singly except for a female and her dependent young. They frequently use the same trails to and from wallowing and feeding sites, which makes them vulnerable to hunters. They are crepuscular, and where hunted they may become almost completely nocturnal (Fragoso 1991).

A single young is born after a thirteen-month gestation period. For the first week after birth it remains in a secluded spot while the mother feeds, and she returns periodically to nurse it. Its striped coat provides concealing coloration when it is at rest. The young ac-

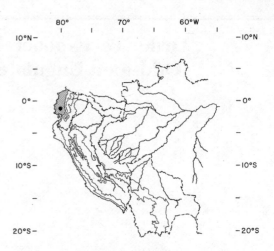

Map 14.1. Distribution of *Tapirus bairdii*.

tively follows the mother at ten days of age. The long gestation and variable interbirth interval, from two to seventeen months, results in a low recruitment rate compared with that of the white-tailed deer (*Odocoileus*). Where hunted, tapirs will take longer to recover than the deer.

Tapirs are selective feeders, eating a wide variety of leaves and fruits (Terwilliger 1978). Janzen has studied the feeding habits of captives and concludes that the tapir is extremely selective, eating a wide range of leaves but taking only small amounts of a given species in a single feeding bout. Although it may crush many seeds of fleshy fruits (e.g., *Enterolobium*), the tapir is an important disperser of seeds, since some (e.g., *Raphia faedigera*) pass through its gut unharmed (Janzen 1981, 1982a,b).

Williams (1984), employing radio telemetry, was able to determine activity ranges for *T. bairdii* in Costa Rica. Diurnal movements could encompass as little as 3 ha or as much as 27 ha. Nocturnal movements were larger, with activity including 38 to 180 ha. While working in Panama, Montgomery and Sunquist (pers. comm.) followed a radio-collared female for twelve days and found her within a 1 km^2 area. She had a nursing young. These data do not give us an idea of total home range or seasonal variations, but they indicate short-term local intensive use of habitat.

Overall (1980) noted a remarkable association between a tapir and a male coati. Tapirs can become infested with ticks. In the case Overall described, the coati fed on the blood-gorged ticks by gleaning them from the tapir's body. The tapir, however, did not reciprocate. I (JFE) have noted the same behavior when a tame tapir and peccary were associated at semiliberty in Panama. The peccary "cleaned" the tapir of ticks, but once again the tapir never reciprocated. Peccaries in groups are known to reciprocally groom each other, as do coatis.

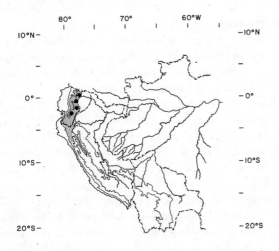

Map 14.2. Distribution of *Tapirus pinchaque.*

Tapirus pinchaque (Roulin, 1829)
Mountain Tapir

Description

This is the smallest of the three South American species of tapir. Head and body length averages 180 cm, and shoulder height 75–80 cm. It is distinguished by its long coat of brown hair and its white lips (plate 18).

Range and Habitat

The species occurs at moderate to high elevations in forested habitat, predominantly in the paramos and dwarf forests of the central and eastern cordilleras of the Andes, extending from Colombia to Ecuador and perhaps to northern Peru (Schauenberg 1969). It has been severely persecuted over much of its range and survives only in fragmented populations (map 14.2).

Natural History

Schauenberg (1969) wrote an early monograph on the behavior, ecology, and distribution of the mountain tapir. Downer (1996a,b) has provided the most recent summary of its status and distribution. In its reproduction and behavior the species resembles *T. bairdii* and *T. terrestris.* Its shaggy coat provides adequate insulation to permit foraging between 2,000 and 4,000 m. It may selectively browse in the paramos. As in the other species, there is little sexual dimorphism. Males may fight severely to gain mating rights with a sexually receptive female. While fighting, males attempt to bite each other on the hind feet. Aside from the mating association or the mother-young unit, the animals move and feed alone (Downer 1996a). The gestation period is thirteen months (Bonney and Crotty 1979).

A study in central Colombia established that mineral licks are an important feature in the ecology of the species. An adult male and female as well as a subadult

male moved within the same range of about 11 km^2. The animals ranged to an altitude of 3,200 m. Seasonal shifts in altitude could be related to availability of forage, but the authors suggest an active attempt to avoid biting flies as a reasonable alternative hypothesis (Acosta, Cavelier, and Londoño, 1996).

In Ecuador Downer (1996a,b) reports that mountain tapirs feed on at least 264 species of vascular plants. In absolute quantity, leaves of *Gynoxys* accounted for 24% of the diet and fern fronds for 19%. In terms of preference ratios, species of *Lupinus* ranked highest. Seeds of over fifty plant species eaten by tapirs germinated from tapir dung piles. These observations suggest that they play an important role as seed dispersers in these high-altitude habitats (Downer 1996a).

Tapirus terrestris (Linnaeus, 1758)
Tapir, Anta, Danta

Description

The tapir is the largest land mammal in most of tropical South America east to the Andes. It is an ungainly looking ungulate with a robust body set on proportionally short, slender legs. Tapirs stand an average of 1,048 mm at the shoulder (range 955–1,128 mm; $n = 6$). The tail is short. The head is large and equipped with the tapir's most distinctive feature, an elongated, flexible nose that resembles a short trunk (plate 18). *T. terrestris* can be differentiated from other tapirs by a more pronounced mane, which is still short, and a more pronounced sagittal crest on the skull. Newborn tapirs have horizontal stripes and dots of yellowish white, a coloration that persists past seven months, disappearing last from the legs. Adults are a dull brownish (Hershkovitz 1954; Wilson and Wilson 1973a,b; pers. obs.).

Distribution

T. terrestris is found from Venezuela south through Paraguay to northern Argentina. Its range has contracted in Argentina owing to hunting and habitat destruction—formerly it ranged south into Tucumán province, but it is now found across the northernmost fringe of the country (Hershkovitz 1954; Ojeda and Mares 1982) (map 14.3).

Life History

The gestation period of this tapir is about 383 days; only a single young is born; the average neonatal weight for four young was 2,200 g. Tapirs in captivity can reproduce for the first time at two years, though three years is the average, and can have young every fourteen months (range 14–28; $n = 4$). Both in captivity and in the wild there appears to be no seasonality

Map 14.3. Distribution of *Tapirus terrestris*.

to reproduction. In captivity one individual lived to thirty-two years (Crespo 1982; Hershkovitz 1954; Mallinson 1969; Wilson and Wilson 1973a,b).

Ecology

Tapirs have usually been thought of as animals of the moist tropical forest, but in southern South America they are common in xeric parts of the Paraguayan and Argentine chaco. They are also found in the tropical forest of northeastern Argentina and at up to 2,000 m. In Mato Grosso, Brazil, they were most abundant in the gallery forests and in low-lying deciduous and secondary forests and were calculated at a crude biomass of 96.4 kg/km^2.

Tapirs are browsers and frugivores. When water is available, they frequently defecate in water, where their large, globular feces are obvious. Away from water tapirs often defecate at fixed points within their home range. Fecal piles with undigested seeds offer definite points for recolonization by seedlings germinating within the fecal mass (Fragoso 1994; Olmos and Galetti 1993).

In captivity tapirs were found to produce four vocalizations: a shrill fluctuating squeal uttered during exploratory behavior, which may also serve as a contact call; a clicking noise; and the characteristic snort that is frequently heard in the wild when an animal has detected the presence of a human and is about to flee. They are solitary (Hunsaker and Hahn 1965; Miller 1930; Olrog 1979; Richter 1966). The most complete review of the biology of *Tapirus terrestris* is included in Padilla and Dowler 1994. Bodmer et al. (1994) assessed populations in Amazonian Peru and concluded that in areas where it is hunted, it is almost invariably overhunted.

References

Acosta, H., J. Cavelier, and S. Londoño. 1996. Aportes al conciemento de la biología de la danta de montaña (*Tapirus pinchaque*) en los Andes centrales de Colombia. *Biotropica* 28:258–66.

Allen, J. A. 1916. Mammals collected on the Roosevelt Brazilian expedition, with field notes by Leo E. Miller. *Bull. Amer. Mus. Nat. Hist.* 34:559–610.

Alvarez del Toro, M. 1966. A note on the breeding of Baird's tapir at Tuxtla Gutierrez Zoo. *Int. Zoo Yearb.* 6:196–97.

Bodmer, R. E., P. Puertas, L. Moya, and T. Fang. 1994. Estado de las poblaciones del tapir en la Amazonia Peruana: En el camino de la extinción. *Bol. Lima* 88:33–42.

Bonney, S., and M. J. Crotty. 1979. Breeding the mountain tapir (*Tapirus pinchaque*) at the Los Angeles Zoo. *Int. Zoo Yearb.* 19:198–200.

Brooks, D. M., R. E. Bodmer, and S. M. Matola, eds. 1997. *Tapir: Status survey and conservation action plan*. Gland, Switzerland: IUCN World Conservation Union, Species Survial Commission.

Crespo, J. A. 1982. Ecología de la comunidad de mamíferos del Parque Nacional Iguazú, Misiones. *Rev. Mus. Argent. Cienc. Nat. "Bernardino Rivadavia," Ecol.* 3 (2): 45–162.

Del Llano, M. 1990. *Los páramos de los Andes.* Bogotá: Montoya y Araugo.

Downer, C. C. 1988. Report on reconnaissance trip to Mountain Tapir Study Site in NE Tirua, Peru. University of Durham, UK.

Downer, C. C. 1996a. The mountain tapir, endangered flagship species of the high Andes. *Oryx* 30:45–58.

Downer, C. C. 1996b. The mountain tapir—flagship species of the northern Andes. In *Tapirs: Status survey and conservation action plan*, ed. D. M. Brooks, R. E. Bodmer, and S. M. Matola. Gland, Switzerland: IUCN World Conservation Union, Species Survial Commission.

Eisenberg, J. F. 1989. *Mammals of the Neotropics.* Vol. 1. *The northern Neotropics: Panama, Colombia, Venezuela, Guyana, Suriname, French Guiana.* Chicago: University of Chicago Press.

Eisenberg, J. F., C. Groves, and K. MacKinnon. 1989. Tapirs. In *Grzimek's encyclopedia: Mammals*, 4:598–608. New York: McGraw-Hill.

Eisenberg, J. F., and R. W. Thorington Jr. 1973. A preliminary analysis of a Neotropical mammal fauna. *Biotropica* 5:150–61.

Enders, R. K. 1935. Mammalian life histories from Barro Colorado Island, Panama. *Bull. Mus. Comp. Zool. (Harvard)* 78 (4): 385–502.

Fragoso, J. M. 1991. The effect of selective logging on Baird's tapir. In *Latin American mammalogy*, ed. M. Mares and D. J. Schmidly. Norman: University of Oklahoma Press.

Fragoso, J. M. 1994. Large mammals and the community dynamics of an Amazonian rain forest. Ph.D. diss., University of Florida, Gainesville.

Hershkovitz, P. 1954. Mammals of northern Colombia, preliminary report no. 7: Tapirs (genus *Tapirus*) with a systematic review of American species. *Proc. U.S. Nat. Mus.* 103:465–96.

Hunsaker, D., II, and T. C. Hahn. 1965. Vocalization of the South American tapir *Tapirus terrestris. Anim. Behav.* 13:69–78.

Janzen, D. 1981. Digestive seed predation by a Costa Rican Baird's tapir, *Tapirus bairdii. Biotropica* 13:59–63.

Janzen, D. H. 1982a. Wild plant acceptability to a captive Costa Rican Baird's tapir. *Brenesia* 19–20:99–128.

Janzen, D. H. 1982b. Seeds in tapir dung in Santa Rosa National Park, Costa Rica. *Brenesia* 19–20:129–35.

Leopold, A. S. 1959. *Wildlife in Mexico.* Berkeley: University of California Press.

Mallinson, J. J. C. 1969. Reproduction and development of Brazilian tapir *Tapirus terrestris. Dodo* 6:47–51.

Miller, F. W. 1930. Notes on some mammals of southern Mato Grosso, Brazil. *J. Mammal.* 11:10–22.

Mondolfi, E. 1971. La danta o tapir. *Defensa Nat.* 1 (4): 13–19.

Naranjo Piñera, E. J. 1995. Hábitos de alimentación del tapir (*Tapirus bairdii*) en bosque tropical humedo en Costa Rica. *Vida Silvestre Neotrop.* 4:32–37.

Ojeda, R. A., and M. A. Mares. 1982. Conservation of South American mammals: Argentina as a paradigm. In *Mammalian biology in South America,* ed. M. A. Mares and H. H. Genoways, 505–22. Pymatuning Symposia in Ecology 6. Special Publication Series. Pittsburgh: Pymatuning Laboratory of Ecology, University of Pittsburgh.

Olmos, M. R. F., and M. Galletti. 1993. Seed dispersal by tapir in southeastern Brazil. *Mammalia* 57 (3): 460–61.

Olrog, C. C. 1979. Los mamíferos de la selva humeda, Cerro Calilegua, Jujuy. *Acta Zool. Lilloana* 33:9–14.

Overall, K. L. 1980. Coatis, tapirs and ticks: A case of mammalian interspecific grooming. *Biotropica* 12:158.

Padilla, M., and R. C. Dowler. 1994. *Tapirus terrestris. Mammal. Species* 481:1–8.

Richter, W. von. 1966. Untersuchung über angeborene Verhaltensweise den Shabrackentapirs und flachland Tapirs. *Zool. Beiträge* (Nuremberg) 1 (12): 67–159.

Schaller, G. B. 1983. Mammals and their biomass on a Brazilian ranch. *Arq. Zool. São Paulo* 31 (1): 1–36.

Schauenberg, P. 1969. Contribution à l'étude du tapir pinchaque *Tapirus pinchaque* Roulin 1829. *Rev. Suisse Zool.* 76:211–56.

Terwilliger, V. J. 1978. Natural history of Baird's tapir on Barro Colorado Island, Panama Canal Zone. *Biotropica* 10 (3): 211–20.

Williams, K. D. 1984. The Central American tapir (*Tapirus bairdii,* Gill) in northwestern Costa Rica. Ph.D. diss., Michigan State University, East Lansing.

Wilson, R., and S. Wilson. 1973a. *Tapirs in captivity, 1973.* Claremont, Calif.: Tapir Research Institute.

Wilson, R., and S. Wilson. 1973b. *Tapirs in the United States.* Claremont, Calif.: Tapir Research Institute.

15   Order Artiodactyla (Even-toed Ungulates)

Diagnosis

The Artiodactyla are especially adapted for feeding on fallen fruit and nuts or on foliage, including grasses and leaves, and occasionally on the reproductive bodies of plants. In New World forms the cheek teeth may be either high crowned or low crowned. In the latter case, the species either is omnivorous or is adapted for browsing (e.g., Tayassuidae) rather than for grazing and browsing (e.g., Cervidae, Camelidae). The premolars are usually simple in structure compared with the molars. The grazing and browsing forms generally exhibit a loss of the upper incisors. The main axis of weight is borne between the third and the fourth digits of the foot. The first, second, and fifth digits are reduced or lost in the more specialized forms. A corresponding reduction in metacarpals and metatarsals to some extent parallels the reduction in toe number. The astragalus of the tarsal bones is modified to articulate at both ends, termed a "double pulley" structure. Enlarged canines or tusks are present in the Suiformes, Tragulidae, and some species of Cervidae. In males and females of many species of this order, bony excrescences develop on the forehead (Cervidae, Antilocapridae, Giraffidae, and Bovidae). Generally there are two types of hornlike structures—either antlers, which lose the epidermal covering and are cast annually, or true horns, which are an extended growth of the frontal bone covered with an epidermally derived cap.

Distribution

The Recent natural distribution is worldwide except for the Australian area, Oceanic islands, and Antarctica. Domesticated and wild members of this order have been introduced into Australia, New Zealand, and many Oceanic islands.

History and Classification

The earliest artiodactyls first appeared in the early Eocene in both North America and Europe. This group rapidly differentiated into three specialized forms that are reflected in the subordinal classification of the Artiodactyla: the Suiformes, represented today by the hippopotamuses, peccaries, and swine; the Tylopoda, represented by the family Camelidae; and the suborder Ruminantia, which today includes the giraffes (Giraffidae), pronghorn antelopes (Antilocapridae), mouse deer (Tragulidae), musk deer (Moschidae), true deer (Cervidae), and bovines (Bovidae).

The artiodactyls had their origin in the northern continents, and since North America and Asia have been in contact off and on, faunal interchange was possible. Faunal interchange between the Northern and Southern Hemispheres increased as Africa became connected to Eurasia. The artiodactyls did not appear in South America until the late Pliocene, when the Panamanian land bridge became complete, linking North and South America. All three suborders are now represented in South America; they have arrived and differentiated only since the Pliocene. The three families in South America at present are the Tayassuidae, or peccaries; the Camelidae, including the vicuna and guanaco (llama); and the Cervidae, including some eleven species currently grouped into six genera. Members of the family Bovidae did not reach South America until Europeans transported them in the sixteenth century. The Cervidae underwent a rapid adaptive radiation in the Pleistocene of South America to fill some niches that are occupied by bovines on the more contiguous continental land masses (Eisenberg 1987, 1989; Eisenberg and McKay 1974).

FAMILY TAYASSUIDAE

Peccaries, Pecaris, Báquiros, Caitatús

Diagnosis

These piglike Artiodactyla have a dental formula of I 2/3, C 1/1, P 3/3, M 3/3. The canines are modified into tusks but are small. The upper canines are directed downward and rub against the lower ones, thus

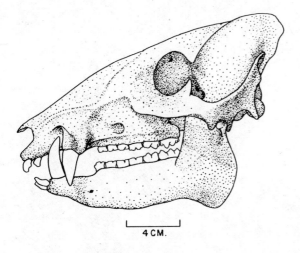

Figure 15.1. Skull of *Tayassu tajacu*.

maintaining a sharp cutting edge (fig. 15.1). The molars are low crowned. Head and body lengths range from 700 to 1,000 mm, the tail averages 20 mm, and weights average about 24 kg. They have four visible toes on the forefeet and three on the hind feet, but only two toes on each foot bear the weight (see ordinal account). Although the stomach is modified into three chambers (Langer 1979), peccaries are not ruminants. The biology of this family is admirably reviewed by Sowls (1984, 1997).

Distribution

Species of this family are distributed from Arizona, New Mexico, and Texas in North America to northern Argentina. There are currently three species, all occuring within the area covered by this volume. The genus *Tayassu* includes *T. tajacu* and *T. pecari,* and the genus *Catagonus* is restricted to Paraguay and adjacent parts of Bolivia and Argentina. Although *Catagonus* had been described as a fossil, it was rediscovered as a living form by Wetzel in 1974 (Wetzel et al. 1975; Wetzel 1977a,b).

Genus *Catagonus* Ameghino, 1904
Catagonus wagneri (Rusconi, 1930)
Chacoan Peccary, Pecari Quimlero, Tagua

Description

Catagonus differs from the other peccaries in the following characters: it is larger, has longer dorsal fur, has no dewclaws, and has a longer, more convex rostrum. The Chacoan peccary is colored much like *T. tajacu,* with a grizzled, brownish gray coat, a dark middorsal stripe, and a faint collar of white hairs that crosses up over the shoulders. The hair on the ears and legs is longer and paler than in *Tayassu.* Long bristles give the animal a shaggy appearance. The head is larger, and the ears, legs, and tail are longer. Males of

Catagonus average 567.8 mm at the shoulder (range 520–690; $n = 22$;) (Mayer and Brandt 1982; Wetzel 1977a,b). Mean weight for eighteen males was 34.8 kg, and for ten females it was 34.6 kg (Redford and Eisenberg 1992).
Chromosome number: $2n = 20$ (Benirschke and Kumamoto 1989; Wetzel et al. 1975).

Distribution

Catagonus wagneri is found in the Gran Chaco of Paraguay, Argentina, and Bolivia. In Paraguay it is found throughout the Chaco, and in Argentina it has been collected in Chaco, Santiago del Estero, and Salta provinces. It occurs only in the extreme southeast of Bolivia (Ludlow, unpubl. data; Olrog and Cajal 1982; Olrog, Ojeda, and Barquez 1976) (map 15.1).

Life History

The average number of embryos per female is 2.7 (range 2–4; mode 2), and the average litter size is 2.5 (range 1–4; mode 2–3). Mating occurs from April to May, and farrowing takes place from early September to early December. Young achieve adult coloration at between three and four months of age. Females have four pairs of mammae (Brooks 1992; Mayer and Brandt 1982).

Ecology

Catagonus is found in the xeric chaco of southern South America in areas of high temperature and low rainfall. It is most prevalent in low to moderate stature thorn forests of the central Paraguayan Chaco, where it can reach densities of 9.24 individuals per square kilometer. Herds are composed of males and females, are stable social units, and average 3.8 members (range 1–10; $n = 148$). During winter *Catagonus* is active di-

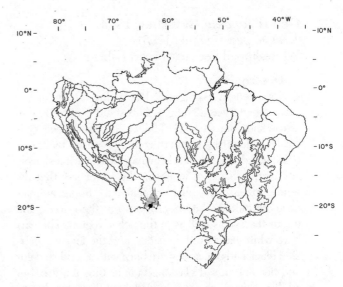

Map 15.1. Distribution of *Catagonus wagneri.*

urnally, between the hours of 0800 and 1100. Animals are inactive at night except during the hottest and coldest weather, when they are active for one to two hours in the middle of the night and are diurnally active from 0800 to 1700 hours.

Like other peccaries, *Catagonus* does a great deal of scent marking using the large dorsal scent gland, from which scent can be rubbed or squirted. Scent marking is most prevalent in areas of high herd use. Members of the herd also frequently scent mark each other. They usually defecate at scat stations.

In the central Paraguayan Chaco, home ranges of three groups of *Catagonus* had a stable core area of about 6 km^2, but with excursions they covered 15 km^2 ($\bar{x} \pm 10.98$ km^2). These ranges are exclusive and probably constitute territories. *Catagonus* may share its range with herds of *T. tajacu* (Taber et al. 1994).

During the winter cacti are the major dietary items, though bromeliads are also eaten. Soil is frequently eaten from the ground, at salt licks, or from leaf-cutter ant (*Atta*) mounds.

Catagonus is heavily hunted throughout its range, for meat and to a limited extent for its hide. Curiosity, use of salt licks, and a tendency to remain near fallen herd members all make it easier to kill this species. The distribution of *Catagonus* appears to be patchy and discontinuous, perhaps owing to the intense hunting that is taking place as the Chaco is developed (Taber et al. 1993, 1994; Mayer and Brandt 1982; Mayer and Wetzel 1986; Sowls 1984; pers. obs.).

Comment

The living Chacoan peccary was unknown to Western scientists until 1974, when it was discovered in the Paraguayan Chaco by Wetzel and his colleagues (Wetzel 1977a,b).

Genus *Tayassu* G. Fisher, 1814
Tayassu pecari (Link, 1795)
White-lipped Peccary, Pecari Labiado

Description

Total length averaged 1.126 m for fifteen males; mean weight was 33.5 kg. Adults stand about 530 mm at the shoulder ($n = 1$). The white-lipped peccary is easily distinguished from the other peccaries by its coloration. Adults are dark brown or black with a sharply contrasting blaze of white on the lower jaw that extends up onto the cheeks (plate 18). Some animals have additional white near the tip of the muzzle and under the eye as well as in the pelvic region. The ears have white hairs on the insides, and the legs are grizzled black and tan. Young are born with a reddish color and become mixed black and tan as they mature, but adult coloration is not achieved until the second year. Because of their strikingly different coloration, imma-

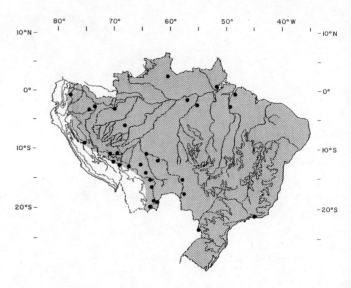

Map 15.2. Distribution of *Tayassu pecari*.

ture animals have been mistaken for another species (Mayer and Brandt 1982; Sowls 1984; March 1991). Chromosome number: $2n = 26$ (Sowls 1984).

Distribution
Tayassu pecari is found from southern Mexico to northern Argentina in lowland tropical forests and savannas (Honacki, Kinman, and Koeppl 1982; Olrog and Lucero 1981; UM) (map 15.2).

Life History
White-lipped peccaries usually produce twins after a gestation period of about 150 days. In captivity there appear to be two annual breeding periods, and females reproduce for the first time at about eighteen months. In the Brazilian Pantanal a female was seen with young in September; in Peru mating was observed between July and August; and in the Paraguayan Chaco two females with two fetuses each were collected in July (Miller 1930; Roots 1966; Sowls 1984).

Ecology
White-lipped peccaries, though usually thought of as animals of the moist tropical forest, are also found in the seasonally xeric parts of southern South America. In the Brazilian Pantanal this species was found at a crude biomass of 44 kg/km^2, and in the Paraguayan Chaco it is found at a density of 1.06/km^2. The first field study of this species, done in the Peruvian rain forest (Kiltie and Terborgh 1983), concluded that if indeed white-lipped peccary herds had home ranges at all they were probably between 60 and 200 km^2.

In Roraima, Brazil, *T. pecari* herds exhibited rather stable home ranges, although seasonal shifts were noted. A large herd of more than 130 individuals utilized 110 km^2 over a two-year period and exhibited site fidelity. A smaller herd of thirty-nine or more individuals had a home range within that of the larger herd and used 22 km² annually (Fragoso 1994).

T. pecari forms the largest groups of any South American terrestrial mammal. There are few good counts of herd size: in Peru the minimum number of individuals in five herds was 106 (range 90–138); in Mato Grasso, Brazil, one herd of at least 200 was counted; in Misiones province, Argentina, one group was estimated at 250; in the Paraguayan Chaco groups averaged much smaller, with 25–60 animals per herd. Work in Peru has demonstrated that these large groups of peccaries move over wide areas in what may be a seasonal pattern, using large patches of mast fruiting palms. Kiltie and Terborgh (1983) have suggested that the large group size may be due to the peccaries' use of patchy food resources and to the ease with which predators can find a herd, so that a large group can be a mutual defense system.

Herds include both males and females. T. pecari has a more elaborate repertoire of vocalizations and a closer affiliation with other herd members than does T. tajacu. In the Paraguayan Chaco white-lipped peccaries eat Prosopis pods and other plant material; in Brazil stomachs have been found to contain figs and palm nuts (n = 3); and in Peru the stomachs of thirty-four animals contained reproductive plant parts, especially palm seeds (61% volume, 97% occurrence); vegetative plant parts (39% volume, 97% occurrence); and animal parts, including snail opercula, tissue from vertebrates, and parts of invertebrate adults and larvae (trace % volume, 82% occurrence). T. pecari is crepuscular and nocturnal in Paraguay and is known to be a prey of pumas. Fragoso (1994) noted that jaguars (Panthera onca) and pumas (Puma concolor) killed many peccaries, and it is one of the most common mammals taken by indigenous hunters throughout its range. Bodmer et al. (1996) review the biology of this species and make specific recommendations for its management (see also Crespo 1982; Davis 1947; Kiltie 1981a,b; Kiltie and Terborgh 1983; Mayer and Brandt 1982; Mayer and Wetzel 1987; Miller 1930; Redford and Robinson 1987; Schaller 1983; Sowls 1984; Yost and Kelley 1983).

Tayassu tajacu (Linnaeus, 1758)
Collared Peccary, Pecari de Collar

Description

Males stand about 450 cm at the shoulder. Mean weight for nine males in Brazil was 19.4 kg, and for females it was 19.7 kg. There is no obvious external sexual dimorphism, though in the United States males are slightly heavier than females. There are four digits on the front feet and three on the hind feet. The coloration of the collared peccary is very similar to that of

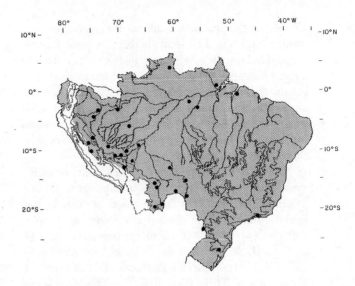

Map 15.3. Distribution of *Tayassu tajacu.*

Catagonus, but T. tajacu is smaller and has shorter hair and darker legs (plate 18). Young animals achieve adult coloration by two to three months. The canine teeth grow continously for the first four to five years. Chromosome number: 2n = 30 (Mayer and Brandt 1982; Sowls 1984).

Distribution

Tayassu tajacu is found from the south-central United States south into northern Argentina. In southern South America it is found in eastern and western Paraguay and at present in the northern Argentine provinces south to at least Tucumán (Lucero 1983; Sowls 1984) (map 15.3).

Life History

In the United States the collared peccary breeds year round, and females exhibit a postpartum estrus. Females conceive at the end of their first year; gestation length is 145 days; litter size varies from one to four with a mode of two. Young are born averaging 665 g (n = 10) and nurse for six to eight weeks. In Paraguay the average litter size is similar (average 1.98; range 1–4). In northeastern Argentina collared peccaries reproduce from August to October and have litters of one or two, though in Amazonian Peru there is no evidence of seasonal breeding. In Brazil one litter was found in a Priodontes burrow. In both wild and captive herds, young will nurse from females other than their own mothers (Byers and Beckoff 1981; Crespo 1982; Lochmiller, Hellgren, and Grant 1984; Mayer and Brandt 1982; Miller 1930; Sowls 1984).

Ecology

Collared peccaries occur in a wide variety of habitats, from moist tropical forest to the xeric Paraguayan

Chaco. In the Brazilian Pantanal they occurred at a crude biomass of 14.1 individuals per square kilometer, and in Peru there was an estimate of 1.2 groups per square kilometer. They can be active at any time during the day or night depending on the weather, season, and availability of food. *T. tajacu* will hide in armadillo burrows or fallen tree trunks and are frequently driven into these shelters by dogs belonging to hunters, who then dig or chop them out.

Mixed-sex bands move along trails, some of which may be traditional. There does not appear to be one leader of the band, and the members are usually dispersed while feeding. Group size varies geographically, with an average of 7.7 in Arizona (U.S.), of 13.6 in Texas (U.S.), of 6.5 in the llanos of Venezuela (n = 65), of 3.3 in the Brazilian Pantanal, of 34 in Roraima, Brazil, and of 5.6 in Peru (n = 18). The basic social unit is a herd of one or more males and several females with young. The herd may divide into smaller groups or may, in some circumstances, temporarily combine with other herds. This latter grouping, termed an aggregation, seems to occur most commonly when there is a great abundance of food, such as during mast fruiting of mesquite or palm trees. Aggregations of more than fifty animals have been seen (Fragoso 1994).

Collared peccary home ranges seem to be fairly small, and in Arizona these home ranges are defended against neighboring herds. In the Venezuelan llanos, dry-season home ranges varied from 38 to 45 ha and wet-season home ranges from 100 to 126 ha. The core area of home ranges contained a permanent defecation station. In Costa Rica, dry-season home ranges were larger than those during the wet season. Wallows and mineral licks are frequently used. As in other peccaries, there is a great deal of mutual rubbing within a herd, usually oriented toward the dorsal scent gland. Herd cohesiveness is maintained through auditory and olfactory signals. In the forests of Roraima, Brazil, Fragoso (1994) determined that collard peccary herds had core home ranges of 2.1 to 5.9 km². The ranges were considerable smaller than those of the sympatric white-lipped peccaries, which had core home ranges of 17.7 to 59.3 km² that overlapped the ranges of the collared peccaries. Home-range size varies seasonally and covaries positively with herd size.

T. tajacu eats fruits, tubers, green grass, shoots of plants, fruit and stems of cacti, and growing points of bromeliads and agave. These peccaries will dig shallow holes in search of underground plant parts and buried seeds. In Peru a study of seventeen stomachs revealed reproductive plant parts (71% volume, 100% occurrence), vegetative plant parts (29% volume, 94% occurrence), and animal parts (trace % volume, 82% occurrence) (see also Byers and Beckoff 1981; Castellanos 1983, 1986; Crespo 1982; Díaz 1978; Eddy

1961; Green and Grant 1984; Kiltie 1981a; Kiltie and Terborgh 1983; Mayer and Brandt 1982; McCoy and Vaughan 1986; Miller 1930; Robinson and Eisenberg 1985; Schaller 1983; Schweinsburg 1971; Smythe 1970; Sowls 1984).

Comment

The genus *Tayassu* is retained here in spite of the recommendation by Grubb (1993) to use *Pecari*.

FAMILY CAMELIDAE

Diagnosis

These artiodactyls have a selenodont dentition, and the upper incisors and canines are retained. The dental formula for the genus *Lama* is 1/3, 1/1, 3/1–2, 3/3 (fig. 15.2). They are ruminants with three-chambered stomachs. The toes bear nails on their tips rather than hooves. In the extant species the head bears no hornlike structures.

Distribution

The family originated in North America but has survived to the present time in the Old World (Africa and Asia) and in South America. The two extant wild species in South America occur in Peru, Bolivia, Chile, and Argentina. Cardozo (1975) outlines the phylogenetic history.

Comment on the Domestic
New World Camelids

We have not mapped the locality records for the llama and alpaca; both of these domestic forms are widespread in the altiplano and paramos of Peru, Bolivia, Argentina, and Ecuador. The llama, *Lama glama*, is used as a pack animal, and the alpaca, *Lama pacos*, is raised for its "wool." Grubb (1993) offers the Latin binomials for the domestic forms, and the use of sepa-

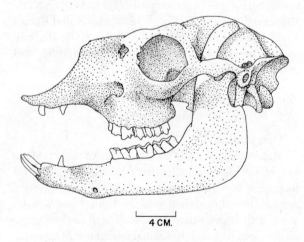

Figure 15.2. Skull of *Lama guanicoe*.

rate binomials for domestic and wild stock has had a long history, but the origin of the two domestic species of the genus *Lama* has provoked controversy. The llama is clearly a domesticated form of *Lama guanicoe,* but the origin of the alpaca has been in dispute. The question can be phrased thus: Is *Lama pacos* a domestic derivative of *Vicugna vicugna*? Or is *L. pacos* a domestic derivative of a hybrid produced by the pre-Inca peoples through crossing an early domestic *L. lama* with *V. vicugna*? The second hypothesis is gaining currency (see Hemmer 1990; Wheeler, Russel, and Stanley 1992). Hemmer (1990) has revived the notion that the alpaca originated as a hybrid between *L. guanicoe* and *V. vicugna,* and Stanley, Adwell, and Wheeler (1994) add biochemical evidence supporting this view.

Genus *Lama* G. Cuvier, 1800

Description
See family account.

Lama guanicoe (Muller, 1776)
Guanaco

Description
Guanacos are large ungulates with slender necks, long legs with broad, modified hooves, and thick, long pelage. They stand between 110 and 120 cm at the shoulder. The head is camel-like, with long, pointed ears and deeply cleft and highly mobile lips. Males have much longer canines than females. The general color of guanacos is reddish brown, with a grayish head and white underparts. Guanacos are distinguishable from vicunas by being larger, by having callosities on the inner sides of the forelimbs, and by lacking the vicuna's characteristic white or yellowish bib (plate 18) (Franklin 1982; Miller, Rottman, and Taber 1973; Osgood 1943; Raedeke 1979).

Distribution
The guanaco was originally found from northern Peru, and perhaps even southern Colombia, south to the southern tip of Chile and throughout Argentina. Now it is found from southern Peru south along the Andean zone in far western Paraguay, Bolivia, Chile, and Argentina to Tierra del Fuego and Navarino Island (Gilmore 1950; Honacki, Kinman, and Koeppl 1982; Olrog and Lucero 1981; Tamayo and Frassinetti 1980; Tonni and Politis 1980) (map 15.4).

Life History
In southern Argentina there is a restricted breeding season, with the peak in February. Females are apparently induced ovulators. The birth season is in the spring months of December to mid-February after a gestation period of about eleven months. The single

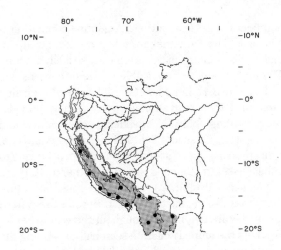

Map 15.4. Distribution of *Lama guanicoe.*

young is born weighing 8–15 kg, and lactation lasts eleven to fifteen months. With good nutrition females can breed in their second year, though full adult size is not achieved until about three years (Franklin 1983; Raedeke 1979).

Ecology
The historical range of guanacos was much larger than the current range. Hunting, habitat destruction, and apparently climatic change all contributed to driving the herds off the lower pampas and into foothills and mountains. The major exception is at the southern tip of the continent, where guanacos are still fairly common. Guanacos inhabit both warm and cold grasslands and shrublands from sea level to over 4,000 m. In some areas they live in forests during the winter. The vicuna, confined to the altiplano above 3,500 m, is much less catholic in its habitat requirements than the guanaco, which can occupy drier habitat because it can go for long periods without drinking. Guanacos are more versatile foragers because they both browse and graze, whereas vicunas only graze. In southern Argentina 61.5% of the food guanacos ate was grass and 15.4% was browse.

Guanacos are found in three principal types of social groups: family bands, male troops, and solitary males. A family band consists of a single breeding male with several females and their young. In southern Argentina the average band consisted of 5.5 females, and the largest had 18 females and a total of 25 individuals. Band males limit the size of their groups by driving off females who try to join and expelling young males at six to twelve months of age. Each family group occupies a territory defended by the male. Family bands occupy the best available habitat and use communal defecation piles that can be quite large and serve to demarcate their territory.

Herd males compose only about 18% of the total

number of males in a southern Argentine population of guanacos. The rest either are solitary or form troops of males. The solitary males are mostly sexually mature. The male troops are not stable in size or composition and occupy peripheral habitat. Males spend their first three to four years in all-male groups, where they develop fighting ability by practicing "chest rams" and play fighting. Guanacos fight much the way equids do; fights proceed unpredictably, with opponents trying to unbalance each other and bite one another on the legs.

Guanacos exhibit a very flexible social structure that varies from habitat to habitat. In some areas fixed territories are defended throughout the year, whereas in others the animals migrate seasonally, spending the winter in more sheltered areas. In San Juan province, Argentina, territories are defended from December to April; outside this time animals combine into larger groups that move into protected areas during the winter.

Young guanacos are particularly vulnerable to predation by pumas. In southern Argentina, where the puma has been exterminated, starvation is a major source of mortality. Before European contact guanacos were very important to several of the indigenous peoples, who based their lives on pursuit of the large herds (Cajal 1983b, 1986; Franklin 1982, 1983; Miller, Rottman, and Taber 1973; Raedeke 1979; Tonni and Politis 1980; Torres 1985; Wilson 1984; Wilson and Franklin 1985).

Genus *Vicugna* Lesson, 1842

Description
See family account.

Vicugna (Lama) vicugna (Molina, 1782)
Vicuna, Vicuña

Description
The lower incisors are very long and slender and have open roots. This is the only ungulate reputed to have continuously growing incisors. Adults weigh from 45–55 kg. This camelid is similar to the guanaco but is smaller, with finer pelage and a conspicuous white or yellowish "apron." The body color is a uniform rich cinnamon with or without a long white chest bib (depending on age and sex) and with a white venter (plate 18) (Franklin 1980, 1983; Osgood 1943).

Distribution
The vicuna is found from southern Peru to northern Chile and northwestern Argentina (Boswall 1972; Cajal 1983b,d; Miller, Rottman, and Taber 1973; Pefaur et al. 1968) (map 15.5).

Life History
Vicunas are seasonal breeders with births taking place during the middle of the season of plant pro-

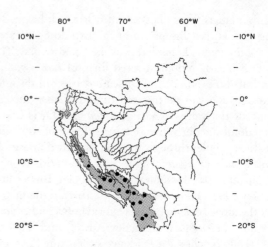

Map 15.5. Distribution of *Vicugna vicugna*.

ductivity. A single young is born after a gestation of 330–50 days, and females breed within two weeks of giving birth. Lactation lasts six to eight months. The age of first reproduction for females is twelve to fourteen months, though most do not breed until they are two years old (Franklin 1980, 1983).

Ecology
Vicunas are found only on the altiplano above about 3,500 m. They inhabit areas of open vegetation and on the best range can reach densities of twenty-one per square kilometer. Unlike guanacos, vicunas are never migratory and exhibit year-round territorial defense, although there are reports of animals' moving off their territories to graze on other pastures. Sleeping territories, in sheltered areas, are separate from feeding territories.

Vicunas are polygynous and are found in family groups (one male and several females with offspring), as solitary males, or in all-male groups. The average family group is composed of a single breeding male, three females, and two young. The largest group recorded by Franklin consisted of nineteen animals. The breeding group occupies a feeding territory that usually includes water and is defended by the male against other individuals. Some family groups occupy peripheral habitat with no water and may attempt to intrude on established territories. Family group size is strongly related to forage production on the territory. Both young males and females are expelled from family groups before they are one year old. There is a very constant group size and composition, and females show little independent behavior.

Solo males are sexually mature and are attempting to establish territories. Male groups are usually composed of five to ten animals but may have more than one hundred; 50% to 80% of the males in a population are in male groups. Males generally become resident at three to four years and are able to defend a territory for at least six years.

Vicunas are grazers, feeding mainly on grass, and their continuously growing incisors seem to be an adaptation for close cropping of small forbs and perennial grasses. In one study of a Peruvian population the diet consisted of two-thirds monocots and one-third dicots. Vicunas of both sexes, all ages, and in all social groups defecate only on traditional dung piles.

Many unsuccessful attempts have been made to domesticate the vicuna because of the potential value of its wool. The Incas reserved this wool for royalty and high officials and managed the wild vicunas by huge roundups that sometimes covered an area 30 km in diameter. Old females and males were killed for meat and hides, while young adults were shorn and released. Since the Incas, the range of the vicuna has decreased because of hunting for the meat and the very valuable wool (Bosch and Svendsen 1987; Cajal et al. 1981; Cajal and Sánchez 1983; Franklin 1980, 1983; Gilmore 1950; Koford 1957; Ménard 1982, 1984; Miller, Rottman, and Taber 1973; Miller et al. 1983).

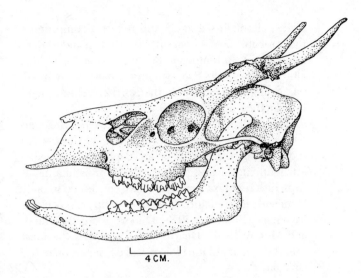

Figure 15.3. Skull of *Mazama americana* (male).

FAMILY CERVIDAE
Deer Family, Venado

Diagnosis

For deer found in South America, the dental formula is I 0/3, C 0/1, P 3/3, M 3/3. The upper incisors are lacking (fig. 15.3). Deer have a four-chambered stomach and are specialized for browsing. They are true ruminants; after feeding they lie up in a sheltered area, eructate, and remasticate the rumen contents. The first, second, and fifth digits are very reduced. Males develop antlers that are usually cast and renewed annually (but see species accounts and figs 15.6 and 15.7). Adults show a uniform dorsal pelage color, but the newborns are usually spotted (plate 19).

Distribution

The Recent distribution includes North and South America, Europe, and Asia; deer were present in Africa only north of the Sahara and are absent from Australia and New Zealand except where introduced by man.

Natural History

Males of the species comprising the family Cervidae typically bear antlers that usually are shed annually and grown anew. The only exceptions are the musk deer (*Moschus*), often placed in their own family, and the Chinese water deer (*Hydropotes*) of Asia, which are antlerless. Most species are adapted for browsing or for mixed browsing and grazing. Deer originated in Europe and Asia and spread to North America in the early Miocene. They then passed to South America in the late Pliocene. Their spread was rapid, and on en-

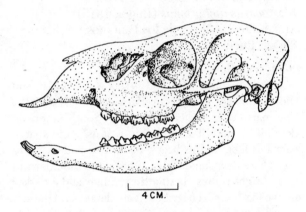

Figure 15.4. Skull of *Mazama americana* (female).

tering South America and they rapidly underwent an adaptive radiation.

Cervids evolved in the high latitudes of the Northern Hemisphere, and reproduction was highly synchronized to ensure that fawns would be born at a time of the year that would favor their survival. In the north, antler growth of the males, breeding, and the timing of birth tend to be highly seasonal. When deer adapt to the equatorial latitudes, births tend to be less sharply seasonal, as does the time of antler casting in males (Brock 1965). Most studies of deer living near the equator have offered the following data: Males still cast their antlers annually, although in any given area the population may show weak synchrony (Brokx 1972a; Branan and Marchinton 1987; Blouch 1987). Conception and birth of fawns, though not tightly synchronized, clearly are timed so that the birth of the fawn favors its survival. A sharp birth peak may not be evident because the determining survival factor is plant productivity, which in turn is tied to rainfall rather than day length. In Suriname there are two peaks in rainfall. In both Venezuela and Suriname births, although scattered, still reflect a compromise that probably favors fawn survival (Branan and Marchinton 1987). In temperate South America, or at high altitudes, reproductive synchrony may be marked, paralleling the tendency in the Northern Hemisphere (Merkt 1987; Polvilitis 1979).

The small deer (*Pudu* and some species of *Mazama*) all live in dense underbrush. Their small stature and reduced antler size could be an adaptation for moving efficiently in thick vegetation. Herbivorous mammals from different taxa worldwide have converged in body form and antipredator behavior to occupy this ecological niche (Bourlière 1973; Dubost 1968; Eisenberg and McKay 1974). Duarte (1997) provides the most useful summary of the biology and conservation of the Brazilian forms.

Genus *Blastocerus* Gray, 1850
Blastocerus dichotomus (Illiger, 1815)
Marsh Deer, Ciervo de los Pantanos,
Guazú Pucú

Description

Measurements: HB 153–91 cm; T 12–16 cm; shoulder height 110–27 cm (Pinder and Grosse 1991). The marsh deer is the largest of the South American deer and is unmistakable because of its size, its large ears and feet, and its heavy, multitined antlers (plate 19). These deer stand 100–120 cm at the shoulder and are a bright rufous chestnut in summer and a darker brown in winter. The lower legs are darker, and there is a black band on the muzzle. The ears are conspicuously large and contain white hairs on the inside. The dorsal pelage is a reddish brown. contrasting with a white venter. The fawn is spotted at birth. Mature males bear robust antlers with a dichotomous branching pattern (figs. 15.5, 15.7) that are very thick for their length (about 60 cm long) and normally have five points per side (Almeida 1976; Hofmann, Ponce del Prado, and Otte 1975–76; Whitehead 1972). As in all artiodactyls, the weight is borne on the two large toes of each foot. The hooves can spread to more than 10 cm while the deer are "strolling" in marshland.

Distribution

Blastocerus is found from southern Amazonian Peru to central Brazil and south through Bolivia and Paraguay to northern Argentina (González Romero, Moreno Ortiz, and Calabrese 1978; Hofmann, Ponce del Prado, and Otte 1975–76; Honacki, Kinman, and Koeppl 1982; G. Schaller and A. Tarak, unpubl. data; Ximénez, Langguth, and Praderi 1972). Pinder and Grosse (1991) imply that the species formerly existed in all major southern drainage systems of the Amazon and Paraná in South America. Currently the species survives in restricted areas of Brazil, Argentina,

Paraguay, and Bolivia (Pinder and Grosse 1991) (map 15.6).

Life History

In captivity does exhibit a postpartum estrus and stags may keep their antlers up to twenty-one months. In the Pantanal of Brazil the mating season is apparently very long: bucks with fully developed antlers were collected through most of the year, and the birth season is apparently from at least May to September. Gestation has been estimated at 271 days. Fawns associate with their mothers for almost one year and may continue the association as yearlings (Almeida 1976; Frädrich 1987; Miller 1930; Schaller and Vasconcelos 1978).

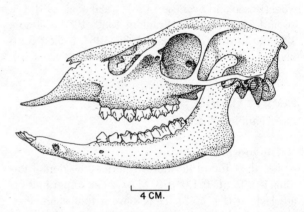

Figure 15.5. Skull of *Blastocerus dichotomus* (female).

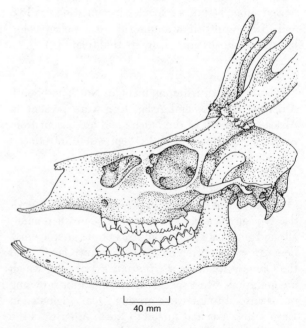

Figure 15.6. Skull of *Odocoileus virginianus* (male).

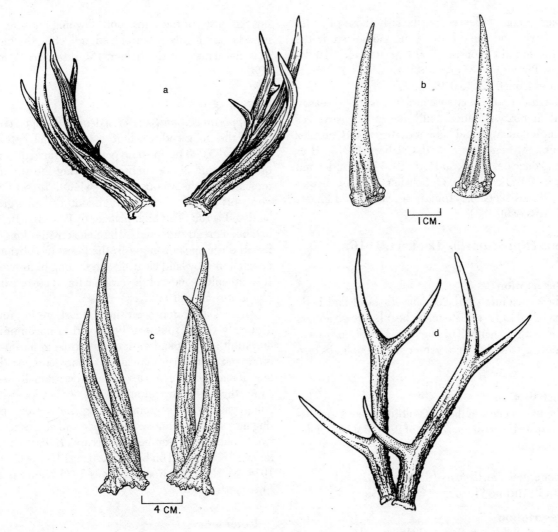

Figure 15.7. Antler forms displayed by males from four genera of South American cervids: (a) *Blastocerus;* (b) *Pudu;* (c) *Hippocamelus;* (d) *Ozotoceros.*

Ecology

Blastocerus, true to its English common name, is found only in areas with standing water and dense cover, though it may move away from these areas to feed. In the Brazilian Pantanal it prefers habitats of reeds or bunchgrass in or near water not over 60 cm deep. As the water levels change seasonally the deer move to find their preferred habitat. In the wet season they are widely dispersed, whereas in the dry season they are concentrated near the remaining water. In Formosa province, Argentina, marsh deer are found in dense stands of *Cyperus* 2 m or more tall during the dry season and in surrounding grassland during the annual floods.

In different parts of its range, *Blastocerus* is reported to be diurnal, crepuscular, and nocturnal, probably changing with season and extent of hunting. This species is often solitary, though in some places small family groups are reported. Stomachs from five

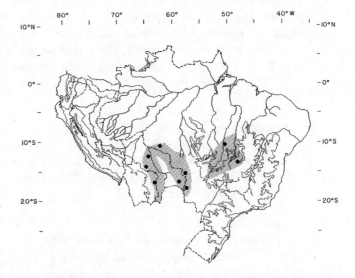

Map 15.6. Distribution of *Blastocerus dichotomus* (present distribution, not historical range).

animals contained leaves, particularly those of water lilies, grass, and some browse. In another study the diet was found to consist of 50% grasses and 31% legumes. Pinder (1993) in Mato Grosso do Sul reports densities from 0.524 to 0.483 km^2.

Marsh deer are very susceptible to cattle diseases, and their rarity is undoubtedly due to this as well as to hunting pressure and habitat destruction (González Romero, Moreno Ortiz, and Calabrese 1978; Hofmann, Ponce del Prado, and Otte 1975–76; Pinder and Grosse 1991; Schaller 1983; Schaller and Vasconcelos 1978; Tomas 1986a,b,c; Tomas, Beccaceci, and Pinder 1997; Voss et al. 1981).

Genus *Hippocamelus* Leukart, 1816
Huemul

Description
These medium-sized deer have a head and body length of 1.4 to 1.6 m. The males bear bifurcate antlers that are cast annually. The dorsum is medium brown, contrasting with the white venter and white limbs (see fig. 15.7).

Distribution
The two species of this high-altitude genus are disjunctly distributed in the Andes of Peru, Bolivia, Chile, and Argentina.

Hippocamelus antisensis (d'Orbigny, 1834)
Huemul Andino, Taruca

Description
Hippocamelus antisensis is slightly smaller and paler than the southern species *H. bisculus,* standing about 73 cm at the shoulder. It is generally a sandy gray (plate 19), and most individuals have a dark Y-shaped facial pattern. The throat and neck are whitish, as are the fronts of the forelegs. The tail has a dark patch at the base and is white on its underside. The antlers branch once to form two points per side and are 20–30 cm long (fig. 15.7) (Pearson 1951; Roe and Rees 1976; Whitehead 1972).

Distribution
Hippocamelus antisensis is found in the Andes from Ecuador to northwestern Argentina (map 15.7) (Cajal 1983a,d; Honacki, Kinman, and Koeppl 1982; Pine, Miller, and Schamberger 1979).

Life History
In southern Peru, young are born from February to April, the end of the rainy season, after an apparent gestation period of 240 days. Mating is most commonly observed in June, during the dry season. Males shed their antlers between September and October, dur-

ing the onset of the rains. Both fawning and the antler cycle are highly synchronized, reflecting the highly seasonal nature of the habitat (Merkt 1987; Pearson 1951).

Ecology
Hippocamelus antisensis is a deer of the open country of the high Andes, where it is found at between 2,500 and 5,000 m. In some parts of its range it is frequently found above timberline during the summer, moving into the woods during winter. In southern Peru this species occurs in high-altitude open grassland and scrub. The high plateau of Peru and Bolivia had been greatly affected by human activities long before the European conquests; the forest has obviously been altered by land clearing and cutting of firewood. It is difficult to guess at the original forest cover where *Hippocamelus* still survives.

Animals are rarely seen singly; most are in groups averaging 6.5 in southern Peru, with a maximum of 40. Adult males spend most of their time in mixed-sex groups except during fawning. Groups composed of only females or only males are also occasionally seen. In northwestern Argentina this species is reported to congregate in large groups during the winter, splitting into smaller groups of one male and an average of two females in September. In this area *H. antisensis* is known to be preyed on by pumas (Cajal 1983a; Crespo 1974; Merkt 1987; Roe and Rees 1976; Tamayo and Frassinetti 1980).

Comment
The biology of *Hippocamelus bisuculus* is reviewed in Redford and Eisenberg (1992) and Diaz (1990).

Genus *Odocoileus* Rafinesque, 1832
Odocoileus virginianus (Zimmermann, 1780)
White-tailed Deer, Venado

Description
Height at the shoulder is about 950 mm; head and body length, 1,500 to 2,200 mm; tail, 100 to 250 mm. Weight ranges from 50 to 120 kg. Adult males bear forked antlers with a variable number of tines, immediately distinguishing them from adult males of *Mazama,* which have spike antlers. The lateral to forward curve of the main antler beam from which all added "points" derive is distinctive. The basic color is a light brown dorsum, seasonally turning to gray, contrasting with a white venter. The underside of the tail is white, and its dorsal portion either is all brown or has a bit of black at the tip. A black band generally extends from between the eyes to the snout. There may be varying degrees of white adjacent to the rhinarium (plate 19). The fawn is spotted at birth.

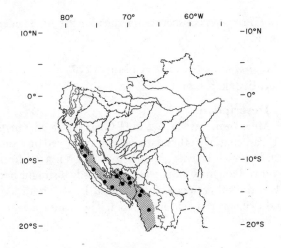

Map 15.7. Distribution of *Hippocamelus antisensis*.

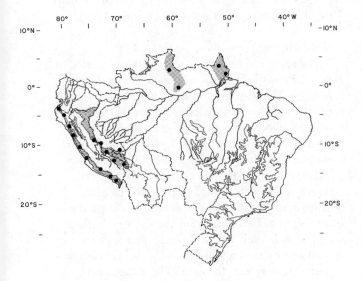

Map 15.8. Distribution of *Odocoileus virginianus*.

Range and Habitat

Odocoileus virginianus is distributed from southern Canada to western Brazil and may occur in central Bolivia (map 15.8). It is broadly tolerant of a variety of habitat types, but the interface between savanna and forest appears to be especially favorable, since it allows the larger white-tailed deer to graze as well as browse. The species extends to moderate elevations in the Andes of Peru and Ecuador.

Natural History

The white-tailed deer is one of the best-studied species of the Artiodactyla (Taylor 1956; Smith 1991; Halls and House 1984). Most research on this species has been carried out in the United States and Canada. The white-tailed deer was studied in Venezuela by Brokx (1972a,b, 1984), in Colombia by Blouch

(1987), in Guyana by Brock (1965), and in Suriname by Branan and Marchinton (1987) and Branan, Werkhoven, and Marchinton (1985). In Venezuela the white-tailed deer is a mixed browser and grazer and is most abundant in the forest-edge habitats of the llanos. When palm fruits are shed in the llanos, mainly during the early wet season (May–June), the deer may feed heavily on *Copernicia* fruits (Brokx and Anderssen 1970). Mean group size in Venezuela is 2.7. The most common association is a mother and her nearly grown offspring or a courting male and female, but in favorable habitats larger aggregations may be observed grazing. In protected areas they may be active in the late afternoon and early morning. Generally, in the llanos the deer rest in a shaded area at midday. Densities can reach 2.2 per km^2 in Venezuela (Eisenberg, O'Connell, and August 1979).

In Venezuela males shed their antlers annually, but the males in a population are not completely synchronized. The season of reproduction seems to be controlled by the pattern of rainfall. There is generally a birth peak, but tight reproductive synchrony is not characteristic of the tropical latitudes. Although northern races of the white-tailed deer frequently twin, twins are exceedingly uncommon in the subspecies of whitetail occurring in northern South America. The spotted fawn is born after seven months' gestation, and for the first ten days of its life it rests in a sheltered area and does not accompany its mother. The mother returns to feed it at least twice a day. At times of flooding in the lowland llanos of Venezuela and Colombia, large aggregations of whitetails may be trapped on islands of high ground, where they are extremely vulnerable to hunting.

Smith (1991) provides the most succinct and useful summary of the biology of *O. virginianus* within its northern distribution.

Genus *Mazama* Rafinesque, 1817
Brocket, Corzuela

Description

These small deer have a shoulder height between 370 mm and 710 mm. There are six species; three are dwarfs (*M. rufina*, *M. nana*, and *M. chunyi*) and may possibly be confused with pudus. *M. chunyi* averages about 380 mm in shoulder height but does not co-occur with any known species of pudu. *M. rufina* averages about 450 mm at the shoulder (plate 19). Members of the genus *Pudu* have a fusion between the cuneiform bone and the navicular-cuboid. In the genus *Mazama* there is no such fusion (Hershkovitz 1982). Brockets are reddish brown and can be distinguished from *Pudu mephistophiles* by their coat color and their greater size. *M. americana* is reddish brown and the largest species, some 710 mm at the shoulder. It can co-occur with the second of the two larger species, *M. gouazoubira*. Over most of its range, *M. gouazoubira* is smaller than *M. americana;* approximately 610 mm would be an average in the northern part of South America (Hershkovitz 1982). Its coat tends to be gray to gray brown. Males of all species bear spike antlers that are usually cast annually but may be retained into a second year (figs. 15.3 and 15.4). As far as can be determined from captive studies, fawns of all species of *Mazama* have spots at birth (Frädrich 1974, 1975). Czernay (1987) recognizes six species, and we follow his designations (see also description for *Pudu*).

Distribution

The genus *Mazama* is distributed from southern Mexico to central Argentina. It occurs in a variety of habitats including montane forests, lowland rain forests, lowland rain forests, and tree savannas.

Natural History

These small deer are adapted to habitats with suitable cover. In savanna situations they can hide effectively in tall grass, but they are absent from shortgrass areas unless they have access to gallery forests. Forest-edge habitats seem to be excellent in providing shrubs for browsing and shelter. These browsers and frugivores occupy a wide variety of elevations and forest types. They do not form large aggregations and are typically seen as single animals or courting pairs. After a gestation period of seven months the newborn spotted fawn is concealed in a sheltered locus, and the female returns at intervals to nurse it. The fawn does not begin to follow the female until it is several weeks old (Frädrich 1974, 1975; Thomas 1975).

Deer of the genus *Mazama* have been studied by Western scientists off and on for the past four centuries, yet we know little about their natural history. Bodmer (1989a,b) notes that these small deer are catholic in their diet. During the fruiting season in lowland Peru they are strongly frugivorous. He suggests that the rumen (forestomach) may be efficient at detoxifying secondary plant compounds before actual digestion of the pericarp and seeds (if small or triturated).

Although *M. americana* and *M. gouazoubira* co-occur over vast regions in South America, to some extent they are segregated by habitat. *M. americana* is often restricted to forests. In southern Brazil *M. nana* may occur with the two more common species. *M. nana* is usually the least abundant of the three. For example, Schneider and Barbosa de Oliviera (1996) report these densities: *M. americana*, 3.7/km^2; *M. gouazoubira*, 5.7/km^2; and *M. nana*, 0.3/km^2 in mixed forests of *Araucaria* and subtropical hardwoods. Bodmer

(1997) reviews conservation issues for the Amazonian cervids.

Mazama americana (Erxleben, 1777)
Red Brocket, Corzuela Roja

Description

Measurements: TL 1.2 m; T 0.15; Wt 28.9 kg ($n = 11$, Paraguay). In other parts of their range red brockets can be much heavier than the weights reported from eastern Paraguay. *M. americana* is easily distinguished from the other species of *Mazama* by its larger size and brilliant reddish brown coat. The venter is white or cream.

Distribution

M. americana is found from southern Mexico south through eastern Paraguay to northern Argentina (map 15.9) (Eisenberg 1989; Lucero 1983; Olrog 1979).

Life History

In captivity red brockets have a gestation period of about 225 days and produce a single young weighing 510–67 g. The young are born spotted and lose their spots after two to three months. In zoos in the Northern Hemisphere males do not show a seasonal rut and irregularly shed and regrow antlers, in some cases retaining hard antlers for over a year. Males can breed while carrying either velvet or hard antlers.

Does in the wild exhibit a postpartum estrus. In northeastern Argentina *M. americana* reproduces from August to October. In Suriname, two of eighteen does had twin embryos, and fawning took place over at least seven months, whereas in Guyana fawning was reported throughout the year. The antler cycle was

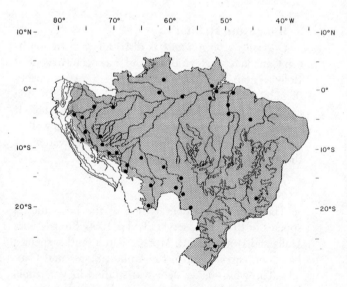

Map 15.9. Distribution of *Mazama americana*.

one year, and the quantity of sperm present was not related to antler condition. The youngest reproducing doe was thirteen months and the youngest stag twelve months or less (Branan and Marchinton 1987; Brock 1965; Crespo 1982; Gardner 1971; MacNamara and Eldridge 1987; Thomas 1975).

Ecology

The red brocket deer is apparently an animal of thick forest, seeming to prefer moist areas. It is a good swimmer and readily takes to the water to escape predators. In Suriname this species was found at a density of approximately one per square kilometer and fed on over sixty plant species. Fungi formed an important component of the diet during the wet season. Fruit was taken preferentially when available, with leaves predominating at the end of the wet season when fruit was scarce (Branan and Marchinton 1987; Branan, Werkhoven, and Marchinton 1985; Crespo 1982; Miller 1930; Schaller 1983).

Mazama bricenii Thomas, 1908

Description

This small species has a skull length of 159 mm, with antlers from 5 to 5.5 cm long. The basic coloration is a reddish chestnut with dark brown cheeks; the underparts are a lighter brown, but the ventral surface of the tail is white. This deer is very similar to *M. rufina* in size and color and is often subsumed under it. Czernay (1987) suggests this form may be a full species.

Distribution

The type locality was the Páramo de la Culata, Mérida, Venezuela. According to Czernay (1987), the specimens in the Andes of western Venezuela and adjacent Colombia belong to *M. bricenii* and the name *M. rufina* should be confined to the population in the premontane forests of southern Colombia and adjacent Ecuador (not mapped).

Mazama chunyi Hershkovitz, 1959

Description

The head and body length is 70 cm, height at the shoulder is about 38 cm, and the antlers are only 3.5 cm long. Weight averages 11 kg (Czernay 1987). This deer is similar to *Pudu puda* in size but is paler, with white markings on the ear margins and the tip of the muzzle. *M. rufina* has a darker brown face (see Hershkovitz 1959).

Distribution

The species is found in the eastern premontane forests of Andean Peru and Bolivia from 1,500 to 3,200 m (map 15.10).

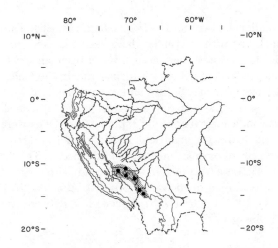

Map 15.10. Distribution of *Mazama chunyi*.

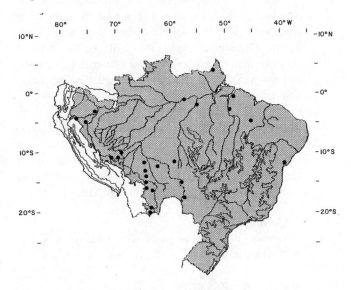

Map 15.11. Distribution of *Mazama gouazoubira*.

Mazama gouazoubira (G. Fischer, 1814)
Gray Brocket, Corzuela Parda

Description

Measurements: TL 1.03 m; T 0.11 m; Wt 16.3 kg ($n = 20$, Paraguay). This deer is smaller and slighter than *M. americana* and larger than *M. rufina*. Its spike antlers are 7–10 cm long (Almeida 1976; Redford and Eisenerg 1992). *M gouazoubira* is easily distinguished from the other species of *Mazama* by its gray to gray brown coat.

Distribution

The gray brocket, as currently defined, is found from Panama south to northern Argentina (Eisenberg 1989; Redford and Eisenberg 1992) (map. 15.11).

Life History

In captivity, the gestation period is 271 days, there is an immediate postpartum estrus, and a stag can keep its antlers for two and a half years. In the Brazilian Pantanal stags with hard antlers have been collected from May to December and those in velvet from January to June, indicating a long or irregular rut. In the Chaco of Paraguay males in hard antlers were similarly seen throughout the year. Spotted fawns were also seen all year, indicating year-round reproduction. Does were found to be pregnant and lactating, indicating the presence of postpartum estrus in this species. The litter size was always one. In Guyana a similar pattern of aseasonal reproduction was observed (Almeida 1976; Brock 1965; Frädrich 1987; Miller 1930; Redford and Eisenberg 1992).

Ecology

Unlike the red brocket, *M. gouazoubira* is not confined to thick forest. It is common in more open areas such as the thorn scrub of the Paraguayan Chaco and the Gran Sabana in Venezuela. It can be found in very dry areas. The gray brocket is a grazer, browser, and frugivore. In the Chaco of Paraguay an examination of twenty-three rumens showed that this species tracks fruit availability as diet varies seasonally, consuming particularly fruit of *Zizyphus* and *Caesalpina*. It feeds heavily on hard, dry fruit in the dry season and soft, fleshy fruit during the wet season and also takes leaves, buds, flowers, twigs, and roots (Pinder and Leeuwenberg 1997; Schaller 1983; Stallings 1984).

Comment

Grubb (1993) recommeds *M. gouazoupira* as the original spelling.

Mazama nana (Lesson, 1842)

Description

Measurements: TL 853; T 78; E 83; Wt 15 kg. Old males may weigh 20 kg, but see the comment at the end of this section. Height at the shoulder is less than 0.50 m. The dorsal pelage is a striking chestnut brown while the venter is snow white. The males bear spike antlers.

Distribution

We have chosen to consider this form a distinct species, although it is clearly related to *M. rufina* and *M. chunyi*. This small cervid is confined to the moist, forested areas that remain in eastern Brazil, northeastern Argentina, and adjacent Paraguay. It is often found in hilly or mountainous terrain (Czernay 1987) (see map 15.12).

Natural History

In northeastern Argentina a single spotted fawn is born between September and February. A male with antlers in velvet was collected in July in eastern

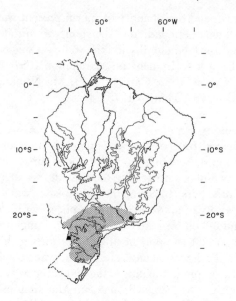

Map 15.12. Distribution of *Mazama nana*.

Paraguay (Crespo 1982), and the deer has been taken there in moist forest with a bamboo understory. It produces a wheezing alarm call when startled (Czernay 1987).

Comment

This deer is sometimes included within *M. rufina*, but Czernay (1987) separates this population into its own species (see also Redford and Eisenberg 1992).

Mazama rufina (Bourcier and Pucheran, 1852) Corzuela Enana

Description

For this small species of *Mazama*, average measurements are TL 853; T 78; E 83; Wt 8.2 kg. It averages about 450 mm at the shoulder. The dorsum is a rich chestnut brown. Although the color is variable to some extent, its coat resembles that of *M. americana* rather than *M. gouazoubira*.

Distribution

In this volume, following Czernay (1987), *M. rufina* will be restricted to those populations of small *Mazama* found in the western Andes of Ecuador (map 15.13). This species is adapted to the dense undergrowth of premontane to montane forests at elevations of 1,500 to 3,500 m. It may range to the paramo.

Comment

Mazama bororo was described by Duarte (1992). The male specimen was captured at Cape Bonito in the state of São Paulo. Slightly larger than *M. nana*, it weighed 22 kg. The diploid autosome number is 36, low compared with that of other known karyotypes for

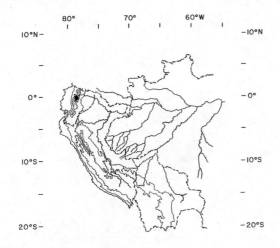

Map 15.13. Distribution of *Mazama rufina*.

Mazama. *M. gouazoubira* exhibits a polymorphic karyotype over its range ($2n = 70$ to 36). Duarte describes the variable chromosome numbers for *Mazama americana*. The diploid number of autosomes ranges from 50 in Mexico to 52 in Brazil. *M. nana* has a chromosome number of $2n = 36$, but the banding pattern differs from that of *M. bororo*. The problems of karyotypic variation and its significance for taxonomy requires further study (Duarte and Merino 1997).

Genus *Ozotoceros* Ameghino, 1891
Ozotoceros bezoarticus (Linnaeus, 1758)
Pampas Deer, Ciervo de la Pampa, Venado

Description
This medium-sized deer measures HB 1.10–1.3 m, with a shoulder height of 0.70 to 0.75 m (Jackson 1987). The pampas deer resembles the white-tailed deer (*Odocoileus*): it is longer-legged than deer of the genus *Mazama* (standing 70–75 cm at the shoulder) and smaller and more slender than the marsh deer, *Blastocerus* (weighing up to 35 kg) (plate 19). The branched antlers are moderate in size and usually bear three points per side, a brow tine and a simple upper bifurcation (fig. 15.7). However, up to eight tines per side have been reported (Almeida 1976; Jackson and Langguth 1987; pers. obs.). The coat is smooth and short-haired, yellowish gray with white underparts, white on the insides of the ears and the sides of the muzzle, and a conspicuous white eye-ring. The single fawn is spotted at birth.
Chromosome number: $2n = 68$; FN = 74 (Spotorno, Brum, and Di Tomaso 1987).

Distribution
Ozotoceros formerly occupied the open savannas of what is now termed the caatinga, cerrado, and chaco of Brazil, Bolivia, Uruguay, Paraguay, and Argentina (map 15.14). This species is now found from central

Brazil south into Bolivia, Paraguay, Uruguay and northeastern and north-central Argentina. Its range has contracted greatly owing to habitat destruction and hunting (Jackson 1978; Jackson, Landa, and Langguth 1980; Ojeda and Mares 1982).

Life History
The gestation period of *Ozotoceros* is a little longer than seven months, after which a single fawn weighing 2,100 g is born. In central Brazil births occur from July to December with a peak in October to November. In Argentina the reproductive activity is from December to February, with most births occurring between September and November. Fawns are spotted until about one month of age. Stags are in hard antlers from November to July. In the Brazilian Pantanal, all observed stags had antlers in velvet between June and August. The general pattern of antler growth is for one-year-old males to have spikes, two-year-olds to have a total of four tines, and males three years old and up to have a total of six tines (Frädrich 1981; Jackson and Langguth 1987; Redford 1987; Tomas 1988a).

Ecology
The pampas deer is an animal of the open vegetation formations of central and southern South America. It was once found in most of the natural grasslands of eastern South America below the equator, but it has been exterminated from the vast majority of its range.

In the Pantanal *Ozotoceros* exhibits seasonal feeding on grasses, but the annual diet is strongly biased toward herbs. *Melochia simplex* (Sterauliacae) predominates in the dry-season diet. Browsing is of secondary importance (Pinder 1997). On the other hand, grazing is the most important foraging activity of pampas deer at Samborombón Bay, San Luis province, Argentina (Jackson and Giuletti 1988).

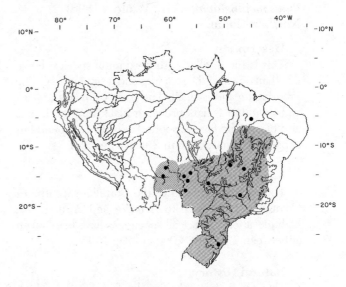

Map 15.14. Distribution of *Ozotoceros bezoarticus*.

In central Brazil the mean group size was 1.36, though on better pasture and during the rainy season groups were considerably larger. In Argentina groups averaged five or six animals and included both males and females, but group size and composition were both fluid. In the Brazilian Pantanal, densities ranged from 1.6 individuals per square kilometer in cerrado to 4.6 in open grassland. Historical documents indicate that even when they were numerous pampas deer lived in small mixed-sex groups rarely exceeding five individuals, except when congregating on feeding grounds. These deer produce a pungent odor from the interdigital gland on the hind feet that is strongest in adult males.

During the late nineteenth century there was intensive commercial hunting of the pampas deer, and this combined with conversion of grassland to cattle pasture and agricultural fields was responsible for the remarkable decrease in population. In 1880 there were 61,401 pampas deer skins exported from Buenos Aires (Jackson 1978, 1985, 1986, 1987; Jackson and Langguth 1987; Langguth and Jackson 1980; Miller 1930; Redford 1987; Tomas 1988b; Pinder 1992).

Genus *Pudu* Gray, 1852

Description

This genus contains two Recent species, *Pudu puda* and *P. mephistophiles*. Only the latter occurs within the range of this book. These deer may be distinguished from all other South American deer because one of the carpal bones, the cuneiform, is fused with the navicular-cuboid. They are the smallest deer in South America, less than 380 mm high at the shoulder. The males bear a spike antler that is cast annually. The genus has been reviewed by Hershkovitz (1982).

Pudu mephistophiles (De Winton, 1896)
Northern Pudu

Description

This is the smaller of the two pudu species, being less than 35 cm at the shoulder and weighing 5 kg. The preorbital gland is small. The basic color of the back and shoulders is dark brown or black, and the face is also black. The fawn is spotted at birth in *P. puda* but apparently is unspotted in *P. mephistophiles*.

Range and Habitat

This deer is confined to the Cordillera Central of Colombia, and thence to Ecuador and Peru. It is a high-elevation species; all specimens have been taken at between 7,700 and 4,000 m (map 15.15).

Natural History

This small deer inhabits the forests of the high Andes. A single fawn is born after a gestation period

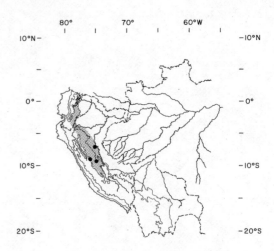

Map 15.15. Distribution of *Pudu mephistophiles*.

of approximately seven months (Hick 1967). The standard reference for this genus is Hershkovitz (1982). Details on captive propagation are given in Frädrich (1974).

Final Comments on Frugivory and Forest Ungulates

Bodmer (1989a) developed an elegant set of hypotheses concerning the consequences of ungulate adaptation to rain forests. Equipped with his ideas, he set off for Peru to study *Mazama*, *Tayassu*, and *Tapirus*. Since seasonal flooding of the Amazon basin is a prime factor in the lives of Neotropical forest ungulates, he attempted to assess the effects of inundation on the dietary habits of the forest ungulate community (Bodmer 1990). In short, he found that *Tapirus* and *Tayassu pecari* are little influenced by flooding, but *Tayassu tajacu* and *Mazama* are affected by the deluge. The latter species shift their exploitation strategies.

Ungulate biomass levels in forests are understandably lower than in savanna or mosaic habitats (Eisenberg 1980), yet forest ungulates living on the fallen fruits from the canopy can achieve a respectable biomass (Bodmer 1989a,c). To be sure, in times of fruit scarcity the ungulates can browse on the saplings of the future forest (Bodmer, 1989b, 1990). During the phase of intensive feeding on fallen fruit, *Mazama* and *Tayassu* can be considered seed predators. Only the tapir disperses a considerable amount of seed after fruits have passed through its gut (Bodmer 1991). Foregut fermentation by *Mazama* of fruits with a fibrous, protective sheath suggests that species of this genus may have a significant impact on seed survival. The members of the genus *Tayassu* "crack" seeds and have long been known as active seed predators, but see Fragoso (1994).

The situation is much more complex than the preceding paragraphs imply. Working in Roraima, Brazil, Fragoso (1994) concentrated on the palm *Maximiliana maripa* and the mammals that utilize its fallen fruits. Seed and seedling mortality was high when fruit fall was studied under the parent tree. *Mazama, Odocoileus,* and *Tayassu tajacu* ate the pulp but spit out the seeds near the parent tree. Such exposed seeds were preyed on by the larvae of bruchid beetles. *Tayassu pecari* was an important seed predator because it cracks many of the seeds with the thinnest endocarps (those containing the most endosperm). *Tapirus terrestris,* on the other hand, swallows the fruit and digests the pulp but defecates the seed at latrines up to 2 km from the parent clump of trees. This brief outline highlights the complexity of plant-animal interactions when only one fruiting tree, *Maximiliana,* is considered, let alone the whole forest (see also chapter 14).

References

• References used in preparing distribution maps

• Allen, J. A. 1916. Mammals collected on the Roosevelt Brazilian expedition, with field notes by Leo E. Miller. *Bull. Amer. Mus. Nat. Hist.* 34:559–610.

Almeida, A. de. 1976. *Jaguar hunting in the Mato Grosso.* England: Stanwill Press.

Ayres, J. M., and C. Ayres. 1979. Aspectos da caça no alto Rio Aripuanã. *Acta Amazonica* 9 (2): 287–98.

• Barlow, J. C. 1965. Land mammals from Uruguay: Ecology and zoogeography. Ph.D. diss., University of Kansas.

Benirschke, K., and A. T. Kumamoto. 1989. Further studies on the chromosomes of three species of peccary. In *Advances in Neotropical mammalogy,* ed. K. H. Redford and J. F. Eisenberg, 309–16. Gainesville, Fla.: Sandhill Crane Press.

• Bisbal, F. J. 1991. Distribución y taxonomía del venado matacán (*Mazama*) en Venezuela. *Acta Biol. Venez.* 13 (1–2): 89–104.

Blouch, R. A. 1987. Reproductive seasonality of the white-tailed deer on the Colombian llanos. In *Biology and management of the Cervidae,* ed. C. Wemmer, 339–43. Washington, D.C.: Smithsonian Institution Press.

Bodmer, R. E. 1989a. Frugivory in Amazon ungulates. Ph.D. diss., Cambridge University.

———. 1989b. Ungulate frugivores and the browser-grazer continuum. *Oikos* 57:319–25.

———. 1989c. Ungulate biomass in relation to feeding strategy within Amazonian forests. *Oecologia* 81:547–50.

———. 1990. Responses of ungulates to seasonal inundations in the Amazonian floodplain. *J. Trop. Ecol.* 6:191–201.

———. 1991. Strategies of seed dispersal and seed predation in Amazonian ungulates. *Biotropica* 23: 255–61.

———. 1997. Ecologia e conservação veados mateiro e catinguero na Amazonia. In *Biologia e conservação de cervídeos sul-americanos:* Blastocerus, Ozotoceros e Mazama, ed. J. M. B. Duarte, 70–77. São Paulo: FUNEP.

Bodmer, R. E., R. Aquino, P. Puertas, C. Reyes, T. Fang, and N. Gottdenker. 1996. *Evaluano el uso sostenible de pecaries en el nor-oriente del Peru.* Iquitos, Peru: TCD and University of Florida.

Bosch, P. C., and G. E. Svendsen. 1987. Behavior of male and female vicuna (*Vicugna vicugna* Molina 1782) as it relates to reproductive effort. *J. Mammal.* 68 (2): 425–29.

Boswall, J. 1972. Vicuna in Argentina. *Oryx* 11 (6): 449–53.

Bourlière, F. 1973. The comparative ecology of rain forest mammals in Africa and tropical America. In *Tropical forest ecosystems in Africa and South America: A comparative review,* ed. B. J. Meggers, E. S. Ayensu, and W. D. Duckworth, 279–92. Washington, D.C.: Smithsonian Institution Press.

Branan, W. V., and R. L. Marchinton. 1987. Reproductive ecology of white-tailed and red brocket deer in Suriname. In *Biology and management of the Cervidae,* ed. C. Wemmer, 344–51. Washington, D.C.: Smithsonian Institution Press.

Branan, W. V., M. C. M. Werkhoven, and R. L. Marchinton. 1985. Food habits of brocket and white-tailed deer in Suriname. *J. Wildl. Manage.* 49 (4): 972–76.

Brock, S. E. 1965. The deer of British Guiana. *J. British Guiana Mus. Zoo Royal Agric. Comm. Soc.* 40: 18–25.

Brokx, P. A. J. 1972a. A study of the biology of Venezuelan white-tailed deer (*Odocoileus virginianus gymnotis* Wiegmann, 1833), with a hypothesis on the origin of South American cervids. Ph.D. diss., University of Waterloo, Ontario.

———. 1972b. Age determination of Venezuelan white-tailed deer. *J. Wildl. Manage.* 36 (4): 1060–67.

———. 1984. South America. In *White-tailed deer: Ecology and management,* ed. L. K. Halls and C. House, 525–46. Harrisburg, Pa.: Stackpole.

Brokx, P. A. J., and F. M. Anderssen. 1970. Analisis estomacales del venado caramerudo de los llanos venezolanos. *Bol. Soc. Venez. Cien. Nat.* 27: 330–53.

Brooks, D. M. 1992. Reproductive behavior and development of the young of the chacoan peccary in the Paraguayan Chaco. *Z. Säugetierk.* 57:316–17.

Bruzone, J. H. 1984. Estación de recría *Pudu pudu,* Isla Victoria, Parque Nacional Nahuel Huapi. *Rev. Mus. Argent. Cienc. "Bernardino Rivadavia," Zool.* 13 (1–60): 281–92.

Byers, J. A., and M. Beckoff. 1981. Social, spacing, and cooperative behavior of the collared peccary, *Tayassu tajacu*. *J. Mammal.* 62 (4): 767–85.

• Cabrera, A. 1943. Sobre la sistemática del venado y su varición individual y geográfica. *Rev. Mus. La Plata*, n.s., 3 (18): 5–41.

• ———. 1960. *Catálogo de los mamíferos de América del Sur.* Vol. 2. Buenos Aires: Museo Argentino de Ciencias Naturales "Bernardino Rivadavia."

Cajal, J. L. 1983a. Über den Bestand des Nord-Andenhirsches (*Hippocamelus antisensis*) in der argentinischen Provinz La Rioja. *Bongo* (Berlin) 7:83–90.

———. 1983b. La vicuña en Argentina: Pautas para su Manejo. *Interciencia* 8 (1): 19–22.

———. 1983c. *La situación del taruca en la provincia de La Rioja república Argentina.* Buenos Aires: Subsecretaría de Ciencia y Tecnología, Programa Nacional de Recursos Naturales Renovables.

———. 1983d. *Estructura social y area de acción del guanaco* (Lama guanicoe) *en la Reserva de San Guillermo* (provincia de San Juan). Buenos Aires: Subsecretaría de Ciencia y Tecnología, Programa Nacional de Recursos Naturales Renovables.

———. 1986. El recurso fauna en la Argentina: Antecedentes y cuadro d situación actual. Conferencia presentada en las Primeras Jornadas Ambientales Sanjuaninas 16 de agosto de 1984, Fundación Ambientalista Sanjuanina. Ministerio de Educación y Justicia, Secretaría de Ciencia y Técnica, Programa Nacional de Recursos Naturales Renovables, Buenos Aires.

Cajal, J. L., N. E. López, A. Reca, and J. C. Pujalte. 1981. *La estrategia de la conservación de los camélidos en Argentina con especial referencia a la vicuña.* Buenos Aires: Subsecretaría de Ciencia y Tecnología.

Cajal, J. L., and E. Sánchez. 1983. *Marcha de los censos de vicuña* (Vicugna vicugna), *guanaco* (Lama guanicoe), *y ñandú cordillerano* (Pterocnemia pennata), *en la Reserva de San Guillermo, provincia de San Juan* (2/1977-2/1979). Buenos Aires: Subsecretaría de Ciencia y Tecnología, Programa National de Recursos Naturales Renovables.

Cardozo, Armando. 1975. *Orígen y filogenía de los camélidos sudamericanos.* La Paz: Academia Nacional de Ciencias de Bolivia.

Castellanos, H. G. 1983. Aspectos de la organización social del báquiro de collar, *Tayassu tajacu* L., en el estado Guárico-Venezuela. *Acta Biol. Venez.* 11 (4): 127–43.

———. 1986. Home range size and habitat selection of the collared peccary in the state of Guárico, Venezuela. In *Proceedings of the peccary workshop,* ed. R. A. Ockenfels, G. I. Day, and V. C. Sup-

plee, 50. Phoenix: Arizona Game and Fish Department.

Courtin, S. L., N. V. Pacheco, and W. D. Eldridge. 1980. Observaciones de alimentación, movimientos y preferencias de hábitat del puma, en el Islote Rupanco. *Medio Ambiente* (Valdivia, Chile) 4 (2): 50–55.

• Crespo, J. A. 1950. Nota sobre mamíferos de Misiones nuevos para Argentina. *Comun. Inst. Nac. Invest. Cienc. Nat., Mus. Argent. Cienc. Nat. "Bernardino Rivadavia," Zool.* 1 (14): 1–14.

• ———. 1974. Comentarios sobre neuvas localidades para mamíferos de Argentina y de Bolivia. *Rev. Mus. Argent. Cienc. Nat. "Bernardino Rivadavia," Zool.* 11 (1): 1–31.

———. 1982. Ecología de la comunidad de mamíferos del Parque Nacional Iguazú, Misiones. *Rev. Mus. Argent. Cienc. Nat. "Bernardino Rivadavia," Ecol.* 3 (2): 45–162.

Czernay, S. 1987. *Die Spiesshirsche und Pudus.* Wittenberg: A. Ziemsen.

Davis, D. E. 1947. Notes on the life histories of some Brazilian mammals. *Bol. Mus. Nac., Zool.* 76:1–8.

Díaz, G. A. C. 1978. Social behavior of the collared peccary (*Tayassu tajacu*) in captivity. *CEIBA* 22 (2): 73–126.

Diaz, N. I. 1990. *El huemal.* Buenos Aires: Talleros Gráficos.

Drouilly, P. 1983. *Recopilación de antecedentes biológicos y ecológicos del huemul chileno y consideraciones sobre su manejo.* Boletín Técnico 5. Santiago: Corporación Nacional Forestal.

Duarte, J. M. B. 1992. Aspectos taxonômicos e citogenéticos de algumas espécies de cervidos brasilieros. Diss. de mestrado, Faculdade de Ciências Agrárias e Veterinârias, Universidad Paulist, São Paulo, Brazil.

———, ed. 1997. *Biologia e conservação de cervideos sul-americanos:* Blastocerus, Ozotoceros e Mazama. São Paulo: FUNEP.

Duarte, J. M. B., and M. L. Merino. 1997. Taxonomia e ecolução. In *Biologia e conservação de cervideos sul-americanos:* Blastocerus, Ozotoceros e Mazama, ed. J. M. B. Duarte, 2–21. São Paulo: FUNEP.

Dubost, G. 1968. Les niches écologiques des forêts tropicales sud-américaines et africaines, sources de convergences remarquables entre ronqeurs et artiodactyles. *Terre et Vie* 1:3–28.

Eddy, T. A. 1961. Foods and feeding patterns of the collared peccary in southern Arizona. *J. Wildl. Manage.* 25 (3): 248–57.

Eisenberg, J. F. 1980. The density and biomass of tropical mammals. In *Conservation biology: An evolutionary and ecological perspective,* ed. M. E.

Soulé and B. A. Wilcox, 35–55. Sunderland, Mass.: Sinauer.

———. 1987. Evolutionary history of the Cervidae with special reference to the South American radiation. In *Biology and management of the Cervidae*, ed. C. Wemmer, 60–64. Washington, D.C.: Smithsonian Institution Press.

———. 1989. *Mammals of the Neotropics.* Vol. 1. *The northern Neotropics: Panama, Colombia, Venezuela, Guyana, Suriname, French Guiana.* Chicago: University of Chicago Press.

Eisenberg, J. F., and G. M. McKay. 1974. Comparison of ungulate adaptations in the New World and Old World tropical forests with special reference to Ceylon and the rainforests of Central America. In *The behavior of ungulates and its relation to management*, ed. V. Geist and F. Walther, 2:585–602. Publications, n.s., 24. Morges, Switzerland: IUCN.

Eisenberg, J. F., M. A. O'Connell, and P. V. August. 1979. Density, productivity and distribution of mammals in two Venezuelan habitats. In *Vertebrate ecology in the northern Neotropics*, ed. J. F. Eisenberg, 187–207. Washington, D.C.: Smithsonian Institution Press.

Eldridge, W. D., M. M. MacNamara, and N. V. Pacheco. 1987. Activity patterns and habitat utilization of pudus (*Pudu pudu*) in south-central Chile. In *Biology and management of the Cervidae*, ed. C. Wemmer, 352–70. Washington, D.C.: Smithsonian Institution Press.

Feer, F. 1984. Observations éthologiques sur *Pudu pudu* (Molina, 1782) en captivité. *Zool. Gart.*, n.s., 54 (1–2): 1–27.

Fonseca, G. A. B. da, A. B. Rylands, C. M. R. Costa, R. B. Machado, and Y. L. R. Leite, eds. 1994. *Livro vermelho dos mammíferes brasileiros ameaçados de extinção.* Belo Horizonte: Fundação Biodiversitas.

Frädrich, H. 1974. Notizen über seltener gehaltene Cerviden. Part 1. *Zool. Gart.*, n.s., 44 (4): 189–200.

———. 1975. Notizen über seltener gehaltene Cerviden. Part 2. *Zool. Gart.*, n.s., 45 (1): 67–77.

———. 1981. Beobachtungen am Pampashirsch, *Blastocerus bezoarticus* (L., 1758). *Zool. Gart.*, n.s., 51:7–32.

———. 1987. The husbandry of tropical and temperate cervids in the West Berlin Zoo. In *Biology and management of the Cervidae*, ed. C. Wemmer, 422–28. Washington, D.C.: Smithsonian Institution Press.

Franklin, W. L. 1980. Territorial marking behavior by the South American vicuna. In *Chemical signals: Vertebrates and aquatic invertebrates*, ed. D. Muller-Schwarze and R. M. Silverstein, 53–66. New York: Plenum Press.

———. 1982. Biology, ecology, and relationship to man of the South American camelids. In *Mammalian biology in South America*, ed. M. A. Mares and H. H. Genoways, 457–89. Pymatuning Symposia in Ecology 6. Special Publication Series. Pittsburgh: Pymatuning Laboratory of Ecology, University of Pittsburgh.

———. 1983. Contrasting socioecologies of South America's wild camelids the vicuna and the guanaco. In *Advances in the study of mammalian behavior*, ed. J. F. Eisenberg and D. G. Kleiman, 573–629. Special Publication 7. Shippensburg, Pa.: American Society of Mammalogists.

Fragoso, J. M. 1994. Large mammals and the community dynamics of an Amazonian rain forest. Ph.D. diss., University of Florida, Gainesville.

Gardner, A. L. 1971. Postpartum estrus in a red brocket deer, *Mazama americana*, from Peru. *J. Mammal.* 52 (3): 623–24.

Gilmore, R. 1950. Fauna and ethnozoology of South America. In *Handbook of South American Indians*, vol. 6, ed. J. Steward, 264–345. Washington, D.C.: Government Printing Office.

González Romero, N., H. Moreno Ortiz, and P. Calabrese. 1978. El ciervo de los pantanos o guazú pucú (*Blastocerus dichotomus*) en el Paraguay. *Informes Cient.* 1 (1).

Green, G. E., and W. E. Grant. 1984. Variability of observed group sizes within collared peccary herds. *J. Wildl. Manage.* 48 (1): 244–48.

Grubb, P. 1993. Artiodactyla. In *Mammal species of the world*, ed. D. E. Wilson and D. M. Reeder, 377–414. Washington, D.C.: Smithsonian Institution Press.

Halls, L. K., and C. House, eds. 1984. *White-tailed deer: Ecology and management.* Harrisburg, Pa.: Stackpole.

Hemmer, H. 1990. *Domestication: The decline of environmental appreciation.* Cambridge: Cambridge University Press.

Hershkovitz, P. 1959. A new species of South American brocket, genus *Mazama* (Cervidae). *Proc. Biol. Soc. Washington* 72:45–54.

———. 1982. Neotropical deer (Cervidae). Part 1. Pudus, genus *Pudu* Gray. *Fieldiana: Zool.*, n.s., 11:1–86.

Hick, U. 1967. Geglückte Aufzucht eines Pudus (*Pudu pudu* Mol.). *Freunde Kölner Zoo* 4:111–18.

Hofmann, R. K., C. F. Ponce del Prado, and K. C. Otte. 1975–76. Registro de dos nuevas especies de mamíferos para el Perú, *Odocoileus dichotomus* (Illiger—1811) y *Chrysocyon brachyurus* (Illiger—1811) con notas sobre su hábitat. *Rev. Forest. Perú* 6:61–81.

Honacki, J. H., K. E. Kinman, and J. W. Koeppl, eds. 1982. *Mammal species of the world.* Lawrence,

Kans.: Allen Press and Association of Systematics Collections.

Husson, A. M. 1978. *The mammals of Suriname.* Leiden: E. J. Brill.

Jackson, J. E. 1978. The IUCN threatened deer programme. 1. Species close to extinction in the wild: The Argentinian pampas deer or venado (*Ozotoceros bezoarticus celer*). In *Threatened deer,* 33–48. Morges, Switzerland: IUCN.

———. 1985. Behavioural observations on the Argentinian pampas deer (*Ozotoceros bezoarticus celer* Cabrera, 1943). Z. *Säugetierk.* 50:107–16.

———. 1986. Antler cycle in pampas deer (*Ozotoceros bezoarticus*) from San Luis, Argentina. *J. Mammal.* 67:175–76.

———. 1987. *Ozotoceros bezoarticus. Mammal. Species* 295:1–5.

Jackson, J. E., and J. D. Giuletti. 1988. The food habits of pampas deer *Ozotoceros bezoarticus celer* in relation to its conservation in a relict natural grassland in Argentina. *Biol. Conserv.* 45:1–10.

Jackson, J. E., P. Landa, and A. Langguth. 1980. Pampas deer in Uruguay. *Oryx* 15:267–72.

Jackson, J. E., and A. Langguth. 1987. Ecology and status of the pampas deer in the Argentinian pampas and Uruguay. In *Biology and management of the Cervidae,* ed. C. Wemmer, 402–9. Washington, D.C.: Smithsonian Institution Press.

Jungius, H. 1974. Beobachtungen am Weisweldelhirsch und an anderen Cerviden in Bolivia. Z. *Säugetierk.* 39 (6): 373–83.

Kiltie, R. A. 1981a. Stomach contents of rain forest peccaries (*Tayassu tajacu* and *T. pecari*). *Biotropica* 13:234–36.

———. 1981b. The function of interlocking canines in rain forest peccaries (Tayassuidae). *J. Mammal.* 62:459–69.

Kiltie, R. A., and J. Terborgh. 1983. Observations on the behavior of rain forest peccaries in Peru: Why do white-lipped peccaries form herds? Z. *Tierpsychol.* 62:241–55.

Koford, C. B. 1957. The vicuna and the puna. *Ecol. Monog.* 27:153–219.

Langer, P. 1979. Adaptational significance of the forestomach of the collared peccary, *Dicotyles tajacu* (L., 1758) (Mammalia: Artiodactyla). *Mammalia* 43 (2): 235–45.

Langguth, A., and J. Jackson. 1980. Cutaneous scent glands in pampas deer *Blastocerus bezoarticus* (L., 1758). Z. *Säugetierk.* 45:82–90.

Lochmiller, R. L., E. C. Hellgren, and W. E. Grant. 1984. Selected aspects of collared peccary (*Dicotyles tajacu*) reproductive biology in a captive Texas herd. *Zoo Biol.* 3:145–49.

• Lucero, M. M. 1983. Lista y distribución de aves y mamíferos de la provincia de Tucumán. Ministerio de Cultura y Educación, Fundación Miguel Lillo, *Miscelánea* 75:5–53.

McCoy, M. B., and C. Vaughan. 1986. Movement, activity, and diet of collared peccaries in Costa Rican dry forest. In *Proceedings of the peccary workshop,* ed. R. A. Ockenfels, G. I. Day, and V. C. Supplee, 52–53. Phoenix: Arizona Game and Fish Department.

McCoy, M. B., C. Vaughan, and V. Villalobos. 1983. An interesting feeding habit for the collared peccary (*Tayassu tajacu* Bangs) in Costa Rica. *Brenesia* 21:456–57.

MacNamara, M., and W. D. Eldridge. 1987. Behavior and reproduction in captive pudu (*Pudu puda*) and red brocket (*Mazama americana*), a descriptive and comparative analysis. In *Biology and management of the Cervidae,* ed. C. M. Wemmer, 371–87. Washington, D.C.: Smithsonian Institution Press.

March, M. I. J. 1991. Monographiedes Weissbartpekaris (*Tayassu pecari*) *Bongo* (Berlin) 18:151–70.

Mayer, J. J., and P. N. Brandt. 1982. Identity, distribution, and natural history of the peccaries, Tayassuidae. In *Mammalian biology in South America,* ed. M. A. Mares and H. H. Genoways, 433–56. Pymatuning Symposia in Ecology 6. Special Publication Series. Pittsburgh: Pymatuning Laboratory of Ecology, University of Pittsburgh.

Mayer, J. J., and R. M. Wetzel. 1986. *Catagonus wagneri. Mammal. Species* 259:1–5.

———. 1987. *Tayassu pecari. Mammal. Species* 293:1–7.

Ménard, N. 1982. Quelques aspects de la socioécologie de la vigogne *Lama vicugna. Rev. Ecol. (Terre et Vie)* 36:15–35.

Ménard, N. 1984. Le régime alimentaire des vigognes (*Lama vicugna*) pendant une période de sécheresse. *Mammalia* 48 (4): 529–39.

Merkt, J. R. 1987. Reproductive seasonality and grouping patterns of the North Andean deer or taruca (*Hippocamelus antisensis*) in southern Peru. In *Biology and management of the Cervidae,* ed. C. M. Wemmer, 388–401. Washington, D.C.: Smithsonian Institution Press.

Miller, F. W. 1930. Notes on some mammals of southern Mato Grosso, Brazil. *J. Mammal.* 11:10–22.

Miller, S. D., J. Rottman, K. J. Raedeke, and R. D. Taber. 1983. Endangered mammals of Chile: Status and conservation. *Biol. Conserv.* 25:335–52.

Miller, S. D., J. Rottman, and R. D. Taber. 1973. Dwindling and endangered ungulates of Chile: Vicugna, llama, hippocamelus, and pudu. *Trans. North Amer. Wildl. Conf.* 38:55–68.

Miranda-Ribeiro, A. de. 1919. Os veados do Brasil,

segundo as coleções rondon e de vários museus nacionais e estrangeiros. *Rev. Mus. Paulista* 11: 1–99.

Ojeda, R. A., and M. A. Mares. 1982. Conservation of South American mammals: Argentina as a paradigm. In *Mammalian biology in South America,* ed. M. A. Mares and H. H. Genoways, 505–22. Pymatuning Symposia in Ecology 6. Special Publication Series. Pittsburgh: Pymatuning Laboratory of Ecology, University of Pittsburgh.

Olrog, C. C. 1979. Los mamíferos de la selva húmeda, Cerro Calilegua, Jujuy. *Acta Zool. Lilloana* 33: 9–14.

Olrog, C. C., and M. M. Lucero. 1981. *Guía de los mamíferos argentinos.* Tucumán, Argentina: Ministerio de Cultura y Educación Fundación Miguel Lillo.

Olrog, C. C., R. A. Ojeda, and R. M. Barquez. 1976. *Catagonus wagneri* (Rusconi) en el noroeste argentino (Mammalia, Tayassuidae). *Neotrópica* 22: 53–56.

Osgood, W. H. 1943. The mammals of Chile. *Field Mus. Nat. Hist., Zool. Ser.* 30:1–268.

Pearson, O. P. 1951. Mammals in the highlands of southern Peru. *Bull. Mus. Comp. Zool.* 106 (3): 117–74.

Pefaur, J., W. Hermosilla, F. Di Castri, R. González, and F. Salinas. 1968. Estudio preliminar de mamíferos silvestres chilenos: Su distribución, valor económico e importancia zoonótica. *Rev. Soc. Med. Vet.* (Chile) 18 (1–4): 3–15.

Pinder, L. 1992. Comportamento social e reproductivo clos venados campeiro e catinguero. *Anais Etol.* 10:167–73.

———. 1993. Estimativa populacional de cerro de porto *Themaga engenharia.* Manuscript, São Paulo.

———. 1997. Niche overlap among brown brocket deer, pampas deer, and cattle in the Pantanal of Brazil. Ph.D. diss., University of Florida, Gainesville.

Pinder, L., and A. P. Grosse. 1991. *Blastocerus dichotomus. Mammal. Species* 380:1–4.

Pinder, L., and F. Leeuwenberg. 1997. Veado-catinguero (*Mazama gouazxoubira* Fisher, 1814). In *Biologia e conservação de cervideos sul-americanos: Blastocerus, Ozotoceros e Mazama,* ed. J. M. B. Duarte, 60–68. São Paulo: FUNEP.

Pine, R. H., S. D. Miller, and M. L. Schamberger. 1979. Contributions to the mammalogy of Chile. *Mammalia* 43:339–76.

Povilitis, A. J. 1979. The Chilean huemul project (1975–1976): Huemul ecology and conservation. Ph.D. diss., Colorado State University, Fort Collins.

———. 1982. The huemul in Chile: National symbol in jeopardy? *Oryx* 17:34–40.

———. 1983. Social organization and mating strategy of the huemul (*Hippocamelus bisulcus*). *J. Mammal.* 64:156–58.

Prichard, H. 1902. Field-notes upon some of the larger mammals of Patagonia, made between September 1900 and June 1901. *Proc. Zool. Soc. London* 1902:272–77.

Raedeke, K. J. 1979. Population dynamics and socioecology of the guanaco (*Lama guanicoe*) of Magallanes, Chile. Ph.D. diss., University of Washington.

Rau, J. 1980. Movimiento, hábitat y velocidad del huemul del sur (*Hippocamelus bisulcus*) (Artiodactyla, Cervidae). *Not. Mens. Mus. Nac. Hist. Nat.* (Santiago) 24 (281–82): 7–9.

• ———. 1982. Situación de la bibliografía e información relativa a mamíferos chilenos. *Mus. Nac. Hist. Nat. Publ. Ocas.* 38:29–51.

Redford, K. H. 1987. The pampas deer (*Ozotoceros bezoarticus*) in central Brazil. In *Biology and management of the Cervidae,* ed. C. M. Wemmer, 410–16. Washington, D.C.: Smithsonian Institution Press.

Redford, K. H., and J. F. Eisenberg. 1992. *Mammals of the Neotropics.* Vol. 2. *The southern cone: Chile, Argentina, Uruguay, Paraguay.* Chicago: University of Chicago Press.

Redford, K. H., and J. G. Robinson. 1987. The game of choice. Patterns of Indian and colonist hunting in the Neotropics. *Amer. Anthropol.* 89:650–67.

Robinson, J. G., and J. F. Eisenberg. 1985. Group size and foraging habits of the collared peccary *Tayassu tajacu. J. Mammal.* 66:153–55.

Roe, N. A., and W. E. Rees. 1976. Preliminary observations of the taruca (*Hippocamelus antisensis:* Cervidae) in southern Peru. *J. Mammal.* 57:722–30.

Roots, C. G. 1966. Notes on the breeding of white-lipped peccaries *Tayassu albirostris* at Dudley Zoo. *Int. Zoo Yearb.* 6:198–99.

Schaller, G. B. 1983. Mammals and their biomass on a Brazilian ranch. *Arq. Zool. São Paulo* 31 (1): 1–36.

Schaller, G. B., and J. M. C. Vasconcelos. 1978. A marsh deer census in Brazil. *Oryx* 14 (4): 341–51.

Schneider, M., and L. F. Barbosa de Oliviera. 1996. Densidade populacional de *Mazama* sp. Estaciao Ecologica de Aracuri, Rio Grande do Sul. *Resumos do XXI Congresso Brasileiro de Zoologia* (Pôrto Alegre), 1210.

Schweinsburg, R. E. 1971. Home range, movements, and herd integrity of the collared peccary. *J. Wildl. Manage.* 35:455–60.

Smith, W. P. 1991. *Odocoileus virginianus. Mammal. Species* 388:1–13.

Smythe, N. D. E. 1970. Ecology and behavior of the

agouti (*Dasyprocta punctata*) and related species on Barro Colorado Island, Panama. Ph.D. diss., University of Maryland.

Sowls, L. K. 1984. *The peccaries.* Tucson: University of Arizona Press.

———. 1997. *Javelinas and other peccaries.* College Station: Texas A&M University Press.

Spotorno O., A. E., N. Brum, and M. Di Tomaso. 1987. Comparative cytogenetics of South American deer. In *Studies in Neotropical mammalogy: Essays in honor of Philip Hershkovitz,* ed. B. D. Patterson and R. M. Timm, 473–83. *Fieldiana: Zool.,* n.s., 39:1–506.

Spotorno O., A. E., and R. Fernández-Donoso. 1975. The chromosomes of the Chilean dwarf-deer "pudu" *Pudu pudu* (Molina). *Mammal. Chrom. Newsl.* 16 (1): 17.

Stallings, J. R. 1984. Notes on feeding habits of *Mazama gouazoubira* in the Chaco Boreal of Paraguay. *Biotropica* 16 (2): 155–57.

Stanley, H. F., M. K. Adwell, and J. C. Wheeler. 1994. Molecular evolution of the Camelidae: A mitochondrial DNA study. *Proc. Roy. Soc. London,* ser. B, 256:467–75.

Taber, A. B., P. Doncaster, N. N. Neris, and F. H. Colman. 1993. Ranging behavior and population dynamics of the Chacoan peccary, *Catagonus wagneri. J. Mammal.* 74:443–54.

———. 1994. Ranging behavior and activity patterns of two sympatric peccaries, *Catagonus wagneri* and *Tayassu tajacu,* in the Paraguayan Chaco. *Mammalia* 58 (1): 61–71.

• Tamayo, M., and D. Frassinetti. 1980. Catálogo de los mamíferos fósiles y vivientes de Chile. *Mus. Nac. Hist. Nat., Chile* 37:323–99.

Taylor, W. P. 1956. *The deer of North America.* Harrisburg, Pa.: Stackpole.

• Texera, W. A. 1974. Algunos aspectos de la biología del huemul (*Hippocamelus bisulcus*) (Mammalia: Ariodactyla, Cervidae) en cautividad. *Anal. Inst. Patagonia, Punta Arenas* 5 (1–2): 155–88.

Thomas, W. D. 1975. Observations on captive brockets *Mazama americana* and *M. gouazoubira. Int. Zoo Yearb.* 15:77–78.

Tomas, W. M. 1986a. Hábitos alimentares do cervo-do-pantanal *Blastocerus dichotomus* (Illiger, 1811) (Cervidae) no Pantanal matogrossense. *Resumos do XIII Congresso Brasileiro de Zoologia,* Poconé-Mt.

———. 1986b. Considerações sobre os ambientes frequentados pelo cervo-do-pantanal *Blastocerus dichotomus* (Illiger, 1811) (Cervidae) no Pantanal motogrossense. *Resumos do XIII Congresso Brasileiro de Zoologia,* Poconé-Mt.

———. 1986c. Padrão diáro de atividade e estrutura

de grupos do cervo-do-pantanal *Blastocerus dichotomus* (Illiger, 1811) (Cervidae) no Pantanal matogrossense. *Resumos do XIII Congresso Brasileiro de Zoologia,* Poconé-Mt.

———. 1988a. Observações preliminares sobre densidades e estrutura de grupos de veado campeiro no Pantanal da Nhecolândia. *Resumos do XV Congresso Brasileiro de Zoologia,* Corumbá.

———. 1988b. Nota sobre a troca das galhadas pelo cervo-do-pantanal (*Blastocerus dichotomus*) e pelo veado camp iro (*Ozotoceros bezoarticus*). *Resumos do XV Congresso Brasileiro de Zoologia,* Corumbá.

Tomas, W. M., M. D. Beccaceci, and L. Pinder. 1997. Cerro-do-pantanal (*Blastocerus dichotomus*). In *Biologia e conservação de cervideos sul-americanos: Blastocerus, Ozotoceros e Mazama,* ed. J. M. B. Duarte, 24–40. São Paulo: FUNEP.

• Tonni, E. P., and G. G. Politis. 1980. La distribución del guanaco (Mammalia, Camelidae) en la provincia de Buenos Aires durante el Pleistoceno tardío y Holoceno: Los factores climáticos como causas de su retracción. *Ameghiniana* 17 (1): 53–66.

Torres, H. 1985. Distribution and conservation of the guanaco (*Lama guanicoe*). International Union for Conservation of Nature and Natural Resources, Species Survival Commission, South American Camelid Specialist Group, Special Report 2:1–37. Gland, Switzerland: IUCN.

Vanoli, T. 1967. Beobachtungen an Pudus, *Mazama pudu* (Molina, 1782). *Säugetierk. Mitt.* 15:155–63.

Voss, W. A., F. R. dos Santos Breyer, G. C. Mattes, and H. G. Konrad. 1981. Constatação e observação de uma população residual de *Blastocerus dichotomus* (Illiger, 1811) (Mammalia, Cervidae). *Iheringia,* ser. zool. 59:25–36.

Wetterberg, G. B. 1972. *Pudu* in a Chilean national park. *Oryx* 11 (5): 347–51.

Wetzel, R. M. 1977a. The extinction of peccaries and a new case of survival. *Ann. New York Acad. Sci.* 288:538–44.

———. 1977b. The Chacoan peccary *Catagonus wagneri* (Rusconi). *Bull. Carnegie Mus. Nat, Hist.* 3:1–36.

Wetzel, R. M., R. E. Dubos, R. L. Martin, and P. Myers. 1975. *Catagonus,* an "extinct" peccary, alive in Paraguay. *Science* 189:379–81.

Wheeler, J. C., A. J. Russel, and H. F. Stanley. 1992. A measure of loss: Prehispanic llama and alpaca breeds. *Archos Zootecnia* 41:467–75.

Whitehead, G. K. 1972. *Deer of the world.* New York: Viking Press.

Wilson, P. 1984. Puma predation on guanacos in Torres del Paine National Park, Chile. *Mammalia* 48 (4): 515–22.

Wilson, P., and W. L. Franklin. 1985. Male group dynamics and intermale aggression of guanacos in southern Chile. *Z. Tierpsychol.* 69:305–28.

Wolffsohn, J. A. 1923. Medidas máximas y mínimas de algunos mamíferos chilenos colectados entre los años 1896 y 1917. *Rev. Chil. Hist. Nat.* 27:159–67.

• Ximénez, A., A. Langguth, and R. Praderi. 1972. Lista sistemática de los mamíferos del Uruguay. *Anal. Mus. Nac. Hist. Nat. Montevideo* 7 (5): 1–45.

Yost, J. A., and P. M. Kelley. 1983. Shotguns, blowguns and spears. In *Adaptive responses of native Amazonians,* ed. R. B. Hames and W. T. Vickers, 189–224. New York: Academic Press.

16   Order Rodentia (Rodents, Roedores)

Diagnosis and Comment

The dental formula for the Rodentia is distinctive: a single pair of upper and lower incisors; no canines; no more than two pairs of premolars per side; no more than three molars per side of the upper and lower jaws. There is a distinct gap or diastema between the incisors and the cheek teeth (fig. 16.1). The incisors are ever-growing and have enamel on the anterior surface. Premolar and molar patterns of cusps are diverse and are indicators of both phylogenetic relationship and adaptation for particular feeding strategies. Keys for the Sigmodontinae are offered in tables 16.1 and 16.2. Anderson (1997) offers a useful key for the rodents of Bolivia.

Distribution

Rodents are distributed over the entire world except Antarctica. They were originally absent from the Oceanic islands and New Zealand but have subsequently been introduced by man. In South America the rodents derive from two separate major colonizations. The hystricognath or caviomorph rodents are well represented from the late Eocene to the present. The sigmodontine rodents (Muridae, Sciuridae, and Geomyoidea) entered South America with a series of later invasions, possibly very late in the Miocene, but

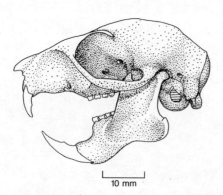

10 mm

Figure 16.1. Skull of *Sciurus aestuans*.

accelerating in the Pliocene. Hershkovitz (1994, 39), however, reiterates his belief that some of the sigmodontine stocks may well have rafted from Africa to South America much the way the caviomorph rodents and platyrrhine monkeys did. He feels firmly that the often repeated scenario of colonization of South America by North American forms may have to be revised in part. Of special concern are colonization events believed to have occurred from the Eocene to the Oligocene, such as those by primates and by caviomorph rodents.

Suborder Sciurognathi

SUPERFAMILY SCIUROIDEA

FAMILY SCIURIDAE

Squirrels and Marmots; Ardillas

Diagnosis

The dental formula is I 1/1, C 0/0, P 1-2/1, M 3/3 (fig. 16.1). The cheek teeth usually show prominent cusps and ridges arranged in a triangular pattern in the upper molars (fig. 16.3). The infraorbital foramen is minute, and no muscle passes through it. There are five digits on the hind feet and four on the forefeet, and each toe bears a long claw. The tail is always haired and may look bushy (plate 11).

Distribution

Species of this family were naturally distributed on all continents except Antarctica and Australia. They have been widely introduced.

Genus *Sciurus* Linnaeus, 1758

Tree Squirrel, Ardilla, Esquilo

Description

The dental formula is I 1/1, C 0/0, P 1-2/1, M 3/3. The dorsal color is highly variable; in the tropics, red, dark brown, or orange generally predominates on the dorsum, although to the north in the temperate zone

Table 16.1   Key to the Genera of Neotropical Sigmodontinae

When Ellerman prepared this key, he removed the Ichthyomyini based on a subjective assessment of an aquatic adaptation as related below:

External form highly modified for aquatic life. Skull much specialized toward aquatic life (with much flattened brain-case, enlarged infraorbital foramen, narrowed zygomatic plate, etc., its general appearance *Hydromys*-like).
  Nasal slanting upward anteriorly, the nasal opening much heightened.
    Lower incisors extremely compressed, but the upper pair less so.
    No ear conch . . . . *Anotomys*
  Nasal not slanting upward anteriorly, nasal opening not specially heightened. Functional ear conch present.
    Upper and lower incisors broad . . . . *Ichthyomys*
    Upper and lower incisors narrow . . . . *Rheomys*

Ellerman then removed the "mole mice" based on a rather subjective description of extreme fossorial adaptation as follows:

External form never extremely modified for aquatic life. Skull without pecularities mentioned above.
External form much modified for fossorial life (either by strong reduction of ears and eyes, or by great enlargement of foreclaws).
  Claws not lengthened. Infraorbital foramen unusually large . . . . *Blarinomys*
Claws greatly lengthened. Infraorbital moderate.
  Posterior portion of palate with deep pits each side, upper incisors grooved . . . . *Chelemyscus*
  Posterior portion of palate not abnormal. Upper incisors plain . . . . *Notiomys*

At the beginning of the dichotomous key the first three genera under 1 (2, 2′, 3, 3′) are also fossorial (but less so than the mole mice). From this point on the key becomes more refined:

External form never excessively modified for fossorial life.
1   Zygomatic plate much narrowed, usually slanting gradually backward from lower to upper border, lower border
     always considerably in front of upper border; infraorbital foramen usually large, well open (skull normally with
     little interorbital constriction, lengthened rostrum, heavy braincase.) 2
1′   Zygomatic plate not narrowed or less narrowed and tilted more strongly upward (in genera in which it is narrowed,
     it is always tilted upward) (skull usually without pecularities mentioned above.) 4
2   Foreclaws prominent and considerably lengthened. . . . . *Oxymycterus*
2′   Foreclaws normal. 3
3   Nasals extending posteriorly beyond front portion of orbit; interparietal large; interorbital constriction scarcely
     apparent . . . . *Lenoxus*
3′   Nasals not extending posteriorly beyond portion of orbit; interparietal small, much reduced; interorbital con-
     striction more marked . . . . *Microxus*
4   Upper cheek teeth specialized, their normal cuspidate pattern not apparent at any time and obliterated (in most
     genera included in this section, cutting teeth have been examined; *Neotomodon* may be an exception to the state-
     ment above); the pattern of the molars prismatic and flat crowned 5
4′   Upper cheek teeth not specialized or less so, their cuspidate pattern as a rule not obliterated, and apparent at
     least at some time of life; usually the molars are not flat crowned. In a few genera, such as *Phyllotis*, *Graomys*,
     Eligmodontia, the molars may have a prismatic appearance, but the molars are never with all folds compressed,
     deep and narrow (compare *Sigmodon*, *Holochilus*, etc.) and are less microtine in appearance and less angular than
     in such forms as *Neotoma*, *Andinomys*, and *Chinchillula*. If the crowns are flat, there are clear traces of subsidary
     ridges in the outer main folds of the upper series (which do not occur in those genera above), and the folds
     isolate as islands on crown surface 16
5   Laminae of the upper molars plain, straight, and equal sized, separated by inner and outer folds that are oppo-
     site each other, about equal in depth, and almost meeting in middle line of teeth . . . . *Irenomys*
5′   Laminae of the upper molars never as just described; inner and outer folds alternating, general effect less simple. 6
6   Third upper molar simplified, without inner folds. . . . . *Nelsonia*
6′   Third upper molar always with inner fold present. 7
7   Upper incisors clearly grooved 8
7′   Upper incisors plain. 11
8   Third upper molar enlarged, complex; groove of upper incisors placed at outer side of tooth; nasals abnormally
     expanded anteriorly (palate with deep pits in posterior portion) . . . . *Neotomys*
8′   Third upper molar not complex or enlarged; groove of upper incisors placed more centrally; nasals not abnor-
     mally expanded anteriorly 9
9   Zygoma robust; pits in posterior portion of palate shallow . . . . *Sigmomys*
9′   Zygoma slender; pits in posterior portion of palate deep 10
10   Zygomatic plate with strongly marked forwardly projecting process on upper border; M_2 not S-shaped . . . . *Reithrodon*
10′ Zygomatic plate without forward-projecting process on upper border; M_2 S-shaped. . . . . *Euneomys*
11   Folds of upper molars in moderately young adult animals are wide open; general dental pattern reminiscent of
     that of Microtinae 12
11′ Folds of upper molars in moderately young adult animals are not wide open, and dentine spaces are less sharply
     projecting, so that appearance of the molars is compressed and not reminiscent of Microtinae (folds typically
     deep but very narrow). 15

Continued on next page

Table 16.1   (*continued*)

12   Palate extending posteriorly to end of tooth rows or slighly behind that level 13

12′   Palate extending posteriorly only about to level of hind part of M_2 or front part of M_3 . . . . *Neotoma*

13   Zygomatic plate with interior border concave and sharply cut back above (folds of upper molars alternating; first
      lower molar exceptionally complex originally; compare *Chinchillula*) . . . . *Andinomys*

13′   Zygomatic plate with anterior border straight 14

14   Braincase broad; rostrum broad; two well-marked inner folds in M_1 (folds of upper molars nearly opposite;
      first lower molar not exceptionally complex; compare *Andinomys*) . . . . *Chinchillula*

14′   Braincase narrow; rostrum narrow; one well-marked inner fold in M_1. (The position of this genus must be
      accepted as provisional, since only one specimen with much-worn teeth has been examined) . . . . *Neotomodon*

15   M_3 is more complex (usually the pattern of upper molars is more angular, with anterior transverse loop of
      M_2 and M_3 straight and with closed triangles more in evidence) . . . . *Holochilus*

15′ M_3 less complex and smaller (usually but not always the pattern of upper molars is less angular, with anterior
      transverse loop of M_2 and M_3 curved to a certain degree, and closed triangles of molars less well marked) . . . . *Sigmodon*

16   Upper incisors clearly grooved (cheek teeth strongly cuspidate; compare *Phyllotis*) . . . . *Reithrodontomys*

16′   Upper incisors plain or scarcely grooved (only one subgenus of *Phyllotis* of those that follow may have weakly
      grooved incisors) 17

17   Cheek teeth complex, with clear subsidiary ridges normally present in main outer folds of first and second
      upper molars; teeth always cuspidate originally, and in most cases through life. (In the genus *Peromyscus*
      there is a tendency for the subsidiary ridges to be lost, and in one subgenus this is a fairly constant character,
      though it contains forms in which these ridges are traceable, and there are intermediate forms between these
      and typical *Peromyscus*. Members of this group without the subsidiary ridges will stand nearest to *Baiomys* in
      the key below, from which they differ in the nonreduced coronoid process and the nonshortened tail) 18

17′   Cheek teeth not excessively complex; subsidiary ridges in the main outer folds of upper molars in M_1 and M_2
      not traceable. 30

18   Dentition weak, molars rather narrow, folds usually not approaching each other and not well marked; sub-
      sidiary ridges present originally but may be lost in adult; molars tend to become more hypsodont. *Akodon*

18′   Dentition stronger, as a rule with relatively broader molars with well-marked folds; subsidiary ridges usually
      retained throughout life; molars generally brachyodont. Cusps of upper molars opposite or nearly so; antero-
      internal cusp not obliterated, except in *Nesomys*; normally M_3 not strongly reduced. 19

19   Upper incisors strongly proodont; braincase much enlarged but not ridged . . . . *Chilomys*

19′   Upper incisors not proodont; braincase when enlarged is strongly ridged. 20

20   Fur composed of bristles or spines . . . . *Neacomys*

20′   Fur not spiny. M_2 smaller than M_1 or more reduced in elements (anterointernal cusp in M_1 retained). 21

21   Anterointernal cusp of M_1 much reduced. 22

21′   Anterointernal cusp of M_1 not much reduced. 23

22   Feet more modified for arboreal life. Skull specialized, with widened frontals, and heavy braincase, as seen in
      specialized arboreal genera . . . . *Nyctomys*

22′   Feet not much modified for arboreal life. Skull generalized, with narrow frontals and moderate braincase . . . . *Phaenomys*

23   Palate not reaching posterior part of tooth rows and without lateral pits in posterior portion 24

23′   Palate reaching behind posterior part of tooth rows and, except in *Scapteromys*, with well-marked lateral pits
      in posterior portion. 27

24   Skull specialized, of arboreal type, with much widened frontals and large, heavy braincase. M_3 similar in
      elements to M_2. Subsidiary ridges of upper molars reduced (tail naked, feet arboreal) 25

24′   Skull not much modified. M_3 more reduced than M_2. Subsidiary ridges in upper molars not reduced or less
      so (tail usually moderately or well haired) 26

25   Pits between cusps of upper molars unusually well developed. Bullae enlarged . . . . *Ototylomys*

25′   Pits between cusps of upper molars not specially developed. Bullae small . . . . *Tylomys*

26   Feet modified for arboreal life. Folds of upper cheek teeth never specially widened . . . . *Rhipidomys*

26′   Feet usually not modified for arboreal life; but if so, folds of upper cheek teeth conspicuously widened . . . . *Thomasomys (Delomys)*

27   Cheek teeth tending to become more or less flat crowned early; outer folds of upper molars isolated, or
      practically so, as islands on crown surface early in life. 28

27′   Cheek teeth not tending to become flat crowned until late in life; cusps usually traceable throughout life; less
      tendency for isolating of outer folds as enamel islands 29

28   Outer folds of upper molars isolate as broad islands; general dental pattern simple; M_1 and M_2 with not more
      than two isolated islands each . . . . *Scapteromys*

28′   Outer folds of upper molars isolated as narrow islands; general dental pattern more complex; M_1 and M_2 with
      three or four isolated islands each. . . . . *Nectomys*

29   Braincase scarcely wider than rostrum (giant form) . . . . *Megalomys*

29′   Braincase clearly wider than rostrum . . . . *Oryzomys*
      (see supplemental key, table 16.2)

30   Some part of backward prolongation of outer folds in the upper molars definitely isolated as deep and
      conspicuous pits between main cusps of upper molars in moderately young animal. Inner and outer folds of
      upper molars weak; cheek teeth narrowed; tail clearly more than half head and body length (averaging about
      60% in those examined). . . . . *Scotinomys*

Table 16.1   (*continued*)

30′	No part of backward prolongation of outer folds of upper molars definitely isolated as conspicuous pits between main cusps.. 31
31	Cusps of upper molars unusually raised and heightened, cheek teeth narrowed (this feature usually apparent even in comparatively old specimens) (M_3 strongly reduced; tail considerably shortened; form thick-set; plantar pads reduced) .... *Onychomys*
31′	Cusps of upper molars not unusually raised up and heightened.. 32
32	Feet apparently considerably specialized for arboreal life (a little known form) .... *Rhagomys*
32′	Feet not specialized for arboreal life 33
33	In most cases upper molars have cusps more marked, and there is less tendency for teeth to take on a prismatic appearance; if molars become more or less a series of transverse plates, general dentition is much weaker, and less angular in appearance 34
33′	In most cases upper molars have cusps not well marked, and pattern is more or less prismatic in appearance or may tend to become a series of transverse plates....................................... 35
34	Folds of upper molars nearly straight; dentition lighter; no tendency for folds to cut cusp areas into partially closed triangles .... *Zygodontomys*
34′	Folds of upper molars considerably curved; dentition heavier; tendency present for folds to cut cusp areas into partially closed triangles .... *Hesperomys (Calomys)*
35	Palms and soles with pads situated on hairy outgrowths (Thomas) .... *Eligmodontia*
35′	Palms and soles usually naked, without abnormalities.. 36
36	Frontals relatively broad, little constricted, with interorbital region evenly divergent backward .... *Graomys*
36′	Frontals strongly narrowed, never with interorbital region evenly divergent backward .... *Phyllotis*

Source: Adapted from Ellerman 1940.

Table 16.2   Key to Some of the Oryzomyini

1	Cheek teeth tending to become flat crowned early; outer folds of upper molars isolated, or nearly so, as islands on crown surface early in life .... *Sigmodontomys*
1′	Cheek teeth not tending to become flat crowned early; cusps usually present throughout life....................... 2
2	Tail considerably less than head and body length; size small.......................... *Microryzomys*
2′	Tail equal to or longer than head and body 3
3	Tail slightly penicillate; zygomatic plate straight anteriorly........................... *Oecomys*
3′	Tail not penicillate; zygomatic plate not straight anteriorly... 4
4	Lachrymal articulating approximately equally with frontal and maxilla anteriorly; tail approximately equal to or longer than head and body 5
4′	Lachrymal articulating almost entirely with maxilla anteriorly; tail approximately three-quarters of head and body length.. *Melanomys*
5	M_2 with central enamel island normally elongate or absent; supraorbital and temporal ridges present; hind foot usually more than 25 mm .... *Oryzomys*
5′	M_2 with central enamel island usually circular; supraorbital and temporal ridges absent; hind foot usually less than 25 mm .... *Oligoryzomys*

Source: Adapted from Goldman 1918.

of North America many species are varying forms of gray and gray brown (plate 11).

Distribution

The genus has speciated widely, and the fifty-five named species occur in Europe and Asia and from Canada south to northern Argentina.

Natural History

These active diurnal, scansorial rodents are striking components of the forest fauna in South America. Although strongly adapted for arboreal life, they come to the ground to forage, and many species scatter-hoard fruits and nuts by burying them in discrete locations within the home range. Nests are constructed in hollow portions of trees, or a leaf nest may be built in the

fork of a branch. The gestation period varies with adult body size, but thirty-eight to forty-five days covers the range. Where there is a favorable food distribution, two litters may be produced in a single year. Litter size ranges from two to five. The young generally do not open their eyes until they are approximately thirty days old; thus there is a rather long developmental period.

Sciurus aestuans Linnaeus, 1776
Brazilian Squirrel, Ardilla Gris, Esquilo

Description

Average measurements: HB 202; T 180; HF 51; E 22; Wt 300 g. This squirrel is medium brown to agouti dark brown dorsally and lighter on the venter, ranging from gray to orange. The chin often has a white patch. The dorsal pelage is individually variable in color. The

Map 16.1. Distribution of *Sciurus aestuans*.

long tail is well furred and may have a prominent terminal tuft. The ends of some of the tail hairs are often orange, creating an impression of indistinct orange bands (see plate 11).

Distribution

Sciurus aestuans ranges from Venezuela to the Guianas and south to Argentina, where it is found in Misiones province. It may also be found on the Paraguayan side of the Río Paraná and across southern Brazil to the Atlantic coast (map 16.1).

Ecology

In addition to feeding on fruits and nuts (Galletti, Paschoal, and Pedroni 1992), this species preys on the eggs and young of birds (Husson 1978). Heavy feeding on palm fruits (91% of diet) of the genera *Astrocaryum*, *Polyandrococos*, and *Attalea* has been seasonally recorded in southern Brazil (Palma 1996). In southern Brazil this species is recorded as the occasional prey of *Cebus apella* (Galletti 1990).

Comment

Sciurus aestuans includes *S. ingrami* Thomas, 1901. *S. ingrami* described from southeastern Brazil is considered a full species by many workers.

Sciurus flammifer Thomas, 1904

Description

The mean measurements for ten Venezuelan specimens were TL 580; HB 274; T 310; HF 66.5 (Allen 1915). In the type specimen the head and ears were red and the upperparts were grizzled yellow and black, darker on the rump. The chin was yellow to orange, with the venter white. The basal third of the tail was black with the rest orange. This species is color poly-

morphic, and all-black specimens are not uncommon (Tate 1939).

Range and Habitat

The species occurs south of the Orinoco in Venezuela and is known from a few localities in the state of Bolívar, but it may extend to adjacent portions of Brazil (not mapped).

Sciurus gilvigularis Wagner, 1842

Description

Five Venezuelan specimens of this medium-sized squirrel averaged TL 342; HB 166; T 173; HF 43. Measurements of one specimen from Brazil were TL 363; T 195; HF 35; E 15. It is similar in coloration to *S. estuans* but paler; the dorsum is a grizzled, reddish buff, there is a buff eye-ring, and the venter is reddish orange. The tail may exhibit a faint banding. Cabrera (1960) considered this a subspecies of *S. aestuans*.

Range and Habitat

This species is broadly distributed in northern Brazil and penetrates southern Venezuela and parts of Guyana. A specimen referable to this species was described by Tate (1939) from Guyana (map 16.2).

Sciurus granatensis Humbolt, 1811
Neotropical Red Squirrel

Description

The dental formula is I 1/1, C 0/0, P 1/1, M 3/3. Total length ranges from 330 to 520 mm; T 140–280; HF 40–65; E 16–36; Wt 228–520 g (Nitikman 1985). The color of the dorsal pelage varies widely. The upperparts tend to be rusty brown to almost black, though they may show some mixing of yellow hair. The

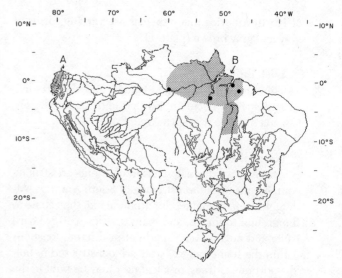

Map 16.2. Distribution of (a) *Sciurus granatensis*; (b) *Sciurus gilvigularis*.

venter tends to be light yellow to dark reddish brown, often with a white chest patch, but it usually contrasts with the dorsum. The range of size and color in northern South America has been analyzed by Hershkovitz (1947). There are thirty-three named subspecies (Nitikman 1985).

Range and Habitat
This species extends from western Costa Rica south through the Isthmus across northern Colombia, south to Ecuador, and east into northern Venezuela (map 16.2).

Natural History
We include a full discussion of this species, although it occurs marginally within the range considered in this volume. This is one of the best studied of the South American squirrels. In Panama it forages both terrestrially and arboreally, frequently eating portions of over twenty-one species of native plants. Most food items are large seeds and fruits, but young leaves, mushrooms, bark, and flowers are also taken. The fruit of the palm *Scheelea* is used heavily when it produces. The squirrels cache food both in trees and in the ground (Heaney 1983).

Litter size varies from one to three, with a mode of two. In seasonal habits such as Panama these squirrels begin breeding when the dry season starts in late December. Births take place in February and early March, and lactation probably continues for nearly two months. Home-range areas for adult females overlap only slightly and average about 0.64 ha. Home ranges of adult males overlap with each other and with those of adult females, averaging about 1.5 ha. In good habitat densities may be extremely high, up to four squirrels per hectare, but in habitats with lower productivity density may fall to 0.5 per hectare. At times of high fruit productivity squirrels may transgress on one another's home ranges and form temporary feeding aggregations (Heaney and Thorington 1978; Glanz et al. 1982; Nitikman 1985).

Sciurus ignitus (Gray, 1867)

Description
Measurements of two specimens from Peru were TL 330–70; T 150–60; HF 46–50; E 20–23 (LSU). This medium-sized squirrel has an agouti gray brown dorsum and a pale orange venter. The tails of some specimens have orange-tipped hairs, but such shading is often organized into "bands." The ear is sparsely haired (see Redford and Eisenberg 1992).

Distribution
This species occurs from lowland to premontane forests in Peru and Bolivia. It has been taken as high

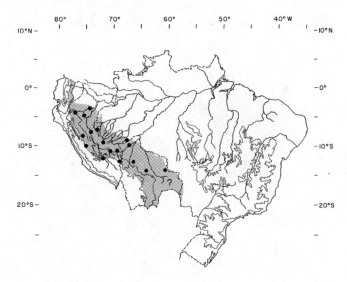

Map 16.3. Distribution of *Sciurus ignitus.*

as 2,000 m elevation. It may extend its range to northwestern Argentina (map 16.3).

Sciurus igniventris Wagner, 1842

Description
Four specimens from Peru exhibited the following range in measurements: TL 460–560; T 240–90; HF 54–78; E 25–30. Weights ranged from 480 to 665 g (LSU). This large squirrel has a medium to dark brown dorsum. The venter is orange, but in some specimens the venter may have white in the midline extending to the groin and axillae. The proximal portion of the tail is brown, but the distal half may have an orange cast dorsally; if not completely orange, the lateral fringes of the tail usually have an orange cast (see Redford and Eisenberg 1992). The species is distinguishable from *S. spadiceus* by the absence of orange patches behind the ears and the lack of orange on the top of the feet (Patton 1984).

Distribution
This species tolerates a wide altitudinal range and has been collected at over 1,800 m in Peru. Typically it occurs in tropical rain forests of Peru, Ecuador, western Brazil, and the southern portions of Colombia and Venezuela (map 16.4).

Sciurus pyrrhinus Thomas, 1898

Description
Two males from Peru measured TL 421–90; T 240; HF 68; E 23–33 (LSU). The dorsum is agouti orange brown, usually darker on the head. The tail is dark brown at the base with an orange fringe on the distal two-thirds. The venter is white. This squirrel looks very similar to *S. spadiceus* but is smaller (plate 11).

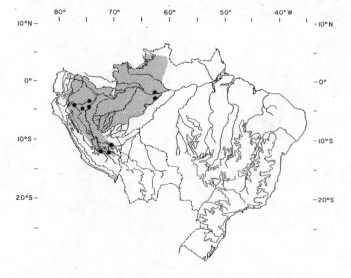

Map 16.4. Distribution of *Sciurus igniventris*.

Map 16.6. Distribution of *Sciurus sanborni*.

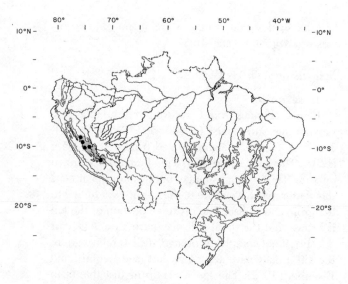

Map 16.5. Distribution of *Sciurus pyrrhinus*.

Distribution

This species is apparently confined to the foothills of the eastern Andean cordillera of Peru (map 16.5).

Sciurus sanborni Osgood, 1944

Description

Measurements: HB 155–75; T 160–80; E 20. This small squirrel is allied to the *S. granatensis* and *S. aestuans* group. The dorsum is olive brown with a pale patch behind each ear, and the venter is orange. The tail may exhibit a faint banding. A faint pale eye-ring is noticeable.

Distribution

The species is found in the Madre de Dios department of lowland Peru (map 16.6).

Sciurus spadiceus Olfers, 1818
Ardilla Roja

Description

Measurements of two males were TL 503–22; T 266–67; HF 67–71; E 33–34; Wt 570–660 g (MVZ). Five specimens from Peru measured TL 525–628; T 250–340; HF 65–70; E 32–34 (LSU; see also Patton 1984). This squirrel is easily separated from *S. aestuans* by its much larger size, but it resembles *S. igniventris* (see below). Its color pattern is variable, but typically it has a reddish brown dorsum that is darker in the midline, a white venter, and a tail that is dark brown at the base with varying amounts of orange on the distal two-thirds. The orange on the tail is especially prominent laterally (plate 11). In the sample viewed at LSU two specimens were melanistic; thus a black "morph" occurs in this species.

Distribution

The species apparently occurs in the lowland forests of southern Colombia and in Venezuela, then extends to Peru, western Brazil, Ecuador, and Bolivia (map 16.7).

Comment

Sciurus duida is a composite and invalid as a species (Lawrence 1988).

Sciurus stramineus Eydoux and Souleyet, 1841

Description

Four specimens from Peru exhibited the following measurements: TL 530–90; T 292–300; HF 62–63; E 32–39. Weight of one adult was noted at 470 g (LSU). This species has an extremely variable color

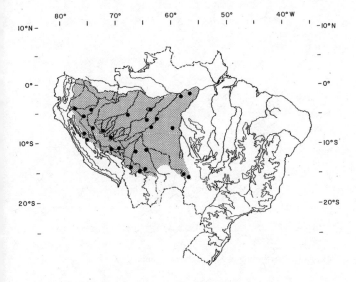

Map 16.7. Distribution of *Sciurus spadiceus*.

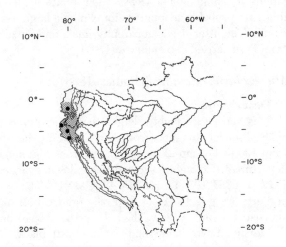

Map 16.8. Distribution of *Sciurus stramineus*.

pattern. The common "morph" has dorsal hairs with black to orange bases and white tips, yielding an agouti gray head and dorsum grading to orange brown on the rump. There is a buff collar on the back of the neck. The tips of the hairs on the tail are white, but the dark basal portion of the hairs is prominent. The venter ranges from dirty white to gray (plate 11). Variations in color pattern include the following: the collar may be almost white; the reddish cast to the rump may be suppressed; and finally, I (JFE) examined a "black" morph where the venter and dorsum were black, contrasting with a white dorsal collar and a few white spots on the anterior dorsum.

Distribution

This squirrel apparently ranges to 300 m elevation on the western face of the Andes in Peru and Ecuador (map 16.8).

Figure 16.2. Skull of *Sciurillus pusillus* (after Ellerman 1940). CB-= 24 mm.

Genus *Sciurillus* Thomas, 1914
Sciurillus pusillus (Desmarest, 1817)
Pygmy Squirrel

Description

Three specimens from Peru measured TL 220–40; T 120–35; HF 27–29; E 10–12; Wt 38–48 g (LSU). This is the smallest squirrel in the Western Hemisphere (fig. 16.2). The dorsum is gray, often grizzled because the hair tips tend to be yellow or gray in contrast to the darker base of the hair. The venter does not contrast with the dorsum.

Distribution

This species seems to occur in the lowland rain forest of the southern Guianas, western Brazil, and Peru (map 16.9).

Natural History

As a denizen of the rain forest canopy, this diminutive squirrel is not often observed. Thorington and Thorington (1989) have compared its limb proportions with those of *Microsciurus* and conclude that the elongate forelimbs are an adaptation to permit a small mammal to climb large tree trunks. Litter size ranges from two to three (see also Eisenberg 1989).

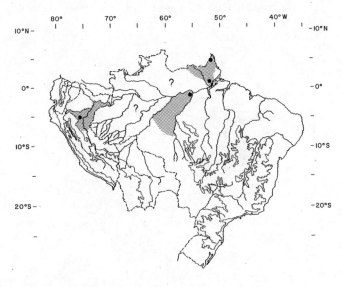

Map 16.9. Distribution of *Sciurillus pusillus*.

Genus *Microsciurus* J. A. Allen, 1895
Neotropical Dwarf Squirrel, Ardilla Menor

Description
The dental formula is I 1/1, C 0/0, P 2/1, M 3/3. Some seventeen species have been described in the literature. The head and body length averages 120–60 mm, larger than *Sciurillus*. The fur varies with geographic location. The ears are well haired, contrasting with the sparsely haired ears of *Sciurus pucheranii*, which is similar in size to *Microsciurus*.

Distribution
Species of this genus occur from southern Nicaragua south through the Isthmus to the Amazonian portion of South America. Many are associated with elevations above 1,800 m.

Microsciurus flaviventer (Gray, 1867)

Description
Five adults from Peru exhibited the following measurements: TL 275–330; T 155–80; HF 39–43; E 16–17; WT 99–132 g (LSU). This small squirrel has an agouti brown dorsum (quite dark in some specimens). The venter is yellow to light orange. The ears are haired, and the tail is faintly banded with orange; the rings are incomplete in some specimens (plate 11).

Distribution
This species seems to occur in the lowlands and in the foothills of the Andes, ranging to 2,000 m in eastern Colombia, Ecuador, western Brazil, and Amazonian Peru (map 16.10).

Natural History
Based on a morphometric analysis, Thorington and Thorington (1989) suggest that the small squirrels

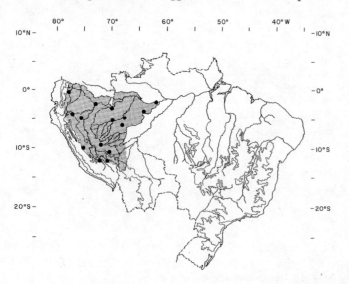

Map 16.10. Distribution of *Microsciurus flaviventer*.

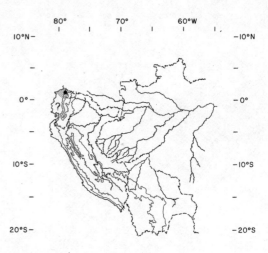

Map 16.11. Distribution of *Microsciurus mimulus*.

offer many challenges concerning their varied adaptations to climbing tall trees. Their relatively long limbs are probably adaptations for climbing large tree trunks. The litter size is apparently small: one female from Peru carried two embryos (LSU).

Microsciurus mimulus (Thomas, 1898)

Description
This species is slightly larger than *Microsciurus alfari*. Total length ranges from 235 to 268 mm, the tail from 100 to 116 mm. Five specimens from Antioquia department, Colombia, showed the following range: TL 250–63; T 105–14; HF 36–42 (CMNH). The upperparts are grizzled black and pale orange buff or, alternatively, buffy yellow. The tail is grizzled black on the dorsum, tawny below, and bordered with black.

Range and Habitat
Microsciurus mimulus generally occurs at higher elevations than *M. alfari* where their ranges overlap. It is found in Panama in montane regions, south through Colombia to northern Ecuador. The limits of its distribution in Colombia are poorly understood. Apparently this small species is replaced by *Sciurus pucheranii* in the high montane regions of Colombia (map 16.11).

SUPERFAMILY GEOMYOIDEA

FAMILY HETEROMYIDAE

Diagnosis
The dental formula is I 1/1, C 0/0, P 1/1, M 3/3. In common with the pocket gophers, this family has externally opening fur-lined cheek pouches. A wide variety of morphological forms are displayed, ranging from the bipedal kangaroo rats to the quadrupedal spiny pocket mice. In no case are members of this family as specialized for a fossorial life as are the pocket

gophers. The evolutionary history of the family was covered in monographs by Wood (1935) and Wahlert (1993).

Distribution

Members of this family are distributed from the plains of Saskatchewan, Canada, to northern South America.

Natural History

There is an enduring trend toward an adaptation for arid habitats within this family that culminates in specialized bipedal forms such as the kangaroo rats (*Dipodomys*) and the kangaroo mice (*Microdipodops*) (see Genoways and Brown 1993). Many of the pocket mice belonging to the genera *Perognathus* and *Chaetodipus* are similarly adapted for xeric habitats but are not bipedal. Two genera, *Liomys* and *Heteromys*, have adapted to moister subtropical to tropical habitats. They are referred to as spiny pocket mice or forest spiny pocket mice. They construct burrows but forage on the surface, gathering seeds, nuts, and fruits in their capacious cheek pouches and transporting them back to the burrow system for storage. *Liomys* may be one of the most important seed predators in the tropical dry, deciduous forests of Central America (Janzen 1982). The behavior of the family was reviewed by Eisenberg (1963) and Jones (1993).

Genus *Heteromys* Desmarest, 1817
Forest Spiny Pocket Mouse

Description

Head and body length ranges from 125 to 160 mm and the tail from 130 to 200 mm. The tail tends to be longer than the head and body; in this *Heteromys* differs from *Liomys*. The dorsal pelage is generally dark brown to almost black, contrasting sharply with the white venter. In some species flattened hairs give the fur a spiny texture.

Distribution

The animals occur from southern Veracruz, Mexico, south to northern South America at low elevations.

Natural History

Species of *Heteromys* are nocturnal. They construct elaborate burrow systems but forage on the surface, gathering seeds, fruits, and nuts in their capacious cheek pouches and carrying them to the burrow system for storage (Vandermeer 1979). Litter size averages about four, and a female may produce several litters each year depending on local availability of food, which is ultimately controlled by rainfall. The biology of *Heteromys* is reviewed by Sanchez-Cordero and Fleming (1993).

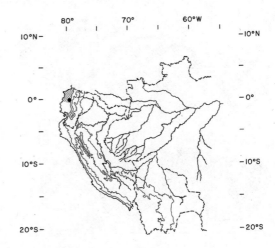

Map 16.12. Distribution of *Heteromys australis*.

Heteromys australis Thomas, 1901

Description

Total length ranges from 240 to 260 mm, tail length from 120 to 133 mm, and the hind foot from 32 to 33.5 mm. The dorsum is grizzled slate black, the underparts are usually white, and the tail is brown above and light below.

Range and Habitat

This species is distributed primarily to the west of the central Andean cordillera in Colombia. It extends from Ecuador through western Colombia to the extreme westernmost part of Panama (map 16.12). Throughout the rest of Panama it is replaced by *Heteromys desmarestianus*.

SUPERFAMILY MUROIDEA
FAMILY MURIDAE
SUBFAMILY SIGMODONTINAE
New World Rats and Mice

Diagnosis

The dental formula I 1/1, C 0/0, P 0/0, M 3/3 distinguishes this group from the Sciuridae and Geomyoidea (see figs. 16.3 and 16.4). *Neusticomys oyapocki* is exceptional with two upper and lower molars per side. The cheek teeth are variable in cusp pattern and may be laminate, prismatic, or cuspidate. When cusps are present they are arranged in two longitudinal rows in both upper and lower molars (see fig. 16.6). The infraorbital foramen is rather small; generally a strip of masseter muscle passes through it to attach to the rostrum. The bulk of the masseter attaches on the zygomatic arch and the rostrum. There is no postorbital process on the frontal bone.

We follow Carleton and Musser (1984) and Musser and Carleton (1993) in their classification and termi-

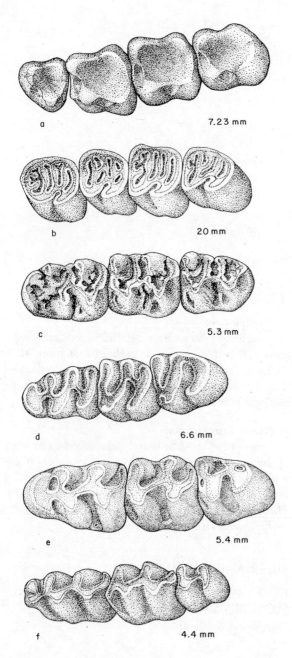

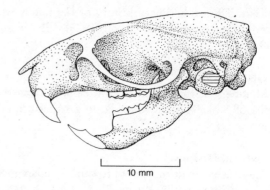

Figure 16.4. Skull of *Oryzomys* sp.

Figure 16.3. Some molar teeth of rodents; figures in millimeters indicate actual length of tooth row: (a) *Sciurus aestuans;* (b)-*Dasyprocta;* (c) *Nyctomys sumichrasti;* (d) *Sigmodon hispidus;* (e) *Zygodontomys brevicauda;* (f) *Akodon* sp. Dorsal tooth row; anterior, left; posterior, right.

nology. A key is given in tables 16.1 and 16.2. Data on reproduction are given in table 16.3.

Distribution

Sigmodontine rodents are found from northern Canada throughout North America to Patagonia in South America. The Sigmodontinae is a New World subfamily.

History and Classification

The sigmodontine rodents are defined in this volume as a subfamily of the family Muridae (Carleton and Musser 1984; Reig 1980). They are frequently referred to as the New World cricetines. For the purpose of this account we consider the gerbils to be within their own subfamily, Gerbillinae, and the voles to be in their own subfamily, Arvicolinae. The New World Sigmodontinae have as their nearest relatives in the Old World the hamsters (Cricetinae) of Europe and Asia and the pouched rats (Cricetomyinae) of Africa. Cricetine rodents appear first in the Oligocene of Europe and by the mid-Oligocene are represented in North America. Rodents that can be referred to the subfamily Sigmodontinae are identifiable in the Miocene of North America.

The classification of these rodents above the generic level into tribes has been plagued with difficulty. It seems fair to say that they were originally adapted to forested environments, but with the increasing trends of drying during the Miocene and the creation of extensive grasslands, the original stock began to adapt to more open grassland and xeric habitats (Hershkovitz 1962). How often this happened during the early divergence of the ancestral stock is unknown.

If we consider only the terrestrial mammals of South America, then rodents constitute 42% of the species described to date. Approximately half of these rodent species may be assigned to the subfamily Hesperomyinae of the Cricetidae or to the subfamily Sigmodontinae of the Muridae. Reig (1986) has proposed that the South American cricetids be included in the subfamily Sigmodontinae and be subdivided into seven tribes: Oryzomyini, Ichthyomyini, Akodontini, Scapteromyini, Wiedomyini, Phyllotini, and Sigmodontini. This grouping does not include the Central American sigmodontines that enter parts of northern South America from Panama. In addition, four South American genera cannot be placed with certainty in any of the seven tribes. These genera are

Table 16.3 Reproduction Data for Some South American Rodents

Taxon	Litter Size	Age at First Reproduction	Gestation Period
Sciuridae			
Sciurus granatensis	mode 2, max 3	—	—
Geomyidae			
Orthogeomys hispidus	2	—	—
Heteromyidae			
Heteromys australis	2.83	—	—
Muridae (Sigmodontinae)			
Akodon boliviensis	5	—	—
Akodon azare	5.0 (3–7)	♀ 75 d; ♂ 83.7 d	24.5 d
Akodon longipilis	3.7 (2–5)	—	—
Akodon molinae	3.65 (4–5)	—	—
Akodon olivaceus	5.08 (4–8)	—	—
Akodon urichi	5.0 (4–6)	2.7 mo ♀ and ♂	—
Andinomys edax	3	—	—
Anotomys leander	1	—	—
Anotomys trichotis	1	—	—
Auliscomys bolivianus	3–4	—	—
Auliscomys micropus	4.07	—	—
Blarinomys breviceps	1.25	—	—
Bolomys lasiurus	3.62 (1–6) mode 4	♀ 40.9 d	21 d~23 d
Bolomys obscurus	6.2	—	—
Calomys callosus	4.3	♀ 50 d	21.8 d
Calomys laucha	5.7 (4–8)	—	—
Calomys lepidus	4.4	—	—
Calomys musculinus	5.4 (1–15); 7.8 (embryos)	♀ 72.5 d	20.96 d
Eligmodontia typus	5.9 (3–9) embryos	♀ 42–64 d	18 d
Graomys griseoflavus	6.0 (3–8)	—	—
Holochilus brasiliensis	1–5	♀ 2–4 mo	28.4 d
Holochilus sciureus	3.12 (1–8)	—	—
Lundomys magnus	3 (placental scars)	—	—
Ichthyomys tweedi	2 embryos	—	—
Irenomys tarsalis	3.5	—	—
Neacomys tenuipes	3.8 (2–5)	♀ 30 d	—
Nectomys squamipes	2.5 (2–3) (4–5)	—	—
Neusticomys monticolus	2	—	—
Oecomys bicolor	3.5	—	—
Oecomys concolor	3.60 (2–5)	♀ 90 d	—
Oligoryzomys delicatus	3.0	—	—
Oligoryzomys flavescens	5.2	—	—
Oligoryzomys fulvescens	4.0 (2–6)	—	—
Oligoryzomys longicaudatus	4.9 (3–9)	—	—
Oligoryzomys nigripes	3.0 (1–5)	♀ 57.6 d	—
Oryzomys albigularis	3–4 embryos	—	—
Oryzomus alfaroi	4.7 (4–6)	—	—
Oryzomys bombycinus	3.33	—	—
Oryzomys caliginosus	1.5	—	—
Oryzomys capito	3.68	♀ 50 d	27.6 d
Oryzomys delticola	3.4	—	—
Oryzomys melanotis	4.6 (3–6)	—	—
Oryzomys subflavus	4.1 (3–6)	—	—
Oxymycterus rutilans	3.1	♀ 90 d	—
Phyllotis darwini	3.9 (2–7)	♀ 60 d	33.6 d? (from pairing)
Phyllotis osilae	4.38 (2–6)	—	—
Punomys lemminus	2	—	—
Reithrodontomys mexicanus	3	—	—
Rhipidomys mastacalis	3.8 (3–5)	♀ 90 d	—
Sigmodon alstoni	5.0 (2–6)	♀ 60 d	—
Wiedomys pyrrhonotus	3.8	—	—
Zygodontomys brevicauda	4.6	♀ 50 d	25.1 d

Zygodontomys, Rhagomys, Punomys, and *Abrawayaomys* (Reig 1986).

Reig (1986) summarizes his previous work and that of others concerning the origin of the tribes of sigmodontine rodents. He concludes that some sigmodontines crossed into South America as early as the Miocene. The four largest tribes in South America are the Oryzomyini, Akodontini, Phyllotini, and the Sigmodontini. Reig believes these taxa originated in South America and radiated into many genera before the Pleistocene. Some genera of the Sigmodontini and the Oryzomyini subsequently reinvaded Central America and North America via the Panamanian land bridge.

Earlier workers considered the neotomine-peromyscine group of the Hesperomyinae or Sigmodontinae to be North American in origin and a clearly defined taxonomic unit. The remaining sigmodontines were thought to have evolved primarily in the Isthmus and the southern parts of the United States (Hooper and Musser 1964). The synonymy of the fossil *Bensonomys* of North America with the extant South American *Calomys* suggests that the major stocks of the South American sigmodontines (cricetines) may have differentiated in North America before a Pliocene transfer to South America (Baskin 1978). However, this conclusion has been challenged.

In an effort to better understand the relationships among the sigmodontine rodents, numerous characters have been evaluated. Early workers such as Cabrera (1960), Moojen (1952a), Hershkovitz (1962, 1966), and Pearson (1958) undertook important descriptions and attempts at classifications. Penile morphology has had important implications (Hooper and Musser 1964; Spotorno 1992). Teeth have proved useful for delineating some groups but have provided equivocal evidence of relationship for others. Studies have been done comparing blood proteins, and extensive work has been done on chromosomal evolution, especially by Reig, Olivo, and Kiblisky (1971), Pearson and Patton (1976), and Gardner and Patton (1976).

Vorontsov (1979) provides a useful overview of the evolutionary trends by examining the digestive system for both New World and Old World murids. Some of his conclusions are reviewed in a contribution by Carleton (1973) on the stomach morphology of New World sigmodonts. Gyldenstolpe (1932) provides an exceedingly useful compilation of the descriptions and measurements for all type specimens of sigmondontine rodents described before 1931. Recent molecular research offers promise in resolving the problems of rodent phylogeny. Brownell (1983) and Catzeflis et al. (1993) have pointed in fruitful directions and offered resolution of some points concerning divergence times and the structure of monophyletic groups. The evolu-

tionary divergence creating distinct clades within the Oryzomini was very rapid and may have occurred five to six million years ago.

TRIBE PEROMYSCINI
White-footed Mice and Allies

Systematics
This assemblage includes *Peromyscus* and allied genera as well as *Reithrodontomys, Onychomys,* and *Ochrotomys.* These rodents are highly diversified and distributed primarily in North America. One species from one genus occurs within the range covered by this volume.

Genus *Reithrodontomys* Giglioli, 1874
Harvest Mouse, Grooved-toothed Mouse

Description
The dental formula is I 1/1, C 0/0, P 0/0, M 3/3. The upper incisors bear a longitudinal groove, which distinguishes these mice from all similar-sized forms within the range covered here. Head and body length ranges from 50 to 145 mm, and the tail from 65 to 95 mm. In most species the pelage is brown above and white to cream below. The genus has been reviewed by Hershkovitz (1941) and Hooper (1952).

Distribution
The genus is distributed from southwestern Canada through Mexico to Ecuador west of the central cordillera of the Andes.

Natural History
Very little is known about the natural history of the southern species of the genus *Reithrodontomys.* The northern species have been well studied in North America. These are omnivorous rodents preferring edge habitats. All species can climb well. Gestation in the northern species is twenty-three to twenty-four days; litter size averages about four. The young open their eyes at approximately eight days of age, and the female nurses them for approximately eighteen days, then they disperse. It is assumed that the southern species exhibit similar life history traits, but the details need much work (see Jones and Genoways 1970).

Reithrodontomys mexicanus (Saussure, 1860)

Description
Head and body length ranges from 68 to 77 mm, tail 92 to 126 mm, hind foot 17 to 21 mm, ear 14.5 to 18 mm. The dorsum is tawny to orange cinnamon, and the venter is variable, ranging from white to cinnamon.

Range and Habitat

This species is distributed from Tamaulipas in eastern Mexico and Oaxaca in western Mexico south through the Isthmus, through western Colombia to Ecuador. It occurs at moderate elevations up to 2,000 m (not mapped).

Natural History

In Central America this species breeds from June to August. Embryo counts range from three to five (Jones and Genoways 1970).

TRIBE TYLOMYINI
Central American Climbing Rats

Genus *Tylomys* Peters, 1866

Description

The tylomyine rodents include two genera, *Tylomys* and *Ototylomys*. *Ototylomys* includes four species distributed from southern Mexico to Costa Rica and does not occur within the range covered by this volume. *Tylomys* includes seven species. The dental formula is I 1/1, C 0/0, P 0/0, M 3/3. Head and body length ranges from 170 to 240 mm; the almost naked tail is longer than the head and body, ranging from 200 to 250 mm. The ears are large and naked.

Distribution

Tylomys is primarily Central American in its distribution, ranging into northwestern South America.

Tylomys mirae Thomas, 1899

Description

This species possibly includes *Tylomys fulviventer*. Total length ranges from 330 to 400 mm: HB 160–200; T 170–210; HF 30–40. The vibrissae are stout and long. The dorsum is gray brown, the venter buff to gray. There is generally a russet median stripe on the dorsum. The tail is black, with the terminal portion flesh colored; the ventral side of the tail tends to be gray.

Range and Habitat

The species ranges from central Colombia south to northern Ecuador. If it is considered conspecific with *T. fulviventer*, then it extends to extreme eastern Panama (not mapped).

Natural History

These nocturnal, arboreally adapted rodents appear to be primarily frugivorous. In captivity females of *T. nudicaudus* gave birth after a gestation period of thirty-two to thirty-three days to litters ranging from one to three (mean 1.6) (Baker and Petersen 1965; Helm 1973, 1975).

TRIBE THOMASOMYINI
South American Climbing Rats

Description

The molars are complex (pentalophodont) (figs. 16.5 and 16.6), and M_3 is large (1/2 to 3/4 the length of M_2). The tail is longer than the head and body, and these rodents have a relatively long hind foot and moderate to long ears. They possess a complex baculum. They may be confused with oryzomyine rodents over much of their range but are usually distinguishable by their size and relatively long tail (see species accounts and Voss 1993 for an interim assessment of the tribe). Although we have included *Delomys* and *Wilfredomys* within the thomasomyine rodents, these eastern genera may bear no special relation to the Andean forms and to *Rhipidomys*, especially if the Thomasomyini prove to be polyphyletic (see also Voss 1993).

Distribution

Predominately a South American group, the Thomasomyini are replaced ecologically in the north by Central American climbing rats of the genus *Tylomys*.

Genus *Aepeomys* Thomas, 1898

Description

The dental formula is I 1/1, C 0/0, P 0/0, M 3/3. These small mice have a head and body length of approximately 113 to 126 mm. The tail is longer than the head and body, averaging approximately 118 to 127 mm. The dorsum is gray brown to brown, and the venter is paler. The genus is considered distinct from *Thomasomys* by Gardner and Patton (1976).

Distribution

The genus is distributed in the Andean portions of Venezuela, Colombia, and Ecuador.

Aepeomys lugens (Thomas, 1896)

Description

Total length ranges from 232 to 263 mm, tail length from 115 to 127 mm, and hind foot from 27 to 28 mm. Weight ranges from 32 to 42 g, and males are about 6% heavier than females. The dorsum is gray, grading to a lighter shade on the venter.

Range and Habitat

The species is distributed in the eastern cordillera of the Andes from western Venezuela south to Andean

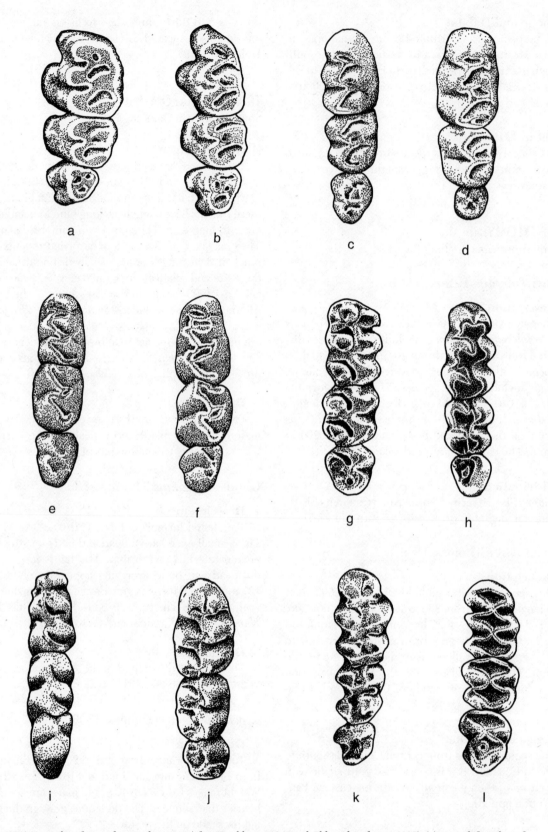

Figure 16.5. Examples of sigmodontine dentitions (after Hershkovitz 1944 and Olds and Anderson 1989): (a) upper left molars of *Sigmodontomys alfari* (worn); (b) upper left molars of *Nectomys* (worn); (c) upper left molars of *Nectomys* (unworn); (d) upper left molars of *Sigmodontomys alfari* (unworn); (e) lower left molars of *Nectomys squamipes* (unworn); (f) lower left molars of *Sigmodontomys alfari* (unworn); (g) upper tooth row of *Thomasomys aureus*; (h) upper tooth row of *Oxymycterus paramensis*; (i) upper tooth row of *Akodon toba*; (j) upper tooth row of *Calomys sorellus*; (k) upper tooth row of *Punomys lemminus*; (l) upper tooth row of *Irenomys tarsalis*.

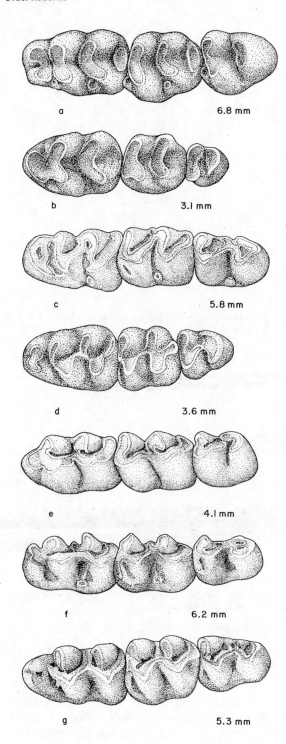

Figure 16.6. Some molar teeth of rodents; figures in millimeters indicate actual length of tooth row: (a) *Rattus rattus;* (b) *Mus musculus;* (c) *Oryzomys capito;* (d) *Calomys laucha;* (e) *Scotinomys xerampelinus;* (f) *Isthmomys (Peromyscus) flavidus;* (g) *Rhipidomys leucodactylus.* Dorsal tooth row: anterior, left, posterior, right.

a   6.8 mm

b   3.1 mm

c   5.8 mm

d   3.6 mm

e   4.1 mm

f   6.2 mm

g   5.3 mm

Ecuador (not mapped). It is strongly associated with moist habitats and cloud forest. In western Venezuela it was taken at between 1,900 and 3,560 m elevation.

Natural History

Aepeomys lugens breeds in the wet season and is almost entirely terrestrial (Aagaard 1982).

Genus *Chilomys* Thomas, 1897
Chilomys instans (Thomas, 1895)
Colombian Forest Mouse

Description

The dental formula is I 1/1, C 0/0, P 0/0, M 3/3. Average measurements are TL 219; T 123; HF 24; E 16. Head and body length ranges from 86 to 99 mm, and the tail from 105 to 130 mm. Weight ranges from 15 to 18 g. The dorsum is gray brown to very dark brown, and the venter usually matches it. There may be a buffy pectoral spot on the venter, and some specimens have a white stripe from the throat to the middle of the stomach (Osgood 1912). The tip of the tail is often white.

Range and Habitat

The species is distributed in the eastern cordillera of the Andes of Venezuela and Colombia. It may extend to Ecuador. It is strongly associated with cloud forest, and in Venezuela it occurs at between 2,405 and 2,700 m elevation (not mapped).

Natural History

Very little is known about this mouse other than that it prefers moist cloud forest and co-occurs with the larger *Thomasomys hylophilus* (Aagaard 1982), which it resembles.

Genus *Delomys* Thomas, 1917

Description and Taxonomy

Voss (1993) has recently reviewed the genus. Although *Delomys* is often listed under *Thomasomys* Coues, 1884, he points out that it exhibits the pentalophodont molar pattern characteristic of the more conservative genera of the Sigmodontinae. But since this is a primitive character, it does not necessarily convincingly imply a relationship with *Thomasomys.* The two recent species share a unique combination of morphological characters that distinguish them at the generic level. The complex of characters includes the incipient lophodont condition of the molars and a notch in the zygomatic arch, both of which are found in *Delomys* but not in *Oryzomys palustris* and *Thomasomys cinereus* (see Voss 1993, table 3). Zanchin et al. (1992) have determined that the karyotypes of the two species are distinct. In *D. sublineatus* $2n = 72$, whereas in *D. dorsalis* $2n = 82$.

Delomys dorsalis (Henzel, 1872)

Description

Measurements: HB 125; T 129; HF 32; E 20 (Rio). The ear is short (see plate 12). The dorsum is a soft gray brown, and the venter is a dirty white. The pelage is dense, while the tail is sparsely haired and faintly bicolored.

Distribution

The species occurs in the Atlantic coastal region of Brazil (map 16.13).

Delomys sublineatus (Thomas, 1903)

Description

Measurements: HB 140; T 110; HF 30; E 22. A second specimen measured HB 120; T 130; E 20 (Rio). I (JFE) examined specimens from Rio Terezopolis (Faz Boa Fé, Brazil). These specimens differ from those described under *D. dorsalis;* the ear is relatively larger. The pelage is soft and medium brown dorsally; the venter is gray; the tail is faintly bicolor; the median dorsal area is dark, suggesting a stripe. The mystacial vibrissae are moderately long.

Distribution

The species occurs in the Atlantic coastal region of southeastern Brazil (map 16.14).

Genus *Rhipidomys* Tschudi, 1844
Climbing Rat

Description

This genus is closely related to *Thomasomys,* with which it shares many of the features listed in the tribal description. *Rhipidomys* has several characters that

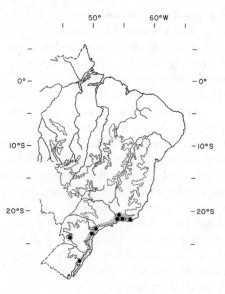

Map 16.13. Distribution of *Delomys dorsalis.*

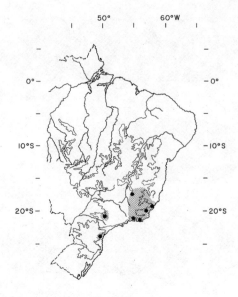

Map 16.14. Distribution of *Delomys sublineatus.*

Figure 16.7. Tail tips and degrees of hairiness: top, haired to tip, *Oecomys;* middle, strongly penicillate, *Rhipidomys;* bottom, extremely well furred, *Nyctomys.*

distinguish it from *Thomasomys.* In *Rhipidomys* the eye is relatively large; the dorsal color is usually sharply set off from the ventral color; the tail tip is penicillate, with a set of elongate hairs (plate 12 and fig. 16.7). The interorbital region of the skull of *Rhipidomys* has a distinct ridge. Reviews of the genus include Zanchin, Langguth, and Mattevi (1992) and Voss (1993). Zanchin, Langguth, and Mattevi (1992) examined the karyotypes of four Brazilian species of *Rhipidomys.* All four species have a diploid number of 44. The fundamental numbers are different: in *R. leucodactylus* $2n = 48$; in *R. mastacalis* $2n = 74$; and an as yet underscribed species exhibited a full number of 50.

Distribution

The genus is typically found in the lowland, forested regions of tropical South America.

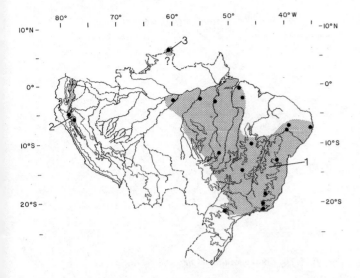

Map 16.15. Distribution of (1) *Rhipidomys mastacalis;* (2) *R. latimanus;* (3) *R. macconnelli.*

Rhipidomys couesi (J. A. Allen and Chapman, 1893)

Distribution
Some museum records indicate that *Rhipidomys couesi* occurs in southeast Peru (not mapped).

Comment
The taxon is here regarded as and included in *Rhipidomys sclateri.*

Rhipidomys latimanus (Tomes, 1860)

Description
Average measurements: TL 250; T 137; HF 23; E 18. The dorsum is reddish brown and the snow white venter contrasts sharply. The penicillate tail is not noticeably bicolored.

Range and Habitat
This species is broadly distributed at high elevations from western Venezuela across Colombia and south to Ecuador. In Venezuela it ranges from 1,160 to 2,422 m elevation. This rat is broadly tolerant of dry habitats but apparently prefers moist areas. It is found in both multistratal tropical evergreen forest and dry deciduous forest near rivers (map 16.15).

Natural History
Rhipidomys latimanus has been studied in Colombia in a cloud forest at 3,000 m elevation. High populations occurred during periods of maximum precipitation. Reproduction was continual throughout the annual cycle. Home ranges averaged 0.21 ha per individual. Stomach content analyses demonstrated a high degree of frugivory, with the diet consisting of 53% pericarp, 18% leaves, 11% seeds, and 6% arthropods.

Populations of two to nine adults inhabited an area of 1.2 ha during an annual cycle (Montenegro-Diaz, Lopez-Arévalo, and Cadena 1991).

Rhipidomys leucodactylus (Tshudi, 1843)

Description
Average measurements: HB 149–61; T 176–209; HF 30–36; E 23 (Argentina). In Peru measurements were HB 136–90; T 187–230; HF 34–38. This mouse is identifiable by its large size and very long, penicillate tail. Dorsally it is light reddish brown to yellowish brown mixed with black hairs, with more yellowish sides and a white to yellowish gray venter. The vibrissae are very long and stout (Gyldenstolpe 1932; pers. obs.).

Range and Habitat
The species occurs from southern Venezuela to northwestern and eastern Ecuador and south through eastern Peru to western Brazil and northern Argentina (map 16.16). This is a lowland form; in Venezuela it occurs at 135 to 145 m elevation and prefers multistratal tropical evergreen forest (Handley 1976). In Argentina it has been trapped in lower montane forest and transitional forest (Ojeda and Mares 1989).

Rhipidomys macconnelli De Winton, 1900

Description
Average measurements: TL 258; T 149; HF 27. In this small species of *Rhipidomys* the dorsum is brown, contrasting sharply with the gray venter, the tail is bicolored, and the feet are reddish brown.

Range and Habitat
The species is found in southeastern Venezuela to Brazil and Guyana in premontane forest. It ranges from 750 to 2,900 m elevation (map 16.15).

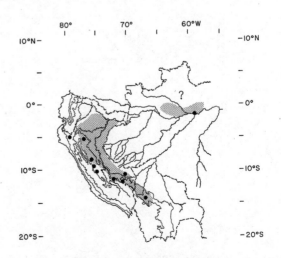

Map 16.16. Distribution of *Rhipidomys leucodactylus.*

Natural History

Females from Mount Duida, Venezuela, contained one or two embryos (AMNH).

Rhipidomys mastacalis (Lund, 1840)

Description

Average measurements: TL 231; T 161; HF 27; E 18. Adults weigh about 100 g. The dorsum is gray brown, and the venter is pale gray to cream. The tail is not bicolored, and the penicillate tip is present but not always accentuated.

Range and Habitat

The subspecies *Rhipidomys mastacalis venezuelae* occurs in northern Venezuela to the Guianas. The subspecies *R. m. tenicauda* occurs in southern Venezuela, the southern Guianas, and to eastern Brazil. This is a lowland species preferring moist habitats, but it can occur in dry deciduous tropical forest. In Venezuela it has been taken at elevations from 13 to 1,500 m (map 16.15).

Natural History

This nocturnal form is the common arboreal rat caught in the mountains of northern Venezuela. It feeds on fruits, seeds, leaves, fungi, and adult insects (O'Connell 1981). There are 3.8 young in an average litter, and adults are rather long-lived compared with semiarboreal rice rats such as *Oryzomys concolor*.

Rhipidomys nitela Thomas, 1901

Description

Measurements: HB 103–19; T 128–42; HF 21–26; E 14–17; Wt 110 g. The dorsum is grayish fawn, varying from gray to chestnut brown. The bases of the dor-

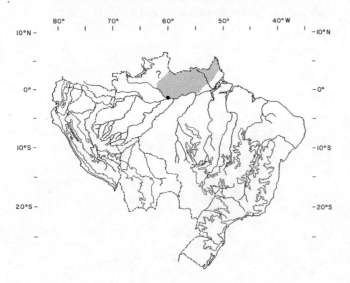

Map 16.17. Distribution of *Rhipidomys nitela*.

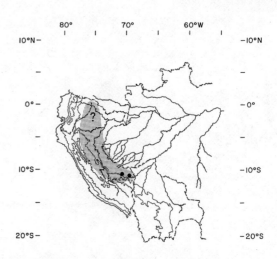

Map 16.18. Distribution of *Rhipidomys sclateri* (includes *R. couesi*).

sal hairs are black, and the tips are fawn. The ventral surface of the head and body is white, and the tops of the feet are yellowish.

Distribution

The species is found in southern Venezuela to the Guianas and south to Brazil (map 16.17).

Rhipidomys ochrogaster J. A. Allen, 1901

Comment

In this volume the species is included in *Rhipidomys leucodactylus*.

Rhipidomys sclateri (Thomas, 1887)

Description

This description includes *R. couesi*. In this extremely large climbing rat, total length ranges from 319 to 410 mm; T 161–209; HF 22–26; E 21–28. Its size distinguishes it from other species of *Rhipidomys*. The dorsum is light brown, the venter is cream, and the tail is not bicolored.

Range and Habitat

The species is distributed in southeastern Venezuela, through the Guianas, and south to Peru. It is broadly tolerant of both moist and dry forests and occurs at elevations of 1 to 1,400 m (map 16.18).

Genus *Thomasomys* Coues, 1884

Description

The molars are complex (pentalophodont), and M_3 is elongated relative to M_2 (fig. 16.5). These medium-sized ratlike rodents have rather long tails. The baculum is complex in structure. The hind feet are rather long, and the pinnae are conspicuous. The eye is reduced in size. The dorsal color is variable, ranging from gray brown to dark brown. The venter is paler than the dorsum, but there is no sharp demarcation.

Distribution

The genus is found throughout most of the Andean slopes of South America. Most of the species occur in northern South America, where many range to 4,000 m.

Natural History

These rodents occur at high altitudes in moist forests extending to elfin forests. Carleton (1973) reviews the stomach morphology of several species of *Thomasomys*. He notes that this is the only South American genus where there is appreciable interspecific gastric differentiation. He points out specifically that *T. aureus* is distinctly different from the other six species he examined. The co-occurrence of several species of *Thomasomys* in the same microhabitat suggests niche separation by specialization for different diets.

Thomasomys apeco M. Leo L. and A. L. Gardner, 1993

Description

Measurements: TL 469–548; T 279–329; HF 50–59; E 27–31; Wt 164–335. This is among the largest sigmodontine rodents from South America. The dorsal fur is long, with pronounced underfur. The dorsum is a bright reddish brown, darker in the middorsum and grading to a reddish buff on the venter (see Leo L. and Gardner 1993).

Distribution

The species is known from the department of San Martín, Peru, at 7°45′ S, 77°15′ W, from 1,000 to 4,000 m (not mapped).

Thomasomys aureus (Tomes, 1860)

Description

Average measurements: TL 366; T 209; HF 37; E 23. Head and body length ranges from 142 to 166 mm. The dorsum is brown, but the bases of the dorsal hairs are black; the venter is pale orange. The tail is uniformly colored and sparsely haired.

Range and Habitat

The species ranges from the western and central cordilleras of the Andes in Colombia and the eastern cordillera of the Andes in Venezuela south to Ecuador and eastern Peru (map 16.19). In Venezuela specimens were taken at 2,400 m in association with cloud forest (Handley 1976), and it has been taken at up to 3,400 m elevation in Colombia. In Peru J. L. Patton took specimens at 1,500–3,000 m (pers. comm.).

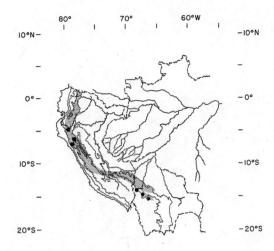

Map 16.19. Distribution of *Thomasomys aureus*.

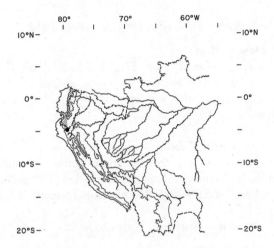

Map 16.20. Distribution of *Thomasomys baeops*.

Natural History

This species is known to feed on fruits. It is an excellent climber and builds nests in trees. Embryo counts of two and three have been recorded from Peru (MVZ).

Thomasomys baeops (Thomas, 1899)

Description

This small species has a total length of 250 mm, of which 142 mm is the tail. The same specimen had a hind foot of 25 mm and an ear of 17 mm and weighed 35 g (USNMNH). The dorsum was a light brown (somewhat darker in the midline) with a venter of dirty white (tips of venter hairs white and bases gray) (locality 3°58′ S, 79°21′ W, Ecuador).

Distribution

The species was found in Chilla valley, Río Pita, El Oro province, Ecuador, at 3,500 m (Wilson and Reeder 1993) (map 16.20).

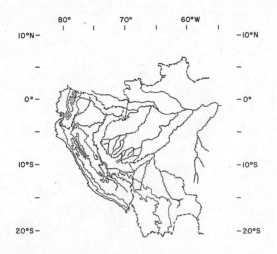

Map 16.21. Distribution of *Thomasomys cinereiventer.*

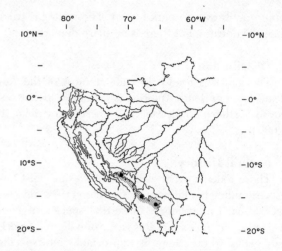

Map 16.23. Distribution of *Thomasomys daphne.*

Thomasomys cinereiventer J. A. Allen, 1912

Description

Average measurements: TL 282; T 153; HF 35. The dorsum is gray brown to dark brown, and the venter is yellow. The tail is uniformly colored and sparsely haired.

Range and Habitat

The species is distributed from central Colombia (Huila and Popayán, Cauca department) to Ecuador. It is found at 2,000–3,500 m elevation (map 16.21).

Thomasomys cinereus (Thomas, 1882)

Description

This description includes *T. caudivarius* Anthony, 1923. Average measurements: TL 275; T 161; HF 30. The extreme tips of the dorsal hairs are brown; the rest of the hair is blackish. The venter is brown, but the dark bases of the hairs lend a dark cast. The tail is not bicolored, but the distal 20% is white. The tail is

sparsely haired, and the scales are evident. The hairs at the very bases of the claws are white.

Distribution

The species is found in the mountains of southwest Ecuador, extending to the mountains of adjacent Peru. It has been taken with *T. baeops* and *T. pyrrhonotus* at 3,000 m in Chilla valley, El Oro province, Ecuador (map 16.22).

Thomasomys daphne Thomas, 1917

Description

This is a small species with measurements HB 92; T 133; HF 25.3; E 16.

Distribution

The type locality is the Ocobamba valley, Peru, at 9,100 ft. The species is distributed from south-central Peru to Boliva (map 16.23).

Thomasomys eleusis Thomas, 1926

Description

Measurements: HB 126; T 141; HF 28.5; E 21. The fur is very long and lax. The dorsum is brown, contrasting with the gray venter.

Distribution

The type locality is Tambo Jemes, Peru, at about 3,800 m elevation (not mapped).

Thomasomys gracilis Thomas, 1917

Description

This account includes *Thomasomys hudsoni* Anthony, 1923. Measurements: TL 213; T 120; HF 23. The dorsum is brown, darker in the midline, and the venter is buff. The claws have whitish hairs at the base. The tail is uniformly colored with no white tip.

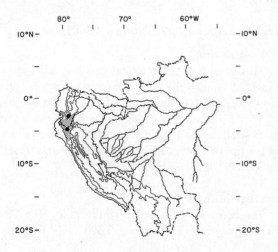

Map 16.22. Distribution of *Thomasomys cinereus.*

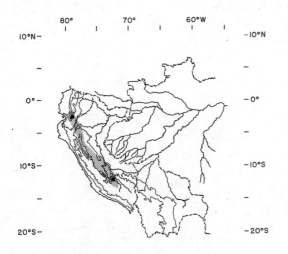

Map 16.24. Distribution of *Thomasomys gracilis* (includes *T.-cinnameus*).

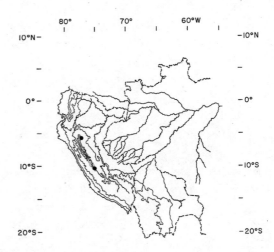

Map 16.26. Distribution of *Thomasomys ischyurus*.

Distribution and Habitat

The species is found in the Andes of Ecuador and adjacent Peru. It can co-occur with *T. baeops, T. aureus,* and *T. daphne.* It has been trapped along streams with dense scrubby undergrowth broken by montane meadows (map 16.24).

Thomasomys incanus (Thomas, 1894)

Description

Measurements: HB 114; T 125; HF 25.5; E 17.5. This form is similar to *T. aureus.* The tail is sparsely haired and not conspicuously bicolored; the venter is gray brown, contrasting with the medium brown dorsum.

Distribution

The type locality is Junín, Peru (map 16.25).

Thomasomys ischyurus Osgood, 1914

Description

Measurements: TL 261; T 151; HF 30. This form is similar in appearance to *T. rhoadsi.* The dorsal hairs

are slate colored for most of their length, but the tips are brown. The transition from the dorsal color to the lighter venter is gradual. The tail is extremely long and almost naked; it is brown dorsally but paler on the underside.

Distribution

The species is found in the western Andes of northern Ecuador (map 16.26).

Thomasomys kalinowskii (Thomas, 1894)

Description

Measurements of the type: HB 140; T 155; HF 32.8. The ears are quite long. The general color is a grizzled brownish gray, with the toes white. The chromosomes of this form were studied by Gardner and Patton (1976). The pattern is similar to that of *T. notatus* and *T. taczanowski,* which it resembles. Chromosome number: $2n = 44$; FN = 44.

Distribution

The type locality is the valley of Vitoc, Junín department, Peru. It is assumed to be confined to the Andes of central Peru (map 16.27).

Thomasomys ladewi Anthony, 1926

Description

Measurements: TL 294; HB 136; T 158; HF 33. The skull of this large species is characterized by a heavy build with low supraorbital ridges and broad zygomatic arches. The pelage is long, with individual hairs of 15–16 mm. The coat is dark gray above and below, and the tip of the tail (20 mm) is whitish.

Distribution

The type locality is Río Aceramarca, northeast of La Paz, Bolivia, at 10,800 ft (map 16.28).

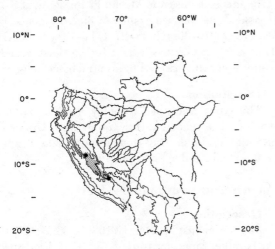

Map 16.25. Distribution of *Thomasomys incanus.*

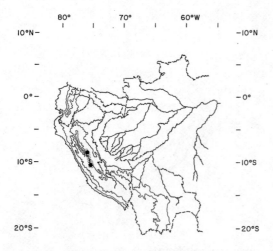

Map 16.27. Distribution of *Thomasomys kalinowskii*.

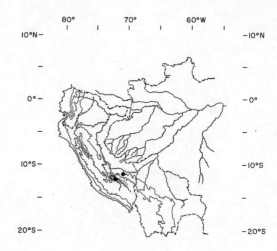

Map 16.29. Distribution of *Thomasomys notatus*.

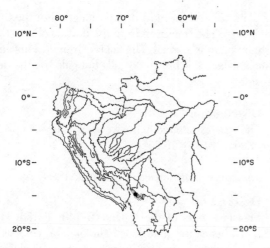

Map 16.28. Distribution of *Thomasomys ladewi*.

Thomasomys macrotis Gardner and Romo R., 1993

Description

Measurements: TL 345–87; T 193–219; HF 44–48; E 28–33; Wt 64–166 g (*n* = 4). Gardner and Romo R. (1993) describe it as "a large bodied, big-eared thomasomyine with a long hind foot." The nasal bones of *T. macrotis* are distinctive when compared with those of *T. aureus* (see Gardner and Romo R. 1993, fig. 6). The dorsum is dark brown, and the tail is white at the tip.

Distribution

The species is described from Río Abiseo national park, 7°45′ S, 77°15′ W, on the Amazonian slope of the Andes in Peru (not mapped).

Comment

The species co-occurs with *Thomasomys apeco*.

Thomasomys monochromus Bangs, 1900

Description

Measurements: TL 230; T 124; HF 25; E 19; Wt 36 g (USNMNH). The dorsal hairs have a light brown distal portion contrasting with an almost black basal segment, giving a faint salt-and-pepper look. The overall impression is of a very dark brown mouse with a venter only slightly paler than the dorsum.

Distribution

The species is found in northeastern Colombia (Wilson and Reeder 1993) (not mapped).

Comment

In the original description from Colombia, Bangs considered the form a species. Cabrera (1960) included it in *T. laniger*; Gardner and Patton (1976) treated it as a species.

Thomasomys notatus Thomas, 1917

Description

Four specimens from Machu Pichu, Peru, at 8,000 ft yielded the following range of measurements: HB 78–112; T 105–30; HF 25–27; E 17–18 (USNMNH). This small species has a light brown dorsum and a white venter (the bases of the ventral hairs are gray).

Distribution

The type locality is "Torontoy," Cuzco, Peru, at 9,500 ft. The species is known only from Urubamba and Alto Madre de Dios drainages at 1,500–2,500 m (Wilson and Reeder 1993) (map 16.29).

Thomasomys oreas Anthony, 1926

Description

Measurements: TL 244; HB 108; T 136; HF 25. The ear is rather large, measuring 18.7 in a dried speci-

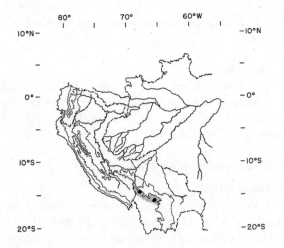

Map 16.30. Distribution of *Thomasomys oreas*.

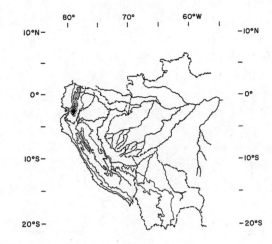

Map 16.31. Distribution of *Thomasomys paramorum*.

men, larger than in *T. daphne*. The hairs of the dorsum are long (12.5 mm) and dark gray at the base; there is a scattering of light-tipped hairs on the back, giving the impression of a dark brown dorsum sprinkled with black. The venter is buff. The area around the eye is lighter, the hind feet are pale brown dorsally, and the tail is faintly bicolored.

Distribution

The type locality is Cocopunco, La Paz department, Bolivia, at 10,000 ft (map 16.30).

Thomasomys paramorum Thomas, 1898

Description

A specimen from the Río Pita, Ecuador, collected at 11,500 ft, measured TL 220; HB 100; T 120; HF 25 (USNMNH). The dorsum was brown grading to a gray venter, and the tail was not noticeably bicolored.

Distribution

The species is found in the Andes of Ecuador (Wilson and Reeder 1993) (map 16.31).

Thomasomys pyrrhonotus Thomas, 1886

Description

Average measurements: TL 345; T 190; HF 32. One specimen collected at El Tembron, Peru, at 79°31′ W, 5°21′ N at 3,100 m measured HB 129; T 166; HF 20; E 26. The dorsum is tawny (reddish brown), but the black hair tips yield a salt-and-pepper effect. The venter is buff to white. A tuft of hair behind the ear is buff, and the tail is uniformly colored.

Distribution

The species is found in the Andes of southern Ecuador and northwestern Peru (map 16.32).

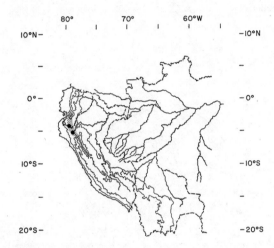

Map 16.32. Distribution of *Thomasomys pyrrhonotus*.

Thomasomys rhoadsi Stone, 1914

Description

Measurements of a male from Río Condor, Huachana, Ecuador, at 10,650 ft elevation were TL 250; T 121; HF 29; E 19; Wt 46 g. The dorsum is very dark brown (almost black), grading to a reddish gray venter.

Distribution

The species is found in the Andes of Ecuador (Wilson and Reeder 1993) (map 16.33).

Thomasomys rosalinda Thomas and St. Leger, 1926

Description

Measurements: HB 135; T 170; HF 24.5; E 20.5 (BMNH). This form is similar to *T. aureas*.

Distribution

The type locality is Goncha, Amazonas department, Peru, at 2,700 m (map 16.33).

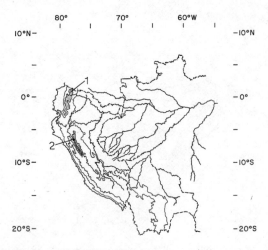

Map 16.33. Distribution of (1) *Thomasomys rhoadsi;* (2) *T. rosalinda.*

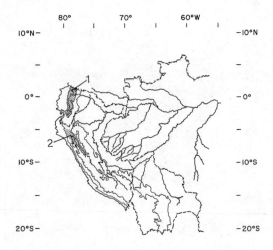

Map 16.34. Distribution of (1) *Thomasomys silvestris;* (2) *T. taczanowskii.*

Thomasomys silvestris Anthony, 1924

Description
A female weighing 35 g, caught at 6,400 ft elevation at Zapadores, Río Salaya, Santo Domingo province, Ecuador (USNMNH), measured TL 255; T 148; HF 28; E 18. The dorsum is very dark brown, almost black; the venter is paler and may be called dark gray.

Distribution
The species is found in the western Andes of Ecuador (Wilson and Reeder 1993) (map 16.34).

Thomasomys taczanowskii (Thomas, 1882)

Description
Measurements of one specimen from the Occahamba valley, Peru, at 9,100 ft elevation were HB 95; T 138; HF 25; E 20 (USNMNH). This rather small

species has a reddish brown dorsum grading to a gray venter.

Distribution
The type locality is Río Malleta, Tambillo, Cajamarca department, Peru, at 1,800 m (map 16.34). The species is found in the Andes of northwestern Peru (Wilson and Reeder 1993).

Thomasomys vestitus Thomas, 1898

Description
A single male specimen measured TL 310, T 169, HF 33, E 21 and weighed 76.5 g (USNMNH). The dorsum is ochraceous brown and the venter a buff-washed gray. The tail is sparsely haired and slightly penicillate.

Distribution and Habitat
Specimens have been taken in the Andes in western Venezuela at elevations over 1,630 m. In common with other species of *Thomasomys*, it is associated with moist habitats and cloud forest (not mapped).

Comment
The genus *Thomasomys* badly needs a major revision.

Genus *Wilfredomys* Avila-Pires, 1960

Wilfredomys has often been subsumed in *Thomasomys.* Pine (1980) reviews the history of the classification of *W. oenax* and argues that *W. pictipes* is closely related to *W. oenax.* We follow Musser and Carleton (1993) in recognizing *Wilfredomys.*

Wilfredomys oenax (Thomas, 1928)

Description
Average measurements: HB 127; T 174; HF 32; E 19; Wt 47 g. This species resembles a large *Oryzomys* except for the extremely long tail. The dorsum is pale gray to agouti brown with a yellowish cast, and the venter is buffy white. The most striking feature is the color of the nose: the tip of the muzzle and the lightly haired ears are bright ochraceous, and there is another ochraceous patch on the rump. The long tail is haired and uniformly buffy (Barlow 1969; Pine 1980; pers. obs.).

Distribution
Wilfredomys oenax is found in southern Brazil and Uruguay (Pine 1980) (map 16.35).

Ecology
In Uruguay this species is apparently limited to patches of dense subtropical woodland. Specimens

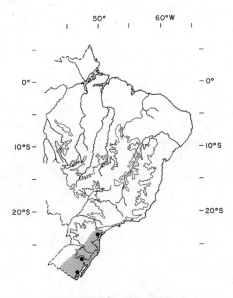

Map 16.35. Distribution of *Wilfredomys oenax*.

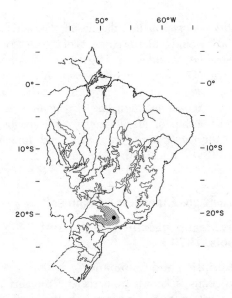

Map 16.36. Distribution of *Wilfredomys pictipes*.

have been caught in trees at night (Barlow 1965, 1969; Vaz-Perreira 1958).

Wilfredomys pictipes (Osgood, 1933)
Ratón Gris

Description
Average measurements: HB 87; T 95; HF 21; E 14. *W. pictipes* is the smallest of the Thomasomyini in southern South America. The tail is slightly shorter than the head and body length, unhaired, bicolored, and slightly penicillate. It is reddish brown to brownish orange dorsally and a warm tan on the sides, with a white venter. In some individuals the feet are distinctly bicolored (Osgood 1933; Pine 1980; pers. obs.).

Distribution
Wilfredomys pictipes is found in southeastern Brazil and in Misiones province of northeastern Argentina (Pine 1980) (map 16.36).

Ecology
In southeastern Brazil two individuals were found nesting off the ground in bamboo and in a bromeliad (Pine 1980).

TRIBE ORYZOMYINI VORONTSOV, 1959
Rice Rats

Description
The molars are often not complex (tetralophodont) (but see below), and the third molar is usually large (half to three-quarters the length of M_2; figs. 16.5 and 16.6). Like the thomasomyine rodents, the genera usu-

ally have tails longer than the head and body. The ears are modest to rather large, and the tail is seldom penicillate. Some genera (*Neacomys* and *Scolomys*) have semispinescent pelage.

Distribution
This tribe has members extending from coastal New Jersey in the United States to southern Argentina.

Natural History
These rodents usually occupy mesic habitats. Some species are highly arboreal; others may exhibit aquatic specializations.

Comment
We include the following note according to Voss and Carleton (1993). Members of the tribe show the following traits:

1. Paired pectoral mammae with a count of eight or more (*Scolomys* has six)
2. A long palate with prominent posterolateral pits (reversed in *Holochilus sciureus*)
3. No alisphenoid strut separating the buccinator-masticatory and accessory oval foramina (reversed in *Holochilus sciureus*)
4. No posterior suspensory process of the squamosal attached to the tegmen tympani (secondary overlap occurs in *Lundomys* [as diagnosed by Voss and Carleton 1993] and some species of *Oecomys*)
5. No gallbladder

Pentalolophodonty is not a unifying character as formerly believed, since this use of the character then bases the "recognition" on a symplesiomorphy.

With *Lundomys molitor* diagnosed, Voss and Carleton (1993) would limit the contents of the Oryzomyini to *Holochilus*, *Lundomys*, *Megalomys*, *Melanomys*, *Microryzomys*, *Neacomys*, *Nectomys*, *Nesoryzomys*, *Oecomys*, *Oligoryzomys*, *Oryzomys*, *Pseudoryzomys*, *Scolomys*, *Sigmodontomys*, and *Zygodontomys*. The true relationship of *Megaoryzomys* (Galápagos, extinct), *Abrawayaomys*, *Phaenomys*, and *Rhagomys* (eastern Brazil) remains problematical.

The Galápagos Endemics: A Comment

Genus *Megaoryzomys* Lenglet and Coppois, 1979

Description and Comment
The genus is known from skeletal remains from cave deposits on Santa Cruz and Isabela Islands in the Galápagos. It is one of the largest species groups of the Sigmodontinae, and Lenglet and Coppois placed the new forms in their own genus and speculated on a close relationship with *Oryzomys*. Steadman and Ray (1982) placed the species of *Megaoryzomys* nearer to *Thomasomys* and *Rhipidomys* than to *Oryzomys*. Skull (condylar-basal length) exceeds 750 mm.

Distribution
See above. One species has been described so far, *M. curioi* from Santa Cruz Island (not mapped).

Genus *Nesoryzomys* Heller, 1904

Description
Measurements: HB 100–200; T 79–140. This genus is clearly allied to the oryzomyines. The snout is long and narrow (see chapter 18).

Distribution
The genus is confined to the Galápagos Islands (not mapped). It will be discussed in chapter 18.

Genus *Abrawayaomys* Cunha and Cruz, 1979

Abrawayaomys ruschii Cunha and Cruz, 1979

Description
Measurements of the type specimen are HB 201; T 85; Wt 46 g. The tail is shorter than the head and body. This small cricetine has stiff, spinescent hairs on the dorsum. The overall impression is of a large *Neacomys*. The brown dorsum is classic agouti brown shading to a lighter yellowish cast.

Distribution
This species was first trapped in the state of Espírito Santo, Brazil, but it has subsequently been collected

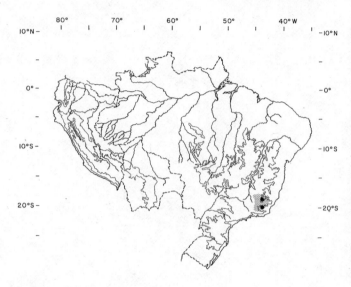

Map 16.37. Distribution of *Abrawayaomys ruschii*.

in Parque Rio Doce, Minas Gerais, Brazil (Stallings 1988a,b) (map 16.37).

Genus *Oryzomys* Baird, 1858 (sensu lato)
Rice Rat

Description
The genus, when considered in the broad sense, is highly variable in size, color, and tail length. The M_3 ranges from one-half to three-fourths the length of the M_2, and the molar structure may be simple (tetralophodont) or complex (pentalophodont). In terrestrial forms, the sparsely haired tail is equal to or slightly shorter than the head and body length. The baculum is complex in structure, and the pinnae are conspicuous. The dorsum is usually some shade of brown with a contrasting ventral color, generally white to cream.

Distribution
This genus ranges from the southeastern United States through Central and South America to northern Chile and Argentina.

Comment on the Taxonomy of *Oryzomys*
As early as 1906, Thomas recognized the distinctive characters of *Oryzomys bicolor* and proposed the generic name *Oecomys*. *Oecomys* was adopted by Cabrera (1960) and others. It was not employed as a generic category by Honacki, Kinman, and Koeppl (1982), but Wilson and Reeder (1993) have recognized the genus as well as other subdivisions of the original genus. Within the genus *Oryzomys*, other subgeneric categories are *Oryzomys* (sensu stricto), *Oligoryzomys*, *Microryzomys*, *Melanomys*, and *Sigmodontomys*. In consideration of the strong arguments made by

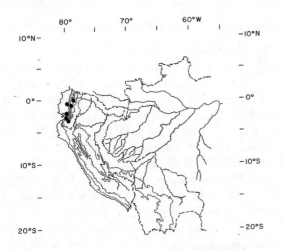

Map 16.38. Distribution of *Melanomys caliginosus*.

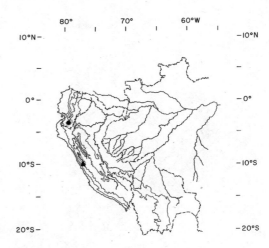

Map 16.39. Distribution of two species of *Melanomys*: circle, *M. robustulus*; triangle, *M. zunigae*.

Carleton and Musser (1989), *Oligoryzomys* and *Microryzomys* will be recognized together with *Oecomys* as full genera in the following accounts. *Melanomys* and *Sigmodontomys* will also be recognized as full genera.

Genus *Melanomys* Thomas, 1902

Description
See account for *Melanomys caliginosus*.

Melanomys caliginosus (Tomes, 1860)
(= *Oryzomys caliginosus* Tomes, 1860)

Description
Cusps are present on adult molars. Total length ranges from 196 to 240 mm; T 85–105; HF 25–27; E 15–16; Wt 47–60 g. The tail is not penicillate and is much shorter than the head and body. The dorsum is very dark brown, almost black; the underparts are paler, the tail is not bicolor, and the tops of the feet are black. One female was recorded with four embryos (MVZ).

Distribution
The species is distributed from Honduras south through Panama west of the Andes to southern Ecuador (map 16.38).

Melanomys robustulus Thomas, 1914

Description
One specimen measured HB 122; T 71; HF 27; E 16. This species has an extremely dark brown dorsum contrasting with an orange gray venter.

Distribution
The species is known only from southeastern Ecuador at high altitudes (map 16.39).

Melanomys zunigae (Sanborn, 1949)

Description
Measurements: TL 206–41; T 81–105; HF 23–28. The tail is markedly shorter than the head and body. The dorsum is grizzled brown, paler on the sides, and the individual hairs are either dark based with a brown tip or wholly gray black from base to tip. The venter is brown with gray showing through. The feet are brown, as is the tail.

Distribution
The type locality is Lomas de Atoconge, Lima department, Peru. The species is found in west-central Peru (map 16.39).

Genus *Microryzomys*

Description
These small oryzomyine rodents have relatively long tails and rather long, narrow hind feet. They are superficially similar to some species of *Oligoryzomys* but are distinguishable by a complex of characters (Carleton and Musser 1989). See also species accounts.

Distribution
The genus is found mainly in the Andes of South America but extends to the north coast ranges of Venezuela.

Comment
According to Carleton and Musser (1989) this genus is distinct from *Oligoryzomys*. They suggest that the two genera may be sister taxa. The two taxa are separable based on clusters of morphological traits rather than any single character. As defined by Carleton and Musser (1989), the genus *Microryzomys* in-

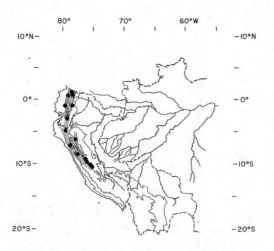

Map 16.40. Distribution of *Microryzomys altissimus*.

cludes two species: *M. altissimus* and *M. minutus,* both small montane forms that may occur in sympatry as well as parapatry.

Microryzomys altissimus (Osgood, 1933)

Description

This small oryzomyine rodent has an extremely long tail (see *M. minutus* below, which it resembles). The tail is relatively longer in *M. minutus* (145% of head and body length compared with 137% in *M. altissimus*) The palatine foramina extend posteriorly beyond the level of the molars in *altissimus,* whereas in *minutus* they terminate anterior to the level of the molars (Carleton and Musser 1989).

Distribution

The species is found at high elevations in the Andes of Colombia, Ecuador, and Peru. Although it can occur sympatrically with *M. minutus,* it tends to range at higher elevations. *M. altissimus* records dominate within the 2,500 to 4,000 m range (map 16.40).

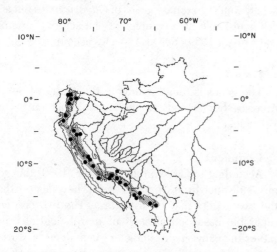

Map 16.41. Distribution of *Microryzomys minutus*.

Microryzomys minutus (Tomes, 1860)

Description

Cusps are present on adult molars. Six specimens from "Bato" on Zapelacha Carmen in Piura department, Peru, at 2,250 m measured TL 190–210; T 110–26; HF 22–23; E 15–16; Wt 10–24 g. The heaviest was a female with three embryos. At Canchaque, Piura department, Peru, at 1,900 m, four specimens measured TL 173–95; T 111–20; HF 22–25; E 14; Wt 10–18 g (LSU). This small mouse has a reddish brown dorsum and a pale orange venter. The tail is not distinctly bicolored and not penicillate.

Distribution

The species is confined to high elevations in the Andes of Ecuador and Peru (map 16.41).

Genus *Oecomys* Thomas, 1906
Arboreal Rice Rat

Description

The anterior margin of the maxillary plate when viewed dorsally does not extend as far anterior to the rest of the maxilla as is the normal condition in *Oryzomys.* The tail exceeds the head and body length in length. The hind feet are wide, suggesting arboreal habits (fig. 16.8), and the fifth toe of the hind foot is rather long. The dorsum is reddish brown, but black hair tips may be conspicuous. The venter varies from white to orange, and the tail is not strongly bicolored.

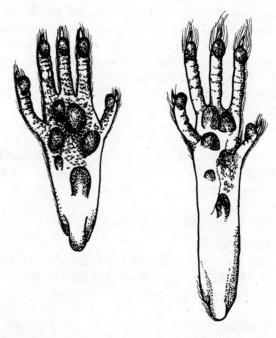

Figure 16.8. Oryzomyine hind feet: (a) arboreal; (b) terrestrial (after Hershkovitz 1960).

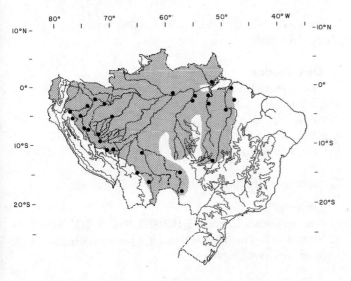

Map 16.42. Distribution of *Oecomys bicolor.*

Distribution

These rodents are found from Costa Rica southward throughout tropical South America.

Comment

Members of this genus have often been included within *Oryzomys*, but the karyological studies of Gardner and Patton (1976) reinforced the original suspicion of Thomas that a true genus distinct from the classical *Oryzomys* should be considered. This generic status was adopted by Carleton and Musser (1984).

Oecomys bicolor (Tomes, 1860)

Description

This species is smaller than *O. concolor*. Seven specimens from Suriname showed the following range in measurements: TL 172–223; T 89–111; HF 21–23; E 13–15. The dorsum is cinnamon brown infused with black, the venter is white, and the tail is not noticeably bicolored. Juveniles have slightly darker pelage (see plate 12).

Range and Habitat

This species is distributed from Panama south to northwestern Bolivia, east to the Guianas, and thence south to south-central Brazil. It tolerates both dry deciduous forest and multistratal tropical evergreen forest (map 16.42).

Natural History

This arboreally adapted species can exist in the Venezuelan llanos by using gallery forest or palms in seasonally inundated areas. In suitable habitat, adult density can range seasonally from 0.4 to 1.25 individuals per hectare. The litter size is 4.5, and the young be-

come sexually mature at approximately three months of age (O'Connell 1981).

Oecomys cleberi Locks, 1981

Distribution

The species is known only from the type locality at 45°54′ W, 15°57′ S, Brasília, Brazil (not mapped).

Comment

This taxon is allied to *Oecomys bicolor* or *O. paricola*.

Oecomys concolor (Wagner, 1845)

Description

For this account *O. trinitatus* is included within *O. concolor*. Average measurements are TL 297; T 63; HF 30; E 19. This species tends to be slightly larger than *O. bicolor* and has slightly different habitat preferences. The dorsal pelage is colored as in *O. bicolor* and gives the effect of a medium agouti brown, but the venter is gray to buff and the tail is not strongly bicolored.

Distribution and Habitat

The species is found from northern South America including the Guianas south to central Brazil, and in the west it is distributed broadly through Peru and Bolivia to northern Argentina (map 16.43). It can range to 2,230 m elevation in northern Venezuela and prefers multistratal tropical evergreen forest, although it tolerates man-made clearings.

Natural History

The common arboreal rice rat in lowland forest and premontane forest is less tolerant of xeric habitats than

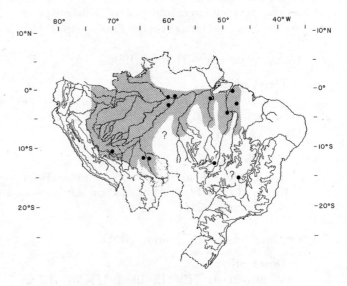

Map 16.43. Distribution of *Oecomys concolor.*

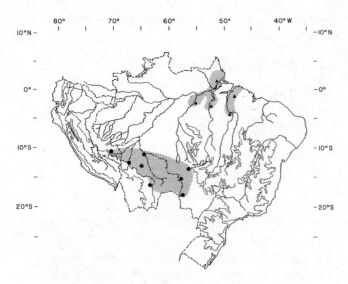

Map 16.44. Distribution of two species of *Oecomys*: circle, *O. mamorae*; triangle, *O. paricola*.

is *O. bicolor*. Litter size ranges from two to four. Density ranges from 0.5 to 2.5 individuals per hectare. The species co-occurs over much of its range with other species of *Oecomys* (O'Connell 1981).

Oecomys mamorae (Thomas, 1906)

Description

Measurements for two adults were TL 333, 338; HB 149, 147; T 184, 181; HF 27, 27. A brown dorsum grades to orange on the sides, which then extends, slightly paler, to the ventral midline. The throat, axillae, and inguinal regions are white. The venter becomes paler with age, and in an old female it was almost white.

Distribution

The species is found from eastern Bolivia to northern Paraguay and then to central Brazil (map 16.44).

Oecomys paricola (Thomas, 1904)

Description

Average measurements for nineteen individuals from Para were HB 106.5; T 122.8; HF 24.5; E 16; Wt 37.4 g (São Paulo).

Distribution

The species is found mainly in east-central Brazil, but it could extend to the Guianas and southeastern Venezuela (map 16.44).

Oecomys phaeotis (Thomas, 1901)

Description

Measurements: HB 112–19; T 111–30; HF 25; E 14–18. The dorsum is ochraceous to tawny brown

finely mixed with dark brown. The venter is buffy and sharply defined.

Distribution

The species is distributed from 200 to 2,000 m in the departments of Cuzco, Puno, and Madre de Dios, Peru, and thence to the departments of La Paz, Cochabamba, and El Beni in northwestern Bolivia (Hershkovitz 1960) (map 16.45).

Oecomys rex Thomas, 1910

Description

One specimen measured HB 160; T 150; HF 26; E 16 (BMNH). The dorsum is reddish brown contrasting with a dirty white venter.

Distribution

The species' range extends from eastern Venezuela across the Guianas to extreme northeastern Brazil (map 16.45).

Oecomys roberti (Thomas, 1904)

Description

Measurements of the type specimens: HB 110; T 145; HF 26.5; E 16. The upperparts are buffy to tawny mixed with dark brown while the underparts are white and sharply defined from the sides. The chest may have a buffy wash.

Distribution

The type locality is Santa Ana de Chapada, Mato Grosso state, Brazil, but the species is broadly distributed in the Amazonian region of Brazil, Peru, and Ecuador (map 16.45).

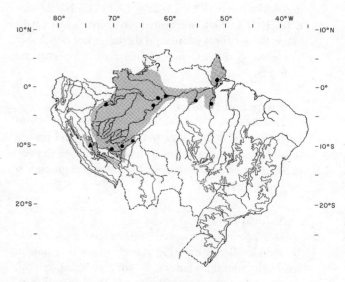

Map 16.45. Distribution of three species of *Oecomys*: circle, *O. roberti*; triangle, *O. phaeotis*; square, *O. rex*.

Oecomys superans Thomas, 1911

Description

This description includes *O. melleus*. Measurements: HB 140; T 156; HF 28. The pelage is long and soft. The bases of the hairs are slate colored and the tips are brown, giving the impression of a brown coat with a sprinkling of black. The venter is honey yellow to cream.

Distribution

The species is found in the Andean foothills to the lowlands of Ecuador and Peru and thence to western Brazil. It is known from Río Bobeneca, Pastaza province, Ecuador, at 2,100 ft elevation (map 16.46).

Oecomys trinitatus (J. A. Allen and Chapman, 1893)

Description

Measurements: TL 296.5; T 163.2; HF 29.8; E 18.8 (*n* = 11). This species has a dorsal pelage of medium agouti brown. The venter is gray to buff, and the tail is *not* strongly bicolored.

Distribution

The species' range extends from southwestern Costa Rica and thence southeast of the Andes to Peru and east across Colombia and Venezuela to Brazil (map 16.47).

Natural History

This is the common arboreal rice rat over its range. It appears to be confined to moist habitats. Litter size ranges from two to four.

Comment

This species is often included in *Oecomys concolor*.

Genus *Oligoryzomys* Bangs, 1900

Description and History

Carleton and Musser (1989) offer an amended diagnosis for this group. These are small mice with the tail exceeding the head and body in length. There are six plantar tubercles on the hind foot as in *Microryzomys*, but the pads are smaller. The molars are brachydont, contrasting with the higher-crowned molars of *Oryzomys*. The third molar is noticeably smaller than the second molar. The carotid circulatory pattern as revealed by size and shape of the relevant foramina differs among *Microryzomys*, *Oligoryzomys*, and *Oryzomys* (see Carleton and Musser 1989 for details).

Carleton and Musser suggest a provisional grouping into five groups:

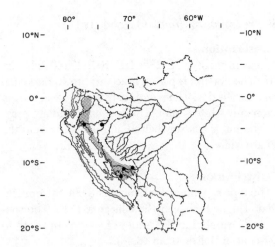

Map 16.46. Distribution of *Oecomys superans*.

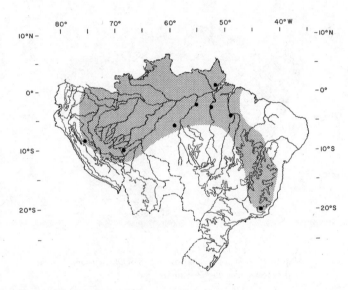

Map 16.47. Distribution of *Oecomys trinitatus*.

1. *Fulvescens* group: *O. fulvescens*, *O. arenalis*, and *O. vegetus*
2. *Microtis* group: *O. microtis*, *O. andinus*, and *O. chacoensis*
3. *Flavescens* group: *O. flavescens* and three as yet unnamed species
4. *Nigripes* group: *O. nigripes*, *O. eliurus*, *O. destructor*, *O. delticola*, and *O. longicaudatus*

Cusps are present on the molars. The long tail is not penicillate. The body is rather small, with the hind foot less than 28 mm. Long hairs are present on the toes, protruding beyond the claws.

Distribution

Mice of this genus extend from Mexico south to Argentina and Chile.

Oligoryzomys andinus (Osgood, 1914)

Description
Measurements: TL 235; HB 104; T 140; HF 26; E 16. The dorsum is pale ochraceous brown darkened by numerous dusky lines and sharply demarcated from the creamy white venter. The face and head are grayish. The tail is sparsely haired and bicolored, and the feet are white.

Distribution
The type specimen was caught at 6,000 ft from Hacienda Hagueda, Upper Río Chicama, Peru. The species is distributed in the Andes of west Peru south to west-central Bolivia (map 16.48).

Map 16.48. Distribution of two species of *Oligoryzomys:* circle, *O. flavescens;* triangle, *O. andinus.*

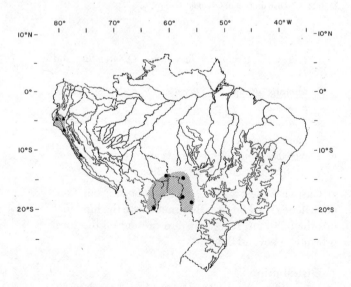

Map 16.49. Distribution of two species of *Oligoryzomys:* circle, *O. chacoensis;* triangle, *O. arenalis.*

Oligoryzomys arenalis (Thomas, 1913)

Description
Measurements: HB 80; T 100; HF 21; E 15. This rather small, pale species has a clay-colored dorsum. The sides are buffy, and the venter is creamy white. The feet are white.

Distribution
The species is found in the arid coastal areas of northwestern Peru. The type is from Eten, Lambayeque department, Peru, at sea level (map 16.49).

Oligoryzomys chacoensis Myers and Carleton, 1981

Description
This medium-sized species of *Oligoryzomys* is unique in its whitish underside with hair white to the base of the chin and throat, relatively long ears with heavy hair on the inner surfaces and short or absent dark basal bands, and small but distinctive tufts of orangish hair in front of the ears. The dorsum is clay colored, heavily lined with black, darkest over the shoulders. The head is grayer, and the cheeks are brown or gray, occasionally with some orange. The tail is weakly bicolored, averaging 131% of the head and body length, and the feet average 25% of the head and body length (Myers and Carleton 1981). Chromosome number: $2n = 58$ (Myers 1982).

Distribution
Oligoryzomys chacoensis is found in the Chaco of Paraguay, Bolivia, and Argentina (map 16.49).

Life History
In the Paraguayan Chaco ten females were found to contain an average of 4.6 embryos (range 2–5) (Myers and Carleton 1981).

Ecology
In the northern and western Chaco of Paraguay this species is common in thrown scrub and grassland, reaching its highest densities in grassland. In the lower, wetter chaco it is found in forest and dry grassland but not in wet marshes, where *O. fornesi* occurs (Myers and Carleton 1981).

Oligoryzomys delticola (Thomas, 1917)
Colilargo Isleno

Description
Average measurements: HB 103; T 137; HF 26; E 18; Wt 29 g. This medium-sized *Oryzomys* closely resembles *O. longicaudatus.* In adults the dorsum is grayish brown to reddish brown, and the venter is

whitish or grayish and never ochraceous. The sides are sometimes washed with yellow. The tail is long and bicolored. *O. delticola* is distinguishable from *O. flavescens* by its longer tail, smaller hind feet, and smaller ears (Barlow 1965; Massoia 1973b; Thomas 1917).

Distribution

Oligoryzomys delticola, as the name suggests, is found in a restricted range centering on the delta of the Río Paraná and including all of southeastern Brazil, Uruguay, and the Argentine provinces of Entre Rios, Misiones, and Buenos Aires (Barlow 1965) (map 16.50).

Life History

In Uruguay *O. delticola* breeds in late summer and early autumn, and females have two to four embryos (*n* = 5) (Barlow 1969).

Ecology

This is the common woodland mouse of Uruguay, found in mesic subtropical forest with a closed canopy and a scanty understory as well as in more xeric, open thorn woods. It is primarily nocturnal, climbs well, and is primarily herbivorous, though it eats some insect material (10%, *n* = 4; Barlow 1969).

Oligoryzomys destructor (Tschudi, 1844)

Description

This description includes *O. stoltzmani.* Measurements: HB 75–87; T 109–30; HF 20–23; E 10–12 (*n* = 5; Colombia, FMNH). The dorsum is medium brown, contrasting with a tan venter. The tail is faintly bicolored.

Distribution

The species occurs in the Andes of southern Colombia and thence south through Ecuador and Peru to west-central Bolivia (map 16.51).

Oligoryzomys eliurus (Wagner, 1845)

Description

This taxon could be conspecific with *O. nigripes.* See that species account.

Distribution

Specimens labeled as *Oligoryzomys eliurus* have been collected in eastern Brazil (map 16.51).

Oligoryzomys flavescens (Waterhouse, 1837)
Colilargo Chico

Description

Average measurements: TL 202; T 112; HF 27; E 14; Wt 32 g. The tail is never longer than 138 mm.

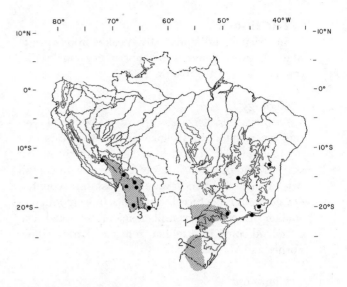

Map 16.50. Distribution of (1) *Oligoryzomys nigripes;* (2) *O. delticola;* (3) *O. longicaudatus.*

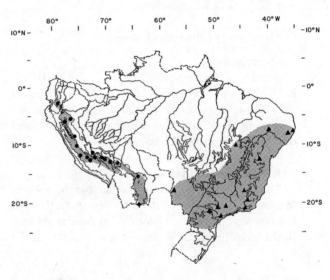

Map 16.51. Distribution of two species of *Oligoryzomys:* circle, *O. destructor;* triangle, *O. eliurus.*

Dorsally this mouse is brown mixed with yellowish and tinged with black. The venter is whitish yellow, sometimes mixed with a little gray. The sides and rump are sometimes heavily washed with orange (plate 12). Some individuals are mottled yellowish brown (Barlow 1965; Gyldenstolpe 1932; Massoia 1973b).

Distribution

Oligoryzomys flavescens ranges from southern Brazil south through Uruguay to northern and central Argentina to at least Buenos Aires province (Contreras and Berry 1983; Contreras and Rosi 1980c) (map 16.48).

Life History

In Uruguay the reproductive peak is from April to May. The average number of embryos per female is 5.1 (range 3–7; $n = 10$) (Barlow 1969).

Ecology

Oligoryzomys flavescens appears to be found typically near water. In Uruguay it has been captured in stands of tall grass in marshes, along rivers and streams, and occasionally away from water. In central Argentina it is found near water in brushy arid country. *O. flavescens* is nocturnal. Ten stomachs were found to contain principally green plant material, though five contained some invertebrate remains (Barlow 1969; Fornes and Massoia 1965).

Comment

Anderson (1997) defines an *O. flavescens* (species group B) in the foothills of the Bolivian Andes as distinct from *O. andinus*. This distribution is plotted in map 16.50 as *O. longicaudatus*.

Oligoryzomys fulvescens (Saussure, 1860)

Description

Average measurements: TL 176; T 100; HF 21; E 14; Wt 10 g. This species is distinctive for its small size. The upperparts are reddish brown with an admixture of black hairs, the underparts are white, and the tail is not noticeably bicolored.

Range and Habitat

The species ranges from Tamaulipas in eastern Mexico and Nayarit in western Mexico south through Panama and northern Colombia to Venezuela and Brazil (map 16.52). In Venezuela it is common in the

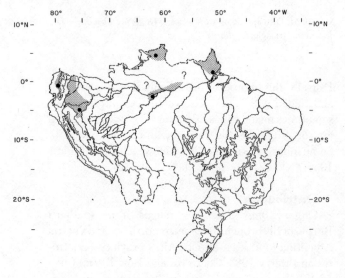

Map 16.52. Distribution of *Oligoryzomys fulvescens*.

lowlands below 1,500 m elevation and prefers multistratal tropical evergreen forest.

Natural History

This is the common low-elevation small rice rat over much of Central America and northern South America.

Oligoryzomys longicaudatus (Bennett, 1832)
Colilargo Común

Description

Measurements: TL 222.5; T 127.3; HF 27; E 15.5; Wt 23.8 g. The tail is strongly bicolored and sparsely haired and decreases in relative length with decreasing latitude. The ears are small and also sparsely haired. The dorsum is buffy with fine lines of blackish to light brown, with the gray at the bases of the hairs occasionally showing through. The venter is gray white. In Argentina the dorsum is often washed with ochraceous, which can be prominent along the sides (Mann 1978; Osgood 1943; pers. obs.).

Distribution

The species is distributed in Bolivia from approximately 25° south latitude to the islands off Tierra del Fuego. It is found in Argentina from the northwesternmost province of Jujy south along the Andes to about Santa Cruz province, where it appears to occur across the whole width of the continent (Olrog and Lucero 1981; Tamayo and Frassinetti 1980; Thomas 1898a). (See map 16.50.)

Life History

In Argentina the average number of fetuses per female was 4.9 (range 2–11; $n = 44$). In the former Malleco province, Chile, females reproduce from November to February and average 4.9 embryos (range 3–9; $n = 43$). In Chile females start reproducing when only a few months old, have litters of three to five, and can breed three times a year (Greer 1966; Mann 1978; Pearson 1983).

Ecology

Oligoryzomys longicaudatus shows a remarkable flexibility in habitat choice, although it requires free water. In one area in north-central Chile it was caught in cloud forest, in brush forest transition, and in open brushy habitat, though it was by far most common in the cloud forest, where it represented 59% of all small captures. In the Patagonian forest of Argentina it occurs occasionally in dense forest but seems to prefer brushy areas, edges of clearings, and roadsides with blackberry and wild rose tangles. In Salta province, Argentina, this species was captured in some areas only near permanent water, while in other areas it was most common in dense second growth, though occasionally taken in mature subtropical forest away from standing water. In general *O. longicaudatus* seems to prefer moister areas with abundant cover.

As suggested by its long tail and long hind feet, *O. longicaudatus* is a good climber and has been captured 3 m off the ground. It builds its nests in bushes and trees and will use abandoned bird's nests. It is also a good jumper.

In one area *O. longicaudatus* was strongly granivorous during the dry season (seeds 72.7% by volume), and in the wet season it specialized on flowers, pollen, and foliage of *Chenopodium* (flowers and pollen 53% by volume; foliage 29.4%). Arthropods were of very little significance in the diet. This species' fondness for seeds applies in forests as well as in brushy habitats, but in one area insects formed 14.7% of the diet measured year round. When bamboos undergo mass flowering, populations of *O. longicaudatus* have been reported to undergo dramatic increases (Contreras 1972a; Glanz 1977a; Greer 1966; Mann 1978; Mares 1973, 1977a; Meserve 1981a,b; Murúa and González 1979, 1981, 1982; Murúa, González, and Jofre 1982; Ojeda 1979; Pearson 1983; Pearson and Pearson 1982; Pefaur, Yáñez, and Jaksić 1979; Pereira 1941; Schamberger and Fulk 1974).

Comment

Some investigators believe that the southernmost members of *Oligoryzomys longicaudatus* belong to a separate species, *O. magellanicus.*

Oligoryzomys microtis (Allen, 1916)

Description

Average measurements: TL 188; T 106; HF 24; E 14; Wt 18 g. This is the smallest *Oligoryzomys* The tail is equal to about 128% of the head and body length, and the hind feet are relatively long (28.5% of head and body length).*O. nigripes* is absent or indistinct. The throat is variably white to gray, and the venter is whitish or grayish, occasionally washed with buffy. The cheeks are orangish brown or gray; the tail is weakly bicolored. The pinnae are densely furred inside (Massoia 1973b).

Chromosome number: $2n = 62-66$ (Myers and Carleton 1981).

Distribution

Oligorysomys microtis is found from lowland Peru and Bolivia east to southern Brazil and south to northern Argentina in the province of Corrientes. It occurs throughout eastern Paraguay and into the Chaco (Contreras 1982; Honacki, Kinman, and Koeppl 1982; Myers and Carleton 1981) (map 16.53).

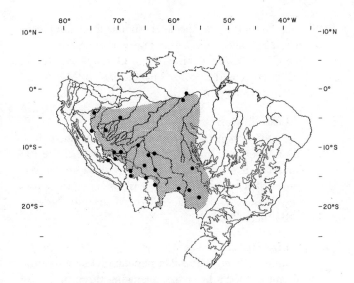

Map 16.53. Distribution of *Oligoryzomys microtis.*

Ecology

Oligorysomys microtis is strongly associated with marshes and wet grasslands in Paraguay and Argentina. In central Brazil it is found most often on the edge of gallery forests and in second-growth habitats. In the Paraguayan Chaco, it has been taken only in marshy habitats both wet and dry. Some individuals have been caught on floating mats of grass (Dietz 1983; Massoia 1973b; Myers 1982; Myers and Carleton 1981).

Comment

Olds and Anderson (1987) state that *O. fornesi* (Massoia, 1973b) is a junior synonym of *O. microtis.*

Oligoryzomys nigripes (Olfers, 1818)

Description

This description includes *O. eliurus*. Average measurements: HB 92; T 126; HF 25; E 17; Wt 24 g. The tail is equal to about 126% of the head and body length, and the relatively small feet equal about 24.5% of the head and body length. The ears are small and sparsely haired. This medium-sized species is characterized by a grayish white venter with white-tipped hairs, often with an orangish pectoral band. It has a light brown dorsum, a lightly bicolored tail, and tan feet. In Paraguay *O. nigripes* can be distinguished from *O. chacoensis*, which has a paler, yellowish dorsal stripe and a white venter and never has a pectoral band. Also, *O. chacoensis* has hairs on the throat that are usually white to the base, and there is a complete or almost complete absence of a dark basal band on the hairs inside the ears (prominent in *O. nigripes*). Some individuals have a pronounced V of black hairs originating at the nose and passing over the eyes.

In Argentina the species is similar to *O. flavescens* in size but always has a light gray venter and dark ears like *O. delticola.* It differs from *O. longicaudatus* in pelage color and texture (Massoia 1973b; Myers and Carleton 1981; pers. obs.).

Distribution

Oligoryzomys nigripes is found from eastern Brazil to Argentina. In Paraguay it occurs east of the Río Paraguay; it occurs throughout Uruguay; and in Argentina it is found south to at least Buenos Aires province (Massoia and Fornes 1967b; Myers and Carleton 1981) (map 16.50).

Life History

Animals from a Brazilian population taken into the laboratory had a litter size averaging three ($n = 18$); neonatal weight averaged 3.3 g; the estrous cycle was ten days; and females reproduced for the first time at fifty-eight days of age. In southern Brazil animals reproduced all year with two peaks in September to November and February to April. The gestation period was twenty-five days; litters were produced every two and a half to three months, and litter size was three to four (Dalby 1975; Mello 1978a; Myers and Carleton 1981; Veiga-Borgeaud 1982).

Ecology

In southern Brazil this species was found in cultivated fields, forest borders, and areas of second growth. In Minas Gerais, Brazil, it was reasonably common in forested areas and was a good climber, making extensive use of low shrubs, where it built nests about 1 m above the ground.

Animals are nocturnal and are caught on the ground as well as in trees. They build ball nests, usually in trees, although they are also reported to burrow under logs. Stomachs were found to contain leaves, grass seeds, fungal hyphae, and insects. In southeastern Brazil, insects were found in abundance in the stomachs examined (Contreras and Berry 1983; Crespo 1982a; Dalby 1975; Fonseca 1988; Mares, Ojeda, and Kosco 1981; Myers and Carleton 1981; Stallings 1988b; Vaz-Ferreira 1958; Veiga-Borrgeaud 1982; Villafañe et al. 1973).

Genus *Oryzomys* Baird, 1858 (sensu stricto)
Rice Rat, Ratón Arrocero

Description

Cusps are present on the molars. The tail is not penicillate and tends to be equal to or slightly shorter than the head and body. The hind foot exceeds 25 mm, thus distinguishing this species assemblage from the subgenera *Oligoryzomys* and *Microryzomys,* which include most of the smaller species.

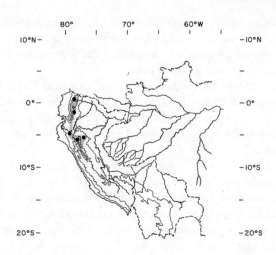

Map 16.54. Distribution of *Oryzomys albigularis.*

Oryzomys albigularis (Tomes, 1860)

Description

Average measurements: TL 296; T 160; HF 32; E 23. The dorsal pelage is dark tawny mixed with black, while the throat is usually white and the venter ranges from cream to gray. In the typical form the white is restricted to the throat, but it may extend to the entire venter (Gardner and Patton 1976).

Range and Habitat

This species is distributed from Costa Rica south through the Andean portion of Colombia to northern Peru (map 16.54). It ranges to the east in Venezuela in montane habitats, being adapted to cloud forest and multistratal evergreen premontane forest at elevations over 1,000 m.

Natural History

Aagaard (1982) found this species at up to 2,600 m elevation in the Venezuelan Andes. The rats were strongly associated with wet forests. They are primarily terrestrial and were sometimes active during the day.

Oryzomys alfaroi (J. A. Allen, 1891)

Description

Average measurements: TL 219; T 109; HF 25; E 17; Wt 33 g. The dorsum is dark reddish brown to tawny mixed with black, which is pronounced in the middorsal region. The venter is grayish white.

Range and Habitat

The group ranges from southern Tamaulipas, Mexico, south to Panama and in western Colombia to Ecuador (map 16.55).

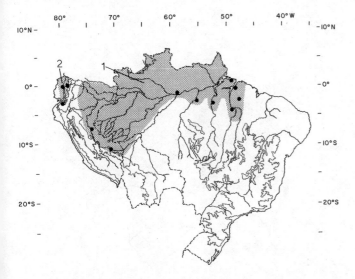

Map 16.55. Distribution of (1) *Oryzomys macconnelli*; (2) *O. alfaroi*.

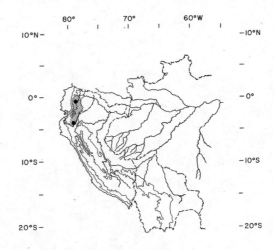

Map 16.56. Distribution of *Oryzomys balneator*.

Comment

Oryzomys alfaroi as currently defined includes at least three unnamed species in Central America. Only *O. alfaroi* is South American in its distribution (Musser and Williams 1985).

Oryzomys auriventer Thomas, 1890

Description

Three specimens from Huanhuachay, Ayacucho department, Peru, at 1,660 m elevation measured TL 304–64; T 183–91; HF 37–39; E 27–28; Wt 115–20 g (LSU). The dorsum is light brown while the venter is gray brown to pale orange. The tail is indistinctly bicolored; the hind feet are tan on the top, and the ears are sparsely haired.

Distribution

The species is recorded from 1,500 to 1,600 m elevation in the Andes of Ecuador and Peru (not mapped).

Oryzomys balneator Thomas, 1900

Description

Measurements: TL 221; HB 94; T 127; HF 27. The ears are small and thinly haired. The dorsum is dark brown mixed with black flecks especially in the middorsal region. The feet are clay colored, and the tail is faintly bicolored. The venter is ivory yellow, and the dorsal and ventral hairs are dark gray at the base.

Distribution

The type locality is the upper Río Pastaza, Mirador, Napo-Pastaza provinces, Ecuador, at 1,500 m elevation. The species is found in the mountains of southeastern Ecuador to northern Peru (map 16.56).

Oryzomys bolivaris J. A. Allen, 1901

Description

Total length ranges from 198 to 267 mm; tail, 90–130 mm; hind foot, 27–35 mm. One specimen from Costa Rica measured TL 265; T 125; HF 32; E 20. The dorsum is cinnamon brown, very dark in the dorsal midline, contrasting strongly with the grayish white venter. The tail is faintly bicolored. The vibrissae are extremely long and conspicuous, distinguishing it from most sympatric congeners.

Note: Musser et al. 1998) redefine the limits of *Oryzomys bolivaris,* previously included in *O. bombycinus* as a full species.

Range and Habitat

The species is distributed from Nicaragua south to Panama and in western Colombia to northwestern Ecuador (not mapped).

Oryzomys buccinatus (Olfers, 1818)
Colilargo Rojizo

Description

Average measurements: HB 164; T 195; HF 37; E 21; Wt 100 g. This is a large *Oryzomys* with a long tail, smaller than *O. ratticeps* with more dark flecking, less tan, and a dark belly. Dorsally it is tan agouti with some black and a tendency to darken on the head, the rump, and along the middorsum. Some individuals have more orange than others. The dorsal color lightens on the sides and grades into a grayish white venter. The tail is slightly bicolored, gray on the top and whitish below except for the proximal few centimeters, which are often yellowish.

Note: Musser et al. (1998) include *O. buccinatus* in *O. angouya* (Fisher, 1814) (see *O. ratticeps* below).

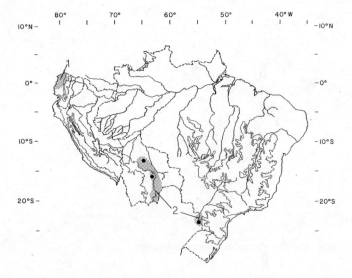

Map 16.57. Distribution of (1) *Oryzomys bolivaris;* (2) *O. buccinatus.*

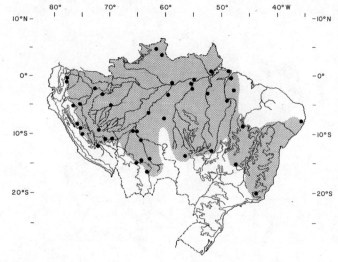

Map 16.58. Distribution of *Oryzomys capito.*

Distribution

Oryzomys buccinatus is distributed from eastern Paraguay and northeastern Argentina to southern Bolivia (map 16.57).

Ecology

In Paraguay this species was almost always caught under bromeliads at the ecotone between fields and forests (UM).

Comment

Oryzomys buccinatus is very similar to *O. subflavus* of the forests and forest edges of southeastern Brazil (Myers 1982). *O. ratticeps* was included in *O. buccinatus* by Hershkovitz (1959), but see Musser and Carleton (1993).

Oryzomys capito (Olfers, 1818)
Colilargo Acanelado

Description

Average measurements: HB 134; T 129; HF 31; E 22; Wt 66 g. This is a smaller, stouter-bodied *Oryzomys* with a proportionally much shorter, thinner tail. The dorsum is chestnut, lighter on the sides and sharply demarcated from a grayish white belly. The face is grayer, with whitish gray at the base of the vibrissae and a white chin. The tail is not bicolored (plate 12).

Note: Musser et al. (1998) have reinstated *Oryzomys megacephalus* (Fisher, 1814) as the senior synonym of *O. capito* (Olfers, 1818).

Distribution

Oryzomys capito is found from the Amazonian portions of Venezuela, Columbia, and the Guianas south to Amazonian Ecuador, Peru, Bolivia, Brazil, and Paraguay (Carleton 1980; 1982; Massoia 1980a) (map 16.58).

Life History

In the laboratory *O. capito* has a gestation period of twenty-five days, an average litter size of 3.5 (range 1–6), and birthweights of 3.7 to 4.0 g. The eyes of the young open after seven or eight days; young mice reproduce at fifty days and live three years. In captivity breeding is continuous (Worth 1967).

Ecology

In Misiones province, Argentina, mice of this species were caught in tropical forest. In Paraguay they were caught on and off the ground in the forest. They are omnivorous, feeding on fruits, seeds, adult insects, insect larvae, and fungi. Densities can vary between 0.5 and 5.0 individuals per hectare in Venezuela. In French Guiana they are strongly frugivorous in second-growth forests. They do not burrow but build leaf nests within natural crevices (Guillotin 1982; Massoia 1976b; O'Connell 1981; UM).

Oryzomys galapagoensis (Waterhouse, 1839)

Description

The type specimen measured HB 152; T 120; HF 29.6 (see chapter 18). The dorsum is brown mixed with black and pale yellow, and the venter is white tinged with yellow. The bases of the hairs are gray.

Distribution

The species is currently found on San Cristóbal and Santa Fe Islands of the Galápagos archipelago (not mapped).

Oryzomys hammondi (Thomas, 1913)

Description
Measurements: HB 203; T 251; HF 41; E 18. This large species has a long tail and broad hind feet probably adapted for an arboreal or semiarboreal way of life. It is a dark brown form resembling *Oecomys concolor* but much larger.

Distribution
The type locality is Mindo, Pichincha province, Ecuador, at 4,213 ft. It is assumed to be confined to northwestern Ecuador (not mapped).

Comment
The form is included in *Nectomys* as *incertae sedis* by Hershkovitz (1944).

Oryzomys intermedius (Leche, 1886)

Description
Measurements: TL 305–35; HB 150–62; T 155–74; HF 32–34. The dorsum is yellow brown with black-tipped hairs more abundant in the dorsal midline. The sides are more yellowish. The area around the eyes and at the base of the tail is blackish. The venter is grayish white, often with a yellow tinge on the breast. The ears are large, with fine blackish hairs on the surface, the feet are gray brown, and the tail is bicolored.
Note: Musser et al. (1998) include *O. intermedius* in *O. russatus* (Wagner, 1848).

Distribution
The species is found from southeastern Brazil to Rio Grande do Sul (map 16.59).

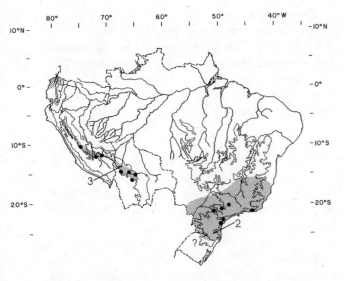

Map 16.59. Distribution of (1) *Oryzomys hammondi;* (2) *O. intermedius;* (3) *O. keaysi.*

Natural History
Adult males average 87.5 g while adult females average 77.6 g. Females become sexually mature at 115 days of age and can bear six litters a year. Maximum longevity in the field may reach eighteen months. The home range for both sexes is about 0.45 ha (Bergallo 1995).

Comment
Gardner and Patton (1976) suggest a synonymy with *Oryzomys nitidus*, but Musser et al. (1988) would reinstate *O. russatus* in their preliminary consideration of the *"nitidus* group."

Oryzomys keaysi J. A. Allen, 1900

Description
Measurements: TL 305–35; HB 150–62; T 155–74; HF 32–34. The dorsum is yellow brown with black tipped hairs more abundant in the dorsal midline, and the sides are more yellowish. The area around the eyes and at the base of the tail is blackish. The venter is grayish white, often with a yellow tinge on the breast. The ears are large, with fine blackish hairs on the surface, the feet are gray brown, and the tail is bicolored.

Distribution
The type locality is the valley of the upper Río Inambari, Inca mines, in Puno department, Peru, at 2,000 ft. It extends through the mountains of central Peru to Bolivia at middle elevations, being replaced higher by *O. levipes* (Patton, Myers, and Smith 1990) (map 16.59).

Oryzomys kelloggi Avila-Pires, 1959

Description
Measurements: HB 145–60; T 145–72; HF 33; Wt 75–115 g. The dorsum is chestnut brown grading to yellowish on the sides. The venter is white and sharply demarcated from the dorsal color. The superficial hairs of the feet are white.
Note: Musser et al. (1998) recommend that *O. kelloggi* be included in *O. russatus* (Wagner, 1848).

Distribution
The type locality is Fazenda São Geraldo, Além Paraíba, Minas Gerais, Brazil. The limits of the species' distribution in southeastern Brazil are poorly known (map 16.60).

Comment
The species' relationship to *O. intermedius* needs to be addressed.

Oryzomys lamia Thomas, 1901

Description

The measurements of the type specimen were HB 145; T 153; HF 33; E 23. This species is rather large but is smaller than *O. subflavus*, with which it co-occurs. The dorsal color is ochraceous buff, extending to the cheeks and flanks. The white venter is sharply demarcated from the sides. The bases of the hairs are slate. The upper surface of the forefeet is white, that of the hind feet is buff, and the bicolored tail is nearly naked, with short, fine hairs.

Note: Musser et al. (1998) recommend that *O. lamia* be included in *O. russatus* (Wagner, 1848).

Distribution

The type locality is Rio Jordas, a tributary of the Rio Paranaíba, Miras Gerais, Brazil, at 700–900 m. The species is assumed to be confined to southeastern Brazil (not mapped).

Comment

This species is assumed to be allied to *Oryzomys intermedius*.

Oryzomys legatus Thomas, 1925

Considered synonymous with *O. nitidus* Thomas, 1884 (map 16.60) (see *Oryzomys nitidus*).

Note: Musser et al. (1998) recommend *O. legatus* be included in *O. russatus* (Wagner, 1848).

Oryzomys levipes Thomas, 1902

Description

One specimen measured HB 145; TL 53; HF 33; E 23 (BMNH). This is a very dark brown species with a gray white venter.

Distribution

The type locality is Limbane, Puno department, Peru, at 2,700 m. The species occurs from southeastern Peru to adjacent Bolivia at about 2,000 m elevation (map 16.60).

Comment

This species is distinct from *Oryzomys albigularis* and *O. keaysi* (Patton 1990) (see Musser and Carleton 1993).

Oryzomys macconnelli Thomas 1910

Description

Average measurements: TL 286; T 152; HF 33; E 21; Wt 58 g. The dorsum is ochraceous tawny interspersed with black, and the sides are reddish. The

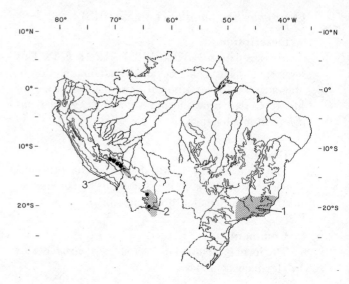

Map 16.60. Distribution of (1) *Oryzomys kelloggi;* (2) *O. legatus;* (3) *O. levipes.*

venter is grayish white, the tail bicolored. The juvenile pelage is dark brown with little or no reddish cast, and the venter is dirty white. The long dorsal hair in the adult distinguishes this species from *O. capito* where they co-occur.

Range and Natural History

The species is distributed in Guyana, Suriname, southern Venezuela, and southern Colombia to eastern Ecuador and Peru and extends to northern Brazil (map 16.55). It prefers moist forest. One embryo count of two indicates a possible small litter size (MVZ).

Oryzomys nitidus Thomas, 1884

Description

Average measurements: TL 274; T 149; HF 33; E 25; Wt 55 g. This large *Oryzomys* has an orangish brown dorsum, lighter on the sides, with a sharply contrasting white or gray white venter. The tail is quite long and bicolored, and the ears are large and sparsely haired (pers. obs.).

Distribution

Oryzomys nitidus is distributed from Ecuador south to northwestern Argentina, where it occurs in Salta province. To the east it extends to the Rio Xingu in Brazil (Mares, Ojeda, and Kosco 1981; Musser and Carleton 1993) (map 16.61).

Ecology

In Salta province *O. nitidus* occurs only in mesic and transitional forests. It is not abundant in any area

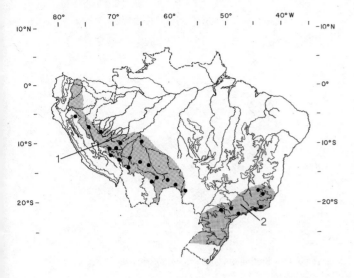

Map 16.61. Distribution of (1) *Oryzomys nitidus;* (2) *O. ratticeps.*

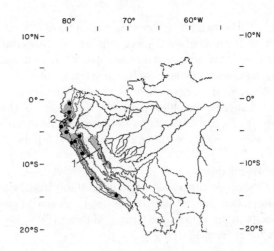

Map 16.63. Distribution of (1) *Oryzomys polius;* (2) *O. xantheolus.*

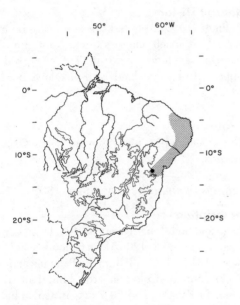

Map 16.62. Distribution of *Oryzomys oniscus.*

but is most common in forests with little undergrowth. Individuals of this species have been taken in dense second growth along streams and roads. They inhabit burrows in the forest floor. Stomachs have been found to contain berries, unidentified plant remains, and insects (Mares, Ojeda, and Kosco 1981; Ojeda and Mares 1989).

Comment
Here considered synonymous with *O legatus.*

Oryzomys oniscus Thomas, 1904

Description
Measurements: HB 140; T 145; HF 31; E 24. The ears are rather large, and the tail is nearly hairless and

finely scaled. The dorsum is gray brown, darker in the midline. The venter is gray white (the bases of the hairs are slaty, the tips white).

Note: Musser et al. (1998) include *O. oniscus* in *O. laticeps* (Lund, 1840).

Distribution
The type locality is São Torenzo, Pernambuco state, Brazil. The total distribution is assumed to extend to northern Bahia state, Brazil (map 16.62).

Oryzomys polius Osgood, 1913

Description
Measurements: HB 157; T 180; HF 30. The type specimen of *Oryzomys polius* had smoky gray dorsal pelage with fawn brown in the middorsum. The venter is grayish white.

Distribution
The type locality is the mountains east of Balsas, Tambo Carrizal, Amazonas department, Peru, at 5,000 ft. The species is assumed to be confined to north-central Peru (map 16.63).

Comment
Osgood described this species from Peru and compares it with *Oryzomys xantheolus.* Its exact status remains to be clarified pending further study.

Oryzomys ratticeps

Description
Average measurements: HB 185; T 209; HF 38; E 25; Wt 144 g. This very large *Oryzomys* has an orangish brown dorsum with a quite sharply demarcated cream-colored venter (see account for *O. bucci-*

natus). The head is darker gray, with a wash of white around the nose of some specimens. The feet are whitish above. The ears are large and sparsely haired, and the tail is very long and not bicolored (pers. obs.).

Note: In their revision of *Oryzomys*, Musser et al. (1998) suggest the oldest name for *O. ratticeps* is *Mus angouya* Fisher, 1814. Thus *O. angouya* would include not only *O. ratticeps* but *O. buccinatus* and others.

Distribution

Oryzomys ratticeps is found in southern Brazil and eastern Paraguay, although it has been collected at one locality in the Paraguayan Chaco (Massoia 1980a; Myers 1982) (map 16.61).

Ecology

This is a forest-dwelling species, though it has also been caught at the ecotones between forests and fields (Myers 1982; UM).

Oryzomys subflavus (Wagner, 1842)

Description

The type specimen measured HB 157; T 159; HF 29.4. Twenty-three specimens had the following average measurements: HB 143.4; T 176.8; HF 33.3; E 21.9; Wt 69.6 g (São Paulo). This is one of the larger species. The dorsum is a yellow brown flecked with black, especially in the middorsal region. The venter is white, but the bases of the hairs are gray.

Distribution

The species is broadly distributed in eastern Brazil, extending to dry forests and transition zones in the Brazilian Atlantic rain forest (map 16.64).

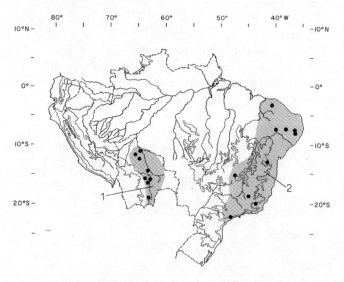

Map 16.64. Distribution of *Oryzomys subflavus*: (1) Bolivian component; (2) Brazilian component.

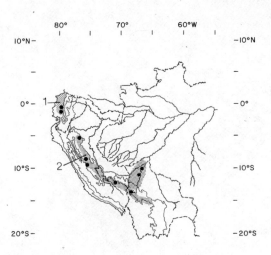

Map 16.65. Distribution of (1) *Oryzomys talamancae*; (2) *O. yunganus*.

Natural History

Stallings (1988a,b) found this species to be mainly terrestrial. Stomach analyses demonstrated a diet of herbaceous vegetation and fruit. The preferred habitat appears to be grasslands and meadows bordering forests.

Comment

This species is possibly synonymous with *Oryzomys buccinatus*.

Oryzomys talamancae J. A. Allen, 1899

Description

Head and body length averages 124.4 (101–36) mm in Panama and 135.2 (120–51) in Colombia; tail length averages 124.7 (110–43) mm in Panama and 125.2 (114–40) mm in Colombia; the hind foot averges 29.2 mm in Panama and 29.5 mm in Colombia; and the ear averages 21.1 mm in Panama and 19.0 mm in Colombia (Musser and Williams 1985). The short dorsal pelage is russet brown mixed with blackish brown, "passing gradually into clear, yellow-brown on the sides" (Allen as quoted in Musser and Williams 1985). The venter is grayish white, and the tail is distinctly bicolored.

Range and Habitat

This rat is distributed from eastern Costa Rica through western Colombia south to Ecuador and east to northern Venezuela (map 16.65). It ranges from sea level to 1,500 m elevation.

Comment

Musser and Williams (1985) note that *Oryzomys villosus* is invalid, since the holotype is a composite of an *O. talamancae* skin and an *O. albigularis* skull.

Oryzomys xantheolus Thomas, 1894

Description

Four specimens from Los Junitas, Lambayeque department, Peru, at about 300 m elevation measured TL 244–82; T 139–77; HF 27–33; E 21–26; Wt 65–96 g (LSU). The dorsum is a grayish brown, somewhat darker in the midline, the venter is cream, and the tail is bicolored. The hind feet are whitish on the dorsal surface, and the ears are sparsely haired.

Distribution

The species occurs in western Peru, possibly extending to Ecuador and Peru (map 16.63).

Oryzomys yunganus Thomas, 1902

Description

Two males from central Columbia measured TL 214–50; T 94–112; HF 29–31; E 18–19 (FMNH). The dorsum is very dark brown, contrasting with a gray venter. The tail is noticeably bicolored. The juneville pelage is gray.

Note: Musser et al. (1998) have named and described two new species of *Oryzomys*. The publication was too recent to integrate with the text and maps of the current volume. The new species are *O. tatei* from central eastern Ecuador and *O. emmonsi* from the middle Xingu River, Brazil.

Distribution

The type locality is Río Secure, Charuplaya, Cochabamba department, Peru. The species extends from southern Venezuela and the adjacent Guianas through southern Colombia to Peru and possibly Bolivia and Brazil (map 16.65).

Comment

The species was formerly included in *O. capito*. Gardner and Patton (1976) discussed the validity of its status (see also Musser et al. 1998).

Genus *Phaenomys* Thomas, 1917
Phaenomys ferrugineus (Thomas, 1894)
Rio de Janeiro Rice Rat

Description

Measurements from BMNH (Rio) are HB 150; T 190. The hind foot is broad, with moderately long claws. The soft dorsal pelage is reddish, but the hairs are gray at the base; the venter is pale red. The tail is not bicolored and resembles the dorsum in color. The vibrissae are prominent and dark brown. Females have eight mammae.

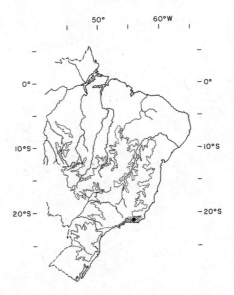

Map 16.66. Distribution of *Phaenomys ferrugineus*.

Distribution

The species is known from the Atlantic rain forests of Rio de Janeiro and Bahia states, Brazil (map 16.66).

Genus *Pseudoryzomys* Hershkovitz, 1962
Rato do Mato

Description

This medium-sized rat has a tail equal to or only slightly longer than the head and body: HB 94–140; T 106–40. The dorsum is yellow brown and the venter is dirty white. Although it was formerly included under *Oryzomys*, its molars are distinctive (Voss and Myers 1991).

Comment

Two species are recorded in the literature, *Pseudoryzomys warrini* and *P. simplex*. Voss and Myers (1991) believe and offer evidence that only a single species is valid, *P. simplex*. Langguth and Neto (1993) examined the penile morphology and suggest affinities with the Oryzomyini. Although such affinities are demonstrable, this genus could be considered Sigmodontinae *incertae sedis*.

Pseudoryzomys simplex (Winge, 1887)

Description

See the genus account. However, Voss and Myers (1991) offer the following measurements for Bolivian specimens: HB 117; T 128; HF 31; E 17.

Distribution

The species is apparently confined to lowland alluvial areas of strong seasonal rainfall in Bolivia, Paraguay,

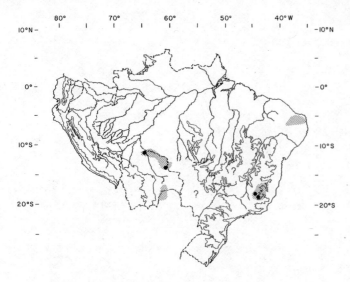

Map 16.67. Distribution of *Pseudoryzomys simplex.*

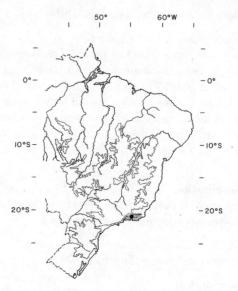

Map 16.68. Distribution of *Rhagomys rufescens.*

Argentina, and Brazil. Such habitats are characterized by palm savannas and thorn scrub (map 16.67).

Ecology
The sparse data available suggest that this rat is terrestrial and favors grassland habitats that may be seasonally inundated.

Rhagomys *Thomas*, 1917
Rhagomys rufescens (Thomas, 1886)
Brazilian Arboreal Mouse

Description
Two specimens in the museum at Rio de Janeiro derived from Mina Gerais state. The measurements were HB 162–50; T 163–55; HF 37–36; E 29–23. On the other hand, specimens from BMNH had head and body length of about 94 mm with the tail almost equal to the head and body; these specimens derived from Rio de Janeiro state. The dorsum is rather pale reddish brown, and the venter is similar. The tail resembles the dorsum and is not bicolored; it has elongated hairs at the tip.

Distribution
The species is known only from the states of Rio de Janeiro and Minas Gerais, Brazil (map 16.68).

Genus *Scolomys* Anthony, 1924

Description
This genus, which resembles *Neacomys*, was reviewed by Patton and da Silva (1995). Measurements: HB = 90; T = 70. The dorsal pelage is spinescent; its color is variable but is some shade of agouti brown. The venter is gray. The tops of the hind feet are gray, but the toes are paler. The tail is faintly bicolored.

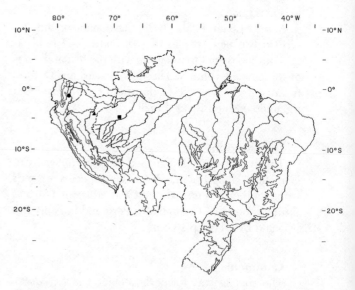

Map 16.69. Distribution of three species of *Scolomys:* circle, *S.-melanops;* triangle, *S. ucayalensis;* square, *S. juruaense.*

Distribution
The genus was originally known only from eastern Ecuador, at 1,150 m (Nowak 1991), but it was subsequently taken in Peru and Brazil (map 16.69).

Scolomys juruaense Patton, Nazareth, and de Silva, 1995

Description
Measurements: TL 160; HB 86; T 76; HF 21; E 16. This spiny mouse nearly exceeds *Neacomys* in size.

Distribution

The type locality is the Rio Juruá of western Brazil at 70°51′ W, 6°45′ S (map 16.69).

Comment

While describing the new species, Patton and de Silva (1995) offer a total review of the genus.

Scolomys melanops Anthony, 1924

Description

Measurements: TL 158; HB 90; T 68; HF 20. This small mouse is superficially similar to *Neacomys*. The rostrum is short and the molars are small. The tail is less than half the total length, and the ears are moderately large. The dorsum is sooty black flecked with brown, and flattened spines tipped with black are mixed among the dorsal hairs. The venter is gray.

Distribution

The type locality is Mera in eastern Ecuador at 1,275 ft (map 16.19).

Scolomys ucayalensis Pacheco, 1991

Description

Measurements: TL 144; T 60; HF 18; E 13. The ears are small and the tail is short; both are sparsely haired, and the tail has no terminal tuft. This small spiny mouse has dorsal pelage composed entirely of flattened spines, black interspersed with grayish brown. The venter is a clear gray, and the feet are whitish.

Distribution

The type locality is the right bank of the Río Ucayali at 79°30′ W, 4°52′ S, 2.8 km east of Jenaro, Herrera, Loreto department, Peru (map 16.69).

Genus *Sigmodontomys* J. A. Allen, 1897
Sigmodontomys alfari

Description

Sigmodontomys alfari has cusps on its molars when extremely young, but they rapidly wear down to give a flat-crowned appearance (fig. 16.5). The lack of cusps in the adult is diagnostic for the species but is shared with the closely related *Nectomys squamipes*. Head and body length ranges from 115 to 152 mm, while tail length ranges from 149 to 190 mm and the hind foot from 33.8 to 40 mm. Three specimens from Panama measured TL 300–312; T 161–80; HF 33–35; E 19. The dorsum is reddish brown, the venter varies from gray to buff, and the tail is not bicolored. Subspecies and their characters are reviewed by Hershkovitz (1944).

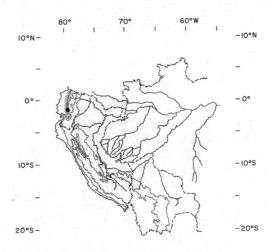

Map 16.70. Distribution of *Sigmodontomys alfari*.

Range and Habitat

This species is found in the lowlands from eastern Honduras to Panama and thence to northern Ecuador (map 16.70).

Natural History

Sigmondontomys alfari is not nearly as specialized for an aquatic life as is *Nectomys squamipes*. There is no webbing between the toes. This is a lowland form confined to moist habitats, but its natural history is poorly known.

Genus *Neacomys* Thomas, 1900
Spiny Mouse

Description

The dental formula is I 1/1, C 0/0, P 0/0, M 3/3. In this small rice rat, head and body length ranges from 64 to 100 mm, and the tail is approximately the same length. The dorsal pelage is composed of hairs interspersed with bristles or spinelike hairs, easily distinguishing it from other small rice rats. The spines have alternating dark and light bands, giving a salt-and-pepper appearance. The underparts are white or cream, in contrast to the dorsum, which ranges from reddish to dark yellow brown. The tail is sparsely haired, brown above and lighter below.

Distribution

The genus occurs from Panama across northern South America to southwestern Brazil. In general it is confined to lower elevations. New forms are currently being described.

Natural History

These small, terrestrial rice rats feed on seeds, insects, and small fruits and are nocturnal. Litter size ranges from two to four. This genus may show great

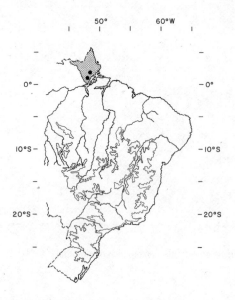

Map 16.71. Distribution of *Neacomys guianae*.

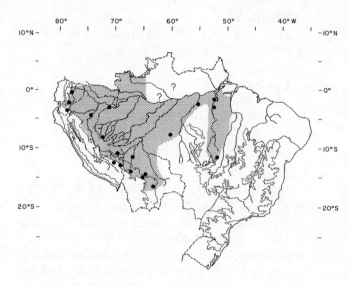

Map 16.72. Distribution of *Neacomys spinosus*.

fluctuations in abundance throughout the annual cycle (O'Connell 1981).

Neacomys guianae Thomas, 1905

Description
Average measurements: TL 154; T 76; HF 20; E 14; Wt 14. This is an extremely small species. The dorsum is reddish agouti brown, dark in the midline, and the sides may be orangish and the venter white. The tail is not noticeably bicolored.

Range and Habitat
The species is distributed in the Guianas and adjacent parts of northern Brazil (map 16.71). It prefers dense, humid forests (Husson 1978).

Neacomys spinosus

Description
Six specimens from Río Curanja, Loreto department, Peru, measured TL 158–80; T 80–101; HF 22–27; E 14–17; Wt 19–32 g (LSU). This species is similar in appearance to *N. guianae* but larger, having a reddish agouti brown dorsum grading to a light orange on the sides, contrasting with a white venter. The tail is not noticeably bicolored. The dorsal hairs are stiff to the touch.

Distribution
The species is found in the lowlands of southwestern Brazil, Peru, Bolivia, and Ecuador (map 16.72).

Neacomys tenuipes Thomas, 1900

Description
Total length ranges from 158 to 163 mm, the tail from 83 to 87 mm, and the hind foot from 20.5 to

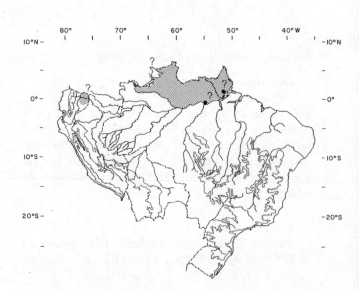

Map 16.73. Distribution of *Neacomys tenuipes*.

21 mm. Average weight is about 16 g. The dorsum is reddish brown mixed with black, the underparts are white to cream, and the tail is faintly bicolor.

Range and Habitat
The species occurs from eastern Panama, south to Colombia and eastern Ecuador and east to Venezuela, thence extending to Brazil (map 16.73). The content of this species is currently under study, and the range will undoubtedly be revised. It is strongly associated with moist habitats and multistratal tropical evergreen forest to cloud forest and ranges up to 1,655 m elevation in northern Venezuela. It prefers dense cover of herbs, ferns, and shrubs (Handley 1976).

Genus *Nectomys* Peters, 1861

Description
See *Nectomys sqamipes*.

Nectomys parvipes Petter, 1979
Neotropical Water Rat, Rata Nadadora

Description
Measurements: HB 135; T 152; HF 37; E 17 (Petter 1979). This small water rat was recently described from French Guiana (14°35′ N, 52°28′ W) and is known only from the type locality. It is similar in appearance to *N. squamipes* but much smaller (not mapped).

Nectomys squamipes (Brants, 1827)
Neotropical Water Rat, Rata Nadadora

Description
Average measurements: HB 207; T 223; HF 50; E 22; Wt 255 g. This large rat is unmistakable because of its long, glossy coat, dark brown dorsum, and grayish venter. The sides are lighter and often show an orangish wash. *Nectomys* can be confused with rats of the genus *Holochilus*, but *Nectomys* has smaller hind feet. The webbing of the hind feet is less well developed than in *Holochilus* and the hind foot "scales" are granular (fig. 16.9). The teeth of *Nectomys* are cuspidate, in contrast to the planed teeth of *Holochilus* (fig. 16.5). Also, in general the coat is short and sleek and the vibrissae are longer (Hershkovitz 1944; Massoia 1976a; pers. obs.).

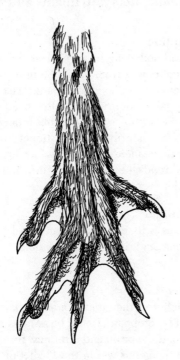

Figure 16.9. Hind foot of *Nectomys squamipes* (after Hershkovitz 1944).

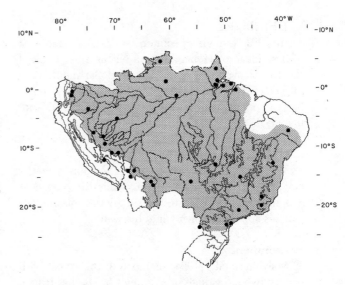

Map 16.74. Distribution of *Nectomys squamipes*.

Distribution
Nectomys squamipes is distributed from the Guianas to Paraguay and northeastern Argentina. In southern South America it is found in eastern Paraguay and northeastern Argentina (Ernest 1986; Hershkovitz 1944; Massoia 1976a; UM) (map 16.74).

Ecology
In Misiones province, Argentina, this species reproduces from October to November with a litter size of four or five (Crespo 1982b). In general this semiaquatic rat is confined to streams within forests but is occasionally found in other moist areas. In central Brazil, home ranges were calculated as 2,200 m². This nocturnal rodent spends considerable time foraging in the water and eats fungi, plant matter, fruits, seeds, invertebrates, and vertebrates. It builds nests under old logs or brush heaps or in tangled roots. When released from a trap individuals almost always flee to the water (Ernest 1986; Ernest and Mares 1986; Hershkovitz 1944).

Comment
Nectomys squamipes is probably a composite species containing at least five named forms (Reig 1984, 1986).

Genus *Amphinectomys*
Amphinectomys savamis Malygin, 1994

Description
This species is superficially similar to *Nectomys squamipes*, but it differs in that the brown dorsal hairs are monochromatic and the webbing between the toes of the hind foot extends to the claws.
Chromosome number: $2n = 52$; FN = 66.

Distribution

In 1991 a single specimen was taken 7 km east of the village of Henaro Errera on the right bank of the Río Ucayali, Requena, Loreto department, Peru (4°55′ S, 73°45′ W). Michael H. Valqui (pers. comm.) took a specimen near this locality in July 1997 (not mapped).

Habitat

The original type was found in the forest at the edge of a stream. M. H. Valqui, while camping near a laguna, was attracted to the site by the splashing of animals apparently diving into the water.

Comment

Malygin et al. (1994) compared this type with twenty-two individuals of *Nectomys squamipes* from a marshy site 5 km from the type locality. The measurement data are discussed in the original publication. To fieldworkers the monochromatic hair and fully webbed hind feet are diagnostic.

Genus *Nesoryzomys* Heller, 1904
Galápagos Rice Rat

Description

Measurements: HB 100–200; T 79–140. The type of *N. indefessus* measured HB 135; T 97; HF 30. The tail is shorter than the head and body. The three species will be described separately in chapter 18.

Comment

Gardner and Patton (1976) consider these insular forms to be at the generic level given their long isolation on this Ecuadorian archipelago. Hutterer and Hirsch (1979), and Patton and Hafner (1983), concur that there are three species: *Nesoryzomys darwini, N. fernandinae,* and *N. indefessus.*

TRIBE PHYLLOTINI

The tribe Phyllotini, though not formally diagnosed, was first proposed as a taxon by Vorontsov (1960). Olds and Anderson (1989) have presented a diagnosis for the group and offer the following caveats. There is no one character that is unique for the tribe. The genera *Reithrodon, Neotomys,* and *Euneomys* form a distinct subgroup within this tribe. The tribe exhibits the following character states: the heel of the hind foot tends to be haired, the pinnae are moderately long to rather long; the incisive foramina are long; the zygomatic notch is deeply excised; the molars are tetralophodont; M_3 is more than half the length of M_2 (this latter character distinguishes the phyllotines from most akodonts) (see fig. 16.5). This diagnosis does not include

Pseudoryzomys (see previous section). The genera *Reithrodon, Neotomys,* and *Euneomys* have grooved upper incisors. In *Euneomys* the molars approximate an S shape, reminiscent of the Sigmodontini (Olds and Anderson 1989), a condition exhibited to a lesser extent by *Reithrodon* and *Neotomys.* The publication by Braun (1993) essentially confirms the phylogenetic unity of this tribe. Walker, Spotorno, and Sans (1991) and Walker and Spotorno (1992) add credence to the conclusions of Braun, but as usual they pose certain cautions.

Steppan (1993, 1995), using ninety-six characters and employing Wagner parsimony analysis, confirms that *Calomys* is the most basal genus and may be paraphyletic. The altiplano endemics *Auliscomys, Galenomys,* and *Chinchillula* form a group, while the genera *Graomys, Andalgalomys,* and *Eligmodontia* seem similarly clumped. *Phyllotis* may be paraphyletic. *Reithrodon, Euneomys,* and *Neotomys* form a subgrouping, supporting Olds and Anderson (1989). The current content of *Phyllotis* will probably be altered.

Braun (1993) has made several taxonomic recommendations in her monograph on the systematics of the phyllotines. She suggests that *Loxodontomys* Osgood, 1947, be raised from subgeneric status to include only *Auliscomys micropus.* Two species of *Phyllotis, P. gerbillus* and *P. amicus,* could be included in the genus *Paralomys* Thomas, 1926. She creates a new genus *Maresomys* to include the former *Auliscomys boliviensis.*

Genus *Andalgalomys* Williams and Mares, 1978

Description

Head and body length ranges from 86 to 115 mm, while the tail ranges from 97 to 130 mm. These phyllotine rodents have moderately large ears and a tail roughly equal to or exceeding the head and body length. The eyes are not noticeably reduced in size. The feet are relatively small, distinguishing it from the similar *Eligmodontia.* The tail is bicolored, sparsely haired, and reddish brown to gray brown, contrasting with the white venter and feet.

Distribution

The genus is found in xeric and grassland habitats of Argentina and Paraguay, extending to southern Bolivia.

Comment

The validity of the genus *Andalgalomys* has been challenged by Steppan (1995), but Mares and Braun (1996) defend the taxon and offer new insights on the position of the genus within the Phyllotini. They also add the new species *A. roigi* to the Argentine fauna.

Andalgalomys pearsoni (Myers, 1977)

Description

Average measurements: HB 97; T 108; HF 25; E 18; Wt 25 g. The tail is longer than the head and body. *A. pearsoni* is larger than the sympatric *Calomys* and smaller than *Graomys griseoflavus*, with proportionally smaller ears and a shorter, nonpenicillate tail (Myers 1977; Williams and Mares 1978; pers. obs.). This moderately small mouse has a yellowish brown dorsum washed with black-tipped hairs. The tail is bicolored, brownish dorsally, and sparsely haired. The hair on the venter and feet is white; the face is brownish and not noticeably lighter than the dorsum, but without the black wash. The sides are often washed with orange.

Distribution

Andalgalomys pearsoni is limited to the Chaco of western Paraguay, extending to southeastern Bolivia (Williams and Mares 1978) (map 16.75).

Ecology

This mouse inhabits dry grasslands that occur as islands in the Chaco of western Paraguay. The surrounding thorn scrub is inhabited by *Graomys griseoflavus* (Myers 1977).

Genus *Andinomys* Thomas, 1902

Andinomys edax Thomas, 1902
Rata Andina

Description

Average measurements: HB 141; T 24; HF 28; E 24; Wt 58 g. This stout-bodied rat varies from uniformly dark buffy to uniformly brown dorsally and gray ventrally. Its very dense fur is long, soft, and fine; the ears are fairly large; and the tail is bicolored and not penicillate (Gyldenstolpe 1932; Pearson 1951; pers. obs.).

Distribution

Andinomys edax is found in high-altitude regions of Peru, Bolivia, northwestern Argentina, and northern Chile. In Argentina it is found along the Andes south to La Rioja province, and in Chile it is found in Arica department, Tarapacá province (map 16.76).

Ecology

In southern Peru, one female was trapped with three embryos (Pearson 1951). This is a high-altitude animal, having been caught at between 1,800 and 5,100 m. Throughout this range it appears to be found mainly in dense vegetation along water courses, although it has also been trapped in rocky terrain (Fo-

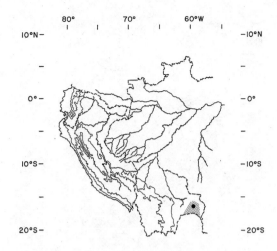

Map 16.75. Distribution of *Andalgalomys pearsoni*.

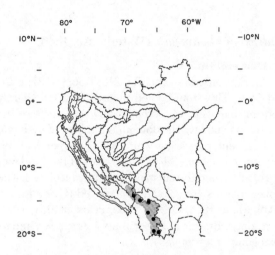

Map 16.76. Distribution of *Andinomys edax*.

nollat 1984; Olrog 1979; Pearson 1951; Pine, Miller, and Schamberger 1979; Tamayo and Frassinetti 1980; Thomas 1919a).

Genus *Auliscomys* Osgood, 1915

Description

Head and body length ranges from 96 to 127 mm, and tail length from 53 to 85 mm. These phyllotine rodents have a complex baculum, a tail shorter than the head and body, relatively short hind feet, a large external ear, a relatively large M_3, and a simplified or tetralophodont molar. The vibrissae are long; the tail tends to be bicolor; and the dorsum is some shade of brown, contrasting with the white to gray venter. The juveniles of some species have gray dorsal pelage.

Distribution

The genus is found in Peru, Bolivia, Argentina, and Chile. It occurs at high elevations, up to 6,000 m.

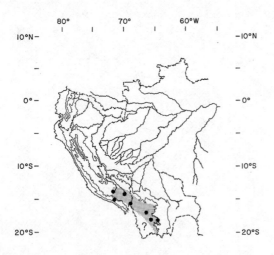

Map 16.77. Distribution of *Auliscomys boliviensis.*

Auliscomys boliviensis (Waterhouse, 1846)

Description
Average measurements: TL 213; T 86; HF 27; E 27; Wt 55 g. This stout-bodied mouse has a tail slightly shorter than its head and body and very large ears. Its fur is long and lax. The dorsum is buffy lined with tan, and the venter is creamy white. There is a patch of yellow fur in front of each ear. The soles of the hind feet are blackish and the tail may be slightly bicolored. The upper incisors are pale and ungrooved (Gyldenstolpe 1932; Mann 1978; Osgood 1943; Pearson 1951, 1958; pers. obs.). Braun (1993) separates this species into its own genus *Maresomys.*

Distribution
Auliscomys boliviensis is found at high elevations in western Bolivia, northern Chile, and southern Peru. In Chile it is found in Tarapacá province (Redford and Eisenberg 1992) (map 16.77).

Life History
In Peru females have three to four embryos. In captivity females will produce three litters a year with three to five young per litter (Pearson 1951; Pine, Miller, and Schamberger 1979).

Ecology
Auliscomys boliviensis is a high-altitude animal and has been trapped at up to almost 6,000 m. It lives is open country and is commonly found on boulder-strewn slopes with sparse vegetation, rock slides, and stone walls, as well as in *Ctenomys* burrows and viscacha colonies. It is diurnal and has a varied diet (Mann 1978; Pearson 1951, 1958; Pine, Miller, and Schamberger 1979; Tamayo and Frassinetti 1980).

Auliscomys micropus (Waterhouse, 1837)

Description
Average measurements: HB 127; T 96; HF 29; E 20; Wt 73 g. This medium-sized, robust species has an unfurred, bicolored tail shorter than its head and body, relatively small ears, and broad, ungrooved incisors. The dorsum varies from gray to brownish, and the venter is lighter, often washed with white (Mann 1978; Osgood 1943; Pearson 1958; pers. obs.).

Distribution
Auliscomys micropus is found in southern Chile and Argentina. In Chile it ranges along the Andes from about the former Talca province south to Magallanes province, and in Argentina north to Tucumán province (Honacki, Kinman, and Koeppl 1982; Mann 1978; Tamayo and Frassetti 1980; CMNH) (not mapped).

Life History
In Chile the average number of embryos per female was 4.8 (range 4–5; *n* = 5); in Argentina the average was 4.1 (range 1–7; *n* = 27) (Greer 1966; Pearson 1983; Pine, Miller, and Schamberger 1979).

Ecology
Auliscomys micropus has been caught in a wide variety of habitats from precordilleran steppe to forests. Its presence seems tied to sufficient cover; in forests, it occurs only where there is well-developed ground cover, and in grasslands it occurs only where there are many bushes. In the Patagonian forests of Argentina it reaches densities of 5.1 per hectare. *A. micropus* ranges from sea level to 3,000 m, is nocturnal, and burrows but also climbs well. It is primarily herbivorous, and stomachs have been found to contain fungus, grass seeds, and fruit (Greer 1966; Mann 1978; Murúa, González, and Jofre 1982; Pine, Miller, and Schamberger 1979; Pearson and Pearson 1982; Reise and Venegas 1974; Tamayo and Frassinetti 1980).

Auliscomys pictus (Thomas, 1884)

Description
Measurements: TL 178–216; T 83–102; HF 25–23; E 17–18 (includes *A. decoloratus*). This form is one of the smaller species. It is similar to *Auliscomys boliviensis*, but the ears are smaller. The dorsum is buff brown, with the head and shoulders somewhat gray. The feet are white.

Distribution
The species occurs in the high Andes of central Peru south to Bolivia (map 16.78).

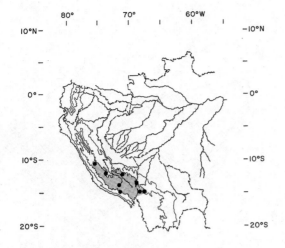

Map 16.78. Distribution of *Auliscomys pictus.*

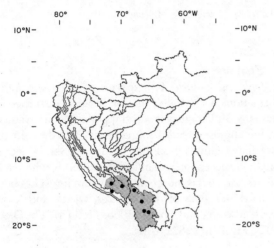

Map 16.79. Distribution of *Auliscomys sublimis.*

Auliscomys sublimis Thomas, 1900
Pericote Andino

Description
Average measurements: HB 111; T 53; HF 23; E 21; Wt 39 g. This is a volelike species of medium size with a short tail and comparatively small ears. There is a faint groove on the anterior face of the upper incisors. It has long, thick fur, and the dorsum is dull grayish fawn to tan flecked with gray, sometimes sharply demarcated from the white venter. Its feet and tail are pale above. Individuals from some populations may have considerable orange on the sides and rump (Gyldenstolpe 1932; Mares, Ojeda, and Kosco 1981; Pearson 1951; pers. obs.).

Distribution
Auliscomys sublimis is found on the altiplano from southern Peru to northeastern Chile and northern Argentina (Honacki, Kinman, and Koeppl 1982; Pearson 1958) (map 16.79).

Ecology
This species is found at high altitudes, from 4,000 to 6,000 m, one of the highest altitude records for a mammal in the Western Hemisphere. *A. sublimis* is an animal of open habitats, preferring areas with bunchgrass and rocks. It shelters under rocks or rock walls or in *Ctenomys* burrows. It is apparently gregarious and is often caught together with other individuals of the same species. In southern Peru it disappeared from October to December, perhaps hibernating (Koford 1955; Mares, Ojeda, and Kosco 1981; Pearson 1951, 1958; Sanborn 1950; Tamayo and Frassinetti 1980).

Genus *Calomys* Waterhouse, 1837
Laucha

Description
The molars are tetralophodont, and M_3 ranges from one-half to three-fourths the length of M_2. Head and body length ranges from 62 to 123 mm, while the tail ranges from 31 to 93 mm. The tail is equal to or shorter than the head and body; the external ear is moderate but not visibly reduced in size. Males have a complex baculum. The dorsum ranges from light brown to gray brown and the venter from white to gray. Species of *Calomys* superficially resemble the introduced urban *Mus musculus*, but figures 16.6 and 16.10 clearly separate these forms. There are seven species recorded in this genus that typically differ in size when sympatric.

Distribution
The genus is found predominately in southern South America (Peru, Bolivia, Chile, Argentina, Paraguay, and southeastern Brazil), but it also occurs in isolated "pockets" of Colombia and Venezuela.

Calomys boliviae (Thomas, 1901)

Description
See species description for *Calomys callosus*. Typical measurements from Bolivia are HB 98–106; T 73–82; HF 20–21; E 19.

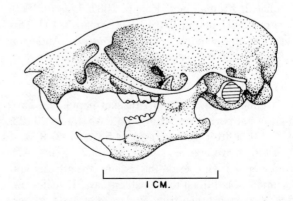

Figure 16.10. Skull of *Calomys laucha.*

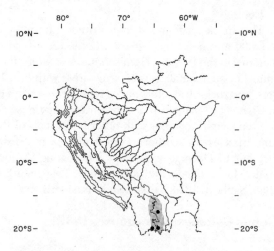

Map 16.80. Distribution of *Calomys boliviae*.

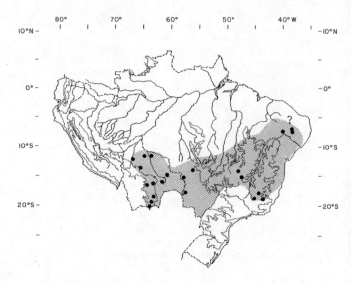

Map 16.81. Distribution of *Calomys callosus*.

Distribution

The species is found in western Bolivia (map 16.80).

Comment

Calomys boliviae is often listed as a subspecies of *C. callosus* and is so treated in Anderson (1997).

Calomys callosus (Rengger, 1830)
Laucha Grande

Description

Average measurements: HB 98; T 87; HF 21; E 20; Wt 31 g. This is the largest species of *Calomys*. It is dark gray brown dorsally and gray ventrally, its sparsely haired tail is bicolored, and its feet are tan (pers. obs.). Chromosome number: $2n = 36$; FN = 48 (Reig 1986).

Distribution

Calomys callosus is found in eastern and southwestern Brazil, Bolivia, Paraguay, and northern Argentina. In southern South America it is found in the southern and western portions of the Paraguayan Chaco and in Argentina across the northern provinces (Honacki, Kinman, and Koeppl 1982; Lucero 1983; Myers 1982; Thomas 1916, 1919b) (map 16.81). This distribution includes *Calomys venustus*. Anderson (1997) separates *C. venustus* from *C. callosus*.

Life History

Animals from a central Brazilian population have a gestation period averaging 21.8 days (range 20–23; $n = 43$), a litter size averaging 4.5 (range 2–9; $n = 13$), and an average neonatal weight of 2.3 g ($n = 42$). Young open their eyes on the sixth or seventh day, and lactation lasts fifteen to seventeen days. Females undergo a postpartum estrus (Hodara et al. 1984; Mares, Ojeda, and Kosco 1981; Mello 1978b).

Ecology

Calomys callosus is a species of open vegetation formations, especially common in areas disturbed by humans. In Paraguay it was trapped in palm savannas, in bunchgrass meadows, and in dry marshes. In the Brazilian caatinga it is found in the later seral stages of old field succession and in low thorn scrub, whereas in central Brazil it is most common where forest has been cleared and grass planted (Mares, Ojeda, and Kosco 1981; Massoia and Fornes 1965; Mello 1978b; Ojeda 1979; Streilein 1982a; UM).

Calomys laucha (Olfers, 1818)
Laucha Chica

Description

Average measurements: TL 127; T 55; HF 17; E 13; Wt 13 g. The tail is only about 40% of the total length. This very small rodent looks like a house mouse, with a white or grayish white venter and a light brown to tawny dorsum mixed with blackish. The distinguishing characteristic is a small white patch behind each ear (Barlow 1965; Mares, Ojeda, and Kosco 1981; Massoia et al. 1968; pers. obs.).
Chromosome number: $2n = 62$; FN = 82 (Reig 1986).

Distribution

Calomys laucha is found in southeastern Brazil, Paraguay, Uruguay, central Argentina, and southern Bolivia. In southern South America it is found in the south and west of the Paraguayan Chaco, through much of Uruguay, and in Argentina south to about Río Negro province (Barlow 1965; Myers 1982; Olrog and Lucero 1981; Vallejo and Gudynas 1981) (map 16.82).

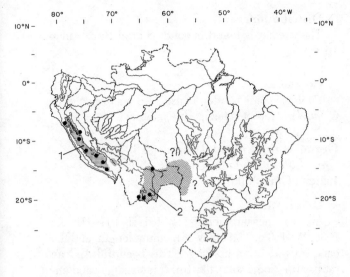

Map 16.82. Distribution of (1) *Calomys sorellus;* (2) *C. laucha.*

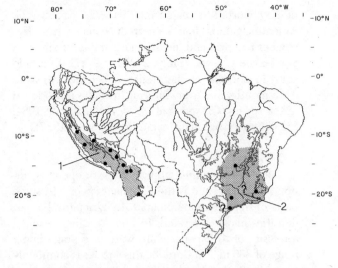

Map 16.83. Distribution of (1) *Calomys lepidus;* (2) *C. tener.*

Ecology

This little cricetine is common in many of the savanna and grassland areas of southern South America. It is one of the few Argentine rodents so far tested that can live for extended periods with no free water. It has been caught in small shrubs near coastal areas and rivers, in old fields, on rocky hillsides, and near dwellings. It has been reported at a maximum density of eighty-seven per hectare. Nests have been found under boards and rocks, in crevices in the ground, and even in trees. Typically it uses subterranean galleries or depressions beneath objects to build its grass nest. *C. laucha* is primarily herbivorous, climbs well, and experiences periodic dramatic population increases (Barlow 1965, 1969; Kravetz and Villafañe 1981; Lucero 1983; Mares 1973; Mares, Ojeda, and Kosco 1981; Myers 1982; Vallejo and Gudynas 1981).

Calomys lepidus (Thomas, 1884)
Laucha Andina

Description

Average measurements: TL 125; T 40; HF 18. This small mouse has a short tail, less than 30% of the total length, which distinguishes it from the longer-tailed *Eligmodontia*. It has dense fur; the grizzled tan and brown dorsum is darker along the spine, grading to lighter brown on the sides, and the venter is white or gray. The cheeks are white or gray, and the tail is entirely tan (Mann 1978; pers. obs.).
Chromosome number: $2n = 36$; FN = 68 (Reig 1986).

Distribution

Calomys lepidus is distributed at high altitudes along the Andes from western Bolivia south to the northernmost Chilean province of Antofagasta and the northwestern Argentine province of Jujuy (Honacki, Kinman, and Koeppl 1982; Tamayo and Frassinetti 1980) (map 16.83).

Ecology

This is a species of the high-altitude grasslands from 3,300 to 5,000 m (Dorst 1971; Koford 1955; Tamayo and Frassinetti 1980).

Calomys musculinus (Thomas, 1913)
Laucha Bimaculada

Description

Average measurements: HB 79; T 73; HF 18; E 13; Wt 18 g. The tail is about 50% of the total length. This small mouse is slightly larger than a house mouse and longer than the conspecific *C. laucha.* It is agouti brown dorsally, washed with orange on the sides and cheeks, and contrasting grayish white on the venter. Its tail is bicolored, and there is a white stripe behind each ear (Massoia et al. 1968; pers. obs.).

Distribution

Calomys musculinus is found in most of the Argentine provinces north of Mendoza. Given the proximity to Bolivia, it is included here (Olrog and Lucero 1981; Contreras and Justo 1974) (not mapped). Anderson (1997) plots a distribution for *C. musculinus* in the Andean foothills of Bolivia.

Life History

In the laboratory gestation ranges from an average of 21 to 24.5 days; average litter size is 5.4 (range 1–15); and for females the average age of first reproduction is reported to range from 37 to 72.5 days. In the Argentine pampas females reproduce mainly from November to April, and the average number of embryos per female is seven (range 2–11; $n = 37$). In a field study in Córdoba maximum life expectancy was six to eight months, and overwintering juveniles reproduced in the spring (Crespo et al. 1970; Hodara et al. 1984; Villafañe 1981b; Villafañe and Bonaventura 1987).

Ecology

Calomys musculinus appears to favor open vegetation formations and is the dominant species in some parts of the Argentine pampas. In agricultural areas, densities may be elevated. In Mendoza they reached a density of 37.6 per hectare with an average home range of 366 m². In Córdoba this species is commonly found in sorghum and alfalfa fields and is a reservoir for Argentine hemorrhagic fever.

This species can tolerate a water-free diet in the laboratory but loses weight. Despite a physiological capability for a xeric existence, in the wild it seems to avoid xeric microhabitats (Contreras and Rosi 1980b; Crespo et al. 1970; Kravetz and Villafañe 1981; Mares 1977d; Villafañe 1981b; Villafañe and Bonaventura 1987).

Comment

Reig (1986) has separated a segment of *Calomys musculinus* and placed it in a separate species, *C. murillus*. It is here considered a subspecies of *C. laucha.*

Calomys sorellus (Thomas, 1900)

Description

Measurements: HB 77–91; T 65–75; HF 16–18; E 17–20. This is the largest species of *Calomys* in Peru. The tail is variable, shorter or longer than the head and body. The pelage is long and soft, with thick underfur. The dorsum is reddish brown mixed with black, and the underparts are gray. The tail is bicolored except at the tip, and there are white tufts of hair behind the ears.

Distribution

Calomys sorellus has been considered a subspecies of *C. lepidus*. If it is considered a species, its range (map 16.82) is confined to the highlands of south-central Peru above 2,000 m, from La Libertad department to Puno.

Calomys tener (Winge, 1887)

Description

A series of twenty-nine specimens from Mato Grosso do Sul and São Paulo, Brazil, yielded the following average measurements: HB 80.17; T 59.85; HF 17.72; E 14.48 (see also genus account).

Distribution

The species is found in south-central Brazil (map 16.83).

Comment

Caolmys tener is often considered a subspecies of *C. laucha.* Here it is mapped separately.

Genus *Chinchillula* Thomas, 1898
Chinchillula sahamae Thomas, 1898
Laucha Chinchilla

Description

Average measurements: HB 161; T 104; HF 35; E 36; Wt 140 g. The tail is of medium length, and the ears, eyes and feet are large. This beautiful Andean rat has very long, soft, silky fur. It is strikingly and unmistakably marked, with a white venter and a dorsum of ashy brown, washed with blackish and intermixed with very long hairs that are black-tipped on the back and white-tipped on the front. The black markings are particularly evident along the dorsum and on the hips. The hips and rump are also white, with a conspicuous band of black and a blaze of white. The tail is well haired and distinctly penicillate (Gyldenstolpe 1932; Pearson 1951; pers. obs.).

Distribution

Calomys sahamae is distributed at high elevations from southern Peru to northern Chile and northwestern Argentina (Tamayo and Frassinetti 1980) (map 16.84).

Ecology

This is an animal of the altiplano, ranging from 4,000 to 4,600 m. It is usually caught among boulders and along stony walls and is often found in association with the viscacha, *Lagidium. Chinchillula* is nocturnal and

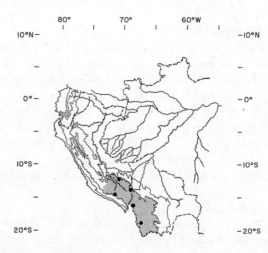

Map 16.84. Distribution of *Chinchillula sahamae*.

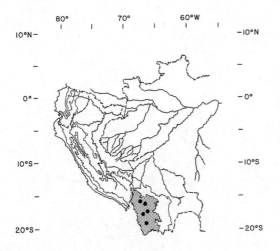

Map 16.85. Distribution of *Eligmodontia puerulus.*

herbivorous; stomachs usually contain leaves, seeds, and a few insects. It has been trapped for its beautiful skin (Dorst 1971, 1972; Mann 1978; Pearson 1951; Tamayo and Frassinetti 1980).

Genus *Eligmodontia* F. Cuvier, 1837
Eligmodontia puerulus

Comment

Eligmodontia puerulus Has often been considered a subspecies of *E. typus.* Here it is mapped as a separate species extending from extreme south-central Peru to Bolivia (map 16.85). Kelt et al. (1991) have established that *Eligmodontia* from Peru and Bolivia have a karyotype of $2n = 50$, distinct from the $2n = 44$ established for southern Argentina, but the boundary of the species is undetermined (see account for *E. typus*).

Eligmodontia typus F. Cuvier, 1837
Laucha Colilarga Bayo

Description

Average measurements: HB 81; T 86; HF 22; E 16; Wt 21 g. Females are larger than males. This is a small, pale, soft-pelaged mouse with the venter wholly or partly white. It has long hind legs and spade-shaped hind feet with distinctive hairy cushions on the soles. Animals from highland populations are much larger than those from the lowlands and have shorter tails and longer, laxer pelage. There is considerable geographic variation in coloration, tail length, and karyotype (Gyldenstolpe 1932; Mann 1978; Mares 1973, 1977b; Osgood 1943; Pearson 1951; Pearson, Martin, and Bellati 1987).

Distribution

Eligmodontia typus is found in western Bolivia, southern Peru, northern Chile, and Argentina. In northern Chile it is found on the altiplano south to the province of Antofagasta, while in the southern part of the country it is found in the former Malleco, Aisén, and Ultima Esperanza provinces. In Argentina this species is distributed along the Andes from at least Salta province south to La Pampa and Buenos Aires provinces and then across the whole country south to the Straits of Magellan (Contreras and Justo 1974; Honacki, Kinman, and Koeppl 1982; Mares, Ojeda, and Kosco 1981; Rau, Yáñez, and Jaksić 1978; Tamayo and Frassinetti 1980; BA) (not mapped).

Life History

In Argentina, near the town of Bariloche, Río Negro province, reproduction begins in October and lasts until the end of April. Males and females breed in the same season they are born. The age of first reproduction is six to eight weeks, the number of embryos averages 5.9 ($n = 20$), and longevity is less than one year (Pearson, Martin, and Bellati 1987).

Ecology

Eligmondontia typus is an animal of sandy soils and cool temperatures, occupying habitats that on other continents are filled by kangaroo rats and gerbils. In the xeric areas where it is found it is usually the most abundant rodent species.

In one study in Argentina, *E. typus* reached a density of 3.5 per hectare in a grazed area, though in seasons when it was scarce the density could fall to 0.4 per hectare; home-range diameter averaged 31 m. In this area it is nocturnal, nests and retreats underground, and is primarily granivorous, feeding on *Berberis* seeds and *Prosopis* (Daciuk 1974; Greer 1966; Mann 1978; Mares 1973, 1977b: Mares, Ojeda, and Kosco 1981; Pearson 1951; Pearson, Martin, and Bellati 1987).

Comment

Eligmodontia has been included in *Phyllotis*, and *E. typus* includes *E. hypogaeus, E. puerulus,* and *E. elegans. E. typus* undoubtedly contains two or more distinct species (Ojeda and Mares 1989; Oliver P. Pearson, pers. comm.).

Genus *Galenomys* Thomas, 1916
Galenomys garleppi (Thomas, 1898)

Description

Average measurements: TL 158; T 41; HF 24; E 22; Wt 59 g. This attractive mouse is unmistakable because of its stout body, very short white tail, and large ears. The incisors are very slender, ungrooved, and have anterior faces of pale yellow. It has close, thick fur; the dorsal color is grayish buff finely lined with black, becoming clear buff on the rump and sides; the venter is white and sharply demarcated from the sides.

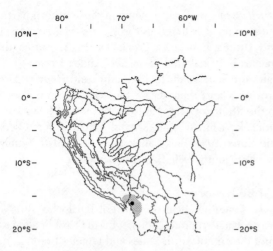

Map 16.86. Distribution of *Galenomys garleppi*.

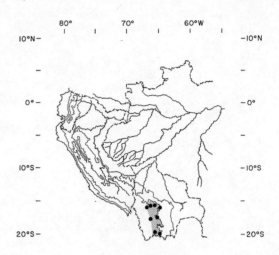

Map 16.87. Distribution of *Graomys domorum*.

The ears, though large, are smaller than in *Phyllotis*. The soles of the hind feet are hairy for the posterior two-thirds. Females of this species have eight mammae (Gyldenstolpe 1932; Thomas 1898b).

Distribution

Galenomys garleppi is found at high altitudes from southern Peru to western Bolivia to northern Chile (Tamayo and Frassinetti 1980) (map 16.86).

Genus *Graomys* Thomas, 1916

Description

These phyllotine rodents share many general characters with *Phyllotis* (see p. ●●●); however, in *Graomys* the tail is longer than the head and body. They are ratlike in size, with total lengths ranging from 235 to 327 mm. The dorsum is usually some shade of brown, and the venter is paler. The tail is frequently bicolored and is covered with fine, short hairs. The tail tip is usually penicillate.

Distribution

The genus is recorded from Bolivia, Paraguay, and Argentina.

Grayomys domorum (Thomas, 1902)
Pericote Palido

Description

Average measurements: HB 142; T 160; HF 30; E 27; Wt 73 g. This large, stout-bodied *Graomys* is brown dorsally, lightening to tan on the sides, and whitish gray ventrally. The tail is long, penicillate and strongly bicolored. The face has long vibrissae and no distinct markings. The feet are white.

Distribution

Graomys domorum is found in the Andes of Bolivia and northwestern Argentina. Its southernmost distri-

bution occurs in Salta province, Argentina (Honacki, Kinman, and Koeppl 1982; Ojeda and Mares 1989) (map 16.87).

Ecology

In Salta province, Argentina, this species is limited to transitional forest, occurring on very low mountains in the central part of the province, but it also may occur in drier portions of the subtropical forests. It was trapped in thick grass along road cuts and from areas of second growth (Mares, Ojeda, and Kosco 1981).

Graomys griseoflavus (Waterhouse, 1837)
Pericote Común

Description

Average measurements: TL 278; T 144; HF 30; E 24; Wt 62 g. This large, long-tailed rat resembles a small North American wood rat (*Neotoma*). It has long, soft fur with a brown or grayish dorsum, lighter on the sides and white on the venter, a strongly bicolored, penicillate, long tail, and large ears (Gyldenstolpe 1932; Mares 1973; pers. obs.).

Distribution

Graomys griseoflavus is found in Argentina, Bolivia, Paraguay, and perhaps southwestern Brazil. In southern South America it is found in the south and west of the Paraguayan Chaco and in Argentina, in at least the following provinces: Tucumán, Chaco, Salta, La Pampa, Chubut, Buenos Aires, and Mendoza (Allen 1901; Contreras 1972b, 1982a,b; Contreras and Justo 1974; Daciuk 1974; Lucero 1983; Honacki, Kinman, and Koeppl 1982; Mares, Ojeda, and Kosco 1981; Myers 1982; Rosi 1983) (map 16.88).

Ecology

Graomys griseoflavus is found in diverse locations that are typically xeric but that include cultivated

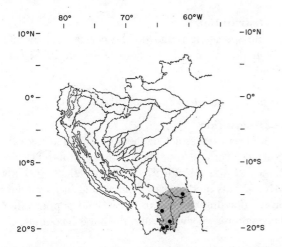

Map 16.88. Distribution of *Graomys griseoflavus*.

fields, semiarid sandy areas with rocks, hillsides with boulders, riverbanks with associated gallery forest, and orchards. Its grass nests have been found in large cacti, thorn bushes, and in *Microcavia* burrows.

These mice are reasonably good at conserving water and appear to be almost strictly herbivorous. They are good climbers and have been seen foraging up to 15 m high in trees. Groups of three to six individuals have been seen foraging together. Individuals are strong and aggressive and bite readily (Mares 1973, 1977c; Mares, Ojeda, and Kosco 1981; Rosi 1983).

Genus *Neotomys* Thomas, 1894
Neotomys ebriosus Thomas, 1894
Ratón Ebrio

Description
Measurements: HB 122.8; T 75.9; HF 23; E 18.5; Wt 64.5 g. This mouse is unmistakable because of its

grizzled gray brown dorsum and the bright cinnamon orange tip of its muzzle. The venter is grayish white to gray, the rump is washed with reddish, and the short, well-haired tail is bicolored (Barquez 1983; Gyldenstolpe 1932; Sanborn 1947b; pers. obs.).

Distribution
The species is found at high altitudes from Peru to northern Chile and northwestern Argentina. In Chile it has been trapped in Arica department, Tarapacá province, and in Argentina it has been found in the provinces of San Juan, Jujuy, Catamarca, and Tucumán (Barquez 1983; Pine, Miller, and Schamberger 1979; Tamayo and Frassinetti 1980) (map 16.89).

Life History
Embryo counts from Peruvian specimens ranged from one to two (MVZ).

Ecology
Neotomys ebriosus is found at altitudes of 3,300 to 5,000 m in dense grasslands, along streams with dense cover, and in marshes (Barquez 1983; Pearson 1951; Pine, Miller, and Schamberger 1979; Tamayo and Frassinetti 1980).

Genus *Phyllotis* Waterhouse, 1837

Description
For a full diagnosis of the tribe see above (p. 404). Species of *Phyllotis* have a very large external ear, a tail roughly equal to or shorter than the head and body, a relatively short hind foot, and a simplified molar dentition (tetralophodont). The last upper molar, M_3, tends to be from one-half to three-fourths of M_2 (figs. 16.5 and 16.11). Pearson (1958) and Hershkovitz (1962) contain useful summaries of the genus.

Head and body length ranges from 70 to 153 mm, and the tail from 45 to 150 mm. The dorsal pelage varies from gray brown to reddish brown, and the venter may be dirty white to buff.

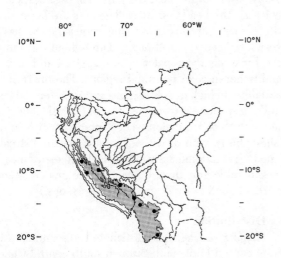

Map 16.89. Distribution of *Neotomys ebriosus*.

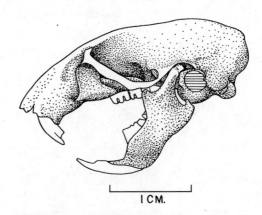

Figure 16.11. Skull of *Phyllotis darwini*.

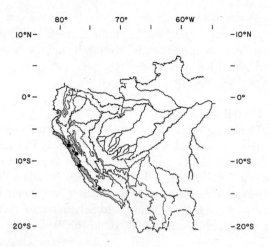

Map 16.90. Distribution of *Phyllotis amicus.*

Distribution

The genus occurs from Ecuador to Argentina, including Bolivia, Chile, and Peru; it is usually found at moderate to high elevations (sea level to 5,000 m).

Natural History

These rodents are mainly nocturnal. They are extremely omnivorous; seeds, arthropods, lichens, and forbs have been recorded in their diets (see species accounts). In spite of the variety in diet, most species are strongly herbivorous, in contrast to the often sympatric species of *Akodon,* to which insects are important.

Phyllotis (Paralomys) amicus Thomas, 1900

Description

Measurements: HB 80–85; T 100–105; HF 22; E 23 (*n* = 2, USNMNH) Tolon, Peru. This small, large-eared phyllotine has a medium brown dorsum contrasting with a white venter.

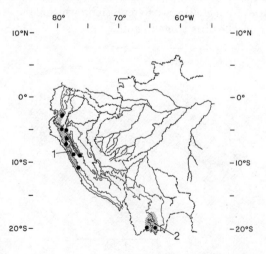

Map 16.91. Distribution of (1) *Phyllotis andium;* (2) *P. caprinus.*

Distribution

The species is found on the coastal and low elevation slopes of western Peru (Pearson 1958) (map 16.90).

Phyllotis andinum Thomas, 1912

Description

Two males taken at Chanocay Lomado Lachay, Lima province, Peru, at roughly 400 m measured TL 204–26; T 101–12; HF 26–27; E 23; Wt 30–32 g (USNMNH). This medium-sized phyllotine has a medium brown dorsum with the usual salt-and-pepper effect. The venter is gray white, and the tail is bicolored.

Distribution

The species is found from the eastern and western slopes of the Andes in central Ecuador to central Peru (Pearson 1958) (map 16.91).

Phyllotis caprinus Pearson, 1958
Pericote Anaranjado

Description

Average measurements: HB 118; T 135; HF 26; E 25. This large, relatively short-furred *Phyllotis* is sympatric only with *P. darwini.*

Range and Habitat

Phyllotis caprinus is found from southern Bolivia to northern Argentina on the eastern slope of the Andes above 2,400 m in brush and thorn scrub (Pearson 1958) (map 16.91).

Phyllotis darwini (Waterhouse, 1837)
Pericote Panza Gris

Description

Average measurements: HB 128; T 124; HF 28; E 27; Wt 58 g. The head and body length is almost always over 90 mm, and the tail is almost always longer than the head and body. The ears are large, and the vibrissae are luxuriant. This robust mouse has a very large geographical range. Throughout this range the fur tends to be paler in arid regions and darker and richer in color in humid regions. The fur is thick and fluffy in high or cold regions and is flatter and less fluffy in specimens from the coast of central Chile and in relatively warm regions. The venter is whitish, grayish, or buffy, with or without a buffy pectoral streak. The fur is predominantly in tones of coffee, often with considerable gray though juveniles are grayer (Mann 1978; Osgood 1943; Pearson 1958; pers. obs.).

Distribution

Phyllotis darwini is distributed discontinuously from central Peru south through south-central Chile,

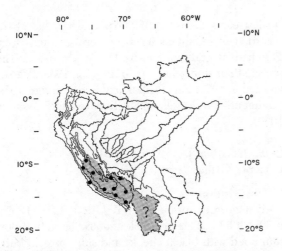

Map 16.92. Distribution of *Phyllotis darwini* (but see map 16.95 and text).

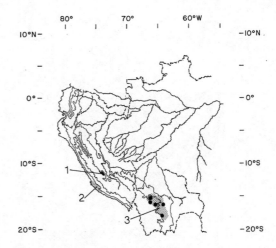

Map 16.93. Distribution of (1) *Phyllotis definitus;* (2) *P. gerbillus;* (3) *P. wolffsohni.*

and in Argentina along the eastern side of the Andes south to Mendoza province, then east to Buenos Aires province; from there it extends across the width of Argentina south to Tierra del Fuego, where it may enter Chile again (Honacki, Kinman, and Koeppl 1982; Mann 1978; Tamayo and Frassinetti 1980) (map 16.92).

Life History

In Peru the breeding season coincides largely with the wet season, but it is poorly defined since there is no clear peak in breeding activity. The average number of embryos per female is 3.7. In some years populations of this species can undergo dramatic increases, in part owing to increased reproduction following successive years of high rainfall (Mann 1978; Meserve and Le Boulengé 1987; Pearson 1975).

Ecology

Over its entire range, this species is found in a very large variety of habitats from near sea level to 4,500 m and from marsh edges to thorn brush to araucaria forests.

P. darwini are good climbers, and nests have been found aboveground in cacti and shrubs, though also in rocks. They are nocturnal and very flexible in their feeding habits. In Chile they have an herbivorous diet in the spring and a largely granivorous diet in the summer, although the extent of granivory varies with vegetation type. In Peru the diet was found to consist of 13% grass, 37% forbs, 11% seeds, and 39% insects ($n = 42$) (Contreras and Rosi 1980a; Fulk 1975; Glanz 1977a; Greer 1966; Mann 1978; Mares 1973, 1977c; Mares, Ojeda, and Kosco 1981; Meserve 1978, 1981b; Pearson 1951, 1958, 1975; Pizzimenti and de Salle 1980; Reise and Venegas 1974; Schamberger and Fulk 1974; Schneider 1946).

Comment

Oliver P. Pearson (pers. comm.) comments that the present taxonomic situation, as promoted by Spotorno, is that *Phyllotis darwini* sensu stricto is a species of the central coast of Chile, that it meets *P. xanthopygus* near Santiago without intergrading, and that they do not interbreed in the lab. This means that many, if not most or all, of the forms that were formerly considered subspecies of *P. darwini* would now be listed as subspecies of *P. xanthopygus* (Spotorno and Walker 1983; Walker, Spotorno, and Arrau 1984).

Phyllotis definitus Osgood, 1915

Description

Measurements of the type: TL 263; HB 131; T 132; HF 28; E 22. This large species has a thick pelage. The dorsum is cinnamon brown, grayer on the rump and cheeks, and the venter is an orange buff, paler on the throat.

Distribution

The type locality is Macate, Ancash department, Peru, at 9,000 ft elevation (map 16.93).

Comment

This species is often included in *P. darwini* or *P. magister,* but the karyotype is unique.

Phyllotis (Paralomys) gerbillus Thomas, 1900

Description

Five specimens averaged HB 74; T 73; E 19; HF 17 (Ratón, Peru, USNMNH). The hind foot is relatively long, suggesting saltatorial locomotion. Braun (1993) suggests a close relationship with *P. amicus.* The dorsal pelage is extremely pale, almost cream, the venter is white, and the tail is only faintly bicolored.

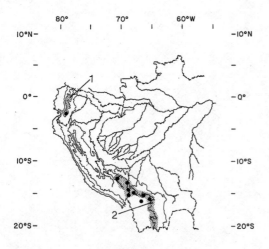

Map 16.94. Distribution of (1) *Phyllotis haggardi;* (2) *P. osilae.*

small streams. Densities of 2.06 per hectare have been recorded in mountain scrub of Peru. Animals from Peru had 7% grass, 60% forbs, 11% seeds, and 22% insects in their stomachs ($n = 5$) (Pearson 1958; Pearson and Ralph 1978; Pine, Miller, and Schamberger 1979; Pizzimenti and de Salle 1980; Tamayo and Frassinetti 1980).

Phyllotis osilae J. A. Allen, 1901
Pericote Osilae

Description
Average measurements: HB 116; T 121; HF 26; E 22; Wt 57 g. This long-tailed mouse has a buff dorsum lined with black, cheeks and sides of pale buffy orange, and a chest with a median longitudinal streak of buff on a white or gray white venter. The fur is thick and rich. It is very similar to *P. darwini* in some areas, separable only by skull characters (Gyldenstolpe 1932; Pearson 1958; pers. obs.).

Distribution
Phyllotis osilae is found from southeastern Peru to northern Argentina, where it occurs in Jujuy, Salta, Tucumán, and Catamarca provinces (Mares 1977c; Olrog and Lucero 1981; Pearson 1958; FM) (map 16.94).

Ecology
This is a species of the high Andean bunchgrass areas from 3,000 to 4,500 m, though it can occur at lower elevations in a few localities. In Peru this species was found at a density of 1.47 per hectare. *P. osilae* is strictly nocturnal, confined to more mesic habitats because it is a poor conserver of water and eats mostly plant material. Eleven stomachs from Peru contained 31% grass, 22% forbs, 26% seeds, and 20% insects (Dorst 1971, 1972; Lucero 1983; Mares 1977c; Pearson 1951, 1958; Pearson and Ralph 1978; Pizzimenti and de Salle 1980).

Phyllotis wolffsohni Thomas, 1902

Description
Measurements; HB 133; T 137; E 25.8. The tail is unusally long, and the pelage is coarse.

Range and Hatibat
The species occurs on the eastern slope of the Andes in Bolivia and in central Peru, at 7,200–11,000 ft, usually in brushy habitat (map 16.93). Where it co-occurs with *P. osilae*, the latter species prefers grassy habitats (Pearson 1958).

Comment
This species is sometimes synonymized with *P. darwini.*

Distribution
The species occurs in the Sechura Desert of northwestern Peru (Pearson 1958) (map 16.93).

Phyllotis haggardi Thomas, 1908

Description
Measurements of one specimen from 1.5 km north of Urbina, Chimborazo province, Ecuador, at 3,609 m were TL 172; T 71; HF 23; E 22; Wt 22 g (USNMNH). These rats have a very dark brown dorsum (the bicolored nature of the dorsal hairs is somewhat obscure). The venter is gray, and the tail is noticeably bicolored.

Distribution
The species is found in the Andes of north-central Ecuador (Wilson and Reeder 1993) (map 16.94).

Phyllotis magister Thomas, 1912

Description
Average measurements: HB 132; T 131; HF 29; E 26; Wt 69 g. This species is distinguishable from *P. darwini,* which it closely resembles, by its larger size and larger feet. Dorsally it is light brown to grizzled gray slightly suffused with buff, and ventrally it is whitish gray. The tail is bicolored, and the ears are sparsely haired (Gyldenstolpe 1932; pers. obs.).

Distribution
Phyllotis magister is found in the Andes of southern Peru and northern Chile from sea level (Honacki, Kinman, and Koeppl 1982; Pearson 1975; Tamayo and Frassinetti 1980) (not mapped).

Ecology
Brushy areas from 2,000 to 4,200 m elevation are the preferred habitat of this species. It has been trapped among rocks, in stone walls, and along the banks of

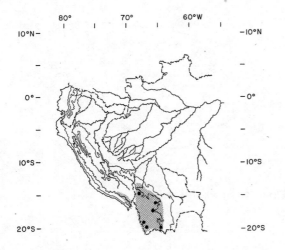

Map 16.95. Distribution of *Phyllotis xanthopygus*.

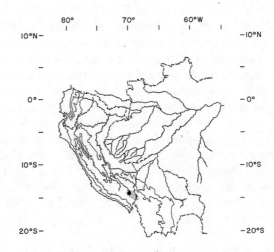

Map 16.96. Distribution of *Punomys lemminus*.

Phyllotis xanthopygus (Gervais, 1841)

Description
Measurements: TL 260–90; T 139–56; HF 26–30; E 22–24; Wt 45–56 g.

Distribution
Map 16.95 portrays the distribution from Anderson (1997) for Bolivia. Map 16.92 portrays the distribution of *P. darwini* that includes Bolivian specimens. The problem of revision needs attention.

Comment
This species was formerly assigned to *P. darwini*, but it is treated separately by Anderson (1997).

Genus *Punomys* Osgood, 1943
Punomys kofordi Pacheco and Patton, 1995

Description
Measurements: TL 203.3; T 69.2; HF 27.2; E 24.7 (*n* = 6). This rodent is similar in color to *P. lemminus* but is slightly larger and darker.

Distribution
The species is distributed to the east of Lake Titicaca in Peru, between 16° and 17°S. It is known only from the high elevations of Abra Aricema and the adjacent Limbani valley above 4,500 m in the Cordillera de Carabaya of northern Puno department, southern Peru (Pacheco and Patton 1995) (not mapped).

Punomys lemminus Osgood, 1943

Description
Four specimens from Puno department, Peru, measured TL 198–223; T 61–74; HF 28–29; E 23–28; Wt 70–92 g (MVZ). The tail is short, and the ear is only moderate in size. Dorsal pelage varies from dark to light gray brown. The venter does not contrast sharply with the dorsum, and the tail is faintly bicolored.

Distribution
The species occurs in the montane portions of southern Peru west of Lake Titicaca at elevations of 4,400–5,200 m (map 16.96).

Natural History
This rodent shelters in rock crevices. It is diurnal and forages on the *Senecio* and *Werneria* plants within this treeless habitat. Females have been recorded with only two embryos (MVZ).

TRIBE AKODONTINI

The Akodontini are small to medium-sized rodents with general adaptations for terrestrial life; some are semifossorial. Thus the tail is usually shorter than the head and body, the ears and eyes are of modest size, and the claws are long (plate 12 and fig. 16.13). The digestive system is relatively unspecialized, reflecting an omnivorous or insectivorous diet. The molar teeth are not high crowned. The molar cusps are crested or secondarily planed (see figs. 16.3 and 16.5). The molar mesoloph and mesolophid are usually small or vestigial. The karyotype does not exceed $2n = 54$ (see Reig 1987). Although the akodonts usually exhibit fossorial adaptations with short ears, short tail, and short hind feet, there are many exceptions; for example, *Lenoxus* has a long tail. The baculum is usually complex, but the structure is simple in *Abrothrix* (see also Spotorno 1992).

Many rodents tend to be flexible in their diets, and Landry (1970) has made an excellent case for omnivory as an underlying factor in their adaptive radiation. The high level of arthropod predation, as evinced from the stomach contents of many akodonts, suggests that the tribe is preadapted to invade insectivorous niches. If the members of the genera *Oxymycterus*, *Blarinomys*, and *Notiomys* (among others) derive from the ancestor of the extant genus *Akodon*, we may well view this evolutionary development as the ultimate occupancy of the semifossorial insectivore niche in South America by the rodent invasion after the formation of the Pliocene land bridge. Smith and Patton (1993) present data suggesting that although *Akodon* and *Oxymycterus* are related, the long-clawed mice may represent a separate line.

Genus *Microakodontomys* Hershkovitz, 1993

Description
These long-tailed sigmodontine rodents resemble *Oligoryzomys*, but the molars lack a mesoloph. Most sigmodontines lacking a mesoloph on the molars are short tailed.

Distribution
The genus occupies the fringe zone habitats between forests and savannas in the cerrado of Brazil.

Microakodontomys transitorius Hershkovitz, 1993

Description
Measurements: HB 70; T 93; HF 23; E 13 (Hershkovitz 1993). This small mouse has a soft, lax pelage. The dorsum is orange brown with a blackish effect, since the longest hairs are brown at the base, and the eyes are rimmed with blackish hairs.

Distribution
The type locality is the vicinity of the federal district of Brasília, Brazil (not mapped).

Comment
Hershkovitz (1993) believes this genus has characteristics that are transitional between *Akodon* and *Oxymycterus*.

Genus *Akodon* Meyen, 1833

Description
Members of the genus *Akodon* are distributed throughout South America. The baculum is usually complex in structure, the tail is shorter than the head and body, the hind foot is relatively short, and the ear is relatively small. The molar teeth tend to be somewhat simplified or tetralophodont, and the M_3 tends to be reduced in size (see figs. 16.5 and 16.12; Reig

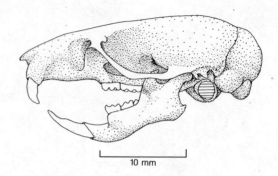

Figure 16.12. Skull of *Akodon* sp.

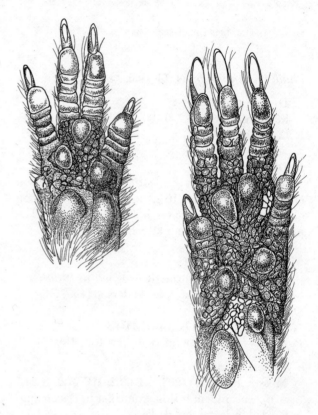

Figure 16.13. Forefeet and hind feet of *Akodon* (after Myers 1989).

1987). Head and body length of adults ranges from 72 to 141 mm, and the tail is 50 to 100 mm. The claws are long (fig. 16.13). These rodents remind a north temperate zone worker of voles, but their dietary habits are quite different. The dorsum varies from gray to brown, and the venter from white to gray.

Comment
Patton and Smith (1992b) note that the content of *Akodon* is poorly understood. There are currently fifteen species recognized for Peru alone, and the list is certain to increase. They attempt to analyze the *"boliviensis"* group and plot its distribution in Peru.

Table 16.4   Content of *Akodon* and *Bolomys*

Taxon	Number of Species	Comment
Genus *Akodon*	54	
Subgenus *Akodon*[a]	31	
varius group	5	
boliviensis group	2	
Subgenus *Chalcomys*	1	*Chalcomys aerosus*
Subgenus *Abrothrix*	6	*Abrothrix illeutus*, *A. lanosus, A. longipilis, A. mansoensis, A. sanborni, A. xanthorhinus,* and *A. siberiae* (?)
Subgenus *Hypsimys*	2	*Hypsimys budini* and *H. siberiae*
Subgenus *Microxus*	3 (2)	*Microxus bogotensis* is an *Akodon* and distinct from *A. latebricola* and *A. mimus*
Subgenus *Deltamys*	1	*Deltamys kempi*
Subgenus *Thaptomys*	1	*Thaptomys nigrita*
Genus *Bolomys*	6	
Unassigned species	5	

Note: Content as outlined by Musser and Carleton 1993. They do *not* advocate a taxon but only note the history.

[a] Subgenus as listed in comments, not necessarily accepted usage.

They then consider the other eight species and their geographical subdivision as distributed on the east and west slopes of the Andes. Musser and Carleton (1993) relegate *Microxus* (considered a full genus in Eisenberg 1989) to a subgenus of *Akodon*. In their phylogenetic assessment of *Akodon*, Smith and Patton (1993) studied the type species of *Microxus, M. mimus.* They conclude that *M. mimus* is clearly an *Akodon* in the strict sense. *M. bogotensis* and *M. latebricola* need further study, since neither is closely related to *M. mimus.* Musser and Carleton (1993) take note of the problem. *Bolomys,* although related to *Akodon,* stands as a full genus (see also Anderson and Olds 1989). *Hypsimys, Abrothrix, Deltamys,* and *Thaptomys* are considered subgenera of *Akodon* by Musser and Carleton (1993). González and Massoia (1995) reassert the validity of *Deltamys* as a full genus. *Thalpomys* is currently recognized as a genus (Hershkovitz 1990). Voss (1991a,b), in his exhaustive reviews of *Zygodontomys,* clarified its content and its distinctiveness from *Bolomys* (see table 16.4). (Important recent papers dealing with the systematics of the Akodontini include Reig 1987; Myers, Patton, and Smith 1990; Myers and Patton 1989a,b; Barros et al. 1990; Reig 1989; Apfelbaum and Reig 1989; Hershkovitz 1990.) In his discussion of *Thalpomys,* Hershkovitz (1990) considers his original hypothesis of metachromism as discussed in chapter 9 of this volume as applicable to an interpretation of coat color variation in rodents (see Hershkovitz 1990, 9–11). Spotorno (1992) considers the problem of parallel evolution in the ontogeny of penile morphology.

Distribution

The genus *Akodon* as here defined is confined to South America from Colombia to Argentina. Species range from sea level to 5,000 m.

Natural History and Systematics

Depending on the authority, some twenty-six to forty-five species may be included in this genus. The confusion depends in part on the subgroupings of the genus, some of which may be raised to generic status (see comments on the definition of the Akodontini). Natural history data that are available suggest that the species of *Akodon* are omnivorous and eat fruits, seeds, insects, and some vegetation.

Akodon aerosus Thomas, 1913

Description

Measurements of the type of *A. a. baliolus:* TL 190; T 83; HF 25. Additional measurements (*n* = 15–17): TL 189.3; T 83.2; HF 23.8 (Myers and Patton 1989a). This species is smaller than *A. boliviensis,* with a very robust skull. The dorsum is very dark, almost black, darkest on the rump. The venter scarcely contrasts with the dorsum but is slightly paler.

Range and Habitat

The species occurs in the eastern Andes from southeastern Ecuador to Peru and northwestern Bolivia (map 16.97). It is found in montane forest at the middle elevations, 1,200–2,400 m.

Akodon albiventer Thomas, 1897
Ratón Ventriblanco

Description

Average measurements: HB 93; T 69; HF 22; E 14; Wt 22 g. The fur is dense, and the short ears are well

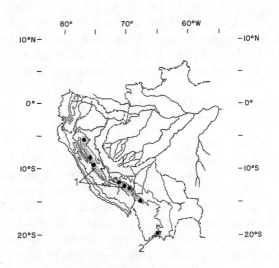

Map 16.97. Distribution of (1) *Akodon aerosus;* (2) *A. azarae.*

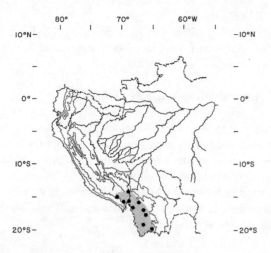

Map 16.98. Distribution of *Akodon albiventer.*

haired. These mice are dark gray through olive brown to light brown dorsally, with the throat, venter, and feet a strongly contrasting white to dirty white. There is sometimes a line of buff separating the venter from the dorsum. The sparsely haired tail is darker on top and white on the bottom (Mann 1978; pers. obs.).

Distribution

Akodon albiventer is found in southern and western Bolivia, northwestern Argentina, northern Chile, and southeastern Peru (Honacki, Kinman, and Koeppl 1982; Tamayo and Frassinetti 1980) (map 16.98).

Ecology

This species, taken at 400 to 3,000 m elevation, is typically an animal of open habitats. It has been found in meadows with dense grass and rock walls, near small marshes, and on high dry slopes. It seems to prefer more mesic areas but is not confined to them. In Peru densities ranged from 3.53 per hectare in quenua habitat to 0.14 in tola habitat. *A. albiventer* is diurnal and has long claws that it reportedly uses for digging tunnels and finding insect larvae (Koford 1955; Mann 1978; Mares, Ojeda, and Kosco 1981; Pearson and Ralph 1978; Pine, Miller, and Schamberger 1979).

Akodon azarae (Fisher, 1829)
Ratón de Azarae

Description

Average measurements: TL 179; T 77; HF 21; E 15; Wt 35 g. This typically volelike *Akodon* is of moderate size. The harsh olive brownish pelage is washed with yellow on the sides and is more strongly yellowish and grayish on the venter. In some populations the venter is sharply demarcated from the dorsum. The hind feet are tan, and there is a faint wash of reddish brown on the nose and shoulders and a faint eye-ring. In Paraguay *A. azarae* is easily distinguishable from the sympatric *A. varius* by its smaller size and its color (Barlow 1965; Myers and Wetzel 1979; pers. obs.).

Distribution

Akodon azarae is found from southern Bolivia and Brazil south to central Argentina. It occurs in both eastern and western Paraguay, throughout Uruguay, and in Argentina along the eastern part of the country south to Buenos Aires and La Pampa provinces (Barlow 1965; Contreras and Justo 1974; Honacki, Kinman, and Koeppl 1982; Massoia 1971a; UM) (map 16.97).

Life History

In Uruguay this species breeds from October to May, and females havethree to seven embryos (mean 4; $n = 12$). In the Argentine pampas, *A. azarae* breeds from October to April, and the number of embryos ranges, on average, from 4.6 to 5.7 (range 3–10). The gestation period is 24.5 days, and neonates weigh 2.2 g.

In the lab males reach sexual maturity at eighty-four days and females at seventy-five days; in the wild maturity can be reached at sixty days. In the wild 50% of females reach sexual maturity by 22 g, though some can begin reproducing at 12 g. Longevity in the wild is ten to twelve months for young born in the fall and seven to eight months for those born in the spring. Young born near the beginning of the season reproduced when only two months old, whereas those born near the end of the season did not breed until they were seven months old (Barlow 1969; Crespo 1966; Crespo et al. 1970; Dalby 1975; Pearson 1967; Villafañe 1981a).

Ecology

Akodon azarae prefers open vegetation formations. During periods of high density these mice can occur at densities of over two hundred per hectare, though averages of about fifty per hectare are more common. Where it occurs, *A. azarae* is usually the dominant small rodent.

These mice live in shallow holes and will occasionally burrow. They can be nocturnal or diurnal. In Uruguay the stomachs of eleven individuals contained 25% plant material and 75% invertebrate material, most commonly Coleoptera, Orthoptera, and Hymenoptera (Barlow 1969; Crespo 1966; Crespo et al. 1970; Dalby 1975; Fornes and Massoia 1965; Myers 1982; Villafañe et al. 1973).

Akodon boliviensis Meyen, 1833
Ratón Plomizo

Description

Myers, Patton, and Smith (1990) reviewed and identified *Akodon boliviensis.* Average measurements:

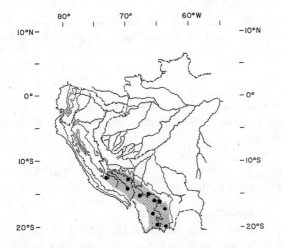

Map 16.99. Distribution of *Akodon boliviensis*.

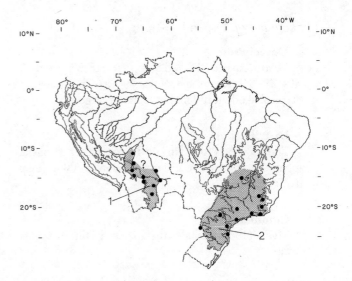

Map 16.100. Distribution of (1) *Akodon dayi;* (2) *A. cursor.*

TB 107; T 76; HF 22; E 17; Wt 33 g. In this medium-sized *Akodon* the dorsum ranges from agouti brown to darker shades of brown. The sides are often washed with tan, and the venter is grayish tan to buffy gray. The ears are lightly furred, and in some populations the nose and cheeks have a slight orange cast. The tail is bicolored (Pearson 1951).

Chromosome number: $2n = 40$ (Barquez et al. 1980).

Distribution

Akodon boliviensis is found from southern Peru through Bolivia (map 16.99).

Life History

In captivity the litter size is three to four. Nests have been found under tree trunks and in holes in the ground (Fonollat 1984).

Ecology

In southern Peru the species is found at up to 4,000 m elevation. It is the common mouse of grassy places, and some were caught in homes.

This terrestrial species is active both day and night. Its diet consists primarily of invertebrates, particularly Coleoptera larvae. Twenty-six stomachs from thirteen sites in Peru contained 8% grass, 6% forbs, 8% seeds, and 78% insects (Dorst 1971, 1972; Fonollat 1984; Pearson 1951; Pizzimenti and de Salle 1980).

Comment

The species includes *A. subfuscus,* but see Anderson (1997).

Akodon cursor (Winge, 1887)

Description

This description includes *A. momtensis.* Average measurements: HB 110; T 87; HF 25; E 19; Wt 40 g.

Akodon cursor is a medium-sized mouse, reddish brown to olive brown dorsally, grading to more tan on the sides, gradually becoming a reddish tan to gray washed with orange on the venter. The tail is sparsely haired and weakly bicolored, the feet are tan, and the fur on the face shows some blackish hairs.

Distribution

Akodon cursor is distributed in southeastern and central Brazil, Uruguay, eastern Paraguay, and northeastern Argentina (Crespo 1982b; Honacki, Kinman, and Koeppl 1982; Massoia 1980a; UM) (map 16.100).

Life History

In northeastern Argentina the litter size is three and the breeding season is September to March (Crespo 1982b).

Ecology

In eastern Paraguay this is one of the most common species in the forests and forest-grassland ecotones. In Misiones province, Argentina, it is found in most habitats but prefers flatter drier areas. *A. cursor* undergoes large population cycles, and stomachs of this species have been found to contain plant material, seeds, and adult and larval Coleoptera, Lepidoptera, and Diptera (Crespo 1982b; Myers 1982).

Comment

Rieger et al. (1995) consider variation in gene frequencies in four populations from Rio Grande do Sul, Brazil. Rieger, Langguth, and Weimer (1995) further explore the divergence times for relatives of the *"cursor"* group.

Akodon dayi Osgood, 1916

Description
Measurements: TL 196.8; T 79.0; HF 26.1; E 18.7 (*n* = 21–34) (Myers and Patton 1989a). This species is a typical akodont rodent but is rather large.

Distribution
The species is found in central to southern Bolivia, apparently from above 270 m elevation to 700 m (map 16.100).

Comment
Myers (1989) includes the species in the *A. varius* group.

Akodon fumeus Thomas, 1902

Description
Measurements: TL 173; T 77.3; HF 21.4; E 15.2 (*n* = 10) (Myers and Patton 1989a,b). *Akodon fumeus* is similar in size to *A. boliviensis,* but the dorsal pelage is darker.

Distribution
The species is found on the eastern Andean slopes of Peru and Bolivia (map 16.101).

Akodon juninensis Myers, Patton, and Smith, 1990

Description
Measurements: TL 149.4; T 61.8; HF 19.5; E 12.2 (*n* = 68–70). This species is clearly a member of the *A. bolviensis* group. According to Myers, Patton, and Smith (1990), the *A. boliviensis* group includes, in ad-

dition to *A. juninensis, A. boliviensis, A. spegazzinni, A. puer,* and *A. subfuscus* (see description for *A. boliviensis*).

Distribution
The species occurs above 2,700 m elevation in central Peru (see Myers, Patton, and Smith 1990) (map 16.101).

Akodon (Deltamys) kempi (Thomas, 1917)
Ratón del Delta

Description
Measurements: TL 178.3; HB 96.3; T 82; HF 20.9; E 13 (*n* = 10); Wt 26.4 g (*n* = 4). The eyes are rather small, without eye-rings, and the ears are short and well haired. *A. kempi* has a tail proportionally longer than the sympatric *A. azarae.* The dorsum of *A. kempi* is blackish brown, inconspicuously washed on the head and sides with olivaceous. The venter is dull brownish gray, the feet are tan, and the tail is only faintly bicolored (Massoia 1964; Thomas 1917).

Distribution
Akodon kempi is known from the islands of the Río Paraná estuary in both Argentina and Uruguay and from Rio Grande do Sul, Brazil (not mapped) (Massoia 1963c; González and Massoia 1995).

Life History
One female with four embryos was captured (MVZ).

Ecology
This is a species of the wet grassy areas of the Río Paraná delta, where it is found with *Holochilus brasiliensis* and *Scapteromys.* It is omnivorous, and one grass nest was found in a log (Massoia 1964).

Comment
The species is listed as *Akodon* in accordance with Musser and Carleton (1993), but see p. 419.

Akodon kofordi Myers and Patton, 1989

Description
Measurements: TL 173.9; T 77.6; HF 22.5; E 15.3. This is a medium-sized *Akodon* with a short tail, similar to *A. fumeus.*

Distribution
Known from Cusco and Puno departments of southeastern Peru at between 2,750 and 2,900 m (map 16.102).

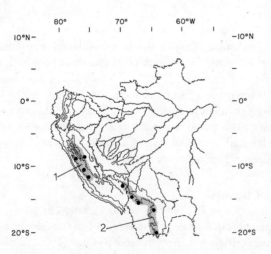

Map 16.101. Distribution of (1) *Akodon juninensis;* (2) *A. fumeus.*

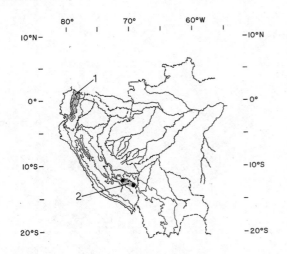

Map 16.102. Distribution of (1) *Akodon latebricola;* (2) *A. kofordi.*

Akodon latebricola (Anthony, 1924)

Description

Measurements of the type: TL 163; HB 83; T 80; HF 22. This rather small species is almost wholly black. Some of the dorsal hairs have white tips, but the overall effect is of a dark mouse. The tail is almost unicolor, only somewhat purer below.

Distribution

The type locality is Hacienda San Francisco, east of Ambato on the Río Cusutagua, Ecuador, at 8,000 ft (map 16.102).

Comment

The species was formerly assigned to *Microxus,* but see p. 419.

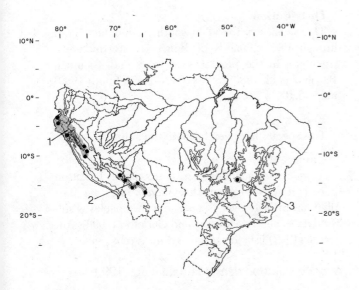

Map 16.103. Distribution of (1) *Akodon mollis;* (2) *A. mimus;* (3) *A. lindberghi.*

Akodon lindberghi Hershkovitz, 1990

Description

Measurements: HB 92; T 63; HF 18.7; E 13.3 (*n* = 6). The dorsum is dark brown; the venter is buff and weakly defined from the dorsal pelage. A reddish eye-ring is visible (Hershkovitz 1990).

Distribution

The species is found near the federal district of Brasília, Brazil (map 16.103).

Comment

In his description, Hershkovitz (1990) allies the species to his *Akodon boliviensis* group. Hershkovitz also elevates the former *Plectomys paludicola* to a full species of *Akodon.*

Akodon mimus (Thomas, 1901)

Description

Formerly *Microxus mimus.* This large akodont measures TL 190.9; HF 24.0; E 18.3 (*n* = 28) (Myers and Patton 1989b).

Distribution

The species is found on the slopes of the Andes of southeastern Peru and thence to central Bolivia (map 16.103).

Akodon mollis Thomas, 1894

Description

Measurements: TL 169.8; T 72; HF 20.4; E 11.

Distribution

The species is found from the Andes of Ecuador to north coastal Peru (map 16.103).

Comment

This group is considered a distinct species by Myers and Patton (1989b).

Akodon (Thaptomys) nigrita (Lichtenstein, 1829)
Ratón Subterráneo

Description

Average measurements: HB 92; T 47; HF 19; E 12; Wt 20 g. This small, dark *Akodon* is glistening olive brown to reddish brown, finely grizzled with ochraceous on the dorsum and dull ochraceous washed with yellowish to gray brown on the venter. The tail is not bicolored. It is smaller than the sympatric *A. cursor,* and its tail is half the length of *A. cursor*'s (Gyldenstolpe 1932; Myers and Wetzel 1979; pers. obs.).

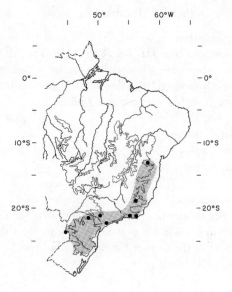

Map 16.104. Distribution of *Akodon nigrita*.

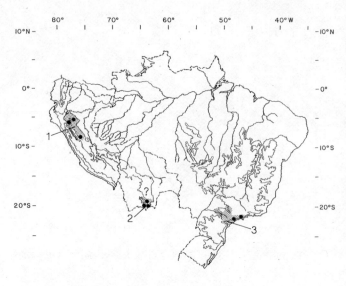

Map 16.105. Distribution of (1) *Akodon orophilius;* (2) *A. simulator;* (3) *A. sanctipaulensis.*

Distribution

Akodon nigrita is found in eastern Paraguay, southeastern Brazil, and northeastern Argentina (Massoia 1963a; Myers 1982) (map 16.104).

Life History

In Brazil females have four pairs of mammae, and three females were found to contain three, four, and five embryos (Davis 1947).

Ecology

Akodon nigrita is apparently confined to moist tropical forests, following the rain forests from southeastern Brazil into eastern Paraguay and Misiones province, Argentina. In Brazil they were found under logs and tree roots. This species will make tunnels in the leaf litter. It is aggressive, strongly terrestrial, and diurnal (Davis 1947; Myers 1982; Myers and Wetzel 1979; UM).

Comment

Akodon nigrita has been included in the genus *Thaptomys* by some authors, but Musser and Carleton (1993) list it as *Akodon*.

Akodon orophilus Osgood, 1913

Description

Akodon orophilus was originally described as a subspecies of *A. mollis*. It is very similar to *A. mollis*, but it has a narrower zygomatic plate (Patton and Smith 1992a,b). Total length averages 168 mm.

Range and Habitat

The species is found in the mountains of northern Peru. The type locality is Leimabamba, Alto Uta-

camba, Amazonas department, Peru, at 2,400 m (map 16.105).

Akodon puer Thomas, 1902

Description

Average measurements: HB 89; T 64; HF 19; E 14; Wt 21 g. The dorsum of this small *Akodon* is brownish olive to light reddish brown, and the venter is slightly brown or washed with pale yellow and not sharply demarcated from the dorsum. The tail is weakly bicolored (Gyldenstolpe 1932; pers. obs.; BA).
Chromosome number: $2n = 24$ for *A. lutescens caenosus* and $2n = 40$ for *A. l. lutescens* (Myers, Patton, and Smith 1990).

Distribution

The species is found in southern montane Peru through the highlands of Bolivia to northwestern Argentina, in the provinces of Jujuy and Tucumán (Barquez et al. 1980; Myers, Patton, and Smith 1990) (map 16.106).

Ecology

This species has been taken in the altiplano.

Comment

Akodon puer includes *A. caenosus.* James L. Patton (pers. comm.) informs us that *Akodon lutescens* J. A. Allen has priority over *A. puer,* but the species is still listed as *A. puer* by Musser and Carleton (1993). Anderson (1997) lists this taxon as *A. lutescens puer.*

Akodon sanctipaulensis Hershkovitz, 1990

Description

Measurements: HB 93; T 72; HF 23.8; E 15 ($n = 4$). In this long-tailed species of *Akodon* the tail averages

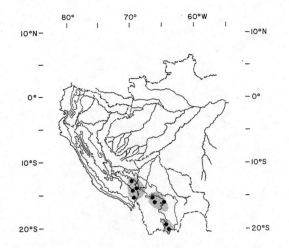

Map 16.106. Distribution of *Akodon puer* (Anderson 1997 lists as *A. lutescens*).

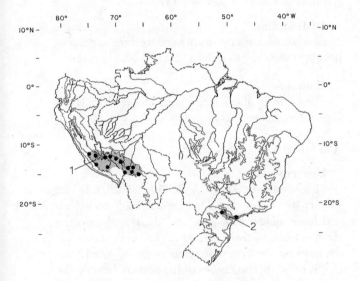

Map 16.107. Distribution of (1) *Akodon subfuscus;* (2) *A. serrensis.*

77% of the head and body length. In his original description Hershkovitz (1990) notes that the expanded braincase combined with the long nasal bones and the truncated interorbital region render this species distinct from other members of the genus *Akodon*.

Distribution

The species is found from the north bank of the Rio Juquia and Rio Etá between 24° to 24°25′ S and 47°35′ to 48°10′ W on the Atlantic coastal plain of southern São Paulo state, Brazil (map 16.105).

Akodon serrensis Thomas, 1902

Description

Average measurements: HB 95; T 81; HF 24; E 18. The fur of this *Akodon* may be thick and woolly, a grizzled olivaceous above, darker along the spine,

paling on the sides and ochraceous below. The venter is not sharply demarcated from the dorsum; the anal region is strikingly ochraceous. The ears are dark brown, slightly darker than the body, and the tail is very thinly haired, almost naked (Justo and de Santis 1977; Thomas 1902; pers. obs.).

Distribution

Akodon serrensis is known from southeastern Brazil and northern Argentina and perhaps into eastern Paraguay (Honacki, Kinman, and Koeppl 1982) (map 16.107).

Ecology

In Argentina an individual of this species was taken in an araucaria forest (Justo and de Santis 1977).

Akodon siberiae Myers and Patton, 1989

Description

Measurements: TL 191.1; T 89.1; HF 24.1; E 18.5. The tail almost equals the head and body in length. The dorsum is a very dark brown. The dorsal hairs in the midline are dark brown at the tip and gray at the base. The transition from the dorsal color to the venter is gradual, from slightly paler sides to a venter with hairs gray at the base but buff at the tip. The tail may be all one color or vaguely bicolored.

Distribution

The species is described from the cloud forests of eastern Cochabamba department, Bolivia (map 16.108).

Comment

Myers and Patton (1989a, 24) offer a phenogram suggesting that *A. siberiae* represents an early branch-

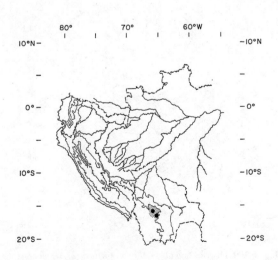

Map 16.108. Distribution of *Akodon siberiae.*

ing. It is related to *A. budini,* which is the type species for the subgenus *Hypsomys.*

Akodon simulator Thomas, 1916

Distribution
The species is found from the eastern Andean foothills of south-central Bolivia to Argentina (map 16.105).

Comment
Akodon simulator is frequently placed as a subspecies of *A. varius,* but see Myers (1989).

Akodon subfuscus Osgood, 1944

Description
Measurements: TL 160.3; T 70.3; HF 20.9; E 14.4 (*n* = 13–19) (Myers and Patton 1989a). This form is similar to specimens of the *A. "bolviensis"* group. The dorsum varies from olivaceous to grayish brown, heavily lined with black. There is no sharp boundary between the dorsal and ventral pelage; the hairs of the venter are gray at the bases and buff or whitish at the tips. Buffy hairs around the eye form a faint ring, and the tail is sharply bicolored (Myers, Patton, and Smith 1990).

Distribution
The species is found on the western and eastern slopes of the Andes from southern Peru to Bolivia (map 16.107).

Akodon surdus Thomas, 1921

Comment
Akodon surdus is a species of uncertain status (see Myers 1989; Patton and Smith 1992a), known from south-central Peru (map 16.109).

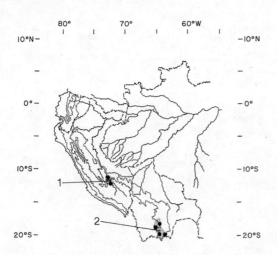

Map 16.109. Distribution of (1) *Akodon surdus;* (2) *A. sylvanus* (included in *A. simulator* by Anderson 1997).

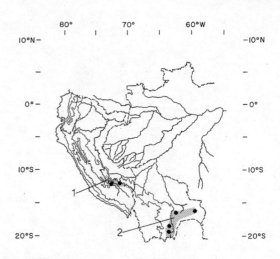

Map 16.110. Distribution of (1) *Akodon torques;* (2) *A. toba.*

Akodon sylvanus Thomas 1921

Distribution
Akodon sylvanus occurs in northwestern Argentina, possibly extending to Bolivia (map 16.109).

Comment
The form is allied to *Akodon azarae,* but Myers (1989) recognized it as a species.

Akodon toba Thomas, 1921

Description
Measurements: TL 203.7; T 85.1; HF 25.5; E 19.2 (Myers 1989). The dorsal pelage is strongly olivaceous and finely ticked with gray, and the dorsal hairs are short (11 mm in the midrump). The venter is sharply demarcated and is grayish white or buffy. A buff eye-ring is present. The insides of the ears are haired. The tops of the feet are white.

Distribution
The species occurs from eastern Paraguay to Bolivia and northern Argentina (map 16.110).

Comment
Akodon toba is often considered a subspecies of *A. varius,* but it is separated by Myers (1989).

Akodon torques (Thomas, 1917)

Description
This large species has a total length of 195.3 mm; the nearly naked tail is about 87% of the head and body length. The dorsum and venter are a similar gray brown. There is no distinctive eye-ring (Patton and Smith 1992a).

Distribution
The species is found in the Andes of southeastern Peru in elfin forest at 2,000–3,500 m (map 16.110).

Akodon torques can co-occur with *A. subfuscus*, but it prefers forested habitat whereas *A. sufuscus* occurs in the bunchgrass.

Comment

Akodon torques is usually considered a subspecies of *A. orophilus*, but it could be more closely related to *A. mollis* (Patton, Myers, and Smith 1989).

Akodon urichi J. A. Allen and Chapman, 1897

Description

Total length averages about 190 mm, of which the tail is 70 mm. A female from Venezuela measured TL 188; T 65; HF 24; E 15. Weight averages 39 g for males; females are lighter, averaging 29 g. The dorsum is dark brown, the venter gray to reddish brown.

Range and Habitat

The species is distributed in northern Venezuela and adjacent portions of Colombia. It recurs in southern Venezuela and from there to central Brazil (not mapped). It prefers moist habitats and is found at from 240 to 2,232 m elevation in northern Venezuela. Though it is predominantly associated with multistratal tropical evergreen forest, it tolerates second-growth forest and man-made clearings. In the Andes of Venezuela it can occur up to 3,020 m.

Natural History

This species of *Akodon* is active both diurnally and nocturnally and appears to be completely terrestrial. Its diet includes fruits, seeds, grass stems, fungi, and insects (O'Connell 1981). It appears to breed during the wet season.

Comment

Akodon urichi could be related to *Bolomys* rather than *Akodon* (Smith and Patton 1993). Note that this species may include three related forms (Reig, Olivo, and Kiblisky 1971).

Akodon varius Thomas, 1902
Ratón Variado

Description

Average measurements: HB 108; T 85; HF 25; E 21; Wt 39 g. As portrayed here, the data are probably derived from a composite (see comment). This large mouse is heavy bodied, with a dark gray to light chestnut dorsum grading laterally to gray; the venter is lighter gray. The feet are white or gray, and some individuals have a distinctive white chin spot. There can be markedly different color in different subspecies: one is entirely gray. The tail is bicolored (Barquez et al.

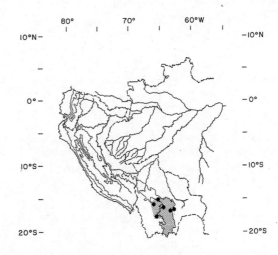

Map 16.111. Distribution of *Akodon varius*.

1980; Gyldenstolpe 1932; Olrog and Lucero 1981; pers. obs.).
Chromosome number: $2n = 41$ (Barquez et al. 1980).

Distribution

Akodon varius is found in western Bolivia on the eastern Andean slopes at 2,000–3,000 m (Myers 1982, 1989) (map 16.111).

Comment

In a recent study, several of the subspecies of *A. varius* that were recognized have been separated out as species. These included *A. neocenus,* found in central and southern Argentina below 1,000 m; *A. simulator,* found in Salta, Santiago del Estero, and Catamarca provinces, Argentina; and *A. toba,* found in the Chaco of western Paraguay, eastern Bolivia, and northwestern Argentina. Under the new revision *A. varius* would be found only in the Bolivian Andes between 2,000 and 3,000 m (Barquez et al. 1980; Myers 1989).

Genus *Blarinomys* Thomas, 1896
Blarinomys breviceps (Winge, 1887)

History and Description

This species was described by Winge from Pleistocene fossil remains collected by Lund. As the generic name implies, the species resembles a shrew in form and presumed function (e.g., *Blarina*, the North American short-tailed shrew). Matson and Abravaya (1977) offer the following data: The molar dentition is reduced in size while the incisors present a typical rodent pattern, although the lower incisors are rather long. Measurements: TL 142.4; T 40.4; HF 18.2; E 9.6. The general appearance is of a soft-furred, short-tailed mammal with foreclaws adapted for digging. The eyes are very small, the fur is velvetlike, and the basic dorsal color is slate gray (plate 12).

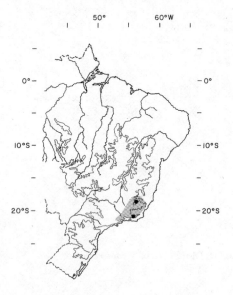

Map 16.112. Distribution of *Blarinomys breviceps.*

Distribution

The species is known from Pleistocene fossil deposits and Recent localities including the states of Rio de Janeiro, Espírito Santo, Minas Gerais, and Bahia, Brazil. It appears to be a form adapted to the higher altitudes within its present range (map 16.112).

Life History

Apparently this species forages in the soil for arthropods and other invertebrates. Fragmentary data suggest it breeds from September to January. The litter size is one or two, with one as the mode (Matson and Abravaya 1977; Abravaya and Matson 1975).

Genus *Bolomys* Thomas 1916

Description

Bolomys is related to the akodont rodents and thus exhibits the major characters of that tribe (see Akodontini, p. 417). According to Reig (1987, 352, 354), the key features of *Bolomys* when contrasted with *Akodon* include:

Braincase broad and deep; occipital region short; rostrum rather short and markedly tapering forward in lateral view; upper profile of skull gradually sloping forward from the middle of parietals; nasals short, with anterior borders well posterior to the level of the anterior border of incisors; frontals long, always longer than nasals; parietals short, less than half the length of frontals and extending forward anteriolaterally by means of narrow spines penetrating between frontals and temporals; interparietal noticeably reduced anteroposteriorly and transversely; occiput short and truncated; interorbital area with well-formed, anteriorly convergent borders; posterior palate moderately long and wide, the median posterior border of M^3; zygomatic plate broad and strong with

anterior border straight or slightly concave, perpendicular to diastema; upper incisors orthodont or protodont; molars mesodont, terraced with moderate wear, broad and robust; upper molars with lophs almost completely transverse, and mesoloph usually completely coalesced with paraloph; procingulum of M^1 simple with anteromedian flexus absent or only slightly developed; lower molars with lingual cusps somewhat anterior to the labial ones, with mesolophid remnants and mesotylids usually absent.

Anderson and Olds (1989) offer the most recent review of the genus and its distribution.

Distribution

The genus includes five species distributed in Peru, Brazil, Paraguay, Bolivia, Uruguay, and Argentina.

Comment

Cabreramys is included with *Bolomys.*

Bolomys amoenus (Thomas, 1900)

Description

The upper incisors are whitish, as in other *Bolomys* from Bolivia. See genus account. Measurements: TL 160–65; T 63–70; HF 20–22; E 10–13 ($n = 5$).

Distribution

The species is found in the highlands of Peru and west-central Bolivia (Anderson and Olds 1989) (map 16.113).

Bolomys lactens (Thomas, 1918)
Ratón Ventrirufo

Description

One specimen measured HB 101; T 67; HF 22; E 16. The dorsal color of this species is mixed blackish

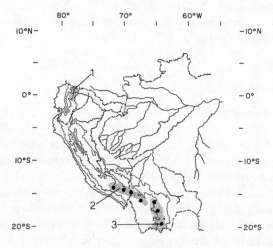

Map 16.113. Distribution of (1) *Bolomys amoenus;* (2) *B. punctulatus;* (3) *B. lactens.*

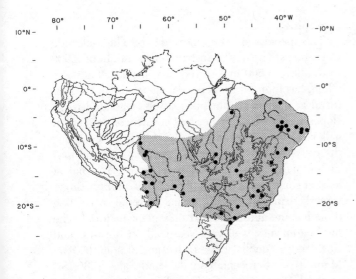

Map 16.114. Distribution of *Bolomys lasiurus* (see text for comments on *B. lenguarum*).

and buffy, with the head grayer and the rump more buffy. The sides and belly are distinctly more buffy, and the chin is white. The ears are about the color of the head, and the claws are rather long (Thomas 1918).

Distribution

Bolomys lactens is known from the northwestern Argentine provinces of Jujuy and Tucumán and extends to Bolivia (Anderson and Olds 1989; Lucero 1983; Thomas 1918) (map 16.113).

Bolomys lasiurus (Lund, 1814)
Ratón Selvatico

Description

Average measurements: HB 103; T 75; HF 25; E 16; Wt 35 g. The tail is much shorter than the head and body. This olivaceous gray mouse has a gray to grayish white venter. It has no distinctive facial markings. The tail is lightly haired and uniformly colored (Macedo and Mares 1987; pers. obs.) (see plate 12).

Distribution

Bolomys lasiurus is found in eastern Brazil and Paraguay and into the Argentine province of Misiones. It occurs patchily throughout eastern and western Paraguay (Macedo and Mares 1987; Myers 1982) (map 16.114).

Life History

In central Brazil, six females had an average of 4.2 embryos per female (Dietz 1983).

Ecology

Bolomys lasiurus is a species of grassland and cerrado, though it is occasionally trapped in forest. In sa-

vanna habitats in northern Brazil it achieves maximum densities of 4.2 individuals per hectare. Male home ranges overlap, but adult female home ranges are exclusive (Magnusson, de Lima Francisco, and Sanaiotti 1995). In central Brazil activity peaks at twilight, just after sunset and just before dawn (Vieira and Baumgarten 1995). In the Brazilian cerrado it is almost exclusively terrestrial and mostly diurnal; it is trapped at the forest edge and in bamboo near gallery forest. The species is seasonally rare, becoming the most common rat during the rainy season. It has been found at an average density of 11.8 per hectare. In the Brazilian caatinga it is found only in cultivated and abandoned fields and appears to be dependent on these areas. It builds nests of grass and leaves and lives in burrows with two to five openings. *B. lasiurus* must have free water in its food. In central Brazil its diet consists primarily of seeds (82%, $n = 32$) although in some localities it also eats significant quantities of invertebrates (Alho and de Souza 1982; Borchert and Hansen 1983; Dietz 1983; Macedo and Mares 1987; Nitikman and Mares 1987; Myers 1982; Streilein 1982a,b,c; UM).

Comment

Bolomys lasiurus usually includes *B. lasiotus, Zygodontomys lasiurus, Akodon arviculoides,* and *A. lenguarum* (Myers 1982; Voss 1991a,b), but Anderson and Olds (1989) separate *A. lenguarum.*

Bolomys lenguarum (Thomas, 1989)

Description

Bolomys lenguarum differs from all other species of *Bolomys* found in Bolivia in that its upper incisors are yellow and the zygomatic breadth of skull is less than 55% of the occiptonasal length. Measurements: Beni, TL 190–212; T 77–86; HF 20.5–26; E 16–17 ($n = 6$); Santa Cruz, TL 172–206; T 70–86; HF 24–28.5; E 15–18 ($n = 8$).

Distribution

The species is found in central Bolivia in the foothills and adjacent lowlands (Anderson and Olds 1989) (not mapped).

Bolomys punctulatus (Thomas, 1894)

Comment

Bolomys punctulatus is a species of uncertain status (see Voss 1991a). Voss (1991b) removes *punctulatus* from *Zygodontomys* and provisionally assigns it to *Bolomys.* The specimens are fragments and lie well outside the range for other species of *Bolomys.* The specimens assigned to this species derive from Ecuador and Colombia (not mapped).

Genus *Chroeomys* Thomas 1916

Status and Description

The genus *Chroeomys* was created by Thomas in 1916 to include *Akodon pulcherrimus* and *A. jelskii*, but it is often considered a subgenus of *Akodon*. Smith and Patton (1991) recognize *Chroeomys* as a genus and include *A. jelskii* and *A. andinus* within it, based on mt DNA data. The external appearance of the two species is quite different (see below).

Chroeomys andinus (Philippi, 1858)
Ratón Andino

Description

Measurements: TL 154; T 59; HF 20.6; E 14.4; Wt 21.5 g. This small *Chroeomys* has very dense, soft fur, light brown or buffy to agouti gray dorsally with a gray or gray white venter. Behind the ears is a pale patch, and in some populations the ears themselves are covered with white hairs. The lips and chin are also frequently white (Contreras and Rosi 1981b; Mann 1978; Osgood 1943; Pearson 1951; pers. obs.).

Distribution

Chroeomys andinus is distributed along the higher part of the Andes from Peru south to Chile and Argentina, though in places it is found as low as 950 m. In Chile it occurs south to about 34°, and in Argentina it is found south to Mendoza province (Honacki, Kinman, and Koeppl 1982; Iriarte and Simonetti 1986; Mann 1978; Roig 1965) (map 16.115).

Life History

In Salta province, Argentina, a lactating female was caught in March, and in Peru a female with six embryos was captured (Ojeda and Mares 1989; MVZ).

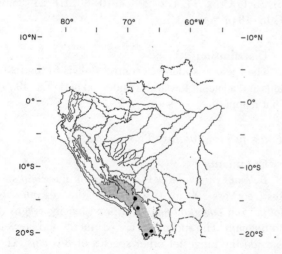

Map 16.115. Distribution of *Chroeomys andinus*.

Ecology

This species has been reported to be a high-altitude *Chroeomys*, apparently distributed from about 2,500 m to above 4,500 m, but recent work near Santiago, Chile, has found it as an occasional visitor in the matorral at 950 m. It inhabits sparsely vegetated rocky slopes and is a good digger that constructs a system of galleries about 5 cm deep in the soil, winding among the rocks. These mice can be very common and may undergo dramatic population increases. Diurnal at least at some times during the year, *C. andinus* shows a marked preference for animal material in its diet, though some authors report only plant material. The food eaten may well vary seasonally (Contreras and Rosi 1981b; Fonollat 1984; Iriarte and Simonetti 1986; Mann 1978; Scrocchi, Fonollat, and Salas 1986; Simonetti, Fuentes, and Otaiza 1985.

Chroeomys jelskii (Thomas, 1894)
Ratón Tricolor

Description

Average measurements: HB 99; T 79; HF 23; E 19. Among the southern subspecies this is the most striking of the akodonts, and its coloring makes identification easy. The unique features of color alone resulted in its original description as a separate genus. In some populations the nose, face, and ears are red, while in others only the nose is orange. The dorsal pelage is gray brown to chocolate brown, and the venter is sharply contrasting white that may extend onto the sides and the cheeks. In some populations a sharply contrasting white patch behind each ear may connect with the white on the cheeks. The feet are reddish tan, the tail is not bicolored, and the fur is dense and long (Pearson 1951; pers. obs.).

Distribution

Chroeomys jelskii is distributed from central Peru to Argentina, where it is found in Salta and Jujuy provinces (BA) (map 16.116).

Life History

Two females from Peru had three embryos each (MVZ).

Ecology

This high-altitude akodont ranges from about 2,800 to 5,600 m. It occupies a variety of habitats in southern Peru, such as grassy places, rocks, vacant huts, and occupied houses. It is active day and night; and seventeen stomachs were found to contain 19% grass, 42% forbs, 4% seeds, and 35% insects (Dorst 1971; Pearson 1951; Pizzimenti and de Salle 1980; Sanborn 1947a).

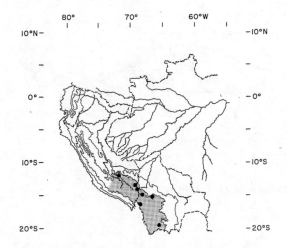

Map 16.116. Distribution of *Chroeomys jelskii*.

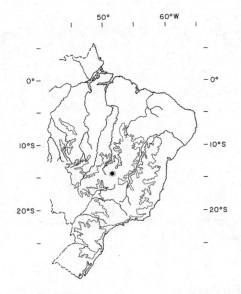

Map 16.117. Distribution of *Juscelinomys candango*.

Genus *Juscelinomys* Moojen, 1965
Juscelinomys candango Moojen, 1965

Description
Measurement of the type: HB 136; T 57; HF 26; E 15. Although *Juscelinomys* is clearly related to *Oxymycterus*, members of this genus does do not display the long rostrum characteristic of its near relative. *Juscelinomys candango* differs from *Oxymycterus* by its rather short dorsal pelage, its robust cranium, and its rather thick tail, which is well haired. The dorsum is agouti brown, and the venter is orange.

Distribution
The species is known from the type locality, Brasília, Brazil (map 16.117).

Natural History
This rodent lives in burrows, and the stomach contents of one individual contained vegetable material and ants (Moojen 1965).

Juscelinomys talpinus (Winge, 1887)

Comment
This Pleistocene form was originally assigned to *Oxymycterus* by Winge. Moojen (1965) assigned *talpinus* to *Juscelinomys*. Voss and Myers (1991) point out that further research is necessary to confirm relationships and explore the possibility of its continued survival, since out of twenty-five fossil murids Winge described from Lagoa Santa in 1887 eighteen have survived to the present. "*J. vulpinus*" in Wilson and Reeder (1993, 706) is an obvious spelling error (not mapped).

Genus *Lenoxus* Thomas, 1909
Lenoxus apicalis (J. A. Allen, 1906)

Description
Measurements: HB 150–70; T 150–90. The dorsum is a grayish black grading to a paler color on the sides and a grayish white venter. The tail is not bicolored, but the tip is whitish.

Distribution
This high-altitude species is taken in montane and elfin forests at between 1,000 and 3,300 m in southeastern Peru and adjacent Bolivia (map 16.118).

Genus *Oxymycterus* Waterhouse, 1837

Description
The genus *Oxymycterus* is affiliated with *Akodon* (Reig 1987; Smith and Patton 1993). Some twelve spe-

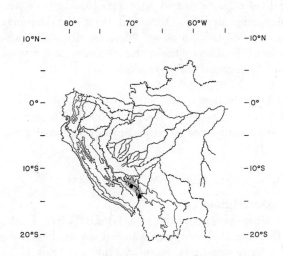

Map 16.118. Distribution of *Lenoxus apicalis*.

cies have been recorded in the literature. All have the following characters: the baculum is complex in structure; the hind feet are relatively short, as is the tail; the external ear is relatively short; the third molar is small; and the molars exhibit a simplified (tetralophodont) pattern. The rostrum of the skull is long and delicate (fig. 16.14), and the claws on the forefeet are long. The general appearance is of an animal adapted for a semifossorial existence. Head and body length ranges from 91 to 170 mm, and the tail from 63 to 153 mm. The dorsal pelage is usually some shade of brown, and the venter is buff to grayish white. Hershkovitz (1994) has reviewed the entire genus and compiled all described forms.

Distribution

South American in its distribution, the genus has been recorded from Brazil, Bolivia, Argentina, Paraguay, and Uruguay. According to Hershkovitz (1994), the known distribution is peripheral to the Amazon basin and the genus is unknown north or west of the Rio Amazonas–Solimões–Marañón system.

Natural History

These mice are semifossorial and highly insectivorous (Redford 1984). In a sense they are an evolutionary bridge between the generalist *Akodon* and the specialist *Blarinomys.* According to Hershkovitz (1994), where two species co-occur in sympatry one species is distinctly larger. This may not hold true over the entire range, and it also suggests ecological segregation according to prey size. (See the species accounts for more information.)

Oxymycterus amazonicus Hershkovitz, 1994

Description

Measurements: TL 208–19; T 80–93; HF 26–29; E 13–15. There are two pairs of pectoral and one pair of inguinal nipples. In this small species the dorsum is brown with reddish brown sides, the slaty basal portions of the hairs on the venter are more conspicuous, and the feet and tail are blackish.

Distribution

The type locality is 3°40′ S, 55°30′ W. The species is known only from the lower Rio Tapajós, Brazil (map 16.119).

Oxymycterus angularis Thomas, 1909

Description

Measurements of the male type: HB 160; T 100; HF 30; E 21. The very dark brown dorsum has a reddish tinge resulting from the red tint at the tips of the hairs; the venter is paler but has a reddish tinge.

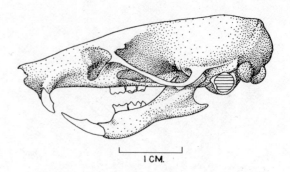

Figure 16.14. Skull of *Oxymycterus* sp.

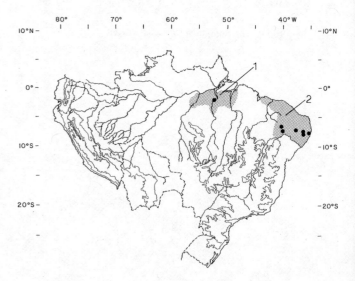

Map 16.119. Distribution of (1) *Oxymycterus amazonicus;* (2) *O. angularis.*

Distribution

The type locality is São Lorenzo, Pernambuco state, Brazil, at about 30 m elevation (map 16.119).

Oxymycterus hiska Hinojosa, Anderson, and Patton, 1987

Description

The holotype measured HB 100; T 77; HF 25; E 16. This is one of the smallest species in the genus. Small individuals overlap in measurements with *O. hucucha.* The dorsal pelage is blackish, the tail and the tops of the hind feet are black, and the chin may have a white spot.

Distribution

The type locality is 14 km west of Yanahuaya, Puna department, Peru, 14°19′ S, 69°21′ W at 2,210 m elevation. The species is found in south-central Peru (map 16.120).

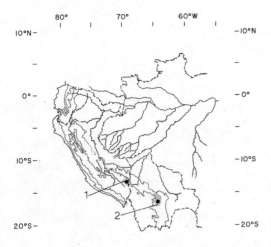

Map 16.120. Distribution of (1) *Oxymycterus hiska;* (2) *O. hucucha.*

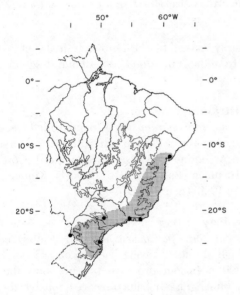

Map 16.121. Distribution of *Oxymycterus hispidus.*

Oxymycterus hispidus Pictet, 1843
Hocicudo Selvático

Description

Average measurements: HB 144; T 120; HF 33; E 22. The dorsum of this large *Oxymycterus* is generally reddish to orangish brown shading into warm yellow brown on the sides and to almost russet on the rump. The venter is buffy gray to orangish gray (Sanborn 1931; pers. obs.).

Distribution

Oxymycterus hispidus is distributed only in eastern Brazil and the northeastern Argentine province of Misiones. It may occur in eastern Paraguay as well (Honacki, Kinman, and Koeppl 1982; Massoia 1980a) (map 16.121).

Oxymycterus hucucha Hinojosa, Anderson, and Patton, 1987

Description

The holotype measured HB 116; T 60; HF 23; E 14. This is one of the smallest species described (but see the slightly larger *O. hiska*). The skull is very narrow. The tips of the dorsal hairs are paler than in *O. hiska* and more reddish. Some hairs on the toes of the hind feet reach to the ends of the claws, and this character helps distinguish it from *O. hiska*.

Distribution

The type locality is Compara, 17°51′ S, 64°40′ W, 28 km west of Santa Cruz, Cochabamba department, Boliva. The species is found in north-central Bolivia (map 16.120)

Oxymycterus iheringi Thomas, 1896
Hocicudo Chico

Description

Average measurements: HB 103; T 86; HF 22; E 18; Wt 43 g. *Oxymycterus iheringi* is a small, slender species with a general uniform color of grizzled brown, only slightly paler below. The ears are fairly large, thinly haired, and brown. The claws are short. This species resembles *Bolomys lasiurus,* with which it is sympatric (Massoia and Fornes 1969; Thomas 1898c).

Distribution

Oxymycterus iheringi is found in southern Brazil (Rio Grande do Sul) and the northeastern Argentine province of Misiones (Honacki, Kinman, and Koeppl 1982; Massoia and Fornes 1969) (map 16.122).

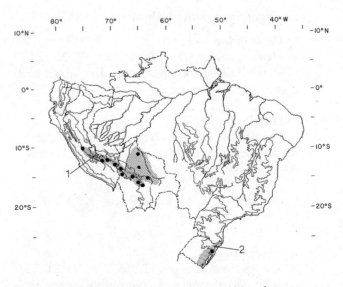

Map 16.122. Distribution of (1) *Oxymycterus inca;* (2) *O. iheringi.*

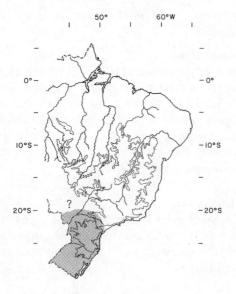

Map 16.123. Distribution of *Oxymycterus nasutus*.

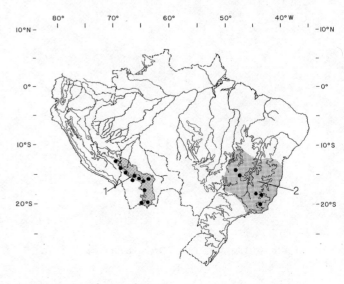

Map 16.124. Distribution of (1) *Oxymycterus paramensis*; (2) *O. roberti*.

Ecology
This species has been caught in forests (Massoia 1963b).

Oxymycterus inca Thomas, 1900

Description
 Measurements of the type: HB 135; T 105; HF 30; E 21 (BMNH). A reddish brown dorsum grades to a paler venter.

Distribution
The type specimen is from Río Perene, Junín department, Peru, at 800 m elevation. The species occurs in the lower montane forests from central and southern Peru to northern Boliva (map 16.122).

Oxymycterus nasutus (Waterhouse, 1837)

Description
 Measurements: HB 135; T 85; HF 25; E 17 (USNMNH). The dorsum is a very dark brown grading to a brown to tan venter.

Distribution
The type locality is Ita Petinnya, São Paulo, Brazil. The species occurs in Uruguay and adjacent southeastern Brazil (Wilson and Reeder 1993) (map 16.123).

Oxymycterus paramensis Thomas, 1902
Hocicudo Parameno

Description
 Average measurements: TL 235; T 90; HF 30; E 21; Wt 42 g. The dorsum and sides of this species are a rich yellow brown, and the venter is ochraceous tawny.

A sharply marked blackish brown frontal spot extends back from the rhinarium about 10 mm (Osgood 1944; pers. obs.).

Distribution
Oxymycterus paramensis is found from the eastern Andes of southern Peru to eastern Bolivia, then south to the Argentine provinces of Salta, Jujuy, Córdoba, and Tucumán (Barquez 1976; Honacki, Kinman, and Koeppl 1982) (map 16.124).

Ecology
Oxymycterus paramensis is found at intermediate and high altitudes, having been taken at from 1,500 to 4,300 m. In southern Peru it occurs above the tree line in bunchgrass. In Salta province it inhabits the forest floor of the northern wet forests and probably the mesic forests to the south. In Tucumán province it has been trapped in brushy areas near watercourses and in forests (Barquez 1976; Lucero 1983; Mares, Ojeda, and Kosco 1981; Osgood 1944; Thomas 1920).

Oxymycterus roberti Thomas, 1901

Description
 One male measured HB 187; T 110; HF 30; E 22. The reddish brown dorsum grades to a gray venter.

Distribution
The species is found in Minas Gerais state, Brazil (map 16.124).

Oxymycterus rufus (Fisher, 1814)

Description
This description includes *O. rutilans*. Average measurements: TL 224; T 98; HF 28; E 18; Wt 92 g. Adult

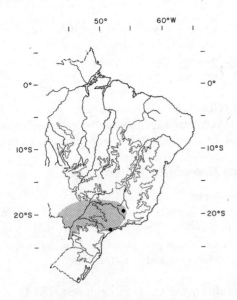

Map 16.125. Distribution of *Oxymycterus rufus.*

dorsal pelage is a grizzled reddish to yellowish black, darker along the spine, becoming ochraceous on the sides, and the ventral pelage is bright ochraceous mixed with gray. The tail is dark, well scaled, and sparsely haired (Barlow 1965; pers. obs.).
Chromosome number: $2n = 54$ (Kajon et al. 1984).

Distribution

Oxymycterus rufus is distributed from Brazil south through Paraguay and Uruguay and along eastern Argentina to Buenos Aires province (Musser and Carleton 1993) (map 16.125).

Life History

In Uruguay the breeding season is protracted, and females were found to have an average of 2.1 embryos (range 1–4; $n = 7$). In Buenos Aires province, Argentina, females are found in breeding condition throughout the year, with most activity from September to late May. There the average litter size is 3.1, and young are weaned at fourteen days (Barlow 1969; Dalby 1975; Kravetz 1972).

Ecology

In Uruguay *O. rufus* is found in wet meadows with stands of bunchgrass, in tall grass adjacent to streams and rivers, and in drier parts of marshes. In Buenos Aires province, Argentina, it is found in moist grassy areas and also in rocky areas of hills, though it is most common where there is good cover. It can reach densities of forty animals per hectare in suitable habitat.

In some areas *O. rufus* is diurnal, and in others it is nocturnal. It is strictly terrestrial, does not swim well, and feeds primarily on invertebrates. In Uruguay, stomachs had 100% occurrence of Coleoptera,

though Formicidae, Diptera, and Hemiptera were also common.

Genus *Podoxymys* Anthony, 1929
Podoxymys roraimae Anthony, 1929

Description

Head and body length averages 101 mm, the tail 94 mm, and the hind foot 23 mm. The claws are long, suggesting a semifossorial adaptation. The pelage is long and soft, with the dorsum gray and black, darker on the rump, and the underparts slightly paler. The feet are brown. Anthony (1929) noted the strong relationship with the akodont rodents.

Range and Habitat

The species is known from Mount Roraima, Venezuela, but it may extend to Brazil (see Perez-Zapata et al. 1992) (not mapped).

Genus *Thalpomys* Thomas, 1916

Description

Thalpomys are small to medium-sized akodonts. The molars exhibit an anteromedial fold of the upper first molar that is absent in the lower. The supraorbital ridges are strongly divergent, and the sides form ridges over the parietal bones to the mastoidal crests (see Hershkovitz 1990 for further details). The dorsum is pale to reddish, in contrast to that of other akodonts.

Distribution

Species of the genus are confined to the cerrado vegetation zone of Brazil.

Comment

Thalpomys has been considered a subgenus of *Akodon* (see Hershkovitz 1990).

Thalpomys cerradensis Hershkovitz, 1990

Description

Measurements: HB 96–102; T 59–63; HF 20; E 15 (Hershkovitz 1990).

Distribution

The species is apparently confined to the cerrado of central Brazil (map 16.126).

Thalpomys lasiotis Thomas, 1916

Description

Measurements: HB 121; T 80; HF 23. In this rather small akodont rodent, the dorsum is reddish brown.

Distribution

The type locality is Logoa Santo, Minas Gerais state, Brazil (map 16.127).

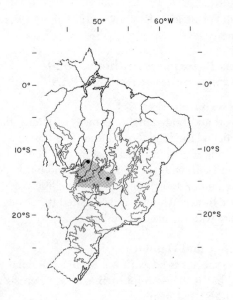

Map 16.126. Distribution of *Thalpomys cerradensis*.

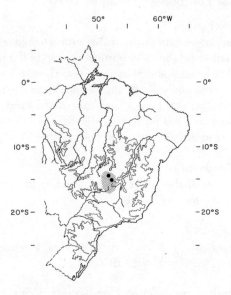

Map 16.127. Distribution of *Thalpomys lasiotus*.

Comment

The species includes *Akodon lasiotus* Ellermann, 1940, and Cabrera, 1961, and Dietz, 1983.

TRIBE SIGMODONTINI

Description

Students of the group should be warned that the common character is an S-shaped molar pattern (see fig. 16.3), but this character is also found in some species of the tribe Phyllotini (p. 404). This may not be a natural grouping, and much more research is necessary to establish its phylogenetic integrity. The grouping includes dry grassland forms (*Sigmodon*

and *Sigmomys*) as well as semiaquatic forms (*Holochilus*). However, Voss and Carleton (1993) remove *Holochilus* from the Sigmodontini and place it near the Oryzomyini.

Distribution

As currently recognized, this "tribe" extends from the southeastern United States to northern Argentina.

Genus *Holochilus* Brandt, 1835

Comment

Although they bear a "sigmodont" dentition, rodents formerly assigned to *Holochilus* may represent a rather unnatural grouping (see Aguilera and Perez-Zapata 1989; Voss and Carleton 1993).

Holochilus brasiliensis (Desmarest, 1819)
South American Water Rat; Rata Nutria o Colorada

Description

Average measurements: TL 405; T 201; HF 57; E 23; Wt 326 g. There is a good deal of variation in size. The webbed hind feet are large, and the tail is sparsely haired with scales clearly visible. This species is similar to the North American muskrat (*Ondatra*). The dorsal pelage is ochraceous tawny mixed with blackish, the sides are paler and more orange, and the venter is white, occasionally orangish. The fur is soft, quite dense and shiny (Barlow 1965; Fornes and Massoia 1965; Massoia 1976a; pers. obs.).

Distribution

Holochilus brasiliensis is found from eastern Brazil into Uruguay, Paraguay, and Argentina. In southern South America it is found throughout eastern Paraguay and in gallery forests into the Chaco, throughout Uruguay, and in Argentina (see Massoia 1976a; Honacki, Kinman, and Koeppl 1982; Vaz-Ferreira 1958) (map 16.128).

Life History

Females with three and four embryos have been examined. Rainfall seems to influence reproductive pattern in this species, and it can undergo eruptive population increases (Barlow 1969; Mares, Ojeda, and Kosco 1981; Veiga-Borgeaud, Lemos, and Bastos 1987).

Ecology

This is a semiaquatic rodent, distributed throughout southern and eastern South America in low, marshy areas. It contructs spherical nests of leaves, sticks, and grasses that average 22 to 27 cm in diameter. These nests are often in tangles of vines and termi-

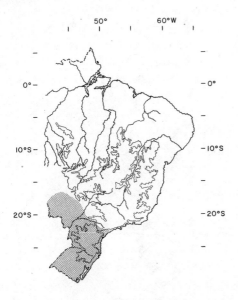

Map 16.128. Distribution of *Holochilus brasiliensis.*

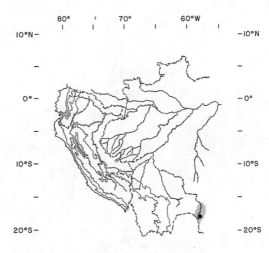

Map 16.129. Distribution of *Holochilus chacarius.*

nal forks of tree branches, but the rats also dig burrows with up to three entrances.

The diet of *H. brasiliensis* consists primarily of the more tender parts of aquatic, riparian, and domestic plants. It can be a pest in sugarcane and rice fields (Barlow 1969; Mares, Ojeda, and Kosco 1981; Massoia 1971b,c, 1976a).

Comment

This species may be a composite of *Holochilus balnearum* and *H. amazonicus* (Reig 1986).

Holochilus chacarius Thomas, 1906

Description

Average measurements: HB 177; T 167; HF 39; E 19; Wt 186 g. The hind foot length (without the claws) never exceeds 45 mm. This is the smallest species of *Holochilus*. The dorsum is a comparatively light orange chestnut, sometimes reddish with a few dark hairs mixed in, and the venter is pure white to grayish white. The interdigital membranes are rudimentary (Massoia 1976a).

Distribution

Holochilus chacarius is found in Paraguay and northeastern Argentina and may extend to adjacent Brazil. In Argentina it occurs in the provinces of Santa Fe, Chaco, Formosa, Corrientes, Jujuy, Salta, Tucumán, and Santiago del Estero (Massoia 1976a) (map 16.129).

Ecology

This semiaquatic rodent swims, dives, and climbs well. It makes grass nests 20–30 cm in diameter in rice

fields, on banana plantations, and along river courses. It is strictly herbivorous and can cause extensive damage to fields of rice, bananas, sugarcane, and other crops. In some places it can be very common (Massoia 1976a).

Holochilus sciureus Wagner, 1842
Marsh Rat, Rata Acuática

Description

Average measurements: TL 334; T 162; HF 38; E 19; Wt 164 g. Head and body length can range from 130 to 220 mm; the tail is nearly equal to the head and body. The toes are partially webbed, and the tail has a fringe of hair on the ventral side. The dorsum is buffy or tawny, usually mixed with black. The sides are paler, and the underparts vary from white to orange.

Range and Habitat

The group is distributed east of the Andes across northern South America and south to Brazil (map 16.130).

Natural History

These rats are semiaquatic. They may construct a nest from pieces of reeds and plant stems, creating a very large mound; nests have been reported with a diameter of 400 mm. In the sugarcane fields of Guyana, nests may be constructed about 1 m off the ground (Twigg 1962). Marsh rats feed on plant stems and seeds and may eat snails. Litter size varies from four to six. This species can reach high densities (713 per hectare) in rice fields and can be a serious agricultural pest (Cartaya and Aguilera 1985). Food habits are elegantly analyzed by Martino and Aguilera M. (1989).

Comment

Holochilus sciureus Wagner, 1842, was included in *H. brasiliensis* by Hershkovitz (1955). Considerable

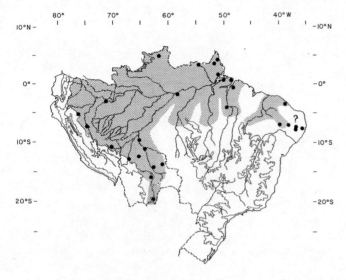

Map 16.130. Distribution of *Holochilus sciureus*.

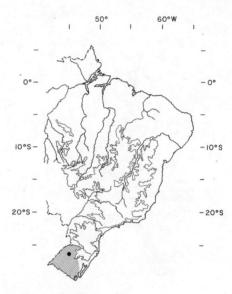

Map 16.131. Distribution of *Lundomys molitor*.

confusion surrounds the taxonomy of this genus, and at least six species have been named. The genus *Holochilus* was revised by Hershkovitz (1955), but Gardner and Patton (1976) believe his *H. brasiliensis* is a composite. The northern form is usually referred to as *H. sciureus,* while the form south of the Amazon is called *H. brasiliensis.* Martino and Aguilera M. (1989) maintain the species name *H. venezuelae,* and we concur that they have a valid case.

Genus *Lundomys* Voss and Carleton, 1993

Description
These large rodents have a tail longer than the head and body. The ears are small and the hind feet are conspicuously webbed. The dorsal pelage is dark brown grading to yellow brown along the sides (see Voss and Carleton 1993).

Comment
From their analysis of *Lundomys,* Voss and Carleton (1993) suggest that *Holochilus, Lundomys, Pseudoryzomys, Zygodontomys,* and the oryzomyines may form a natural grouping. They synonymize *Holochilus molitor* and *H. magnus. H. molitor* then becomes the type and the only valid species.

Lundomys molitor (Winge, 1887)

Description
Average measurements: HB 228; T 267; HF 62; E 25; Wt 239 g. *L. molitor* is one of the largest South American cricetines, with very large feet and a long, strong tail. The dorsum is ochraceous to tawny mixed with grayish or blackish, especially along the midline, and the venter is tawny to ochraceous. The feet are

whitish, with reasonably well developed interdigital membranes (Barlow 1965; Massoia 1976a; pers. obs.).

Distribution
The Recent distribution includes Rio Grande do Sul, Brazil, and Uruguay (map 16.131).

Ecology
In Uruguay this semiaquatic species has been taken along streams in tall bunchgrass and reeds and in marshy areas. Its diet consists of 95% plant material (Barlow 1969).

Genus *Sigmodon* Say and Ord, 1825
Cotton Rat, Rata Afelpada

Description
Head and body length ranges from 125 to 200 mm, tail from 75 to 125 mm, and weight from 70 to 200 g. The dorsum is grayish brown to blackish brown, and the underparts are gray to buff. The molar teeth have an S-shaped folding pattern (fig. 16.3). The species of *Sigmodon* present in South America are reviewed by Voss (1992).

Distribution
The genus occurs from the southeastern United States through Mexico and Central America to northern South America.

Sigmodon (Sigmomys) alstoni (Thomas, 1881)

Description
The measurements of a female from Suriname were TL 252; T 107; HF 30, E 21; Wt 74 g. The species is similar to *S. hispidus* (see below), though smaller, but

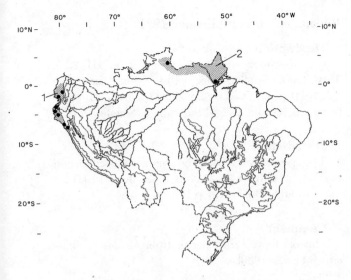

Map 16.132. Distribution of (1) *Sigmodon peruanus;* (2) *S. alstoni.*

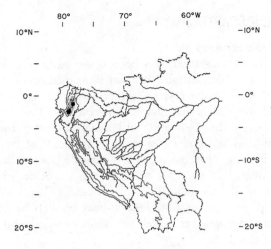

Map 16.133. Distribution of *Sigmodon inopinatus.*

it is characterized by a longitudinal groove in the upper incisors. For this reason it has frequently been placed in a separate genus, *Sigmomys*. The dorsum is agouti brown, and the venter is gray. The eye is ringed with pale yellow hairs.

Range and Habitat

The species is found in northern and eastern Venezuela east to the Guianas and south to northern Brazil (map 16.132). In Venezuela it is associated with low-elevation grasslands. Although it occurs in moist habitats, it is very tolerant of seasonally arid habitats and in general replaces *S. hispidus* in these areas.

Natural History

This species feeds primarily on forbs and green plants as well as seeds. It typically has a litter size of five and may show seasonal fluctuations in numbers depending on rainfall and the abundance of food. It appears to occupy a volelike niche (O'Connell 1982).

Sigmodon hispidus Say and Ord, 1825

Description

Total length ranges from 224 to 365 mm; T 81–166; HF 28–41; E 16–24. The dorsum is brown mixed with blackish hairs, giving a salt-and-pepper effect; the venter is pale gray. The tail is sparsely haired and tends to be a single color, usually some shade of brown.

Range and Habitat

The species is distributed from the southeastern United States through Mexico and across most of northern South America. In Venezuela it was found in moist habitats and at elevations of 200–1,580 m (not mapped).

Natural History

Sigmodon hispidus has been well studied in the northern parts of its range. The gestation period there is twenty-seven days, and up to six young are born. The neonates are furred, and their eyes open within twenty-four hours. The young begin to disperse from the nest at less than ten days of age, and females commence breeding at about six weeks old. Nest construction has been described by Dawson (1973). The nest, built from grass, provides protection from light rainfall and also buffers the external temperature by approximately 5°C.

Sigmodon inopinatus Anthony, 1924

Description

Measurements of the type: TL 259; T 991; HF 31. The dorsal hairs are slate colored for the basal two-thirds; some hairs have white tips, but most are dark tipped. The overall impression is clay color sprinkled with black. The venter is gray.

Distribution

The type locality is Mount Chimborazo, Chimborazo province, Ecuador, at 11,400 ft. The species is found in south-central Ecuador (map 16.133).

Sigmodon peruanus J. A. Allen, 1897

Description

Measurements: TL 245; HB 150; T 95; HF 31; E 18. The ears are rather large. The dorsum is ashy gray, speckled with darker hairs, and the venter is pale yellowish gray. The sides of head are buffy, and the feet are buffy, but paler than the dorsum.

Distribution

The species occurs on the coastal plain of western Ecuador and northwestern Peru (map 16.132).

TRIBE SCAPTEROMYINI

Description

This vexing group has been reviewed by Hershkovitz (1966). Basically, these are semiaquatic rodents with considerable fossorial ability. The skull is robust, and the molars are complex in structure. The ears are short, the eye is small, and the tail is shorter than the head and body. The claws on the forefeet tend to be long. The color pattern is variable (see species accounts). In common with other South American genera, the baculum is complex. Massoia (1979) has proposed that the genus *Bibimys* may be allied to the classical genera that Hershkovitz (1966) included within this group.

Distribution

The tribe as we recognize it is confined to the wetlands of southeastern Brazil and to eastern Paraguay, adjacent portions of Uruguay, and northeastern Argentina.

Genus *Bibimys* Massoia, 1979

Description

Massoia (1979) suggested that *Bibimys* is closely related to the genera *Scapteromys* and *Kunsia* and is clearly adapted for a fossorial life in wetlands. The ear and tail are reduced in length, and the claws are enlarged. The skull is robust, and certain dental characters reinforce the inclusion of the genus *Bibimys* in the Scapteromyini. Head and body length ranges from 84 to 97 mm. The dorsum is some variable shade of brown, and the venter is usually gray or buff. *B. torresi* has a reddish nose.

Distribution

The genus is recorded from southeastern Brazil and adjacent parts of northeastern Argentina.

Bibimys chacoensis (Shamel, 1931)

Description

One specimen measured TL 160; T 66; HF 22.5. This mouse is oblivaceous, with some buff along the sides of the body, and is dark along the back. The venter is whitish with a slight buffy tinge (Shamel 1931).

Distribution

The species is known only from a few localities in north-central Argentina (Myers 1982; Shamel 1931), it may extend to Bolivia (map 16.134).

Comment

Massoia moved this species from *Akodon* to *Bibimys* (Massoia 1980b).

Bibimys labiosus (Winge, 1887)

History and Description

Measurements: HB 77, T 56. This small species was first described as a fossil from Lagoa Santa, Minas Gerais state, Brazil. Hershkovitz (1966) included it in *Akodon;* Massoia (1980b) placed it in *Bibimys.*

Distribution

The species is found in Lagoa Santa, Minas Gerais state, Brazil (map 16.135). It demands further study (Voss and Myers 1991).

Genus *Scapteromys* Waterhouse 1837
Scapteromys tumidus Waterhouse, 1837

Description

Average measurements: HB 181; T 147; HF 36; E-22; Wt 146 g. *Scapteromys* looks similar to *Rattus*

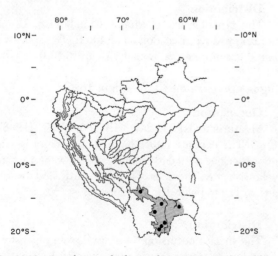

Map 16.134. Distribution of *Bibimys chacoensis* (unconfirmed by Anderson 1997).

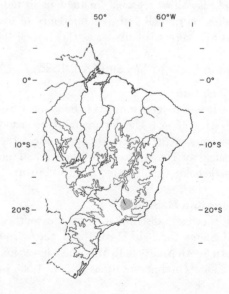

Map 16.135. Distribution of *Bibimys labiosus* (postulated).

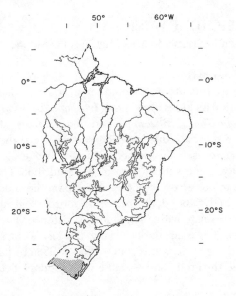

Map 16.136. Distribution of *Scapteromys tumidus*.

but has a relatively larger head, a stouter body, larger feet, and a relatively longer tail, which often has a distinct fringe of stiff hairs. The dorsal pelage is long and glossy, ranging from medium to dark brown, with a grayish wash in some individuals. The sides have a yellowish cast, and the venter is dirty white. Juveniles are dark grayish on the dorsum and grayish white to light gray on the venter. The forefeet are well developed, with fairly long claws. The pollex, as well as all other digits, is equipped with a claw (Barlow 1965; Hershkovitz 1966; pers. obs.).

Distribution
Scapteromys tumidus is distributed in Uruguay, adjacent Brazil, eastern Argentina, and southern Paraguay along the Río Paraguay (Massoia 1964) (map 16.136).

Life History
In Uruguay the average number of embryos or young was four, with a range of three to five; females have eight nipples. Breeding males were found year round (Barlow 1969).

Ecology
Scapteromys tumidus is a semiaquatic species of low, flooded grasslands, including salt marshes. Throughout its range it is found only in or near areas with standing water. *S. tumidus* swims well using its tail with its "swimming fringe," and it has also been observed to dive. It digs for food and is trapped throughout the day and night (Cueto, Cagnon, and Piantanida 1995). Its diet consists of arthropods (85%), mostly beetles, and earthworms. There is no indication that this species digs burrows, and young have been found

in shallow depressions beneath matted grass (Barlow 1969; Massoia and Fornes 1964; Myers 1982).

Comment
Scapteromys tumidus is considered to include *S. aquaticus*, but Reig (1986) reports $2n = 32$ for *aquaticus* and $2n = 24$ for *tumidus*.

Genus *Kunsia* Hershkovitz, 1966
Kunsia fronto (Winge, 1887)
Rata Acuática Grande

Description
Measurements: HB 160; T 75; HF 25; E 17.8. This species is similar in size to *Rattus norvegicus,* but the tail is proportionally shorter. It has small, densely haired ears. The subspecies *K. fronto* from central Brazil is black or blackish above, with sides speckled with white, and grayish white ventrally (Avila-Pires 1972; Hershkovitz 1966).

Distribution
Kunsia fronto is known from Chaco province, Argentina, and from the central Brazilian plateau. Its precise distribution is poorly understood (Hershkovitz 1966) (map 16.137).

Ecology
In central Brazil specimens of this species were captured in a marsh (Avila-Pires 1972; pers. obs.).

Comment
Formerly placed in *Scapteromys, Kunsia fronto* includes *K. chacoensis.*

Kunsia tomentosus (Lichtenstein, 1830)

Description
Measurements: HB 287–70; T 150; HF 41–45; E 33. This is one of the largest species of sigmodontine rodents found in the Neotropics. The tail is much shorter than the head and body, and the ears are reduced in size. The claws of the forefeet are long and curved (fig. 16.15). The dorsal pelage is harsh and dark brown. The hairs of the venter are white tipped but dark gray basally. The upper surfaces of the feet are gray (Hershkovitz 1966).

Distribution
This form has been found in the states of Minas Gerais, Mato Grosso, and Rondônia in Brazil. The range extends to El Beni department of Bolivia (map 16.137).

Natural History
As noted in the description, this species is strongly adapted for burrowing. In areas of seasonal rain it may

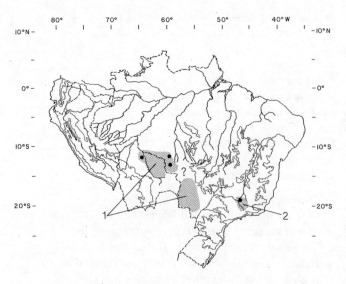

Map 16.137. Distribution of (1) *Kunsia tomentosus;* (2) *K. fronto.*

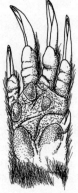

Figure 16.15. Forefeet and hind feet of *Kunsia tomentosus* (after Hershkovitz 1966).

spend most of the day in underground tunnels. With flooding in the wet season it spends more time aboveground. This rat feeds on roots and grasses exposed while burrowing. It is difficult to trap by conventional methods and is poorly represented in museum collections (Hershkovitz 1966).

TRIBE ICHTHYOMYINI
Fish-eating Rats and Mice, Ratones Pescadores

Diagnosis
These rodents are adapted for an aquatic existence. The tribe includes five genera, *Ichthyomys, Anotomys, Chibchanomys, Rheomys,* and *Neusticomys.* These genera show a reduction in the external ear; the toes either are partially webbed or have a fringe of hairs, increasing the surface area of the hind foot. The fur tends to be velvety, and the vibrissae on the snout are stout and rather long. All these morphological features correlate with a semiaquatic existence. The dorsal pelage is usally dark brown to blackish, contrasting with the gray to white venter. Voss (1988a,b) has reviewed the group and concludes it is monophyletic.

Distribution
These species occur from moderate to high elevations in the vicinity of freshwater streams in Panama, Colombia, Venezuela, the Guianas, Ecuador, Peru, and Bolivia.

Natural History
As the common name suggests, some species, especially the genus *Ichthyomys,* may feed on small freshwater fishes in addition to crustaceans. Others, such as *Rheomys,* may eat small crustaceans and snails. Species within the generic assemblage display different degrees of adaptation for a semiaquatic habitat. *Ichthyomys* seems highly specialized in its partially webbed feet; *Anotomys* shows specialization with the loss of the pinnae; *Neusticomys* shows the least specialization for semiaquatic life.

Genus *Anotomys* Thomas, 1906
Anotomys leander Thomas, 1906

Description
Measurements: HB 101–22; T 125–53. The dorsum is a dark slate gray with paler sides, and the venter is pale gray. There are no pinnae, but white hairs surround the slitlike ear openings. The ears lie closed when the rat is swimming.

Distribution
The species is known from the Andes of Pichincha province, Ecuador, and has been collected at 3,600–3,755 m (Voss 1988b) (map 16.138).

Genus *Chibchanomys* Voss, 1988
Chibchanomys trichotis (Thomas, 1897)

Description
The dental formula is I 1/1, C 0/0, P 0/0, M 3/3. Head and body length averages 130 mm, the tail 125–

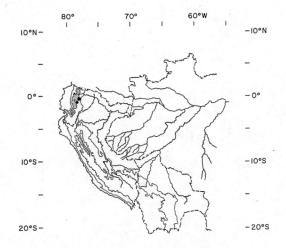

Map 16.138. Distribution of *Anotomys leander*.

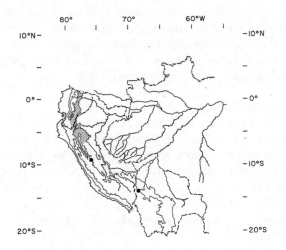

Map 16.139. Distribution of *Chibchanomys trichotis* (dot in Bolivia marks undescribed species; see Anderson 1997).

50 mm, and the hind foot 30–32 mm. This group is immediately distinguishable from the other genera by the extreme reduction of the external ear, which is almost absent. Generally the opening to the ear is surrounded by whitish hairs. The dorsum is blackish gray, and the venter is pale gray. The dorsal surface of the broad feet is gray, and the ventral surface is black.

Range and Habitat

This species is found in the montane portions of Colombia and northern Venezuela south to northern Ecuador. In northern Venezuela it occurs in streams at 2,400 m elevation in association with cloud forest (map 16.139). Barnett (1997) studied an as yet unnamed species in Ecuador. He noted the strong association with streamside habitats and use of aquatic resources.

Comment

Chibchanomys trichotis Voss, 1988, was formerly *Antomys trichotis* (Thomas, 1897). Barnett (1997) describes the diet and ecology of a new species from Ecuador. This species has now been named *C. orcesi* by Jenkins and Barnett (1997). It is found only on the Las Cajos Plateau of Ecuador at 3,100 to 4,000 m.

Genus *Ichthyomys* Thomas, 1893
Rata Pescadora

Description

The dental formula is I 1/1, C 0/0, P 0/0, M 3/3. The upper incisors are V-shaped, apparently an adaptation for seizing aquatic prey. Head and body length ranges from 145 to 190 mm. The eyes are reduced in size, as are the pinnae. The vibrissae are long and stout. The toes of the hind feet are partially webbed, and there is a distinct fringe of stiff, short hairs on the outer sides of the hind feet. The pelage is thick, with the upperparts dark olive brown to blackish and the underparts white. The bicolored tail is covered with fine hairs.

Distribution

The genus is found from Peru north to Ecuador and montane Colombia and Venezuela, with an isolated population in Panama (Voss 1988b).

Ichthyomys hydrobates (Winge, 1891)

Description

One male from FMNH measured HB 182; T 150; HF 35; E 9. The dorsum is very dark agouti brown, the tail is bicolored, and the venter is gray. The hind feet are very broad, and the fringe of hair on the toes is conspicuous.

Range and Habitat

The species is broadly distributed where suitable habitat is available in the foothills of the Andes in Colombia (map 16.140).

Ichthyomys stolzmanni Thomas, 1893

Description

Measurements: HB 110–97. The tail is equal to or shorter than the head and body (T:HB = 0.70:1.15. There are five separate plantar pads on the manus. The pelage is grizzled brownish on the dorsum, contrasting sharply with a silver to white venter.

Distribution

The type locality is Chanchamayo, near Tarma, Junín department, Peru, at 3,000 ft. The species occurs in western Ecuador and Peru (map 16.140).

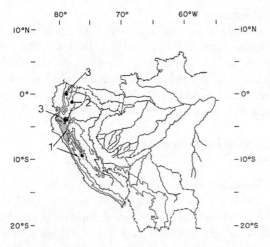

Map 16.140. Distribution of (1) *Ichthyomys stolzmanni;* (2) *I. hydrobates;* (3) *I. tweedii.*

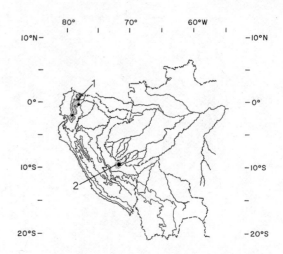

Map 16.141. Distribution of (1) *Neusticomys monticolus;* (2) *N. peruviensis.*

Ichthyomys tweedii Anthony, 1921

Description

Measurements: HB 167; T 150; HF 36; E 9. These rats are large and dark colored; the tail is not bicolored.

Distribution

The species occurs from northwestern Ecuador to Panama (map 16.140). In Ecuador it is found in the southern province of El Oro and the upper Río Esmeraldas drainage of Pichincha province.

Genus *Neusticomys* Anthony, 1921

Description

The dental formula is I 1/1, C 0/0, P 0/0, M 2/2 or 3/3. Head and body length averages 100 to 131 mm, and the tail is 105 to 111 mm, usually shorter than the head and body. There are long guard hairs with a dense underfur. The upperparts are balckish brown, and the venter is paler. The characters of the genus are reviewed by Musser and Gardner (1974) and by Voss (1988b).

Distribution

The genus is known from the northern and southern montane region of Venezuela and lowland Peru and from French Guiana (Musser and Gardner 1974).

Neusticomys monticolus Anthony, 1921

Description

The type specimen (dry) measured: TL 204; T 111; HF 25. This small rodent has dense, soft fur, and Anthony implied that it was the least specialized for aquatic life. It is clearly allied to the other genera of ichthyomyine rodents, but the hallux is greatly reduced. It is superficially similar to *Rheomys* (see Eisenberg 1989). The external ear is not noticeably reduced. The

dorsal pelage is an olive brown with the dorsal hairs dark brown at the base; the venter is grayish, with the tail color paralleling that of the body. Voss (1988b) believes and offers sound evidence that the Ichthyomyini are monophyletic (see Eisenberg 1989).

Distribution

The type locality is Nono Hacienda San Francisco, Pichincha province, Ecuador, at 3,150 m elevation. The species extends to Andean Colombia, where it has been recorded from Risaralda department (map 16.141).

Neusticomys peruviensis (Musser and Gardner, 1974)

Description

The species is known only from a single specimen. Measurements: HB 128; T 108; HF 30; E 12 (LSU 14407). The glossy dorsum is brownish. The eye is minute.

Distribution

Neusticomys peruviensis is known only from the type locality, 10°08′ S, 71°13′ W, Ucayali department, Peru, at 300 m elevation (map 16.141).

Neusticomys venezuelae (Anthony, 1929)

Description

The dental formula is I 1/1, C 0/0, P 0/0, M 3/3. Average measurements are HB 114; T 108; HF 27; E 10. The dorsum is blackish brown, and the venter is gray.

Range and Habitat

This species ranges from 750 to 1,400 m in southern Venezuela in association with montane streams. It is rare throughout its range and could extend to Guyana and thence to Brazil (not mapped).

Comment

Formerly *Daptomys venezuelae*.

TRIBE WIEDOMYINI
Genus *Wiedomys* Hershkovitz, 1959
Wiedomys pyrrhorhinos (Wied-Neuwied, 1821)
Red-nosed Mouse

Description

Measurements: HB 100–128; T 160–205. This long-tailed mouse has a light brown dorsum and a white venter. The nose and face, including the ears, are reddish, as are the forelimbs, and rump (see plate 12). It resembles *Thomasomys* in body proportions, but the dentition is similar to that of *Calomys*.
Chromosome number: $2n = 62$; FN = 86 (Maia and Langguth 1987).

Distribution

This species inhabits dry deciduous and scrub forest in the caatinga and cerrado vegetation formations of eastern Brazil (map 16.142).

Life History and Ecology

This mouse is an able climber. It will construct nests in termite mound cavities and in the nests abandoned by birds. It appears to be granivorous and also takes insects. Several animals, presumably parents and young, can occupy the same nest (Streilen 1982c). A captive colony at the NZP exhibited the following reproductive data: age at first birth (females), eighty-three days; interbirth interval forty days (mean) range thirty-nine to seventy days; litter size 3.8 (mean) (range 3–5; $n = 6$) (Eugene Maliniak, pers. comm.; see also Streilen 1982c).

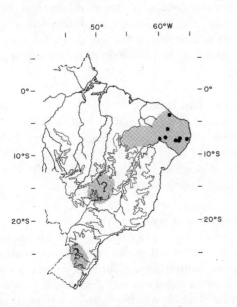

Map 16.142. Distribution of *Wiedomys pyrrhorhinos*.

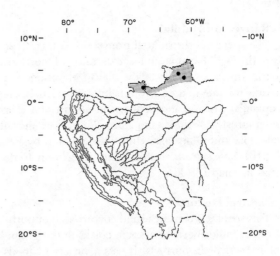

Map 16.143. Distribution of *Zygodontomys brevicauda*.

TRIBE ZYGODONTOMYINI
Genus *Zygodontomys* J. A. Allen, 1897

Introduction

This genus has been assigned to various higher-level taxa, but thanks to Voss (1991b), we can rest and reflect on the premise that the genus is not an "oryzomyine," but just might have its own tribal rank legitimately assigned. Voss and Carleton (1993), after redefining the content of the Oryzomyini and reconsidering the content of *Holochilus*, now suggest including *Zygodontomys* within the Oryzomyini. Voss (1991b) points out that this genus is associated with habitats in South America that are subject to extreme alternating wet and dry seasons. The area in question is a mosaic of grasslands, riverine gallery forests, and seasonally inundated low-stature forests, as well as scrub forest on noninundated areas that may well burn, given lightning strikes and human activities. In short, the ruminations of Voss (1991b) can be read with great profit by neophytes recently indoctrinated with the concept of "rain forest biodiversity."

In short, Voss (1991a,b) has promulgated the notion that, aside from insular forms, there are two mainland species, *Z. brunneus* of Columbia and *Z. brevicauda*, ranging from Central America to Brazil (the range is disjunct). The Brazilian distribution is portrayed in map 16.143.

Zygodontomys brevicauda (J. A. Allen and Chapman, 1893)

Description

Average measurements: TL 236; T 96; HF 27; E 15; Wt 65 g. The upperparts are gray to agouti brown, the underparts white to gray. The tail is scantily haired and tends to be bicolored, whitish below and gray brown above. The pelage of juveniles tends to be gray. The molar teeth are illustrated in figure 16.3

Range and Habitat

The species is distributed from southeastern Costa Rica through Panama and across most of northern South America, usually below 500 m elevation. It is broadly tolerant of a variety of habitat types, including croplands, pasture, and clearings in multistratal evergreen forest. It is the dominant rodent in the llanos of Colombia and Venezuela. Under the revision by Voss (1991b), Z. brevicauda extends into Brazil only to the Amazon (map 16.143).

Natural History

This species is terrestrial and nocturnal. It opportunistically builds nests of grass in cracks in the ground under trees or in burrows. It eats a variety of foods, including insects, seeds, and fruits. In the field, the average litter size is 4.1 and the young attain sexual maturity before two months of age. Reproduction can occur throughout the year, but the timing is frequently controlled by rainfall (O'Connell 1981). In the laboratory, Aguilera (1985) found that gestation averaged twenty-five days. Litter size for primiparous females averaged 3.3, while the value for multiparous females was 5. The sex ratio at birth was 1:1, and in captivity females attain first estrus at 25.6 days of age and males reach sexual maturity at 42.3 days. At high densities home ranges of individuals can show considerable overlap (Vivas and Roca 1983).

Comment

The species name Z. microtinus is often applied to the populations north of the Orinoco in Venezuela (Aguilera 1985). Voss (1991a) does not recognize Z. microtinus. Zygodontomys reigi, described by Trainer (1976) from Cayenne, French Guiana, may be conspecific with Z. brevicauda. Z. borreroi (Hernandez-Camacho, 1957) has a restricted distribution in Santander department, Colombia (see commentary in Voss 1991b).

FAMILY MURIDAE

SUBFAMILY MURINAE

Old World Rats and Mice

Diagnosis

These rodents have a dental formula of I 1/1, C 0/0, P 0/0, M 3/3 and typically have the molar cusps aligned in three longitudinal rows (fig. 16.6), thus differing from the varied sigmodontine forms. Superficially very similar to cricetine rodents in body form and proportions.

Distribution

These rodents are native to the Old World, but species of the subfamily Murinae have been widely intro-duced by man around the world. Three species occur near human settlements in the area covered by this volume: Rattus rattus, Rattus norvegicus, and Mus musculus.

Genus *Rattus* Fischer, 1803

Rattus norvegicus (Berkenhout, 1769)
Brown Rat

Description

Total length ranges from 320 to 480 mm, tail 153 to 218 mm, hind foot 37 to 44 mm. The tail is shorter than the head and body, and this rat is more robust than R. rattus. The dorsal pelage is brown interspersed with black hairs, and the underparts are pale gray.

Habitat and Range

This terrestrial species is associated with large urban centers. It does not penetrate into undisturbed habitats (not mapped).

Natural History

Rattus norvegicus is much more terrestrial than R. rattus. It tunnels actively, and its presence is usually first detected when its burrows are noticed. Up to seven young are born after a twenty-three-day gestation period. The brown rat is an omnivore and is often a serious pest when it exploits stored cereal grains. The ecology and behavior of the brown rat have been summarized by Calhoun (1962).

Rattus rattus (Linnaeus, 1758)
Black Rat

Description

Total length ranges from 327 to 430 mm; tail 160 to 220 mm; hind foot 35.5 mm. A large male may attain 200 g. This species may be distinguished from R. norvegicus by its longer tail. The dorsal color is grayish black to brown, and the venter may be gray or almost black.

Range and Habitat

As a species introduced by man, it tends to be associated with human dwellings. This rat is highly arboreal and may extend into forest regions far from points of original introduction. Generally it occurs near coastal settlements (not mapped).

Natural History

This rat is highly adaptable and an excellent climber. It is omnivorous and readily eats fruits and nuts when occupying natural forests. Three to five young are produced after a twenty-three-day gestation period. The behavior of black rats has been described by Ewer (1971).

Genus *Mus* Linnaeus, 1766

Mus musculus Linnaeus, 1766
House Mouse

Description

The total length ranges from 140 to 205 mm; tail 69 to 85 mm; hind foot 16 to 20 mm. The dorsum is gray brown to brown; the venter is lighter and often buffy.

Range and Habitat

This species is closely associated with human dwellings and farms. It has been widely introduced in coastal cities (not mapped).

Natural History

This mouse is omnivorous and mainly terrestrial. It climbs well but does not nest arboreally. Five to seven young are born after twenty-one to twenty-two days' gestation. These mice are almost always associated with man as a commensal. Their natural history has been summarized for Australia by Newsome (1971) and for England by Chitty and Southern (1954). Since this species has been introduced widely and currently exhibits great genetic variation, it is of great theoretical interest to population geneticists (see Boursot et al. 1993).

HYSTRICOGNATH RODENTS

Introduction

As indicated in the section "Rodentia: History and Classification" (p. 356), the earliest rodents on the South American continent were hystricognathous. The classification employed here follows Woods (1982), who groups the South American hystricognath rodents into four superfamilies. All four occur within the range covered by this volume: the Erethizontoidea, the Cavioidea, the Chinchilloidea, and the Octodontoidea (Woods 1982) (see also table 16.6).

In the New World, the hystrigognath rodents are characterized by greatly enlarged infraorbital foramina. The basic molar cusp pattern is pentatlophodont, but there may be profound variation over a series of species (fig. 16.16). The hystricognath rodents also exhibit a number of unique reproductive characters. Litter sizes tend to be small (one to three), and gestations are long (table 16.3). The young are born precocially. The biology of hystricognath rodents is admirably summarized in the volume edited by Rowlands and Weir (1974). Reproduction has been summarized by Weir (1974b) and by Kleiman, Eisenberg, and Maliniak (1979). Behavior patterns were reviewed by Kleiman (1974), Eisenberg (1974), and Eisenberg and

Table 16.5   Reproductive Data for Caviomorph Rodents (in Order of Increasing Weight)

Species	Adult Weight (g)	Birthweight (g)	Mean Litter Size (range)	Age When 50% Adult Weight Is Achieved weeks	Gestation Period (days)
Ctenomys talarum	150	8.0	4.3 (1–7)		102
Abrocoma cinerea	150	22.0	2.2 (1–3)	10	116
Octodontomys gliroides	190	20.0	2.1 (1–4)	6	104
Octodon degus	200	14.4	4.9 (1–8)	5	90
Microcavia australis	280	30.0	2.8 (1–5)	9	54
Proechimys suariae	305	20.9	2.7 (1–6)	7	63
Trichomys cunicularis	335	29.0	3.4 (1–5)		89
Diplomys daelingei	350	35.0	1.2 (1–2)		
Galea musteloides	400	37.0	2.7 (1–5)	11	53
Hoplomys gymnurus	450	24.3	2.1 (1–3)	12	
Proechimys semispinosus	450	26.5	3.3 (1–5)	14	65
Chinchilla lanigera	500	35.0	2.0		111
Cavia aperea	550	57.6	2.3 (1–5)	11	62
Geocapromys ingrahami	740	85.0	1.0	13	115
Kerodon rupestris rupestris	800	80.0	1.6 (1–2)		77
Myroprocta pratti	950	77.0	1.8 (1–3)	12	99
Plagiodontia aedium	1,267	110.0	1.3 (1–2)	15	
Lagidium peruanum	1,300	226.0	1.0	11	140
Pediolagus salinicola	1,700	199.0	1.5 (1–3)		77
Dasyprocta punctata	3,000		(1–2)	11	
Lagostomus maximus	3,000	193	1.9 (1–3)	17	154
Coendou prehensilis	3,100	390	1.0	18	
Capromys pilorides	5,200	200	1.8 (1–3)	24	125
Myocastor coypu	6,000	225	5.0		129
Agouti paca	8,000	710	1.0	14	118
Dinomys branickii	13,000	900	(1–2)		223–80
Hydrochaeris hydrochaeris	49,000	14,000	3.5 (1–6)	52	120

Source: Kleiman, Eisenberg, and Maliniak 1979.

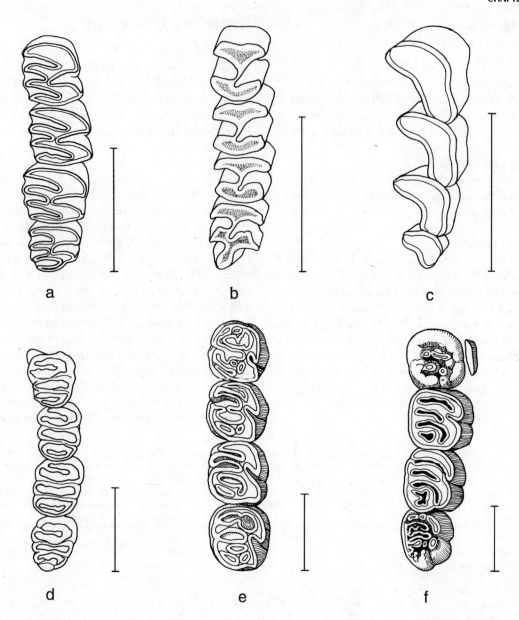

Figure 16.16. Examples of hystricognath rodent dentition (after Ellerman 1940). Upper cheek teeth (labial left; anterior top): (a) *Kannabateomys ambylonyx;* (b) *Abrocoma bennetti;* (c) *Ctenomys* sp.; (d) *Chaetomys subspinosus;* (e) *Coendou prehensilis;* (f) *Dasyprocta punctata.*

Kleiman (1977); distribution and ecology were covered by Mares and Ojeda (1982).

SUPERFAMILY ERETHIZONTOIDEA

FAMILY ERETHIZONTIDAE
Porcupines, Puercos Espinosos

Diagnosis
The dental formula is I 1/1, C 0/0, P 1/1, M 3/3. The genera of this family are adapted for an arboreal life; the thumb is replaced by a broad movable pad, and the sole of the hind foot is wide. The dorsal pelage is modified so that many hairs have become spinelike (plate 11). The tips of the spines have minute barbs. The spines are easily detached and embed themselves in potential predators. Table 16.9 gives a classification and body sizes for the Erethizontodae.

Distribution
Species of this group are distributed from northern Canada and Alaska south to northern Argentina. The northern species belong to the genus *Erethizon.* The species covered in the range of this volume belong to the genera *Echinoprocta, Sphiggurus,* and *Coendou.*

Table 16.6   Classification for Hystricognath Rodents

Suborder	Family	Subfamily
Hystricognathi		
	Bathyergidae	
	Hystricidae	
	Petromuridae	
	Thryonomyidae	
	Erethizontidae	
	Chinchillidae	
	Dinomyidae	
	Caviidae	
		Caviinae
		Dolichotinae
	Hydrochaeridae	
	Dasyproctidae	
	Agoutidae	
	Ctenomyidae	
	Octodontidae	
	Abrocomidae	
	Echimyidae	
		Chaetomyinae
		Dactylomyinae
		Echimyinae
		Eumysopinae
		+ Heteropsomyinae
	Capromyidae	
		Capromyinae
		+ Hexolobodontinae
		+ Isolobodontinae
		Plagiodontinae
	+ Heptaxodontidae	
		+ Clidomyinae
		+ Heptaxodontinae
	Myocastoridae	

Source: Woods 1993.
+ = extinct.

Table 16.7   Nomenclature for the Erethizontidae of South and Central America

Emmons 1989 ("Lowland")	Honacki, Kinman, and Koeppl 1982 and Cabrera 1960
Coendou bicolor	*Coendou bicolor*
Coendou insidiosus	*Sphiggurus insidiosus*
Coendou melanurus	*Sphiggurus insidiosus* (Cabrera)
Coendou mexicanus	*Sphiggurus mexicanus*
Coendou paraguayensis	*Sphiggurus spinosus*
Coendou prehensilis	*Coendou prehensilis*
Coendou pruinosus	*Sphiggurus vestitus*
Coendou quichua (Ecuador)	*Sphiggurus quichua* (Cabrera)
Coendou rothschildi	*Coendou bicolor*
Coendou sneiderni	*Echinoprocta rufescens*
Coendou vestitus	*Sphiggurus vestitus*
Coendou villosus	*Sphiggurus villosus*
Chaetomys subspinosus	*Chaetomys subspinosus*

Table 16.8   Body Proportions of the Erethizontidae Including *Chaetomys*

Taxon	Measurements
Echinoprocta	HB 309–70; T 105–50 (~ 30%)
Coendou	HB 300–600; T 330–485 (~ 80 +%)
Sphiggurus	HB 363–80; T 375–70 (~ 90 +%)
Erethizon	HB 645–80; T 145–300 (< 50%)
Chaetomys	HB 430–57; T 255–80 (59–61%)

Note: % = percentage of head and body length.

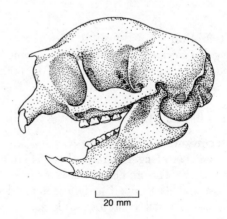

20 mm

Figure 16.17. Skull of *Coendou prehensilis*.

Comment on Taxonomy

Leaving aside the North American genus *Erethizon*, Honacki, Kinman, and Koeppl (1982) recognized four genera from South America: *Chaetomys*, *Coendou*, *Echinoprocta*, and *Sphiggurus*. Emmons and Feer (1990) include *Sphiggurus* within *Coendou* (table 16.7), Woods (1993) includes *Chaetomys* within the Echimyidae. Eisenberg (1989) acknowledged the distinctiveness of *Sphiggurus*. Furthermore, Emmons and Feer (1990) recognize most subspecies of *Coendou* (*Sphiggurus*) deriving from the Atlantic rain forest of Brazil as full species. Handley and Pine (1992) subsume *Sphiggurus* within *Coendou*, but Concepcion and Molinari (1991) offer evidence for separating *Sphiggurus* from *Coendou*. The situation is confused to say the least. Until someone conducts a thorough review of the family, we will remain conservative with respect to the taxonomy and follow Woods (1993); however, where controversial points are raised we will outline the alternatives (table 16.8).

Genus *Coendou* Lacépède, 1799
Prehensile-tailed Porcupine, Puerco-espín

Description

The dental formula is I 1/1, C 0/0, P 1/1, M 3/3 (fig. 16.17). Head and body length ranges from 300 to 600 mm, and the tail is 330 to 450 mm. Adults may weigh over 4 kg. The naked tail is fully prehensile. The dorsal pelage is distinctive; the long dorsal hairs do not cover the prominent spines.

Table 16.9   Classification and Body Sizes for the Erethizontidae of South America

Arrangement according to Woods 1993	Arrangement according to Emmons and Feer 1990	Measurements			Color Pattern and Spinescence	
		HB	T	% T/HB	Condition of Dorsum	Basic Dorsal Color
Coendou bicolor	*Coendou bicolor*	378–493	460–540	110–21	Spiny	Less white on tips, blackish
Coendou koopmani	*Coendou sp. nov.*	319–64	280–326	88–90	Spiny	Black tips on spines (S bank east Amazon)
Coendou prehensilis	*Coendou prehensilis*	444–560	330–578	74–103	Spiny	Tips of spines whitish
Coendou rothschildi	*Coendou rothschildi*	332–438	260–413	78–94	Spiny	Similar to *bicolor* (Panama)
Sphiggurus insidiosus (*melanurus*)	*Coendou melanurus*	280–380	220–363	79–96	Hairy	Black-white tips, black tail (Sur, S Vez)
Sphiggurus insidiosus	*Coendou insidiosus*	290–350	180–222	62–63	Hairy	Smoky brown (Bahia)
Sphiggurus spinosus	*Coendou paraguayensis*	288–340	228–60	79–77	Hairy	Pale, yellow (SE Paraguay)
Sphiggurus vestitus	*Coendou vestitus*	290	130	45	Hairy	Smoky brown (Col Andes)
Sphiggurus vestitus (*pruinosus*)	*Coendou pruinosus*	330–80	190	52	Hairy	Black brown frosted (Vez Andes)
Sphiggurus villosus	*Coendou villosus*	300–538	200–378	66–70	Hairy	Brown, orange tips (SE Brazil)
Echinoprocta rufescens	*Coendou sneideri*	335	125	37	Rump hairy, rest spiny	White blaze on head, tips black brown, spines orange (W Cord Col Andes)
Sphiggurus mexicanus	*Coendou mexicanus*	350–460	200–360	57–78	Hairy	Black–dark brown (Mexico)
Chaetomys subspinosus	*Chaetomys subspinosus*	380–450	260–75	68–61	Spiny	Salt-and-pepper, pale brown (Bahia)

Note: Italic in column % T/HB indicates extremely short tail.

Distribution

The genus *Coendou* as defined in this volume is distributed from eastern Panama south to Argentina and east across Brazil to the Guianas.

Coendou bicolor (Tschudi, 1844)

Description

Head and body length averages about 543 mm, and the tail 481 mm. The prominent spines are pale yellow at the base and black tipped, and those in middorsum are very dark. The tail is fully prehensile; spines extend on the dorsal side for most of its length, while the distal 50 mm is haired. This species is darker than *C. prehensilis*. The venter may be light to dark brown.

Range and Habitat

The species is distributed from northern Colombia to western Ecuador and thence to Amazonian Peru and Bolivia and adjacent parts of Brazil (map 16.144).

Coendou nycthemerae (Olfers, 1818)

Description

Measurements of old adult: TL 678.6; T 327.4; HF 65.2 ($n = 5$–7). The skull is small, with a condylar

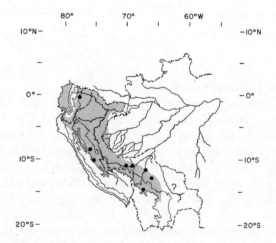

Map 16.144. Distribution of *Coendou bicolor*.

basal length less than 70 mm. *C. nycthemerae* appears smaller than *C. melanurus*. It is sympatric with *C. prehensilis* and much smaller. This small form falls within the size class of *Sphiggurus*, yet it is considered a species of *Coendou*, since Handley and Pine reject the status of *Sphiggurus* based on an elaborate series of cranial measurements that they carefully explain. This small, long-tailed *Coendou* is covered with quills and

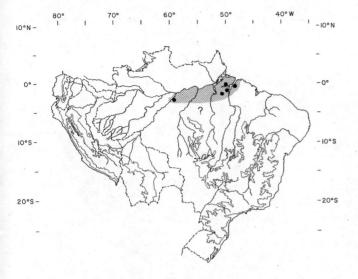

Map 16.145. Distribution of *Coendou koopmani.*

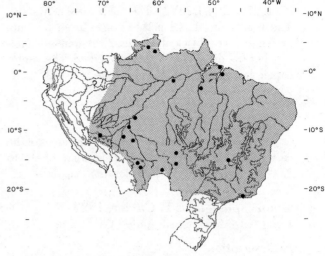

Map 16.146. Distribution of *Coendou prehensilis.*

lacks long, soft hairs as an adult (Handley and Pine 1992).

Note: *Coendou koopmani* (Handley and Pine, 1992) is included in *C. nycthemerae* (see Voss and Angermann 1997).

Distribution

The species is found south of the Amazon in Brazil from Ilha de Marajó, Belém, to Ilha Tayaúna. According to Handley and Pine (1992), it co-occurs with *C. prehensilis* (map 16.145).

Coendou prehensilis (Linnaeus, 1758)
Coendú Grande

Description

Average measurements: HB 524; T 520; HF 83; E 15; Wt 4.9 kg. In this rather large porcupine the quills are conspicuous and not covered by dorsal hairs. The tail almost equals the head and body length and is fully prehensile. The shape of the skull is distinctive (fig. 16.17).

Distribution

Coendou prehensilis is broadly distributed from Venezuela south to northern Argentina. In southern South America it has been collected in Uruguay, eastern Paraguay, and the Argentine provinces of Salta and Misiones (Olrog 1976; Tálice 1969; Yepes 1938) (map 16.146).

Life History

In captivity, the gestation has been determined to last 195–210 days, with a postpartum estrus and a minimum interbirth interval of 203 days. There is no seasonality in reproduction, but rainfall can cause birth

peaks in selected areas. The litter size is one, and the average neonatal weight is 415 g (range 364–447; $n = 8$). Young are born with long red fur that hides the spines. Lactation lasts approximately fifteen weeks. The young is left alone in a sheltered place and nursed at least once a day by the mother. Female age at first reproduction is about nineteen months, and animals can reproduce for more than twelve years (Roberts, Brand, and Maliniak 1985).

Ecology

In Venezuela this porcupine is nocturnal and rests in trees during the day, often sheltering in cavities or remaining inconspicuously high in a tree crown. The home range of this solitary species may be quite large depending on the availability of suitable feeding trees. In the Venezuelan llanos, animals range over an area of 15 to 20 ha.

In Suriname, *C. prehensilis* was exclusively herbivorous, with a large proportion of the diet consisting of buds, bark, fruits, and unripe seeds of large fruit. In captivity these porcupines sleep aboveground and are nocturnal. Adults are primarily solitary but socially tolerant. Both sexes anal rub, and males mark females with urine. Nine vocalization types are produced, including a long moan employed as a contact call between isolated individuals (Charles-Dominique et al. 1981; Montgomery and Lubin 1978; Roberts, Brand, and Maliniak 1985).

Genus *Echinoprocta* Gray, 1865
Echinoprocta rufescens (Gray, 1865)
Short-tailed Porcupine

Description

Three specimens of this small porcupine from Colombia measured HB 309–70; T 105–50; HF 60–

67; E 17 (FMNH, MVZ). The tail is extremely short. The very spiny dorsal pelage ranges from medium brown to almost black; the bases of the spines are yellow, but the distal 20 mm are pigmented. The venter is pale brown and spinescent. There is usually some white on the face.

Range and Habitat

The species is known from the montane areas of the eastern cordillera of the Andes at elevations of 800 to 2,000 m. It may range to Ecuador (not mapped).

Genus *Sphiggurus* F. Cuvier, 1823
Long-haired Prehensile-tailed Porcupine

Description

The dental formula is I 1/1, C 0/0, P 1/1, M 3/3. The dorsum is covered with long hairs that cover and conceal the spines in the midportion, distinguishing *Spiggurus* from *Coendou*. The long hairs are white tipped, the middle part being blackish brown while the basal third is paler. The spines on the dorsum are rather short, about 25 mm long, and are yellowish with brown tips. The tail is fully prehensile and hairless.

Distribution

If one considers *Coendou mexicanus* to be part of *Sphiggurus*, then the genus is distributed from San Luis Potosí, Mexico, south through the Isthmus across much of northern South America and into Amazonian Brazil.

Natural History

Little work has been done with this genus. It is similar to *Coendou* in its behavior.

Comment

Sphiggurus has been proposed as a generic category for two northern species formerly considered within the genus *Coendou* (Husson 1978): *S. insidiosus* and *S. vestitus*. *C. mexicanus* exhibits the long dorsal hairs that are diagnostic for *Sphiggurus*. There is no reason not to include *C. mexicanus* within the genus.

Sphiggurus insidiosus (Lichtenstein, 1818)

Description

Head and body length ranges from 290 to 350 mm; the tail is 180 to 222 mm, the hind foot 62 to 75 mm. The long prehensile tail is black and sparsely haired to the tip. The bases of the dorsal hairs are black, and the tips are yellow. The venter is sparsely covered with a mixture of black and yellow hairs. The spines are hidden by the long dorsal hairs (see also *Sphiggurus melanurus*) (see plate 11).

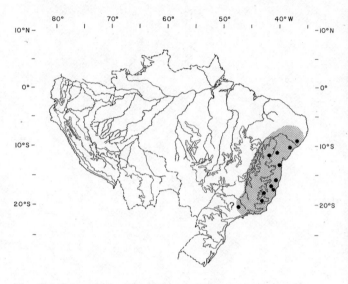

Map 16.147. Distribution of *Sphiggurus insidiosus*.

Range and Habitat

The species ranges from eastern Brazil north to Suriname and perhaps Guyana (map 16.147).

Natural History

Husson (1978) reports that these porcupines feed on fruit, ant pupae, cultivated vegetables, and roots. *Sphiggurus insidiosus* in eastern Brazil can occur in very degraded secondary forests. It can exhibit two color patterns. In southern Bahia nearly white individuals have been collected (Oliver and Santos 1991).

Sphiggurus melanurus (Wagner, 1842)

Description

Sphiggurus melanurus is a rather large species: HB 280–330; T 220–363. It has the black tail described for *S. insidiosus*, but the entire tail is usually melanistic. It is often considered a subspecies of *S. insidiosus*.

Distribution

If considered a full species, *Sphiggurus melanurus* extends from the state of Bolívar, Venezuela, to adjacent portions of Brazil (map 16.148).

Sphiggurus spinosus (Lichtenstein, 1818)

Description

Average measurements: TL 581; HB 329; T 251; HF 49; E 14. *S. spinosus* has gray brown hair; black is always prominent on the distal portion of the body hairs. Quills are scarcely visible anywhere on the body, being obscured by the longer hairs. The tail is relatively short, with no quills on the tip. Young are covered with reddish hair. This species is easily distinguishable form *Coendou prehensilis* by the much

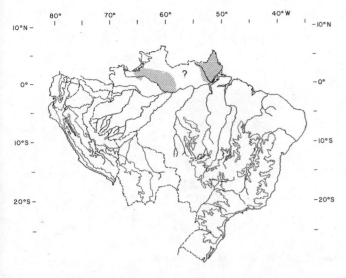

Map 16.148. Distribution of *Sphiggurus melanurus*.

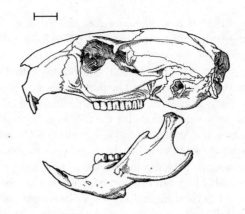

Figure 16.18. Skull of *Chaetomys subspinosus* (after Ellerman 1940). Scale bar = 8.5 mm.

Map 16.149. Distribution of *Sphiggurus spinosus*.

Map 16.150. Distribution of *Chaetomys subspinosus*.

smaller size and longer hair with no quills showing through (Miranda-Ribeiro 1936; pers. obs.).

Distribution

Sphiggurus spinosus is found in southern and eastern Brazil, eastern Paraguay, throughout Uruguay, and in the Argentine province of Misione (Crespo 1974a; Honacki, Kinman, and Koeppl 1982; Ximénez, Langguth, and Praderi 1972) (map 16.149).

Comment

Sphiggurus spinosus is often listed as *Coendou paraguayensis*.

Sphiggurus villosus (F. Cuvier, 1823)

Comment

Some workers consider *Sphiggurus villosus* a subspecies of *S. insidiosus*.

Genus *Chaetomys* (see p. 479)

Comment

Whether systematists finally decide to include *Chaetomys* within the Erethizontidae or to relegate it to the Echimyidae remains to be seen. Nevertheless, the structure of the nose and tail, regardless of the deciduous nature of the premolar (fig. 16.18), strongly suggests a close affinity between *Chaetomys* and the rest of the Erethizontidae (see Voss and Angermann, 1997) (map 16.150).

Comments on the Taxonomy, Ontogeny, and Phylogeny of the Erethizontidae

The condition of the newborn in *Coendou* is not appreciated by many workers (Miranda-Ribeiro 1936). The neonate has extremely soft spines that are completely hidden by a coat of hair ranging from light orange to red. In other words, the adult *Coendou* looks quite different from the juvenile. Indeed, members of the genus *Sphiggurus* carry on into adulthood the characters of the young *Coendou*. This could imply that *Sphiggurus* is neotenic or, contrarywise, that *Coendou* is gerontomorphic. The porcupines are a vexing group badly in need of revision. Table 16.8 gives some important measurements for the family. Note that the tail is long and prehensile in *Sphiggurus* and *Coendou*. In *Echinoprocta* the tail is short and presumably not prehensile, as it is also in *Erethizon*. *Chaetomys* has a reasonably long tail that can coil, but to what extent it is used in climbing has not as yet been adequately described. One thing should be noted: in *Coendou*, *Sphiggurus*, and *Chaetomys* the nose is distinctive; it is not only naked, but bulbous.

FAMILY DINOMYIDAE
Genus *Dinomys* Peters, 1873
Dinomys branickii Peters, 1873
Branick's Giant Rat, Pacarana, Guagua Loba

Systematics
This family of rodents shows great species diversity in the Miocene, but only a single living representative survives. It is considered to be within the superfamily Erethizontoidea in accordance with Grand and Eisenberg (1982).

Diagnosis
The dental formula is I 1/1, C 0/0, P 1/1, M 3/3. The teeth are high crowned (fig. 16.19). Head and body length ranges from 730 to 790 mm, the tail averages 200 mm, and adult weight may exceed 15 kg. The head is massive, and the limbs and ears are short. *D. branickii* has long, powerful claws and a short, stout tail that is haired and not prehensile. The dorsal pelage is brown to black, and on each side are rows of white spots.

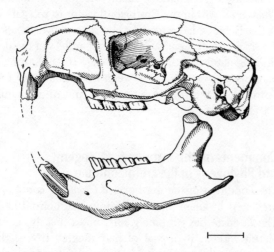

Figure 16.19. Skull of *Dinomys branickii* (after Ellerman 1940). Scale bar = 20 mm.

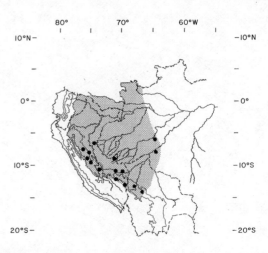

Map 16.151. Distribution of *Dinomys branickii*.

Distribution
The single living species of this family, *D. branickii* is currently found in isolated localities in Amazonian Brazil, Peru, Ecuador, Venezuela, and Colombia (map 16.151) at up to 1,000 m elevation.

Natural History
These large, slow-moving rodents have some climbing ability, especially prominent in young animals. Their behavior in captivity has been summarized by Meritt (1984). They tend to be active at night. Their diet includes fruits, some leaves, and herbaceous vegetation. They shelter in caves or holes at the bases of trees. They produce a wide variety of vocalizations, and males seeking mates produce a complicated, intricate series of calls (Eisenberg 1974). Gestation lies between 222 and 280 days; generally only two young are born. This species is hunted for food and may be on the verge of extinction (Collins and Eisenberg 1972; White and Alberico 1992).

SUPERFAMILY CAVIOIDEA

FAMILY CAVIIDAE

Diagnosis
The dental formula is 1/1, 0/0, 1/1, 3/3. The cheek teeth are ever-growing and have two enamel prisms with sharp folds and angular projections (fig. 16.20). Body form is stout. Reduction in the number of toes is common; four toes are retained on the forefoot and three on the hind foot. The tail is extremely reduced.

Distribution
Disjunct distribution characterizes the northern population, but the genus occurs over most of South America except the easternmost parts of Brazil and the extremely arid portions of southern Peru and northwestern Chile.

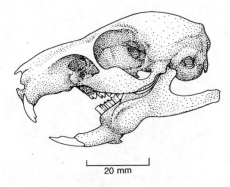

Figure 16.20. Skull of *Cavia aperea*.

Genus *Cavia* Pallas, 1766

Description
Most characters are as noted for the family; however, in the genus *Cavia* the tail is virtually absent (see species accounts).

Cavia aperea Erxleben, 1777
Cavy, Guinea Pig, Cuis Selvatico, Cuis Campestre

Description
This account includes *Cavia nana* and *C. pamparum*. Average measurements: HB 272; T 2.4; HF 47; E 26; Wt 637 g. This species closely resembles the domestic guinea pig but is less robust. It has an extremely short or nonexistent tail. The dorsum is dark olive mixed with brownish and blackish, and the venter is paler, usually a whitish or yellowish gray (Barlow 1965, 1969; pers. obs.).
Chromosome number: $2n = 64$; FN = 116 (Maia 1984).

Distribution
Cavia aperea is found from Colombia south through Peru and Bolivia to Brazil and into Argentina. In southern South America it is found in Uruguay, eastern Paraguay, and in northeastern and east-central Argentina, south to Buenos Aires province (Contreras 1972b, 1980; Massoia 1973a; Ximénez, Langguth, and Praderi 1972) (map 16.152).

Life History
In captivity the gestation is sixty-two days, the litter size averages 2.3 (range 1–5), and neonates average 57.6 g. In Uruguay this species has a protracted breeding season beginning in September and extending until April or May. In Buenos Aires province, Argentina, young are born throughout the year; two neonates weighed 56.6 and 58.7 g; the litter size averages 2.1; and females can have up to five litters per year. The minimum age of first reproduction is thirty days (Barlow 1969; Kleiman, Eisenberg, and Maliniak 1979; Rood 1972).

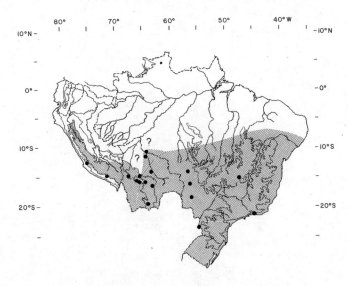

Map 16.152. Distribution of *Cavia aperea* (includes *C. porcellus* in the broad sense. Peruvian and Bolivian specimens also plotted in map 16.155).

Ecology
Cavia aperea is a grassland herbivore, ecologically analogous to some of the North American microtine rodents. It does not excavate burrows but constructs a maze of interconnecting surface tunnels and runs 8–12 cm in diameter. Piles of its characteristic bean-shaped fecal pellets and plant cuttings are found at intervals along these runs. It may forage in the open but remains close to shelter.

C. aperea is a grazer, but it will eat grass inflorescences as well and feeds during evening and early morning. It can occur in high densities; in Buenos Aires province, Argentina, it reaches a maximum of 38.7 individuals per hectare. In this habitat, male home ranges averaged 1,387 m² and female home ranges 1,173 m². This species can depress the biomass of other grazing rodents (Barlow 1969; Dalby 1975; Rood 1972).

Comment
The domesticated form, *Cavia porcellus*, is included within *C. aperea*. The status of *C. magna* is uncertain (see below).

Cavia fulgida Wagler, 1831

Description
Measurements: TL 265; HF 51. This small species is similar to *Cavia aperea*, but it has a reddish brown dorsum. The venter is paler orange brown.

Distribution
Apparently *Cavia fulgida* does not overlap with *C. magna*, but it can co-occur with *C. aperea*. According to Ximénez (1980) the range is the coastal areas of southeastern Brazil (map 16.153).

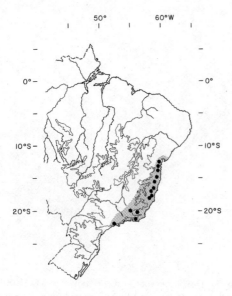

Map 16.153. Distribution of *Cavia fulgida.*

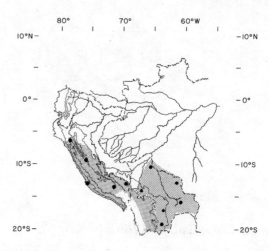

Map 16.155. Distribution of *Cavia tschudii.*

Cavia tschudii Fitzinger, 1857
Cuis Serrano

Description

Average measurements: HB 247; T 4; HF 43; E 28. The color of this species varies throughout its range. In Peru specimens were very dark, almost reddish brown heavily mixed with black and with dark buffy gray underparts. In Chile specimens were pale agouti brown dorsally, and in Bolivia they are agouti olive and brown with a white or cream venter (Sanborn 1949; pers. obs.).

Distribution

Cavia tschudii is distributed from Peru south to Arica in the northern Chilean province of Tarapacá and in northwesternmost Argentina in the province of Tucumán and also probably Jujuy; fossil remains have been found in Salta as well (Honacki, Kinman, and Koeppl 1982; Massoia 1973a; Pine, Miller, and Schamberger 1979; Tonni 1984; Thomas 1926) (map 16.155).

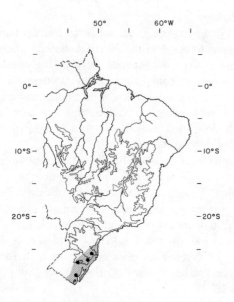

Map 16.154. Distribution of *Cavia magna.*

Cavia magna Ximénez, 1980

Description

Cavia magna is a very large species. The following measurements were recorded by Ximénez (1980): males, TL 310.8; HF 52.6; E 26.4; Wt 635.9 g (*n* = 36); females, TL 303.1; HF 51.4; E 26.0; Wt 536.6 g (*n* = 29). The dorsal pelage is dark agouti brown, and the venter is reddish.

Distribution

The species apparently is found from eastern Uruguay through Rio Grande do Sul to Santa Catarina province, Brazil (map 16.154).

Life History

In captivity the gestation period is 63.3 days, litters average 1.9 (range 1–4), and the average age of first reproduction is two months (Weir 1974b).

Ecology

Cavia tschudii typically is found in moist habitats with scattered rocks between 2,000 and 3,800 m, and specimens have been taken in dense riparian habitats where extensive runways were evident. In Peru the animals live in thick grass in which they make distinct runways. In Argentina this species has been reported to live in burrows with multiple entrances (Lucero 1983; Olrog 1979; Pearson 1957; Pine, Miller, and Schamberger 1979; Tamayo and Frassinetti 1980; Thomas 1926).

Genus *Dolichotis* Desmarest, 1820
Patagonian Cavy, Mara

Description
The description includes *Pediolagus* Marelli, 1927. These are rather large rodents with a head and body length ranging from 430 to 780 mm. The limbs are long, and the feet are digitigrade. The tail is extremely short, and the ears are of modest length (see species accounts). The body form resembles that of a European hare or small artiodactyl, and they can run rather fast. They cannot be confused with any other rodent in the southern cone of South America.

Distribution
The current range of the two species includes the scrub and grassland areas of southern Bolivia, Paraguay, and Argentina.

Dolichotis (Pediolagus) salinicola
(Burmeister, 1876)
Mara Chico, Conejo del Palo

Description
Average measurements: HB 435; T 24; HF 100; E 61; Wt 2 kg. The ears are moderately large, broad, and pointed. This harelike species is agouti brown dorsally, lighter on the sides, and white on the venter and chin. The dorsal coloration extends onto the throat. The cheeks are washed with tan, and there are white patches in front of and behind the eyes. The vibrissae are long and reasonably stout. The legs are very thin (pers. obs.).

Distribution
Dolichotis salinicola is found in extreme southern Bolivia, northwestern Argentina as far south as the province of Córdoba, and in the Paraguayan Chaco (Redford and Eisenberg 1992) (map 16.156).

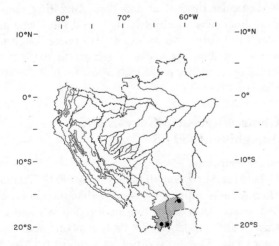

Map 16.156. Distribution of *Dolichotis salinicola*.

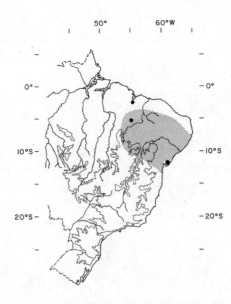

Map 16.157. Distribution of *Galea flavidens*.

Life History
Females have been caught with one and two embryos ($n = 3$). In captivity the gestation period is seventy-seven days, litter size averages 1.5 (range 1–3), and neonates averaged 199 g (Kleiman, Eisenberg, and Maliniak 1979; Mares, Ojeda, and Kosco 1981).

Ecology
Dolichotis salinicola is an animal of the arid chaco of Bolivia, Paraguay, and Argentina and is typically found in dry, low, flat thorn scrub, where it digs large burrows, conspicuous by the extensive piles of dirt outside the entrances (Mares, Ojeda, and Kosco 1981; pers. obs.).

As in *D. patagonica*, the young are integrated into the social unit by various forms of chemical marking and "playful social interactions" (see Wilson and Kleiman 1974).

Genus *Galea* Meyen, 1832
Galea flavidens (Brandt, 1835)

Description
Galea flavidens is considered synonymous with *G. spixii* (Paula Couto 1950). Its status is uncertain.

Distribution
The type locality is possibly the state of Minas Gerais, Brazil (map 16.157).

Galea musteloides Meyen, 1832
Cuis Común

Description
Average measurements: TL 204; HF 38; E 19; Wt 226 g. This is a large, hamster-sized rodent, very

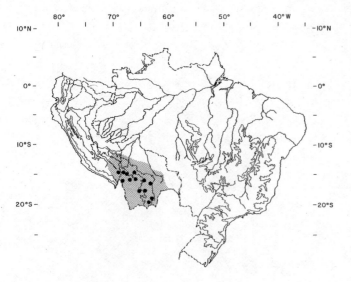

Map 16.158. Distribution of *Galea musteloides*.

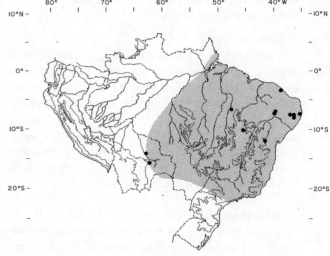

Map 16.159. Distribution of *Galea spixii*.

similar to *Cavia aperea* in coloration but with a slightly whiter belly; it is also less robust and considerably smaller. It ranges from dark agouti flecked with black dorsally, slightly darker on the rump and head, to a much lighter agouti brown. The dorsum is sharply demarcated from the white belly (pers. obs.).

Distribution

Galea musteloides is found from southern Peru south into Chile and Argentina (Honacki, Kinman, and Koeppl 1982; Olrog and Lucero 1981; Tamayo and Frassinetti 1980) (map 16.158).

Life History

In captivity the gestation period is fifty-three days, litter size is 2.7 (range 1–5), and neonates average 37 g. Lactation lasts about three weeks, and the age of first reproduction is one to three months. Females may have up to seven litters per year. Male neonates average 36.4 g, female neonates average 37.6 g; and the minimum age of first reproduction is twenty-eight days (Kleiman, Eisenberg, and Maliniak 1979; Rood 1972; Weir 1974b; UConn).

Ecology

In Peru *Galea* may associate with *Ctenomys*. It uses their burrows and appears to respond to the alarm calls of *Ctenomys* by seeking cover (Sanborn and Pearson 1947). *Galea musteloides* is found through an enormous altitudinal range: in the Andes it occurs up to 5,000 m, whereas in Paraguay it is found in the low Chaco. In Salta province, Argentina, *G. musteloides* is most common in moist areas such as stream edges and croplands. The home range of a female in Buenos Aires province, Argentina, was 4,275 m². Males mark females during courtship using chin gland secre-

tions (Cabrera 1953; Kleiman 1974; Mares, Ojeda, and Kosco 1981; Ojeda 1979; Ojeda and Mares 1989; Rood 1970, 1972).

Galea spixii (Wagler, 1831)

Description

Galea spixii is similar to *G. musteloides*, and they may in fact be conspecific.

Distribution

If this is considered a full species, then it is found within the caatinga of northeastern Brazil (map 16.159).

Natural History

Galea spixii has a litter size of one to five with a mode of three. Young females exhibit a perforate vagina at eighty days of age. The young male exhibits testicular descent at 135 days. Gestation ranges from forty-nine to fifty-two days, and the young are born fully haired and with their eyes open. Young are weaned at about six weeks of age. In the caatinga of Brazil females may reproduce at any time of the year (Lacher 1981).

Genus *Microcavia* Gervais and Ameghino, 1880

Description

Head and body length ranges from 200 to 220 mm. *Microcavia* is appreciably smaller than the closely allied genus *Cavia*. As noted for the family, these are small, tailless rodents with short ears and short but stout claws. The eye is offset by a rim of light hairs, yielding the effect of a ring. The adults lack a submandibular gland, thus distinguishing them from the similar-sized *Galea*.

Distribution

The genus has been recorded form the montane regions of Bolivia, southeastern Peru, and Chile. *M. australis* occurs in lowland Argentina and southern Chile (see species accounts).

Microcavia australis (I. Geoffroy and d'Orbigny, 1833)
Cuis Chico

Description

Average measurements: TL 219; HF 50; E 17; Wt 286 g. The ears are small and rounded, and there is no external tail. The dorsal color ranges from yellowish gray through very light agouti brown to gray. It is lighter on the sides, and the venter is gray to white washed with yellow (Allen 1903; Osgood 1943; pers. obs.).

Distribution

Microcavia australis is distributed in the northernmost Chilean province of Aisén and in Argentina between Jujuy and Santa Cruz provinces. It also occurs in extreme southern Bolivia. This species has also been recovered from archaelogical sites in Magallanes province, Chile (Sielfeld 1979) (map 16.160).

Life History

The average litter size is 2.8 (*n* = 16), born after a gestation period of fifty-three to fifty-five days. Females have a minimum age of first reproduction of eighty-five days and exhibit a postpartum estrus. Neonates average 30 g in weight, and lactation lasts three to four weeks. Females can have up to four lit-

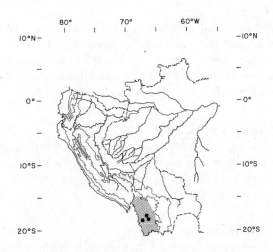

Map 16.161. Distribution of *Microcavia niata*.

ters per year (Kleiman, Eisenberg, and Maliniak 1979; Rood 1970, 1972).

Ecology

In Salta province, Argentina, this species prefers riparian situations or forested areas along gullies or in sandy forested flats. It is particularly abundant along cultivated fields, under rock walls, and in rock piles. In the forested areas burrows are found under shrubs with overhanging branches or under fallen trees.

In Buenos Aires province, Argentina, *M. australis* clears paths between clumps of thorn bushes. Only occasionally will it build burrows, though it frequently uses holes dug by armadillos and viscachas.

Though not a graceful climber, *M. australis* frequently climbs bushes and trees, jumping down when startled. The diet consists of grass, leaves, seed heads, flowers, and berries. They are diurnal and appear not to need free water (Daciuk 1974; Mares 1973; Mares, Ojeda, and Kosco 1981; Rood 1970, 1972).

Microcavia niata (Thomas, 1925)

Distribution

The species occurs in southwestern Bolivia (map 16.161).

Comment

The relation of *Microcavia niata* to *M. australis* is poorly understood; it may be a subspecies of *M. australis*.

Genus *Kerodon* F. Cuvier, 1825
Kerodon rupestris (Wied, 1820)
Moco, Rock Cavy

Description

Kerodon is similar to *Cavia*, but the tail is vestigial or absent. Weights average about 1 kg. The toes are

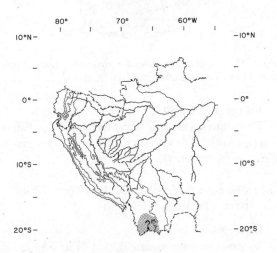

Map 16.160. Distribution of *Microcavia australis*.

short and stout with blunt nails. The dorsum is gray agouti, and the venter is light brown. The throat may be white. The species is well adapted to living among rocks and has converged in form and function with the rock hyrax, *Procavia* (Mares and Lacher 1987).

Range and Habitat

The species is found only in eastern Brazil, from eastern Piauí state to eastern Minas Gerais. This cavy is a specialist inhabiting dry, rocky areas supporting a low scrub vegetation (map 16.162).

Life History

The gestation period averages seventy-five days, and average litter size is 1.5. The young are precocial at birth. Females have a postpartum estrus and can produce three litters annually of two young each. Females can conceive at 151 days of age. In captivity they can give birth throughout the year (Roberts, Maliniak, and Deal 1984; Lacher 1981).

Ecology

Lacher (1981) describes the habitat of this species in some detail. The animals prefer rocky terrain, where they shelter in crevices. The males defend rock piles against other adult males, and several females live within a male's defended area. They feed on the leaves of a variety of scrubby plants growing within the defended area. They are able climbers, springing from shrub to shrub (Lacher 1981).

FAMILY HYDROCHAERIDAE
Genus *Hydrochaeris* Brunnich, 1772
Capybara, Chigüire

Diagnosis

The dental formula is 1/1, 0/0, 1/1, 3/3. The teeth are distinctive. The third molar is longer than the other three molariform teeth (fig. 16.21). The molars consist of transverse lamellae joined with cementum. The skull is massive, and the cheek teeth are vastly modified. This is the largest living rodent and cannot be confused with any other living form. Head and body length ranges from 1 to 1.3 m, height at the shoulder averages 0.5 m, and weight may reach 50 kg. The tail is vestigial. The capybara gives the appearance of an enormous guinea pig (fig. 16.22). There are four toes on the forefeet and three on the hind feet, and the digits are semiwebbed. The upperparts are reddish brown and the underparts are lighter, usually some shade of yellow brown. In the sexually mature male, a bare raised glandular area on top of the snout is conspicuous.

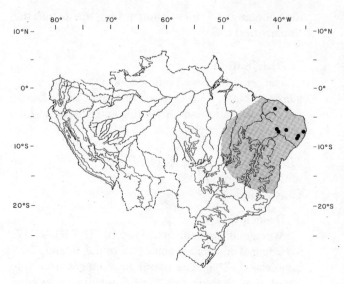

Map 16.162. Distribution of *Kerodon rupestris*.

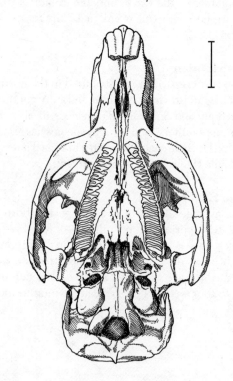

Figure 16.21. Skull of *Hydrochaeris hydrochaeris* showing upper molars (after Ellerman 1940). Scale bar = 34 mm.

Range and Habitat

The genus occurs in lowland wetlands from Panama through lowland Amazonas to northern Argentina and Uruguay (map 16.163).

Hydrochaeris hydrochaeris (Linnaeus, 1766)
Capybara, Carpincho, Chigüire

Description

Average measurements: HB 1,176; HF 223; E 63; Wt 63 kg. This pig-sized rodent is unmistakable

Figure 16.22. The capybara, *Hydrochaeris hydrochaeris*.

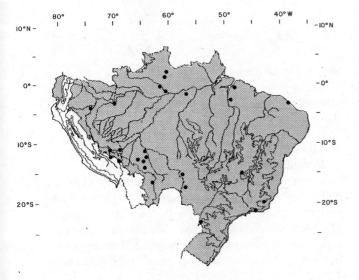

Map 16.163. Distribution of *Hydrochaeris hydrochaeris*.

(figs. 16.21 and 16.22). Its body is covered with coarse hair, chestnut to dull brown dorsally and ventrally. It has a very short tail and a large squared-off muzzle with large nostrils. Adult males can be told by the large black sebaceous gland on top of the muzzle and the bulging testes (Barlow 1965; Massoia 1976a; Schaller and Crawshaw 1981).

Distribution

Hydrochaeris hydrochaeris has an extensive distribution from Panama south into northern Argentina. The distribution of capybaras has been severely affected by hunting, and local populations have been eliminated; its original range may have been more extensive (Barlow 1969; Honacki, Kinman, and Koeppl 1982; Massoia 1976a; Rabinovich et al. 1987) (map 16.163).

Natural History

In Venezuela capybaras breed throughout the year; there is usually only one breeding cycle per year,

though two are possible in excellent habitat. In the Brazilian Pantanal, reproduction is seen year round, with the birth peak in mid-February. Litter size ranges from 1 to 7 with an average of 3.5. The gestation period is 120 days, neonates average 1,400 g in weight, and lactation lasts approximately ten weeks (Alho, Campos, and Gonçalves 1989; Barlow 1969; Crespo 1982b; Ojasti 1973; Kleiman 1974; Kleiman, Eisenberg, and Maliniak 1979; Schaller and Crawshaw 1981). Capybaras are confined to areas with standing or running water and can occur in marshes, in estuaries, and along rivers and streams. Depending on habitat and hunting pressure, they can be found singly or in groups. In the Venezuelan llanos the median group size was 10.5. The normal herd is composed of a dominant male, several adult females with young, and subordinate males as peripheral members. In the Brazilian Panatanal the largest group was 37, with the mean group size varying between 3.6 and 5.8 depending on the season; densities could reach 12.5 animals per hectare.

In the Amazonian floodplain forest of Peru the capybara restricts its activities to the riverine habitat. A home range of 17–22 ha was used by a single group. They fed on over forty species of plants but used the grass *Echinocola polystachya* heavily. The jaguar *Panthera onca* and the anaconda *Eunectes murinus* were the principal predators (Soini and Soini 1992).

In areas with seasonal rainfall, groups are larger during the dry season as animals concentrate around available water holes. Within a single population, it is possible to see several different social structures. Males mark using the large sebaceous gland on the muzzle and have been known to kill young animals.

Capybaras can be diurnal or nocturnal depending on season and hunting pressure. When alarmed, an animal gives a characteristic "woof" alarm bark and flees into the water. Capybaras are excellent swimmers and can remain submerged for several minutes. They are grazers, eating grass as well as aquatic vegetation (Alho, Campos, and Gonçalves 1989).

Capybara leather is valued in southern South America, and from 1976 to 1979 almost 80,000 skins were exported from Argentina. There is also a sizable internal market for these skins (Barlow 1969; Kleiman 1974; Mares and Ojeda 1984; Massoia 1976a; Ojasti 1973; Ojasti and Sosa Burgos 1985; Schaller and Crawshaw 1981).

FAMILY AGOUTIDAE
Genus *Agouti* Lacépède, 1799

Description

The dental formula is 1/1, 0/1, 1/1, 3/3. Head and body length averages 600–795 mm. The tail is vesti-

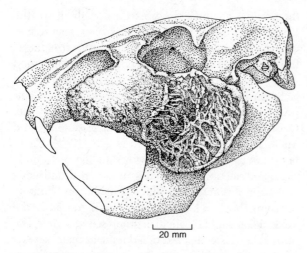

Figure 16.23. Skull of *Agouti paca*.

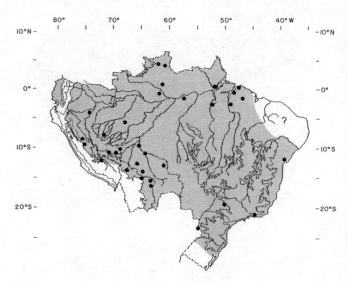

Map 16.164. Distribution of *Agouti paca*.

gial, barely exceeding 20 mm in length. The hind foot averages 188 mm, and the ear 45 mm. Weight can reach 10 kg. The adult male is about 15% larger than the adult female. A small slit on the underside of the zygomatic arch is visible externally. It opens to a blind cavity on each side of the head. The zygomatic arch is massive and distinctively sculpted (fig. 16.23). There are four toes on the forefeet and the hindfeet. The dorsal pelage is brown to almost black, usually with four horizontal lines of white dots on each side of the body. The venter tends to be white to buff.

Distribution

Members of the genus are distributed from southern Mexico to northern Argentina in suitable lowland habitats.

Agouti paca (Linnaeus, 1766)
Paca, Lapa

Description

Average measurements: TL 636; T 19; HF 113; Wt 7.5 kg. This large rodent has short legs and a moderately robust body. The skull is unmistakable, with rugose, greatly enlarged zygomatic arches, more pronounced in males. The square muzzle has large lips, elaborate nostrils, large eyes, and stiff vibrissae as well as a prominent externally opening fur-lined cheek pouch. The coloration is distinctive: the dorsum varies from reddish brown to dark chocolate, with two to seven prominent irregular rows of white spots on the flanks. The venter is white to cream. The eyes shine brilliant yellow when spotlighted (Collett 1981; Smythe 1970; Perez 1992).

Distribution

Agouti paca is widely distributed in mesic habitats from Mexico south through eastern Paraguay and

into the northeastern Argentine province of Misiones (Crespo 1974a) (map 16.164).

Life History

In captivity, animals from Panama have from one to two single births per year. Young are born weighing 650 g and nurse for approximately ninety days. The gestation period is 116 days. In Venezuela animals give birth throughout the year, whereas in Argentina reproduction takes place from August to January (Collett 1981; Crespo 1982b; Matamoros 1982; Mondolfi 1972).

Ecology

Pacas are restricted to forested habitats but occupy a wide range of forest types including mangrove swamps, narrow gallery forests, and dense upland scrub. They are strongly associated with moist habitats and are frequently encountered in gallery forests, near standing water and rivers. They are strictly terrestrial, spending the day in a burrow either of their own construction or built by another animal such as an armadillo. They are excellent swimmers and often take to the water when pursued. The entrance to the burrow is frequently concealed with leaf litter. In favorable habitat in Colombia they occurred at densities ranging from eighty-four to ninety-three per km². Territories are defended by mated pairs.

Pacas are primarily frugivorous, though they also browse. Fruit is frequently eaten at a "midden," and this species may be an important disperser of fruit seeds. Pacas are important game animals throughout their range and are frequently the preferred bush meat (Collett 1981; Crespo 1982b; Mondolfi 1972; Redford and Robinson 1987; Smythe 1970; pers. obs.).

Smythe conducted a long-term study to explore the feasibility of breeding the paca in captivity, and

Map 16.165. Distribution of *Agouti taczanowskii.*

through selection and rearing conditions he derived a domesticated stock. His efforts and success are evaluated in a monograph (Smythe and Brown de Guanti 1995).

Agouti taczanowskii (Stolzmann, 1865)
Mountain Paca

Description
This species is similar in size and color to other members of the genus, but the mountain paca has a dense undercoat.

Range and Habitat
This paca is found in the high cloud forest of the Andean cordilleras of Venezuela and Colombia, and south to Peru (map 16.165). Aagaard (1982) lists its maximum abundance as occurring at 2,000–3,050 m.

FAMILY DASYPROCTIDAE

Diagnosis
The dental formula is I 1/1, C 0/0, P 1/1, M 3/3 (fig. 16.16). These large cursorial rodents are similar to the pacas but are more slender in build and lack the spotted pattern (plate 11).

Distribution
The family ranges from Veracruz in Mexico south to northern Argentina and Uruguay in appropriate moist habitats.

Genus *Dasyprocta* Illiger, 1811
Agouti, Agutí, Picure

Description
Head and body length ranges from 415 to 620 mm; the tail is 10 to 25 mm, the hind foot 125 to 140 mm,

and the ear about 35 mm. The weight may reach 4 kg (see fig. 16.24). The forefoot has four functional digits and a vestigial thumb, and the hind foot has three digits. The upperparts vary from reddish brown to almost black, while the underparts are white, yellow, or gray. In the dark brown form the rump may be reddish.

Distribution
The genus ranges from Veracruz, Mexico, south to Uruguay and northern Argentina at low elevations in both dry deciduous and evergreen forests.

Natural History
The best-studied species of *Dasyprocta* is *D. punctata* (Smythe 1970, 1978). The agouti is diurnal, but where it is extremely persecuted by man it may extend its activity to the hours of darkness. Though adults rarely use burrows, the young are generally born near a crevice or some burrow dug by another animal, to which they retreat for shelter except when the female calls them for nursing. Gestation lasts about 120 days, and litter size is one or two. The young, like all hystricognath rodents, are born furred and with their eyes open (Weir 1974b; Smythe 1978).

The agouti is a significant harvester of seeds and fruits. Within its home range it accumulates stores for times of fruit scarcity, caching food in small pits that it excavates and then covers. Adult females defend parts of their home ranges when food is scarce. Young may be tolerated within the parent's home range. An adult male defends as large an area as he can against other adult males, thereby ensuring the paternity of the young in his range. At high densities the ranges of a male and a female may be coincident, and thus they give the appearance of living in pairs (Smythe 1978). During times of fruit scarcity, the subadults may lose weight. On Barro Colorado Island they appear to be food limited (Smythe, Glanz, and Leigh 1982).

When startled, agoutis produce a barklike warning call to alert family members within the home range to a potential predator, and the long hairs of the rump

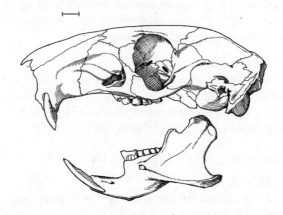

Figure 16.24. Skull of *Dasyprocta punctata* (after Ellerman 1940). Scale bar = 10 mm.

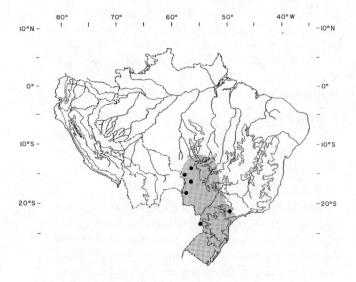

Map 16.166. Distribution of *Dasyprocta azarae*.

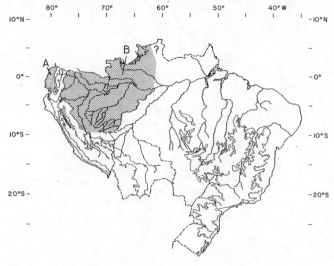

Map 16.167. Distribution of (a) *Dasyprocta punctata;* (b) *D. fuliginosa.* The distribution of *D. fuliginosa* is hypothetical pending a revision. Specimens from this area may bear several different names in museum collections. Hence locations are not plotted. See comments under *D. leporina* and *D. agouti.*

are erected, increasing the animal's apparent size. On encountering boas (*Constrictor constrictor*), they sit at a distance and drum with a hind foot, attracting other family members, who join them in foot drumming until the snake moves off (Kleiman 1974; Eisenberg 1974; Smythe 1978).

Comment

Many species are assigned to the genus *Dasyprocta*. In general, none are sympatric. Since the natural history of all forms is similar, in the species accounts we will merely list them and give their distribution. The genus *Dasyprocta* is in need of taxonomic revision. The names employed here are in accordance with Woods (1993).

Dasyprocta azarae Lichtenstein, 1823
Agouti, Agutí Bayo

Description

Average measurements: HB 494; T 24; HF 113; E 37; Wt 2.7 kg. This species is light to medium agouti brown with a pronounced speckled pattern. It can be very yellow on the sides and venter.

Distribution

Dasyprocta azarae is found from east-central and southern Brazil into eastern Paraguay and the northeastern Argentine province of Misiones (Honacki, Kinman, and Koeppl 1982; Massoia 1980a) (map 16.166).

Dasyprocta fuliginosa Wagler, 1832

Description

Dasyprocta fuliginosa is similar in size and proportion to *D. azarae,* but it lacks the conspicuous orange yellow venter. The dorsum is a grizzled black.

Distribution

The species is distributed in southern Venezuela and adjacent portions of Colombia, southward to west-central Brazil (map 16.167).

Dasyprocta kalinowskii Thomas, 1897

Description

This species is of uncertain status and needs further comparison with *Dasyprocta variegata* and *D. fuliginosa.*

Distribution

The species is found in southeastern Peru (map 16.168).

Dasyprocta leporina (Linnaeus, 1758)

Description

Dasyprocta leporina includes *D. agouti*. This agouti is similar in size to *D. azarae.* The head and foreparts are olive green speckled with gray, the rump patch is reddish orange, and there may be some black on the top of the head.

Distribution

The species has been introduced to the Lesser Antilles. Continental distribution includes Venezuela, the Guianas, and central Brazil (map 16.169).

Dasyprocta prymnolopha Wagler, 1831

Description

Dasyprocta prymnolopha is similar in size to *D. punctata* (see below). The dorsum is speckled (agouti)

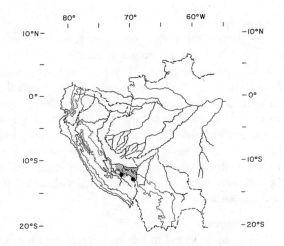

Map 16.168. Distribution of *Dasyprocta kalinowskii*.

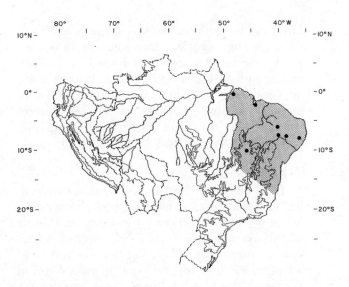

Map 16.170. Distribution of *Dasyprocta prymnolopha*.

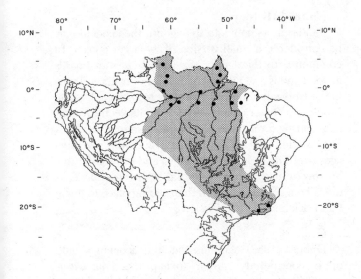

Map 16.169. Distribution of *Dasyprocta leporina* (includes *D. agouti*).

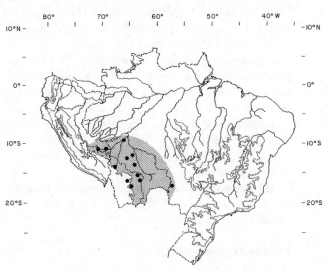

Map 16.171. Distribution of *Dasyprocta variegata* (considered *D. punctata* by some authors).

red. The dark brown of the rump often extends anteriorly in the midline.

Distribution

The species occurs in northeastern Brazil south of the Amazon. This is the common agouti of the caatinga (map 16.170).

Dasyprocta punctata Gray, 1842
Agouti, Agutí Rojizo

Description

Dasyprocta punctata includes *D. variegata*. This agouti is similar in size to *D. azarae;* two specimens weighed 3.1 and 3.6 kg (AMNH). Animals from Colombia weighed 3.5–4 kg. This agouti is a pronounced reddish agouti brown with much less pronounced speckling than *D. azarae*. When aroused or when

fleeing, it erects the long rump hairs (Kleiman 1974; Vergara 1982).

Distribution

Earlier accounts indicate that *D. punctata* is distributed from southern Mexico south to the northwestern Argentine provinces of Jujuy and Salta and to southwestern Brazil (Woods 1993), but see (maps 16.167 and 16.171). In our estimation *D. punctata* should be restricted to the northern form.

Life History

In captivity this species breeds throughout the year, with an interbirth interval of 127 days and a gestation period of somewhat less. Neonates weight 22.7 g (range 200–256; $n = 17$); litter size is two, and two litters can be produced per year. Animals can breed for the first time by at least sixteen months. In Venezuela the normal litter size is two (range 1–3), and animals apparently breed year round. In Panama young are generally born in the vicinity of some burrow or crevice to which the young retreat for shelter except when called by the female (Meritt 1983; Ojasti 1972; Smythe 1978; Vergara 1982).

Ecology

These cursorial, diurnal rodents occur in broadleaf forests up to 1,500 m in Argentina, though they are more common at lower elevations. In Panama mated pairs occupy an area of 2–3 ha that they defend against conspecifics. Adults dig burrows and build a nest where the young are born. In a dry forest in Guatemala they occur at a density no greater than 0.1 per hectare, while in the Brazilian Pantanal they occur at densities ranging from 0.9 to 3.2 per km^2, with the highest densities in gallery forest. In Costa Rica one female had a home range of 3.9 ha, larger than the 1–2 ha ranges reported from Panama.

Agoutis are frugivores and bury fruit that they do not immediately eat. Because of this habit of scatter-hoarding they are important agents for the dispersal of seeds. When fleeing, animals emit a sharp alarm bark (Cant 1977; Eisenberg 1974, 1989; Olrog 1979; Rodriguez 1985; Schaller 1983; Smythe 1970, 1978).

Comment

If *Dasyprocta variegata* is a valid species, then the range of *D. punctata* would be restricted to west of the Andes (see map 16.167).

Genus *Myoprocta*
Acouchi

Taxonomy

Honacki, Kinman, and Koeppl (1982) recognized two species, *Myoprocta acouchy* and *M. exilis*, as does Woods (1993). Emmons and Feer (1997) acknowledge two species as well: *M. acouchy* and *M. pratti*, while Cabrera (1960) includes *M. exilis* in *M. acouchy*. Two color morphs are easily recognized: a reddish agouti brown and a greenish agouti form. For some time it was thought that these color "morphs" had no taxonomic significance, but the consensus now is that we may indeed be dealing with two distinct species. Contrary to the discussion in *Mammals of the Neotropics*, volume 1, we concur that the two-species concept is valid (Eisenberg 1989). Dubost (1988) offers interesting data derived from specimens studied in French Guiana. He refers to his radio-collared specimens as *M. exilis*, whereas Emmons and Feer (1997) refer to the specimens as *M. acouchy*. Kleiman's research (1971, 1972) obviously deals with *M. pratti*, as she has indicated in her references.

Description

Members of this genus are smaller than the agoutis. Head and body length ranges from 250 to 380 mm. The tail is relatively longer than in *Dasyprocta*, some 45 to 70 mm; the hind foot averages 100 mm, and the ear 28 mm. Weight ranges from 880 to 1,200 g. The dorsal pelage varies geographically from olive green (*M. pratti*) to reddish brown (*M. acouchy*) (plate 11). The venter generally contrasts with the dorsum, being some shade of orange to cream, and the tip of the tail may bear a few white hairs, enhancing its conspicuousness.

Distribution

The genus is found in lowland evergreen forest in the Amazonian portion of Colombia, Ecuador, Peru, Brazil, Venezuela, and the Guianas.

Natural History

Similar in its habits to the agouti, the acouchi may be considered a small version of its larger cousin. In common with the agouti, the acouchi scatter-hoards its food and is a significant consumer of fruits and nuts in the tropical rain forest. Entirely terrestrial, like the agouti it is cursorial and depends on its keen eyesight and hearing for avoiding predators (Kleiman 1974).

Gestation lasts ninety-nine days, and litter size averages two. The young are precocial and will shelter in a burrow or natural crevice until they are several weeks old. The mother returns to their hiding place to nurse them (Kleiman 1970, 1972).

Ecology

Dubost (1988) cautions that the acouchi should not be considered a "junior" form of the genus *Dasyprocta*. To be sure, *Myoprocta* caches fruits and seeds in the manner described for *Dasyprocta* (Smythe 1978; Kleiman 1972). Studies in captivity suggest that monogamy could characterize the social system as described by Smythe for *Dasyprocta*. Dubost (1988) suggests that the mating system for *Myoprocta* may be somewhat different in that the activity patterns do not suggest a "strict" demarcation of territoriality.

Comment

The green color morph is often referred to as *M. pratti* and the red phase as *M. acouchy*.

Myoprocta acouchy (Erxleben, 1777)

Description

Myoprocta acouchy includes *M. exilis* (Cabrera 1960). The description is the same as given for the genus. This taxon is usually reddish.

Range and Habitat

The range is given under the genus. This lowland species is strongly tied to multistratal tropical evergreen forest (map 16.172).

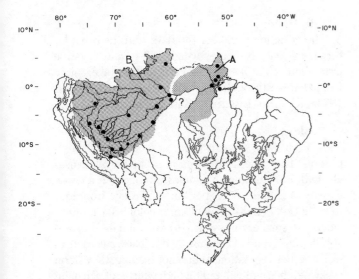

Map 16.172. Distribution of (a) *Myoprocta acouchy;* (b) *M. pratti.*

Myoprocta pratti (Wagler, 1831)

Description
See genus account. The dorsum of this species is usually olive green.

Distribution
See map 16.172.

SUPERFAMILY CHINCHILLOIDEA

FAMILY CHINCHILLIDAE

Diagnosis
The dental formula is I 1/1, C 0/0, P 1/1, M 3/3. The lacrimal bone is large, and the lacrimal canal opens on the side of the rostrum; the angular process is delicate and not strongly hystricognathous; there is no massateric crest on the dentary; the auditory canal is deflected dorsally and has an accessory foramina at the base (Woods 1984). Size ranges from 0.5 kg to over 9 kg. These are stout rodents whose tails are shorter than the head and body and fully furred. They look like a rabbit with short ears. The toes on the hind foot are reduced to a minimum of three in *Lagostomus*. Coloration is highly variable, from a blue gray dorsum with cream underparts (*Chinchilla*) to a brown dorsum with a bold black band across the face (e.g., *Lagostomus*).

Distribution
The family occurs from the highlands of Peru south through the Andes to Chile. *Lagostomus* is found in in the lowlands of Argentina. By and large this is a temperate zone family that has colonized both montane and lowland habitats.

Natural History
Chinchilla is colonial and prefers montane habitats. It has been taken at elevations over 4,500 m. The genus formerly occurred in Chile, Peru, Bolivia, and Argentina, but now wild colonies have been reduced to remnants. *Lagidium* is also colonial and is confined to elevations below 5,000 m. *Lagostomus* is colonial and occurs below 500 m. All species appear to be entirely herbivorous, eating a variety of herbaceous vegetation. In common with other hystricognath rodents, they have relatively long gestation periods and give birth to exceedingly precocial young. In *Lagostomus* the male is far larger than the female, whereas in *Chinchilla* the female is larger than the male. Further discussion will be found in the species accounts.

Genus *Chinchilla* Bennett, 1829

Description
This rodent weighs about 0.5 kg. The tympanic bullae are inflated. It is stout and rabbitlike, but it has a short, bushy tail (fig. 16.25).

Chinchilla brevicaudata Waterhouse, 1848

Description
This species is distinguishable from *C. lanigera* by its shorter tail and smaller ears (Mann 1978). It is basically brownish gray.

Distribution
Chinchilla brevicaudata is found only in the Andes from southern Peru to northern Chile and northwestern Argentina. Severe human persecution has undoubtedly greatly reduced its original range, and none have been recorded from Peru for over fifty years (map 16.173) (Honacki, Kinman, and Koeppl 1982; Tamayo and Frassinetti 1980; BA).

Ecology
Chinchilla brevicaudata lives in mountain grassland and scrub areas, very similar to those occupied by the viscacha, *Lagidium* (Mann 1978).

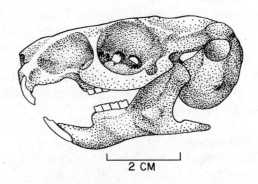

Figure 16.25. Skull of *Chinchilla lanigera.*

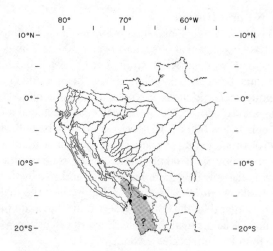

Map 16.173. Distribution of *Chinchilla brevicaudata*.

Comment

Some authors include *C. brevicaudata* in *C. lanigera* (Honacki, Kinman, and Koeppl 1982, but see Woods (1993).

Genus *Lagidium* Meyen, 1833
Viscacha

Description

Head and body length ranges from 300 to 450 mm and the tail from 200 to 400 mm. These are rabbit-sized rodents with long ears and a rather long tail. They look like a larger version of the *Chinchilla*. The thick fur of the dorsum ranges from gray to brown, contrasting with the cream to white venter.

Distribution

This genus is montane in its distribution, contrasting with the closely related plains denizen, *Lagostomus*. *Lagidium* is found from the highlands of Peru and Bolivia, south to central Chile and western Argentina.

Lagidium peruanum Meyen, 1833
Mountain Viscacha

Description

Average measurements: TL 639; T 267; HF 91; E 65; Wt 1.3 kg. The tail is medium in length with a long black tuft. The rabbitlike ears are fairly long, and the vibrissae are very long and stout. This species has short, soft, dense ash brown fur, browner on the shoulders. The sides are washed with yellow, and the belly is washed with white.

Distribution

The species is found in the Andes of central and southern Peru (map 16.174). It may extend into Bolivia.

Life History

The gestation period is 140 days, and the usual litter size is one. Neonates weigh 180 g, and lactation lasts about eight weeks. In Peru mating takes place from October through November (Pearson 1948; Weir 1974a; Kleiman, Eisenberg, and Maliniak 1979).

Ecology

In Peru these animals occur at 3,000–5,000 m elevation. They choose shelters in rocky crevices, usually with one entrance. Colonies can occur at a crude density of 11 individuals per km^2 and range from four to seventy-five individuals. Larger colonies are subdivided into smaller units of two to five animals, presumably family groups (Pearson 1948). They eat cacti as well as other vegetation. When not feeding, they sit on rocky perches and will emit a high whistle of alarm on sighting a potential predator (Dávila et al. 1982).

Lagidium viscacia (Molina, 1782)
Chinchillón, Vizcacha Serrana

Description

Average measurements: HB 396; T 287; HF 97; E 70; Wt 1.5 kg. This large rodent has dense, soft, slightly woolly fur, elongated ears, and a long, heavily penicillate tail covered with long hairs on the dorsal surface and haired to the tip. The tail is held half curled or fully curled when the animal is at rest but carried fully extended when moving. There is a great deal of individual variation in pelage color, since the fur is molted throughout the year and there are pronounced age-related changes in color. But the basic dorsal color is agouti gray and brown with a range of cream to black and, and the venter is pale yellow to tan (Osgood 1943; Mann 1978; Rowlands 1974; pers. obs.).

Distribution

Lagidium viscacia is distributed from southern Peru along the Andes of Bolivia to south-central Chile

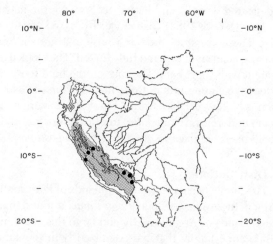

Map 16.174. Distribution of *Lagidium peruanum*.

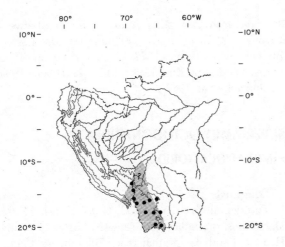

Map 16.175. Distribution of *Lagidium viscacia*.

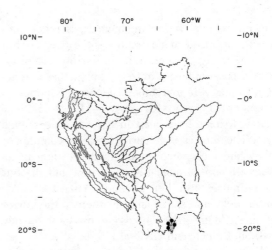

Map 16.176. Distribution of *Lagostomus maximus*.

and Argentina (map 16.175) (Honacki, Kinman, and Koeppl 1982; Olrog and Lucero 1981; Tamayo and Frassinetti 1980).

Ecology

This gregarious species lives in colonies in rocky outcroppings up to 4,000 m. They are diurnal and apparently do not need free water (Mann 1978; Rowlands 1974).

Genus *Lagostomus* Brooks, 1828

Lagostomus maximus (Desmarest, 1817)
Plains Viscacha, Vizcacha

Description

Measurements: male, TL 753; T 189; HF 127; E 59; Wt 6.4 kg; female, TL 641; T 160; HF 113; E 56; Wt 4 kg. The head is large and massive, and the skull is distinctive (see fig. 16.26). Males have considerably larger heads and heavier bodies than the more gracile females. The vibrissae are very long and stout, and females have a set of stout bristles on the cheeks below the eyes. The face is broad and prominently marked by two parallel black stripes, one passing between the eyes and the other over the nose, separated by a white

band. There is additional white on the cheeks, above the eyes, and at the base of the moderately haired ears. The dorsal fur is gray brown, lightening on the sides to a pure white venter (pers. obs.).

Distribution

Lagostomus maximus is found in extreme western Paraguay, probably in southeastern Bolivia, and into Argentina, then south across the country to at least Río Negro province, with at least one specimen recorded from Uruguay. It may be expanding its range toward the north and west into the monte because of human persecution in the preferred grassland habitat. On the other hand, it has been eliminated from large parts of its range owing to intense eradication programs (Rabinovich et al. 1987; Weir 1974b) (map 16.176).

Life History

In central Argentina the birth peak of plains viscachas is from July to September, with one litter a year and two young per litter. In captivity litters of five have been born after a gestation period of 154 days. The birthweight is 200 g, and the age of first reproduction is eight months. Females have four nipples. In northeastern Argentina this species produces two litters a year, one in July to August and the other between February and April.

In La Pampa province, Argentina, the breeding season is from May to June; there may be two breeding peaks, and there is usually one litter a year. Lactation lasts less than one month, and if two young are born usually only one is raised (Crespo 1982a; Hudson 1872; Kleiman, Eisenberg, and Maliniak 1979; Llanos and Crespo 1952; Maik-Siembida 1974; Rabinovich et al. 1987).

Ecology

Lagostomus maximus is characteristically found in open grassy or brushy areas, frequently with little rain.

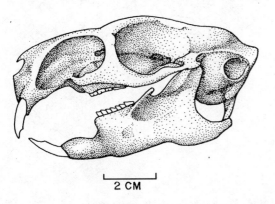

2 CM

Figure 16.26. Skull of *Lagostomus maximus*.

They are colonial, burrowing animals, and their presence is easy to determine because of the extensive burrow systems. By preference these burrows are placed near the bases of bushes or trees, though they may also occur in cracks in limestone. The average number of openings per burrow is 3.6 (range 1–11; $n = 35$). Some burrow systems have been in use for at least eight years. The species is gregarious, with a complex social system incorporating dominant breeding males. There are several animals per burrow, and the burrow entrances are often clumped. Viscachas collect bones, sticks, and stones and pile them at the burrow entrances. Many other animals also use these burrows (Jackson, Branch, and Villareal 1996).

Within a viscacha colony burrow systems are used communally. A social unit consists of several females and their offspring. One or two adult males occupy a colony and defend it against other adult males. Males transfer to new colonies at the onset of the breeding season, and fights among males are common. During the breeding season the displays and fighting among males result in pronounced weight loss (Branch 1993a,b).

In central Argentina densities range from 2.0 to 7.5 individuals per hectare. Plains viscachas are strictly grazing animals, and the areas around burrow entrances are often heavily grazed. They feed nocturnally and appear not to need surface water. They are very vocal and have a rich repertoire. The viscacha in the scrub habitats of central Argentina may exhibit periodic population declines resulting from prolonged drought and predation by *Puma concolor* (Branch, Villareal, and Fowler 1994).

With the increase in cattle grazing, the overall range of *L. maximus* has undoubtedly increased, though local populations have been eliminated because they damage pasture and their meat is prized. One author and gourmet commented that the flesh of young animals was "rather insipid; the old males are tough, but the mature females are excellent." Viscacha skins are exported from Argentina in large numbers; between 1976 and 1979 over 370,000 skins left the port of Buenos Aires (Hudson 1872; Llanos and Crespo 1952; Mares, Ojeda, and Kosco 1981; Mares and Ojeda 1984; Rabinovich et al. 1987).

SUPERFAMILY OCTODONTOIDEA
FAMILY MYOCASTORIDAE

Diagnosis
The dental formula is I 1/1, C 0/0, P 1/1, M 3/3. Deciduous premolars are retained throughout life. Head and body length may reach 700 mm and the tail 450 mm. The sexes differ in body size, with the males some 2–3 kg heavier than the females. The body is heavy. Ear size is reduced, the tail is long and sparsely haired, and the hind feet are webbed (fig. 16.27). The fur is long, and the dense underfur makes the pelt commercially valuable.

Distribution
The Myocastoridae are found in the temperate portions of Chile and Argentina, northward to Bolivia and southern Brazil (map 16.177). They have been introduced to parts of Europe and the United States.

Natural History
These animals are highly adapted for life in water, where they feed on roots and aquatic grasses and are known also to take snails and freshwater mussels. They construct a simple burrow containing a nest chamber, where they feed heavily. Platform nests may be constructed in the portions of the marsh that have been eaten out.

Figure 16.27. The coypu, *Myocastor*.

Map 16.177. Distribution of *Myocastor coypus*.

Genus *Myocastor* Kerr, 1792
Myocastor coypus (Molina, 1782)
Nutria, Coipo

Description

Average measurements: HB 499; T 364; HF 133; E 30; Wt 4.9 kg. This large rodent has glossy pelage with long guard hairs and a long, rounded, tapering, and thinly haired tail. It has a rounded head, short neck, short snout, and large, stout vibrissae. The hind feet are large, and the middle toes are connected by a basal swimming membrane. The mammae are high on the sides rather than on the abdomen. The color, uniform over the body, is highly variable, ranging from light brown to reddish black (Barlow 1965; Mann 1978; Massoia 1976a; Osgood 1943).

Distribution

Myocastor coypus is found in southern Brazil, Paraguay, Uruguay, Bolivia, Argentina, and Chile (map 16.177) (Honacki, Kinman, and Koeppl 1982; Massoia 1976a; Tamayo and Frassinetti 1980).

Life History

For a full account, see Woods et al. (1992). In northern Argentina, nutrias reach sexual maturity at four months or at 2–3 kg weight. The adult females cycle every twenty-three to twenty-six days throughout the year, and the average number of embryos per female is 4.9 (range 2–7). In Chile females give birth in spring and summer and can have up to two litters per year, with litters ranging from two to eleven. The gestation period is 123–50 days; neonates weigh 225 g; and the female has four or five pairs of nipples (Crespo 1974b; Kleiman, Eisenberg, and Maliniak 1979; Mann 1978).

Ecology

Myocastor coypus is a semiaquatic rodent confined to areas with standing water, from swamps and marshes to streams and drainage ditches, but almost always with succulent vegetation in or near the water. The nutria is an excellent swimmer and diver and can remain submerged for up to seven minutes. It eats aquatic vegetation (Barlow 1969; Crespo 1974b; Greer 1966; Henry et al., n.d.; Mares and Ojeda 1984; Massoia 1976a; Woods et al. 1992).

FAMILY OCTODONTIDAE

Diagnosis

The dental formula is I 1/1, C 0/0, P 1/1, M 3/3. The cheek teeth have a simplified figure-eight occlusal pattern. The jugal process is variable, and the lacrimal does not open on the side of the rostrum. The paroccipital process is short and fused to the bulla; the coronoid process of the dentary is very pronounced (Woods 1984). These are rather small, stout-bodied rodents with the tail usually shorter than the head and body. Weight ranges from 200 to 300 g. All are adapted for burrowing; the most extreme adaptation is shown in the genus *Spalacopus*. The fur is soft, and the dorsum ranges from gray to brown. The venter usually contrasts, being cream or white.

Distribution

Octodonts are confined to southern South America, extending from southwestern Peru through the Andes of Bolivia, Chile, and Argentina. Forms such as *Octodontomys* may occur at high elevations. Others, such as *Spalacopus*, are confined to low-lying plains. The biogeography of this group has been reviewed by Contreras, Torres-Mura, and Yáñez (1987).

Natural History

These rodents are to some extent adapted for digging. Most live in burrows of their own construction. They appear to be herbivorous, eating a wide variety of herbaceous vegetation. One can recognize a continuum from strongly fossorial forms, such as *Spalacopus*, to forms that live in burrows but forage extensively on the surface, such as *Octomys*. Within *Octomys*, the postulated subgenus *Tympanoctomys* exhibits tympanic bullae that are inflated, suggesting a convergence toward arid-adapted rodents such as *Dipodomys* and *Salpingotus*. *Tympanoctomys* is considered a genus and is discussed in volume 2 (Redford and Eisenberg 1992). All members of the family have long gestation periods for their body size and produce precocial young.

Genus *Octodontomys* Palmer, 1903
Octodontomys gliroides (Gervais and d'Orbigny, 1844)
Rata Cola Pincel

Description

Average measurements: HB 171; T 143; HF 36; E 29; Wt 158 g. The tympanic bullae are large, but not as extreme as shown by *Tympanoctomys*. This large rat has large ears, a long bicolored tail with a tuft, and soft, dense fur with longer hairs along the spine. The dorsum is light brown, and the dark tips of the dorsal hairs make the coat look darker than that of *Octomys*. The venter and feet are white(Ipinza, Tamayo, and Rottmann 1971; Mann 1978; pers. obs.).

Distribution

Octodontomys gliroides is found in the high Andes of southwestern Bolivia, northernmost Chile, and the northwestern Argentine provinces from Jujuy to La Rioja (Contreras, Torres-Mura, and Yáñez 1987; Ho-

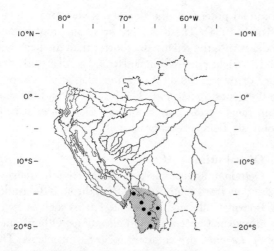

Map 16.178. Distribution of *Octodontomys gliroides*.

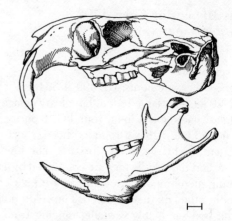

Figure 16.28. Skull of *Ctenomys* sp. (after Ellerman 1940). Scale bar = 5 mm.

nacki, Kinman, and Koeppl 1982; Lucero 1983; Mann 1945; BA) (map 16.178).

Life History

In captivity, gestation lasts between 99 and 104 days, litter size averages two (range 1–4), young weigh 15 g at birth, and lactation lasts five to six weeks (Kleiman 1974; Kleiman, Eisenberg, and Maliniak 1979; Weir 1974b).

These forms dust bathe and have elaborate vocal repertoires. The behavior of the young during maturation has been documented by Wilson and Kleiman (1974).

Ecology

Octodontomys gliroides is an animal of the arid high Andes. It occurs between 2,000 and 5,000 m in open, dry habitats with bushes, cacti, and rocks. It is nocturnal and lives in small galleries among rocks and cactus roots, with superficial trails between burrow mouths. It does not dig and is a good rock climber (Ipinza, Tamayo, and Rottmann 1971; Mann 1978; Tamayo and Frassinetti 1980).

FAMILY CTENOMYIDAE
Tuco-tucos

Diagnosis

This family includes the single genus *Ctenomys*. The dental formula is I 1/1, C 0/0, P 1/1, M 3/3 (fig. 16.28). Skull characters are similar to those of the Octodontidae, except that the cheek teeth are not in the shape of a figure eight but rather are kidney shaped. The jugal process is always present, and the zygoma is very broad. Average weights vary from 100 to 700 g. The tail is extremely short. The foreclaws are very long, and the eyes are reduced in size, as is the external ear. Overall, these animals are clearly adapted for a fossorial life. The fur is thick and is usually some shade of brown.

Distribution

The Ctenomyidae are found from the southern highlands of Peru south to Tierra del Fuego. The distribution extends into eastern Brazil but is discontinuous. Cook, Anderson, and Yates (1990) discuss the situation in Bolivia.

Natural History

These animals spend most of their life underground. Some species appear to lead a solitary existence, with one adult per burrow; others seem more colonial. Altman (1991) describes mating behavior for *Ctenomys*. Species of this family have a long gestation period of up to 130 days. The young are born fully furred and with their eyes open. Litter size varies from two to four (see species accounts). Mating takes place in burrows, though in the solitary species pairs must make contact underground or possibly on the surface. A courtship of thirty to forty minutes precedes copulation. Males are submissive to females. Copulation may last up to five minutes, with a single ejaculation (Camin, n.d.; Altuna 1991; Altuna and Lessa 1985). The long, complex tunnel systems contain chambers for food storage as well as a nesting chamber. Camin, Modery, and Roig (1995) describe burrowing behavior. They feed on roots and grasses and are convergent in habits and behavior with the gophers (Geomyidae) of North America and the mole rats (Spalacidae) of Eurasia. Their vernacular name, tuco-tuco, refers to the call they produce. This call may be made at the entrance to burrows and also underground (Weir 1974a).

Comment

Cook, Anderson, and Yates (1990) have reviewed the *Ctenomys* of Bolivia. The lowland forms are regarded as a group to be dealt with later, but they consider the highland forms in some detail. They confirm the chromosome numbers for *C. frater, C. opimus, C. lewisi,* and *C. leucodon* as 52, 26, 56, and 36, respectively. The highland races of *Ctenomys* in Bolivia are mapped.

This vexing group may have many lessons to teach us concerning the rates and modes of organic evolution. While there appear to be many species of similar morphology, fossorial rodents are prone to speciation because they have very limited means of movement aboveground. Hence localized populations, not too far distant, may exhibit limited exchange of genetic material. Localized populations may also be subjected to extreme environmental selection resulting from profound differences in soil texture, depth, mean low temperatures, and so forth (Nevo 1979; Patton and Yang 1977). Clearly, evolutionary divergence among populations of *Ctenomys* is in a very dynamic state. Some putative species may show extreme genetic divergence, whereas other species may exhibit much less divergence from a presumptive common ancestral form. The application of advanced genetic analysis to these groups is in its infancy, but it offers great promise (Novello and Lessa 1986; Sage et al. 1986; Cook, Anderson, and Yates 1990).

Although chromosomal polymorphisms are common in *Ctenomys,* the actual genome size remains constant; thus chromosomal evolution is independent of DNA content (Ruedas et al. 1993). Comparative sperm morphology is reviewed by Vitullo and Cook (1991). Evolutionary divergence sequences for the Bolivan species are analyzed by Cook and Yates (1994). Cytogenetics, morphology, and distributions are discussed by Freitas and Lessa (1984) and Freitas (1995).

Genus *Ctenomys* Blainville, 1826

Comment on *Ctenomys* in Southeastern Brazil

Freitas (1995) lists four species of *Ctenomys* for southern Rio Grande do Sul. *C. torquatus* extends from Uruguay to Rio Grande do Sul in a band somewhat inland from the coast. On the coastal plain itself three species occur. One is as yet undescribed. *Ctenomys minutus* is narrowly distributed on the strip of land between the Atlantic and Lagoa dos Patos, where it overlaps with *Ctenomys flamarioni* to the east. The status of *C. flamarioni* remains to be clarified.

Ctenomys boliviensis Waterhouse, 1848
Tuco-tuco Boliviano

Description

Average measurements: TL 315; T 82; HF 43; E 9; Wt 421 g. This handsome *Ctenomys* demonstrates

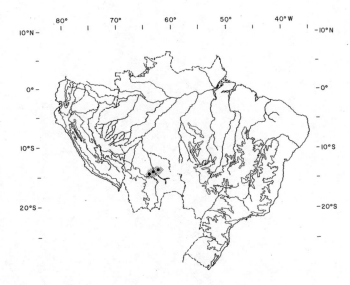

Map 16.179. Distribution of *Ctenomys boliviensis* (includes *C. goodfellowi*).

considerable variation in color, but all specimens have a dark middorsal stripe grading into lighter brown on the flanks and cream to chestnut on the venter. The venter also has varying amounts of white. There is a distinct light stripe extending from the base of the ear to the chin (pers. obs.).

Distribution

The species is described from central Bolivia and recorded in western Paraguay and Formosa province, Argentina (map 16.179) (Honacki, Kinman, and Koeppl 1982; Olrog and Lucero 1981).

Ctenomys brasiliensis Blainville, 1826

Description

The dorsal pelage is reddish brown and the venter is pale red. The tail has short hairs (Moojen 1952a).

Distribution

The species is apparently confined to south-central Brazil. Minas Gerais is listed as the type locality (map 16.180).

Ctenomys conoveri Osgood, 1946

Description

Measurements: TL 403; T 110; HF 58; E 13.4; Wt 900 g. The proodont incisors are very broad and heavy; each of their anterior surfaces has an inner and an outer wide lateral groove and three shallow, narrow median grooves, the last little more than striations and not always fully continuous. This is a very large, stout *Ctenomys.* The dorsum is tan with a faint middorsal line, and the venter is a lighter yellowish. The top of

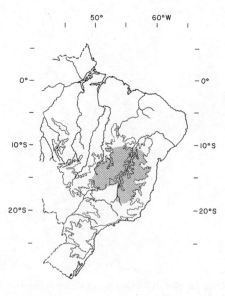

Map 16.180. Distribution of *Ctenomys brasiliensis*.

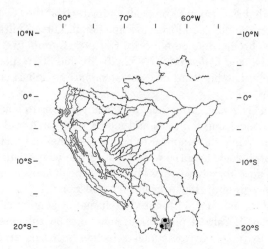

Map 16.181. Distribution of *Ctenomys conoveri*.

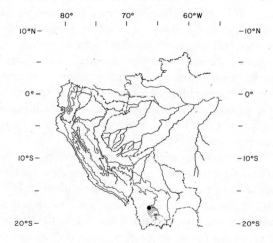

Map 16.182. Distribution of *Ctenomys frater*.

the head is gray brown, lightening to flecking on the back of the head, the cheeks are brownish yellow, and there are touches of white around the mouth and on the chin (Osgood 1946; pers. obs.).

Distribution

Ctenomys conoveri is distributed in the Paraguayan Chaco and across the border into Argentina (Olrog and Lucero 1981) (map 16.181).

Ctenomys frater Thomas, 1902
Tuco-tuco Colorado

Description

Average measurements: HB 176; T 80; HF 35; E 8; Wt 173 g. Individuals of this species exhibit a great range in dorsal color, ranging from rich dark brown to light brown, almost tan. The venter is gray brown to light brown, in most cases not contrasting sharply with the dorsum. In some populations there is white on the venter and along the hind and front legs (pers. obs.).

Distribution

The species is distributed in the Andes from southwestern Bolivia into the Argentine provinces of Jujuy and Salta (Honacki, Kinman, and Koeppl 1982) (map 16.182).

Ecology

Ctenomys frater is an animal of forests and moist meadows at higher elevations. It has been taken in moist forest, in grassy meadows at 2,000 m, and at 4,500 m in Jujuy province, Argentina. In Salta province, Argentina, it inhabits flat areas with deep soil, often near creeks, does not appear to burrow extensively, and is not as vocal as desert *Ctenomys* (Mares, Ojeda, and Kosco 1981; Olrog 1979).

Comment

Ctenomys frater includes *C. budini* and *C. sylvanus*.

Ctenomys leucodon Waterhouse, 1848

Description

This rather small species is almost the same size as *C. frater* but is slightly paler.
Chromosome number: $2n = 36$ (Cook, Anderson, and Yates 1990).

Distribution

The species is found in west-central Bolivia (see map 16.183) (Cook, Anderson, and Yates 1990).

Ctenomys lewisi Thomas, 1926

Description

This species is large for its genus and has a dark dorsum.
Chromosome number: $2n = 56$ (Cook, Anderson, and Yates 1990).

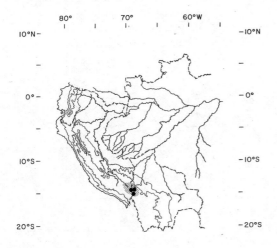

Map 16.183. Distribution of *Ctenomys leucodon*.

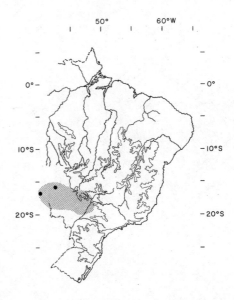

Map 16.185. Distribution of *Ctenomys minutus*.

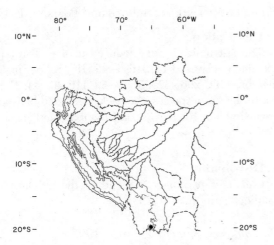

Map 16.184. Distribution of *Ctenomys lewisi*.

Distribution

The species is recorded from Bolivia adjacent to Argentina (Jujuy province) (map 16.184).

Ctenomys minutus Nehring, 1887

Description

Average measurements: HB 175; T 77; HF 32; E 7. *C. minutus* exhibits color variation. Some specimens are black on the dorsum and venter with no markings; others are glossy tan, yellow on the sides and venter and darker on the head, with a dark stripe down the top of the tail (Langguth and Abella 1970; AMNH). Chromosome number: $2n = 50$ (Reig and Kiblisky 1969).

Distribution

This species is found in southern Brazil, Uruguay, and east-central Argentina in the province of Entre Ríos (Langguth and Abella 1970) (map 16.185).

Comment

A subspecies of *Ctenomys minutus*, *C. m. rionegrensis*, is regarded as a full species by Altuna and Lessa (1985).

Ctenomys nattereri Wagner, 1848

Description

Measurements of the type specimen: HB 230; HF 40; T 80. The dorsum is light brown speckled with black. The venter is whitish, and pure white in some individuals.

Distribution

The species is found in the Mato Grosso of Brazil (map 16.186).

Ctenomys opimus Wagner, 1848
Tuco-tuco Tujo

Description

Average measurements: TL 314; T 91; HF 44; E 7; Wt 439 g. Males are much larger than females. This is one of the few mammals of the Peruvian puna to accumulate fat. In contrast to other Chilean *Ctenomys*, this species has long, silky pelage. It is light brown to buffy gray dorsally, lighter on the sides and venter, with a blackish head and sometimes a faint dorsal line. The anterior surface of the incisors is a bright orange yellow (Mann 1978; Osgood 1943; Pearson 1951; Thomas 1900a; pers. obs.). Chromosome number: $2n = 26$ (Gallardo 1979; Reig and Kiblisky 1969).

Distribution

Ctenomys opimus is found from southern Peru south to northern Chile and northwestern Argentina.

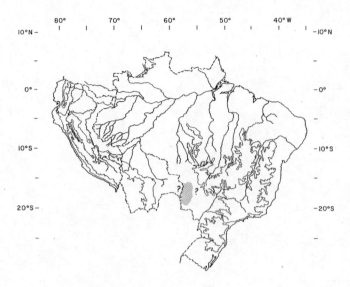

Map 16.186. Distribution of *Ctenomys nattereri*.

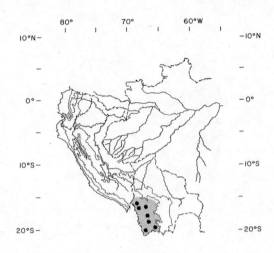

Map 16.187. Distribution of *Ctenomys opimus*.

In Chile it occurs in the northernmost altiplano, and in Argentina it is found at high altitudes in the provinces of Jujuy and Salta (Honacki, Kinman, and Koeppl 1982; Mares, Ojeda, and Kocso 1981; Tamayo and Frassinetti 1980; Thomas 1900a) (map 16.187).

Life History
In southern Peru most births are between October and March; females have three pairs of nipples and average 1.6 embryos (range1–2; *n* = 10). Animals are capable of breeding in the first year (Pearson 1959).

Ecology
Ctenomys opimus is a tuco-tuco of the altiplano, occurring between 2,500 and 5,000 m. It is abundant where vegetation is sparse and the soil is loose. It is found on sandy, gravely, or cindery soils, usually on slopes. It digs burrows by loosening earth with its front feet and sweeping it out of tunnels with the hind feet. Burrow systems usually consist of a single main tunnel from which short lateral branches diverge every few meters and include one or more chambers with stored vegetation and other chambers with nests. Only one animal lives in each burrow system. The burrows are expanded as vegetation is eaten, and *C. opimus* reveals its presence by stripping large areas of natural vegetation. In southern Peru it can occur at densities of one to seventeen individuals per acre.

Feeding and digging usually occur between 6:30 and 11:00 A.M.; animals are not seen aboveground very often. They eat roots, stems, or leaves of most of the available plants. The tuco-tuco digs a tunnel to the food source, then forages no more than two to three body lengths outside its burrow, bringing the food back to the hole to eat (Mares, Ojeda, and Kosco 1981; Pearson 1951, 1959; Pine, Miller, and Schamberger 1979; Tamayo and Frassinetti 1980).

Ctenomys peruanus Sanborn and Pearson, 1947

Description
The last molar is very reduced in size. The type specimen measured TL 316; T 88; HF 41. Four males measured TL 296–325; T 80–89; HF 40–44; E 8 10. In this large species, the back, sides, and belly are cream buff and the feet are dark (Sanborn and Pearson 1947)

Distribution
Ctenomys peruanus is found in the altiplano of southern Peru (map 16.188).

Ctenomys steinbachi Thomas, 1907

Description
The species ranges from 139 to 235 mm in head and body length, rather large yet smaller than *C. conoveri*. There is no sexual dimorphism. The short dorsal pelage is blackish.

Distribution
Ctenomys steinbachi is confined to western Santa Cruz department of Bolivia (map 16.189).

Comment
This species has one of the lowest chromosome numbers for the genus ($2n = 10$; Anderson, Yates, and Cook 1987).

Ctenomys torquatus Lichtenstein, 1830
Tuco-tuco de Collar

Description
Average measurements: TL 267; T 77; HF 38; E 7; Wt 229 g. This tuco-tuco has a dark yellow to mahogany dorsum, sometimes with a darker line along the spine. The flanks and venter are yellowish white, and most individuals have a whitish or yellow-

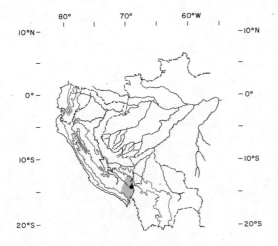

Map 16.188. Distribution of *Ctenomys peruanus.*

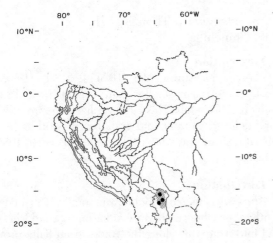

Map 16.189. Distribution of *Ctenomys steinbachi.*

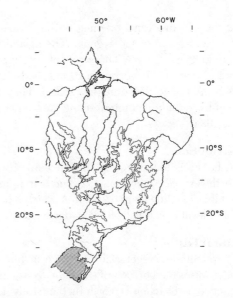

Map 16.190. Distribution of *Ctenomys torquatus.*

ish collar. In the inguinal and axillary regions, many have light patches. The tail is usually a uniform dark brown. There is a great deal of inter- and intrapopulation variation in color (Barlow 1965; Fernandes 1965; Freitas and Lessa 1984; Langguth and Abella 1970; pers. obs.).

Distribution

Ctenomys torquatus is found in extreme southern Brazil through Uruguay and into Argentina in the provinces of Entre Ríos and Corrientes (Reig and Kiblisky 1969). (map 16.190).

Life History

Females are monestrous, breeding between June and October. Young are born in September to December after a gestation period of about 105 days. Females have two or three embryos (Barlow 1969; Tálice, Mosera, and Sprechmann 1972).

Ecology

In Argentina this species prefers high dry areas with sandy soils and is common in open palm savannas.

In Uruguay its distribution is probably limited by soil composition, since *C. torquatus* prefers sandy, rock-free soil and disappears in areas under cultivation. The density of this species depends on the density of the food plants.

C. torquatus lives singly except during the reproductive season, when it is found in pairs. Twice as many females as males were captured in one study ($n = 180$). In six study sites, the male-to-female ratio varied from 1:2.5 to 1:14.5. This species has good vision and is aggressive toward conspecifics.

Burrows nowhere exceed 60 cm in depth and average 4.5 m long and 6–10 cm in diameter. Food is stored in side galleries. *C. torquatus* is diurnal in its surface activity. Animals have never been seen more than two-thirds of the way out of a burrow entrance when foraging or resting. They are strictly herbivorous, feeding primarily at the surface and eating entire plants around the burrow mouth (Barlow 1969; Crespo 1982; Fernandes 1965; Freitas and Lessa 1984; Talice and Momigliano 1954; Tálice, Capario, and Momigliano 1954; Tálice, Mosera, and Sprechmann 1972).

Comment

The chromosome number of this species is reported as $2n = 68$, one of the highest recorded for the genus. However, more recent work has shown that $2n$ varies from 56 to 70 and that *C. torquatus* is probably a superspecies (Altuna and Lessa 1985; Lessa and Altuna 1984; Reig and Kiblisky 1969).

FAMILY ABROCOMIDAE

Diagnosis

The dental formula is I 1/1, C 0/0, P 1/1, M 3/3. Members of this family differ from the Octodontidae in that the lacrimal canal opening on the rostrum is near the dorsal root of the zygoma. The palatal foramina are long and narrow. These rodents are approximately the size of a large rat; head and body length ranges from 150 to 250 mm (fig. 16.29). They have short limbs, a rather pointed snout, and large ears (Woods 1984).

Distribution

The family is found from southern Peru to Chile and northwestern Argentina. *A. bennetti* is found in lowland hills of Chile up to 3,700 m; *A. cinerea* occurs at higher elevations, usually in cold, dry habitats.

Natural History

The rodents are vegetarians and live in burrows. *A. bennetti* uses the burrows of *Octodon degus*, and its dependence on *Octodon* is worth further study. Little is known concerning the natural history of *A. cinerea*, but it has been studied in captivity. In common with other hystricognath rodents, they give birth to precocial young after a lengthy gestation period.

Comment

Glanz and Anderson (1990) review the characters of the family and suggest a close relationship with the family Chinchillidae.

Abrocoma boliviensis Glanz and Anderson, 1990

Description

Measurements: HB 174; T 141; HF 30; E 22.5 (*n* = 2). Among the living species of *Abrocoma*, *A. boliviensis* most resembles *A. bennetti*. The tympanic bullae are small compared with those of *A. cinerea*. The tail is relatively long and hairy. The dorsum is dark brown, and the venter is pale brown.

Distribution

The species is known only from the type locality, Comarapa, Santa Cruz department, Bolivia (not mapped).

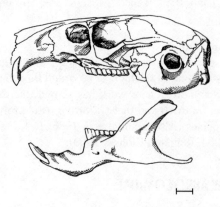

Figure 16.29. Skull of *Abrocoma bennetti* (after Ellerman 1940). Scale bar = 5 mm.

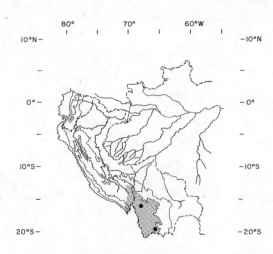

Map 16.191. Distribution of *Abrocoma cinerea*.

Abrocoma cinerea Thomas, 1919
Rata Chinchilla

Description

Average measurements: HB 174; T 61; HF 26; E 24. *A. cinerea* is smaller than *A. bennetti* and has a much shorter tail. This large rat has large, rounded ears, delicate incisors, and granular soles on its hind feet. It is silver gray with a paler venter (Pearson 1951; pers. obs.).

Distribution

Abrocoma cinerea is found from southwestern Peru south to the Chilean province of Parinacota and El Loa and into Argentina along the Andes from Salta province and as far south as Mendoza province (Roig 1965; Tamayo and Frassinetti 1980) (map 16.191).

Life History

In captivity the gestation period is 116 days, litter size averages 2.2 (range 1–3), and neonates weigh 22 g (Kleiman, Eisenberg, and Maliniak 1979).

Ecology

This is a species of arid, rocky zones at elevations up to 5,000 m. In tola habitat in Peru it was found at a density of 0.17 per hectare. It lives in small colonies and digs burrows about 5 cm in diameter at the bases of boulders and shrubs. *A. cinerea* is not as arboreal as *A. bennetti* (Koford 1955; Lucero 1983; Mann 1978; Pearson and Ralph 1978; Tamayo and Frassinetti 1980).

FAMILY ECHIMYIDAE
Spiny Rats, Ratas Espinosas, Casiraguas

Description

The dental formula is I 1/1, C 0/0, P 1/1, M 3/3. These are rat-sized hystricognath rodents, usually with a long tail. As the common name implies, many species, but not all, have spinescent hairs interspersed in their dorsal pelage. The dorsum is highly variable in color but is usually some shade of brown to almost black. The venter often contrasts sharply with the dorsum, being white or yellowish. A character matrix for the genera is given in table 16.10.

Distribution

Echimyid rodents are distributed from Honduras south through the Isthmus over much of South America to northern Argentina.

Natural History

The best-studied echimyid rodents are members of the genus *Proechimys*. These terrestrial forms use hollow logs, natural crevices, or burrows they construct to cache food, bear their young, and seek refuge. Some terrestrial genera of southern South America are poorly known, including *Clyomys*, *Carterodon*, and *Euryzygomatomys*. *Proechimys* may be one of the most abundant rodents in the lowland tropical rain forest of the northern Neotropics. Other spiny rat genera are strongly arboreal. These include the genera *Dactylomys*, *Kannabateomys*, *Echimys*, *Diplomys*, *Isothrix*, *Mesomys*, and *Olallamys* (formerly *Thrinacodus*). The

Table 16.10   Matrix of Characters for Some Echimyid Rodents

Taxon	Heavy Spines	Tail Scaly or Sparsely Haired	Tail Haired	Tail Fully Haired
Mesomys	+ +		+	
Lonchothrix	+ +	+ (but tuft)		
Echimys				
Echimys armatus	+		±	
Echimys chrysurus	+ +		+	
Echimys grandis	+ +			+
Echimys macrourus?				
Echimys occasius	+ +	+		
Echimys rhipidurus	+ +		+	
Echimys saturnus	+ +			+
Echimys semivillosus	+ +	+		
Nelomys				
Nelomys blainvillei	+		+	
Nelomys lamarum	+ +	+		
Nelomys nigrispinis	+ +	+		
Dactylomys	−	+		
Kannabateomys	−		+	
Diplomys	−		+	
Isothrix	−		+	
Nelomys brasiliensis	+ +			
Nelomys thomasi (Ilha São Sebastião, an island subspecies of *N. nigrispinis*)		− (but bristles)		+

arboreal genera have not been studied in as much detail as the terrestrial forms (Emmons 1982).

SUBFAMILY CHAETOMYINAE
Genus *Chaetomys* Gray, 1843
Chaetomys subspinosus (Olfers, 1818)

Description

After examing the microstructure of the incisor enamel, Martin (1994) concludes that *Chaetomys* is an erethizontid (see p. 453). The skull is portrayed in figure 16.18. Measurements are HB 430–57; T 255–80. This very spinescent echimyid superficially resembles a species of *Coendou*, but the "spines" on the dorsum are not as stiff as those of the true porcupines. The tail is relatively hairless and somewhat prehensile. The forefeet and hind feet have four digits, all bearing claws (see plate 11) (Nowak 1991).

Range and Habitat

The species is apparently confined to the fragments of the Brazilian rain forests of Bahia, Espírito Santo, and Rio de Janeiro (Oliver and Santos 1991) (not mapped). *Chaetomys* is tolerant of a wide range of habitats including both primary and secondary forests. It appears to be much more restricted than *Sphiggurus*, with which it may occur in sympatry (Oliver and Santos 1991).

Natural History

Chiarello, Passamani, and Zortea (1997) report on a brief radio telemetry study of *Chaetomys* conducted in an Atlantic rain forest patch in Espírito Santo, Brazil. They determined that the animal is strictly nocturnal and that almost all its activity is arboreal. It feeds about 18% of each night and appears to eat leaves exclusively. They suggest it may be categorized as an arboreal folivore.

SUBFAMILY EUMYSOPINAE (IN PART)
Genus *Clyomys* Thomas, 1916
Clyomys bishopi Avila-Pires and Wutke, 1981

Description

Measurements: HB 206.4; T 77; HF 33.5; E 20.2 ($n = 11$). Cranial measurements and the development of a cranial index led the authors to separate this form from *C. laticeps*, from which it is also distinguished by the yellowish brown dorsum.

Distribution

The type locality is Itapetininga, São Paulo state, Brazil. It is apparently confined to southern São Paulo state.

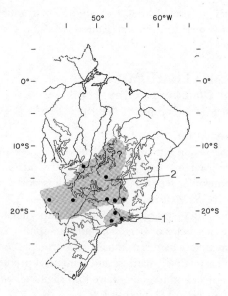

Map 16.192. Distribution of (1) *Clyomys bishopi;* (2) *C. laticeps.*

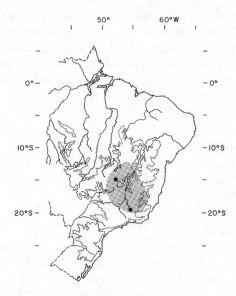

Map 16.193. Distribution of *Carterodon sulcidens.*

Clyomys laticeps (Thomas, 1909)

Description

Average measurements: HB 187; T 73; HF 32. The tympanic bullae are inflated. In this medium-sized spiny echymid the claws are well developed and the ears and tail are short. It is agouti buffy gray brown or grizzled rufous black, blacker on the back and becoming strongly rufous on the rump and head. The venter is dull whitish or buffy, and the tail is dark brown (Bishop 1974; Moojen 1952a; pers. obs.).

Distribution

Clyomys laticeps is found in southern Brazil and into eastern Paraguay (Honacki, Kinman, and Koeppl 1982; Moojen 1952b) (map 16.192).

Life History

In Mato Grosso, Brazil, three females were caught from June to December, each with a single embryo (Bishop 1974).

Ecology

This highly fossorial species is one of the few South American rodents apparently confined to savanna environments. In the Brazilian Pantanal it is a common member of the small mammal fauna, inhabiting edge habitats and frequently caught in gardens. It is colonial (Alho et al. 1987; Bishop 1974).

Genus *Carterodon* Waterhouse, 1848
Carterodon sulcidens (Lund, 1839)

Description

Measurements: HB 155–200; T 68–80. This is a typical short-tailed terrestrial echimyid (plate 13). The tympanic bullae are inflated. The dorsum is a yellow brown and the lower parts of the throat are reddish. The venter is white in the midportion, while laterally it parallels the color of the dorsum.

Distribution

This species was first described by Lund from Pleistocene deposits in Minas Gerais state, Brazil. It apparently occurs in the cerrado of eastern Brazil, an area of seasonal rainfall with a basic vegetative cover of deciduous trees interspersed with grassland (map 16.193).

Ecology and Life History

These rodents apparently dig burrows about 2 m long and construct a grass nest at the end of the excavation. A female was trapped with one embryo (Jao Moojen, pers. comm. to Walker, as cited in Nowak 1991). Owls *Tyto alba* are the principal predators (Moojen 1952a).

Comment

The species' phylogenetic relationship with *Clyomys* remains to be investigated.

Genus *Euryzygomatomys* Goeldi, 1901
Euryzygomatomys spinosus (G. Fischer, 1814)
Rata Guira

Description

Average measurements: HB 195; T 50; E 17; Wt 188 g. This rat is stout and has short ears; it is somewhat less spiny than *Clyomys* and has a shorter tail, a longer tooth row, and broad incisors. Although the bullae are somewhat inflated, they are not as inflated as in

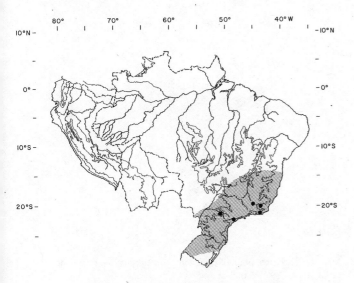

Map 16.194. Distribution of *Euryzygomatomys spinosus*.

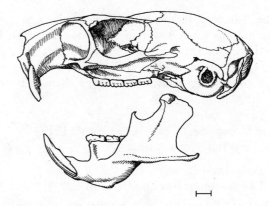

Figure 16.30. Skull of *Kannabateomys amblyonyx* (after Ellerman 1940). Scale bar = 5 mm.

Clyomys. The dorsum is agouti brown to black. The venter is more tan than the dorsum and is marked with distinctive pure white hourglass-shaped blotches. The chin and throat are often heavily washed with tan (Bishop 1974; pers. obs.).

Distribution

Euryzygomatomys spinosus is found in southern and eastern Brazil, in northeastern Argentina, and in Paraguay (map 16.194).

Ecology

In Minas Gerais state, Brazil, this species was caught in wet meadow habitat (Stallings 1988a,b).

SUBFAMILY DACTYLOMYINAE
Genus *Kannabateomys* Jenkins, 1891
Kannabateomys amblyonyx (Wagner, 1845)
Rata Tacuarera

Description

Average measurements: HB 247; T 321; HF 23; Wt 475 g. This large rodent is unmistakable because of its size and the length of its tail (fig. 16.30). The dorsum is light agouti brown along the midline grading to reddish brown on the sides. The venter and the sides of the face are reddish white. The very long tail is well haired along the basal 6 cm, becoming less well haired and ending in a pronounced tuft (pers. obs.).

Distribution

Kannabateomys amblyonyx is found in southern Brazil, into eastern Paraguay, and in the northeastern province of Misiones, Argentina (map 16.195).

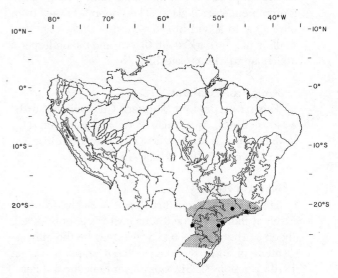

Map 16.195. Distribution of *Kannabateomys amblyonyx*.

Ecology

This nocturnal, arboreal rodent appears to be confined to moist tropical forest, where it lives in thickets of bamboo, especially *Guadua angustifolia* when it grows near water. When alarmed this species makes loud squeaks or shrieks. A variant of these calls may serve a spacing function. In its ecology it seems to be similar to *Dactylomys* (see Emmons 1981). In Brazil home ranges average 1,000 m² or more, and densities can be as high as 1.47 individuals per 1,000 m² (Crespo 1950, 1982b; Stallings, Kierulff, and Silva 1994; Kierulff, Stallings, and Sabato 1991; Olmos et al. 1993).

Genus *Dactylomys* I. Geoffroy, 1838
Dactylomys boliviensis Anthony, 1920

Description and Comment

It is difficult to determine the relation of this species to *Dactylomys peruanus*. I have not examined the

type. Emmons and Feer (1997) recognize one species (*D. dactylinus*) within the genus; however, it
has been confirmed as a species by da Silva and Patton (1993). It has been recorded from southeastern
Peru to northern Bolivia, extending to Acre in Brazil
(map 16.196).

Dactylomys dactylinus (Desmarest, 1817)
Bamboo Rat, Coro-coro

Description

Two specimens measured TL 710, 720; T 410, 415;
HF 50, 55; E 17, 20. Its large size and very long tail distinguish this species from all other spiny rats within its
range. The tail is furred for about 30 mm at the base;
the dorsal pelage is soft and lacks spines. Males have
a prominent sternal gland. The upperparts vary from
olive gray to agouti brown with a reddish tinge; the
midline may be dark agouti brown, and the underparts
are cream to white (plate 13).

Range and Habitat

This rat occurs from southern Colombia to northwestern Brazil, Amazonian Peru, and Ecuador. This
lowland form is strongly associated with multistratal
tropical evergreen forest (map 16.197).

Natural History

This species is nocturnal and strongly arboreal.
At night individuals emit a characteristic pulsed call,
"boop . . . boop . . . boop," which may be the only indication of their presence without trapping (Laval
1976). Emmons (1981), working in Ecuador and Peru,
established that these rodents are highly folivorous.
Gut modifications suggest they can ferment masticated leaves in the cecum. They are arboreal foragers in bamboo clumps. Home ranges vary from 0.23 to
0.43 ha, and they appear to forage in family groups.

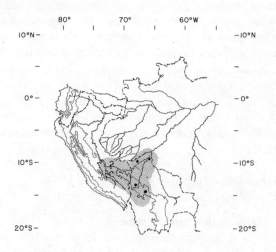

Map 16.196. Distribution of *Dactylomys boliviensis*.

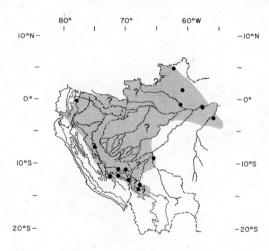

Map 16.197. Distribution of *Dactylomys dactylinus* (see comments
in Anderson 1997 for Bolivia).

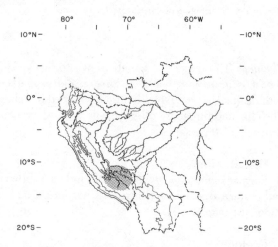

Map 16.198. Distribution of *Dactylomys peruanus* (hypothetical).

Dactylomys peruanus J. A. Allen, 1900

Description

Measurements of the type are TL 560; HB 240;
T 320; HF 51; E 14. This species is distinctly smaller
than *D. dactylinus*, and the tail is much hairier. The
pelage is soft, with long, bristly overhairs. The dorsum
is yellowish brown varied with black-tipped hairs. The
sides of the nose are whitish, and the top is sooty gray.
The venter is white only in the midline, because the
dorsal color extends far down the sides.

Distribution

There are records from cloud forests in southeastern Peru (map 16.198).

Genus *Olallamys* Emmons, 1988 (formerly *Thrinacodus* Gunther, 1879)

Description

Head and body length ranges from 180 to 240 mm.
The tail is longer than the head and body—a diagnos-

tic feature of this genus. These rats are smaller than *Dactylomys*. The fur is soft, without spines or bristles. The upperparts are usually some shade of reddish brown, and the underparts are white to yellow. The tail is sparsely haired, faintly bicolored, and paler at the tip.

Distribution

The genus is known from Colombia and Venezuela at higher elevations. Specimens are generally taken at above 2,000 m. It could extend to Ecuador.

Comment

Emmons (1988) advocates *Olallamys* as a replacement for *Thrinacodus*.

Olallamys albicauda (Gunther, 1879)

Description

Average measurements: TL 554; T 327; HF 49; E 23. See the descripton for the genus. The tail is distinctly bicolored, brown above and light below and often paler at the tip.

Range and Habitat

The species is found in the northwestern and central portions of Colombia to the west of the central cordillera of the Andes. It occurs at up to 3,000 m elevation and is often associated with moist, dense bamboo thickets (not mapped).

Olallamys edax Thomas, 1916

Description

See the description for the genus. The tail is bicolored, and there is a tendency for the distal third to be entirely white.

Range and Habitat

The species occurs at high elevations in the western Venezuelan Andes and adjacent portions of northeastern Colombia and may extend to Ecuador (not mapped).

Natural History

This nocturnal, arboreal form produces a whistle-like call at night.

SUBFAMILY ECHIMYINAE
Genus *Diplomys* Thomas, 1916
Arboreal Soft-furred Spiny Rat, Ratón Marenero

Description

Head and body length ranges from 250 to 480 mm, and the tail from 200 to 280 mm. The fur is very soft, with no spines; the tail is haired and tends to be tufted at the tip. The dorsal pelage is brown mixed with black hairs, and the venter is lighter, ranging from buffy to white.

Distribution

The genus occurs from Panama through Colombia to northern Ecuador, west of the eastern cordillera of the Andes.

Natural History

These highly arboreal rodents nest in tree cavities. Their presence often goes undetected, but by thumping on the side of a tree one can occasionally flush a specimen from its hole. They are nocturnal and feed on fruit, seeds, and nuts (Tesh 1970a). In Panama *D. labilis* has a litter size of one or two.

Diplomys caniceps (Gunther, 1877)

Description

Average measurements: TL 390; T 189; HF 42; E 15. This rat is similar in appearance to *D. labilis,* but the dorsum is not as reddish. The venter is tan. The tail is sparsely haired to the tip and is not noticeably bicolored.

Range and Habitat

The species is found in the central portion of Colombia between the western and central cordilleras of the Andes, perhaps south to Ecuador (map 16.199).

Diplomys labilis (Bangs, 1901)

Description

This account includes *Diplomys darlingi.* The type has a total length of 540 mm; T 200; HF 47; E 15

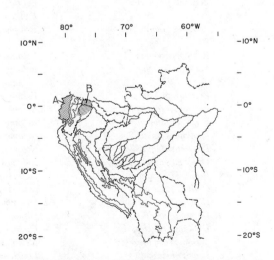

Map 16.199. Distribution of (a) *Diplomys labilis;* (b) *D. caniceps.*

(FMNH). The upperparts are reddish brown, darker at the middorsal region. There is a small yellowish white patch at the base of the mystacial vibrissae, above the eye and behind the ear. The underparts vary from buff to orange. The distal portion of the tail may have white hairs.

Range and Habitat

The species is found from central Panama to the east and then south into Colombia to the west of the western cordillera of the Andes to Ecuador (map 16.199).

Genus *Echimys* G. Cuvier, 1809
Arboreal Spiny Rat, Rata Espinosa

Taxonomy

Emmons and Feer (1997) use the genus *Nelomys* for species of the Echimyidae in southeastern Brazil. We shall hold with the nomenclature as outlined in Woods (1993). With respect to Woods's retention of *Makalata*, we offer the following comments: *Makalata* Husson (1978) was established as a genus to separate *Echimys armatus* from *E. chrysurus* at the generic level. The genus *Makalata* has thus been reserved for *armatus*, which became *M. armata* (I. Geoffroy, 1830). Emmons (1993) points out that *Mesomys didelphoides* (Desmarest, 1817) is actually referable to *M. armata* and has priority; thus *Echimys* (*Makalata*) *didelphoides* is the correct name for *Echimys armatus*. Husson (1978) primarily focused on the spiny rats

of Suriname and adjacent portions of northeastern Brazil. Recent work by da Silva and Patton (1993) implies that the spiny rats at the mouth of the Amazon (both banks) are distinct from the spiny rats of the upper Amazon. Based on these data Louise Emmons (pers. comm.) suggests that *Echimys macrurus* may well apply to forms on the upper Amazon, since *Mesomys* (*Echimys*) *obscurus* is based on a composite (Emmons 1993). *Echimys macrurus* (Wagner, 1842) was not recognized in Emmons and Feer (1990) (see table 16.11).

Description

Head and body length ranges from 170 to 350 mm and tail length from 150 to 300 mm. The dorsum is usually brown to gray brown, but dorsal coloration varies geographically. The underparts are usually buff to white (plate 13).

Distribution

The genus is found over most of northern South America south to Brazil.

Natural History

These nocturnal, arboreal rats nest in hollow trees. Their presence can sometimes be detected at night by their distinctive chattering vocalizations. The female appears to bear only one or two young. Small family groups can use the same nesting hole in a tree.

Table 16.11   Nomenclature for Lowland Rain Forest Echimyidae (Exclusive of *Proechimys* and *Trinomys*)

Moojen 1952 (Brazil Only)	Emmons and Feer 1990	Woods 1993
—	*Hoplomys gymnurus*	*Hoplomys gymnurus*
Phyllomys brasiliensis	?	*Echimys brasiliensis*
Echimys paleaceus	*Echimys chrysurus*	*Echimys chrysurus*
?	*Echimys saturnus*	*Echimys saturnus*
Echimys grandis	*Echimys grandis*	*Echimys grandis*
?	*Echimys rhipidurus*	*Echimys rhipidurus*
Echimys armatus	*Echimys armatus*	*Makalata armata*
Echimys occasius	*Echimys occasius*	*Makalata armata*
—	*Echimys semivillosus*	*Echimys semivillosus*
Echimys pictus	*Echimys pictus*	*Echimys pictus*
Phyllomys nigrispina	*Nelomys nigrispinus*	*Echimys nigrispinus*
Phyllomys thomasi	*Nelomys thomasi*	*Echimys thomasi*
Phyllomys lamarum	*Nelomys lamarum*	*Echimys lamarum*
Phyllomys blainvillei	*Nelomys blainvillei*	*Echimys blainvillei*
Phyllomys dasythrix	*Nelomys dasythrix*	*Echimys dasythrix*
Isothrix bistriata	*Isothrix bistriata*	*Isothrix bistriata*
Isothrix pagurus	*Isothrix pagurus*	*Isothrix pagurus*
Diployms caniceps	*Diployms caniceps*	*Diplomys caniceps*
Diplomys labilis	*Diplomys labilis*	*Diplomys labilis*
Diplomys rufodorsalis	*Diplomys rufodorsalis*	*Diplomys rufodorsalis*
Mesomys hispidus	*Mesomys hispidus*	*Mesomys hispidus*
Lonchothrix emiliae	*Lonchothrix emiliae*	*Lonchothrix emiliae*
Dactylomys dactylinus	*Dactylomys dactylinus*	*Dactylomys dactylinus*
Kannabateomys amblonyx	*Kannabateomys amblonyx*	*Kannbateomys amblonyx*
Echimys macrurus	—	*Echimys macrurus*

Echimys (Makalata) didelphoides
(Desmarest, 1817)

Description

Two specimens measured HB 238, 264; T 195, 220; HF 35, 44; E 18, 19 (FMNH). This species may attain a weight of 300 g. The dorsal pelage is furry but strongly admixed with spines. The tail is furred for about 30 mm at the base, and the distal portion is sparsely haired. The dorsum is dark brown, and the venter is gray brown (plate 13).

Range and Habitat

The species in the older sense of *Echimys didelphoides (armatus)* occurs in northern Ecuador, Peru, Colombia, Venezuela, the Guianas, and Brazil. Distributions in northern Colombia were documented by Hershkovitz (1948), but see *Echimys macrurus*. It is a lowland form preferring seasonally flooded tropical evergreen forest. In northern Venezuela most specimens were taken at below 350 m elevation (map 16.200).

Natural History

Unlike *Echimys semivillosus*, this rat is strongly tied to moist habitats. Charles-Dominique et al. (1981) found it to be a significant predator on immature seeds, particularly *Virola* and *Inga*. The species is strictly nocturnal and nests in hollow trees. The litter size is one or two, and the young are precocial.

Comment

Formerly *Echimys armatus* (I. Geoffroy, 1830). The older "*armatus*" group needs a thorough revision. If *Makalata* is accepted as a genus, then clearly *Echimys macrurus* and *E. occasius*, and possibly *E. rhipidurus*

and *E. saturnus*, are somehow related. The following outline employs the genus *Echimys* but attempts to confine the descriptions regionally (see comments by da Silva and Patton 1993; Patton, da Silva, and Malcolm 1994.

Echimys Species of the Upper Amazon
Echimys macrurus (Wagner, 1842)

Description

Echimys macrurus is heavily spinescent, with hair noticeable on the tail. See also comments under *E. didelphoides*.

Distribution

The species occurs in the upper Amazon drainage of Peru (map 16.200).

Echimys occasius Thomas, 1921

Description

This rodent is heavily spinescent, with the tail sparsely haired. (See account for *E. didelphoides*.)

Distribution

The species is currently thought to occupy Amazonian Ecuador and extend thence to adjacent Amazonian Peru (map 16.200).

Echimys rhipidurus Thomas 1928

Description

In this species the heavily spinescent tail is sparsely haired (see *didelphoides*).

Distribution

The species has a possibly disjunct distribution in eastern, central, and northern Peru (map 16.201).

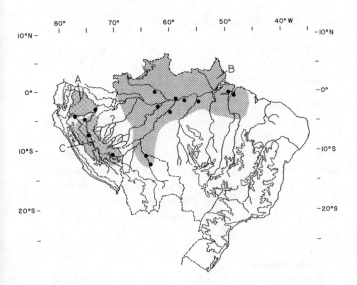

Map 16.200. Distribution of three species of *Echimys*: (a) *E. occasius*; (b) *E. didelphoides* (formerly *Makalata armata;* see text) (c) *E. macrurus.*

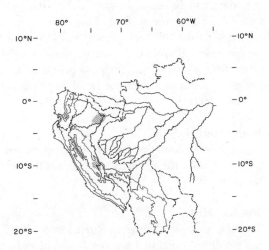

Map 16.201. Distribution of *Echimys rhipidurus.* Specimens from this area have been assigned various names, including *E. armata* and *E. saturnus*. Localities are not marked, pending revision.

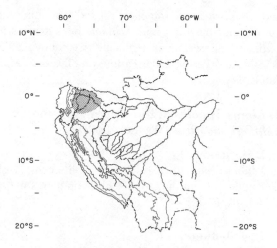

Map 16.202. Distribution of *Echimys saturnus*.

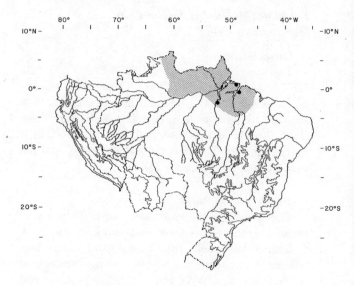

Map 16.203. Distribution of *Echimys chrysurus*.

Echimys saturnus Thomas, 1928

Description
This rodent is heavily spinescent, with a well-haired tail (see comment under *didelphoides*).

Distribution
The species occurs in Amazonian Ecuador (map 16.202).

Echimys Species from the Central Amazon Region
Echimys chrysurus (Zimmermann, 1780)

Description
Head and body length ranges from 230 to 290 mm and the tail from 310 to 335 mm; the hind foot is 51 mm and the ear 20 mm. Weight may reach over 500 g. The dorsal pelage includes very broad, heavy spines interspersed with hairs; the tips of the spines tend to be brown while the shafts are gray. The dorsum is dark rufous brown to gray brown, darkest in the midline. The dorsal surface of the head is usually much darker than the back, almost blackish, and in some specimens there is a distinctive white to yellow median stripe running from the nose over the face to the neck. Specimens with white facial stripes usually have white tail tips.

Range and Habitat
The species occurs in multistratal tropical evergreen forest from the Guianas south to northeastern Brazil (map 16.203).

Natural History
This is a strongly arboreal, nocturnal form. The female bears two young that are born fully furred and with their eyes open (Husson 1978).

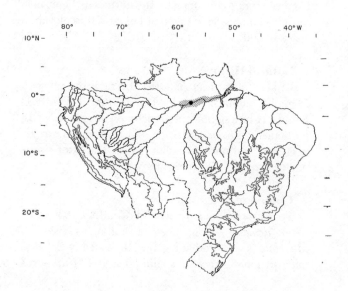

Map 16.204. Distribution of *Echimys grandis*.

Echimys grandis (Wagner, 1845)

Description
Three specimens measured HB 251–345; T 265–88; HF 50–58; E 15–18 (Belém). In this large, heavy-bodied, dark-colored species the dorsum has coarse, agouti banded hairs, almost black in the midline and yielding a reddish cast to the sides. In some specimens there are flecks of cream. The venter is orange to yellow. The tail is haired and ranges from dark brown to almost black.

Distribution
The species is found in the central Amazon of Brazil (map 16.204).

Echimys Species from Brazil Referable to *Nelomys* or *Phyllomys*

General Remarks

These rats are strongly arboreal and exhibit varying degrees of spinescence. Olmos (1997) reports that some species of *Nelomys* construct leaf nests in the branches of trees several meters above the ground. This is in contrast to nesting sites reported for members of the general *Diplomys* and *Echimys*. Species of these genera apparently habitually nest in tree cavities.

Echimys (Nelomys) blainvillei (Cuvier, 1837)

Description

Measurements: HB 213.8; T 223.2; HF 37.2; E 16.0; Wt 407.5 g (*n* = 19; São Paulo). The dorsal pelage is moderately spinescent, and the tail is sparsely haired.

Distribution

The type locality is given as Isla Deas. There are several islands with that name, but if we assume it is the island off north Bahia, then it is possible to assign this form to northeastern Brazil (map 16.205). Emmons and Feer (1997) list this species as interior in its distribution, but they also list an undescribed species, the "rusty-sided Atlantic tree rat," as coastal. This situation will be clarified with further study.

Echimys (Nelomys) brasiliensis (Waterhouse, 1848)

Description

These rodents have heavy spines on the dorsum and hair on the tail, albeit sparse.

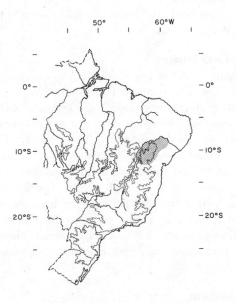

Map 16.205. Distribution of *Echimys blainvillei.*

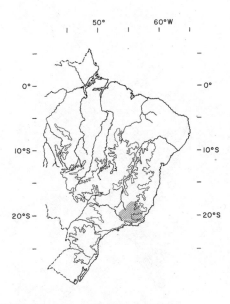

Map 16.206. Distribution of *Echimys brasiliensis.*

Distribution

The species is apparently confined to Rio de Janeiro and Minas Gerais states, Brazil (map 16.206).

Comment

The status of this form is still in question.

Echimys (Nelomys) dasythrix (Hensel, 1872)

Description

Measurements of one female were TL 460; T 230; HF 34; E 19. The dorsal pelage is mildly spinescent. The dorsum is reddish brown, becoming darker in the midline. The venter is white on the throat and in the inguinal region, with the rest dirty white. The tail is moderately haired and has a dark brown tuft at the tip.

Distribution

The species occurs in extreme southeastern Brazil (map 16.207). All reported localities were within 60 km of Pôrto Alegre.

Echimys (Nelomys) lamarum (Thomas, 1916)

Description

Three specimens measured HB 200–205; T 200–230; HF 40–45; E 16–20; Wt 250–90 g. These are spinescent rats with a pale brown back. The tail is sparsely haired, and the tip of the tail is tufted.

Distribution

The species is found in the coastal regions of Bahia, Brazil (map 16.208). The specimens viewed were from the vicinity of Belo Horizonte, Ceccá, and St. Serra

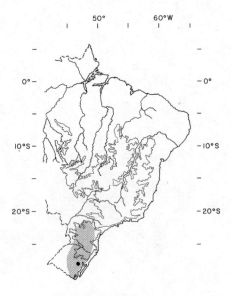

Map 16.207. Distribution of *Echimys dasythrix*.

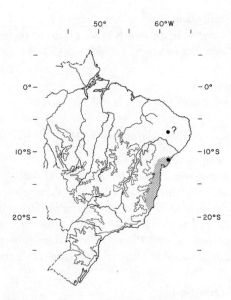

Map 16.208. Distribution of *Echimys lamarum*.

Bechibe in northern Brazil. The extent of the range to the interior is not mapped, but the exact identity of species within this region is unclear (see *Echimys blainvillei*).

Echimys (Nelomys) nigrispinus (Wagner, 1842)

Description

This is a heavily spinescent form with a sparsely haired tail.

Distribution

The species is apparently confined to Rio de Janeiro and São Paulo states, Brazil (map 16.209).

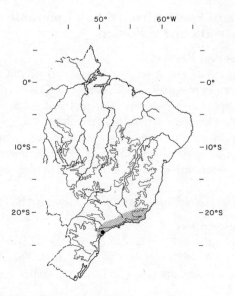

Map 16.209. Distribution of *Echimys nigrispinus*.

Natural History

Olmos (1997) summarizes data indicating that this species may construct an arboreal leaf nest, similar to what he describes for *E. thomasi* (see below).

Echimys (Nelomys) thomasi (Ihering, 1897)

Description

Head and body length is greater than 300 mm, and the tail is nearly equal. One specimen examined from the vicinity of São Paulo was moderately spinescent. The dorsum was light brown, grading to darker brown on the midline. The tail was darker brown, lightly haired, and the tip of the tail was tufted.

Distribution

The type locality is Ilha de São Sebastião, São Paulo state, Brazil (not mapped).

Natural History

Olmos (1997) reviewed what is known concerning the biology of *E. thomasi*. He examined an occupied nest 13 m above the ground in the crown of a *Terminalia catappa*. The domelike nest was constructed of interwoven dried and twisted leaves of *Terminalia*, with the floor supported by a forked branch, and the entrance was on the side. The nest was 38 cm long, 30 cm wide, and 20 cm high. Captive specimens appear to feed on fruit, but not foliage. Olmos also noted that at an apparent feeding platform 3.5 m below the nest there was evidence of feces, discarded plant material, and urine stains, perhaps indicating olfactory marking.

Comment

Echimys thomasi is often considered a subspecies of *E. blainvillei*.

Genus *Isothrix* Wagner, 1845

Description

Head and body length ranges from 220 to 290 mm; the tail is approximately 250 mm. The color pattern is highly variable for the genus, and one species, *I. pictus*, is blotched dark brown broken by a buff patch across the shoulders. The fur is soft and without spines. The tail is very well haired, almost squirrellike (plate 13).

Distribution

The genus is confined to the Amazonian portions of Venezuela, Colombia, Peru, and Brazil. Vié et al. (1996) have described a new *Isothrix* from French Guiana.

Isothrix bistriatus Wagner, 1845

Description

Measurements: TL 505; T 242; HF 47; E 16.5. Head and body length ranges from 220 to 251 mm. This rat is almost squirrellike in appearance. The soft dorsal pelage is brown, somewhat lighter on the sides, the venter is orange, and the forehead is yellow to cream, framed by two brown stripes. The tail is well haired, pale brown at the base grading to black at the tip.

Range and Habitat

The species occurs in southwestern to north-central Brazil, Amazonian Peru, southern Venezuela, and southern Colombia (map 16.210). It is found in moist habitats in multistratal tropical evergreen forests.

Natural History

This soft-furred form is nocturnal and strongly arboreal. It nests in tree cavities. The tail may be semiprehensile in some of the proposed species (Patton and Emmons 1985). The female generally bears only a single young, which is born furred and with its eyes open.

Isothrix pagurus (Wagner, 1845)

Comment

Isothrix pagurus is often considered synonymous with *I. bistriatus,* but it has been confirmed as a species by Patton and Emmons (1985). It is mapped here as a mid-Amazon isolate in Brazil (map 16.211).

Isothrix pictus (Pictet, 1841)
Painted Spiny Rat

Description

One male specimen from Rio de Birasa, Brazil, measured HB 290; T 310; HF 54; E 18; Wt 480 g.

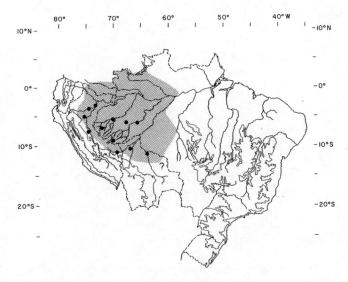

Map 16.210. Distribution of *Isothrix bistriatus.*

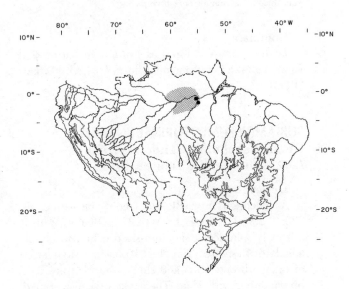

Map 16.211. Distribution of *Isothrix pagurus.*

The hind feet are relatively short but quite broad. This large echimyid rodent has soft pelage, with no obvious spinelike hairs. The tail is furred black proximately with a white tip. The color pattern is unique among the echimyids: white sides, venter, foreshoulder, throat, and face. The top of the head, nape, and back are black or dark brown (see plate 13) (Emmons and Feer 1997; Moojen 1952a).

Distribution

This species appears to be confined to the Atlantic rain forest of Brazil. Specimens have been collected in Bahia (map 16.212).

Ecology

This rat is apparently arboreal. It can survive in partially cleared forest where cacao is grown under the canopy of large native trees (Moojen 1952a).

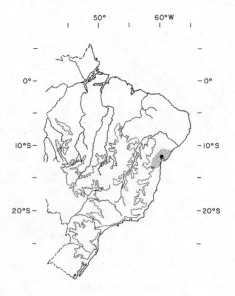

Map 16.212. Distribution of *Isothrix pictus*.

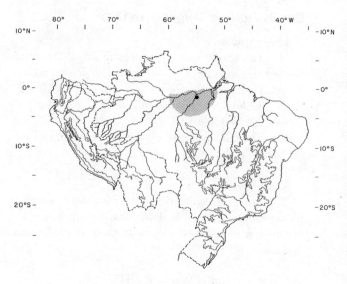

Map 16.213. Distribution of *Lonchothrix emiliae*.

Comment

Isothrix pictus is listed as *Isothrix* in Patton and Emmons (1985) but included in *Echimys* by Woods (1993).

SUBFAMILY EUMYOSOPINAE (IN PART)
Genus *Lonchothrix* Thomas, 1920
Lonchothrix emiliae Thomas, 1920

Description

A specimen from Rio de Janeiro measured TL 380; T 200; HF 35; E 15; Wt 237 g (Belém). This is a heavily spinescent, medium-sized echimyid closely allied to *Mesomys*. The dorsal pelage is a mixture of orange brown hairs and stout spinesthat are light brown tipped with white. The flanks are lightly spined. The central portion of the venter is covered with stiff, cream-colored hairs. The terminal portion of the tail is strongly penicillate (plate 13).

Distribution

The species is found in Amazonian Brazil, south of the Amazon River (map 16.213).

Genus *Mesomys* Wagner, 1845

Description

Head and body length ranges from 150 to 200 mm and the tail from 120 to 220 mm. The dorsum is usually some shade of brown. The spiny hairs on the dorsum are banded, giving a salt-and-pepper effect; when stroked, these spines become much more obvious than in *Proechimys*. The underparts are buff to white. The tail is sparsely haired, slightly penicillate at the tip, and not bicolored (plate 13).

Distribution

Members of the genus are chiefly found in the Amazonian portions of Brazil, Colombia, Peru, Ecuador, the Guianas, and southern Venezuela. They are associated with multistratal tropical evergreen forest at up to 1,950 m elevation. From an analysis of genetic variation, Patton, da Silva, and Malcolm (1996) conclude that *Mesomys* has long occupied the Amazon basin. Populations on the lower Amazon are distinct from populations on the upper reaches.

Natural History

These spiny rats are nocturnal and arboreal but often forage terrestrially. They may nest in vine tangles close to the ground. *M. hispidus* has a relatively short gastrointestinal tract, suggesting that leaves do not make up a significant proportion of its diet (Emmons 1981). Females bear a single young that is born furred and with its eyes open.

Mesomys hispidus (Desmarest, 1817)

Description

Average measurements: TL 354; T 165; HF 33; E 13. This species is larger than *M. stimulax*. See also the description for the genus. *M. ferrugineus* is included in *M. hispidus* (Woods 1993).

Range and Habitat

The species is found in multistratal tropical forest from the lowlands to the foothills of the Andes, in northern Brazil, Peru, Ecuador, Colombia, and Venezuela. It prefers moist habitats and is arboreal, although it frequently forages on the ground (map 16.214).

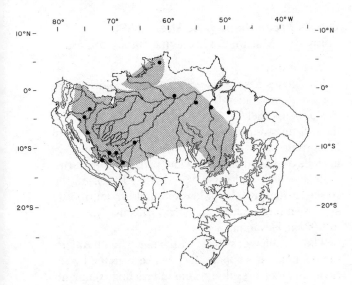

Map 16.214. Distribution of *Mesomys hispidus*.

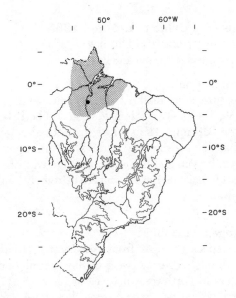

Map 16.215. Distribution of *Mesomys stimulax*.

Mesomys stimulax Thomas, 1911

Description

Measurements of one specimen were HB 158; T 122; HF 29. This form matches the description given for the genus but is slightly smaller than *M. hispidus*.

Range and Habitat

The species occurs in multistratal tropical evergreen forest in northeastern Brazil, extending to Suriname (map 16.215).

Genus *Trichomys* Troessart, 1880
Trichomys aperoides (Lund, 1839)

Description

Average measurements: HB 237; T 203; HF 47; E 25; Wt 339 g. This large, stocky rat is agouti greenish brown with more tan on the sides and a grayish white venter. The tail is long, well pencilled, strongly bicolored, and very well furred. The ears are relatively small. It has remarkably long vibrissae and long vibrissae between the ears and eyes. There are often white spots above the eyes (plate 13).

Distribution

Trichomys aperoides is found in southern Brazil south to Paraguay (Alho et al. 1987; Honacki, Kinman, and Koeppl 1982; UM) (map 16.216).

Life History

In the Brazilian caatinga there is some reproductive activity throughout the year, and two or three litters are produced per year; litters average 3.1 (range 1–6). Gestation lasts eighty-nine days. The young are born very precocial, weighing about 29 g, and start eating solid food on the first day. In captivity, lactation lasts

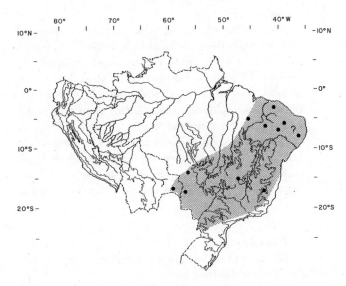

Map 16.216. Distribution of *Trichomys apereoides*.

fifty to sixty days. The age of first reproduction is between seven and nine months (Kleiman, Eisenberg, and Maliniak 1979; Roberts, Thompson, and Cranford 1988; Streilein 1982a,c).

Ecology

Trichomys aperoides is usually associated with rocky outcroppings, though a recent study in the Brazilian Pantanal found it abundant in patches of cerrado with no rocks. It is crepuscular and moderately good at conserving water. As in *Proechimys*, there is zone of fracture in the tail, which is easily detached from the body when grabbed (Alho et al. 1987; Myers 1982; Streilein 1982a,b).

Genus *Hoplomys* J. A. Allen, 1908
Hoplomys gymnurus (Thomas, 1897)
Thick-spined or Armored Rat

Description

Measurements of the type, an adult male, were HB 380; T 170; HF 53; female head and body length is about 320 mm. The sparsely haired tail is generally shorter than the head and body. The upperparts are dark brown, and the underparts are whitish. The spinescent hairs on the dorsum are extremely long, up to 33 mm.

Range and Habitat

The species ranges from east-central Honduras south through Panama, west of the central Andean cordillera to northwestern Ecuador (map 16.217). It prefers lowland multistratal tropical evergreen forest habitats.

Natural History

This nocturnal, terrestrial form constructs elaborate burrows (Buchanan and Howell 1965). Armored rats feed on seeds, fruits, and nuts and are known to cache food in their burrows. Like other burrowing rodents, they tend to use a separate chamber of the burrow system for defecation. Litter size averages 2.1 in Panama; the young are precocial (Tesh 1970b). In captivity males and females may tolerate each other, but males are highly intolerant of other adult males. A complicated series of vocalizations are employed during social encounters, but their precise function has not been determined.

Comment

Hoplomys may be at best a subgenus of *Proechimys* (see Patton and Reig 1989).

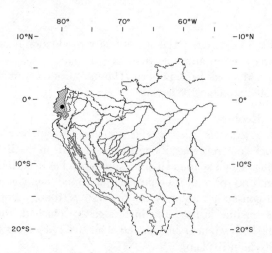

Map 16.217. Distribution of *Hoplomys gymnurus*.

Genus *Proechimys* J. A. Allen, 1899
Spiny Rat, Macangu, Casiragua, Rata Espinosa

Taxonomy

This genus has speciated widely, and some speciation events are evidently quite recent. Many of the species are superficially very similar in appearance, and their distinctiveness cannot be determined by careful measurement of adult specimens. Many characters must be evaluated, including foot pads, color patterns, skull structure, and karyotype (Patton 1987). All the species primarily inhabit lowland multistratal tropical evergreen forest, although some forms may range up to 1,000 m elevation.

The evolution of species within many South American mammals has apparently been accelerated by periodic isolations resulting from alternating continent-wide wet and dry cycles deriving from variations in the extent of polar ice caps (see chapter 19). A combination of geographic isolation and genetic drift can result in populations that are incapable of breeding and producing viable offspring when they subsequently come in contact. Such allopatric speciation events have been well documented (Mayr 1963). Through enormous efforts by Oswaldo Reig and his colleagues, the genus *Proechimys* in Venezuela has been studied from the standpoint of karyotypic variations. Six such variants were identified in Venezuela alone (Reig and Useche 1976; Reig et al. 1980).

Gardner and Emmons (1984), building on Patton and Gardner (1972) and Martin (1970), have provided us with a general review of this vexing genus and have also evaluated morphological characters, including the structure of the tympanic bulla. The septa dividing the bulla into compartments range from a few simple ones to many, apparently varying with how fossorial the species is. Emmons (1982) confirms the contention of Hershkovitz (1948) and Patton and Gardner (1972) that up to five species of *Proechimys* apparently can coexist in sympatry within Amazonian South America, but such sympatry is difficult to document for most of the area covered by this volume (Gardner and Emmons 1984). Guillotin and Ponge (1984) have developed evidence that two species are sympatric in French Guiana.

The basic contention of Gardner and Emmons (1984) is that the genus *Proechimys* falls into four groups of species. The species *P. semispinosus* is the most geographically defined, extending from Nicaragua through the Isthmus and then west of the Andes to Ecuador (Gardner 1983). However, isolated populations in central Colombia show affinities with the *seminspinosus* group. Gardner and Emmons (1984) further define a group of karyotypic variants in northeastern Colombia across northern Venezuela, which they refer to as the *guairae* group. This group, studied by Reig et al. (1980), is divisible into at least six variants based on karyotype alone. As with the *semispinosus* groups, outlying populations are found south of the Orinoco in south-central Venezuela. The third group is referred to as the *brevicauda* complex. It is primarily distributed in the Amazon basin and the Guianas but has a few pockets of apparent relatives in northeastern Colombia. Again this suggests range expansion and contraction paralleling global climate changes (Haffer 1974). The fourth group of species is predictably isolated in southeastern Brazil, a site of extensive endemism. This is the *"Trinomys"* group, here considered as a genus (see Lara, Patton, and da Silva 1996).

Patton (1987) recognizes nine species groups of the subgenus *Proechimys*. Although he agrees with Gardner and Emmons (1984) to separate the southeastern Brazilian spiny rats as a separate genus, *Trinomys*, and concurs with Gardner (1983) on the identity of *P. semispinosus*, he believes that the *brevicauda* group of Gardner and Emmons is too inclusive. Patton divided the *brevicauda* groups of Gardner and Emmons into a number of different groups. Patton presents evidence that three groups are monotypic: *decumanus, canicollis,* and *simonsi*. The other six groups are polytypic, but "the number of species in each remains unclear." He uses baculum, molars, and the palatal foramina as characters to define his groupings. Patton (1987) maps all groups based on museum specimens.

It should be noted that the karyotypic variants of *Proechimys* cannot easily be characterized by external features. Indeed, in the south (e.g., Peru) two or three species may coexist that are quite different in their habits but not easily distinguished by inspection in the hand. Since this volume is only a general guide to what the best experts have defined as true species, we follow Woods (1993) with respect to the species names, with modifications as noted by Gardner and Emmons (1984), Patton (1987), and Emmons and Feer (1997). New species are currently being described (da Silva 1995) (see also table 16.12).

Description
Head and body length ranges from 165 to 290 mm; tail, 121 to 242 mm; hind foot, 35 to 57 mm; ear, 17 to 29 mm. The dorsum is generally some shade of brown, contrasting sharply with a white or cream-colored venter. The tail is very sparsely haired. The dorsal pelage has spinescent hairs, but this is not nearly as pronounced as in some of the other genera (plate 13 and fig. 16.31).

Distribution
The genus is distributed in moist evergreen forest from Honduras south to Peru and central Brazil and east to the Guianas.

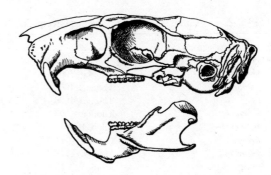

Figure 16.31. Skull of *Proechimys* sp. (after Ellerman 1940). Scale = 1.25X.

Natural History
Proechimys guyannensis (cayennensis) is nocturnal and lives in burrows and hollow logs. Usually only a single adult inhabits a burrow system (Guillotin 1982). In *P. semispinosus* the gestation period is sixty-four to sixty-five days, and the average litter size is 2.8, with a mode of three (Maliniak and Eisenberg 1971).

Proechimys emerges from its burrow at night to forage for seeds and fruits on the forest floor, and fungi are widely used as food. It caches food in its burrow or shelter system. Emmons (1982) reported that three species live in sympatry in Peru, but usually the species may be graded in size. Further work in Peru has demonstrated now that up to five species may be sympatric. Within a 4 km^2 area of lowland forest five species of *Proechimys* may be taken. Although microhabitat segregation is demonstrable, two or three species may be caught on a single sample line (Valqui 1995).

Where represented in the Amazon basin by up to five sympatric species, this genus may be important in dispersing spores of mycorrhizal fungi. Janos, Sahley, and Emmons (1995) report that 60% of *Proechimys* fecal samples contain spores of the genus *Glomus*. *Sclerocystis* spores may be present in 37.5% of the samples.

Proechimys amphichoricus Moojen, 1948

Description
Chromosome number: $2n = 26$; FN = 44.

Distribution
The species is distributed from extreme southern Venezuela and may extend to adjacent Brazil (not mapped).

Proechimys bolivianus Thomas, 1901

Description
Measurements: HB 245; T 195; HF 53; E 27. This species is lighter in color than *P. simonsi*, with spines

Table 16.12   Names Allocated to Species Groups of *Proechimys* (Excluding the subgenus *Trinomys*) by Patton 1987

The *guyannensis* group
- • *guyannensis*   (E. Geoffroy, 1803)
- ? *cherriei*   Thomas, 1901
-   *vacillator*   Thomas, 1903
-   *oris*   Thomas, 1904
-   *warreni*   Thomas, 1905
-   *boimensis*   Allen, 1916
-   *arescens*   Osgood, 1944
-   *riparem*   Moojen, 1948
-   *abrabupa*   Moojen, 1948

The *goeldii* group
- • *goeldii*   Thomas, 1905
-   *steerei*   Goldman, 1911
-   *kermiti*   Allen, 1915
-   *pachita*   Thomas, 1923
-   *hilda*   Thomas, 1924
-   *rattinus*   Thomas, 1926
-   *quadruplicatus*   Hershkovitz, 1948
-   *liminalus*   Moojen, 1948
- • *amphicharicus*   Moojen, 1948
-   *hyleae*   Moojen, 1948
-   *resiotes*   Moojen, 1948
-   *leioprimna*   Moojen, 1948

The *longicaudatus* group
- • *longicaudatus*   (Rengger, 1830)
-   *brevicauda*   (Gunther, 1877)
-   *boliviensis*   Thomas, 1901
-   *securus*   Thomas, 1902
-   *gularis*   Thomas, 1911
-   *leucomystax*   Osgood, 1944
-   *villacauda*   Moojen, 1948
-   *ribeiroi*   Moojen, 1948

The *simonsi* group
- • *simonsi*   Thomas, 1900
-   *hendeei*   Thomas, 1926
-   *rigrifulvus*   Osgood, 1944

The *cuvieri* group
- • *cuvieri*   Petter, 1978

The *trinitatus* group
-   *trinitatus*   (Allen and Chapman, 1893)
-   *chrysaeolus*   (Thomas, 1898)
-   *mincae*   (Allen, 1899)
-   *urichi*   (Allen, 1899)
- • *quairae*   Thomas, 1901
-   *ochraceus*   Osgood, 1912
-   *poliopus*   Osgood, 1914
-   *haplomyoides*   Tate, 1939
-   *magdalenae*   Hershkovitz, 1948

The *semispinosis* group
- • *semispinosus*   (Tomes, 1860)
-   *centralis*   (Thomas, 1896)
-   *rosa*   Thomas, 1900
-   *chiriquiensis*   Thomas, 1900
-   *panamensis*   Thomas, 1900
-   *burrus*   Bangs, 1901
-   *gorgonae*   Bangs, 1905
-   *calidor*   Thomas, 1911
-   *oconnelli*   Allen, 1913
-   *rubellus*   Hollister, 1914
-   *colombianus*   Thomas, 1914
-   *goldmani*   Bole, 1937
-   *ignotus*   Kellogg, 1946

The *canicollis* group
-   *canicollis*   Allen, 1899

The *decumanus* group
-   *decumanus*   (Thomas, 1899)

Note: Many names could be synonyms.
• = species having a good natural history record.

about 19 mm long. The dorsum is dark brown with a noticeably darker patch on the rump. The venter is white and well defined from the dorsal brown. The tail is faintly bicolored and thinly haired. See also the genus description.

Distribution

The type locality is Mapiri, upper Río Beni, northwestern Bolivia, at 1,000 m elevation. The species could be widespread on the upper Amazon (map 16.218).

Comment

Anderson (1997) includes *Proechimys bolivianus* in *P. brevicauda*.

Proechimys brevicauda (Gunther, 1877)

Description

Measurements: HB 235; T 148.8; HF 52.1; E 24.2. This medium-sized species has a reddish brown dor-

sum. The venter has white patches on a red background, and the tail is bicolored (Patton and Gardner 1972).
Chromosome number: $2n = 28$.

Distribution

The species occurs in eastern Amazonian Peru to western Brazil (map 16.218).

Comment

Proechimys brevicauda is often considered within *P. longicaudatus*.

Proechimys cuvieri Petter, 1978

Description

This rat is distributed from the Guianas south to the Amazon region. Guillotin and Ponge (1984) note that *Proechimys cuvieri* is distinct karyotypically and can be discriminated from *P. guyannensis* only by a multi-

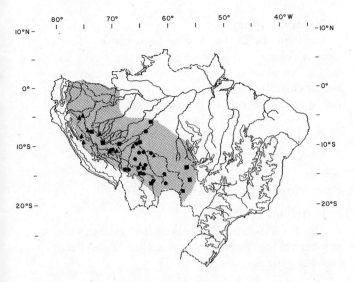

Map 16.218. Distribution of three species of *Proechimys*: circle, *P. bolivianus;* triangle, *P. brevicauda;* square, *P. longicaudatus* (the *P. longicaudatus* group in part).

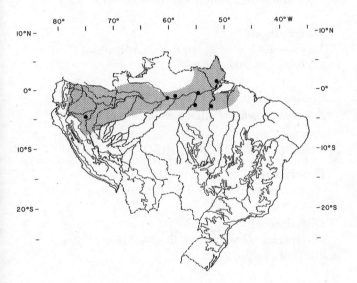

Map 16.219. Distribution of *Proechimys cuvieri*.

variate analysis of several measurements or by examining the shape of the incisiform foramina in the palate. Patton (1987) demonstrated how a variety of cranial and bacular characters could discrimate the species. It co-occurs with several species over its range.
Chromosome number: $2n = 28$.

Distribution
The range extends from the Guianas to northern Brazil (map 16.219).

Proechimys decumanus (Thomas, 1899)

Description
The baculum in this species is long, about 12 mm.

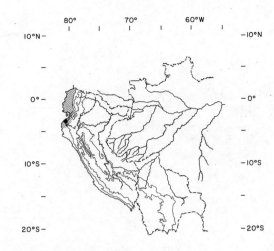

Map 16.220. Distribution of *Proechimys decumanus*.

Distribution
This rat is known from Ecuador south of the Bahía Guayquil and thence to northwestern Peru (map 16.220).

Comment
Patton (1987) considers this species to represent the *decumanus* group. Gardner and Emmons place it within the *brevicauda* group, but Patton separates it based on several characters including the baculum and characteristics of the palate. It can co-occur with *P. semispinosus.*

Proechimys goeldii Thomas, 1905

Description
This orangish brown spiny rat has relatively little spinescence. The venter is white, contrasting sharply with the orange brown dorsum. The tail is distinctly bicolored.

Distribution
The species is known from the Amazonian portion of Brazil (map 16.221).

Proechimys gularis Thomas, 1911

Description
The throat and venter are gray; *Proechimys gularis* resembles *P. brevicauda.*

Distribution
The species occurs in Amazonian Ecuador (not mapped).

Comment
Proechimys gularis may be a local variant of *P. brevicauda.*

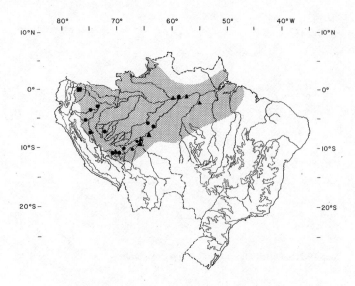

Map 16.221. Distribution of three species of *Proechimys:* circle, *P. steerei;* triangle, *P. goeldii;* square, *P. quadruplicatus* (the *P. goeldii* group in part).

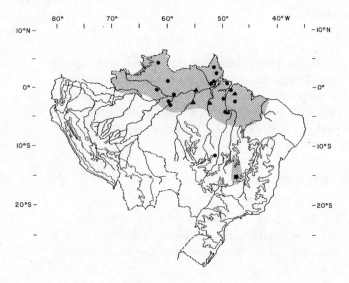

Map 16.222. Distribution of three species of *Proechimys:* circle, *P. guyannensis;* triangle, *P. oris;* square, *P. roberti* (the *P. guyannensis* group in part).

Proechimys guyannensis (E. Geoffroy, 1803)

Description

The chromosome number is $2n = 40$, FN = 54, but both Gardner and Emmons (1984) and Patton (1987) report great variation. Reig, Trainer, and Barros (1979) demonstrated clear karyotypic separation from *P. cuvieri.*

Distribution

This form is distributed from south-central Venezuela to the Guianas and thence to northeastern Brazil (map 16.222).

Natural History

Guillotin (1982) studied *P. guyannensis* in French Guiana. Home ranges average 0.84 for males and 0.31 ha for females. The species is highly frugivorous but eats insects in the dry season. The animals are nocturnal and use a complex burrow system.

Comment

Proechimys guyannensis includes *P. cayennensis.*

Proechimys haplomyoides (Tate, 1939)

Distribution

This form is found in southeastern Venezuela and may extend to adjacent portions of Guyana and Brazil (not mapped).

Proechimys hendeei Thomas, 1926

Description

The chromosome number is $2n = 32$, FN = 52.

Distribution

This species is distributed from northeastern Peru to southern Colombia (map 16.223)

Comment

Proechimys hendeei often includes *P. simonsi.*

Proechimys hilda (Thomas, 1924)

Description

Measurements: TL 245; T 99; HF 39; E 20.

Distribution

The species is known from the Río Mamoré, Beni, Bolivia (not mapped).

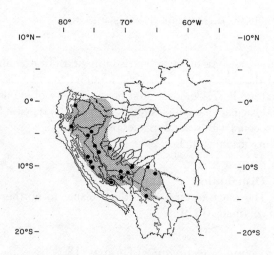

Map 16.223. Distribution of two species of *Proechimys:* closed circle, *P. simonsi;* open circle, *P. hendeei* (the *P. simonsi* group in part).

Comment

Proechimys hilda belongs to the *goeldii* group and is related to *P. steerëi*. Anderson (1997) outlines his reasons for applying the name.

Proechimys longicaudatus (Rengger, 1830)

Description

Average measurements: HB 222; T 152; HF 47; E 21. This large rat has glossy chestnut fur, not heavily bristled. It is more orange on the sides and has a sharply demarcated pure white belly and chin. The relatively short tail is bicolored, lightly haired, and easily broken off (pers. obs.).

Distribution

Proechimys longicaudatus ranges from lowland Bolivia through southwest Brazil and into Paraguay (map 16.218).

Proechimys oris Thomas, 1904

Description

A series of nine specimens from the upper Rio Xingu, Brazil, measured HB 207.3; T 145.7; HF 33; E 24.1. The mean weight of six specimens was 214 g.

Distribution

The species is distributed in eastern Brazil, with isolated populations in gallery forests near Brasília (map 16.222).

Comment

Proechimys oris is probably a synonym of *P. roberti*.

Proechimys quadruplicatus Hershkovitz, 1948

Description

Measurements: HB 224–34; T 152–71; HF 46–51; E 22–26. The upperparts are reddish brown mixed with black, the dorsal midline is black, and the venter is white. The upper surfaces of the feet are brown. The tail is almost naked.

Distribution

The type locality is listed as Llunchi Island in the Río Napo, about 18 km below the mouth of the Río Coca (map 16.221). The species is found in eastern Ecuador and northern Peru.

Comment

Proechimys quadruplicatus is possibly a junior synonym of *P. amphichoricus*. It is found with *P. gularis*, *P. simonsi*, and *P. cuvieri*.

Proechimys semispinosus (Tomes, 1860)

Description

This is a rather large rat with head and body ranging from 261 to 254 mm and the weight from 134 to 438 g.

The dorsum is agouti brown, contrasting sharply with a white venter. The spinescent hairs of the dorsum are conspicuous and most pronounced in the middorsal region. The feet are dark. The tail is easily broken, and "amputation" between the fifth and sixth tail vertebrae is frequent.

Chromosome number: $2n = 30$, FN $= 50$–54.

Distribution

This species ranges from Honduras through Panama to the west of the western cordillera of the Andes, southward to Peru (map 16.224).

Natural History

Fleming (1971) studied *P. semispinosus* in Panama. Median densities ranged form 3.0 to 8.5 per hectare. They breed throughout the year and can have three or four litters a year with a mean litter size of about 2.7. The highest densities (9.7 ha) were recorded by Gliwicz (1973) on small offshore islands in Lake Gatun. Maliniak and Eisenberg (1971) studied this species in captivity, where it was nocturnal and cached food in its nest box. Adults are intolerant of one another, but a male and a female could be housed together in a large cage. During courtship a variety of vocalizations are employed. Males utter a whimper when approaching the female, who responds with a twitter. Both animals may continue calling during copulation. In captivity, gestation was from sixty-three to sixty-five days, and litter sizes ranged from 1 to 5, with an average of 2.8. The young are born fully haired and with their eyes open. Adult proportions are reached at eighty-five to ninety days of age. Earliest age of conception by a captive-born female was one hundred days. Bowdre (1986), comparing the reproductive capacity of two Panamanian populations (Atlantic coast versus Pacific coast), demonstrated significant differences in litter size and age at sexual maturity correlated with differences in plant productivity.

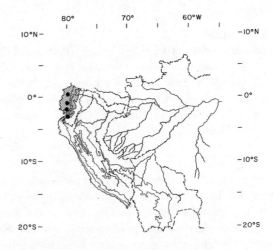

Map 16.224. Distribution of *Proechimys semispinosus*.

Proechimys simonsi Thomas, 1900

Description

Head and body length ranges from 150 to 250 mm (including a subadult with a mean of 193.7 mm). The tail is about 84% of the head and body length. The spines are soft and sparse. The dorsum is a light reddish brown, contrasting with a pure white venter. The tops of the feet are whitish.
Chromosome number: $2n = 32$.

Distribution

The species occurs in Amazonian Ecuador, Peru, and Bolivia and extends to western Brazil (map 16.223).

Comment

Proechimys hendeei is a junior synonym of *P. simonsi.*

Proechimys steerei Goldman, 1911

Description

Measurements: HB 209–72 (mean = 235). The tail is moderately long—about 70% of the head and body length. The spines of this large rat are sparse and soft. The dorsum is red brown, and the venter is snow white. The tops of the hind feet are bicolored, gray and white.
Chromosome number: $2n = 24$.

Distribution

The species is found in Amazonian Peru, Ecuador, and Brazil (map 16.221).

Comment

Proechimys steerei is considered a synonym of *P. goeldii.*

Proechimys (Trinomys) yonenagae da Rocha, 1995

Description

Measurements: HB 159.8; T 190.3; HF 43.8; E 24.0 (*n* = 24). Assigned to the subgenus *Trinomys*, this small species exhibits some unique features. It has a relatively long tail and hind foot. The tympanic bullae of the skull are inflated. The tip of the tail has a well developed tuft.

Range and Habitat

A very localized distribution has been established on the left bank of the Rio São Francisco in the state of Bahia, Brazil (not mapped). This is an area of sand dunes within the caatinga climatic zone. These rodents construct extensive burrows in the sand. This species occurs with several endemic lizards and exhibits some morphological convergences with desert-adapted rodents (Rocha 1995).

Genus *Trinomys* Thomas, 1921

Description

The skull is characterized by the absence of dorsal ridges; the postorbital bar involves only the squamosal; and the posterior edges of the palatal foramina are slightly anterior to or on a plane with the first molars. The tympanic bullae are of average size, but when sectioned they exhibit numerous interior subdivisions deriving from thin-walled bony septa. The tail ranges from 86% to 106% of the head and body length. The dorsal pelage is characterized by a superficial set of long, stiff guard hairs that may be spinescent or soft. The dorsum is reddish brown (Pessôa and dos Reis 1993; Moojen 1948, 1952a; Gardner and Emmons 1984). A key to the species is offered in Pessôa and dos Reis (1993).

Distribution

The genus is confined to the eastern portion of Brazil south of the Amazon River. Moojen (1948) includes excellent distribution maps.

Comment

Trinomys is considered a subgenus of *Proechimys* by Moojen (1948) and Woods (1993) but is confirmed as a genus by Lara, Patton, and da Silva (1996).

Trinomys albispinus (I. Geoffroy, 1838)

Description

Measurements: HB 175–85; T 125–62; HF 37–40. The skull is small, less than 49 mm, and the incisive foramina are long. The dorsum is a reddish light brown; the venter is white. The pale guard hairs average 1.2 mm in length and are white at the base. The tail does not have a white tip (Moojen 1952a).

Distribution

The species is found on the coast and offshore islands of Bahia state, Brazil (Cabrera 1960). Reis and Pessôa (1995) review the Recent distribution and name a new subspecies (map 16.225).

Trinomys dimidiatus (Günter, 1877)

Description

Average measurements: HB 199; T 170; HF 46 (males). The tail is shorter than the head and body. The guard hairs are barely spinescent. The dorsum is reddish buff, and the venter and inner portions of the limbs are white (Pessôa and dos Reis 1991b, 1993).

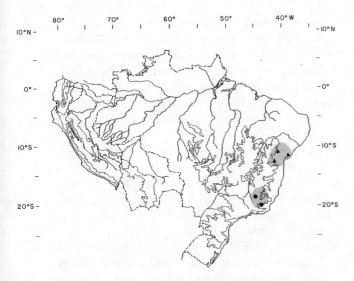

Map 16.225. Distribution of two species of *Trinomys:* circle, *T. setosus;* triangle, *T. albispinus.*

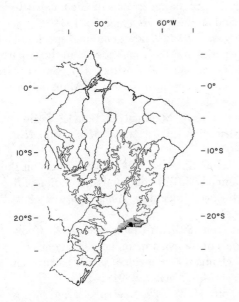

Map 16.226. Distribution of *Trinomys dimidiatus.*

Distribution

The current distribution is confined to coastal eastern Rio de Janeiro. The species occurs from sea level to 1,000 m (map 16.226).

Natural History

The breeding season extends from September to November and from March to May. A female may have two litters per year. Litter size ranges from one to five, but the mode is two. These rodents are adapted to climax forest; they usually shelter in crevices associated with rocks. This species can co-occur with *Trinomys iheringi. T. dimidiatus* is strongly diurnal in captivity (Pessôa and dos Reis 1993).

Trinomys iheringi Thomas, 1911

Description

Adult males weigh an average of 217.4 g, and adult females are about 27 g lighter. *Trinomys iheringi* is superficially similar to *T. dimidiatus,* but the guard hairs of *T. iheringi* are wider and rather stiff. The incisive foramina are elongate. The bacular morphology of the two species is distinct (Pessôa and dos Reis 1990, 1991a, 1992).

Distribution

Although the species is apparently more broadly tolerant of xeric habitats, it can be sympatric with *T. dimidiatus.* Its range extends from the coastal portions of eastern Espírito Santo to eastern São Paulo state, Brazil (Cabrera 1960; Moojen 1948; Pessôa and dos Reis 1996) (map 16.227).

Natural History

Females become sexually mature at 240 days of age and can bear four litters per year. Maximum longevity in the field may reach thirty months. The home range of adult males averages 1.37 ha; females have ranges of 0.86 ha (Bergallo 1995).

Trinomys myosurus (Lichtenstein, 1818)

Comment

Trinomys myosurus is a possible synonym of *T. setosus* (see below).

Trinomys setosus (Thomas, 1921)

Description

Measurements: HB 190; T 190; HF 47. This rat is similar in size and color to *T. albispinus,* but the tail has a white tip (Moojen 1952a).

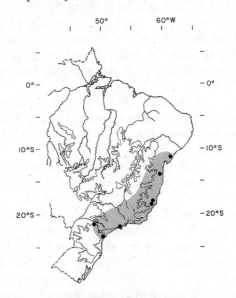

Map 16.227. Distribution of *Trinomys iheringi.*

Distribution

The species is apparently able to occupy forest to the west of the coastal mountain ranges in Bahia and Minas Gerais states, Brazil (map 16.225).

References

• References used in preparing distribution maps

• Aagaard, E. M. 1982. Ecological distribution of mammals in the cloud forests and paramos of the Andes, Mérida, Venezuela. Ph.D. diss., Colorado State University.

Abravaya, J. P., and J. O. Matson. 1975. Notes on a Brazilian mouse, *Blarinomys breviceps* (Winge). *Los Angeles Co. Mus. Nat. Hist. Contrib. Sci.* 270:1–8.

Aguilera, M. 1985. Growth and reproduction in *Zygodontomys microtinus* (Rodentia-Cricetidae) from Venezuela in a laboratory colony. *Mammalia* 49: 58–67.

Aguilera, M., and A. Perez-Zapata. 1989. Cariología de *Holochilus venezuelae* (Rodentia: Cricetidae). *Acta Cient. Venez.* 40:198–207.

Alho, C. J. R. 1982. Brazilian rodents: Their habitats and habits. In *Mammalian biology in South America*, ed. M. A. Mares and H. H. Genoways, 143–66. Pymatuning Symposia in Ecology 6. Special Publication Series. Pittsburgh: Pymatuning Laboratory of Ecolology, University of Pittsburgh.

Alho, C. J. R., Z. M. Campos, and H. C. Gonçalves. 1989. Ecology, social behavior and management of the capybara (*Hydrochaeris hydrochaeris*) in the Pantanal of Brazil. In *Advances in Neotropical mammalogy,* ed. K. H. Redford and J. F. Eisenberg, 163–94. Gainesville, Fla.: Sandhill Crane Press.

Alho, C. J. R., T. E. Lacher Jr., Z. M. S. Campos, and H. C. Gonçalves. 1987. Mamíferos da Fazenda Nhumirim, sub-região de Nhecolândia, Pantanal do Mato Grosso do Sul. 1. Levantamento preliminar de espécies. *Rev. Brasil. Zool., São Paulo* 4 (2): 151–64.

Alho, C. J. R., and M. J. de Souza. 1982. Home range and use of space in *Zygodontomys lasiurus* (Cricetidae, Rodentia) in the cerrado of central Brazil. *Ann. Carnegie Mus.* 51:127–32.

• Allen, J. A. 1901. New South American Muridae and a new *Metachirus*. *Bull. Amer. Mus. Nat. Hist.* 14: 405–12.

• ———. 1903. Descriptions of new rodents from southern Patagonia, with a note on the genus *Euneomys coues,* and an addendum to article IV, on Siberian mammals. *Bull. Amer. Mus. Nat. Hist.* 19:185–96.

• ———. 1912. Mammals from western Colombia. *Bull. Amer. Mus. Nat. Hist.* 31:71–95.

• ———. 1915. Review of the South American Sciuridae. *Bull. Amer. Mus. Nat. Hist.* 34 (8): 147–309.

• ———. 1916a. Mammals collected on the Roosevelt Brazilian expedition, with field notes by Leo E. Miller. *Bull. Amer. Mus. Nat. Hist.* 34:559–610.

• ———. 1916b. New mammals collected on the Roosevelt Brazilian expedition. *Bull. Amer. Mus. Nat. Hist.* 35:523–30.

Altman, C. A. 1991. On the copulatory behavior of *Ctenomys. Mammalia* 55 (3):440–42.

Altuna, C. A. 1983. Sobre la estructura de las construcciones de *Ctenomys pearsoni* Lessa y Langguth, 1983 (Rodentia, Octodontidae). *Res. Com. J. C. Nat., Montevideo* 3:70–72.

Altuna, C. A., G. Francescoli, and G. Izquierdo. 1991. Copulatory patterns of *Ctenomys pearsoni* (Rodentia, Octodontidae) from Balneario Solis, Uruguay. *Mammalia* 55 (3): 440–42.

Altuna, C. A., and E. P. Lessa. 1985. Penial morphology in Uruguayan species of *Ctenomys* (Rodentia: Octodontidae). *J. Mammal.* 66:483–88.

Anderson, S. 1992. A new genus and species of a large phyllotine rodent from Tapecua in the department of Tarija, Bolivia. Abstract 177, Seventy-second meeting, American Society of Mammalogists, Salt Lake City, Utah.

• ———. 1997. Mammals of Bolivia: Taxonomy and distribution. *Bull. Amer. Mus. Nat. Hist.* 231: 1–652.

• Anderson, S., and N. Olds. 1989. Notes on Bolivian mammals. 5. Taxonomy and distribution of *Bolomys* (Muridae, Rodentia). *Amer. Mus. Novitat.* 2935: 1–22.

• Anderson, S., T. Y. Yates, and J. A. Cook. 1987. Notes on Bolivian mammals: The genus *Ctenomys* (Rodentia, Ctenomyidae) in the eastern lowlands. *Amer. Mus. Novitat.* 2891:1–20.

• Anthony, H. E. 1925. New species and subspecies of *Thomasomys. Amer. Mus. Novitat.* 178:1–4.

• ———. 1929. Two new genera of rodents from South America. *Amer. Mus. Novitat.* 383:1–6.

Apfelbaum, L. I., and O. A. Reig. 1989. Allozyme genetic distances and evolutionary relationships in species of akodontine rodents (Cricetidae: Sigmodontinae). *Biol. J. Linn. Soc.* 38:257–80.

• Avila-Pires, F. D. de. 1972. A new subspecies of *Kunsia fronto* (Winge, 1888) from Brazil (Rodentia, Cricetidae). *Rev. Brasil. Biol.* 32 (3): 419–22.

• Avila-Pires, F. D. de, and M. R. C. Wutke. 1981. Taxonomia e evolução de *Clyomys* Thomas, 1916 (Rodentia, Echimyidae). *Rev. Brasil. Biol.* 41 (3): 529–34.

Baker, R. H., and M. K. Petersen. 1965. Notes on a climbing rat, *Tylomys* from Oaxaca, Mexico. *J. Mammal.* 46:694–95.

• Barlow, J. C. 1965. Land mammals from Uruguay: Ecology and zoogeography. Ph.D. diss., University of Kansas.

———. 1969. Observations on the biology of rodents in Uruguay. *Life Sci. Contrib., Roy. Ontario Mus.* 75:1–59.

Barnett, A. A. 1997. The ecology and natural history of a fishing mouse *Chibchanomys* spec. nov. from the Andes of southern Ecuador. *Z. Säugetierk.* 62:43–52.

• Barquez, R. M. 1976. Nuevo registro de distribución de *Oxymycterus paramensis* (Mammalia, Rodentia, Cricetidae). *Neotrópica* 22 (68): 115.

• ———. 1983. La distribución de *Neotomys ebriosus* Thomas en la Argentina y su presencia en la provincia de San Juan (Mammalia, Rodentia, Cricetidae). *Hist. Nat.* 3 (22): 189–91.

• Barquez, R. M., D. F. Williams, M. A. Mares, and H. H. Genoways. 1980. Karyology and morphometrics of three species of *Akodon* (Mammalia: Muridae) from northwestern Argentina. *Ann. Carnegie Mus.* 49:279–403.

Barros, M. A., R. C. Liascovich, L. Gonzalez, M. Lizarralde, and O. A. Reig. 1990. Banding pattern comparisons between *Akodon iniscatus* and *A. puer. Z. Säugetierk.* 55:115–27.

Barros, M. A., and O. A. Reig. 1979. Doble polimorfismo robertsoniano en *Microxus bogotensis* (Rodentia, Cricetidae) del Páramo de Mucubaji (Mérida, Venezuela). *Acta Cient. Venez.* 30 (suppl. 1): 96.

Barros, M. A., O. A. Reig, and A. Perez-Zapata. 1992. Cytogenetics and karyosystematics of South American oryzomyine rodents (Cricetidae, Sigmodontinae). 4. Karyotypes of Venezuelan, Trinidadian, and Argentinian water rats of the genus *Nectomys. Cytogenet. Cell Genet.* 59:34–38.

Baskin, J. A. 1978. *Bensonomys, Calomys,* and the origin of the phyllotine group of Neotropical cricetines (Rodentia: Cricetidae). *J. Mammal.* 59:125–35.

Bergallo, H. G. 1995. Comparative life history characteristics of two species of rats: *Proechimys iheringi* and *Oryzomys intermedius* in an Atlantic rainforest of Brazil. *Mammalia* 59:51–64.

Bianchi, N. O., S. Merani, and M. Lizarralde. 1979. Cytogenetics of the South American *Akodon* rodents (Cricetidae). 6. Polymorphism in *Akodon dolores* (Thomas). *Genetica* 50 (2): 99–104.

Bianchi, N. O., O. A. Reig, C. J. Molina, and G. N. Dulout. 1971. Cytogenetics of the South American akodont rodents (Cricetidae), part 1. *Evolution* 25:724–36.

• Bishop, I. R. 1974. An annotated list of caviomorph rodents collected in northeastern Mato Grasso, Brazil. *Mammalia* 38 (3): 489–502.

• Boher B., S., J. Naveuva S., and L. Escobar M. 1988. First record of *Dinomys branickii* for Venezuela. *J. Mammal.* 69:433.

Borchert, M., and R. L. Hansen. 1983. Effects of flooding and wildfire on valley side wet campo residents in central Brazil. *Rev. Brasil. Biol.* 43 (3): 229–40.

• Borrero-H., J. L., and J. Hernandez-Camacho. 1957. Informe preliminar sobre aves y mamíferos de Santander, Colombia. *Ann. Soc. Biol. Bogotá* 7 (5): 197–230.

Boursot, P., J. C. Auffray, J. Britton-Davidian, and F. Bonhomme. 1993. The evolution of house mice. *Ann. Rev. Ecol. Syst.,* 24:119–52.

Bowdre, L. P. 1986. Comparative life history of *Proechimys semispinosus* in two contrasting environments in the republic of Panama. Ph.D. diss., Michigan State University, East Lansing.

Branch, L. C. 1993a. Social organization and mating system of the plains viscacha (*Lagostomus maximus*). *J. Zool.* (London) 229:473–91.

———. 1993b. Seasonal patterns of activity and body mass in the plains viscacha, *Lagostomus maximus* (family Chinchillidae). *Can. J. Zool.* 71:1041–45.

Branch, L. C., D. Villareal, and G. S. Fowler. 1994. Factors influencing population dynamics of the plains viscacha (*Lagostomus maximus*, Mammalia, Chinchillidae) in scrub habitat of central Argentina. *J. Zool.* (London) 232:383–95.

Braun, J. K. 1993. *Systematic relationships of the tribe Phyllotini (Muridae: Sigmodontinae) of South America.* Special Publications. Norman: Oklahoma Museum of Natural History.

Brownell, E. 1983. DNA/DNA hybridization studies of muroid rodents: Symmetry and rates of molecular evolution. *Evolution* 37:1034–51.

Buchanan. O. M., and T. R. Howell. 1965. Observations on the natural history of the thick spined rat, *Hoplomys gymnuras* in Nicaragua. *Ann. Mag. Nat. Hist.,* ser. 13, 8:549–59.

Cabrera, A. 1953. *Los roedores argentinos de la familia Caviidae.* Publication 6. Buenos Aires: Facultad de Agronomía y Veterinaria, Escuela de Veterinaria.

• ———. 1960. *Catálogo de los mamíferos de América del Sur.* Vol. 2. Buenos Aires: Museo Argentino de Ciencias Naturales "Bernardino Rivadavia."

Calhoun, J. B. 1962. *The ecology and behavior of the Norway rat.* Washington, D.C.: National Institutes of Health.

Camin, S. n.d. Mating behaviour in *Ctenomys mendocinus. J. Zool.* (London). In press.

Camin, S., I. Modery, and V. Roig. 1995. The burrowing behavior of *Ctenomys mendocinus. Mammalia* 59:9–18.

Cant, J. G. H. 1977. A census of the agouti (*Dasyprocta punctata*) in seasonally dry forest at Tikal, Guatemala, with some comments on strip censusing. *J. Mammal.* 58:688–90.

Carleton, M. D. 1973. A survey of gross stomach morphology in New World Cricetinae Rodentia, Muroidea, with comment on functional interpretations. *Misc. Pub. Mus. Zool. Univ. Mich.* 146:1–43.

———. 1980. Phylogenetic relationships in neotomine-peromyscine rodents (Muroidea) and a re-appraisal of the dichotomy with New World Cricetinae. *Misc. Publ. Mus. Zool. Univ. Mich.* 157:1–146.

Carleton, M. D., and G. G. Musser. 1984. Muroid rodents. In *Orders and families of Recent mammals of the world,* ed. S. Anderson and J. Knox Jones Jr., 289–379. New York: John Wiley.

• ———. 1989. Systematic studies of oryzomyine rodents: A synopsis of *Microryzomys. Bull. Amer. Mus. Nat. Hist.* 191:1–83.

Cartaya, E., and M. Aguilera. 1985. Estudio de la comunidad de roedores plaga en un cultivo de arroz. *Acta Cient. Venez.* 36:250–57.

Catzeflis, F. M., A. W. Dickerman, J. Michaux, and J. A. W. Kirsch. 1993. DNA hybridization and rodent phylogeny. In *Mammal phylogeny: Placentals,* ed. F. Szalay, M. J. Novacek, and M. C. McKenna, 159–71. New York: Springer-Verlag.

Charles-Dominique, P., M. Atramentowicz, M. Charles-Dominique, H. Gérard, A. Hladik, C. M. Hladik, and M. F. Prévost. 1981. Les mammifères frugivores arboricoles nocturnes d'une forêt guyanaise: Interrelations plantes-animaux. *Rev. Ecol. (Terre et Vie)* 35:342–435.

Chiarello, A. G., M. Passamani, and M. Zoreta. 1997. Field observations on the thin-spined porcupine, *Chaetomys subspinosus* (Rodentia; Echimyidae). *Mammalia* 61:29–36.

Chitty, D., and H. N. Southern, eds. 1954. *Control of rats and mice.* 3 vols. Oxford: Clarendon Press.

Collett, S. F. 1981. Population characteristics of *Agouti paca* (Rodentia) in Colombia. *Publ. Mus., Michigan State Univ., Biol. Ser.* 5:487–601.

Collins, L. R., and J. F. Eisenberg. 1972. Notes on the behavior and breeding of pacaranas. *Int. Zoo Yearb.* 12:108–14.

Concepcion, J. L., and J. Molinari. 1991. *Sphiggurus vestitus pruinosus* (Mammalia, Rodentia, Erethizontidae): The karyotype and its phylogenetic implications. *Stud. Neotrop. Fauna Environ.* 26:237–41.

Contreras, J. R. 1972a. El home range en una población de *Oryzomys longicaudatus* Philippi (Landbeck) (Rodentia, Cricetidae). *Physis* 31 (83): 353–61.

• ———. 1972b. Nuevos datos acerca de la distribución de algunos roedores en las provincias de Buenos Aires, La Pampa, Entre Ríos, Santa Fe y Chaco. *Neotrópica* 18 (55): 27–30.

———. 1980. Sobre el limite occidental de la distribución geográfica del cuis grande *Cavia aperea pamparum* en la Argentina. *Hist. Nat.* 1 (11): 73–74.

• ———. 1982a. Mamíferos de Corrientes. 1. Nota preliminar sobre la distribución de algunas especies. *Hist. Nat.* 2 (10): 71–72.

• ———. 1982b. *Graomys griseoflavus* (Waterhouse, 1837) en la provincia del Chaco, república Argentina (Rodentia, Cricetidae). *Hist. Nat.* 2 (2): 252.

• Contreras, J. R., and L. M. Berry. 1983. Notas acerca de los roedores del género *Oligoryzomys* de la provincia del Chaco, república Argentina (Rodentia, Cricetidae). *Hist. Nat.* 3 (15): 145–48.

Contreras, J. R., and E. R. Justo. 1974. Aportes a la mastozoología pampeana. 1. Nuevas localidades para roedores Cricetidae (Mammalia, Rodentia). *Neotrópica* 20 (62): 91–96.

Contreras, J. R., and M. I. Rosi. 1980a. Comportamiento territorial y fidelidad al hábitat en una población de roedores del centro de la provincia de Mendoza. *Ecología* 5:17–29.

———. 1980b. El ratón de campo *Calomys musculinus cordovensis* (Thomas) en la provincia de Mendoza. 1. Consideraciones taxonómicas. *Hist. Nat.* 1 (5): 17–28.

———. 1980c. Una nueva subespecie del ratón colilargo para la provincia de Mendoza: *Oligoryzomys flavescens occidentalis* (Mammalia, Rodentia, Cricetidae). *Hist. Nat.* 1 (22): 157–60.

———. 1980d. Acerca de la presencia en la provincia de Mendoza del ratón de campo *Akodon molinae* Contreras, 1968 (Rodentia: Cricetidae). *Hist. Nat.* 1 (26): 181–84.

———. 1981a. Notas sobre los Akodontini argentinos (Rodentia, Cricetidae). 1. *Abrothrix longipilis moerens* Thomas, 1919, en el Parque Nacional Nahuel Huapi. *Hist. Nat.* 1 (30): 209–12.

———. 1981b. Notas sobre los Akodontini argentinos (Rodentia, Cricetidae). 2. *Akodon andinus andinus* (Philippi, 1868) en la provincia de Mendoza. *Hist. Nat.* 1 (832): 233–36.

Contreras, L. C., J. C. Torres-Mura, and J. L. Yáñez. 1987. Biogeography of octodontid rodents: An eco-evolutionary hypothesis. *Fieldiana; Zool.,* n.s., 39:401–11.

• Cook, J. A., S. Anderson, and T. L. Yates. 1990. Notes on Bolivian mammals. 6. The genus *Ctenomys* (Rodentia, Ctenomyidae) in the highlands. *Amer. Mus. Novitat.* 2980:1–27.

• Cook, J. A., and T. L. Yates. 1994. Systematic relationships of the Bolivian tuco-tucos genus *Ctenomys. J. Mammal.* 75 (3): 583–99.

• Crespo, J. A. 1950. Nota sobre mamíferos de Misiones nuevos para Argentina. *Comun. Inst. Nac. Invest. Cienc. Nat., Mus. Argent. Cienc. Nat. "Bernardino Rivadavid," Zool.* 1 (14): 1–14.

• ———. 1966. Ecología de una comunidad de roedores silvestres en el Partido de Rojas, provincia de Buenos Aires. *Rev. Mus. Argent. Cienc. Nat. "Bernardino Rivadavia," Ecol.* 1 (3): 79–134.

———. 1974a. Comentarios sobre nuevas localidades para mamíferos de Argentina y de Bolivia. *Rev. Mus. Argent. Cienc. Nat. "Bernardino Rivadavia," Zool.* 11 (1): 1–31.

———. 1974b. Observaciones sobre la reproducción de la nutria en estado silvestre. *Primer Congreso Argentino de Producción Nutriera,* fasc. 1:60–73.

———. 1982a. Introducción a la ecología de los mamíferos del Parque Nacional el Palmar, Entre Ríos. *Anal. Parques Nac.* 15:1–33.

———. 1982b. Ecología de la comunidad de mamíferos del Parque Nacional Iguazú, Misiones. *Rev. Mus. Argent. Cienc. Nat. "Bernardino Rivadavia," Ecol.* 3 (2): 45–162.

Crespo, J. A., M. S. Sabattini, M. J. Piantanida, and G. de Villafañe. 1970. Estudios ecológicos sobre roedores silvestres, observaciones sobre densidad, reproducción y estructura de comunidades de roedores silvestres en el sur de Córdoba. Buenos Aires: Minsterio de Bienestar Social.

Cueto, V. R., M. Cagnoni, and M. J. Piantanida. 1995. Habitat use of *Scapteromys tumidus* (Rodentia: Cricetidae) in the delta of the Paraná River, Argentina. *Mammalia* 59:25–34.

Daciuk, J. 1974. Notas faunísticas y bioecológicas de Península Valdés y Patagonia. 12. Mamíferos colectados y observados en la Peninsula Valdés y zona litoral de los Golfos San José y Nuevo (provincia de Chubut, república Argentina). *Physis,* sec. C, 33 (86): 23–39.

Dalby, P. L. 1975. Biology of pampa rodents, Balcarce area, Argentina. *Publ. Mus., Michigan State Univ., Biol. Ser.* 5 (3): 153–271.

Dávila, J., E. López, G. Mamani, and P. Jiménez. 1982. Consideraciones ecológicas de algunas poblaciones de vizcachas en Arequipa-Perú. In *Zoología neotropical,* vol. 2, ed. P. J. Salinas, 949–58. Mérida, Venezuela: Congreso Latinamericano de Zoología.

Davis, D. E. 1947. Notes on the life histories of some Brazilian mammals. *Bol. Mus. Nac., Zool.* 76:1–8.

Dawson, G. A. 1973. The functional significance of nest building by a Neotropical rodent (*Sigmodon hispidus*). *Amer. Midl. Nat.* 89:503–9.

• De Santis, L. J. M., and E. R. Justo. 1980. *Akodon (Abrothrix) mansoensis* sp. nov. un nuevo "ratón lanoso" de la provincia de Río Negro, Argentina (Rodentia, Cricetidae). *Neotrópica* 26 (75): 121–27.

Dietz, J. M. 1983. Notes on the natural history of some small mammals in central Brazil. *J. Mammal.* 64:521–23.

Dorst, J. 1971. Nouvelles recherches sur l'écologie des rongeurs des hauts plateaux péruviens. *Mammalia* 35:515–47.

———. 1972. Morphologie de l'estomac et régime alimentaire de quelques rongeurs des hautes Andes du Pérou. *Mammalia* 36:647–56.

Dubost, G. 1988. Ecology and social life of the red acouchy *Myoprocta exilis,* comparison with the orange-rumped agouti *Dasyprocta leporina. J. Zool.* (London) 214:107–23.

Eisenberg, J. F. 1963. The behavior of heteromyid rodents. *Univ. Calif. Publ. Zool.* 69:1–100.

———. 1974. The function and motivational basis of hystricomorph vocalizations. In *The biology of hystricomorph rodents,* ed. I. W. Rowlands and B. J. Weir, 211–44. London: Academic Press.

———. 1984. New World rats and mice. In *The encyclopaedia of mammals,* ed. D. Macdonald, 640–49. New York: Facts on File.

• ———. 1989. *Mammals of the Neotropics.* Vol. 1. *The northern Neotropics: Panama, Colombia, Venezuela, Guyana, Suriname, French Guiana.* Chicago: University of Chicago Press.

Eisenberg, J. F., and D. G. Kleiman. 1977. Communication in lagomorphs and rodents. In *How animals communicate,* ed. T. Sebeok, 634–54. Bloomington: Indiana University Press.

Ellerman, J. R. 1940. *The families and genera of living rodents.* Vol. 1. *Rodents other than Muridae.* London: British Museum of Natural History.

———. 1941. *The families and genera of living rodents.* Vol. 2. *Family Muridae.* London: British Museum of Natural History.

Emmons, L. 1981. Morphological, ecological, and behavioral adaptations for arboreal browsing in *Dactylomys dactylinus* (Rodentia, Echimyidae). *J. Mammal.* 62:183–89.

———. 1982. Ecology of *Proechimys* in southeasthern Peru. *J. Trop. Ecol.* 23:280–90.

———. 1988. Replacement name for a genus of South American rodent (Echimyidae). *J. Mammal.* 69:421.

———. 1993. On the identity of *Echimys didelphoides* Desmarest, 1817. *Proc. Biol. Soc. Wash.* 106:1–4.

• Emmons, L. H., and F. Feer. 1990. *Neotropical rainforest mammals: A field guide.* Chicago: University of Chicago Press.

———. 1997. *Neotropical rainforest mammals: A field guide.* 2d ed. Chicago: University of Chicago Press.

• Ernest, K. A. 1986. *Nectomys squamipes. Mammal. Species* 265:1–5.

Ernest, K. A., and M. A. Mares. 1986. Ecology of *Nectomys squamipes,* the Neotropical water rat in central Brazil. *J. Zool.* (London), ser. A, 210:599–612.

Everard, C. O. R., and E. S. Tikasingh. 1973. Ecology of the rodents *Proechimys guyannensis trinitatis* and *Oryzomys capito velutinus* on Trinidad. *J. Mammal.* 54:875–86.

Ewer, R. F. 1971. The biology and behavior of a free-living population of black rats (*Rattus rattus*). *Anim. Behav. Monogr.* 4 (3): 127–74.

Fernández, R. G. 1965. Tuco-tuco—um roedor nocivo. *Rev. Fac. Agron. Vet. Pôrto Alegre* 7:253–57.

Fleming, T. H. 1971. Population ecology of three species of Neotropical rodents. *Misc. Publ. Mus. Zool. Univ. Michigan* 143:1–77.

Fonollat, A. M. P. de. 1984. Cricetidos de la provincia de Tucumán (Argentina). *Acta Zool. Lilloana* 37 (2): 219–25.

• Fonollat, A. M. P. de, and M. S. de Décima. 1979. Comportamiento en cautividad de *Oryzomys longicaudatus* (Bennett, 1832) y *Akodon varius* (Thomas, 1902) (Rodentia-Chordata). *Acta Zool. Lilloana* 33 (2): 91–94.

Fonseca, G. A. B. da. 1988. Patterns of small mammal species diversity in the Brazilian Atlantic forest. Ph.D. diss., University of Florida, Gainesville.

Fonseca, G. A. B. da, and M. C. M. Kierulff. 1988. Biology and natural history of Brazilian Atlantic forest small mammals. *Bull. Florida State Mus., Biol. Sci.* 34:99–152.

Fonseca, G. A. B. da, A. B. Rylands, C. M. R. Costa, R. B. Machado, and Y. L. R. Leite, eds. 1994. *Livro vermelho dos mammíferos brasileiros ameaçados de extincão.* Belo Horizonte: Fundação Biodiversitas.

Forget, P. M. 1993. Scatter hoarding of *Astrocaryum paramaca* by *Proechimys* sp. in French Guyana: Comparisons with *Myoprocta exilis. Trop. Ecol.* 32:155–67.

Fornes, A., and E. Massoia. 1965. Micromamíferos (Marsupialia y Rodentia) recolectados en la localidad Bonaerense de Miramar. *Physis* 25 (69): 99–108.

Freitas, T. R. O. de. 1995. Geographic distribution and conservation of four species of the genus *Ctenomys* in southern Brazil. *Stud. Neotrop. Fauna Environ.* 30:53–59.

• Freitas, T. R. O. de, and E. P. Lessa. 1984. Cytogenetics and morphology of *Ctenomys torquatus* (Rodentia: Octodonidae). *J. Mammal.* 65 (4): 637–42.

Fulk, G. W. 1975. Population ecology of rodents in the semiarid shrublands of Chile. *Occas. Pap. Mus., Texas Tech Univ.* 33:1–40.

———. 1976a. Notes on the activity, reproduction, and social behavior of *Octodon degus. J. Mammal.* 57 (3): 495–505.

———. 1976b. Owl predation and rodent mortality: A case study. *Mammalia* 40:423–27.

• Gallardo, M. 1979. Las especies chilenas de *Ctenomys* (Rodentia, Octodontidae). 1. Estabilidad cariotípica. *Arch. Biol. Med. Exp.* 12:71–92.

Galletti, M. 1990. Predation on the squirrel *Sciurus aestuans* by capuchin monkeys, *Cebus apella. Mammalia* 54:152–54.

Galletti, M., M. Paschoal, and F. Pedroni. 1992. Predation on palm nuts (*Syagurus romanzoffiana*) by squirrels (*Sciurus ingrami*) in southeast Brazil. *J. Trop. Ecol.* 8:121–23.

• Gardner, A. L. 1983. *Proechimys semispinosus* (Rodentia: Echimyidae): Distribution, type locality, and taxonomic history. *Proc. Biol. Soc. Washington* 96:134–44.

• Gardner, A. L., and L. Emmons. 1984. Species groups in *Proechimys* (Rodentia: Echimyidae) as indicated by karyology and bullar morphology, *J. Mammal.* 65:10–25.

Gardner, A. L., and J. L. Patton. 1976. Karyotypic variation in oryzomyine rodents (Cricetinae) with comments on chromosomal evolution in the Neotropical cricetine complex. *Occas. Pap. Mus. Zool. Louisiana State Univ.* 49:1–48.

Gardner, A. L., and M. Romo R. 1993. A new *Thomasomys* (Mammalia: Rodentia) from the Peruvian Andes. *Proc. Biol. Soc. Washington* 106 (4): 762–74.

Genoways, H., and J. H. Brown, eds. 1993. *Biology of the Heteromyidae.* Special Publication 10. Shippensburg, Pa.: American Society of Mammalogists.

Glanz, W. E. 1977a. Comparative ecology of small mammal communities in California and Chile. Ph.D. diss., University of California, Berkeley.

Glanz, W. E., and S. Anderson. 1990. Notes on Bolivian mammals, no. 7: A new species of *Abrocoma* (Rodentia) and relationships of the Abrocomidae. *Amer. Mus. Novitat.* 2991:1–32.

Glanz, W. E., R. W. Thorington Jr., J. Giacalone-Madden, and L. R. Heaney. 1982. Seasonal food use and demographic trends in *Sciurus granatensis.* In *The ecology of the tropical forest,* ed. E. Leigh Jr., A. Stanley Rand, and D. M. Windsor, 239–52. Washington, D.C.: Smithsonian Institution Press.

Gliwicz, J. 1973. A short-term characteristic of a population of *Proechimys semispinosus,* a rodent species of the tropical rain forest. *Bull. Acad. Polonaise Sci., Biol. Ser.* 21:413–18.

• Goldman, E. S. 1918. *The rice rats of North America (genus Oryzomys).* North American Fauna, vol. 43. Washington, D.C.: U.S. Department of Agriculture.

González, E. M., and E. Massoia. 1995. Revalidación del género *Deltomys* con la descripción de una nueva subespecie de Uruguay sur y del Brasil. *Comm. Zool. Mus. Hist. Nat. Montevideo* 12 (182): 1–8.

Grand, T. H., and J. F. Eisenberg. 1982. On the affinities of the Dinomyidae. *Säugertierk. Mitt.* 30:151–57.

• Greer, J. K. 1966. Mammals of Malleco province, Chile. *Publ. Mus., Michigan State Univ., Biol. Ser.* 3 (2): 49–152.

Guillotin, M. 1982. Rythmes d'activité et régimes alimentaires de *Proechimys cuvieri* et d'*Oryzomys capito velutinus* (Rodentia) en forêt guyanaise. *Rev. Ecol. (Terre et Vie)* 36:337–71.

Guillotin, M., and J. F. Ponge. 1984. Identification de deux espèces de rongeurs de Guyane française *Proechimys cuvieri* et *Proechimys guyannensis*. *Mammalia* 48:287–91.

• Gyldenstolpe, N. 1932. A manual of Neotropical sigmodont rodents. *Kungl. Svensk. Vetenskapsakad. Handl.* 11 (3): 1–164.

• Haffer, J. 1974. *Avian speciation in tropical South America*. Publication 14. Cambridge, Mass.: Nuttall Ornithological Club.

• Handley, C. O., Jr. 1976. Mammals of the Smithsonian Venezuela project. *Brigham Young Univ. Sci. Bull., Biol. Ser.* 20 (5): 1–90.

Handley, C. O., Jr., and R. Pine. 1992. A new species of prehensile-tailed porcupine genus *Coendou* Lacépède from Brazil. *Mammalia* 56 (2): 238–44.

Heaney, L. R. 1983. *Sciurus granatensis*. In *Costa Rican natural history*, ed. D. H. Janzen, 489–90. Chicago: University of Chicago Press.

Heaney, L. R., and R. W. Thorington Jr. 1978. Ecology of Neotropical red-tailed squirrels, *Sciurus granatensis*, in the Panama Canal Zone. *J. Mammal.* 59:846–51.

Helm, J. D. 1973. Reproductive biology of *Tylomys* and *Ototylomys*. Ph.D. diss., Michigan State University, East Lansing.

———. 1975. Reproductive biology of *Ototylomys* (Cricetidae). *J. Mammal.* 56:575–90.

Henry, O. 1994. Saisons de reproduction chez trois rongeurs et un Artiodactyla en Guyane française. *Mammalia* 58 (2): 183–200.

Henry, S., V. Leites, A. Marcos, J. P. Pacheco de Sila, and R. Rodriguez. n.d. La nutria, recurso natural de interés económico. Contribución 30, Programa Investigación Costera. Grupo de Investigación y Conservación, CIPPE.

Hernandez-Camacho, J. 1960. Primitiae mastozoologicae Colombianae. 1. Status taxonomical de *Sciurus pucheranii santanderensis*. *Caldasia* 8:359–68.

• Hershkovitz, P. 1941. The South American harvest mice of the genus *Reithrodontomys*. *Occas. Pap. Mus. Zool. Univ. Michigan* 441:1–7.

• ———. 1944. A systematic review of the Neotropi-

cal water rats of the genus *Nectomys* (Cricetinae). *Misc. Publ. Mus. Zool. Univ. Michigan* 58:1–88.

• ———. 1947. Mammals of northern Colombia, preliminary report no. 1: Squirrels (Sciuridae). *Proc. U.S. Nat. Mus.* 97:1–46.

• ———. 1948. Mammals of northern Colombia, preliminary report no. 2: Spiny rats (Echimyidae), with supplemental notes on related forms. *Proc. U.S. Nat. Mus.* 97:125–40.

• ———. 1955. South American marsh rats, genus *Holochilus*, with a summary of sigmodont rodents. *Fieldiana: Zool.* 37:639–73.

• ———. 1959. Nomenclature of the Neotropical mammals described by Olfers, 1818. *J. Mammal.* 40:337–53.

• ———. 1960. Mammals of northern Colombia, preliminary report no. 8: Arboreal rice rats, a systematic revision of the subgenus *Oecomys* genus *Oryzomys*. *Proc. U.S. Nat. Mus.* 110:513–68.

• ———. 1962. Evolution of Neotropical cricetine rodents (Muridae) with special reference to the phyllotine group. *Fieldiana: Zool.* 46:1–524.

———. 1966. South American swamp and fossorial rats of the scapteromyine group (Cricetinae, Muridae) with comments on the glans penis in murid taxonomy. *Z. Säugetierk.* 31:81–149.

———. 1990. The Brazilian rodent genus *Thalpomys* (Sigmodontinae, Cricetidae) with a description of a new species. *J. Nat. Hist.* 24:763–83.

———. 1993. A new central Brazilian genus and species of Sigmodontine rodent transitional between akodonts and oryzomyines. *Fieldiana: Zool.*, n.s., 75:1–18.

———. 1994. The description of a new speices of South American hocicudo, or long-nose mouse, genus *Oxymycterus* (Sigmodontinae: Muroidea) with a critical review of the generic content. *Fieldiana: Zool.*, n.s., 19:1–43.

Hinojosa, F., S. Anderson, and J. L. Patton. 1987. Two new species of *Oxymycterus* (Rodentia) from Peru and Bolivia. *Amer. Mus. Novitat.* 2898:1–17.

Hodara, V. L., A. E. Kajón, C. Quintans, L. Montoro, and M. S. Merani. 1984. Parámetros métricos y reproductivos de *Calomys musculinus* (Thomas, 1913) y *Calomys callidus*, Thomas 1916 (Rodentia, Cricetidae). *Rev. Mus. Argent. Cienc. Nat. "Bernardino Rivadavia," Zool.* 13 (1–60): 453–59.

• Honacki, J. H., K. E. Kinman, and J. W. Koeppl, eds. 1982. *Mammal species of the world*. Lawrence, Kans.: Allen Press and Association of Systematics Collections.

Hooper, E. T. 1952. A sytematic review of the harvest mice (genus *Reithrodontomys*) of Latin America. *Misc. Publ. Mus. Zool. Univ. Michigan* 77:1–255.

Hooper, E. T., and G. G. Musser. 1964. The glans penis in Neotropical cricetines, with comments on the classification of murid rodents. *Misc. Publ. Mus. Zool. Univ. Michigan* 123:1–57.

Hudson, W. H. 1872. On the habits of the vizcacha (*Lagostomus trichodatylus*). *Proc. Zool. Soc. London* 1872:822–33.

Husson, A. M. 1978. *The mammals of Suriname.* Leiden: E. J. Brill.

Hutterer, R., and U. Hirsch. 1979. Ein neuer *Nesoryzomys* von dem Insel Fernandina, Galapagos. *Bonner Zool. Beitr.* 30:276–83.

Ipinza R., J., M. Tamayo H., and J. Rottmann S. 1971. Octodontidae en Chile. *Not. Mens.* 16 (183): 3–10.

Iriarte, J. A., and J. A. Simonetti. 1986. *Akodon andinus* (Philippi, 1858): Visitante ocasional del matorral esclerófilo centro-chileno. *Not. Mens., Mus. Nac. Hist. Nat.* 311:6–7.

Jackson, J. E., L. Branch, and D. Villareal. 1996. *Lagostomus maximus. Mammal. Species* 543:1–6.

Jaksić, F., and J. Yáñez. 1979. Tamaño corporal de los roedores del distrito mastozoológico santiaguino. *Not. Mens.* 23 (271): 3–4.

Janos, D. P., C. T. Sahley, and L. H. Emmons. 1995. Rodent dispersal of vesicular-arbuscular mycorrhizal fungi in Amazonian Peru. *Ecology* 76:1852–58.

Janzen, D. H. 1982. Attraction of *Liomys* mice to horse dung and the extinction of this response. *Anim. Behav.* 30:483–89.

Jenkins, P. D., and A. A. Barnett. 1997. A new species of water mouse, of the genus *Chibchanomys* (Rodentia, Muridae, Sigmodontinae) from Ecuador. *Bull. Nat. Hist. Mus. Lond. (Zool.)* 63 (2): 123–28.

• Jones, J. Knox, Jr., and H. H. Genoways. 1970. Harvest mice (genus *Reithrodontomys*) of Nicaragua. *Occas. Pap. Western Foundation Vert. Zool.* 2:1–16.

Jones, T. 1993. Social systems of heteromyid rodents. In *Biology of the Heteromyidae,* ed. H. Genoways and J. H. Brown, 575–95. Special Publiction 10. Shippensburg, Pa.: American Society of Mammalogists.

• Justo, E. R., and L. de Santis. 1977. *Akodon serrensis serrensis* Thomas en la Argentina (Rodentia, Cricetidae). *Neotrópica* 23 (69): 47–48.

Kajon, A. E., O. A. Scaglia, C. Horgan, C. Velázquez, M. S. Merani, and O. A. Reig. 1984. Tres nuevos cariotipos de la tribu Akodontini (Rodentia, Cricetidae). *Rev. Mus. Argent. Cienc. Nat. "Bernardino Rivadavia," Zool.* 13 (1–60): 461–69.

Kelt, D. A., R. E. Palma, M. H. Gallardo, and J. A. Cook. 1991. Chromosomal multiformity in *Eligmodontia* and verification of the status of *E. morgani. Z. Saügtierk.* 56:352–58.

Kierulff, M. C., J. R. Stallings, and E. L. Sabato. 1991. A method of capture of the bamboo rat (*Kannabateomys amblyonyx*) in bamboo forests. *Mammalia* 55 (4): 633–36.

Kleiman, D. G. 1970. Reproduction in the female green acouchi, *Myoprocta pratti. J. Reprod. Fertil.* 23:55–65.

———. 1971. The courtship and copulatory behavior of the green acouchi, *Myoprocta pratti. Z. Tierpsychol.* 29:259–78.

———. 1972. Maternal behavior of the green acouchi, *Myoprocta pratti,* the South American caviomorph rodent. *Behaviour* 43:48–84.

———. 1974. Patterns of behavior in hystricomorph rodents. In *The biology of hystricomorph rodents,* ed. T. W. Rolands and B. Weir, 171–209. London: Academic Press.

Kleiman, D. G., J. F. Eisenberg, and E. Maliniak. 1979. Reproductive parameters and productivity of caviomorph rodents. In *Vertebrate ecology in the northern Neotropics,* ed. J. F. Eisenberg, 173–83. Washington, D.C.: Smithsonian Institution Press.

• Koford, C. B. 1955. New rodent records for Chile and for two Chilean provinces. *J. Mammal.* 365: 465–66.

Kolt, D. A., R. E. Palma, M. H. Gallardo, and J. A. Cook. 1991. Chromosomal multiformity in *Eligmodontia* and verification of the status of *E. morgani. Z. Säugetierk.* 56:352–58.

Kravetz, F. O. 1972. Estudio del régimen alimentario, periodos de actividad y otros rasgos ecológicos en una población de "ratón hocicudo" (*Oxymycterus rufus platensis* Thomas) de Punta Lara. *Acta Zool. Lilloana* 29:201–12.

Kravetz, F. O., and G. de Villafañe. 1981. Poblaciones de roedores en cultivo de maíz durante las etapas de madurez y rastrojo. *Hist. Nat.* 1 (31): 213–32.

Lacher, T. E., Jr. 1981. The comparative social behavior of *Kerodon rupestris* and *Galea spixii* and the evolution of behavior in the Caviidae. *Bull. Carnegie Mus. Nat. Hist.* 17:1–71.

Lander, E. 1974. Observaciones preliminares sobre lapas *Agouti paca* (Linne, 1766) (Rodentia, Agoutidae) en Venezuela. Thesis, Universidad Central de Venezuela, Facultad de Agronomía, Instituto de Zoologica Agrícola, Maracay.

Landry, S. 1970. The Rodentia as omnivores. *Quart. Rev. Biol.* 45:351–72.

Langguth, A., and A. Abella. 1970. Las especies uruguayas del género *Ctenomys* (Rodentia-Octodontidae). *Comun. Zool. Mus. Hist. Nat. Montevideo* 10 (129): 1–20.

Langguth, A., and E. J. S. Neto. 1993. Morfologia do pênis em *Pseudoryzomys warrimi* e *Wiedomys pyrrhorhinos. Rev. Nordest. Biol.* 8 (1): 55–59.

• Lara, M. C. 1994. Systematics and phylogeography of the spiny rat genus *Trinomys* (Rodentia: Echimyidae). Ph.D. diss., University of California, Berkeley.

Lara, M. C., J. L. Patton, and M. N. F. da Silva. 1996. Simultaneous diversification of South American echimyid rodents based on complete cytochrome b sequences. *Mol. Phylogen. Evol.* 5:403–13.

Laval, R. K. 1976. Voice and habitat of *Dactylomys dactylinus* (Rodentia: Echimyidae) in Ecuador. *J. Mammal.* 57:402–4.

• Lawrence, B. 1941. A new species of *Octomys* from Argentina. *Proc. New England Zool. Club* 18: 43–46.

Lawrence, M. 1988. The identity of *Sciurus duida* J. A. Allen (Rodentia: Sciuridae). *Amer. Mus. Novitat.* 2919:1–8.

Leo L., M. Gardner, and A. L. Gardner. 1993. A new species of a giant *Thomasomys* (Mammalia: Muridae: Sigmodontinae) from the Andes of north-central Peru. *Proc. Biol. Soc. Washington* 106 (3): 417–28.

Lessa, E. P., and C. A. Altuna. 1984. Estudio comparativo de la morfología del pene en poblaciones uruguayas de *Ctenomys* (Rodentia, Octodontidae). *Rev. Mus. Argent. Cienc. Nat. "Bernardino Rivadavia," Zool.* 13 (1–60): 471–78.

Llanos, A. C., and J. A. Crespo. 1952. Ecología de la vizcacha (*Lagostomus maximus maximus* Blainv.) en el nordeste de la provincia de Entre Ríos. *Rev. Invest. Agríc.* 6 (3–4): 289–378.

Lucero, M. M. 1983. Lista y distribución de aves y mamíferos de la provincia de Tucumán. Ministerio de Cultura y Educación, Fundación Miguel Lillo, *Miscelánea* 75:5–53.

• Macedo, R. H., and M. A. Mares. 1987. Geographic variation in the South American cricetine rodent *Bolomys lasiurus. J. Mammal.* 68:278–94.

Magnusson, W. E., A. de Lima Francisco, and T. A. Sanaiotti. 1995. Home range size and territoriality in *Bolomys lasiurus* (Rodentia, Murida) in the Amazonian savanna. *J. Trop. Ecol.* 11:179–88.

Maia, V. 1984. Karyotypes of three species of Caviinae (Rodentia, Caviidae). *Experientia* 40:564–66.

Maia, V., and A. Langguth. 1987. Chromosomes of the Brazilian cricetid rodent *Wiedomys pyrrhorhinus. Rev. Bras. Genet.* 10 (2): 229–23.

Maik-Siembida, J. 1974. Breeding viscachas *Lagostomus maximus* at Lodz Zoo. *Int. Zoo Yearb.* 14: 116–17.

Maliniak, E., and J. F. Eisenberg. 1971. The breeding of *Proechimys semispinosus* in captivity. *Int. Zoo Yearb.* 11:93–98.

Malygin, V. M., V. M. Aniskin, S. I. Isaev, and A. N. Milishnikov. 1994. *Amphinectomys savamis* Malygin gen. et sp. n., a new genus and a new species of water rat (Cricetidae, Rodentia) from the Peruvian Amazon. *Zool. Zh.* 73:195–208. In Russian.

Mann, G. 1945. Mamíferos de Tarapaca: Observaciones realizadas durante una expedición al alto norte de Chile. *Biológica* 2:23–141.

———. 1978. *Los pequeños mamíferos de Chile.*

Guyana: Zoología 40. Santiago: Universidad de Concepción.

Mares, M. A. 1973. Climates, mammalian communities and desert rodent adaptations: An investigation into evolutionary convergence. Ph.D. diss., University of Texas at Austin.

———. 1977a. Aspects of the water balance of *Oryzomys longicaudatus* from northwestern Argentina. *Comp. Biochem. Physiol.* 57A:237–38.

———. 1977b. Water economy and salt balance in a South American desert rodent, *Eligmodontia typus. Comp. Biochem. Physiol.* 56A:325–32.

———. 1977c. Water balance and other ecological observations on three species of *Phyllotis* in northwestern Argentina. *J. Mammal.* 58:514–20.

———. 1977d. Water independence in a South American non-desert rodent. *J. Mammal.* 58:653–56.

Mares, M., and J. K. Braun. 1996. A new species of phyllotine rodent, genus *Andalgalomys* (Muridae: Sigomdontinae) from Argentina. *J. Mammal.* 77 (4): 928–41.

Mares, M. A., and T. E. Lacher. 1991. Ecological, morphological, and behavioral convergence in rock-dwelling mammals. *Curr. Mammal.* 1:307–45.

Mares, M. A., and R. A. Ojeda. 1982. Patterns of diversity and adaptation in South American hystricognath rodents. In *Mammalian biology in South America*, ed. M. A. Mares and H. H. Genoways, 393–432. Pymatuning Symposia in Ecology 6. Special Publication Series. Pittsburgh: Pymatuning Laboratory of Ecology, University of Pittsburgh.

———. 1984. Faunal commercialization and conservation in South America. *BioScience* 34 (9): 580–84.

Mares, M. A., R. A. Ojeda, and M. P. Kosco. 1981. Observations on the distribution and ecology of the mammals of Salta province, Argentina. *Ann. Carnegie Mus.* 50 (6): 151–206.

Martin, R. E. 1970. Cranial and bacular variation in populations of spiny rats of the genus *Proechimys* (Rodentia: Echimyidae) from South America. *Smithsonian Contrib. Zool.* 35:1–19.

Martin, T. 1994. On the systematic position of *Chaetomys subspinous* (Rodentia: Caviomorph) based on evidence from the incisor enamel microstructure. *J. Mammal. Evol.* 2 (2): 117–23.

Martino, A. M. G., and M. Aguilera M. 1989. Food habits of *Holochilus venezuelae* in rice fields. *Mammalia* 53 (4): 545–61.

• Massoia, E. 1963a. Sobre la posición sistemática y distribución geográfica de *Akodon (Thaptomys) nigrita* (Rodentia-Cricetidae). *Physis* 24 (67): 73–80.

• ———. 1963b. *Oxymycterus iheringi* (Rodentia-Cricetidae) nueva especie para la Argentina. *Physis* 24 (67): 129–36.

• ———. 1963c. Sistemática, distribución geográfica

y rasgos etoecológicas de *Akodon (Deltamys) kempi* (Rodentia, Cricetidae). *Physis* 24 (67): 240.

• ———. 1964. Sistemática, distribución geográfica y rasgos etoecológicos de *Akodon (Deltamys) kempi* (Rodentia-Cricetidae). *Physis* 24 (68): 299–305.

• ———. 1971a. Descripción y rasgos bioecológicas de una nueva subespecie de cricétido: *Akodon azarae bibianae* (Mammalia-Rodentia). *Rev. Invest. Agropecuarias, INTA* (Buenos Aires), ser. 4, *Patolog. Anim.* 8 (5): 131–40.

———. 1971b. Caracteres y rasgos bioecológicos de *Holochilus brasiliensis chacarius* Thomas (rata nutria) de la provincia de Formosa y comparaciones con *Holochilus brasiliensis vulpinus* (Brants) (Mammalia-Rodentia-Cricetidae). *Rev. Invest. Agropecuarias, INTA* (Buenos Aires), ser. 1, *Biol. Prod. Anim.* 8 (1): 13–40.

———. 1971c. Las ratas nutrias argentinas del género *Holochilus* descritas como *Mus brasiliensis* por Waterhouse (Mammalia-Rodentia-Cricetidae). *Rev. Invest. Agropecuarias, INTA* (Buenos Aires), ser. 4, *Patolog. Anim.* 8 (5): 141–48.

• ———. 1973a. Zoogeografía del género *Cavia* en la Argentina con comentarios bioecológicos y sistemáticos (Mammalia-Rodentia-Caviidae). *Rev. Invest. Agropecuarias, INTA* (Buenos Aires), ser. 1, *Biol. Prod. Anim.* 19 (1): 1–11.

• ———. 1973b. Descripción de *Oryzomys fornesi*, nueva especie y nuevos datos sobre algunas especies y subespecies argentinas del subgénero *Oryzomys (Oligoryzomys)* (Mammalia-Rodentia-Cricetidae). *Rev. Invest. Agropecuarias, INTA* (Buenos Aires), ser. 1, *Biol. Prod. Anim.* 10 (1): 21–37.

• ———. 1976a. Mammalia. In *Fauna de agua dulce de la república Argentina,* vol. 44, *Mammalia.* Buenos Aires: Fundación para la Educación, la Ciencia y la Cultura.

• ———. 1976b. Datos sobre un cricétido nuevo para la Argentina: *Oryzomys (Oryzomys) capito intermedius* y sus diferencias con *Oryzomys (Oryzomys) legatus* (Mammalia-Rodentia). *Rev. Invest. Agropecuarias, INTA* (Buenos Aires), ser. 5, 11 (1): 1–7.

• ———. 1979. Descripción de un género y especie nuevos: *Bibimys torresi* (Mammalia-Rodentia-Cricetidae-Sigmodontinae-Scapteromyini). *Physis,* sec. C, 38 (95): 1–7.

———. 1980a. Mammalia de Argentina. 1. Los mamíferos silvestres de la provincia de Misiones. *Iguazú* 1 (1): 15–43.

———. 1980b. El estado sistemático de cuatro especies de cricétidos sudamericanos y comentarios sobre otras congenéricas (Mammalia: Rodentia). *Ameghiana* 17:280–87.

———. 1982. Diagnosis previa de *Cabreramys tem-*

chuki, nueva especie (Rodentia, Cricetidae). *Hist. Nat.* 2 (11): 91–92.

• Massoia, E., and A. Fornes. 1964. Notas sobre el género *Scapteromys* (Rodentia-Cricetidae). 1. Systemática, distribución geográfica y rasgos etoecológicos de *Scapteromys tumidus* (Waterhouse). *Physis* 24 (68): 279–97.

• ———. 1965. Nuevos datos sobre la morfología, distribución geográfica y etoecología de *Calomys callosus callosus* (Rengger) (Rodentia-Cricetidae). *Physis* 25 (70): 325–31.

Massoia, E., and A. Fornes. 1967a. Roedores recolectados en la Capital Federal (Caviidae, Cricetidae y Muridae). *IDIA* 240:47–53.

———. 1967b. El estado sistemático, distribución geográfica y datos etoecológicos de algunos mamíferos neotropicales (Marsupialia y Rodentia) con la descripción de *Cabreramys,* género nuevo (Cricetidae). *Acta Zool. Lilloana* 23:407–30.

———. 1969. Caracteres comunes y distintivos de *Oxymycterus nasutus* (Waterhouse) y *O. iheringi* Thomas (Rodentia, Cricetidae). *Physis* 28 (77): 315–21.

Massoia, E., A. Fornes, R. L. Wainberg, and T. G. de Fronza. 1968. Nuevos aportes al conocimiento de las especies bonaerenses del género *Calomys* (Rodentia-Cricetidae). *Rev. Invest. Agropecuarias, INTA* (Buenos Aires), ser. 1, *Biol. Prod. Anim.* 5 (4): 63–92.

Matamoros H., Y. 1982. Notas sobre la biología del tepezcuinte, *Cuniculus paca,* Brisson (Rodentia: Dasyproctidae) en cautiverio. *Brenesia* 19–20:71–82.

Matson, J. O., and J. P. Abravaya. 1977. *Blarinomys breviceps. Mammal. Species* 74:1–3.

Mayr, E. 1963. *Animal species and evolution.* Cambridge: Harvard University Press.

Mello, D. A. 1978a. Some aspects of the biology of *Oryzomys eliurus* (Wagner, 1845) under laboratory conditions (Rodentia, Cricetidae). *Rev. Brasil. Biol.* 38 (2): 293–95.

———. 1978b. Biology of *Calomys callosus* (Rengger, 1830) under laboratory conditions (Rodentia, Cricetinae). *Rev. Brasil. Biol.* 38 (4): 807–11.

Meritt, D. A., Jr. 1983. Preliminary observations on reproduction in the Central American agouti, *Dasyprocta punctata. Zoo Biol.* 2:127–31.

———. 1984. The pacarana, *Dinomys branickii.* In *One medicine,* ed. O. Ryder and M. L. Byrd, 154–61. Berlin: Springer-Verlag.

Meserve, P. L. 1978. Water dependence in some Chilean arid zone rodents. *J. Mammal.* 59:217–19.

———. 1981a. Resource partitioning in a Chilean semi-arid small mammal community. *J. Anim. Ecol.* 40:747–57.

———. 1981b. Trophic relationships among small

mammals in a Chilean semiarid thorn scrub community. *J. Mammal.* 62:304–14.

Meserve, P. L., and E. B. Le Boulengé. 1987. Population dynamics and ecology of small mammals in the northern Chilean semiarid region. *Fieldiana: Zool.,* n.s., 39:413–31.

Miranda-Ribeiro, A. de. 1936. The new-born of the Brazilian tree-porcupine (*Coendou prehensilis* Linn.) and of the hairy tree-porcupine (*Sphiggurus villosus* F. Cuvier). *Proc. Zool. Soc. London* 1936: 971–74.

Mondolfi, E. 1972. La lapa o paca. *Defensa Nat.* 2 (5): 4–16.

Montenegro-Diaz, O., H. L. Lopez-Arévalo, and A. Cadena. 1991. Aspectos ecológicos del roedor arbicola *Rhipidomys latimanus* Tomes, 1860 (Rodentia: Cricetidae) en el oriente de Cudinamarca, Colombia. *Caldasia* 16 (79): 565–72.

Montgomery, G. G., and Y. Lubin. 1978. Movements of *Coendou prehensilis* in the Venezuelan llanos. *J. Mammal.* 59:887–88.

• Moojen, J. 1948. Speciation in the Brazilian spiny rats (genus *Proechimys,* family Echimydae). *Univ. Kansas Publ. Mus. Nat. Hist.* 1 (19): 301–406.

• ———. 1952a. Os roedores do Brasil. *Bibl. Cient. Brasil.* (Rio de Janeiro), ser. A, 2:1–214.

• ———. 1952b. A new *Clyomys* from Paraguay (Rodentia: Echimyidae). *J. Washington Acad. Sci.* 42 (3): 102.

——— . 1965. Novo género de Cricetidae do Brasil central. *Rev. Brasil. Biol.* 25:281–85.

Moore, J. C. 1959. Relationships among living squirrels of the Sciurinae. *Bull. Amer. Mus. Nat. Hist.* 118 (4): 159–206.

Murúa, R., and L. A. González. 1979. Distribución de roedores silvestres con relación a las caracteristicas del hábitat. *Anal. Mus. Hist. Nat. Valparaiso* 12:69–75.

——— . 1981. Estudios de preferencias y hábitos alimentarios en dos especies de roedores cricétidos. *Med. Amb.* 5 (1–2): 115–24.

——— . 1982. Microhabitat selection in two Chilean cricetid rodents. *Oecologia* (Berlin) 52:12–15.

• Murúa, R., L. González, and C. Jofre. 1982. Estudios ecológicos de roedores silvestres en los bosques templados fríos de Chile. *Mus. Nac. Hist. Nat. Publ. Ocas.* 38:105–16.

Musser, G. G., M. D. Carleton, E. M. Brothers, and A. L. Gardner. 1998. Systematic studies of the oryzomyine rodents (Muridae, Sigmodontinae): Diagnoses and distributions of species formerly assigned to *Oryzomys* "capito." New York: *Bull. Amer. Mus. Nat. Hist.* 236:1–376.

Musser, G. G., and A. L. Gardner. 1974. A new species of ichthyomyine, *Daptomys* from Peru. *Amer. Mus. Novitat.* 2537:1–23.

Musser, G. G., and M. M. Williams. 1985. Systematic studies of oryzomyine rodents (Muridae): Definitions of *Oryzomys villosus* and *Oryzomys telamancae. Amer. Mus. Novitat.* 2810:1–22.

• Myers, P. 1977. A new phyllotine rodent (genus *Graomys*) from Paraguay. *Occas. Pap. Mus. Zool. Univ. Michigan* 676:1–7.

——— . 1982. Origins and affinities of the mammal fauna of Paraguay. In *Mammalian biology in South America,* ed. M. A. Mares and H. H. Genoways, 85–94. Pymatuning Symposia in Ecology 6. Special Publication Series. Pittsburgh: Pymatuning Laboratory of Ecology, University of Pittsburgh.

• ——— . 1989. A preliminary revision of the *varius* group of *Akodon* (A. *dayi, dolores, molinae, neocenus, simulator, toba,* and *varius*). In *Advances in Neotropical mammalogy,* ed. K. H. Redford and J. F. Eisenberg. Gainesville, Fla.: Sandhill Crane Press.

• ——— . 1990. A review of the *boliviensis* group of *Akodon* (Muridae: Sigmodontinae), with emphasis on Peru and Bolivia. *Misc. Publ. Mus. Zool. Univ. Michigan* 177:1–104.

Myers, P., and M. D. Carleton. 1981. The species of *Oryzomys* (*Oligoryzomys*) in Paraguay and the identity of Azare's "rat sixième ou rat tarse noir." *Misc. Publ. Mus. Zool. Univ. Michigan* 161:1–41.

Myers, P., B. Lundrigen, and P. K. Tucker. 1995. Molecular phlogenetics of oryzomyine rodents: The genus *Oligoryzomys. Mol. Phylogen. Ecol.* 4:372–82.

• Myers, P., and J. L. Patton. 1989a. *Akodon* of Peru and Bolivia—revision of the *Fumeus* group (Rodentia: Sigmodontinae). *Occas. Pap. Mus. Zool. Univ. Michigan.* 721:1–35.

——— . 1989b. A new species of *Akodon* from the cloud forests of eastern Cochabamba department, Bolivia (Rodentia: Sigmodontinae). *Occas. Pap. Mus. Zool. Univ. Michigan.* 720:1–28.

Myers, P., J. L. Patton, and M. F. Smith. 1990. A review of the *boliviensis* group of *Akodon* with emphasis on Peru and Bolivia. *Misc. Publ. Mus. Zool. Univ. Mich.* 177:1–104.

Myers, P., and R. M. Wetzel. 1979. New records of mammals from Paraguay. *J. Mammal.* 60:638–41.

Nevo, E. 1979. Adaptive convergence and divergence of subterranean mammals. *Ann. Rev. Ecol. Syst.* 10:269–308.

Newsome, A. E. 1971. The ecology of house mice in cereal haystacks. *J. Anim. Ecol.* 40:116.

Nitikman, L. Z. 1985. *Sciurus granatensis. Mammal. Species* 246:1–8.

Nitikman, L. Z., and M. A. Mares. 1987. Ecology of small mammals in a gallery forest of central Brazil. *Ann. Carnegie Mus.* 56:75–95.

Novello, A. F., and E. P. Lessa. 1986. G-band homology

in two karyomorphs of the *Ctenomys pearsoni* complex (Rodentia: Octodontidae) of Neotropical fossorial rodents. *Z. Säugetierk.* 51:378–80.

Nowak, R. M. 1991. *Walker's "Mammals of the World."* 5th ed. Baltimore: Johns Hopkins University Press.

O'Connell, M. A. 1981. Population ecology of small mammals from northern Venezuela. Ph.D. diss., Texas Tech University.

———. 1982. Population biology of North and South American grassland rodents: A comparative review. In *Mammalian biology in South America*, ed. M. A. Mares and H. H. Genoways, 167–86. Pymatuning Symposia in Ecology 6, Special Publication Series. Pittsburgh: Pymatuning Laboratory of Ecology, University of Pittsburgh.

• Ojasti, J. 1972. Revisión preliminar de los picures o agutís de Venezuela. *Mem. Soc. Cienc. Nat. La Salle* 32 (93): 150–204.

———. 1973. *Estudio biológico del chigüire o capibara.* Caracas: Fundo Nacional de Investigaciones Agropecuarias.

Ojasti, J., and L. M. Sosa Burgos. 1985. Density regulation in populations of capybara. *Acta Zool. Fennica* 173:81–83.

Ojeda, R. A. 1979. Aspetos ecológicos de una comunidad de cricétidos del bosque subtropical de transción. *Neotrópica* 25 (73): 27–35.

• Ojeda, R. A., and M. A. Mares. 1989. *A biogeographic analysis of the mammals of Salta province, Argentina.* Special Publications of the Museum 27. Lubbock: Texas Tech University Press.

Olds, N., and S. Anderson. 1987. Notes on Bolivian mammals: Taxonomy and distribution of rice rats of the subgenus *Oligoryzomys. Fieldiana: Zool.*, n.s., 39:261–82.

———. 1989. A diagnosis of the tribe Phyllotini (Rodentia, Muridae). In *Advances in Neotropical mammalogy*, ed. K. H. Redford and J. F. Eisenberg, 55–74. Gainesville, Fla.: Sandhill Crane Press.

Oliver, W. L., and I. B. Santos. 1991. *Threatened endemic mammals of the Atlantic rainforest of southeast Brazil.* Special Scientific Report 4. Jersey: Jersey Wildlife Preservation Trust.

Olmos, F. 1997. The giant Atlantic forest tree rat *Nelomys thomasi* (Echimyidae): A Brazilian insular endemic. *Mammalia* 61:130–34.

Olmos, F., M. Galletti, M. Pashoal, and S. L. Mendes. 1993. Habits of the southern bamboo rat, *Kannabateomys amblyonyx* (Rodentia, Echimyidae) in southeastern Brazil. *Mammalia* 57 (3): 325–26.

• Olrog, C. C. 1976. Sobre mamíferos del noroeste argentino. *Acta Zool. Lilloana* 32:5–12.

• ———. 1979. Los mamíferos de la selva húmeda, Cerro Calilegua, Jujuy. *Acta Zool. Lilloana* 33:9–14.

• Olrog, C. C., and M. M. Lucero. 1981. *Guía de los*

mamíferos argentinos. Tucumán, Argentina: Ministerio de Cultura y Educación, Fundación Miguel Lillo.

• Osgood, W. H. 1912. Mammals from western Venezuela and eastern Colombia. *Field Mus. Nat. Hist., Zool. Ser.* 10 (5): 32–66.

———. 1925. The long-clawed South American rodents of the genus *Notiomys. Field Mus. Nat. Hist., Zool. Ser.* 12 (9): 113–25.

———. 1933. Two new rodents from Argentina. *Field Mus. Nat. Hist., Zool. Ser.* 20:11–14.

• ———. 1943. The mammals of Chile. *Field Mus. Nat. Hist., Zool. Ser.* 30:1–268.

———. 1944. Nine new South American rodents. *Field Mus. Nat. Hist., Zool. Ser.* 29 (13): 191–204.

• ———. 1946. A new octodont rodent from the Paraguayan Chaco. *Fieldiana: Zool.* 31 (6): 47–49.

Pacheco, V. 1991. A new species of *Scolomys* (Muridae: Sigmodontinae) from Peru. *Publ. Mus. Hist. Nat. Univ. Nac. Mayor San Marcos, Zool.* 37:1.

Pacheco, V., and J. L. Patton. 1995. A new species of puna mouse *Punomys*, Osgood from the southeastern Andes of Peru. *Z. Säugetierk.* 60:85–96.

• Palma, A. R. Torre. 1996. *Sciurus aestuans ingrami:* Historia natural e uso do espaco. In *Resumos do XXI Congresso Brasileiro de Zoologia* (Pôrto Alegre), 230.

Parreira, G. G., and F. M. Cardoso. 1993. Seasonal variation of the spermatogenic activity in *Bolomys lasiurus* (Lund, 1841) from southeastern Brazil. *Mammalia* 57:27–33.

Patterson, B. D., M. H. Gallardo, and K. E. Freas. 1984. Systematics of mice of the subgenus *Akodon* (Rodentia: Cricetidae) in southern South America, with the description of a new species. *Fieldiana: Zool.*, n.s., 23:1–16.

Patton, J. L. 1984. Systematic status of the large squirrels (subgenus *Urosciurus*) of the western Amazon basin. *Stud. Neotrop. Fauna Environ.* 19 (2): 53–72.

• ———. 1987. Species groups of spiny rats, genus *Proechimys* (Rodentia: Echimyidae). In *Studies in Neotropical mammalogy: Essays in honor of Philip Hershkovitz*, ed. B. D. Patterson and R. M. Timm, 305–45. *Fieldiana: Zool.*, n.s., 39:1–506.

———. 1990. Geomyid evolution: The historical, selective and random basis for divergence patterns within and among species. In *Evolution of subterranean mammals at the organismal and molecular levels*, ed. E. Nevo and O. A. Reig, 49–69. New York: A. R. Liss.

Patton, J. L., and L. Emmons. 1985. A review of the genus *Isothrix* (Rodentia, Echimyidae). *Amer. Mus. Novitat.* 2817:1–14.

Patton, J. L., and A. L. Gardner. 1972. Notes on the systematics of *Proechimys* (Rodentia: Echimyidae), with emphasis on Peruvian forms. *Occas. Pap. Mus. Zool. Louisiana State Univ.* 44:1–30.

Patton, J. L., and M. S. Hafner. 1983. Biosystematics of the native rodents of the Galápagos archipelago, Ecuador. In *Patterns of evolution in Galápagos organisms*, ed. R. J. Bowman, M. Benson, and A. E. Leviton, 539–68. San Francisco: American Association for the Advancement of Science, Pacific Division.

Patton, J. L., P. Myers, and M. F. Smith. 1989. Electomorphic variation in selected akodontine rodents (Muridae, Sigmodontinae) with comments on systematic implications. *Z. Säugetierk.* 54:347–59.

• ———. 1990. Vicariant versus gradient models of diversification: The small mammal fauna of eastern Andean slopes of Peru. In *Vertebrates in the tropics*, ed. G. Peters and R. Hutterer, 355–72. Bonn: Museum Alex Koenig.

Patton, J. L., and O. Reig. 1989. Genetic differentiation among echimid rodents, with emphasis on the genus *Proechimys*. In *Advances in Neotropical mammalogy*, ed. K. H. Redford and J. F. Eisenberg, 75–96. Gainesville, Fla.: Sandhill Crane Press.

Patton, J. L., and M. N. F. da Silva. 1995. A review of the spiny mouse *Scolomys* with the description of a new species from western Brazil. *Proc. Biol. Soc. Washington* 108 (2): 319–37.

Patton, J. L., M. N. F. da Silva, and J. R. Malcolm. 1994. Gene genealogy and differentiation among arboreal spiny rats of the Amazon basin: A test of riverine barrier hypothesis. *Evolution* 48:1314–23.

———. 1996. Hierarchical genetic structure and gene flow in three sympatric species of Amazonian rodents. *Mol. Ecol.* 5:229–38.

Patton, J. L., and M. F. Smith. 1992a. mt DNA pylogeny of Andean mice: A test of diversity across ecological gradients. *Evolution* 46:174–83.

———. 1992b. Evolution and systematics of akodontine rodents (Muridae: Sigmodontinae) of Peru, with emphasis on the genus *Akodon*. *Mem. Mus. Hist. Nat.*, UNMSM (Lima) 21:83–103.

Patton, J. L., and S. Y. Yang. 1977. Genetic variation in *Thomomys bottae* pocket gophers: Macrogeographic patterns. *Evolution* 31:697–720.

Paula Couto, C. 1950. Footnote number 242. In *Memórias sobre a Paleontologis brasileira*, ed. P. W. Lund, with notes and comments by C. Paula Couto, 232. Rio de Janeiro: Instituto Nacional do Livro.

Pearson, O. P. 1948. Life history of mountain viscachas in Peru. *J. Mammal.* 29:345–74.

———. 1951. Mammals in the highlands of southern Peru. *Bull. Mus. Comp. Zool.* 106 (3): 117–74.

———. 1957. Additions to the mammalian fauna of Peru and notes on some other Peruvian mammals. *Breviora: Mus. Comp. Zool.* 73:1–7.

• ———. 1958. A taxonomic revision of the rodent genus *Phyllotis*. *Univ. Calif. Publ. Zool.* 56 (4): 391–496.

———. 1959. Biology of the subterranean rodents, *Ctenomys*, in Peru. *Mem. Mus. Hist. Nat. "Javier Prado"* 9:1–56.

———. 1967. La estructura por edades y la dinámica reproductiva en una populación de ratones de campo *Akodon azarae*. *Physis* 27 (74): 53–58.

———. 1975. An outbreak of mice in the coastal desert of Peru. *Mammalia* 39:375–86.

———. 1983. Characteristics of a mammalian fauna from forests in Patagonia, southern Argentina. *J. Mammal.* 64:476–92.

Pearson, O. P., N. Binsztein, L. Boiry, C. Busch, M. Di Pace, G. Gallopin, P. Penchaszadeh, and M. Piantanida. 1968. Estructura social, distribución espacial y composición por edades de una población de tuco-tucos (*Ctenomys talarum*). *Invest. Zool. Chil.* 13:47–80.

Pearson, O. P., S. Martin, and J. Bellati. 1987. Demography and reproduction of the silky desert mouse (*Eligmodontia*) in Argentina. *Fieldiana: Zool.*, n.s., 23:433–46.

Pearson, O. P., and J. L. Patton. 1976. Relationships among South American phyllotine rodents based on chromosome analysis. *J. Mammal.* 57:339–50.

Pearson, O. P., and A. K. Pearson. 1982. Ecology and biogeography of the southern rainforests of Argentina. In *Mammalian biology in South America*, ed. M. A. Mares and H. H. Genoways, 129–42. Pymatuning Symposia in Ecology 6. Special Publication Series. Pittsburgh: Pymatuning Laboratory of Ecology, University of Pittsburgh.

• Pearson, O. P., and C. P. Ralph. 1978. The diversity and abundance of vertebrates along an altitudinal gradient in Peru. *Mem. Mus. Hist. Nat. "Javier Prado"* 18:1–97.

Pefaur, J. E., J. L. Yáñez, and F. M. Jaksić. 1979. Biological and environmental aspects of a mouse outbreak in the semi-arid region of Chile. *Mammalia* 43:313–22.

Pereira, C. 1941. Sobre las "ratadas" no sul do Brasil e o ciclo vegetativo das taquaras. *Aarq. Inst. Biol. São Paulo* 12:175–96.

Perez, E. M. 1992. *Agouti paca*. *Mammal. Species* 404:1–7.

Perez-Zapata, A., D. Lew, M. Aguilera, and O. A. Rcig. 1992. New data on the] systematics and karyology of *Podoxomys roraimae*. *Z. Säugetierk.* 57:216–24.

Pessôa, L. M. 1991. Cranial intraspecific differentiation in *Proechimys iheringi* Thomas (Rodentia: Echimyidae). *Z. Säugetierk.* 56:34–40.

Pessôa, L. M., and S. F. dos Reis. 1990. Componentes de variação craniana intra e interpopulacional em *Proechimys dimidiatus* e *P. iheringi*. *Resumos do XVII Congresso Brasileiro de Zoologia*, 220.

————. 1991a. Analise da variação craniana e diferenção subespecifica em populações de *Proechimys iheringi*. *Resumos do XVIII Congresso Brasileiro de Zoologia*, 413.

————. 1991b. Natural selection, morphological divergence and phenotypic evolution in *Proechimys dimidiatus*. *Rev. Brasil. Gen.* 14 (3): 705–11.

————. 1992. Revisão taxonomica de *Proechimys iheringi* com descrição uma novo subespécie de estado Rio de Janiero. *Resumos do XIX Congresso Brasileiro de Zoologia*, 157.

————. 1993. *Proechimys dimidiatus*. *Mammal. Species* 441:1–3.

• ————. 1996. *Prochimys iheringi*. *Mammal. Species* 536:1–4.

Petter, F. 1979. Une nouvelle espèce de rat d'eau de Guyane française *Nectomys parvipes*. *Mammalia* 43:507–10.

Pine, R. H. 1973. Una nueva especie de *Akodon* (Mammalia: Rodentia: Muridae) de la Isla Wellington, Magallanes, Chile. *Anal. Inst. Patagonia, Punta Arenas* 4 (1–3): 423–26.

————. 1980. Notes on rodents of the genera *Wiedomys* and *Thomasomys* (including *Wilfredomys*). *Mammalia* 44:195–202.

Pine, R. H., S. D. Miller, and M. L. Schamberger. 1979. Contributions to the mammalogy of Chile. *Mammalia* 43:339–76.

Pine, R. H., and R. M. Wetzel. 1975. A new subspecies of *Pseudoryzomys wavrini* (Mammalia: Rodentia: Muridae: Cricetinae) from Bolivia. *Mammalia* 39: 649–55.

Pizzimenti, J. J., and R. de Salle. 1980. Dietary and morphometric variation in some Peruvian rodent communities: The effect of feeding strategy on evolution. *Biol. J. Linnean Soc.* 13:263–85.

Rabinovich, J., A. Capurro, P. Folgarait, T. Kitzberger, G. Kramer, A. Novaro, M. Puppo, and A. Travaini. 1987. Estado del conocimiento de 12 especies de la fauna silvestre argentina de valor comerical. Paper presented at the second workshop on Elaboración de propuestas de investigación orientada al manejo de la fauna silvestre de valor comercial, Buenos Aires.

• Rau, J., J. Yáñez, and F. Jaksić. 1978. Confirmación de *Notiomys macronyx alleni* O. y *Eligmodontia typus typus* C., y primer registro de *Akodon (Abrothrix) lanosus* T. (Rodentia: Cricetidae) en la zona de Ultima Esperanza (XII Region, Magallanes). *Anal. Inst. Patagonia, Punta Arenas* 9:203–4.

Redford, K. H. 1984. Mammalian predation on termites: Tests with the burrowing mouse (*Oxymycterus roberti*) and its prey. *Oecologia* 65:145–52.

• Redford, K. H., and J. F. Eisenberg. 1992. *Mammals of the Neotropics*. Vol. 2. *The southern cone: Chile,* *Argentina, Uruguay, Paraguay.* Chicago: University of Chicago Press.

Redford, K. H., and J. G. Robinson. 1987. The game of choice: Patterns of Indian and colonist hunting in the Neotropics. *Amer. Anthropol.* 89:650–67.

Reig, O. A. 1980. A new fossil genus of South American cricetid rodents allied to *Wiedomys*, with an assessment of the Sigmodontinae. *J. Zool.* (London) 192:257–81.

————. 1984. Significado de los métodos citogenéticos para la distinción y la interpretación de las especies, con especial referencia a los mamíferos. *Rev. Mus. Argent. Cienc. Nat. "Bernardino Rivadavia," Zool.* 13 (1–60): 19–44.

————. 1986. Diversity patterns and differentiation of high Andean rodents. In *High altitude tropical biogeography*, ed. F. Vuilleumier and M. Monasterio, 404–42. New York: Oxford University Press.

————. 1987. An assessment of the systmatics and evolution of the Akodontini, with the description of new fossil species of *Akodon* (Cricetidae: Sigmodontinae). *Fieldiana: Zool.*, n.s., 39:347–99.

————. 1989. Karyotypic patterning as one triggering factor in cases of explosive speciation. In *Evolutionary biology of transient unstable populations*, ed. A. Fontdevila, 246–89. Berlin: Springer-Verlag.

Reig, O. A., M. Aguilera, M. A. Barros, and M. Useche. 1980. Chromosomal speciation in a rassenkreis of Venezuelan spiny rats (genus *Proechimys*, Rodentia, Echimyidae). *Genetica* 52–53:291–312.

Reig, O. A., and P. Kiblisky. 1968. Chromosomes in four species of rodents of the genus *Ctenomys* (Rodentia, Octodontidae) from Argentina. *Experientia* 24:274–76.

————. 1969. Chromosome multiformity in the genus *Ctenomys* (Rodentia, Octodontidae). *Chromosome* (Berlin) 28:211–44.

Reig, O. A., N. Olivo, and P. Kiblisky. 1971. The idiogram of the Venezuelan vole mouse, *Akodon urichi venezuelensis* Allen (Rodentia, Cricetidae). *Cytogenetics* 10:99–114.

Reig, O. A., M. Trainer, and M. A. Barros. 1979. Sur l'identification chromosomique de *Proechimys guyannensis* et de *Proechimys cuvieri*. *Mammalia* 43:501–5.

Reig, O. A., and M. Useche. 1976. Diversidad cariotípica y sistemática en poblaciones venezolanas de *Proechimys* (Rodentia, Echimyidae), con datos adicionales sobre poblaciones de Perú y Colombia. *Acta Cient. Venez.* 27:132–40.

Reis, S. F., and L. M. Pessôa. 1995. *Proechimys albispinus minor*, a new subspecies from the state of Bahia, northeastern Brazil. *Z. Säugetierk.* 60:237–42.

Reise, D., and W. Venegas. 1974. Observaciones sobre el comportamiento de la fauna de micromamíferos en la región de Puerto Ibañez (Lago General Carrera), Aysen, Chile. *Bol. Soc. Biol. Concepción* 47:71–85.

Rieger, T. T., A. Langguth, and T. A. Weimer. 1995. Allozymic characterization and evolution relationships in the Brazilian *Akodon cursor* species group. *Biochem. Genet.* 336 (91): 283–95.

Rieger, T. T., T. A. Weimer, A. Langguth, and L. F. B. Oliviera. 1995. Temporal variation of allele frequencies in four population samples of *Akodon monotensis. Brazil. J. Genet.* 16 (3): 385–90.

Roberts, M., S. Brand, and E. Maliniak. 1985. The biology of captive prehensile-tailed porcupines, *Coendou prehensilis. J. Mammal.* 66 (3): 476–82.

Roberts, M., E. Maliniak, and M. Deal. 1984. The reproductive biology of the rock cavy *Kerodon rupestris* in captivity. *Mammalia* 48:253–66.

Roberts, M. S., K. V. Thompson, and J. Cranford. 1988. Reproduction and growth in captive punare of the Brazilian caatinga with reference to the reproductive strategies of the Echimyidae. *J. Mammal.* 69 (3): 542–51.

Rocha, P. L. B. da. 1995. *Proechimys yonenagae:* A new species of spiny rat (Rodentia: Echimyidae) from fossil sand dunes in the Brazilian caatinga. *Mammalia* 59 (4): 537–49.

Rodriguez, J. M. 1985. Ecología de la guatusa (*Dasyprocta punctata punctata* Gray). In *Investigaciones sobre fauna silvestre de Costa Rica,* 9–22. San José, Costa Rica: Ministerio de Agricultura y Ganadería.

• Roig, V. G. 1965. Elenco sistemático de los mamíferos y aves de la provincia de Mendoza y notas sobre su distribución geográfica. *Bol. Est. Geogr.* 12 (49): 175–222.

Rood, J. P. 1970. Ecology and social behavior of the desert cavy (*Microcavia australis). Amer. Midl. Nat.* 83:415–54.

———. 1972. Ecological and behavioral comparisons in three genera of Argentine cavies. *Anim. Behav. Monogr.* 5:67–83.

Rosi, M. I. 1983. Notas sobre la ecología, distribución y sisetmática de *Graomys griseoflavus griseoflavus* (Waterhouse, 1837) (Rodentia, Cricetidae) en la provincia de Mendoza. *Hist. Nat.* 3 (17): 161–67.

Rowlands, I. W. 1974. Mountain viscacha. *Symp. Zool. Soc. London* 34:131–41.

Rowlands, I. W., and B. J. Weir, eds. 1974. *The biology of hystricomorph rodents.* London: Academic Press.

Ruedas, L., J. A. Cook, T. L. Yates, and J. W. Bickham. 1993. Conservative genome size and rapid chromosomal evolution in the South American tuco-tucos. *Genome* 36:449–58.

Sage, R. D., J. R. Contreras, V. G. Roig, and J. L. Patton. 1986. Genetic variation in the South American burrowing rodents of the genus *Ctenomys* (Rodentia: Ctenomyidae). *Z. Säugetierk.* 51:158–72.

• Sanborn, C. C. 1931. A new *Oxymycterus* from Misiones, Argentina. *Proc. Biol. Soc. Washington* 44:1–2.

• ———. 1947a. Geographical races of the rodent *Akodon jelskii* Thomas. *Fieldiana: Zool.* 31 (17): 133–42.

———. 1947b. The South American rodents of the genus *Neotomys. Fieldiana: Zool.* 31:51–57.

———. 1949. Cavies of southern Peru. *Proc. Biol. Soc. Washington* 63:133–34.

• ———. 1950. Small rodents from Peru and Bolivia. *Publ. Mus. Hist. Nat. "Javier Prado,"* ser. A, *Zool.* 5:1–16.

Sanborn, C. C., and O. P. Pearson. 1947. The tuco-tucos of Peru (genus *Ctenomys). Proc. Biol. Soc. Washington* 60:135–38.

Sanchez-Cordero, V., and T. H. Fleming. 1993. Ecology of tropical heteromyids. In *Biology of the Heteromyidae,* ed. H. Genoways and J. H. Brown, 596–617, Special Publication 10. Shippensburg, Pa.: American Society of Mammalogists.

Schaller, G. B. 1983. Mammals and their biomass on a Brazilian ranch. *Arq. Zool. São Paulo* 31 (2): 1–36.

Schaller, G. B., and P. G. Crawshaw Jr. 1981. Social organization in a capybara population. *Säugetierk. Mitt.* 29:3–16.

Schamberger, M., and G. Fulk. 1974. Mamíferos del Parque Nacional Fray Jorge. *Idesia* (Chile) 3: 167–79.

• Schneider, C. O. 1946. Catálogo de los mamíferos de la provincia de Concepción. *Bol. Soc. Biol. Concepción* 21:67–83.

• Scrocchi, G., A. M. P. de Fonollat, and H. H. Salas. 1986. *Akodon andinus dolichonyx* (Philippi), (Rodentia, Cricetidae) en la provincia de Tucumán, Argentina. *Acta Zool. Lilloana* 38 (2): 113–18.

Shamel, H. H. 1931. *Akodon chacoensis,* a new cricetine rodent from Argentina. *J. Mammal.* 12: 427–29.

Sielfeld, W. H. 1979. Presencia de *Microcavia australis* (G. y d'O.) en Magallanes (Mammalia: Caviidae). *Anal. Inst. Patagonia, Punta Arenas* 20:197–99.

• Silva, M. N. F. da. 1995. Systematics and phlyogeography of Amazonian spiny rats of the genus *Proechimys* (Rodentia: Echimyidae). Ph.D. diss., University of California, Berkeley.

Silva, M. N. F. da, and J. L. Patton. 1993. Amazonian phylogeography: mt DNA sequence variation in arboreal echimyid rodents. *Mol. Phyl. Evol.* 2 (3): 243–55.

Simonetti, J. A., E. R. Fuentes, and R. D. Otaiza. 1985.

Habitat use by two rodent species in the high Andes of central Chile. *Mammalia* 49:19–25.

Simpson, G. G. 1945. The principles of classification and a classification of the mammals. *Bull. Amer. Mus. Nat. Hist.* 85:1–350.

Smith, M. F., and J. L. Patton. 1991. Variation in mitochondrial cytochrome 6 sequence in natural populations of South American akodontine rodents (Muridae: Sigmodontinae). *Mol. Biol. Evol.* 8 (1): 85–103.

———. 1993. The diversification of South American murid rodents: Evidence from mitochondrial DNA sequence data for the akodontine tribes. *Biol. J. Linnean Soc.* 50:149–77.

Smythe, N. 1970. Ecology and behavior of the agouti (*Dasyprocta punctata*) and related species on Barro Colorado Island, Panama. Ph.D. diss., University of Maryland.

———. 1978. The natural history of the central American agouti (*Dasyprocta punctata*). *Smithsonian Contrib. Zool.* 257:1–52.

Smythe, N., and O. Brown de Guanti. 1995. *La domesticación y cría de la paca* (Agouti paca). Guía de Conservación 26. Rome: FAO.

Smythe, N., W. E. Glanz, and E. Leigh. 1982. Population regulation in some terrestrial frugivores. In *The ecology of a tropical forest*, ed. E. Leigh, A. S. Rand, and D. Windsor, 227–38. Washington, D.C.: Smithsonian Institution Press.

Soini, P., and M. Soini. 1992. Ecología del ronsoco o capibara (*Hydrochaeris hydrochaeris*) en la reserva nacional Pacaya-Jamilia, Peru. *Folia Amazonica* 4 (2): 120–33.

Spotorno, A. E. 1992. Parallel evolution and ontogeny of simple penis among New World rodents. *J. Mammal.* 73 (3): 504–14.

Spotorno, A. E., and L. Walker. 1983. Análisis electroforético y biométrico de dos especies de *Phyllotis* en Chile central y sus híbridos experimentales. *Rev. Chil. Hist. Nat.* 56 (1): 15–23.

Stallings, J. R. 1988a. Small mammal communities in an eastern Brazilian park. Ph.D. diss., University of Florida, Gainesville.

———. 1988b. Small mammal inventories in an eastern Brazilian park. *Bull. Florida State Mus.* 34: 153–200.

Stallings, J. R., M. C. M. Kierulff, and L. F. B. M. Silva. 1994. Use of space and activity patterns of Brazilian bamboo rats (*Kannebateomys amblyonyx*) in an exotic habitat. *J. Trop. Ecol.* 10:434–38.

Steadman, D. W., and C. Ray. 1982. The relationships of *Megaoryzomys curioi*. *Smithsonian Contrib. Paleobiol.* 51:1–23.

Steppan, S. 1993. Phylogenetic relationships among Phyllotini (Rodentia: Sigmodontinae) using morphological characters. *J. Mammal. Evol.* 1 (3): 187–213.

———. 1995. Revision of the tribe Phyllotini (Rodentia, Sigmodontinae) with a phlytogenetic hypothesis for the Sigmodontinae. *Fieldiana: Zool.*, n.s., 80:1–112.

Streilein, K. E. 1982a. The ecology of small mammals in the semiarid Brazilian caatinga. 1. Climate and faunal composition. *Ann. Carnegie Mus.* 51 (5): 79–107.

———. 1982b. The ecology of small mammals in the semiarid Brazilian caatinga. 2. Water relations. *Ann. Carnegie Mus.* 51 (6): 109–26.

———. 1982c. The ecology of small mammals in the semiarid Brazilian caatinga. 3. Reproductive biology and population ecology. *Ann. Carnegie Mus.* 51 (13): 251–69.

Suarez, O. V. 1994. Diet and habitat selection of *Oxymycterus rutilans*. *Mammalia* 58 (2): 225–34.

Tálice, R. V. 1969. Mamíferos autóctonos. *Nuestra Tierra* 5:1–68.

Tálice, R. V., R. Caprio, and E. Momigliano. 1954. Distribución geográfica y hábitat de *Ctenomys torquatus*. *Arch. Soc. Biol. Montevideo* 21:133–39.

Tálice, R. V., and E. Momigliano. 1954. Arquitectura y microclima de las tuqueras o moradas de *Ctenomys torquatus*. *Arch. Soc. Biol. Montevideo* 21: 126–33.

Tálice, R. V., S. L. de Mosera, and A. M. S. de Sprechmann. 1972. Problemas de captura y sobrevida en cautividad en *Ctenomys torquatus*. *Rev. Biol. Uruguay* 1 (2): 121–28.

• Tamayo, M., and D. Frassinetti. 1980. Catálogo de los mamíferos fosiles y vivientes de Chile. *Mus. Nac. Hist. Nat. Chile* 37:323–99.

• Tate, G. H. H. 1939. The mammals of the Guiana region. *Bull. Amer. Mus. Nat. Hist.* 76 (5): 151–229.

Tesh, R. B. 1970a. Notes on the reproduction, growth and development of echimyid rodents in Panama. *J. Mammal.* 51:199–202.

———. 1970b. Observations on the natural history of *Diplomys darlingi*. *J. Mammal.* 51:197–99.

Thomas, O. 1898a. On the small mammals collected by Dr. Borelli in Bolivia and northern Argentina. *Boll. Mus. Zool. Anat. Comp.* 13 (315): 1–4.

———. 1898b. On some new mammals from the neighborhood of Mount Sahama, Bolivia. *Ann. Mag. Nat. Hist.*, ser. 7, 1:277–83.

———. 1898c. On new small mammals from the Neotropical region. *Ann. Mag. Nat. Hist.*, ser. 6, 18:301–14.

———. 1900a. Descriptions of new rodents from western South America. *Ann. Mag. Nat. Hist.*, ser. 7, 6:383–87.

———. 1900b. Descriptions of new rodents from

western South America. *Ann. Mag. Nat. Hist.*, ser. 7, 6:294–302.

———. 1902. On mammals from the Serra do Mar of Paraná, collected by Mr. Alphonse Robert. *Ann. Mag. Nat. Hist.*, ser. 7, 4 (9): 59–64.

———. 1906. Notes on South American rodents. 2. On the allocation of certain species hitherto referred respectively to *Oryzomys*, *Thomasomys*, and *Thipidomys. Ann. Mag. Nat. Hist.*, ser. 7, 18:442–48.

———. 1916. Notes on Argentine, Patagonian, and Cape Horn Muridae. *Ann. Mag. Nat. Hist.*, ser. 8, 17:182–87.

———. 1917. On small mammals from the delta of the Paraná. *Ann. Mag. Nat. Hist.*, ser. 8, 20:95–100.

———. 1918. On small mammals from Salta and Jujuy collected by Mr. E. Budin. *Ann. Mag. Nat. Hist.*, ser. 9, 1:186–93.

———. 1919a. On small mammals from "Otro Cerro," north-eastern Rioja, collected by Sr. E. Budin. *Ann. Mag. Nat. Hist.*, ser. 9, 3:489–500.

• ———. 1919b. List of mammals from the highlands of Jujuy, north Argentina, collected by Sr. E. Budin. *Ann. Mag. Nat. Hist.*, ser. 9, 4:128–35.

• ———. 1920. A further collection of mammals from Jujuy. *Ann. Mag. Nat. Hist.*, ser. 9, 5 (26): 188–96.

———. 1921. A new tuco-tuco from Bolivia. *Ann. Mag. Nat. Hist.*, ser. 9, 7 (37): 136–37.

———. 1926. The Spedan Lewis South American exploration. 3. On mammals collected by Sr. Budin in the province of Tucumán. *Ann. Mag. Nat. Hist.*, ser. 9, 17 (101): 602–8.

Thorington, R. W., and E. M. Thorington. 1989. Postcranial porportions of *Microsciurus* and *Sciurillus*, the American pygmy squirrels. In *Advances in Neotropical mammalogy*, ed. K. H. Redford and J. F. Eisenberg, 125–36. Gainesville, Fla.: Sandhill Crane Press.

• Tonni, E. P. 1984. The occurrence of *Cavia tschudi* (Rodentia, Caviidae) in the southwest of Salta province, Argentina. *Stud. Neotrop. Fauna Environ.* 19 (3): 155–58.

Trainer, M. M. 1976. Nouvelles donneés sur l'evolution non parallele du caryotype et de la morphologie chez les phyllotines (Rongeurs, Cricétidés) *C. R. Acad. Sci. Paris*, ser. D, 283:1201–3.

Twigg, G. I. 1962. Notes on *Holochilus sciureus* in British Guiana. *J. Mammal.* 43:369–74.

• Vallejo, S., and E. Gudynas. 1981. Notas sobre la distribución y ecología de *Calomys laucha* en Uruguay (Rodentia: Cricetidae). *C. E. D. Orione Cont. Biol.* 4 (l): 1–16.

Valqui, M. H. 1995. A terrestrial, small rodent community in northeastern Peru. M.S. thesis, Department of Wildlife Ecology and Conservation, University of Florida, Gainesville.

Vandermeer, J. H. 1979. Hoarding behavior of captive *Heteromys desmarestianus* (Rodentia) on the fruits of *Welfa georgii*, a rainforest dominant palm in Costa Rica. *Brenesia* 16:107–16.

Vaz-Ferreira, R. 1958. Nota sobre Cricetinae del Uruguay. *Arch. Soc. Biol. Montevideo* 24:66–75.

Veiga-Borgeaud, T. 1982. Données écologiques sur *Oryzomys nigripes* (Desmarest, 1819) (Rongeurs; Cricétidés) dans le forêt natural de Barracao dos Mendes. *Mammalia* 46; 335–59.

Veiga-Borgeaud, T., D. R. Lemos, and C. B. Bastos. 1987. Étude de la dynamique de population d'*Holochilus brasiliensis* (Rongeurs, Cricétidés), réservoir sauvage de *Schistosoma mansoni* (Baixada du Maranhão, São Luiz, Brésil). *Mammalia* 51 (2): 249–57.

Vergara, S. G. 1982. Cría masiva y explotación del agutí en cautiverio. In *Zoología neotropical*, vol. 2, ed. P. J. Salinas, 1479–96. Mérida, Venezuela: Congreso Latinamericano de Zoología.

Vié, J. C., V, Volobouev, J. L. Patton, and L. Granjon. 1996. A new species of *Isothrix* (Rodentia: Echimyidae) from French Guiana. *Mammalia* 60:393–406.

Vieira, C. 1945. Sobre uma coleção de mamíferos de Mato Grosso. *Arq. Zool. (São Paulo)* 4:395–430.

Vieira, E. M., and L. C. Baumgarten. 1995. Daily activity patterns of small mammals in a cerrado area from central Brazil. *J. Trop. Ecol.* 11:255–62.

Villafañe, G. de. 1981a. Reproducción y crecimiento de *Akodon azarae azarae* (Fischer, 1829). *Hist. Nat.* 1 (28): 193–204.

———. 1981b. Reproducción y crecimiento de *Calomys musculinus murillus* (Thomas, 1916). *Hist. Nat.* 1 (33): 237–56.

Villafañe, G., and S. M. Bonaventura. 1987. Ecological studies in crop fields of the endemic area of Argentine hemorrhagic fever. *Calomys musculinus* movements in relation to habitat and abundance. *Mammalia* 51 (2): 233–48.

Villafañe, G. de, F. O. Kravetz, M. J. Piantanida, and J. A. Crespo. 1973. Dominancia, densidad e invasión en una comunidad de roedores de la localidad de Pergamino (provincia de Buenos Aires). *Physis*, sec. C, 32 (84): 47–59.

Vitullo, A. D., and J. A. Cook. 1991. The role of sperm morphology in the evolution of tuco-tucos *Ctenomys. Z. Säugetierk.* 56:359–64.

Vivas, A., and R. Roca. 1983. Uso del espacio por *Zygodontomys microtinus* (Rodentia, Cricetidae) en el estado Guárico. *Acta Cient. Venez.* 34 (suppl. 1): 146.

Vorontsov, N. N. 1960. The ways of food specialization and evolution of the alimentary system in Muroidea. In *Symposium theriologicum*, ed. J. Kratochvil, 360–77. Brno: Ceskoslovenska Akademic Ved.

———. 1979. *Evolution of the alimentary system in myomorph rodents*. Washington, D.C.: Smithsonian Institution and National Science Foundation. Translated from the Russian; originally published in 1967.

Voss, R. S. 1988a. Comparative morphology and systematics of ichthyomyine rodents (Muroidea). Ph.D. diss., University of Michigan.

———. 1988b. Systematics and ecology of ichthyomyine rodents (Muroidea): Patterns of morphological evolution in a small adaptive radiation. *Bull. Amer. Mus. Nat. Hist.* 188 (2): 259–493.

———. 1991a. On the identity of *"Zygodontomys" punctulatus* (Rodentia: Muroidea). *Amer. Mus. Novitat.* 3026:1–8.

———. 1991b. An introduction to the Neotropical muroid rodent genus *Zygodontomys*. *Bull. Amer. Mus. Nat. Hist.* 210:1–113.

———. 1992. A revision of the South American species of *Sigmodon* (Mammalia: Muridae) with notes on their natural history and biogeography. *Amer. Mus. Novitat.* 3050:1–56.

———. 1993. A revision of the Brazilian rodent genus *Delomys* with remarks on "thomasomyine" characters. *Amer. Mus. Novitat.* 3073:1–44.

Voss, R. S., and R. Angermann. 1997. Revisionary notes on Neotropical porcupines 1: Type material described by Olfers (1818) and Kuhl (1820) in the Berlin Zoological Museum. *Amer. Mus. Novitat.* 3214:1–44.

Voss, R. S., and M. D. Carleton. 1993. A new genus for *Hesperomys molitor* Winge and *Holochilus magnus* Hershkovitz (Mammalia: Muridae) with an analysis of its phylogenetic relationships. *Amer. Mus. Novitat.* 3085:1–39.

Voss, R. S., and A. V. Lindzey. 1981. Comparative gross morphology of male accessory glands among Neotropical Muridae (Mammalia: Rodentia) with comments on systematic implications. *Misc. Publ. Mus. Zool. Univ. Michigan* 159:1–41.

Voss, R. S., and P. Myers. 1991. *Pseudoryzomys simplex* and the significance of Lund's collections from the caves of Lagoa Santa, Brazil. *Bull. Amer. Mus. Nat. Hist.* 206:414–32.

Voss, R. S., J. L. Silva, and J. A. Valdes. 1982. Feeding behavior and diets of Neotropical water rats, genus *Ichthyomys* Thomas, 1893. *Z. Säugertierk.* 47:364–69.

Wahlert, J. H. 1993. The fossil record. In *Biology of the Heteromyidae*, ed. H. Genoways and J. H. Brown, 1–37. Special Publication 10. Shippensburg, Pa.: American Society of Mammalogists.

Walker, L. I., and A. E. Spotorno. 1992. Tandem and centric fusions in the chromosomal evolution of the South American phyllotines of the genus *Auliscomys* (Rodentia, Cricetidae) *Cytogenet. Cell Genet.* 61:135–40.

Walker, L. I., A. E. Spotorno, and J. Arrau. 1984. Cytogenetic and reproductive studies of two nominal subspecies of *Phyllotis darwini* and their experimental hybrids. *J. Mammal.* 65:220–30.

Walker, L. I., A. E. Spotorno, and J. Sans. 1991. Genome size variation and its consequences in *Phyllotis* rodents. *Hereditas* 115:99–107.

Weir, B. J. 1974a. The tuco-tuco and plains viscacha. *Symp. Zool. London* 34:113–30.

———. 1974b. Reproductive characteristics of hystricomorph rodents. In *The biology of hystricomorph rodents,* ed. W. Rowlands and B. J. Weir, 265–99. London: Academic Press.

White, T. G., and M. S. Alberico. 1992. *Dinomys branickii. Mammal. Species* 410:1–5.

Williams, D. F., and M. A. Mares. 1978. A new genus and species of phyllotine rodent (Mammalia: Muridae) from northwestern Argentina. *Ann. Carnegie Mus.* 47 (9): 193–221.

Wilson, D. E., and D. M. Reeder, eds. 1993. *Mammal species of the world*. 2d ed. Washington, D.C.: Smithsonian Institution Press.

Wilson, S. C., and D. G. Kleiman. 1974. Eliciting play: A comparative study. *Amer. Zool.* 14:341–70.

Wolffsohn, J. A. 1923. Medidas máximas y mínimas de algunos mamíferos chilenos colectados entre los años 1896 y 1917. *Rev. Chil. Hist. Nat.* 27:159–67.

Wood, A. E. 1935. Evolution and relationship of the heteromyid rodents with new forms from the Tertiary of western North America. *Ann. Carnegie Mus.* 24:73–262.

———. 1955. A revised classification of the rodents. *J. Mammal.* 36:165–87.

———. 1974. The evolution of the Old World and New World hystricomorphs. In *The biology of hystricomorph rodents,* ed. W. Rowlands and B. J. Weir, 21–54. London: Academic Press.

Woods, C. A. 1982. The history and classification of South American hystricognath rodents: Reflections on the far away and long ago. In *Mammalian biology of South America*, ed. M. A. Mares and H. H. Genoways, 377–92. Pymatuning Symposia in Ecology 6. Special Publication Series. Pittsburgh: Pymatuning Laboratory of Ecology, University of Pittsburgh.

Woods, C. A. 1984. Hystricognath rodents. In *Orders and families of Recent mammals of the world*, ed. S. Anderson and J. Knox-Jones Jr., 389–446. New York: John Wiley.

———. 1993. Suborder Hystricognathi. In *Mammal species of the world,* ed. D. E. Wilson and D. M.

Reeder, 771–806. Washington, D.C.: Smithsonian Institution Press.

Woods, C. A., and D. K. Boraker. 1975. *Octodon degus. Mammal. Species* 67:1–5.

Woods, C. A., L. Contreras, G. Willner-Chapman, and H. P. Whidden. 1992. *Myocastor coypus. Mammal. Species* 398:1–8.

Worth, C. B. 1967. Reproduction, development and behavior of captive *Oryzomys laticeps* and *Zygodontomys brevicauda* in Trinidad. *Lab. Anim. Care* 17 (4): 355–61.

Ximénez, A. 1980. Notas sobre género *Cavia* Pallas con la descripción de *Cavia magna* sp. n. (Mammalia-Caviidae). *Rev. Nordest. Biol.* 3 (special issue): 145–79.

• Ximénez, A., A. Langguth, and R. Praderi. 1972. Lista sistemática de los mamíferos del Uruguay. *Anal. Mus. Nac. Hist. Nat. Montevideo* 7 (5): 1–45.

• Yepes, J. 1938. Disquisiciones zoogeográficas referidas a mamíferos: Comunes a las faunas de Brasil y Argentina. *Anal. Soc. Argent. Estud. Geogr.* 6:37–60.

———. 1944. Comentarios sobre cien localidades nuevas para mamíferos sudamericanos. *Rev. Argent. Zoogeogr.* 4 (162): 59–71.

Zanchin, N. I., M. Galletti, and M. S. Mattevi. 1992. Karyotypes of Brazilian species of *Rhipidomys. J. Mammal.* 73:120–22.

Zanchin, N. I., A. Langguth, and M. S. Mattevi. 1992. Karyotypes of Brazilian species of *Rhipidomys. J. Mammal.* 73:120–22.

Zanchin, N. I., I. J. Sbalqueiro, A. Langguth, R. C. Bassla, E. C. Castro, L. F. Oliviera, and M. Mattevi. 1992. Karyotype and species diversity of the genus *Delomys* in Brazil. *Acta Theriol.* 37:163–69.

Diagnosis

Lagomorphs are small, digitigrade mammals with the tail either extremely short and well furred or virtually absent. The upper lip is cleft but mobile and can be compressed in the midline. The dental formula for the family is variable: I 2/1, C 0/0, P 3/2, M 2-3/3. The first upper incisor has a longitudinal groove down the midline. The second pair of upper incisors is immediately behind the first, and this character is diagnostic (fig. 17.1).

Distribution

Aside from introduction by humans, members of the Lagomorpha are found on all major continents and islands except Australia, New Zealand, and Antarctica. They were introduced by Europeans in Australia and New Zealand, where they have multiplied to become a serious pest. Introductions of exotic lagomorphs (*Oryctolagus* and *Lepus*) have had serious ecological consequences.

History and Classification

There are two Recent families, the Ochotonidae and the Leporidae; both originated in the Oligocene. The oldest fossils believed to belong to this order trace from the Paleocene. Lagomorphs radiated and diversified in Eurasia and North America; they did not enter South America until the completion of the Panamanian land bridge at the beginning of the Pliocene. European hares and rabbits have been widely introduced in southern South America.

FAMILY LEPORIDAE
Rabbits and Hares

Diagnosis

The dental formula is I 2/1, C 0/0, P 3/2, M 3/3. The ears are rather long. The hind limbs are longer than the forelimbs and typify an adaptation for quadrupedal, ricochetal locomotion. There is a strong reduction of the first digit on the forefeet and hind feet. The tail is short but well haired, with a fluffy appearance.

Distribution
See the ordinal account.

Genus *Sylvilagus* Gray, 1867
Cottontail Rabbit, Conejo, Coelho

Description
These rather small rabbits have a head and body length ranging from 250 to 450 mm; the tail is 25 to 60 mm (fig. 17.2). Weight ranges from 400 to 2,300 g. There are thirteen species. The South American forms were reviewed by Tate (1933).

Distribution
The genus is distributed from southern Canada to Argentina, occupying a wide range of habitats from semiarid, brushy areas to forests. The species *S. floridanus* has been studied in Venezuela (Ojeda-Castillo and Keith 1982). One species occurs in the area covered by this volume, *Sylvilagus brasiliensis*.

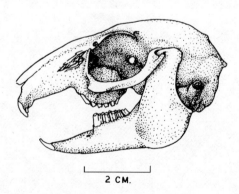

2 CM.

Figure 17.1. Skull of *Sylvilagus brasiliensis*.

Figure 17.2. *Sylvilagus brasiliensis.*

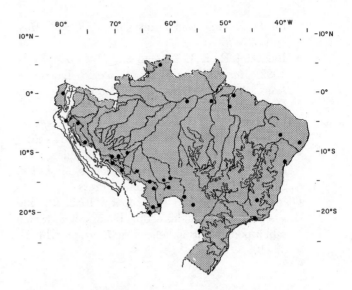

Map 17.1. Distribution of *Sylvilagus brasiliensis.*

Subgenus *Tapeti*
Sylvilagus brasiliensis (Linnaeus, 1758)
Brazilian Cottontail, Tapiti

Description
Average measurements: HB 320; T 21; HF 71; E 54; Wt 934 g. Its small size, delicate form, small tail, warmly colored rump, and six mammae distinguish *S. brasiliensis* from all other Neotropical rabbits. The dorsum is brown, but the black tips of the dorsal hairs give a speckled appearance. There is a rufous spot at the nape of the neck. The chest is rufous, as is the underside of the tail, and the venter is white. There are pale spots above the eye and on the muzzle (Hershkovitz 1950; pers. obs.).

Distribution
Sylvilagus brasiliensis ranges from southern Mexico to northern Argentina. It occurs in Amazonian Peru, Ecuador, and Bolivia and is common in eastern Brazil. In southern South America it is found in both eastern and western Paraguay and in Argentina across the northern provinces and as far south as Tucumán province (Redford and Eisenberg 1992) (map 17.1).

Natural History
The ecology of this species in the paramos of Venezuela was addressed by Durant (1983, 1984), who studied the high-altitude race *Sylvilagus brasiliensis meridensis*. He noted that this subspecies has a long gestation period of 44.9 days, compared with the 27-day gestation period of *S. floridanus*. The litter size of *S. b. meridensis* is small (3.9), and the reproductive cycle has an interval length of 270 days. The recruitment of this subspecies is extremely low. Throughout its habitat the Brazilian cottontail is a crepuscular browser and may show seasonal decline in abundance. It is strongly associated with moist habitats.

In Misiones province, Argentina, females reproduce in September. In Paraguay a female with three embryos was collected (Crespo 1982a,b).

Ecology
In Jujuy province, Argentina, this rabbit is common to about 2,500 m, preferring forest edges and brushy areas. It nests in brush heaps, in hollow trunks of trees, and among roots at the bases of trees. In Paraguay specimens were collected in forest with bamboo understory, in orange groves, and among bromeliads (Hershkovitz 1950; Olrog 1979). In Iguazú National Park this species is commonly sighted in both the Argentine and Brazilian portions of the park (Crespo 1982b).

References
• References used in preparing distribution maps
• Crespo, J. A. 1982a. Introducción a la ecología de los mamíferos del Parque Nacional el Palmar, Entre Ríos. *Anal. Parques Nac.* 15:1–33.
———. 1982b. Ecología de la comunidad de mamíferos del Parque Nacional Iguazú, Misiones. *Rev. Mus. Argent. Cienc. Nat. "Bernardino Rivadavia," Ecol.* 3 (2): 45–162.
Durant, P. 1983. Estudios ecológicos del conejo silvestre *Sylvilagus brasiliensis meridensis* (Lagomorpha: Leporidae) en los páramos de los Andes venezolanos. *Caribbean J. Sci.* 19 (1–2): 21–29.
———. 1984. Estudios ecológicos y nesesidad de protección del conejo del páramo. *Rev. Ecol. Conserv. Ornit. Latinoamer.* 1:35–37.
• Hershkovitz, P. 1950. Mammals of northern Colombia, preliminary report no. 6: Rabbits (Leporidae), with notes on the classification and distribution of

the South American forms. *Proc. U.S. Nat. Mus.* 100:327–75.

• Honacki, J. H., K. E. Kinman, and J. W. Koeppl, eds. 1982. *Mammal species of the world.* Lawrence, Kans.: Allen Press and Association of Systematics Collections.

• Hoogmoed, M. S. 1983. The occurrence of *Sylvilagus brasiliensis* in Surinam. *Lutra* 26 (1): 34–45.

• Mares, M. A., R. A. Ojeda, and R. M. Barquez. 1989. *Guide to the mammals of Salta province, Argentina.* Norman: University of Oklahoma Press.

Ojeda-Castillo, M., and L. B. Keith. 1982. Sex and age composition and breeding biology of cottontail rabbit populations in Venezuela. *Biotropica* 14 (1): 99–107.

• Olrog, C. C. 1979. Los mamíferos de la selva húmeda, Cerro Calilegua, Jujuy. *Acta Zool. Lilloana* 33: 9–14.

• Redford, K. H., and J. F. Eisenberg. 1992. *Mammals of the Neotropics.* Vol. 2. *The southern cone: Chile, Argentina, Uruguay, Paraguay.* Chicago: University of Chicago Press.

Schaller, G. B. 1983. Mammals and their biomass on a Brazilian ranch. *Arq. Zool. São Paulo* 31 (1): 1–36.

Tate, G. H. H. 1933. Taxonomic history of the Neotropical hares of the genus *Sylvilagus. Amer. Mus. Novitat.* 661:1–10.

Thomas, O. 1920. A further collection of mammals from Jujuy. *Ann. Mag. Nat. Hist.*, ser. 9, 5 (26): 188–96.

Part 3

Biogeography of
Land Mammal Faunas

In volume 1 of *Mammals of the Neotropics* (Eisenberg 1989), the nearshore islands, including Trinidad, Tobago, Isla Margarita, Grenada, Bonaire, Curaçao, and Aruba, were considered in some detail. The conclusion was that little specific distinctiveness had evolved but that subspecific differences could be noted. A similar conclusion could be made based on a consideration of the Panamanian islands: Coiba and Islas de las Perlas.

In volume 2 (Redford and Eisenberg 1992) the Falkland Islands were briefly considered, but we neglected to discuss the singular case of *Dusicyon (Pseudalopex) australis* (see appendix), although Isla Chiloé, Tierra del Fuego, and other inshore islands were dealt with in detail.

In this volume we discuss an outstanding group of islands deriving from volcanic processes that lie off the coast of Ecuador. The biology of the Galápagos has been reviewed in numerous publications (see Bowman, Benson, and Leviton 1983 for one of many reviews).

The mammals of the Galápagos are summarized in table 18.1. The species accounts are given in the text. Allen (1942) provides an useful review of status and extinction events for the mid-twentieth century. Figure (18.1) depicts the major islands.

The Galápagos are truly "Oceanic Islands." The marine mammal fauna (Otariidae) clearly have occupied the islands opportunistically as a breeding ground. The hoary bat (*Lasiurus cinereus*) is migratory and must have colonized the area by being blown off course. Note that the same species reached the Hawaiian archipelago. Living specimens have been collected from Santa Cruz and San Cristóbal. *Lasiurus borealis* has been described by Steadman (1986) as a fossil from Isla Floreana and may still survive. The pinniped species are discussed in the main text and listed in table 18.1. Only one species is endemic, *Arctocephalus galapagoensis*.

The oryzomyine rodents (*Nesoryzomys*, *Oryzomys*, and *Megaoryzomys*) present special problems. The

original stocks must have arrived from the continent of South America on floating masses of vegetation. This is an improbable event, but the three genera may represent separate invasions. *Megaoryzomys* could repre-

Table 18.1   Recent Mammals of the Galápagos Archipelago (Excluding Cetacea)

Taxon	Location
Chiroptera	
Vespertilionidae	
Lasiurus cinereus	Santa Cruz
Lasiurus borealis	Santa Cruz, Floreana, San Cristóbal
Pinnipedia	
Otariidae	
• *Arctocephalus australis*	
A. galapagoensis	
Zalophus californicus	
Rodentia	
Muridae	
Sigmodontinae	
Tribe Oryzomyini	
Described species	
Nesoryzomys	
+*N. darwini*	Santa Cruz
N. fernandina	Fernandina
N. indefessus	
+*N. i. indefessus*	Santa Cruz, Baltra
N. i. narboroughi	Fernandina
+*N. i. swarthi*	Santiago
Oryzomys	
O. galapagoensis	
+*O. g. galapagoensis*	San Cristóbal
O. g. bauri	Santa Fe
Megaoryzomys[a]	
+*M. curioi*	Santa Cruz
Undescribed species	
Nesoryzomys (1) (large form)	Isabela
Nesoryzomys (2) (small form)	Isabela
Megaoryzomys	Isabela

[a]Steadman and Ray 1982 suggest that *Megaoryzomys* shows affinities with *Thomasomys*, but see comments in chapter 16.
+ Extinct..
• Does not breed (?).

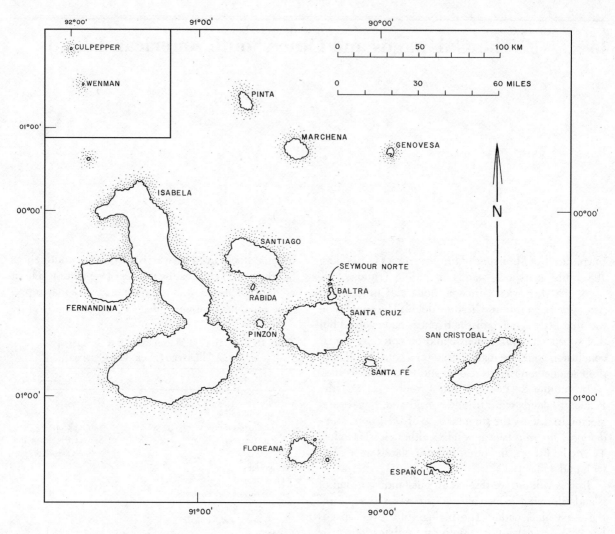

Figure 18.1. Map of the Galápagos Islands (adapted from Steadman and Zousmer 1988).

Table 18.2   Equivalent Names for Some Islands of the
              Galápagos Archipelago

English	Spanish
Narborough	Fernandina
Albermarle	Isabela
Indefatigable	Santa Cruz
James	San Salvador, Santiago
Barrington	Santa Fe
Chatham	San Cristóbal
Charles	Floreana
South Seymour	Baltra

sent an early colonization, followed by *Nesoryzomys*
and subsequently by *Oryzomys galapagoensis* (see
Patton and Hafner 1983; Steadman and Ray 1982).

The genus *Nesoryzomys* has been included as a
subgenus of *Oryzomys*, but most authors now recognize the distinctive nature of this grouping (Patton and
Hafner 1983). *Nesoryzomys narboroughi* from Fer-

nandina and *N. swarthi* from San Salvador have been
subsumed under *N. indefessus*, although they may be
considered subspecies. *N. indefessus* in the strict sense
occurs on Santa Cruz. In the broad sense *N. indefessus*
co-occurred with *N. fernandina* on Fernandina and
with *N. darwini* on Santa Cruz. *N. fernandina* and *N.
darwini* were smaller than *N. indefessus*, suggesting
that selection produced sympatric species adapted for
different microniches when they occurred on the same
island. *N. indefessus narborourghi* is still common, but
the other species are extinct or extremely rare. Data
presented by Steadman et al. (1991) established that
species of *Megaoryzomys* persisted until at least five
hundred years ago, as did four described and three as
yet undescribed species of *Nesoryzomys*. These authors conclude that humans, together with their introductions *Felis, Rattus,* and *Mus,* contributed to the extinction of *Nesoryzomys* and *Megaoryzomys* on Santa
Cruz and Isabela.

Oryzomys galapagoensis occurred on Santa Fe and

San Cristóbal, but it is extinct on the latter island. It apparently represents a recent colonization and did not co-occur with *Nesoryzomys*. The large *Megaoryzomys* formerly occurred on Isabela and Santa Cruz islands. Although it is now extinct, the causes of its demise are poorly understood. That *Nesoryzomys, N. darwini, N. indefessus,* and *Megaoryzomys curioi* may have occurred together in the past is a tantalizing possibility.

Additional Island Species

The Juan Fernández islands were discussed in volume 2 (Redford and Eisenberg, 1992). Although many exotic vertebrates have been introduced, the fur seal *Arctocephalus philippi* is the only endemic mammal other than offshore cetaceans.

Inshore islands that have not been discussed separately include Isla Chiloé, Tierra del Fuego, and Isla Mocha. In addition we should discuss the only terrestrial mammal of the Islas Malvinas (Falkland Islands), *Dusicyon australis*. The mammals of Isla Chiloé are plotted in Redford and Eisenberg (1992), but we did not discuss *Pseudalopex fulvipes*. The large *culpeo* of Tierra del Fuego was not discussed in detail and deserves inclusion, as does the new species of *Octodon* from Isla Mocha.

Dusicyon australis
Falkland Island Fox (Wolf)

Since several reviewers of volume 2 mourned our brief mention of this extinct species, we have decided to include it here. Consider this an apologetic correction. The facts as we understand them are these: The species in question was isolated during the last rise in sea level that separated the Islas Malvinas from the mainland and was exterminated at the close of the nineteenth century (1876 is often cited). Clearly, European settlement and the introduction of sheep brought the fox (rightly or wrongly) into conflict with the human settlers, but a brief trade in its pelt accelerated its demise. Given the confused taxonomic status of the genera of the Canidae (especially in South America) and the debates concerning the status of *Pseudalopex* and *Dusicyon,* and since we lack the specimens to even contemplate the issue, we despair of a taxonomic resolution within our lifetimes. We can say only that this canid taxon was exterminated.

Allen (1942) summarizes the evidence for this tragic case of insular extinction accompanying the transfer of European animal husbandry to an isolated island complex. Lest we become complacent, Massoia and Chebez (1993) sound a similar warning concerning *Dusicyon culpaeus lycoides* on Tierra del Fuego, which is equally threatened.

Octodon pacificus Hutterer, 1994

Description
Measurements: TL 390–92; T 165–70; HF 40–42; E 20; Wt 290 g. This large, heavy form of the genus *Octodon* has fur of a uniform dark brown, lighter on the venter. Dental features suggest that it is an isolate similar to the ancestor of the extant species.

Distribution
The species is confined to Isla Mocha, 38°22′ S, 75°55′ W, at 31 km from the mainland.

Pseudalopex fulvipes (Martin, 1837)
Darwin's Fox

Description
Measurements: TL 790; T 248; HF 123; E 77. This rather small fox has a dark brown grizzled pelage. The hind legs are almost black, the feet are buff, and the throat is whitish.

Distribution
Believing this form was confined to Isla Chiloé, Osgood thought it was distinct from the mainland *P. griseus*. Jaksić et al. (1990) reinforce this view with further evidence for the presence of this form on the adjacent mainland.

Pseudalopex culpeus lycoides (Philippi, 1896)
Fuegan Wolf

Description
This is the largest subspecies of *Pseudalopex* but is clearly related to the mainland *culpeo*. Philippi considered it a distinct species.

Distribution
The species is found only on Tierra del Fuego. The status of this form is discussed by Massoia and Chebez (1993).

Comment
This large canid preys on the guanaco. The puma has apparently never occupied the island.

References

Allen, G. M. 1942. *Extinct and vanishing mammals of the Western Hemisphere.* Special Publication 11. New York: American Committee for International Wild Life Protection.

Bowman, R. J., M. Benson, and A. E. Levition, eds. 1983. *Patterns of evolution in Galápagos organisms.* San Francisco: American Association for the Advancement of Science, Pacific Division.

Hutterer, R. 1994. Island rodents: A new species of *Octodon* from Isla Mocha, Chile. *Z. Säugetierk.* 59:27–41.

Jaksić, F. M., J. E. Jimenez, R. Medel, and P. A. Marquet. 1990. Habitat and diet of Darwin's fox (*Pseudalopex fulvipes*) on the Chilean mainland. *J. Mammal.* 71:246–48.

Massoia, E., and J. C. Chebez. 1993. *Mammíferos silvestres del archipiélago Fueguino*. Buenos Aires: Lezano.

Osgood, W. S. 1943. The mammals of Chile. *Field Mus. Nat. Hist. Zool. Ser.* 30:1–268.

Patton, J. L., and M. S. Hafner. 1983. Biosystematics of the native rodents of the Galápagos archipelago, Ecuador. In *Patterns of evolution in Galápagos organisms*, ed. R. J. Bowman, M. Benson, and A. E. Leviton, 539–68. San Francisco: American Association for the Advancement of Science, Pacific Division.

Steadman, D. W. 1986. Holocene vertebrate fossils from Isla Floreana, Galápagos. *Smithsonian. Contrib. Zool.*, vol. 413.

Steadman, D. W., and C. Ray. 1982. The relationships of *Megaoryzomys curioi*. *Smithsonian Contrib. Paleobiol.* 51:1–23.

Steadman, D. W., T. W. Stafford Jr., D. J. Donahue, and A. J. T. Jul. 1991. *Quat. Res.* 36:126–33.

Steadman, D. W., and S. Zousmer. 1988. *Galápagos: Discovery on Darwin's islands*. Washington, D.C.: Smithsonian Institution Press.

19   Biodiversity Reconsidered

John F. Eisenberg

Contemporary Habitats and Species Richness

Introduction

A continent with the latitudinal span of South America will obviously yield gradients of biodiversity that replicate those found in the northern continents. Vegetation zones in the north are tropical, but the rain shadow of the Andes and the failure of the westerly winds between 10° N and 10° S (the area of the equatorial convergence), as well as the mixed and variable air and ocean currents until 22° south latitude, combine to offer the student a kaleidoscope of rainfall patterns, soil fertility, and subsequent vegetation cover (Eisenberg and Redford 1982).

The savanna areas of the Neotropics include the llanos, pantanal, and palm forests, while the seasonal dry areas of woody vegetation include the cerrado, caatinga, and chaco (see chapter 1). Evergreen tropical rain forest areas are disjunct and are broadly divisible

into the Amazonas, Guyanan, and Atlantic rain forests. This division does not consider the temperate areas of the southern portion of the continent, including the pampas grasslands, the Patagonian steppe, the sub-Antarctic beech forests, and the semidesert monte. The Andes Mountains provide the ultimate climatic zonation, dominating the western rim of the continent and at times exceeding 7,000 m elevation (Eisenberg 1989; Hershkovitz 1972).

The tropical rain forests of Amazonas are not a homogeneous habitat; they exhibit both vertical and horizontal heterogeneity. Vertical heterogeneity is best exemplified by forest strata. Seasonal flooding determines the presence or rarity of terrestrial forms and imposes horizontal heterogeneity (fig. 19.1). The dry, deciduous tropical forests and savannas also exhibit pronounced horizontal heterogeneity, since gallery forests near permanent streams serve as "ribbons of life" for species adapted to evergreen tropical forests

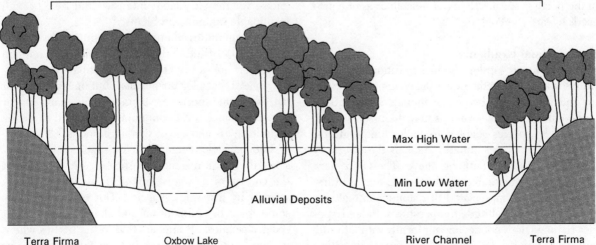

Figure 19.1. Generalized diagram of a multistratal tropical evergreen forest in western Brazil. Here we have great vertical heterogeneity in the vegetation form but, in addition, an annual disturbance effect deriving from fluctuations in the flow of rivers. In areas that are flooded for long periods, the terrestrial mammal fauna may be sparse even during periods of low water.

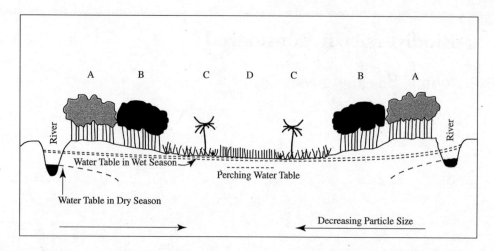

Figure 19.2. Profile of vegetation types in a seasonally dry savanna. In this region with a pronounced dry and wet season, the vegetation forms away from permanent water may include low-stature trees and have an open woodland appearance. Near permanent streams, trees of different strata and closed canopies may be apparent. The figure emphasizes lateral heterogeneity of the vegetation: (a) evergreen forest; (b) deciduous forest; (c) shrubs and palms; (d) wet grassland. (Redrawn from Sarmiento 1984.)

(Eisenberg 1990). The presence of tropical forests along river courses in seasonally arid areas offers a genuine challenge to students of zoogeography (fig. 19.2). August (1983) demonstrated that species richness correlates positively with habitat complexity.

Latitudinal Gradients

South America tapers as one approaches the South Pole, and the southern 30% of the continent resembles a huge peninsula. Eisenberg and Redford (1982), reinforced in Redford and Eisenberg (1992), demonstrated that the niches occupied by mammals in southern South America were equivalent; however, the number of species per niche appeared to be depauperate when compared with North America, yielding a peninsular effect. Given this caveat, mammalian fauna of the north—species richness—declines as one proceeds toward the Pole (fig. 19.3).

Altitudinal Gradients

As one moves "upslope" on an elevational gradient from sea level within the topics (all edaphic factors being equal), species richness of mammals declines. Species common at low elevations may disappear and be replaced by species adapted to cooler temperatures. This trend in species number parallels the effects of latitudinal gradients outlined above. The Chiroptera are especially vulnerable to changes in temperature, but it should be appreciated that some mammals are adapted to higher-altitude temperatures. Indeed, species richness does not decline uniformly with altitude, and many taxa exhibit replacement, especially within the order Rodentia (Eisenberg 1989). Thus where altitudinal gradients occur, *total* species number will be high for areas including both lowland and highland forms (fig. 19.4).

Species Richness and Rarity

Some mammal species may be broadly distributed geographically yet be numerically rare. Rarity is often a case of trophic adaptation; for example, carnivores are numerically less abundant than herbivores (Elton 1927; Eisenberg, O'Connell, and August 1979).

Species richness, or the number of species tabulated for a defined geographical area, is not easy to determine, but calculating diversity (H') is even more taxing (Pielou 1969). To calculate diversity requires estimating numerical abundance for each species, a task that has been approximated for only a few areas. Even then the accuracy is questionable (see chapters by Emmons, Janson, Glanz, and Malcolm in Gentry 1990). Density estimates are critical for conservation planning and for understanding microhabitat preferences, yet they are labor intensive and fraught with formidable sampling problems.

In communities where density estimates have been performed, it is clear that some species are common and others are rare. The few sites that have been studied suggest that a log normal distribution is a recurrent pattern. Some species are abundant, but many appear very rare, since most communities sampled in reality include an assemblage of subdistributions.

Ecological dominance by a few species is not necessarily the result of random processes but may reflect the control of a large segment of a community's resources by just a few species. This concept implies some form of competition and thereby can restrict species richness. It appears that in the tropics where mammal communities have been analyzed (albeit imperfectly) the log normal distribution describes the species abundance curves. This is especially true for complex communities exhibiting great vertical and horizontal heterogeneity. In simpler communities the

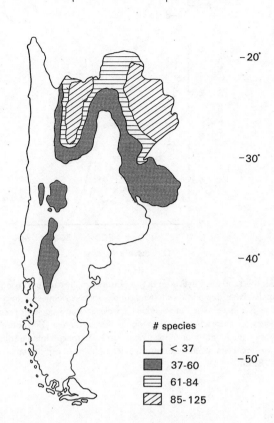

75°   65°   55°

−20°

−30°

−40°

−50°

species

< 37

37-60

61-84

85-125

Figure 19.3. Species richness for the southern portion of South America. The map shows the political boundaries of Chile, Argentina, Paraguay, and Uruguay. The rain shadow of the Andes to the west can profoundly influence species richness to the east. Regions of high species richness include the tropical portions and areas associated with permanent bodies of water. Note the rather high species richness in the Andes between 32° and 42° S. This area of endemism contributes to the higher species richness in the Andean region. (Resolution, 1/10°; compiled from Redford and Eisenberg 1992.)

not only is often limiting but is uncommon in the environment.

Great emphasis has been placed on the supposed stability of the tropics in terms of climate and productivity; yet much of the tropics is anything but stable in a temporal or spatial sense. Although plant communities often exhibit high diversity (H'), animal communities may not exactly parallel this trend. Most mammal communities in the Neotropics exhibit a log normal distribution (Peterson, Roberts, and Pinheiro 1981). That is, some species are very abundant, many are very rare, and most have intermediate values (figs. 19.5–19.8). The skew toward a few very abundant species is accentuated in disturbed habitats or those showing pronounced seasonal productivity. These trends are exhibited by both chiropteran and nonchiropteran communities (figs. 19.9 and 19.10).

Low species richness for a given area may be the result of primary productivity as well as Recent catastrophic events or geographic isolation. Both temperate Australia and temperate South America show lowered species richness when compared with equivalent areas in the Northern Hemisphere. Massive Pleistocene extinctions occurred in these areas. Isla Chiloé and the Península Valdés in South America show near equivalent species richness when compared with Tasmania, but all three areas have lower values than comparable areas in North America and Asia. This reflects massive extinction of the megafauna within the past twenty thousand years.

The Community Ecology of Small Mammals in South America

Introduction

To answer questions concerning community ecology, we must have a species inventory; in short, we must know how many species are present. In addition to this, we need to know the relative abundance of each species. Do some species predominate while others are rather rare, or is there evenness in numbers across all species categories? To sample a mammalian community, it is customary to trap. The sample can be gathered by live trapping with mark-and-release or by trapping and removal. In a complex lowland tropical forest, there is vertical as well as horizontal heterogeneity. Thus, what species are sampled and in what numbers may be subject to bias with respect to the microhabitat. Of course, in sampling small mammals by trapping, all things being equal, forms that are easy to trap might be overrepresented. Some animals that are not trapable will have to be collected by other means.

Establishing species richness or number of species for a study area might seem trivial, but Voss and

dominance of a few species is more exaggerated, and something like ecological dominance may be an appropriate description. Such simple communities in the Neotropics include the tropical grasslands or semiarid scrub characteristic of the llanos or caatinga formations of Venezuela and northeastern Brazil.

Scaling factors are also involved, since within a trophic class smaller species are usually more abundant than large species. Many life history traits such as longevity, litter size, reproductive potential, and generation interval scale predictably with body size when taxa are examined from the standpoint of a trophic classification (Eisenberg 1981). Of course, phylogenetic inertia is also a factor (Kinnaird and Eisenberg 1989; Robinson and Redford 1986). Given the caveats above, some taxa are apparently rare for reasons that are not readily apparent and must be related to some key specialization for exploiting a resource that

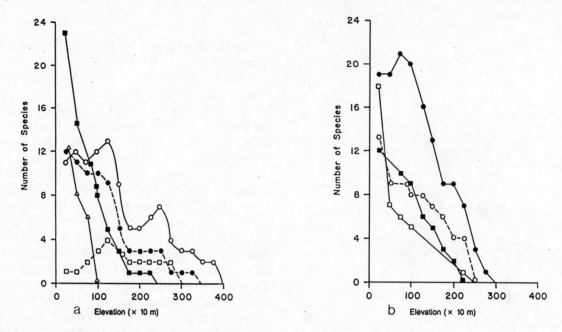

Figure 19.4. (a) Elevational distributions of some mammals in Venezuela: solid square, Phyllostominae; open square, Sturnirinae; triangle, Emballonurinae; open circle, sigmodontine rodents; solid circle, Marsupalia. Note that all taxa show maximum species richness at lower elevations, but that rodents have evolved a group of species adapted for higher elevations. For most bats, species richness declines precipitously above 1,000 m elevation, but the Sturnirinae seem to achieve greatest species richness in premontane forested areas (Eisenberg 1989). (b) Elevational distributions of some bats in northern Venezuela: solid circle, Stenoderminae; open circle, Vespertilionidae; solid square, Glossophaginae; open square, Molossidae. Note that the Stenoderminae can reach maximum species richness in premontane forests. The Molossidae decline rapidly in species richness above 500 m (Eisenberg 1989).

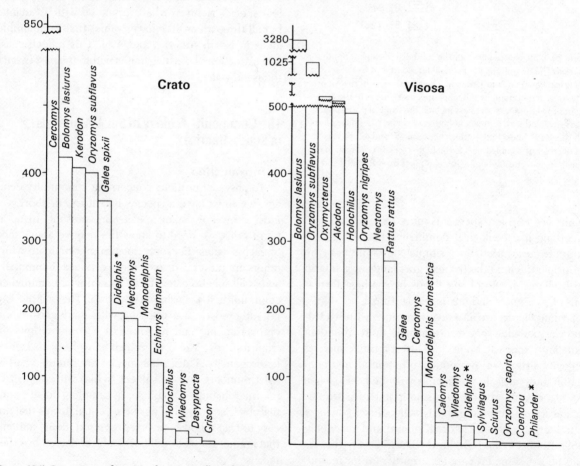

Figure 19.5. Comparison of two Brazilian seasonally arid sites intensively collected (rank order of abundance). Crato is an area of the caatinga with streams and scrub vegetation. Visosa is a cerrado grassland with streams and fringing trees. Note the numerical dominance of a few species and the exaggerated abundance of *Bolomys* at Visosa (from Peterson, Roberts, and Pinheiro 1981).

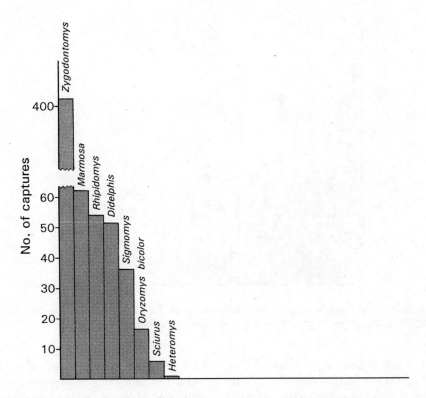

Figure 19.6. A Venezuelan site with eight small mammal species. The habitat is grassland and open woodland. Note the extreme dominance of one species, *Zygodontomys brevicauda* (O'Connell 1981). The graph represents one year of trapping on 4 ha.

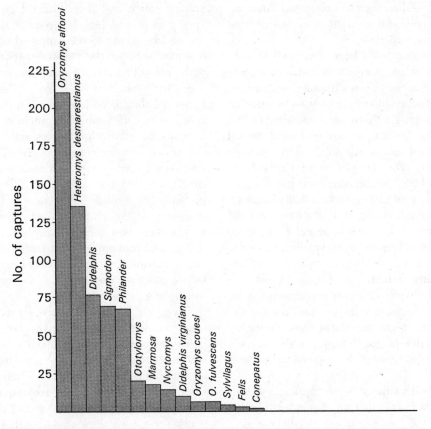

Figure 19.7. Trapping results for small mammals from a locality in southern Mexico. Although numerical dominance by a few species is still evident, the pronounced effect shown in figure 19.6 is not duplicated (Medellin 1992).

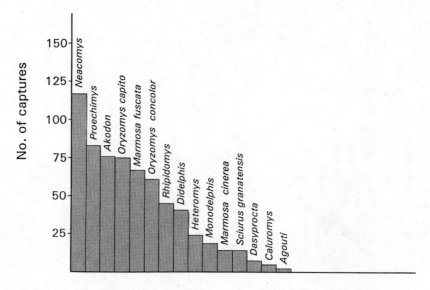

Figure 19.8. A premontane forest site in northern Venezuela. Note the more even distribution of numbers despite the numerical dominance of *Neacomys*. The graph represents one year of trapping on 4 ha. *Source:* O'Connell 1981.

Emmons (1996) carefully document how difficult it is to develop such lists. Only a handful of study areas in the lowland Neotropics have been studied in sufficient detail to warrant an assertion of a complete inventory. Population data for each species within a defined community have been achieved only for some segments of a community, seldom for a complete vertebrate assemblage. Yet in spite of these limitations, we can draw some tentative conclusions.

A species inventory for a large area, such as a national park, may include many forms that are not readily censused by trapping. Thus a trapping program, in fact, may represent a subsample of those forms that are actually present. Furthermore, depending on the ease of trapping, the sample may be biased toward those forms that are most easily obtained by a specific trapping technique. For example, what is trapped is in part a function of the size and construction of the trap itself. If one relies on bait, then trapability depends on the bait employed. The forms that are trapped and are representative of their true abundance will in no small measure be influenced by the placement of the trap with respect to microhabitat or, in the case of forests, the verticality of the trapping station. Finally, an aspect frequently overlooked is site preparation for the trap and the animal's familiarity with the area and the trap site. Thus long-term trapping at the same station can yield very different results from short-term trapping with quick placement of the trap and subsequent removal.

Since logistics limit the style of trapping and trap placement, it is fair to say that the comparisons in this section should involve animals of a moderate body size. It is often impossible to replicate a trapping procedure

for specimens weighing more and less than a kilogram within the same designated spacing of a trapping grid. At present I believe we have enough knowledge of the taxonomy and distribution of most South American mammals so we can make certain generalizations.

It is probable that genera that have a wide geographic range and that exhibit high karyotypic diversity as well as high heterosis at the species level will also be shown to be composed of "superspecies" that may be broken into several potential species. Second, genera that contain many species probably have a greater probability of being represented across a range of habitats. Within areas of high diversity, then, specious genera probably contribute greatly to the diversity by subdividing niches within a community. When two or more congeners co-occur, they usually segregate according to size, microhabitat, diet, or some combination of these. Dietary differences, in fact, probably often reflect differences in the body size of congeners. Although genera with many species may well be represented across many habitats, it is also true that genera containing few species may have not only broad distributions, but probably broad trophic adaptations and thus can co-occur in many communities, usually as a single species.

Given the limitations above, let us look at several small mammal trapping programs carried out over the past two decades. In particular I will focus on those studies that have attempted to examine vertical as well as horizontal heterogeneity and have used a diversity of trap sizes and baiting techniques to guarantee a rather fair sampling of the extant community. For comparison, I will draw on data gathered by Medellin (1992, 1994) in Chiapas, Mexico, which represents a

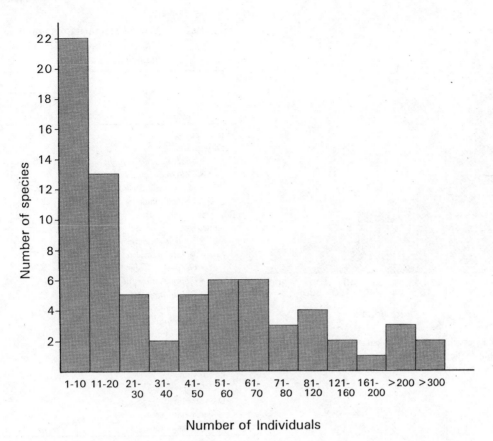

Figure 19.9. Plot of species number against number of individuals for small nonvolant mammals from lowland Neotropical sites. (Includes summaries of data from Medellin 1992; Malcolm 1990; Eisenberg, O'Connell, and August 1979; Fonseca and Kierulff 1988; and Stallings 1988a,b.)

northern extreme for Neotropical forms. I will also use the data that Stallings (1988a,b) and Fonseca (1988) derived from the Atlantic rain forest of Brazil, which for our purposes could represent a southern extreme. Within any of these regions, there is microhabitat heterogeneity caused by variations in rainfall, drainage, and other edaphic factors. Thus data from grassland and savanna communities in Brazil gathered by Mares and associates (see Mares 1992) will be compared with data from Venezuela developed by O'Connell (1989) and August (1983) in the llanos. The true forested habitats at low elevations in continental South America will include those studied by O'Connell (1981, 1989) in Guatopo National Park, Venezuela; Malcolm (1991) in the vicinity of Manaus, Brazil; Valqui (1995) in the upper Amazon of Peru and Woodman, Slade, and Timm (1995) in southeastern Peru. In addition, I will refer to species lists developed from inventories in Peru by Hutterer et al. (1995); Jansen and Emmons (1990): Patton and Gardner (1972); and Patton, Myers, and Smith (1990). Data for Panama are from Fleming (1971, 1975).

As August (1983) has demonstrated, the species richness of small mammal communities varies posi-

tively with habitat complexity, usually a combination of horizontal as well as vertical heterogeneity (fig. 19.11). This generalization has been amply confirmed by studies of deep gallery forest along major rivers as they intersect through areas that normally would climax in grasslands or dry deciduous forests (see also Mares 1992; Mares and Ernest 1992). Table 19.1, derived from a group effort extending from 1976 to 1984, shows both the absolute number of animals trapped in the savanna woodlands and the species known to be present in a gallery forest but not estimated with respect to density. Nine species are noted for the savanna woodlands and thirteen for the gallery forests. Table 19.2 is based on trapping by Streilein in the semiarid caatinga of Brazil. Twelve small mammal species were taken or observed. Table 19.3 presents some trapping results from the cerrado, where six species were taken in the grasslands (campos) and nine in the gallery forests. Thus, in the cerrado of Brazil, grasslands may support a community of five small rodent species dominated by two genera, *Bolomys* and *Oxymycterus*, whereas the gallery forests may contain six rodent species dominated by *Bolomys* and *Oligoryzomys*. Correspondingly, the marsupial communities

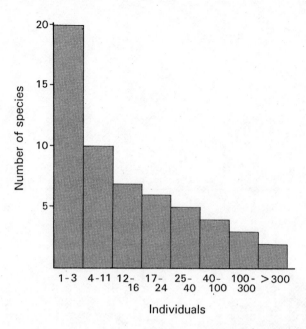

Figure 19.10. Plot of the Chiroptera taken at La Selva, Costa Rica. Once again, the dominance of a small number of species in terms of numerical superiority is evident (Wilson 1990).

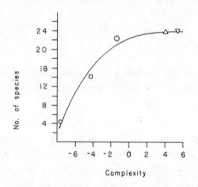

Figure 19.11. Number of mammalian species plotted against an index of habitat complexity (vertical plus horizontal heterogeneity) (data from August 1981).

Table 19.1   A Small Mammal Community from the Llanos of Venezuela, Hato Masaguaral, Venezuela

	Habitat	
Taxon	°Savanna-Woodland	Gallery Forest and Swamp
Marsupialia		
Didelphidae		
• *Didelphis marsupialis*	53	+
• *Caluromys philander*		+
• *Marmosa robinsoni*	65	+
Total Marsupialia	118	
Rodentia		
Sciuridae		
• *Sciurus granatensis*	1	+
Muridae (Sigmodontinae)		
• *Oecomys bicolor*	57	+
• *Oligoryzomys fulvescens*	+	+
• *Rhipidomys* sp.	57	+
• *Sigmomys alstoni*	37	+
Holochilus sciureus	−	+
• *Zygodontomys brevicauda*	404	+
Heteromyidae		
• *Heteromys anomalus*	17	+
Echimyidae		
Echimys semivillosus	−	+
Dasyproctidae		
Dasyprocta sp.	+	+
Total Rodentia	573	

Source: Data from O'Connell, August, Sunquist, and Ludlow.
• Species trapped by O'Connell on 4 ha grid.
+ Occurs, but not tabulated.
− Absent.
° Number of individuals.

in the grasslands may fall to two species, whereas in the gallery forest they increase to three. It is the addition of the vertical heterogeneity that promotes the extra species in the gallery as opposed to the grassland. In the llanos of Venezuela, O'Connell was able to collect five species of rodents and two species of marsupials in an open woodlands savanna (table 19.1). The rodents were dominated numerically by a single genus, *Zygodontomys*. Marsupials and rodents characteristic of multistratal tropical evergreen forest may be found in the seasonally arid habitats dominated by open woodlands and grassland where multistratal evergreen forest is found associated with large permanent rivers. For example, in the llanos of Venezuela nine small mammal species were present in the savanna woodlands, but an additional four species were found together with the savanna woodlands species in the gallery forests. The semiarid savanna woodlands of

South America typically contain three species of marsupials and eight or nine species of rodents. Although species composition differs between eastern Brazil and northern Venezuela, the community structures are remarkably similar (compare table 19.1 and 19.2). Once again, when the woodland savannas are compared with gallery forest, certain species seem well adapted to the grasslands. Others may be found in both grasslands and gallery forests, and still a third group of species appears to be confined solely to gallery forests (see table 19.3).

It is in the tropical forest areas where true multistratal tropical evergreen forest occurs that small mammal diversity increases dramatically. Stallings and Fonseca in southeastern Brazil recorded twelve species of rodents and seven species of marsupials. in Manaus, Brazil, Malcolm recorded eight species of marsupials and twelve species of rodents. The contribution of various species to the enhanced diversity will be considered in the next two sections.

Marsupial Diversity

At the northern limits of the tropics, such as Chiapas, Mexico, and the republic of Panama, the potential

Table 19.2   A Small Mammal Community from
               Northeastern Brazil

Taxon	Individuals Trapped (Grid 1)
Marsupialia	
• Monodelphis domestica	38
Thylamys pusilla (karimii)	+
• Didelphis albiventris	32
Total Marsupialia	70
Rodentia	
Muridae (Sigmodontinae)	
Oligoryzomys eliurus	+
Oryzomys subflavus	+
• Bolomys lasiurus	6
• Calomys callosus	6
• Wiedomys pyrrhorhinos	3
Caviidae	
• Kerodon rupestris	5
Galea spixii	58
Echimyidae	
• Trichomys aperoides	59
Dasyproctidae	
Dasyprocta prymnolopha	+
Total Rodentia	137

Source: Data from Streilein 1982a,b,c (caatinga).
+ Present but not trapped regularly.
• Trapped regularly.

Table 19.3   A Small Mammal Community from the
               "Cerrado"/Gallery Forest of Central Brazil

Taxon	Grassland (Campos)		Gallery	
Marsupialia				
Didelphidae				
Didelphis albiventris	+		+	
Gracilinanus agilis	−		192	(85)
Monodelphis americana	?		34	(15)
Total Marsupialia			226	
Rodentia				
Muridae (Sigmodontinae)				
Akodon cursor (?)	11	(8)	52	(14)
Bolomys lasiurus	81	(57)	141	(37)
Calomys callosus	4	(3)	−	−
Oligoryzomys nigripes	7	(5)	133	(35)
Oxymycterus roberti	40	(30)	−	−
Oecomys bicolor	−	−	22	(16)
Oryzomys capito	−	−	68	(18)
Oecomys concolor trinitatus	−	−	68	(18)
Total Rodentia	143		384	

Note: Caviomorpha not trapped.
Source: Alho 1981; Lacher, Mares, and Alho 1989; Nitikman and Mares 1987.
(?) *Thalpomys lasiotis.*
+   Present but not trapped regularly.
−   Absent.

maximum species richness of marsupials for any given locality is about five, because the number of species of the smaller forms diminishes as one proceeds north. In northern Venezuela where O'Connell studied, there were six easily trapped species of marsupials for any given elevation, but because of the geographical isolation of the north coast range of Venezuela, at least two genera, *Metachirus* and *Philander*, were absent. This absence is due to biogeographical processes rather than to other causes such as available habitat (see table 19.4).

In the lowland tropical forest of continental South America, nine genera of marsupials are usually represented. A tenth, *Chironectes*, may often occur but must be sampled by special trapping techniques. The genus *Didelphis* is found in most habitats and usually constitutes from 15% to 45% of the marsupials trapped. If arboreal traps are set, the genus *Micoureus* is well represented, with densities ranging from 10% to 33% of the sample. The genus *Marmosops* tends to be rather terrestrial but is well represented in most areas sampled. The genus *Marmosa* is scansorial and usually is represented in most areas. *Metachirus* is highly terrestrial and in some respects is a microhabitat specialist, apparently favoring early successional areas. In addition, it is not uniformly distributed over South America, being absent from the llanos and drier areas of Brazil. Its density can range from as low as 1% up

to 21% of the trapped marsupials. *Caluromys* is highly arboreal, and trapping success depends on attention to this dimension. Where efforts have been made, it has represented from 3% to 24% of the total marsupial catch. *Philander* is rather terrestrial, although it climbs some. In common with *Metachirus*, it has a disjunct distribution and thus is absent from certain sampled areas. Nevertheless, where found, it may dominate and can average about 21% of the marsupial catch. *Monodelphis* occurs broadly in continental South America but may require special trapping techniques. It extends well into the lowland dry zone of Brazil following gallery forests and can represent as much as 13% of the total small marsupials trapped. The genus *Gracilinanus* is clearly somewhat specialized. It is extremely small and highly arboreal, and where it occurs it appears to forage at the extreme terminal branches of shrubs and trees. The genus is mainly distributed either at moderate elevations in northern South America or at lower elevations in southeastern Brazil. Here it may be taken in moist forest as well as gallery forest and dry deciduous areas. It tends thus to have a scattered distribution and is not uniformly sampled in the six core areas that have been cited, including Guatopo, Venezuela, Manaus, Brazil, and southeastern Brazil. Supplementary data from efforts by Valqui and by Woodman et al. in Peru support the contentions above (see also table 19.5). When the old genus *Marmosa*

Table 19.4   Basic Studies for Seven Lowland Tropical Forests and Masaguaral Gallery Forest for Marsupialia (Percentage of All Marsupials, Number of Individuals, First Capture)

Taxon	Chiapas, Mexico	Guatopo, Venezuela	Manaus, Brazil	Hato Masaguaral, Venezuela	Panama	SE Brazil	[Peru][a]	Range	Mean
Didelphis	32	29	25	45	40	15	—	15–45	31
Micoureus	—	10	12	—	—	33	51	10–33	18
Marmosops	—	45	15	—	—·	24	32	15–45	28
Marmosa	< 4	?	6	66	39	4	4	4–66	29
Metachirus	2	—	6	—	< 1	21	6	1–21	7.5
Caluromys	—	3	24	—	—·	4		3–24	10
Philander	62	—	< 1	—	20	< 2	—	1–62	21.3
Monodelphis	—	13	13	—	—	< 1	—	< 1–13	9
Gracilinanus	—	—	—	—	—	< 1			

[a]Woodman et al. 1991 data from Peru for comparison only; not calculated in final two columns.

(sensu Tate) is broken into the new generic categories such as *Micoureus, Marmosops, Gracilinanus,* and *Thylamys,* much of the former confusion in understanding community dynamics for the marsupials vanishes. In most of the areas studied a single species for each genus dominates, suggesting microhabitat specialization. The case is quite different, however, when the genera of smaller rodents are considered.

The Small Rodents of Lowland Tropical South America

Small rodents exhibit high diversity within the lowland tropical forest of South America. Over a range of habitats one can see that the same genera are usually represented by at least one species; but some genera are highly speciose, and in areas with a high number of rodent species the increase in species number usually comes from some key genera that are in themselves highly speciose, with two or more species at an alpha (habitat) level of diversity (see table 19.6). The same trend can be noted to a lesser degree for the Marsupialia (see table 19.5). In areas with high small mammal biodiversity, fourteen species of rodents can coexist with eight species of marsupials at an alpha diversity level. Clearly, some of this diversity can be accounted for by fine subdivision of niches by congeners. At the generic level, potential niche separation is easily demonstrable when activity rhythms, substrate choice, and diet are considered (see table 19.7).

If one considers nondeciduous or evergreen lowland tropical forest (at or below 700 m), then over much of South America the communities of small rodents fall out as follows. Table 19.8 presents a series of small rodent species caught at nine study sites in the lowland tropics. Five families and twenty-two genera are represented. There may be from one to three species of squirrels (*Sciurus*), one to three species of *Oecomys,* one species of *Oligoryzomys,* one to four species of *Oryzomys,* one or two species of *Neacomys,* one

species of *Rhipidomys,* and in some special cases one species of *Akodon* and one species of *Oxymycterus.* Returning to the Echimyidae, there may be one species of *Echimys,* one to five species of *Proechimys,* one species of *Isothrix,* and one species of *Mesomys. Isothrix* and *Mesomys* are found only in the complex rain forest of Amazonas (see table 19.6). Of particular interest to us in this discussion are *Proechimys* and the old genus *Oryzomys,* which now has been broken into *Oryzomys* in the strict sense, *Oecomys,* and *Oligoryzomys* (genera germane to the lowlands).

The genus *Oecomys* is characterized by arboreally adapted rice rats. Some, such as *O. bicolor,* are highly arboreal. Others such as *O. concolor* are more scansorial. But all of them show adaptations for using the vertical aspect of the forest. Depending on the sampling procedure, they may be moderately abundant to quite rare, ranging from 1% to 15% of the total small rodent catch. *Oligoryzomys* is usually represented, if at all, by one species. This small rice rat is rather terrestrial and is usually associated with disturbed or ruderal habitats. Generally they compose a small percentage of the total catch, less than 4%. The genus *Oryzomys* may contain up to four species at one locus. These are predominantly terrestrial and apparently segregate in multispecies communities according to microhabitat preference. They may dominate under certain conditions and achieve as high a level as 45% of the total small mammal catch. *Neacomys,* the spiny mice, are usually represented by a single species in any given community. They may show dramatic population fluctuations in some habitats; although they are often a minor component, in "population highs" they can achieve 24% (O'Connell 1981). *Rhipidomys* is usually represented by a single species. This highly arboreal rat is not a dominant and rarely achieves over 4% of the numerical total for a given site. The genus *Akodon* is one of the most speciose genera in South America, with roughly forty-five named forms. On the other

Table 19.5   Presence of Marsupial Species at Ten Neotropical Sites

Taxon	Woodman et al., Peru	Valqui, Peru	Stallings, Brazil	Fonseca, Brazil	Malcolm, Brazil	Hutterer, Peru	Janson and Emmons, Peru	O'Connell, Venezuela	Medellin, Chiapas, Mexico	Charles-Dominique, French Guiana
Didelphis albiventris										
D. aurita			+	+						
D. marsupialis	+	+			+	+	+	+	+	+
Micoureus cinerea (demararae)		+	+	+	+		+	+		+
M. phaea						+				
M. regina	+					+				
Marmosops fuscatus								+		
M. impavidus		+?								
M. incanus			+	+						
M. noctivagus	+	+					+	+		
M. parvidens	+	+			+					
Marmosa lepida						+				
M. mexicana									+	
M. murina	+	+			+			+		+
Metachirus nudicaudatus	+	+	+	+	+	+	+		+	
Gracilinanus agilis				+						
G. microtarsus			+	+						
Caluromys derbianus									+	
C. lanatus					+	+	+			
C. philander			+	+	+			+		+
Monodelphis americana				+						
M. brevicaudata					+			+		
Philander mcilhennyi						+				
P. opossum				+	+	+	+		+	+
Chironectes						+		+	+	
Glironia							+			
Caluromysiops							+			
Total	6	7	6	9	9	10	8	7	6	5

hand, it is most speciose in the south temperate part of the continent and extends up the foothills to rather high elevations in the Andes all the way to the north of Peru. Where it occurs in lowland tropical habitats, it is usually represented by a single species. It can achieve high numerical densities only in southeastern Brazil and Venezuela, where it constitutes 55% of the total small mammal catch. *Oxymycterus* has a widespread distribution south and at the eastern portion of the Amazon basin, and although it extends into the moist tropics, it is most speciose in lowland dry to temperate zones of southeastern South America and thus

does not figure prominently our general pattern (see table 19.8).

Within the family Echimyidae most of the multistratal tropical rain forest has one species of *Echimys*. In western Amazonas two other genera are represented, usually by a single species: *Mesomys* and *Isothrix*. Wherever they are trapped, they are strongly arboreal and represent only a small fraction, less than 3%, of the total small rodents trapped. The genus *Proechimys*, however, is quite a different story and may include up to 80% of the trapped small mammals. This is an extreme value, though, and 20% would be a

Table 19.6 Relative Abundance of Rodents under 500 Grams in South American Lowland Tropical Forests: Basic Studies

| | | | | | | Genus Only | |
| | | | | | | Number of Species Co-occurring | Range of Percentage Contribution to Total |
Taxon	Valqui, Peru	Woodman, Peru	Malcolm, Brazil[a]	O'Connell, Venezuela[a]	Stallings and Fonseca, SE Brazil[a]		
Muridae (Sigmodontinae)							
•Oecomys bicolor	1%	14%	3%	—	—	1–2	1–15%
Oecomys trinitatus/concolor	—	—	—	13%			
Oecomys paricola/tapajinus	—	3%	7%	—	—		
Oecomys superans	—	—	—				
•Oligoryzomys microtis	4%	—	—	—	—	0–1	0–4%
Oligoryomys nigripes	—	—	—	—	1%		
•Oryzomys capito	0.4%	45%	20%	16%	3%	1–3	1.7–45%
Oryzomys macconnelli	2%	—	20%	—	—		
Oryzomys nitidus	—	22%	—	—	—		
Oryzomys subflavus	—	—	—	—	6%		
Oryzomys yunganus	6%	1.7%	—	—	—		
•Neacomys guianae/tenuipes	—	—	<1%°	24%	—		
Neacomys spinosus	<1%	1.5%	—	—	—	1–2?	0–24%
•Nectomys sqamipes	<3%	—	—	—	4%	0–1	0–4%
•Rhipidomys sp.	—	3%	4%	4%	1%	0–1	1–4%
•Akodon cursor	—	—	—	—	55%	0–1	0–55%
Adodon urichi	—	—	—	16%	—		
•Oxymycterus sp.	—	—	—	—	1%	0–1	0–0.2%
•Calomys laucha	—	—	—	—	0.2%	0–1	0–0.2%
Echimyidae							
•Echimys sp.	<1%	—	<1%	<1%	<0.1%	0–1	0–<1%
•Proechimys sp.	Σ ~80%	7%	49%	17%	14%	1–4 +	<7–80%
Proechimys goeldii group	<1%	—	—	—	—		
Proechimys guayanensis group	—	—	42%	—	—		
Proechimys longicaudatus group	<4%	—	—	—	—		
Proechimys simonsi group	6%						
Proechimys trinitatus group	—	—	—	17%	—		
Proechimys trinomys group	—	—	—	—	14%		
Nov. sp.	62%	—	—	—	—		
•Mesomys	<1%	3%	<1%	—	—	0–1	0–3%
•Isothrix	<1%	<1%	<1%	—	—	0–1	0–1%
Heteromyidae							
•Heteromys	—	—	—	5%	—	0–1	0–5%

Note: Sciuridae are omitted.
[a] Represents a rather complete inventory.
• Major genus.
° Could range from ~1% to high of 5%.

Table 19.7   Niche Partitioning among Selected Rodent Genera

Taxon	Activity			Substrate				Diet					Insects	
	D	"C"	N	T	Sc	A₁	A₂	Grass	Leaf	Fruit	Seed	Fungi	Larvae	Adult
Sciuridae														
Microsciurus	⊙				⊙					⊙	⊙			+
Sciurus	⊙					⊙				⊙	⊙			+
Muridae														
(Sigmodontinae)														
Oecomys			⊙		⊙					⊙	⊙			+
Oligoryzomys			⊙	⊙						⊙	⊙			+
Oryzomys			⊙	⊙						⊙	⊙			+
Neacomys			⊙	⊙						⊙	⊙			⊙
Oxymycteris		⊙		⊙									+	+
Rhipidomys			⊙			⊙				⊙	⊙			+
Akodon		⊙		⊙				+	+	+		?	+	+
Echimyidae														
Echimys			⊙				⊙			⊙	⊙			
Proechimys			⊙	⊙						⊙	⊙	⊙		
Isothrix			⊙				⊙			⊙	⊙			
Mesomys			⊙				⊙							

D = diurnal, "C" = crepuscular, N = nocturnal, T = terrestrial, Sc = scansorial, A₁ = arboreal, A₂ = strongly arboreal, + = marginal, ⊙ = strong association.

more reasonable average. The genus *Proechimys* may contain some thirty-two species. According to Patton (1987), it is possible to divide this genus into eight species groups. Some of these groups eventually may be elevated to full genera, as has occurred with *Trinomys*. The terrestrial spiny rat, *Proechimys*, is usually found in most moist tropical rain forest habitats of lowland South America. It may exist as a single species or with as many as five sympatric species. When multispecies assemblages occur, each individual species almost invariably falls into one of the eight proposed species groups, suggesting that Patton's groupings reflect morphological characters that further reflect specialization for the subdivision of niches. At the extreme geographic extent of the genus *Proechimys*, it is usually represented by a single species. Within the heartland of Amazonian Peru and adjacent portions of Brazil and Ecuador, multispecies assemblages are demonstrable (Patton and Gardner 1972; Silva 1995; Valqui 1995). The exact way the niches are subdivided remains to be ascertained, but the work of Emmons (1982), Valqui (1995), and Malcolm (1991) suggests that microhabitat segregation is at least in part responsible (see also p. 532).

Small Mammals and Altitudinal Gradients

As indicated in the introduction to this section, certain small mammal taxa are highly associated with the southern part of South America, the temperate area. These taxa however, extend northward by following altitudinal gradients, thus effectively excluding themselves from the lowland tropics.

The high plateaus of Peru tend to be somewhat semiarid. Adjacent foothills, however, can exhibit considerable heterogeneity in the vegetative cover as well as drainage and slope. Pearson and Pearson Ralph (1978) analyzed vertebrate community structure along an altitudinal gradient in Peru. The transect stations sampled the west face of the Andes over a region dominated by a low annual rainfall. Commencing near sea level at 18° S, the Pearsons passed through the arid "lifeless zone" from sea level until about 2,500 m elevation. Sampling from 3,000 to 4,300 m allowed them to separate in part the effects of rainfall and altitude (fig. 19.12a).

The west face of the Andes throughout much of the tropical zone is dominated by low rainfall. Indeed, high rainfall in the north is characteristic only of the Chocó region of Colombia and northwest Ecuador. This feature contrasts sharply with the structurally complex east face of the Peruvian Andes, where rainfall is higher and the vegetational complexity reaches exceedingly high levels, with a correspondingly high diversity in the vertebrate faunal component (fig. 19.12b).

The sampling scheme the Pearsons used sought to develop density estimates; hence not all species were found on a sample area, but they noted species caught near their formal trapping grids (table 19.9). At low elevations (< 2,000 m) only two species were recorded: *Mus musculus* and *Phyllotis darwini*. At 3,000 to 4,000 m the number of species increased to seven. At higher elevations small mammal species increased to ten, and this increase was not solely a correlate of vegetational

Table 19.8   Presence of Rodent Species at Nine Neotropical Sites

	Peru				Brazil			Venezuela	
Taxon	Valqui	Hutterer	Janson	Woodman et al.	Malcolm	Stallings	Fonseca	Guatopo	Masaguaral
Sciuridae									
⊙ *Microsciurus flaviventer*	+	+							
Sciurus gilvigularis					+				
Sciurus ignitus/granatensis			+	+				+	+
Sciurus igniventris	?	+							
Sciurus sanborni/aestuans			+			+	+		
Sciurus spadiceus	?	+	+						
Muridae (Sigmodontinae)									
⊙ *Oecomys bicolor/concolor*	+	+	+	+				+	+
Oecomys paricola/tapajinus				+	+				
Oecomys superans			+	+					
Oecomys trinitatus	?			+	+				
Oligoryzomys microtis	+		+						
Oligoryzomys nigripes/fulvescens						+	+		+
⊙ *Oryzomys capito*	+	+	+	+	+	+	+	+	
Oryzomys macconnelli	+		+		+				
Oryzomys nitidus			+	+					
Oryzomys subflavus						+	+		
Oryzomys yunganus	+			+					
Neacomys spinosus	+			+					
Neacomys guianae/tenuipes				+	+			+	
Neacomys sp.		+							
Abrawayaomys ruschii						+	+		
Nectomys squamipes	+	+	+			+	+		
⊙ *Rhipidomys couesii*			+	+					
Rhipidomys mastacalis/sp.					+	+	+		
Rhipidomys venezuelae								+	+
Akodon cursor						+	+		
Akodon urichi								+	
Holochilus sciureus									+
Oxymycteris sp.			+				+		
Oxymycteris roberti						+			
Zygodontomys sp.									+
Calomys laucha						+			
Sigmomys alstoni									+
Scolomys ucayalensis	+								
Echimyidae									
Echimys rhipidurus	+								
Echimys sp.			+		+		+		
Echimys semivillosus									+
Echimys didelphoides (Makalata)								+?	
⊙ *Proechimys brevicauda*	+	+	+						
Proechimys cuvieri	+				+				
Proechimys simonsi	+	+	+						
Proechimys steerei	+		+						
Proechimys sp. nov.	+								
Proechimys guyannensis, cayennensis				+					
Proechimys guairae								+	
Proechimys sp. +				+					
Trinomys setosus						+?	+		
Isothrix bistriatus	+	+		+					
Isothrix pagurus					+				
Mesomys hispidus	+	+	+	+	+				
Euryzygomatomys spinosus						+			
Dactylomys dactylinus	+		+						
Caviidae									
Cavia fulgida						+			
Heteromyidae									
Heteromys anomalus								+	+

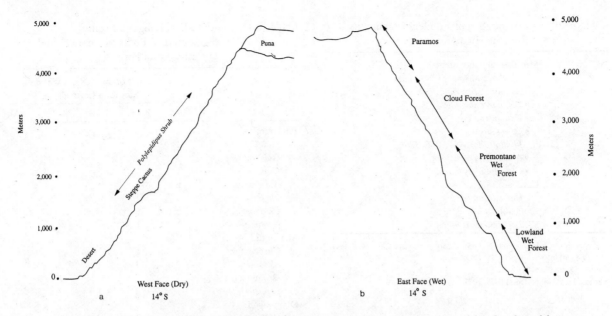

Figure 19.12. (a) West face of the Andes and plateau demonstrating altitudinal zonation of the vegetation of the "dry" face of the range. (b) East face of the Andes and edge of the plateau demonstrating the "wet" side at 18° south latitude and resulting vegetational zonation. (Adapted from Pearson and Pearson Ralph 1978.)

Table 19.9   Mammal Diversity along a West Andean Altitudinal Gradient

Location and Elevation	Vegetation Form	Species Trapped on Grid	Species Trapped or Noted in Vicinity
Loma (near sea level)	Cloud-dependent chaparral	*Phyllotis darwini, Mus musculus*	
Tillandsia (2,000 m)	Bromeliad cover	*Mus musculus*	*Lagidium*
Desert scrub (3,000 m)	Scrub	*Phyllotis darwini, P. magister*	*Thylamys elegans, Akodon berlepschii*
Mountain scrub (3,500 m)	Scrub	*Phyllotis darwini, P. magister, Akodon boliviensis, A. berlepschii*	
Quenus (3,900 m)	Scrub forest	*Phyllotis darwini, Akodon berlepschii*	*Lagidium, Phyllotis magister, Thylamys elegans, Chinchillula sahamae, Abrocoma cinerea*
Yareta (4,500 m)	Cushion plant community	*Phyllotis darwini, Akodon andinus*	*Lagidium sp., Auliscomys boliviensis*
Tola (4,300 m)	Sagebrush	*Eligmodontia typus, Auliscomys sublimus, Akodon andinus, A. berlepschii, Abrocoma cinerea*	*Phyllotis darwini, Calomys lepidus, Auliscosmy boliviensis, Ctenomys opimus*
Ichu (3,900 m)	Bunchgrass	*Phyllotis osilae, Akodon berlepschii, Calomys lepidus*	*Neotomys ebriosus, Akodon boliviensis, A. jelskii, Auliscomys pictus, Chinchillula sahamae, Phyllotis darwini, Galea musteloides*

Source: Pearson and Pearson-Ralph 1978.

complexity. For example, at 4,300 m (tola) five species were trapped on a sample grid, and four more species were noted near the grid in a sagebrush-dominated habitat, in contrast to four species trapped and noted at 3,500 m in the structurally complex mountain scrub habitat.

At high plateau sites such as Ichu, dominated by bunchgrass, three species were trapped, but seven others were noted nearby. The high plateau diversity is apparently correlated with great horizontal heterogeneity in the physical habitat, in combination with subtle shifts in vegetational microhabitats (see tables 19.9 and 19.10). Up to nine species of small mammals may be found in such habitats, with rodents dominat-

Table 19.10   Composition of a Rodent Community from the High Andes

Muridae (Sigmodontinae)
- • *Akodon boliviensis* (?), *A. puer*
- • *Akodon (Chroeomys) jelskii*
- • *Bolomys amoenus*
- • *Calomys sorella*
- • *C. lepidus*
  - *Phyllotis darwini* (?), *P. osilae*
- • *Auliscomys pictus*
- • *Chinchillula sahamae*

Caviidae
-    *Cavia tschudii*

Chinchillidae
-    *Lagidium peruanum*

Note: Study area was dry puna and rock escarpments near Checayani, Peru, at over 4,000 m.
Source: Dorst 1958.
• Studied in detail by Dorst 1958, 1971.

ing and marsupials having fallen out at lower altitudes. Rodents can show differences in community structure as a function of altitude, and even high-altitude communities can be exceedingly complex, especially at the beta diversity level. Eleven species of rodents may be found at between 4,000 and 5,000 m elevation, which compares favorably with communities of approximately the same number at lower elevations (see table 19.10). At lower elevations the community structure of small mammals becomes more complicated with the addition of marsupials that are absent at high elevations, where rodents dominate within the small mammal community.

We can consider the data presented by Patton, Myers, and Smith (1990) as instructive for the east face. They segregate rodents according to altitude considering a gradient on the eastern face of the Andes of Peru (table 19.11). If we consider only the sigmodontine rodents, we can see that such genera as *Punomys, Galenomys, Chinchillula, Auliscomys, Phyllotis, Calomys,* and *Neotomys* are more or less confined to grasslands at elevations exceeding 3,000 m (see table 19.11). The genus *Akodon* has species generally occurring above 1,000 m. The genus *Microryzomys* has representatives between 2,000 and 4,000 m elevation. What stands out is the dominance of the genera *Oryzomys, Neacomys, Oecomys, Nectomys,* and *Rhipidomys,* which become speciose at elevations below 1,000 m, although some representatives of *Oryzomys* extend to tree line elevations over 3,000 m (table 19.11).

The marsupial genera rarely extend above 3,000 m and are most speciose at lower elevations, but *Lestoros* at 3,500 m is an exception. The caviomorph rodents, including *Cavia* and *Lagidium,* do extend above 4,000 m. At 4,000 m elevation some species of *Ctenomys* can occur, although the genus is reaching the limits of its geographical range.

Table 19.11   Altitudinal Ranges for some Sigmodontine Rodent Genera Found in East-Central Peru

Taxon	Number of Species at 1,000 m Elevation Intervals				
	5–4	4–3	3–2	2–1	1–0
Phyllotini					
Punomys	1				
Galenomys	1				
Chinchillula	1				
Auliscomys	2	1			
Phyllotis	2	1			
Calomys	2	2			
Neotomys	1	1			
Akodontini					
Chroeomys	1	1			
Akodon	2	3	1	2	
Oxymycterus		1		1	
Lenoxus			1		
Oryzomyini					
• Oligoryzomys		2	2	1	1
Microryzomys		1	1		
• Oryzomys			1	2	4
Neacomys				1	2
• Oecomys				1	2
• Nectomys					1
Thomasomyini					
Thomasomys		4	4	1	
• Rhipidomys					1

Source: Derived in part from Patton, Myers, and Smith 1990.
• Genera contributing to the species richness of the lowland tropics.

Thus it is important to consider elevational gradients in defining small mammal communities. It is clear that certain taxa of rodents and marsupials dominate at low elevations, but as one proceeds upslope over an altitudinal gradient, species present at lower elevations are replaced by new species within the same genus, or new genera manifest themselves at the extreme elevations. These latter taxa are the montane endemics and contribute considerably to the species richness of any defined geographical area that contains considerable elevational gradients.

The Community Ecology of Small Mammals in Semiarid Habitats of the Tropics

Of course semiarid areas occur at high altitudes, and small mammal communities of the arid highlands of the Andes have been studied by Pearson (1957) and Dorst (1958). But the lowland dry zones, including grasslands and dry deciduous tropical forests, cover a considerable area, including the llanos of Venezuela and Colombia and, above all, the caatinga, cerrado, and chaco, extending across central Brazil southward to Paraguay and adjacent Bolivia.

As noted by numerous authors, including Alho (1981) and Nitikman and Mares (1987), dry deciduous forests contain many tracts of tropical evergreen forests in association with river systems. This latter forest type, termed "gallery forest," allows penetration of the lowland dry topics by some forms that would normally be associated with lowland wet forests. One can often see marked contrasts between inhabitants of grasslands or campos and those of gallery forest. For example, in central Brazil marsupials such as *Monodelphis* may be entirely associated with gallery forests and absent from grassland. Other species that are more broadly tolerant, such as *Didelphis albiventris*, may be taken in both gallery forests and grasslands. Among the rodents, *Akodon* may be taken in both grassland and gallery forests, but most species dominate in the gallery forest, whereas *Bolomys* dominates in the grasslands and is less abundant in the gallery forests. Semiarboreal rodents such as *Oecomys* may be confined to the gallery forest and absent from the grassland. Mares and Ernest (1995) noted the high species richness attained by small mammals in the gallery forests of central Brazil, where four marsupials, one lagomorph, and ten rodent species were taken.

The caatinga is especially interesting because of the unpredictability of rainfall in the area, coupled with a low annual rainfall in general; the great variation from year to year can create a temporal heterogeneity in plant productivity. In such a community, small mammals may be highly segregated with respect to microhabitat. Rocky outcroppings provide shelter for such forms as *Kerodon* and *Trichomys*, while the adjacent grassland and shrubs harbor rather depauperate communities of marsupials and small rodents. These communities are discussed in detail by Streilein (1982a, b,c) (see tables 19.2 and 19.3).

Open woodland savannas, such as the area studied by O'Connell, August, Sunquist, and Ludlow in Venezuela, often exhibit a rather rich fauna of small rodents and marsupials, especially if the open woodland connects with a riverine drainage system and associated gallery forest (Eisenberg, O'Connell, and August 1979; Sunquist, Sunquist, and Daneke 1989).

The Chiroptera

The community ecology of bats in the lowland Neotropics was discussed in volume 1 of *Mammals of the Neotropics* and in numerous publications cited there. In addition, the Temperate Zone bat communities of southern South America were considered in volume 2 (Redford and Eisenberg 1992). Patterson, Pacheco, and Solari (1996) sampled bats over an elevational gradient in the Andes of southeastern Peru. The bat species richness of 112 species in the lowlands (below 500 m) fell to about 65 species at 1,000 m. In the highlands at 3,000 to 3,500 meters only 15 to 18 species of bats are to found. Altitudinal gradients are critical in interpreting patterns of bat distribution. The lowland dry zone area of the caatinga studied by Willig (1983) must be mentioned at this point, however. In spite

of the unpredictable seasonal aridity of the caatinga, bat communities can be astonishingly rich if the animals have access to appropriate roosting sites. Areas of rock outcroppings with associated springs and seeps not only support some evergreen vegetation but are often dissected with crevices and caverns. Such localities provide roosting sites with sufficient humidity and temperature constancy to allow a great variety of bats to survive. They may then sally forth at night and, given their high mobility, disperse over wide areas, seeking out appropriate feeding niches. The richness of the bat communities in the caatinga is comparable to the rather rich bat faunas often associated with the arid southwestern United States, such as Arizona and New Mexico, as well as the adjacent states of the republic of Mexico, including Durango and Chihuahua. The analysis of the structure of such communities is expanded greatly in the review by Findley (1993).

A Reconsideration of Data from Nontraditional Sources

A Comment on Mammalian Community Structure in the Lowland Tropics of South America Derived from Epidemiological Studies

If one seeks structural rules for niche occupancy by nonvolant mammals in the lowlands of the tropical regions, South America is instructive. Topographical relief lies in the Andes to the west. Old, degraded upthrusts compose the Guiana and Brazilian shields in the east (see Eisenberg 1989, 9–12). If one were to examine trapping records for Brazil and make allowances for bias and pseudoreplication, some interesting hypotheses could be derived. Consider the efforts of public health teams who collect specimens (by whatever means) in order to determine possible reservoirs of microorganisms transmissible to humans with a potentially negative influence (in short, disease involving pathogens). Results of trapping by epidemiologists have been looked on with suspicion by taxonomists because species identifications are sometimes suspect. Additionally, the collectors' failure to maintain "voucher" specimens in museums has contributed to the controversy. Of course the instability of our revered Linnaean nomenclature is certainly in part to blame. The results are still instructive.

Trapping techniques vary. In the old days a mixed approach was traditional, with steel traps, mouse traps, shotguns, and rifles as standard equipment. It became recognized that rodents and bats constituted not only the dominant component of the mammal fauna but also the key reservoirs for some of the parasites and microbes that afflicted humans. The bats require several approaches to trapping and sampling (Kunz 1988).

Rodents and marsupials are easily trapped, but they may use the vertical component of a forest so that it takes a versatile trapping technique to reflect their true abundance (Malcolm 1990).

Basically, if one is livetrapping or killtrapping (terrestrially) small mammals in the lowland forest of South America to the east of the Andean cordilleras, then regardless of the species' names, the composition of the "ecotypes" is basically similar. There will be herbivores, omnivores, and insectivores adapted to terrestrial, scansorial, and arboreal niches. Within any trophic-substrate class, species may segregate according to size (reflecting prey size?) or activity pattern. There will also be reflected different tolerances for human-induced habitat alterations. Some adaptations by individual species or the persistence of species may reflect an ability to survive in disturbed habitats, whether the disturbance is human induced or the result of natural processes. One will find a group of mammalian genera that could differentially group themselves according to the major habitat type in which they are found and the presence of human disturbance. First, there are areas undisturbed by human activity: forest, gallery forest, dry deciduous forest, and savannas subject to fire or flooding or both. Then there are human-disturbed habitats across the former categories. The small, trapable mammals will also fall into two categories: abundant, versatile survivors and rare specialists. And of course there is the third category of species that are "intermediate."

Let us consider the publication by Peterson, Roberts, and Pinheiro (1981). The authors were epidemiologists with an excellent eye for habitats and human disturbance effects on the sampled habitats in eastern Brazil, south of the Amazon. We will grant that arboreal trapping was *not* a priority and limit our examination to the results of terrestrial trapping. Among the marsupials, the forested areas favor *Didelphis marsupialis*, *Philander opossum*, and *Marmosa cinerea*. Cultivated areas and areas of secondary vegetation favor *Monodelphis brevicaudata*. Among the rodents, *Oryzomys capito*, *O. macconnelli*, *Neacomys guianae*, and *Proechimys oris* predominate in the forests, with *Oligoryzomys fulvescens* and *Zygodontomys* dominant within successional habitats (see figs. 19.5 and 19.6). Now, leaving aside the arboreal rodents and marsupials, we only have a few families numerically contributing to the communities in the South American lowland tropics.

Consider a rather mature forest with obvious patches of second growth—for example, Parque Nacional, Guatopo, Venezuela. For the Marsupialia we find *Didelphis marsupialis*, *Marmosops fuscatus*, and *Monodelphis brevicaudata* numerically dominant. Among the rodents one can enumerate *Oryzomys*

capito, *O. concolor*, *Proechimys guyannensis*, and *Neacomys tenuipes*. In the llanos of Venezuela *Zygodontomys brevicauda*, *Oecomys bicolor*, *Didelphis marsupialis*, and *Marmosa robinsoni* predominate (Eisenberg, O'Connell, and August 1979). An inspection of figures 19.7 to 19.9 will complete the discussion. We conclude that there is an underlying structure of mammalian communities paralleling vegetational heterogeneity (August 1983; Peterson, Roberts, and Pinheiro 1981).

When we are forced to consider the rough-and-ready collectors with jaw traps, shotguns, rifles, snares, and "kill traps"; their brothers and sisters in the epidemiological service; and the trap-mark-release artists of the mid-twentieth century seeking insights into life history strategies—and when we then discover concordance among the results of their efforts—we come to believe that some basic patterns of the way mammals exploit their environment must reflect the energy limits imposed by the "producers" in an ecosystem.

Concerning the Density of Carnivores within Neotropical Habitats

INTRODUCTION

When ranked by mean body size, mammalian densities generally scale negatively. That is to say, larger forms exist at lower densities than smaller forms. When trophic adaptations are taken into consideration, however, carnivores (in a trophic sense) exhibit a similar negative scaling but have the lowest densities of any trophic class (Elton 1927; Eisenberg and Thorington 1973; Robinson and Redford 1986). On the other hand, there are great variations in recorded densities within the same species and wide variations within the same size class. One source of variation may be differential hunting pressures. If a particular size class of carnivores is persecuted for furs or as pest animals, then where they co-occur with humans their densities may be depressed below carrying capacity. The way predator densities are estimated may in no small measure contribute to the variation. Indeed, methods of collection have changed through the years even when we consider museum collectors. Jaw traps were in common use around the turn of the century, but in many areas today their use is illegal. Museum collectors may have also relied on purchases and donations, so assemblages within the world's museums may be riddled with biases.

Of course the best density estimates would derive from long-term field studies using individually marked animals, perhaps in conjunction with radiotelemetry, to derive an absolute count of adults within the appropriate habitat categories. When carnivores are treated as an order, some additional biases may enter in that some of the variation can derive from dietary specialization. The omnivorous canids often can exist at higher densities than the more truly carnivorous mustelids and felids. Even within a taxonomic category such as the Felidae, variation in density estimates may derive from selective hunting pressures. For example, spotted cats have a higher market value for their pelts than solid-colored felids, whose pelts are often not sought after for the fur trade.

Another source of variation derives from the selection of a base area. For example, if the politically defined area of Venezuela is taken as a base, then those species that have very restricted habitat requirements, such as Andean forms, will appear to exist at a far lower density than if only the Andean biogeographic province were taken as a base reference. Furthermore, there are biogeographic considerations. The Orinoco River approximately bisects Venezuela as it meanders to the west, and the fauna south of the river is quite distinct from that to the north. Thus some forms do not extend north of the Orinoco, so their overall density when Venezuela is considered as a base area will be lower than could have been attained if the species in question ranged on both sides of the Orinoco.

Given the reality of the llanos, forest-adapted forms may be underrepresented north of the Orinoco. Furthermore, considering the case of Venezuela, the north coast range may in fact have a fair sample of forest-adapted forms. These species may in turn range along the major rivers draining the north coast range within gallery forest but, if they are forest adapted, not extend their range into the more open llanos habitat that so typifies much of north-central Venezuela. Again, their relative densities will appear to be low mainly because of the restricted nature of their preferred habitat.

This brings us to the final point—that habitat specificity by certain species may account for their apparent rarity when their density is averaged over a large area not containing optimum habitat. Some carnivores that are large are almost landscape animals in the sense that they range over a wide variety of habitat types. Other, often smaller forms, such as the little spotted cat, *Felis tigrina,* may be quite specific with respect to their preferred habitat.

A POTENTIAL DATABASE FOR
ESTIMATING CARNIVORE DENSITIES

If we know that two carnivore species inhabit the same biogeographic region and have similar habitat requirements, and if we know the density of one, it is possible to estimate the density of the other because certain databases are available that indicate rough ratios of abundance, such as furs exported. If we assume that the species in question are hunted with the same intensity and that ease of capture is approximately equal, then pelt export data might give some indication of

relative abundance. With the same set of assumptions, specimens present in collections may also reflect relative abundance, and ratios could be calculated from these data. This assumes that the inherent biases already outlined in the introduction to this section have been corrected for.

Rough field estimates are often useful indicators of relative abundance, but by far the best data are those that derive from long-term radiotelemetric studies where known individuals and their ranges have been mapped. We propose to examine some of these possible sources of data for each of the major carnivore families. We fully recognize that hunting efforts for museum or the pelt trade vary widely among the different families of carnivores. Spotted cats are highly prized and thus will be overrepresented compared with canids, especially those from the tropics, whose pelts are not particularly desirable. Because of the extensive collections assembled from there, Venezuela will be the geographic source for most of these data, although pelt export data from Peru will also be consulted. The data derive from Bisbal (1986) and Grimwood (1969).

Exports of skins from Peru vary considerably from year to year, but Grimwood (1969) offers data that in general show an increase in exports from 1958 to 1968. This undoubtedly reflects the increasing demand for pelts by the fur trade. If we pick years of maximum support, then ocelot skins are fourteen times more abundant than jaguar pelts and three times as abundant as the combined total for skins of the little spotted cat and the margay. These spotted cats are hunted assiduously. We do not know the catchment area they were taken from, but assuming that the habitat is reasonably uniform in Amazonian Peru, it is safe to say that the vast majority came from lowland tropical rain forest.

Bisbal (1986) offers totals for spotted carnivores from museums holding collections deriving from Venezuela. Looking at the gross totals for such carnivore specimens, Bisbal's data reflect the following: The crab-eating fox is the most frequent, with 312 out of 1,233 specimens; next in rank is the kinkajou with 286; and third is the ocelot with 108. Unless corrected for habitat bias, these data are not particularly meaningful. For this reason we considered Bisbal's data set from the standpoint of two major biogeographical provinces: the north coastal range, which consists of lowland tropical rain forest, premontane tropical forest with very little savanna habitat, and the state of Guárico, which in the main is savanna, dry deciduous forest, and strips of gallery forest. Within the north coastal range, the crab-eating fox is the carnivore most frequently encountered in museum collections, closely followed by the crab-eating raccoon, the kinkajou, and the tayra. In the savanna habitats of the state of Guárico the crab-eating fox again is the specimen most frequently encountered, followed by the ocelot and the jaguarundi. Perhaps the grain of these biogeographic provinces is a bit too rough. If we take Venezuela as a whole and examine specimen numbers from dry deciduous forest and savanna habitats, the crab-eating fox leads with 187 specimens, followed by the ocelot with 50, the crab-eating raccoon with 31, and the jaguarundi and the hog-nosed skunk with 24 each. If we look at humid tropical forest, then the specimen most frequently encountered in collections is the kinkajou with 162, followed by the crab-eating fox with 48, the ocelot with 34, the tayra with 20, and the olingo with 17. If we turn to very humid tropical forest, the species most frequently encountered in collections is again the kinkajou with 25, followed by the long-tailed weasel with 20, the crab-eating fox with 13, and the ocelot with 6.

Clearly the sample effort has not been equal within the different habitat types, nor has it been equal between biogeographical provinces. Nevertheless, the data are suggestive. No matter what habitat type we look at, after correcting for biogeographical factors it is clear that the little spotted cat is exceedingly uncommon in collections, as is the margay. Mustelids are not common in collections, although the hog-nosed skunk is most frequently encountered. The ocelot occurs with reasonable abundance in collections, and the crab-eating fox is very abundant.

These data may be compared and contrasted with density estimates based on radiotelemetric studies. Sunquist, Sunquist, and Daneke (1989) have made density estimates for the more common carnivores found in the llanos of Venezuela. Their estimates are not out of line with the relative abundances as developed from museum and pelt export data. This lends credence to our idea that some carnivore species are rather rare compared with others. It is not surprising that omnivorous forms such as the crab-eating fox are relatively more abundant than the more strictly carnivorous species. But why the little spotted cat and the margay should be rarer than the ocelot remains somewhat of a mystery.

We can form several hypotheses concerning the abundance of organisms. For example, organisms that appear to be rare by our methods of enumeration may actually be abundant because we have a poorly developed sampling technique. On the other hand, so-called rare organisms may in fact exist at low densities regardless of our data collection methods. How are we to resolve this dilemma? Fortunately, at the close of the twentieth century we have data available from paleontologists, ecologists, museum collectors, and naturalists. The consensus is that some species are *not* numerically abundant.

References

• References used in preparing distribution maps

Aagaard, E. M. J. 1982. Ecological distribution of mammals in the cloud forests and paramos of the Andes, Mérida, Venezuela. Ph.D. diss., Colorado State University.

Alho, C. J. R. 1981. Small mammal populations of Brazilian cerrado: The dependence of abundance and diversity on habitat complexity. *Rev. Brasil. Biol.* 41 (1): 223–30.

August, P. V. 1981. Population and community ecology of small mammals in northern Venezuela. Ph.D. diss., Boston University.

———. 1983. The role of habitat complexity and heterogeneity in structuring mammalian communities. *Ecology* 64:1495–1507.

Bisbal, F. 1986. Food habits of some Neotropical carnivores in Venezuela. *Mammalia* 50:329–39.

Bisbal, F. 1989. Distribution and habitat association of the carnivore community in Venezuela. In *Advances in Neotropical mammalogy,* ed. K. H. Redford and J. F. Eisenberg, 339–62. Gainesville, Fla.: Sandhill Crane Press.

Dorst, J. 1958. Contribucion a l'étude écologique de rongeurs des hauts plateaux du Perou méridional. *Mammalia* 22:547–65.

———. 1971. Nouvelles recherches sur l'écologie des rongeurs des hauts plateaux péruviens. *Mammalia* 35:515–47.

Eisenberg, J. F., ed. 1979. *Vertebrate ecology in the northern Neotropics.* Washington, D.C.: Smithsonian Institution Press.

———. 1981. *The mammalian radiations: An analysis of trends in evolution, adaptation, and behavior.* Chicago: University of Chicago Press.

———. 1989. *Mammals of the Neotropics.* Vol. 1. *The northern Neotropics: Panama, Colombia, Venezuela, Guyana, Suriname, French Guiana.* Chicago: University of Chicago Press.

———. 1990. Neotropical mammal communities. In *Four Neotropical rainforests,* ed. A. H. Gentry, 358–68. New Haven: Yale University Press.

Eisenberg, J. F., M. A. O'Connell, and P. V. August. 1979. Density, productivity and distribution of mammals in two Venezuelan habitats. In *Vertebrate ecology in the northern Neotropics,* ed. J. F. Eisenberg, 187–207. Washington, D.C.: Smithsonian Institution Press.

Eisenberg, J. F., and K. H. Redford. 1982. Comparative niche structure and evolution of mammals of the Nearctic and southern South America. In *Mammalian biology in South America,* ed. M. A. Mares and H. H. Genoways, 77–84. Pymatuning Symposia in Ecology 6. Special Publication Series. Pittsburgh: Pymatuning Laboratory of Ecology, University of Pittsburgh.

Eisenberg, J. F., and R. W. Thorington Jr. 1973. A preliminary analysis of a Neotropical mammal fauna. *Biotropica* 5:150–61.

Elton, C. S. 1927. *Animal ecology.* London: Sidgwich and Jackson.

Emmons, L. 1982. Ecology of *Proechimys* in southeastern Peru. *J. Trop. Ecol.* 23:280–90.

Findley, J. S. 1993. *Bats: A community perspective.* Cambridge: Cambridge University Press.

Fleming, T. H. 1971. Population ecology of three species of Neotropical rodents. *Misc. Publ. Mus. Zool. Univ. Michigan* 143:1–77.

———. 1975. The role of small mammals in tropical ecosystems. In *Small mammals: Their productivity and population dynamics,* ed. G. B. Golly, K. Petrusewicz, and L. Ryszkowski, 269–98. Cambridge: Cambridge University Press.

Fonseca, G. A. B. da 1988. Patterns of small mammal species diversity in the Brazilian Atlantic forest. Ph.D. diss., University of Florida, Gainesville.

Fonseca, G. A. B. da, and M. C. M. Kierulff. 1988. Biology and natural history of Brazilian Atlantic forest small mammals. *Bull. Florida State. Mus., Biol. Sci.* 34:99–152.

Gardner, A., and J. L. Patton. 1976. Karotypic variation in oryzomyine rodents (Cricetinae) with comments on chromosomal evolution in the Neotropical cricetine complex. *Occas. Pap. Mus. Zool. Lousiana State University* 49:1–48.

Gentry, A. H., ed. 1990. *Four Neotropical rainforests.* New Haven: Yale University Press.

Gentry, A. H., and L. H. Emmons. 1987. Geographical variation in fertility, phenology and composition of the understory of Neotropical forests. *Biotropica* 19 (3): 216–27.

Grimwood, I. R. 1969. *Notes on the distribution and status of some Peruvian mammals.* Special Publication 21. New York: American Committee for Wild Life Protection.

Hershkovitz, P. 1972. The Recent mammals of the Neotropical region: A zoogeographic and ecological review. In *Evolution, mammals, and southern continents,* ed. A. Keast, F. C. Erk, and B. Glass, 311–431. Albany: State University of New York Press.

Hutterer, R., M. Verhaagh, J. Diller, and R. Podloucky. 1995. An inventory of mammals observed at Panguana Biological Station, Amazonian Peru. *Ecotropica* 1:3–20.

Janson, C. H., and L. H. Emmons. 1990. Ecological structure of the non-flying mammal community at Cocha Cashu Biological Station, Manu National Park, Peru. In *Four Neotropical rainforests,* ed. A. H. Gentry, 314–38. New Haven: Yale University Press.

Kinnaird, M., and J. F. Eisenberg. 1989. A consideration of body size, diet and population biomass for Neotropical mammals. In *Advances in Neotropical mammalogy,* ed. K. H. Redford and J. F. Eisenberg, 595–604. Gainesville, Fla.: Sandhill Crane Press.

Kunz, T. H., ed. 1988. *Ecological and behavioral methods for the study of bats.* Washington, D.C.: Smithsonian Institution Press.

Lacher, T. E., Jr. 1981. The comparative social behavior of *Kerodon rupestris* and *Galea spixii* and the evolution of behavior in the Caviidae. *Bull. Carnegie Mus. Nat. Hist.* 17:1–71.

Lacher, T. E., Jr., M. A. Mares, and C. J. R. Alho. 1989. The structure of a small mammal community in a central Brazilian savanna. In *Advances in Neotropical mammalogy,* ed. K. H. Redford and J. F. Eisenberg, 137–62. Gainesville, Fla.: Sandhill Crane Press.

Malcolm, J. R. 1990. Estimation of mammalian densities in a continuous forest north of Manaus. In *Four Neotropical rainforests,* ed. A. H. Gentry, 339–57. New Haven: Yale University Press.

———. 1991. The small mammals of Amazonian forest fragments: Pattern and process. Ph.D. diss., University of Florida, Gainesville.

Mares, M. A. 1992. Neotropical mammals and the myth of Amazonian biodiversity. *Science* 255:976–79.

Mares, M. A., and K. A. Ernest. 1995. Population and community ecology of small mammals in a gallery forest of central Brazil. *J. Mammal.* 76:750–68.

Medellin, R. A. 1992. Community ecology and conservation of mammals in a Mayan tropical rainforest and abandoned agricultural fields. Ph.D. diss., University of Florida, Gainesville.

———. 1994. Mammal diversity and conservation in the Selva Lacandona, Chiapas, Mexico. *Conserv. Biol.* 8:780–99.

Nitikman, L. Z., and M. A. Mares. 1987. Ecology of small mammals in a gallery forest of central Brazil. *Ann. Carnegie Mus.* 56:75–95.

O'Connell, M. A. 1979. Ecology of didelphid marsupials from northern Venezuela. In *Vertebrate ecology in the northern Neotropics,* ed. J. F. Eisenberg, 73–87. Washington, D.C.: Smithsonian Institution Press.

———. 1981. Population ecology of small mammals in northern Venezuela. Ph.D. diss., Texas Tech University.

———. 1989. Population dynamics of Neotropical small mammals in seasonal habitats. *J. Mammal.* 70:532–48.

Oliviera, T. G. de. 1994. *Neotropical cats: Ecology and conservation.* São Luis, Brazil: EDUFAMA.

Pacheco, V., B. D. Patterson, J. L. Patton, L. H. Emmons, S. Solari, and C. F. Ascorra. 1993. List of mammal species known to occur in Manu Biosphere Preserve. *Publ. Mus. Hist. Nat. Univ. Nac. Mayor San Marcos, Zool.* 44:1–12.

Patterson, B. D., V. Pacheco, and S. Solari. 1996. Distributions of bats along an elevational gradient in the Andes of southeastern Peru. *J. Zool.* (Lond.) 240:637–58.

Patton, J. L. 1987. Species groups of spiny rats genus *Proechimys* (Rodentia, Echimyidae). In *Studies in Neotropical mammalogy: Essays in honor of Philip Hershkovitz,* ed. B. D. Patterson and R. M. Timm, 305–45. *Fieldiana: Zool.,* n.s., 39:1–506.

Patton, J. L., and A. L. Gardner. 1972. Notes on the systematics of *Proechimys* (Rodentia: Echimyidae), with emphasis on Peruvian forms. *Occas. Pap. Mus. Zool. Louisiana State University* 44:1–30.

Patton, J. L., P. Myers, and M. F. Smith. 1990. Vicariant versus gradient models of diversification: The small mammal fauna of eastern Andean slopes of Peru. In *Vertebrates in the tropics,* ed. G. Peters and R. Hutterer, 355–72. Bonn: Museum Alex Koenig.

Pearson, O. P. 1951. Mammals in the highlands of southern Peru. *Bull. Mus. Comp. Zool.* 106 (3): 117–74.

———. 1957. Additions to the mammalian fauna of Peru and notes on some other Peruvian mammals. *Breviora: Mus. Comp. Zool.* 73:1–7.

Pearson, O. P., and C. Pearson Ralph. 1978. The diversity and abundance of vertebrates along an altitudinal gradient in Peru. *Mem. Mus. Hist. Nat. "Javier Prado"* 18:1–97.

Peterson, N. E., D. R. Roberts, and F. P. Pinheiro. 1981. Programa multidisciplinario de vigilancia de las enfermedades infecciosas en zonas colindantes con carretera transamazonica en Brasil. 3. Estudio de los mamíferos. *Bol. Sanit. Panama* 91 (4): 324–36.

Pielou, E. C. 1969. *An introduction to mathematical ecology.* New York: John Wiley.

Redford, K. H., and J. F. Eisenberg. 1992. *Mammals of the Neotropics.* Vol. 2. *The southern cone: Chile, Argentina, Uruguay, Paraguay.* Chicago: University of Chicago Press.

Robbins, L. W., M. P. Moulton, and R. J. Baker. 1983. Extent of geographic range and magnitude of chromsomal evolution. *J. Biogeogr.* 10:533–41.

Robinson, J., and K. Redford. 1986. Body size, diet and density of Neotropical forest mammals. *Amer. Nat.* 128:665–80.

Sarmiento, G. 1984. *The ecology of Neotropical savannas.* Cambridge: Harvard University Press.

Silva, M. N. F. da. 1995. Systematics and phylogeography of Amazonian spiny rats of the genus *Proechimys* (Rodentia: Echimyidae). Ph.D. diss., University of California, Berkeley.

Stallings, J. R. 1988a. Small mammal communities in an eastern Brazilian park. Ph.D. diss., University of Florida, Gainesville.

———. 1988b. Small mammal communities in an eastern Brazilian park. *Bull. Florida Mus. Nat. Hist.* 34:159–200.

Stallings, J. R., G. A. B. de Fonseca, L. P. Souza Pinto, L. M. de Souza Aguiar, and E. Lima Sábato. 1991. Mamíferos de Parque Florestal Estadual do Rio Doce, Minas Gerais, Brasil. *Rev. Bras. Zool.* 7: 663–77.

Streilein, K. E. 1982a. The ecology of small mammals in the semiarid Brazilian caatinga. 1. Climate and faunal composition. *Ann. Carnegie Mus.* 51:79–107.

———. 1982b. The ecology of small mammals in the semi-arid Brazilian caatinga. 3. Reproductive biology and population ecology. *Ann. Carnegie Mus.* 51:251–69.

———. 1982c. Ecology of small mammals in the semi-arid Brazilian caatinga. 4. *Ann. Carnegie Mus.* 51: 331–43.

Sunquist, M. E., F. Sunquist, and D. E. Daneke. 1989. Ecological separation of carnivores in a Venezuelan llanos carnivore community. In *Advances in Neotropical mammalogy,* ed. K. H. Redford and J. F. Eisenberg, 197–232. Gainesville, Fla.: Sandhill Crane Press.

Valqui, M. H. 1995. A terrestrial, small rodent community in northeastern Peru. M.S. thesis, Department of Wildlife Ecology and Conservation, University of Florida, Gainesville.

Voss, R. S., and L. H. Emmons. 1996. Mammalian diversity in Neotropical lowland rainforests: A preliminary assessment. *Bull. Amer. Mus. Nat. Hist.* 230:1–115.

Willig, M. R. 1983. Composition, microgeographic variation, and sexual dimorphism in caatinga and cerrado bat communities from northeastern Brazil. *Bull. Carnegie Mus. Nat. Hist.* 23:1–131.

Wilson, D. E. 1990. Mammals of La Selva, Costa Rica. In *Four Neotropical rainforests,* ed. A. H. Gentry, 273–86. New Haven: Yale University Press.

Woodman, N., N. A. Slade, and R. M. Timm. 1995. Mammalian community structure in lowland, tropical Peru, as determined by removal trapping. *Zool. J. Linnean Soc.* 113:1–20.

Woodman, N., R. M. Timm, R. Arana C., V. Pacheco, C. A. Schmidt, E. D. Hooper, and C. Pacheco A. 1991. Annotated checklist of the mammals of Cuzco, Amazonica, Peru. *Occas. Pap. Mus. Nat. Hist. Univ. Kansas.* 145:1–12.

20 Macrogeography of Brazilian Mammals

Gustavo A. B. da Fonseca, Gisela Herrmann,
and Yuri L. R. Leite

Introduction

With an area of 8,511,996 km², Brazil accounts for 48% of South America. Four of the five major Brazilian biomes, the Atlantic Forest, the Cerrado, the Caatinga, and the Pantanal, are virtually confined to the nation's territory. Fully 62% of Amazonia lies within Brazil. These superlatives have given rise to high levels of animal and plant diversity, particularly for some groups. Worldwide, Brazil has the highest number of primates (68 species), pscittacine birds (70 species), amphibians (518 species), terrestrial vertebrates (3,022 species), freshwater fish (ca. 3,000 species) and flowering plants (ca. 55,000 species), making it the top country in terrestrial biodiversity (Mittermeier et al. 1992).

As good-quality species-level continental and regional databases pertaining to better studied groups, such as birds and mammals, are progressively becoming more available and easier to handle, some interesting macroecological and macrogeographical analyses of fauna assemblages have been produced (e.g., Fonseca and Redford 1984; Redford and Fonseca 1986; Brown and Maurer 1989; Redford, Taber, and Simonetti 1990; Mares 1992). These studies are helping to shed light on biome-level faunal assemblages as well as on biogeographic processes that operate on the regional and landscape levels, searching for recurrent patterns that could reveal large-scale phenomena (Ricklefs 1987; Brown and Maurer 1989; Ricklefs and Schluter 1993).

Using a similar approach in the mid-1980s, researchers presented evidence that the lack of significant endemicity for the Brazilian Cerrado meant savanna mammal fauna may have been derived from Pleistocene climatic events that favored versatile species and selected against pastoral, open-habitat mammals (Fonseca and Redford 1984; Redford and Fonseca 1986; see also Cartelle, chapter 4 of this volume). Furthermore, these events seem to have been evolutionarily neutral in relation to forest-adapted species.

This chapter reinforces the previous suggestion that the bulk of Brazilian mammals are either forest adapted or versatile in habits. It lays out the basic descriptive statistics for the entire mammalian fauna of the country, using as units of analysis its five major biomes. We have included the neighboring Gran Chaco, which represents the southwestern limit of the belt of open vegetation formations bisecting the two major forest biomes east of the Andes. In addition, mammals of extra-Brazilian Amazonia were also incorporated in the analysis.

Focusing primarily on Brazilian land mammals, this chapter also examines area size and distribution patterns, faunal affinity parameters, body weight frequency distributions, and feeding and locomotor niche relationships at a biome level, advancing suggestions concerning the history of this fauna. The database we compiled includes the South American continent as a whole, except for distribution patterns pertaining to non-Brazilian habitats and biomes.

Considering 904 South American land mammals belonging to 13 orders, 47 families, and 249 genera, our latest account indicates there are 457 species in Brazil. This makes it the top country in the Western Hemisphere in mammalian diversity, followed closely by Mexico with 449 (Ceballos and Navarro 1991). Mammal species diversity in Brazil, totaling 51% of that of all of South America, is distributed among 11 orders, 38 families, and 180 genera, representing, respectively, 85%, 81%, and 72% of the totals for the continent.

Four orders (Rodentia, Chiroptera, Primates, and Didelphimorphia, with respectively 33%, 31%, 15%, and 9% of the total) account for close to 88% of all species in Brazil, exactly the same proportion found for these orders over the entire South American mammalian fauna. Carnivores are responsible for another 5% of species diversity in both Brazil and South America, while edentates are particularly well represented in the former. Close to 25% of all Brazilian mammals are endemic to the nation's territory.

Despite these high levels of diversity and endemism, Brazil is also a leader in number of species under threat. Conservative figures, contained in the official list of Brazilian species threatened with extinction (Bernardes, Machado, and Rylands 1989), indicate that 58 mammals that occur in the country could be included in one of IUCN's categories of threat. It is believed that at least twice that number, or close to 25% of all Brazilian mammals, are currently threatened (Fonseca et al. 1994b).

Methods

Over a period of five years, we collected distribution data on land mammals from various sources (see Fonseca et al. 1994a,b for a more comprehensive listing of sources). The database is composed mostly of published information, but it was also complemented with numerous unpublished reports and abstracts of congresses as well as personal communications from specialists. Data on geographical distribution, body weight, and basic ecology of South American mammals were compiled, checked, stored, and processed at the Biodiversity Conservation Data Center (CDCB), at the Fundação Biodiversitas (Belo Horizonte, Brazil). The database is linked with a Geographical Information System (CISIG, a PC-based GIS developed by Conservation International), which aided in the analyses of species distribution.

Except for more recent reviews of particular groups, the latest mammalian taxonomic arrangement of the CDCB database closely follows the two previous volumes of *Mammals of the Neotropics* and Wilson and Reeder (1993). Feeding and locomotor niche categories follow Eisenberg (1981). Species were also classified according to their distribution patterns, borrowing from Hershkovitz's terminology (1969), and divided into three categories: sylvan, versatile, and pastoral forms. Sylvan forms occur only in forest biomes (Amazonia or Atlantic Forest or both), while pastoral species are restricted to semiarid and open biomes (Cerrado, Chaco, Caatinga, or a combination). Versatile forms are usually of widespread distribution, occurring in both biome types.

Major Brazilian Biomes

For the purposes of this study, the biogeographical division of Brazil, having its five major biomes as basic units (fig. 20.1), is that of IBGE (1993), with minor modifications. The limits of the non-Brazilian Chaco, as well as those of the extraterritorial Amazonia and of the southwesternmost borders of the Atlantic Forest, were adapted from Hueck (1972).

Although minor enclaves of other habitat types can be found within all Brazilian biomes (e.g., Amazonian savannas), ecologically we have considered the whole of Amazonia and the Atlantic Forest as major tropical forest major habitats. The Cerrado, Caatinga, Chaco, and the Pantanal were considered open habitat formations. The GIS analysis of the map used for this paper indicated the area of Brazil, Amazonia, Cerrado, Atlantic Forest, Gran Chaco, Caatinga, and the Pantanal as, respectively, 8,511,996 km^2, 6,355,643 km^2, 2,052,533 km^2, 1,272,539 km^2, 1,085,393 km^2, 770,442 km^2, 483,305 km^2, and 143,046 km^2. For a physiognomic description of each major biome, see Hueck (1972).

Species/Area Relationships and Diversity Comparisons

Among the six major units analyzed, Brazilian Amazonia, the largest in area, comes out as the most speciose biome, with 350 species, followed by the Atlantic Forest, Cerrado, Pantanal, Chaco, and Caatinga with, respectively, 229, 159, 124, 113 and 102 species (table 20.1). There is a statistically significant relation between area size and the number of mammal species (fig. 20.2a). All semiarid biomes (Cerrado, Chaco, and

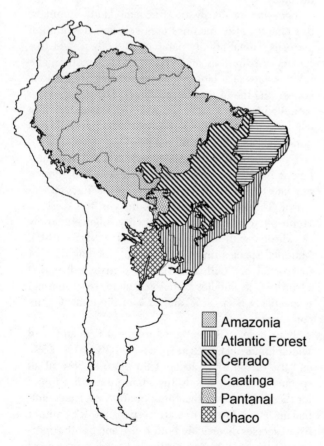

Amazonia
Atlantic Forest
Cerrado
Caatinga
Pantanal
Chaco

Figure 20.1. Approximate geographical limits of Amazonia, Cerrado, Caatinga, Atlantic Forest, Pantanal, and Gran Chaco.

Table 20.1 Number of Orders, Families, Genera, and Species of Mammals in Brazil, South America, and Each Biome Analyzed, and Biome-Level Endemism of Species and Genera (in Parentheses)

Taxon	Amazonia	Atlantic Forest	Cerrado	Caatinga	Pantanal	Chaco	Brazil	South America
Orders	11	9	9	8	9	9	11	13
Families	36	31	28	26	28	26	38	47
Genera	145 (41)	127 (10)	109 (4)	74 (0)	92 (0)	84 (7)	180 (13)	249
Species	350 (205)	229 (73)	159 (23)	102 (0)	124 (3)	113 (19)	457 (107)	904

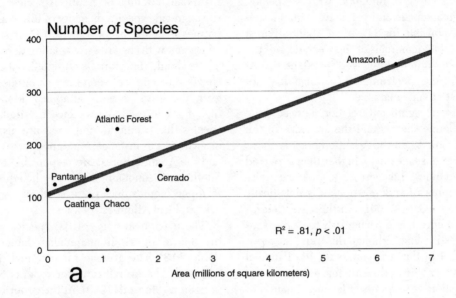

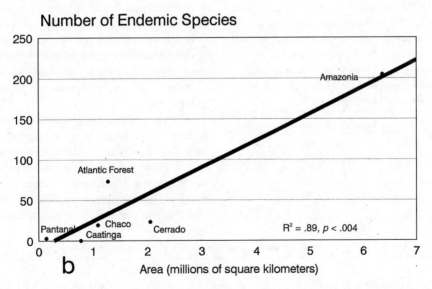

Figure 20.2. Regression analyses of richness versus area size of the biome: (a) number of species; (b) number of endemic species.

Caatinga) are below the regression line. The Atlantic Forest, on the other hand, stands out as the most diverse for its size. At the generic level, a similar area size relationship is observed in the order of diversity from the highest to the lowest, even though not statistically significant ($R^2 = .62, p < .06$). However, the Cerrado and the Pantanal are slightly above the line, while the

Atlantic Forest clearly shows higher generic diversity for its size.

As expected, area size effects should weaken at higher taxonomic levels. Even though Amazonia and the Atlantic Forest harbor a larger number of families (table 20.1), followed by the Cerrado and the Pantanal and by the Caatinga and the Chaco, the relationship is not significant, the same being true for the number of mammalian orders in each biome.

Amazonia stands out in degree of endemicity, with 205 species restricted to that biome, or 59% of the total (table 20.1). Rodents, bats, and primates (65, 61, and 55 endemic species) account for 88% of all endemic species of this region. The Atlantic Forest also displays a high level of endemism, with 73 endemics out of 229 species (32%). Rodentia, with 39 species, together with marsupials and primates, with 11 species each, is responsible for 84% of all Atlantic Forest endemics. Endemic mammals of the Cerrado and the Chaco are mostly rodents, followed by edentates. At the other end of the spectrum, the Caatinga does not harbor a single endemic mammal.

There is also a significant relation between the number of endemic species and the area size of the biomes (fig. 20.2b). Again the Atlantic Forest is above the regression line, suggesting a higher than expected number of endemics. Likewise, the area size relation for the number of endemic genera is statistically significant ($R^2 = .93$, $p < .001$). In this case Amazonia, the Atlantic Forest, and particularly the Chaco are above the regression line, although this trend may represent an artifact of the zero counts for the Pantanal and Caatinga. Over 50% of the endemic genera for the sample in question are rodents. Eleven primate and ten bat genera (all Amazonian) account for another 34% of endemics. Three carnivore, two marsupial, and two edentate genera, together with *Catagonus* and

Inia, make up the total universe of endemic genera for the sample of the six biomes examined.

For the database composed of the six biome samples investigated, Amazonia and the Atlantic Forest together harbor 455 species of mammals. Conversely, the four open vegetation samples are home to no more than 210 species (see table 20.2). These figures represent, respectively, 50% and 23% of the mammalian fauna of South America. Nonetheless, the aggregate species diversity of forest and open area biomes is nearly proportional to their summed area, consistent with the area size relation described earlier. There should therefore be no significant differences in species diversity between the two biome types. However, the level of zoogeographical identity is much higher for forest biomes, since the bulk of mammal species that occur in open areas are also distributed elsewhere. For instance, endemic species of Amazonia and the Atlantic Forest are responsible for 31% of all South American mammals, while the equivalent figure for open biomes does not exceed 5% (see also "Zoogeographical Affinities" below).

There is an average of 3.6 species per genus in the South American mammalian fauna as a whole (table 20.3). The average for mammals in general is 3.2, while the overall ratio for rodents is 4.3 species per genus (Stenseth 1989). At the generic level, South American primates are the most speciose group, with an average of 5.6, followed by rodents (4.6) and, surprisingly enough, by marsupials (4.1), one of the two ancient South American mammal groups. Bats have an average of 3.1 species per genus, while the remaining three most diverse groups have 2 species or fewer per genus.

The ratio of number of species to number of genera is fairly constant across open vegetation samples, and with the exception of Chiroptera, is ranges from 1.0 to 1.7, being remarkably similar across all other taxonomic groups. Significant deviations are found for the two forested biomes, especially Amazonia. Amazonia and the Atlantic Forest have, respectively, 2.4 and 1.8

Table 20.2   Distribution of Species according to Sylvan, Pastoral, and Versatile Habits, in Each of the Six Biomes Analyzed

Biome	Sylvan	Pastoral	Versatile	Endemic
Amazonia	224		126	205
Atlantic Forest	92		137	73
Cerrado		28	131	23
Caatinga		6	96	0
Pantanal		13	111	3
Chaco		31	82	19

Table 20.3   Number of Species per Genus for South America and for Each Biome Analyzed

Order	Amazonia	Atlantic Forest	Cerrado	Caatinga	Pantanal	Chaco	South America
Didelphimorphia	1.8	1.9	1.2	1.0	1.3	1.0	4.1
Chiroptera	2.6	1.9	1.8	1.8	1.7	1.9	3.1
Rodentia	2.6	1.9	1.4	1.0	1.1	1.3	4.6
Carnivora	1.2	1.2	1.2	1.1	1.1	1.1	1.7
Artiodactyla	1.3	2.0	1.2	1.3	1.2	1.2	2.0
Xenarthra	1.4	1.2	1.2	1.0	1.2	1.2	2.2
Primates	4.4	2.5	1.7	1.5	1.0	1.2	5.6
Ratios for South American biomes	2.4	1.8	1.5	1.4	1.3	1.3	3.6

species per genus. As such, these figures indicate that for Amazonia there is only one additional species per genus compared with the other five vegetation formations. The region thus owes a large fraction of its mammal species richness to a higher generic diversity and to a significant subgeneric diversity concentrated on primates (4.4 species per genus) and to a lesser extent on bats and rodents (both with a ratio of 2.6). The higher species richness of the Atlantic Forest, compared with open-area biomes, is more diluted among genera pertaining to the seven more diverse orders, but there is still a noticeable contribution of primates. Sylvan forms taken together show the same trend of one additional species per genus in relation to pastoral mammals (respectively 2.4 and 1.6).

Biome-Level Assemblages

Taxonomic and Body Weight Distribution

In all cases, bats make up the largest proportion of the mammalian subfamily taxa of each biome, followed by rodents (fig. 20.3), a trend long noticed for the Neotropics (Fleming 1973; Eisenberg 1981). Except for these two orders, the taxonomic dominance at the species level does not show a clear-cut pattern among all the biomes analyzed, even though some trends can be identified. For instance, primates, marsupials, and carnivores are the next most speciose groups in Amazonia. The same is true for the Atlantic Forest, but in this case marsupials and carnivores outnumber the primates. Edentates and artiodactylans are the next most representative order in both biomes. In the four open or semiarid biomes, Carnivora is the third most abundant order, except in the Caatinga, where Edentata predominates. The fourth and the fifth most speciose orders in these biomes alternate between edentates and marsupials.

At the generic level, bats and rodents predominate in all biomes. Only in the Chaco do Rodentia outnumber Chiroptera. Carnivora is the next order in number of genera in every biome examined, with marsupials coming in fourth or fifth place. Primates are the fourth most represented order in number of genera in Amazonia.

In virtually every biome, species of small body weight predominate. For the whole of South America, close to 90% of the mammalian fauna is composed of species with body weight less than 3,000 g (fig. 20.4). This figure varies, according to the particular biome, from 85% for Amazonia and 83% for the Atlantic Forest to 78%, 73%, and 71%, respectively, in the Cerrado, Pantanal, and Chaco. The less diverse Caatinga lacks the relatively more significant representation of higher body weight categories found for the other

open and semiarid biomes, and 83% of its species fall below 3,000 g.

At the species level, the numerical dominance of bats and rodents reflects the distribution of body weights at the continental and biome scales. The frequency distribution of the number of species plotted on a logarithmic scale of body weight classes shows the often recurrent distribution highly skewed toward smaller size classes (Ceballos and Navarro 1991; see also Brown and Maurer 1989), for South America and for each biome analyzed (fig. 20.4). The mode for the South American aggregate data, as well as for the Cerrado and for the Chaco, can be found at about 25 g, while for the rest of the biomes it is the next smaller size class or approximately 16 g.

As expected, the distribution is much flatter at the biome level than that for the South American continent as a whole (see fig. 20.3). However, some interesting patterns also emerged from these relationships. First, in five out of six biomes, as well as for South America as a whole, there is a sharp decrease in the number of species weighing little more than 100 g. This reflects the fact that the dominant taxa in all cases are represented by small-bodied bats and rodents and by marsupials to a minor extent. Second, although both forest biomes show a somewhat flatter version of the distribution for the whole continent, all four semiarid biomes have a bimodal distribution with a second peak between 3,000 and 6,000 g. This recurrent pattern may occur because medium- and large-bodied species are usually wide ranging, common to most biomes.

Feeding and Locomotor Niche Distribution

For South America as a whole, 80% of the species are distributed in just five out of the fifteen feeding niche categories used for this analysis (fig. 20.5). The first three—frugivore-omnivore, frugivore-granivore and insectivore-omnivore—are responsible for respectively 22%, 19%, and 19% of all South American mammals. Aerial insectivores and herbivore-grazers make up another 12% and 8% of the species.

This relationship does not necessarily follow at the biome level. While frugivore-omnivores represent close to 30% of Amazonian mammals (fig. 20.6), aerial insectivores are the second most abundant niche category, followed by frugivore-granivores and frugivore-herbivores. These first four categories embrace 70% of all Amazonian species as well as all Atlantic Forest species. For the latter biome, however, aerial insectivores and insectivore-omnivores are at the top of two niche categories, followed by frugivore-omnivores and frugivore-granivores.

For the open vegetation formations, the niche distribution resembles that for the Atlantic Forest. Aerial

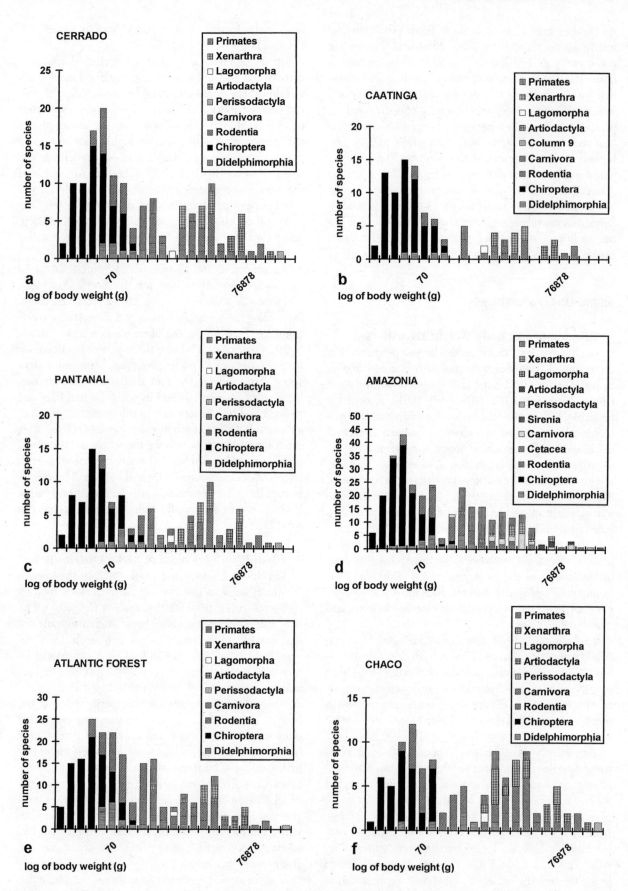

Figure 20.3. Frequency distribution of body weights for mammalian fauna: (a) Cerrado; (b) Caatinga; (c) Pantanal; (d) Amazonia; (e) Atlantic Forest; (f) Chaco.

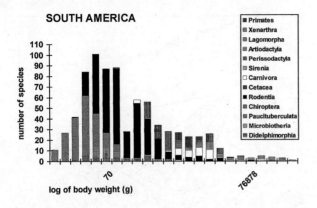

Figure 20.4. Frequency distribution of body weights for the South American mammalian fauna and distribution of mammalian orders according to body weight classes.

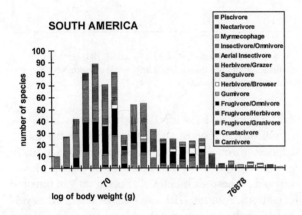

Figure 20.5. Distribution of feeding niche categories according to body weight classes for the South American mammalian fauna.

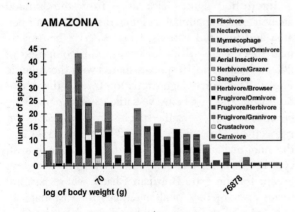

Figure 20.6. Distribution of feeding niche categories according to body weight classes for the Amazonian mammalian fauna.

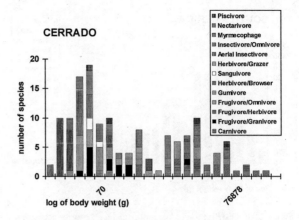

Figure 20.7. Distribution of feeding niche categories according to body weight classes for the Cerrado mammalian fauna.

are the fourth most abundant class of mammals in the latter, accounting for 10% of all species in that biome. The least diverse Caatinga has a concentration of 73% of its mammals in the top four feeding niches.

On the continental scale, smaller body weight classes are dominated by volant, scansorial and terrestrial species (fig. 20.8). Arboreal mammals commence having a more representative contribution in body weight classes ranging from 70 to 2,000 g. This trend holds for Amazonia (fig. 20.9), but not so much for the Atlantic Forest, the other forested biome analyzed (fig. 20.10). In the latter, the intermediate body weight classes are considerably more segregated among arboreal, scansorial, and terrestrial niche categories. For open-area vegetation formations, this body weight range is dominated by terrestrial, semifossorial, and scansorial species.

Zoogeographical Affinities

Simpson's similarity indexes yielded a number of interesting results (fig. 20.11). The Cerrado and Pantanal come up as basically one major zoogeographical unit, sharing 64% of their species. Since the Pantanal has only three endemic species, its mammal fauna represents a more depauperate subset of that of the Cerrado. This same relationship is also true when the Cerrado and the Caatinga are compared. The affinities start diluting at the level of the Chaco, with the influence emanating from southern cone faunal elements, even though a high proportion of its mammals are shared with the neighboring Pantanal and Cerrado. Therefore there is a significant zoogeographical affinity among the biomes that make up the open-area belt of vegetation formations separating the Atlantic Forest and the Amazon basin.

The higher endemicity of Amazonia (50%) makes it the biome with the least zoogeographical affinity

insectivores, frugivore-omnivores, insectivore-omnivores, and frugivore-granivores, in that order, account for 65% of the feeding niches of Cerrado mammals (fig. 20.7). Percentages for the top four categories in the Chaco and Pantanal are also very close to this figure, with the noticeable distinction that carnivores

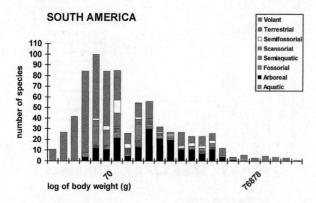

Figure 20.8. Distribution of locomotor niche categories according to body weight classes for the South American mammalian fauna.

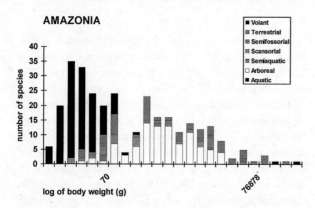

Figure 20.9. Distribution of locomotor niche categories according to body weight classes for the Amazonian mammalian fauna.

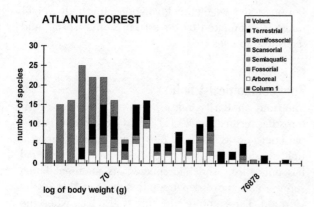

Figure 20.10. Distribution of locomotor niche categories according to body weight classes for the Atlantic Forest mammalian fauna.

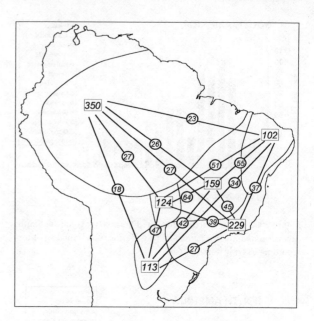

Figure 20.11. Simpson's similarity indexes (× 100) between the different biomes (in circles). Large numbers within biomes indicate species richness.

The analysis is reinforced when we examine the overall distribution of species included in the entire database throughout the six vegetation formations in question (table 20.3). Out of a total of 507 species in the pool of these six biomes, 297 occur only in forested formations. Another 31% can be characterized as versatile in habits, occurring in at least one forested and one open area biome. Only 10% are typical of open or semiarid regions.

Interesting figures also arise from the lumped analysis of the mammalian fauna distribution in open area and forested biomes. Close to 64% of Amazonian species are sylvan forms, a group composed of 205 endemics and 19 species shared with the Atlantic Forest. This percentage drops to 40% for the Atlantic Forest, the rest being versatile, wide-ranging species. Conversely, versatile species are responsible for, respectively, 36%, 60%, 73%, 82%, 90%, and 94% of the mammalian fauna of Amazonia, Atlantic Forest, Chaco, Cerrado, Pantanal and Caatinga. Therefore, except for the extra-Brazilian Chaco, where pastoral forms make up 39% of all mammals, the vast majority of species of the open-area belt formed by the Caatinga, Cerrado, and Pantanal are versatile in habits, with a reduced number of elements typical of open and semiarid regions.

Discussion

Despite the higher absolute number of species found for Amazonia and the Atlantic Forest, this analysis somewhat substantiates the recent claims that there is

among the six analyzed (table 20.3). The Atlantic Forest, on the other hand, shares 121 out of a total of 267 species with the neighboring Cerrado and has a degree of endemicity of 32%. At the species level, therefore, the two major forest biomes occurring within the Brazilian territory display a higher degree of faunal identity than the open vegetation formations.

nothing unusual in regionwide tropical forest mammalian diversity in South America (Redford, Taber, and Simonetti 1990; Mares 1992) other than what would be expected from species/area relationships. However, it has also become startlingly clear that the two forest biomes that occur in Brazil, particularly Amazonia, have a more distinctive mammal fauna than the open vegetation formations, which can be partially translated through their higher degrees of endemicity. The open biomes are chiefly composed of more tolerant, widely distributed versatile species. The formations that compose the open vegetation belt also display high levels of similarity in their pool of species. Therefore the previous suggestion advanced for the Cerrado (Fonseca and Redford 1984; Redford and Fonseca 1986), indicating that the bulk of its vertebrate fauna is composed of mesically derived versatile forms, also holds for the entire belt of open formations, from northeastern Brazil south to the Gran Chaco, with a diluting trend in the latter. The same tendency has been shown by Vanzolini (1976) for the lizard fauna of the Caatinga.

These conclusions diverge from those of Mares (1992), a much cited study, owing primarily to the different physiognomic classification of South America adopted in the two analyses, resulting in disparate mammal diversity figures for forested and open biomes. For instance, the area given for the Atlantic Forest (0.19 million km^2) is only one-sixth the figure considered here. Furthermore, Mares (1992) includes vast areas of subtropical drylands in his database of open and semiarid biomes.

It is also worth noticing that, since habitat (alpha) diversity is not independent of regional diversity (Ricklefs and Schluter 1993; Lawton, Lewinsohn, and Compton 1993), species richness of local communities should be higher for the two forested biomes of the sample examined. In fact, Peres (chapter 21 of this volume) has advanced data indicating an average of sixty-eight nonvolant species for five Amazonian lowland sites. Bat diversity is more difficult to incorporate and compare because of incomplete sampling across communities and biomes. Stallings et al. (1991) found forty-five nonvolant species in a 35,000 ha Atlantic Forest park. Conversely, data from localized sampling of open vegetation yielded the much lower counts of twenty-six species for the Caatinga and thirty-two for the Pantanal (Lacher and Mares 1986; Alho et al. 1988). These figures suggest that local mammalian species richness is about a third of that total, closer to the proportions found for bracken herbivores in global heterogeneous samples (Lawton, Lewinsohn, and Compton 1993).

There are many possible interpretations arising from these comparisons, two of which we will explore

here. In Amazonia, mammalian species' richness has been shown to vary considerably in response to physical and biological parameters (Emmons 1984), and heterogeneity in general across regions can be pronounced (Endler 1982). Thus, for the forest biomes the regional pool available to build up any given local community may consist of a smaller subset of the presumed diversity for the whole biome. However, this question cannot be resolved until better resolution data on distribution of Amazonian species become available. For the open vegetation formations, the higher ratios of local to regional diversity are consistent with the more widely distributed nature of its mammalian fauna and suggest less community identity than in the forest biomes.

These trends have repercussions for the understanding of the Recent history of the South American mammalian fauna. During the past twenty-five years, the traditional idea that Neotropical diversity has been produced, shaped, and maintained by long periods of environmental stability (e.g., Connell and Orias 1964), or that higher evolutionary rates result from favorable tropical climatic conditions (Stehli, Douglas, and Newell 1969), was gradually reworked with the recognition that the region has undergone a number of climatic cycles, which in turn led to a widespread dynamic process of forest expansion and retraction. Although these cycles have occurred many times in geological history, the Pleistocene events are assumed to have been most effective in producing modern distribution conformations (Haffer 1969, 1982; Vanzolini 1976; Vanzolini and Williams 1970; Simpson and Haffer 1978). Many times during the Pleistocene, populations of forest-adapted species became isolated in a variable number of refugia, resulting in genetic divergence that ultimately may have led to allopatric speciation. Conversely, during periods of rainfall maxima, it is reasonable to assume that xeric refugia would also have been formed (Granville 1982; Mares, Willig, and Lacher 1985). Many researchers think this process produced a large fraction of the high subgeneric diversity observed in present-day South American forest biomes (Haffer 1969; Vuilleumier 1971; Simpson and Haffer 1978; Brown and Ab'Saber 1979; Prance 1987).

As more distribution data become available, there have been suggestions that refugia might actually have caused more extinction than speciation (for a discussion, see Brown 1987). The magnitude of both processes remains unknown (Simpson and Haffer 1978). Leading specialists in some groups also maintain that the bulk of Neotropical species diversity is Tertiary (Hershkovitz 1969; Heyer and Maxson 1982; Cracraft and Prum 1988), while at least one recent study of sequences of the mitochondrial cytochrome-b DNA gene of Amazonian arboreal echimyids places speciation events in pre-Pleistocene periods (Silva and Patton 1993). Within this framework, it is unlikely that the short period since the Pleistocene has been sufficient to produce the large number of species observed today. Furthermore, island biogeographical processes suggest that species saturation in refugia may have led to a decrease in species richness measured over a large area, even if the time span involved was long enough for speciation to occur. In fact, the equilibrium model (MacArthur and Wilson 1967) has previously been invoked to explain the magnitude of South American mammal generic diversity after the closure of the Panamanian land bridge some three million years ago (Marshall et al. 1982).

If these arguments hold, we may argue that the Pleistocene climatic cycles have brought a net loss of Tertiary mammalian diversity, even if populations of some groups were able to diverge to subspecific and more rarely to specific taxonomic levels during the Pleistocene. Therefore, even though the number of extant mammal species in South America can be seen as high, it may once have been higher. For instance, Brazil as a whole has a few more mammal species than Mexico, which spans a region for which species-producing refugia for vertebrates have not been as widely proposed. However, Mexico is 23% the size of Brazil and spans some fourteen fewer degrees of latitude. If the Brazilian country sample can be considered representative of the South American continent (equivalent to 48% of its land surface), its mammalian fauna has to be judged relatively less rich. Therefore, alathough Pleistocene refugia indeed have shaped some present South American mammal distribution patterns (e.g., Kinzey 1982), they may also have driven a large number of groups to extinction and possibly halted differentiation of others already under way. It is reasonable to assume that the taxa most affected by stochastic extinction in constricted refugia would be homeotherm vertebrates of restricted vagility, especially those of large body size (Robinson and Redford 1989; Arita, Robinson, and Redford 1990). Interestingly, the number of forest refugia postulated for Neotropical mammals is indeed low and is restricted to a few primate groups (Kinzey 1982).

There is mounting evidence that a large number of South American pastoral forms, both endemic and of allochthonous origin, went extinct during the Pleistocene (Webb 1976, 1978; Cartelle, chapter 4 of this volume; Rancy, chapter 3 of this volume), despite the logical assumption that open-area or xeric refugia must also have been formed. These forms represented the characteristic mammalian fauna of the area that is now occupied by the open-area vegetation belt that transverses the Brazilian territory (Cartelle, this volume). These trends can be observed at present in the form of low endemicity in the mammalian fauna of the Cerrado (Redford and Fonseca 1986), Caatinga, Pantanal, and Chaco (this study), and they run against the suggestion that the Pleistocene witnessed bursts of radiation catalyzed by cyclic fluctuations in climate (Simpson and Haffer 1978), at least for Neotropical mammals. Lovejoy (1982) rightly pointed out that the number of functioning refugia should be inversely related to the body size of the taxa in question, and pastoral species are usually larger than sylvan forms (see Eisenberg 1981). As Dial and Marzluff (1988) also suggest, under fluctuating environmental conditions smaller organisms should be favored, whereas the reverse is expected during long stable periods.

The episodes of forest expansion and retraction, which were accelerated especially since the beginning of the Pleistocene, would have resulted in demographic stress to specialized pastoral groups of large body size and to a lesser extent to those that, although smaller, were forest specialists. This is especially true considering that the intervals between dry and arid phases could be as short as one thousand to two thousand years (Cerqueira 1982), halting incipient local adaptations that may have developed during these intervals. Given that large mammals should require large areas to maintain genetic and demographically viable populations (Robinson and Ramirez 1982), extinction should be expected. That approximately 90% of the present South American mammalian diversity is made up of species weighing less than 3 kg, despite the many ecological advantages of large body size (Brown and Maurer 1986), corroborates this interpretation. Versatile forms, on the other hand, should have been less affected by these events and should be overrepresented in the modern fauna. Again, the data support this contention.

Comparing some examples of pairs of extinct and extant related species, the trend was for elimination of the larger member of the pair and persistence of the form with smaller body size (Cartelle, this volume). For instance, among giant armadillos, *Priodontes* persisted and *Holmesina* perished. The same can be observed between extant *Hydrochaeris* and extinct *Neochoerus* and between *Blastocerus* and *Morenelaphus* (Cartelle, this volume). Another interesting example is the extinction of the specialized and strictly carnivorous *Protocyon troglodites* and the persistence of the omnivorous maned wolf, *Chrysocyon brachyurus*.

As Vanzolini (1976) suggested in a model he called "turbulent mixing," if changes are rapid and widespread, homogenizing of characters rather than divergence is to be expected. This mechanism, coupled with the relatively long generation times for medium-sized and large species, may have meant that the number of open-area or xeric refugia for the endemic pastoral mammals, and of forest refugia for exclusively sylvan forms, may have been minimal. In other words, refugia proposed for plants and other animals were to a large extent evolutionarily functionless, at least for mammals. Comparing similarity indexes and distribution data for extant mammals shows that the forest expansion/retraction dynamics may have functioned to homogenize the fauna over a large part of the South American continent.

In relation to forest biomes, the data have demonstrated that only Amazonia harbors a larger number of exclusive sylvan forms: in the Atlantic Forest versatile mammals predominate. But even in Amazonia the higher degree of faunal identity is concentrated on rodents, bats, and primates, while the higher number of species per genus follows a similar pattern and primates greatly outnumber the other two orders in the ratio of number of species to number of genera. Although largely speculative, coupled with the poor Amazonian fossil record, the data suggest that extinction of Pleistocene mammals in response to forest contraction may also have occurred and that speciation events were not sufficient to counteract species extirpation. The higher absolute mammalian diversity of this region in relation to neighboring biomes should be found at the generic rather than at the species level.

Conservation Status of the Brazilian Mammal Fauna

The environmental problems facing Brazil are historically and geographically dissimilar, depending on the different ecological and socioeconomic settings. For instance, the Brazilian Amazon remains in a fairly pristine stage, with about 12% of the region altered by development programs, as recent history has demonstrated. Case studies are the widespread eradication of forests in the states of Pará and Rondônia (Mahar 1989). The Cerrado of central Brazil lingers unstudied, while export-oriented cash crops and pastures begin dominating the landscape, and protected areas account for only 1.5% of the total area of the region (Dias 1992).

The last remaining populations of several highly endangered species of the semiarid lands of the northeastern Caatinga are holding poorly to the few available portions of relatively unaltered habitats, in a sad mixture of human tragedy and overexploitation of biodiversity (Santos et al. 1994). Billions of dollars of development funds are just about to burst from the bilateral and multilateral pipelines to radically change the two-hundred-year-old sustainable economic activity of the Pantanal wetlands, based on low-density cattle ranching (Lourival and Fonseca, n.d.). Finally, the biologically rich Atlantic Forest, the first to be colonized in Brazil, is barely surviving as a biotic complex, with millions of years of evolutionary history now compounded onto the few isolated and fragile remaining

fragments, totaling no more than 10% of the original area (Fonseca 1985; Câmara 1991).

Therefore, although Brazil is one of the richest countries in mammalian diversity, owing to habitat degradation, in addition to overexploitation of particular groups (mostly primates, carnivores, and edentates) for meat, skins, and the pet trade, a significant fraction of this fauna is currently facing a high degree of threat. The number of species and subspecies of mammals officially recognized as threatened with extinction has grown from twenty-nine in 1973 to fifty-eight in 1992 (Fonseca et al. 1994c), a figure that represents 12% of all Brazilian mammal species.

But it is almost certain that there are many more threatened mammals than this, reflecting the relatively poor knowledge of most species, the bias toward better-studied groups, and the caution with which the Brazilian specialist group responsible for elaborating this list has proceeded with the threat analysis (Fonseca et al. 1994c). It is worth noticing that this group recognized an additional twenty-one species and subspecies as insufficiently known, but presumably with poor conservation status based on distribution data, as well as eight genera that may include species facing threat. Some of these, such as *Proechimys*, *Echimys*, and *Marmosa*, are highly speciose.

Corroborating this argument, the 1989 official list included no bats, which make up the most diverse mammal group in Brazil, no doubt as a consequence of the limited number of field studies. Only seven rodents are listed, six of them added in 1989. In addition, over 40% of all mammals recognized as threatened are primates. Although this group is particularly susceptible to habitat disturbance, primatology is one of the better-developed branches of field zoology in Brazil. As the number of field studies increases, so do data indicating the decline of wild primate populations in many parts of Brazil (Fonseca et al., 1994a). Indeed, of the fifty-two primate species of Brazilian Amazonia, 28% are considered threatened. If this trend can be applied to other groups and regions as well, there are clear indications that the status of wild mammal populations of much more degraded biomes, such as the Atlantic Forest, is potentially very poor. For instance, 60% of its primates are threatened, including two endemic genera (*Brachyteles* and *Leontopithecus*). In addition, five Atlantic Forest rodents, four belonging to monotypic genera (*Abrawayaomys*, *Chaetomys*, *Rhagomys*, and *Phaenomys*), have poor conservation status.

Last, owing to the structure of tropical mammalian communities, characterized by few dominant groups, the new species that are constantly being described are usually of restricted geographical distribution or locally rare. These newly described species, especially those of the highly degraded Atlantic Forest or of

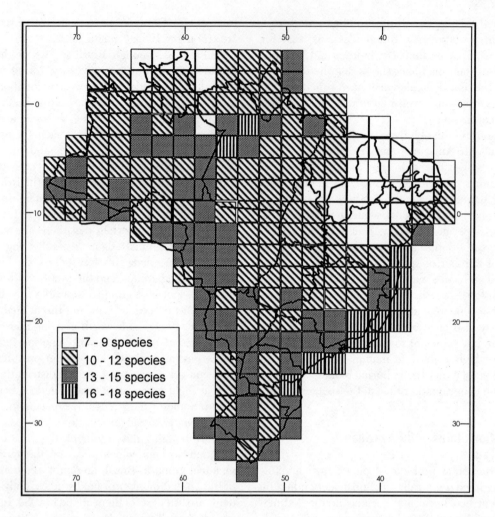

Figure 20.12. Density of threatened species of Brazilian mammals in quadrats of 220 km by 220 km. (From Fonseca et al. 1994c.)

the increasingly altered Cerrado, are also potentially threatened. Examples are the black-faced lion tamarin (*Leontopithecus caissara*), described in 1990 in one of the most developed and populous regions of Brazil (Lorini and Persson 1990), and the terrestrial Cerrado akodontine *Akodon lindberghi* (Hershkovitz, 1990).

Taking into account the biases previously described, Amazonia stands out with the highest total number of threatened species, which include seventeen endemics (Fonseca et al. 1994c). The Atlantic Forest follows with fourteen endemic species threatened, and the Cerrado and Caatinga appear on the list with one endemic mammal each. Of the forty-nine land species listed, thirty-three are endemic to their respective biomes (Fonseca et al. 1994c) and 50% are restricted to Brazil. In relation to the Cerrado, the few species included on the list do not reflect the actual degree of threat facing the region's mammalian communities. Examples of plant and animal species with limited distribution, such as *Juscelinomys candango,* are apparently not uncommon (Dias 1992).

Figure 20.12 shows the absolute density of threatened Brazilian mammal species in quadrats of 220 km by 220 km, or close to 48,400 km^2. The Atlantic Forest has the highest number of quadrats with sixteen to eighteen species, particularly in the southern portion of the state of Bahia, the states of Espírito Santo and Rio de Janeiro, and the coastal areas of the states of São Paulo and Paraná (Fonseca et al. 1994a). The remaining area of this biome is virtually characterized by quadrats in the next higher class of threatened mammals, ranging from thirteen to fifteen species. In Amazonia, two areas of western Pará that are high in primate diversity are also included in the top threatened class, and nearly half of the quadrats in the thirteen- to fifteen-species range are in this region. The Cerrado, perhaps the least surveyed Brazilian biome, is mostly characterized by quadrats with ten to twelve species. Caatinga has the fewest species at risk owing to its significantly lower diversity.

Sadly, there are rising indications of deterioration in the status of a great many Brazilian mammals, particularly those endemic to the biomes where they occur. Although all major habitats should be given appropriate attention, forest biomes, especially the Brazilian Atlantic Forest, stand to lose a great deal of highly distinctive mammalian biodiversity. As Emmons and Feer (1997) have noted, 60% of all forest mammals belong to endemic South American families, a significant proportion of which are ancient dwellers on the continent. On the other hand, the data available from field studies are fragmentary and insufficient even to document the process of declining wild populations. Therefore direct measures to counteract the current erosion of mammalian diversity in Brazilian native habitats are highly unlikely to be generated in the short term, except for the few best-studied species. At present, landscape ecology approaches are thus more prone to be productive and cost effective than species-specific strategies, such as those exemplified by species action plans previously advocated by IUCN's Species Specialists' Groups.

Acknowledgments

We thank John Eisenberg for the invitation to contribute to the third volume of this most important work, which significantly advances our understanding of Neotropical mammals. We also acknowledge the thorough review of this chapter by Anthony Rylands and Castor Cartelle and thank Kent Redford for their stimulus and for reviewing an earlier draft. This is a contribution of the research program in Ecology, Conservation, and Wildlife Management, Federal University of Minas Gerais, Brazil.

References

Alho, C. J. R., T. E. Lacher Jr., Z. M. S. Campos, and H. C. Gonçalves. 1988. Mamíferos de Fazenda Nhumirim sub-região de Nhecolândia, Pantanal do Mato Grosso do Sul: Levantamento preliminar de espécies. *Rev. Brasil. Biol.* 48 (2): 213–25.

Arita, H. T., J. G. Robinson, and K. H. Redford. 1990. Rarity in Neotropical forest mammals and its ecological correlates. *Conserv. Biol.* 4:181–92.

Bernardes, A. T., A. B. M. Machado, and A. B. Rylands. 1989. *Fauna brasileira ameaçada de extinção.* Belo Horizonte: Fundação Biodiversitas.

Brown, J. H., and B. A. Maurer. 1986. Body size, ecological dominance and Cope's rule. *Nature* 324: 248–50.

———. 1989. Macroecology: The division of food and space among species on continents. *Science* 243: 1145–50.

Brown, K. S., Jr. 1987. Conclusions, synthesis, and alternative hypotheses. In *Biogeography and Quater-

ary history in tropical America*, ed. T. C. Whitmore and G. T. Prance, 175–91. Oxford: Oxford University Press.

Brown, K. S., Jr., and A. N Ab'Saber. 1979. Ice-age refuges and evolution in the Neotropics: Correlation of paleoclimatological, geomorphological and pedological data with modern biological endemism. *Paleoclimas* 5:1–30.

Câmara, I. G. 1991. *Plano de ação para a Mata Atlântica.* Fundação SOS Mata Atlântica–Sociedade Brasileira de Proteção Ambiental–World Wildlife Fund U.S.

Ceballos, G., and D. Navarro L. 1991. Diversity and conservation of Mexican mammals. In *Latin American mammalogy: History, biodiversity, and conservation*, ed. M. A. Mares and D. J. Schmidly, 167–98. Norman: University of Oklahoma Press.

Cerqueira, R. 1982. South American landscapes and their mammals. In *Mammalian biology in South America*, ed. M. A. Mares and H. H. Genoways, 53–57. Pymatuning Symposia in Ecology 6. Special Publication Series. Pittsburgh: Pymatuning Laboratory of Ecology, University of Pittsburgh.

Connell, J. H., and E. Orias. 1964. The ecological regulation of species diversity. *Amer. Nat.* 98:399–414.

Cracraft, J., and R. O. Prum. 1988. Patterns and processes of diversification: Speciation and historical congruence in some Neotropical birds. *Evolution* 42:603–20.

Dial, K. P., and J. M. Marzluff. 1988. Are the smallest organisms the most diverse? *Ecology* 69:1620–24.

Dias, B. F. S. 1992. *Alternativas de desenvolvimento dos cerrados: Manejo e conservação dos recursos naturais.* Brasília: Fundação Pró-Natureza (Funatura) e Ibama.

Eisenberg, J. F. 1981. *The mammalian radiations: An analysis of trends in evolution, adaptation, and behavior.* Chicago: University of Chicago Press.

Emmons, L. H. 1984. Geographic variation in densities and diversities of non-flying mammals in Amazonia. *Biotropica* 16:210–22.

Emmons, L. H., and F. Feer. 1997. *Neotropical rainforest mammals: A field guide.* 2d ed. Chicago: University of Chicago Press.

Endler, J. A. 1982. Pleistocene forest refuges: Fact or fancy? In *Biological diversification in the tropics*, ed. G. T. Prance, 641–57. New York: Columbia University Press.

Fleming, T. H. 1973. Numbers of mammal species in north and central forest communities. *Ecology* 54:555–63.

Fonseca, G. A. B. da. 1985. The vanishing Brazilian Atlantic forest. *Biol. Conserv.* 34:17–34.

Fonseca, G. A. B. da, and K. H. Redford. 1984. The

mammals of IBGE's Ecological Reserve Brasília, and an analysis of the role of gallery forests in increasing diversity. *Rev. Brasil. Biol.* 44 (4): 517–23.

Fonseca, G. A. B. da, C. M. R. Costa, R. B. Machado, Y. L. R. Leite and C. Furlani, eds. 1994a. *Mamíferos brasileiros: Uma coletânea bibliográfica.* Belo Horizonte: Fundação Biodiversitas.

Fonseca, G. A. B. da, A. B. Rylands, C. M. R. Costa, R. B. Machado, and Y. L. R. Leite, eds. 1994b. *Livro vermelho dos mamíferos brasileiros ameaçados de extinção.* Belo Horizonte: Fundação Biodiversitas.

Fonseca, G. A. B. da, A. B. Rylands, C. M. R. Costa, R. B. Machado, and Y. L. R. Leite. 1994c. Mamíferos brasileiros sobre ameaça. In *Livro vermelho dos mamíferos brasileiros ameaçados de extinção,* ed. G. A. B. da Fonseca, A. B. Rylands, C. M. R. Costa, R. B. Machado, and Y. L. R. Leite, 1–10. Belo Horizonte: Fundação Biodiversitas.

Granville, J. J. de. 1982. Rain forest and xeric flora refuges in French Guiana. In *Biological diversification in the tropics,* ed. G. T. Prance, 159–81. New York: Columbia University Press.

Haffer, J. 1969. Speciation in Amazonian forest birds. *Science* 165:131–37.

———. 1982. General aspects of the refuge theory. In *Biological diversification in the tropics,* ed. G. T. Prance, 6–24. New York: Columbia University Press.

Hershkovitz, P. 1969. The evolution of mammals on southern continents. 4. The Recent mammals of the Neotropical region: A zoogeographical and ecological review. *Quart. Rev. Biol.* 44:1–70.

———. 1990. Mice of the *Akodon boliviensis* size class (Sigmodontinae, Cricetidae) with the description of two new species from Brazil. *Fieldiana: Zool.,* n.s., 57:1–35.

Heyer, W. R., and L. R. Maxson. 1982. Distributions, relationships and zoogeography of lowland frogs: The *Leptodactylus* complex in South America, with special reference to Amazonia. In *Biological diversification in the tropics,* ed. G. T. Prance, 375–88. New York: Columbia University Press.

Hueck, K. 1972. *Las florestas da América do Sul.* São Paulo: Editora Polígono.

IBGE. 1993. *Mapa de vegetação do Brasil.* Brasília: Fundação Instituto Brasileiro de Geografía e Estatística.

Kinzey, W. G. 1982. Distribution of primates and forest refuges. In *Biological diversification in the tropics,* ed. G. T. Prance, 455–82. New York: Columbia University Press.

Lacher, T. E., Jr., and M. A. Mares. 1986. The structure of Neotropical mammal communities: An appraisal of current knowledge. *Rev. Chil. Hist. Nat.* 59:121–34.

Lawton, J. H., T. M. Lewinsohn, and S. G. Compton. 1993. Patterns of diversity for the insect herbivores on bracken. In *Species diversity in ecological communities,* ed. R. E. Ricklefs and D. Schluter, 178–84. Chicago: University of Chicago Press.

Lorini, M. L., and V. G. Persson. 1990. Nova espécie de *Leontopithecus* Lesson, 1840, do sul do Brasil (Primates, Callitrichidae). *Bol. Mus. Nac.,* n.s., 338: 1–14.

Lourival, R. F. F., and G. A. B. da Fonseca. n.d. A caça na regiço da Nhecolândia, Mato Grosso do Sul. In *Manejo de fauna na America Latina,* ed. C. V. Pádua and R. Bodmer. Belém: CNPq–Sociedade Civil Mamirauá. In press.

Lovejoy, T. E. 1982. Designing refugia for tomorrow. In *Biological diversification in the tropics,* ed. G. T. Prance, 455–82. New York: Columbia University Press.

MacArthur, R. H., and E. O. Wilson. 1967. *The theory of island biogeography.* Princeton: Princeton University Press.

Mahar, D. J. 1989. *Government policies and deforestation in Brazil's Amazon region.* Washington, D.C.: World Bank.

Mares, M. A. 1992. Neotropical mammals and the myth of Amazonian biodiversity. *Science* 255: 976–79.

Mares, M. A., M. R. Willig, and T. E. Lacher Jr. 1985. The Brazilian caatinga in South American zoogeography: Tropical mammals in a dry region. *J. Biogeogr.* 12:57–69.

Marshall, L. G., S. D. Webb, J. J. Sepkoski, and D. M. Raupp. 1982. Mammalian evolution and the Great American interchange. *Science* 215:1351–57.

Mittermeier, R. A., J. M. Ayres, T. Werner, and G. A. B. da Fonseca. 1992. O país da megadiversidade. *Ciênc. Hoje* 14 (81): 20–27.

Prance, G. T. 1987. Biogeography of Neotropical plants. In *Biogeography and Quaternary history in tropical America,* ed. T. C. Whitmore and G. T. Prance, 46–65. Oxford: Oxford University Press.

Redford, K. H., and G. A. B. da Fonseca. 1986. The role of gallery forests in the zoogeography of the cerrado's non-volant mammalian fauna. *Biotropica* 18 (2): 126–35.

Redford, K. H., A. Taber, and J. A. Simonetti. 1990. There is more to biodiversity than the tropical rain forests. *Conserv. Biol.* 4:328–40.

Ricklefs, R. E. 1987. Community diversity: Relative roles of local and regional processes. *Science* 235: 167–71.

Ricklefs, R. E., and D. Schluter. 1993. Species diversity: Regional and historical influences. In *Species diversity in ecological communities,* ed. R. E. Ricklefs and D. Schluter, 350–63. Chicago: University of Chicago Press.

Robinson, J. G., and J. Ramirez. 1982. Conservation biology of Neotropical primates. In *Mammalian biology in South America,* ed. M. A. Mares and H. H. Genoways, 329–44. Pymatuning Symposia in Ecology 6. Special Publication Series. Pittsburgh: Pymatuning Laboratory of Ecology, University Pittsburgh.

Robinson, J. G., and K. H. Redford. 1989. Body size, diet, and population variation in Neotropical forest mammal species: Predictors of local extinction? In *Advances in Neotropical mammalogy,* ed. K. H. Redford and J. F. Eisenberg, 567–94. Gainesville, Fla.: Sandhill Crane Press.

Santos, I. B., G. A. B. da Fonseca, S. E. Rigueira, and R. B. Machado. 1994. The rediscovery of the Brazilian three-banded armadillo and notes on its conservation status. *Edentata* 1 (1): 11–15.

Silva, M. N. F. da, and J. L. Patton. 1993. Amazonian phylogeography: mtDNA sequence variation in arboreal echimyid rodents (Caviomorpha). *Mol. Phyl. Evol.* 2:243–55.

Simpson, B. B., and J. Haffer. 1978. Speciation patterns in the Amazonian forest biota. *Ann. Rev. Ecol. Syst.* 9:497–518.

Stallings, J. R., G. A. B. da Fonseca, L. P. S. Pinto, L. M. S. Aguiar, and E. L. Sábato. 1991. Mamíferos do Parque Estadual do Rio Doce, Minas Gerais, Brasil. *Rev. Brasil. Zool.* 7 (4): 663–77.

Stehli, F. G., R. G. Douglas, and N. D. Newell. 1969. Generation and maintenance of gradients of taxonomic diversity. *Science* 164:947–49.

Stenseth, N. C. 1989. On the evolutionary ecology of mammalian communities. In *Patterns in the structure of mammalian communities,* ed. D. W. Morris, Z. Abramsky, B. J. Fox, and M. R. Willig, 219–28. Lubbock: Texas Tech University Press.

Vanzolini, P. E. 1976. On the lizards of a cerrado-caatinga contact: Evolutionary and zoogeographical implications (Sauria). *Pap. Avul. Zool. (São Paulo)* 29 (16): 111–19.

Vanzolini, P. E., and E. E. Williams. 1970. South American anoles: The geographical differentiation and evolution of the *Anolis chrysolepis* species group (Sauria, Iguanidae). *Arq. Zool., São Paulo* 19:1–298.

Vuilleumier, B. S. 1971. Pleistocene changes in the fauna and flora of South America. *Science* 173: 771–80.

Webb, S. D. 1976. Mammalian faunal dynamics of the great American interchange. *Paleobiology* 2: 216–34.

———. 1978. A history of savanna vertebrates in the New World. 2. South America and the great interchange. *Ann. Rev. Ecol. System.* 9:393–426.

Wilson, D. E., and D. M. Reeder, eds. 1993. *Mammal species of the world.* 2d ed. Washington, D.C.: Smithsonian Institution Press.

21 The Structure of Nonvolant Mammal Communities in Different Amazonian Forest Types

Carlos A. Peres

Introduction

From a historical perspective Amazonian mammalogy has yet to mature far beyond the stage of the great expeditions led by European travelers since the eighteenth century.[1] Comprehensive museum collections have not been assembled from most major interfluvial basins, particularly in the case of the least known and highly speciose small-bodied taxa, such as the Chiroptera, Rodentia, and Didelphimorphia. A complete inventory of forest mammals has yet to be conducted at any Amazonian site, and even the most exhaustive catalogs and long-term studies have failed to saturate local species richness curves as a function of sampling effort (but see Patton, Berlin, and Berlin 1982; Terborgh, Fitzpatrick, and Emmons 1984; Pacheco et al. 1993; and Woodman et al. 1991 for nearly complete inventories). Moreover, whereas previous geographic sampling has been largely restricted to more accessible areas near large, navigable rivers, most of the unbroken horizons of Amazonian forests are in relatively remote and still poorly known interfluvial areas.

Documented mammal communities in the region are therefore very few and far between. For instance, at least three new chromatic forms of diurnal primates, clearly some of the most visible mammals in tropical forests anywhere, have been described from newly discovered specimens from east-central Amazonia in the past three years (Mittermeier, Schwarz, and Ayres 1992; Ferrari and Lopes 1992; Queiroz 1992). This situation is magnified when mammals are considered from an ecologist's perspective rather than a systematist's. Autoecological studies of Amazonian mammals are largely restricted to perhaps eighteen species of primates, five ungulates, and a handful of rodents, marsupials, and carnivores, whereas even anecdotal ecological data are still lacking for most other forest and nonforest taxa.

Our present knowledge of mammal community ecology in the region thus remains at best glaringly incomplete. Not surprisingly, there have been few attempts to examine the structure of mammal communities in relation to regional differences in forest types. Yet the clear physiognomic distinctions in the macrohabitats of Amazonia (e.g., Pires 1972; Pires and Prance 1985) suggest that terrestrial and arboreal mammal communities could diverge in a predictable fashion depending on key determinants of habitat complexity and heterogeneity, resource spectrum, and primary productivity. At present, however, a detailed analysis of this topic would at best be preliminary, given the paucity of quantitative mammal surveys providing density estimates and even presence/absence data from species checklists.

Here I attempt to piece together and compare existing data on mammal community organization and guild structure within different Amazonian forest types. As in previous communitywide studies of Neotropical mammal faunas in both forest habitats (Eisenberg and Thorington 1973; Emmons 1984; Janson and Emmons 1990; Malcolm 1990) and semiopen habitats (Eisenberg, O'Connell, and August 1979; August 1983; Schaller 1983), bats are largely excluded from this analysis, for most available data are confined to nonvolant species. Bats constitute the most poorly sampled taxon in these studies because of the inherent difficulty of capture techniques at different forest levels and the extremely high species diversity of Neotropical forest chiropterans.

From the outset, this comparison is deliberately intended to avoid any potential effects of different forms of anthropogenic disturbance such as subsistence hunting (Peres 1990, 1996, n.d.), selective logging (Johns 1986), and habitat fragmentation (Malcolm 1991a) on the patterns of mammal abundance and diversity at any one site. Critical data for this preliminary undertaking have recently been made available from communitywide studies at three undisturbed sites from which species abundance data can be compiled across the class Mammalia: (1) Cocha Cashu, Madre de Dios, southern Peru (Terborgh 1983, 1986; Janson and Emmons 1990); (2) Lago Teiú, Amazonas, Brazil (Ayres 1986); and (3) Urucu, Amazonas, Brazil

(Peres 1993a, this chapter). Fortunately, these three sites are appropriately distributed along what may turn out to be important environmental gradients determining the structure of Amazonian vertebrate communities: distance from large rivers and flooding periodicity.

I begin by presenting data on densities and biomass of nonvolant mammals from a recently surveyed un-flooded (hereafter terra firma) forest near the head-waters of the Rio Urucu. I then compile data from a mammal capture operation after the closing of a hy-droelectric dam in eastern Amazonia, briefly consid-ering how such rescue attempts can approach exhaus-tive mammal inventories. Next I compare the Urucu mammal community with those of forest types un-der the direct influence of major white-water rivers.[2] Finally, I consider how regional differences in habitat structure, diversity, and productivity might affect the abundance of nonvolant mammals, particularly that of primary consumers, and examine how Amazonian mammal communities diverge from those elsewhere in the American tropics.

Typology of Amazonian Vegetation

Amazonian mammal macrohabitats can be sorted into two basic categories: those consisting of closed-can-opy forests—the hylaea—and those in nonforest or semiopen habitats, including relatively large expanses of open savannas, savanna woodlands, and seasonally inundated savannas. Embedded in the closed-canopy forest matrix, however, one may find a number of smaller or transitional vegetation subtypes, including palm swamps, liana forests, bamboo forests at differ-ent stages of their flowering cycle, and variable-sized patches of blowdowns at different stages of regenera-tion, as well as small enclaves of cerrado, campinas, and campos rupestres,[3] all of which may provide natu-ral canopy breaks for mammals in an otherwise rela-tively monotonic forest landscape.

Amazonia is a macromosaic of forest types in-tersected by different types of watersheds, in which natural fluvial dynamics and disturbance are likely to play an important role in maintaining habitat diversity (Salo et al. 1986; Terborgh and Petren 1991; Puhakka et al. 1993; fig. 21.1). Amazonian closed-canopy for-ests have traditionally been distinguished (by both na-tive inhabitants and phytogeographers) according to drainage type and inundation regime. This essentially reflects the geographic position of a given site relative to large rivers and proximity to headwaters. Floodplain forests can thus be termed igapós in the case of black- or clear-water rivers, and várzeas in the case of white-water rivers (Prance 1979). Of particular interest, this distinction is closely tied to the underlying soil nutrient regime a given forest is subject to (Furch and Klinge

1989). Clear- and black-water floodplains, like terra firma forests, have not been subject to the seasonal input of suspended inorganic sediments and solutes even near major rivers, whereas forests near white-water rivers typically receive an annual (or supra-annual) influx of nutrient-rich alluvial sediments. This can profoundly affect the amount of macronutrients available to plants in different forest types and, in turn, the overall productivity of vegetative and repro-ductive plant parts harvested by primary consumers. Mammal community density or biomass should thus be expected to be low in terra firma forests far re-moved from the influence of nearby rivers, merely as a reflection of low nutrient availability. Conversely, over-all densities and biomass of mammals, particularly her-bivores, should be expected to be high in typical várzea and alluvial sites abutting major white-water rivers, should nutrient input be an important determinant of primary productivity.

Proximity to major rivers and tributaries may thus become a decisive factor in the structure and beta di-versity of faunal assemblages. Clearly, however, a vast proportion of Amazonia (about 94% of the basin) con-sists of terra firma systems well beyond the influence of large rivers. These areas roughly match the distri-bution of the long, folded crystalline rocks of Precam-brian origin forming the Guiana and central Brazilian (or Guaporé) shields, which have received no sedi-ment deposits since the Triassic or lower Cretaceous (Putzer 1984). As a result, only 6% of Amazonian soils have no major limitations of one key nutrient or an-other to plant photosynthesis (Sanchez 1981). Terra firma forests have thus been deprived of recent alluvial nutrients and could be described as truly oligotrophic systems, perhaps comparable only to remote nonvolca-nic regions of central Borneo and parts of the Congo basin.

This condition is in stark contrast to forest sites close to white-water rivers of great discharge and transport capacity (e.g., the Madeira, Purus, Juruá, Japurá Rivers), which have received enormous sedi-ment loads of mud and silt. This is particularly the case with the young valleys and meander belts of Brazilian Amazonia, which often become inundated for four to six months of the year. Riverine sites in the headwaters region of the same rivers, on the other hand, tend to be subject to considerably more ephemeral floods that may not occur every year, yet they too have been pro-foundly shaped by alluvial deposits (Terborgh and Pe-tren 1991; Kalliola et al. 1991).

Terra Firma, Alluvial, and Várzea Forests

According to this brief overviw of Amazonian geo-chemistry, three forest sites that are affected to vary-ing degrees by fluvial activity are considered in this

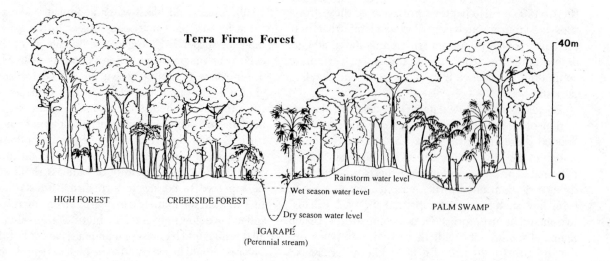

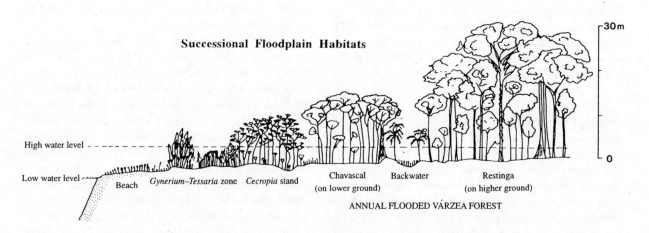

Figure 21.1. Cross section of a typical terra firma forest (Urucu forest) and floodplain habitats of west-central Amazonia. Maximum variation in vegetation structure at the Urucu forest (above) occurred along a perennial black-water forest stream, where high terra firma forest was replaced by creekside forest and small palm swamps. Várzea forests on lower or higher ground (below), on the other hand, can represent late successional stages of white-water floodplain vegetation, whose structure is affected by age, depth and duration of annual floods, and degree of fluvial disturbance.

chapter (table 21.1). The first site, the Urucu forest, is thought to be a typical undisturbed terra firma forest representative of the yet vast pristine areas of remote interfluvial basins of Amazonia (Peres 1991a). The mammal community at this site should thus serve as a base for comparison with those near large rivers, which may otherwise be depicted as the rule rather than the exception for the entire Amazon. The second site, Cocha Cashu, is assumed to represent a typical alluvial forest (Terborgh 1983). Cashu is characterized by soils formed by rich alluvial deposits, but only rarely is it actually flooded by unusually high water levels of the Rio Manu, and that for relatively ephemeral periods (Emmons 1987; J. Terborgh, pers. comm.). Shallow standing water may cover 25% of even the high-ground forest for part of the year, however, thus

reducing habitat availability for some strictly terrestrial mammals (Janson and Emmons 1990). Finally, a site in the lower Rio Japurá (Lago Teiú), which typically remains under 1 to 6 m of standing water for four months of the year (Ayres 1986), was selected to represent a Holocene várzea forest.

Forest structure at these sites remains entirely or almost entirely undisturbed by humans, and recent selective hunting has not taken place (Urucu: Peres 1990; Cocha Cashu: Terborgh 1983) or has been restricted to very light harvest of large-bodied game species (Lago Teiú: Ayres 1986; Ayres, pers. comm.). The structure of mammal communities there is thus largely assumed to be a response to intrinsic features of the habitat, such as forest structure and composition, flooding regime, and soil fertility, rather than to differ-

Table 21.1   Physical Profile of Three Undisturbed Amazonian Forest Sites

	Urucu, Brazil	Cocha Cashu, Peru	Lago Teiú, Brazil
Geographic position and elevation	4°50′S, 65°16′W 53–71 m	11°54′S, 71°22′W ca. 400 m	2°48′S, 65°10′W ca. 50 m
Predominant forest type	Remote terra firme forest	"Mature" floodplain forest	Seasonally flooded (várzea) forest
Flooding periodicity	Never flooded	Supra-annual, irregular	Annual, regular
Duration of floods	Never flooded	Ephemeral (few days)	4–6 months/year
Type of drainage system	Black-water	White-water	White-water
Predominant soil type	Yellow clay latosols	Young alluvial fluvisols	Very young alluvial fluvisols
Influence of nearest white-water river	None: 181 km inland	Moderate: abutting river oxbow lake	Heavy: within system of channels and levées
Human disturbance	None	None	Very low
Habitat heterogeneity (and main source)	Extremely low (small stream)	High (extensive lacustrine succession)	Moderate (natural flood disturbance)
Continuity to upland habitats	Complete	Partial	None (isolated in a várzea archipelago)
Annual rainfall (N) and strength of dry season[a]	2,882 mm (5 yrs) 3 months, 9.5%	2,028 mm (4 yr) 4 months, 7.2%	2,850 mm (1 yr) 3 months, 10.9%

Note: These sites from which data are available on the nonvolant mammal community are subject to widely divergent conditions imposed by proximity to major white-water rivers and watershed position.

Sources: Urucu: Peres 1991a; Cocha Cashu: Terborgh 1983; Janson and Emmons 1990; Lago Teiú: Ayres 1986, 1993.

[a] Measured in terms of duration (number of consecutive months with substantially lower rainfall) and intensity (proportion of the annual rainfall in the three driest months of the year).

ences in anthropogenic disturbance. A more detailed description of the physical setting of each of these sites can be found in the references above.

The Urucu Forest Mammal Community

Situated 180 km from the nearest major white-water river (Rio Solimões), the Urucu forest (4°50′ S, 65°16′ W) is considered representative of most remote terra firma forests south of the Amazon, where lack of access has prevented selective hunting and habitat disturbance (Peres 1990, 1991a). The nonvolant mammal fauna at Urucu was studied by at least three observers operating simultaneously for about twenty-five days a month between February 1988 and October 1989 after a pilot survey in April and May 1987. The study area was accessible only by small aircraft subcontracted to an oil company and consisted of a 900 ha polygon 4 km or more from a relatively small black-water river (Rio Urucu). Approximately 93% of this area consisted of high terra firma forest of low structural heterogeneity; the rest was made up of a few minor habitats including creekside forest and small palm swamps along the only perennial stream intersecting the study plot (fig. 21.1). A more detailed description of the climate, habitat types, forest structure and composition, and plant phenology at this site can be found elsewhere (Peres 1991a, 1993a, 1994a,b).

Mammals were studied using a mixture of several techniques, including diurnal and nocturnal line transect censuses (Peres 1993a, unpubl. data), systematic observations of habituated groups of five primate species (e.g., Peres 1993b, 1994b), and thanks to the efforts of J. Malcolm, J. Patton, and M. N. da Silva, trapping of small rodents and marsupials using both terrestrial and arboreal traplines at trap heights over 18 m (sensu Malcolm 1991b). Understory mist netting of bats was carried out by C. Peres and R. Gribel but was too occasional to merit inclusion in this analysis. All voucher specimens collected at Urucu are deposited in mammal collections of the Museu Goeldi, Belém, Instituto Nacional de Pesquisa da Amazônia, Manaus, and Museum of Vertebrate Zoology, University of California, Berkeley.

Primate densities and biomass reported here are from Peres (1993a). Those reported for all other diurnal taxa are based on a similar repeat line transect census method carried out every month over a two-year period. Censuses covered a cumulative walking distance of 359.4 km on a 4.5 km transect at speeds of 1 to 1.5 km h^{-1}. Nocturnal mammals ameanable to line transect counts (e.g., *Aotus nigriceps*, *Potos flavus*, and several didelphid and echimyid species) were censused at a slower pace (0.5–0.6 km h^{-1}) for a total of 50 km, but otherwise according to the method used by Emmons (1984). Reliable population densities could not be estimated for most carnivores (other than the most abundant, frugivorous species) and several edentates and small-bodied rodents because of their natural rarity and low frequency of detection at this site. Despite my thorough familiarity with the Urucu forest and its fauna at the end of the study, we attempted no "guesses" at population densities of rare species—as once suggested by Caughley (1977)—other than coarse abundance indexes. Several species of nocturnal rodents and marsupials could not be reliably separated in the field during night surveys based on either eyeshine pattern or external morphology and were thus lumped under pooled estimates. Analysis of line transect data was based on Fourier series estimators of strip width (sensu Burnham, Anderson, and Laake 1980) using perpendicular distances to animals or groups of animals sighted (see Peres 1993a).

The Urucu nonvolant mammal fauna consisted of at least 68 nonvolant species (18 rodents, 15 carnivores, 13 primates, 9 marsupials, 8 edentates, and 5 ungulates) recorded over the entire study (see chapter appendix). At least 26 species of chiropterans (20 phyllostomids, 2 emballonurids, 1 noctilionid, 1 vespertilionid, 1 thyropterid, and 1 molossid) were also found at the study area, but the bat checklist could have been at least doubled had we been able to assemble a more complete list based on a vertically stratified capture program. Total mammal richness in the Urucu could thus in theory exceed 120 species. Such species packing is known to fluctuate seasonally, since not all Urucu mammal species are year-round residents. Residency status of several species appeared to be affected by the low availability of soil nutrients and by strong seasonality in fruit production (Peres 1994a), which is probably typical of most Amazonian terra firma forests. This was the case with several vagrant species living in large groups, including a few seminomadic primates (*Saimiri sciureus, Cebus albifrons, Lagothrix lagothricha*: Peres 1993a) and one nomadic ungulate (*Tayassu pecari*: Peres 1996). Long-distance seasonal movements are most likely also exercised by other highly mobile frugivores, including canopy bats, toucans (*Ramphastos cuvieri*), large cotingas (e.g., *Gymnoderus foetidus*), large parrots (*Amazona* spp.), and macaws (*Ara* spp.) (Peres and Whittaker 1991). Indeed, the extent to which Amazonian terra firma mammals exhibit lateral seasonal movements from and to riverine habitats is perhaps best explained by the staggering of mature fruiting peaks between adjacent forest types subject to variable flooding (Peres 1994a; Kubitzki and Ziburski 1994).

Total nonvolant mammal biomass at Urucu, as compiled from the chapter appendix, is 891 kg/km². This is a minimum figure, since no data are available on many uncommon or rare species, which nevertheless are unlikely to substantially inflate the overall community biomass. In contrast to many Neotropical forest sites where sloths (*Bradypus* and *Choloepus*) can represent over half the biomass of nonflying mammals (Eisenberg and Thorington 1973), the figure for the Urucu is unlikely to be severely underestimated by the absence of these arboreal folivores in the biomass tally (see notes to chapter appendix). The highly folivorous red howler monkey (*Alouatta seniculus*), for instance, was surprisingly rare even though the low detection rate of these vocally conspicuous monogastric primates was clearly not an artifact of hunting pressure at this site (Peres 1990, n.d.). The three groups of howlers known to use the study area were all restricted to within 200 m of the perennial stream, the area corresponding to creekside forest on humic or sandy soils. The only other significant contributor to nonvolant mammal biomass in other Neotropical forests missing in this tally is the white-lipped peccary (*Tayassu pecari*), but only once was a large herd of this easily detectable species observed crossing our study plot during the study. The small-group-living collared peccary (*T. tajacu*), in contrast, was common and could be observed almost every day.

Considering different trophic guilds, the community biomass was dominated by arboreal consumers of fruit pulp (42%), followed by terrestrial seed eaters (32%), terrestrial browsers (18%), arboreal seed eaters (5%), and arboreal folivores with only 4% of the biomass (see table 21.3 for taxa considered within each of these categories). The arboreal component of the mammal biomass (444 kg/km²), however, was almost equivalent to that of terrestrial species (432 kg/km²). Density and biomass estimates for most nonvolant species at Urucu are presented in the chapter appendix. Primates were the top-ranking order in terms of biomass (391 kg/km²), followed by artiodactyls (275 kg/km²), rodents (110 kg/km²), perissodactyls (66 kg/km²), carnivores (32+ kg/km²), marsupials (16+ kg/km²), and edentates (1+ kg/km²). The greatest density and biomass underestimates are to be found in the edentates, carnivores, and rodents, but these are more a function of natural rarity than an artifact of sampling. For instance, the surprisingly low density of both sloth species could perhaps be attributed to the inherent problems of censusing these highly cryptic canopy species using standard line transect techniques. However, even though over four thousand hours of primate observations were logged by three observers at this site, we detected only three sloths during the entire study, and two of them were from skeletal remains (one *Bradypus* and one *Choloepus*) presumably left by a natural predator.

Nonvolant Mammals at the Tucuruí Dam

Ethical considerations aside, rescue operations after the closing of hydroelectric dams, if well designed and systematically carried out, can provide a unique opportunity for near exhaustive counts of forest mammals over large areas (see Eisenberg and Thorington's 1973 analysis of 8,261 mammals rescued at the Afobaka Dam, Suriname; see also Walsh and Gannon 1967). Such rescue programs often involve systematic captures and counts of mammals stranded in isolated treetops from boats operating in the submerged area. This census method is particularly valid in the case of nocturnal or behaviorally inconspicuous species, such as sloths and pygmy anteaters, which cannot be reliably censused otherwise. Such sampling opportunity was provided once again by the filling of the Tucuruí Dam reservoir (September 1984 to April 1985), which flooded about 2,430 km^2 of mostly low-lying forest running along 170 km of the Rio Tocantins, eastern Amazonia. The large-scale capture operation that ensued, equivalent to 470 people based in six field stations working for eight months, yielded a total of

101,326 individuals (including 1,613 "smaller rodents") of forty-eight or more nonvolant mammal species (Mascarenhas and Puorto 1988).

The biomass of thirty-six nonvolant mammal species at Tucuruí, calculated from 99,713 capture records reported by Mascarenhas and Puorto (1988), is shown in figure 21.2. This analysis excludes a number of species of "smaller rodents" (e.g., *Oryzomys, Proechimys, Echimys, Rattus*) that were not identified to species and whose numbers (1,613 captures) have been drastically underestimated. Only one species identification from the original capture records appears to be questionable: the second species of long-nosed armadillos occurring at Tucuruí is probably *Dasypus septemcinctus* rather than the reported *D. kappleri* (Wetzel and Mondolfi 1979). The occurrence of *Cerdocyon thous* may seem surprising for such a forest habitat, but these foxes can occur in many fringing areas of the Amazonian hylaea (pers. obs.).

The most important arboreal folivores in Amazonian mammal faunas—the two species of sloths and howler monkeys—were by far the top-ranking species

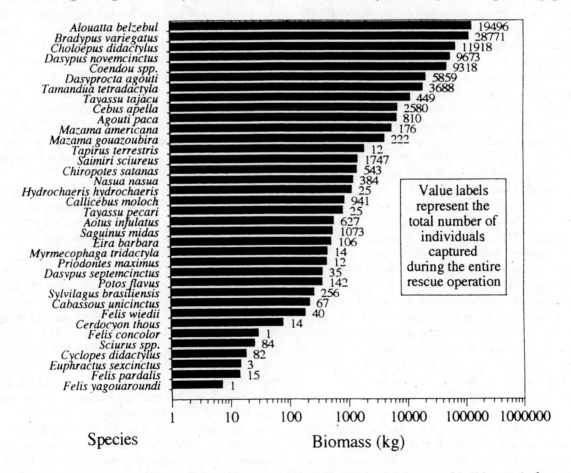

Figure 21.2. Number of individuals and crude biomass (on a log scale) of nonvolant mammal species rescued from the 2,430 km^2 reservoir area flooded by a 170 km section of the Rio Tocantins, upriver of the Tucuruí Hydroelectric Dam, eastern Amazonia (adapted from Mascarenhas and Puorto 1988).

both in numbers and in biomass. These three species combined accounted for 60,185 individuals captured (60%) and 62% of the 489,434 kg of nonvolant mammals rescued. Red-handed howlers (*Alouatta belzebul*), three-toed sloths (*Bradypus variegatus*), and two-toed sloths (*Choloepus didactylus*) represented 26%, 23%, and 13% of the biomass, respectively. These figures are considerably higher than those calculated for the same species (or ecological equivalents) from the Afobaka Dam in Suriname, which first showed that folivores as inconspicuous as sloths could be the greatest contributors to mammalian biomass in Neotropical forests (Eisenberg and Thorington 1973). This conclusion was also reached by G. Montgomery and M. Sunquist based on their study of sloths at Barro Colorado Island, Panama, although their early, highly inflated density estimates have been revised to more realistic figures of 123 individuals per km^2 (*Bradypus infuscatus*) and 25 individuals per km^2 (*Choloepus hoffmanni*) (Montgomery and Sunquist 1975).

The ubiquitous nine-banded armadillo (*Dasypus novemcinctus*), whose density appears to be highly variable across Amazonia, accounted for almost 10,000 captures (9.7%) at Tucuruí, despite its poor climbing ability: animals that escaped drowning tended to crowd into the many unflooded islets on higher ground formed by the rising water level on uneven topography. This was also the case for agoutis (*Dasyprocta agouti*), which accounted for nearly 6,000 captures, and a number of other terrestrial species.

Not surprisingly, several arboreal species were also captured in large numbers. Brazilian porcupines (*Coendou prehensilis*) accounted for 9,318 captures and ranked fifth in the overall tally, resulting in an estimated density of at least 3.8 individuals per km^2. Two species of *Coendou* were also frequently captured at the Afobaka Dam in Suriname, accounting for 17.4% of all captures, and were the second most frequently captured mammal genus after *Bradypus* (Eisenberg and Thorington 1973). This suggests a strong tendency toward folivory for these otherwise poorly known and mostly nocturnal arboreal seed predators. Other arboreal species captured in considerable numbers include collared anteaters (*Tamandua tetradactyla*), brown capuchins (*Cebus apella*), squirrel monkeys (*Saimiri sciureus*), and tamarins (*Saguinus midas*).

Such cumulative rescue counts, however, are not necessarily good predictors of each species' total population size within the reservoir perimeter, even for arboreal and scansorial taxa. This is because there are species-specific differences in survivorship and mortality curves during the dramatic rising of the water level that are a function of body size (and basal metabolic rate), degree of arboreality, swimming ability, degree of folivory, and ultimately the ability to avoid confinement to isolated treetops or to endure prolonged fasts there (Peres and Johns 1991). Nevertheless, discrepancies in postflooding mortality can be corrected to at least some extent if reliable density estimates from censuses conducted before flooding are available (Johns 1986) for comparison with counts of animals rescued from the reservoir area.

Comparisons with Other Neotropical Faunas

Patterns of Species Richness

Amazonian forests contain more mammal species than even the best-sampled sites elsewhere in the Neotropics, in that they can sustain the greatest recorded alpha diversity. This may not necessarily be true for bats, however, although comparisons of bat species richness are prohibitively complicated by the enormous heterogeneity in sampling effort that has been allocated to them. Numbers of nonvolant species recorded in Amazonian terra firma and alluvial forests (sixty-five to seventy-one species) are unrivaled by those of other closed-canopy habitats of well-studied northern Neotropical sites, such as Guatopo, Barro Colorado Island, La Selva, and Montes Azules (forty to fifty species; see table 21.2). However, existing data suggest a positive east-west diversity gradient for forest mammals across Amazonian interfluvial basins, a geographic trend that may be explained by ecological and biogeographic factors, including rainfall and plant diversity gradients, as well as by dispersal opportunities. Further evidence for this pattern has been provided by recent large-scale mammal collecting and surveying expeditions along the Juruá, Tarauacá, and upper Purus Rivers of western Brazilian Amazonia (Malcolm et al. 1992; Peres 1993c, unpubl. data). In these terms, the richest mammal communities would straddle western Amazonian watersheds right up to lowland forests along the eastern flanks of the Andes, whereas the poorest would be found in the drier eastern limits of the region, toward the Pará-Maranhão border. This may not seem obvious from the single eastern Amazonian site listed in table 21.2 (Belém: Handley 1967; Pine 1973), but species richness at this site may have been overestimated because it is based on both field surveys and presumed geographic distributions. Indeed, western Amazonian sites contain the richest known microsympatric assemblages of primates (14 species at Lago da Fortuna, Amazonas, Brazil: Peres 1988), and mammals of all orders (122 species at Balta, Ucayali, Peru: Patton, Berlin, and Berlin 1982; 135 species[4] at Reserva Cuzco Amazónico, Peru: Woodman et al. 1991; 138 species at Cocha Cashu,[5] Manu National Park, Peru: Pacheco et al. 1993) for tropical forests anywhere. The eastward de-

Table 21.2   Number of Mammal Species at Neotropical Forest Sites within and outside Lowland Amazonia

Taxon	Amazonian Lowland Forests[a]							Northern Neotropical Forests			
	Narañón, Peru	Balta, Peru	Cashu,[b] Peru	RCA, Peru	Urucu, Brazil	Manaus, Brazil	Belém, Brazil	Guatopo, Venezuela	BCI, Panama	La Selva, Costa Rica	Mts. Azules, Mexico
Marsupialia	10	9	12	9 (10)[c]	9	9	9	7	6	5	7
Chiroptera	38+	57	61	42 (54)	26+	?	45	29	56	65	64
Primates	6	9	13	7 (13)	13	6	6	3	5	3	2
Xenarthra	9	6	7	5 (9)	8	8	11	5	6	7	4
Lagomorpha	1	1	1	1 (1)	0	0	1	1	1	1	1
Rodentia	26	24	26	24 (27)	18	16	18	13	14	15	17
Carnivora	14	12	13	9 (15)	15	7	17	8	12	14	12
Ungulates	5	4	5	3 (6)	5	5	5	3	5	5	5
Nonvolant species	71	65	77	58 (81)	68	51	67	40	49	50	48
All species	109+	122	138	100 (135)	94+	—	112	69	105	115	112

Sources: Alto Marañón: Patton, Berlin, and Berlin 1982; Balta: Patton, Berlin, and Berlin 1982; Cocha Cashu: Janson and Emmons 1990; Ascorra, Wilson, and Romo 1991; Pacheco et al. 1993; RCA (Reserva Cuzco Amazónico): Woodman et al. 1991; Urucu: this study; Manaus: Malcolm 1990; Belém: Pine 1973; Handley 1967; Guatopo: Eisenberg, O'Connell, and August 1979; BCI: Glanz 1990 (all recent extinctions of original fauna also included) Handley, Wilson, and Gardner 1991; La Selva: Wilson 1990; Mts. Azules: Medellin 1994.

[a] Ordered longitudinally from west to east.

[b] Four marsupial, three rodent and thirty-three bat species that are not known to Cocha Cashu Biological Station but have been collected at a nearby locality 12 km downriver (Pakitza) are also included here.

[c] Numbers in parentheses include additional species of "probable" occurrence but that were not actually sighted or captured at this site.

cline in local species richness is partly accounted for by dropouts of arboreal nonvolant taxa rather than of chiropterans. For example, several primates, rodents, and carnivores typical of eastern Peru and westernmost Brazil (state of Acre and western Amazonas) including Goeldi's monkeys (*Callimico goeldii*), pygmy marmosets (*Cebuella pygmaea*), pacaranas (*Dinomys branickii*), and olingos (*Bassaricyon gabbii*) fail to occur at some point farther east and are not replaced parapatrically by ecologically similar congeners. In addition, a faunal inventory in the northeastern Brazilian caatinga[6] (Mares et al. 1981) found only 23 species of nonvolant mammals but 51 species of bats, suggesting that the relative contribution of bats to mammal faunas actually increases eastward.

Despite the record-breaking local richness of western Amazonian mammal communities, drier macrohabitats bordering Amazonia can be equally important from a conservation viewpoint. Indeed, these can be almost as impressive in terms of occurrence of endemic species (e.g., the "drylands" of Mares 1992, which include pooled figures for the llanos, cerrado, caatinga, chaco, pampas, paramos, and deserts of South America), and richness of large-bodied species (e.g., Paraguayan Chaco: Redford, Taber, and Simonetti 1990). Comparisons of crude numbers of species alone should not belittle the importance of mammal faunas in the east-central Amazon or in Neotropical ecosystems elsewhere.

Within Amazonia, the greatest number of syntopic mammal species (excluding bats) is found in terra firma and alluvial forests, such as the Urucu (sixty-eight or more species) and Cocha Cashu (seventy-

seven species). Adjacent alluvial and terra firma sites qualitatively share much the same fauna once biogeographic effects of regional source faunas are removed (Peres 1997a, unpubl. data). The minimum number of sympatric mammal species (including bats) at Urucu could thus conceivably rival that of western Amazonian sites where chiropterans have been more exhaustively sampled (Patton, Berlin, and Berlin 1982; Woodman et al. 1991; Pacheco et al. 1993). This is in striking contrast with mammal faunas at annually flooded forests such as Teiú, which are relatively impoverished compared with adjacent unflooded forests. The low species richness of typical várzea forests has now also been fairly well established based on multiple-site comparisons for primates (Peres 1997a), rodents and marsupials (Malcolm et al. 1992), and other vertebrate taxa (birds: Remsen and Parker 1983; Peres and Whittaker, unpubl. data; amphibians: C. Gascon, unpubl. data), as well as overstory trees (Balslev et al. 1987; Junk 1989). Indeed, nonvolant mammal communities in typical Amazonian várzea forests tend to be poorer than those in non-Amazonian forests of the northern Neotropics.

Patterns of Community Biomass

We now turn to a cursory examination of how an Amazonian mammal community removed from the effects of large white-water rivers might diverge from those directly affected by either river channel migration within broad meander belts or by pronounced annual flooding. Biomass estimates presented in table 21.3 show that the overall mammal biomass in either alluvial or flooded forests is considerably greater than

Table 21.3   Primary Consumer Biomass (and Percentage of Totals) of the Nonvolant Mammal Fauna in Different Trophic Guilds at Amazonian Terra Firme, Alluvial, and Várzea Forests

Foraging Position and Trophic Guild	Mammalian Taxa Included[a]	Crude Biomass (kg/km^2)		
		Urucu (Terra firme)	Cocha Cashu (Alluvial)	Lago Teiú (Várzea)
Arboreal folivores	*Bradypus, Choloepus Dactylomys, Alouatta*	35 (4)	> 198[b] (15)	> 735[c] (68)
Arboreal frugivores (mostly ripe pulp)	Small didelphids, most primates, arboreal procyonids	365 (42)	542 (40)	> 111 (10)
Arboreal granivore/ frugivores	Pitheciine primates, sciurids, *Coendou* spp., arboreal echimyids	44 (5)	> 15[d] (1)	> 61[e] (6)
Total (arboreal guilds)		444 (51)	755 (55)	907 (83)
Terrestrial/semiaquatic grazer/browsers	*Hydrochaeris, Sylvilagus*	0 (0)	72 (5)	> 180[6] (17)
Terrestrial granivore/ frugivores	*Oryzomys, Proechimys, Myoprocta Dasyprocta, Tayassu* spp.	277 (32)	351 (26)	0 (0)
Terrestrial frugivore/ browsers	*Agouti paca, Mazama* spp., *Tapirus*	155 (18)	186 (14)	0 (0)
Total (terrestrial guilds)		432 (49)	609 (45)	180 (17)
Total biomass of the taxa above		875	> 1364	> 1087
Minimum estimate of nonvolant mammal biomass		891	> 1416	> 1087

Sources: Urucu: this study; Cocha Cashu: Janson and Emmons 1990; Lago Teiú: Ayres 1986. Numbers preceded by a "greater than" (>) sign represent minimum estimates in that biomass figures were unavailable and were calculated based on conservative density estimates for taxa reported to be common.

[a] Unlisted genera included in the above pooled estimates as following: "small didelphids" = *Micoureus, Marmosa, Marmosops, Caluromysiops,* and *Caluromys;* "most primates" = *Cebuella, Saguinus, Aotus, Callicebus, Saimiri, Cebus, Lagothrix,* and *Ateles;* "arboreal procyonids" = *Potos* and *Bassaricyon;* "pitheciine primates" = *Pithecia* and *Cacajao;* "sciurids" = *Sciurus* and *Microsciurus;* "arboreal echimyids" = *Mesomys* and *Isothrix.*

[b] Biomass of the bamboo rat *Dactylomys dactylinus* ("0.09 encounters/km walked": Janson and Emmons 1990) conservatively calculated as 30 individuals/km^2 × 0.6 kg = 18 kg/km^2.

[c] Sloths (*Bradypus variegatus* and *Choloepus didactylus*) were reported as very common and could be "observed almost daily" (Ayres 1986, 92, 108). Because biomass estimates were not provided, figures herein are based on the most conservative biomass estimates for these genera at BCI, Panama (Eisenberg 1980). However, available evidence suggests that sloth abundance is greater at Teiú than at BCI.

[d] Biomass of *Sciurus spadiceus* recalculated from Janson and Emmons 1990 as 25 individuals/km^2 x 0.6 kg = 15 kg/km^2 (rather than the reported "5.0"). The biomass of the porcupine *Coendou prehensilis* is not available for this site even though it may be considerably more common than suggested by these authors.

[e] Biomass of giant Amazonian red squirrels (*Sciurus* cf. *igniventris*) and capybaras (*H. hydrochaeris*), which are reported to be common at Teiú (Ayres 1986), are derived from conservative density estimates for the same species or ecological equivalents from typical várzea sites along the Rio Juruá (C. Peres, unpubl. data).

that of terra firma forests. This becames obvious once the two best-sampled sites—Cashu and Urucu—are examined separately. Differences between these sites are even greater should one consider, first, that previous estimates of mammalian biomass at Cashu (1,705 kg/km^2; Terborgh 1986) are substantially greater than Janson and Emmons's more conservative figure of 1,400+ kg/km^2 and, second, that mammal biomass at Urucu, which might appear to be low for a truly pristine tropical forest site, can actually be significantly greater than that of many other distant terra firma sites across Amazonia. To further test this impression, I calculated the nonvolant mammal biomass from species density estimates obtained at a second terra firma forest—a central Amazonian site north of Manaus—from studies by Malcolm (1990), Emmons (1984), and Rylands and Keuroghlian (1988). Mammal biomass at this site, excluding only the contribution of a few potentially important diurnal species such as *Dasyprocta leporina* and *Tayassu tajacu* (diurnal surveys are yet

to be conducted for most taxa other than primates), was a mere 206+ kg/km^2 (table 21.4). Although this is a minimum estimate, nonvolant mammal biomass at Manaus could hardly exceed 400 kg/km^2, even if 190 kg were added to allow for the few taxa that have not been censused. This suggests that the overall mammalian biomass at Urucu could be nearly three times that of Manaus, a ratio that largely reflects that estimated for primates (391 vs. 96 kg/km^2: Peres 1988, 1993a). Malcolm (1990, 1991a,b), after his herculean twenty-five-month trapping effort targeting small mammals, generally agreed with Emmons's (1984) earlier impression that mammal densities north of Manaus are the lowest yet recorded from any Neotropical forest site. Such enormous discrepancies in mammal biomass may be explained by the well-known poor yellow latosols of many forest sites of the Guiana shield, and by the fact that fruit productivity at Manaus is markedly lower than at other Amazonian sites (Gentry and Emmons 1987).

Table 21.4   Comparison of Nonvolant Mammal Crude Biomass in Different Neotropical Habitats

Sites	Dominant Vegetation Types	Biomass (kg/km^2)	Sources
Closed-canopy forests			
Manaus, Brazilian Amazonia	Evergreen primary terra firme forest	206+	1
Urucu, Brazilian Amazonia	Evergreen primary terra firme forest	891	2
Cashu, Peruvian Amazonia	Evergreen late-successional alluvial forest	1416	3
Teiú, Brazilian Amazonia	Evergreen annually flooded várzea forest	1987	4
Guatopo, Venezuela	Secondary (30 yrs old) submontane forest	946	5
Barro Colorado Is., Panama	Evergreen late-secondary moist forest	2264	6
Open or semiopen savannas			
Acurizal, Pantanal, Brazil	Deciduous forest, wet grassland	380 (3750)[a]	7
W. Masaguaral, Venezuela	Mixed palm forest, llano grassland	479 (7600)	5
E. Masaguaral, Venezuela	Semideciduous forest, llano grassland	1086 (7600)	5
Hato El Frio, Venezuela	Open grasslands in wet llanos	3730 (18504)	6

Sources: Compiled and calculated from (1) Malcolm 1990; Emmons 1984; and Rylands and Keuroghlian 1988; (2) this study; (3) Janson and Emmons 1990; (4) Ayres 1986; (5) Eisenberg, O'Connell, and August 1979; (6) Eisenberg 1980; (7) Schaller 1983.

[a] Figures in parentheses represent the estimated crude biomass of domestic livestock (savanna sites only) not included in that of native mammals.

There appears to be no further increase in mammal community biomass from the alluvial site to the várzea site, as might have been expected from differences in system productivity alone. Any change, however, might be seriously masked because mammal sampling was less complete at Lago Teiú, and because its biomass of arboreal folivores was most likely severely underestimated (see notes to table 21.3). Obligate folivores, such as three-toed sloths, indeed account for much of the variation in Neotropical mammal biomass. In any case the mammal community at Teiú, as in many other typical várzeas, appears to be dramatically simplified in part by the complete lack of a ground substrate for much of the year, and hence of suitable habitat for strictly terrestrial species, because of the long, prohibitive inundation period dominated by deep standing water. The low floristic diversity of várzea forests (Ayres 1993) could also be detrimental to mammal diversity, although this relationship has yet to be established beyond a mere correlation. Fewer mammal species thus contributed to the overall biomass pool. Only a handful of large species appears to occur at Teiú at any one time (four primates, five edentates, three or more rodents, four carnivores, and two or more marsupials: Ayres 1986). This obviously excludes two co-occurring cetaceans (*Inia geoffrensis* and *Sotalia fluviatilis*) and one sirenian (*Trichechus inunguis*), although these aquatic species are an equally intricate part of várzea systems. Strictly terrestrial mammals, on the other hand, (e.g., large caviomorph rodents, peccaries, brocket deer), were conspicuously absent, as is often the case for many isolated várzea forests not contiguous with upland habitats. This does not take place in floodplain sites interdigitating with terra firma forests, which can then serve as a seasonal source of migrant stocks of terrestrial ungulates moving back into flooded forests during the low-water season (e.g., Bodmer 1990). This absence of terrestrial

mammals in flooded forests is partly compensated for because many of the fallen seeds and whole fruits are eaten by characid fish and other aquatic seed predators and frugivores during the high-water season (e.g., Goulding 1980; Kubitzki and Ziburski 1994), and to a lesser extent by arboreal taxa descending to the forest floor to forage on residual seeds and fruit pulp during the low-water season (e.g., *Cacajao calvus*: Ayres 1986).

In terms of guild structure, the greatest discrepancies between mammal communities at different forest types are accounted for by arboreal taxa such as sloths and primates, which harvest vegetative and reproductive plant parts directly at the source. Considerably smaller differences are observed for ground-dwelling frugivores and granivores, which may be explained if the seasonality of residual fruit and seed fall not consumed by arboreal vertebrates is even stronger than that for terrestrial taxa, thus imposing greater population thresholds for the latter. Arboreal taxa, on the other hand, may enjoy more relaxed fluctuations in resource availability, since they have unencumbered access to the standing fruit production during periods of scarcity (Terborgh 1983; Peres 1994a).

One of the most puzzling differences in mammal biomass at Amazonian sites sampled to date is the conspicuous natural rarity of arboreal folivores in terra firma forests, where their density pales next to that of floodplain sites (table 21.3). Indeed, the abundance of habitual leaf eaters, such as sloths, howler monkeys, and the Amazon bamboo rat (*Dactylomys dactylinus*),[7] becomes drastically reduced beyond a critical distance from rivers throughout western Amazonia (Peres 1997b). This pattern could be extended to yet other poorly known species pending dietary studies that may be common in floodplains but very rare in terra firma forests, such as the tree rat *Echimys (Makalata) armatus*. In birds and reptiles this effect is also clearly paralleled by the few specialized folivores that manage to thrive in Amazonian forests—for example, hoatzins (*Opisthocomus hoazin*), horned screamers (*Anhima cornuta*), and iguanid lizards (*Iguana iguana*)—which can also be common near rivers but completely drop out in upland forests (pers. obs.). Indeed, there appears to be a positive abundance gradient for both arboreal and terrestrial folivores from terra firma to alluvial forests, and from alluvial to várzea forests. This finds support in the fact that terrestrial grazers and browsers with little propensity toward frugivory (e.g., tapitis[8] and capybaras) were conspicuously absent in the Urucu, whereas they may become increasingly abundant near rivers (pers. obs.). This may simply reflect intrinsic differences in natural disturbance regimes in that, compared with floodplains, terra firma forests tend to have smaller and fewer canopy gaps, little herbaceous cover (e.g., Poaceae, Marantaceae, Cyperaceae), and fewer shade-intolerant seedlings (Peres and Malcolm, unpubl. data).

Apart from the pervasive effects of baseline system productivity and soil fertility, patterns of Amazonian mammal abundance under closed-canopy core hylaean forests may diverge appreciably from those near the phytogeographic limits of the region, particularly those on the outskirts of forest savanna boundary zones. Southern Amazon areas in the vicinity of cerrado enclaves of the upper Rio Xingu, for instance, appear to sustain atypically high densities of large herbivores such as tapirs, peccaries, and brocket deer (Peres 1996, unpubl. data), which are unheard of in unbroken terra firma forest elsewhere in the region. Visibly high densities of ungulates have also been noted, albeit anecdotally, at the edge of the Roraiman grasslands and Gran Savanna of southern Venezuela (pers. obs.), in southeastern Rondônia (P. Develley, pers. comm.), in northern Mato Grosso (A. Whittaker, pers. comm.), and in other transitional areas into the Brazilian cerrado scrublands. In seasonally flooded savannas amid gallery forests, such as those in eastern Marajó Island, Pará, ungulates are replaced by capybaras, but their biomass can be even more impressive (pers. obs.). This is presumably a result of increased primary productivity for browsers and grazers owing to canopy openness, natural edges, and greater forest structural heterogeneity, if not a more simplified stand composition consisting of fewer and possibly more favorable food species, Indeed, mammal communities in these areas, which may well prove to be less species rich independent of biogeographical factors, should theoretically sustain a greater offtake by subsistence hunters and thus be more amenable to game management.

Species richness is poorly correlated with overall community biomass both within and outside Amazonia. The Venezuelan llanos (Eisenberg, O'Connell, and August 1979; Eisenberg 1980) and Brazilian Pan-tanal wetlands (Schaller 1983) have relatively species-poor mammal faunas but sustain much greater crude biomasses should one take into account even a fraction of the grazing livestock than those of even the least hunted Amazonian forests (table 21.4). This confirms a clear worldwide pattern of high mammal biomasses in tropical savannas and grasslands compared with those in closed-canopy forests (e.g., Bodmer 1989; Prins and Reitsma 1989). An extreme case from the northern Neotropics comes from a wet llano site, Hato El Frio, which supports a mammal biomass almost one hundred times greater than that of terra firma forests north of Manaus (table 21.4). Mammal populations in terra firma forests such as Manaus and Urucu are typically characterized by natural rarity a function of low ecological densities, rather than of habitat specificity and patchy distributions even in the absence of any form of disturbance. But this condition often results in extremely high levels of within-habitat (rather than between-habitat) diversity. This difference sets the structurally monotonous terra firma sites apart from alluvial sites, which often contain far greater levels of habitat diversity owing to vegetation succession driven by active fluvial dynamics (fig. 21.1).

Exploring Divergences in Community Structure

The greater mammalian biomass in areas under the immediate influence of white-water rivers cannot be interpreted as an artifact of differential hunting pressure, for the opposite pattern would be expected should that be the case: hunting near navigable rivers where most tribal and nontribal Amazonians are concentrated is far more intensive than in roadless inland areas (Peres 1990). The most plausible explanation for such discrepancies seems to be the dramatic differences in edaphic conditions and nutrient input between forest types. Such differences, which can be intuitively obvious to the naked eye, given the generally kaolinitic to white-sand soil texture of terra firma forests, have been clearly confirmed by biogeochemical studies of forests under contrasting inundation regimes. Chemical analyses of soil macronutrients important to plants (N, P, K, Na, Ca, and Mg) have consistently shown that várzea floodplain soils of Amazonia have elemental concentrations unquestionably greater than those of terra firma forests (Irion 1978). This is particularly true for phosphorus and potassium (19 and 6 times greater), and calcium (as much as 190 times greater: Furch and Klinge 1989), which have the greatest effects on tropical forest productivity (Vitousek and Sanford 1986). These patterns are clearly supported by data from the Peruvian alluvial site discussed here: soils at Cocha Cashu under typical mature floodplain forest contain between 1.5 and 3 times as much carbon, nitrogen, and potassium, and 2 to 60 times as much calcium and magnesium, as soils in adjacent areas supporting typical terra firma forests (M. P. Riley, pers. comm.). Recent laboratory analyses of topsoil samples from a number of forest sites of western Brazilian Amazonia, although not yet fully available, appear to show much the same trends (C. Peres, unpubl. data).

The nature of soils in different forest types, in turn, could affect plant communities in three ways important to primary consumers, which clearly make up the bulk of mammal community biomass in tropical forests (Brockie and Moeed 1986; White 1994; this study). First, foliage phytochemistry in rich soils could be substantially more favorable (or less limiting) to folivores, because plants in these conditions may be more prone to replace vegetative growth lost to herbivores than to invest in chemical deterrants (Janzen 1974), which may include a wide arsenal of toxic compounds and digestibility reducers. This hypothesis is supported by data presented here in that both arboreal and terrestrial folivores were the least prominent trophic guilds in the poor terra firma site, compared with nutrient-rich floodplain forests.

Second, overall fleshy fruit production (e.g., standing and residual fruit biomass ha^{-1} year^{-1}), which is a function of both fruit crop size and fruit crop density, could be considerably greater in floodplain habitats simply in response to more fertile soils, thus favoring most generalist frugivores. Ripe fruit output in terra firma forests such as Urucu (Peres 1994a) and Manaus (Gentry and Emmons 1987), by contrast, appears to be low even during the most pronounced fruiting peak compared with what few data are available from floodplain sites (Terborgh 1983; Janson and Emmons 1990). Third, the dynamics of flooding and fluvial disturbance generate an active process of vegetation succession and a greater spectrum of habitat types, which tend to

have staggered fruiting peaks throughout the year. Mobile, frugivorous vertebrates are then able to overcome locally depressed food supplies during periods of greatest fruit scarcity by shifting between different habitats. This option is often unavailable to vertebrates in the structurally monotonic terra firma forests, where habitat diversity tends to be extremely low. Frugivorous mammals in these forests that are unable to move to more favorable habitats far afield are thus more likely to suffer the brunt of the dry-season fruit crunch, which may set the stage for more severe limits to population density.

Finally, there may be yet poorly explored but consistent relations between soil fertility and the effects of floristic composition on fruiting potential of Amazonian forests. Stands dominated by several tree families producing mostly dry and sclerocarpic fruits that are not the mainstay of generalist frugivores (e.g., Lecythidaceae, Chrysobalanaceae, Vochysiaceae, Caryocaraceae) may be associated with poor, often sandy soils, whereas those dominated by at least certain tree families important to frugivores (e.g., Moraceae, Arecaceae, Meliaceae) may occur primarily in rich soils of recent alluvial origin (Gentry 1982, 1988; Kahn and Granville 1992). Obviously there will be exceptions to this pattern in that several tree families important to certain canopy frugivores, such as the Sapotaceae (primates) and Myristicaceae (several families of avian canopy frugivores, including toucans and trogons), can be common in terra firma forests. Patterns of abundance in Amazonian mammal communities could thus be shaped in part by the synergistic effects of forest stand composition and per capita fruit productivity. This possibility remains to be tested from comparisons of multiple sampling sites where data on nonvolant mammal community structure, stand composition, and soil fertility are obtained concomitantly.

Appendix

Appendix   Use of Space, Trophic Status, Population Density, and Biomass of Nonvolant Mammals Censused in the Urucu Forest, Amazonas, Brazil

Species (Documentation)[a]	Habitat[b]	Forest Level[c]	Trophic Status[d]	Body Mass (kg)[e]	Density (ind/km^2)[f]	Biomass (kg/km^2)
Marsupialia						
Didelphidae						
Mouse opossums[g] (V)	H,C,I	U,Sc	In/Fr/N	0.10	38.4	3.8
Caluromys lanatus[h] (V)	H,C	Sc,C	Fr/In/N	0.35	13.3	4.6
Philander mcilhennyi (V)	H,C	U,T	In/Fr	0.41	15.5	6.4
Metachirus nudicaudatus (V)	H,C	T	In/Fr	—	R[i]	—
Didelphis marsupialis (V)	H,C	T,U	In/Fr/Vp	1.05	1.3	1.4
Primates						
Callitrichidae						
Cebuella pygmaea (S)	E,G,C,I	Sc,U	In/Ex	0.15	2.8	0.4
Saguinus mystax pileatus (V)	H,C,I,E	Sc,U	Fr/In	0.52	15.3	7.9
Saguinus fuscicollis avilapiresi (V)	H,C,I,E	U,Sc	Fr/In	0.39	9.8	3.8

Continued on next page

Species (Documentation)[a]	Habitat[b]	Forest Level[c]	Trophic Status[d]	Body Mass (kg)[e]	Density (ind/km^2)[f]	Biomass (kg/km^2)
Primates (continued)						
Cebidae						
Saimiri sciureus (S)	I,C	Sc,C	Fr/In	0.9	10.2 (V)	9.2
Aotus nigriceps (S)	C,I,H,E	Sc,C	Fr/In	1.0	8.8	8.8
Callicebus cupreus (S)	E,G,I,C	U,Sc	Fr/Fo/Sp	1.05	5.1	5.4
Callicebus torquatus (S)	C,I	Sc,U	Fr/In	1.4	10.1	14.1
Pithecia albicans (S)	H,C,I	Sc,C	Sp/Fr	2.5	9.8	24.5
Cebus apella (S)	H,C,I,E	Sc,C,U	Fr/Sp/H/Vp	2.6	32.3	84.0
Cebus albifrons (S)	I,C	Sc,T	Fr/Sp/H/Vp	2.4	16.5 (V)	39.6
Alouatta seniculus (S)	I,C	C,Sc	Fo/Fr	6.0	5.7	34.2
Lagothrix lagothricha cana (S)	H,C,I	C,Sc	Fr/Fo	8.25	19.3 (V)	159.2
Ateles paniscus chamek (S)	I	C,Sc	Fr	—	U[j]	—
Xenarthra						
Myrmecophagidae						
Cyclopes didactylus (V)	H,I	U,Sc	A/In	0.22	0.7+	0.15
Tamandua tetradactyla[k] (P)	H,I	Sc,U,T	A/In	4.6	0.1	0.46
Myrmecophaga tridactyla (S)	H,C	T	A/In	—	R[l]	—
Bradypodidae						
Bradypus variegatus (C)	I,C	Sc,C	Fo	—	R[l]	—
Megalonychidae						
Choloepus didactylus (C)	H,C	Sc,C	Fo/Fr	—	R[l]	—
Dasypodidae						
Dasypus novemcinctus (S)	H,C	T	In	5.5	0.10	0.55
Dasypus kappleri (C)	H,C	T	In	—	R	—
Priodontes maximus (C)	H,C	T	In	—	R	—
Rodentia						
Sciuridae						
Sciurus cf. *spadiceus* (V)	Pw,C	U,T	Sp/Fr	0.6	2.2	1.3
Sciurus ignitus (V)	H,C	Sc,U	Sp/Fr	0.22	2.1	0.46
Microsciurus flaviventer (V)	H,C	U,Sc	Sp/In/Fu	0.10	3.1	0.31
Erethizontidae						
Coendou prehensilis (S)	H,C	Sc,U	Sp/Fr/Fo	4.5	2.0	9.0
Muridae (Cricetini)						
Murids[m] (V)	—	T	Sp/Fr/In	—	—	—
Echimyidae						
Proechimys spp.[n] (V)	H,C	T	Sp/Fu/Fr	0.20	145.9	28.9
Arboreal echimyids[o] (V)	H,C	Sc,C?	Fr/Sp?	0.31	27.5	8.5
Dasyproctidae						
Agouti paca (V)	C,I	T	Sp/Fr	8.0	4.5	36.0
Dasyprocta fuliginosa (S)	H,C	T	Sp/Fr	4.6	5.1	23.5
Myoprocta pratti (C,S)	H,C	T	Sp/Fr	0.75	2.2	1.7
Carnivora						
Procyonidae						
Nasua nasua (S)	H,C	T,U	In/Vp/In	3.1	0.10	0.31
Potos flavus (S)	H,C	Sc,C	Fr/In	2.5	9.5	23.8
Procyon cancrivorous (F)	R,S	W,T	In/Vp/Fi	—	R[l]	—
Mustelidae						
Eira barbara (S)	H,C	T,U	Fr/In/Vp	4.8	1.6	7.7
Galictis vittata (S)	H,C	T	In/Vp/H	—	R	—
Mustela africana[p] (H)	H?	T	Vp/In?	—	R	—
Lutra longicaudis (S)	S	W	Fp/In	—	R	—
Pteronura brasiliensis (S)	R	W	Fp	—	R	—
Canidae						
Speothos venaticus[q] (S)	C,H	T	Vp	—	R	—
Atelocynus microtis[q]	H,C	T	Vp/Fr?	—	R[j]	—
Felidae						
Felis wiedii (S)	H,C,I	U,Sc,T	Vp	—	R	—
Felis yagouaroundi (S)	H,C,I	T	Vp	—	R	—
Felis pardalis (S)	H,C,I	T,U	Vp	—	R	—
Felis concolor (S)	H,C,I	T	Vp	—	R	—
Panthera onca (S)	H,C,I	T	Vp	—	R	—

Appendix   (*continued*)

Species (Documentation)[a]	Habitat[b]	Forest Level[c]	Trophic Status[d]	Body Mass (kg)[e]	Density (ind/km²)[f]	Biomass (kg/km²)
Perissodactyla						
Tapiridae						
Tapirus terrestris (S)	Pw,C,I	T	Fr/Fo	150.0	0.4	66.0
Artiodactyla						
Tayassuidae						
Tayassu tajacu (S,C)	H,C,I	T	Sp/Fr/Vp	25.0	8.9	222.5
Tayassu pecari (S)	Pw,I,H	T	Sp/Fr	—	V[r]	—
Cervidae						
Mazama americana (S)	C,I,H	T	Fr/Fo	30.0	0.87	26.1
Mazama gouazoubira (S)	H,C,I	T	Fr/Fo	18.0	1.5	26.6

[a]*Refers to the most reliable documentation available:* (V) captured: vouchers desposited in museum collections; (C) carcass or skeletal remains found, presumably left by a natural predator; (P) photographed but not captured; (S) confirmed sightings; (F) footprints only; (H) hypothetical.

[b]*Habitats* are ordered according to preferences as based on record frequencies. H: high forest; I: igapó forest; C: creekside forest; Pw: *Mauritia* palm swamp; E: natural forest edge; G: tree-fall gaps; S: border of perennial streams; R: river edge.

[c]*Forest levels:* T: terrestrial; U: understory; Sc: subcanopy; C: canopy; W: over or near water.

[d]Fr: frugivore (mostly ripe fruit pulp); Fo: folivore; Sp: seed predator; Ex: exudate feeder; Fu: fungus; In: invertebrate predators (mostly nonsocial insects); A: ants; H: hymenopteran colonies (usually wasp and bee larvae); Vp: terrestrial vertebrate predator; Fp: fish predator.

[e]Body weights represent means of all individuals (of different age-sex classes) weighed and are derived from animals captured at the study site or adjacent river basins or from fresh carcasses obtained by hunters elsewhere.

[f]Density estimates are based on repeated line transect censuses, given a sampling effort equivalent to a cumulative walking distance of 359.4 km and 50.0 km for diurnal and nocturnal censuses. A crude abundance index (C: common; U: uncommon; R: rare; V: vagrant), reflecting the combined detection rate by three observers over a twenty-one-month period, is listed when density estimates could not be derived from censuses. Vagrant species are those that may go undetected depending on time of year they are sampled, since they were decidedly not year-round residents at the study plot.

[g]In the Urucu this group comprised the following four species not readily distinguishable during night surveys: *Micoureus demerarae, Marmosa murina, Marmosops noctivagus,* and *Marmosops parvidens* (J. Malcolm, J. Patton, and M. N. da Silva, pers. comm.).

[h]Sightings of woolly opossum (*Caluromys lanatus*) could not be reliably distinguished from those of the presumably rarer bushy-tailed opossum (*Glironia venusta*), which was also collected on one occasion at this site (J. Malcolm, J. Patton, and M. N. da Silva, pers. comm.).

[i]Only two individuals were obtained from both terrestrial and arboreal traplines operated during over 5,200 trap-nights (J. Malcolm, J. Patton, and M. N. da Silva, pers. comm.). None were sighted during censuses.

[j]Almost entirely restricted to the strip of igapó forest along the Urucu river (Peres 1993a); it did not occur in the 900 ha terra firme forest plot 4 km or farther from the river.

[k]Those found in the Urucu are nonvested and uniform fuliginous black, probably corresponding to a melanistic phase of this species (see Wetzel 1975).

[l]Only one or two individuals of these species were sighted in over sixty-three observer-months of study.

[m]Five species of murids are known to the Urucu—*Nectomys squamipes, Oecomys bicolor, Oecomys paricola, Oryzomys yunganus, Oryzomys macconnelli* (J. Malcolm, J. Patton, and M. N. da Silva, pers. comm.)—but no density estimates from line transect censuses are available for them.

[n]The four sympatric species of *Proechimys*—*P. simonsi, P. steerei,* and two as yet undescribed species—collected within the study plot (J. Malcolm, J. Patton, and M. N. da Silva, pers. comm.) could not be easily distinguished in the field.

[o]Represented here by *Isothrix bistriata, Mesomys hispidus,* and another as yet undescribed species of arboreal spiny rat (*Mesomys* sp. 1). I was unable to differentiate between these species during night surveys.

[p]Occurrence of Amazon weasel at the study plot remains hypothetical. Despite the spotty record localities of this species throughout Amazonia (Emmons and Feer 1997), one individual has been unmistakably sighted at a terra firme forest 32 km northwest of the Urucu (Oleoduto site, C. Peres, pers. obs.).

[q]See Peres (1991b) for further information on these extremely rare forest canids.

[r]Only one large herd of 150+ white-lipped peccaries was sighted crossing the study plot on a single occasion in two years of study. Indeed data from a number of unhunted sites (Peres 1996) suggest that this species is largely nomadic in vast expanses of Amazonian terra firme forest.

Notes

Original data presented here resulted from field studies funded by the Wildlife Conservation Society (1991–94) and the World Wildlife Fund U.S. (1987–89). I thank these organizations for critical financial support. I am grateful to Jay Malcolm, Jim Patton, Maria Nazareth F. da Silva, and Rogério Gribel for kindly making available their unpublished identifications of small mammals collected at the Urucu forest. Many thanks to Raimundo Nonato and Luís Pereira Lopes for field assistance during line transect surveys. Access to the Urucu and logistical support there were provided by the Brazilian oil company (Petrobras): I thank all the employees of this company and subcontracted firms for helping me to organize air transportation. I am grateful to John Eisenberg, Jim Patton, and Anina Carkeek for their editorial improvement of the manuscript, and to the Brazilian Science and Technology Council (CNPq) for partial funding.

1. A large proportion of Amazonian mammals were described from specimens collected during expeditions and sojourns led by Alexandre Rodrigues Ferreira (1783–92), Johann von Spix, Carl von Martius (1817–20), Johann Natterer (1825–35), Henry Bates (1848–59), Alfred Wallace (1848–51), and more recently the Ollala brothers (1934–38). See Wallace (1853), Bates (1892), and Hershkovitz (1987).

2. White-water rivers here refer to those carrying turbid, sediment-loaded waters (usually first- or second-order meandering tributaries of the upper-

central Amazon) draining the eastern slopes of the Andes.

3. Campinas here refers to a stunted form of sclerophyllous forest growing on white-sand soils, whereas campos rupestres refers to an unusual savannalike vegetation on rocky soils (Pires 1972).

4. Includes thirty-two "probable" species that are likely to occur but not actually recorded at Reserva Cuzco Amazónico and three species found only at a nearby locality (Woodman et al. 1991).

5. Recent mammal sampling at Manu National Park includes several "new" species added to previous checklists (e.g., Terborgh Fitzpatrick, and Emmons 1984) based on specimens collected at Pakitza (gateway to the park) and Rio Panagua (below 490 m elevation), which most likely also occur at Cocha Cashu Biological Station (Pacheco et al. 1993).

6. Caatinga here refers to the dominant semiarid vegetation of the drought polygon scrublands of northeastern Brazil.

7. This relation is somewhat more complicated in the case of folivores by a strong predilection toward stands dominated by bamboo (*Guadua* spp.) and other bamboo specialists, because the distribution of bamboo forests in Amazonia is not necessarily congruent with that of rivers (B. Nelson, pers. comm.).

8. Tapitis (*Sylvilagus brasiliensis,* Lagomorpha) are rather common near human settlements close to large rivers of western Brazilian Amazonia (pers. obs.), despite their large geographic gap from collecting localities in the region (Emmons and Feer 1997).

References

Ascorra, C. F., D. E. Wilson, and M. Romo. 1991. Lista anotada de los quirópteros del Parque Nacional Manu, Peru. *Publ. Mus. Hist. Nat., Univ. Nac. Mayor San Marcos, Zool.* 42:1–14.

August, P. V. 1983. The role of habitat complexity and heterogeneity in structuring mammalian communities. *Ecology* 64:1495–1507.

Ayres, J. M. 1986. Uakaris and Amazonian flooded forest. Ph.D. diss., Cambridge University.

———. 1993. As matas de várzea do Mamirauá: Médio Rio Solimões. MCT-CNPq Programa do Trópico Húmido. Belém: Sociedade Civil Mamiruá.

Balslev, H., J. Luteyn, B. Ollgaard, and L. B. Holm-Nielsen. 1987. Composition and structure of adjacent unflooded and floodplain forest in Amazonian Ecuador. *Opera Bot.* 92:37–57.

Bates, H. W. 1892. *The naturalist on the river Amazon.* London: John Murray.

Bodmer, R. E. 1989. Ungulate biomass in relation to feeding strategy within Amazonian forests. *Oecologia* 81:547–50.

———. 1990. Responses of ungulates to seasonal inundations in the Amazon floodplain. *J. Trop. Ecol.* 6:191–201.

Brockie, R. E., and A. Moeed. 1986. Animal biomass in a New Zealand forest compared with other parts of the world. *Oecologia* 70:24–34.

Burnham, K. P., D. R. Anderson, and J. L. Laake. 1980. Estimation of density from line transect sampling of biological populations. *Wildl. Monogr.* 72:1–202.

Caughley, G. 1977. *Analysis of vertebrate populations.* London: John Wiley.

Eisenberg, J. F. 1980. The density and biomass of tropical mammals. In *Conservation biology: An evolutionary-ecological perspective,* ed. M. E. Soulé and B. A. Wilcox, 35–55. Sunderland, Mass.: Sinauer.

Eisenberg, J. F., M. A. O'Connell, and P. V. August. 1979. Density, productivity, and distribution of mammals in two Venezuelan habitats. In *Vertebrate ecology in the northern Neotropics,* ed. J. F. Eisenberg, 187–207. Washington D.C.: Smithsonian Institution Press.

Eisenberg, J. F., and R. W. Thorington Jr. 1973. A preliminary analysis of a Neotropical mammal fauna. *Biotropica* 5:150–61.

Emmons, L. H. 1984. Geographic variation in densities and diversities of non-flying mammals in Amazonia. *Biotropica* 16:210–22.

———. 1987. Comparative feeding ecology of felids in a Neotropical rainforest. *Behav. Ecol. Sociobiol.* 20:271–83.

Emmons, L. H., and F. Feer. 1997. *Neotropical rainforest mammals: A field guide.* 2d ed. Chicago: University of Chicago Press.

Ferrari, S. F., and M. A. Lopes. 1992. A new species of marmoset, genus *Callithrix* Erxleben, 1777 (Callitrichidae, Primates) from western Brazilian Amazonia. *Goeldiana Zool.* 11:1–13.

Furch, K., and H. Klinge. 1989. Chemical relationships between vegetation, soil and water in contrasting inundation areas of Amazonia. In *Mineral nutrients in tropical forest and savanna ecosystems,* ed. J. Proctor, 189–204. Oxford: Blackwell Scientific.

Gentry, A. H. 1982. Patterns of Neotropical plant species diversity. *Evol. Biol.* 15:1–84.

———. 1988. Changes in plant community diversity and floristic composition on environmental and geographical gradients. *Ann. Missouri Bot. Gard.* 75:1–34.

Gentry, A. H., and L. H. Emmons. 1987. Geographical variation in fertility, phenology, and composition of the understory of Neotropical forests. *Biotropica* 19:216–27.

Glanz, W. E. 1990. Neotropical mammal densities: How unusual is the community on Barro Colorado

Island, Panama? In *Four Neotropical rainforests,* ed. A. Gentry, 287–311. New Haven: Yale University Press.

Goulding, M. 1980. *The fishes and the forest: Explorations in Amazonian natural history.* Los Angeles: University of California Press.

Handley, C. O., Jr. 1967. Bats of the canopy of an Amazonian forest. *Atas Simp. Biota Amazônica, Zool.* 5:211–15.

Handley, C. O., Jr., D. E. Wilson, and A. L. Gardner, eds. 1991. Demography and natural history of the common fruit bat, *Artibeus jamaicensis,* on Barro Colorado Island, Panama. *Smithsonian Contrib. Zool.* 511:1–173.

Hershkovitz, P. 1987. A history of the recent mammalogy of the Neotropical region from 1492 to 1850. *Fieldiana: Zool.,* n.s., 39:11–98.

Irion, G. 1978. Soil infertility in the Amazonian rain forest. *Naturwissenschaften* 65:515–19.

Janson, C. H., and L. H. Emmons. 1990. Ecological structure of the non-flying mammal community at the Cocha Cashu Biological Station, Manu National Park, Peru. In *Four Neotropical rainforests,* ed. A. Gentry, 314–38. New Haven: Yale University Press.

Janzen, D. 1974. Tropical black water rivers, animals, and mast fruiting by the Dipterocarpaceae. *Biotropica* 6:69–103.

Johns, A. D. 1986. Effects of habitat disturbance on rainforest wildlife in Brazilian Amazonia. Final Report to the World Wildlife Fund, Washington, D.C.

Junk, W. J. 1989. Flood tolerance and tree distribution in central Amazonian floodplains. In *Tropical forests: Botanical dynamics, speciation, and diversity,* ed. L. B. Holm-Nielsen, I. C. Nielsen, and H. Balslev, 47–64. London: Academic Press.

Kahn, F. and J.-J. de Granville. 1992. *Palms in forest ecosystems of Amazonia.* Berlin: Springer-Verlag.

Kalliola, R., J. Salo, M. Puhakka, and M. Rajasilta. 1991. New site formation and colonizing vegetation in primary succession on the western Amazon floodplains. *J. Ecol.* 79:877–901.

Kubitzki, K., and A. Ziburski. 1994. Seed dispersal in floodplain forests of Amazonia. *Biotropica* 26:30–43.

Malcolm, J. R. 1990. Estimation of mammalian densities in continuous forest north of Manaus. In *Four Neotropical rainforests,* ed. A. H. Gentry, 339–57. New Haven: Yale University Press.

———. 1991a. The small mammals of Amazonian forest fragments: Pattern and process. Ph.D. diss., University of Florida, Gainesville.

———. 1991b. Comparative abundances of Neotropical small mammals by trap height. *J. Mammal.* 72 (1): 188–92.

Malcolm, J. R., C. A. Peres, J. L. Patton, M. N. F. da Silva, and C. Gascon. 1992. Ecological and evolutionary significance of Amazonian river barriers. Unpubl. Report to the Wildlife Conservation Society, New York.

Mares, M. A. 1992. Neotropical mammals and the myth of Amazonian biodiversity. *Science* 255:976–79.

Mares, M. A., M. R. Willig, K. E. Streilein, and T. E. Lacher Jr. 1981. The mammals of northeastern Brazil, a preliminary assessment (list). *Ann. Carnegie Mus.* 50:81–137.

Mascarenhas, B. M., and G. Puorto. 1988. Nonvolant mammals rescued at the Tucuruí Dam in the Brazilian Amazon. *Primate Conserv.* 9:91–93.

Medellín, R. A. 1994. Mammal diversity and conservation in the Selva Lacandona, Chiapas, Mexico. *Conserv. Biol.* 8:780–99.

Mittermeier, R. A., M. Schwarz, and J. M. Ayres. 1992. A new species of marmoset, genus *Callithrix* Erxleben, 1777 (Callitrichidae, Primates) from the Rio Maués region, state of Amazonas, Central Brazilian Amazonia. *Goeldiana Zool.* 14:1–17.

Montgomery, G. G., and M. E. Sunquist. 1975. Impact of sloths on Neotropical forest energy flow and nutrient cycling. In *Tropical ecological systems: Trends in terrestrial and aquatic research,* ed. F. B. Golley and E. Medina, 69–98. New York: Springer-Verlag.

Pacheco, V., B. D. Patterson, J. L. Patton, L. H. Emmons, S. Solari, and C. F. Ascorra. 1993. List of mammal species known to occur in Manu Biosphere Reserve, Peru. *Publ. Mus. Hist. Nat. Univ. Nac. Mayor San Marcos, Zool.* 44:1–12.

Patton, J. L., B. Berlin, and E. A. Berlin. 1982. Aboriginal perspectives of a mammal community in Amazonian Peru: Knowledge and utilization patterns among the Aguaruna Jívaro. In *Mammalian biology in South America,* ed. M. A. Mares and H. H. Genoways, 111–28. Pymatuning Symposia in Ecology 6. Special Publication Series. Pittsburgh: Pymatuning Laboratory of Ecology, University of Pittsburgh.

Peres, C. A. 1988. Primate community structure in western Brazilian Amazonia. *Primate Conserv.* 9:83–87.

———. 1990. Effects of hunting on western Amazonian primate communities. *Biol. Conserv.* 54:47–59.

———. 1991a. Ecology of mixed-species groups of tamarins in Amazonian terra firme forests. Ph.D. diss., Cambridge University.

———. 1991b. Observations on hunting by small-eared (*Atelocynus microtis*) and bush dogs (*Speothos venaticus*) in central-western Amazonia. *Mammalia* 55 (4): 635–39.

————. 1993a. Structure and spatial organization of an Amazonian terra firme forest primate community. *J. Trop. Ecol.* 9:259–76.

————. 1993b. Diet and feeding ecology of saddleback and moustached tamarins in an Amazonian terra firme forest. *J. Zool.* (London) 230:567–92.

————. 1993c. Biodiversity conservation by native Amazonians: A pilot study in the Kaxinawá Indigenous Reserve of Rio Jordão, Acre, Brazil. Final Report to the World Wildlife Fund, Washington, D.C.

————. 1994a. Primate responses to phenological changes in an Amazonian terra firme forest. *Biotropica* 26:98–112.

————. 1994b. Diet and feeding ecology of gray woolly monkeys (*Lagothrix lagotricha cana*) in central Amazonia: Comparisons with other atelines. *Int. J. Primatol.* 15 (3): 333–72.

————. 1996. Population status of white-lipped and collared peccaries in hunted and unhunted Amazonian forests. *Biol. Conserv.* 77:115–23.

————. 1997a. Primate community structure at twenty Amazonian flooded and unflooded forests. *J. Trop. Ecol.* 13:381–405.

————. 1997b. Effects of habitat quality and hunting pressure on arboreal folivore densities in Neotropical forests: A case study of howler monkeys (*Alouatta* spp.). *Folia Primatol.* 68:199–222.

Peres, C. A., and A. D. Johns. 1991. Patterns of primate mortality in a drowning forest: Lessons from the Tucuruí Dam, eastern Amazonia. *Primate Conserv.* 12:7–10.

Peres, C. A., and A. Whittaker. 1991. Annotated checklist of bird species of the upper Rio Urucu, Amazonas, Brazil. *Bull. Brit. Ornithol. Club* 111 (3): 156–71.

Pine, R. H. 1973. Mammals (exclusive of bats) of Belém, Pará, Brazil. *Acta Amazonica* 3 (2): 47–79.

Pires, J. M. 1972. Tipos de vegetação da Amazônia. *Publ. Avul. Mus. Goeldi* 20:179–202.

Pires, J. M., and G. T. Prance. 1985. The vegetation types of the Brazilian Amazon. In *Key environments: Amazonia,* ed. G. T. Prance and T. E. Lovejoy, 109–45. Oxford: Pergamon Press.

Prance, G. T. 1979. Notes on the vegetation of Amazonia. 3. The terminology of Amazonian forest types subject to inundation. *Brittonia* 31 (1): 26–38.

Prins, H. H. T., and J. M. Reitsma. 1989. Mammalian biomass in an African equatorial rain forest. *J. Anim. Ecol.* 58:851–61.

Puhakka, M., R. Kalliola, M. Rajasilta, and J. Salo. 1993. River types, site evolution and successional vegetation patterns in Peruvian Amazonia. *J. Biogeogr.* 19:651–65.

Putzer, H. 1984. The geological evolution of the Amazon basin and its mineral resources. In *The Amazon: Limnology and landscape ecology of a mighty tropical river and its basin,* ed. H. Sioli, 15–46. Dordrecht: Junk.

Queiroz, H. L. 1992. A new species of capuchin monkey, genus *Cebus* Erxleben, 1777 (Cebidae: Primates) from eastern Brazilian Amazonia. *Goeldiana Zool.* 15:1–13.

Redford, K. H., A. Taber, and J. A. Simonetti. 1990. There is more to biodiversity than the tropical rain forests. *Conserv. Biol.* 4:328–30.

Remsen, J. V., Jr., and T. A. Parker III. 1983. Contribution of river-created habitats to bird species richness in Amazonia. *Biotropica* 15:223–31.

Rylands, A. B., and A. Keuroghlian. 1988. Primate populations in continuous forest and forest fragments in central Amazonia. *Acta Amazonica* 18:291–307.

Salo, J., R. Kalliola, I. Häkkinen, Y. Mäkinen, P. Niemelä, M. Puhakka, and P. D. Coley. 1986. River dynamics and the diversity of Amazon lowland forest. *Nature* 322:254–58.

Sanchez, P. A. 1981. Soils of the humid tropics. In *Blowing in the wind: Deforestation and long-term implications,* 347–410. Williamsburg, Va.: College of William and Mary, Department of Anthropology.

Schaller, G. B. 1983. Mammals and their biomass on a Brazilian ranch. *Arq. Zool. São Paulo* 31 (1): 1–36.

Terborgh, J. 1983. *Five New World primates: A study in comparative ecology.* Princeton: Princeton University Press.

————. 1986. Community aspects of frugivory in tropical forests. In *Frugivores and seed dispersal,* ed. A. Estrada and T. H. Fleming, 371–84. Dordrecht: Junk.

Terborgh, J., J. W. Fitzpatrick, and L. Emmons. 1984. Annotated checklist of bird and mammal species of Cocha Cashu Biological Station, Manu National Park. *Fieldiana: Zool.,* n.s., 21:1–29.

Terborgh, J., and K. Petren. 1991. Development of habitat structure through succession in an Amazonian floodplain forest. In *Habitat structure: The physical arrangements of objects in space,* ed. S. S. Bell, E. D. McCoy, and H. R. Mushinsky, 28–44. London: Chapman-Hall.

Vitousek, P. M., and R. L. Sanford. 1986. Nutrient cycling in moist tropical forest. *Ann. Rev. Ecol. System.* 17:137–67.

Wallace, A. F. 1853. *Travels on the Amazon and Rio Negro.* London: Ward, Lock.

Walsh, J., and R. Gannon. 1967. *Time is short and the water rises.* Camden, N.J.: Thomas Nelson.

Wetzel, R. M. 1975. The species of *Tamandua* Gray (Edentata, Myrmecophagidae). *Proc. Biol. Soc. Washington* 88:95–112.

Wetzel, R. M., and E. Mondolfi. 1979. The subgenera and species of long-nosed armadillos, genus *Dasypus* L. In *Vertebrate ecology in the northern Neotropics,* ed. J. F. Eisenberg, 43–63. Washington, D.C.: Smithsonian Institution Press.

White, L. J. T. 1994. Biomass of rain forest mammals in the Lopé Reserve, Gabon. *J. Anim. Ecol.* 63: 499–512.

Wilson, D. E. 1990. Mammals of La Selva, Costa Rica. In *Four Neotropical rainforests,* ed. A. H. Gentry, 273–86. New Haven: Yale University Press.

Woodman, N., R. M. Timm, R. Arana C., V. Pacheco, C. A. Schmidt, E. D. Hooper, and C. Pacheco A. 1991. Annotated checklist of the mammals of Cuzco, Amazonico, Peru. *Occas. Pap. Mus. Nat. Hist. Univ. Kansas* 145:1–12.

22   The Contemporary Mammalian Fauna of South America

John F. Eisenberg

Introduction and Historical Background

Given the history of mammalian evolution and the long isolation of South America after the breakup of Gondwanaland, the distribution of tropical mammals in the Western Hemisphere is remarkably complicated. Autochthonous South American mammals extend into North America, and Central America is the current "bridge." One could argue about the contemporary limits of the "Neotropical realm," but for this chapter I will confine most of my remarks to the continent of South America.

South America was surely conjoined with Antarctica, India, Africa, and Australia until about one hundred million years ago, the middle of the Cretaceous. By the end of the Cretaceous, South America was isolated from the other continental land masses and was truly an island continent. The original terrestrial mammalian fauna, at the beginning of the Cenezoic, included both eutherians and marsupials. The marsupials came to occupy the omnivore-carnivore niches, including both small and large forms. The eutherians came to fill the herbivorous, terrestrial niches and included Xenungulata, Pantodonta, Notoungulata, Litopterna, and Pyrotheria. The Xenarthra are eutherians in a sense, but some workers consider them the "sister group" for all other eutherian mammals (Simpson 1980; Webb 1976, 1985). They filled an insectivorous semifossorial niche at the outset but soon diversified into other adaptive zones.

Some inter-American exchange of mammalian fauna may well have occurred in the late Paleocene and the Eocene. The caviomorph rodents and ceboid primates were established by the late Oligocene, and probably somewhat earlier. Their exact origins are disputed. In the late Miocene, procyonid carnivores moved south and some ground sloths moved north to North America: an archipelago that was to become Panama is implicated. In the late Pliocene, the Panamanian uplift was completed and isolation came to an end. Thus the great American interchange between North and South America was in full swing two to

three million years ago. Although extinctions occurred in the Pliocene, the interchange had disastrous consequences for most southern forms; but the Xenarthra, ceboid primates, and caviomorph rodents persisted. Although many northern invaders to the south did not establish themselves, enough ordinal taxa did so that currently the small- to large-carnivore niches are filled by the eutherian Carnivora, the medium- to large-herbivore niches are occupied by the Perissodactyla and Artiodactyla, and finally, the northern-derived rodent stocks (noncaviomorph) obviously colonized in several waves from the late Miocene to the Recent (Simpson 1980; Webb 1976).

Some workers argue strongly for contact and exchange between Africa and South America at the close of the Cretaceous, but later exchanges with Africa were more probable. Certainly the presence of hystricognath rodents and primates in both areas reinforces the notion. The discovery of *Eurotamandua* in the Eocene Messel beds of Germany is provocative (Storch 1990).

The formation of the Andes Mountains commencing in the Miocene had a profound influence on the climate of the continent. Drainage patterns were altered, and the orogenic activities to the west were to continue to the present while the older areas of the Guiana and Brazilian shields quietly eroded.

The megafauna of South America, including the glyptodonts, giant ground sloths, and "Notoungulata," waned commencing in the Pliocene, but it was not a continuous decline and was completed only at the close of the Pleistocene. Clearly the Pleistocene glaciation episodes have had a profound effect, yet some workers argue that *Homo sapiens* was the decisive influence. A similar case could be made for extinctions in Australia and other areas (Martin and Klein 1984).

A Brief History of the "Older" Mammalian Fauna

Monotremes appeared in the Cretaceous but did not persist (Pascual et al. 1992). The Multituberculata also put in an appearance (Kraus, Kielan-Jaworowska, and Bonaparte 1992). The Marsupialia (Metatheria) appeared in the Cretaceous of South America, although older finds are noted in North America. The first fossil forms of the Metatheria may be assigned to the Didelphidae, which persist to this day. The Caenolestidae first appeared in the Eocene and survive as relicts in the Andes and the southern, temperate zone of the continent. The equally ancient Microbiotheriidae persist as a single genus and species, *Dromiciops australis,* in southwestern Chile and adjacent Argentina. The other families referable to the Marsupialia—the Polydolopidae, Argyrolagidae, Necrolestidae, Groeberiidae, and Borhyaenidae—all appeared in the Paleocene to Oligocene and disappeared by the end of the Pliocene (Hershkovitz 1972; Simpson 1980).

The Xenarthra appeared in the Eocene and rapidly differentiated to form nine families, of which four remain in the contemporary fauna: the Bradypodidae, Megalonychidae, Myrmecophagidae, and Dasypodidae (Hershkovitz 1972).

The "ungulates" such as the Condylarthra, Litopterna, Notoungulata, and Toxodonta appeared in the Paleocene. Whereas such taxa as the Pyrotheria and Astropotheria appeared in the Eocene and disappeared by the Miocene, representatives of the Litopterna and Toxodonta remained until the Pleistocene and co-occurred with the Perissodactyla and Artiodactyla, which moved into South America during the Pliocene.

Exactly when the Sirenia and Cetacea arrived in the coastal waters of South America is conjectural. Fossil Chiroptera have been detected in the Miocene but were surely present long before.

The caviomorph rodents appeared in the late Eocene, and primates appeared first in the Deseaden (Oligocene). Their origins have generated considerable debate, since their nearest contemporary relatives occur in Africa. Various hypotheses have been evaluated in the volume edited by Ciochon and Chiarelli (1980).

Northern Invaders with Special Reference to the Rodentia

Synopsis

The noncaviomorph rodents probably began to enter South America in the Miocene at the same time as the first Carnivora (Procyonidae). A more consistent exchange between the Northern and Southern Hemispheres occurred in the Pliocene. Of the northern invaders, some ordinal taxa such as the Proboscidea (Gomphotheriidae) did not persist, whereas other ordinal taxa persisted and radiated (Artiodactyla, Perissodactyla, and Carnivora). Some ordinal taxa appear to be rather recent entrants into South America (Lagomorpha, Insectivora). The extinction of the megafauna including both autochthonous and allochthonous forms during the late Pleistocene may have derived from climatic changes coupled with the recent entry of *Homo sapiens* (Martin and Klein 1984; Webb 1976, 1985).

The Special Case of the Rodentia
THE HISTORICAL FACTS CONCERNING THE SIGMODONTINAE

The Cucarachua fauna was uncovered during the construction of the Panama canal. It is dated as middle Miocene, and its composition clearly resembles that of the fauna of the state of Texas in North America. The sigmodontine (cricetine) rodents identifiable from the fauna are peromyscine, and one genus has been given the name *Texomys* (Whitmore and Stewart 1965). Webb and Perrigo (1984) have examined Pliocene assemblages from Honduras and El Salvador. Again the evidence is overwhelming that the faunal affinities were North American, and comparable representatives are demonstrable in Texas and Florida. Thus the North American faunas of five million years ago were well represented in Central America and were distinctively North American in character. Yet rodent faunas that have been examined in Argentina indicate the existence of some sigmodontine rodents in the Monte Hermosan or the immediately overlying Chapadmalan in the Pliocene, either just before the Great American interchange or at its onset (Marshall 1985). These include the genera *Auliscomys* and *Bolomys*. Further biochemical evidence suggests that the Akodontini and Phyllotini are derived when compared with the Peromyscine rodents (Catzeflis et al. 1993). It is quite clear, however, that sigmodontine rodents become much more numerous in the Pliocene. when four genera can be recognized in the Chapadmalan strata (Savage and Russell 1983). The question before the house then is, How many invasions of sigmodontine rodents occurred from North America to South America between the Miocene and the present? Of course, the companion question is, How many stocks that evolved in South America reemigrated to North America?

What do contemporary distributions of rodents have to offer? Let us consider the case of Costa Rica. From extensive studies in Costa Rica McPherson (1985) offers the following conclusions. Contemporary rodent species are closely tied to specific microclimatic conditions. Hystricognath rodents have obviously dispersed from South America to the north since the Pliocene (Ahearn and Lance 1980). Sigmodontine rodents, on the other hand, have made at least two movements from Central America to South America (Reig 1980). The heteromyid rodents evidently have been late entrants into South America, as have the geomyid rodents. The sciurids clearly have had a mixed history, and there may have been more than one movement from North America to South America (see also Moore 1959).

We know that if questions of competition are set aside mammal invasions can take place quite rapidly. Contemporary studies of introductions of exotics indicate that large mammals can advance at a rate of 10 km per year and smaller mammals at a maximum of 50 km per year. It has even been proposed that during human invasions movement may proceed at a rate of 16 km per year. Thus 10 to 15 km per year is not an unreasonable estimate. Since we are dealing in terms of millions of years, and since by all odds the greatest invasion of South America occurred 2.5 million years ago, the question of rate of movement is a bit academic.

A CONSIDERATION OF THE NONMURID RODENT TAXA

The hystricognath rodents have been in South America for a considerable span of time (Flynn et al. 1991). They may have entered South America from Central America via the "proto-Antilles" by island hopping late in the Eocene (Wood 1974), or they may have derived from Africa (Landry 1957). They may have been a taxon at the time of the separation of Gondwanaland. The hystricognath rodents had diverged widely, as evinced by the diversity of forms from the Salla beds of Bolivia in the late Oligocene. Some were able to reinvade North America after the completion of the Pliocene land bridge. These included the capybara, *Hydrochaerus,* and the porcupine, *Erethizon* (Ahearn and Lance 1980). *Erethizon* was exceedingly successful and persists today throughout North America in suitable habitats (Frazier 1981).

The sciuromorph rodents had a more complex history. Moore (1959) suggested that *Sciurillus* may have been an early (Miocene) entrant. Clearly it is an aberrant genus, but it is well established as a canopy dweller over much of northern South America. The other genera of squirrels may well have entered later. The genus *Microsciurus* is adapted to montane habitats and has expanded its distribution by taking advantage of the Andean chain. The various subgenera of the genus *Sciurus,* including *Guerlinguatus, Leptosciurus,* and *Hadrosciurus,* are probably late entrants. It is worth noting that whereas the squirrels have been able to penetrate the tropical portions of South America, they have as yet been unable to colonize temperate South America. Thus the niches one would expect to be occupied in the temperate portions of Chile and Argentina have no squirrellike counterparts.

Turning to the Geomyoidea we have two families, the Geomyidae and the Heteromyidae. By and large fossorial rodents have not been active colonizers of the lowland tropics in either the New World or the Old World because of adverse soil conditions. Thus it is not surprising that representatives of the Geomyidae (*Orthogeomys*) have penetrated only marginal portions of Colombia from Panama in premontane areas. The Heteromyidae are predominantly dwellers of the arid portions of North America, but two genera, *Heteromys* and *Liomys,* have successfully colonized tropical habitats. Whereas *Liomys* penetrates parts of western Colombia, *Heteromys* has been much more successful in moving around the eastern cordillera of the Andes and colonizing northern Venezuela (Eisenberg 1989).

THE SIGMODONTINAE RECONSIDERED

The genuine puzzle, then, concerns the sigmodontine rodents, hereafter defined in accordance with Reig (1980). If we look at Central America we find that the tribe Neotomini is highly speciose in Mexico. Notwithstanding its long evolutionary history in North America, it has been unable to penetrate truly tropical habitats. The Peromyscini, widely speciose in Central America and extending to the Arctic Circle in North Ameria, have been only partially successful in colonizing the lowland tropics (Carleton 1980). Depending on one's definition of a peromyscine tribe, one can say that only two species have penetrated Panama. The genus *Reithrodontomys* has penetrated to Peru, but this is an exception to the rule. In any event, the genus has not exhibited speciation. The Baiomyini have had a long history of occupancy in the southwestern United States and Mexico and again have been partially successful in colonizing Central America, mostly in premontane habitats. The Tylomyini are Central American and, although they have passed into Colombia, cannot be considered serious contenders for occupancy of the South American continent.

The Ichthyomyini are of special interest. While moderately speciose in Central America, they occupy a scattered and fragmented range in northern South America. Voss (1988) has proposed that the group is monophyletic and thus a natural one. If this is true, then it is an example of an early penetration to South America by northern stocks only to have them become fragmented and isolated owing to Pleistocene climatic cycles. The Nyctomyini are clearly North American in origin, a very conservative stock, and still confined to the Central American area (Carleton 1980).

The Sigmodontini currently show the greatest species richness in Mexico. They may in fact be an artificial grouping. Their diagnostic characters are based on a peculiar folding of the molars that is not unique to what are generally considered sigmodontines, since some phyllotines show the same trait. If *Holochilus* is unrelated to *Sigmodon,* then clearly the Sigmodontini as we understand them are recent invaders that have not penetrated fully to the continent of South America. This problem remains to be resolved.

The remaining sigmodontine rodents are by and large South American in their distribution. The thomasomyine rodents are clearly South American, with only one genus, *Rhipidomys,* penetrating to Central America. The Akodontini are South American. The Zygodontomyini are predominantly South American, with one species penetrating to Central America. The Phyllotini are entirely South American, as are the Scapteromyini (Reig 1987; Olds and Anderson 1989).

The Oryzomyini present special problems in that they are the most speciose group in the South American continent; yet clearly several species distribute themselves coastally in Central America and on into North America as far north as New Jersey. If they are a natural grouping this could well be an example of a sigmodontine taxon that differentiated in South America and subsequently reinvaded North America. After all the evidence has been weighed, the genus *Oryzomys* has been split into five genera (Wilson and Reeder 1993).

Clearly, the principal interchange may have occurred at about 3×10^6 years before the present. But the lingering question remains: How many movements occurred before this date and how many occurred subsequently? A further problem must also be addressed: there are no consistent diagnoses for the tribes of the Sigmodontinae. Reig (1987) attempted to diagnose the akodonts, and Olds and Anderson (1989) tried to diagnose the phyllotines. Yet in neither case were the other tribes diagnosed. This poses a difficult question: How can you diagnose a single tribe without

a diagnosis for all the others? Clearly the authors have stated the problem and understand it, and they have attempted to make—for want of a better term—what one could call a partial diagnosis. This compounds the difficulty in researching the fossil record. For example, Baskin (1978) assigns *Bensonomys* to the phyllotines, but how can one with any confidence assign a genus to a tribe when the tribe itself has not been properly diagnosed? This is especially problematic if one works mainly (or only) with teeth. I submit that until we have an adequate set of diagnoses for the tribes and some sort of reasonable natural classification, then the assignment of fossil forms remains questionable. The present evidence indicates that there was more than one invasion and that traffic through the isthmus has proceeded in both directions. Nevertheless, many well-established taxa from North America have been unable to penetrate South America in Recent times, probably owing to specialization for microclimatic conditions that prevent their free movement through the lowland tropics.

A Summary of the Contemporary Mammalian Fauna of South America Including Panama

The Metatheria include 272 species; South America contains 69 species, or 25% of the world's total.[1] Of the extant South American species, 6 genera include 85% of all South American marsupials: the genera *Marmosa, Marmosops, Micoureus, Gracilinanus,* and *Thylamys* include 42 species, and *Monodelphis* has 17 species.

Table 22.1   South American Mammals (Including Panama)

Taxon	World		South America		
	Number of Genera	Number of Species	Number of Genera	Number of Species	Percentage of Species
Metatheria	91	272	19	69	25
Didelphimorpha					
Didelphidae			15	63	
Paucituberculata					
Caenolestidae			3	5	
Microbiotheria			1	1	
Xenarthra	13	29	13	29	100
Bradypodidae			1	3	
Megalonychidae			1	2	
Dasypodidae			8	20	
Myrmecophagidae			3	4	
Insectivora	66	428	1	6	1.4
Soricidae			1	6	
Chiroptera	177	925	67	190	21
Emballonuridae			8	16	
Noctilionidae			1	2	
Mormoopidae			2	5	
Phyllostomidae			38	105	
Natalidae			1	2	

Continued on next page

Table 22.1   (*continued*)

Taxon	World		South America		
	Number of Genera	Number of Species	Number of Genera	Number of Species	Percentage of Species
Furipteridae			2	2	
Thyropteridae			1	2	
Vespertilionidae			6	29	
Molossidae			8	27	
Primates	60	233	16	83	36
Callitrichidae			5	26	
Cebidae			11	57	
Carnivora	69	162	27	47	29
Mustelidae			7	13	
Procyonidae			6	11	
Ursidae			1	1	
Canidae			7	12	
Felidae			6	10	
Pinnipedia	18	34	4	6	18
Otariidae			3	5	
Phocidae			1	1	
Cetacea	41	78	25	37	47
Balaenidae			1	1	
Balaenopteridae			2	6	
Neobalaenidae			1	1	
Delphinidae			11	15	
Phocoenidae			1	2	
Physeteridae			2	3	
Platanistidae			2	2	
Ziphiidae			5	7	
Sirenia	3	5	1	2	40
Trichechidae			1	2	
Perissodactyla	6	18	1	3	17
Tapiridae			1	3	
Artiodactyla	81	220	10	18	8
Tayassuidae			2	3	
Camelidae			2	2	
Cervidae			6	13	
Rodentia	443	2,021	110	454	23
Sciuridae			3	11	
Geomyidae			1	1	
Heteromyidae			2	3	
Muridae	281	1,326	61	270	20
Sigmodontinae	79	423	61	168	63
Hystricognathi		197	43	168	85
Erethizontidae			4	11	
Chinchillidae			3	6	
Dinomyidae			1	1	
Caviidae			5	14	
Hydrochoeridae			1	1	
Dasyproctidae			2	11	
Agoutidae			2	2	
Ctenomyidae			1	38	
Octodontidae			6	9	
Abrocomidae			1	3	
Echimyidae			16	71	
Myocastoridae			1	1	
Lagomorpha	13	80	1	2	2.5
Leporidae			1	2	

Note: Figures in far right column are percentages of total world's species.
Source: Wilson and Reeder 1993 (in part).

The extant species of the Xenarthra are all represented in South America and include 29 species, of which 20 belong to the Dasypodidae.

The order Insectivora includes 66 genera and 428 species worldwide, but only 1 genus with 6 species occur in South America.

The order Chiroptera includes a world total of 925 species, and 21% of the world's species (190) are found in South America; 53% of them (105) belong to the family Phyllostomidae and 29% (62) belong to the Vespertilionoidea.

The order Primates contains a world total of 233 species, and 36% or 83 species are to be found in South America. There are 162 species of terrestrial Carnivora (Fissipedia), and 47 species or 29% of the world total are found in South America. The Pinnipedia number 34 species worldwide, and 6 species or 18% are found off the coasts of South America.

The Cetacea include 78 species worldwide, and 37 or 47% have been noted in the rivers or offshore waters of South America. The order Sirenia includes 5 extant species, of which 2 are found in South America. The Perissodactyla includes 18 living species worldwide; 3 of them are found in South America. The Artiodactyla contains 220 species worldwide, of which only 18 (8%) occur in South America. The order Lagomorpha contains 80 species worldwide, with only are to be2 of them found in South America.

The Rodentia are represented globally by 2,021 species subsumed under 443 genera. South America includes 23% of the world total. Only 15 species of the South American rodents are Sciuromorpha. Of the Myomorpha South America includes 63% (270 species) of the world total for the Muridae and Sigmodontinae. Of the South American sigmodontine rodents, *Akodon* (16%) and *Oryzomys, Oecomys, Oligoryzomys,* and *Microryzomys* (19.6%) together include 35.6% of all sigmodontines. Runner-ups are *Thomasomys,* 25 species (9%); *Rhipidomys,* 14 (5%); *Phyllotis,* 13 (5%); and *Oxymycterus,* 13 (4%). In total the preceding genera include 159 species, or 59% of all South American Sigmodontinae.

The hystricognath rodents include 197 species, but South America retains 168 species, or 85% of the world total. Three South America genera—*Proechimys* (40), *Ctenomys* (38), and *Echimys* (11)—contain 89 named taxa, or 53% of the named South American species (Wilson and Reeder 1993).

The total number of mammalian species recorded for continental South America including Panama but *not* including the Cetacea is 909, or 19.9% of the world total, of which 21% are bats and 50% rodents (Wilson and Reeder 1993; Corbet and Hill 1991). The South American totals compare favorably with the world ratios (44% rodents and 20% bats) from a grand total of 4,629 species, of which 78 are Cetacea.

Totals and percentages for the Indo-Malayan region, which is similar to South America in size and geographical position, are 867 species, of which rodents and bats constitute 64% (not including 30 species of Cetacea) (Corbet and Hill 1992).

I conclude that in 1998 our knowledge of the South American mammal fauna, though incomplete, is probably as complete as our knowledge of the mammals of the Old World.

Species Richness and Political Units

Countries in South and Central America are rapidly attempting to inventory their floras and faunas. Mammalian species richness shows the following trends.

Any tropical nation of South America that contains high mountains as well as lowlands will have greater species richness. This derives from coupling the rich highland rodent fauna with the rich lowland bat fauna. By the same token, when a country spans both temperate and tropical zones it will have a combined fauna of great diversity. Given the area, topographic relief, and latitudinal span, it should be no surprise that for its size and latitudethe political unit of Mexico exhibits a high species richness (Fa and Morales 1993) (see table 22.2). Brazil, with its vast area, has 483 species recorded to date (Fonseca et al. 1996). Peru, with 460 species (Pacheco et al. 1995), is very rich given its smaller area, as is Colombia with 423 species (Rodriguez-Mahecha et al. 1995). These numbers will no doubt increase with further discovery, but they illuminate two interesting trends. Ecuador, Peru, and Bolivia contain portions of the Andes with extreme elevational gradients and shifts in vegetation, although they lie within the tropics. The vertebrates on the

Table 22.2   Mammalian Species Richness (Exclusive of Cetacea) according to Realm and Some Subregions

Geographical Area	Number of Species	Source
Nearctic realm	643	Wilson and Reeder 1993
Mexico (transitional with Neotropical)	449	Fa and Morales 1993
Neotropical realm	1,096	Wilson and Reeder 1993
Panama	203	Eisenberg 1989
Argentina	267	Redford and Eisenberg 1992
South America (and Panama)	909	Wilson and Reeder 1993
Palearctic realm	843	Wilson and Reeder 1993
Ethiopean realm	1,045	Wilson and Reeder 1993
Oriental realm	995	Wilson and Reeder 1993
Indo-Malaya	867	Corbet and Hill 1992
Thailand	251	Corbet and Hill 1992
Australian realm	472	Wilson and Reeder 1993
New Guinea	188	Flannery 1990
Australia	232	Strahan 1983
Tasmania	38	Strahan 1983

slopes of the Andes exhibit specific adaptations and endemism, not to mention the fauna of the high central plateaus. The same could be said for Chile and Argentina; these two countries lie mainly in the temperate zone and demonstrate lower species richness than do the tropical countries.

Peru shares the rich upper Amazon region with Brazil, and they also share species. Brazil has true centers of mammalian endemism when one considers the cerrado-caatinga and the eastern Atlantic rain forest. This latter region, with the adjoining cerrado, offers great opportunity for further study but is one of the most threatened ecosystems in the Neotropics. The smaller areas of montane habitats are equally imperiled and cause grave concern for all biologists. The time is short (Dinerstein et al. 1995).

Conservation and Management

Chapters 2 to 4 of this volume document the massive Pleistocene extinctions of the South American megafauna. The causes are surely complex. The event was not caused by early humans acting alone. Up to A.D. 1500 only the Andean cultures were practicing plant and animal husbandry on a large scale. The transport of European agriculture to the New World after 1500 brought large-scale landscape modification, especially in temperate South America, as discussed in volume 2 of this work (Redford and Eisenberg 1992).

At present the boundaries of the taxonomic divisions are still being determined, and the distributions of species are still uncertain. Much work is still in progress, and this volume offers a guide to future efforts. Mammal species of concern are still attracting funding in proportion to their impact on human ecology. Rodents and bats that are reservoirs of human pathogens are the domain of epidemiologists. Rodents as agricultural pests promote research on rodent population control. Primates widely used in biomedical research must be studied to ensure appropriate data for either sustained wild harvest or sustained captive maintenance.

The larger mammal species have always been harvested for human needs. Hides, furs, and meat from the ungulates and primates are today a part of the economy of both indigenous peoples and *colonistas.* Commercial trade in such products has for the most part been regulated by international treaties. The conversion of landscape for commercial agriculture or animal husbandry persists.

The nonconsumptive use of the landscape for wildlife viewing by cash-paying tourists is relatively new, and ecotourism has entered the marketplace. This development has promoted the notion that ecosystem conservation may have economic value.

Beyond strict economic issues is the growing notion that healthy ecosystems may have a value that transcends political boundaries. It is possible to retain some portions of the globe that function as true ecosystems? Can some parts of the world be used for sustained-yield consumption by humans? Clearly we must distinguish between short-term gains and long-term losses. The Rio Conference of 1992 tried to focus on these issues. One outcome was the notion that it is neither necessary nor desirable for the entire world to attempt to replicate the consumer economy of the "Western World."

A working knowledge of what species constitute an ecosystem and how they interact is a first step in planning for either management or preservation of fragile systems. At the close of the twentieth century we may have the minimum of information for some educated guesses. Ultimately, conservation decisions are political decisions. The future of many ecosystems and their component species is still undetermined.

Notes

This chapter is based in part on a lecture presented 3 May 1994 at the Alexander Koenig Museum in Bonn.

1. Excluding Antarctica, South America represents 12% of the earth's surface (tabulations from Wilson and Reeder 1993; Eisenberg 1989; Redford and Eisenberg 1992; see also table 22.1).

References

Ahearn, M. E., and J. F. Lance. 1980. A new species of *Neochoerus. Proc. Biol. Wash.* 93:435–42.

Alberico, M. 1990. Systematics and distribution of the genus *Vampyrops* (Chiroptera: Phyllostomidae) in northwestern South America. In *Vertebrates in the tropics,* ed. G. Peters and R. Hutterer, 345–54. Bonn: Museum Alexander Koenig.

Anderson, S., and J. Knox Jones Jr. 1984. *Orders and families of Recent mammals of the world.* New York: John Wiley.

Arita, H. T. 1993. Riqueza de especies de la masto fauna de México. In *Avances en el estudio de los mamíferos de México,* ed. R. Medellin y Gerardo and G. Ceballos, 1:109–28. Mexico City: A. C. Centro Ecológica UNAM.

August, P. V. 1983. The role of habitat complexity and heterogeneity in structuring mammalian communities. *Ecology* 64:1495–1507.

Baker, R. H. 1991. The classification of Neotropical mammals: A historical resumé. In *Latin American mammalogy: History, biodiversity and conservation,* ed. M. A. Mares and D. J. Schmidly, 7–31. Norman: University of Oklahoma Press.

Baskin, J. A. 1978. *Bensonomys, Calomys* and the origin of the phyllotine group of Neotropical cricetines. *J. Mammal.* 59 (1): 125–35.

Bisbal, F. 1989. Distribution and habitat association of the carnivore community in Venezuela. In *Advances in Neotropical mammalogy*, ed. K. H. Redford and J. F. Eisenberg, 339–62. Gainesville, Fla.: Sandhill Crane Press.

Cabrera, A. 1958. *Catálogo de los mamíferos de América del Sur.* Vol. 1. Buenos Aires: Museo Argentino de Ciencias Naturales "Bernardino Rivadavia."

———. 1960. *Catálogo de los mamíferos de América del Sur.* Vol. 2. Buenos Aires: Museo Argentino de Ciencias Naturales "Bernardino Rivadavia."

Carleton, M. D. 1980. Phylogenetic relationships in neotomine-peromyscine rodents (Muroidea) and a reappraisal of the dichotomy with New World Cricetinae. *Misc. Publ. Mus. Zool. Univ. Michigan,* 157:1–146.

Catzeflis, F. M., A. W. Dickerman, J. Michaux, and J. A. W. Kirsch. 1993. DNA hybridization and rodent phylogeny. In *Mammal phylogeny: Placentals,* ed. F. Szalay, M. J. Novacek, and M. C. McKenna, 159–71. New York: Springer-Verlag.

Ciochon, P. L., and A. B. Chiarelli, eds. 1980. *Evolutionary biology of the New World monkeys and continental drift.* New York: Plenum Press.

Corbet, G. B. 1978. *The mammals of the palearctic region.* London: British Museum (Natural History).

Corbet, G. B., and J. E. Hill. 1991. *A world list of mammalian species.* 3d ed. Oxford: Oxford University Press.

———. 1992. *The mammals of the Indomalayan region: A systematic review.* Oxford: Oxford University Press.

Dinerstein, E., D. M. Olson, D. J. Graham, A. L. Webster, S. A. Primm, M. P. Bookbinder, and G. Leder. 1995. *A conservation assessment of the terrestrial ecoregions of Latin America and the Caribbean.* Washington, D.C.: World Wildlife Fund and World Bank.

Eisenberg, J. F., ed. 1979. *Vertebrate ecology in the northern Neotropics.* Washington, D.C.: Smithsonian Institution Press.

Eisenberg, J. F. 1981. *The mammalian radiations: An analysis of trends in evolution, adaptation, and behavior.* Chicago: University of Chicago Press.

———. 1989. *Mammals of the Neotropics.* Vol. 1. *The northern Neotropics: Panama, Colombia, Venezuela, Guyana, Suriname, French Guiana.* Chicago: University of Chicago Press.

———. 1990. Neotropical mammal communities. In *Four Neotropical rainforests,* ed. A. H. Gentry, 358–68. New Haven: Yale University Press.

Eisenberg, J. F., M. A. O'Connell, and P. V. August. 1979. Density, productivity and distribution of mammals in two Venezuelan habitats. In *Vertebrate ecology in the northern Neotropics*, ed. J. F. Eisenberg, 187–207. Washington, D.C.: Smithsonian Institution Press.

Eisenberg, J. F., and K. H. Redford. 1979. A biogeographic analysis of the mammalian fauna of Venezuela. In *Vertebrate ecology in the northern Neotropics,* ed. J. F. Eisenberg, 31–38. Washington, D.C.: Smithsonian Institution Press.

———. 1982. Comparative niche structure and evolution of mammals of the Nearctic and southern South America. In *Mammalian biology in South America,* ed. M. A. Mares and H. H. Genoways, 77–84. Pymatuning Symposia in Ecology 6. Special Publication Series. Pittsburgh: Pymatuning Laboratory of Ecology, University of Pittsburgh.

Eisenberg, J. F., and R. W. Thorington Jr. 1973. A preliminary analysis of a Neotropical mammal fauna. *Biotropica* 5:150–61.

Emmons, L. H. 1984. Geographic variation in densities and diversities of non-flying mammals in Amazonia. *Biotropica* 16:210–22.

———. 1988. A field study of ocelots (*Felis pardalis*) in Peru. *Rev. Ecol. (Terre et Vie),* 43:133–57.

Emmons, L. H., and F. Feer. 1997. *Neotropical rainforest mammals: A field guide.* 2d ed. Chicago: University of Chicago Press.

Fa, J. E., and L. M. Morales. 1993. Patterns of mammalian diversity in Mexico. In *Biological diversity of Mexico,* ed. T. P. Ramamoorthy, R. Bye, A. Lot, and J. E. Fa, 319–61. Oxford: Oxford University Press.

Flannery, T. 1990. *Mammals of New Guinea.* Carina, Queensland: Robert Brown.

Flynn, J. J., M. A. Norell, C. C. Swisher III, and A. R. Wyss. 1991. Pre-Deseadan, post-Musterean mammals from Central Chile. *J. Vert. Paleontol.,* no. 3, suppl., 29A.

Fonseca, G. A. B. da, G. Herrmann, Y. L. R. Leite, R. A. Mittermeier, A. B. Rylands, and J. L. Patton. 1996. *Lista anotada dos mamíferos do Brasil.* Occasional Papers in Conservation Biology 4. Washington, D.C.: Conservation International.

Fonseca, G. A. B. da, and M. C. M. Kierulff. 1988. Biology and natural history of Brazilian Atlantic forest small mammals. *Bull. Florida State Mus., Biol. Sci.* 34:99–152.

Fragoso, J. M. V. 1991. The effect of hunting on tapirs in Belize. In *Neotropical wildlife use and conservation,* ed. J. G. Robinson and K. H. Redford, 154–62. Chicago: University of Chicago Press.

Frazier, M. 1981. *Erethizon. Bull. Florida State Mus.* 27:1–76.

Gentry, A. H., ed. 1990. *Four Neotropical rainforests.* New Haven: Yale University Press.

Glanz, W. E. 1990. Neotropical mammalian densities: How unusual is the community on Barro Colorado Island, Panama? In *Four Neotropical rainforests,* ed. A. H. Gentry, 287–311. New Haven: Yale University Press.

Hall, E. R. 1981. *The mammals of North America.* 2d ed., 2 vols. New York: John Wiley.

Hershkovitz, P. 1972. The Recent mammals of the Neotropical region: A zoogeographic and ecological review. In *Evolution, mammals, and southern continents,* ed. A. Keast, F. C. Erk, and B. Glass, 311–431. Albany: State University of New York Press.

———. 1987. A history of the recent mammalogy of the Neotropical region from 1492 to 1850. *Fieldiana: Zool.,* n.s., 39:11–98.

Jacobs, L. L., and E. H. Lindsay. 1984. Holuretic radiation of Neogene muroid rodents and origin of South American cricetids. *J. Vert. Paleontol.* 4 (2): 265–72.

Janson, C. H., and L. H. Emmons. 1990. Ecological structure of the non-flying mammal community at Cocha Cashu Biological Station, Manu National Park, Peru. In *Four Neotropical rainforests,* ed. A. H. Gentry, 314–38. New Haven: Yale University Press.

Kinnaird, M., and J. F. Eisenberg. 1989. A consideration of body size, diet, and population biomass for Neotropical mammals. In *Advances in Neotropical mammalogy,* ed. K. H. Redford and J. F. Eisenberg, 595–604. Gainesville, Fla.: Sandhill Crane Press.

Kleiman, D. G., J. F. Eisenberg, and E. Maliniak. 1979. Reproductive parameters and productivity of caviomorph rodents. In *Vertebrate ecology in the northern Neotropics,* ed. J. F. Eisenberg, 173–83. Washington, D.C.: Smithsonian Institution Press.

Konecny, M. J. 1989. Movement patterns and food habits of four sympatric carnivore species in Belize, Central America. In *Advances in Neotropical mammalogy,* ed. K. H. Redford and J. F. Eisenberg, 243–64. Gainesville, Fla.: Sandhill Crane Press.

Krause, D. W., Z. Kielan-Jaworowska, and J. F. Bonaparte. 1992. *Ferugliotherium* Bonaparte, the first known multituberculate from South America. *J. Vert. Paleontol.* 12 (3): 351–76.

Landry, S. 1957. The interrelationships of New and Old World hystricomorph rodents. *Univ. Calif. Publ. Zool.* 56 (1): 1–118.

Malcolm, J. R. 1990. Estimation of mammalian densities in continuous forests north of Manaus. In *Four Neotropical rainforests,* ed. A. H. Gentry, 339–57. New Haven: Yale University Press.

Mares, M. A., and H. H. Genoways, eds. 1982. *Mammalian biology in South America.* Pymatuning Symposia in Ecology 6. Special Publication Series. Pittsburgh: Pymatuning Laboratory of Ecology, University of Pittsburgh.

Mares, M. A., and D. J. Schmidly, eds. 1992. *Latin American mammalogy: History, biodiversity and conservation.* Norman: University of Oklahoma Press.

Marshall, L. G. 1985. Geochronology and land mammal biochronology of the trans-American faunal interchange. In *The great American biotic interchange,* ed. F. G. Stehli and S. D. Webb, 49–85. New York: Plenum Press.

Martin, P. S., and R. G. Klein, eds. 1984. *Quaternary extinctions.* Tucson: University of Arizona Press.

McPherson, A. B. 1985. A biogeographical analysis of factors influencing the distribution of Costa Rican rodents. *Brenesia* 23:97–273.

Medellin, R. A. 1992. Community ecology and conservation of mammals in a Mayan tropical rainforest and abandoned agricultural fields. Ph.D. diss., University of Florida, Gainesville.

Meester, J., and H. W. Setzer. 1961. *The mammals of Africa: An identification manual.* Washington, D.C.: Smithsonian Institution Press.

Moore, J. C. 1959. Relationships among living squirrels of the Sciurinae. *Bull. Amer. Mus. Nat. Hist.* 118 (4): 159–206.

O'Connell, M. A. 1989. Population dynamics of Neotropical small mammals in seasonal habitats. *J. Mammal.* 70:532–48.

Olds, N., and S. Anderson. 1989. A diagnosis of the tribe Phyllotini (Rodentia, Muridae). In *Advances in Neotropical mammalogy,* ed. K. H. Redford and J. F. Eisenberg, 55–74. Gainesville, Fla.: Sandhill Crane Press.

Pacheco, V. H., H. de Macedo, E. Vivar, C. Ascorra, R. Arana-Cardó, and S. Solari. 1995. *Lista anotada de los mamíferos peruanos.* Occasional Papers in Conservation Biology. Washington, D.C.: Conservation International.

Pascual, R., M. Archer, E. O. Jaurequizar, J. L. Prado, H. Godthelp, and S. J. Hand. 1992. First non-Australian monotreme. In *Platypus and echidnas,* ed. M. L. Augee, 1–14. Mosman, New South Wales: Royal Zoological Society.

Patton, J. L., P. Myers, and M. Smith. 1990. Vicarient versus gradient models of diversification: The small mammal fauna of eastern Andean slopes of Peru. In *Vertebrates in the tropics,* ed. G. Peters and R. Hutterer, 355–72. Bonn: Museum Alexander Koenig.

Peterson, N. E., D. R. Roberts, and F. P. Pinheiro. 1981. Programa multidisciplinario de vigilancia de las enfermedades infecciosas en zonas colindantes con carretera transamazonica en Brasil. 3.

Estudio de los mamíferos. *Bol. Sanit. Panama* 91 (4): 324–36.

Rabinowitz, A., and B. G. Nottingham. 1989. Mammal species richness and relative abundance of small mammals in a subtropical wet forest of Central America. *Mammalia* 53:217–26.

Redford, K. H., and J. F. Eisenberg. 1992. *Mammals of the Neotropics*. Vol. 2. *The southern cone: Chile, Argentina, Uruguay, Paraguay*. Chicago: University of Chicago Press.

Reig, O. A. 1980. A new fossil genus of South American cricetid rodents allied to *Wiedomys*, with an assessment of the Sigmodontinae. *J. Zool.* (London) 192:257–81.

———. 1987. An assessment of the systematics and evolution of the Akodontini, with the description of new fossil species of *Akodon* (Cricetidae: Sigmodontinae). *Fieldiana: Zool.*, n.s., 39:347–99.

Robinson, J. G., and K. H. Redford. 1989. Body size, diet, and population variation in Neotropical mammal species: Predictors of local extinction? In *Advances in Neotropical mammalogy*, ed. K. H. Redford and J. F. Eisenberg, 567–94. Gainesville, Fla.: Sandhill Crane Press.

———, eds. 1991. *Neotropical wildlife use and conservation*. Chicago: University of Chicago Press.

Rodriguez-Mahecha, J. V., J. I. Hernández-Camacho, T. R. Defler, M. Alberico, R. B. Mast, R. A. Mittermeier, and A. Cadena. 1995. Mamíferos colombianos: Sus nombres comunes e indígenas. *Occas. Pap. Conserv. Biol.* 3:1–56.

Savage, D. E., and D. E. Russell. 1983. *Mammalian paleofaunas of the world*. London: Addison-Wesley.

Schaller, G. B. 1983. Mammals and their biomass on a Brazilian ranch. *Arq. Zool. São Paulo* 31 (1): 1–36.

Scott, W. B. 1913. *A history of land mammals in the Western Hemisphere*. New York: Macmillan.

Shoshani, J., M. Goodman, J. Czelusniak, and G. Braunitzer. 1985. A phylogeny of Rodentia and other eutherian orders: Parsimony analysis utilizing amino acid sequences of alpha and beta hemoglobin chains. In *Evolutionary relationships among rodents*, ed. W. P. Luckett and J.-L. Hartenberger, 191–210. New York: Plenum Press.

Simpson, G. G. 1980. *Splendid isolation: The curious history of South American mammals*. New Haven: Yale University Press.

Smythe, N. 1978. The ecology and behavior of the agouti (*Dasyprocta punctata*) in Panama. *Smithsonian Contrib. Zool.* 257:1–52.

Stallings, J. R. 1988. Small mammal inventories in an eastern Brazilian park. *Bull. Florida State Mus.* 34:153–200.

Stehli, F. G., and S. D. Webb, eds. 1985. *The great American biotic interchange*. New York: Plenum Press.

Storch, G. 1990. The Eocene mammal fauna from Messel—a palaeogeographical jigsaw puzzle. In *Vertebrates in the tropics*, ed. G. Peters and R. Hutterer, 23–32. Bonn: Museum Alexander Koenig.

Strahan, R. 1983. *The complete book of Australian mammals*. Sydney: Angus and Robertson.

Sunquist, M. E., F. Sunquist, and D. E. Daneke. 1989. Ecological separation in a Venezuelan llanas carnivore community. In *Advances in Neotropical mammalogy*, ed. K. H. Redford and J. F. Eisenberg, 197–232. Gainesville, Fla.: Sandhill Crane Press.

Voss, R. S. 1988. Systematics and ecology of the ichthyomyine rodents (Muroidea): Patterns of morphological evolution in a small adaptive radiation. *Bull. Amer. Mus. Nat. Hist.* 188 (2): 259–493.

Webb, S. D. 1976. Mammalian faunal dynamics of the great American interchange. *Paleobiology* 2:216–34.

———. 1985. Late Cenozoic mammal dispersals between the Americas. In *The great American biotic interchange*, ed. F. G. Stehli and S. D. Webb, 357–86. New York: Plenum Press.

Webb, S. D., and S. C. Perrigo. 1984. Late Cenozoic vertebrates from Honduras and El Salvador. *J. Vert. Paleontol.* 4 (2): 237–54.

Wetzel, R. 1977. The Chacoan peccary *Catagonus wagneri* (Rusconi). *Bull. Carnegie Mus. Nat. Hist.* 3:1–36.

Whitmore, F. C., and R. H. Stewart. 1965. Miocene mammals and Central American seaways. *Science* 148:180–85.

Wilson, D. E. 1990. Mammals of La Selva, Costa Rica. In *Four Neotropical rainforests*, ed. A. H. Gentry, 273–86. New Haven: Yale University Press.

Wilson, D. E., and D. M. Reeder, eds. 1993. *Mammal species of the world*. 2d ed. Washington, D.C.: Smithsonian Institution Press.

Wilson, E. O., ed. 1988. *Biodiversity*. Washington D.C.: National Academy Press.

Wood, A. E. 1974. The evolution of the Old World and New World hystricomorphs. In *The biology of hystricomorph rodents*, ed. W. Rowlands and B. J. Weir, 21–54. London: Academic Press.

Woods, C. A., and J. F. Eisenberg. 1989. Land mammals of Madagascar and the Greater Antilles: Comparisons and analyses. In *Biogeography of the West Indies*. ed. C. A. Woods, 799–826. Gainesville, Fla.: Sandhill Crane Press.

Appendix

In Eisenberg (1989) and in Redford and Eisenberg (1992) we used Honacki, Kinman, and Koeppl (1982) as a primary guide for taxonomic usage. The publication of Wilson and Reeder (1993) has set a revised standard for nomenclature and is the primary reference for the current volume. To aid the owners of the previous two volumes of *Mammals of the Neotropics*, we offer a guide to the major taxonomic changes and some important additions to the literature since 1989 and 1992.

Volume 1, *The Northern Neotropics: Panama, Colombia, Guyana, Suriname, French Guiana*

Guides to the mammals of Venezuela have been published by Badillo et al. (1988) and Linares (1998).

Marsupialia

A well-illustrated book on the marsupials of Venezuela was published in 1993 by Perez-Hernandez, Soriano, and Lew.

Ordinal status has been recognized for the Didelphimorphia (Didelphidae in part), Paucituberculata (Caenolestidae, in part), and Microbiotheria (Microbiotheriidae) (Gardner 1993).

Full generic status has been acknowledged for the former subgenera *Marmosops*, *Micoureus*, *Thylamys*, and a new genus *Gracilinanus* was created for the smaller, long-tailed forms. *Marmosa* is employed in a restricted sense.

Marmosa cinerea is now recognized as *Micoureus demerarae*.

The status of *Philander andersoni* has been clarified, and the content of *P. opossum* has been restricted (see this volume).

Monodelphis orinoci is considered a valid species in the llanos of northern Venezuela (Peres-Hernandez, Soriano, and Lew 1993; Ventura, Perez-Hernandez, and Lopez-Fuster 1998).

Insectivora

Cryptotis has been reexamined, and new species have been described. The definition of *C. thomasi* has been refined, and *C. columbiana* has been described as a new species (see *Cryptotis* in this volume).

Chiroptera

The bat fauna of French Guiana has been completely reviewed with many new localities recorded (Brosset and Charles-Dominique 1990).

Within the Glossophaginae the distinctiveness of *Lonchophylla* is recognized, and a new subfamily has been designated as Lonchophyllinae. *Nyctinomops*, formerly designated a subgenus, is now considered a full genus. The South American forms recognized as *Lasiurus borealis* may be assigned to *L. blossevillii* (see this volume). *Vampyrops* has been changed to *Platyrrhinus* (see this volume). Alberico and Velasco-A. (1994) have named a new species of *Platyrrhinus* from the Chocó region of Colombia *P. chocensis*.

Primates

The subspecies of *Aotus* and *Callicebus* have in many cases been raised to full species. New species have been named (see chapter 9 in this volume).

Carnivora

New data on the ecology of *Tremarctos* in Venezuela have been published by Goldstein (1991). Reproductive data for the Procyonidae have been reviewed by Poglayan-Neuwall (1996), and a complete review of the biology of *Bassariscus sumichrasti* is included in Poglayen-Neuwall and Poglayen-Neuwall (1995).

Artiodactyla

The small *Mazama* from the mountains of western Venezuela is now considered a full species, *M. bricenii*, and the name *M. rufina* is restricted to the species in the Cordillera Central of southern Colombia and Ecuador (Czernay 1987).

Sirenia

The record for *Trichechus inunguis* from the Río Negro at the Venezuelan boundary remains unconfirmed.

Rodentia

Mendez (1993) has published a comprehensive book on the rodents of Panama.

Geomyidae

Alberico (1990) recorded the first specimens of *Orthogeomys* from the northern portion of the Andean cordillera in Colombia. The species was named *O. thaeleri*.

Muridae

SIGMODONTINAE

Thomasomyini

Gomez-Laverde et al. (1997) separate *Thomasomys niveipes* from *T. laniger* as a valid species.

Ichthyomyini

Voss (1988) has renamed *Anatomys trichotis* as *Chibchanomys trichotis*. The genus *Daptomys* is now included within *Neusticomys*. Ochoa and Soriano (1991) named *Neusticomys mussoi* from the state of Táchira, Venezuela.

Oryzomyini

The subgenera of *Oryzomys*, *Oecomys*, *Microryzomys*, *Sigmodontomys*, *Melanomys*, and *Oligoryzomys* are now recognized as full genera. *Oecomys trinitatis* is now separated as a full species from *O. concolor* as the species of *Oecomys* in the Guianas and Trinidad as well as the east-central Andes of Colombia and Peru. *Oecomys speciosus* is recognized as a full species found in the savannas of central Colombia and northern Venezuela. *Oryzomys albigularis* is apparently a superspecies complex (Aguilera, Perez-Zapata, and Martino 1995). Forms assigned to *Oryzomys albigularis* from the north coastal range of Venezuela differ in karyotype from other locations in Venezuela. Aguilera, Perez-Zapata, and Martino (1995) propose that the name *O. caracolus* be applied to the form from the north coastal range. *Oligoryzomys griseolus* in the Táchira Andes of Venezuela is distinct from *O. fulvescens*; it may extend to the Cordillera Oriental of Colombia.

Oryzomys capito was employed in volume 1 for the lowland and intermediate elevation forms of *O. capito*-like rice rats. Although *O. talamancae* was plotted, we did not emphasize that the forms from intermediate elevations evidently were all *O. talamancae*.

Musser et al. (1998) have reinstated *Oryzomys megacephalus* (Fisher, 1814) as the senior synonym of *O. capito* (Olfers, 1818). This newly defined taxon includes the lowland larger rice rat ranging from eastern Venezuela through the Guianas and eastern Brazil to Paraguay. *Oryzomys yunganus* is endemic to the true lowland wet rain forests of Amazonia in the classical sense. *Oryzomys talamancae* ranges in northwestern Venezuela to western Colombia and thence south to western Ecuador.

Akodontini

Microxus bogotensis is considered a member of *Akodon*, but this assignment is provisional and controversial (see *Akodon* in this volume).

Erethizontidae

The small, hairy porcupines assigned to *Sphiggurus* are under revision. The form listed as *S. vestitus* in the Andes of western Venezuela could be assigned to *S. pruinosus* instead. The latter taxon could be restricted to the Cordillera Central of Colombia. *S. melanurus* may be more appropriate for those forms in southern Venezuela than *S. insidiosus* (see p. 452 in this volume).

Dasyproctidae

The species names for *Dasyprocta* in the northern Neotropics are still under review.

Echimyidae

The genus *Thrinacodus* has been replaced by *Ollalomys*. The content of *Echimys armatus* is under review (see chapter 16 in this volume).

The *Proechimys guairae* complex of northern Venezuela has been studied continuously by Aguilera and associates. In addition the *P. trinitatus* group has been closely examined. *P. trinitatus* is elevated to a full species as well as *P. urichi* in the state of Sucre. A new species *P. barinas* has been described from west-central Venezuela (see Aguilera and Corti 1994; Perez-Zapata, Aguilera, and Reig 1992; Corti and Aguilera 1995).

Volume 2, *The Southern Cone: Chile, Argentina, Uruguay, Paraguay*

Barquez, Mares, and Ojeda (1991) have published a guide to the mammals of Tucumán Province, Argentina.

Marsupialia

Ordinal status has been recognized for the Didelphimorphia (Didelphidae); Paucituberculata (Caenolestidae), and Microbiotheria (Microbiotheriidae). Full generic status has been acknowledged for the former subgenera *Marmosops*, *Micoureus*, and *Thylamys*. The content of *Marmosa* is restricted. *Gracilinanus* is reserved for the smaller long-tailed forms. *Marmosa cinerea* is now recognized as *Micoureus demerarae*. The content of *Philander opossum* has been restricted (see this volume).

Chiroptera
Barquez, Giannini, and Mares (1993) have published a guide to the bats of Argentina. *Lasiurus borealis* as portrayed in volume 2 may actually reflect in part the distribution of *L. blossevillii*. *Vampyrops* has been changed to *Platyrrhinus*. *Nyctinomops* is recognized as a full genus.

Carnivora
Wozencraft (1993) has raised the former subgenera of *Felis* to full generic status (see chapter 10 in this volume). *Lontra* is now generally accepted as the generic name for South American forms assigned to *Lutra*. *Dusicyon* is restricted to *D. australis*, and *Pseudalopex* is recommended for all species assigned to *Dusicyon* in Redford and Eisenberg (1992). *Pseudalopex fulvipes* is recognized as a valid species (see p. 525 in this volume).

Artiodactyla
The small brocket deer of eastern Paraguay and northeastern Argentina is more properly referred to as *Mazama nana*.

Rodentia
Pearson (1995) has published a key to the small mammals found in the moist temperate rain forests of Argentina and adjacent Chile.

Muridae
SIGMODONTINAE
Thomasomyini
Delomys is a recognized as the appropriate genus for *Thomasomys sublineatus* and *T. dorsalis*. *Wilfredomys* is recognized as the appropriate genus for *Thomasomys oenax* and *T. pictipes* (see chapter 16 in this volume).

Oryzomyini
The former subgenera of *Oryzomys* are now recognized as full genera. For volume 2 this affects *Oryzomys*, *Oecomys*, and *Oligoryzomys*. The appropriate names for the forms subsumed under *Oryzomys* are

now *Oryzomys buccinatus* and *O. nitidus*. *O. ratticeps* remains the same, but forms labeled *O. capito* may be subject to a revision. Members of *Oligoryzomys* include *Oligoryzomys chacoensis*; *O. delticola*; *O. flavescens*; *O. longicaudatus*; *O. nigripes*, and *O. microtis*. *Oligoryzomys fornesi* is now included in *O. microtis*.

Musser et al. (1998) have reinstated *Oryzomys megacephalus* (Fisher, 1814) as the senior synonym of *O. capito* (Olfers, 1818). This newly defined taxon includes the lowland larger rice rat ranging from eastern Venezuela through the Guianas and eastern Brazil to Paraguay. *Oryzomys nitidus* ranges from eastern Peru through Bolivia south to Paraguay and northern Argentina.

Musser et al. (1998) recommend that *Oryzomys intermedius* be included in *O. russatus*.

Akodontini
Patterson (1992) has described a new genus and species *Pearsonomys annectens* from the Chilean temperate forests.

The map for *Akodon varius* in volume 2 (map 11.69) should read "*Akodon varius* group." The distribution reflects three now recognized species as explained in part on page 427 in this volume.

The subgenus *Abrothrix* has been raised to a full genus by some authors (see p. 419 in this volume).

Anderson (1997) points out that *Akodon lutescens* is conspecific with *A. puer*. Further, *A. puer caenosus* Thomas, 1918 was considered by Reig (1986) to be included in *A. puer*, but *Akodon* from Carapari, Bolivia, assigned by Thomas to *A. caenosus*, are now assigned to *A. boliviensis*. Anderson concludes that *A. boliviensis* is a valid species that co-occurs over much of Bolivia with *A. lutescens*, which now includes the following as subspecies: *A. lutescens caenosus* (includes *A. caenosus* of Cabrera and *A. puer caenosus*); *A. lutescens lutescens* (includes *A. puer lutescens*); and *A. lutescens puer* (includes *A. puer* of Tate, *A. puer puer* of Myers, and *A. fumeus* of Myers and Patton). However, *A. fumeus* as defined by Thomas, 1902 stands as a valid species.

Phyllotini
Braun (1993) recommends that *Auliscomys micropus* be set aside in its own genus *Loxodontomys* (formerly a subgenus). She further assigns *A. boliviensis* to a new genus *Maresomys*. Mares and Braun (1966) name a new *Andalgalomys*, *A. roigi* from central Argentina.

A new genus and species of phyllotine from north-central Argentina, *Salinomys delicatus*, was described by Braun and Mares (1995).

The description of the genus *Euneomys* in volume 2 recognized a single species but introduced the possibility of a second species (2:284). Musser and Carleton (1993) list four species and further discuss the problems of assigning appropriate names. Pearson and Christie (1991) discuss the occurrence of sympatry of range for two species in the genus. The problem needs attention.

Sigmodontini

Holochilus magnus as discussed in volume 2 is now placed in its own genus *Lundomys*. The integrity of the old content of Sigmondontini in South America is challenged (see *Sigmodon* and *Holochilus* in this volume).

Caviidae

Two natural history studies amplifying the conclusions under *Dolichotis patagonum* have been published by Taber and MacDonald (1992a,b).

Abrocomidae

A revision with the naming of a new species *Abrocoma vaccarum* validates the status of former subspecies (see chapter 16 in this volume). Braun and Mares (1996) describe new morphological characters for the species *A. vaccarum.*

Chinchillidae

The recent work by Branch and associates with *Lagostomus maximus* is included in this volume (see pp. 469–70). Maps 11.110 and 11.111 were transposed in the first printing of volume 2.

Octodontidae

A new species of *Octodon, O. pacificus,* was named by Hutterer (see chapter 18 in this volume).

Dasyproctidae

The use of *Dasyprocta punctata* for the species in northern Argentina should be changed to *D. variegata.*

References

Aguilera, M., And M. Corti. 1994. Craniometric differentiation and chromosomal speciation of the genus *Proechimys. Z. Säugetierk.* 59:366–77.

Aguilera, M., A. Perez-Zapata, and A. Martino. 1995. Cytogenetics and karyosystematics of *Oryzomys albigularis* from Venezuela. *Cytogenet. Cell Genet.* 69:44–49.

Alberico, M. 1990. A new species of pocket gopher (Rodentia, Geomyidae) from South America and its biogeographic significance. In *Vertebrates in the tropics,* ed. G. Peters and R. Hutterer, 103–11. Bonn: Museum Alexander Koenig.

Alberico, M., and E. Velasco-A. 1994. Extended description of *Platyrrhinus chocoensis* from the Pacific lowlands of Colombia. *Trianea* 5:343–51.

Anderson, S. 1997. Mammals of Bolivia: Taxonomy and distribution. *Bull. Amer. Mus. Nat. Hist.* 231: 1–652.

Badillo F., A., R. Guerrero, R. Lord, J. Ochoa, and G. Ulloa. 1988. *Mamíferos en Venezuela.* Caracas: Central de Venezuela.

Barquez, R. M., N. P. Giannini, and M. A. Mares. 1993. *Guide to the bats of Argentina.* Norman: University of Oklahoma Press.

Barquez, R. M., M. Mares, and R. A. Ojeda. 1991. *Mammals of Tucumán.* Norman: University of Oklahoma Press.

Braun, J. K. 1993. *Systematic relationships of the tribe Phyllotini (Muridae, Sigmodontinae) of South America.* Special Publications. Norman: Oklahoma Museum of Natural History.

Braun, J. K., and M. Mares. 1995. A new genus and species of phyllotine rodent from South America. *J. Mammal.* 76:504–21.

———. 1996. Unusual morphological and behavioral traits in *Abrocoma* from Argentina. *J. Mammal.* 77:891–97.

Brosset, A., and P. Charles-Dominique. 1990. The bats from French Guiana: A taxonomic, faunistic, and ecological approach. *Mammalia* 54:509–60.

Corti, M., and M. Aguilera. 1995. Allometry and chromosomal speciations of the casiraguas *Proechimys. J. Zool. Syst. Evol. Res.* 33:109–15.

Czernay, S. 1987. *Die Spiesshirsche und Pudus.* Wittenberg: A. Ziemsen.

Eisenberg, J. F. 1989. *Mammals of the Neotropics.* Vol. 1. *The northern Neotropics: Panama, Colombia, Venezuela, Guyana, Suriname, French Guiana.* Chicago: University of Chicago Press.

Gardner, A. L. 1993. Didelphimorphia. In *Mammal species of the world,* ed. D. E. Wilson and D. M. Reeder, 15–23. Washington, D.C.: Smithsonian Institution Press.

Goldstein, I. 1991. Spectacled bear predation and feeding behavior on livestock in Venezuela. *Stud. Neotrop. Fauna Environ.* 26:231–35.

Gómez-Laverde, M., O. Montenegro-Díaz, H. López-Arévalo, A. Cadena, and M. L. Bueno. 1997. Karyology, morphology, and ecology of *Thomasomys laniger* and *T. niveipes* (Rodentia) in Colombia. *J. Mammal.* 78 (4): 1282–89.

Honacki, J. H., K. E. Kinman, and J. W. Koeppl, eds. 1982. *Mammal species of the world.* Lawrence, Kans.: Allen Press and Association of Systematic Collections.

Hutterer, R. 1994. Island rodents: A new species of *Octodon* from Isla Mocha, Chile. *Z. Säugetierk.* 59:27–41.

Linares, O. J. 1998. *Mamíferos de Venezuela.* Caracas: Sociedad Conservacionista Audubon de Venezuela.

Mares, M., and J. K. Braun. 1996. A new species of phyllotine rodent, genus *Andalgalomys* (Muridae:

Sigmodontinae) from Argentina. *J. Mammal.* 77 (4): 928–41.

Medel, R. G., J. Jiménez, and F. Jaksić. 1990. Discovery of a continental population of the rare Darwin's fox *Dusicyon fulvipes* in Chile. *Biol. Conserv.* 51:71–77.

Mendez, F. 1993. *Los roedores de Panama.* Panama City: Privately published.

Musser, G., and M. D. Carleton. 1993. Family Muridae. In *Mammal species of the world,* ed. D. E. Wilson and D. M. Reeder, 501–756. Washington, D.C.: Smithsonian Institution Press.

Musser, G. G., M. D. Carleton, E. M. Brothers, and A. L. Gardner. 1998. Systematic studies of the Oryzomyini rodents (Muridae, Sigmodontinae): Diagnoses and distributions of species formerly assigned to *Oryzomys "capito." Bull. Amer. Mus. Nat. Hist.* 236:1–367.

Ochoa G., J., and P. Soriano. 1991. A new species of water rat, genus *Neusticomys* Anthony, from the Andes of Venezuela. *J. Mammal.* 72:97–103.

Patterson. B. D. 1992. A new genus and species of long-clawed mouse (Rodentia: Muridae) from temperate rainforests of Chile. *Zool. J. Linn. Soc.* 106: 127–45.

Pearson, O. P. 1995. Annotated keys for identifying small mammals living in or near Nahuel Huapi National Park or Lamin National Park, southern Argentina. *Mastozool. Neotrop.* 2:98–148.

Pearson, O. P., and M. I. Christie. 1991. Sympatric species of *Euneomys* (Rodentia: Cricetidae). *Stud. Neotrop. Fauna Environ.* 2:121–27.

Perez-Hernandez, R., P. Soriano, and D. Lew. 1993. *Marsupiales de Venezuela.* Lagoven, S. A.: Cuadernos Lagoven.

Perez-Zapata, A., M. Aquilera, and O. Reig. 1992. An allopatric karyomorph of the *Proechimys guairae* complex in eastern Venezuela. *Interciencia* 17:235–40.

Poglayen-Neuwall, I. 1995. Developmental data on *Bassariscus sumichrasti. Zool. Gart.* 65:391–96.

———. 1996. A comprehensive table of reproduction and longevity data of Procyonidae (Mammalia, Carnivora). *Zool. Gart.* 66:261–63.

Poglayen-Neuwall, I., and I. Poglayen-Neuwall. 1995. Observations on the ethology and biology of the Central American cacomixtle *Bassariscus sumichrasti. Zool. Gart.* 65:11–49.

Redford, K., and J. F. Eisenberg. 1992. *Mammals of the Neotropics.* Vol. 2. *The southern cone: Chile, Argentina, Uruguay, Paraguay.* Chicago: University of Chicago Press.

Reid, F. A. 1997. *A field guide to the mammals of Central America and southeast Mexico.* New York: Oxford University Press.

Reig, O. A. 1986. Diversity patterns and differentiation of high Andean rodents. In *High altitude tropical biogeography,* ed. F. Vuilleumier and M. Monasterio, 404–42. New York: Oxford University Press.

Taber, A. B., and D. W. MacDonald. 1992a. Communal breeding in the mara, *Dolichotis patagonum. J. Zool.* (London) 227:439–52.

———. 1992b. Spatial organization and monogamy in the mara. *J. Zool.* (London) 227:417–38.

Ventura, J., R. Perez-Hernandez, and M. J. Lopez-Fuster. 1998. Morphometric assessment of the *Monodelphis brevicaudata* group in Venezuela. *J. Mammal.* 79:104–17.

Voss, R. S. 1988. Systematics and ecology of the ichthyomyine rodents. *Bull. Amer. Mus. Nat. Hist.* 188: 259–493.

Wilson, D. E., and D. M. Reeder, eds. 1993. *Mammal species of the world.* 2d ed. Washington, D.C.: Smithsonian Institution Press.

Wozencraft, W. C. 1993. Carnivora. In *Mammal species of the world,* ed. D. E. Wilson and D. M. Reeder, 279–348. Washington, D.C.: Smithsonian Institution Press.

Index of Scientific Names

Abrawayaomys, 382
  ruschii, 382
Abrocoma, 478–79
  cinerea, 479
Abrocomidae, 477–78
Abrothrix, 419, 595
Aepeomys, 369–70
  lugens, 369–70
Agouti, 461–63
  paca, 462
  taczanowskii, 463
Agoutidae, 461–63
Akodon, 418–28
  aerosus, 419
  albiventer, 419–20
  azarae, 420
  boliviensis, 420–21
  cursor, 421
  dayi, 422
  fumeus, 422
  juninensis, 422
  kempi, 422
  kofordi, 422
  latebricola, 423
  lindberghi, 423
  mimus, 423
  mollis, 423
  nigrita, 423
  orophilus, 424
  puer, 424
  sanctipaulensis, 424
  serrensis, 425
  siberiae, 425–26
  subfuscus, 426
  surdus, 426
  toba, 426
  torques, 426
  urichi, 427
  varius, 427
Akodontini, 417–36
Alouatta, 262–66
  belzebul, 262
  caraya, 263–64
  fusca, 264

  palliata, 264–65
  sara, 265
  seniculus, 265–66
Ametrida, 185–86
  centurio, 185–86
Amorphochilus, 190
  schnablii, 190
Amphinectomys, 403–4
Andalgalomys, 404–5
  pearsoni, 404–5
Andinomys, 405
  edax, 405
Anotomys, 442
  leander, 442
Anoura, 159–61
  caudifer, 159–60
  culturata, 160
  geoffroyi, 160–61
  latidens, 161
Aotus, 247–49
  azarae, 248
  infulatus, 248
  lemurinus, 247–48
  micronax, 249
  nancymaae, 249
  nigriceps, 249
  trivirgatus, 248
  vociferans, 248
Arctocephalus, 314–15
  australis, 314
  galapagoensis, 314–15
  tropicalis, 315
Artibeus, 179–85
  amplus, 179–80
  andersoni, 180
  cinereus, 180
  concolor, 180–81
  fimbriatus, 181
  fraterculus, 181
  fuliginosus, 181
  glaucus, 181–82
  hartii, 182
  jamaicensis, 182–84
  lituratus, 184

  obscurus, 184
  phaeotis, 184
  planirostris, 186
  toltecus, 186
Artiodactyla, 332–55
Ateles, 267–69
  belzebuth, 267–68
  chamek, 268
  fusciceps, 268
  marginatus, 268–69
  paniscus, 269
Atelocynus, 280–81
  microtis, 280–81
Auliscomys, 405–7
  boliviensis, 406
  micropus, 406
  pictus, 406–7
  sublimus, 407

Balaenidae, 319
Balaenoptera, 319–20
Balaenopteridae, 319–20
Balantiopteryx, 129
  infusca, 129
Bartionycteris, 141–42
Bassaricyon, 287–88
  alleni, 287–88
  beddardi, 287
  gabbi, 287–88
Berardius, 319–20
Bibimys, 440
  chacoensis, 440
Blarinomys, 427–28
  breviceps, 427–28
Blastocerus, 340–42
  dichotomus, 340–42
Bolomys, 428–29
  amoenus, 428
  lactens, 428–29
  lasiurus, 429
  lenguarum, 429
  punctulatus, 429
Brachyteles, 269–70
  arachnoides, 269–70

Isothrix (continued)
   paguras, 489
   picta, 489–90

Juscelinomys, 431
   cadango, 431

Kannabateomys, 481–82
   amblonyx, 481–82
Kerodon, 459–60
   rupestris, 459–60
Kogia, 319–20
Koopmania, 180
Kunsia, 441–42
   fronto, 441
   tomentosa, 441–42

Lagenorhynchus, 319–20
Lagidium, 468–69
Lagomorpha, 518–20
Lagostomus, 469–70
   maximus, 469–70
Lagothrix, 266–67
   flavicauda, 266
   lagothricha, 266–67
Lama, 337–39
   glama, 337
   guanacoe, 337–38
   pacos, 337
   vicugna, 338–39
Lasiurus, 201–3
   blossevillii, 201
   borealis, 201
   cinereus, 202
   ebenus, 202
   ega, 202
   egreguis, 202
Lenoxus, 431
Leontideus. See *Leontopithecus*
Leontopithecus, 243–45
   caissara, 243
   chrysomelas, 244
   chysopygus, 244–45
   rosalia, 244
Leopardus, 299–302
Leporidae, 518–19
Lestoros, 83
   inca, 83
Lichonycteris, 161
   obscura, 161
Lionycteris, 156
   spurrelli, 156
Lonchophylla, 156–58
   bokermanni, 157
   dekeyseri, 157
   handleyi, 157
   hesperia, 157
   mordax, 158
   robusta, 158
   thomasi, 158

Lonchophyllinae, 154
Lonchorhina, 145–46
   aurita, 145–46
   marinkellei, 146
   orinocensis, 146
Lontra, 294–95
   felina, 294
   longicaudis, 294–95
   platensis, 295
Loxodontomys, 404, 595
Lundomys, 438
Lutra. See *Lontra*
Lutreolina, 54–55
   crassicaudata, 54–55
Lutrinae, 294–96
Lycalopex, 285, 280
Lynchailurus, 298

Macrophyllum, 146
   macrophyllum, 146
Makalata, 484
Maresomys, 404, 406, 595
Marmosa, 60–62
   andersoni, 61
   lepida, 61
   murina, 61–62
   robinsoni, 62
   rubra, 62
Marmosini, 57–76
Marmosops, 62–65
   dorothea, 63
   impavidus, 63
   incanus, 63–64
   neblina, 64
   noctivagus, 64
   parvidens, 64–65
   paulensis, 65
Marsupialia, 49–89
Mazama, 343–47
   americana, 344–45
   bororo, 346–47
   bricenii, 345
   chunyi, 345
   gouazoubira, 345–46
   nana, 345
   rufina, 346
Megachiroptera, 118
Megalonychidae, 96–97
Megaoryzomys, 382
Megaptera, 319–20
Melanomys, 383
   caliginosus, 383
   robustulus, 383
   zunigae, 383
Mesomys, 490–91
   hispidus, 490–91
Mesophylla, 178–79
   macconnelli, 178–79
Mesoplodon, 319–20
Metachirini, 77–78

Metachirops. See *Philander*
Metachirus, 77–78
   nudicaudatus, 77–78
Micoureus, 65–67
   cinerea (see *Micoureus demerarae*)
   constantiae, 65–66
   demerarae, 66–67
   regina, 67
Microakodontomys, 418
Microbiotheria, 49
Microcavia, 458–59
   australis, 459
   niata, 459
Microchiroptera, 121–220
Micronycteris, 140–45
   behnii, 141
   brachyotis, 141
   daviesi, 141–42
   hirsuta, 142
   megalotis, 142–43
   minuta, 143
   nicefori, 143
   pusilla, 143–44
   sanborni, 144
   schmidtorum, 144
   sylvestris, 144–45
Microryzomys, 383–84
   altissimus, 384
   minutus, 384
Microsciurus, 364
   flaviventer, 364
   mimulus, 364
Microxus, 419
Mimon, 148–49
   bennettii, 149
   crenulatum, 149
Mirounga, 313
Molossidae, 203–14
Molossops, 204–7
   abrasus, 204–5
   aequatorianus, 205
   greenhalli, 205
   mattogrossensis, 205–6
   neglectus, 206
   planirostris, 206
   temminckii, 206–7
Molossus, 211–13
   ater, 211–12
   bondae, 212
   molossus, 212–13
Monodelphis, 67–74
   americana, 69
   brevicaudata, 69–70
   dimidiata, 70
   domestica, 70–71
   emiliae, 71
   henseli, 71
   iherengi, 71–72
   kunsi, 72
   maraxina, 72

Index of Common Names